PREFACE

The book continues the tradition of providing a firm foundation in chemical calculations. I have included quite a good number of problems with hints wherever necessary. The challenge to me has always been to present a solid understanding of the basic facts and principles of chemistry. I am fully confident that the present book will help the students prepare for the JEE and other engineering and medical entrance examinations.

The design of question papers, particularly of the JEE, changes from time to time. Remember that chemistry has not changed. I feel strongly that if you understand the subject well, you can solve any question in any form. I have also noticed that questions are now being asked from all chapters including the mole concept, stoichiometric calculations, etc. Previously, it used to be selective. Another important fact is that the questions are sometimes too difficult for students of +2 level. In that case you may consult my book 'Modern Approach to Physical Chemistry, volumes I and II' for an advanced knowledge of physical chemistry. Basically, read every chapter very carefully till you have a vivid picture of the subject in your mind. Finally, follow Swami Vivekananda's saying,

Arise, Awake and Stop Not
till the goal is achieved.

Ramendra C Mukerjee

PREFACE TO THE EIGHTH EDITION

I take immense pleasure in presenting you the eighth edition of this book. On the basis of helpful suggestions given by learned teachers and students from all over the country, and also in view of the recent changes in the syllabus and in the examination pattern of IIT-JEE, all the chapters have been critically reviewed. Some changes in the presentation of the text and the necessary additions, both in the text and in the problems, have been carried out. The revision chapters 21 and 22 have been updated by including the questions asked in IIT-JEE of the recent past years. My goal in this revision is to provide students with the best possible tool for learning numerical chemistry, by incorporating and amplifying features that enhance their understanding of concepts. Here are some of the new features:

The presentation of thermodynamics has been reorganized. The consistent presentation helps to consolidate the concepts of thermodynamics. As the central concepts of chemical change are best understood in the sequence of chemical thermodynamics (Is the forward or the reverse reaction favoured?), chemical thermodynamics is now placed as Chapter 14, followed by chemical equilibrium (Chapter 15), ionic equilibrium (Chapter 16—How far the reaction would go?) and then by chemical kinetics (Chapter 17—How fast the reaction would go?).

Though the oxidation and reduction processes have been discussed in chapters 6, 7 and 18, a new chapter (19) on the concepts of oxidation number and its application to balance redox reactions has been added. Chapter 20 on solid state has been extended by including the liquid state. Chapters 21 and 22 contain only problems for the purpose of revision.

Many of my colleagues have helped me write this book, to all of whom I am sincerely thankful. I especially thank Mr D Kumar, former Head, Chemistry Department of our college, with whom I have discussed many problems, and Dr G K Verma, C M P Degree College, for offering me many useful suggestions. I am greatly indebted to Prof. S S Shukla, Lamar University, Texas, USA, for going through the manuscript of the first edition and making invaluable suggestions. I wish to express my appreciation to all the students and teachers who have been kind enough to write letters with helpful suggestions for the improvement of the book. Special thanks are due to my own students for pointing out a number of errors. Finally, I thank all the members of the Bharati Bhawan family who were involved in the production of this edition.

I believe this edition will continue to prove useful to the students and interesting to the teachers.

132 B/2A, Alopibagh
Allahabad 211 006

Ramendra C Mukerjee

MODERN APPROACH TO Chemical Calculations

AN INTRODUCTION TO THE MOLE CONCEPT

Ramendra C Mukerjee, MSc, PhD
Department of Chemistry
Ewing Christian College, Allahabad
(An Autonomous College of Allahabad University)

Bharati Bhawan
PUBLISHERS & DISTRIBUTORS

Published by

BHARATI BHAWAN (Publishers & Distributors)

www.bharatibhawan.in   email: editorial@bbpd.in

4271/3 Ansari Road, Daryaganj, NEW DELHI 110 002, Phone: 23286557
A-61 B/2 Sector 63, NOIDA 201 307, Phone: 4757400
Thakurbari Road, PATNA 800 003, Phone: 2670325
10 Raja Subodh Mallick Square, KOLKATA 700 013, Phone: 22250651
No. 98 Sirsi Circle, Mysore Road, BENGALURU 560 018, Phone: 26740560

First edition 1989
Eighth edition 2019
2023 print

Modern Approach to Chemical Calculations
Printed at Rajkamal Electric Press, Kundli, Haryana-131 028

A Message

Dear Students,

This is a textbook on chemical calculations primarily intended for students who are preparing for the entrance tests of IIT and other engineering colleges. This text is equally useful to the students of Intermediate(+2) and BSc classes of Indian universities.

Chemistry is just one of the subjects you have to study and, therefore, you have limited time for each subject. Keeping in mind your limited time and requirements, I have tried my best to write this book in such a way that it fulfils your requirements. The special features of the book are:

1. This book has been written in the language of the mole, as the concept of mole is a basic tool in understanding chemistry. Try to have fun with the mole; I did.

2. Sufficient fundamental principles have been provided before introducing problems. These principles, in the form of rules or equations, have been referred to throughout the book while solving problems. This will help you in following each step of the solutions of these problems.

3. A wide variety of problems have been selected; at the same time repetitions of similar types of problems have been avoided. This allowed me to give maximum coverage in minimum space.

4. Mixed system of units including SI units have been used in this text in accordance with the latest trend of the entrance tests of IIT and most of the other engineering colleges. Moreover, a selected list of basic and derived SI units and conversions of selected nonsystem units to SI units have been provided at the beginning of the book to help you solve problems using different systems of units.

Finally, I would like to suggest two important points so that you can take the maximum advantage from this book.

- Read Chapter 1 and Chapter 14 very carefully as both these chapters have wide applications in other chapters.
- Do not jump directly to solving the problems until you have thoroughly gone through the text provided at the beginning of each chapter.

Ramendra C Mukerjee

Selected SI Units (Systeme International d' Unites)

Quantity	Unit	
	Name	*Symbol*
Basic Units:		
Length	metre	m
Mass	kilogram	kg
Time	second	s
Electric current	ampere	A
Temperature	kelvin	K
Luminous intensity	candela	cd
Amount of substance	mole	mol
Derived Units:		
Volume	cubic metre	m^3
Density	kilogram per cubic metre	kg/m^3
Force	newton	N ($kg{\cdot}m/s^2$)
Pressure	pascal	Pa (N/m^2)
Energy, work, heat	joule	J (N·m)
Power	watt	W (J/s)
Quantity of electricity	coulomb	C (A·s)
Electric voltage, electrical potential, emf	volt	V (W/A)
Resistance	ohm	Ω (V/A)
Conductance	siemens	Ω^{-1}
Frequency	hertz	Hz (s^{-1})

❑

Conversion of Selected Nonsystem Units to SI Units

Quantity	*Unit*	*Conversion factor to* SI
Length	centimetre	1×10^{-2} m
	micrometre (μm)	1×10^{-6} m
	nanometre (nm)	1×10^{-9} m
	angstrom (Å)	1×10^{-10} m
	inch	$2{\cdot}54 \times 10^{-2}$ m
Mass	gram	1×10^{-3} kg
Volume	litre	1×10^{-3} m^3 = dm^3
Pressure	bar	10^5 Pa
	atmosphere (atm)	$1{\cdot}01325 \times 10^5$ Pa
	millimetre of Hg or torr	133·322 Pa
Energy, work, amount of heat	electronvolt (eV)	$1{\cdot}60219 \times 10^{-19}$ J
	calorie (cal)	4·1868 J
	kilocalorie (kcal)	4186·8 J
	erg	10^{-7} J

❑

Values of Selected Fundamental Physical Constants

Quantity	*Symbol*	*Value*
Planck constant	h	$= 6{\cdot}6262 \times 10^{-27}$ erg · s
		$= 6{\cdot}6262 \times 10^{-34}$ J · s (SI)
Avogadro number (Avogadro constant)	N	$= 6{\cdot}022 \times 10^{23}$ mol^{-1} (SI)
Boltzmann constant	k	$= 1{\cdot}3807 \times 10^{-16}$ erg deg^{-1} molecule^{-1}
		$= 1{\cdot}3807 \times 10^{-23}$ JK^{-1} (SI)
Gas constant	R	$= 8{\cdot}314 \times 10^{7}$ erg deg^{-1} mole^{-1}
		$= 8{\cdot}314$ JK^{-1} mol^{-1} (SI)
		$= 0{\cdot}0821$ L · atm deg^{-1} mole^{-1}
		$= 1{\cdot}987$ cal deg^{-1} mole^{-1}
Molar volume of gas at NTP	V	$= 22{\cdot}414$ L mole^{-1}
		$= 22{\cdot}414 \times 10^{-3}$ m^{3} mole^{-1} (SI)
Velocity of light	c	$= 2{\cdot}998 \times 10^{10}$ cm s^{-1}
		$= 2{\cdot}998 \times 10^{8}$ m s^{-1} (SI)
Electron mass	m_e	$= 9{\cdot}109 \times 10^{-28}$ g
		$= 9{\cdot}109 \times 10^{-31}$ kg (SI)
Electron charge	e	$= 4{\cdot}803 \times 10^{-10}$ esu
		$= 1{\cdot}602 \times 10^{-19}$ coulomb (SI)
Faraday constant	F	$= 96485$ coulomb eq^{-1}
		$= 96484$ coulomb mol^{-1} (SI)
Acceleration due to gravity	g	$= 980{\cdot}665$ cm s^{-2}
		$= 9{\cdot}80665$ m s^{-2} (SI)
Atomic mass unit	**amu**	$= 1{\cdot}66 \times 10^{-27}$ kg (SI)

❑

A Few Abbreviations Used in This Book

Principle of Atom Conservation	POAC
Weight	Wt.
Atomic weight	At. wt.
Molecular weight	Mol. wt.
Equivalent weight	Eq. wt.
Gram Equivalent or Equivalent	Eq.
Equation	Eqn.
Equilibrium	Eqb.
Milli-equivalent	m.e.
Molar	M
Molal	*m*
Normal	N
Demal	D

❑

CONTENTS

❑

CHAPTER ONE

ELEMENTARY PROBLEMS BASED ON DEFINITION OF MOLE: THE MOLE CONCEPT

The mole concept is an essential tool for the study of the fundamentals of chemical calculations. This concept is simple but its application in problems requires a thorough practice.

The Mole

The 14th General Conference on Weights and Measures (1971) adopted the mole (mol) as the basic SI unit of the amount of a substance.

There are many ways of measuring the amount of a substance, weight and volume being the most common. But the basic unit of chemistry is the atom or a molecule and to measure the number of atoms or molecules is, therefore, of foremost importance.

Mole in Latin means heap or mass or pile. A mole of atoms is a collection of atoms whose total weight is the number of grams equal to the atomic weight. As equal numbers of moles of different elements contain equal numbers of atoms, it is convenient to express amounts of the elements in terms of moles. Just as a dozen means twelve objects, a score means twenty objects, chemists have defined a mole as a 'definite number' of particles, viz., atoms, molecules, ions or electrons, etc. This 'definite number' is called the Avogadro constant, equal to $6{\cdot}022 \times 10^{23}$, in honour of Amedeo Avogadro. However, for many years scientists have made use of the concept of a mole without knowing the value of the Avogadro constant. Thus, a mole of hydrogen atoms or a mole of hydrogen molecules or a mole of hydrogen ions or a mole of electrons means the Avogadro constant of hydrogen atoms, hydrogen molecules, hydrogen ions or electrons respectively.

The value of the Avogadro constant depends on the atomic-weight scale. At present **the mole is defined as the amount of a substance containing as many atoms, molecules, ions, electrons or other elementary entities as there are carbon atoms in exactly 12 g of ^{12}C**. The value of the Avogadro constant was changed by a very small amount in 1961 when the basis of the atomic-weight scale was changed from the naturally occurring mixture of oxygen isotopes at 16 amu to ^{12}C, which put oxygen at 15.9994 amu.

It is quite interesting and surprising to know that the mole is such a big number that it will take 10^{16} years to count just one mole at the rate of one count per second, and the world population would be only of the order of 10^{-14} mole in chemical terminology.

In modern practice, a gram-molecule and a gram-atom are termed as a mole of molecules and a mole of atoms respectively, e.g., 1 gram-molecule of oxygen and 1 gram-atom of oxygen are expressed as 1 mole of O_2 and 1 mole of O respectively. Similarly, gram-equivalent shall be written as equivalent.

The number of moles of a substance can be calculated by various means. The choice of method depends on the data provided.

RULES IN BRIEF

The following are the definitions of 'mole' represented in the form of equations:

(1) $\text{Number of moles of molecules} = \dfrac{\text{weight in g}}{\text{molecular weight}}$

(2) $\text{Number of moles of atoms} = \dfrac{\text{weight in g}}{\text{atomic weight}}$

(3) $\text{Number of moles of gases} = \dfrac{\text{volume at NTP}}{\text{standard molar volume}}$

(Standard molar volume is the volume occupied by 1 mole of any gas at NTP, which is equal to 22·4 litres.)

(4) Number of moles of atoms / molecules / ions / electrons

$$= \frac{\text{no. of atoms / molecules / ions / electrons}}{\text{Avogadro constant}}$$

(5) $\text{Number of moles of solute} = \text{molarity} \times \text{volume of solution in litres}$

or $\text{no. of millimoles} = \text{molarity} \times \text{volume in mL.}$

$$\frac{\text{Millimoles}}{1000} = \text{moles}$$

(6) For a compound $M_x N_y$, x moles of N = y moles of M

These rules are very important and have been frequently applied throughout the book. In the beginning, the rules have been mentioned wherever they are applied in solving problems. Before discussing the mole concept in further detail, the understanding of a balanced chemical equation is necessary.

Significance of Chemical Equations

A chemical equation describes the chemical process both qualitatively and quantitatively. The stoichiometric coefficients in the chemical equation give the quantitative information of the chemical process. These coefficients represent the relative number of molecules or moles of the reactants and products, e.g.,

	$2\,KClO_3$ (s)	$\longrightarrow$	$2\,KCl$ (s)	+	$3\,O_2$ (g)
	2 molecules		2 molecules		3 molecules
or	$2\,N$ molecules		$2\,N$ molecules		$3\,N$ molecules
or	2 moles		2 moles		3 moles
					(if N is Av. const.)

Again, Avogadro's principle states that under the same conditions of temperature and pressure, equal volumes of gases contain the same number of molecules. Thus, for homogeneous gaseous reactions, the stoichiometric coefficients of the chemical equation also signify the relative volumes of each reactant and product under the same conditions of temperature and pressure, e.g.,

	H_2 (g)	+	I_2 (g)	$\rightarrow$	$2\,HI$ (g)
	1 molecule		1 molecule		2 molecules
or	1 mole		1 mole		2 moles
or	1 volume		1 volume		2 volumes (T & p constant)
or	1 pressure		1 pressure		2 pressures (T & V constant)

The coefficients, however, do not represent the relative weights of reactants and products, which is due to the fact that **1 mole is a fixed number of particles but not a fixed weight.** It is to be noted that the coefficients do not even represent the relative number of equivalents (eq.) or milli-equivalents (m.e.) of reactants and products. In any reaction the number of equivalents or milli-equivalents of each reactant and product is the same, e.g.,

	$2\,KClO_3$	$\rightarrow$	$2\,KCl$	+	$3\,O_2$
	no. of equivalents of $KClO_3$	=	no. of equivalents of KCl	=	no. of equivalents of O_2
or	no. of m.e. of $KClO_3$	=	no. of m.e. of KCl	=	no. of m.e. of O_2

Principle of Atom Conservation (POAC)

The principle of conservation of mass, expressed in the concepts of atomic theory, means the conservation of atoms. And if atoms are conserved, moles of atoms shall also be conserved. This is known as the principle of atom conservation. This principle is in fact the basis of the mole concept.

In order to solve problems of nearly all stoichiometric calculations, let us first see how this principle works. Choose an example,

$$KClO_3\ (s)\ \rightarrow\ KCl\ (s) + O_2\ (g)$$

Apply the principle of atom conservation (POAC) for K atoms.

Moles of K atoms in reactant = moles of K atoms in products

or moles of K atoms in $KClO_3$ = moles of K atoms in KCl.

Now, since 1 molecule of $KClO_3$ contains 1 atom of K

or 1 mole of $KClO_3$ contains 1 mole of K, similarly, 1 mole of KCl contains 1 mole of K.

Thus, moles of K atoms in $KClO_3 = 1 \times$ moles of $KClO_3$

and moles of K atoms in KCl $= 1 \times$ moles of KCl.

$$\therefore \quad \text{moles of } KClO_3 = \text{moles of KCl}$$

$$\text{or} \quad \frac{\text{wt. of } KClO_3 \text{ in g}}{\text{mol. wt. of } KClO_3} = \frac{\text{wt. of KCl in g}}{\text{mol. wt. of KCl}}. \qquad \text{(Rule 1)}$$

The above equation gives the weight relationship between $KClO_3$ and KCl which is important in stoichiometric calculations.

Again, applying the principle of atom conservation for O atoms,

moles of O in $KClO_3$ = moles of O in O_2.

But since 1 mole of $KClO_3$ contains 3 moles of O and 1 mole of O_2 contains 2 moles of O,

thus, moles of O in $KClO_3 = 3 \times$ moles of $KClO_3$

moles of O in $O_2 = 2 \times$ moles of O_2

$$\therefore \quad 3 \times \text{moles of } KClO_3 = 2 \times \text{moles of } O_2$$

$$\text{or} \quad 3 \times \frac{\text{wt. of } KClO_3}{\text{mol. wt. of } KClO_3} = 2 \times \frac{\text{vol. of } O_2 \text{ at NTP}}{\text{standard molar vol.}}.$$

(Rules 1 and 3)

The above equation thus gives the weight–volume relationship of reactants and products.

From the above discussion we see that the procedure to use the mole method is first to set up an equation based on the principle of atom conservation and then apply rules 1 to 6 to it.

Advantages of the Mole Method over other Methods

The advantages of the Mole Method are as under:

(1) Balancing of chemical equations is not required in the majority of problems as the method of balancing the chemical equation is based on the principle of atom conservation.

(2) Number of reactions and their sequence, leading from reactants to products, need not be given.

(3) It is a general method, applicable in solving many types of problems as may be seen in different chapters.

[**Note:** The students should carefully note that POAC should be applied for only those atoms which remain conserved in a chemical reaction.]

Let us first solve some elementary problems based on the definition of mole.

Elementary Chemical Calculations

Ex. 1. *Calculate the weight of* $6{\cdot}022 \times 10^{23}$ *molecules of* $CaCO_3$.

Solution : No. of moles of $CaCO_3 = \dfrac{\text{no. of molecules}}{\text{Av. cons.}}$ (Rule 4)

$$= \frac{6{\cdot}022 \times 10^{23}}{6{\cdot}022 \times 10^{23}} = 1.$$

Weight of $CaCO_3$ = no. of moles × molecular wt. (Rule 1)

$$= 1 \times 100 = 100 \text{ g}.$$

Ex. 2. *Calculate the weight of* $12{\cdot}044 \times 10^{23}$ *atoms of carbon.*

Solution : No. of moles of $C = \dfrac{12{\cdot}044 \times 10^{23}}{6{\cdot}022 \times 10^{23}} = 2.$ (Rule 4)

Wt. of C atoms = no. of moles × at. wt. (Rule 2)

$$= 2 \times 12 = 24 \text{ g}.$$

Ex. 3. *What will be the number of oxygen atoms in* 1 *mole of* O_2*?*

Solution : 1 mole of O_2 contains $6{\cdot}022 \times 10^{23}$ molecules of oxygen, and since the oxygen molecule is diatomic, i.e., 1 molecule contains 2 atoms, the no. of oxygen atoms in 1 mole of O_2 is equal to $2 \times 6{\cdot}022 \times 10^{23}$.

Ex. 4. *A piece of* Cu *weighs* 0·635 g. *How many atoms of* Cu *does it contain?*

Solution : No. of moles of $Cu = \dfrac{\text{wt. of Cu}}{\text{at. wt. of Cu}}$ (Rule 2)

$$= \frac{0{\cdot}635}{63{\cdot}5} = 0{\cdot}01.$$

No. of atoms of Cu = no. of moles of atoms × Av. const. (Rule 4)

$$= 0{\cdot}01 \times 6{\cdot}022 \times 10^{23} = 6{\cdot}022 \times 10^{21}.$$

Ex. 5. *Calculate the number of molecules in* 11·2 *litres of* SO_2 *gas at* NTP.

Solution : No. of moles of $SO_2 = \dfrac{\text{vol. at NTP (litres)}}{\text{std. molar volume (litres)}}$ (Rule 3)

$$= \frac{11{\cdot}2}{22{\cdot}4} = 0{\cdot}5.$$

No. of molecules of SO_2 = no. of moles × Av. const. (Rule 4)

$$= 0{\cdot}5 \times 6{\cdot}022 \times 10^{23}$$
$$= 3{\cdot}011 \times 10^{23}.$$

Ex. 6. *One atom of an element* X *weighs* $6{\cdot}644 \times 10^{-23}$ g. *Calculate the number of gram-atoms in* 40 kg *of it.*

Solution : Wt. of 1 mole of atoms of X = wt. of 1 atom × Av. const.

$$= 6{\cdot}644 \times 10^{-23} \times 6{\cdot}022 \times 10^{23}$$

$$= 40 \text{ g.}$$

Thus the at. wt. of X = 40.

No. of moles (or gram-atoms) of $X = \dfrac{\text{weight in g}}{\text{at. wt.}}$ (Rule 2)

$$= \frac{40 \times 1000}{40} = 1000.$$

Ex. 7. *From* 200 mg *of* CO_2, 10^{21} *molecules are removed. How many moles of* CO_2 *are left?*

Solution : Total no. of moles of $CO_2 = \dfrac{\text{wt. in g}}{\text{mol. wt.}}$ (Rule 1)

$$= \frac{0{\cdot}2}{44} = 0{\cdot}00454.$$

No. of moles removed $= \dfrac{10^{21}}{6{\cdot}022 \times 10^{23}} = 0{\cdot}00166.$ (Rule 4)

No. of moles of CO_2 left $= 0{\cdot}00454 - 0{\cdot}00166 = 0{\cdot}00288.$

Ex. 8. *What will be the volume occupied by* 1 *mole atom of a (i) monoatomic gas, and (ii) diatomic gas at* NTP?

Solution : For monoatomic gases,

1 mole atom occupies 22·4 litres at NTP, and for diatomic gases,

1 mole atom occupies 11·2 litres at NTP, as 1 mole of O_2 contains 2 moles of O.

Ex. 9. *Calculate the volume of* 20 g *of hydrogen gas at* NTP.

Solution : Moles of hydrogen gas $= \dfrac{20}{2} = 10.$ (Rule 1)

Volume of the gas at NTP = no. of moles × 22·4

$$= 10 \times 22{\cdot}4 = 224 \text{ litres.}$$

Ex. 10. *What volume shall be occupied by* $6{\cdot}022 \times 10^{23}$ *molecules of any gas at* NTP?

Solution : $6{\cdot}022 \times 10^{23}$ molecules are contained in 1 mole, and 1 mole, of any gas at NTP occupies a volume of 22·4 litres.

Ex. 11. *Calculate the number of atoms present in* 5·6 *litres of a (i) monoatomic, and (ii) diatomic gas at* NTP.

Solution : No. of moles of the gas at NTP $= \dfrac{5{\cdot}6}{22{\cdot}4} = \dfrac{1}{4}\cdot$ (Rule 3)

$\therefore$ no. of molecules of the gas $= \frac{1}{4} \times 6{\cdot}022 \times 10^{23}$ (Rule 4)

$= 1{\cdot}5 \times 10^{23}$.

Now, if the gas is monoatomic, the no. of atoms of the gas

= no. of molecules = $1{\cdot}5 \times 10^{23}$.

And if the gas is diatomic,

no. of atoms $= 2 \times$ no. of molecules $= 2 \times 1{\cdot}5 \times 10^{23}$

$= 3{\cdot}0 \times 10^{23}$.

Ex. 12. *Calculate the number of sulphate ions in* 100 mL *of* 0·001 M H_2SO_4 *solution.*

Solution : No. of moles = molarity × volume in litres (Rule 5)

$= 0{\cdot}001 \times 0{\cdot}1 = 0{\cdot}0001$.

Now,

1 molecule of H_2SO_4 contains 1 SO_4^{2-} ion.

$\therefore$ 1 mole of H_2SO_4 contains 1 mole of SO_4^{2-}.

$\therefore$ 0·0001 mole of H_2SO_4 contains 0·0001 mole of SO_4^{2-}.

$\therefore$ number of sulphate ions = moles of ions × Av. const.

$= 0{\cdot}0001 \times 6{\cdot}022 \times 10^{23} = 6{\cdot}022 \times 10^{19}$. (Rule 4)

Ex. 13. *How many atoms are there in* 100 amu *of* He?

Solution : We know that, 1 amu $= \frac{1}{12} \times$ weight of one ^{12}C atom

or weight of one ^{12}C atom = 12 amu (at. wt. of C = 12 amu).

Similarly, as the atomic weight of He is 4 amu,

weight of one He atom = 4 amu.

Thus, the number of atoms in 100 amu of He $= \frac{100}{4} = 25$.

Ex. 14. *If a mole were to contain* 1×10^{24} *particles, what would be the mass of (i) one mole of oxygen, and (ii) a single oxygen molecule?*

Solution : Mass of 1 mole of O_2 is its molecular weight in g, i.e., 32 g.

$\therefore$ mass of a single O_2 molecule $= \frac{\text{mass of 1 mole}}{\text{no. of molecules in 1 mole}}$

$= \frac{32}{1 \times 10^{24}} = 3{\cdot}2 \times 10^{-23}$ g.

Ex. 15. *The density of* O_2 *at* NTP *is* 1·429 g/L. *Calculate standard molar volume of the gas.*

Solution : Standard molar volume is the volume occupied by 1 mole of the gas at NTP. Now,

1·429 g of O_2 at NTP occupies a volume of 1 litre.

$$\therefore \quad 32 \text{ g (i.e., 1 mole) of } O_2 \text{ occupies a volume} = \frac{32}{1{\cdot}429} \text{ L}$$
$$= 22{\cdot}39 \text{ litres.}$$

Ex. 16. *The measured density of* He *at* NTP *is* 0·1784 g/L. *What is the weight of* 1 *mole of it?*

Solution : Weight of 1 mole of He = density (g/L) at NTP × standard molar volume (litres)

$$= (0{\cdot}1784 \times 22{\cdot}4) \text{ g}$$
$$\approx 4 \text{ g.}$$

[**Note**: Weight of 1 mole is the molecular weight in g.]

Ex. 17. *A metal* M *of atomic weight* 54·94 *has a density of* 7·42 g/cc. *Calculate the apparent volume occupied by one atom of the metal.*

Solution : 7·42 g of metal M occupies a volume of 1 cc.

$$\therefore \quad 54{\cdot}94 \text{ g occupies a volume of} = \frac{54{\cdot}94}{7{\cdot}42} = 7{\cdot}404 \text{ cc.}$$

Since the weight of 1 mole of atoms is the atomic weight and 1 mole of atoms contain the Avogadro constant of atoms,

$$\text{volume occupied by 1 atom} = \frac{\text{vol. of 1 mole}}{\text{Av. const.}}$$
$$= \frac{7{\cdot}404}{6{\cdot}022 \times 10^{23}} = 1{\cdot}23 \times 10^{-23} \text{ cc.}$$

Ex. 18. *Calculate the number of moles, and the number of atoms of* H, S *and* O *in* 5 *moles of* H_2SO_4.

Solution : 1 molecule of H_2SO_4 contains 2 atoms of H

or 1 mole of H_2SO_4 contains 2 moles of H

or 5 moles of H_2SO_4 contain 10 moles of H

1 mole of H_2SO_4 contains 1 mole of S

or 5 moles of H_2SO_4 contain 5 moles of S

and again,

1 mole of H_2SO_4 contains 4 moles of O

or 5 moles of H_2SO_4 contain 20 moles of O.

$$\left.\begin{aligned} &\text{No. of atoms of H} = 10 \times 6{\cdot}022 \times 10^{23} \\ &\text{No. of atoms of S} = 5 \times 6{\cdot}022 \times 10^{23} \\ &\text{No. of atoms of O} = 20 \times 6{\cdot}022 \times 10^{23} \end{aligned}\right\} \quad \text{(Rule 4)}$$

Ex. 19. *Calculate the number of oxygen atoms and its weight in* 50 g *of* $CaCO_3$.

Solution : No. of moles of $CaCO_3 = \frac{50}{100} = 0{\cdot}5.$ (Rule 1)

(mol. wt. of $CaCO_3 = 100$)

Now,

1 molecule of $CaCO_3$ contains 3 atoms of O

or 1 mole of $CaCO_3$ contains 3 moles of O

or 0·5 mole of $CaCO_3$ contains 1·5 moles of O.

No. of atoms of $O = 1{\cdot}5 \times 6{\cdot}022 \times 10^{23}$ (Rule 4)

$= 9{\cdot}033 \times 10^{23}.$

Wt. of atoms of O = no. of moles × at. wt. of O (Rule 2)

$= 1{\cdot}5 \times 16$

$= 24$ g.

Ex. 20. *Calculate the number of atoms of each element present in* 122·5 g *of* $KClO_3$.

Solution : No. of moles of $KClO_3 = \frac{122{\cdot}5}{122{\cdot}5} = 1.$

(mol. wt. of $KClO_3 = 122{\cdot}5$)

From the formula $KClO_3$, we know that 1 mole of $KClO_3$ contains 1 mole of K atoms, 1 mole of Cl atoms and 3 moles of O atoms.

$$\left.\begin{aligned} \therefore \text{ no. of atoms of } K &= 1 \times 6{\cdot}022 \times 10^{23} \\ \text{no. of atoms of } Cl &= 1 \times 6{\cdot}022 \times 10^{23} \\ \text{no. of atoms of } O &= 3 \times 6{\cdot}022 \times 10^{23} \end{aligned}\right\} \quad \text{(Rule 4)}$$

Ex. 21. *Calculate the total number of electrons present in* 1·6 g *of* CH_4.

Solution : No. of moles of $CH_4 = \frac{1{\cdot}6}{16} = 0{\cdot}1.$ (Rule 1)

No. of molecules in 1·6 g of $CH_4 = 0{\cdot}1 \times 6{\cdot}022 \times 10^{23}$ (Rule 4)

$= 6{\cdot}022 \times 10^{22}$ molecules

$\because$ 1 molecule of CH_4 has 10 electrons.

$\therefore$ $6{\cdot}022 \times 10^{22}$ molecules of CH_4 has $10 \times 6{\cdot}022 \times 10^{22}$ electrons

$= 6{\cdot}022 \times 10^{23}$ electrons.

Ex. 22. *Find the charge of* 1 g-ion *of* N^{3-} *in coulombs.*

Solution : $\because$ 1 ion of N^{3-} carries a charge of $3 \times 1{\cdot}602 \times 10^{-19}$ coulombs

(charge on an electron $= 1{\cdot}602 \times 10^{-19}$ coulombs)

$\therefore$ $6{\cdot}022 \times 10^{23}$ ions (1 g-ion or 1 mole) carry a charge of

$$3 \times 1{\cdot}602 \times 10^{-19} \times 6{\cdot}022 \times 10^{23} \text{ coulombs}$$
$$= 2{\cdot}894 \times 10^{5} \text{ coulombs.}$$

Ex 23. *Find the charge of 27 g of* Al^{3+} *ions in coulombs.*

Solution : One Al^{3+} ion has the charge of 3 protons, and a proton has the same magnitude of charge as that on an electron.

$$\text{No. of moles of } Al^{3+} \text{ ions} = \frac{\text{wt. in g}}{\text{at. wt.}} \quad \text{(Rule 1)}$$
$$= \frac{27}{27} = 1.$$

$$\text{No. of } Al^{3+} \text{ ions in 27 g} = \text{no. of moles} \times \text{Av. const.} \quad \text{(Rule 4)}$$
$$= 1 \times 6{\cdot}022 \times 10^{23}.$$

$$\text{Charge of 27 g of } Al^{3+} \text{ ions} = 3 \times \text{charge of a proton} \times \text{no. of } Al^{3+} \text{ ions}$$
$$= 3 \times 1{\cdot}602 \times 10^{-19} \times 6{\cdot}022 \times 10^{23}$$
$$= 2{\cdot}894 \times 10^{5} \text{ coulombs.}$$

Ex. 24. *Equal masses of oxygen, hydrogen and methane are taken in a container in identical conditions. Find the ratio of the volumes of the gases.*

Solution : Suppose each gas has a mass of X g.

Therefore,	O_2 : H_2 : CH_4	
Weight –	X X X	
No. of moles –	$\frac{X}{32}$ $\frac{X}{2}$ $\frac{X}{16}$	(Rule 1)
Volume ratio –	$\frac{X}{32} : \frac{X}{2} : \frac{X}{16}$	(Avogadro's principle)

Hence, $O_2 : H_2 : CH_4 = 1 : 16 : 2$.

Ex. 25. *If the components of air are* N_2, 78%; O_2, 21%; Ar, 0·9% *and* CO_2, 0·1% *by volume, what would be the molecular weight of air?*

Solution : The volume ratio of the gases will be the same as their mole ratio (Avogadro's principle)

$$\therefore \quad \text{mol. wt. of air} = \frac{78 \times 28 + 21 \times 32 + 0{\cdot}9 \times 40 + 0{\cdot}1 \times 44}{78 + 21 + 0{\cdot}9 + 0{\cdot}1}$$

(wt. in g per mole)

$$= 28{\cdot}964.$$

$(N_2 = 28,\ O_2 = 32,\ Ar = 40 \text{ and } CO_2 = 44)$

Ex. 26. *The atomic weights of two elements* (A *and* B) *are* 20 *and* 40 *respectively. If x* g *of* A *contains y atoms, how many atoms are present in* 2x g *of* B?

Solution : Number of moles of $A = \frac{x}{20}$. (Rule 2)

Number of atoms of $A = \frac{x}{20} \times N$. (Rule 4)

(N is the Av. const.)

$\therefore \quad y = \frac{x}{20} \times N; \quad x = \frac{20y}{N}$.

Further,

number of moles of $B = \frac{2x}{40}$,

number of atoms of $B = \frac{2x}{40} \times N$

$$= \frac{2N}{40} \times \frac{20y}{N}$$

$$= y.$$

Ex. 27. *Oxygen is present in a 1-litre flask at a pressure of* $7{\cdot}6 \times 10^{-10}$ mmHg *at* 0°C. *Calculate the number of oxygen molecules in the flask.*

Solution : Pressure $= 7{\cdot}6 \times 10^{-10}$ mm

$$= 0{\cdot}76 \times 10^{-10} \text{ cm}$$

$$= \frac{0{\cdot}76 \times 10^{-10}}{76} \text{ atm} \quad (1 \text{ atm} = 76 \text{ cm})$$

$$= 10^{-12} \text{ atm.}$$

Volume = 1 litre, $R = 0{\cdot}0821$ lit. atm/K/mole, temperature = 273 K.

We know that $pV = nRT$ or $n = \frac{pV}{RT}$

$$n = \frac{10^{-12} \times 1}{0{\cdot}0821 \times 273} = 0{\cdot}44 \times 10^{-13}.$$

No. of molecules = no. of moles × Av. const. (Rule 4)

$$= 0{\cdot}44 \times 10^{-13} \times 6{\cdot}022 \times 10^{23} = 2{\cdot}65 \times 10^{10}.$$

Ex. 28. *The density of mercury is* 13·6 g/cc. *Calculate approximately the diameter of an atom of mercury, assuming that each atom is occupying a cube of edge length equal to the diameter of the mercury atom.*

Solution : Suppose the length of the side of the cube is X cm, i.e., the diameter of one Hg atom.

$\therefore$ volume occupied by 1 Hg atom $= X^3$ cc

and mass of one Hg atom $= 13{\cdot}6 \times X^3$ g.

Mass of one Hg atom $= \frac{\text{at. wt.}}{\text{Av. const.}} = \frac{200}{6{\cdot}022 \times 10^{23}}$ g.

($\because$ mass of 1 mole of atoms is the atomic weight in g, and 1 mole contains the Av. const. of atoms)

$$\text{Hence,}\quad 13{\cdot}6 \times X^3 = \frac{200}{6{\cdot}022 \times 10^{23}}$$

$$X^3 = \frac{200}{13{\cdot}6 \times 6{\cdot}022 \times 10^{23}} = 2{\cdot}44 \times 10^{-23}.$$

$$X = 2{\cdot}9 \times 10^{-8} \text{ cm.}$$

Ex. 29. *The density of a particular crystal of* LiF *is* 2·65 g/cc. *X-ray analysis shows that* Li^+ *and* F^- *ions are arranged in a cubic array at a spacing of* 2·01 Å. *From these data calculate the apparent Avogadro constant.*
[Li = 6·939, F = 18·998 (1 Å = 10^{-8} cm.)]

Solution : In this problem, we have to calculate Av. const., i.e., the number of LiF molecules present in 1 mole. As 1 mole of LiF weighs 25·937 g (LiF = 6·939 + 18·998),

$$\text{volume of 1 mole} = \frac{\text{wt. of 1 mole}}{\text{wt. per cc (density)}} = \frac{25{\cdot}937}{2{\cdot}65}$$

$$= 9{\cdot}78 \text{ cc.}$$

Since this volume is supposed to be of a cube,

$$\text{the length of each edge of the cube} = \sqrt[3]{9{\cdot}78}$$

$$= 2{\cdot}138 \text{ cm.}$$

$$\text{No. of ions present in one edge} = \frac{2{\cdot}138}{2{\cdot}01 \times 10^{-8}}$$

$$= 1{\cdot}063 \times 10^{8}.$$

$$\text{No. of ions } (Li^+ + F^-) \text{ present in the cube} = (1{\cdot}063 \times 10^8)^3$$

$$= 1{\cdot}201 \times 10^{24}.$$

$$\text{No. of LiF molecules per mole (i.e., Avogadro constant)} = \frac{1{\cdot}201 \times 10^{24}}{2}$$

$$= 6{\cdot}01 \times 10^{23}.$$

Ex. 30. *To* 50 mL of 0·5 M H_2SO_4, 75 mL *of* 0·25 M H_2SO_4 *is added. What is the concentration of the final solution if its volume is* 125 mL?

Solution : No. of moles in 0·05 litre of H_2SO_4 = M × V (litre) (Rule 5)

$= 0{\cdot}5 \times 0{\cdot}05 = 0{\cdot}025.$

No. of moles in 0·075 litre of H_2SO_4 added = $0{\cdot}25 \times 0{\cdot}075$

$= 0{\cdot}01875.$ (Rule 5)

Total no. of moles in 0·125 litre of H_2SO_4 = 0·025 + 0·01875 (Rule 5)

$= 0{\cdot}04375.$

$\therefore$ molarity of $H_2SO_4 = \dfrac{0{\cdot}04375}{0{\cdot}125} = 0{\cdot}35$ M.

Ex. 31. *It is found that in* 11·2 *litres of any gaseous compound of phosphorus at* NTP, *there is never less than* 15·5 g *of* P. *Also, this volume of the vapour of phosphorus itself at* NTP *weighs* 62 g. *What should be the atomic weight and molecular weight of phosphorus?*

Solution : Since 1 mole of any phosphorus compound contains at least 1 mole of P atoms, and also 22·4 litres (at NTP) of the gaseous compound contains 31 g of phosphorus, therefore, 1 mole of phosphorus weighs 31 g, i.e., 31 is the atomic weight of phosphorus. Similarly, the molecular weight of phosphorus is 124 as 1 mole of its vapour weighs 124 g.

Ex. 32. *A polystyrene, having the formula* $Br_3C_6H_3(C_3H_8)_n$, *was prepared by heating styrene with tribromobenzoyl peroxide in the absence of air. If it was found to contain* 10·46% *bromine by weight, find the value of n.*

Solution : Let the weight of polystyrene prepared be 100 g.

$\therefore$ no. of moles of Br in 100 g of polystyrene $= \dfrac{10{\cdot}46}{79{\cdot}9} = 0{\cdot}1309$.

From the formula of polystyrene, we have,

no. of moles of Br $= 3 \times$ moles of $Br_3C_6H_3(C_3H_8)_n$

or $\quad 0{\cdot}1309 = 3 \times \dfrac{\text{wt.}}{\text{mol. wt.}} = \dfrac{3 \times 100}{314{\cdot}7 + 44n}$.

$\therefore \quad n = 44.9 \approx 45.$

Ex. 33. *It has been estimated that* 93% *of all atoms in the entire universe are hydrogen and that the vast majority of those remaining are helium. Based on only these two elements, estimate the mass percentage composition of the universe.*

Solution : Given that out of 100 atoms of H and He, 93 atoms are of H and 7 atoms are of He, that is, the number of moles of H and He atoms, out of 100 moles, are 93 and 7 respectively.

Mass of H $= 93 \times 1 = 93$ g.

Mass of He $= 7 \times 4 = 28$ g.

$\therefore$ mass percentage of H $= \dfrac{93}{93 + 28} \times 100 = 76.86\%$.

$\therefore$ mass percentage of He $= 23.14\%$.

Ex. 34. *The molecular weight of haemoglobin is about* 65,000 g/mol. *Haemoglobin contains* 0.35% Fe *by mass. How many iron atoms are there in a haemoglobin molecule?*

Solution : 100 g of haemoglobin contain 0.35 g of iron

or, $\frac{100}{65000}$ mole of haemoglobin contains $\frac{0.35}{56}$ mole of Fe.

$\therefore$ 1 mole of haemoglobin contains $\frac{0.35}{56} \times \frac{65000}{100}$ moles of Fe.

$= 4.06$ moles of Fe.

Thus one molecule of haemoglobin contains four iron atoms.

Ex. 35. *At room temperature, the density of water is* 1.0 g/mL *and the density of ethanol is* 0.789 g/mL. *What volume of ethanol contains the same number of molecules as are present in* 175 mL *of water?*

Solution : Let the volume of ethanol containing the same number of molecules as are present in 175 mL of H_2O be v mL. As given,

moles of C_2H_5OH in v mL = moles of H_2O in 175 mL

Now, $$\frac{\text{wt. of } C_2H_5OH}{\text{mol. wt. of } C_2H_5OH} = \frac{\text{wt. of } H_2O}{\text{mol. wt. of } H_2O}$$

or $$\frac{0.789 \times v}{46} = \frac{1.0 \times 175}{18}.$$

$\therefore$ $$v = 566.82 \text{ mL}.$$

Ex. 36. *A sample of ethane has the same mass as* 10.0 *million molecules of methane. How many* C_2H_6 *molecules does the sample contain?*

Solution : Let the number of C_2H_6 molecules in the sample be n. As given,

mass of C_2H_6 = mass of 10^7 molecules of CH_4

$$\frac{n}{\text{Av. const.}} \times \text{mol. wt. of } C_2H_6 = \frac{10^7}{\text{Av. const.}} \times \text{mol. wt. of } CH_4$$

$$\frac{n \times 30}{\text{Av. const.}} = \frac{10^7 \times 16}{\text{Av. const.}}.$$

$\therefore$ $$n = 5.34 \times 10^6.$$

Ex. 37. *How many grams of* $CaWO_4$ *would contain the same mass of tungsten that is present in* 569 g *of* $FeWO_4$? (W = 184)

Solution : Let the mass of $CaWO_4$ be w g. As given,

mass of W in w g of $CaWO_4$ = mass of W in 569 g of $FeWO_4$.

Moles of W in $CaWO_4 \times$ at. wt. of W

= moles of W in $FeWO_4 \times$ at. wt. of W.

As both $CaWO_4$ and $FeWO_4$ contain 1 atom of W each,

$\therefore$ moles of $CaWO_4 \times$ at. wt. of W = moles of $FeWO_4 \times$ at. wt. of W

$$\frac{w}{288} \times 184 = \frac{569}{304} \times 184$$

$$w = 539.05 \text{ g.}$$

Ex. 38. 0.75 *mole of solid 'A_4' and 2 moles of gaseous O_2 are heated in a sealed vessel, completely using up the reactants and producing only one compound. It is found that when the temperature is reduced to the initial temperature, the contents of the vessel exhibit a pressure equal to half the original pressure. What conclusions can be drawn from these data about the product of the reaction?*

***Solution* : First Method :** As both A_4 and O_2 are consumed

$$\frac{\text{moles of } A_4}{\text{moles of } O_2} = \frac{0.75}{2.0}$$

$$\frac{(\text{moles of A})/4}{(\text{moles of O})/2} = \frac{0.75}{2.0}$$

$$\frac{\text{moles of A}}{\text{moles of O}} = \frac{3}{4}.$$

Thus, the empirical formula of the product is A_3O_4. Further, as 2 moles of O_2 give 1 mole of A_3O_4 (for gases, pressure $\propto$ mole at constant temperature and volume), A_3O_4 is also the molecular formula of the product.

Second Method : $\underset{0.75 \text{ mole}}{A_4(s)} + \underset{2.0 \text{ moles}}{O_2(g)} \rightarrow \underset{1.0 \text{ mole}}{A_xO_y(g)}$

Applying POAC for A atoms,

$$4 \times \text{moles of } A_4 = x \times \text{moles of } A_xO_y$$

$$4 \times 0.75 = x \times 1; \; x = 3$$

Applying POAC for O atoms,

$$2 \times \text{moles of } O_2 = y \times \text{moles of } A_xO_y$$

$$2 \times 2 = y \times 1; \; y = 4.$$

In the following chapters, we shall apply the principle of atom conservation (POAC) along with the said rules in tackling the various problems encountered in chemical practice.

PROBLEMS

(Answers bracketed with questions)

1. Find the number of atoms in 48 g of ozone at NTP. ($1{\cdot}8066 \times 10^{24}$)

2. What is the ratio of the volumes occupied by 1 mole of O_2 and 1 mole of O_3 in identical conditions? (1 : 1)

3. Calculate the mass of 5 moles of $CaCO_3$ in grams. (500 g)

4. The vapour density of a gas is 11·2. Calculate the volume occupied by 11·2 g of the gas at NTP.
[Hint: Mol. wt. = $2 \times$ VD] (11·2 litres)

5. Calculate the number of oxygen atoms in 0·2 mole of $Na_2CO_3 \cdot 10H_2O$. ($1{\cdot}56 \times 10^{24}$)

6. Calculate the number of moles of $CuSO_4$ contained in 100 mL of 1 M $CuSO_4$ solution. Also, find the number of SO_4^{2-} ions in it. (0·1 mole, 0.6022×10^{23})

7. Find the total number of nucleons present in 12 g of ^{12}C atoms. ($12 \times 6{\cdot}022 \times 10^{23}$)

8. Find (i) the total number of neutrons, and (ii) the total mass of neutrons in 7 mg of ^{14}C. (Assume that the mass of a neutron = mass of a hydrogen atom)
[**Hint:** 1 ^{14}C atom contains 8 neutrons.] ($24{\cdot}088 \times 10^{20}$, 0·004 g)

9. How many moles are there in 1 metre^3 of any gas at NTP? (1 m^3 = 10^3 litres) (44·6 moles)

10. 3 g of a salt of molecular weight 30 is dissolved in 250 g of water. Calculate the molality of the solution. (0·4 *m*)

11. Calculate the volume occupied by 5·25 g of nitrogen at 26°C and 74·2 cm of pressure. (4·71 litres)

12. Find the ratio of the number of molecules contained in 1 g of NH_3 and 1 g of N_2. (28 : 17)

13. How many molecules of CO_2 are contained in one litre of air if the volume content of CO_2 is 0·03 % at NTP? ($8{\cdot}06 \times 10^{18}$)

14. Is the number of molecules in 1 kg of H_2 and 1 kg of O_2 the same ? What is the ratio of weights of H_2 and O_2, the mixture of which contains equal number of molecules of each gas? (no, 1 : 16)

15. The measured density at NTP of a gaseous sample of a compound was found to be 1·78 g/L. What is the weight of 1 mole of the gaseous sample? (39·9 g)

16. If the concentration of a solution is 2 M calculate the number of millimoles present in 2 litres of the solution.
[**Hint:** Follow Rule 5.] (4000)

17. How many moles of oxygen are contained in one litre of air if its volume content is 21% at NTP? (0·0093)

18. How many atoms do mercury vapour molecules consist of if the density of mercury vapour relative to air is 6·92? (Hg = 200). The average mass of air is 29 g/mole. (One)

19. Calculate the total number of atoms in 0·5 mole of $K_2Cr_2O_7$.
[**Hint:** Follow Example 19] ($3{\cdot}31 \times 10^{24}$)

20. What is the volume of 6 g of hydrogen at 1 atm and 0°C? (67·2 litres)

21. What is the density of oxygen gas at NTP?
[**Hint:** See Example 14] (1·429 g/L)

22. Calculate the total number of electrons present in 18 mL of water. ($10 \times 6{\cdot}022 \times 10^{23}$)

23. Calculate the number of electrons, protons and neutrons in 1 mole of $^{16}O^{-2}$ ions. ($10 \times 6{\cdot}022 \times 10^{23}$, $8 \times 6{\cdot}022 \times 10^{23}$, $8 \times 6{\cdot}022 \times 10^{23}$)

24. Find the mass of the nitrogen contained in 1 kg of (i) KNO_3, (ii) NH_4NO_3, and (iii) $(NH_4)_2$ HPO_4. [(i) 138·5 g (ii) 350 g and (iii) 212 g]

25. Find the mass of each element in 7·84 g of $FeSO_4 \cdot (NH_4)_2SO_4 \cdot 6H_2O$. What will be the volume of O_2 at NTP in this sample? (3·136 litres)

26. The density of solid AgCl is 5·56 g/cc. The solid is made up of a cubic array of alternate Ag^+ and Cl^- ions at a spacing of 2·773 Å between centres. From these data calculate the Avogadro constant. $(6{\cdot}04 \times 10^{23})$

27. Three atoms of magnesium combine with 2 atoms of nitrogen. What will be the weight of magnesium which combines with 1·86 g of nitrogen?
[Hint: Use Rule 6] (4·86 g)

28. 600 mL of a mixture of O_3 and O_2 weighs 1 g at NTP. Calculate the volume of ozone in the mixture. (200 mL)

29. The vapour density (hydrogen = 1) of a mixture consisting of NO_2 and N_2O_4 is 38·3 at 26·7°C. Calculate the number of moles of NO_2 in 100 g of the mixture.
[Hint: Wt. of $NO_2 = x$ g.

$$\therefore \text{ obs. mol. wt. (wt./mole)} = \frac{\text{wt. in g}}{\text{total moles}} = \frac{100}{\left(\frac{x}{46} + \frac{100 - x}{92}\right)} = 2 \times 38{\cdot}3.\Big]$$

(0·437 mole)

30. A nugget of gold and quartz weighs 100 g. Sp. gr. of gold, quartz and the nugget are 19·3, 2·6 and 6·4 respectively. Calculate the weight of gold in the nugget.

[Hint: $\frac{x}{19{\cdot}3} + \frac{100 - x}{2{\cdot}6} = \frac{100}{6{\cdot}4}$; x = wt. of gold] (68·6 g)

31. The nucleus of an atom of X is supposed to be a sphere with a radius of 5×10^{-13} cm. Find the density of the matter in the atomic nucleus if the atomic weight of X is 19.

$$\left[\textbf{Hint: } \text{Density} = \frac{\text{mass of 1 mole (i.e., at. wt.)}}{\text{vol. of 1 mole}}\right]$$

$(6{\cdot}02 \times 10^{13}$ g/mL)

32. Copper forms two oxides. For the same amount of copper, twice as much oxygen was used to form the first oxide than to form the second one. What is the ratio of the valencies of copper in the first and second oxides?
[Hint: Assume that the oxides are Cu_2O_x and Cu_2O_y and apply Rule 6] (2 : 1)

33. 105 mL of pure water (4°C) is saturated with NH_3 gas, producing a solution of density 0·9 g/mL. If this solution contains 30% of NH_3 by weight, calculate its volume.

[Hint: Density $= \frac{\text{total mass}}{\text{total volume}} = \frac{105\,(H_2O) + 45\,(NH_3)}{V}$] (166·67 mL)

34. How many iron atoms are present in a stainless steel ball bearing having a radius of 0·1 inch (1 inch = 2·54 cm)? The stainless steel contains 85·6% Fe by weight and has a density of 7·75 g/cc. $(4{\cdot}91 \times 10^{21})$

35. How many litres of liquid CCl_4 ($d = 1{\cdot}5$ g/cc) must be measured out to contain 1×10^{25} CCl_4 molecules? (1·61 L)

36. A sample of potato starch was ground in a ball mill to give a starchlike molecule of lower molecular weight. The product analysed 0·086% phosphorus. If each molecule is assumed to contain one atom of phosphorus, what is the molecular weight of the material? ($3{\cdot}6 \times 10^4$ amu)

37. The dot at the end of this sentence has a mass of about one microgram. Assuming that the black stuff is carbon, calculate the approximate number of atoms of carbon needed to make such a dot. (1 microgram $= 1 \times 10^{-6}$ g) (5×10^{16} atoms)

38. To what volume must 50 mL of 3·50 M H_2SO_4 be diluted in order to make 2 M H_2SO_4?
[**Hint:** Use Rule 5] (87·5 mL)

39. Sulphur molecules exist under various conditions as S_8, S_6, S_4, S_2 and S.
(a) Is the mass of one mole of each of these molecules the same?
(b) Is the number of molecules in one mole of each of these molecules the same?
(c) Is the mass of sulphur in one mole of each of these molecules the same?
(d) Is the number of atoms of sulphur in one mole of each of these molecules the same? (No, Yes, No, No)

40. Two minerals that contain Cu are $CuFeS_2$ and Cu_2S. What mass of Cu_2S would contain the same mass of Cu as is contained in 125 lb of $CuFeS_2$? (54.2 lb)

41. What is the maximum number of moles of CO_2 that could be obtained from the carbon in 4 moles of $Ru_2(CO_3)_3$? (12 mol)

42. What mass of NaCl would contain the same total number of ions as 245 g of $MgCl_2$? (225 g)

43. An unknown sample weighing 1.5 g was found to contain only Mn and S. The sample was completely reacted with oxygen and it produced 1.22 g of Mn(II) oxide and 1.38 g of SO_3. What is the simplest formula for this compound? (MnS)

44. The two sources of Zn, that is, $ZnSO_4$ and $Zn(CH_3COO)_2 \cdot 2H_2O$, can be purchased at the same price per kilogram of compound. Which is the most economical source of Zn and by how much? ($ZnSO_4$, 35.9%)

45. How many moles of H_2O form when 25.0 mL of 0.10 M HNO_3 solution is completely neutralised by NaOH? ($2{\cdot}5 \times 10^{-3}$ mole)

46. Which would be larger: an atomic mass unit based on the current standard or one based on the mass of a Be-9 atom set at exactly 9 amu? (latter)

47. The enzyme carbonic anhydrase catalyses the hydration of CO_2. This reaction: $CO_2 + H_2O \rightarrow H_2CO_3$, is involved in the transfer of CO_2 from tissues to the lungs via the bloodstream. One enzyme molecule hydrates 10^6 molecules of CO_2 per second. How many kg of CO_2 are hydrated in one hour in one litre by 1×10^{-6} M enzyme? (0.1584 kg)

48. An oxybromo compound, $KBrO_x$, where x is unknown, is analysed and found to contain 52.92% Br. What is the value of x? (2)

49. Radium disintegrates at an average rate of 2.24×10^{13} α-particles per minute. Each α-particle takes up two electrons from the air and becomes a neutral helium atom. After 420 days, helium gas collected was 0.5 mL, measured at 27°C and 750 mmHg. Calculate the Avogadro constant. (6.7×10^{23})

50. If the value of Avogadro number is $6{\cdot}023 \times 10^{23}$ mol^{-1} and the value of Boltzmann constant is $1{\cdot}380 \times 10^{-23}$ JK^{-1}, then the number of significant digits in the calculated value of the universal gas constant is ... . **(IIT 2014 Adv.)** (4)

[Hint: Since k and N, both have four significant figures, the value of R is also rounded off up to four significant figures.]

51. Three moles of B_2H_6 are completely reacted with methanol. The number of moles of boron containing product formed is ... . **(IIT 2015 Adv.)** (2)

Objective Problems

1. The density of chlorine relative to air is

(a) 2·44 (b) 3
(c) found only experimentally (d) 4

2. A gaseous oxide contains 30·4% of nitrogen, one molecule of which contains one nitrogen atom. The density of the oxide relative to oxygen is

(a) 0·94 (b) 1·44 (c) 1·50 (d) 3·0

3. The mass of an oxygen atom is half that of a sulphur atom. Can we decide on this basis that the density of sulphur vapour relative to oxygen is 2?

(a) Yes (b) No

4. Density of air is 0·001293 g/cc. Its vapour density is

(a) 0·001293 (b) 1·293 (c) 14·48 (d) cannot be calculated

5. 5·6 litres of oxygen at NTP is equivalent to

(a) 1 mole (b) $\frac{1}{2}$ mole (c) $\frac{1}{4}$ mole (d) $\frac{1}{8}$ mole

6. 22·4 litres of water vapour at NTP, when condensed to water, occupies an approximate volume of

(a) 18 litres (b) 1 litre (c) 1 mL (d) 18 mL

7. Which of the following has the highest mass?

(a) 1 g-atom of C (b) $\frac{1}{2}$ mole of CH_4
(c) 10 mL of water (d) $3{\cdot}011 \times 10^{23}$ atoms of oxygen

8. $6{\cdot}022 \times 10^{22}$ molecules of N_2 at NTP will occupy a volume of

(a) 22·4 litres (b) 2·24 litres (c) 6·02 litres (d) 6·02 mL

9. How many grams are contained in 1 gram-atom of Na?

(a) 13 g (b) 23 g (c) 1 g (d) $\frac{1}{23}$ g

10. The weight of 350 mL of a diatomic gas at 0°C and 2 atm pressure is 1 g. The wt. of one atom is

(a) $\frac{16}{N}$ (b) $\frac{32}{N}$

(c) 16 N (d) 32 N (N is the Av. const.)

11. The number of atoms present in 16 g of oxygen is

(a) $6{\cdot}02 \times 10^{11{\cdot}5}$ (b) $3{\cdot}01 \times 10^{23}$ (c) $3{\cdot}01 \times 10^{11{\cdot}5}$ (d) $6{\cdot}02 \times 10^{23}$

12. 1 mole of a compound contains 1 mole of C and 2 moles of O. The molecular weight of the compound is

(a) 3 (b) 12 (c) 32 (d) 44

13. The volume of a gas at 0°C and 700 mm pressure is 760 cc. The number of molecules present in this volume is

(a) $1{\cdot}88 \times 10^{22}$ (b) $6{\cdot}022 \times 10^{23}$ (c) $18{\cdot}8 \times 10^{23}$ (d) $18{\cdot}8 \times 10^{22}$

14. 1 mole of a diatomic element X_2 contains 34 and 40 moles of electrons and neutrons respectively. The isotopic formula of the element is

(a) ${}^{74}_{34}X$ (b) ${}^{37}_{17}X$ (c) ${}^{40}_{34}X$ (d) ${}^{40}_{20}X$

15. 2 moles of H atoms at NTP occupy a volume of

(a) 11·2 litres (b) 44·8 litres (c) 2 litres (d) 22·4 litres

16. No. of electrons in 1·8 mL of H_2O (l) is

(a) $6{\cdot}02 \times 10^{23}$ (b) $3{\cdot}011 \times 10^{23}$ (c) $0{\cdot}6022 \times 10^{23}$ (d) $60{\cdot}22 \times 10^{23}$

17. Molecular weight of a gas, 11·2 litres of which at NTP weighs 14 g, is

(a) 14 (b) 28 (c) $\frac{14}{11{\cdot}2}$ (d) $14 \times 11{\cdot}2$

18. The weight of 1 mole of a gas of density 0·1784 g/L at NTP is

(a) 0·1784 g (b) 1 g (c) 4 g (d) cannot be calculated

19. Number of HCl molecules present in 10 mL of 0·1 N HCl solution is

(a) $6{\cdot}022 \times 10^{23}$ (b) $6{\cdot}022 \times 10^{22}$ (c) $6{\cdot}022 \times 10^{21}$ (d) $6{\cdot}022 \times 10^{20}$

20. Number of atoms in 12 g of ${}^{12}_{6}C$ is

(a) 6 (b) 12 (c) $6{\cdot}022 \times 10^{23}$ (d) $12 \times 6{\cdot}022 \times 10^{23}$

21. 5 moles of a gas in a closed vessel was heated from 300 K to 600 K. The pressure of the gas doubled. The number of moles of the gas will be

(a) 5 (b) 2·5 (c) 10 (d) 20

22. Which of the following contains the greatest number of oxygen atoms?

(a) 1 g of O (b) 1 g of O_2

(c) 1 g of O_3 (d) all have the same number of atoms

23. If the atomic weight of carbon were set at 24 amu, the value of the Avogadro constant would be

(a) $6{\cdot}022 \times 10^{23}$ (b) $12{\cdot}044 \times 10^{23}$ (c) $3{\cdot}011 \times 10^{23}$ (d) none of these

24. If 32 g of O_2 contain $6{\cdot}022 \times 10^{23}$ molecules at NTP then 32 g of S, under the same conditions, will contain,

(a) $6{\cdot}022 \times 10^{23}$S (b) $3{\cdot}011 \times 10^{23}$S (c) $12{\cdot}044 \times 10^{23}$S (d) 1×10^{23}S

25. How many moles of electrons weigh one kilogram?

(a) 6.022×10^{23} (b) $\frac{1}{9.108} \times 10^{31}$ (c) $\frac{6.022}{9.108} \times 10^{54}$ (d) $\frac{1}{9.108 \times 6.022} \times 10^{8}$

(IIT 2002)

26. Experimentally it was found that a metal oxide has formula $M_{0.98}O$. Metal M present as M^{2+} and M^{3+} in its oxide. Fraction of the metal which exists as M^{3+} would be

(a) 7.01% (b) 4.08% (c) 6.05% (d) 5.08%

[**Hint:** $M_{0.98}O$ → MO (x, x mole) and M_2O_3 (y, $\frac{3}{2}$y mole) ; $\frac{x+y}{x+\frac{3y}{2}} = 0.98$]

(IIT 2013 Main)

27. The ratio of masses of oxygen and nitrogen of a particular gaseous mixture is 1:4. The ratio of number of molecules is

(a) $\frac{1}{4}$ (b) $\frac{7}{32}$ (c) $\frac{1}{8}$ (d) $\frac{3}{16}$

(IIT 2014 Main)

28. The most abundant amounts by mass in the body of a healthy human adult are oxygen (61.4%), carbon (22.96%) hydrogen (10.0%) and nitrogen (2.6%). The weight which a 75 kg person would gain if all ^1H atoms are replaced by ^2H atoms is

(a) 15 kg (b) 37.5 kg (c) 7.5 kg (d) 10 kg

[**Hint:** Only the mass of H atoms doubles] **(IIT 2017 Main)**

29. 1 g of a carbonate (M_2CO_3) on treatment with excess HCl produces 0.01186 mole of CO_2. The molar mass of M_2O_3 is g mol^{-1} is

(a) 1186 (b) 84.3 (c) 118.6 (d) 11.86

[**Hint:** Apply POAC for C, mole of M_2CO_3 = mol of CO_2] **(IIT 2017 Main)**

Answers

1-a, 2-b, 3-b, 4-c, 5-c, 6-d, 7-a, 8-b, 9-b, 10-a, 11-d, 12-d, 13-a, 14-b, 15-d, 16-a, 17-b, 18-c, 19-d, 20-c, 21-a, 22-d, 23-b, 24-a, 25-d, 26-b, 27-b, 28-c, 29-b.

❑

CHAPTER TWO

PROBLEMS BASED ON EQUATIONS: STOICHIOMETRY

The word 'stoichiometry' is derived from the Greek words *stoicheion,* which means element, and *metrein,* which means to measure. The numerals used to balance a chemical equation are known as stoichiometric coefficients. These numbers are essential for solving problems based on chemical equations. Hence, such problems are also called stoichiometric calculations. The mole method is very useful in such calculations.

For stoichiometric calculations, the mole relationships between different reactants and products are required, as from them, the mass–mass, mass–volume and volume–volume relationships between different reactants and products can be obtained.

For a given balanced equation say,

$$2\,KClO_3 = 2\,KCl + 3\,O_2$$

we can get such relationships directly from the stoichiometric coefficients, e.g.,

$$3 \times \text{moles of } KClO_3 = 2 \times \text{moles of } O_2$$

$$2 \times \text{moles of } KClO_3 = 2 \times \text{moles of } KCl$$

and, $3 \times \text{moles of } KCl = 2 \times \text{moles of } O_2$

For balanced chemical equations, one can also apply the **Factor-Label Method** (Ex. 38 and 39).

The above equations can also be obtained **from an unbalanced equation say**

$$KClO_3 = KCl + O_2$$

by applying POAC for different atoms as explained in Chapter 1.

Another important method used for solving problems based on chemical equations is the **equivalent (or milli-equivalent) method.** This method is based on the fact that for the different amounts of reactants and products involved in the reaction, the number of equivalents (or milli-equivalents) of each reactant and each product are equal.

For the reaction,

$$VO + Fe_2O_3 = FeO + V_2O_5$$

eq. of VO = eq. of Fe_2O_3 = eq. of FeO = eq. of V_2O_5

or, m.e. of VO = m.e. of Fe_2O_3 = m.e. of FeO = m.e. of V_2O_5.

In this method too, balancing of chemical equations is not required. This method is generally applied in volumetric stoichiometric calculations.

Concept of Limiting Reagent

In single-reactant reactions, the calculations are carried out with only that amount of the reactant which has converted to the product. This is done in all the methods mentioned above.

In the reactions where more than one reactant is involved, one has to first identify the **limiting reactant, i.e., the reactant which is completely consumed.** All calculations are to be carried out with the amount of the limiting reactant only. **Now the question is how to identify the limiting reactant?** The procedure is simple as illustrated below:

Initially...	5 moles		12 moles		0 moles
	A	+	2 B	$\rightarrow$	4 C

If A is the limiting reactant: moles of C produced = 20.

If B is the limiting reactant: moles of C produced = 24.

The reactant producing the least number of moles of the product is the limiting reactant and hence A is the limiting reactant. Thus,

Initially ...	5 moles		12 moles		0 moles
	A	+	2 B	$\rightarrow$	4 C
Finally ...	0 moles		2 moles		20 moles

The limiting reactant can also be ascertained by knowing the initial number of equivalents (or milli-equivalents) of each reactant. **The reactant with the least number of equivalents (or milli-equivalents) is the limiting reactant.** The equivalent method to identify the limiting reactant can be applied even if the chemical equation is not balanced.

However, in the following examples we shall mainly follow the POAC method, just for the practice of this modern concept.

[**Note:** In problems involving complicated reactions in which their balanced chemical equations are given, one should prefer to apply the mole method rather than the m.e. method as followed in Example 50 in Ch. 7.]

EXAMPLES

Ex. 1. *What amount of* CaO *will be produced by* 1 g *of calcium?*

Solution : Ca $\rightarrow$ CaO

Applying POAC for Ca atoms as Ca atoms are conserved,

moles of Ca atoms in the reactant

= moles of Ca atoms in the product

= moles of Ca atoms in CaO = $1 \times$ moles of CaO

($\because$ 1 mole of CaO contains 1 mole of Ca atoms)

$$\underset{\text{(Rule 2)}}{\frac{\text{wt. of Ca}}{\text{at. wt. of Ca}}} = \underset{\text{(Rule 1)}}{\frac{\text{wt. of CaO}}{\text{mol. wt. of CaO}}}$$

$$\frac{1}{40}=\frac{\text{wt. of CaO}}{56};\ \text{wt. of CaO}=\frac{56}{40}=1{\cdot}4\text{ g.}$$

Ex. 2. *What weight of oxygen will react with* 1 g *of calcium?* (Ca = 40)

***Solution* : First Method**

Since all the atoms of Ca have changed into CaO, the amount of Ca in CaO is 1 g. Now from the formula of CaO, we have,

$$\text{moles of Ca}=\text{moles of O} \qquad \text{(Rule 6)}$$

Now,
$$\frac{\text{wt. of Ca}}{\text{at. wt. of Ca}}=\frac{\text{wt. of O}}{\text{at. wt. of O}}$$

$$\therefore \quad \text{wt. of oxygen}=\frac{1}{40}\times 16=0{\cdot}4\text{ g.}$$

Second Method

$Ca + O_2 \rightarrow CaO$ (balancing of the equation is not required)

1 g $\quad$ x g (say)

Applying POAC for Ca atoms,

moles of Ca in the reactant = moles of Ca in CaO

$$\frac{1}{40}\text{ (Rule 2)}=1\times\text{moles of CaO}$$

($\because$ 1 mole of CaO contains 1 mole of Ca atom)

$$\therefore \quad \text{moles of CaO}=\frac{1}{40} \qquad \dots (1)$$

Again applying POAC for oxygen atoms,

moles of O in O_2 = moles of O in CaO

$$2\times\text{moles of } O_2=1\times\text{moles of CaO} \qquad \dots (2)$$

($\because$ 1 moles O_2 contains 2 moles of O and 1 mole of CaO contains 1 mole of O).

From eqns. (1) and (2), eliminating moles of CaO, we have,

$$\text{moles of } O_2=\frac{1}{2\times 40}=\frac{1}{80};\quad \text{or} \quad \frac{\text{wt. of } O_2}{32}=\frac{1}{80}. \qquad \text{(Rule 1)}$$

$$\therefore \quad \text{wt. of } O_2=\frac{1}{80}\times 32=0{\cdot}4\text{ g.}$$

[**Note:** The chemical equation of the above given problem is simple, i.e., easy to balance. But in complicated reactions (Ex. 5, 6, etc.) where the balancing is not very easy, the student can apply POAC without balancing the equation. This is where the mole method has its importance.]

Ex. 3(a) *Calculate the volume occupied by* 1 *mole of* He, H *and* O *atoms at* NTP.

Solution : As He is monoatomic, 1 mole of it will occupy 22·4 litres at NTP.

But hydrogen and oxygen being diatomic gases, 1 mole of their respective atoms will occupy a volume of 11·2 litres at NTP.

(b) *What volume of oxygen* (NTP) *will be required to react with* 1 g *of* Ca?

Solution : As Ca combines with O_2 to give CaO, from the formula of CaO, we have,

moles of Ca = moles of O (atoms)

$$\frac{\text{wt. of Ca}}{\text{at. wt.}} = \frac{\text{vol. of O atoms at NTP}}{\text{vol. occupied by 1 mole of O atoms at NTP}}$$

$$\frac{1}{40} = \frac{\text{vol. of O}}{11{\cdot}2}.$$

$$\therefore \text{ vol. of O} = \frac{11{\cdot}2}{40} = 0{\cdot}28 \text{ litre.}$$

Ex. 4. *A sample of* $KClO_3$ *on decomposition yielded* 448 mL *of oxygen gas at* NTP. *Calculate*

(i) *weight of oxygen produced,*

(ii) *weight of* $KClO_3$ *originally taken, and*

(iii) *weight of* KCl *produced.*

(K = 39, Cl = 35·5 and O = 16)

Solution : (i) Mole of oxygen $= \frac{448}{22400} = 0{\cdot}02.$ (Rule 3)

Wt. of oxygen $= 0{\cdot}02 \times 32 = 0{\cdot}64$ g. (Rule 1)

(ii) $KClO_3 \rightarrow KCl + O_2$

Applying POAC for O atoms,

moles of O atoms in $KClO_3$ = moles of O atoms in O_2

$3 \times$ moles of $KClO_3 = 2 \times$ moles of O_2.

(1 mole of $KClO_3$ contains 3 moles of O and 1 mole of O_2 contains 2 moles of O)

$$3 \times \underset{\text{(Rule 1)}}{\frac{\text{wt. of } KClO_3}{\text{mol. wt. of } KClO_3}} = 2 \times \underset{\text{(Rule 3)}}{\frac{\text{vol. at NTP (litres)}}{22{\cdot}4}}$$

$$3 \times \frac{\text{wt. of } KClO_3}{122{\cdot}5} = 2 \times \frac{0{\cdot}448}{22{\cdot}4}.$$

Wt. of $KClO_3 = 1{\cdot}634$ g.

(iii) Again applying POAC for K atoms,

moles of K atoms in $KClO_3$ = moles of K atoms in KCl

or $1 \times$ moles of $KClO_3 = 1 \times$ moles of KCl

(1 mole of $KClO_3$ contains 1 mole of K and 1 mole of KCl contains 1 mole of K)

$$1 \times \frac{\text{wt. of } KClO_3}{\text{mol. wt. of } KClO_3} = 1 \times \frac{\text{wt. of KCl}}{\text{mol. wt. of KCl}}$$

$$\frac{1{\cdot}634}{122{\cdot}5} = \frac{\text{wt. of KCl}}{74{\cdot}5}.$$

$$\text{Wt. of KCl} = 0{\cdot}9937 \text{ g.}$$

Ex. 5. 27·6 g *of* K_2CO_3 *was treated by a series of reagents so as to convert all of its carbon to* $K_2Zn_3[Fe(CN)_6]_2$. *Calculate the weight of the product.*

Solution : $K_2CO_3 \xrightarrow[\text{steps}]{\text{several}} K_2Zn_3[Fe(CN)_6]_2$

Since C atoms are conserved, applying POAC for C atoms,

moles of C in K_2CO_3 = moles of C in $K_2Zn_3[Fe(CN)_6]_2$

$1 \times$ moles of $K_2CO_3 = 12 \times$ moles of $K_2Zn_3[Fe(CN)_6]_2$

($\because$ 1 mole of K_2CO_3 contains 1 mole of C and 1 mole of $K_2Zn_3[Fe(CN)_6]_2$ contains 12 moles of C)

$$\frac{\text{wt. of } K_2CO_3}{\text{mol. wt. of } K_2CO_3} = 12 \times \frac{\text{wt. of the product}}{\text{mol. wt. of product}}$$

$$\text{wt. of } K_2Zn_3[Fe(CN)_6]_2 = \frac{27{\cdot}6}{138} \times \frac{698}{12} = 11{\cdot}6 \text{ g.}$$

[mol. wt. of K_2CO_3 = 138 and mol. wt. of $K_2Zn_3[Fe(CN)_6]_2 = 698$]

Ex. 6. *In a gravimetric determination of* P, *an aqueous solution of dihydrogen phosphate ion* $H_2PO_4^-$ *is treated with a mixture of ammonium and magnesium ions to precipitate magnesium ammonium phosphate,* $Mg(NH_4)PO_4 \cdot 6H_2O$. *This is heated and decomposed to magnesium pyrophosphate,* $Mg_2P_2O_7$, *which is weighed. A solution of* $H_2PO_4^-$ *yielded* 1·054 g *of* $Mg_2P_2O_7$. *What weight of* NaH_2PO_4 *was present originally?*
(Na = 23, H = 1, P = 31, O = 16, Mg = 24)

Solution : $NaH_2PO_4 + Mg^{2+} + NH_4^+ \rightarrow Mg(NH_4)PO_4 \cdot 6H_2O$

$$\xrightarrow{\text{heated}} Mg_2P_2O_7$$

Since P atoms are conserved, applying POAC for P atoms,

moles of P in NaH_2PO_4 = moles of P in $Mg_2P_2O_7$

$1 \times$ moles of $NaH_2PO_4 = 2 \times$ moles of $Mg_2P_2O_7$

($\because$ 1 mole of NaH_2PO_4 contains 1 mole of P and 1 mole of $Mg_2P_2O_7$ contains 2 moles of P)

$$\frac{\text{wt. of } NaH_2PO_4}{\text{mol. wt. of } NaH_2PO_4} = 2 \times \frac{\text{wt. of } Mg_2P_2O_7}{\text{mol. wt. of } Mg_2P_2O_7}$$

$$\frac{\text{wt. of } NaH_2PO_4}{120} = 2 \times \frac{1{\cdot}054}{222}.$$

$$\text{Wt. of } NaH_2PO_4 = 1{\cdot}14 \text{ g.}$$

[**Note:** In Ex. 5 and Ex. 6 the students should note that if in any reaction, a particular atom is conserved, the principle of atom conservation (POAC) with respect to that atom can be applied regardless of the number of steps of the reaction and their sequence.]

Ex. 7. *What weight of* AgCl *will be precipitated when a solution containing* 4·77 g *of* NaCl *is added to a solution of* 5·77 g *of* $AgNO_3$? (Na = 23, Cl = 35·5, Ag = 108, N = 14 *and* O = 16)

Solution : $NaCl + AgNO_3 \rightarrow NaNO_3 + AgCl$

$$\text{No. of moles of NaCl} = \frac{4{\cdot}77}{58{\cdot}5} = 0{\cdot}08154.$$

$$\text{No. of moles of } AgNO_3 = \frac{5{\cdot}77}{170} = 0{\cdot}03394.$$

Since no. of moles of $AgNO_3$ is less than that of NaCl, the whole of $AgNO_3$ shall convert into AgCl ($AgNO_3$ is a limiting reagent). Applying thus the POAC for Ag atoms as the Ag atoms are conserved,

$$\text{moles of Ag in } AgNO_3 = \text{moles of Ag in AgCl}$$

$$1 \times \text{moles of } AgNO_3 = 1 \times \text{moles of AgCl}$$

$$\frac{\text{wt. of } AgNO_3}{\text{mol. wt. of } AgNO_3} = \frac{\text{wt. of AgCl}}{\text{mol. wt. of AgCl}}$$

$$\text{wt. of AgCl} = 0{\cdot}03394 \times 143{\cdot}5 = 4{\cdot}87 \text{ g.}$$

Ex. 8. *What is the number of moles of* $Fe(OH)_3$ (s) *that can be produced by allowing* 1 *mole of* Fe_2S_3, 2 *moles of* H_2O *and* 3 *moles of* O_2 *to react?*

$$2Fe_2S_3\,(s) + 6H_2O\,(l) + 3O_2\,(g) \rightarrow 4Fe(OH)_3\,(s) + 6S\,(s)$$

Solution : From the given mole ratio of Fe_2S_3, H_2O and O_2 (i.e., 1 : 2 : 3) and mole ratio in the reaction, (i.e., 2 : 6 : 3) it is clear that H_2O is the limiting reagent as H_2O shall be fully consumed in the reaction. Thus,

$$\text{moles of } Fe(OH)_3 \text{ produced by 2 moles of } H_2O = \frac{4}{6} \times 2 = 1{\cdot}34.$$

Ex. 9. *Equal weights of phosphorus and oxygen are heated in a closed vessel producing* P_2O_3 *and* P_2O_5 *in a* 1 : 1 *mole ratio. If the limiting component is exhausted, find which component and also what fraction of it is left over.*

Solution : As 1 mole each of P_2O_3 and P_2O_5 is formed, the mole ratio of the reacting P and O atoms is 4 : 8 or 1 : 2.

As the given weights of phosphorus and oxygen are same, say w g, their moles of atoms are $\frac{w}{31}$ and $\frac{w}{16}$ respectively.

$\therefore$ 2 moles of O combine with 1 mole of P.

$\therefore$ $\frac{w}{16}$ moles of O combine with $\frac{w}{32}$ moles of P.

So phosphorous is left over and

$$\text{moles of P remaining} = \frac{w}{31} - \frac{w}{32}.$$

$$\therefore \text{ fraction of P left over} = \frac{\left(\frac{w}{31} - \frac{w}{32}\right)}{\frac{w}{31}} = \frac{1}{32} \text{ or } 3{\cdot}125\%.$$

Ex. 10. *In a process for producing acetic acid, oxygen gas is bubbled into acetaldehyde containing manganese(II) acetate (catalyst) under pressure at 60°C.*

$$2CH_3CHO + O_2 \rightarrow 2CH_3COOH$$

In a laboratory test of this reaction, 20 g of CH_3CHO and 10 g of O_2 were put into a reaction vessel.

(a) How many grams of CH_3COOH can be produced?

(b) How many grams of the excess reactant remain after the reaction is complete?

Solution :

(20/44)	(10/32)		0	... initial mole
$2\ CH_3CHO$	$+ O_2$	$\rightarrow$	$2\ CH_3COOH$	
0	$(0{\cdot}3125 - 0{\cdot}2272)$ $= 0{\cdot}0853$		$0{\cdot}4545$	... final mole

(a) CH_3CHO is the limiting reagent and hence

$$\text{amount of } CH_3COOH \text{ produced} = 0{\cdot}4545 \times 60 \text{ g} = 27{\cdot}27 \text{ g.}$$

(b) $$\text{Amount of } O_2 \text{ left} = 0{\cdot}0853 \times 32 = 2{\cdot}73 \text{ g.}$$

Ex. 11. $0{\cdot}75$ *mole of solid 'A_4' and 2 moles of gaseous O_2 are heated in a sealed vessel, completely using up the reactants and producing only one compound. It is found that when the temperature is reduced to the initial temperature, the contents of the vessel exhibit a pressure equal to half the original pressure. What conclusions can be drawn from these data about the product of the reaction?*

Solution : As both the reactants are limiting reagents, the ratio of stoichiometric coefficients of the reactants would be $0{\cdot}75 : 2$ or $3 : 8$ for A_4 and O_2. Thus,

$$3\,A_4 + 8\,O_2 \rightarrow \text{product}$$

As the product exerts considerable pressure, it has to be a gas.

Further, as the final pressure is half that of oxygen initially present, the stoichiometric coefficient of the product must be half that of O_2, i.e., 4. Obviously, the molecular formula would be A_3O_4 to balance the equation.

$$3\,A_4 + 8\,O_2 \rightarrow 4\,A_3O_4$$

Ex. 12. *Calculate the weight of lime* (CaO) *that can be prepared by heating* 200 kg *of limestone* ($CaCO_3$) *which is 95% pure.*

Solution : $CaCO_3 \rightarrow CaO + CO_2$

Actual amount of $CaCO_3 = \dfrac{95}{100} \times 200000\text{ g} = 190000\text{ g}.$

Since Ca atoms are conserved, applying POAC for Ca atoms,

moles of Ca in $CaCO_3$ = moles of Ca in CaO

$1 \times$ moles of $CaCO_3 = 1 \times$ moles of CaO

$$\frac{190000}{100} = \frac{\text{wt. of CaO}}{56}. \qquad \text{(Rule 1, Chapter 1)}$$

Wt. of CaO = 106400 g = 106·4 kg.

Ex. 13. 1 g-*atom of* Ca *was burnt in excess of oxygen and the oxide was dissolved in water to make up a one-litre solution. Calculate the normality of the alkaline solution.*

***Solution :* First Method**

$$Ca + O_2 \rightarrow CaO \xrightarrow{H_2O} Ca(OH)_2$$

Ca atoms are conserved. ∴ applying POAC for Ca atoms,

moles of Ca atoms in reactant = moles of Ca in $Ca(OH)_2$

$1 = 1 \times$ moles of $Ca(OH)_2$.

(∵ 1 mole of $Ca(OH)_2$ contains 1 mole of Ca)

∴ moles of $Ca(OH)_2 = 1$.

Since the solution is made up to 1 litre,

molarity of $Ca(OH)_2 = 1$ M

and normality of $Ca(OH)_2 = 2$ N. [Eqn. 6 (i), Chapter 7]

Second Method Please see the first method, Ex. 31, Chapter 7.

[**Note:** In the above problems, the rules applied have been referred to wherever necessary just to make every step easy to follow. It may, thus, be expected that the students must now have been well-acquainted with the rules and their applications, and so in the following problems rules have been rarely mentioned except where they are at all required.]

Ex. 14. 10 cc *of* H_2O_2 *solution when reacted with* KI *solution produced* 0·5 g *of iodine. Calculate the percentage purity of* H_2O_2. (I = 127)

Solution : $H_2O_2 + KI \rightarrow KOH + I_2$

First, applying POAC for H atoms to calculate moles of KOH, then applying POAC for K atoms to calculate moles of KI, and then finally applying POAC for I atoms to calculate moles of iodine.

POAC for H atoms:

moles of H in H_2O_2 = moles of H in KOH

$$2 \times \text{moles of } H_2O_2 = 1 \times \text{moles of KOH}. \quad \text{... (i)}$$

Applying POAC for K atoms,

moles of K atoms in KI = moles of K atoms in KOH

$$1 \times \text{moles of KI} = 1 \times \text{moles of KOH}$$

$$= 2 \times \text{moles of } H_2O_2 \quad \text{... (ii)}$$

[from the eqn. (i)].

Applying POAC for I atoms,

moles of I atoms in KI = moles of I atoms in I_2

$$1 \times \text{moles of KI} = 2 \times \text{moles of } I_2$$

or $$\text{moles of } I_2 = \frac{1}{2} \times \text{moles of KI}$$

$$= \frac{1}{2} \times 2 \text{ moles of } H_2O_2. \quad \text{[from the eqn. (ii)]}$$

Now, $$\frac{\text{wt. of } I_2}{\text{mol. wt. of } I_2} = \frac{\text{wt. of } H_2O_2}{\text{mol. wt. of } H_2O_2}.$$

Suppose x is the wt. of H_2O_2. Then,

$$\frac{0{\cdot}5}{254} = \frac{x}{34}$$

$$x = \frac{34 \times 0{\cdot}5}{254} = 0{\cdot}0669 \text{ g}.$$

$$\% \text{ of } H_2O_2 = \frac{0{\cdot}0669}{10} \times 100 = 0{\cdot}669\%.$$

Ex. 15. 0·32 *mole of* $LiAlH_4$ *in ether solution was placed in a flask and* 74 g (1 *mole*) *of t-butyl alcohol was added. The product* $LiAlHC_{12}H_{27}O_3$ *weighed* 81·28 g. *Is the claim that the yield is 'quantitative' a valid one if* Li *atoms are conserved?* (Li = 7, Al = 27, H = 1, C = 12, O = 16)

Solution : $$\underset{0{\cdot}32 \text{ mole}}{LiAlH_4} \xrightarrow[\text{alcohol}]{t\text{–butyl}} \underset{81{\cdot}28 \text{ g}}{LiAlHC_{12}H_{27}O_3}$$

Applying POAC for Li atoms,

moles of Li in $LiAlH_4$ = moles of Li in $LiAlHC_{12}H_{27}O_3$

$1 \times$ moles of $LiAlH_4 = 1 \times$ moles of $LiAlHC_{12}H_{27}O_3$.

Thus, 0·32 mole of $LiAlH_4$ should produce 0·32 mole of $LiAlHC_{12}H_{27}O_3$ if the yield is quantitative, i.e., 100%.

Moles of $LiAlHC_{12}H_{27}O_3 = \frac{81 \cdot 28}{254} = 0 \cdot 32$ mole (Rule 1, Chapter 1)

(mol. wt. of $LiAlHC_{12}H_{27}O_3 = 254$).

Thus the yield is 100%.

Ex. 16. 52·5 *millimoles of* $LiAlH_4$ *was treated with* 15·6 g *(210 millimoles) of t-butyl alcohol. A total of* 157·5 *millimoles of hydrogen was evolved for the reaction*

$$LiAlH_4 + 3\,(CH_3)_3\,COH \rightarrow 3H_2 + Li\,[(CH_3)_3O]_3\,AlH$$

The addition of an excess of another alcohol, methanol, to the above reaction mixture caused the fourth H *atom of the* $LiAlH_4$ *to be replaced according to the equation*

$$Li\,[(CH_3)_3\,O]_3\,AlH + CH_3OH \rightarrow H_2 + Li\,[(CH_3)_3O\,]_3\,[CH_3O]\,Al$$

How much H_2 *was evolved due to the addition of* CH_3OH?

Solution **:** According to the given equations,

1 mole of $Li\,[(CH_3)_3O]_3$ AlH produces 1 mole of H_2 and

1 mole of $Li\,[(CH_3)_3O]_3$ AlH is produced by 1 mole of $LiAlH_4$.

$\therefore$ 1 mole of $LiAlH_4$ should produce 1 mole of H_2 (by CH_3OH)

or 1 millimole of $LiAlH_4$ should produce 1 millimole of H_2 (by CH_3OH)

or 52·2 millimoles of $LiAlH_4$ should produce 52·5 millimoles of H_2. (by CH_3OH)

Ex. 17. *Equal weights of* Zn *metal and iodine are mixed together and the iodine is completely converted to* ZnI_2. *What fraction by weight of the original zinc remains unreacted?* (Zn = 65, I = 127)

Solution **:** Let x g be the initial weight of the Zn metal and iodine each. Since I_2 is completely converted to ZnI_2, we have,

	Zn	+	I_2	$\rightarrow$	ZnI_2
initial no. of moles:	$\frac{x}{65}$		$\frac{x}{254}$		0
No. of moles at the end of the reaction:	$\left(\frac{x}{65} - \frac{x}{254}\right)$		0		$\frac{x}{254}$

$$\therefore \text{ fraction of Zn remained unreacted} = \frac{\left(\frac{x}{65} - \frac{x}{254}\right)}{\frac{x}{65}} = 0{\cdot}74.$$

Ex. 18. 1·0 g *of an alloy of* Al *and* Mg *when treated with excess of dil.* HCl *forms* $MgCl_2$, $AlCl_3$ *and hydrogen. The evolved hydrogen, collected over* Hg *at* 0°C, *has a volume of* 1·20 *litres at* 0·92 atm *pressure. Calculate the composition of the alloy.* (Al = 27 *and* Mg = 24)

Solution : The equations,

$$Al + 3H^+ \rightarrow Al^{3+} + \frac{3}{2}H_2$$

$$Mg + 2H^+ \rightarrow Mg^{2+} + H_2$$

show that 1 mole of Al produces $\frac{3}{2}$ moles of hydrogen and 1 mole of Mg produces 1 mole of hydrogen.

Thus the mole equation is,

$\frac{3}{2}$ moles of Al + 1 mole of Mg = 1 mole of H_2.

Let the weight of Al be x g.

$\therefore$ wt. of Mg = $(1 - x)$ g

$$\frac{3}{2} \times \frac{x}{27} + \frac{1-x}{24} = \frac{\text{vol. of } H_2 \text{ at NTP}}{22{\cdot}4}.$$

$$\text{Volume of } H_2 \text{ at NTP} = \frac{1{\cdot}2 \times 0{\cdot}92}{273} \times \frac{273}{1} = 1{\cdot}104 \text{ litres}$$

$$\left(\text{using } \frac{p_1 V_1}{T_1} = \frac{p_2 V_2}{T_2}\right).$$

$$\therefore \quad \frac{3}{2} \times \frac{x}{27} + \frac{1-x}{24} = \frac{1{\cdot}104}{22{\cdot}4}; \quad x = 0{\cdot}55$$

Wt. of Al = 0·55 g; wt. of Mg = 0·45 g.

$$\text{Thus, \% of Al} = \frac{0{\cdot}55}{1} \times 100 = 55\%$$

$$\text{and \% of Mg} = \frac{0{\cdot}45}{1} \times 100 = 45\%.$$

Ex. 19. *A* 1-g *mixture of cuprous oxide and cupric oxide was quantitatively reduced to* 0·839 g *of metallic copper. What was the weight of cupric oxide in the original sample?* (Cu = 63·5, O = 16)

Solution : Let the weight of CuO be x g. The weight of Cu_2O will be $(1 - x)$ g. As the Cu atoms are conserved, applying POAC for Cu atoms, moles of Cu in CuO + moles of Cu in Cu_2O

= moles of Cu in the product

$1 \times$ moles of CuO + $2 \times$ moles of Cu_2O = moles of Cu

$$\frac{x}{79\cdot5} + 2 \times \frac{1-x}{143} = \frac{0\cdot839}{63\cdot5} \quad \begin{bmatrix} CuO = 79\cdot5 \\ Cu_2O = 143 \end{bmatrix}$$

$$x = 0\cdot55 \text{ g.}$$

Ex. 20. *A mixture of* FeO *and* Fe_3O_4 *when heated in air to a constant weight, gains 5% in weight. Find the composition of the initial mixture.*

Solution : When FeO and Fe_3O_4 are heated, both change to Fe_2O_3. Let the weights of FeO and Fe_3O_4 be x g and y g respectively.

$\therefore$ total weight of reactant = $(x + y)$ g.

Since weight increases 5% on heating,

$\therefore$ when FeO and Fe_3O_4 change completely to Fe_2O_3,

the wt. of $Fe_2O_3 = \frac{105}{100} \times (x + y) = 1\cdot05\,(x + y)$ g.

Now,

$$\begin{array}{ccccc} FeO & + & Fe_3O_4 & \rightarrow & Fe_2O_3 \\ x \text{ g} & & y \text{ g} & & 1\cdot05(x+y) \text{ g.} \end{array}$$

Applying POAC for Fe atoms,

moles of Fe in FeO + moles of Fe in Fe_3O_4 = moles of Fe in Fe_2O_3

$1 \times$ moles of FeO + $3 \times$ moles of Fe_3O_4 = $2 \times$ moles of Fe_2O_3

$$\frac{x}{72} + \frac{3y}{232} = \frac{2 \times 1\cdot05\,(x+y)}{160}. \quad \begin{bmatrix} FeO = 72 \\ Fe_3O_4 = 232 \\ Fe_2O_3 = 160 \end{bmatrix}$$

Dividing by y, we get

$$\frac{1}{72} \times \frac{x}{y} + \frac{3}{232} = \frac{2 \times 1\cdot05}{160} \times \frac{x}{y} + \frac{2 \times 1\cdot05}{160}$$

$$\frac{x}{y}\left(\frac{1}{72} - \frac{2\cdot1}{160}\right) = \frac{2\cdot1}{160} - \frac{3}{232}$$

$$\frac{x}{y} = \frac{81}{319}.$$

$\therefore$ $\quad$ % of FeO $= \frac{81}{(81+319)} \times 100 = 20\cdot02\%$

and $\quad$ % of $Fe_3O_4 = \frac{319}{(81+319)} \times 100 = 79\cdot98\%$.

Ex. 21. *What weight of* CO *is required to form* $Re_2(CO)_{10}$ *from* 2·50 g *of* Re_2O_7 *according to the unbalanced reaction:*

$$Re_2O_7 + CO \rightarrow Re_2(CO)_{10} + CO_2$$

($Re = 186{\cdot}2$, $C = 12$ and $O = 16$)

Solution : Suppose the relative moles of each reactant and product are as follows (just for convenience)

$$\begin{array}{ccccccc} Re_2O_7 & + & CO & \rightarrow & Re_2(CO)_{10} & + & CO_2 \\ a \text{ moles} & & b \text{ moles} & & c \text{ moles} & & d \text{ moles} \end{array}$$

Applying POAC for Re atoms,

moles of Re in Re_2O_7 = moles of Re in $Re_2(CO)_{10}$

$$2 \times \text{moles of } Re_2O_7 = 2 \times \text{moles of } Re_2(CO)_{10}$$

$$2a = 2c$$

or $$a = c. \qquad \text{... (i)}$$

Applying POAC for C atoms,

moles of C atoms in CO = moles of C in $Re_2(CO)_{10}$ + moles of C in CO_2.

$$1 \times \text{moles of CO} = 10 \times \text{moles of } Re_2(CO)_{10} + 1 \times \text{moles of } CO_2$$

or $$b = 10c + d. \qquad \text{... (ii)}$$

Applying POAC for O atoms,

moles of O in Re_2O_7 + moles of O in CO = moles of O in $Re_2(CO)_{10}$ + moles of O in CO_2.

$$7 \times \text{moles of } Re_2O_7 + 1 \times \text{moles of CO} = 10 \times \text{moles of } Re_2(CO)_{10} + 2 \times \text{moles of } CO_2$$

or $$7a + b = 10c + 2d. \qquad \text{... (iii)}$$

From the eqns. (i), (ii) and (iii), we get,

$$17a = b$$

i.e., $$17 \times \text{moles of } Re_2O_7 = \text{moles of CO}$$

$$17 \times \frac{2{\cdot}50}{484{\cdot}4} = \frac{\text{wt. of CO in g}}{28}. \quad \left[\begin{array}{c} \text{mol. wt. of } Re_2O_7 = 484{\cdot}4 \\ \text{mol. wt. of CO} = 28 \end{array}\right]$$

Wt. of CO = 2·46 g.

Ex. 22. *A 1-g sample of* $KClO_3$ *was heated under such conditions that a part of it decomposed according to the equation:*

(i) $2KClO_3 = 2KCl + 3O_2$

and the remaining underwent change according to the equation:

(ii) $4KClO_3 = 3KClO_4 + KCl$

If the amount of O_2 *evolved was* 146·8 mL *at* NTP, *calculate the percentage by weight of* $KClO_4$ *in the residue.*

Solution : $KClO_3 \rightarrow KCl + O_2$

Applying POAC for O atoms in the eqn. (i),

moles of O in $KClO_3$ = moles of O in O_2

$$3 \times \text{moles of } KClO_3 = 2 \times \text{moles of } O_2$$

$$3 \times \frac{\text{wt. of } KClO_3}{\text{mol. wt. of } KClO_3} = 2 \times \frac{\text{volume at NTP (mL)}}{22400}.$$

$$\text{Wt. of } KClO_3 = \frac{2 \times 146{\cdot}8 \times 122{\cdot}5}{3 \times 22400}$$

$$= 0{\cdot}5358 \text{ g.}$$

Again applying POAC for K atoms,

moles of K atoms in $KClO_3$ = moles of K atoms in KCl

$$1 \times \text{moles of } KClO_3 = 1 \times \text{moles of KCl}$$

$$\frac{\text{wt. of } KClO_3}{\text{mol. wt. of } KClO_3} = \frac{\text{wt. of KCl}}{\text{mol. wt. of KCl}}.$$

$$\text{Wt. of KCl} = \frac{0{\cdot}5358}{122{\cdot}5} \times 74{\cdot}5 = 0{\cdot}3260 \text{ g.} \qquad \text{... (i)}$$

In the second reaction:

the amount of $KClO_3$ left = 1 − 0·5358 = 0·4642 g.

We have,

$$\underset{0{\cdot}4642 \text{ g.}}{KClO_3} \rightarrow KClO_4 + KCl$$

Applying POAC for O atoms,

moles of O in $KClO_3$ = moles of O in $KClO_4$

$$3 \times \text{moles of } KClO_3 = 4 \times \text{moles of } KClO_4$$

$$3 \times \frac{\text{wt. of } KClO_3}{\text{mol. wt. of } KClO_3} = 4 \times \frac{\text{wt. of } KClO_4}{\text{mol. wt. of } KClO_4}.$$

$$\text{Wt. of } KClO_4 = \frac{3 \times 0{\cdot}4642 \times 138{\cdot}5}{122{\cdot}5 \times 4}$$

$$= 0{\cdot}3937 \text{ g.} \qquad \text{... (ii)}$$

Wt. of KCl produced by second reaction

$$= \text{wt. of } KClO_3 - \text{wt. of } KClO_4$$

$$= 0{\cdot}4642 - 0{\cdot}3937 = 0{\cdot}0705 \text{ g.} \qquad \text{... (iii)}$$

Now since on heating $KClO_3$, O_2 shall escape out, the substances as residue are KCl produced by the reactions (i) and (ii) and $KClO_4$.

Wt. of residue = (i) + (ii) + (iii)

$$= 0{\cdot}3260 + 0{\cdot}3937 + 0{\cdot}0705$$

$$= 0{\cdot}7902 \text{ g.}$$

$\therefore$ % of $KClO_4$ in the residue $= \dfrac{0{\cdot}3937}{0{\cdot}7902} \times 100$

$= 49{\cdot}8\%.$

Ex. 23. *A mixture of* KBr *and* NaBr *weighing* 0·560 g *was treated with aqueous* Ag^+ *and all the bromide ion was recovered as* 0·970 g *of pure* AgBr. *What was the fraction by weight of* KBr *in the sample?*

(K = 39, Br = 80, Ag = 108, Na = 23)

Solution : $KBr + NaBr + Ag^+ \rightarrow AgBr$

x g $\quad (0{\cdot}56 - x)$ g $\quad$ 0·97 g

Since Br atoms are conserved, applying POAC for Br atoms,

moles of Br in KBr + moles of Br in NaBr = moles of Br in AgBr

or $1 \times$ moles of KBr $+ 1 \times$ moles of NaBr $= 1 \times$ moles of AgBr

$$\frac{x}{119} + \frac{0{\cdot}56 - x}{103} = \frac{0{\cdot}97}{188} \qquad \{KBr = 119,\ NaBr = 103,\ AgBr = 188\}$$

$$x = 0{\cdot}1332 \text{ g.}$$

Fraction of KBr in the sample $= \dfrac{0{\cdot}1332}{0{\cdot}560} = 0{\cdot}2378.$

Ex. 24. *Igniting* MnO_2 *in air converts it quantitatively to* Mn_3O_4. *A sample of pyrolusite is of the following composition:* MnO_2 – 80%, SiO_2 *and other constituents* – 15%, *rest being water. The sample is ignited in air to constant weight. What is the percentage of* Mn *in the ignited sample?*
(Mn = 55, O = 16)

Solution : Suppose the weight of pyrolusite is x g.

$\therefore$ wt. of $MnO_2 = \dfrac{80}{100} \times x = 0{\cdot}8\,x.$

Wt. of SiO_2, etc. $= \dfrac{15}{100} \times x = 0{\cdot}15\,x.$

Wt. of water $= \dfrac{5}{100} \times x = 0{\cdot}05\,x.$

When pyrolusite is ignited, MnO_2 changes to Mn_3O_4 and H_2O evaporates. The residue contains, therefore, SiO_2, etc., and Mn_3O_4.

Now, we know,

$$MnO_2 \rightarrow Mn_3O_4$$

$0{\cdot}8x$ g

Applying POAC for Mn atoms,

moles of Mn in MnO_2 = moles of Mn in Mn_3O_4

$1 \times$ moles of $MnO_2 = 3 \times$ moles of Mn_3O_4 ... (i)

$$\frac{0.8\,x}{87} = 3 \times \frac{\text{wt. of } Mn_3O_4}{229}. \quad \begin{bmatrix} MnO_2 = 87 \\ Mn_3O_4 = 229 \end{bmatrix}$$

Wt. of $Mn_3O_4 = 0.702\,x$ g.

$\therefore$ wt. of the residue = wt. of Mn_3O_4 + wt. of SiO_2, etc.

$$= 0.702\,x + 0.15\,x = 0.852x \text{ g.}$$

Now, since Mn atoms are conserved,

moles of Mn = moles of Mn in Mn_3O_4

= moles of Mn in MnO_2

$= 1 \times$ moles of MnO_2

$= \frac{0.8x}{87}$ ($MnO_2 = 87$).

$\therefore$ wt. of Mn = moles of Mn $\times$ at. wt. of Mn

$= \frac{0.8x}{87} \times 55$ g.

$$\% \text{ of Mn in residue} = \frac{\text{wt. of Mn}}{\text{wt. of residue}} \times 100$$

$$= \frac{0.8x \times 55}{87} \times \frac{100}{0.852x} = 59.37\%.$$

Ex. 25. 1·84 g *of a mixture of* $CaCO_3$ *and* $MgCO_3$ *was heated to a constant weight. The constant weight of the residue was found to be* 0·96 g. *Calculate the percentage composition of the mixture.* (Ca = 40, Mg = 24, C = 12, O = 16)

Solution : On heating $CaCO_3$ and $MgCO_3$, one of the products, CO_2, escapes out.

We have,

$$\underset{\substack{x \text{ g} \\ \text{(say)}}}{CaCO_3} + \underset{(1.84 - x)\text{ g}}{MgCO_3} \rightarrow \underset{\substack{y \text{ g} \\ \text{(say)}}}{CaO} + \underset{(0.96-y)\text{ g}}{MgO} + CO_2\uparrow$$

Applying POAC for Ca atoms,

moles of Ca atoms in $CaCO_3$ = moles of Ca atoms in CaO

$1 \times$ moles of $CaCO_3 = 1 \times$ moles of CaO

$$\frac{x}{100} = \frac{y}{56}. \quad \begin{bmatrix} CaCO_3 = 100 \\ CaO = 56 \end{bmatrix} \quad \ldots \text{(i)}$$

Again applying POAC for Mg atoms,

moles of Mg in $MgCO_3$ = moles of Mg in MgO

$1 \times$ moles of $MgCO_3 = 1 \times$ moles of MgO

$$\frac{1.84-x}{84}=\frac{0.96-y}{40}\quad\begin{bmatrix}MgCO_3=84\\ MgO=40\end{bmatrix}\qquad \text{... (ii)}$$

From eqns. (i) and (ii), we get $x=1$ g, $y=0.84$ g

$$\% \text{ of } CaCO_3=\frac{1}{1.84}\times 100=54.34\%$$

and $$\% \text{ of } MgCO_3=45.66\%\,.$$

Second Method Apply POAC for C atoms.

Ex. 26. *A sample of a mixture of* $CaCl_2$ *and* NaCl *weighing* 4·22 g *was treated to precipitate all the* Ca *as* $CaCO_3$, *which was then heated and quantitatively converted to* 0·959 g *of* CaO. *Calculate the percentage of* $CaCl_2$ *in the mixture.*

(Ca = 40, O = 16, C = 12 and Cl = 35·5)

Solution **:** We have,

$$\underset{x\text{ g (say)}}{CaCl_2} \rightarrow CaCO_3 \rightarrow \underset{0.959\text{ g}}{CaO}$$

Since Ca atoms are conserved, applying POAC for Ca atoms,

moles of Ca in $CaCl_2$ = moles of Ca in CaO

$1\times$ moles of $CaCl_2 = 1\times$ moles of CaO

$$\frac{x}{111}=\frac{0.959}{56}\qquad\begin{bmatrix}CaCl_2=111\\ CaO=56\end{bmatrix}$$

$$x=1.901.$$

Percentage of $CaCl_2 = 1.901\times\frac{100}{4.22} = 45.04\%$.

Ex. 27. *A 2-g sample containing* Na_2CO_3 *and* $NaHCO_3$ *loses* 0·248 g *when heated to* 300°C, *the temperature at which* $NaHCO_3$ *decomposes to* Na_2CO_3, CO_2 *and* H_2O. *What is the percentage of* Na_2CO_3 *in the given mixture?*

(Na = 23, C = 12, O = 16 and H = 1)

Solution **:** On heating Na_2CO_3 and $NaHCO_3$, Na_2CO_3 remains unchanged while $NaHCO_3$ changes into Na_2CO_3, CO_2 and H_2O. The loss in weight is due to removal of CO_2 and H_2O which escape out on heating.

$\therefore$ wt. of Na_2CO_3 in the product = 2·00 – 0·248 = 1·752 g.

Let the weight of Na_2CO_3 in the mixture be x g.

$\therefore$ wt. of $NaHCO_3$ = (2·00 – x) g.

Since Na_2CO_3 in the products contains x g of unchanged reactant Na_2CO_3 and rest produced from $NaHCO_3$,

the wt. of Na_2CO_3 produced by $NaHCO_3$ only = (1·752 – x) g.

Now, we have,

$$NaHCO_3 \rightarrow Na_2CO_3 + (H_2O + CO_2)\uparrow$$
$$(2{\cdot}0 - x)\ g \quad (1{\cdot}752 - x)\ g$$

Applying POAC for Na atoms,

moles of Na in $NaHCO_3$ = moles of Na in Na_2CO_3

$$1 \times \text{moles of } NaHCO_3 = 2 \times \text{moles of } Na_2CO_3$$

$$\frac{2{\cdot}0 - x}{84} = 2 \times \frac{1{\cdot}752 - x}{106} \qquad \begin{bmatrix} NaHCO_3 = 84 \\ Na_2CO_3 = 106 \end{bmatrix}$$

$$x = \frac{82}{62} = 1{\cdot}328\ g.$$

$$\therefore \quad \%\ \text{of } Na_2CO_3 = \frac{1{\cdot}328}{2{\cdot}0} \times 100 = 66{\cdot}4\%.$$

Ex. 28. *(i) A solution of lead nitrate prepared by dissolving 2·07 g of pure lead in nitric acid was treated with HCl, Cl_2 gas and NH_4Cl. What will be the maximum weight of $(NH_4)_2PbCl_6$ so produced?*

$$(N = 14,\ H = 1,\ Pb = 207,\ Cl = 35{\cdot}5)$$

(ii) If on performing the experiment, 2·28 g of $(NH_4)_2PbCl_6$ was produced, calculate the percentage yield of $(NH_4)_2PbCl_6$.

Solution : (i) We have,

$$\underset{2{\cdot}07\ g}{Pb} \rightarrow Pb(NO_3)_2 \xrightarrow[Cl_2,\ NH_4Cl]{HCl} (NH_4)_2PbCl_6$$

Now, for maximum yield of $(NH_4)_2PbCl_6$ (i.e., 100% yield), Pb has to be conserved.

Let us apply POAC for Pb atoms,

moles of Pb in the reactant = moles of Pb in $(NH_4)_2PbCl_6$

$$= 1 \times \text{moles of } (NH_4)_2PbCl_6$$

$$\text{or} \quad \frac{2{\cdot}07}{207} = \frac{\text{maximum wt. of } (NH_4)_2PbCl_6}{456}. \qquad [(NH_4)_2PbCl_6 = 456]$$

$$\therefore \quad \text{maximum wt. of } (NH_4)_2PbCl_6 = 4{\cdot}56\ g.$$

$$\text{(ii) \% yield} = \frac{\text{amount produced experimentally}}{\text{amount for 100\% yield}} \times 100$$

$$= \frac{2{\cdot}28}{4{\cdot}56} \times 100 = 50\%.$$

Ex. 29. *Three different brands of 'liquid chlorine' are available in the market for use in purifying water of swimming pools. All are sold at the same rate of Rs 10 per litre and all are water solutions. Brand A contains 10% hypochlorite (ClO) (wt./vol.), brand B contains 7% available chlorine (Cl) and brand C contains 14% sodium hypochlorite (NaClO). Which of the three would you buy?*

Solution : Let each of the solutions be 100 mL.

Then, for brand A, wt. of ClO = 10 g
for brand B, wt. of Cl = 7 g
for brand C, wt. of NaClO = 14 g.

Let us now calculate wt. of Cl in each brand.

(i) Moles of Cl in ClO in brand A = 1 × moles of ClO

$$= 1 \times \frac{10}{51 \cdot 5}. \qquad (\text{ClO} = 51 \cdot 5)$$

$$\text{Wt. of Cl in brand A} = \frac{10}{51 \cdot 5} \times 35 \cdot 5 = 6 \cdot 89 \text{ g.}$$

(ii) Wt. of Cl in brand B = 7 g (given).

(iii) Moles of Cl in NaClO in brand C = 1 × moles of NaClO

$$= 1 \times \frac{14}{74 \cdot 5}. \qquad (\text{NaClO} = 74 \cdot 5)$$

$$\text{Wt. of Cl in brand C} = \frac{14}{74 \cdot 5} \times 35 \cdot 5 = 6 \cdot 67 \text{ g.}$$

As brand B contains the highest weight of Cl, it should be bought, being the cheapest.

Ex. 30. *In the analysis of a* 0·50-g *sample of feldspar, a mixture of the chlorides of* Na *and* K *is obtained, which weighs* 0·1180 g. *Subsequent treatment of the mixed chlorides with silver nitrate gives* 0·2451 g *of* AgCl. *What is the percentage of sodium oxide and potassium oxide in feldspar?*

Solution : Suppose the weight of NaCl is x g.

$$\underset{\text{(in feldspar)}}{Na_2O + K_2O} \xrightarrow{\text{Step I}} \underset{x\text{ g}}{NaCl} + \underset{(0 \cdot 1180 - x)\text{ g}}{KCl} \xrightarrow[\text{AgNO}_3]{\text{Step II}} \underset{0 \cdot 2451\text{ g}}{AgCl}$$

Applying POAC for Cl atoms in Step II to calculate x,

moles of Cl in NaCl + moles of Cl in KCl = moles of Cl in AgCl

1 × moles of NaCl + 1 × moles of KCl = 1 × moles of AgCl

$$\frac{x}{58 \cdot 5} + \frac{0 \cdot 1180 - x}{74 \cdot 5} = \frac{0 \cdot 2451}{143 \cdot 5}; \quad x = 0 \cdot 0343 \text{ g.}$$

(NaCl = 58·5, KCl = 74·5 and AgCl = 143·5)

∴ wt. of NaCl = 0·0343 g,

wt. of KCl = 0·0837 g.

Again, applying POAC for Na and K atoms to calculate the weight of Na_2O and K_2O respectively,

we get,

$$2 \times \text{moles of } Na_2O = \text{moles of NaCl}$$

and $\quad 2 \times \text{moles of } K_2O = \text{moles of KCl}.$

$$\therefore \quad \text{wt. of } Na_2O = \frac{1}{2} \times \frac{\text{wt. of NaCl}}{\text{mol. wt. of NaCl}} \times \text{mol. wt. of } Na_2O$$

$$\text{and} \quad \text{wt. of } K_2O = \frac{1}{2} \times \frac{\text{wt. of KCl}}{\text{mol. wt. of KCl}} \times \text{mol. wt. of } K_2O.$$

$$\therefore \quad \text{wt. of } Na_2O = \frac{1}{2} \times \frac{0{\cdot}0343}{58{\cdot}5} \times 62 = 0{\cdot}018 \text{ g}$$

$$\text{and} \quad \text{wt. of } K_2O = \frac{1}{2} \times \frac{0{\cdot}0837}{74{\cdot}5} \times 94 = 0{\cdot}053 \text{ g}.$$

$$\therefore \quad \% \text{ of } Na_2O = \frac{0{\cdot}018}{0{\cdot}50} \times 100 = 3{\cdot}6\%$$

$$\text{and} \quad \% \text{ of } K_2O = \frac{0{\cdot}053}{0{\cdot}50} \times 100 = 10{\cdot}6\%.$$

Ex. 31. 4·08 g *of a mixture of* BaO *and an unknown carbonate* MCO_3 *was heated strongly. The residue weighed* 3·64 g. *This was dissolved in* 100 mL *of* 1·0 N HCl. *The excess acid required* 16 mL *of* 2·5 N NaOH *solution for complete neutralisation. Identify the metal* M.

Solution : (Please see solved ex. 54, Chapter 7)

Ex. 32. *A mixture contains* NaCl *and an unknown chloride* MCl.

(i) 1 g *of this is dissolved in water. Excess of acidified* $AgNO_3$ *solution is added to it.* 2·567 g *of a white precipitate is formed.*

(ii) 1·0 g *of the original mixture is heated to* 300°C. *Some vapours come out which are absorbed in acidified* $AgNO_3$ *solution.* 1·341 g *of a white precipitate is obtained. Find the molecular weight of the unknown chloride.*

Solution : Suppose the molecular weight of MCl is *M*.

Given that:

$$\underset{(1-x)\text{ g}}{NaCl} + \underset{x \text{ g (say)}}{MCl} \xrightarrow{AgNO_3} \underset{2{\cdot}567 \text{ g}}{AgCl}$$

Applying POAC for Cl atoms,

moles of Cl in NaCl + moles of Cl in MCl = moles of Cl in AgCl

$$1 \times \text{moles of NaCl} + 1 \times \text{moles of MCl} = 1 \times \text{moles of AgCl}$$

$$\frac{1-x}{58{\cdot}5} + \frac{x}{M} = \frac{2{\cdot}567}{143{\cdot}5}. \qquad \dots \text{(i)}$$

Now, further at 300°C MCl is supposed to undergo sublimation while NaCl does not.

We have,

$$MCl \xrightarrow[300\,°C]{AgNO_3} AgCl$$

$$x\text{ g} \qquad\qquad 1{\cdot}341\text{ g}$$

Applying POAC for Cl atoms,

moles of Cl in MCl = moles of Cl in AgCl

1 × moles of MCl = 1 × moles of AgCl

$$\frac{x}{M} = \frac{1{\cdot}341}{143{\cdot}5}. \qquad \text{... (ii)}$$

From equations (i) and (ii), we get,

$$M = 53{\cdot}5.$$

$\therefore$ mol. wt. of $MCl = 53{\cdot}5$.

Ex. 33. *A solid mixture (5 g) consisting of lead nitrate and sodium nitrate was heated below 600°C until the weight of the residue is constant. If the loss in weight is 28%, find the amount of the lead nitrate and sodium nitrate in the mixture.*

Solution : The loss in weight (28% of 5 g), i.e., 1·4 g, is due to the formation of the gases NO_2 and O_2 which escape out.

$$\underset{x\text{ g (say)}}{Pb(NO_3)_2} \rightarrow \underset{(x-y)\text{ g}}{PbO} + \underbrace{NO_2 + O_2}_{y\text{ g (say)}}$$

$$\underset{(5-x)}{NaNO_3} \rightarrow \underset{(3{\cdot}6-x+y)}{NaNO_2} + \underset{(1{\cdot}4-y)}{O_2}$$

Applying POAC for Pb and Na atoms, we get respectively,

moles of $Pb(NO_3)_2$ = moles of PbO

$$\frac{x}{331} = \frac{x-y}{223} \qquad \{Pb(NO_3)_2 = 331,\ PbO = 223\}$$

and moles of $NaNO_3$ = moles of $NaNO_2$ $\{NaNO_3 = 85, NaNO_2 = 69\}$

from which, we get,

$$x = 3{\cdot}3246\text{ g}$$

$$\frac{5-x}{85} = \frac{3{\cdot}6-x+y}{69}. \ (NaNO_3 = 85,\ NaNO_2 = 69)$$

Thus, wt. of $Pb(NO_3)_2 = 3{\cdot}3246$ g,

wt. of $NaNO_3 = 5 - 3{\cdot}3246 = 1{\cdot}6754$ g.

Ex. 34. *In a particular experiment, 272 g of phosphorus, P_4, reacted with excess of oxygen to form P_4O_{10} in 89.5% yield. In the second step of the reaction, a 97.8% yield of H_3PO_4 was obtained. What mass of H_3PO_4 was obtained?*

Solution : Apply POAC for P atoms for the following steps.

$$P_4 \xrightarrow[\text{yield}]{89.5\%} P_4O_{10} \xrightarrow[\text{yield}]{97.8\%} H_3PO_4$$

In the first step, for 89.5% yield,

$$\text{moles of } P_4O_{10} \text{ produced} = \text{moles of } P_4 \times 0.895$$

$$= \frac{272}{124} \times 0.895 = 1.9632.$$

In the second step, for 97.8% yield,

$$\text{moles of } H_3PO_4 \text{ produced} = 4 \times \text{moles of } P_4O_{10} \times 0.978$$

$$= 4 \times 1.9632 \times 0.978 = 7.680.$$

$\therefore$ wt. of H_3PO_4 produced $= 7.680 \times 98$ g

$= 752.65$ g.

Ex. 35. *For the production of equal amounts of hydrogen from the following reactions, which metal, Zn or Al, is less expensive if Zn costs about half as much as Al on a mass basis and by how much?*

$$Zn + 2HCl \rightarrow ZnCl_2 + H_2$$

$$2Al + 6HCl \rightarrow 2AlCl_3 + 3H_2$$

Solution : 1 mole of H_2 is produced by 1 mole, i.e., 65 g of Zn and

1 mole of H_2 is produced by $\frac{2}{3}$ mole, i.e., $\frac{2 \times 27}{3}$ g of Al = 18 g of Al.

Now that Zn costs about half as much as Al, to purchase Zn and Al to produce the same amount of H_2, the cost ratio of Zn and Al will be 65 : 36.

Al is thus less expensive by $\frac{65 - 36}{65} \times 100$, i.e., 44.61%.

Ex. 36. 20.2 mL *of* CH_3COOH *reacts with* 20.1 mL *of* C_2H_5OH *to form* $CH_3COOC_2H_5$ ($d = 0.902$ g/mL) *by the following reaction.*

$$CH_3COOH + C_2H_5OH \rightarrow CH_3COOC_2H_5 + H_2O$$

(a) Which compound is the limiting reagent?

(b) If 27.5 mL *of pure ethyl acetate is produced, what is the per cent yield? Densities of* CH_3COOH *and* C_2H_5OH *are* 1.05 g/mL *and* 0.789 g/mL *respectively.*

Solution : (a) $CH_3COOH + C_2H_5OH \rightarrow CH_3COOC_2H_5 + H_2O$

$$\text{Mole of } CH_3COOH = \frac{20.2 \times 1.05}{60} = 0.3535.$$

$$\text{Mole of } C_2H_5OH = \frac{20.1 \times 0.789}{46} = 0.3447.$$

As CH_3COOH reacts with C_2H_5OH in a 1 : 1 mole ratio and mole of C_2H_5OH is less than that of CH_3COOH, C_2H_5OH is the limiting reagent.

(b) As C_2H_5OH is the limiting reagent, mole of $CH_3COOC_2H_5$ to be produced theoretically

$$= 0.3447 \text{ mole.}$$

But experimental yield of $CH_3COOC_2H_5 = \dfrac{27.5 \times 0.902}{84}$

$$= 0.2953 \text{ mole.}$$

$\therefore$ per cent yield of ethyl acetate $= \dfrac{0.2953}{0.3447} \times 100$

$$= 85.66\%.$$

Ex. 37. *A mixture of pure* AgCl *and pure* AgBr *is found to contain* 60.94% Ag *by mass. What are mass percentages of* Cl *and* Br *in the mixture?*

(Ag = 108, Cl = 35.5, Br = 80)

Solution : Let us first calculate the weight ratio of AgCl and AgBr, their wts. being supposed to be x and y g respectively in the mixture. Apply POAC for Ag atoms.

$$AgCl + AgBr \rightarrow Ag$$

$$1 \times \text{moles of AgCl} + 1 \times \text{moles AgBr} = \text{moles of Ag}$$

$$\frac{x}{143.5} + \frac{y}{188} = \text{moles of Ag.}$$

$\therefore$ wt. of Ag in the mixture $= \left(\dfrac{x}{143.5} + \dfrac{y}{188}\right) \times 108.$ g

As given,

$$\frac{\left(\dfrac{x}{143.5} + \dfrac{y}{188}\right) \times 108}{x + y} = 0.6094$$

$$\frac{x}{y} = \frac{0.035}{0.1432}.$$

$\therefore$ % of AgCl $= \dfrac{x \times 100}{x + y} = \dfrac{0.035}{0.035 + 0.1432} \times 100 = 19.64$

and % of AgBr = 80.36.

Thus, a 100-g mixture contains 19.64 g of AgCl and 80.36 g of AgBr. Now calculate the amount of Cl in AgCl and the amount of Br in AgBr in 100 g of the mixture.

Wt. of Cl in 19.64 g of AgCl $= \dfrac{35.5}{143.5} \times 19.64 = 4.85$ g.

Wt. of Br in 80.36 g of AgBr $= \dfrac{80}{188} \times 80.36 = 34.19$ g.

Thus 100 g of the mixture contains 4.85 g of Cl and 34.19 g of Br.

$\therefore$ percentage of Cl = 4.85%

and percentage of Br = 34.19%.

Factor-Label Method (Dimensional Analysis)

This method is applied to stoichiometric calculations, provided chemical equations are balanced. In this method, moles of one reagent are converted to moles of another, using a ratio of the stoichiometric coefficients of the balanced chemical equation. The use of this method may be clearly understood by the following examples.

Ex. 38. *From the following reaction sequence,*

$$CaC_2 + H_2O \rightarrow CaO + C_2H_2$$

$$C_2H_2 + H_2 \rightarrow C_2H_4$$

$$n\,C_2H_4 \rightarrow (C_2H_4)_n$$

calculate the mass of polyethylene which can be produced from 10 kg *of* CaC_2.

Solution : Wt. of $(C_2H_4)_n = (10000\text{ g } CaC_2) \times \left(\frac{1 \text{ mole } CaC_2}{64\text{ g } CaC_2}\right)$

$$\times \left(\frac{1 \text{ mole } C_2H_2}{1 \text{ mole } CaC_2}\right) \times \left(\frac{1 \text{ mole } C_2H_4}{1 \text{ mole } C_2H_2}\right)$$

$$\times \left(\frac{1 \text{ mole } (C_2H_4)_n}{n \text{ mole } C_2H_4}\right) \times \left(\frac{28n \text{ g } (C_2H_4)_n}{1 \text{ mole } (C_2H_4)_n}\right)$$

$$= 4375 \text{ g.}$$

(note that n cancels)

Second Method Apply POAC for C atoms.

Ex. 39. *From the following series of reactions,*

$$Cl_2 + 2KOH \rightarrow KCl + KClO + H_2O$$

$$3KClO \rightarrow 2KCl + KClO_3$$

$$4KClO_3 \rightarrow 3KClO_4 + KCl$$

calculate the mass of chlorine needed to produce 100 g *of* $KClO_4$.

Solution : Wt. of $Cl_2 = 100\text{ g } KClO_4 \times \left(\frac{1 \text{ mole } KClO_4}{139\text{ g } KClO_4}\right)$

$$\times \left(\frac{4 \text{ mole } KClO_3}{3 \text{ mole } KClO_4}\right) \times \left(\frac{3 \text{ mole } KClO}{1 \text{ mole } KClO_3}\right)$$

$$\times \left(\frac{1 \text{ mole } Cl_2}{1 \text{ mole } KClO}\right) \times \left(\frac{71\text{ g } Cl_2}{1 \text{ mole } Cl_2}\right)$$

$$= 204{\cdot}5 \text{ g.}$$

PROBLEMS

(Answers bracketed with questions)

1. Calculate orally
(i) How many moles of $CaCO_3$ shall be produced from 5 moles of Ca atoms?
(ii) How many moles of $BaSO_4$ shall be formed from 5 moles of $BaCl_2$?
(iii) How many moles of Na_2O shall be produced from 5 moles of Na atoms?
[(i) 5 (ii) 5 (iii) 2·5]

2. What weight of oxygen will react with 40 g of Ca? (16 g)

3. Calculate, without balancing the following equation, the volume of chlorine at NTP produced from 50 g of sodium chloride.

$$NaCl + MnO_2 + H_2SO_4 \rightarrow NaHSO_4 + MnSO_4 + Cl_2 + H_2O$$ (9·575 litres)

4. Two tonnes of an iron ore containing 94% of Fe_2O_3 produces iron in pure state. Calculate the weight of iron. (1315 kg)

5. Calculate the volume of acetylene at NTP produced by 100 g of CaC_2 with water. (35 litres)

6. How many litres of detonating gas will be produced at NTP in the decomposition of 0·1 mole of water by an electric current? (3·36 litres)

7. Find the mass of $Cu(NO_3)_2 \cdot 3H_2O$ produced by dissolving 10 g of copper in nitric acid and then evaporating the solution. (38 g)

8. 4·90 g of $KClO_3$, on heating, shows a weight loss of 0·384 g. What per cent of the original $KClO_3$ has decomposed? (20%)

9. When the mixture of $MgCO_3$ and $CaCO_3$ was heated for a long time, the weight decreased by 50%. Calculate the percentage composition of the mixture. ($MgCO_3$: 71.59%; $CaCO_3$: 28.41%)

10. How many moles of $Zn(FeS_2)_2$ can be made from 2 g of Zn, 3 g of Fe and 4 g of S? (0·0269 mole)

11. Calculate the weight of V_2O_5 produced from 2 g of VO and 5·75 g of Fe_2O_3.

$$VO + Fe_2O_3 \rightarrow FeO + V_2O_5.$$ (2·18 g)

[**Hint:** Fe_2O_3 is the limiting reagent.]

12. Equal weights of mercury and iodine are allowed to react completely to form a mixture of mercurous and mercuric iodide leaving none of the reactants. Calculate the ratio by weight of Hg_2I_2 and HgI_2 formed. (0·532 : 1)

13. 5·5 g of a mixture of $FeSO_4 \cdot 7H_2O$ and $Fe_2(SO_4)_3 \cdot 9H_2O$ requires 5·4 mL of 0·1 N $KMnO_4$ solution for complete oxidation. Calculate the number of moles of hydrated ferric sulphate in the mixture. (0·0095 mole)

14. Anhydrous sodium sulphate can absorb water vapour and be converted to the decahydrate. By how many grams would the mass of a 1-g sample of the thoroughly dried Na_2SO_4 increase if exposed to sufficient water vapour to be converted to the decahydrate? (1·27 g)

15. A partially dried clay sample contained 50% of silica and 7% of water. The original clay contained 12% of water. Find the percentage of silica in the original sample. (47·3%)

16. 1 g of a sample containing NaCl, NaBr and an inert material, with excess of $AgNO_3$, produces 0·526 g of precipitate of AgCl and AgBr. By heating this precipitate in a current of chlorine, AgBr converted to AgCl and the precipitate then weighed 0·426 g. Find the percentage of NaCl and NaBr in the sample. (NaCl: 4·25%; NaBr: 23·2%)

17. 3·90 g of a mixture of Al and Al_2O_3, when reacted with a solution of sodium hydroxide, produced 840 mL of a gas at NTP. Find the composition of the mixture. (Al: 17·3%)

[Hint: Only Al produces H_2.]

18. To determine the NaCl content in commercial NaOH, 2 g of the latter was dissolved in water and an excess amount of an $AgNO_3$ solution was added to this solution. The precipitate formed was washed and dried. Its mass was 0·287 g. Find the mass of NaCl in the initial sample. (0·117 g)

19. One litre of an acidified solution of $KMnO_4$ containing 15·8 g of $KMnO_4$ is decolourised by passing sufficient amount of SO_2. If SO_2 is produced by roasting of iron pyrites (FeS_2), what will be the amount of pyrites required to produce the necessary amount of SO_2? (15 g)

20. When a mixture of NaBr and NaCl is repeatedly digested with sulphuric acid, all the halogens are expelled and Na_2SO_4 is formed quantitatively. With a particular mixture, it was found that the weight of Na_2SO_4 obtained was precisely the same as the weight of NaBr–NaCl mixture taken. Calculate the ratio of the weights of NaCl and NaBr in the mixture. (1·454 : 1)

21. 25·4 g of iodine and 14·2 g of chlorine are made to react completely to yield a mixture of ICl and ICl_3. Calculate the number of moles of ICl and ICl_3 formed. (0·1 mole, 0·1 mole)

22. An alloy of aluminium and copper was treated with aqueous HCl. The aluminium dissolved according to the reaction:

$$Al + 3H^+ \rightarrow Al^{3+} + \frac{3}{2}H_2,$$

but the copper remained as pure metal. A 0·350-g sample of the alloy gave 415 cc of H_2 measured at 273 K and 1 atm pressure. What is the weight percentage of Al in the alloy? (95·3%)

23. 1 g of dry green algae absorbs 4.7×10^{-3} mole of CO_2 per hour by photosynthesis. If the fixed carbon atoms were all stored after photosynthesis as starch, $(C_6H_{10}O_5)_n$, how long would it take for the algae to double their own weight assuming photosynthesis takes place at a constant rate? (7·88 hours)

[Hint: Wt. of $(C_6H_{10}O_5)_n = 2 - 1 = 1$ g. Apply POAC for C in

$$CO_2 \rightarrow (C_6H_{10}O_5)_n$$

$$\therefore \text{ time} = \frac{\text{moles of } CO_2}{\text{rate of absorption of } CO_2}.$$]

24. Crude calcium carbide is made in an electric furnace by the following reaction:

$$CaO + 3C \rightarrow CaC_2 + CO$$

The product contains 85% of CaC_2 and 15% of unreacted CaO.

(a) How much CaO is to be added to the furnace charge for each 1000 kg of CaC_2 (pure) produced?

(b) How much CaO is to be added to the furnace charge for each 1000 kg of crude product? [(a) 1051·5 kg; (b) 893·7 kg]

25. 2·5 g of a mixture of BaO and CaO when treated with an excess of H_2SO_4, produced 4·713 g of the mixed sulphates. Find the percentage of BaO present in the mixture. (60%)

26. A mixture of NaI and NaCl, when heated with H_2SO_4, produced the same weight of sodium sulphate as that of the original mixture. Calculate percentage of NaI in the mixture. (28·85%)

27. 7·46 g of KCl was heated with excess of MnO_2 and H_2SO_4. The gas so produced was then passed through a solution of KI. Calculate the weight of iodine.

$$\left[\textbf{Hint: } KCl \xrightarrow[H_2SO_4]{MnO_2} Cl_2 \xrightarrow{KI} I_2; \text{ (moles of } Cl_2 = \text{moles of } I_2)\right]$$

(12·7 g)

28. Carnalite is a double chloride of potassium and magnesium containing 38·86% of water. 0·458 g of it gave 0·71 g of AgCl and 0·666 g of it gave 0·27 g $Mg_2P_2O_7$. Find the percentage of KCl in the carnalite. (26·46%)

29. What volume of hydrogen at NTP is needed to reduce 125 g of MoO_3 to the metal? (58·3 litres)

30. How much gas (in litres) will be produced at 0°C and 760 mm of pressure when 10 g of oxalic acid was heated with concentrated sulphuric acid? (4·97 litres)

[**Hint:** $(COOH)_2 \xrightarrow{H_2SO_4} CO + CO_2 + H_2O$. Apply POAC for C atoms.]

31. A natural gas sample contains 84% (by volume) of CH_4, 10% of C_2H_6, 3% of C_3H_8 and 3% of N_2. If a series of catalytic reactions could be used for converting all the carbon atoms of the gas into butadiene, C_4H_6, with 100% efficiency, how much butadiene could be prepared from 100 g of the natural gas? (82 g)

32. What weights of P_4O_6 and P_4O_{10} will be produced by the combustion of 2 g of P_4 in 2 g of oxygen leaving no P_4 and O_2? (1·996 g, 2·004 g)

33. From the following reactions, $2CoF_2 + F_2 \rightarrow 2CoF_3$

$$(CH_2)_n + 4n\,CoF_3 \rightarrow (CF_2)_n + 2n\,HF + 4n\,CoF_2$$

calculate how much F_2 will be consumed to produce 1 kg of $(CF_2)_n$. (1·52 kg)

34. A mixture containing $KClO_3$, $KHCO_3$, K_2CO_3 and KCl was heated, producing CO_2, O_2 and H_2O gases according to the following equations:

$$2KClO_3(s) \rightarrow 2KCl(s) + 3O_2$$

$$2KHCO_3(s) \rightarrow K_2O(s) + H_2O(g) + 2CO_2(g)$$

$$K_2CO_3(s) \rightarrow K_2O(s) + CO_2(g)$$

The KCl does not react under the conditions of the reaction. If 100.0 g of the mixture produces 1.80 g of H_2O, 13.20 g of CO_2 and 4.0 g of O_2, what was the composition of the original mixture? ($KClO_3$: 10.2 g, $KHCO_3$: 20 g, K_2CO_3: 13.8 g)

35. The ammonia prepared by treating ammonium sulphate with calcium hydroxide is completely used by $NiCl_2 \cdot 6H_2O$ to form a stable coordination compound. Assume that both the reactions are 100% complete. If 1584 g of ammonium sulphate and 952 g of $NiCl_2 \cdot 6H_2O$ are used in the preparation, the combined weight (in gram) of gypsum and the nickel ammonia coordination compound thus produced is ... .

(H = 1, N = 14, O = 16, S = 32, Cl = 35.5, Ca = 40, Ni = 59) (2992 g)

[Hint: $(NH_4)_2SO_4 + Ca(OH)_2 \rightarrow 2NH_3 + CaSO_4 \cdot 2H_2O$ **(IIT 2018 Adv.)**

$\frac{1584}{132} = 12$ mol 24 mol 12 mol

$6NH_3 + [Ni(H_2O)_6]Cl_2 \rightarrow [Ni(NH_3)_6]Cl_2 + 6H_2O$

24 mol 4 mol 4 mol

As weight in gram = no. of mol × Mol. wt.

Wt. of $CaSO_4 \cdot 2H_2O$(gypsum) + $[Ni(NH_3)_6]Cl_2$

$= (12 \times 172 + 4 \times 232)$ g]

36. Galena (an ore) is partially oxidised by passing air through it at high temperature. After some time, the passage of air is stopped, but the heating is continued in a closed furance such that the contents undergo self-reduction. The weight (in kg) of Pb produced per kg of O_2 consumed is ... .

(O = 16, S =32, Pb = 207) (6.47 kg)

[Hint: $PbS + O_2 \rightarrow Pb + SO_2$ **(IIT 2018 Adv.)**

mol of O_2 = mol of Pb

$\frac{1000\text{ (g)}}{32} = \frac{\text{Wt. of Pb(g)}}{207}$]

Objective Problems

1. In a gaseous reaction of the type, $aA + bB \rightarrow cC + dD$, which statement is wrong?

(a) *a* litres of A combine with *b* litres of B to give C and D

(b) *a* moles of A combine with *b* moles of B to give C and D

(c) *a* g of A combine with *b* g of B to give C and D

(d) *a* molecules of A combine with *b* molecules of B to give C and D

2. The equation $2Al\,(s) + \frac{3}{2}O_2\,(g) \rightarrow Al_2O_3\,(s)$ shows that

(a) 2 moles of Al react with $\frac{3}{2}$ moles of O_2 to produce $\frac{7}{2}$ moles of Al_2O_3

(b) 2 g of Al react with $\frac{3}{2}$ g of O_2 to produce one mole of Al_2O_3

(c) 2 g of Al react with $\frac{3}{2}$ litre of O_2 to produce 1 mole of Al_2O_3

(d) 2 moles of Al react with $\frac{3}{2}$ moles of O_2 to produce 1 mole of Al_2O_3

3. If 5 litres of H_2O_2 produce 50 litres of O_2 at NTP, H_2O_2 is
(a) '50V' (b) '10V' (c) '5V' (d) '250V'

4. 2·76 g of silver carbonate on being strongly heated yields a residue weighing
(a) 2·16 g (b) 2·48 g (c) 2·32 g (d) 2·64 g

5. Assuming that petrol is octane (C_8H_{18}) and has density 0·8 g/mL, 1·425 litres of petrol on complete combustion will consume
(a) 50 moles of O_2 (b) 100 moles of O_2
(c) 125 moles of O_2 (d) 200 moles of O_2

6. 12 g of Mg will react completely with an acid to give
(a) 1 mole of O_2 (b) $\frac{1}{2}$ mole of H_2 (c) 1 mole of H_2 (d) 2 moles of H_2

7. 10 mL of gaseous hydrocarbon on combustion gives 40 mL of CO_2(g) and 50 mL of H_2O (vap.). The hydrocarbon is
(a) C_4H_5 (b) C_8H_{10} (c) C_4H_8 (d) C_4H_{10}

8. For complete oxidation of 4 litres of CO at NTP, the required volume of O_2 at NTP is
(a) 4 litres (b) 8 litres (c) 2 litres (d) 1 litre

9. The minimum quantity in grams of H_2S needed to precipitate 63·5 g of Cu^{2+} will be nearly
(a) 63·5 g (b) 31·75 g (c) 34 g (d) 20 g

10. If 0·5 mole of $BaCl_2$ is mixed with 0·2 mole of Na_3PO_4, the maximum number of mole of $Ba_3(PO_4)_2$ that can be formed is
(a) 0·7 (b) 0·5 (c) 0·30 (d) 0·1

11. For the reaction $A + 2B \rightarrow C$, 5 moles of A and 8 moles of B will produce
(a) 5 moles of C (b) 4 moles of C
(c) 8 moles of C (d) 13 moles of C

12. A mixture of N_2 and H_2 is caused to react in a closed container to form NH_3. The reaction ceases before either reactant has been totally consumed. At this stage, 2.0 moles each of N_2, H_2 and NH_3 are present. The moles of N_2 and H_2 present originally were respectively,
(a) 4 and 4 moles (b) 3 and 5 moles
(c) 3 and 4 moles (d) 4 and 5 moles

13. A compound with molar mass 180 is acylated with CH_3COCl to get a compound with molar mass 390. The number of amino groups present per molecule of the former compound is
(a) 2 (b) 5 (c) 4 (d) 6

(IIT 2013 Main)

14. The molecular formula of a commercial resin used for exchanging ions in water softening is $C_8H_7SO_3Na$ (mol. wt. = 206). What would be the maximum uptake of Ca^{2+} ions by the resin when expressed in mole per gram resin? **(IIT 2015 Main)**
(a) $\frac{1}{103}$ (b) $\frac{1}{206}$ (c) $\frac{1}{309}$ (d) $\frac{1}{412}$

[**Hint:** $2C_8H_7SO_3Na + Ca^{2+} \rightarrow (C_8H_7SO_3)_2Ca + 2Na$]

15. A sample of $NaClO_3$ is converted by heat to NaCl with a loss of 0.16 g of oxygen. The residue is dissolved in water and precipitated as AgCl. The mass of AgCl (in g) obtained will be ... . (AgCl = 143.5 g mol^{-1})

(a) 0.54 (b) 0.41 (c) 0.48 (d) 0.35

(IIT 2018 Main)

[**Hint:** $NaClO_3 \rightarrow NaCl + \underset{0.16\text{ g}}{O_2} \rightarrow AgCl$; Apply POAC]

Answers

1-c, **2**-d, **3**-b, **4**-a, **5**-c, **6**-b, **7**-d, **8**-c, **9**-c, **10**-d, **11**-b, **12**-b, **13**-b, **14**-d, **15**-c.

❑

CHAPTER THREE

EUDIOMETRY OR GAS ANALYSIS

Gaseous reactions are carried out in a special type of tube known as an eudiometer tube. The tube is graduated in millimetres for volume measurement. The reacting gases taken in the eudiometer tube are exploded by sparks, produced by passing electricity through the platinum terminals provided in the tube. The volumes of the products of a gaseous explosion are determined by absorbing them in suitable reagents, e.g., CO_2 and SO_2 are absorbed in KOH solution, O_2 is absorbed in a solution of alkaline pyrogallol, and CO is absorbed in a solution of ammoniacal cuprous chloride. Since H_2O vapour produced during the reaction changes to liquid on cooling, the volume of water is neglected, but while applying POAC, moles of H_2O produced cannot be neglected.

Eudiometry is mainly based on Avogadro's law, which states that equal volumes of all gases under similar conditions of temperature and pressure contain equal number of molecules. Two gases having equal number of molecules, also have equal number of moles. The mole concept may be applied in solving the problems of this chapter, keeping in mind that in a gaseous reaction the relative volumes (measured under identical conditions) of each reactant and product represent their relative numbers of moles.

A(g) +	B(g) →	C(g) +	D(g)
a volumes	*b* volumes	*c* volumes	*d* volumes
a moles	*b* moles	*c* moles	*d* moles

In the following problems we shall again see that the balancing of gaseous reactions are not required if solved by the POAC method.

EXAMPLES

Ex. 1. *What volume of oxygen will be required for the complete combustion of* 18·2 *litres of propane at* NTP?

Solution : $C_3H_8 + O_2 \rightarrow CO_2 + H_2O$

Applying POAC for C atoms, we have,

moles of C in C_3H_8 = moles of C in CO_2

$3 \times \text{moles of } C_3H_8 = 1 \times \text{moles of } CO_2$. ... (1)

Similarly, applying POAC for H and O atoms,

$8 \times \text{moles of } C_3H_8 = 2 \times \text{moles of } H_2O$ ... (2)

$$2 \times \text{moles of } O_2 = 2 \times \text{moles of } CO_2 + 1 \times \text{ moles of } H_2O. \quad \dots (3)$$

Substituting moles of CO_2 and moles of H_2O from eqns. (1) and (2) in eqn. (3),

$$2 \times \text{moles of } O_2 = 2 \times 3 \times \text{moles of } C_3H_8 + 1 \times 4 \times \text{moles of } C_3H_8$$

$$\text{moles of } \quad O_2 = 5 \times \text{moles of } C_3H_8$$

$$\frac{\text{volume of } O_2 \text{ at NTP (lit.)}}{22{\cdot}4} = 5 \times \frac{\text{volume of } C_3H_8 \text{ at NTP (lit.)}}{22{\cdot}4}.$$

(Rule 3, Chapter 1)

Volume of O_2 at NTP $= 5 \times 18{\cdot}2 = 91$ litres.

Ex. 2. 20 mL *of* CO *was mixed with* 50 mL *of oxygen and the mixture was exploded. On cooling, the resulting mixture was shaken with* KOH. *Find the volume of the gas that is left.*

Solution :

$$CO + O_2 \rightarrow CO_2$$

Applying POAC for C and O atoms, we get respectively

$$\text{moles of } CO = \text{moles of } CO_2 \quad \dots (1)$$

$$\text{and moles of } CO + 2 \times \text{ moles of } O_2 = 2 \times \text{moles of } CO_2. \quad \dots (2)$$

From eqns. (1) and (2),

$$\text{moles of } CO = 2 \times \text{moles of } O_2$$

or $$\frac{\text{moles of } CO}{\text{moles of } O_2} = 2$$

or $$\frac{\text{volume of } CO}{\text{volume of } O_2} = 2. \quad \dots \text{(Avogadro's law)}$$

Since the volume of CO is twice that of O_2, for 20 mL of CO the volume of O_2 will be 10 mL, which will react with 20 mL of CO. Thus, the volume of O_2 remaining unreacted is 40 mL. Further, as KOH absorbs CO_2 produced in the reaction, the only gas left is O_2, the volume of which is 40 mL.

Ex. 3. 10 mL *of a gaseous hydrocarbon was burnt completely in* 80 mL *of* O_2 *at* NTP. *The remaining gas occupied* 70 mL *at* NTP. *This volume became* 50 mL *on treatment with* KOH *solution. What is the formula of the hydrocarbon?*

Solution : Let the hydrocarbon be C_xH_y.

We know that the hydrocarbon on combustion gives CO_2 and H_2O. As said in the beginning, since the volume occupied by water is neglected, we have,

volume of CO_2 produced + unreacted O_2 = 70 mL and

volume of unreacted O_2 (CO_2 absorbed by KOH) = 50 mL.

$\therefore$ volume of O_2 reacted with 10 mL of hydrocarbon

$$= (80 - 50) \text{ mL} = 30 \text{ mL}$$

and volume of CO_2 produced by 10 mL of hydrocarbon and 30 mL of $O_2 = (70 - 50)$ mL $= 20$ mL.

Now the reaction is

$$
\begin{array}{lllll}
 & C_xH_y + O_2 & \rightarrow & CO_2 + H_2O \\
 & 10 \text{ mL} \quad 30 \text{ mL} & & 20 \text{ mL} \\
\text{or} & 10 \text{ moles} \quad 30 \text{ moles} & & 20 \text{ moles}
\end{array}
$$

Applying POAC for O atoms, we get,

$$2 \times \text{moles of } O_2 = 2 \times \text{moles of } CO_2 + 1 \times \text{moles of } H_2O$$

$$2 \times 30 = 2 \times 20 + \text{moles of } H_2O.$$

$\therefore$ moles of $H_2O = 20$.

Applying POAC for C atoms, we get,

$$x \times \text{moles of } C_xH_y = 1 \times \text{moles of } CO_2$$

$$x \times 10 = 1 \times 20. \quad \therefore \; x = 2.$$

Applying POAC for H atoms,

$$y \times \text{moles of } C_xH_y = 2 \times \text{moles of } H_2O$$

$$y \times 10 = 2 \times 20. \quad \therefore \; y = 4.$$

The formula of the hydrocarbon is C_2H_4.

Ex. 4. 16 mL *of a hydrocarbon gas was exploded with excess of oxygen. On cooling, the volume of the resulting gaseous mixture was reduced by* 48 mL. *When* KOH *was added, there was a further decrease of* 48 mL *in the volume. Find the molecular formula of the compound.*

Solution : Since water vapour condenses to practically zero volume of water, the decrease in volume on cooling is the volume of water vapour. CO_2 is absorbed by KOH and so volume of CO_2 is equal to 48 mL.

Thus,

$$
\begin{array}{lllll}
 & C_xH_y \text{ (say)} + O_2 & \rightarrow & CO_2 + H_2O \\
 & 16 \text{ mL} & & 48 \text{ mL} \quad 48 \text{ mL} \\
\text{or} & 16 \text{ moles} & & 48 \text{ moles} \quad 48 \text{ moles}
\end{array}
$$

Since C and H are completely converted to CO_2 and H_2O respectively, applying POAC for C and H atoms we get respectively,

$$x \times \text{moles of } C_xH_y = 1 \times \text{moles of } CO_2$$

$$16x = 48. \qquad \therefore \; x = 3.$$

$$y \times \text{moles of } C_xH_y = 2 \times \text{moles of } H_2O$$

$$16y = 2 \times 48. \qquad \therefore \; y = 6.$$

Hence, the formula of the hydrocarbon is C_3H_6.

Ex. 5. 7·5 mL *of a gaseous hydrocarbon was exploded with* 36 mL *of oxygen. The volume of gases on cooling was found to be* 28·5 mL, 15 mL *of which was absorbed by* KOH *and the rest was absorbed in a solution of alkaline pyrogallol. If all volumes are measured under the same conditions, deduce the formula of the hydrocarbon.*

Solution : Volume of CO_2 produced = 15 mL. (absorbed by KOH)

Volume of unused $O_2 = (28{\cdot}5 - 15)$ mL = 13·5 mL. (absorbed by pyrogallol)

∴ volume of O_2 reacted with 7·5 mL of hydrocarbon

$$= (36{\cdot}0 - 13{\cdot}5)\text{ mL} = 22{\cdot}5\text{ mL}.$$

Thus,

	C_xH_y (say)	+ O_2	→ CO_2	+ H_2O
	7·5 mL	22·5 mL	15 mL	
or	7·5 moles	22·5 moles	15 moles	

Applying POAC for C, H and O atoms, we get respectively,

$x \times$ moles of $C_xH_y = 1 \times$ moles of CO_2 ... (1)

$7{\cdot}5x = 15\,;\quad x = 2$

$y \times$ moles of $C_xH_y = 2 \times$ moles of H_2O ... (2)

$2 \times$ moles of $O_2 = 2 \times$ moles of $CO_2 + 1 \times$ moles of H_2O. ... (3)

From eqns. (2) and (3), eliminating moles of H_2O, we get

$$2 \times \text{moles of } O_2 = 2 \times \text{moles of } CO_2 + 1 \times \frac{y \times \text{moles of } C_xH_y}{2}$$

$$2 \times 22{\cdot}5 = 2 \times 15 + \frac{7{\cdot}5}{2}y.$$

$$\therefore \quad y = 4.$$

Hence, the formula of the hydrocarbon is C_2H_4.

Ex. 6. *A sample of a gaseous hydrocarbon occupying* 1·12 *litres at* NTP *when completely burnt in air produced* 2·2 g *of* CO_2 *and* 1·8 g *of* H_2O. *Calculate the weight of the compound taken and the volume of* O_2 *at* NTP *required for its burning. Find the molecular formula of the hydrocarbon.*

Solution : Mole of hydrocarbon $= \frac{1{\cdot}12}{22{\cdot}4} = \frac{1}{20}\cdot$ (Rule 3, Chapter 1)

Mole of $CO_2 = \frac{2{\cdot}2}{44} = \frac{1}{20}\cdot$ (Rule 1, Chapter 1)

Mole of $H_2O = \frac{1{\cdot}8}{18} = \frac{1}{10}\cdot$ (Rule 1, Chapter 1)

Thus,

$$C_xH_y \text{ (suppose)} + O_2 \rightarrow CO_2 + H_2O$$

$$\frac{1}{20}\text{ mole} \qquad \frac{1}{20}\text{ mole} \quad \frac{1}{10}\text{ mole}$$

Applying POAC for C and H atoms we get respectively,

$$x \times \text{moles of } C_xH_y = 1 \times \text{moles of } CO_2$$

$$x \times \frac{1}{20} = 1 \times \frac{1}{20};\ x = 1$$

and $y \times$ moles of $C_xH_y = 2 \times$ moles of H_2O

$$y \times \frac{1}{20} = 2 \times \frac{1}{10};\ y = 4.$$

Thus, formula of the hydrocarbon is CH_4.

$$\text{Wt. of } CH_4 = \text{moles of } CH_4 \times \text{mol. wt.}$$

$$= \frac{1}{20} \times 16 = 0{\cdot}8 \text{ g.}$$

Now, again applying POAC for O atoms,

$$2 \times \text{moles of } O_2 = 2 \times \text{moles of } CO_2 + 1 \times \text{moles of } H_2O$$

$$2 \times \frac{\text{volume of } O_2 \text{ (NTP) in lit.}}{22{\cdot}4} = 2 \times \frac{1}{20} + \frac{1}{10} = \frac{1}{5}.$$

$\therefore$ vol. of O_2 at NTP $= 2{\cdot}24$ litres.

Ex. 7. 7·5 mL *of a hydrocarbon gas was exploded with excess of oxygen. On cooling, it was found to have undergone a contraction of* 15 mL. *If the vapour density of the hydrocarbon is* 14, *determine its molecular formula.* (C = 12, H = 1)

Solution : A contraction in volume on cooling shows the volume of H_2O vapour produced.

Thus,

$$C_xH_y \text{ (suppose)} + O_2 \rightarrow CO_2 + H_2O \text{ (vapour)}$$

7·5 mL	15 mL
7·5 moles	15 moles

Applying POAC for H atoms,

$y \times$ moles of $C_xH_y = 2 \times$ moles of H_2O

$$y \times 7{\cdot}5 = 2 \times 15;\quad y = 4.$$

Since molecular weight of hydrocarbon = 2 × vapour density

$$= 2 \times 14 = 28$$

and the formula is C_xH_4, we have,

$$12 \times x + 4 = 28;\ x = 2.$$

Hence, the formula of the hydrocarbon is C_2H_4.

Ex. 8. 10 mL *of a gaseous organic compound containing* C, H *and* O *only was mixed with* 100 mL *of oxygen and exploded under conditions which allowed the water formed to condense. The volume of the gas after explosion was* 90 mL. *On treatment with potash solution, a further contraction of* 20 mL *in volume was observed. Given that the vapour density of the compound is* 23, *deduce the molecular formula. All volume measurements were carried out under the same conditions.*

Solution : Volume of unreacted O_2 + vol. of CO_2 = 90 mL.
(H_2O vapour condensed)

Volume of CO_2 (absorbed by KOH) = 20 mL.

∴ volume of unreacted O_2 = (90 – 20) mL = 70 mL.

Volume of O_2 reacted with 10 mL of compound
= (100 – 70) mL = 30 mL.

Thus,

	$C_xH_yO_z$ (suppose) + O_2	→	CO_2 + H_2O
	10 mL	30 mL	20 mL
or	10 moles	30 moles	20 moles

Applying POAC for C atoms,

$$x \times \text{moles of } C_xH_yO_z = 1 \times \text{moles of } CO_2$$

$$x \times 10 = 1 \times 20\,; \quad x = 2.$$

Again applying POAC for H and O atoms, we get respectively,

$$y \times \text{moles of } C_xH_yO_z = 2 \times \text{moles of } H_2O$$

$$10y = 2 \times \text{moles of } H_2O \qquad \dots (1)$$

and $z \times \text{moles of } C_xH_yO_z + 2 \times \text{moles of } O_2$

$$= 2 \times \text{moles of } CO_2 + 1 \times \text{moles of } H_2O$$

$$10z + 2 \times 30 = 2 \times 20 + \text{moles of } H_2O. \qquad \dots (2)$$

Eliminating moles of H_2O from eqns. (1) and (2), we get

$$y - 2z = 4. \qquad \dots (3)$$

Now,
molecular wt. of the compound = 2 × 23 = 46.

∴ for the compound $C_xH_yO_z$,

$$2 \times 12 + y \times 1 + z \times 16 = 46$$

$$y + 16z = 22. \qquad \dots (4)$$

From eqns. (3) and (4), we get,

$$y = 6 \text{ and } z = 1.$$

∴ formula of the compound is C_2H_6O.

Ex. 9. *9 volumes of a gaseous mixture consisting of a gaseous organic compound A and just sufficient amount of oxygen required for complete combustion, yielded on burning, 4 volumes of* CO_2, *6 volumes of water vapour and 2 volumes of* N_2, *all volumes measured at the same temperature and pressure. If the compound A contained only C, H and N, (i) how many volumes of oxygen are required for complete combustion, and (ii) what is the molecular formula of the compound A?*

Solution : Suppose the compound A is $C_xH_yN_z$

(i)	$C_xH_yN_z$	+ O_2	→ CO_2	+ H_2O (vapour)	+ N_2
	$(9-v)$ vol.	v vol. (suppose)	4 vol.	6 vol.	2 vol.
or	$(9-v)$ moles	v moles	4 moles	6 moles	2 moles

Applying POAC for O atoms,

$$2 \times \text{moles of } O_2 = 2 \times \text{moles of } CO_2 + 1 \times \text{moles of } H_2O$$

$$2v = 2 \times 4 + 1 \times 6 = 14\ ;\ v = 7 \text{ volumes.}$$

(ii) The reaction is

	$C_xH_yN_z$	+ O_2	→ CO_2	+ H_2O (vapour)	+ N_2
	2 vol.	7 vol.	4 vol.	6 vol.	2 vol.
or	2 moles	7 moles	4 moles	6 moles	2 moles

Applying POAC for C, H and N, we get respectively,

$x \times \text{moles of } C_xH_yN_z = 1 \times \text{moles of } CO_2$

$$x \times 2 = 1 \times 4\ ;\ x = 2$$

$y \times \text{moles of } C_xH_yN_z = 2 \times \text{moles of } H_2O \text{ (vapour)}$

$$y \times 2 = 2 \times 6\ ;\ y = 6$$

$z \times \text{moles of } C_xH_yN_z = 2 \times \text{moles of } N_2$

$$z \times 2 = 2 \times 2;\ z = 2.$$

Hence, the compound is $C_2H_6N_2$.

Ex. 10. *An organic compound* $C_xH_{2y}O_y$ *was burnt with twice the amount of oxygen needed for complete combustion to* CO_2 *and* H_2O*. The hot gases when cooled to 0°C and 1 atm pressure, measured 2·24 litres. The water collected during cooling weighed 0·9 g. The vapour pressure of pure water at 20°C is 17·5 mmHg and is lowered by 0·104 mm when 50 g of organic compound is dissolved in 1000 g of water. Give the molecular formula of the organic compound.*

Solution : Suppose the number of moles of $C_xH_{2y}O_y$ is m and total volume of O_2 used is $2v$ litres at 0°C and 1 atm pressure (NTP).

$\therefore$ volume of O_2 reacted = v lit.

and volume of O_2 remaining = v lit.

As given,

volume of O_2 not reacted + volume of CO_2 formed= 2·24 lit.

$\therefore$ volume of CO_2 formed = $(2{\cdot}24 - v)$ lit.

Thus we have,

$$
\begin{array}{ccccccc}
C_xH_{2y}O_y & + & O_2 & \rightarrow & H_2O & + & CO_2 \\
m \text{ moles} & & v \text{ lit.} & & 0{\cdot}9 \text{ g} & & (2{\cdot}24 - v) \text{ lit.} \\
m \text{ moles} & & \frac{v}{22{\cdot}4} \text{ moles} & & \frac{0{\cdot}9}{18} = 0{\cdot}05 \text{ mole} & & \frac{(2{\cdot}24 - v)}{22{\cdot}4} \text{ moles} \\
 & & \text{(Rule 3)} & & \text{(Rule 1)} & & \text{(Rule 3)} \\
 & & \text{Chapter 1} & & \text{Chapter 1} & & \text{Chapter 1}
\end{array}
$$

Applying POAC for C, H and O we get respectively,

$x \times$ moles of $C_xH_{2y}O_y$ = moles of CO_2

or $$x \times m = \frac{2{\cdot}24 - v}{22{\cdot}4}. \quad \ldots (1)$$

$2y \times$ moles of $C_xH_{2y}O_y = 2 \times$ mole of H_2O

$$2y \times m = 2 \times 0{\cdot}05 = 0{\cdot}1 \quad \ldots (2)$$

$y \times$ mole of $C_xH_{2y}O_y + 2 \times$ mole of O_2

$$= 1 \times \text{mole of } H_2O + 2 \times \text{moles of } CO_2$$

$$y \times m + 2 \times \frac{v}{22{\cdot}4} = 0{\cdot}05 + 2 \times \frac{(2{\cdot}24 - v)}{(22{\cdot}4)}. \quad \ldots (3)$$

Substituting $(y \times m)$ from Eqn. (2) in (3) we get,

$$v = 1{\cdot}12.$$

Putting the value of v in Eqn. (1), we get,

$$x \times m = 0{\cdot}05 \quad \ldots (4)$$

Dividing Eqn. (4) by Eqn. (2)

$$\frac{x}{y} = 1. \quad \ldots (5)$$

Now, from the vapour pressure data, we can calculate the molecular weight of $C_xH_{2y}O_y$.

From Raoult's law, we know

$$\frac{\text{decrease in VP}}{\text{VP of solvent}} = \frac{\text{moles of solute}}{\text{moles of solute} + \text{moles of solvent}}$$

(See dilute solutions, Chapter 13)

$$\frac{0{\cdot}104}{17{\cdot}5} = \frac{50/M}{50/M + 1000/18} \quad (M = \text{mol. wt. of } C_xH_{2y}O_y)$$

$\therefore$ $M = 150.$

Thus, we have another equation from the formula $C_xH_{2y}O_y$ and its molecular weight,

$$12x + 18y = 150. \quad \ldots (6)$$

From eqns. (5) and (6),

$$x = 5, \qquad y = 5.$$

Hence, the molecular formula of the organic compound is $C_5H_{10}O_5$.

Ex. 11. 100 mL *of a gas at* NTP *was heated with tin. Tin converted into stannous sulphide and hydrogen was left. This hydrogen when passed over hot* CuO, *produced* 0·081 g *of water. If the vapour density of the gas is 17, find its formula.*

Solution : In this problem it is clear that the gas contains H and S only.

Let the formula of the gas be H_xS_y.

Thus,

$$H_xS_y + Sn \rightarrow SnS + H_2 \xrightarrow{(H_2 \text{ passed over hot CuO})} H_2O$$

Since all the H of the gas converted into H_2O,

applying POAC for H atom, we have,

$$x \times \text{moles of } H_xS_y = 2 \times \text{moles of } H_2O$$

$$x \times \frac{100}{22400} = 2 \times \frac{0{\cdot}081}{18}; \qquad x = 2.$$

(Rule 3) Chapter 1   (Rule 1) Chapter 1

Hence, the formula of the gas is H_2S_y, the molecular weight of which is 34. (mol., wt. $= 2 \times$ vapour density)

$$= 2 \times 17 = 34.$$

$$\therefore \quad 2 \times 1 + y \times 32 = 34$$

$$y = 1.$$

Thus, the formula of the gas is H_2S.

Ex. 12. *Determine the formula of ammonia from the following data:*

Volume of ammonia = 25 mL

Volume on addition of O_2 *after explosion* = 71·2 mL

Volume after explosion with O_2 *(on cooling)* = 14·95 mL

Volume after being absorbed by alkaline pyrogallol = 12·5 mL

Solution : Since ammonia on explosion produces N_2 and H_2,

volume of $(N_2 + H_2 + \text{total } O_2) = 71{\cdot}2$ mL.

Volume of $(N_2 + \text{unused } O_2) = 14{\cdot}95$ mL.

($\because$ on further explosion with O_2, H_2 changes to H_2O, the volume of which on cooling is zero)

Volume of N_2 = 12·5 mL.

($\because$ unused O_2 is absorbed by alkaline pyrogallol)

$\therefore$ volume of $(H_2 + \text{total } O_2) = (71{\cdot}2 - 12{\cdot}5)$ mL $= 58{\cdot}7$ mL.

$\therefore$ volume of unused $O_2 = (14{\cdot}95 - 12{\cdot}5)$ mL $= 2{\cdot}45$ mL.

$\therefore$ volume of $H_2 + \text{used } O_2 = (58{\cdot}7 - 2{\cdot}45)$ mL $= 56{\cdot}25$ mL.

Now, let the volume of H_2 be x' mL.

$$\begin{array}{lccc} & H_2 & + \; O_2 & \rightarrow \; H_2O \\ & x' \text{ mL} & (56{\cdot}25 - x') \text{ mL} & \\ \text{or} & x' \text{ moles} & (56{\cdot}25 - x') \text{ moles} & \end{array}$$

Applying POAC for H atoms and O atoms,

$$2 \times \text{moles of } H_2 = 2 \times \text{moles of } H_2O \quad \ldots (1)$$

$$2 \times \text{moles of } O_2 = 1 \times \text{moles of } H_2O. \quad \ldots (2)$$

From eqns. (1) and (2), we have

$$\text{moles of } H_2 = 2 \times \text{moles of } O_2$$

$$x' = 2 \times (56{\cdot}25 - x')$$

or $$x' = 37{\cdot}50 \text{ mL}.$$

Now, let the formula of ammonia be N_xH_y.

$$\begin{array}{lcccc} & N_xH_y & \xrightarrow{\text{explosion}} & N_2 \; + & H_2 \\ & 25 \text{ mL (given)} & & 12{\cdot}5 \text{ mL (given)} & 37{\cdot}5 \text{ mL (calculated)} \\ \text{or} & 25 \text{ moles} & & 12{\cdot}5 \text{ moles} & 37{\cdot}5 \text{ moles} \end{array}$$

Applying POAC for N atoms,

$x \times$ moles of $N_xH_y = 2 \times$ moles of N_2

$$x \times 25 = 2 \times 12{\cdot}5; \; x = 1.$$

Again applying POAC for H atoms,

$$y \times \text{moles of } N_xH_y = 2 \times \text{moles of } H_2$$

$$y \times 25 = 2 \times 37{\cdot}50; \quad y = 3.$$

Hence, the formula of ammonia is NH_3.

Ex. 13. *At a high temperature the compound S_4N_4 decomposes completely into N_2 and sulphur vapour. If all measurements are made under the same conditions of temperature and pressure, it is found that for each volume of S_4N_4 decomposed, 2·5 volumes of gaseous products are formed. What is the molecular formula of sulphur?*

Solution : Let the molecular formula of sulphur be S_x.

$$\begin{array}{lccc} & S_4N_4 & \rightarrow \; S_x & + \; N_2 \\ & \text{I vol.} & v \text{ vol.} & (2{\cdot}5 - v) \text{ vol.} \\ & & \text{(suppose)} & \\ \text{or} & 1 \text{ mole} & v \text{ moles} & (2{\cdot}5 - v) \text{ moles} \end{array}$$

Applying POAC for S and N atoms, we get respectively,

$$4 \times \text{moles of } S_4N_4 = x \times \text{moles of } S_x$$

$$4 = x \times v \quad \ldots (1)$$

and $4 \times \text{moles of } S_4N_4 = 2 \times \text{moles of } N_2$

$$4 = 2 \times (2{\cdot}5 - v) \quad \dots (2)$$

From the eqns. (1) and (2), we get $x = 8$.
Thus the molecular formula of sulphur is S_8.

Ex. 14. 50 mL *of pure and dry oxygen was subjected to silent electric discharge and on cooling to the original temperature, the volume of ozonised oxygen was found to be* 47 mL. *The gas was then brought in contact with turpentine oil, when after the absorption of ozone, the remaining gas occupied a volume of* 41 mL. *Find the molecular formula of ozone.*

Solution : Total volume of O_2 = 50 mL.

Volume of O_2 (not converted to O_x) + ozone produced = 47 mL.

Volume of O_2 (not converted to O_x) = 41 mL.

(ozone absorbed in turpentine oil)

$\therefore$ volume of O_2 converted to $O_x = 50 - 41 = 9$ mL.

Volume of ozone produced = 47 − 41 = 6 mL.

Let the formula of ozone be O_x.

$$O_2 \rightarrow O_x$$

9 mL 6 mL

or 9 moles 6 moles

Applying POAC for the O atom, we get,

$$2 \times \text{moles of } O_2 = x \times \text{moles of } O_x$$

$$2 \times 9 = x \times 6.$$

$$\therefore \quad x = 3.$$

Hence, the formula of ozone is O_3.

Ex. 15. 50 mL *of a mixture of* CO *and* CH_4 *was exploded with* 85 mL *of* O_2. *The volume of* CO_2 *produced was* 50 mL. *Calculate the percentage composition of the gaseous mixture if all volumes are measured under the same conditions, and the given volume of* O_2 *is just sufficient for the combustion of* 50 mL *of the mixture of* CO *and* CH_4.

Solution : $CO + CH_4 + O_2 \rightarrow CO_2 + H_2O$

(50 − x) mL x mL (suppose) 85 mL 50 mL

or (50 − x) moles x moles 85 moles 50 moles

Applying POAC for H and O atoms, we get respectively,

$$4 \times \text{moles of } CH_4 = 2 \times \text{moles of } H_2O. \quad \dots (1)$$

and $1 \times \text{moles of } CO + 2 \times \text{moles of } O_2$

$$= 2 \times \text{moles of } CO_2 + 1 \times \text{moles of } H_2O. \quad \dots (2)$$

From (1) and (2), eliminating moles of H_2O, we get,

$$\text{moles of } CO + 2\times\text{moles of } O_2$$
$$= 2\times\text{moles of } CO_2 + 2\times\text{moles of } CH_4$$

$$(50-x) + 2\times 85 = 2\times 50 + 2\times x$$

$\therefore$ $x = 40.$

Hence volume of CH_4 = 40 mL

and volume of CO = (50 − 40) = 10 mL.

$\therefore$ percentage of CH_4 in the mixture $= \frac{40}{50}\times 100 = 80\%$

and percentage of CO = 20%.

Ex. 16. *A* 40-mL *sample of a mixture of* H_2 *and* O_2 *was placed in a gas burette at* 18°C *and* 1 atm *pressure. A spark was applied so that the formation of* H_2O *was complete. The remaining pure gas had a volume of* 10 mL *at* 18°C *and* 1 atm *pressure. If the remaining gas was hydrogen, what was the initial mole per cent of hydrogen in the mixture?*

Solution : Total volume of H_2 and O_2 which reacted with each other

= (40 − 10) mL = 30 mL.

Let the volume of H_2 which reacted be x mL.

$$H_2 \quad + \quad O_2 \quad \rightarrow \quad H_2O$$

x mL $\quad (30-x)$ mL $\quad$ under similar conditions
x moles $\quad (30-x)$ moles $\quad$ of temperature and pressure

Applying POAC for H and O atoms,

$$2\times\text{moles of } H_2 = 2\times\text{moles of } H_2O \quad \ldots (1)$$
$$2\times\text{moles of } O_2 = 1\times\text{moles of } H_2O \quad \ldots (2)$$

From the eqns. (1) and (2),

$$\text{moles of } H_2 = 2\times\text{moles of } O_2$$
$$x = 2\times(30-x).$$

$\therefore$ $x = 20.$

Total volume of hydrogen

= vol. of H_2 undergoing reaction + vol. of H_2 remained

= 20 mL + 10 mL

= 30 mL.

Now, total volume of gas is 40 mL.

$\therefore$ volume percentage of H_2 in the mixture

$$= \frac{30}{40}\times 100 = 75\%$$

and from Avogadro's law, mole per cent of H_2 is also 75%.

Ex. 17. *A sample of coal gas contained* H_2, CH_4 *and* CO. 20 mL *of this mixture was exploded with* 80 mL *of oxygen. On cooling, the volume of gases was* 68 mL. *There was a contraction of* 10 mL *when treated with* KOH. *Find the composition of the original mixture.*

Solution : Volume of total O_2 = 80 mL.

Volume of CO_2 + unreacted O_2 = 68 mL.

Volume of CO_2 = 10 mL. (absorbed in KOH)

$\therefore$ volume of unreacted $O_2 = (68 - 10)$ mL = 58 mL.

Volume of O_2 used in explosion = (80 – 58) mL = 22 mL.

Let the volume of H_2 and CO in the mixture be x mL and y mL respectively. Thus,

H_2 +	CO +	CH_4 +	O_2 →	CO_2 +	H_2O
x mL	y mL	$(20 - x - y)$ mL	22 mL	10 mL	
x moles	y moles	$(20 - x - y)$ moles	22 moles	10 moles	

Applying POAC for H, O and C atoms, we get respectively,

$$2 \times \text{moles of } H_2 + 4 \times \text{moles of } CH_4 = 2 \times \text{moles of } H_2O$$

$$2x + 4(20 - x - y) = 2 \times \text{moles of } H_2O \quad \ldots (1)$$

$$1 \times \text{moles of } CO + 2 \times \text{moles of } O_2 = 2 \times \text{moles of } CO_2 + 1 \times \text{moles of } H_2O$$

$$y + 2 \times 22 = 2 \times 10 + \text{moles of } H_2O \quad \ldots (2)$$

and $1 \times \text{moles of } CO + 1 \times \text{moles of } CH_4 = 1 \times \text{moles of } CO_2$

$$y + (20 - x - y) = 10. \quad \ldots (3)$$

$$\therefore \quad x = 10 \text{ mL.}$$

From the eqns. (1) and (2), eliminating moles of H_2O we get

$$y + 44 = 20 + x + 2(20 - x - y)$$

Substituting the value of x,

$$y + 44 = 20 + 10 + 2(20 - 10 - y)$$

$$3y = 6$$

$$y = 2 \text{ mL}$$

$\therefore$ volume of $CH_4 = 20 - (10 + 2) = 8$ mL.

$\therefore$ percentage volume of H_2 in the mixture $= \frac{10}{20} \times 100 = 50\%$

percentage volume of CO in the mixture $= \frac{2}{20} \times 100 = 10\%$

percentage volume of CH_4 in the mixture $= \frac{8}{20} \times 100 = 40\%$.

Ex. 18. 1 *litre of a mixture of* CO *and* CO_2 *is taken. This mixture is passed through a tube containing red hot charcoal. The volume now becomes* 1·6 *litres. The volumes are measured under the same condition. Find the composition of the mixture by volume.*

Solution : When the mixture of CO and CO_2 is passed through red hot charcoal, CO_2 changes to CO.

Let the volume of CO_2 in the mixture be x litres

i.e., $CO_2 = x$ litres, $CO = (1 - x)$ litres in the mixture.

On passing through red hot charcoal,

$$\underset{x \text{ litre}}{CO_2} \xrightarrow{C} CO$$

Since O atoms are conserved, applying POAC for O atoms,

$$2 \times \text{moles of } CO_2 = 1 \times \text{moles of } CO$$

or $\quad 2 \times \text{volume of } CO_2 = 1 \times \text{volume of } CO$

$$2 \times x = \text{volume of } CO$$

or $\quad \text{volume of } CO = 2x.$

$\therefore$ total volume of $CO = (1 - x) + 2x = (1 + x)$

and $\quad 1 + x = 1{\cdot}6$, (given)

$\therefore \quad x = 0{\cdot}6$ litre.

$\therefore$ volume of CO_2 in the mixture $= 0{\cdot}6$ litre.

Volume of CO in the mixture $= 1 - 0{\cdot}6 = 0{\cdot}4$ litre.

Ex. 19. 10 mL *of a mixture of* CO, CH_4 *and* N_2, *exploded with excess of oxygen, gave a contraction of* 6·5 mL. *There was a further contraction of* 7 mL *when the residual gas was treated with* KOH. *What is the composition of the original mixture?*

Solution : In the explosion, N_2 does not take part in the reaction, while CO and CH_4 change to CO_2 and H_2O, volume of H_2O (water) being zero.

Let the volume of $CO = x$ mL

$$CH_4 = y \text{ mL}$$

$\therefore \quad N_2 = (10 - x - y)$ mL.

The explosion reaction is

$$\underset{\substack{x \text{ mL} \\ x \text{ moles}}}{CO} + \underset{\substack{y \text{ mL} \\ y \text{ moles}}}{CH_4} + O_2 \rightarrow \underset{\substack{7 \text{ mL} \\ 7 \text{ moles}}}{CO_2} + H_2O$$

(or: x moles, y moles, 7 moles)

Applying POAC for C, O and H atoms, we get respectively,

moles of CO + moles of CH_4 = moles of CO_2

$$x + y = 7 \qquad \ldots (1)$$

moles of CO + 2 × moles of O_2 = 2 × moles of CO_2 + moles of H_2O

$$x + 2 \times \text{moles of } O_2 = 14 + \text{moles of } H_2O \quad \dots (2)$$

and $$4 \times \text{moles of } CH_4 = 2 \times \text{moles of } H_2O \quad \dots (3)$$

$$4y = 2 \times \text{moles of } H_2O.$$

From eqns. (2) and (3), we get

$$\text{moles of } O_2 = \left(7 + y - \frac{x}{2}\right) \cdot \quad \text{(used in explosion)}$$

Now, again we consider the reaction,

CO	+	CH_4	+	O_2	→	CO_2	+	H_2O
x moles		y moles		$[7 + y - (x/2)]$ moles		7 moles		
or x mL		y mL		$\left(7 + y - \frac{x}{2}\right)$ mL		7 mL		

As given in the problem,

volume of reactants – volume of products = 6·5 mL

volume of ($CO + CH_4 + O_2$) – vol. of CO_2 (vol. of water = 0) = 6·5 mL

(N_2 has not been included as it would be in both reactants and products)

$$x + y + \left(7 + y - \frac{x}{2}\right) - 7 = 6{\cdot}5$$

$$x + 4y = 13. \quad \dots (4)$$

From eqns. (1) and (4), we get,

$$x = 5;\ y = 2$$

$$\therefore \begin{cases} \text{vol. of } CO = 5 \text{ mL} \\ \text{vol. of } CH_4 = 2 \text{ mL} \\ \text{vol. of } N_2 = 10 - 5 - 2 = 3 \text{ mL} \end{cases}$$

Ex. 20. 10 mL *of a mixture of* CH_4, C_2H_4 *and* CO_2 *was exploded with excess of air. After explosion there was a contraction of* 17 mL *and after treatment with* KOH, *there was a further reduction of* 14 mL. *What was the composition of the mixture?*

Solution : In the explosion the reactant CO_2 does not change, while CH_4 and C_2H_4 change to CO_2 and H_2O. The volume of H_2O is taken to be zero.

∴ vol. of CO_2 in the reactant + CO_2 produced = 14 mL.

Let the volume of CH_4 and C_2H_4 in the mixture be respectively x and y mL.

∴ volume of CO_2 in the mixture = $(10 - x - y)$

and vol. of CO_2 produced on explosion = $14 - (10 - x - y)$

$$= (4 + x + y) \text{ mL}.$$

Now, we know,

CH_4	+	C_2H_4	+	O_2	$\rightarrow$	CO_2	+	H_2O
x mL		y mL				$(4+x+y)$ mL		
x moles		y moles				$(4+x+y)$ moles		

Applying POAC for C, H and O atoms, we get respectively,

$$1 \times \text{moles of } CH_4 + 2 \times \text{moles of } C_2H_4 = 1 \times \text{moles of } CO_2,$$

$$x + 2y = 4 + x + y; \quad y = 4 \qquad \dots (1)$$

$$4 \times \text{moles of } CH_4 + 4 \times \text{moles of } C_2H_4 = 2 \times \text{moles of } H_2O$$

$$4x + 4y = 2 \times \text{moles of } H_2O \qquad \dots (2)$$

$$2 \times \text{moles of } O_2 = 2 \times \text{moles of } CO_2 + 1 \times \text{moles of } H_2O$$

$$2 \times \text{moles of } O_2 = 2\,(4 + x + y) + \text{moles of } H_2O \qquad \dots (3)$$

From eqns. (2) and (3), eliminating moles of H_2O, we get

moles of $O_2 = (4 + 2x + 2y)$. (used in explosion)

Now, again consider the explosion reaction,

CH_4	+	C_2H_4	+	O_2	$\rightarrow$	CO_2	+	H_2O
x moles		y moles		$(4+2x+2y)$ moles		$(4+x+y)$ moles		
x mL		y mL		$(4+2x+2y)$ mL		$(4+x+y)$ mL		

$\because$ vol. of reactants – vol. of products = 17 mL (as given)

$\therefore \; x + y + (4 + 2x + 2y) - (4 + x + y) = 17$

or $$2x + 2y = 17. \qquad \dots (4)$$

From eqns. (1) and (4), we get,

$$x = 4{\cdot}5$$

$$\therefore \begin{cases} \text{vol. of } CH_4 = 4{\cdot}5 \text{ mL} \\ \text{vol. of } C_2H_4 = 4{\cdot}0 \text{ mL} \\ \text{vol. of } CO_2 = (10 - 4{\cdot}5 - 4) \text{ mL} = 1{\cdot}5 \text{ mL} \end{cases}$$

Ex. 21. 60 mL *of a mixture of nitrous oxide and nitric oxide was exploded with excess of hydrogen. If* 38 mL *of* N_2 *was formed, calculate the volume of each gas in the mixture.*

Solution : Let the volume of NO in the mixture be x mL.

	NO	+	N_2O	+	H_2	$\rightarrow$	N_2	+	H_2O
	x mL		$(60-x)$ mL				38 mL		
or	x moles		$(60-x)$ moles				38 moles		

Since N atoms are conserved, applying POAC for N atoms, we get,

$$1 \times \text{moles of NO} + 2 \times \text{moles of } N_2O = 2 \times \text{moles of } N_2$$

$$x + 2\,(60 - x) = 2 \times 38$$

$$x = 44.$$

Hence, volume of NO = 44 mL

and volume of $N_2O = (60 - 44)$ mL = 16 mL

Ex. 22. *A mixture of formic acid and oxalic acid is heated with concentrated* H_2SO_4. *The gas produced is collected and on its treatment with* KOH *solution the volume of the gas decreased by one-sixth. Calculate the molar ratio of the two acids in the original mixture.*

Solution : The decomposition of the acids takes place as follows:

HCOOH	→	H_2O	+	CO	
a moles (say)		a moles		a moles	after decomposition
		a vol.		a vol.	

$H_2C_2O_4$	→	H_2O	+	CO	+	CO_2	
b moles (say)		b moles		b moles		b moles	after decomposition
		b vol.		b vol.		b vol.	

H_2O is absorbed by H_2SO_4 and CO_2 is absorbed by KOH.

Thus, as given,

$$\frac{\text{volume of } CO_2}{\text{total volume of } (CO + CO_2)} = \frac{b}{a+b+b} = \frac{1}{6}$$

or
$$\frac{a}{b} = 4.$$

The molar ratio of HCOOH and $H_2C_2O_4 = 4 : 1$.

PROBLEMS

(Answers bracketed with questions)

1. Calculate the volume of CO_2 produced by the combustion of 40 mL of acetone in the presence of excess of oxygen. (120 mL)

2. What volume of air will be required to oxidise 210 mL of sulphur dioxide to sulphur trioxide, if the air contains 21% of oxygen? (500 mL)

3. What volume of CO_2 is obtained in the combustion of 2 litres of butane? (8 litres)

4. If a mixture containing 12 litres of hydrogen and 11·2 litres of chlorine is exploded in an eudiometer tube, what will be the composition of the resulting mixture by volume? (HCl = 22·4 litres, H_2 = 0·8 litre)

5. What volume of oxygen is required for complete combustion of 2·2 g of propane at NTP? (5·6 litres)

6. 500 mL of a hydrocarbon gas, burnt in excess of oxygen, yields 2500 mL of CO_2 and 3 litres of water vapour, all volumes being measured at the same temperature and pressure. What is the formula of the hydrocarbon? (C_5H_{12})

7. When 0·02 litre of a mixture of hydrogen and oxygen was exploded, 0·003 litre of oxygen remained. Calculate the initial composition of the mixture in per cent by volume. (O_2 : 44·0%; H_2 : 56%)

8. 12 mL of a gaseous hydrocarbon was exploded with 50 mL of oxygen. The volume measured after explosion was 32 mL. After treatment with KOH the volume diminished to 8 mL. Determine the formula of the hydrocarbon. (C_2H_6)

9. 15 mL of a gaseous hydrocarbon was required for complete combustion in 357 mL of air (21% of oxygen by volume) and the gaseous products occupied 327 mL (all volumes being measured at NTP). What is the formula of the hydrocarbon? (C_3H_8)

10. 0·90 g of a solid organic compound (molecular weight = 90) containing C, H and O was heated with oxygen corresponding to a volume of 224 mL at STP. After the combustion the total volume of the gases was 560 mL at STP. On treatment with KOH the volume decreased to 112 mL. Determine the molecular formula of the compound. ($C_2H_2O_4$)

11. The explosion of a mixture consisting of one volume of a gas being studied and one volume of H_2 yielded one volume of water vapour and one volume of N_2, the volumes being measured under identical conditions. Find the formula of the gas being studied. (N_2O)

12. 5 mL of a gas containing C and H was mixed with an excess of oxygen (30 mL) and the mixture exploded by means of an electric spark. After the explosion the volume of the mixed gases remaining was 25 mL. On adding a concentrated solution of KOH, the volume further diminished to 15 mL, the residual gas being pure oxygen. All volumes have been reduced to NTP. Calculate the molecular formula of the hydrocarbon gas. (C_2H_4)

13. 40 mL of ammonia gas taken in an eudiometer tube was subjected to sparks till the volume did not change any further. The volume was found to increase by 40 mL. 40 mL of oxygen was then mixed and the mixture was further exploded. The gases remained were 30 mL. Deduce the formula of ammonia. (NH_3)

14. 20 mL of a gas containing H and S was heated with tin. When the reaction was over, there was no change in volume. The residual gas was hydrogen. If the molecular weight of the gas is 34, calculate the molecular formula. (H_2S)

15. When a certain quantity of oxygen was ozonised in a suitable apparatus, the volume decreased by 4 mL. On addition of turpentine the volume further decreased by 8 mL. All volumes were measured at the same temperature and pressure. From these data, establish the formula of ozone. (O_3)

16. 1 litre of a sample of ozonised oxygen weighs 1·5 g at 0°C and one atm pressure. 100 mL of this sample reduced to 90 mL when treated with turpentine under the same conditions. Find the molecular weight of ozone. (48)

17. 280 mL of sulphur vapour at NTP weighs 3·2 g. Determine the molecular formula of the sulphur vapour. (S_8)

18. 1 litre of oxygen and 1 litre of hydrogen are taken in a vessel of 2-litre capacity at NTP. The gases are made to combine by applying electric sparks. Assume that water is formed quantitatively. How many grams of water are formed? What is the other component present in the vessel and in what weight? If the vessel is now heated to 100°C, what will be the pressure inside the vessel in mmHg?

(0·8036 g; O_2; 0·7143 g; 778 mm)

19. 20 mL of a mixture of C_2H_2 and CO was exploded with 30 mL of oxygen. The gases after the reaction had a volume of 34 mL. On treatment with KOH, 8 mL of oxygen remained. Calculate the composition of the mixture.

(C_2H_2 : 6 mL; CO : 14 mL)

20. On passing 25 mL of a gaseous mixture of N_2 and NO over heated copper, 20 mL of the gas remained. Calculate the percentage of each in the mixture.

(N_2 : 60%; NO : 40%)

21. 40 mL of a mixture of hydrogen, CH_4 and N_2 was exploded with 10 mL of oxygen. On cooling, the gases occupied 36·5 mL. After treatment with KOH, the volume reduced by 3 mL and again on treatment with alkaline pyrogallol, the volume further decreased by 1·5 mL. Determine the composition of the mixture.

(H_2 : 12·50%; CH_4 : 7·50%; N_2 : 80%)

22. What volume of air is needed for the combustion of 1 metre^3 of a gas having the following composition in percentage volume:
50% of H_2, 35% of CH_4, 8% of CO, 2% of C_2H_4 and 5% of noncombustible admixture? The air contains 21% (by volume) of oxygen. (5 cubic metre)

23. 38 mL of a mixture of CO and H_2 was exploded with 31 mL of O_2. The volume after the explosion was 29 mL which reduced to 12 mL when shaken with KOH. Find the percentage of CO and H_2 in the mixture. (CO = 44·7% ; H_2 = 55·3%)

24. A mixture of CH_4 and C_2H_2 occupied a certain volume at a total pressure of 63 mm. The sample was burnt to CO_2 and H_2O and the CO_2 alone was collected and its pressure was found to be 69 mm in the same volume and at the same temperature as the original mixture. What fraction of the mixture was methane? (0·90)

25. A mixture of 20 mL of CO, CH_4 and N_2 was burnt in an excess of oxygen, resulting in reduction of 13 mL of volume. The residual gas was treated with KOH solution when there was a further reduction of 14 mL in volume. Calculate the volumes of CO and CH_4 in the given mixture. (10 mL, 4 mL)

26. A mixture of oxygen and hydrogen is analysed by passing it over hot copper oxide and through a drying tube. Hydrogen reduces the CuO according to the equation, $CuO + H_2 \rightarrow Cu + H_2O$; oxygen then oxidises the copper formed:

$$Cu + \frac{1}{2}O_2 \rightarrow CuO$$

100 cm^3 of the mixture measured at 25°C and 750 mm yields 84·5 cm^3 of dry oxygen measured at 25°C and 750 mm after passing over CuO and a drying agent. What is the mole per cent of H_2 in the mixture? (10·3%)

[**Hint:** $\underset{x}{CuO + H_2} \rightarrow \underset{x}{Cu + H_2O}\,;\, \underset{x}{Cu} + \underset{x/2}{\frac{1}{2}O_2} \rightarrow CuO$

If x is moles of H_2 or vol. of H_2 in the mixture, $100 - \left(x + \frac{x}{2}\right) = 84{\cdot}5.$]

27. When a mixture consisting of 10 moles of SO_2 and 15 moles of O_2 was passed over a catalyst, 8 moles of SO_3 was formed. How many moles of SO_2 and O_2 did not enter into the reaction?

$$\begin{bmatrix} SO_2: 2 \text{ moles} \\ O_2: 11 \text{ moles} \end{bmatrix}$$

28. When 100 mL of an O_2–O_3 mixture was passed through turpentine, there was reduction of volume by 20 mL. If 100 mL of such a mixture is heated, what will be the increase in volume? (10 mL)

[Hint: O_3 is absorbed by turpentine.]

29. One volume of a compound of carbon, hydrogen and oxygen was exploded with 2·5 volumes of oxygen. The resultant mixture contained 2 volumes of water vapour and 2 volumes of carbon dioxide. All volumes were measured in identical conditions. Determine the formula of the compound. (C_2H_4O)

30. $5{\cdot}22 \times 10^{-4}$ mole of a gas containing H_2, O_2 and N_2 exerted a pressure of 67·4 mm in a certain standard volume. The gas was passed over a hot platinum filament which combined H_2 and O_2 into H_2O which was frozen out. When the gas was returned to the same volume, the pressure was 14·3 mm. Extra oxygen was added to increase the pressure to 44·3 mm. The combustion was repeated, after which the pressure read 32·9 mm. What was the mole fraction of H_2, O_2 and N_2 in the gas sample? (0·638, 0·262, 0·1)

[Hint: 2 volumes of H_2 combine with 1 vol. of O_2. In the first combustion H_2 is in excess, while in the second one, O_2 is in excess.]

31. At 300 K and 1 atm, 15 mL of gaseous hydrocarbon requires 375 mL air containing 20% O_2 by volume for complete combustion. After combustion, the gases occupy 345 mL. Assuming that water formed is in liquid form and the volumes were measured at the same temperatrue and pressure, the formula of the hydrocarbon is

(a) C_3H_8 (b) C_4H_8 (c) C_4H_{10} (d) C_2H_6

[Hint: Vol. of $O_2 = 75$ mL, vol. of air = 300 mL and vol. of $CO_2 = 45$ mL. Apply POAC for C, H and O atoms.]

❑

CHAPTER FOUR

ATOMIC WEIGHT

The atomic weight of an element is defined as the average weight of the atoms of the element relative to a carbon atom, taken as exactly 12. **Atomic weight in grams is, in fact, the weight of one mole of atoms,** e.g., the atomic weight of oxygen is 16 and so 16 grams is the weight of 1 mole of oxygen atoms. Mathematically,

$$\text{atomic weight} = \frac{\text{weight of atoms in grams}}{\text{number of moles of atoms}}.$$

Atomic weight is measured in atomic mass unit (amu). Atomic mass unit is defined as $\frac{1}{12}$ of the mass of the ^{12}C isotope (1 amu = $1{\cdot}66 \times 10^{-24}$ g). One amu is also called one **dalton**.

We shall see here how the mole method can be applied in solving the problems on atomic weight by on different methods, viz., from Dulong and Petit's law, vapour density of chloride of the elements, law of isomorphism, Cannizzaro's method, etc.

EXAMPLES

Ex. 1. *A sample of pure* Ca *metal weighing* 1·35 *grams was quantitatively converted to* 1·88 *grams of pure* CaO. *What is the atomic weight of* Ca? (O = 16)

Solution : From the formula of CaO,

we know, moles of Ca = moles of O

$$\frac{\text{weight of Ca}}{\text{atomic wt. of Ca}} = \frac{\text{weight of O}}{\text{atomic wt. of O}} \qquad \text{(Rule 2, Chapter 1)}$$

$$\frac{1{\cdot}35}{\text{at. wt. of Ca}} = \frac{1{\cdot}88 - 1{\cdot}35}{16}.$$

Atomic weight of Ca = 40·75.

Ex. 2. *A compound contains* 28% *of nitrogen and* 72% *of a metal by weight.* 3 *atoms of the metal combine with* 2 *atoms of nitrogen. Find the atomic weight of the metal.* (N = 14)

Solution : The formula of the compound is M_3N_2 (M representing the metal)

$2 \times$ moles of M $= 3 \times$ moles of N. (Rule 6, Chapter 1)

Now if the weight of the compound is 1 gram

then weight of M = 0·72 g and weight of N = 0·28 g.

$$\therefore \qquad 2\times\frac{0{\cdot}72}{\text{atomic weight of M}}=3\times\frac{0{\cdot}28}{14}.$$

Atomic weight of the metal = 24.

Ex. 3. 1·00 g *of* $EuCl_2$ *is treated with excess of aqueous* $AgNO_3$ *and all the chlorine is recovered as* 1·29 g *of* AgCl. *Calculate the atomic weight of* Eu. (Cl = 35·5, Ag = 108)

Solution : $EuCl_2 + AgNO_3 \rightarrow AgCl$

1 g 1·29 g

Since Cl atoms are conserved, applying POAC for Cl atoms,

$$\text{moles of Cl in } EuCl_2 = \text{moles of Cl in AgCl}$$

$$2\times\text{moles of } EuCl_2 = 1\times\text{moles of AgCl}$$

$$2\times\frac{1}{(x+35{\cdot}5\times 2)}=1\times\frac{1{\cdot}29}{(108+35{\cdot}5)}.$$

(where $x \equiv$ atomic weight of Eu)

$$\therefore \qquad x = 152{\cdot}48.$$

Ex. 4. *Two elements* A (*at. wt.* = 75) *and* B (*at. wt.* = 16) *combine to give a compound having* 75·8% *of* A. *What is the formula of the compound?*

Solution : 100 g of the compound should contain 75·8 g of A and 24·2 g of B.

$$\left.\begin{aligned}\text{Moles of A} &= \frac{\text{weight of A}}{\text{atomic wt. of A}}=\frac{75{\cdot}8}{75}\cong 1\\ \text{Moles of B} &= \frac{\text{weight of B}}{\text{atomic wt. of B}}=\frac{24{\cdot}2}{16}=1{\cdot}5\end{aligned}\right\}\quad\text{(Rule 2, Chapter 1)}$$

Thus, molar ratio of A and B in whole numbers is 2 : 3, and the formula is, therefore, A_2B_3.

From Cannizzaro's Method

Ex. 5. *The following data were obtained from the analysis of five volatile phosphorous compounds. The percentages of phosphorous in these compounds having the molecular weights* 222, 154, 140, 300 *and* 126 *are respectively* 27·9%, 20·2%, 22·5%, 43·7% *and* 24·6%. *Calculate the atomic weight of phosphorous* (*to the nearest whole number*).

Solution : The smallest weight of an element present in 1 mole of its various compounds gives either the atomic weight of the element or a simple multiple of the atomic weight.

(i) Weight of P in 1 mole (222 g) $=\frac{27{\cdot}9}{100}\times 222 = 61{\cdot}94.$

(ii) Weight of P in 1 mole (154 g) $=\frac{20{\cdot}2}{100}\times 154 = 31{\cdot}08.$

(iii) Weight of P in 1 mole (140 g) $= \frac{22.5}{100} \times 140 = 31.50.$

(iv) Weight of P in 1 mole (300 g) $= \frac{43.7}{100} \times 300 = 131.10.$

(v) Weight of P in 1 mole (126 g) $= \frac{24.6}{100} \times 126 = 30.99.$

The smallest weight of P being 31 (to the nearest whole number) may be its atomic weight.

From Dulong and Petit's Law

Ex. 6. 7·38 g *of a sample of a metal oxide is quantitatively reduced to* 6·84 g *of pure metal. If the specific heat of the metal is* 0·0332 cal/g, *calculate the valency and the accurate atomic weight of the metal.*

Solution : Let the formula of the oxide be M_2O_x, where x is the valency of the metal M. We have, therefore,

$x \times$ moles of M = 2 × moles of O. (Rule 6, Chapter 1) ... (1)

We know that, atomic weight × specific heat ≈ 6·4 (Dulong and Petit's law)

$$\therefore \text{ approximate atomic weight} = \frac{6.4}{0.0332} = 193$$

From eqn. (1), $x \times \frac{6.84}{193} = 2 \times \frac{(7.38 - 6.84)}{16}$ (Rule 2, Chapter 1)

or $x = 1.9.$

But valency is always a whole number and so the valency of the metal is 2. Now, to calculate the accurate value of the atomic weight of the metal, substitute the value of x as 2 in equation (1) again,

or $$2 \times \frac{6.84}{\text{atomic wt.}} = 2 \times \frac{(7.38 - 6.84)}{16}.$$

Atomic weight (accurate) = 202·67.

Ex. 7. *On dissolving* 2·0 g *of a metal in sulphuric acid,* 4·51 *grams of the metal sulphate was formed. The specific heat of the metal is* 0·057 cal/g. *What is the valency of the metal and its exact atomic weight?*

Solution : Let the formula of the metal sulphate be $M_2(SO_4)_x$ (x being the valency of the metal M)

$$\therefore \quad x \times \text{moles of M} = 2 \times \text{moles of } SO_4^{2-}$$

or $$x \times \frac{\text{wt. of M}}{\text{atomic weight of M}} = 2 \times \frac{\text{wt. of sulphate ion}}{\text{ionic wt. of } SO_4^{2-}} \quad \ldots (1)$$

$$x \times \frac{2}{112} = 2 \times \frac{(4.51 - 2)}{96}.$$

$$x = 2{\cdot}85 \approx 3.$$

$$\left\{ \text{at. wt. of M} \approx \frac{6{\cdot}4}{0{\cdot}057} \approx 112 \text{ (approx.)} \right\}$$

The valency is, therefore, 3.

Now, substitute the valency of x as 3 in eqn. (1) again to calculate the accurate value of the atomic weight of the metal.

$$3 \times \frac{2}{\text{atomic weight}} = 2 \times \frac{(4{\cdot}51 - 2)}{96}.$$

Accurate atomic weight = 114·7.

From Vapour Density of the Chloride of the Element

Ex. 8. 1·00 g *of a chloride of an element contains* 0·835 g *of chlorine. If the vapour density of the chloride is* 85, *find the atomic weight of the element and its valency.*

Solution : Let the formula of the chloride be MCl_x, where x is the valency of the element M.

$$x \times \text{moles of M} = 1 \times \text{moles of Cl} \qquad \text{(Rule 6, Chapter 1)}$$

or $$x \times \frac{1 - 0{\cdot}835}{\text{at. wt. of M}} = \frac{0{\cdot}835}{35{\cdot}5}. \qquad \dots (1)$$

Now, we know,

molecular weight = 2 × vapour density

= 2 × 85 = 170.

For MCl_x,

mol. wt. of MCl_x = at. wt. of M + x × at. wt. of Cl

or $$170 = \text{at. wt. of M} + 35{\cdot}5\,x.$$

∴ $$\text{at. wt. of M} = 170 - 35{\cdot}5\,x$$

or $$x = \frac{170 - \text{at. wt. of M}}{35{\cdot}5}. \qquad \dots (2)$$

From (1) and (2), we get, at. wt. of M = 28·05.

Substituting at. wt. of M in the eqn. (1) we get, $x = 4$.

Ex. 9. *The percentage of chlorine in the chloride of an element is* 44·71%. 158·5 g *of this chloride on vaporisation occupies a volume of* 22·4 *litres at* NTP. *Calculate the atomic weight and valency of the element.*

Solution : Since 22·4 litres are occupied by 1 mole at NTP and wt. of 1 mole is the mol. wt. in grams,

∴ mol. wt. of chloride = 158·5.

Let the formula of the chloride be MCl_x, where x is the valency of the element M.

$$x \times \text{moles of M} = 1 \times \text{moles of Cl}$$

or $$x \times \frac{\text{wt. of M}}{\text{at. wt. of M}} = 1 \times \frac{\text{wt. of Cl}}{\text{at. wt. of Cl}}. \quad \dots (1)$$

Now, mol. wt. of MCl_x = at. wt. of M + 35·5x

$$= 158·5.$$

$\therefore$ at. wt. of M = 158·5 − 35·5 x.

Again, if the wt. of the chloride is 1 g,

$$\text{wt. of Cl} = 0·4471 \text{ g}.$$

$\therefore$ $$\text{wt. of M} = 1 - 0·4471 = 0·5529 \text{ g}.$$

Equation (1) becomes,

$$x \times \frac{0·5529}{158·5 - 35·5\,x} = \frac{0·4471}{35·5}$$

$$x = 1·996 \approx 2.$$

$\therefore$ valency = 2.

From Law of Isomorphism

Ex. 10. *The equivalent weight of an element is* 13·16. *It forms a salt, isomorphous with* K_2SO_4. *Deduce the atomic weight of the element.*

Solution **:** Isomorphous substances have the same formula as they have the same number of atoms similarly arranged, e.g., in this problem the salt of the element X (say) will have the formula X_2SO_4 of the type of K_2SO_4. The element X will have, therefore, the valency equal to that of K, i.e., 1.

Atomic weight = equivalent weight × valency

$$= 13·16 \times 1 = 13·16.$$

Ex. 11. Cu_2S *and* Ag_2S *are isomorphous in which percentages of sulphur are* 20·14% *and* 12·94% *respectively. Calculate the atomic weight of silver.* (Cu = 63·5)

Solution **:** Suppose the atomic weight of S and Ag are x and y respectively.

Now, for Cu_2S,

$$1 \times \text{moles of Cu} = 2 \times \text{moles of S} \quad \text{(Rule 6, Chapter 1)}$$

or $$1 \times \frac{0·7986}{63·5} = 2 \times \frac{0·2014}{x}. \quad \dots (1)$$

(1 g of Cu_2S contains 0·7986 g and 0·2014 g of Cu and S respectively)

And, for Ag_2S,

$$1 \times \text{moles of Ag} = 2 \times \text{moles of S}$$

$$1 \times \frac{0{\cdot}8706}{y} = 2 \times \frac{0{\cdot}1294}{x}. \qquad \text{... (2)}$$

(1 g of Ag_2S contains 0·8706 g and 0·1294 g of Ag and S respectively)

From eqns. (1) and (2), we get,

$$y = 107{\cdot}7.$$

∴ atomic weight of silver is 107·70.

Ex. 12. 4·08 g *of a mixture of* BaO *and an unknown carbonate* MCO_3 *was heated strongly. The residue weighed* 3·64 g. *This was dissolved in* 100 mL *of* N HCl. *The excess acid required* 16 mL *of* 2·5 N NaOH *solution for complete neutralisation. Identify the metal* M.

(See Example 54, Chapter 7)

Ex. 13. 2·180 g *of a sample contains a mixture of* XO *and* X_2O_3 *which are completely oxidised to* XO_4^- *by* 0·015 mole *of* $K_2Cr_2O_7$. *Calculate the atomic weight of* X *if* 0·0187 mole *of* XO_4^- *is formed.*

Solution **:** Let the atomic weight of X be A.

$$\underset{+2}{XO} \rightarrow \underset{+7}{XO_4^-}$$

$$\therefore \text{ equivalent weight of XO} = \frac{\text{mol. wt.}}{\text{change in ON}} = \frac{A+16}{5}.$$

$$\underset{+6}{X_2O_3} \rightarrow \underset{+14}{2XO_4^-} \quad \therefore \text{ eq. wt. of } X_2O_3 = \frac{2A+48}{8}$$

$$\underset{+12}{K_2Cr_2O_7} \rightarrow \underset{+6}{2Cr^{3+}}$$

$$\therefore \text{ no. of eq. of } K_2Cr_2O_7 = \text{no. of moles of } K_2Cr_2O_7 \times \text{change in ON}$$
$$= 0{\cdot}015 \times 6 = 0{\cdot}09.$$

As both XO and X_2O_3 are oxidised to XO_4^- by $K_2Cr_2O_7$

$$\therefore \qquad \text{eq. of XO} + \text{eq. of } X_2O_3 = \text{eq. of } K_2Cr_2O_7$$

$$\frac{\text{wt. of XO}}{\text{eq. wt. of XO}} + \frac{\text{wt. of } X_2O_3}{\text{eq. wt. of } X_2O_3} = \text{eq. of } K_2Cr_2O_7$$

Let the weight of XO be x grams.

$$\therefore \qquad \frac{x}{(A+16)/5} + \frac{2{\cdot}18 - x}{(2A+48)/8} = 0{\cdot}09. \qquad \text{... (1)}$$

Further,

$$XO + X_2O_3 \rightarrow XO_4^-$$

Applying POAC for X atoms,

moles of X in XO + moles of X in X_2O_3 = moles of X in XO_4^-.

$1 \times$ moles of XO $+ 2 \times$ moles of $X_2O_3 = 1 \times$ moles of XO_4^-

or $$\frac{x}{A+16} + \frac{2(2{\cdot}18 - x)}{2A+48} = 0{\cdot}0187. \quad \dots (2)$$

On solving eqns. (1) and (2), we get

$$A = 99.$$

$\therefore$ atomic weight of X is 99.

Ex. 14. *Two elements* P *and* Q *form the compounds* P_2Q_3 *and* PQ_2. *If* 0·15 *mole of* P_2Q_3 *weighs* 15·9 g *and* 0·15 *mole of* PQ_2 *weighs* 9·3 g, *find the atomic weights of* P *and* Q.

Solution **:** Molecular weight of P_2Q_3 = wt. of 1 mole

$$= \frac{15{\cdot}9 \times 1}{0{\cdot}15} = 106.$$

Molecular weight of PQ_2 = wt. of 1 mole $= \frac{9{\cdot}3 \times 1}{0{\cdot}15} = 62.$

Suppose that the atomic weights of P and Q are respectively x and y. Thus,

$$2x + 3y = 106 \quad (P_2Q_3 = 106)$$

and $$x + 2y = 62. \quad (PQ_2 = 62)$$

On solving: $x = 26$

$y = 18.$

Ex. 15. *A compound which contains one atom of* X *and two atoms of* Y *for each three atoms of* Z *is made by mixing* 5·00 g *of* X, $1{\cdot}15 \times 10^{23}$ *atoms of* Y *and* 0·03 mole *of* Z *atoms. Given that only* 4·40 g *of the compound is formed. Calculate the atomic weight of* Y *if the atomic weights of* X *and* Z *are* 60 *and* 80 amu *respectively.*

Solution **:** Moles of X $= \frac{5}{60} = 0{\cdot}08$ mole. (Rule 2, Chapter 1)

Moles of Y $= \frac{1{\cdot}15 \times 10^{23}}{6{\cdot}022 \times 10^{23}} = 0{\cdot}19$ mole (Rule 4, Chapter 1)

Moles of Z = 0·03 mole.

Since the formula of the compound is XY_2Z_3,

moles of X : moles of Y : moles of Z = 1 : 2 : 3 = 0·01 : 0·02 : 0·03.

Comparing these values with the moles of X, Y and Z calculated above, we find that moles of X and Y are in excess and therefore, moles of X and Y associated with 0·03 mole of Z are 0·01 and 0·02 mole respectively. Now,

$$\text{wt. of Z + wt. of Y + wt. of Z = wt. of } XY_2Z_3$$

or $0{\cdot}01 \times 60 + 0{\cdot}02 \times \text{at. wt. of } Y + 0{\cdot}03 \times 80 = 4{\cdot}4$ (Rule 2, Chapter 1)

$\therefore$ at. wt. of $Y = 70$ amu.

PROBLEMS

(Answers bracketed with questions)

1. 1·5276 grams of $CdCl_2$ was found to contain 0·9367 gram of Cd. Calculate the atomic weight of Cd. (112·4 amu)

2. In air, element X is oxidised to compound XO_2 . If 1·0 gram of X reacts with 0·696 g of oxygen, what is the atomic weight of X? (46 amu)

3. When $BaBr_2$ is heated in a stream of chlorine, it is completely converted to $BaCl_2$. From 1·50 g of $BaBr_2$, just 1·05 g of $BaCl_2$ is obtained. Calculate the atomic weight of Ba from this data. (137·5)

4. Naturally occurring boron consists of two isotopes whose atomic weights are 10·01 and 11·01. The atomic weight of natural boron is 10·81. Calculate the percentage of each isotope in natural boron. (20%; 80%)

5. In an experiment pure carbon monoxide was passed over red hot copper oxide. CO_2, so produced, weighed 0·88 g and the weight of copper oxide was reduced by 0·3232 g. Calculate the atomic weight of carbon. (12·0)

6. 4·008 g of pure $KClO_3$ was quantitatively decomposed to give 2·438 g of KCl and oxygen. KCl was dissolved in water and treated with $AgNO_3$ solution. The result was a precipitate of AgCl weighing 4·687 g. Under further treatment, AgCl was found to contain 3·531 g of Ag. What are the atomic weights of Ag, Cl and K relative to O = 16? (108·0, 35·5, 39·0)

7. From the following data calculate the atomic weight of carbon.

Compound	*Vapour density*	*% of C*
CO	14	42·8
CS_2	38	15·8
C_2H_2	14	85·7
C_3H_4	22	81·4
C_6H_6	39	92·3

(12·0)

8. The vapour density of a volatile chloride of a metal is 74·6. If the specific heat of the metal is 0·55, calculate the exact atomic weight of the metal and the formula of its chloride. (7·2, MCl_4)

9. A sample of a metal chloride weighing 0·22 g required 0·51 g of $AgNO_3$ to precipitate the chloride completely. The specific heat of the metal is 0·057. Find the molecular formula of the chloride if the metal is M. (MCl_3)

10. A hydrated sulphate of a metal contained 8·1% of the metal and 43·2% of SO_4^{2-} by weight. Assuming the specific heat of the metal to be 0·242, determine the formula of the hydrated sulphate. [$M_2(SO_4)_3 \cdot 18\,H_2O$]

11. 0·096 g of stannic chloride gave 25·84 mL of its vapour at 120°C and 350 mm pressure. If the chloride contains 54·6% of chlorine and tin has the valency equal to 4, what will be the atomic weight of tin? (118)

12. A metal forms three volatile chlorides containing 23·6%, 38·2% and 48·3% of chlorine respectively. The vapour densities of the chlorides (H = 1) are 74·6, 92·9 and 110·6 respectively. The specific heat of the metal is 0·055. Find the exact atomic weight of the metal and formulae of its chlorides.

(114·9; MCl; MCl_2 ; MCl_3)

13. Cu_2S and Ag_2S are isomorphous. The percentages of sulphur in these compounds are 20·14% and 12·94% respectively. If the atomic weight of Cu is 63·5, calculate the atomic weight of Ag. (107·9)

14. The natural titanium oxide, known as rutile and containing 39·95% of oxygen, is isomorphous with SnO_2, known as cassiterite. Calculate the atomic weight of titanium. (48·1)

15. The atomic weight of sulphur was determined by decomposing 6·2984 g of Na_2CO_3 with H_2SO_4 and weighing the resultant Na_2SO_4 formed. The weight was found to be 8·4380 g. Taking the atomic weights of C, O and Na as 12·011, 15·999 and 22·990 respectively, calculate the atomic weight of sulphur. (32·019)

16. 12·5843 grams of a sample of $ZrBr_4$ was dissolved and after several chemical steps, all of the combined bromine was precipitated as AgBr. The silver content of the AgBr was found to be 13·216 g. Calculate the atomic weight of Zr from this experiment. (Ag = 107·868, Br = 79·904) (91·32)

17. In a chemical determination of the atomic weight of vanadium, 2·8934 g of pure $VOCl_3$ was allowed to undergo a set of reactions as a result of which all the chlorine contained in this compound was converted to AgCl which weighed 7·1801 g. Calculate the atomic weight of vanadium. (Ag = 107·868, Cl = 35·453) (50·91)

❑

CHAPTER FIVE

MOLECULAR WEIGHT

Molecular weight of a compound is defined as the weight of a molecule of the compound relative to a carbon atom, the atomic weight of which is supposed to be exactly 12. The molecular weight when expressed in grams is called gram molecular weight. **The molecular weight in grams is, in fact, the weight of 1 mole of molecules, e.g.,** molecular weight of O_2 is 32 and 32 g is the weight of 1 mole of oxygen molecules. Mathematically,

$$\text{molecular weight} = \frac{\text{weight of molecules in grams}}{\text{number of moles of molecules}}.$$

Molecular weight is measured in atomic mass unit (amu). Atomic mass unit is defined as 1/12 of the mass of the ^{12}C isotope (1 amu $= 1{\cdot}66 \times 10^{-24}$ g). There are various methods to determine the molecular weight of compounds viz., vapour-density method (e.g., Regnault's method, Victor Meyer's method, Dumas method and Hofmann method), depression-in-freezing-point method, elevation-in-boiling-point method, gravimetric method, (silver-salt method of organic acids and platinichloride method for organic bases), volumetric method, etc. The problems on molecular weight based on depression-in-freezing-point and elevation-in-boiling-point methods shall be discussed in Chapter 13 (Dilute Solution and Colligative Propertries).

The mole method is found to be very useful in tackling the problems on molecular weight based on the aforesaid methods.

EXAMPLES

Ex. 1. 96·00 g *of a gas occupies the same volume as* 84 g *of nitrogen under similar conditions of temperature and pressure. Find the molecular weight of the gas.* (N = 14)

Solution : According to Avogadro's principle, equal volumes of all gases under identical conditions of temperature and pressure contain equal numbers of molecules or moles.

∴ number of moles of N_2 = number of moles of the gas

$$\frac{84}{28} = \frac{96}{M} \quad \left\{ \begin{array}{l} \text{say } M = \text{mol. wt. of the gas} \\ \quad N_2 = 28 \end{array} \right\}$$

$M = 32.$ (Rule 1, Chapter 1)

Ex. 2. 5 *litres of a gas weigh* 14·4 g *at* NTP. *Calculate its molecular weight.*

Solution : Number of moles of the gas $= \frac{5}{22 \cdot 4}$. (Rule 3, Chapter 1)

Since the weight of 1 mole in grams is numerically equal to mol. wt.,

$$\therefore \text{ molecular weight} = 14 \cdot 4 \times \frac{22 \cdot 4}{5}$$

$$= 64 \cdot 51.$$

Ex. 3. *It was found that* 380 mL *of a gas at* 27°C *and* 800 mmHg *weighed* 0·455 g. *Find the molecular weight of the gas.*

Solution : Volume of the gas at NTP $= \frac{800 \times 380}{300} \times \frac{273}{760}$ mL

$$= 364 \text{ mL}.$$

Number of moles of the gas $= \frac{364}{22400}$. (Rule 3, Chapter 1)

Since weight of 1 mole of the gas is numerically equal to mol. weight,

$$\text{molecular weight} = 0 \cdot 455 \times \frac{22400}{364} = 28.$$

Ex. 4. *When* 2 g *of a gas 'A' is introduced into an evacuated flask kept at* 25°C, *the pressure is found to be one atmosphere. If* 3 g *of another gas 'B' is then added to the same flask, the total pressure becomes* 1·5 atm. *Assuming ideal gas behaviour, calculate the ratio of the molecular weights,* $M_A : M_B$.

Solution : See Example 20, Chapter 12.

Ex. 5. $3 \cdot 011 \times 10^{23}$ *molecules of a gas weigh* 14 g. *Calculate the molecular weight of the gas.*

Solution : Number of moles $= \frac{3 \cdot 011 \times 10^{23}}{6 \cdot 022 \times 10^{23}} = 0 \cdot 5.$ (Rule 4, Chapter 1)

$\therefore$ molecular weight $= \frac{14}{0 \cdot 5} = 28.$ (Rule 1, Chapter 1)

Ex. 6. 100 mL *of a* 2 M *solution contains* 11·7 g *of a substance. Calculate the molecular weight of the compound.*

Solution : Number of moles of the solute = volume (litre) × molarity

$= 0 \cdot 1 \times 2 = 0 \cdot 2.$ (Rule 5, Chapter 1)

$\therefore$ molecular weight $= \frac{11 \cdot 7}{0 \cdot 2} = 58 \cdot 5.$ (Rule 1, Chapter 1)

Ex. 7. 116 g *of* Fe_3O_4 *has* 1·5 *moles of* Fe. *Calculate the molecular weight of* Fe_3O_4 *without using atomic weights of* Fe *and* O.

Solution : Since 1 mole of Fe_3O_4 contains 3 moles of Fe,

moles of Fe in Fe_3O_4 = 3 × moles of Fe_3O_4

$$1{\cdot}5 = 3 \times \frac{116}{\text{mol. wt. of } Fe_3O_4}. \qquad \text{(Rule 1, Chapter 1)}$$

$$\therefore \quad \text{mol. wt. of } Fe_3O_4 = \frac{3 \times 116}{1{\cdot}5} = 232.$$

Ex. 8. 3·7 g *of a gas at* 25°C *occupied the same volume as* 0·184 g *of hydrogen at* 17°C *and at the same pressure. What is the molecular weight of the gas?*

Solution : Volume of 3·7 g of the gas at 25°C

= volume of 184 g of H_2 at 17°C.

Since pressure and volume are same for both the gases,

$$n_1 R T_1 = n_2 R T_2 \qquad (pV = nRT)$$

or

$$n_1 T_1 = n_2 T_2.$$

Let the molecular weight of the gas be M.

$$\frac{3{\cdot}7}{M} \times 298 = \frac{0{\cdot}184}{2} \times 290$$

$$M = 41{\cdot}32.$$

Density, Vapour Density and Relative Density (or, Sp. Gr.)

$$\text{Absolute density} = \frac{\text{mass}}{\text{volume}}$$

$$\text{Vapour density} = \frac{\text{mass of 1 mole of vapour}}{\text{mass of 1 mole of hydrogen}}$$

$$= \frac{\text{molecular weight}}{2}$$

$$\text{Relative density (or, sp. gravity)} = \frac{\text{absolute density}}{\text{density of pure water at 4°C}}$$

$$= \text{absolute density}$$

Further,

$$\text{vapour density} = \frac{\text{mass of the vapour per mL at NTP}}{\text{mass of hydrogen per mL at NTP}}$$

$$= \frac{\text{absolute density (or, sp. gravity or rel. density)}}{1/11200}$$

Vapour density = specific gravity × 11200. (for gases)

(1 mole of hydrogen, i.e., 2 g at NTP occupies 22400 mL)

From Vapour-Density Method

Ex. 9. *In a Victor Meyer apparatus* 0·168 g *of a volatile compound displaced* 49·4 mL *of air measured over water at* 20°C *and* 740 mm *of pressure. Calculate the molecular weight of the compound.* (*Aqueous tension at* 20°C = 18 mm)

Solution : Pressure due to dry air only = (740 – 18) mm.

Volume of vapour of 0·168 g of compound at NTP = volume of air displaced by 0·168 g at NTP

$$= \frac{(740-18)\times 49{\cdot}4\times 273}{293\times 760}$$

$$= 43{\cdot}72 \text{ mL.}$$

$\therefore$ mole of vapour $= \dfrac{43{\cdot}72}{22400}$. (Rule 3, Chapter 1)

Molecular weight (Wt. of 1 mole) $= \dfrac{\text{wt. in grams}}{\text{no. of moles}} = 0{\cdot}168 \times \dfrac{22400}{43{\cdot}72}$

$$= 86{\cdot}06.$$

Ex. 10. *A Dumas bulb was filled with* 0·4524 g *of vapour at* 136°C *and* 758 mm *of pressure. Calculate molecular weight of the vapour if the capacity of the bulb is* 127 mL.

Solution : Volume of vapour at NTP $= \dfrac{758\times 127\times 273}{409\times 760}$

$$= 84{\cdot}57 \text{ mL.}$$

Mole of vapour $= \dfrac{84{\cdot}57}{22400}$. (Rule 3, Chapter 1)

$\therefore$ molecular weight $= 0{\cdot}4524 \times \dfrac{22400}{84{\cdot}57}$ (Rule 1, Chapter 1)

$$= 119{\cdot}8.$$

Ex. 11. 0·3151 g *of a substance when introduced into a Hofmann's tube generated* 128·5 mL *of vapour at* 30°C, *the level of* Hg *inside being* 430 mm *higher than outside the tube, and the barometer reading was* 758 mm. *Calculate the molecular weight of the substance.*

Solution : Pressure of vapour in Hofmann's tube = (758 – 430) mm = 328 mm.

Volume of vapour at NTP $= \dfrac{328\times 128{\cdot}5\times 273}{303\times 760} = 50$ mL.

Mole of vapour $= \dfrac{50}{22400}$.

Molecular weight $= 0{\cdot}3151 \times \dfrac{22400}{50}$

$$= 141{\cdot}16.$$

From Silver-Salt Method (for organic acids)

The salt of the acid is quantitatively reduced to pure silver. The molecular weight of the salt is first determined from which the molecular weight of the acid is calculated as given below. The mole method is applied by using POAC for Ag atoms which remain conserved.

Ex. 12. 0·701 g *of the silver salt of a dibasic acid on ignition yielded* 0·497 g *of metallic silver. Calculate the molecular weight of the acid.* (Ag = 108)

Solution : Suppose the dibasic acid is H_2A.

$$\underset{}{H_2A} \rightarrow \underset{\substack{\text{salt} \\ 0{\cdot}701\text{ g}}}{Ag_2A} \rightarrow \underset{0{\cdot}497\text{ g}}{Ag}$$

Since Ag atoms are conserved, applying POAC for Ag atoms in the second step,

moles of Ag atoms in Ag_2A = moles of Ag atoms in the product

$$2 \times \text{moles of } Ag_2A = \text{moles of Ag}$$

(POAC, page 3, Chapter 1)

$$2 \times \frac{0{\cdot}701}{\text{mol. wt. of } Ag_2A} = \frac{0{\cdot}497}{108}.$$

$$\therefore \quad \text{mol. wt. of } Ag_2A = 304{\cdot}7.$$

Now since the Ag salt is formed from the dibasic acid by the replacement of 2 H atoms by 2 Ag atoms,

$$\begin{aligned}\text{molecular weight of the acid} &= \text{molecular weight of } Ag_2A \\ &\quad - 2 \times \text{at. wt. of Ag} + 2 \times \text{at. wt. of H} \\ &= 304{\cdot}7 - 216 + 2 \\ &= 90{\cdot}7.\end{aligned}$$

Ex. 13. 0·607 g *of a silver salt of a tribasic organic acid was quantitatively reduced to* 0·370 g *of pure silver. Calculate the molecular weight of the acid.* (Ag = 108)

Solution : Suppose the tribasic acid is H_3A.

$$\underset{\text{acid}}{H_3A} \rightarrow \underset{\substack{\text{salt} \\ 0{\cdot}607\text{ g}}}{Ag_3A} \rightarrow \underset{0{\cdot}37\text{ g}}{Ag}$$

Since Ag atoms are conserved, applying POAC for Ag atoms,

moles of Ag atoms in Ag_3A = moles of Ag atoms in the product

$$3 \times \text{moles of } Ag_3A = \text{moles of Ag in the product}$$

$$3 \times \frac{0{\cdot}607}{\text{mol. wt. of } Ag_3A} = \frac{0{\cdot}37}{108} \qquad (\text{Ag} = 108)$$

$$\text{mol. wt. of } Ag_3A = 531.$$

∴ mol. weight of tribasic acid (H_3A)

$$= \text{mol. wt. of the salt } (Ag_3A) - 3 \times \text{at. wt. of Ag} + 3 \times \text{at. wt. of H}$$

$$= 531 - 324 + 3 = 210.$$

From Platinichloride Method (for organic bases)

Organic bases combine with chloroplatinic acid, H_2PtCl_6, to form insoluble anhydrous chloroplatinates (platinichlorides), which on ignition leave a residue of metallic Pt. For a base, say B

formula of its chloroplatinate is $B_2H_2PtCl_6$ – for monoacid base

formula of its chloroplatinate is BH_2PtCl_6 – for diacid base

Molecular weight of the base is determined from the molecular weight of platinichloride as given below.

Ex. 14. 0·80 g *of the chloroplatinate of a monoacid base on ignition gave* 0·262 g *of* Pt. *Calculate the molecular weight of the base.* (Pt = 195)

Solution : Suppose the monoacid base is B.

$$\underset{\substack{\text{base}\\\text{(monoacid)}}}{B} + \underset{\text{acid}}{H_2PtCl_6} \rightarrow \underset{\substack{\text{chloroplatinate}\\0{\cdot}80\text{ g}}}{B_2H_2PtCl_6} \rightarrow \underset{0{\cdot}262\text{ g}}{Pt}$$

Since Pt atoms are conserved, applying POAC for Pt atoms,

moles of Pt atoms in $B_2H_2PtCl_6$ = moles of Pt in the product

$1 \times$ moles of $B_2H_2PtCl_6$ = moles of Pt in the product

$$\frac{0{\cdot}80}{\text{mol. wt. of } B_2H_2PtCl_6} = \frac{0{\cdot}262}{195}.$$

∴ mol. wt. of $B_2H_2PtCl_6 = 595{\cdot}4.$

Now from the formula $B_2H_2PtCl_6$, we know

$$\text{mol. wt. of B} = \frac{\text{mol. wt. of } B_2H_2PtCl_6 - \text{mol. wt. of } H_2PtCl_6}{2}$$

$$= \frac{595{\cdot}40 - 410{\cdot}00}{2}$$

$$= 92{\cdot}70.$$

Ex. 15. 0·532 g *of the chloroplatinate of a diacid base on ignition left* 0·195 g *of residue of* Pt. *Calculate molecular weight of the base.* (Pt = 195)

Solution : Suppose the diacid base is B.

$$\underset{\substack{\text{diacid}\\\text{base}}}{B} + \underset{\text{acid}}{H_2PtCl_6} \rightarrow \underset{\substack{\text{chloroplatinate}\\0{\cdot}532\text{ g}}}{BH_2PtCl_6} \rightarrow \underset{0{\cdot}195\text{ g}}{Pt}$$

Since Pt atoms are conserved, applying POAC for Pt atoms,

moles of Pt atoms in BH_2PtCl_6 = moles of Pt atoms in the product

$$1 \times \text{ moles of } BH_2PtCl_6 = \text{moles of Pt in the product}$$

$$\frac{0{\cdot}532}{\text{mol. wt. of } BH_2PtCl_6} = \frac{0{\cdot}195}{195}.$$

$\therefore$ mol. wt. of $BH_2PtCl_6 = 532.$

From the formula BH_2PtCl_6, we get,

mol. wt. of B = mol. wt. of BH_2PtCl_6 – mol. wt. of H_2PtCl_6

$$= 532 - 410 = 122.$$

From Volumetric Method

Ex. 16. 0·45 g *of a dibasic acid required* 200 mL *of* 0·05 N NaOH *solution for neutralisation. Calculate molecular weight of the organic acid.*

Solution **:** (The students are advised to solve this problem after doing volumetric problems.)

m.e. of NaOH solution $= 0{\cdot}05 \times 200 = 10.$ ... (Eqn. 1, Chapter 7)

$\therefore$ m.e. of 0·45 g of dibasic acid = 10. ... (1) ... (Eqn. 2, Chapter 7)

Now eq. of the acid $= \frac{0{\cdot}45}{E}$ (E being the eq. wt. of the acid)

... (Eqn. 4, Chapter 7)

m.e. of the acid $= \frac{0{\cdot}45}{E} \times 1000$ ... (Eqn. 3, Chapter 7)

$$= \frac{450}{E}. \quad \text{... (2)}$$

From eqns. (1) and (2), we have,

$$\frac{450}{E} = 10.$$

$$\therefore \quad E = 45.$$

$$\frac{\text{mol. wt.}}{\text{basicity}} = \text{eq. weight.}$$

$\therefore$ molecular weight $= 2 \times 45 = 90.$

Ex. 17. *A mixture contains* NaCl *and an unknown chloride* MCl.

(i) 1 g *of this is dissolved in water. Excess of acidified* $AgNO_3$ *solution is added to it.* 2·567 g *of a white precipitate is formed.*

(ii) 1 g *of the original mixture is heated to* 300°C. *Some vapours come out which are absorbed in acidified* $AgNO_3$ *solution.* 1·341 g *of a white precipitate is obtained.*

Find the molecular weight of the unknown chloride.

Solution **:** (See Example 32, Chapter 2)

Ex. 18. *When* 2·96 g *of mercuric chloride is vaporised in a* 1-*litre bulb at* 680 K, *the pressure is* 458 mm. *What is the molecular weight and molecular formula of mercuric chloride vapour?* (Hg = 200·6)

Solution : Let M be the molecular weight of mercuric chloride.

No. of moles of the mercuric chloride vapour $= \frac{2{\cdot}96}{M}$.

Again from the gas equation $pV = nRT$,

$$\text{no. of moles} = n = \frac{pV}{RT} = \frac{\left(\frac{458}{760}\right) \times 1}{0{\cdot}0821 \times 680},$$

where $p = \frac{458}{760}$ atm, $V = 1$ litre, $R = 0{\cdot}0821$ lit. atm K^{-1} mol^{-1} and $T = 680$ K.

Hence,
$$\frac{2{\cdot}96}{M} = \frac{\frac{458}{760} \times 1}{0{\cdot}0821 \times 680}$$
$$M = 274.$$

Since $HgCl_2$ has a molecular weight of $(200{\cdot}6 + 2 \times 35{\cdot}5) = 271{\cdot}6$ which is nearly equal to the calculated value of 274, $HgCl_2$ is the molecular formula for mercuric chloride.

PROBLEMS

(Answers bracketed with questions)

1. What is the volume of 6 g of hydrogen at 1·5 atm and 273°C? (89·6 litres)

2. A certain gas occupies 0·418 litre at 27°C and 740 mmHg. (i) What is its volume at STP? (ii) If the same gas weighs 3·00 g, what is its molecular weight? (0·3704 litre. 181·4)

3. 33·6 cc of phosphorous vapour weighs 0·0625 g at 546°C and 76 cmHg pressure. What is the molecular weight of phosphorous? How many atoms are there in one molecule of phosphorous vapour? (125, 4)

4. The mass of a sulphur atom is double that of an oxygen atom. Can we decide on this basis that the density of sulphur vapour relative to oxygen is two? (No)

5. In a Victor Meyer determination, 0·0926 g of a liquid gave 28·9 mL of gas collected over water and measured at 16°C and 753·6 mm pressure. Calculate molecular weight and vapour density of the substance. (Aq. tension at 16°C = 13·6 mm) (78, 39)

[**Hint:** Find weight of 22400 mL (NTP) of the gas.]

6. A Dumas bulb of capacity 200 mL weighs 22·567 g at 120°C. Filled with vapour of a substance at 120°C and 755 mmHg pressure, it weighs 22·8617 g. Find the molecular weight of the substance. (47·8)

7. 0·2704 g of a substance when introduced into a Hofmann tube generated 110 mL of vapours at 99·6°C and 747 mmHg pressure. The height of mercury inside the tube was 285·2 mm. Calculate the molecular weight of the substance. (122·6)

8. 0·607 g of the silver salt of a tribasic acid on combustion deposited 0·37 g of pure silver. Calculate the molecular weight of the acid. (210·16)

9. 0·304 g of a silver salt of a dibasic acid left 0·216 g of silver on ignition. Calculate its molecular weight. (90)

10. 0·66 g of platinichloride of a monoacid base left 0·150 g of platinum. Calculate its molecular weight. (Pt = 195) (221)

11. The chloroplatinate of a diacid base contains 39% platinum. What is the molecular weight of the base? (Pt = 195) (90)

12. A solution containing 3·00 g of a monobasic organic acid was just neutralised by 40 mL of 0·5 N NaOH solution. Calculate the molecular weight of the acid. (150)

13. 0·366 g of an organic base required 15 mL of $\frac{N}{5}$ HCl for exact neutralisation. If the molecular weight of the base is 122, find its acidity. (1)

❑

CHAPTER SIX

CHEMICAL EQUIVALENCE

Significance of Equivalent Weight

An equivalent of a substance is defined as the amount of it which combines with 1 mole of hydrogen atoms or replaces the same number of hydrogen atoms in a chemical reaction. The weight in grams of 1 equivalent is called the equivalent weight in grams. For example, in the compounds HBr, H_2O and NH_3; one mole of H combines with one mole of Br, ½ mole of O and ⅓ mole of N respectively. Hence the equivalent weights in grams of Br, O and N are the weights of 1 mole of Br, ½ mole of O, ⅓ mole of N respectively. In other words, 1 mole each of Br, O and N contains their 1 equivalent, 2 equivalents and 3 equivalents respectively.

Thus: eq. wt. of Br $= 1 \times 79{\cdot}9 = 79{\cdot}9$

$$\text{eq. wt. of O} = \frac{1}{2} \times 16 = 8{\cdot}0$$

$$\text{eq. wt. of N} = \frac{1}{3} \times 14 = 4{\cdot}67$$

(Atomic weights of Br, O and N are 79·9, 16 and 14 respectively.)

To determine the equivalent weight of an element, it is not necessary to proceed from its hydrogen compound only. Equivalent weight of an element can be calculated using the composition of the compound of the given element with any other element, whose equivalent weight is known by the knowledge of the **Law of Equivalence.** The law states that **one equivalent of an element combines with one equivalent of the other.** Accordingly, the equivalent weight of an element is the weight of its mole combining with one equivalent of another element. It can be further illustrated by the following example:

The equivalent weight of Al in Al_2O_3 can be calculated if it is known that 1 mole of O contains 2 equivalents of it. From the composition of Al_2O_3, we write, 3 moles of O combine with 2 moles of Al, or 6 equivalents of O combine with 2 moles of Al or 1 equivalent of O combines with ⅓ mole of Al.

$$\therefore \text{ equivalent weight of Al} = \frac{1}{3} \times 27 = 9$$

(at. wt. of Al = wt. of 1 mole of Al = 27)

An element can have more than one value of equivalent weight, e.g., the equivalent weights of Fe in FeO and Fe_2O_3 are 56/2 and 56/3 respectively.

The equivalent weight (in grams) of a compound taking part in a reaction is the weight of the compound which combines with 1 equivalent of another compound. Thus knowing that 1 mole of HCl is equal to 1 equivalent of it, equivalent weight of Na_2CO_3 can be calculated from the following equation.

$$Na_2CO_3 + 2HCl \rightarrow 2NaCl + CO_2 + H_2O$$

2 moles of HCl combine with one mole of Na_2CO_3

or 2 equivalents of HCl combine with 1 mole of Na_2CO_3

$\therefore$ 1 equivalent of HCl combines with $\frac{1}{2}$ mole of Na_2CO_3.

$\therefore$ eq. wt. of $Na_2CO_3 = \frac{1}{2} \times 106 = 53$.

(mol. wt. of Na_2CO_3 = wt. of 1 mole = 106)

But if the reaction takes place as:

$$Na_2CO_3 + HCl \rightarrow NaHCO_3 + NaCl$$

1 equivalent of HCl combines with 1 mole of Na_2CO_3.

$\therefore$ eq. wt. of $Na_2CO_3 = 1 \times 106 = 106$.

Thus a compound may have different values of equivalent weights.

To calculate the equivalent weight of any substance taking part in a reaction we generally use the following expression of the law of equivalence. For a reaction

$$aA + bB \rightarrow mM + nN,$$

eq. of A = eq. of B = eq. of M = eq. of N ... (1)

or $\frac{\text{wt. of A}}{E_A} = \frac{\text{wt. of B}}{E_B} = \frac{\text{wt. of M}}{E_M} = \frac{\text{wt. of N}}{E_N}$,

where E stands for equivalent weight.

In an acid–base neutralisation reaction the equivalent weight (in grams) of an acid is that portion of the weight of 1 mole of the acid which can furnish 1 mole of H^+, and the equivalent weight of a base is the portion of weight of one mole of the base which can furnish 1 mole of OH^- or accept 1 mole of H^+. The equivalent weights of HCl, H_2SO_4 and H_3PO_4 are therefore 36·5, 98/2 and 98/3 respectively and the equivalent weights of NaOH and $Ca(OH)_2$ are 40 and 74/2 respectively. One equivalent of an acid just reacts with one equivalent of any base.

The law of equivalence allows us to derive the following formulae for calculating equivalent weights of compounds.

For acids: $E = \frac{\text{mol. wt. of acid (i.e., wt. of 1 mole)}}{\text{basicity of acid (mole of } H^+ \text{ furnished)}}$ ... (2)

For bases: $E = \frac{\text{mol. wt. of base}}{\text{acidity of base (moles of } OH^- \text{ furnished or moles of } H^+ \text{ accepted)}}$ ... (3)

For salts:

$$E = \frac{\text{mol. wt. of salt}}{\text{moles of H equivalent to total no. of cations or anions}} \quad \dots (4\text{ i})$$

$$E = \frac{\text{mol. wt. of salt}}{\text{moles of metal atoms} \times \text{valency of metal}} \quad \dots (4\text{ ii})$$

For oxides: $$E = \frac{\text{mol. wt. of oxide}}{\text{moles of element atoms} \times \text{valency of element}} \quad \dots (5)$$

However, the equivalent weight of a compound taking part in a reaction should be determined from the chemical equation, as equivalent weight of a compound depends on the stoichiometry of the reaction.

In redox reactions, the equivalent weight (in grams) of an oxidising or a reducing agent is the portion of the weight of 1 mole of substance that picks up or releases 1 mole of electrons respectively, e.g., when $\overset{+7}{MnO_4^-}$ is reduced to Mn^{2+}, the oxidation number changes by 5 (i.e., from +7 to +2), the equivalent weight (in grams) of MnO_4^- is thus 1/5 of the weight of 1 mole (molecular weight) since 5 moles of electrons are needed for the reduction of 1 mole of MnO_4^-. When Fe^{2+} is oxidised to Fe^{3+}, the number of electrons required is 1 mole and so the equivalent weight of Fe^{2+} or Fe^{3+} is the same as its ionic weight. Thus for oxidising or reducing agents we can write:

$$E = \frac{\text{mol. weight of oxidising or reducing agent}}{\text{change in oxd. no. per mole of the oxidising or reducing agent}} \quad \dots (6)$$

The oxidation number of ions taking part in the precipitation reaction does not change. For such ions the equivalent weight is the weight of 1 mole of ions (i.e., ionic weight) per mole of the charge on it, e.g.,

$$E_{Fe^{2+}} = \frac{56}{2} = 28$$

$$E_{Fe^{3+}} = \frac{56}{3} = 18{\cdot}67$$

$$E_{SO_4^{2-}} = \frac{96}{2} = 48.$$

Again for electrolytes, i.e., a substance undergoing complete ionisation,

$$E_{(\text{electrolyte})} = E_{\text{cations}} + E_{\text{anions}}. \quad \dots (7)$$

The equations (1 to 7) are useful in solving problems based on volumetric analysis.

In this chapter we shall solve problems on equivalent weight using the concept of the law of equivalence.

EXAMPLES

Ex. 1. *If W_1 and W_2 are the weights of two reactants in any reaction, having their equivalent weights E_1 and E_2 respectively, which of the following equations represents the law of equivalence correctly?*

(a) $W_1E_1 = W_2E_2$ (b) $W_1E_2 = W_2E_1$ (c) $W_1W_2 = E_1E_2$

Solution : The answer is (b), because number of equivalents

$$= W_1 / E_1 = W_2 / E_2 .$$

Ex. 2. 12 g *of an element combines with* 32 g *of oxygen. What is the equivalent weight of the element if the equivalent weight of oxygen is 8?*

Solution : Suppose the element is M.

No. of eq. of M = no. of eq. of O

$$\frac{\text{wt. of M}}{E_M} = \frac{\text{wt. of oxygen}}{E_O}$$

$$\frac{12}{E_M} = \frac{32}{8}$$

$$E_M = \frac{12 \times 8}{32} = 3.$$

Ex. 3. *What are the equivalent volumes of hydrogen, oxygen and nitrogen?*

Solution : Equivalent volume is the volume occupied by one equivalent of any gas at NTP.

We know, 1 mole of molecules of any gas at NTP occupies 22·4 litres and 1 mole of atoms of any diatomic gas at NTP occupies 11·2 litres.

Now,

1 mole of H contains 1 equivalent of H,

1 mole of O contains 2 equivalents of O,

1 mole of N contains 3 equivalents of N.

Thus at NTP,

1 equivalent of H occupies 11·2 litres,

1 equivalent of O occupies $\frac{11·2}{2} = 5·6$ litres,

1 equivalent of N occupies $\frac{11·2}{3} = 3·73$ litres.

Ex. 4. *How many molecules are present in* 1 *equivalent each of hydrogen, oxygen and nitrogen?*

Solution : 1 equivalent of H contains 1 mole of H.

∴ 1 equivalent of H contains 1/2 mole of H_2.

$\therefore$ 1 equivalent of H contains $\frac{6{\cdot}022\times10^{23}}{2}$ H_2 molecules.

... (Rule 4, Chapter 1)

1 equivalent of O contains $\frac{1}{2}$ mole of O.

1 equivalent of O contains $\frac{1}{4}$ mole of O_2.

1 equivalent of O contains $\frac{6{\cdot}022\times10^{23}}{4}$ O_2 molecules.

... (Rule 4, Chapter 1)

1 equivalent of N contains $\frac{1}{3}$ mole of N.

1 equivalent of N contains $\frac{1}{6}$ mole of N_2.

1 equivalent of N contains $\frac{6{\cdot}022\times10^{23}}{6}$ N_2 molecules.

... (Rule 4, Chapter 1)

Ex. 5. *If* 0·5 *equivalent of* $H_2C_2O_4$ *undergoes complete dissociation, what will be the number of equivalents of* H^+ *and* $C_2O_4^{2-}$ *ions?*

Solution : Equivalents of H^+ = equivalents of $C_2O_4^{2-}$

$= \text{eq. of } H_2C_2O_4 = 0{\cdot}5.$ ... (Eqn. 1)

Ex. 6. *Calculate the number of equivalents in* 1 mole *of oxygen molecules.*

Solution : We know, 1 mole of O_2 at NTP occupies 22·4 litres and 1 equivalent of oxygen at NTP occupies 5·6 litres.

$\therefore$ no. of equivalents of oxygen $= \frac{22{\cdot}4}{5{\cdot}6} = 4.$

Ex. 7. *The two acids* H_2SO_4 *and* H_3PO_4 *are neutralised separately by the same amount of an alkali when sulphate and dihydrogen orthophosphate are formed respectively. Find the ratio of the masses of* H_2SO_4 *and* H_3PO_4*.*

Solution : $H_2SO_4 \rightarrow 2H^+ + SO_4^{2-}$

$H_3PO_4 \rightarrow H^+ + H_2PO_4^-$

The basicity of H_2SO_4 is 2 and that of H_3PO_4 is 1.

Eq. of alkali = eq. of H_2SO_4 = eq. of H_3PO_4.

$$\therefore \quad \frac{\text{wt. of } H_2SO_4}{\text{eq. wt. of } H_2SO_4} = \frac{\text{wt. of } H_3PO_4}{\text{eq. wt. of } H_3PO_4}.$$

$$\therefore \quad \text{wt. of } H_2SO_4/\text{wt. of } H_3PO_4 = \frac{49}{98} = \frac{1}{2}.$$

Ex. 8. *Find the equivalent weight of* H_3PO_4 *in the reaction*

$$Ca(OH)_2 + H_3PO_4 = CaHPO_4 + 2H_2O$$

Solution : Since 1 mole of H_3PO_4 replaces 2 moles of H^+,

$$\text{equivalent weight of } H_3PO_4 = \frac{\text{mol. weight}}{2} = \frac{98}{2} = 49.$$

Ex. 9. *Equivalent weight of sulphur in* SCl_2 *is 16, what is the equivalent weight of S in* S_2Cl_2? (S = 32, Cl = 35·5)

Solution : From the data we conclude that 1 mole of S contains 2 equivalents as the weight of 1 mole is 32 and that of 1 equivalent is 16. In SCl_2,

1 mole of S combines with 2 moles of Cl

or 2 equivalents of S combine with 2 moles of Cl.

∴ 1 equivalent of S combines with 1 mole of Cl.

∴ equivalent weight of $Cl = 1 \times 35{\cdot}5 = 35{\cdot}5$. ... (Eqn. 1)

Thus for Cl, 1 mole = 1 equivalent.

Now, in S_2Cl_2,

2 moles of Cl combine with 2 moles of S

or 2 equivalent of Cl combine with 2 moles of S

∴ 1 equivalent of Cl combines with 1 mole of S

∴ equivalent weight of S = 1 × 32 = 32. ... (Eqn. 1)

Ex. 10. *The equivalent weight of a metal is double that of oxygen. How many times is the weight of its oxide greater than the weight of the metal?*

Solution : Equivalent of metal = equivalent of oxygen.

$$\therefore \quad \frac{\text{weight of metal}}{\text{eq. wt. of metal}} = \frac{\text{weight of oxygen}}{\text{eq. wt. of oxygen}}$$

$$\therefore \quad \frac{\text{wt. of oxygen}}{\text{wt. of metal}} = \frac{\text{eq. wt. of oxygen}}{\text{eq. wt. of metal}}$$

$$= \frac{1}{2} \text{ (given).}$$

$$\frac{\text{weight of oxygen}}{\text{weight of metal}} + 1 = \frac{1}{2} + 1 = \frac{3}{2} = 1{\cdot}5.$$

$$\frac{\text{weight of oxygen + weight of metal}}{\text{weight of metal}} = 1{\cdot}5.$$

$$\therefore \quad \frac{\text{weight of oxide}}{\text{weight of metal}} = 1{\cdot}5.$$

Ex. 11. 1·60 g *of* Ca *and* 2·60 g *of* Zn *when treated with an acid in excess separately, produced the same amount of hydrogen. If the equivalent weight of* Zn *is* 32·6, *what is the equivalent weight of* Ca?

Solution : Since Ca and Zn produce the same amount of hydrogen with an acid separately,

equivalent of Ca = equivalent of Zn = equivalent of hydrogen.

$$\therefore \quad \frac{1{\cdot}60}{\text{eq. wt. of Ca}} = \frac{2{\cdot}60}{32{\cdot}6}.$$

∴ equivalent weight of Ca = 20.

Ex. 12. *One gram of an alloy of* Al *and* Mg *when treated with excess of dilute* HCl *forms* $MgCl_2$, $AlCl_3$ *and hydrogen. The evolved hydrogen collected over* Hg *at* 0°C *has a volume of* 1·20 *litres at* 0·92 atm *pressure. Calculate the composition of the alloy.* (Al = 27, Mg = 24)

Solution : Volume of H_2 produced at NTP $= \frac{1{\cdot}2 \times 0{\cdot}92}{273} \times \frac{273}{1}$

$$\left(\frac{p_1V_1}{T_1} = \frac{p_2V_2}{T_2}\right)$$

$$= 1{\cdot}104 \text{ litres.}$$

Since both Al and Mg produce 1·104 litres of H_2,

equivalent of Al + equivalent of Mg

= equivalent of hydrogen.

$$\therefore \quad \frac{\text{weight of Al}}{\text{eq. wt. of Al}} + \frac{\text{weight of Mg}}{\text{eq. wt. of Mg}} = \text{eq. of hydrogen}$$

$$= \frac{\text{volume of hydrogen (NTP)}}{\text{equivalent volume of hydrogen}}.$$

Let the weight of Al be x g.

$$\therefore \quad \frac{x}{27/3} + \frac{1-x}{24/2} = \frac{1{\cdot}104}{11{\cdot}2} \qquad \left(E_{Al} = \frac{27}{3}\right) \left(E_{Mg} = \frac{24}{2}\right)$$

$$x = 0{\cdot}55$$

$$(1-x) = 0{\cdot}45.$$

Thus, weight of Al = 0·55 g

and weight of Mg = 0·45 g.

[**Note:** See the mole method in Chapter 2, Ex. 18]

Ex. 13. 0·376 g *of aluminium reacted with an acid to displace* 0·468 litre *of hydrogen at* NTP. *Find the equivalent volume of hydrogen if the equivalent weight of* Al *is* 9.

Solution : Equivalent of Al = equivalent of hydrogen.

$$\therefore \quad \frac{\text{weight of Al}}{\text{eq. weight of Al}} = \frac{\text{volume of hydrogen at NTP}}{\text{equivalent volume of hydrogen}}$$

$$\frac{0.376}{9} = \frac{0.468}{V\text{ (litre)}}$$

$$\therefore \qquad V = 11.2 \text{ litres.}$$

Ex. 14. *One litre of oxygen at* NTP *weighs* 1·46 g. *How many litres of oxygen are needed for the combustion of* 21·0 g *of* Mg *whose equivalent weight is* $\frac{1}{2}$ mole?

Solution : The volume occupied by 1 equivalent at NTP is equivalent volume.

∴ equivalent volume of oxygen

= volume of 8·0 g of oxygen at NTP (eq. wt. of O = 8)

$= \frac{8}{1.46} = 5.48$ litres. (1.46 g oxygen contains 1 L at NTP)

Given that the equivalent weight of $\text{Mg} = \frac{1}{2} \times 24 = 12.$

(wt. of 1 mole of Mg = 24 g)

Now, in the combustion,

equivalent of Mg = equivalent of oxygen.

$$\therefore \quad \frac{\text{weight of Mg}}{\text{eq. weight of Mg}} = \frac{\text{volume of oxygen at NTP}}{\text{eq. volume of oxygen}}$$

$$\frac{21}{12} = \frac{\text{volume of oxygen}}{5.48}.$$

$$\therefore \quad \text{volume of oxygen} = 9.59 \text{ litres.}$$

Ex. 15. *Dry hydrogen was passed over* 1·58 g *of red hot copper oxide till all of it completely reduced to* 1·26 g *of copper* (Cu). *If in this process* 0·36 g *of* H_2O *is formed, what will be the equivalent weight of* Cu *and* O? (H = 1)

Solution : Copper oxide + H_2 $\rightarrow$ Cu + H_2O

∴ equivalent of copper oxide = equivalent of Cu

= equivalent of H_2O.

$$\therefore \quad \frac{\text{weight of copper oxide}}{\text{eq. weight of copper oxide}} = \frac{\text{weight of Cu}}{\text{eq. wt. of Cu}} = \frac{\text{weight of } H_2O}{\text{eq. wt. of } H_2O}.$$

Suppose eq. wt. of Cu and O are x and y respectively and since,

eq. wt. of H is 1, we have,

equivalent weight of copper oxide = $x + y$

equivalent weight of copper = x

equivalent weight of $H_2O = 1 + y$ $(E_{H_2O} = E_H + E_O)$

$$\therefore \qquad \frac{1{\cdot}58}{x+y}=\frac{1{\cdot}26}{x}=\frac{0{\cdot}36}{1+y}.$$

$$\therefore \qquad x = 31{\cdot}5$$

and $$y = 8.$$

Ex. 16. *The equivalent weight of chlorine is* 35·5 *and the equivalent weight of copper chloride is* 99·5. *Find the formula of copper chloride if the atomic weight of* Cu *is* 63·5.

Solution : The equivalent weight of chlorine is given as equal to 35·5 which is also its atomic weight. The valency of Cl is thus 1.

Equivalent weight of copper

= eq. wt. of copper chloride – eq. wt. of chlorine

= 99·5 – 35·5 = 64.

The equivalent weight of Cu is 64 which is approximately equal to its given atomic weight, i.e., 63·5, showing that valency of Cu is also 1. Hence the formula for copper chloride is CuCl.

Ex. 17. 4·215 g *of a metallic carbonate was heated in a hard glass tube and the* CO_2 *evolved was found to measure* 1336 mL *at* 27°C *and* 700 mm *pressure. What is the equivalent weight of the metal?*

Solution : Volume of CO_2 at NTP $= \frac{1336 \times 273}{300} \times \frac{700}{760}$

$= 1120$ mL.

Suppose the equivalent weight of the metal is E.

$\therefore$ equivalent weight of the metal carbonate = $(E + 30)$

($\because$ eq. wt. of $CO_3^{2-} = \frac{60}{2} = 30$).

Now, equivalent of metallic carbonate = equivalent of CO_2

$$\frac{4{\cdot}215}{E+30} = \frac{1120 \text{ (vol. at NTP)}}{11200 \text{ (vol. of 1 eq. at NTP)}}.$$

$$\therefore \qquad E = 12{\cdot}15.$$

Ex. 18. 0·324 g *of copper was dissolved in nitric acid and the copper nitrate so produced was burnt till all copper nitrate converted to* 0·406 g *of copper oxide. Calculate the equivalent weight of copper.*

Solution : $Cu + HNO_3 \rightarrow$ copper nitrate $\rightarrow$ copper oxide

0·324 g 0·406 g

Suppose the equivalent weight of copper (Cu) is E_{Cu}.

Equivalent of Cu = equivalent of copper oxide

$$\frac{0{\cdot}324}{E_{Cu}} = \frac{0{\cdot}406}{\text{eq. wt. of copper oxide}}$$

$$\frac{0{\cdot}324}{E_{Cu}} = \frac{0{\cdot}406}{\text{eq. wt. of Cu} + \text{eq. wt. of O}} = \frac{0{\cdot}406}{E_{Cu} + 8}$$

$$E_{Cu} = 31{\cdot}60.$$

Hence the equivalent weight of Cu is 31·60.

Ex. 19. 0·13 g *of* Cu, *when treated with* $AgNO_3$ *solution, displaced* 0·433 g *of* Ag. 0·13 g *of* Al, *when treated with* $CuSO_4$ *solution displaced* 0·47 g *of* Cu. 1·17 g *of* Al *displaces* 0·13 g *of hydrogen from an acid. Find the equivalent weight of* Ag *if equivalent weights of* Cu *and* Al *are not known.*

Solution :

$$\underset{0{\cdot}13\text{ g}}{Cu} + AgNO_3 \rightarrow \underset{0{\cdot}433\text{ g}}{Ag} \quad \text{... (1)}$$

$$\underset{0{\cdot}13\text{ g}}{Al} + CuSO_4 \rightarrow \underset{0{\cdot}47\text{ g}}{Cu} \quad \text{... (2)}$$

$$\underset{1{\cdot}17\text{ g}}{Al} + \text{Acid} \rightarrow \underset{0{\cdot}13\text{ g}}{H_2} \quad \text{... (3)}$$

For Eqn. (3) : eq. of Al = eq. of hydrogen

$$\therefore \quad \frac{1{\cdot}17}{\text{eq. wt. of Al}} = \frac{0{\cdot}13}{1}; \text{ eq. wt. of Al is 9.}$$

For Eqn. (2) : eq. of Al = eq. of Cu

$$\therefore \quad \frac{0{\cdot}13}{9} = \frac{0{\cdot}47}{\text{eq. wt. of Cu}}; \text{ eq. wt. of Cu} = 32{\cdot}5.$$

For Eqn. (1) : eq. of Cu = eq. of Ag

$$\therefore \quad \frac{0{\cdot}13}{32{\cdot}5} = \frac{0{\cdot}433}{\text{eq. wt. of Ag}}$$

$$\text{eq. wt. of Ag} = 108{\cdot}25.$$

Ex. 20. 1·0 g *of* Mg *is burnt in a closed vessel which contains* 0·5 g *of* O_2.

(i) Which reactant is left in excess?

(ii) Find the weight of excess reactant.

(iii) How many millilitres of 0·5 N H_2SO_4 *will dissolve the residue in the vessel?*

Solution : $Mg + O_2 \rightarrow MgO$

$$\text{Equivalent of Mg} = \frac{1}{12} = 0{\cdot}0833 \qquad \left(\text{eq. wt. of Mg} = \frac{\text{at. wt.}}{\text{valency}} = \frac{24}{2}\right)$$

$$\text{Equivalent of oxygen } \frac{0{\cdot}5}{8} = 0{\cdot}0625 \qquad (\text{eq. wt. of O} = 8)$$

(i) Mg is in excess because its eq. is greater than that of oxygen.

(ii) Equivalent of Mg in excess $= 0{\cdot}0833 - 0{\cdot}0625$

$$= 0{\cdot}0208.$$

$\therefore$ weight of Mg in excess = equivalent × equivalent weight

$$= 0{\cdot}0208 \times 12 = 0{\cdot}25 \text{ g.}$$

(iii) The residue contains MgO and Mg which has not taken part in the reaction. Suppose v mL of H_2SO_4 is required to dissolve the residue.

$\therefore$ m.e. of H_2SO_4 = m.e. of Mg + m.e. of MgO (not reacted)

$\therefore$ $0{\cdot}5\,v$ = eq. of Mg × 1000 + eq. of MgO × 1000

(Eqn. 1, Chapter 7) ... (Eqn. 3, Chapter 7)

$$= 1000 \text{ (eq. of Mg + eq. of MgO)}$$
$$= 1000 \text{ (eq. of Mg + eq. of O)}$$
$$= 1000\,(0{\cdot}0208 + 0{\cdot}0625)$$
$$= 83{\cdot}3.$$

$\therefore$ $v = 166{\cdot}6$ mL.

Ex. 21. 448 mL *of* SO_2 *at* NTP *is passed through* 100 mL *of a* 0·2 N *solution of* NaOH. *Find the weight of the salt formed.*

Solution : $SO_2 + NaOH \rightarrow NaHSO_3$

m.e. of NaOH = 0·2 × 100 = 20. ... (Eqn. 1, Chapter 7)

Equivalent of NaOH $= \frac{20}{1000} = 0{\cdot}02.$ ... (Eqn. 3, Chapter 7)

Since 1 equivalent of NaOH combines with 1 mole of SO_2 according to the above reaction,

$\therefore$ for SO_2, 1 mole = 1 equivalent, i.e., 1 equivalent of SO_2 will occupy 22·4 litres at NTP.

Equivalent of $SO_2 = \frac{448}{22400}$

$$= 0{\cdot}02.$$

We see that number of equivalents of SO_2 and that of NaOH are equal. Number of eq. of $NaHSO_3$ will also be 0·02 by the law of equivalence.

$\therefore$ weight of $NaHSO_3$ = equivalent of SO_2 × equivalent weight

$$= 0{\cdot}02 \times 104$$
$$= 2{\cdot}08 \text{ g.}$$

(According to the given reaction, eq. wt. of $NaHSO_3$ = mol. wt. = 104.)

Ex. 22. *Calculate the equivalent weight of the following:*

(i) $KMnO_4$ *in acidic medium*

(ii) $KMnO_4$ *in alkaline medium*

(iii) $FeSO_4(NH_4)_2SO_4 \cdot 6H_2O$ *(converting to* Fe^{3+}*)*

(iv) $K_2Cr_2O_7$ *in acidic medium*

(v) $H_2C_2O_4$ *(converting to* CO_2*)*

(vi) $Na_2S_2O_3 \cdot 5H_2O$ *(reacting with* I_2*)*

$$\underset{1\ \text{mole}}{} $$

Solution : (i) $\underset{+7}{\overset{1\ \text{mole}}{KMnO_4}} \rightarrow \underset{+2}{Mn^{2+}}$ (acidic medium)

$$\text{Equivalent weight of } KMnO_4 = \frac{\text{mol. weight of } KMnO_4}{\text{change in ON per mole}}$$

$$= \frac{158}{5} = 31{\cdot}6.$$

(ii) $\underset{+7}{\overset{1\ \text{mole}}{KMnO_4}} \rightarrow \underset{+6}{MnO_4^{2-}}$ (alkaline medium)

$$\text{Equivalent weight of } KMnO_4 = \frac{158}{1} = 158.$$

(iii) $Fe^{2+} \rightarrow Fe^{3+}$

$$\text{Equivalent weight of } FeSO_4(NH_4)_2SO_4 \cdot 6H_2O = \frac{392}{1} = 392.$$

(iv) $\underset{+12}{\overset{1\ \text{mole}}{K_2Cr_2O_7}} \rightarrow \underset{+6}{2Cr^{3+}}$

$$\text{Equivalent weight of } K_2Cr_2O_7 = \frac{294{\cdot}2}{6} = 49{\cdot}03.$$

(v) $\underset{+6}{\overset{1\ \text{mole}}{H_2C_2O_4}} \rightarrow \underset{+8}{2CO_2}$

$$\text{Equivalent weight of } H_2C_2O_4 = \frac{90}{2} = 45.$$

(vi) $\overset{1\ \text{mole}}{Na_2S_2O_3 \cdot 5H_2O} + I_2 \rightarrow Na_2S_4O_6 + 2I^-$

or $\underset{+4}{S_2O_3^{2-}} \rightarrow \underset{+5}{\tfrac{1}{2}S_4O_6^{2-}}$

$$\text{Equivalent weight of } Na_2S_2O_3 \cdot 5H_2O = \frac{248}{1} = 248.$$

Ex. 23. *The equivalent weight of* $MnSO_4$ *is half its molecular weight when it is converted to* (a) Mn_2O_3 (b) MnO_2 (c) MnO_4^- (d) MnO_4^{2-}. *Indicate the correct answer.* **(IIT 1988)**

Solution : In (b),

$$\underset{+2}{MnSO_4} \rightarrow \underset{+4}{MnO_2}$$

Change in ON of Mn per mole of $MnSO_4 = 2$

$$\therefore \text{ eq. wt. of } MnSO_4 = \frac{\text{Mol. wt.}}{2}$$

Hence the answer is (b).

PROBLEMS

(Answers bracketed with questions)

1. Fe forms two chlorides, $FeCl_2$ and $FeCl_3$. Does Fe have the same value of equivalent weight in its compounds? (No)

2. The equivalent weight of a metal is 12. What is the equivalent weight of its oxide? (20)

3. 0·2 g of oxygen and 3·17 g of a halogen combine separately with the same amount of a metal. What is the equivalent weight of the halogen? (127)

4. Arsenic forms two oxides, one of which contains 65·2% and the other, 75·7% of the element. Determine the equivalent weights of arsenic in both cases. (15, 24·9)

5. 1·80 g of a metal oxide required 833 mL of hydrogen at NTP to be reduced to its metal. Find the equivalent weights of the oxide and the metal. (24·2, 16·2)

6. A certain amount of a metal whose equivalent weight is 28, displaces 0·7 litre of hydrogen measured at NTP from an acid. Calculate the weight of the metal. (1·75 g)

7. 9·44 g of a metal oxide is formed by the combination of 5 g of the metal. Calculate the equivalent weight of the metal. (9·01)

8. 14·7 g of sulphuric acid was needed to dissolve 16·8 g of a metal. Calculate the equivalent weight of the metal and the volume of hydrogen liberated at NTP. (56, 3·36 litres)

9. The salt Na_2HPO_4 is formed when orthophosphoric acid is reacted with an alkali. Find the equivalent weight of orthophosphoric acid. (49)

10. 0·501 g of silver was dissolved in nitric acid and HCl was added to this solution, AgCl so formed weighed 0·6655 g. If the equivalent weight of chlorine is 35·5, what will be the equivalent weight of silver? (108)

11. 5 g of zinc displaced 4·846 g of copper from a copper sulphate solution. If zinc has an equivalent weight of 32·5, find the equivalent weight of copper. (31·5)

12. (i) What is the weight of sodium bromate and molarity of solution necessary to prepare 85·5 mL of 0·672 N solution when the half-cell reaction is

$$BrO_3^- + 6H^+ + 6e \rightarrow Br^- + 3H_2O?$$

(ii) What would be the weight as well as molarity if the cell reaction is

$$2BrO_3^- + 12H^+ + 10e \rightarrow Br_2 + 6H_2O?$$ **(IIT 1987)**

[(i) 1·4479 g, 0·112 M (ii) 1·7236 g, 0·134 M]

13. 2 g of a metal when dissolved in nitric acid converted to its nitrate. The nitrate was then precipitated to 2·66 g of the metal chloride. Find the equivalent weight of the metal. (107·57)

14. 2 g of anhydrous $BaCl_2$, present in a solution, was quantitatively converted to 2·25 g of $BaSO_4$. Find the equivalent weight of Ba. (64·5)

15. The chloride of a metal 'M' contains 47·23% of the metal. 1 g of this metal displaced from a compound 0·88 g of another metal 'N'. Find the equivalent weight of 'M' and 'N' respectively. (31·77 ; 27·96)

16. What weight of $CuSO_4 \cdot 5H_2O$ must be taken to make 0·5 litre of 0·01 M Cu(II) ion solution? (1·248 g)

17. How many litres of SO_2 taken at NTP have to be passed through a solution of $HClO_3$ to reduce 16·9 g of it to HCl? (13·44 litres)

18. How many grams of H_2S will react with 6·32 g of $KMnO_4$ to produce K_2SO_4 and MnO_2? (0·511 g)

[**Hint:** ON change for Mn, + 7 to + 4 and for S, – 2 to + 6]

19. One gram of the acid $C_6H_{10}O_4$ requires 0·768 g of KOH for complete neutralisation. How many neutralisable hydrogen atoms are in this molecule? (2)

20. What is the weight of 1 gram-equivalent of the oxidising and the reducing agent in the following reaction?

$$5Zn + V_2O_5 \rightarrow 5ZnO + 2V \quad (V = 50{\cdot}94, Zn = 65{\cdot}38 \text{ and } O = 16)$$

[**Hint:** Zn is reducing agent and its ON change = 2; V_2O_5 is oxidising agent and its ON change = 10] (32·69 g, 18·2 g)

21. Write the following oxidising agents in the increasing order of equivalent weight:

$$KMnO_4 \rightarrow Mn^{2+}$$
$$K_2Cr_2O_7 \rightarrow Cr^{3+}$$
$$KMnO_4 \rightarrow MnO_2$$
$$KIO_3 \rightarrow I^-$$
$$KClO_3 \rightarrow Cl^-$$

($KClO_3$, $KMnO_4$, KIO_3, $K_2Cr_2O_7$, $KMnO_4$)

22. Find the equivalent weight of H_3PO_4 in each of the following reactions:

$$H_3PO_4 + OH^- \rightarrow H_2PO_4^- + H_2O$$
$$H_3PO_4 + 2OH^- \rightarrow HPO_4^{2-} + 2H_2O$$
$$H_3PO_4 + 3OH^- \rightarrow PO_4^{3-} + 3H_2O$$

(98, 49, 32·67)

23. Calculate the equivalent weight of SO_2 in the following reactions:

(a) $SO_2 + 2H_2S = 3S + 2H_2O$

(b) $5SO_2 + 2KMnO_4 + 2H_2O = K_2SO_4 + 2MnSO_4 + 2H_2SO_4$ (16, 32)

24. How many equivalents per mole of H_2S are there in its oxidation to SO_2? (6)

❑

CHAPTER SEVEN

VOLUMETRIC CALCULATIONS

The quantitative analysis in chemistry is primarily carried out by two methods, viz., volumetric analysis and gravimetric analysis. In the first method the mass of a chemical species is measured by measurement of volume, whereas in the second method it is determined by taking the weight.

In the volumetric analysis, the process of determination of strength of a solution by another solution of known strength under volumetric conditions is known as *titration.* Titrations are of various types, viz., acid and base titration, oxidation–reduction titration, iodine titration, etc. The fundamental basis of all titrations is the **law of equivalence** which states that at the end point of a titration the volumes of the two titrants reacted have the same number of equivalents or milliequivalents.

The strength of a solution in volumetric analysis is generally expressed in terms of normality, i.e., number of equivalents per litre but since the volume in the volumetric analysis is generally taken in millilitres (mL), the normality is expressed by milliequivalents per millilitre.

Milliequivalents (m.e.)

From the definition of normality we know,

$$\text{normality of a solution} = \frac{\text{number of equivalents}}{\text{volume in litres}}.$$

$\therefore$ number of equivalents = normality × volume in litres.

If the volume is taken in mL,

number of milliequivalents (m.e.) = normality × vol. in mL.

$$\therefore \quad \frac{\text{number of milliequivalents}}{1000} = \text{number of equivalents}$$

$$\left(\frac{\text{m.e.}}{1000} = \text{equivalents}\right).$$

For a given solution, number of equivalents per litre is the same as the number of milliequivalents per mL.

Milliequivalents and Chemical Reactions

It is important at this stage to mention that for any given reaction, say,

$$A + 2B = 2C + 3D,$$

the stoichiometric coefficients of A, B, C and D represent the molar ratio of A, B, C and D, i.e., 1 mole of A combines with 2 moles of B to produce 2 moles of C and 3 moles of D. But these coefficients do not represent the equivalent ratio of the reactants and products at all. However, the number of equivalents or milliequivalents of each reactant and product is the same.

It is to be noted that if an acid and a base react with each other, the milliequivalents of one which is in excess in the reacting mixture will be equal to the difference between milliequivalents of the acid and that of the base but if the two acids or two bases are mixed, the milliequivalents of mixture will be the sum of milliequivalents of the two acids or the two bases because the two acids or the two bases do not react with each other.

Further, in volumetric calculations one should carefully note that the number of milliequivalents of solute in a solution does not change on its dilution.

Titrations with Two Indicators

When a reaction goes in two steps in a titration, two indicators are needed to detect the end point of each step. For example, when Na_2CO_3 solution is titrated with HCl or H_2SO_4, the reaction

$$Na_2CO_3 + 2\,HCl = 2\,NaCl + H_2O + CO_2$$

takes place in the following two steps,

$$Na_2CO_3 + HCl = NaHCO_3 + NaCl;\ (E_{Na_2CO_3} = 106)$$

$$NaHCO_3 + HCl = NaCl + H_2O + CO_2;\ (E_{NaHCO_3} = 84)$$

requiring respectively the indicators phenolphthalein and methyl orange to detect the end points.

HCl is taken in a burette while Na_2CO_3 solution is taken in a conical flask. First, phenolphthalein is added to the Na_2CO_3 solution and HCl is added from the burette till the first end point is reached. Now the second indicator, methyl orange, is added and the burette is again run till the second end point is reached. Thus we have the following equations.

For the first reaction with phenolphthalein,

m.e. of Na_2CO_3 converted to $NaHCO_3$ = m.e. of HCl added till the first end point is reached

And for the second reaction with methyl orange,

m.e. of $NaHCO_3$ produced by the first reaction = m.e. of HCl added between the first and second end points

If the sample contains both Na_2CO_3 and $NaHCO_3$, the same procedure is followed and we have the following equations:

For the first step,

m.e. of Na_2CO_3 converted to $NaHCO_3$ = m.e. of HCl added till the first end point is reached

And for the second step,

m.e. of $NaHCO_3$ produced by the first reaction + m.e. of $NaHCO_3$ given in the sample = m.e. of HCl added between the first and second end points.

Further, if the sample contains NaOH or KOH besides Na_2CO_3 and $NaHCO_3$, we have, for the first step,

m.e. of NaOH + m.e. of Na_2CO_3 converted to $NaHCO_3$	=	m.e. of HCl added till the first end point is reached

The second step is the same as in the previous one.

Now using the above equations, one can calculate the amounts of Na_2CO_3, $NaHCO_3$ and NaOH, whatever is given in the sample, as illustrated by the following examples (Ex. 44 to Ex. 47).

EDTA Titrations

The estimation of many metal ions like Ca^{2+}, Mg^{2+}, Zn^{2+}, etc., may be carried out by **complexometric titrations.** These metallic ions form complexes with ethylenediaminetetraacetic acid (EDTA). EDTA is a tetrabasic acid, represented by H_4X and its disodium salt is represented by Na_2H_2X

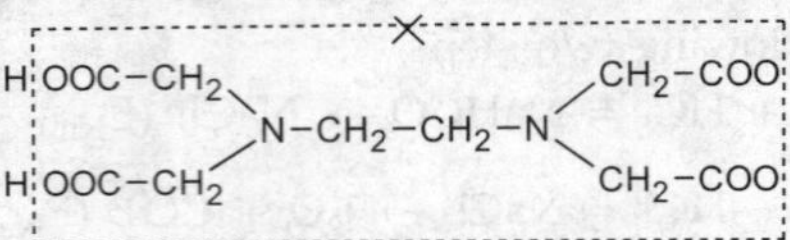

EDTA has an advantage of forming very stable complexes so that small amounts of cations can be detected. Disodium salt affords the complex-forming ions, H_2X^{2-} in aqueous medium. These ions form a complex with dipositive metal ions on a one-to-one basis. In this titration the pH is controlled by using suitable buffers. The end point is detected by a suitable metal-ion indicator which forms a complex with specific metal ions having a different colour from the free indicator. Ex 73 is just one of the vast number of applications of this type of titrations to analytical problems.

Useful Formulae for Volumetric Calculations

1. m.e. = normality × volume in millilitres.

2. At the end point of titration, the two titrants, say 1 and 2, have the same number of milliequivalents, i.e., $N_1V_1 = N_2V_2$, volumes being in mL.

3. No. of equivalents $= \dfrac{\textbf{m.e.}}{\textbf{1000}}$.

4. (i) No. of equivalents $= \dfrac{\textbf{weight in grams}}{\textbf{equivalent weight}}$.

(ii) No. of equivalents for a gas

$$= \frac{\textbf{volume at NTP}}{\textbf{equivalent volume (vol. of 1 eq. at NTP)}}.$$

5. Grams per litre = normality × equivalent weight.

6. (i) Normality = molarity × factor relating mol. wt. and eq. wt.

(ii) No. of equivalents = no. of moles × factor relating mol. wt. and eq. wt.

The factors relating mol. wt. and eq. wt. in case of acids and bases are respectively the basicity and acidity; in oxidants and reductants, it is the change in oxidation number per mole, and so on.

7. (i) In a given reaction $aA + bB \rightarrow mM + nN$,

eq. of A = eq. of B = eq. of M = eq. of N

or m.e. of A = m.e. of B = m.e. of M = m.e. of N.

(ii) In a compound M_xN_y,

equivalents of M_xN_y = equivalents of M = equivalents of N

or m.e. of M_xN_y = m.e. of M = m.e. of N.

EXAMPLES

Ex. 1. *Calculate the number of* m.e. *of* H_2SO_4 *present in* 10 mL *of* N/2 H_2SO_4 *solution.*

Solution : Number of m.e. = normality × volume in mL ... (Eqn. 1)

$$= \frac{1}{2} \times 10 = 5.$$

Ex. 2. *Calculate the number of* m.e. *and equivalents of* NaOH *present in* 1 *litre of* N/10 NaOH *solution.*

Solution : Number of m.e. = normality × volume in mL ... (Eqn. 1)

$$= \frac{1}{10} \times 1000 = 100.$$

Number of equivalents $= \frac{\text{no. of m.e.}}{1000}$ ... (Eqn. 3)

$$= \frac{100}{1000} = 0{\cdot}10.$$

Ex. 3. *Calculate the weight of* NaOH *in grams in Example* 2.

Solution : Weight in g = equivalents × eq. wt. ... (Eqn. 4)

$$= 0{\cdot}1 \times 40 = 4 \text{ g}.$$

Ex. 4. *Calculate number of* m.e. *of the acids present in*

(i) 100 mL *of* 0·5 M *oxalic acid solution,*

(ii) 50 mL *of* 0·1 M *sulphuric acid solution.*

Solution : Normality = molarity × factor relating mol. wt. and eq. wt. ... (Eqn. 6)

(i) Normality of oxalic acid = $0{\cdot}5 \times 2 = 1$ N

m.e. of oxalic acid = normality × vol. in mL = $1 \times 100 = 100$.

(ii) Normality of sulphuric acid = $0{\cdot}1 \times 2 = 0{\cdot}2$ N

m.e. of sulphuric acid = $0{\cdot}2 \times 50 = 10$.

Ex. 5. *A* 100 mL *solution of* KOH *contains* 10 *milliequivalents of* KOH. *Calculate its strength in normality and grams/litre.*

Solution : $\text{Normality} = \dfrac{\text{no. of m.e.}}{\text{volume in mL}}$ ... (Eqn. 1)

$$= \frac{10}{100} = 0{\cdot}1.$$

∴ strength of the solution = N/10. Again,

strength in grams/litre = normality × eq. wt. $= \dfrac{1}{10} \times 56 = 5{\cdot}6$ grams/litre.

$$\left(\text{eq. wt. of KOH} = \frac{\text{molecular wt.}}{\text{acidity}} = \frac{56}{1} = 56\right)$$

Ex. 6. *Calculate the normality of a solution of* $FeSO_4 \cdot 7H_2O$ *containing* 2·4 g/100 mL (Fe = 56, S = 32, O = 16, H = 1) *which converts to ferric form in a reaction.*

Solution : Weight per 100 mL = 2·4 g.

$$\text{Equivalents/100 mL} = \frac{2{\cdot}4}{278} = 0{\cdot}0086. \quad (\text{eq. wt. of } FeSO_4 \cdot 7H_2O = 278)$$

$$\left\{\begin{array}{l} \text{As } Fe^{2+} \rightarrow Fe^{3+} \\ \text{eq. wt. of } FeSO_4 \cdot 7H_2O = \dfrac{\text{molecular wt.}}{\text{change in ON}} = \dfrac{278}{1} \end{array}\right\}$$

Thus m.e. per 100 mL = $0{\cdot}0086 \times 1000 = 8{\cdot}6$. ... (Eqn. 3)

$\text{Normality of solution} = \dfrac{\text{m.e.}}{\text{volume in mL}}$ ... (Eqn. 1)

$$= \frac{8{\cdot}6}{100}$$

$$= 0{\cdot}086 \text{ N}.$$

Ex. 7. *Calculate the number of milliequivalents, gram eq., weight in grams and number of moles contained in* 10 litres *of* 0·5 M $Ba(OH)_2$ *solution.* (Ba = 137)

Solution : Normality of $Ba(OH)_2$ solution = molarity × acidity ... (Eqn. 6)

$$= 0{\cdot}5 \times 2$$

$= 1\text{ N}.$

m.e. of the solution = normality × vol. in mL

$= 1 \times 10000$

$= 10000.$

Equivalent of the solution $= \dfrac{\text{m.e.}}{1000}$ ... (Eqn. 3)

$= \dfrac{10000}{1000}$

$= 10.$

Weight of $Ba(OH)_2$ in solution = equivalents × eq. wt. ... (Eqn. 4)

$= 10 \times 85{\cdot}5$

$= 855\text{ g}.$

$\left\{\text{equivalent wt. of } Ba(OH)_2 = \dfrac{\text{mol. wt.}}{\text{acidity}} = \dfrac{171}{2} = 85{\cdot}5\right\}$

Moles of $Ba(OH)_2$ solution $= \dfrac{\text{wt. in grams}}{\text{mol. wt.}}$

$= \dfrac{855}{171} = 5{\cdot}00$ moles.

Ex. 8. *What is the strength in grams/litre of a solution of* H_2SO_4, 12 cc *of which neutralises* 15 cc of $\dfrac{N}{10}$ NaOH *solution?*

Solution : m.e. of NaOH solution $= \dfrac{1}{10} \times 15 = 1{\cdot}5$ ... (Eqn. 1)

m.e. of 12 cc of $H_2SO_4 = 1{\cdot}5$ ... (Eqn. 2)

∴ normality of $H_2SO_4 = \dfrac{1{\cdot}5}{12}$ ... (Eqn. 1)

Strength in grams/litre = normality × eq. wt. ... (Eqn. 5)

$= \dfrac{1{\cdot}5}{12} \times 49$ grams/litre

$= 6{\cdot}12$ grams/litre.

$\left\{\text{eq. wt. of } H_2SO_4 = \dfrac{\text{mol. wt.}}{\text{basicity}} = \dfrac{98}{2} = 49\right\}$

Ex. 9. *What weight of* $KMnO_4$ *will be required to prepare* 250 mL *of its* $\dfrac{N}{10}$ *solution if eq. wt. of* $KMnO_4$ *is* 31·6?

Solution : m.e. of $KMnO_4$ solution $= \dfrac{1}{10} \times 250 = 25$ ... (Eqn. 1)

$$\text{Equivalent of } KMnO_4 = \frac{25}{1000} = 0{\cdot}025. \qquad \text{... (Eqn. 3)}$$

Wt. of $KMnO_4$ solution required to prepare

$$250 \text{ mL of } \frac{N}{10} \text{ solution} = \text{eq.} \times \text{eq. wt.} \qquad \text{... (Eqn. 4i)}$$

$$= 0{\cdot}025 \times 31{\cdot}6$$

$$= 0{\cdot}79 \text{ g.}$$

Ex. 10. 100 mL *of* $0{\cdot}6$ N H_2SO_4 *and* 200 mL *of* $0{\cdot}3$ N HCl *were mixed together. What will be the normality of the resulting solution?*

Solution : m.e. of H_2SO_4 solution $= 0{\cdot}6 \times 100 = 60$

m.e. of HCl solution $= 0{\cdot}3 \times 200 = 60$

$\therefore$ m.e. of 300 mL (100 + 200) of acid mixture $= 60 + 60 = 120.$

$$\text{Normality of the resulting solution} = \frac{\text{m.e.}}{\text{total vol.}} \qquad \text{... (Eqn. 1)}$$

$$= \frac{120}{300} = \frac{2}{5} \text{ N.}$$

Ex. 11. *What will be the normality of the resulting solution of Example* 10 *if it is diluted to* 600 mL?

Solution : In this problem it should be remembered that the m.e. of a solution does not change on dilution.

Thus, m.e. of 600 mL of the resulting solution = 120 and therefore, normality of this diluted solution

$$= \frac{120}{600} = \frac{N}{5}. \qquad \text{... (Eqn. 1)}$$

Ex. 12. *A sample of* $Na_2CO_3 \cdot H_2O$ *weighing* $0{\cdot}62$ g *is added to* 100 mL *of* $0{\cdot}1$ N H_2SO_4. *Will the resulting solution be acidic, basic or neutral?*

Solution : Equivalents of $Na_2CO_3 \cdot H_2O = \frac{0{\cdot}62}{62} = 0{\cdot}01.$ ... (Eqn. 4i)

$$\left(\text{eq. wt. of } Na_2CO_3 \cdot H_2O = \frac{124}{2} = 62\right)$$

m.e. of $Na_2CO_3 \cdot H_2O = 0{\cdot}01 \times 1000 = 10$ ... (Eqn. 3)

m.e. of $H_2SO_4 = 0{\cdot}1 \times 100 = 10.$ ... (Eqn. 1)

Since the m.e. of $Na_2CO_3 \cdot H_2O$ is equal to that of H_2SO_4, the resulting solution will be neutral.

Ex. 13. *How many millilitres of a* 0·05 M $KMnO_4$ *solution are required to oxidise* 2·0 g *of* $FeSO_4$ *in a dilute acid solution?* (Fe = 56, S = 32, O = 16)

Solution : Normality of $KMnO_4$ solution = 0·05 × 5 = 0·25 N ... (Eqn. 6)

$$\left\{\begin{array}{l}\text{factor relating mol. wt. and} \\ \text{eq. wt., i.e., the change in ON} \\ \text{is 5 from } \underset{+7}{MnO_4^-} \rightarrow \underset{+2}{Mn^{2+}}\end{array}\right\}$$

Let the volume of $KMnO_4$ solution be v mL.

Thus m.e. of $KMnO_4 = 0{\cdot}25\,v$.

$$\text{Equivalents of } FeSO_4 = \frac{2}{152} \quad \text{... (Eqn. 1)}$$

$$\left\{\begin{array}{l}Fe^{2+} \rightarrow Fe^{3+} \\ \text{In the above reaction,} \\ \text{eq. weight of } FeSO_4 = \dfrac{\text{molecular wt.}}{\text{change in ON}} = \dfrac{152}{1} = 152\end{array}\right\}$$

$$\text{m.e. of } FeSO_4 = \frac{2}{152} \times 1000 \quad \text{... (Eqn. 3)}$$

Now,

$$\text{m.e. of } KMnO_4 = \text{m.e. of } FeSO_4 \quad \text{... (Eqn. 2)}$$

$$0{\cdot}25\,v = \frac{2}{152} \times 1000.$$

$$\therefore \quad v = 52{\cdot}63 \text{ mL.}$$

Ex. 14. *Hydroxylamine reduces iron*(III) *according to the equation:*

$$2NH_2OH + 4Fe^{3+} \rightarrow N_2O\,(g)\uparrow + H_2O + 4Fe^{2+} + 4H^+$$

Iron (II) *thus produced is estimated by titration with a standard solution of permanganate. The reaction is:*

$$MnO_4^- + 5Fe^{2+} + 8H^+ \rightarrow Mn^{2+} + 5Fe^{3+} + 4H_2O$$

A 10 mL *sample of hydroxylamine solution was diluted to* 1 *litre.* 50 mL *of this diluted solution was boiled with an excess of iron*(III) *solution. The resulting solution required* 12 mL *of* 0·02 M $KMnO_4$ *solution for complete oxidation of iron*(II). *Calculate the weight of hydroxylamine in one litre of the original solution.*

Solution : m.e. of 50 mL of the diluted solution of NH_2OH

= m.e. of Fe^{3+}

= m.e. of Fe^{2+} produced

= m.e. of $KMnO_4$ solution

= 0·1 × 12 = 1·2 ... (Eqn. 1)

$$\left\{\begin{array}{l} \underset{+7}{MnO_4^-} \rightarrow \underset{+2}{Mn^{+2}} \text{ change in ON} = 5 \\ \therefore \text{ normality} = 5 \times \text{molarity} = 5 \times 0{\cdot}02 = 0{\cdot}1 \end{array}\right\}$$

$\therefore$ m.e. of 1000 mL of diluted solution of $NH_2OH = \frac{1{\cdot}2}{50} \times 1000 = 24.$

$\therefore$ m.e. of 10 mL of original solution of $NH_2OH = 24.$

$\therefore$ m.e. of 1000 mL of original solution of $NH_2OH = 2400.$

$\therefore$ number of equivalents per litre $= \frac{2400}{1000}.$ ... (Eqn. 3)

$\therefore$ strength of NH_2OH in g/L $= \frac{2400}{1000} \times 16{\cdot}5 = 39{\cdot}6.$ ... (Eqn. 4i)

$$\left\{ \text{eq. wt. of } NH_2OH \text{ from the given reaction} = \frac{33}{2} = 16{\cdot}5 \right\}$$

Ex. 15. *In a quantitative determination of iron in an ore, an analyst converted* 0·42 g *of the ore into its ferrous form. This required* 42·00 mL *of* 0·1 N *solution of* $KMnO_4$ *for titration.*

(i) *How many milliequivalents of* $KMnO_4$ *does* 42·00 mL *of* 0·1 N *solution represent?*

(ii) *How many equivalents of iron were present in the sample of the ore taken for analysis?*

(iii) *How many grams of iron were present in the sample?*

(iv) *What is the percentage of iron in the ore?*

(v) *What is the molarity of* $KMnO_4$ *solution used?*

(vi) *How many moles of* $KMnO_4$ *were used for titration?* (Fe = 56)

Solution : In this problem Fe^{2+} is oxidised to Fe^{3+} by $KMnO_4$.

(i) m.e. of $KMnO_4$ solution $= 0{\cdot}1 \times 42 = 4{\cdot}2.$ ... (Eqn. 1)

(ii) Equivalent of Fe present in the sample

$= $ equivalent of $KMnO_4$ solution ... (Eqn. 7)

$= \frac{\text{m.e. of } KMnO_4 \text{ solution}}{1000}$ ... (Eqn. 3)

$= \frac{4{\cdot}2}{1000} = 0{\cdot}0042.$

(iii) Wt. of iron = equivalent × eq. wt. ... (Eqn. 4i)

$= 0{\cdot}0042 \times 56 = 0{\cdot}2352$ g.

$$\left\{ Fe^{2+} \rightarrow Fe^{3+} \quad \therefore \text{ eq. wt. of Fe} = \frac{\text{at. wt.}}{\text{change in ON}} = \frac{56}{1} = 56 \right\}$$

(iv) % of iron $= \dfrac{0{\cdot}2352}{0{\cdot}42} \times 100 = 56{\cdot}00\%$.

(v) Molarity of $KMnO_4 = \dfrac{\text{normality}}{\text{factor relating mol. wt. and eq. wt.}}$

i.e., change in ON ... (Eqn. 6i)

$$= \frac{0{\cdot}1}{5} = 0{\cdot}02 \text{ M}.$$

$$\begin{Bmatrix} MnO_4^- \rightarrow Mn^{2+} \\ +7 \qquad +2 \end{Bmatrix}$$

(vi) Moles of $KMnO_4 = \dfrac{\text{equivalents}}{\text{factor relating mol. wt. and eq. wt.}}$

... (Eqn. 6ii)

$$= \frac{0{\cdot}0042}{5} = 0{\cdot}00084 \text{ mole}.$$

[**Note:** Thus we see how the equations 1 to 7 have been put into use in the problems given above. However, application of these equations and the ability to use them come only from practice. The students are advised to apply the said rules in as many problems as possible.]

Ex. 16. *A* 0·5-g *sample of an iron-containing mineral, mainly in the form of* $CuFeS_2$, *was reduced suitably to convert all the ferric iron into ferrous form and was obtained as a solution. In the absence of any interfering matter, the solution required* 42 mL *of* 0·01 M $K_2Cr_2O_7$ *solution for titration. Calculate the percentage of* $CuFeS_2$ *in the mineral.* (Cu = 63·5, Fe = 55·8, S = 32, O = 16)

Solution : In this problem Fe^{2+} is oxidised by $K_2Cr_2O_7$ to Fe^{3+}.

Normality of $K_2Cr_2O_7$ solution = molarity × ON change per mole

... (Eqn. 6i)

$$= 0{\cdot}01 \times 6 = 0{\cdot}06$$

$$\begin{Bmatrix} Cr_2O_7^{2-} \rightarrow 2Cr^{3+}\text{; change in ON} = 6 \\ +12 \qquad\quad +6 \end{Bmatrix}$$

m.e. of $K_2Cr_2O_7 = 0{\cdot}06 \times 42 = 2{\cdot}52$. ... (Eqn. 1)

∴ m.e. of $CuFeS_2$ = m.e. of Fe^{2+} salt

= m.e. of $K_2Cr_2O_7$

= 2·52. ... (Eqn. 2)

Equivalents of $CuFeS_2 = \dfrac{2{\cdot}52}{1000}$. ... (Eqn. 3)

Weight of $CuFeS_2$ in the mineral $= \frac{2.52}{1000} \times 183.3$ ... (Eqn. 4i)

$= 0.4619$ g.

$\left\{ \begin{array}{l} \text{eq. weight of } CuFeS_2 \text{ is equal to its molecular weight, i.e., 183.3} \\ \text{because the change in ON of Fe is one } (Fe^{3+} \rightarrow Fe^{2+}) \end{array} \right\}$

Percentage of $CuFeS_2$ in the mineral $= \frac{0.4619 \times 100}{0.5} = 92.83\%$.

Ex. 17. 25 *grams of a sample of ferrous sulphate was dissolved in dilute sulphuric acid and water and its volume was made up to* 1 *litre.* 25 mL *of this solution required* 20 mL *of* $\frac{N}{10}$ $KMnO_4$ *solution for complete oxidation. Calculate the percentage of* $FeSO_4 \cdot 7H_2O$ *in the sample.*

Solution : m.e. of $KMnO_4$ solution $= \frac{1}{10} \times 20 = 2.$ ... (Eqn. 1)

$\therefore$ m.e. of 25 mL of $FeSO_4 \cdot 7H_2O$ solution = 2.

$\therefore$ m.e. of 1000 mL of $FeSO_4 \cdot 7H_2O$ solution $= \frac{2}{25} \times 1000 = 80.$

Equivalents of $FeSO_4 \cdot 7H_2O = \frac{80}{1000}$. ... (Eqn. 3)

$\therefore$ weight of $FeSO_4 \cdot 7H_2O$ = equivalent × eq. wt.

$= \frac{80}{1000} \times 278 = 22.24$ g.

$\left\{ \text{As } Fe^{2+} \rightarrow Fe^{3+}, \text{ eq. wt. of } FeSO_4 \cdot 7H_2O = \frac{\text{mol. wt.}}{\text{change in ON}} = \frac{278}{1}. \right\}$

Thus the percentage of $FeSO_4 \cdot 7H_2O$ in the sample $= \frac{22.24}{25} \times 100$

$= 88.96\%$.

Ex. 18. 5·39 g *of a mixture of* $FeSO_4 \cdot 7H_2O$ *and anhydrous ferric sulphate requires* 80 mL *of* 0·125 N *permanganate solution for complete conversion to ferric sulphate. Calculate the individual weights of each component of the original mixture.*

Solution : Ferrous sulphate present in the mixture is oxidised to ferric sulphate by permanganate solution. Let the weight of ferrous sulphate be x gram.

m.e. of $FeSO_4 \cdot 7H_2O$ = m.e. of permanganate solution

$= 0.125 \times 80 = 10.$

$\therefore$ eq. of $FeSO_4 \cdot 7H_2O = \frac{10}{1000}$. ... (Eqn. 3)

$\therefore$ wt. of $FeSO_4 \cdot 7H_2O$ = eq. × eq. wt.

$$= \frac{10}{1000} \times 278$$

$$= 2{\cdot}78 \text{ g}$$

and wt. of $Fe_2(SO_4)_3 = (5{\cdot}39 - 2{\cdot}78)$ g

$$= 2{\cdot}61 \text{ g.}$$

Ex. 19. 100 g *of a sample of* HCl *solution of relative density* 1·17 *contains* 33·4 g *of* HCl. *What volume of this* HCl *solution will be required to neutralise exactly* 5 *litres of* $\frac{N}{10}$ NaOH *solution?*

Solution : Volume of HCl solution $= \frac{100}{1{\cdot}17}$ mL.

$$\left(\text{density} = \frac{\text{mass}}{\text{volume}}\right).$$

$$\text{Equivalents of HCl} = \frac{33{\cdot}4}{36{\cdot}5} \quad \text{... (Eqn. 4i)}$$

(eq. wt. of HCl = 36·5)

$$\text{m.e. of HCl} = \frac{33{\cdot}4}{36{\cdot}5} \times 1000. \quad \text{... (Eqn. 3)}$$

$$\text{Normality of HCl} = \frac{\text{m.e.}}{\text{volume in mL}} \quad \text{... (Eqn. 1)}$$

$$= \frac{33{\cdot}4}{36{\cdot}5} \times 1000 \times \frac{1{\cdot}17}{100}$$

$$= 10{\cdot}7 \text{ N.}$$

Now let the volume of HCl, of normality calculated above, required to neutralise exactly the given NaOH solution be v mL.

m.e. of HCl = m.e. of NaOH

$$10{\cdot}7 \times v = \frac{1}{10} \times 5000$$

$$10{\cdot}7 \times v = 500.$$

$$\therefore \quad v = 46{\cdot}7 \text{ mL.}$$

Ex. 20. *Derive a formula to calculate the normality of an acid of sp. gr. 'd' containing x% by weight. The eq. wt. of the acid is E.*

Solution : 100 g of the acid solution contains x g of the acid

or $\frac{100}{d}$ mL of the acid solution contains $\frac{x}{E}$ eq. of the acid ... (Eqn. 4i)

or $\frac{100}{d}$ mL solution contains $\frac{x}{E} \times 1000$ m.e. of the acid ... (Eqn. 3)

$$\text{Normality of acid} = \frac{\text{m.e.}}{\text{volume in mL}} \quad \text{... (Eqn. 1)}$$

$$= \frac{x}{E} \times 1000 \times \frac{d}{100}.$$

$$\therefore \quad \text{normality} = \frac{10 \times x \times d}{E}.$$

Ex. 21. *A bottle of commercial sulphuric acid (density* 1·787 g/mL) *is labelled as* 86% *by weight. What is the molarity of the acid? What volume of the acid has to be used to make* 1 *litre of* 0·2 M H_2SO_4?

Solution : ∵ 100 g of H_2SO_4 solution contains 86 g of H_2SO_4.

$$\therefore \quad \frac{100}{1{\cdot}787} \text{ mL } H_2SO_4 \text{ solution contains } \frac{86}{98} \text{ mole.}$$

$$\therefore \quad 1000 \text{ mL } H_2SO_4 \text{ solution contains } \frac{86}{98} \times \frac{1{\cdot}787}{100} \times 1000.$$

$$= 15{\cdot}68 \text{ M.}$$

$$\therefore \quad \text{normality of } H_2SO_4 = (2 \times 15{\cdot}68) \text{ N} = 31{\cdot}36 \text{ N.}$$

(basicity of $H_2SO_4 = 2$)

Suppose that v mL of 31·36 N H_2SO_4 is to be used to make 1000 mL of 0·2 M (i.e., 0·4 N) H_2SO_4.

∴ m.e. of v mL of 31·36 N H_2SO_4 = m.e. of 1000 mL of 0·4 N H_2SO_4

$$31{\cdot}36 \times v = 0{\cdot}4 \times 1000$$

$$\therefore \quad v = \frac{0{\cdot}4 \times 1000}{31{\cdot}36} = 12{\cdot}75 \text{ mL.}$$

Ex. 22. *How many millilitres of concentrated sulphuric acid of sp. gr.* 1·84 *containing* 98% H_2SO_4 *by weight are required to prepare* 200 mL *of* 0·50 N solution?

Solution : 98% of H_2SO_4 by weight means 100 g H_2SO_4 solution contains 98 g of H_2SO_4.

$$\text{Volume of 100 g of } H_2SO_4 = \frac{100}{1{\cdot}84} \text{ mL}$$

i.e., $\frac{100}{1{\cdot}84}$ mL of H_2SO_4 solution contains 98·00 g of H_2SO_4.

$$\text{Equivalents of } H_2SO_4 = \frac{98}{49} = 2 \quad \text{... (Eqn. 4i)}$$

(eq. wt. of $H_2SO_4 = 49$)

$$\text{m.e. of } H_2SO_4 = 2 \times 1000 = 2000. \quad \text{... (Eqn. 3)}$$

$$\text{Normality of } H_2SO_4 \text{ solution} = \frac{\text{m.e.}}{\text{volume in mL}} \quad \text{... (Eqn. 1)}$$

$$= \frac{2000}{100/1{\cdot}84} = 36{\cdot}8 \text{ N.}$$

Let the volume of H_2SO_4 of normality calculated above to prepare 200 mL of 0·5 N H_2SO_4 solution be v mL.

$\therefore$ m.e. of v mL of H_2SO_4 of normality 36·8 N = 36·8 v ... (Eqn. 1)

and m.e. of 200 mL of H_2SO_4 of normality 0·5 N = 0·5 × 200

= 100.

Since both the solutions of H_2SO_4 should have the same number of m.e. we have,

$$36{\cdot}8\, v = 100.$$

$$\therefore \quad v = \frac{100}{36{\cdot}8} = 2{\cdot}72 \text{ mL.}$$

Ex. 23. *A piece of aluminium weighing* 2·7 g *is heated with* 75·0 mL *of sulphuric acid (sp. gr.* 1·18 *containing* 24·7% H_2SO_4 *by weight). After the metal is carefully dissolved the solution is diluted to* 400 mL. *Calculate the molarity of the free* H_2SO_4 *in the resulting solution.*

Solution : Normality of the given H_2SO_4 = 5·95 N

(calculated as in Example 20 or 21)

$\therefore$ m.e. of 75 mL of H_2SO_4 = 5·95 × 75 ... (Eqn. 1)

= 446·25.

$$\text{Equivalent of Al} = \frac{2{\cdot}7}{9} = 0{\cdot}3. \quad \text{... (Eqn. 4i)}$$

$$\left\{ \text{eq. wt. of Al} = \frac{\text{at . wt.}}{\text{valency}} = \frac{27}{3} = 9 \right\}$$

$\therefore$ m.e. of Al = 0·3 × 1000 = 300. ... (Eqn. 3)

Since 300 m.e. of Al will react with 300 m.e. of H_2SO_4,

m.e. of free H_2SO_4 = total m.e. of H_2SO_4 − 300

= 446·25 − 300

= 146·25.

Now the free H_2SO_4 is diluted to 400 mL and we know that the m.e. of H_2SO_4 does not change on dilution.

$\therefore$ normality of the diluted free H_2SO_4 in the resulting solution

$$= \frac{\text{m.e. of free } H_2SO_4}{\text{volume (mL)}}$$

$$= \frac{146{\cdot}25}{400} = 0{\cdot}366 \text{ N.}$$

$$\therefore \quad \text{molarity} = \frac{0{\cdot}366}{2} = 0{\cdot}183 \text{ M.} \qquad \text{... (Eqn. 6i)}$$

(basicity of $H_2SO_4 = 2$)

Ex. 24. *Two litres of ammonia at* 30°C *and* 0·90 *atmospheric pressure neutralised* 134 mL *of a solution of sulphuric acid. Calculate the normality of the acid.*

Solution : Volume of NH_3 at NTP $= \frac{0{\cdot}9 \times 2 \times 273}{303}$ lit. = 1·622 lit.

$$\left(\because \frac{p_1 V_1}{T_1} = \frac{p_2 V_2}{T_2} \right)$$

$$\therefore \text{ number of moles of } NH_3 = \frac{\text{volume at NTP (litres)}}{22{\cdot}4}$$

$$= \frac{1{\cdot}622}{22{\cdot}4} = 0{\cdot}0724.$$

Number of equivalents of NH_3

= number of moles × factor relating mol. wt. and eq. wt. ... (Eqn. 6ii)

$= 0{\cdot}0724.$

$\left\{ \because \text{ according to the reaction } 2NH_3 + H_2SO_4 \rightarrow (NH_4)_2SO_4, \text{ eq. wt. of } NH_3 \text{ is equal to its mol. wt.} \right\}$

$\therefore$ m.e. of NH_3 = equivalent × 1000

$= 0{\cdot}0724 \times 1000 = 72{\cdot}4.$

Let the normality of H_2SO_4 be N

$\therefore$ m.e. of $H_2SO_4 = N \times 134.$

Now m.e. of H_2SO_4 = m.e. of NH_3 ... (Eqn. 2)

$134 \times N = 72{\cdot}4$

$$\therefore \text{ normality of } H_2SO_4 = \frac{72{\cdot}4}{134}$$

$= 0{\cdot}54$ N.

Ex. 25. *(a) Calculate the strength of '20 V' of H_2O_2 in terms of:*
(i) normality (ii) grams per litre (iii) molarity and (iv) percentage.
(b) Calculate the volume strength of 3·58 N H_2O_2 *solution.*

Solution : (a) The strength of H_2O_2 as '20 V' means 1 volume of H_2O_2 on decomposition gives 20 volumes of oxygen at NTP or 1 litre of H_2O_2 gives 20 litres of oxygen at NTP.

(i) $2H_2O_2 \rightarrow 2H_2O + O_2$

1 lit. 20 lit. at NTP

1 lit. $\frac{20}{5 \cdot 6}$ eq. ... (Eqn. 4ii)

$$\left\{\begin{array}{l} \because \text{ 1 mole of oxygen (32 g) occupies a vol. of 22·4 lit. at NTP} \\ \therefore \text{ 1 eq. of oxygen (8 g) shall occupy } \frac{22 \cdot 4}{4} \text{ lit. at NTP} = 5 \cdot 6 \text{ lit.} \end{array}\right\}$$

$\therefore$ equivalent in 1 lit. of H_2O_2 = eq. of oxygen produced

$$= \frac{20}{5 \cdot 6} = 3 \cdot 58.$$

$\because$ equivalent per litre represents normality.

$\therefore$ normality of '20 V' $H_2O_2 = 3 \cdot 58$ N

(ii) From the above reaction,

$$\text{eq. wt. of } H_2O_2 = \frac{\text{mol. wt.}}{2} = \frac{34}{2} = 17.$$

$\therefore$ strength of '20 V' H_2O_2 = normality × eq. wt.

$$= 3 \cdot 58 \times 17$$

$$= 60 \cdot 86 \text{ g/lit.}$$

(iii) $\text{Molarity} = \frac{\text{normality}}{\text{factor relating mol. wt. and eq. wt.}}$ ... (Eqn. 6i)

$$= \frac{3 \cdot 58}{2} = 1 \cdot 79 \text{ M.}$$

(iv) Strength in percentage is grams per 100 mL so from part (ii) of the solution,

strength in percentage of '20 V' H_2O_2 = 6·086% (g/100 mL).

(b) The strength of H_2O_2 is 3·58 equivalents per litre.

$\therefore$ the equivalent of oxygen = 3·58.

The volume of O_2 (lit.) at NTP = $3 \cdot 58 \times 5 \cdot 6 = 20$ litres. ... (Eqn. 4ii)

(1 eq. of oxygen occupies 5·6 litres.)

Thus 1 litre of H_2O_2 at NTP gives a volume of 20 litres

That is to say— strength of H_2O_2 is '20 V'.

Ex. 26. *Calculate the percentage of free* SO_3 *in an oleum (considered as a solution of* SO_3 *in* H_2SO_4*) that is labelled* '109% H_2SO_4'.

Solution : '109% H_2SO_4' refers to the total mass of pure H_2SO_4, i.e., 109 g that will be formed when 100 g of oleum is diluted by 9 g of H_2O which (H_2O) combines with all the free SO_3 present in oleum to form H_2SO_4

$$H_2O + SO_3 \rightarrow H_2SO_4$$

1 mole of H_2O combines with 1 mole of SO_3

or 18 g of H_2O combines with 80 g of SO_3

or 9 g of H_2O combines with 40 g of SO_3.

Thus, 100 g of oleum contains 40 g of SO_3 or oleum contains 40% of free SO_3.

Ex. 27. *A solution of a* 0·4-g *sample of* H_2O_2 *reacted with* 0·632 g *of* $KMnO_4$ *in the presence of sulphuric acid. Calculate the percentage purity of the sample of* H_2O_2.

Solution :

$$\underset{+7}{2KMnO_4} + 3H_2SO_4 + 5H_2O_2 \rightarrow K_2SO_4 + \underset{+2}{2MnSO_4} + 8H_2O + 5O_2$$

$$\text{Eq. wt. of } KMnO_4 = \frac{\text{mol. wt.}}{\text{change in ON per mole}} = \frac{158}{5} = 31{\cdot}6.$$

Again from the above reaction we see that

2 moles of $KMnO_4$ combine with 5 moles of H_2O_2

or 10 equivalents (2 moles) of $KMnO_4$ combine with 5 moles of H_2O_2

or 1 equivalent of $KMnO_4$ combines with $\frac{1}{2}$ moles of H_2O_2.

$$\therefore \text{ equivalent wt. of } H_2O_2 = \frac{\text{mol. wt.}}{2} = \frac{34}{2} = 17.$$

Now, m.e. of H_2O_2 = m.e. of $KMnO_4$

or eq. of H_2O_2 = eq. of $KMnO_4$

$$\frac{x}{17} = \frac{0{\cdot}632}{31{\cdot}6} \qquad \text{... (Eqn. 4i)}$$

x being the amount of H_2O_2 in grams,

$$x = 0{\cdot}34 \text{ g}.$$

$$\text{Percentage of } H_2O_2 \text{ in the sample} = \frac{0{\cdot}34}{0{\cdot}4} \times 100 = 85\%.$$

Ex. 28. *A* 1·00-g *sample of* H_2O_2 *solution containing* x *per cent* H_2O_2 *by weight requires* x mL *of a* $KMnO_4$ *solution for complete oxidation under acidic conditions. Calculate the normality of the* $KMnO_4$ *solution.*

Solution :

$$\underset{(+7)}{2MnO_4^-} + 5H_2O_2 + 6H^+ \rightarrow 5O_2 + \underset{(+2)}{2Mn^{2+}} + 8H_2O$$

From the equation, we see that

change in oxidation number of Mn = 7 − 2 = 5.

∴ 1 mole of $KMnO_4$ = 5 eq. of $KMnO_4$.

∴ 10 eq. of $KMnO_4$ combines with 5 moles of H_2O_2.

$\therefore$ 1 eq. of $KMnO_4$ combines with $\frac{1}{2}$ mole of H_2O_2.

$\therefore$ eq. wt. of $H_2O_2 = \frac{34}{2} = 17$.

Now,

$\because$ 100 g of H_2O_2 solution contains x g of H_2O_2

$\therefore$ 100 g of H_2O_2 contains $\frac{x}{17}$ equivalent of H_2O_2.

$\therefore$ 1 g of H_2O_2 contains $\frac{x}{17 \times 100}$ eq. of H_2O_2

$\therefore$ number of m.e. of H_2O_2 in 1 g solution $= \frac{x}{17 \times 100} \times 1000$

$= \frac{10\,x}{17}$. ... (Eqn. 3)

m.e. of H_2O_2 = m.e. of $KMnO_4$

$\frac{10\,x}{17} = x\,N$. ... (Eqn. 1)

(N – normality of $KMnO_4$)

$\therefore$ normality of $KMnO_4$ solution $= \frac{10}{17}$ eq. lit.$^{-1}$

Ex. 29. *In a 50-mL solution of H_2O_2 an excess of KI and dilute H_2SO_4 were added. The I_2 so liberated required 20 mL of 0·1 N $Na_2S_2O_3$ for complete reaction. Calculate the strength of H_2O_2 in grams per litre.*

Solution : $2KI + H_2SO_4 + H_2O_2 \rightarrow K_2SO_4 + 2H_2O + I_2$

$2Na_2S_2O_3 + I_2 \rightarrow Na_2S_4O_6 + 2NaI$

m.e. of H_2O_2 in 50 mL = m.e. of I_2 = m.e. of $Na_2S_2O_3$

$\therefore$ m.e. of H_2O_2 in 50 mL $= 0{\cdot}1 \times 20 = 2$. ... (Eqn. 1)

$\therefore$ m.e. of H_2O_2 in 1000 mL $= \frac{2}{50} \times 1000 = 40$.

Equivalents per litre $= \frac{40}{1000}$. ... (Eqn. 3)

Grams per litre of $H_2O_2 = \frac{40}{1000} \times 17$ ... (Eqn. 4)

$= 0{\cdot}68$ g / litre.

(eq. wt. of $H_2O_2 = \frac{34}{2} = 17$)

Ex. 30. *Calculate the percentage amount of oxalate in a given sample of oxalate salt when* 0·3 g *of salt was dissolved in* 100 mL *and* 10 mL *of which required* 8 mL *of* $\frac{N}{20}$ $KMnO_4$ *solution.*

Solution : Let the amount of oxalate in the sample be x g.

Now,

$$\text{m.e. of } KMnO_4 = \frac{1}{20} \times 8 = 0{\cdot}4. \qquad \text{... (Eqn. 1)}$$

$\therefore$ m.e. of 10 mL of oxalate salt solution = 0·4. ... (Eqn. 2)

$\therefore$ m.e. of 100 mL of oxalate solution = 4·0

but m.e. of oxalate = m.e. of oxalate salt = 4. ... (Eqn. 7)

$$\therefore \text{ equivalent of oxalate} = \frac{4}{1000} = 0{\cdot}004. \qquad \text{... (Eqn. 3)}$$

$$\text{Wt. of oxalate} = (0{\cdot}004 \times 44)\text{ g} \qquad \text{... (Eqn. 4i)}$$

$$= 0{\cdot}176 \text{ g}.$$

$$\left\{ \text{eq. wt. of } C_2O_4^{2-} = \frac{88}{2} = 44. \right\}$$

$$\text{Percentage amount of oxalate} = \frac{0{\cdot}176}{0{\cdot}3} \times 100 = 58{\cdot}67\%.$$

Ex. 31. *One gram-atom of* Ca *was burnt in excess of oxygen and the oxide was dissolved in water to make up a* 1-*litre solution. Calculate the normality of the alkaline solution.*

Solution : **First Method**

$$Ca \xrightarrow[\text{burnt}]{O_2} CaO \xrightarrow{\text{dissolved in water}} Ca(OH)_2$$

m.e. of Ca = m.e. of CaO = m.e. of 1000 mL of $Ca(OH)_2$ solution ... (Eqn. 7)

$\therefore$ m.e. of 1000 mL of $Ca(OH)_2$ solution = m.e. of Ca

$$= \text{eq.} \times 1000 \qquad \text{... (Eqn. 3)}$$

$$= (\text{gram-atom} \times \text{valency of Ca}) \times 1000$$

$$= 1 \times 2 \times 1000$$

$$= 2000.$$

$$\therefore \text{ normality of } Ca(OH)_2 \text{ solution} = \frac{\text{m.e.}}{\text{vol. in mL}}$$

$$= \frac{2000}{1000} = 2\,\text{N}.$$

Second Method See the first method in Example 13, Chapter 2.

Ex. 32. *How much* AgCl *will be formed by adding* 1·70 g *of* $AgNO_3$ in 200 mL *of* 5 N HCl *solution?* (Ag = 108, N = 14, O = 16)

Solution : Equivalent of $AgNO_3 = \frac{1{\cdot}70}{170} = 0{\cdot}01.$ ... (Eqn. 4i)

(eq. wt. of $AgNO_3 = 170$)

m. e. of HCl solution $= 5 \times 200 = 1000.$ ... (Eqn. 1)

$\therefore$ equivalent of HCl solution $= \frac{1000}{1000} = 1.$ ... (Eqn. 3)

Since equivalent of $AgNO_3$ is less than the eq. of HCl,

equivalent of AgCl = eq. of $AgNO_3 = 0{\cdot}01.$

$\therefore$ wt. of AgCl $= 0{\cdot}01 \times 143{\cdot}5$ ... (Eqn. 4i)

$= 1{\cdot}435$ g.

Ex. 33. *A mixture of aluminium and zinc weighing* 1·67 g *was completely dissolved in acid and evolved* 1·69 *litres of hydrogen at* NTP. *What was the weight of aluminium in the original mixture?* (Al = 27, Zn = 65·4)

Solution : Since H_2 is formed by both Al and Zn,

eq. of Al + eq. of Zn = eq. of H_2.

Let w be the mass in grams of Al in the mixture.

$$\therefore \quad \frac{w}{\text{eq. wt. of Al}} + \frac{(1{\cdot}67 - w)}{\text{eq. wt. of Zn}} = \frac{1{\cdot}69}{\text{vol. of 1 eq. of } H_2 \text{ at NTP in lit.}}$$

$$\frac{w}{27/3} + \frac{1{\cdot}67 - w}{65{\cdot}4/2} = \frac{1{\cdot}69}{11{\cdot}2} \quad \left\{ \begin{array}{l} \text{eq. wt. of Al} = \frac{27}{3} \\ \text{eq. wt. of Zn} = \frac{65{\cdot}4}{2} \end{array} \right\}$$

$$\therefore \quad w = 1{\cdot}24 \text{ g.}$$

Ex. 34. *A metal weighing* 0·43 g *was dissolved in* 50 mL *of* N H_2SO_4. *The unreacted* H_2SO_4 *required* 14·2 mL *of* N NaOH *for neutralisation. Find out the equivalent weight of the metal.*

Solution : Eq. of metal $= \frac{0{\cdot}43}{E} \cdot$ (eq. wt. of metal $\equiv E$)

$\therefore$ m.e. of metal $= \frac{0{\cdot}43}{E} \times 1000 = \frac{430}{E}$ ... (Eqn. 3)

m.e. of total H_2SO_4 solution $= 1 \times 50 = 50$ ... (Eqn. 1)

But m.e. of H_2SO_4 reacted with metal = m.e. of the metal

$= \frac{430}{E}.$

$\therefore$ m.e. of unreacted $H_2SO_4 = \left(50 - \frac{430}{E}\right)$.

Again m.e. of unreacted H_2SO_4 = m.e. of NaOH. ... (Eqn. 2)

$$\therefore \quad 50 - \frac{430}{E} = 1 \times 14{\cdot}2 \quad \text{... (Eqn. 1)}$$

$$\therefore \quad E = 12{\cdot}01.$$

Ex. 35. 7·35 g *of a dibasic acid was dissolved in water and diluted to* 250 mL. 25 mL *of this solution was neutralised by* 15 mL *of* N NaOH *solution. Calculate eq. wt. and mol. wt. of the acid.*

Solution : Let the equivalent weight of the acid be E.

$$\text{Equivalent of acid} = \frac{7{\cdot}35}{E}. \quad \text{... (Eqn. 4i)}$$

$$\text{m.e. of the acid} = \frac{7{\cdot}35}{E} \times 1000 = \frac{7350}{E}. \quad \text{... (Eqn. 3)}$$

Now,

$$250 \text{ mL of the acid contains } \frac{7350}{E} \text{ m.e.}$$

$$\therefore \quad 25 \text{ mL of the acid contains } \frac{735}{E} \text{ m.e.}$$

Again, m.e. of 25 mL of the acid = m.e. of NaOH ... (Eqn. 2)

$$\frac{735}{E} = 1 \times 15$$

$$E = \frac{735}{15} = 49.$$

$$\therefore \quad \text{eq. wt. of acid} = 49.$$

$$\therefore \quad \text{molecular weight of the acid} = \text{eq. wt.} \times \text{basicity}$$
$$= 49 \times 2$$
$$= 98.$$

Ex. 36. 2·0 g *of a mixture of carbonate, bicarbonate and chloride of sodium, on heating, produced* 56 mL *of* CO_2 *at* NTP. 1·6 g *of the same mixture required* 25 mL *of* N HCl *solution for neutralisation. Calculate the percentage of* Na_2CO_3, $NaHCO_3$ *and* NaCl *in the mixture from the given data.*

Solution : On heating the given mixture, only $NaHCO_3$ decomposes as

$$2NaHCO_3 \rightarrow Na_2CO_3 + H_2O + CO_2$$

$$\therefore \quad \text{eq. of } NaHCO_3 = \text{eq. of } CO_2 = \frac{56}{11200}. \quad \text{... (Eqn. 4ii)}$$

(1 eq. of CO_2 occupies 11200 mL at NTP)

$\therefore$ wt. of $NaHCO_3 = \dfrac{56}{11200} \times 84 = 0{\cdot}42$ g. ... (Eqn. 4i)

(eq. wt. of $NaHCO_3 = 84$)

$\therefore$ % of $NaHCO_3 = \dfrac{0{\cdot}42}{2} \times 100 = 21\%$.

Now, if x is the weight of NaCl in 1·6 g of the mixture

then wt. of $NaHCO_3 = 0{\cdot}336$ g (i.e., 21% of 1·6 g)

and wt. of $Na_2CO_3 = 1{\cdot}6 - 0{\cdot}336 - x = (1{\cdot}264 - x)$ g.

Since Na_2CO_3 and $NaHCO_3$ are neutralised by HCl solution as:

$$Na_2CO_3 + 2HCl \rightarrow 2NaCl + H_2O + CO_2$$
$$NaHCO_3 + HCl \rightarrow NaCl + H_2O + CO_2$$

m.e. of Na_2CO_3 + m.e. of $NaHCO_3$ = m.e. of HCl

or eq. of $Na_2CO_3 \times 1000$ + eq. of $NaHCO_3 \times 1000$ = m.e. of HCl ... (Eqn. 3)

$$\frac{1{\cdot}264 - x}{53} \times 1000 + \frac{0{\cdot}336}{84} \times 1000 = 1 \times 25$$

$$x = 0{\cdot}151 \text{ g.}$$

$\therefore$ % of NaCl $= \dfrac{0{\cdot}151}{1{\cdot}6} \times 100 = 9{\cdot}42\%$

and % of $Na_2CO_3 = 100 - (21 + 9{\cdot}42) = 69{\cdot}58\%$.

Thus, $\begin{cases} Na_2CO_3 = 69{\cdot}58\% \\ NaHCO_3 = 21{\cdot}00\% \\ NaCl = 9{\cdot}42\%. \end{cases}$

Ex. 37. *Find out the volume in* mL *of* 0·1 N HCl *solution required to react completely with* 1·0 g *of a mixture of* Na_2CO_3 *and* $NaHCO_3$ *containing equimolar amounts of the two compounds.*

Solution : Let the amount of Na_2CO_3 in 1 g of mixture be x g and since the Na_2CO_3 and $NaHCO_3$ are in equimolar amounts,

$$\frac{x}{106} = \frac{1-x}{84}, \text{ where } \begin{Bmatrix} \text{mol. wt. of } Na_2CO_3 = 106 \\ \text{mol. wt. of } NaHCO_3 = 84 \end{Bmatrix}$$

$\therefore$ $x = 0{\cdot}558$ g.

Thus wt. of $Na_2CO_3 = 0{\cdot}558$ g

and wt. of $NaHCO_3 = 1 - 0{\cdot}558 = 0{\cdot}442$ g.

Now,

m.e. of HCl = m.e. of Na_2CO_3 + m.e. of $NaHCO_3$

m.e. of HCl = eq. of $Na_2CO_3 \times 1000$ + eq. of $NaHCO_3 \times 1000$. ... (Eqn. 3)

If v is the volume of HCl in mL

then $$0{\cdot}1 \times v = \frac{0{\cdot}558}{53} \times 1000 + \frac{0{\cdot}442}{84} \times 1000. \quad \text{... (Eqn. 4i)}$$

$$\therefore \quad v = 157{\cdot}9 \text{ mL}.$$

Ex. 38. 5 mL *of* 8 N *nitric acid,* 4·8 mL *of* 5 N HCl *and a certain volume of* 17 M *sulphuric acid are mixed together and made up to 2 litres.* 30 mL *of this mixture exactly neutralises* 42·9 mL *of sodium carbonate solution containing* 1 g *of* $Na_2CO_3 \cdot 10H_2O$ *in* 100 mL *of water. Calculate the amount in grams of the sulphate ions in the solution.* **(IIT 1985)**

Solution : Let the volume of 17 M (i.e., 34 N) H_2SO_4 solution be v mL.

$\therefore$ total m.e. of the acid mixture $= 8 \times 5 + 5 \times 4{\cdot}8 + 34v$ ... (Eqn. 1)

$$= (64 + 34v).$$

$$\therefore \text{ normality of the mixture} = \frac{\text{m.e.}}{\text{total volume (mL)}} \quad \text{... (Eqn. 1)}$$

$$= \frac{64 + 34v}{2000}.$$

$$\text{m.e. of 30 mL of this acid mixture} = \frac{64 + 34v}{2000} \times 30. \quad \text{... (Eqn. 1)}$$

$$\text{Now, normality of } Na_2CO_3 \cdot 10H_2O \text{ solution} = \frac{\text{g/litre}}{\text{eq. wt.}}$$

$$= \frac{10}{143}.$$

$$\left\{ \begin{array}{l} \because \text{ grams/litre of } Na_2CO_3 \cdot 10H_2O = 10 \\ \qquad \text{and eq. wt.} = \dfrac{\text{mol. wt.}}{2} = \dfrac{286}{2} \\ \qquad\qquad\qquad = 143 \end{array} \right\}$$

$$\therefore \text{ m.e. of 42·9 mL of } Na_2CO_3 \,.\, 10H_2O \text{ solution} = \frac{10}{143} \times 42{\cdot}9.$$

Thus, m.e. of 30 mL of acid mixture

= m.e. of 42·9 mL of $Na_2CO_3 \cdot 10H_2O$ solution ... (Eqn. 2)

$$\therefore \quad \frac{64 + 34\,v}{2000} \times 30 = \frac{10}{143} \times 42{\cdot}9$$

$$v = \frac{68}{17}.$$

$$\therefore \text{ m.e. of 34 N (i.e., 17 M) } H_2SO_4 = 34 \times \frac{68}{17} \quad \text{... (Eqn. 1)}$$

$$= 136.$$

$$\therefore \quad \text{equivalent of } H_2SO_4 = \frac{136}{1000} = 0{\cdot}136. \quad \text{... (Eqn. 3)}$$

$$\therefore \quad \text{equivalent of } SO_4^{2-} = 0{\cdot}136 \quad \text{... (Eqn. 7ii)}$$

$$\begin{aligned} \text{weight of } SO_4^{2-} &= \text{eq.} \times \text{eq. wt. of } SO_4^{2-} \quad \text{... (Eqn. 4i)} \\ &= 0{\cdot}136 \times 48 \\ &= 6{\cdot}528 \text{ g.} \end{aligned}$$

$$\left(\text{eq. wt. of } SO_4^{2-} = \frac{\text{ionic wt.}}{\text{valency}} = \frac{96}{2} = 48\right)$$

Ex. 39. 1·42 g *of a mixture of* $CaCO_3$ *and* $MgCO_3$ *was dissolved in* 200 mL *of* 0·2 N HCl *solution which was then diluted to* 250 mL. 10 mL *of this solution was neutralised by* 12 mL *of* (N/30) Na_2CO_3 . *Find out the percentage of each in the mixture.* (Ca = 40, Mg = 24, C = 12, O = 16)

Solution : Let the wt. of $CaCO_3$ be x g.

$\therefore$ wt. of $MgCO_3 = (1{\cdot}42 - x)$ g.

$$\therefore \text{ eq. of } CaCO_3 = \frac{x}{50} \text{ and eq. of } MgCO_3 = \frac{1{\cdot}42 - x}{42}. \quad \text{... (Eqn. 4)}$$

$$\left(\text{eq. wt. of } CaCO_3 = \frac{100}{2} = 50 \text{ ; eq. wt. of } MgCO_3 = \frac{84}{2} = 42\right)$$

$$\text{Total m.e. of } CaCO_3 \text{ and } MgCO_3 = \frac{x}{50} \times 1000 + \frac{1{\cdot}42 - x}{42} \times 1000 \quad \text{... (Eqn. 3)}$$

$$\text{m.e. of HCl} = 0{\cdot}2 \times 200 = 40. \quad \text{... (Eqn. 1)}$$

From the given question it is clear that m.e. of HCl is greater than those of $CaCO_3$ and $MgCO_3$.

$\because$ m.e. of excess HCl = m.e. of HCl – m.e. of $CaCO_3$ and $MgCO_3$

$$= 40 - \left\{1000\left(\frac{x}{50} + \frac{1{\cdot}42 - x}{42}\right)\right\}.$$

$\therefore$ the m.e. of the resulting solution does not change on dilution.

$\therefore$ normality of excess HCl in the diluted resulting solution

$$= \frac{\text{m.e.}}{250} = \frac{40 - \left\{1000\left(\frac{x}{50} + \frac{1{\cdot}42 - x}{42}\right)\right\}}{250}.$$

$\therefore$ m.e. of 10 mL of the resulting solution

$$= \frac{40 - \left\{1000\left(\frac{x}{50} + \frac{1{\cdot}42 - x}{42}\right)\right\}}{250} \times 10. \quad \text{... (Eqn. 1)}$$

$$\text{m.e. of } Na_2CO_3 \text{ solution} = \frac{1}{30} \times 12. \quad \text{... (Eqn. 1)}$$

$$\therefore \quad \frac{40 - \left\{1000\left(\frac{x}{50} + \frac{1{\cdot}42 - x}{42}\right)\right\}}{250} \times 10 = \frac{1}{30} \times 12. \qquad \text{... (Eqn. 2)}$$

$\therefore \quad x = 1.$

$$\therefore \quad \% \text{ of } CaCO_3 = \frac{1}{1{\cdot}42} \times 100 = 70{\cdot}4\%,$$

$$\% \text{ of } MgCO_3 = 100 - 70{\cdot}4 = 29{\cdot}6\%.$$

Ex. 40. 3·68 g *of a mixture of* $CaCO_3$ *and* $MgCO_3$, *on reaction with* 1000 mL *of* N/10 HCl *solution produced* 1·76 g *of* CO_2. *Calculate the percentage of each in the mixture.*

Solution :

$$CaCO_3 + 2HCl \rightarrow CaCl_2 + H_2O + CO_2$$
$$MgCO_3 + 2HCl \rightarrow MgCl_2 + H_2O + CO_2$$

From the equation, 2 eq. of HCl produces 1 mole of CO_2.

$\therefore$ 1 eq. of HCl produces 1/2 mole of CO_2.

$$\therefore \text{ eq. wt. of } CO_2 = \frac{44}{2} = 22.$$

$$\therefore \quad \text{eq. of } CO_2 = \frac{1{\cdot}76}{22} = 0{\cdot}08. \qquad \text{... (Eqn. 4i)}$$

From the given problem it is clear that the number of equivalents of the mixture of $CaCO_3$ and $MgCO_3$ is lesser than that of HCl solution. And so the CO_2 is produced by the complete consumption of $CaCO_3$ and $MgCO_3$.

$\therefore$ eq. of $CaCO_3$ + eq. of $MgCO_3$ = eq. of CO_2

$$\frac{x}{50} + \frac{3{\cdot}68 - x}{42} = 0{\cdot}08 \quad (x = \text{amount of } CaCO_3)$$

$$x = 2{\cdot}0 \text{ g}.$$

$$\therefore \quad \% \text{ of } CaCO_3 = \frac{2}{3{\cdot}68} \times 100 = 54{\cdot}34\%,$$

$$\% \text{ of } MgCO_3 = 100 - 54{\cdot}34 = 45{\cdot}66\%.$$

Ex. 41. 1·25 g *of a mixture of* Na_2CO_3 *and* Na_2SO_4 *was dissolved in* 250 mL *of water.* 25 mL *of this solution required* 20 mL *of* 0·1 N H_2SO_4 *solution for exact neutralisation. Calculate the percentage of* Na_2CO_3 *in the mixture.*

Solution : In this problem only Na_2CO_3 is neutralised by H_2SO_4.

Let the amount of Na_2CO_3 be x g.

$$\therefore \text{ equivalent of } Na_2CO_3 = \frac{x}{53} \qquad \text{... (Eqn. 4i)}$$

$$\text{m.e. of } Na_2CO_3 = \frac{x}{53} \times 1000 \qquad \text{... (Eqn. 3)}$$

$\because$ m.e. of Na_2CO_3 in 250 mL of the mixture solution $= \frac{1000x}{53}$.

$\therefore$ m.e. of Na_2CO_3 in 25 mL of the mixture solution $= \frac{100x}{53}$.

Now m.e. of 25 mL of mixture solution

$$= \text{m.e. of 20 mL of } 0{\cdot}1 \text{ N } H_2SO_4 \qquad \text{... (Eqn. 2)}$$

$$\frac{100x}{53} = 0{\cdot}1 \times 20$$

$$x = 1{\cdot}06 \text{ g.}$$

$$\therefore \quad \% \text{ of } Na_2CO_3 = \frac{1{\cdot}06}{1{\cdot}25} \times 100 = 84{\cdot}8\%.$$

Ex. 42. 4·35 g *of a mixture of* NaCl *and* Na_2CO_3 *was dissolved in* 100 mL *of water,* 20 mL *of which was exactly neutralised by* 75·5 mL *of* N/10 *solution of* H_2SO_4. *Calculate the percentage of* NaCl *and* Na_2CO_3 *in the mixture.*

Solution : In this problem, Na_2CO_3 is neutralised by H_2SO_4 solution. Let the amount of Na_2CO_3 be x g.

$\therefore$ equivalent of Na_2CO_3 in 100 mL of solution $= \frac{x}{53}$. ... (Eqn. 4i)

$\therefore$ m.e. of Na_2CO_3 in 100 mL of solution $= \frac{x}{53} \times 1000$. ... (Eqn. 3)

$\therefore$ m.e. of Na_2CO_3 in 20 mL of solution $= \frac{1000x}{53} \times \frac{20}{100} = \frac{200x}{53}$.

Now, m.e. of H_2SO_4 solution $= \frac{1}{10} \times 75{\cdot}50$. ... (Eqn. 1)

$\because$ m.e. of Na_2CO_3 = m.e. of H_2SO_4 ... (Eqn. 2)

$$\therefore \quad \frac{200x}{53} = 7{\cdot}55$$

$$x = 2.$$

$$\therefore \quad \% \text{ of } Na_2CO_3 = \frac{2}{4{\cdot}35} \times 100 = 45{\cdot}99\%,$$

$$\% \text{ of NaCl} = 100 - 45{\cdot}99 = 54{\cdot}01\%.$$

Ex. 43. 1·216 g *of a sample of* $(NH_4)_2SO_4$ *was boiled with excess of* NaOH *and the ammonia gas so produced was absorbed in* 100 mL *of* N H_2SO_4 *solution. The unreacted* H_2SO_4 *required* 81·6 mL *of normal solution of a base for exact neutralisation. Calculate percentage amount of ammonia in ammonium sulphate.*

Solution : m.e. of unreacted H_2SO_4 = m.e. of the base

$$= 1 \times 81{\cdot}6 = 81{\cdot}6.$$

m.e. of NH_3 = m.e. of H_2SO_4 reacted with ammonia

= m.e. of total H_2SO_4 – m.e. of unreacted H_2SO_4

$$= 1 \times 100 - 81{\cdot}6 = 18{\cdot}4.$$

$$\therefore \quad \text{equivalent of } NH_3 = \frac{18{\cdot}4}{1000}. \qquad \text{... (Eqn. 3)}$$

$$\text{Wt. of } NH_3 = \frac{18{\cdot}4}{1000} \times 17 \qquad \text{... (Eqn. 4i)}$$

$$= 0{\cdot}3128 \text{ g.}$$

$$\therefore \quad \% \text{ of } NH_3 \text{ in } (NH_4)_2SO_4 = \frac{0{\cdot}3128}{1{\cdot}216} \times 100$$

$$= 25{\cdot}72\%.$$

Ex. 44. *A solution contains* Na_2CO_3 *and* $NaHCO_3$. 10 mL *of the solution requires* 2·5 mL *of* 0·1 M H_2SO_4 *for neutralisation using phenolphthalein as an indicator. Methyl orange is added when a further* 2·5 mL *of* 0·2 M H_2SO_4 *was required. Calculate the amount of* Na_2CO_3 *and* $NaHCO_3$ *in one litre of the solution.*

Solution : The neutralisation reactions are

$$2Na_2CO_3 + H_2SO_4 \rightarrow 2NaHCO_3 + Na_2SO_4$$

$$2NaHCO_3 + H_2SO_4 \rightarrow Na_2SO_4 + H_2O + CO_2$$

The volume of H_2SO_4 (2·5 mL), used while using phenolphthalein corresponds to the volume required for conversion of Na_2CO_3 to $NaHCO_3$ while volume of H_2SO_4 (2·5 mL) further added corresponds to the volume required for conversion of $NaHCO_3$ to Na_2SO_4. Thus at the end point with phenolphthalein, we have,

m.e. of 2·5 mL of 0·1 M (i.e., 0·2 N) H_2SO_4 = m.e. of Na_2CO_3

or $\quad 2{\cdot}5 \times 0{\cdot}2$ = m.e. of Na_2CO_3

or $\quad$ m.e. of $Na_2CO_3 = 0{\cdot}5$.

$$\text{Equivalent of } Na_2CO_3 = \frac{0{\cdot}5}{1000}.$$

$$\text{Wt. of } Na_2CO_3/10 \text{ mL} = \frac{0{\cdot}5}{1000} \times 106 = 0{\cdot}053 \text{ g.}$$

*(equivalent wt. of Na_2CO_3 is 106 according to given reaction)

$\therefore$ wt. of Na_2CO_3 per litre = 5·3 g.

Again with methyl orange, we have,

m.e. of 2·5 mL of 0·2 M (i.e., 0·4 N) H_2SO_4 solution

---

* Read page 91, Chapter 6.

= m.e. of $NaHCO_3$ produced from Na_2CO_3
+ m.e. of $NaHCO_3$ originally present.

Since m.e. of $NaHCO_3$ (produced) = m.e. of Na_2CO_3.

$\therefore$ $2{\cdot}5 \times 4$ = m.e. of Na_2CO_3 + m.e. of $NaHCO_3$ originally present.

$\therefore$ $1 = 0{\cdot}5$ + m.e. of $NaHCO_3$ originally present.

$\therefore$ m.e. of $NaHCO_3$ originally present $= 1 - 0{\cdot}5 = 0{\cdot}5$.

$$\therefore \quad \text{equivalent of } NaHCO_3 = \frac{0{\cdot}5}{1000}.$$

$$\therefore \quad \text{wt. of } NaHCO_3 \text{ per 10 mL} = \frac{0{\cdot}5}{1000} \times 84 = 0{\cdot}042 \text{ g}.$$

(eq. wt. of $NaHCO_3 = 84$ according to given reaction)

Wt. of $NaHCO_3$ per litre = 4·2 g.

Ex. 45. *A solution contained* Na_2CO_3 *and* $NaHCO_3$. 25 mL *of this solution required* 5 mL *of* 0·1 N HCl *for titration with phenolphthalein as indicator. The titration was repeated with the same volume of the solution but with methyl orange.* 12·5 mL *of* 0·1 N HCl *was required this time. Calculate the amount of* Na_2CO_3 *and* $NaHCO_3$ *in the solution.*

Solution : Neutralisation reaction with phenolphthalein is

$$Na_2CO_3 + HCl \rightarrow NaHCO_3 + NaCl$$

while with methyl orange, the reactions are,

$$Na_2CO_3 + HCl \rightarrow NaHCO_3 + NaCl$$

$$\underset{\text{(produced)}}{NaHCO_3} + HCl \rightarrow NaCl + H_2O + CO_2$$

and

$$\underset{\text{(originally present)}}{NaHCO_3} + HCl \rightarrow NaCl + H_2O + CO_2$$

Thus,

we have with phenolphthalein,

m.e. of Na_2CO_3 = m.e. of 5 mL of 0·1 N HCl

$= 0{\cdot}1 \times 5 = 0{\cdot}5$.

$$\therefore \quad \text{eq. of } Na_2CO_3 = \frac{0{\cdot}5}{1000} = 0{\cdot}0005$$

$\therefore$ wt. of $Na_2CO_3 = (0{\cdot}0005 \times 106)$ g

$= 0{\cdot}053$ g.

{Eq. wt. of Na_2CO_3 in the given reaction is 106}

And with methyl orange,

m.e. of Na_2CO_3 + m.e. of $NaHCO_3$ (produced) + m.e. of $NaHCO_3$ (originally present)

= m.e. of 12·5 mL of 0·1 N HCl

or $\quad 0{\cdot}5 + 0{\cdot}5$ + m.e. of $NaHCO_3 = 0{\cdot}1 \times 12{\cdot}5 = 1{\cdot}25$

or $\quad$ m.e. of $NaHCO_3 = 0{\cdot}25$

$\therefore \quad$ wt. of $NaHCO_3 = \dfrac{0{\cdot}25}{1000} \times 84 = 0{\cdot}021$ g.

(eq. wt. of $NaHCO_3 = 84$)

Ex. 46. *A mixed solution of* KOH *and* Na_2CO_3 *required* 20 mL *of* N/20 HCl *solution when titrated with phenolphthalein as indicator. But the same amount of solution when titrated with methyl orange as indicator required* 30 mL *of the same acid. Calculate the amount of* KOH *and* Na_2CO_3.

Solution : Neutralisation reactions are

$$\left.\begin{aligned} KOH + HCl &\rightarrow KCl + H_2O \\ Na_2CO_3 + HCl &\rightarrow NaHCO_3 + NaCl \end{aligned}\right\} \text{phenolphthalein is used}$$

$NaHCO_3 + HCl \rightarrow NaCl + H_2O + CO_2$; methyl orange is used.

As discussed in the previous example we have with phenolphthalein,

m.e. of 20 mL of N/20 HCl = m.e. of KOH + m.e. of Na_2CO_3

or m.e. of KOH + m.e. of $Na_2CO_3 = 20 \times \dfrac{1}{20} = 1.$ ... (1)

Now, with methyl orange,

m.e. of 30 mL of N/20 HCl

= m.e. of KOH + m.e. of Na_2CO_3 + m.e. of $NaHCO_3$ produced.

Since m.e. of Na_2CO_3 = m.e. of $NaHCO_3$ produced.

$\therefore \quad 30 \times \dfrac{1}{20}$ = m.e. of KOH + m.e. of Na_2CO_3 + m.e. of Na_2CO_3

or m.e of KOH + 2 × m.e. of $Na_2CO_3 = 1{\cdot}5.$ ... (2)

Subtracting Eqn. (1) from Eqn. (2), we get,

m.e. of $Na_2CO_3 = 1{\cdot}5 - 1 = 0{\cdot}5$ ... (3)

$\therefore \quad$ equivalent of $Na_2CO_3 = \dfrac{0{\cdot}5}{1000}$.

$\therefore \quad$ wt. of $Na_2CO_3 = \dfrac{0{\cdot}5}{1000} \times 106 = 0{\cdot}053$ g.

(eq. wt. of $Na_2CO_3 = 106$)

From eqns. (3) and (1),

m.e. of KOH = $1 - 0{\cdot}5 = 0{\cdot}5$.

Equivalent of KOH = $\dfrac{0{\cdot}5}{1000}$.

Weight of KOH = $\dfrac{0{\cdot}5}{1000} \times 56 = 0{\cdot}028$ g. (eq. wt. of KOH = 56.)

Ex. 47. 50 mL *of a solution, containing* 1 g *each of* Na_2CO_3, $NaHCO_3$ *and* NaOH, *was titrated with* N HCl. *What will be the titre readings if*

(a) *only phenolphthalein is used as indicator?*

(b) *only methyl orange is used as indicator from the very beginning?*

(c) *methyl orange is added after the first end point with phenolphthalein?*

Solution : (a) The titration reactions in this case are

$$Na_2CO_3 + HCl \rightarrow NaHCO_3 + NaCl$$

and $$NaOH + HCl \rightarrow NaCl + H_2O$$

Thus, we have,

m.e. of Na_2CO_3 + m.e. of NaOH = m.e. of v_1 mL (say) of N HCl

$$\frac{1}{106} \times 1000 + \frac{1}{40} \times 1000 = 1 \times v_1 ; \quad \therefore \quad v_1 = 34{\cdot}4 \text{ mL.}$$

(b) The reactions in this case are,

$$Na_2CO_3 + HCl \rightarrow NaHCO_3 + NaCl$$

$$\underset{\text{(produced)}}{NaHCO_3} + HCl \rightarrow NaCl + H_2O + CO_2$$

$$\underset{\text{(originally present)}}{NaHCO_3} + HCl \rightarrow NaCl + H_2O + CO_2$$

and $$NaOH + HCl \rightarrow NaCl + H_2O$$

Thus, we have,

m.e. of Na_2CO_3 + m.e. of $NaHCO_3$ (produced) + m.e. of $NaHCO_3$ (originally present) + m.e. of NaOH

= m.e. of v_2 mL (say) of N HCl

$$\frac{1}{106} \times 1000 + \frac{1}{106} \times 1000 + \frac{1}{84} \times 1000 + \frac{1}{40} \times 1000 = 1 \times v_2$$

$$\therefore \qquad v_2 = 55{\cdot}8 \text{ mL.}$$

(c) The reactions in this case are,

$$NaHCO_3 \text{ (produced)} + HCl \rightarrow NaCl + H_2O + CO_2$$

and $$NaHCO_3 \text{ (originally present)} + HCl \rightarrow NaCl + H_2O + CO_2$$

Thus we have,

m.e. of $NaHCO_3$ (produced) + m.e. of $NaHCO_3$ (originally present) = m.e. of v_3 mL (say) of N HCl

or m.e. of Na_2CO_3 + m.e. of $NaHCO_3$ = m.e. of v_3 mL (say) of N HCl

$$\frac{1}{106} \times 1000 + \frac{1}{84} \times 1000 = 1 \times v_3.$$

$$\therefore \qquad v_3 = 21{\cdot}3 \text{ mL.}$$

Ex. 48. 1·245 g *of* $CuSO_4 \cdot x\,H_2O$ *was dissolved in water and* H_2S *was passed into it till* CuS *was completely precipitated. The* H_2SO_4 *produced in the filtrate required* 10 mL *of* N NaOH *solution. Calculate x.*

Solution : $CuSO_4 + H_2S \rightarrow CuS + H_2SO_4$

m.e. of $CuSO_4 \cdot x\,H_2O$ solution = m.e. of H_2SO_4

= m.e. of 10 mL of N NaOH

$= 1 \times 10 = 10.$

$\therefore$ number of equivalent of $CuSO_4 \cdot x\,H_2O$ solution $= \dfrac{10}{1000}$.

Weight of $CuSO_4 \cdot x\,H_2O$ = equivalent × eq. wt.

$$= \frac{10}{1000} \times \frac{159{\cdot}5 + 18x}{2}.$$

$$\left\{\text{eq. wt. of } CuSO_4 \cdot x\,H_2O = \frac{159{\cdot}5 + 18x}{2}\right\}$$

Thus, $\dfrac{10}{1000} \times \dfrac{159{\cdot}5 + 18x}{2} = 1{\cdot}245$ (given)

$$18x = 89{\cdot}5$$

$$x \approx 5.$$

Ex. 49. *A* 10 mL *of* $K_2Cr_2O_7$ *solution liberated iodine from* KI *solution. The liberated iodine was titrated by* 16 mL *of* M/25 *sodium thiosulphate solution. Calculate the concentration of* $K_2Cr_2O_7$ *solution in grams per litre.*

Solution : $Cr_2O_7^{2-} + 14H^+ + 6I^- = 2Cr^{3+} + 7H_2O + 3I_2$

$+12$ $\qquad\qquad\qquad\qquad +6$

$$S_2O_3^{2-} + \frac{1}{2}I_2 = \frac{1}{2}S_4O_6^{2-} + I^-$$

$+4$ $\qquad\qquad\qquad +5$

Eq. wt. of $K_2Cr_2O_7 = \dfrac{\text{mol. wt.}}{\text{change in ON per mole}}$

$$= \frac{294{\cdot}18}{6}$$

$$= 49{\cdot}03.$$

m.e. of 10 mL of $K_2Cr_2O_7$ solution = m.e. of iodine

= m.e. of sodium thiosulphate

$$= \frac{1}{25} \times 16 = 0{\cdot}64.$$

$\therefore$ equivalent of 10 mL of $K_2Cr_2O_7$ solution $= \dfrac{0{\cdot}64}{1000} = 0{\cdot}00064.$

$\therefore$ weight per 10 mL $= 0{\cdot}00064 \times 49{\cdot}03$
$= 0{\cdot}0313$ g.

$\therefore$ concentration of $K_2Cr_2O_7$ in grams per litre $= 0{\cdot}0313 \times 100$
$= 3{\cdot}13$ g/L.

Ex. 50. *Mercuric iodate* $[Hg_5(IO_6)_2]$ *reacts with a mixture of* KI *and* HCl *according to the following equation:*

$$Hg_5(IO_6)_2 + 34KI + 24HCl \rightarrow 5K_2HgI_4 + 8I_2 + 24KCl + 12H_2O$$

The liberated iodine is treated with $Na_2S_2O_3$ *solution,* 1 mL *of which is equivalent to* 0·0499 g *of* $CuSO_4 \cdot 5H_2O$. *What volume (in* mL*) of* $Na_2S_2O_3$ *solution will be required to react with iodine liberated from* 0·7245 g *of* $[Hg_5(IO_6)_2]$? (Hg = 200·5; Cu = 63·5; I = 127)

Solution : From the stoichiometry of the given equations,

$$Hg_5(IO_6)_2 + 34KI + 24HCl \rightarrow 5K_2HgI_4 + 8I_2 + 24KCl + 12H_2O$$

and $$2CuSO_4 \cdot 5H_2O + 4KI \rightarrow Cu_2I_2 + 2K_2SO_4 + I_2,$$

we have,

$$8 \times \text{moles of } Hg_5(IO_6)_2 = \text{moles of } I_2$$

and $$\text{moles of } CuSO_4 \cdot 5H_2O = 2 \times \text{moles of } I_2.$$

$$\therefore \text{ moles of } I_2 = \frac{1}{2} \times \text{ moles of } CuSO_4 \cdot 5H_2O$$
$$= 8 \times \text{moles of } Hg_5(IO_6)_2$$

or $$\text{moles of } CuSO_4 \cdot 5H_2O = 16 \times \text{moles of } Hg(IO_6)_2$$

$$\frac{\text{wt. of } CuSO_4 \cdot 5H_2O}{249{\cdot}5} = 16 \times \frac{0{\cdot}7245}{1448{\cdot}5}. \quad \left[\begin{array}{r} CuSO_4 \cdot 5H_2O = 249{\cdot}5 \\ Hg_5(IO_6)_2 = 1448{\cdot}5 \end{array}\right]$$

$\therefore$ wt. of $CuSO_4 \cdot 5H_2O = 1{\cdot}9967$ g.

Since 0·0499 g of $CuSO_4 \cdot 5H_2O = 1$ mL of $Na_2S_2O_3$

$$\therefore \quad 1{\cdot}9967 \text{ g of } CuSO_4 . 5H_2O = \frac{1{\cdot}9967}{0{\cdot}0499} = 40 \text{ mL}.$$

Ex. 51. *Calculate the concentration of* $Na_2S_2O_3 \cdot 5H_2O$ *solution in grams per litre,* 10 mL *of which just decolourised* 15 mL *of* N/20 *iodine solution.*

Solution :
$$\underset{+4}{S_2O_3^{2-}} + I_2 \rightarrow \underset{+5}{\frac{1}{2}S_4O_6^{2-}} + I^-$$

$$\text{Eq. wt. of } Na_2S_2O_3 \cdot 5H_2O = \frac{\text{mol. wt.}}{\text{change in ON per mole}}$$

$$= \frac{248.2}{1}.$$

m.e. of 10 mL of $Na_2S_2O_3 \cdot 5H_2O$ solution = m.e. of iodine solution

$$= \frac{1}{20} \times 15 = 0.75.$$

Equivalent of 10 mL of $Na_2S_2O_3 \cdot 5H_2O = \frac{0.75}{1000}$.

Weight per 10 mL = $\frac{0.75}{1000} \times 248 = 0.186$ g.

Concentration of hypo in grams per litre $= 0.186 \times \frac{1000}{10}$ g/L

$$= 18.6 \text{ g/L}.$$

Ex. 52. 25 mL *of* 0.017 M $H_2SO_3^-$, *in strongly acidic solution required the addition of* 16·9 mL *of* 0·01 M MnO_4^- *for its complete oxidation to* SO_4^{2-} *or* HSO_4^-. *In neutral solution it required* 28·6 mL. *Assign oxidation numbers to* Mn *in each of the products.*

Solution :

$$\underset{+4}{HSO_3^-} \xrightarrow{\text{change in ON} = 2} \underset{+6}{SO_4^{2-}} \text{ or } \underset{+6}{HSO_4^-}$$

$\therefore$ 0·017 M $HSO_3^- \equiv 2 \times 0.017$ N ... (Eqn. 6i)

$$= 0.034 \text{ N}.$$

In the first case suppose the ON of Mn in the product is X

$\therefore$ 0·01 M $MnO_4^- = 0.01(7 - X)$ N MnO_4^- ... (Eqn. 6i)

m.e. of HSO_3^- = m.e. of $\underset{+7}{MnO_4^-}$

$$0.034 \times 25 = 0.01\,(7 - X)\,16.9$$

$$\therefore \quad 7 - X = \frac{0.034 \times 25}{16.9 \times 0.01} = 5.00$$

or $X = 2$.

Now in the second titration, suppose the ON of Mn in the product is Y.

$\therefore$ 0·01 M $MnO_4^- = 0.01\,(7 - Y)$ N MnO_4^-

$$0.034 \times 25 = 0.01\,(7 - Y) \times 28.6$$

$$7 - Y = \frac{0.034 \times 25}{0.01 \times 28.6} = 3$$

$\therefore$ $Y = 4$.

Ex. 53. *A polyvalent metal weighing* 0·1 g *and having atomic weight* 51 *reacted with dilute* H_2SO_4 *to give* 43·9 mL *of hydrogen at* STP. *The solution containing the metal in this lower oxidation state, was found to require* 58·8 mL *of* 0·1 N *permanganate for complete oxidation. What are the valencies of the metal?*

Solution : Suppose the lower oxidation number of the metal is X.

Given that:

$$\begin{array}{lcl} \text{metal} + H_2SO_4 & \rightarrow & H_2 \\ 0{\cdot}1 \text{ g} & & 43{\cdot}9 \text{ mL at STP} \\ \dfrac{0{\cdot}1}{51/X}\text{ eq.} & & \dfrac{43{\cdot}9}{11200}\text{ eq.} \end{array}$$

(eq. wt. of the metal = $\dfrac{51}{X}$ and volume occupied by 1 eq. of hydrogen at NTP = 11200 mL.)

Now, eq. of the metal = eq. of hydrogen

$$\frac{0{\cdot}1}{51/X} = \frac{43{\cdot}9}{11200}; \quad X = 2.$$

Further, the metal is changing from lower oxidation number 2 to higher oxidation number, say, Y.

$$\therefore \quad \text{eq. wt. of the metal} = \frac{51}{\text{change in ON}} = \frac{51}{Y-2}.$$

$$\text{Eq. of metal} = \text{eq. of } KMnO_4 = \frac{\text{m.e. of } KMnO_4}{1000}.$$

$$\therefore \quad \frac{0{\cdot}1}{51/(Y-2)} = \frac{0{\cdot}1 \times 58{\cdot}8}{1000}.$$

$$\therefore \quad Y = 5.$$

Ex. 54. 4·08 g *of a mixture of* BaO *and an unknown carbonate* MCO_3 *was heated strongly. The residue weighed* 3·64 g. *This was dissolved in* 100 mL *of* 1 N HCl. *The excess of acid required* 16 mL *of* 2·5 N NaOH *for complete neutralisation. Identify the metal* M.

Solution :

$$\begin{array}{ccc} 4{\cdot}08 \text{ g} & & 3{\cdot}64 \text{ g} \quad (4{\cdot}08 - 3{\cdot}64) = 0{\cdot}44 \text{ g} \\ \underbrace{BaO + MCO_3} & \rightarrow & \underbrace{BaO + MO} + CO_2 \end{array}$$

BaO does not change on heating.

Suppose the weight of MCO_3 is x g and at. wt. of M is y

$$\begin{array}{lll} MCO_3 \rightarrow & MO \quad + & CO_2 \\ x \text{ g} & [3{\cdot}64 - (4{\cdot}08 - x)] & 0{\cdot}44 \text{ g} \\ & = (x - 0{\cdot}44) \text{ g.} & \end{array}$$

Applying POAC for M and C atoms, we get, (p. 4, Chapter 1)

$$\frac{x}{y+60} = \frac{(x-0{\cdot}44)}{y+16} \qquad \text{... (1)}$$

and $$\frac{x}{y+60} = \frac{0{\cdot}44}{44} = 0{\cdot}01. \qquad \text{... (2)}$$

From eqns. (1) and (2), we have,

$$\frac{x-0{\cdot}44}{y+16} = 0{\cdot}01. \qquad \text{... (3)}$$

Now, m.e. of NaOH $= 2{\cdot}5 \times 16 = 40$. ... (Eqn. 1)

$\therefore$ m.e. of excess acid = 40.

$\therefore$ m.e. of the acid used to neutralise BaO and MO

= m.e. of total acid – m.e. of excess acid

$= 1 \times 100 - 40 = 60.$

$\therefore$ eq. of the acid $= \dfrac{60}{1000} = 0{\cdot}06 =$ eq. of BaO + eq. of MO

or $$\frac{4{\cdot}08 - x}{154/2} + \frac{(x-0{\cdot}44)}{(y+16)/2} = 0{\cdot}06. \qquad \text{... (4)}$$

$\left(\text{eq. wt. of BaO} = \dfrac{154}{2}\text{; eq. wt. of MO} = \dfrac{y+16}{2}\right)$

Substituting the value of $\left(\dfrac{x-0{\cdot}44}{y+16}\right)$ from Eqn. (3) in Eqn. (4),

we get, $x = 1$ and $y = 40$.

Hence, the metal M must be Ca.

Ex. 55. (a) (i) *A sample of* $MnSO_4 \cdot 4H_2O$ *is strongly heated in air. The residue is* Mn_3O_4.

(ii) *The residue is dissolved in* 100 mL *of* 0·1 N $FeSO_4$ *containing dilute* H_2SO_4.

(iii) *The solution reacts completely with* 50 mL *of* $KMnO_4$ *solution.*

(iv) 25 mL *of the* $KMnO_4$ *solution used in step* (ii) *requires* 30 mL *of* 0·1 N $FeSO_4$ *solution for complete reaction.*

Find the amount of $MnSO_4 \cdot 4H_2O$ *present in the sample.*

(b) 0·5 g *of fuming* H_2SO_4 *(oleum) is diluted with water. The solution is completely neutralised by* 26·7 mL *of* 0·4 N NaOH. *Find the percentage of free* SO_3 *in the sample of oleum.*

Solution : (a) m.e. of 25 mL of $KMnO_4$ solution

= m.e. of 30 mL of 0·1 N $FeSO_4$ solution

$= 0{\cdot}1 \times 30 = 3$

$\therefore$ m.e. of 50 mL of $KMnO_4$ solution $= 2 \times 3 = 6.$

$\therefore$ m.e. of $FeSO_4$ (remained which did not react with Mn_3O_4) = 6.

Now, m.e. of total $FeSO_4$ solution $= 0{\cdot}1 \times 100 = 10$.

m.e. of $FeSO_4$ oxidised by $Mn_3O_4 = 10 - 6 = 4$

$\therefore$ m.e. of $Mn_3O_4 = 4$

$\therefore$ eq. of $Mn_3O_4 = \frac{4}{1000}$

From the redox reaction

$$\underset{+8}{Mn_3O_4} + Fe^{2+} \rightarrow \underset{+6}{3Mn^{2+}} + Fe^{3+}$$

$$\text{Equivalent wt. of } Mn_3O_4 = \frac{\text{mol. wt.}}{\text{change in ON per mole}}$$

$$= \frac{229}{2} = 114{\cdot}50$$

$$\therefore \quad \text{wt. of } Mn_3O_4 = \text{equivalent} \times \text{eq. wt.}$$

$$= \frac{4}{1000} \times 114{\cdot}5 = 0{\cdot}458 \text{ g.}$$

As given in the problem, Mn_3O_4 is obtained by heating $MnSO_4 \cdot 4H_2O$.

$$MnSO_4 \cdot 4H_2O \xrightarrow{\Delta} Mn_3O_4$$

Applying POAC for Mn atoms, we get, (p. 4, Chapter 1)

$1 \times$ moles of $MnSO_4 \cdot 4H_2O = 3 \times$ moles of Mn_3O_4

$$\text{or} \quad \frac{\text{wt. of } MnSO_4 \cdot 4H_2O}{\text{mol. wt. of } MnSO_4 \cdot 4H_2O} = \frac{3 \times 0{\cdot}458}{229}.$$

$$\therefore \quad \text{wt. of } MnSO_4 \cdot 4H_2O = \frac{3 \times 0{\cdot}458}{229} \times 223$$

$$= 1{\cdot}338 \text{ g.}$$

(b) Reactions involved are

$$SO_3 + 2NaOH \rightarrow Na_2SO_4 + H_2O$$

$$H_2SO_4 + 2NaOH \rightarrow Na_2SO_4 + 2H_2O$$

Eq. wt. of $SO_3 = \frac{80}{2} = 40$.

Eq. wt. of $H_2SO_4 = \frac{98}{2} = 49$.

Now, m.e. of SO_3 + m.e. of H_2SO_4 = m.e. of NaOH.

$\therefore$ eq. of $SO_3 \times 1000$ + eq. of $H_2SO_4 \times 1000 = 0{\cdot}4 \times 26{\cdot}7$.

Suppose the wt. of SO_3 is x g.

$\therefore$ wt. of $H_2SO_4 = (0{\cdot}5 - x)$ g.

$$\therefore \quad \frac{x}{40} + \frac{(0{\cdot}5 - x)}{49} = 0{\cdot}4 \times 26{\cdot}7; x = 0{\cdot}1036 \text{ g.}$$

$$\therefore \quad \% \text{ of } SO_3 = \frac{0{\cdot}1036}{0{\cdot}5} \times 100 = 20{\cdot}67\%.$$

Ex. 56. *A mixture of* $H_2C_2O_4$ *(oxalic acid) and* $NaHC_2O_4$ *weighing* 2·02 g *was dissolved in water and the solution made up to one litre.* 10 mL *of the solution required* 3 mL *of* 0·1 N NaOH *solution for complete neutralisation. In another experiment,* 10 mL *of the same solution, in hot dilute* H_2SO_4 *medium, required* 4 mL *of* 0·1 N $KMnO_4$ *solution for complete reaction. Calculate the amount of* $H_2C_2O_4$ *and* $NaHC_2O_4$ *in the mixture.* **(IIT 1990)**

Solution : Let the wt. of $H_2C_2O_4$ in 10 mL of the solution be x g. The weight of $NaHC_2O_4$ in 10 mL will be $(0{\cdot}0202 - x)$ g.

In the first experiment, $H_2C_2O_4$ and $NaHC_2O_4$ are neutralised by NaOH changing into $Na_2C_2O_4$. The eq. wt. of $H_2C_2O_4$ and $NaHC_2O_4$ will, therefore, be 90/2 and 112 respectively.

Thus,

m.e. of $H_2C_2O_4$ + m.e. of $NaHC_2O_4$ = m.e. of NaOH

$$\frac{x}{90/2} \times 1000 + \frac{(0{\cdot}0202 - x)}{112} \times 1000 = 0{\cdot}1 \times 3. \quad \ldots (1)$$

In the second experiment, both $H_2C_2O_4$ and $NaHC_2O_4$ are oxidised to CO_2 by $KMnO_4$. The equivalent weight of $H_2C_2O_4$ and $NaHC_2O_4$ will, therefore, be 90/2 and 112/2 respectively $\left(\underset{+6}{C_2O_4^{2-}} \rightarrow \underset{+8}{2CO_2}\right)$.

Thus,

m.e. of $H_2C_2O_4$ + m.e. of $NaHC_2O_4$ = m.e. of $KMnO_4$

$$\frac{x}{90/2} \times 1000 + \frac{(0{\cdot}0202 - x)}{112/2} \times 1000 = 0{\cdot}1 \times 4. \quad \ldots (2)$$

Subtracting (1) from (2), we get,

$$\frac{0{\cdot}0202 - x}{112} = \frac{0{\cdot}1}{1000}.$$

$$\therefore \quad x = 0{\cdot}009 \text{ g/10 mL of solution.}$$

The 1000 mL of solution contains

$$H_2C_2O_4 = 0{\cdot}9 \text{ g}$$

and $$NaHC_2O_4 = 2{\cdot}02 - 0{\cdot}9 = 1{\cdot}12 \text{ g.}$$

[**Note:** This question may be solved by the data of the first experiment only.]

Ex. 57. *A solution of* 0·2 g *of a compound containing* Cu^{2+} *and* $C_2O_4^{2-}$ *ions on titration with* 0·02 M $KMnO_4$ *in presence of* H_2SO_4 *consumes* 22·6 mL *of the* *oxidant. The resultant solution is neutralised with* Na_2CO_3, *acidified with dilute acetic acid and treated with excess* KI. *The liberated* I_2 *required* 11·3 mL *of* 0·05 M $Na_2S_2O_3$ *solution for complete reduction. Find out the mole ratio of* Cu^{2+} *to* $C_2O_4^{2-}$ *in the compound. Write down the balanced redox reactions involved in the above titrations.* **(IIT 1991)**

Solution : At the first stage, $C_2O_4^{2-}$ is oxidised to CO_2 by the oxidant $KMnO_4$.

$$5C_2O_4^{2-} + 2\underset{+7}{MnO_4^-} + 16H^+ \rightarrow 10CO_2 + 2\underset{+2}{Mn^{2+}} + 8H_2O$$

$\therefore$ normality of $KMnO_4$ solution $= 0{\cdot}02 \times (7-2) = 0{\cdot}1$ N.

At the second stage, Cu^{2+} liberates I_2 from KI and this liberated I_2 requires 11·3 mL of 0·05 M $Na_2S_2O_3$ solution for complete reaction.

$$2Cu^{2+} + 2I^- \rightarrow 2Cu^+ + I_2$$

$$2\underset{+4}{S_2O_3^{2-}} + I_2 \rightarrow \underset{+5}{S_4O_6^{2-}} + 2I^-$$

Normality of $Na_2S_2O_3$ solution $= 0{\cdot}05(5-4) = 0{\cdot}05$ N.

Now, $$\frac{\text{m.e. of } Cu^{2+}}{\text{m.e. of } C_2O_4^{2-}} = \frac{\text{m.e. of } Na_2S_2O_3}{\text{m.e. of } KMnO_4} = \frac{0{\cdot}05 \times 11{\cdot}3}{0{\cdot}1 \times 22{\cdot}3} = \frac{1}{4}.$$

As $$\underset{+2}{Cu^{2+}} \rightarrow \underset{+1}{Cu^+} \text{ and } \underset{+6}{C_2O_4^{2-}} \rightarrow \underset{+8}{2CO_2}$$

$$\therefore \quad \frac{\text{mmol of } Cu^{2+} \times 1}{\text{mmol of } C_2O_4^{2-} \times 2} = \frac{1}{4}$$

or $$\frac{\text{mole of } Cu^{2+}}{\text{mole of } C_2O_4^{2-}} = \frac{1}{2}.$$

Ex. 58. 1·25 g *of a sample of bleaching powder is dissolved in* 100 mL *of water and* 25 mL *of which are treated with* KI *solution. The iodine so liberated required* 12·5 mL *of* N/25 *hypo solution in titration. Find the percentage of chlorine available from the sample of bleaching powder.*

Solution : m.e. of available chlorine in 25 mL of bleaching powder solution

= m.e. of iodine liberated

= m.e. of hypo solution

$$= \frac{1}{25} \times 12{\cdot}5 = 0{\cdot}5.$$

$\therefore$ m.e. of available chlorine in 100 mL $= \dfrac{0{\cdot}5 \times 100}{25} = 2.$

$\therefore$ eq. of available chlorine $= \dfrac{2}{1000}.$

wt. of available chlorine in 1·25 grams of bleaching powder

$$= \frac{0{\cdot}2}{1000} \times 35{\cdot}5 = 0{\cdot}071 \text{ g.} \quad (\text{eq. wt. of chlorine} = 35{\cdot}5)$$

$\therefore$ percentage of available chlorine $= \frac{0{\cdot}071}{1{\cdot}25} \times 100$

$= 5{\cdot}68\%.$

Ex. 59. *An equal volume of a reducing agent is titrated separately with* 1 M $KMnO_4$ *in acid, neutral and alkaline media. The volumes of* $KMnO_4$ *required are* 20 mL *in acid,* 33·4 mL *in neutral and* 100 mL *in alkaline media. Find out the oxidation state of manganese in each reduction product. Give the balanced equations for all the three half reactions. Find out the volume of* 1 M $K_2Cr_2O_7$ *consumed, if the same volume of the reducing agent is titrated in acid medium.* **(IIT 1989)**

Solution : Given that:

$$KMnO_4 + \text{reducing agent} \begin{cases} \xrightarrow{\text{acidic}} Mn^{+x_1} \\ \xrightarrow{\text{neutral}} Mn^{+x_2} \\ \xrightarrow{\text{alkaline}} Mn^{+x_3} \end{cases}$$

where x_1, x_2 and x_3 are the oxidation states of Mn in the product in acidic, neutral and alkaline media respectively. Since equal volumes of the reducing agent is used in each titration,

$\therefore$ m.e. of reducing agent = m.e. of $KMnO_4$ in acidic medium

= m.e. of $KMnO_4$ in neutral medium

= m.e. of $KMnO_4$ in alkaline medium

or $\quad 1 \times (7 - x_1) \times 20 = 1 \times (7 - x_2) \times 33{\cdot}4$

$= 1 \times (7 - x_3) \times 100$

[m.e. $= N \times V$ (mL) ; $N = M \times$ change in ON]

or $\quad \frac{7 - x_1}{5} = \frac{7 - x_2}{3} = \frac{7 - x_3}{1}.$

On inspection, we see that the equality exists for

$x_1 = +2$, $x_2 = +4$ and $x_3 = +6$ as x_1, x_2 and x_3 can never be greater than 7.

The balanced chemical equations of all the three half reactions are

$$MnO_4^- + 8H^+ + 5e \rightarrow Mn^{2+} + 4H_2O \quad (\text{acidic medium})$$

$$MnO_4^- + 2H_2O + 3e \rightarrow \underset{+4}{MnO_2} + 4OH^- \quad (\text{neutral medium})$$

$$MnO_4^- + e \rightarrow \underset{+6}{MnO_4^{2-}} \quad (\text{alkaline medium})$$

Further, in acidic medium,

$$\underset{+12}{Cr_2O_7^{2-}} \rightarrow \underset{+6}{2Cr^{3+}}; \text{ change in ON} = 6$$

$\therefore$ normality of $K_2Cr_2O_7$ solution $= 1 \times 6 = 6$ N.

Let the volume of $K_2Cr_2O_7$ solution be v mL.

$\therefore$ m.e. of $K_2Cr_2O_7$ = m.e. of $KMnO_4$ in acidic medium

$$6 \times v = 5 \times 20 \quad (\text{normality of } KMnO_4 = 5 \text{ N})$$

$$v = 16{\cdot}67 \text{ mL.}$$

Ex. 60. *A* 1·0-g *sample of* Fe_2O_3 *solid of* 55·2 *per cent purity is dissolved in acid and reduced by heating the solution with zinc dust. The resultant solution is cooled and made up to* 100 mL. *An aliquot of* 25 mL *of this solution requires* 17 mL *of* 0·0167 M *solution of an oxidant for titration. Calculate the number of electrons taken up by the oxidant in the reaction of the above titration.*

(IIT 1991)

Solution : Weight of $Fe_2O_3 = 0{\cdot}552$ g.

$$\text{Number of eq. of } Fe_2O_3 = \frac{0{\cdot}552}{80}. \quad \left\{ \begin{array}{l} \underset{+6}{Fe_2O_3} \rightarrow \underset{+4}{2FeO} \\ \text{eq. wt. of } Fe_2O_3 = \dfrac{160}{2} = 80 \end{array} \right\}$$

Let the number of electrons taken up by the oxidant in the reaction be n (i.e., the change in oxidation number).

$\therefore$ normality of the oxidant $= 0{\cdot}0167\, n$ N.

$\therefore$ m.e. of the oxidant $= 0{\cdot}0167\, n \times 17$.

$\therefore$ m.e. of 25 mL of Fe^{2+} solution $= 0{\cdot}0167\, n \times 17$

$\therefore$ m.e. of 100 mL of Fe^{2+} solution $= 4 \times 0{\cdot}0167\, n \times 17$

$\therefore$ eq. of 100 mL of Fe^{2+} solution $= \dfrac{68 \times 0{\cdot}0167\, n}{1000}$.

eq. of Fe_2O_3 = eq. of FeO

$$\frac{0{\cdot}552}{80} = \frac{68 \times 0{\cdot}0167n}{1000}; \; n = 6.$$

Ex. 61. 1.6 g *of pyrolusite ore was treated with* 50 cc *of* 1.0 N *oxalic acid and some sulphuric acid. The oxalic acid left undecomposed was raised to* 250 cc *in a flask.* 25 cc *of this solution when titrated with* 0·1 N $KMnO_4$ *required* 32 cc *of the solution. Find out the percentage of pure* MnO_2 *in the sample and also the percentage of available oxygen.*

Solution : m.e. of the remaining 25 mL of oxalic acid = m.e. of $KMnO_4$
$= 0{\cdot}1 \times 32 = 3{\cdot}2$

m.e. of 250 mL of oxalic acid = 32.

$\therefore$ m.e. of MnO_2 = m.e. of oxalic acid reacted $= 1 \times 50 - 32 = 18$.

$$\text{wt. of } MnO_2 = \frac{18}{1000} \times 43{\cdot}5 \qquad \left\{\begin{array}{l} Mn^{4+} \longrightarrow Mn^{2+} \\ E_{MnO_2} = \frac{87}{2} = 43{\cdot}5 \end{array}\right\}$$

$$= 0{\cdot}783 \text{ g.}$$

$$\% \; MnO_2 = \frac{0{\cdot}783}{1{\cdot}6} \times 100 = 48{\cdot}9\%.$$

$$\text{Mole of O in } MnO_2 = 2 \times \text{mole of } MnO_2 = 2 \times \frac{0{\cdot}783}{87} = 0{\cdot}018.$$

$$\text{Wt. of O} = 0{\cdot}018 \times 16 = 0{\cdot}288 \text{ g.}$$

$$\% \text{ of available oxygen} = \frac{0{\cdot}288}{1{\cdot}6} \times 100 = 18\%.$$

Ex. 62. *The iodide content of a solution was determined by titration with cerium* (IV) *sulphate in the presence of* HCl, *in which* I^- *is converted to* ICl. *A* 250 mL *sample of the solution required* 20 mL *of* 0·05 N Ce^{4+} *solution. What is the iodide concentration in the original solution in* g/L?

Solution : m.e. of iodide = m.e. of Ce^{4+}

$$= 20 \times 0{\cdot}05 = 1.$$

$$\therefore \text{ normality of iodide} = \frac{1}{250} \text{ N}$$

For, $\underset{-1}{I^-} \longrightarrow \underset{+1}{ICl}$

change in ON of I^- = 2.

$$\text{Eq. wt. of } I^- = \frac{127}{2}.$$

$$\therefore \text{ iodide concentration} = \frac{1}{250} \times \frac{127}{2} \text{ g/L}$$

$$= 0{\cdot}254 \text{ g/L.}$$

Ex. 63. *To a* 25 mL H_2O_2 *solution, excess of acidified solution of* KI *was added. The iodine liberated required* 20 mL *of* 0·3 N $Na_2S_2O_3$ *solution. Calculate the volume strength of* H_2O_2 *solution.* **(IIT 1997)**

Solution : m.e. of 25 mL of H_2O_2 = m.e. of I_2 = m.e. of $Na_2S_2O_3$

$$= 20 \times 0{\cdot}3 = 6.$$

$$\therefore \quad \text{normality of } H_2O_2 = \frac{6}{25} \text{ eq/L}$$

Let the volume strength of H_2O_2 be '*xV*'. Thus,

1 litre of H_2O_2 produces *x* litres of O_2 at NTP.

$\therefore$ equivalent of H_2O_2/L = eq. of O_2 = $\dfrac{V\,(\text{NTP})}{\text{eq. vol. of } O_2} = \dfrac{x}{5{\cdot}6}$.

$$\therefore \frac{x}{5{\cdot}6} = \frac{6}{25}$$

$$x = 1{\cdot}344.$$

The volume strength of H_2O_2 is '1·344*V*'.

Ex. 64. *One litre of a mixture of* O_2 *and* O_3 *at* NTP *was allowed to react with an excess of acidified solution of* KI. *The iodine liberated required* 40 mL *of* M/10 *sodium thiosulphate solution for titration. What is the weight per cent of ozone in the mixture? Ultraviolet radiation of wavelength* 300 nm *can decompose ozone. Assuming that one photon can decompose one ozone molecule, how many photons would have been required for the complete decomposition of ozone in the original mixture?* **(IIT 1997)**

Solution : The reactions involved are

$$2\,KI + H_2O + O_3 = 2\,KOH + I_2 + O_2$$

$$I_2 + 2\,Na_2S_2O_3 = Na_2S_4O_6 + 2\,NaI$$

From the stoichiometry of the equations, we get,

$$\text{mole of } O_3 = \text{mole of } I_2 = \frac{1}{2} \text{ mole of } Na_2S_2O_3$$

$$= \frac{1}{2} \times \frac{40}{1000} \times \frac{1}{10} = \frac{1}{500}.$$

$$\text{Wt. of } O_3 = \frac{48}{500} = 0{\cdot}096 \text{ g.}$$

$$\text{Volume of } O_3 \text{ at NTP} = \frac{22.4}{500} = 0{\cdot}0448 \text{ litre.}$$

$$\text{Volume of } O_2 = 1 - 0{\cdot}0448 = 0{\cdot}9552 \text{ litre.}$$

$$\text{Wt. of } O_2 = \frac{0{\cdot}9552}{22{\cdot}4} \times 32 = 1{\cdot}3654 \text{ g.}$$

$$\text{Wt. \% of } O_3 = \frac{0{\cdot}096}{0{\cdot}096 + 1{\cdot}3654} \times 100 = 6{\cdot}57\,\%.$$

Further,

$$\text{number of photons} = \text{no. of } O_3 \text{ molecules} = \frac{6{\cdot}022 \times 10^{23}}{500}$$

$$= 1{\cdot}2 \times 10^{21}.$$

Ex. 65. *A* 3-g *sample containing* Fe_3O_4, Fe_2O_3 *and an inert impure substance is treated with excess of* KI *solution in presence of dilute* H_2SO_4. *The entire iron is converted into* Fe^{2+} *along with the liberation of iodine. The resulting solution is diluted to* 100 mL. 20 mL *of the diluted solution requires* 11 mL *of* 0·5 M

$Na_2S_2O_3$ *solution to reduce the iodine present.* 50 mL *of the diluted solution, after complete extraction of the iodine requires* 12·8 mL *of* 0·25 M $KMnO_4$ *solution in dilute* H_2SO_4 *medium for the oxidation of* Fe^{2+}. *Calculate the percentages of* Fe_2O_3 *and* Fe_3O_4 *in the original sample.* **(IIT 1996)**

Solution **:** Let the amount of Fe_3O_4 and Fe_2O_3 in the sample be x and y moles respectively. Thus the mixture contains x moles of FeO and $(x + y)$ moles of Fe_2O_3.

In the first titration KI is oxidised to I_2 and so Fe_2O_3 is reduced. ($\overset{+6}{Fe_2O_3} \rightarrow \overset{+4}{2\,FeO}$; change in ON = 2)

m.e. of Fe_2O_3 = m.e. of I_2 = m.e. of $Na_2S_2O_3$

$$(x + y) \times 2 \times 1000 = 5 \times 11 \times 0{\cdot}5 = 27{\cdot}5. \qquad \ldots (1)$$

After the first titration, all the iron is in Fe^{2+} state, i.e., FeO, which is now titrated with $KMnO_4$ to oxidise Fe^{2+} to Fe^{3+}.

Total moles of FeO = moles of FeO (originally present) + moles of FeO (produced by red. of Fe_2O_3)

$= \{x + 2(x + y)\}$

Thus,

m.e. of FeO = m.e. of $KMnO_4$

$$\{x + 2\,(x + y)\} \times 1000 = 2 \times 12{\cdot}8 \times 0{\cdot}25 \times 5 = 32. \qquad \ldots (2)$$

Solving (1) and (2), we get

$x = 0{\cdot}0045.$

$\therefore$ wt. of Fe_3O_4 = $0{\cdot}0045 \times 232 = 1{\cdot}044$ g.

Percentage of $Fe_3O_4 = \dfrac{1{\cdot}044}{3} \times 100 = 34{\cdot}8$ %.

Substituting the value of x in equation (1), we get,

$y = 0{\cdot}00925.$

$\therefore$ wt. of Fe_2O_3 = $0{\cdot}00925 \times 160 = 1{\cdot}49$ g.

Percentage of $Fe_2O_3 = \dfrac{1{\cdot}49}{3} \times 100 = 49{\cdot}66$ %.

Ex. 66. *A sample of hard water contains* 96 ppm *of* SO_4^{2-} *and* 183 ppm *of* HCO_3^-, *with* Ca^{2+} *as the only cation. How many moles of* CaO *will be required to remove* HCO_3^- *from* 1000 kg *of this water? If* 1000 kg *of this water is treated with the amount of* CaO *calculated above, what will be the concentration (in ppm) of residual* Ca^{2+} *ions? (Assume* $CaCO_3$ *to be completely insoluble in water.) If* Ca^{2+} *ions in one litre of the treated water are completely exchanged with hydrogen ions, what will be its* pH*? (One* ppm *means one part of the substance in one million parts of water, weight/weight.)* **(IIT 1997)**

Solution **:** $Ca(HCO_3)_2 + CaO = 2CaCO_3 + H_2O$

Mole of CaO = mole of $Ca(HCO_3)_2$ in 10^6 g of solution

$$= \frac{1}{2} \times \text{mole of } HCO_3^-$$

$$= \frac{1}{2} \times \frac{183}{61} = 1{\cdot}5.$$

As $CaCO_3$ is assumed to be completely insoluble in water, Ca^{2+} ions left are, therefore, those associated only with SO_4^{2-} ion (96 ppm).

For $CaSO_4$, we have,

$$\text{mole of } Ca^{2+}/10^6 \text{ g} = \text{mole of } SO_4^{2-}/10^6 \text{ g} = \frac{96}{96} \text{ mole}/10^6 \text{ g}$$

$$= 1 \text{ mole}/10^6 \text{ g}.$$

Wt. of Ca^{2+} in g/10^6 g = 40 g/10^6 g = 40 ppm.

Now, assuming density of solution to be 1 g/mL,

mole of Ca^{2+} per litre = mole of SO_4^{2-} per litre

$$= \frac{96}{10^6} \times \frac{10^3}{96} = 10^{-3} \text{ M}.$$

If Ca^{2+} is replaced by H^+,

$$[H^+] = 2 \times 10^{-3} \text{ M}$$

$$pH = -\log(2 \times 10^{-3}) = 2{\cdot}7.$$

Ex. 67. *A sample of* Mg *was burnt in air to give a mixture of* MgO *and* Mg_3N_2. *The ash was dissolved in* 60 m.e. *of* HCl *and the resulting solution back titrated with* NaOH. 12 m.e. *of* NaOH *were required to reach the end point. An excess of* NaOH *was then added and the solution distilled. The ammonia released was then trapped in* 10 m.e. *of second acid solution. Back titration of this solution required* 6 m.e. *of the base. Calculate the percentage of* Mg *burnt to nitride.*

Solution :

First Method : m.e. method

m.e. of MgO + m.e. of Mg_3N_2 = m.e. of HCl reacted

= m.e. of total HCl − m.e. of NaOH

= 60 − 12 = 48.

In the dissolution of ash, HCl reacts with total Mg in Mg_3N_2 and in MgO and also with N in Mg_3N_2.

∴ m.e. of total Mg + m.e. of N = 48

or m.e. of total Mg + m.e. of NH_3 = 48

m.e. of total Mg = 48 − 4 = 44.

Further, Mg converted to Mg_3N_2 whose N converts to NH_4Cl (or NH_3),

∴ m.e. of Mg converted to Mg_3N_2 = 3 × m.e. of NH_3

$$= 3 \times (10 - 6)$$
$$= 12.$$

$\therefore$ percentage of Mg converted to $Mg_3N_2 = \frac{12}{44} \times 100$

$$= 27.27\%.$$

Second Method : Mole Method

The reactions involved are

$$\underset{x\text{ mmol (say)}}{Mg} \rightarrow \underset{x\text{ mmol}}{MgO}\ ;\ \underset{x\text{ mmol}}{MgO} + 2HCl = MgCl_2 + H_2O$$

$$\underset{y\text{ mmol (say)}}{Mg} \rightarrow \underset{\frac{y}{3}\text{ mmol}}{Mg_3N_2}\ ;\ \underset{\frac{y}{3}\text{ mmol}}{Mg_3N_2} + 8HCl = 3MgCl_2 + \underset{\frac{2y}{3}\text{ mmol}}{2NH_4Cl}$$

$\therefore$ $2x$ mmol of HCl + $\frac{8y}{3}$ mmol of HCl

= total mmol of HCl – mmol of NaOH

$= 60 - 12 = 48$

$$2x + \frac{8y}{3} = 48 \qquad \text{... (1)}$$

Further, mmol of NH_4Cl = mmol of $NH_3 = (10 - 6)$

or $\frac{2y}{3} = 4.$ ... (2)

From eqns. (1) and (2), one can calculate: $x = 16$ and $y = 6$.

$\therefore$ percentage of Mg converted to $Mg_3N_2 = \frac{y}{x+y} = 27.27\%.$

(**Note:** Mole method is more convenient to equivalent method.)

Ex. 68. *Hydrogen peroxide solution* (20 mL) *reacts quantitatively with a solution of* $KMnO_4$ (20 mL) *acidified with dilute* H_2SO_4. *The same volume of the* $KMnO_4$ *solution is just decolourised by* 10 mL *of* $MnSO_4$ *in neutral medium forming a dark brown precipitate of hydrated* MnO_2. *The brown precipitate is dissolved in* 10 mL *of* 0.2 M *sodium oxalate under boiling condition in the presence of dilute* H_2SO_4. *Write the balanced equations involved in the reactions and calculate the molarity of* H_2O_2. **(IIT 2001)**

Solution : m.e. of H_2O_2 in 20 mL solution

= m.e. of 20 mL of $KMnO_4$

= m.e. of 20 mL of $MnSO_4$

= m.e. of MnO_2

= m.e. of $Na_2C_2O_4$

$= 0.4 \times 10$ (normality = molarity × change in ON)

$= 4.$

Eq. of H_2O_2 in 20 mL solution

$$= \frac{4}{1000}.$$

$$\text{Mole of } H_2O_2/20 \text{ mL} = \frac{4}{1000} \times \frac{1}{2}.$$

$$\text{Molarity (mole/L)} = \frac{4}{2000} \times \frac{1000}{20}.$$

$$= 0.1 \text{ M}.$$

The reactions involved are,

$$5H_2O_2 + 2KMnO_4 + 3H_2SO_4 = 2MnSO_4 + K_2SO_4 + 8H_2O + 5O_2$$

$$2KMnO_4 + 3MnSO_4 + 2H_2O = 5MnO_2 + K_2SO_4 + 2H_2O$$

$$MnO_2 + Na_2C_2O_4 + 2H_2SO_4 = MnSO_4 + Na_2SO_4 + CO_2 + 2H_2O$$

Ex. 69. *You are given a* 2.198-g *sample containing a mixture of* XO *and* X_2O_3. *It requires* 0.015 mol *of* $K_2Cr_2O_7$ *to oxidise the sample completely to form* XO_4^- *and* Cr^{3+}. *If* 0.0187 *mole of* XO_4^- *is formed, what is the atomic mass of X?*

Solution :

change in ON = 6

$$\underset{+2}{\overset{m\text{g}}{XO}} + \underset{+6}{\overset{(2.198 - m)\text{ g}}{X_2O_3}} + \underset{+12}{K_2Cr_2O_7} \rightarrow \underset{+14}{2XO_4^-} + \underset{+7}{XO_4^-} + \underset{+6}{2Cr^{3+}}$$

change in ON = 8

change in ON = 5

Applying POAC for X atoms of atomic mass, say x,

mole of XO + 2 × mole of X_2O_3 = mole of XO_4^-

$$\frac{m}{x + 16} + \frac{2(2.198 - m)}{2x + 48} = 0.0187. \qquad \text{... (1)}$$

Applying law of equivalence,

eq. of XO + eq. of X_2O_3 = eq. of $K_2Cr_2O_7$

$$\frac{m}{x + 16} \times 5 + \frac{(2.198 - m)}{2x + 48} \times 8 = 0.015 \times 6 \qquad \text{... (2)}$$

Solving eqns. (1) and (2),

$$x = 100.04.$$

Ex. 70. *The arsenic in a* 1.22-g *sample of a pesticide was converted to* AsO_4^{3-} *by suitable chemical treatment. It was then titrated using* Ag^+ *to form* Ag_3AsO_4

as a precipitate. If it took 25 mL *of* 0.102 M Ag^+ *to reach to equivalence point in this titration, what is the percentage of arsenic in the pesticide?* (As = 75)

Solution : The reaction

$$3Ag^+ + AsO_4^{3-} \rightarrow Ag_3AsO_4^{3-}$$

is not a redox reaction but a precipitation reaction. The equivalent weights of AsO_4^{3-} and Ag^+ are given by $E_{AsO_4^{3-}} = \frac{139}{3}$ and $E_{Ag^+} = \frac{108}{1}$

Now, m.e. of AsO_4^{3-} = m.e. of Ag^+ = mmol of Ag^+

$$= 0.102 \times 25 = 2.55.$$

Eq. of $AsO_4^{3-} = \frac{2.55}{1000} = 0.00255.$

Mole of $AsO_4^{3-} = \frac{0.00255}{3} = 0.00085.$

Mole of As = 0.00085.

Wt. of As $= 0.00085 \times 75 = 0.06375$ g.

Percentage of arsenic in the pesticide

$$= \frac{0.06375}{1.22} \times 100$$

$$= 5.22\%.$$

Ex. 71. *A mixture containing* As_2O_3 *and* As_2O_5 *required* 20 mL *of* 0.05 N *iodine for titration. The resulting solution is then acidified and excess of* KI *was added. The liberated iodine required* 1.0 g *of* $Na_2S_2O_3 \cdot 5H_2O$ *for complete reaction. Calculate the weight of the mixture.* (As = 75, O = 16, S = 32, Na = 23)

Solution : m.e. of As_2O_3 = m.e. of iodine

$$= 20 \times 0.05 = 1.0.$$

Wt. of $As_2O_3 = \frac{1.0}{1000} \times \frac{198}{4}$ g = 0.0495 g

$$\left(\underset{+6}{As_2O_3} + I_2 \rightarrow \underset{+10}{As_2O_5} + I^-; \; E_{As_2O_3} = \frac{198}{4} \right)$$

m.e. of As_2O_5 (produced by As_2O_3) + m.e. of As_2O_5 (in the mixture) = m.e. of I_2

= m.e. of $Na_2S_2O_3$

or m.e. of As_2O_3 + m.e. of As_2O_5 = m.e. of $Na_2S_2O_3$

$$1.0 + \text{m.e. of } As_2O_5 = \frac{1}{248/1} \times 1000$$

m.e. of As_2O_5 = 4.032.

$$\text{Wt. of } As_2O_5 \text{ in the mixture} = \frac{4.032}{1000} \times \frac{230}{4} \text{ g}$$
$$= 0.232 \text{ g.}$$

$$\left(\underset{+10}{As_2O_5} + I^- \rightarrow \underset{+6}{As_2O_3} + I_2;\ E_{As_2O_5} = \frac{230}{4} \right)$$

$$\therefore \text{ wt. of } (As_2O_3 + As_2O_5) = 0.0495 + 0.232 \text{ g} = 0.2815 \text{ g.}$$

Ex. 72. *A sample of* 0.15 g *of the compound* $[Pt(NH_3)_xBr_y]^{z+} \cdot z\,Br^-$, *ignited and heated to decomposition produced* 0.0502 g *of* Pt. *A second* 0.15-g *sample was dissolved in water and titrated rapidly with* 0.01 M $AgNO_3$ *solution.* 51.50 mL *was required to precipitate all the ionic bromide. A third* 0.15-g *sample was heated for two hours on a steam bath in a solution to which* 0.2 *mole of* $AgNO_3$ *has been added. This precipitated all the bromide (not just the free ionic* Br^-*) as* AgBr. *The weight of the precipitate thus produced was* 0.20 g. *Find x, y and z.* (Pt = 195, Ag = 108, Br = 80, N = 14 and H = 1)

Solution : $[Pt(NH_3)_xBr_y]^{z+} \cdot zBr^- \xrightarrow{\Delta} Pt$

Applying POAC for Pt atoms,

1 × mole of the compound = 1 × mole of Pt

$$\frac{0.15}{M} = \frac{0.0502}{195}, \qquad \dots (1)$$

where $M = 195 + 17x + 80y + 80z.$ ... (2)

For 0.15 g of the second sample containing z Br atoms per molecule (only the ionic bromide), one molecule of the compound shall combine with z molecules of $AgNO_3$ to give z molecules of AgBr.

$$\{Pt(NH_3)_xBr_y\}^{z+} \cdot zBr^- + zAgNO_3 \rightarrow zAgBr$$

Applying mole ratio method to reactants

z × mole of the compound = 1 × mole of $AgNO_3$

$$z \times \frac{0.15}{M} = \frac{0.01 \times 51.5}{1000}. \qquad \dots (3)$$

For the 0.15 g of the third sample, all bromine atoms $(y + z)$ in the compound combine with $AgNO_3$ to give $(y + z)$ molecules of AgBr,

$$[Pt\,(NH_3)_xBr_y]^{z+} \cdot z\,Br^- + (y + z)\,AgNO_3 \rightarrow (y + z)\,AgBr$$

Applying mole ratio method,

$(y + z)$ × mole of the compound = 1 × mole of AgBr

$$(y + z) \times \frac{0.15}{M} = 1 \times \frac{0.20}{188}. \qquad \dots (4)$$

Solving equations (1), (2), (3) and (4), we get,

$x = 4,\ y = 2$ and $z = 2.$

Ex. 73. 10 mL *of tap water containing* Ca^{2+} *and* Mg^{2+} *in the presence of* HCO_3^- *was properly buffered and the indicator murexide added. The sample was diluted and heated to* 60°C. *Titration with* 0.01 M *EDTA solution changed the indicator colour at* 7.50 mL. *This complexed* Ca^{2+} *only.*

A second 10-mL *sample was made basic and Erio T indicator added. Titration with* 0.01 M *EDTA solution changed the indicator colour at* 13.02 mL. *Under these conditions both* Ca^{2+} *and* Mg^{2+} *are complexed.*

If the 10 mL *of water sample were to be evaporated to dryness, what weight of* $CaCO_3 + MgCO_3$ *would be formed?*

Solution : All EDTA complexes are formed on a one-to-one basis with dipositive ions.

Mole of $Ca^{2+} + Mg^{2+}$ = mole of $CaCO_3$ + mole of $MgCO_3$

$$= \frac{0.01 \times 13.02}{1000} = 13 \times 10^{-5}$$

Mole of Ca^{2+} = mole of $CaCO_3$

$$= \frac{0.01 \times 7.50}{1000} = 7.50 \times 10^{-5}.$$

$\therefore$ mole of $MgCO_3 = 13 \times 10^{-5} - 7.50 \times 10^{-5} = 5.5 \times 10^{-5}$.

$\therefore$ weight of $CaCO_3 + MgCO_3 = 7.50 \times 10^{-5} \times 100 + 5.5 \times 10^{-5} \times 84$

$= 1.21 \times 10^{-2}$ g.

($CaCO_3 = 100$, $MgCO_3 = 84$)

PROBLEMS

(Answers bracketed with questions)

1. Calculate the strength in g/L of 3 N HCl and $\frac{N}{2}$ H_2SO_4 solutions. (109·5, 24·5)

2. How many mL of 1 M sulphuric acid is required to neutralise 10 mL of 1 M sodium hydroxide solution? (5 mL)

3. 2 litres of ammonia at 13°C and 0·90 atmospheric pressure is neutralised by 134 mL of H_2SO_4 solution. Find the normality of the acid. (0·57 N)

4. What weight of $CuSO_4 \cdot 5H_2O$ must be taken to make 0·5 litre of 0·01 M copper (II) ion solution? (1·248 g)

5. (a) Calculate the molarity of hydrogen chloride in a solution when 0·365 g of it has been dissolved in 100 mL of the solution.

(b) 3 g of a salt of molecular weight 30 is dissolved in 250 g of water. The molality of the solution is … . [(a) 0·1 M (b) 0·4 *m*]

6. Derive a formula for the volume of water, V_2, which must be added to V_1 mL of concentrated solution of molarity M_1 to give a solution of molarity M_2.

$$\left[V_2 = \frac{V_1(M_1 - M_2)}{M_2}\right]$$

7. Find the equivalent weight of H_3PO_4 in the reaction

$[Ca(OH)_2 + H_3PO_4 = CaHPO_4 + 2H_2O]$ (49)

8. A 250-mL sample of 0·20 M hydrochloric acid is to be made by diluting the approximate amount of the concentrated reagent 11·7 M. What volume of the latter should be used? (4·27 mL)

9. How many mL of each of two hydrochloric acids of strengths 12 N and 3 N are to be mixed to make one litre of 6 N solution? (333·33 mL, 666·67 mL)

10. What volumes of 2 M and 6 M solutions of HCl have to be mixed to prepare 500 mL of a 3 M solution? Disregard the change in the volume in mixing. (375 mL, 125 mL)

11. 1 litre of a solution contains 18·9 g of HNO_3 and 1 litre of another solution contains 3·2 g of NaOH. In what volume ratio must these solutions be mixed to obtain a solution having a neutral reaction? (1 : 3·75)

12. 10 mL of sulphuric acid solution (sp. gr. = 1·84) contains 98% by weight of pure acid. Calculate the volume of 2·5 N NaOH solution required to just neutralise the acid. (147·2 mL)

13. What is the molarity and molality of a 13% solution (by weight) of H_2SO_4 solution? Its density is 1·090 g/mL. To what volume should 100 mL of this acid be diluted in order to prepare 1·5 N solution? (1·45 M, 1·52 *m*, 193·3 mL)

14. How many mL of concentrated sulphuric acid of sp. gr. 1·84 containing 98% H_2SO_4 solution by weight is required to prepare 200 mL of 0·5 N solution? (2·71 mL)

15. 26 mL of a 1 N Na_2CO_3 solution is neutralised by the acids *A* and *B* in different experiments. The volumes of the acids *A* and *B* required were 10 mL and 40 mL respectively. How many volumes of *A* and *B* are to be mixed in order to prepare 1 litre of normal acid solution? (179·4, 820·6)

16. 25 mL of a solution of Fe^{2+} ions was titrated with a solution of the oxidising agent $Cr_2O_7^{2-}$. 32·45 mL of 0·0153 M $K_2Cr_2O_7$ solution was required. What is the molarity of the Fe^{2+} solution? (0·1192 M)

17. Upon heating a litre of a $\frac{N}{2}$ HCl solution, 2·675 g of hydrogen chloride is lost and the volume of the solution shrinks to 750 mL. Calculate (i) the normality of the resultant solution (ii) the number of milliequivalents of HCl in 100 mL of the original solution. (0·569 N, 50)

18. The reaction $Zn + CuSO_4 = Cu + ZnSO_4$ goes completely to the right. In one experiment 10 g of metallic zinc was added to 200 mL of copper sulphate solution.

After all copper is precipitated it was found that not all the zinc had dissolved. After filtration the total solid at the end of the reaction was 9·810 g. Calculate (i) the weight of copper deposited and (ii) molarity of copper sulphate in the original solution. (Cu = 63·5, Zn = 65·4) (6·35 g, 0·5 M)

19. 0·108 g of finely-divided copper was treated with an excess of ferric sulphate solution until copper was completely dissolved. The solution after the addition of excess dilute sulphuric acid required 33·7 mL of 0·1 N $KMnO_4$ for complete oxidation. Find the equation which represents the reaction between metallic copper and ferric sulphate solution. (Cu = 63·6, Fe = 56)

$$[Cu + Fe_2(SO_4)_3 = CuSO_4 + 2FeSO_4]$$

20. A commercial sample (2·013 g) of NaOH containing Na_2CO_3 as an impurity was dissolved to give 250 mL of solution. A 10 mL portion of the solution required 20 mL of 0·1 N H_2SO_4 solution for complete neutralisation. Calculate the percentage by weight of Na_2CO_3 in the sample. (2·29%)

21. (i) A solution of a mixture of KCl and KOH was neutralised with 120 mL of 0·12 N HCl. Calculate the amount of KOH in the mixture.

(ii) After titration, the resultant solution was made acidic with HNO_3. Then excess of $AgNO_3$ solution was added to precipitate the AgCl which weighed 3·7 g after drying. Calculate percentage of KOH in the original mixture. (0·806 g, 48·7 %)

22. 10·03 g of vinegar was diluted to 100 mL and a 25 mL sample was titrated with the 0·0176 M $Ba(OH)_2$ solution. 34·30 mL was required for equivalence. What is the percentage of acetic acid in the vinegar? (2·90%)

23. Zinc can be determined volumetrically by the precipitation reaction

$$3\,Zn^{2+} + 2\,[4K^+, Fe(CN)_6^{4+}] \longrightarrow K_2Zn_3[Fe(CN)_6]_2 + 6\,K^+$$

A sample of zinc ore weighing 1·5432 g was prepared for reaction and required 34·68 mL of 0·1043 M $K_4Fe(CN)_6$ for titration. What is the percentage of zinc in the ore? (23%)

24. 5·5 g of a mixutre of $FeSO_4 \cdot 7H_2O$ and $Fe_2(SO_4)_3 \cdot 9H_2O$ requires 5·4 mL of 0·1 N $KMnO_4$ solution for complete oxidation. Calculate the number of mole of hydrated ferric sulphate in the mixture. (0·0077 mole)

25. A chemist is preparing to analyse samples that will contain no more than 0·5 g of uranium. His procedure calls for preparing the uranium as U^{4+} ion and oxidising it by MnO_4^- in acid solution.

$$5U^{4+} + 2MnO_4^- + 6H_2O \rightarrow 5UO_2^{2+} + 2Mn^{2+} + 4H_3O^+$$

If he wants to react the total U^{4+} sample with a maximum of 50 mL of $KMnO_4$ solution, what concentration should he choose? (0·0168 M)

26. For the standardisation of a $Ba(OH)_2$ solution, 0·2 g of potassium acid phthalate (wt. of one mole = 204·2 g) was weighed which was then titrated with $Ba(OH)_2$ solution. The titration indicated equivalence at 27·80 mL of $Ba(OH)_2$ solution. What is the molarity of the base? The equation for reaction is

$$2KHC_8H_4O_4 + Ba(OH)_2 \rightarrow 2H_2O + 2K^+ + 2C_8H_4O_4^{2-} + Ba^{2+}$$ (0·0176 M)

27. A sample of magnesium metal containing some magnesium oxide as impurity was dissolved in 125 mL of 0·1 N H_2SO_4. The volume of hydrogen evolved at 27·3°C and 1 atm was 120·1 mL. The resulting solution was found to be 0·02 N with respect to H_2SO_4. Calculate (i) the weight of the sample dissolved and (ii) the percentage by weight of magnesium in the sample. Neglect any change in the volume of the solution. (0·1235 g; 95·95%)

28. A piece of aluminium weighing 2·7 g is treated with 75 mL of H_2SO_4 (sp. gr. 1·18 containing 24·7% H_2SO_4 by weight). After the metal is completely dissolved, the solution is diluted to 400 mL. Calculate the molarity of free sulphuric acid in the resulting solution. (0·177 M)

29. 4·00 g of mixture of NaCl and Na_2CO_3 was dissolved in water and the volume made up to 250 mL; 25 mL of this solution required 50 mL of N/10 HCl for complete neutralisation. Calculate percentage composition of the original mixture. (33·75% ; 66·25%)

30. 1·00 g of a mixture consisting of equal number of moles of carbonates of the two univalent metals, required 44·4 mL of 0·5 N HCl for complete reaction. If the atomic weight of one of the metals is 7, find the atomic weight of the other metal. What will be the total amount of sulphate formed on gravimetric conversion of 1 g of the mixture of sulphates? (23·1 ; 1·4 g)

31. A mixture containing only Na_2CO_3 and K_2CO_3 weighing 1·22 g was dissolved in water to form 100 mL of a solution. 20 mL of this solution required 40 mL of 0·1 N HCl for neutralisation. Calculate the weight of Na_2CO_3 in the mixture. If another 20 mL portion of the solution is heated with excess of $BaCl_2$, what will be the weight of the precipitate obtained? (0·53 g ; 0·3946 g)

32. 1·00 g of a moist sample of a mixture of KCl and $KClO_3$ was dissolved in water and made up to 250 mL. 25 mL of this solution was treated with SO_2 to reduce the chlorate to chloride and excess SO_2 was removed by boiling. The total chloride was precipitated as silver chloride. The weight of the precipitate was 0·1435 g. In another experiment, 25 mL of the original solution was heated with 30 mL of 0·2 N solution of ferrous sulphate and unreacted ferrous sulphate required 37·5 mL of 0·08 N solution of an oxidising agent for complete oxidation. Calculate the molar ratio of chlorate to chloride in the given mixture. Fe^{2+} reacts with ClO_3^- according to the reaction

$$ClO_3^- + 6Fe^{2+} + 6H^+ \rightarrow Cl^- + 6Fe^{3+} + 3H_2O$$

($0·5 \times 10^{-3}$ mole ; $0·5 \times 10^{-3}$ mole)

33. 0·6 g of a sample of pyrolusite was boiled with 200 mL of N/10 oxalic acid and excess of dilute H_2SO_4. The liquid was filtered and the residue washed. The filtrate and washings were mixed and made up to 500 mL. 100 mL of this solution required 50 mL of N/30 $KMnO_4$ solution. Calculate the percentage of MnO_2 in the given sample of pyrolusite. (84·58%)

34. 25 mL of a 0·107 M H_3PO_4 was titrated with a 0·115 M solution of a NaOH solution to the end point identified by the colour change of the indicator, bromocresol green. This required 23·1 mL. The titration was repeated using phenolphthalein indicator. This time 25 mL of 0·107 M H_3PO_4 required 46·8 mL of the 0·115 M NaOH. What is the coefficient n in the equation,
$H_3PO_4 + nOH^- \rightarrow nH_2O + [H_{(3-n)}PO_4]^{n-}$ for each reaction? (1; 2)

35. 9·8 g of $FeSO_4(NH_4)_2SO_4 \cdot xH_2O$ was dissolved in 250 mL of its solution. 20 mL of this solution required 20 mL of $KMnO_4$ solution containing 3·53 g of 90% pure $KMnO_4$ dissolved per litre. Calculate x. (6)

36. 10 mL of H_2O_2 weighs 10·205 g. The solution was diluted to 250 mL, 25 mL of which required 35·8 mL of a decinormal solution of $KMnO_4$. Calculate the weight in grams of H_2O_2 in 100 mL and also the volume strength of the solution.
(6·086 g; 20 V)

37. 50 mL of H_2O_2 was treated with excess of KI in presence of dilute sulphuric acid. I_2 so liberated was titrated with 20 mL of 0·1 N hypo solution. Calculate the strength of H_2O_2 in grams per litre. (0·68)

38. Calculate the percentage of available chlorine in a sample of 3·546 g of bleaching powder which was dissolved in 100 mL of water; 25 mL of this solution, on treatment with KI and dilute acid, required 20 mL of 0·125 N sodium thiosulphate solution. (10·01%)

39. (a) A zinc rod weighing 25·00 g was kept in 100 mL of 1 M $CuSO_4$ solution. After a certain time molarity of Cu^{2+} in solution was 0·8 M. What was the molarity of the sulphate ion (SO_4^{2-})? What was the weight of the zinc rod after cleaning?

(b) If the above experiment was done with a copper rod of weight 25 g and 50 mL of 2M zinc sulphate ($ZnSO_4$) solution, what would be the molarity of Zn^{2+} at the end of the same interval? [(a) 1 M, 23·6926 g; (b) 2 M]

40. 25 mL of a mixed solution of Na_2CO_3 and $NaHCO_3$ required 12 mL of N/20 HCl when titrated using phenolphthalein as an indicator. But 25 mL of the same, when titrated separately with methyl orange required 30 mL of N/20 HCl. Calculate the amount of Na_2CO_3 and $NaHCO_3$ in grams per litre. (2·544 g; 1·008 g)

41. 20 mL of a mixed solution of Na_2CO_3 and NaOH required 17·5 mL of N/10 HCl when titrated with phenolphthalein as indicator. But when methyl orange was added, a second end point was observed on the further addition of 2·5 mL of HCl. Calculate the amount of Na_2CO_3 and NaOH in the solution. (0·02648 g, 0·06 g)

42. 0·2 g of a chloride of an element was dissolved in water and then treated with excess of silver nitrate solution resulting in the formation of 0·47 g of silver chloride. Find the equivalent weight of the element. (25·56)

43. A sample of hydrazene sulphate ($N_2H_4SO_4$) was dissolved in 100 mL of water. 10 mL of this solution was reacted with excess of ferric chloride solution and warmed to complete the reaction. Ferrous ion formed was estimated and it required 20 mL of M/50 potassium permanganate solution. Estimate the amount of hydragene sulphate in one litre of the solution. **(IIT 1988)**

Reactions:

$$4Fe^{+++} + N_2H_4 \rightarrow N_2 + 4Fe^{++} + 4H^+$$

$$MnO_4^- + 5Fe^{++} + 8H^+ \rightarrow Mn^{++} + 5Fe^{+++} + 4H_2O$$

(6·5 g/L)

$$\left[\textbf{Hint: } \text{Eq. wt. of } N_2H_4 = \frac{\text{mol. wt.}}{4}\right]$$

44. 5 g of bleaching powder was suspended in water and volume made up to half a litre. 20 mL of this suspension when acidified with acetic acid and treated with excess of KI solution, liberated I_2 which required 20 mL of a N/10 hypo solution for titration. Calculate the percentage of available chlorine in bleaching powder.

[**Hint:** See Example 58.]

45. A 10-g mixture of Cu_2S and CuS was treated with 200 mL of 0·75 M MnO_4^- in acid solution producing SO_2, Cu^{2+} and Mn^{2+}. The SO_2 was boiled off and the excess MnO_4^- was titrated with 175 mL of 1 M Fe^{2+} solution. Calculate the percentage of CuS in the original mixture. (57·4%)

$$\left[\textbf{Hint: } \underset{+2\ \ -2}{Cu_2\,S} \rightarrow \underset{+4}{2Cu^{2+}} + \underset{+4}{SO_2}\text{; eq. wt. of } Cu_2S = \frac{\text{mol. wt.}}{8}\right]$$

46. A 0·518-g sample of limestone is dissolved and then Ca is precipitated as CaC_2O_4. After filtering and washing the precipitate, it requires 40 mL of 0·25 N $KMnO_4$ solution acidified with H_2SO_4 to titrate it. What is the percentage of CaO in limestone?

$$MnO_4^- + CaC_2O_4 + H_2SO_4 \rightarrow CaSO_4 + Mn^{2+} + CO_2 + H_2O$$

(54·1%)

47. 20 mL of M/60 solution of $KBrO_3$ was added to a certain volume of SeO_3^{2-} solution. Br_2 evolved was removed by boiling and the excess of $KBrO_3$ was back titrated with 5.1 mL of $\frac{M}{25}$ solution of $NaAsO_2$. Calculate the weight of SeO_3^{2-} in the solution. (0.084 g)

$$\left[\begin{array}{l}\textbf{Hint: } \underset{+4}{SeO_3^{2-}} + \underset{+5}{BrO_3^-} + H^+ \rightarrow \underset{+6}{SeO_4^{2-}} + \underset{0}{Br_2} + H_2O \\ \qquad\ \ \underset{+5}{BrO_3^-} + \underset{+3}{AsO_2^-} + H_2O \rightarrow \underset{-1}{Br^-} + \underset{+5}{AsO_4^{3-}} + H^+\end{array}\right]$$

48. Concentrated HCl solution is 37.0% HCl and has a density of 1.19 g/mL. A dilute solution of HCl is prepared by diluting 4.50 mL of this concentrated HCl solution to 100 mL with water. Then 10 mL of this dilute HCl solution reacts with an $AgNO_3$ solution. Calculate the volume of 0.108 M $AgNO_3$ solution required to precipitate all the chloride as AgCl(s).

49. 0.9546 g of a Rochelle salt, $NaKC_4H_4O_6 \cdot 4H_2O$, on ignition, gave $NaKCO_3$, which was treated with 41.72 mL of 0.1307 N H_2SO_4. The unreacted H_2SO_4 was then neutralised by 1.91 mL of 0.1297 N NaOH. Find the percentage purity of the Rochelle salt in the sample. (76.87%)

50. A mixture of $KMnO_4$ and $K_2Cr_2O_7$ weighing 0.24 g on being treated with KI in acid solution liberates just sufficient I_2 to react with 60 mL of 0.1 N $Na_2S_2O_3$ solution. Calculate percentage of Cr and Mn in the mixture. (20.83%, 14.17%)

[**Hint:** $\underset{+7}{KMnO_4} \rightarrow \underset{+2}{Mn^{2+}}$ and $\underset{+12}{K_2Cr_2O_7} \rightarrow \underset{+6}{2Cr^{3+}}$]

51. Federal regulations set an upper limit of 50 parts per million (ppm) of NH_3 in the air in a work environment (that is, 50 mL NH_3 per 10^6 mL of air). The density of $NH_3(g)$ at room temperature is 0.771 g/L. Air from a manufacturing operation was drawn through a solution containing 100 mL of 0.0105 M HCl. The NH_3 reacts with HCl as follows:

$$NH_3(aq) + HCl(aq) \rightarrow NH_4Cl(aq)$$

After drawing air through the acid solution for 10 minutes at a rate of 10 litres/min, the acid was titrated. The remaining acid required 13.1 mL of 0.0588 M NaOH to reach the equivalence point.

(a) How many grams of NH_3 were drawn into the acid solution?

(b) How many ppm of NH_3 were in the air?

(c) Is this manufacturer in compliance with regulations?

[(a) 0.00475 g (b) 61.6 ppm (c) No]

52. A compound H_2X with molar weight 80 g is dissolved in a solvent having density 0.4 g mL^{-1}. Assuming no change in volume upon dissolution, the molality of a 3.2 Molar solution is .... . **(IIT 2014 Adv.)** (8)

53. To measure the quantity of $MnCl_2$ dissolved in an aqueous solution, it was completely converted to $KMnO_4$ using the reaction:

$$MnCl_2 + K_2S_2O_8 + H_2O \rightarrow KMnO_4 + H_2SO_4 + HCl \text{ (equation not balanced)}$$

Few drops of concentrated HCl were added to this solution and gently warmed. Further oxalic acid (225 mg) was added in portions till the colour of permanganate ion disappeared. The quantity of $MnCl_2$ (in mg) present in the initial solution is ... .

(Mn = 55, Cl=35.5) **(IIT 2018 Adv)** (126 mg)

[**Hint:** m.e. of $MnCl_2$ = m.e. of $KMnO_4$ = m.e. of oxalic acid

$$= \left(\frac{0.225(g)}{\frac{90}{2}} \times 1000 \right)$$

(Eq. wt. of oxalic acid = 90/2)

$$\text{Weight of } MnCl_2 = \left(\frac{0.225\text{ (g)}}{\frac{90}{2}} \times 1000 \right) \times \frac{1}{1000} \times \frac{126}{5} = 0.126 \text{ g}$$

{Eq. wt. of $MnCl_2 = \frac{126}{5}$ $(\underset{+2}{MnCl_2} \rightarrow \underset{+7}{KMnO_4})$}]

Objective Problems

1. Normality of a solution is defined as
(a) number of eq./litre of solution (b) number of eq./litre of solvent
(c) number of mole/kg of solvent (d) number of mole/kg of solution

2. In the reaction $2S_2O_3^{2-} + I_2 \rightarrow S_4O_6^{2-} + 2I^-$, the eq. wt. of $Na_2S_2O_3$ is equal to its
(a) mol. wt. (b) mol. wt./2 (c) 2 × mol. wt. (d) mol. wt./6

3. In the reaction $VO + Fe_2O_3 \rightarrow FeO + V_2O_5$, the eq. wt. of V_2O_5 is equal to its
(a) mol. wt. (b) mol. wt./8 (c) mol. wt./6 (d) none of these

4. In the reaction $Na_2CO_3 + HCl \rightarrow NaHCO_3 + NaCl$, the eq. wt. of Na_2CO_3 is
(a) 53 (b) 106 (c) 10·6 (d) 5·3

5. The eq. wt. of iodine in $I_2 + 2S_2O_3^{2-} \rightarrow 2I^- + S_4O_6^{2-}$ is equal to its
(a) mol. wt. (b) mol. wt./2 (c) mol. wt./4 (d) none of these

6. The eq. wt. of K_2CrO_4 as an oxidising agent in acid medium is
(a) mol. wt./2 (b) $\frac{2 \times \text{mol. wt.}}{3}$ (c) $\frac{\text{mol. wt.}}{3}$ (d) $\frac{\text{mol. wt.}}{6}$

7. In alkaline condition $KMnO_4$ reacts as follows:

$$2KMnO_4 + 2KOH \rightarrow 2K_2MnO_4 + H_2O + O,$$

the eq. wt. of $KMnO_4$ is
(a) 52·7 (b) 158 (c) 31·6 (d) 79

8. 0·126 g of an acid requires 20 mL of 0·1 N NaOH for complete neutralisation. Eq. wt. of the acid is
(a) 45 (b) 53 (c) 40 (d) 63

9. H_3PO_4 is a tribasic acid and one of its salt is NaH_2PO_4. What volume of 1 M NaOH solution should be added to 12 g of NaH_2PO_4 to convert it into Na_3PO_4?
(a) 100 mL (b) 200 mL (c) 80 mL (d) 300 mL

10. 2 g of a base whose eq. wt. is 40 reacts with 3 g of an acid. The eq. wt. of the acid is
(a) 40 (b) 60 (c) 10 (d) 80

11. In a reaction 4 moles of electrons are transferred to one mole of HNO_3 when acted as an oxidant. The possible reduction product is
(a) (1/2) mole of N_2 (b) (1/2) mole of N_2O
(c) 1 mole of NO_2 (d) 1 mole of NH_3

12. Normality of 1% H_2SO_4 solution is nearly
(a) 2·5 (b) 0·1 (c) 0·2 (d) 1

13. What volume of 0·1N HNO_3 solution can be prepared from 6·3 g of HNO_3?
(a) 1 litre (b) 2 litres (c) 0·5 litre (d) 5 litres

14. The volume of water to be added to 200 mL of seminormal HCl solution to make it decinormal is
(a) 200 mL (b) 400 mL (c) 600 mL (d) 800 mL

15. 0·2 g of a sample of H_2O_2 required 10 mL of N $KMnO_4$ in a titration in the presence of H_2SO_4. Purity of H_2O_2 is
(a) 25% (b) 85% (c) 65% (d) 95%

16. 100 mL of 0·5 N NaOH solution is added to 10 mL of 3 N H_2SO_4 solution and 20 mL of 1 N HCl solution. The mixture is
(a) acidic (b) alkaline (c) neutral (d) none of these

17. Which of the following has the highest normality?
(a) 1 M H_2SO_4 (b) 1 M H_3PO_3 (c) 1 M H_3PO_4 (d) 1 M HNO_3

18. Eq. wt. of a metal, *x* g of which reacts with 1 eq. of an acid, is
(a) 1 (b) $x/2$ (c) $2x$ (d) x

19. The molarity of 98% H_2SO_4 ($d = 1{\cdot}8$ g/mL) by wt. is
(a) 6 M (b) 18 M (c) 10 M (d) 4 M

20. 0·7 g of $Na_2CO_3 \cdot x\,H_2O$ is dissolved in 100 mL of water, 20 mL of which required 19·8 mL of 0·1 N HCl. The value of *x* is
(a) 4 (b) 3 (c) 2 (d) 1

21. The normality of 10 mL of a '20 V' H_2O_2 is
(a) 1·79 (b) 3·58 (c) 60·86 (d) 6·086

22. If 8·3 mL of a sample of H_2SO_4 (36 N) is diluted by 991·7 mL of water, the approximate normality of the resulting solution is
(a) 0·4 (b) 0·2 (c) 0·1 (d) 0·3

23. 10 mL of an HCl solution gave 0·1435 g of AgCl when treated with excess of $AgNO_3$. The normality of the resulting solution is
(a) 0·1 (b) 3 (c) 0·3 (d) 0·2

24. 500 mL of a 0·1 N solution of $AgNO_3$ is added to 500 mL of a 0·1 N KCl solution. The concentration of nitrate in the resulting mixture is
(a) 0·1 N (b) 0·05 N (c) 0·01 N (d) 0·2 N

25. The ratio of amounts of H_2S needed to precipitate all the metal ions from 100 mL of 1 M $AgNO_3$ and 100 mL of 1 M $CuSO_4$ is
(a) 1 : 2 (b) 2 : 1 (c) zero (d) infinite

26. If 0·5 mole of $BaCl_2$ is mixed with 0·2 mole of Na_3PO_4, the maximum number of mole of $Ba_3(PO_4)_2$ that can be formed is
(a) 0·7 (b) 0·5 (c) 0·30 (d) 0·1

27. 0·45 g of an acid of molecular weight 90 was neutralised by 20 mL of 0·5 N caustic potash. The basicity of the acid is
(a) 1 (b) 2 (c) 3 (d) 4

28. 1 litre of 18 molar H_2SO_4 has been diluted to 100 litres. The normality of the resulting solution is
(a) 0·09 N (b) 0·18 N (c) 1800 N (d) 0·36 N

29. The best indicator for detection of end point in titration of a weak acid and a strong base is
(a) methyl orange (3 to 4) (b) methyl red (4 to 6)

(c) bromothymol blue (6 to 7·5) (d) phenolphthalein (8 to 9·6)

NB Figures in the brackets show the pH range of the indicator. **(IIT 1985)**

30. In a compound A_xB_y,

(a) moles of A = moles of B = moles of A_xB_y

(b) eq. of A = eq. of B = eq. of A_xB_y

(c) $y \times$ moles of A = $y \times$ moles of B = $(x + y) \times$ moles of A_xB_y

(d) $y \times$ moles of A = $y \times$ moles of B

31. The volume strength of 1·5 N H_2O_2 solution is

(a) 4·8 (b) 8·4 (c) 3·0 (d) 8·0

32. The number of moles of $KMnO_4$ that will be needed to react completely with one mole of ferrous oxalate in acidic solution is

(a) 3/5 (b) 2/5 (c) 4/5 (d) 1 **(IIT 1997)**

33. The number of moles of $KMnO_4$ that will be needed to react with one mole of sulphite ions in acidic solution is

(a) 2/5 (b) 3/5 (c) 4/5 (d) 1 **(IIT 1997)**

34. Which of the following reactions is not a disproportionation reaction?

(a) $2NO_2 + H_2O = HNO_3 + HNO_2$

(b) $Cl_2 + H_2O = HCl + HClO$

(c) $3K_2MnO_4 + 2H_2O = 2KMnO_4 + MnO_2 + 4KOH$

(d) $2FeSO_4 + 2H_2O = Fe_2(OH)_2SO_4 + H_2SO_4$

35. Dissolving 120 g urea (mol. wt. = 60) in 1000 g of water gives a solution of density 1.15 g/mL. The molarity of the solution is

(a) 1.78 M (b) 2.00 M (c) 2.05 M (d) 2.22 M **(IIT 2011)**

[Hint : Volume of solution = mass of solution × density]

36. The molarity of a solution obtained by mixing 750 mL of 0.5 M HCl with 250 mL of 2 M HCl will be

(a) 0.875 M (b) 1.00 M

(c) 1.75 M (d) 0.0975 M **(IIT 2013 Main)**

37. 3 g of activated charcoal was added to 50 mL of acetic acid solution (0.06 N) in a flask. After an hour it was filtered and the strength of the filtrate was found to be 0.042 N. The amount of acetic acid adsorbed (per gram of charcoal) is

(a) 18 mg (b) 36 mg

(c) 42 mg (d) 54 mg **(IIT 2015 Main)**

[Hint: m.e. of CH_3COOH adsorbed = 50(0.06 – 0.042)]

Answers

1-a, **2**-a, **3**-c, **4**-b, **5**-b, **6**-c, **7**-b, **8**-d, **9**-b, **10**-b, **11**-b, **12**-c, **13**-a, **14**-d, **15**-b, **16**-c, **17**-c, **18**-d, **19**-b, **20**-c, **21**-b, **22**-d, **23**-a, **24**-b, **25**-a, **26**-d, **27**-b, **28**-d, **29**-d, **30**-b, **31**-b, **32**-a, **33**-a, **34**-d, **35**-c, **36**-a, **37**-a.

❑

CHAPTER EIGHT

ELECTROLYSIS AND ELECTROLYTIC CONDUCTANCE

The electrical and chemical concepts are interdependent. A flow of electricity through a substance may produce a chemical reaction, and also, a chemical reaction may cause a flow of electricity through some external circuit. The former involves the study of electrolysis and conductance, while the latter, the measurement of electromotive force. In this chapter we shall deal with the phenomena of electrolysis and conductance.

ELECTROLYSIS

Faraday's Laws

The quantitative relationship between the amount of electricity passed through a cell and the amount of substances discharged at the electrodes was systematised by Michael Faraday in the form of the following laws:

First law: The amount of substance discharged (deposited or dissolved) at an electrode is proportional to the quantity of the electricity passing through the electrolyte.

Mathematically:

$$\left.\begin{aligned} &w \propto q \\ &w \propto I.t \quad (q = I.t) \\ &w = z.I.t \end{aligned}\right\}, \qquad \ldots (1)$$

where w is the weight of the substance discharged at an electrode in grams; q is the charge in coulombs, t is the time of flow of electricity in seconds, I is the current in amperes and z is a constant known as the electrochemical equivalent which is defined as the number of grams of the substance deposited or dissolved by one coulomb of electricity.

Second law: When the same quantity of electricity is passed through different solutions, the amounts of different substances deposited or dissolved at the electrodes in different electrolytic cells are proportional to their equivalent weights, and in an electrolytic cell, chemically equivalent amounts of substances are discharged at both the electrodes.

Interpretation of Faraday's Second Law

Let us now interpret the second law of Faraday in a simple manner.

1 electron reduces and deposits 1 M^+ ion at an electrode (i.e., $M^+ + e \rightarrow M$)

∴ 1 mole of electrons* shall reduce and deposit 1 mole of M^+ ions.

If the ion has a valency of n,

n mole of electrons shall reduce 1 mole of M^{n+} ions.

∴ 1 mole of electrons shall reduce $\frac{1}{n}$ mole of M^{n+} ions.

For example: 1 mole of electrons reduces or deposits 1 mole of Ag^+ or $\frac{1}{2}$ mole of Cu^{2+} or $\frac{1}{3}$ mole or Al^{3+}.

Now, that (number of moles × valency) represents number of equivalents

∴ 1 mole of electrons shall reduce or deposit 1 equivalent of Ag^+ or Cu^{2+} or Al^{3+}. In general, 1 mole of electricity (electrons) liberates 1 equivalent of matter.

Again we know

charge of 1 mole of electrons = charge of an electron × Av. const.

$$= 1{\cdot}6021 \times 10^{-19} \times 6{\cdot}022 \times 10^{23} \text{ coulombs}$$
$$= 96487 \text{ coulombs}$$
$$\approx 96500 \text{ coulombs}$$
$$= 26{\cdot}8 \text{ ampere-hour per equivalent}$$
$$= 1 \text{ faraday.}$$

Thus the essential content of Faraday's second law is that 1 faraday, which corresponds to 1 mole of electrons, liberates 1 equivalent of matter.

In redox reactions, the amount of the reactant, corresponding to 1 mole of electrons, is thus its equivalent mass.

Electrochemical Equivalent and Equivalent Weight

The weight in grams of a substance liberated by 1 coulomb of electricity is called electrochemical equivalent, whereas the weight in grams liberated by 96500 coulombs (or 1 faraday or 1 mole of electrons) is called gram equivalent weight of the substance.

From Faraday's law, we can deduce the relationship between the electrochemical equivalent and equivalent weight.

From the first law we have $w = z \,.\, q$;

dividing both sides by eq. wt.,

$$\frac{w}{\text{eq. wt.}} = \frac{z}{\text{eq. wt.}} \cdot q$$

or $$\text{number of equivalents} = \frac{z}{\text{eq. wt.}} \cdot q.$$

Now, from the second law we know that 96500 coulombs liberate 1 equivalent of matter, i.e., when q = 96500 coulombs, no. of eq. = 1.

* 1 mole of electrons corresponds to Av. const. of electrons.

$$\therefore \quad 1 = \frac{z}{\text{eq. wt.}} \times 96500$$

$$z = \frac{\text{equivalent weight}}{96500}.$$

$$\text{Equivalent weight} = \text{electrochemical weight} \times 96500. \quad \ldots (2)$$

If, E_1 and E_2 are the equivalent weights of two substances whose electrochemical equivalents are z_1 and z_2 respectively then

$$\frac{E_1}{E_2} = \frac{z_1}{z_2}.$$

Now, if the same amount of current is passed for the same interval of time through two different electrolytes then from the first law, i.e., $w = z \cdot I \cdot t$, we get,

$$\frac{E_1}{E_2} = \frac{z_1}{z_2} = \frac{w_1}{w_2}, \quad \ldots (3)$$

where w_1 and w_2 are the weights of the substances deposited from the two electrolytes, the electrochemical equivalents of which are respectively z_1 and z_2.

Further, if an incandescent bulb of W watts or a resistance of 'R' ohms is connected in series with an electrolytic cell, the following equations should be remembered to calculate the current I in amperes.

$$V = IR \quad \ldots (4)$$

$$W = IV \text{ (}V\text{ is the voltage in volt).} \quad \ldots (5)$$

[**Note:** If two or more different cations or anions are present in the solution, the decreasing order of preferential discharge of ions is

cations: Ag^+, Hg^{2+}, Cu^{2+}, H^+, Fe^{3+}, Zn^{2+}, Al^{3+}, Mg^{2+}, Ca^{2+}, Na^+, K^+

anions: I^-, Br^-, Cl^-, OH^-, NO_3^-, SO_4^{2-}, F^-]

ELECTROLYTIC CONDUCTANCE

The passage of a current through an electrolyte involves the movement of ions carrying an electric charge and so the study of electrolytic conduction may supply useful chemical information.

The magnitude of the conductance, i.e., the reciprocal of resistance, depends mainly on three factors: the number of ions, magnitude of charge on each ion and the ionic mobility.*

The conductance of an electrolyte may be measured in terms of molar conductance, that is, the conductance due to one mole of an ionic solute and secondly, the equivalent conductance, that is, the conducting power of all ions produced by one equivalent of the electrolyte in the given solution. But to

* Ionic mobility or absolute ionic velocity is defined as the distance travelled by an ion per second under a potential gradient of one volt per cm.

compare the conductance of two solutions, equivalent conductance is considered because one equivalent of different electrolytes involves the same number of electrons (i.e., Avogadro constant of electrons) in accordance with Faraday's second law of electrolysis while one mole of different electrolytes may or may not involve the same number of electrons. In other words, **the solutions, each containing one equivalent of different electrolytes, are equivalent in terms of moles of electrons being carried.**

Conductance, Specific Conductance, Equivalent Conductance and Molar Conductance

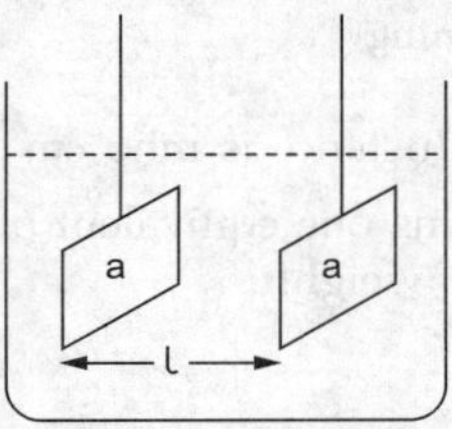

The reciprocal of resistance offered by an electrolyte to the flow of electricity through it is known as conductance. The unit of conductance is ohm^{-1} or mho.

$$\text{Conductance} = \frac{1}{\text{resistance}}.$$

Consider the portion of the electrolyte lying between the two parallel and equal electrodes as shown in the diagram; the area of each electrode being 'a' sq cm and the distance between them is 'l' cm. Since the electrolytes, like metallic conductors, also offer resistance to the flow of electricity, Ohm's law can be applied to the electrolytic conductor.

Ohm's law relates the resistance 'R' offered by a medium with the applied voltage 'V' and the passing current 'I'

$$V = IR. \qquad \ldots (6)$$

The resistance 'R' is related to the dimensions of the conductor as:

$$R \propto l$$

and $$R \propto \frac{1}{a}.$$

$\therefore$ $$R \propto \frac{l}{a};\ R = \rho\frac{l}{a}$$

where R = resistance, ρ = specific resistance

or $$\frac{1}{\rho} = \frac{1}{R} \times \frac{l}{a}.$$

$$\therefore \text{ specific conductance} = \text{conductance} \times \frac{l}{a}. \qquad \ldots (7a)$$

Just as $\frac{1}{R}$ is called conductance, $\frac{1}{\rho}$ is called specific conductance, denoted by 'K'. The unit of specific conductance is mho per cm. The SI unit of conductivity is the siemens (S) which is equal to ohm^{-1} or mho.

The solution under study is filled in a conductivity cell made of pyrex glass having two platinum electrodes fixed parallel to each other. For a given cell, $\frac{l}{a}$ is known as cell constant. Thus,

specific conductance = conductance × cell constant. ... (7b)

From the definition of equivalent conductance and molar conductance mentioned above, we get

equivalent conductance = specific conductance × vol. in cc containing 1 eq. of the electrolyte

or $\Lambda = KV$. ... (8)

Similarly,

molar conductance = sp. conductance × vol. in cc containing 1 mole of electrolyte.

The unit of equivalent conductance and molar conductance is mho cm^2.

If 1 cc of an electrolytic solution theoretically contains one equivalent of the electrolyte whose equivalent weight is equal to mol. weight;

conductance = specific conductance
= equivalent conductance
= molar conductance.

Equivalent Conductance at ∞ Dilution

Equivalent conductance increases with the increase in dilution but after a limit it becomes constant and does not further increase. The maximum value of equivalent conductance is known as equivalent conductance at ∞ dilution, denoted by Λ_∞. If this increase in equivalent conductance with dilution is only due to the increase in the degree of dissociation of the electrolyte, we can write,

$$\text{degree of dissociation } \alpha = \frac{\Lambda_c}{\Lambda_\infty}, \qquad \text{... (9)}$$

where Λ_c is equivalent conductance and α is the degree of dissociation at a concentration c of the weak electrolyte.

Kohlrausch's Law

At infinite dilution an ionic species (cation or anion) contributes a fixed value, at a given temperature, towards equivalent conductance of the electrolyte irrespective of the other ionic species in combination with it. These contributions are known as equivalent ionic conductance at ∞ dilution, denoted by $\overset{\circ}{\lambda}_c$ and $\overset{\circ}{\lambda}_a$ for cations and anions respectively.

The equivalent conductance of an electrolyte at ∞ dilution, following the Kohlrausch's law, is then equal to the sum of equivalent ionic conductances of cations and anions, $\Lambda^\circ = \overset{\circ}{\lambda}_c + \overset{\circ}{\lambda}_a$. ... (10)

The equivalent ionic conductance (λ) and ionic mobility (U) are not the same thing but are proportional to each other.

$$\left.\begin{aligned}\lambda^{\circ}_{c} \propto U^{\circ}_{c} \quad \text{or} \quad \lambda^{\circ}_{c} = KU^{\circ}_{c} \\ \lambda^{\circ}_{a} \propto U^{\circ}_{a} \quad \text{or} \quad \lambda^{\circ}_{a} = KU^{\circ}_{a}\end{aligned}\right\} \qquad \ldots (11)$$

The proportionality constant K may be shown to be equal to 96,500 coulombs. Ionic mobility is measured in cm per second for a field of one volt per cm.

Mathematically,

$$\text{ionic mobility} = \frac{\text{ionic velocity}}{\text{potential gradient}} \qquad \ldots (12)$$

$$= \frac{\text{ionic velocity (cm/s)}}{\text{pot. difference (volt)/distance between the electrodes (cm)}}$$

The unit of ionic mobility is thus $\text{cm}^2\ \text{volt}^{-1}\ \text{s}^{-1}$.

[**Note:** The following combinations of electrical units may be expressed as a single unit.]

volt · ampere	... watt
ampere · second	... coulomb
volt/ampere	... ohm
joule/volt	... coulomb
watt/ampere · ohm	... ampere
joule/second	... watt
joule/ampere · second	... volt
joule/ampere^2 · second	... ohm

EXAMPLES ON ELECTROLYSIS

Ex. 1. *From the following data calculate the value of the Avogadro constant. Charge of an electron* $= 1{\cdot}6 \times 10^{-19}$ *coulomb.* (1 *faraday* = 96500 *coulombs*)

Solution : Since 1 faraday, i.e., 96500 coulombs, is the charge of 1 mole of electrons, i.e., Av. no. of electrons.

$$\therefore \quad \text{Av. constant} = \frac{\text{charge of 1 mole of electrons}}{\text{charge of one electron}}$$

$$= \frac{96500}{1{\cdot}6 \times 10^{-19}} = 6{\cdot}03 \times 10^{23}.$$

Ex. 2. *From the values of* 1 *faraday and Avogadro constant, show that* 1 *faraday may be called* 1 *mole of electricity.*

Solution : Since 1 faraday = 96500 coulombs

$$\therefore \quad \frac{1\text{ F}}{\text{Av. no.}} = \frac{96500}{6{\cdot}022 \times 10^{23}}$$

$$= 1{\cdot}6 \times 10^{-19} \text{ coulomb}$$

$$= \text{charge of an electron.}$$

$$\therefore \quad 1\text{ F} = \text{charge of an electron} \times \text{Av. no.}$$

Since 1 F of electricity is the charge of Av. no. of electrons, 1 faraday may be called 1 mole of electricity.

Ex. 3. *A metal wire carries a current of* 1 *ampere. How many electrons pass a point in the wire in* 1 *second?*

Solution : Charge in coulomb = current in ampere × time(s)

$$= 1 \times 1 = 1.$$

Since 1 F (96,500 coulombs) of electricity is carried out by 1 mole of electrons, i.e., $6{\cdot}022 \times 10^{23}$ electrons, therefore, 1 coulomb shall involve $6{\cdot}022 \times 10^{23}/96500$, i.e., $6{\cdot}24 \times 10^{18}$ electrons. Thus, $6{\cdot}24 \times 10^{18}$ electrons pass a point in the wire in 1 second.

Ex. 4. *How many electrons are lost and gained by* 2 g *of* Cl^- *ions and* 1 g *of* Zn^{2+} *ions as the result of electrolysis respectively?* (Cl = 35·5; Zn = 65)

Solution : Number of equivalents of Cl^- ions $= \dfrac{\text{weight}}{\text{eq. wt.}}$

$$= \frac{2}{35{\cdot}5}.$$

Number of equivalent of Zn^{2+} ions $= \dfrac{1}{65/2} = \dfrac{2}{65}$.

Now, 1 mole (1 F) of electric charge discharges 1 equivalent of matter.

Mole of electric charge involved in case of $Cl^- = \dfrac{2}{35{\cdot}5}$.

Mole of electric charge involved in case of $Zn^{2+} = \dfrac{2}{65}$.

As 1 mole of electric charge corresponds to Av. no. of electrons, thus,

no. of electrons lost by $Cl^- = \dfrac{2}{35{\cdot}5} \times 6{\cdot}022 \times 10^{23}$

and no. of electrons gained by $Zn^{2+} = \dfrac{2}{65} \times 6{\cdot}022 \times 10^{23}$.

Ex. 5. *Calculate the quantity of ferrous and ferric ions that would be deposited by* 1 *faraday.* (Fe = 56)

Solution : 1 mole of electricity (i.e., 1 faraday) produces 1 eq. of matter.

Mole of electric charge = 1 faraday (given).

∴ no. of equivalent of $Fe^{2+} = 1$

and no. of equivalent of $Fe^{3+} = 1$.

∴ wt. of $Fe^{2+} = 1 \times \frac{56}{2} g = 28 g$ (wt. = eq. × eq. wt.)

and wt. of $Fe^{3+} = 1 \times \frac{56}{3} g = 18{\cdot}6 g$. [Eqn. 4(i), Chap. 7]

Ex. 6. 0·2864 g *of* Cu *was deposited on passage of a current of* 0·5 *ampere for* 30 *minutes through a solution of copper sulphate. What is the electrochemical equivalent of copper?* (1 F = 96500 coulombs)

Solution : Electrochemical equivalent (z) is defined as the weight of the substance deposited by the passage of 1 coulomb of electricity.

$$\therefore \quad z = \frac{\text{wt. of Cu deposited}}{\text{charge in coulombs}}$$

$$= \frac{\text{wt. of Cu}}{\text{current in ampere} \times \text{time in seconds}}$$

$$= \frac{0{\cdot}2864}{0{\cdot}5 \times 30 \times 60}$$

$$= 0{\cdot}00032 \text{ g/coulomb.}$$

Ex. 7. *Calculate how much current is necessary to produce hydrogen gas at the rate of* 1 cc *per second under standard conditions.*

Solution : ∵ 1 equivalent of substance is produced by 1 mole of electric charge

∴ 11200 cc (i.e., 1 equivalent) of hydrogen is produced by 96500 coulombs

∴ 11200 cc of hydrogen is produced per second by 96500 amperes ($q = I.t$)

∴ 1 cc of hydrogen is produced per second by $\frac{96500}{11200}$ amp

= 8·61 amp.

Ex. 8. *In an industrial electrolytic cell it is desired to produce* 36 kg *of magnesium metal per hour. Calculate the current required.* (Mg = 24)

Solution : 1 mole of electric charge (1 F) produces 1 eq. of Mg or 1 mole of electric charge (1 F) produces 12 g of Mg. (eq. wt. $= \frac{24}{2} = 12$)

No. of equivalents of Mg (in 36 kg) $= \frac{36000}{12} = 3000$

[Eqn. 4(i), Chapter 7]

$\therefore$ moles of electric charge to produce 36000 g of Mg = 3000 faradays

$= 3000 \times 96500$ coulombs

$$\therefore \text{ current required per hour} = \frac{\text{charge in coulombs}}{\text{time in seconds}}$$

$$= \frac{3000 \times 96500}{60 \times 60} \text{ amp}$$

$$= 8{\cdot}04 \times 10^4 \text{ amp.}$$

Ex. 9. *How much time is required for the complete decomposition of 2 moles of water using a current of 2 amp?*

Solution : Electrolysis of water takes place according to the equation:

$$2H_2O \rightarrow 2H_2 + O_2$$

$\because$ 1 equivalent of hydrogen is produced by $\frac{1}{2}$ mole of water

$\therefore$ no. of equivalents of H_2O decomposed

$= 2 \times$ no. of moles of water

$= 2 \times 2 = 4.$ [Eqn. 6(ii), Chapter 7]

$\because$ 1 equivalent of H_2O will be decomposed by 1 mole of electricity (1 F)

$\therefore$ mole of electric charge = 4 faradays $= 4 \times 96500$ coulombs

$$\therefore \text{ duration of electrolysis} = \frac{\text{charge in coulombs}}{\text{current in amperes}}$$

$$= \frac{4 \times 96500}{2} \text{ seconds}$$

$$= \frac{4 \times 96500}{2 \times 60 \times 60} \text{ hours}$$

$$= 53{\cdot}61 \text{ hours.}$$

Ex. 10. *In the electrolysis of an aqueous* $SnCl_2$ *solution, 4·48 litres of chlorine (in standard conditions) is liberated at the anode. Find the mass of tin deposited at the cathode.* (Sn = 119; Cl = 35·5)

Solution : Number of equivalents of chlorine liberated at anode = number of equivalents of Sn^{2+} deposited at cathode.

$$\therefore \frac{\text{volume of chlorine (NTP)}}{\text{vol. occupied by 1 eq.}} = \frac{\text{wt. of } Sn^{2+} \text{ deposited}}{\text{eq. wt. of } Sn^{2+}}$$

... [Eqns. 4(i) & 4(ii), Chapter 7]

$$\therefore \quad \frac{4{\cdot}48}{11{\cdot}2} = \frac{\text{wt. of } Sn^{2+} \text{ deposited}}{119/2} \qquad \left(\text{eq. wt. of Sn} = \frac{119}{2}\right)$$

$\therefore$ wt. of Sn deposited = 23·80 g.

Ex. 11. *In the electrolysis of an aqueous solution of* NaOH, 2·8 *litres of oxygen was liberated at the anode at* NTP. *How much hydrogen was liberated at the cathode?*

Solution : Eq. of O_2 liberated

$$= \frac{\text{vol. of } O_2 \text{ liberated at NTP}}{\text{vol. occupied by 1 eq. of } O_2 \text{ at NTP}}$$

$$= \frac{2{\cdot}8 \text{ lit.}}{5{\cdot}6 \text{ lit.}} = \frac{1}{2}.$$

Since same number of equivalent of hydrogen liberates at the cathode, equivalent of hydrogen liberated at the cathode $= \frac{1}{2}$.

$\therefore$ volume of hydrogen = no. of equivalents × vol. occupied by 1 eq. of hydrogen at NTP

$$= \frac{1}{2} \times 11{\cdot}2$$

$$= 5{\cdot}6 \text{ litres.}$$

Ex. 12. 30 mL *of* 0·13 M $NiSO_4$ *is electrolysed using a current of* 360 *milliamperes for* 35·3 *minutes. How much of the metal would have been plated out if current efficiency was only* 60%? (Ni = 58·7)

Solution : Charge in coulombs = current in amperes × time in seconds

$$= 0{\cdot}36 \times 35{\cdot}3 \times 60$$

$$= \frac{0{\cdot}36 \times 35{\cdot}3 \times 60}{96500} \text{ F}$$

$$= 0{\cdot}0079 \text{ F.}$$

$\because$ 1 mole of electric charge (1 F) produces 1 equivalent of the substance.

$\therefore$ eq. of Ni which should be deposited for 100% current efficiency

$$= 0{\cdot}0079$$

$$= \frac{0{\cdot}0079 \times 58{\cdot}7}{2} \text{ g}$$

$$= 0{\cdot}2318 \text{ g.} \qquad \left(\text{eq. wt. of Ni} = \frac{58{\cdot}7}{2}\right)$$

But since the current efficiency is 60%,

$$\text{actual amount of Ni deposited} = 0{\cdot}2318 \times \frac{60}{100}$$

$$= 0{\cdot}1391 \text{ g.}$$

Ex. 13. *Calculate the quantity of electricity that would be required to reduce* 12·3 g *of nitrobenzene to aniline, if the current efficiency for the process is* 50%. *If the potential drop across the cell is* 3 *volts, how much energy will be consumed?*

(IIT 1990)

Solution : No. of eq. of nitrobenzene to aniline

$$= \frac{\text{wt. in g}}{\text{eq. wt. of } C_6H_5NO_2}$$

$$= \frac{12{\cdot}3}{123/6} = 0{\cdot}6.$$

$$\left\{ \begin{array}{l} \underset{+3}{C_6H_5NO_2} \rightarrow \underset{-3}{C_6H_5NH_2}\text{ ; change in ON} = 6 \\ \therefore \text{ eq. wt. of } C_6H_5NO_2 = \dfrac{\text{mol. wt.}}{\text{change in ON}} = \dfrac{123}{6} \end{array} \right\}$$

∴ mole of electricity for 100% current efficiency = 0·6 F.

But the current efficiency is 50%.

∴ mole of electricity used $= 0{\cdot}6 \times 2 = 1{\cdot}2$ F

$= 1{\cdot}2 \times 96500$ coulombs

= 115800 coulombs.

*The energy consumed = electricity in coulomb × pot. drop in volt

$= 115800 \times 3$ J

= 347400 J

= 347·40 kJ.

Ex. 14. *Potassium chlorate is prepared by electrolysis of* KCl *in basic solution:*

$$6OH^- + Cl^- \rightarrow ClO_3^- + 3H_2O + 6\,e^-$$

If only 60% *of the current is utilised in the reaction, what time will be required to produce* 10 g *of* $KClO_3$ *using a current of* 2 amp?

Solution : According to the given equation,

$$\text{eq. wt. of } KClO_3 = \frac{\text{mol. wt.}}{6} = \frac{122{\cdot}5}{6} = 20{\cdot}4.$$

$$\therefore \quad \text{equivalent of } KClO_3 \text{ to be produced} = \frac{10}{20{\cdot}4}.$$

Since current efficiency is 60%; hence,

1 faraday (i.e., 96500 coulombs) shall produce

0·6 equivalent instead of 1 equivalent.

∴ production of $\frac{10}{20{\cdot}4}$ eq. of $KClO_3$ shall require

---

* 1 joule = 1 volt × 1 coulomb

$$\frac{96500}{0{\cdot}6} \times \frac{10}{20{\cdot}4} \text{ i.e., } 78839{\cdot}86 \text{ coulombs}$$

$$\therefore \text{ time required} = \frac{\text{no. of coulombs}}{\text{current in amp}}$$

$$= \frac{78839{\cdot}86}{2} = 39419{\cdot}93 \text{ seconds}$$

$$\approx 39420 \text{ seconds}$$

$$= \frac{39420}{60 \times 60} \text{ hours}$$

$$= 10{\cdot}95 \text{ hours.}$$

Ex. 15. *After electrolysis of a sodium chloride* (NaCl) *solution with inert electrodes for a certain period of time,* 600 mL *of the solution was left which was found to be* 1 N *in sodium hydroxide. During the same time* 31·8 g *of* Cu *was deposited in a* Cu *voltameter in series with the electrolytic cell. Calculate the percentage of theoretical yield of the sodium hydroxide obtained.* (Cu = 63·5)

Solution : *No. of equivalent of NaOH that can be produced theoretically for 100% current efficiency

= no. of eq. of NaCl decomposed

= no. of eq. of Cu deposited (Faraday's 2nd law)

$$= \frac{\text{wt. of Cu}}{\text{eq. wt. of Cu}} = \frac{31{\cdot}8}{63{\cdot}5/2} = 1. \qquad \text{[Eqn. 4(i), Chapter 7]}$$

Now,

no. of eq. of NaOH produced experimentally

$$= \frac{\text{m.e. of NaOH}}{1000} \qquad \text{(Eqn. 3, Chapter 7)}$$

$$= \frac{\text{normality} \times \text{volume}}{1000} = \frac{1 \times 600}{1000} = 0{\cdot}6. \qquad \text{(Eqn. 1, Chapter 7)}$$

$$\therefore \text{ percentage yield} = \frac{0{\cdot}6}{1} \times 100 = 60\%.$$

Ex. 16. *The density of* Cu *is* 8·94 g per mL. *Find out the charge in coulombs needed to plate an area* 10 cm × 10 cm *to a thickness of* 10^{-2} cm *using* $CuSO_4$ *solution as electrolyte.*

Solution : Wt. of Cu to be coated = volume of the deposit × density

$$= 10 \times 10 \times 10^{-2} \times 8{\cdot}94$$

$$= 8{\cdot}94 \text{ g.}$$

* On electrolysis of NaCl solution, NaOH is produced.

$$\therefore \quad \text{eq. of Cu} = \frac{8\cdot94}{31\cdot8}\left(\text{eq. wt. of Cu} = \frac{63\cdot5}{2} = 31\cdot8\right)$$
$$= 0\cdot281.$$

$\therefore$ mole of electric charge = 0·281 faraday

= 0·281 × 96500 coulombs

= 27116·5 coulombs.

Ex. 17. *How long a current of* 3 amp *has to be passed through a solution of* $AgNO_3$ *to coat a metal surface of* 80 cm^2 *with* 0·005 mm *thick layer? Density of* Ag is 10·5 g/cm^3. **(IIT 1985)**

Solution : No. of eq. of Ag deposited

$$= \frac{\text{wt. of Ag deposited in grams}}{\text{equivalent wt. of Ag}} \quad \text{[Eqn. 4(i), Chapter 7]}$$
$$= \frac{\text{volume} \times \text{density}}{\text{equivalent weight}}$$
$$= \frac{\text{surface area} \times \text{thickness} \times \text{density}}{\text{equivalent weight}}$$
$$= \frac{80 \times 0\cdot0005 \times 10\cdot5}{108}$$
$$= \frac{7}{1800} = 0\cdot0039.$$

Mole of electric charge = 0·0039 faraday

= 0·0039 × 96500 coulombs

= 376·35 coulombs.

$$\therefore \text{ duration of electrolysis} = \frac{\text{charge in coulombs}}{\text{current in amp}}$$
$$= \frac{376\cdot35}{3}\text{ s}$$
$$\approx 125 \text{ s}.$$

Ex. 18. *A* 100-*watt,* 110-*volt incandescent lamp is connected in series with an electrolytic cell containing cadmium sulphate solution. What weight of cadmium will be deposited by the current flowing for* 10 *hours?* **(IIT 1987)**

Solution : We know

$$\text{current in amperes} = \frac{\text{wattage}}{\text{voltage}} = \frac{100}{110}. \quad \text{... (Eqn. 5)}$$

Further, charge in coulombs = current in amperes × time in seconds

$$= \frac{100}{110} \times 10 \times 60 \times 60 = 32727\cdot27.$$

$$\therefore \quad \text{charge in faraday} = \frac{32727 \cdot 27}{96500} \text{ F} = 0 \cdot 34 \text{ F.}$$

$$\therefore \quad \text{mole of electricity} = 0 \cdot 34 \text{ F.}$$

$$\begin{aligned} \text{Amount of Cd deposited} &= 0 \cdot 34 \text{ eq.} \\ &= (0 \cdot 34 \times \text{eq. wt.}) \text{ g} \\ &= 0 \cdot 34 \times \frac{112 \cdot 4}{2} = 19 \cdot 11 \text{ g.} \end{aligned}$$

$$\left(\text{eq. wt. of cadmium} = \frac{112 \cdot 4}{2}\right)$$

Ex. 19. *A mixture of hydrogen and oxygen is evolved when a dilute solution of* NaOH *is electrolysed. How many moles of each gas would be liberated by a current which deposited* 20·942 g *of* Ag?

Solution : 1 mole (faraday) electric charge yields 1 equivalent of matter.

$$\therefore \quad \text{equivalent of Ag} = \frac{20 \cdot 942}{108} \cdot \qquad \text{[Eqn. 4 (i), Chapter 7]}$$

$$\therefore \text{ mole of electric charge} = \frac{20 \cdot 942}{108} \text{ faraday.}$$

Since the same amount of electricity is passed through dilute NaOH solution,

$$\text{equivalent of hydrogen liberated} = \frac{20 \cdot 942}{108}$$

$$\text{and} \quad \text{equivalent of oxygen liberated} = \frac{20 \cdot 942}{108} \cdot$$

$$\left.\begin{aligned} \therefore \text{ mole of } H_2 \text{ liberated} &= \frac{20 \cdot 942}{108} \times \frac{1}{2} = 0 \cdot 097 \\ \text{mole of } O_2 \text{ liberated} &= \frac{20 \cdot 942}{108} \times \frac{1}{4} = 0 \cdot 0485 \end{aligned}\right\} \text{[Eqn. 6(ii), Chapter 7]}$$

Ex. 20. *A current of* 3·7 amp *is passed for* 6 *hours between inert electrodes in* 0·5 *litre of a* 2 M *solution of* $Ni(NO_3)_2$. *What will be the molarity of the solution at the end of electrolysis?*

Solution :

$$\begin{aligned} \text{Charge in coulombs} &= \text{current in amp} \times \text{time in s} \\ &= 3 \cdot 7 \times 6 \times 60 \times 60 \\ &= 79920. \end{aligned}$$

$$\begin{aligned} \therefore \text{ mole (faraday) of electric charge} &= \frac{79920}{96500} \\ &= 0 \cdot 8283 \text{ F.} \end{aligned}$$

Eq. of $Ni(NO_3)_2$ decomposed on electrolysis = 0·8283.

$\therefore$ m.e. of $Ni(NO_3)_2$ decomposed $= 0{\cdot}8283 \times 1000 = 828{\cdot}3$

(Eqn. 3, Chapter 7)

Now, m.e. of $Ni(NO_3)_2$ solution before electrolysis

$= \text{normality} \times \text{volume (mL)}$ (Eqn. 1, Chapter 7)

$= 4 \times 500 = 2000$

{normality $= 2 \times$ molarity $= 2 \times 2 = 4$ [Eqn. 6(i), Chapter 7]}

m.e. of $Ni(NO_3)_2$ after electrolysis $= 2000 - 828{\cdot}3 = 1171{\cdot}7$.

$\therefore$ normality of $Ni(NO_3)_2$ after electrolysis $= \dfrac{1171{\cdot}7}{500} = 2{\cdot}34$ N.

(Eqn. 1, Chapter 7)

$\therefore$ molarity of $Ni(NO_3)_2$ solution after electrolysis $= \dfrac{2{\cdot}34}{2} = 1{\cdot}17$ M.

[Eqn. 6(i), Chapter 7]

Ex. 21. *A current of* 1·70 A *is passed through* 300 mL *of a* 0·16 M *solution of* $ZnSO_4$ *for* 230 s *with a current efficiency of* 90%. *Find out molarity of* Zn^{2+} *after the deposition of* Zn. *Assume the volume of the solution to remain constant during the electrolysis.* **(IIT 1991)**

Solution : Mole of electric charge $= \dfrac{1{\cdot}70 \times 230}{96500} = 0{\cdot}004052$ F.

$\therefore$ eq. of Zn to be deposited for 100% current efficiency $= 0{\cdot}004052$

or mole of Zn to be deposited for 100% current efficiency

$= 0{\cdot}004052/2 = 0{\cdot}002026$

or mole of Zn to be deposited for 90% current efficiency

$= 0{\cdot}9 \times 0{\cdot}002026 = 0{\cdot}0018234$.

Initial mole of Zn (or $ZnSO_4$) $= 0{\cdot}16 \times 0{\cdot}3 = 0{\cdot}048$.

Mole of Zn remained undeposited $= 0{\cdot}048 - 0{\cdot}0018234$

$= 0{\cdot}0461766$.

Molarity after electrolysis $= \dfrac{0{\cdot}0461766}{0{\cdot}3} = 0{\cdot}154$ M.

Ex. 22. *A current of* 0·0965 amp *is passed for* 1000 *seconds through* 50 mL *of* 0·1 M *of* NaCl. *If the only reactions are reduction of* H_2O *to* H_2 *at the cathode and oxidation of* Cl^- *to* Cl_2 *at the anode, what will be the average concentration of* OH^- *in the final solution?*

Solution : Mole of electric charge $= \dfrac{0{\cdot}0965 \times 1000}{96500} = 0{\cdot}001$ F.

$\therefore$ equivalent of OH^- liberated $= 0{\cdot}001$.

Mole of OH^- liberated $= 0{\cdot}001$.

$\therefore$ concentration of OH^- in mole per litre $= \dfrac{0{\cdot}001 \text{(mole)}}{0{\cdot}05 \text{ (litre)}}$

$= 0{\cdot}02$ M.

Ex. 23. *When a current of* 1·5 amp *was passed for* 30 *minutes through a solution of a salt of a trivalent metal,* 1·071 g *of the metal was deposited at the cathode. Calculate the atomic weight of the metal.*

Solution : Charge passed through the solution $= 1{\cdot}5 \times 30 \times 60$

$= 2700$ coulombs.

Mole of electric charge passed $= \dfrac{2700}{96500}$ faraday $= \dfrac{27}{965}$ F.

$\therefore$ equivalent of the metal deposited $= \dfrac{27}{965}$.

Let the at. wt. of the metal be x.

$\therefore$ eq. wt. of the metal $= \dfrac{x}{3}$. ($\because$ valency of the metal = 3)

$\therefore$ weight of the metal deposited = no. of eq. $\times$ eq. wt.

$= \dfrac{27}{965} \times \dfrac{x}{3}$ g.

$\therefore$ as given, $\dfrac{27 \times x}{965 \times 3} = 1{\cdot}071$

$x = 114{\cdot}8$ amu.

Ex. 24. 19 g *of molten* $SnCl_2$ *is electrolysed for some time, using inert electrodes.* 0·119 g *of* Sn *is deposited at the cathode. No substance is lost during the electrolysis. Find the ratio of the weight of* $SnCl_2$ *and* $SnCl_4$ *after electrolysis.* (Sn = 119)

Solution : According to this problem $SnCl_2$ should first ionise into the ions of tin and chloride. Sn gets deposited at the cathode and as Cl_2, produced at anode, is not lost, it combines with the remaining $SnCl_2$ to give $SnCl_4$.

$$SnCl_2 \rightarrow Sn + Cl_2$$

$$SnCl_2 + Cl_2 \rightarrow SnCl_4$$

Equivalent of Sn deposited $= \dfrac{0{\cdot}119}{119/2} = 0{\cdot}002$. $\left(\text{eq. wt. of Sn } = \dfrac{119}{2}\right)$

$\therefore$ equivalent of $SnCl_2$ decomposed $= 0{\cdot}002$.

Equivalent of chlorine produced $= 0{\cdot}002$.

Equivalent of $SnCl_4$ formed $= 0{\cdot}002$.

$\therefore$ weight of $SnCl_4$ formed = no. of eq. $\times$ eq. wt.

$$= 0{\cdot}002 \times \frac{261}{2} = 0{\cdot}261 \text{ gram.}$$

$$\left(\text{eq. wt. of } SnCl_4 = \frac{261}{2}, \text{ in } \underset{+2}{SnCl_2} + Cl_2 \rightarrow \underset{+4}{SnCl_4}\right)$$

Further, eq. of chlorine produced

$= $ eq. of $SnCl_2$ which combined with $Cl_2 = 0{\cdot}002$.

$\therefore$ weight of $SnCl_2$ remained

= initial wt. of $SnCl_2$ – (wt. of $SnCl_2$ decomposed

+ wt. of $SnCl_2$ combined with Cl_2 to form $SnCl_4$)

= 19 – (eq. of $SnCl_2$ decomposed $\times$ eq. wt. of $SnCl_2$

+ eq. of $SnCl_2$ which combined with $Cl_2 \times$ eq. wt. of $SnCl_2$).

$$= 19 - \left[0{\cdot}002 \times \frac{190}{2} + 0{\cdot}002 \times \frac{190}{2}\right]$$

$$= 19 - 0{\cdot}38$$

$$= 18{\cdot}62 \text{ g.}$$

$\therefore$ $SnCl_2 : SnCl_4 = 18{\cdot}62 : 0{\cdot}261$. (weight ratio)

Ex. 25. *Ten grams of a fairly concentrated solution of cupric sulphate is electrolysed using* 0·01 *faraday of electricity. Calculate* (i) *the weight of the resulting solution and* (ii) *the number of equivalent of acid or alkali in the solution.* (Cu = 63·5, F = 96500 *coulombs*)

Solution : (i) Since mole of electric charge used = 0·01 F

$\therefore$ eq. of Cu deposited = 0·01

$\therefore$ wt. of Cu deposited $= 0{\cdot}01 \times 31{\cdot}75 = 0{\cdot}3175$ g.

$$\left(\text{eq. wt. of Cu} = \frac{63{\cdot}5}{2} = 31{\cdot}75\right)$$

Eq. of oxygen discharged = 0·01.

$\therefore$ wt. of oxygen discharged $= 0{\cdot}01 \times 8 = 0{\cdot}08$ g.

$$\left(\text{eq. wt. of O} = \frac{16}{2} = 8\right)$$

Total loss in weight due to electrolysis $= 0{\cdot}3175 + 0{\cdot}08$

$= 0{\cdot}3975$ g.

$\therefore$ the weight of the resulting solution $= 10 - 0{\cdot}3975 = 9{\cdot}6025$ g.

(ii) Electrolysis of $CuSO_4$ solution follows through

$$CuSO_4 = Cu^{2+} + SO_4^{2-}$$

At cathode: $Cu^{2+} + 2e^- \rightarrow Cu$

At anode: $SO_4^{2-} + H_2O \rightarrow H_2SO_4 + \frac{1}{2}O_2 \uparrow + 2e^-$

Thus H_2SO_4 is produced during electrolysis.

$\therefore$ eq. of H_2SO_4 produced $= 0{\cdot}01$. (due to passage of 0·01 faraday)

Ex. 26. *An acidic solution of* Cu^{2+} *salt containing* 0·4 g *of* Cu^{2+} *is electrolysed until all the copper is deposited. The electrolysis is continued for seven more minutes with the volume of the solution kept at* 100 mL *and the current at* 1·2 amp. *Calculate the volume of gases evolved at* NTP *during the entire electrolysis.* **(IIT 1989)**

Solution : Let us suppose that the acidic solution of Cu^{2+} salt contains H_2SO_4.

In the beginning of electrolysis, Cu will be deposited at the cathode and O_2 will discharge at anode.

$$Cu^{2+} + 2e \rightarrow Cu \qquad \text{(at cathode)}$$

$$\left.\begin{aligned} 2OH^- &\rightarrow 2OH + 2e \\ 2OH &\rightarrow H_2O + \frac{1}{2}O_2 \end{aligned}\right\} \qquad \text{(at anode)}$$

$\therefore$ equivalent of oxygen evolved = eq. of Cu deposited

$$= \frac{0{\cdot}4}{31{\cdot}8}.$$

$\therefore$ volume of O_2 (at NTP) evolved $= \frac{0{\cdot}4}{31{\cdot}8} \times 5600 = 70{\cdot}44$ mL.

(1 eq. of oxygen at NTP = 5600 mL)

During another seven minutes (420 s) of electrolysis, H_2 and O_2 will evolve at cathode and anode respectively.

Now, charge $= 1{\cdot}2 \times 420 = 504$ coulombs $= \frac{504}{96500}$ F.

$\therefore$ eq. of H_2 evolved $= \frac{504}{96500}\cdot$ (at cathode)

Volume of H_2 evolved at NTP $= \frac{504}{96500} \times 11200 = 58{\cdot}49$ mL.

(1 eq. of H_2 at NTP = 1200 mL)

Eq. of O_2 evolved $= \frac{504}{96500}\cdot$ (at anode)

$\therefore$ volume of O_2 evolved at NTP $= \frac{504}{96500} \times 5600 = 29{\cdot}24$ mL.

Thus, during the entire electrolysis,

$$H_2 \text{ evolved} = 58{\cdot}49 \text{ mL}$$
$$O_2 \text{ evolved} = (70{\cdot}44 + 29{\cdot}24) \text{ mL}$$
$$= 99{\cdot}68 \text{ mL}.$$

Ex. 27. *The electrode reactions for charging of a lead storage battery are:*

$$PbSO_4 + 2e^- = Pb + SO_4^{2-}$$
$$PbSO_4 + 2H_2O = PbO_2 + SO_4^{2-} + 4H^+ + 2e^-$$

The electrolyte in the battery is an aqueous solution of sulphuric acid. Before charging, the specific gravity of the liquid was found to be 1·11 (15·7% H_2SO_4 *by weight). After charging for* 100 *hours, the sp. gr. of the liquid was found to be* 1·28 (36·9 % H_2SO_4 *by wt.). If the battery contained two litres of the liquid, calculate the average current used for charging the battery.*

NB. (e^-) *denotes electrons. Assume that the volume of the battery liquid remained constant during charging.* 1 *faraday* = 96500 *coulombs.*

Solution : The overall battery reaction is

$$2PbSO_4 + 2H_2O = Pb + PbO_2 + 2H_2SO_4$$

∵ two moles of electrons are involved in the production of two moles of H_2SO_4

∴ equivalent wt. of H_2SO_4 = molecular wt. of H_2SO_4 = 98.

Before electrolysis;

∵ 100 g of H_2SO_4 solution contains 15·7 g of H_2SO_4

or $\frac{100}{1{\cdot}11}$ mL of H_2SO_4 solution contains $\frac{15{\cdot}7}{98}$ eq. of H_2SO_4.

∴ 2000 mL H_2SO_4 solution contains $\frac{15{\cdot}7}{98} \times \frac{1{\cdot}11}{100} \times 2000 = 3{\cdot}556$ eq.

Similarly, we get

number of equivalent of H_2SO_4 after electrolysis

$$= \frac{36{\cdot}9}{98} \times \frac{1{\cdot}28}{100} \times 2000 = 9{\cdot}6387.$$

∴ number of eq. of H_2SO_4 produced = (9·6387 − 3·5560) = 6·0827

∴ moles of electric charge used = 6·0827 faradays

$$\text{Average current used} = \frac{\text{charge in coulombs}}{\text{time in seconds}}$$
$$= \frac{6{\cdot}0827 \times 96500}{100 \times 60 \times 60} \text{ amp} = 1{\cdot}629 \text{ amp}.$$

Ex. 28. *During the discharge of a lead storage battery, the density of sulphuric acid fell from* 1·294 *to* 1·139 g/mL. *Sulphuric acid of density* 1·294 g/mL *is* 39%

H_2SO_4 *by wt. and that of density* 1·139 g/mL *is* 20% *by weight. The battery holds* 3·5 *litres of the acid and the volume remained practically constant during the discharge. Calculate the number of ampere-hours for which the battery must have been used. The charging and discharging reactions are*

$$Pb + SO_4^{2-} = PbSO_4 + 2e \quad (\textit{charging})$$

$$PbO_2 + 4H^+ + SO_4^{2-} + 2e^- = PbSO_4 + 2H_2O \ (\textit{discharging})$$ **(IIT 1986)**

Solution : The overall battery reaction is

$$Pb + PbO_2 + 2H_2SO_4 = 2PbSO_4 + 2H_2O$$

$\because$ two moles of electrons are involved for the reaction of two moles of H_2SO_4,

$\therefore$ eq. wt. of H_2SO_4 = mol. wt. of $H_2SO_4 = 98$.

Following in the same way as in Example 27 we get

no. of eq. of H_2SO_4 present in 3·5 litres of solution of a charged battery

$$= \frac{39}{98} \times \frac{1{\cdot}294}{100} \times 3500$$

$$= 18{\cdot}0235.$$

No. of equivalents of H_2SO_4 present in 3·5 litres of solution after getting discharged $= \frac{20}{98} \times \frac{1{\cdot}139}{100} \times 3500$

$$= 8{\cdot}1357.$$

Number of eq. of H_2SO_4 lost $= 18{\cdot}0235 - 8{\cdot}1357$

$$= 9{\cdot}8878.$$

$\therefore$ moles of electric charge produced by the battery $= 9{\cdot}8878$ F

$$= 9{\cdot}8878 \times 96500 \text{ coulombs}$$

$$= 9{\cdot}8878 \times 96500 \text{ amp-seconds}$$

$$= \frac{9{\cdot}8878 \times 96500}{60 \times 60} \text{ amp-hours}$$

$$= 265 \text{ amp-hours.}$$

Ex. 29. *A lead storage battery has initially* 200 g *of lead and* 200 g *of* PbO_2, *plus excess* H_2SO_4. *Theoretically, how long could this cell deliver a current of* 10 amp., *without recharging, if it were possible to operate it so that the reaction goes to completion?*

Solution : Discharging of battery takes place through the reaction

$$\underset{\frac{200}{207}\text{ mole}}{Pb} + \underset{\frac{200}{239}\text{ mole}}{PbO_2} + \underset{\text{(excess)}}{2H_2SO_4} = 2PbSO_4 + 2H_2O$$

As mole of PbO_2 is less than that of Pb, PbO_2 is the limiting reactant which shall be totally consumed.

No. of faraday delivered by the battery

$$= \text{no. of eq. of } PbO_2 \text{ lost}$$

$$= \frac{200}{239} \times 2.$$

$$\therefore \quad \text{charge} = \frac{400}{239} \times 96500 \text{ coulombs}$$

$$\therefore \quad \text{time of discharge} = \frac{400 \times 96500}{239} \times \frac{1}{10} \text{ seconds}$$

$$= \frac{400 \times 96500}{239 \times 10} \times \frac{1}{60 \times 60} \text{ h}$$

$$= 4.486 \text{ h.}$$

Ex. 30. *Suppose* 250 mL *of a* 0.433 M *solution of* $CuCl_2$ *is electrolysed. How long will a current of* 0.75 A *have to run in order to reduce the concentration of* Cu^{2+} *to* 0.167 M? *What mass of* Cu(s) *will be deposited on the cathode during this time?*

Solution : Mole of Cu^{2+} reduced $= 0.433 \times 0.250 - 0.167 \times 0.250$

$$= 0.0665.$$

Wt. of Cu deposited $= 0.0665 \times 63.5$ g

$$= 4.22 \text{ g.}$$

Let the time to deposit this amount of Cu be t seconds

$\therefore$ no. of F = no. of eq.

$$\frac{0.75 \times t}{96500} = 0.0665 \times 2$$

$$t = 1.71 \times 10^4 \text{ seconds.}$$

Ex. 31. *A current of* 10 A *is applied for* 1 h *to* 1 *litre of a solution containing* 1 *mole of* HCl. *Calculate the* pH *of the solution at the end of this time.*

Solution : At the cathode, H_2 is discharged. At the anode Cl_2 is preferentially discharged over O_2. Thus during electrolysis, HCl is lost from the solution.

Eq. of HCl lost = number of faradays used

$$= \frac{10 \times 60 \times 60}{96500} = 0.373$$

or, mole of HCl lost = 0.373.

Mole of HCl remained/litre $= 1.0 - 0.373 = 0.627.$

$\therefore$ $\text{pH} = -\log(0.627) = 0.20.$

Ex. 32. *Calculate the minimum number of kilowatt-hours of electricity required to produce* 1000 kg *of* Al *by electrolysis of* Al^{3+} *if the required emf is* 4.50 V.

Solution : The no. of faradays required for electrolysis

= number of eq. of Al deposited

$$= \frac{10^6}{27/3} F = \frac{10^6}{9} F.$$

$$\text{Charge} = \frac{10^6}{9} \times 96500 \text{ coulombs.}$$

$$= 1.07 \times 10^{10} \text{ C.}$$

$$\therefore \text{ electric energy} = 1.07 \times 10^{10} \times 4.50 \text{ J} = 4.815 \times 10^{10} \text{ J} \qquad (\because \text{J} = \text{C} \times \text{V})$$

$$\text{kilowatt-hours} = \frac{4.815 \times 10^{10}}{3.6 \times 10^6} \text{ kWh} \qquad (\because 1 \text{ kWh} = 3.6 \times 10^6 \text{ J})$$

$$= 1.34 \times 10^4 \text{ kWh.}$$

EXAMPLES ON CONDUCTANCE

Ex. 33. *The resistance of an* N/10 KCl *solution is* 245 *ohms. Calculate the specific conductance and the equivalent conductance of the solution if the electrodes in the cell are* 4 cm *apart and each having an area of* 7·0 sq cm.

Solution : Cell constant $= \frac{l}{a} = \frac{4}{7} \text{ cm}^{-1}$.

Specific conductance = conductance × cell constant

$$= \frac{1}{\text{resistance}} \times \frac{l}{a}$$

$$= \frac{1}{245} \times \frac{4}{7} = 2{\cdot}332 \times 10^{-3} \text{mho cm}^{-1}.$$

Equivalent conductance = sp. cond. × volume (cc)

(containing 1 eq. of the substance)

$$= 2{\cdot}332 \times 10^{-3} \times 10000 \qquad (\text{for N/10 solution, } v = 10{,}000 \text{ cc})$$

$$= 23{\cdot}32 \text{ mho cm}^2.$$

Ex. 34. *The resistance of a solution A is* 50 *ohms and that of solution B is* 100 *ohms, both solutions being taken in the same conductivity cell. If equal volumes of solution A and B are mixed, what will be the resistance of the mixture using the same cell? (Assume that there is no increase in the degree of dissociation of A and B on mixing.)*

Solution : Let K_1 and K_2 be the specific conductances of the solutions A and B respectively and the constant of the cell be x.

∴ for solution A: sp. conductance = conductance × cell constant

$$K_1 = \frac{1}{50} \times x \qquad \dots (1)$$

For solution B: sp. conductance $K_2 = \frac{1}{100} \times x.$ ... (2)

When equal volumes of A and B are mixed, both the solutions get doubly diluted; hence their individual contribution towards the sp. conductance of the mixture will be $\frac{K_1}{2}$ and $\frac{K_2}{2}$ respectively and the sp. conductance of the mixture will be $\frac{1}{2}(K_1 + K_2)$.

$\therefore$ for the mixture: $\frac{1}{2}(K_1 + K_2) = \frac{1}{R} \times x.$ ... (3)

(R is the resistance of mixture)

From eqns. (1), (2) and (3), we get $R = 66{\cdot}67$ ohms.

Ex. 35. *In a conductivity cell the two platinum electrodes, each of area* 10 sq cm, *are fixed* 1·5 cm *apart. The cell contained* 0·05 N *solution of a salt. If the two electrodes are just half dipped into the solution which has a resistance of* 50 *ohms, find equivalent conductance of the salt solution.*

Solution : Since the electrodes of the cell are just half dipped, the effective area will be 5 sq cm.

$$\text{Cell constant} = \frac{l}{a} = \frac{1{\cdot}5}{5} = 0{\cdot}3 \text{ cm}^{-1}.$$

$$\text{Specific conductance} = \text{conductance} \times \text{cell constant}$$

$$= \frac{1}{\text{resistance}} \times \text{cell constant}$$

$$= \frac{1}{50} \times 0{\cdot}3 = \frac{3}{500} \text{ mho cm}^{-1}.$$

Equivalent conductance = specific conductance × volume ... (Eqn. 8)

$$= \frac{3}{500} \times 20000 = 120 \text{ mho cm}^2.$$

(0·05 N = N/20 $\therefore$ V = 20,000 cc)

Ex. 36. *A big, irregular-shaped vessel contained water, the sp. conductance of which was* $2{\cdot}56 \times 10^{-5}$ mho cm^{-1}. 500 g *of* NaCl *was then added to the water and the sp. cond. after the addition of* NaCl, *was found to be* $3{\cdot}10 \times 10^{-5}$ mho cm^{-1}. *Find the capacity of the vessel if it is fully filled with water.* ($\Lambda^\circ_{NaCl} = 149{\cdot}9$)

Solution : Let the volume of the vessel be V cc.

$$\text{Number of equivalent of NaCl} = \frac{\text{wt. in grams}}{\text{eq. weight}}$$

$$= \frac{500}{58 \cdot 5} = 8 \cdot 547.$$

$\therefore$ volume of water (cc) containing 1 eq. of NaCl $= \dfrac{V}{8 \cdot 547}$.

The sp. cond. of the NaCl solution (only due to presence of Na^+ and Cl^- ions)

$$= 3 \cdot 10 \times 10^{-5} - 2 \cdot 56 \times 10^{-5}$$

$$= 0 \cdot 54 \times 10^{-5}.$$

$$\therefore \qquad \Lambda_{NaCl} = 0 \cdot 54 \times 10^{-5} \times \frac{V}{8 \cdot 547}. \qquad \text{... (Eqn. 8)}$$

Since the vessel is big, the resulting solution may be supposed to be dilute.

$$\therefore \qquad \Lambda_{NaCl} = \Lambda^{\circ}_{NaCl}$$

$$0 \cdot 54 \times 10^{-5} \times \frac{V}{8 \cdot 547} = 149 \cdot 9$$

$$V = 2 \cdot 37 \times 10^{8} \text{ cc.}$$

Ex. 37. *The equivalent conductance of* 0·10 N *solution of* $MgCl_2$ *is* 97·1 mho cm^2 $eq.^{-1}$ *at* 25°C. *A cell with electrodes that are* 1·50 cm^2 *in surface area and* 0·50 cm *apart is filled with* 0·1 N $MgCl_2$ *solution. How much current will flow when the potential difference between the electrodes is* 5 *volts?*

Solution : Cell constant $= \dfrac{0 \cdot 50}{1 \cdot 50} = \dfrac{1}{3}$.

$$\text{Specific conductance} = \frac{\text{equivalent conductance}}{\text{volume (cc) containing 1 eq.}} \qquad \text{... (Eqn. 8)}$$

$$= \frac{97 \cdot 1}{10{,}000} \qquad \text{(for 0·1 N solution } V = 10{,}000 \text{ cc)}$$

$$= 0 \cdot 00971 \text{ mho cm}^{-1}.$$

Conductance = specific conductance/cell constant ... (Eqn. 7)

$$= \frac{0 \cdot 00971}{1/3} = 0 \cdot 02913 \text{ mho.}$$

$$\therefore \qquad \text{resistance} = \frac{1}{0 \cdot 02913} \text{ ohm.}$$

$$\therefore \qquad \text{current in amp} = \frac{\text{potential difference (volt)}}{\text{resistance (ohm)}} \qquad \text{(Ohm's law)}$$

$$= \frac{5}{1/0 \cdot 02913} = 0 \cdot 1456 \text{ ampere.}$$

Ex. 38. *A* 0·01 D* *solution of* KCl *has a specific conductance value of* 0·00141 mho cm^{-1}. *A cell filled with this solution has a resistance of* 4·2156 *ohms.*

(a) What is the cell constant?

(b) The same cell filled with a solution of HCl *has a resistance of* 1·0326 *ohms. What is specific conductivity of the* HCl *solution?*

Solution : (a) Cell constant $= \dfrac{\text{specific conductance}}{\text{conductance}}$

$$= \frac{\text{specific conductance}}{1/\text{resistance}}$$

$$= \frac{0{\cdot}00141}{1/4{\cdot}2156} = 0{\cdot}00594 \text{ cm}^{-1}.$$

(b) Conductance of HCl solution $= \dfrac{1}{1{\cdot}0326}$ mho.

Since the same cell is used, we shall take the same value of cell constant in this case.

Sp. conductance = cell constant × conductance

$$= 0{\cdot}00594 \times \frac{1}{1{\cdot}0326}$$

$$= 0{\cdot}00575 \text{ mho cm}^{-1}.$$

Ex. 39. *The equivalent conductance at infinite dilution are sodium benzoate* = 82·4; *hydrochloric acid* = 426·2 *and sodium chloride* = 126·5 mho cm^2. *Calculate* $\Lambda°$ *for benzoic acid.*

Solution : From Kohlrausch's law, we get,

$$\Lambda°_{\text{benzoic acid}} = \Lambda°_{\text{sod. benzoate}} + \Lambda°_{\text{hydrochloric acid}} - \Lambda°_{\text{sodium chloride}}$$

$$= 82{\cdot}4 + 426{\cdot}2 - 126{\cdot}5$$

$$= 382{\cdot}1 \text{ mho cm}^2.$$

Ex. 40. *At* 18°C *the mobilities of* NH_4^+ *and* ClO_4^- *ions are* $6{\cdot}6 \times 10^{-4}$ *and* $5{\cdot}7 \times 10^{-4}$ cm^2 $volt^{-1}$ s^{-1} *at infinite dilution. Calculate equivalent conductance of ammonium chlorate solution.*

Solution : From Kohlrausch's law, we have,

$$\Lambda°_{NH_4ClO_4} = \lambda°_{NH_4^+} + \lambda°_{ClO_4^-}$$

$$= FU°_{NH_4^+} + FU°_{ClO_4^-} \qquad \text{... (Eqn. 11)}$$

$$= F(U°_{NH_4^+} + U°_{ClO_4^-})$$

$$= 96500\,(6{\cdot}6 \times 10^{-4} + 5{\cdot}7 \times 10^{-4})$$

$$= 118{\cdot}67 \text{ mho cm}^2.$$

---

* A one demal (1 D) solution contains 1 mole of salt in 1 litre solution at 0°C.

Ex. 41. *For* H^+ *and* Na^+ *the values of* $\lambda°$ *are* 349·8 *and* 50·11 *respectively. Calculate the mobilities of these ions and their velocities if they are in a cell in which the electrodes are* 5 cm *apart and to which a potential of* 2 volts *is applied.*

Solution : We have

$$U°_{H^+} = \frac{\lambda°_{H^+}}{F} = \frac{349 \cdot 8}{96500}$$

$$= 3 \cdot 62 \times 10^{-3} \text{ cm}^2 \text{ volt}^{-1} \text{ s}^{-1}$$

$$U°_{Na^+} = \frac{\lambda°_{Na^+}}{F} = \frac{50 \cdot 11}{96500}$$

$$= 5 \cdot 20 \times 10^{-4} \text{ cm}^2 \text{ volt}^{-1} \text{ s}^{-1}$$

Further, we know that

$$U° = \frac{\text{ionic velocity (cm/s)}}{\text{pot. diff. (volt) / distance between the electrodes (cm)}} \quad \text{... (Eqn. 12)}$$

$$\therefore \text{ velocity of H}^+ = 3 \cdot 62 \times 10^{-3} \times \frac{2}{5}$$

$$= 1 \cdot 45 \times 10^{-3} \text{ cm s}^{-1}.$$

$$\text{Velocity of Na}^+ = 5 \cdot 20 \times 10^{-4} \times \frac{2}{5}$$

$$= 2 \cdot 08 \times 10^{-4} \text{ cm s}^{-1}.$$

Ex. 42. *The equivalent conductance of an infinitely dilute solution of* NH_4Cl *is* 150 *and the ionic conductances of* OH^- *and* Cl^- *ions are* 198 *and* 76 *respectively. What will be the equivalent conductance of the solution of* NH_4OH *at infinite dilution. If the equivalent conductance of a* 0·01 N *solution of* NH_4OH *is* 9·6, *what will be its degree of dissociation?*

Solution : $\Lambda°_{NH_4Cl} = \lambda°_{NH_4^+} + \lambda°_{Cl^-}$

$$\therefore \quad \lambda°_{NH_4^+} = \Lambda°_{NH_4Cl} - \lambda°_{Cl^-} = 150 - 76 = 74$$

$$\therefore \quad \Lambda°_{NH_4OH} = \lambda°_{NH_4^+} + \lambda°_{OH^-} = 74 + 198 = 272$$

Further,

$$\text{degree of dissociation} = \frac{\Lambda_c}{\Lambda°} \quad \text{... (Eqn. 9)}$$

$$= \frac{9 \cdot 6}{272} = 0 \cdot 0353.$$

Ex. 43. *Calculate the dissociation constant of water at 25°C from the following data:*
Specific conductance of $H_2O = 5 \cdot 8 \times 10^{-8}$ mho cm^{-1}
$\lambda°_{H^+} = 350 \cdot 0$ *and* $\lambda°_{OH^-} = 198 \cdot 0$ mho cm^2

Solution : Suppose water contains X moles per litre (or X eq./L) of H^+ ions (or OH^- ions).

$\because$ X equivalents of H^+ ions are produced from X eq. of water

$\therefore$ volume (cc) containing 1 eq. of water which dissociated into its ions

$$= \frac{1000}{X}.$$

$\therefore$ eq. conductance of water = sp. cond. $\times V$

$$= 5{\cdot}8 \times 10^{-8} \times \frac{1000}{X}.$$

Since water dissociates feebly, i.e., water may be considered to be a dilute solution of H^+ and OH^- ions,

$$\Lambda_{H_2O} = \Lambda^{\circ}_{H_2O} = \lambda^{\circ}_{H^+} + \lambda^{\circ}_{OH^-}.$$

$$\therefore\ 5{\cdot}8 \times 10^{-8} \times \frac{1000}{X} = 350 + 198 = 548.$$

$$\therefore\ X = 1{\cdot}0 \times 10^{-7}.$$

$$\therefore\ [H^+] = [OH^-] = 1 \times 10^{-7}.$$

For the equilibrium,

$$H_2O = H^+ + OH^-$$

$$\text{Equilibrium constant } (K) = \frac{[H^+][OH^-]}{[H_2O]}$$

$$K_w = K \times [H_2O] = [H^+][OH^-]$$
$$= 1{\cdot}0 \times 10^{-7} \times 1{\cdot}0 \times 10^{-7} = 1 \times 10^{-14}.$$

$$\therefore\ K = \frac{K_w}{[H_2O]} = \frac{1 \times 10^{-14}}{55{\cdot}5} = 1{\cdot}8 \times 10^{-16} \text{ mole/litre.}$$

$$\left([H_2O] = \frac{1000}{18} = 55{\cdot}5 \text{ moles/litre} \right)$$

Ex. 44. *Calculate K_a of acetic acid if its* 0·05 N *solution has equivalent conductance of* 7·36 mho cm^2 *at* 25°C. ($\Lambda^{\circ}_{CH_3COOH} = 390{\cdot}7$)

Solution : Degree of dissociation $(x) = \dfrac{\Lambda_c}{\Lambda_0} = \dfrac{7{\cdot}36}{390{\cdot}7} = 0{\cdot}0188.$

For the equilibrium

0·05		0		0	Initial concn. (moles/litre)
CH_3COOH	=	CH_3COO^-	+	H^+	
0·05 $(1 - x)$		0·05 x		0·05 x	Equilibrium concentration

(for CH_3COOH, 0·05 N = 0·05 M)

$$K_a = \frac{0{\cdot}05x \times 0{\cdot}05x}{0{\cdot}05(1 - x)}$$

Since x is very small,

$$K_a = 0{\cdot}05x^2 = 0{\cdot}05 \times (0{\cdot}0188)^2$$

$$= 1{\cdot}76 \times 10^{-5} \text{ mole/litre.}$$

Ex. 45. *The specific conductance of a saturated solution of* AgCl *at* 25°C *after subtracting the specific conductance of conductivity of water is* $2{\cdot}28 \times 10^{-6}$ mho cm^{-1}. *Find the solubility product of* AgCl *at* 25°C.
($\Lambda°_{AgCl} = 138{\cdot}3$ mho cm^2)

Solution : For equilibrium,

$$AgCl = Ag^+ + Cl^-$$

$$K_{sp} = [Ag^+][Cl^-]$$

If the solubility of AgCl in water is, say, x moles/litre or x eq./L,

$$K_{sp} = x \cdot x = x^2.$$

$\therefore$ volume containing 1 eq. of AgCl $= \dfrac{1000}{x}$

$$\Lambda_{AgCl} = \text{sp. cond.} \times V$$

$$= 2{\cdot}28 \times 10^{-6} \times \frac{1000}{x}.$$

Since AgCl is sparingly soluble in water, $\Lambda_{AgCl} = \Lambda°_{AgCl} = 138{\cdot}3$.

$$\therefore \; 2{\cdot}28 \times 10^{-6} \times \frac{1000}{x} = 138{\cdot}3$$

or $\quad x = 1{\cdot}644 \times 10^{-5}$ eq./litre or mole/litre.

$$K_{sp} = x^2 = (1{\cdot}644 \times 10^{-5})^2$$

$$= 2{\cdot}70 \times 10^{-10} \text{ (mole/litre)}^2.$$

PROBLEMS

(Answers bracketed with questions)

1. Molten $AlCl_3$ is electrolysed with a current of 0·5 amp to produce 27·0 g Al. (i) How many g eq. of Al were produced? (ii) How many gram-atoms of Al were produced? (iii) How many atoms of Al were produced? (iv) How many electrons were required? (v) What is the no. of faradays of electricity consumed? (vi) How long did the electrolysis take place? (vii) How many litres of Cl_2 at NTP were produced?

{(i) 3 g eq (ii) 1 (iii) Av. constant (iv) 3 × Av. constant (v) 3 F (vi) 160 h 50 min (vii) 33·6 litres}

2. In the electrolysis of an aqueous $Cr_2(SO_4)_3$ solution using a current of 2 amp, the mass of cathode is increased by 8 g. How long was electrolysis conducted?

(6·19 h)

3. A spoon used as a cathode is dipped in $AgNO_3$ solution and a current of 0·2 amp is passed for one hour. Calculate
(a) how much silver plating has occurred?
(b) how many electrons were involved in the process?
(c) what amount of copper would have been plated under similar conditions?
($0{\cdot}805$ g, $4{\cdot}5 \times 10^{21}$, 0·237 g)

4. A steady current passing through a solution of $AgNO_3$ solution deposits 0·50 g of Ag in 1 h. Calculate the number of coulombs. What volume of hydrogen at 27°C and 750 mm pressure would the same current liberate in one hour?
(446·7 coulombs, 57·7 mL)

5. At the Nangal fertilizer plant in Punjab, hydrogen is produced by the electrolysis of water. The hydrogen is used for the production of ammonia and nitric acid (by oxidation of ammonia). If the average production of ammonium nitrate is 5000 kg/day, estimate the daily consumption of electricity per day.
($2{\cdot}8 \times 10^5$ amp/day)

6. In an electrolysis experiment, current was passed for 5 hours through two cells connected in series. The first cell contains a solution of gold and second contains $CuSO_4$ solution. 9·85 g of gold was deposited in the first cell. If the oxidation number of gold is +3, find the amount of Cu deposited on the cathode of the second cell. Also calculate the magnitude of the current in ampere. (I F = 96500 coulomb) (Au = 197, Zn = 65·4)
(4·765 g, 0·8037 amp)

7. A constant current flowed for 2 hours through a potassium iodide solution oxidising the iodide ion to iodine ($2I^- \rightarrow I_2 + 2e^-$).
At the end of the experiment, the iodine was titrated with 21·75 mL of 0·0831 M sodium thiosulphate solution

$$(I_2 + 2S_2O_3^{2-} \rightarrow 2I^- + S_4O_6^{2-}).$$

What was the average rate of current flow in amperes? (0·0242 amp)

8. During the electrolysis of $CrCl_3$, chlorine gas is evolved at the anode and chromium is deposited at the cathode. How many grams of Cr and how many litres of chlorine (at NTP) are produced, when a current of 6 amperes is passed for one hour?
(3·88 g, 2·507 litres)

9. A current passes through two cells containing respectively—(i) $CuSO_4$ solution between Cu electrode and (ii) $AgNO_3$ solution between Pt electrodes. Calculate the loss or gain in weight of the different electrodes in the time in which 40 cc of oxygen at NTP collects in the second cell.
(Ag = 108 ; Cu = 63·5)

First cell: { 0·228 g Cu deposited at cathode. / 0·228 g Cu dissolved out from anode. }

Second cell: { 0·778 g of Ag deposited at cathode. / Wt. of anode does not change. }

10. What weight of water will be decomposed by a current of 100 amp in 12 h?
(403 g)

11. How many grams of H_2 and O_2 are produced during the electrolysis of water under a 1·30 amp of current for 5 hours? What volumes of dry gases are produced at NTP?

$$\left\{\begin{array}{l} H_2 - 0{\cdot}245\text{ g}, O_2 - 1{\cdot}94\text{ g} \\ H_2 - 2{\cdot}72\text{ L}, O_2 - 1{\cdot}36\text{ L} \end{array}\right\}$$

12. The water is electrolysed in a cell, hydrogen is liberated at one electrode and oxygen is simultaneously liberated at the other. In a particular experiment hydrogen and oxygen so produced were collected together and the total volume measured 16·8 mL at NTP. How many coulombs were passed through the cell in the experiment? (96·5 coulombs)

13. Ag is electrodeposited on a metallic vessel of surface area 800 cm^2 by passing a current of 0·2 amp for three hours. Calculate the thickness of silver deposited, given its density is 10·47 g/cc. ($2{\cdot}88 \times 10^{-4}$ cm)

14. 50 mL of hydrogen gas was collected over at 23°C, 740 mmHg barometric pressure. H_2 was produced by the electrolysis of water. The voltage was constant at 2·1 volts, the current averaged 0·50 amp for 12 minutes and 20 seconds. Calculate Avogadro constant. ($5{\cdot}85 \times 10^{23}$)

15. A current of 0·5 amp is sent through a solution of $CuSO_4$ for 20 minutes using Pt electrodes.

(a) Calculate the weight of Cu deposited.

(b) Find out the total number of copper atoms deposited.

(Cu = 63·57, Av. constant = $6{\cdot}022 \times 10^{23}$, 1 F = 96500 coulombs) (0·1976 g; $1{\cdot}87 \times 10^{21}$)

16. What current is required to pass 1 mole of electrons per hour through an electrolytic bath? How many grams of Al and Cd will be liberated by 1 mole of electrons? (26·8 amp, Al – 8·99 g, Cd – 56·2 g)

17. How many hours are required for a current of 3 amp to decompose electrolytically 18 g of water? (18 hours)

18. 50 mL of a 0·1 M $CuSO_4$ solution is electrolysed for 12 minutes at a current of 0·06 amp. If Cu is produced at one electrode and O_2 at the other, what will be the pH of the final solution? For $HSO_4^- = H^+ + SO_4^{2-}$, $K_{diss} = 1{\cdot}3 \times 10^{-2}$. (2·95)

19. For the electrolytic production of $NaClO_4$ from $NaClO_3$ as per reactions:

$$ClO_3^- + H_2O \rightarrow ClO_4^- + 2H^+ + 2e^-$$

(i) How many faradays of electricity would be required to produce 1 mole of $NaClO_4$?

(ii) What volume of H_2 at STP would be liberated at the cathode in the time that it takes to form 12·25 g of $NaClO_4$? (2 F, 2·24 litres)

20. Electric current is passed through two cells 'A' and 'B' in series. Cell 'A' contains an aqueous solution of Ag_2SO_4 and platinum electrodes. The cell 'B' contains aqueous solution of $CuSO_4$ and Cu electrodes. The current is passed till 1·6 g of oxygen is liberated at the anode of cell 'A'.

(i) Give equations for the reaction taking place at each electrode.

(ii) Calculate the quantities of substances deposited at the cathodes of the two cells. (Ag = 108, Cu = 63·5)

(i) In cell 'A' ; $2H_2O \rightarrow O_2(g) + 4H^+ + 4e$ at anode

$Ag^+ + e \rightarrow Ag$ at cathode

In cell 'B' ; $Cu \rightarrow Cu^{2+} + 2e$ at anode

$Cu^{2+} + 2e \rightarrow Cu$ at cathode

(ii) Ag = 21·6 g; Cu = 6·35 g

21. One hundred millilitres of 0·8 M copper sulphate is electrolysed for 30 minutes by passing a current of 5 amp. Calculate the amount of copper sulphate in grams in the solution. (5·32 g)

22. Anthracene can be oxidised anodically to anthraquinone. What weight of anthraquinone can be produced by the passage of a current of 1 amp for 60 minutes if the current efficiency is 100%? (1·2932 g)

[**Hint:** $C_{14}H_{10} + 3O \rightarrow C_{14}H_8O_2 + H_2O$

$$\text{Eq. wt. of } C_{14}H_8O_2 = \frac{\text{mol. wt.}}{6}$$]

23. To reduce nitrobenzene to aniline, 20 g of $C_6H_5NO_2$, 30 cc of an alcohol, 250 cc of water, 11 g of HCl and 1 g of $SnCl_2 \cdot 2H_2O$ were placed in the cathode space. After passing current at a rate of 26·5 amp-hour through the lead cathode electrolytic cell, 12·76 g of aniline was produced. Determine the current yield. (84·38%)

[**Hint:** See Example 13]

24. 0·2964 g of Cu was deposited on passage of a current of 0·5 amp for 30 min through a solution of copper sulphate. Calculate the atomic weight of Cu. (63·56)

25. Most of the copper used to make wire has been electrically refined by depositing it from copper salts solution (divalent) on to a cathode. What is the cost of electrical energy required per kg of copper if the cost of electricity is Rs. 0·25 per kWh and the cell operates at 0·2 volt? The electrochemical equivalent of copper is 0·00033 g/coulomb. (4·2 p)

[**Hint:** No. of eq. in 1 kg of Cu × 0·25/no. of eq. of Cu deposited by 1 kWh = cost in rupees.]

26. How long should a current of 0·5 amp be passed through 50 mL of a 0·10 M NaCl solution in order to make its pH 12, assuming no volume change? (97 s)

27. Lake Cayuga has a volume of water estimated to be $8{\cdot}2 \times 10^{12}$ litres. A power station not so far above Cayuga's waters produces electricity at the rate of $1{\cdot}5 \times 10^6$ coulombs per second at an appropriate voltage. How long would it take to electrolyse the lake? (1·9 million years)

28. A 200-watt, 110-volt incandescent lamp is connected in series with an electrolytic cell of negligible resistance containing a solution of zinc chloride. What weight of zinc will be deposited from the solution on passing the current for 30 minutes? (Zn = 63·5; 1 faraday = 96500 coulombs) (1·075 g)

29. 40 mL of 0·125 M of $NiSO_4$ solution is electrolysed by a current of 0·05 amp for 40 minutes. (i) Write the equation for the reactions occurring at each electrode. (ii) How many coulombs of electricity passed through the solution? (iii) How many grams of the product deposited on the cathode? (iv) How long will the same current have to pass through the solution to remove completely the metal ions from the solution? (v) At the end of electrolysis how many grams of the product would appear at the anode? [120, 0·037 g, 19300 s, 0·08 g]

30. An electric current is passed through a solution of (i) silver nitrate, (ii) solution of 10 g of copper sulphate ($CuSO_4 \cdot 5H_2O$) crystals in 500 mL of water, platinum electrodes being used in each case. After 30 minutes it was found that 1·307 g of silver has been deposited. What was the concentration of copper, expressed as grams of copper per litre in the copper sulphate solution after electrolysis? (4·32 g/litre)

31. In a fuel cell hydrogen and oxygen react to produce electricity. In the process hydrogen gas is oxidised at the anode and oxygen at the cathode. If 67·2 litres of H_2 at STP react in 15 minutes, what is the average current produced? If the entire current is used for electrodeposition of copper from Cu (II) solution, how many grams of copper will be deposited?

Anode reaction: $H_2 + 2OH^- \rightarrow 2H_2O + 2e^-$

Cathode reaction: $O_2 + 2H_2O + 4\,e^- \rightarrow 4\,OH^-$ **(IIT 1988)** (643·34 amp, 190·5 g)

[**Hint:** Eq. of hydrogen = no. of faradays of electricity = eq. of Cu deposited]

32. 3 amp of current was passed through an aqueous solution of an unknown salt of Pd for 1 hour. 2·977 g of Pd^{n+} was deposited at the cathode. Find n. (Pd = 106·4) (4)

33. A total of 69500 C of electricity was required to reduce 37.7 g of M^{3+} to the metal. What is M? (M = 157)

34. A solution containing Cu(I), Ni and Zn cyanide complexes was electrolysed and a deposit of 0.175 g was obtained. The deposit contained 72.8% Cu, 4.3% Ni and 22.9% Zn. No other element was released. Calculate the number of coulombs passed through the solution. (335·9 C)

35. Calculate the minimum number of kWh of electricity required to produce 1.0 kg of Mg from electrolysis of molten $MgCl_2$ if the applied emf is 5.0 V. (1 kWh = 3.6×10^6 J) (11.0 kWh)

36. A sample of Al_2O_3 dissolved in a molten fluoride bath is electrolysed using a current of 1.20 A. What is the rate of production of Al in g/hour? The oxygen liberated at the positive carbon electrode reacts with the carbon to form CO_2. What mass of CO_2 is produced per hour? (0·403, 0·4924 g/h)

37. The specific conductance of an N/10 KCl solution at 18°C is $1{\cdot}12 \times 10^{-2}$ mho cm^{-1}. The resistance of the solution contained in the cell is found to be 65 ohms. Calculate the cell constant. (0·728 cm^{-1})

38. When a solution of conductance 1·342 mho $metre^{-1}$ was placed in a conductivity cell with parallel electrodes, the resistance was found to be 170·5 ohm. The area of the electrodes is $1{\cdot}86 \times 10^{-4}$ sq metre. Calculate the distance between the two electrodes in metres. ($4{\cdot}25 \times 10^{-2}$ metres)

39. The resistances of two electrolytes, X and Y, were found to be 45 and 100 ohms respectively when equal volumes of both the solutions were taken in the same cell in two different experiments. If equal volumes of these solutions are mixed in the same cell, what will be the conductance of the mixture?

[**Hint:** See Example 34] (0·016 mho)

40. The resistance of an aqueous solution containing 0·624 g of $CuSO_4 \cdot 5H_2O$ per 100 cm^3 of the solution in a conductance cell of cell constant 153·7 per metre is 520 ohms at 298 K. Calculate the molar conductivity.

($CuSO_4 \cdot 5H_2O = 249{\cdot}5$) (118·2 mho cm^2)

41. An aqueous solution of 0·02 M KCl solution is filled in a 25-cm-long capillary tube of internal radius 0·01 cm. The solution was found to have a specific conductance of 0·0027 mho cm^{-1}. What will be the current in amp when a potential of 2 volts is applied across the capillary tube? ($6{\cdot}78 \times 10^{-8}$ amp)

42. Given the equivalent conductance of sodium butyrate, sodium chloride and hydrogen chloride as 83, 127 and 426 mho cm^2 at 25°C respectively. Calculate the equivalent conductance of butyric acid at infinite dilution. (382 mho cm^2)

43. A dilute solution of KCl was placed between two platinum electrodes 10 cm apart, across which a potential of 6 volts was applied. How far would the K^+ ion move in 2 hours at 25°C? Ionic conductance of K^+ ion at infinite dilution at 25°C is 73·52 mho cm^2. (3·29 cm)

44. For 0·0128 N solution of acetic acid at 25°C, equivalent conductance of the solution is 1·4 mho cm^2 eq^{-1} and $\Lambda_0 = 391$ mho cm^2 eq^{-1}. Calculate dissociation constant (K_a) of acetic acid. ($1{\cdot}6 \times 10^{-7}$)

45. The specific conductance at 25°C of a saturated solution of AgCl in water is $1{\cdot}826 \times 10^{-6}$ mho cm^{-1}. If Λ°_{AgCl} is equal to 138·26 mho cm^2, find out the solubility of AgCl in water in grams per litre. ($1{\cdot}89 \times 10^{-3}$ g/L)

46. Specific conductance of pure water at 25°C is $0{\cdot}58 \times 10^{-7}$ mho cm^{-1}. Calculate ionic product of water (K_w) if ionic conductances of H^+ and OH^- ions at infinite dilution are 350 and 198 mho cm^2 respectively at 25°C. (1×10^{-14}) $(mole/litre)^2$

47. The molar conductivity of solution of a weak acid HX (0.01 M) is ten times smaller than the molar conductivity of a solution of a weak acid HY (0.1 M). If $\lambda^\circ_{x^-} = \lambda^\circ_{y^-}$, the difference in their pK_a values, {pK_a(HX) − pK_a(HY)} is ... .

(Consider degree of dissociation of both acids to be << 1. **(IIT 2015 Adv.)** (3)

[**Hint:** $\dfrac{\Lambda_{HX}}{\Lambda_{HY}} = \dfrac{1}{10} = \dfrac{\alpha_{HX}/\Lambda^{\circ}_{HX}}{\alpha_{HY}/\Lambda^{\circ}_{HY}} \quad \because \quad \alpha = \dfrac{\Lambda_c}{\Lambda^{\circ}}$

As $\overset{\circ}{\lambda}_{x^-} = \overset{\circ}{\lambda}_{y^-}$, $\overset{\circ}{\Lambda}_{HX} = \overset{\circ}{\Lambda}_{HY} \quad \therefore \quad \dfrac{\alpha_{HX}}{\alpha_{HY}} = \dfrac{1}{10}$

$\dfrac{(K_a)_{HX}}{(K_a)_{HY}} = \dfrac{0.01\alpha^2_{HX}}{0.1\alpha^2_{HY}}$]

48. The conductance of a 0.0015 M aqueous solution of a weak monobasic acid was determined by using a conductivity cell consisting of Pt electrodes. The distance between the electrods is 120 cm with an area of cross section of 1 cm^2. The conductance of this solution was found to be 5×10^{-7} S. The pH of the solution is 4. The value of limiting molar conductivity (Λ°) of this monobasic acid in aqueous solution is $Z \times 10^2$ S cm^{-1} mol^{-1}. The value of Z is ... . **(IIT 2017 Adv.)** (6)

Objective Problems

1. The number of electrons involved in the reaction when one faraday of electricity is passed through an electrolyte is

(a) 12×10^{46} (b) 96,500 (c) 6×10^{23} (d) 8×10^{16}

2. Number of electrons involved in the electrodeposition of 63·5 g of Cu from a solution of $CuSO_4$ is

(a) $6{\cdot}022 \times 10^{23}$ (b) $3{\cdot}011 \times 10^{23}$
(c) $12{\cdot}044 \times 10^{23}$ (d) $6{\cdot}022 \times 10^{22}$

3. Faraday's laws of electrolysis are related to the
(a) atomic number of the cation
(b) atomic number of the anion
(c) equivalent weight of the electrolyte
(d) speed of the cation **(IIT 1985)**

4. The electric charge for electrodeposition of 1 eq. of a substance is
(a) one ampere per second (b) 96500 coulombs per second
(c) one ampere for one hour (d) charge on one mole of electrons

5. 1 coulomb of electricity produces m kg of a substance X. Electrochemical equivalent of X is

(a) m (b) $m \times 10^3$ (c) $m \times 10^{-3}$ (d) all wrong

6. Electrochemical equivalent of a substance is 0·0006735; its eq. wt. is
(a) 65 (b) 67·35
(c) 130 (d) cannot be calculated

7. When electricity is passed through a solution of $AlCl_3$, 13·5 g of Al is deposited. The number of faradays must be
(a) 1·0 (b) 1·5 (c) 0·5 (d) 2·0

8. 3·17 g of a substance was deposited by the flow of 0·1 mole of electrons. The equivalent weight of the substance is
(a) 3·17 (b) 0·317 (c) 317 (d) 31·7

9. A current of 0·5 ampere when passed through $AgNO_3$ solution for 193 seconds deposited 0·108 g of Ag. The equivalent weight of Ag is
(a) 10·8 (b) 108 (c) 54 (d) 1

10. In the electrolysis of an aqueous solution of a salt, the pH in the space near one of the electrodes increased. A solution of which of the following salts was electrolysed?
(a) $Cu(NO_3)_2$ (b) $CuCl_2$ (c) KCl (d) none of these

11. In the electrolysis of $CuCl_2$ solution (aq) with Cu electrodes, the weight of cathode increased by 3·2 g. In the anode,
(a) 0·05 mole of Cu^{2+} will go into the solution
(b) 560 mL of O_2 will be liberated
(c) 112 mL of Cl_2 will be liberated
(d) 3·2 mole of Cu^{2+} will go into the solution

12. The current required to displace 0·1 g of H_2 in 10 seconds will be
(a) 9·65 amp (b) 1·988 amp (c) 198 amp (d) 965 amp

13. The charge of an electron is $1{\cdot}6 \times 10^{-19}$ coulomb. How many electrons per second pass through a cross section of a Cu wire carrying 10^{-16} amp?
(a) 300 (b) 1800 (c) 1200 (d) 600

14. 96500 coulombs deposit 107·9 g of Ag from its solution. If $e = 1{\cdot}6 \times 10^{-19}$ coulomb, calculate the number of electrons per mole of electrons
(a) 96500 (b) $1{\cdot}6 \times 10^{-19}$ (c) $6{\cdot}02 \times 10^{23}$ (d) $6{\cdot}02 \times 10^{-23}$

15. A current of 2 amp passing for 5 hours through a molten tin salt deposits 22·2 g of tin. The oxidation state of the tin in the salt is
(a) + 4 (b) + 3 (c) + 2 (d) + 1

16. The cost at 5 paise/kWh of operating an electric motor for 8 hours which takes 15 amp at 110 V is
(a) Rs 66 (b) 66 paise (c) 37 paise (d) Rs 6·60
[Hint: $W = I.V$**]**

17. One faraday of current was passed through the electrolytic cells placed in series containing solutions of Ag^+, Ni^{++} and Cr^{+++} respectively. The amounts of Ag (at. wt. = 108), Ni (at. wt. = 59) and Cr (at. wt. = 52) deposited will be

	Ag	Ni	Cr
(a)	108 g	29·5 g	17·4 g
(b)	108 g	59·0 g	52·0 g
(c)	108 g	108·0 g	108·0 g
(d)	108 g	117·5 g	166·0 g

18. The time required to remove electrolytically one-fourth of Ag from 0·2 litre of 0·1 M $AgNO_3$ solution by a current of 0·1 amp is
(a) 320 min (b) 160 min (c) 80 min (d) 40 min

19. In the electrolysis of H_2SO_4, 9·72 litres and 2·35 litres of H_2 and O_2 were liberated. Number of equivalent of persulphuric acid ($H_2S_2O_8$) produced is
(a) 0·448 (b) 0·224 (c) 0·868 (d) 0·42

[Hint: Reactions are $2H_2SO_4 \rightarrow H_2S_2O_8 + H_2$

$$2H_2O \rightarrow 2H_2 + O_2\ ]$$

20. In the electrolysis of H_2O, 11·2 litres of H_2 was liberated at cathode at NTP. How much O_2 will be liberated at anode under the same conditions?
(a) 11·2 litres (b) 22·4 litres (c) 32 g (d) 5·6 litres

21. A galvanic cell was operated under almost ideally reversible conditions at a current of 10^{-16} amp. How long would it take to deliver 1 mole of electrons?
(a) $9{\cdot}65 \times 10^{20}$ s (b) 96,500 s
(c) $9{\cdot}65 \times 10^{-12}$ s (d) none of these

22. The time required for a current of 3 amp to decompose electrolytically 18 g of H_2O is
(a) 18 hours (b) 36 hours (c) 9 hours (d) 18 seconds

23. A constant current of 1.50 amp is passed through an electrolytic cell containing 0.10 N solution of $AgNO_3$ and a silver anode and a platinum cathode are used. After some time, the concentration of the $AgNO_3$ solution may be
(a) equal to 0.10 M (b) less than 0.10 M
(c) greater than 0.10 M (d) none of these

24. Which of the following processes is used in the extraction metallurgy of Mg?
(a) Fused salt electrolysis (b) Self-reduction
(c) Aqueous solution electrolysis (d) Thermite reduction **(IIT 2002)**

25. In which of the following aqueous solutions, H_2 and O_2 are not liberated at cathode and anode respectively on electrolysis using inert electrodes?
(a) H_2SO_4 solution (b) NaOH solution
(c) Na_2SO_4 solution (d) $AgNO_3$ solution

[Hint: Read text for preferential discharge of ions.)

26. The aqueous solutions of the following substances were electrolysed. In which case, the pH of the solution does not change if inert electrodes are used.
(a) $AgNO_3$ (b) $CuSO_4$ (c) NaCl (d) K_2SO_4

27. Electrolytic conduction is due to the movement of
(a) molecules (b) atoms (c) ions (d) electrons

28. Which of the following equations is correct?
(a) Cond. = sp. cond. × cell constant
(b) Eq. cond. = sp. cond. × cell constant
(c) Cond. = eq. cond. × cell constant
(d) Cell constant = sp. cond./cond.

29. Which of the following solutions of KCl has the lowest value of specific conductance?
(a) 1 M (b) 0·1 M (c) 0·01 M (d) 0·001 M

30. Which of the following solutions of KCl has the lowest value of equivalent conductance?
(a) 1 M (b) 0·1 M (c) 0·01 M (d) 0·001 M

31. Under which of the following conditions, conductance, sp. conductance and eq. conductance are all equal?
(a) 1000 cc of the solution contains 1 eq. of the electrolyte
(b) 100 cc of the solution contains 1 eq. of the electrolyte
(c) 10 cc of the solution contains 1 eq. of the electrolyte
(d) 1 cc of the solution contains 1 eq. of the electrolyte

32. Which of the following is the unit of eq. conductance?
(a) mho (b) mho cm^2 (c) mho cm^{-1} (d) mho cm^{-2}

33. If V, in the equation $\Lambda = \text{sp. cond.} \times V$, is the volume in cc containing 1 eq. of the electrolyte; V for a $\frac{N}{10}$ solution will be
(a) 10 cc (b) 100 cc (c) 1000 cc (d) 10,000 cc

34. At infinite dilution, the eq. conductances of CH_3COONa, HCl and CH_3COOH are 91, 426 and 391 mho cm^2 respectively at 25°C. The eq. conductance of NaCl at infinite dilution will be
(a) 126 (b) 209 (c) 391 (d) 908

35. An aqueous solution of X is added slowly to an aqueous solution of Y as shown in List I. The variation in conductivity of these reactions is given in List II. Match List I with List II and select the correct answer using the code given below the Lists.

List I	List II
P. $(C_2H_5)_3N + CH_3COOH$ (X, Y)	1. Conductivity decreases and then increases.
Q. KI(0.1 M) + $AgNO_3$(0.01 M) (X, Y)	2. Conductivity decreases and then does not change much.
R. CH_3COOH + KOH (X, Y)	3. Conductivity increases and then does not change much.
S. NaOH + HI (X, Y)	4. Conductivity does not change much and then increases.

Codes

	P	Q	R	S		P	Q	R	S
(a)	3	4	2	1	(b)	4	3	2	1
(c)	2	3	4	1	(d)	1	4	3	2

[**Hint:** See Mod. App. to Phy. Chemistry, Vol. I Ch. 7.] **(IIT 2013 Adv.)**

36. Resistance of 0.2 M solution of an electrolyte is 50 Ω. The specific conductance of the solution of 0.5 M solution of same electrolyte is 1.4 S m^{-1} and resistance of same solution of the same electrolyte is 280 Ω. The molar conductivity of 0.5 M solution of the electrolyte is 5 m^2 mol^{-1} is
(a) 5×10^{-4} (b) 5×10^{-3}
(c) 5×10^3 (d) 5×10^2 **(IIT 2014 Main)**

37. The equivalent conductance of NaCl at concentration, C and at infinite dilution are Λ_c and Λ_∞ respectively. The correct relationship between Λ_c and Λ_∞ is given by (where the constant B is positive)

(a) $\Lambda_c = \Lambda_\infty + (B)C$ (b) $\Lambda_c = \Lambda_\infty - (B)C$ (c) $\Lambda_c = \Lambda_\infty - (B)\sqrt{C}$ (d) $\Lambda_c = \Lambda_\infty + (B)\sqrt{C}$

[**Hint:** Read Mod. App. to Phy. Chemistry vol I, Ch.7] **(IIT 2014 Main)**

38. Two faraday of electricity is passed through a solution of $CuSO_4$. The mass of Cu deposited at the cathode is (Atomic mass of Cu = 63.5 u)

(a) 0 g (b) 63.5 g (c) 2 g (d) 127 g

(IIT 2015 Main)

39. How long (approximate) water be electrolysed by passing through 100 amperes current so that the oxygen released can completely burn 27.66 g of diborane? (B = 10.8 u)

(a) 1.6 hours (b) 6.4 hours (c) 0.8 hours (d) 3.2 hours

[**Hint:** $B_2H_6 + 3O_2 \rightarrow B_2O_3 + 3H_2O$] **(IIT 2018 Main)**

40. When an electric current is passed through acidified water, 112 mL of hydrogen gas at NTP was collected at the cathode in 965 seconds. The current passed, in ampere, is

(a) 0.5 (b) 0.1 (c) 1.0 (d) 2.0

(IIT 2018 Main)

Answers

1-c, **2**-c, **3**-c, **4**-d, **5**-b, **6**-a, **7**-b, **8**-d, **9**-b, **10**-c, **11**-a, **12**-d, **13**-d, **14**-c, **15**-c, **16**-b, **17**-a, **18**-c, **19**-a, **20**-d, **21**-a, **22**-a, **23**-a, **24**-a, **25**-d, **26**-d, **27**-c, **28**-d, **29**-d, **30**-a, **31**-d, **32**-b, **33**-d, **34**-a, **35**-a, **36**-a, **37**-c, **38**-b, **39**-d, **40**-c.

❑

CHAPTER NINE

ESTIMATION OF ELEMENTS IN ORGANIC COMPOUNDS

In the estimation of elements in organic compounds, the atoms of the element to be estimated, present in the organic compound are quantitatively converted to another compound and then the number of moles of atoms of the element in that compound is calculated by the mole method, e.g., C and H in the organic compound are converted to CO_2 and H_2O respectively. The moles of C in CO_2 and the moles of H in H_2O are calculated which give the moles of C and H present in the organic compound. The same method is applied to the estimation of other elements, the moles of which are calculated as given below:

$$C \xrightarrow{\text{combustion}} CO_2\text{; moles of C in } CO_2 = 1 \times \text{moles of } CO_2$$

(∵ 1 mole of CO_2 contains 1 mole of C)

$$H \longrightarrow H_2O\text{; moles of H in } H_2O = 2 \times \text{moles of } H_2O$$

(∵ 1 mole of H_2O contains 2 moles of H)

$$\text{Halogen (X)} \longrightarrow AgX\text{; moles of X in } AgX = 1 \times \text{moles of } AgX$$

(Carius method)

(∵ 1 mole of AgX contains 1 mole of X)

$$S \xrightarrow[\text{(Carius method)}]{} BaSO_4\text{;}$$

$$\text{moles of S in } BaSO_4 = 1 \times \text{moles of } BaSO_4$$

(∵ 1 mole of $BaSO_4$ contains 1 mole of S)

$$P \xrightarrow[\text{(Carius method)}]{} Mg_2P_2O_7\text{; moles of P in } Mg_2P_2O_7$$

$$= 2 \times \text{moles of } Mg_2P_2O_7$$

(∵ 1 mole of $Mg_2P_2O_7$ contains 2 moles of P)

$$N \xrightarrow[\text{(Dumas method)}]{} N_2 \text{ gas at NTP;}$$

$$\text{moles of N} = 2 \times \text{moles of } N_2 = \frac{2 \times \text{volume (NTP)}}{22400}$$

(∵ 1 mole of N_2 contains 2 moles of N)

$$N \xrightarrow[\text{(Kjeldahl method)}]{} NH_3\text{; moles of N in } NH_3 = 1 \times \text{moles of } NH_3$$

(∵ 1 mole of NH_3 contains 1 mole of N)

Oxygen atoms are not estimated directly. The percentage of all the elements except oxygen are determined and their sum when subtracted from 100 gives the percentage of oxygen.

EXAMPLES
Carbon and Hydrogen

Ex. 1. 0·16 g *of an organic compound, on complete combustion, produced* 0·44 g *of* CO_2 *and* 0·18 g *of* H_2O. *Calculate the percentage of carbon and hydrogen in the organic compound.*

Solution : Moles of C in $CO_2 = 1 \times$ moles of CO_2

$$= 1 \times \frac{0{\cdot}44}{44} = 0{\cdot}01 \qquad \text{(Rule 1, Chapter 1)}$$

Weight of C = moles of C × at. wt. of C (Rule 2, Chapter 1)

$$= 0{\cdot}01 \times 12 = 0{\cdot}12 \text{ g.}$$

Moles of H in $H_2O = 2 \times$ moles of H_2O

$$= 2 \times \frac{0{\cdot}18}{18} = 0{\cdot}02. \qquad \text{(Rule 1, Chapter 1)}$$

weight of H = moles of H × at. wt. of H (Rule 2, Chapter 1)

$$= 0{\cdot}02 \times 1 = 0{\cdot}02 \text{ g.}$$

$$\therefore \qquad \% \text{ of C} = \frac{0{\cdot}12}{0{\cdot}16} \times 100 = 75\%$$

and

$$\% \text{ of H} = \frac{0{\cdot}02}{0{\cdot}16} \times 100 = 12{\cdot}5\%.$$

Oxygen

Ex. 2. 0·66 g *of an organic compound containing* C, H *and* O *gave on combustion* 0·968 g *of* CO_2 *and* 0·792 g *of* H_2O. *Calculate the percentage of* O *in the compound.* (C = 12, H = 1, O = 16)

Solution : Moles of C in $CO_2 = 1 \times$ moles of CO_2 (Rule 1, Chapter 1)

$$= 1 \times \frac{0{\cdot}968}{44} = 0{\cdot}022.$$

$$\text{Wt. of C} = 0{\cdot}022 \times 12 = 0{\cdot}264 \text{ g.} \qquad \text{(Rule 2, Chapter 1)}$$

Moles of H in $H_2O = 2 \times$ moles of H_2O

$$= 2 \times \frac{0{\cdot}792}{18} = 0{\cdot}088.$$

$$\text{Weight of H} = 0{\cdot}088 \times 1 = 0{\cdot}088 \text{ g.}$$

Total weight of C and H in the compound $= (0{\cdot}264 + 0{\cdot}088)$ g $= 0{\cdot}352$ g.

$\therefore$ weight of O in the compound $= (0{\cdot}66 - 0{\cdot}352)$ g $= 0{\cdot}308$ g.

$\therefore$ % of O in the compound $= \dfrac{0{\cdot}308}{0{\cdot}66} \times 100 = 46{\cdot}67\%$.

Nitrogen (From Dumas Method)

Ex. 3. 0·2033 g *of an organic compound in Dumas method gave* 31·7 mL *of moist nitrogen at* 14°C *and* 758 mm *pressure. Calculate the percentage of nitrogen in the compound.* (*Aqueous tension at* 14°C = 14 mm, N = 14)

Solution : Pressure due to nitrogen only $= 758 - 14 = 744$ mm.

$$\text{Volume of nitrogen at NTP} = \frac{744 \times 31{\cdot}7 \times 273}{287 \times 760} = 29{\cdot}52 \text{ mL.}$$

$$\text{Mole of nitrogen} = \frac{29{\cdot}52}{22400}. \qquad \text{(Rule 3, Chapter 1)}$$

$$\text{Weight of nitrogen } (N_2) = \frac{29{\cdot}52}{22400} \times 28 = 0{\cdot}0369 \text{ g.} \qquad \text{(Rule 1, Chapter 1)}$$

$$\%\text{ of nitrogen} = \frac{0{\cdot}0369}{0{\cdot}2033} \times 100 = 18{\cdot}16\%.$$

From Kjeldahl Method

Ex. 4. 1·525 g *of an organic compound was Kjeldahlised and the ammonia so produced was passed into* 30 mL *of* N HCl *solution. The remaining* HCl *was further neutralised by* 120 mL *of* $\frac{N}{10}$ NaOH *solution. Calculate the percentage of nitrogen in the compound.*

{ *The students are advised to solve this problem after doing the volumetric problems.* }

Solution : m.e. of NaOH $= \dfrac{1}{10} \times 120 = 12.$ (Eqn. 1, Chapter 7)

m.e. of remaining HCl = 12.
(neutralised by NaOH) (Eqn. 2, Chapter 7)

m.e. of HCl (total) $= 1 \times 30 = 30.$ (Eqn. 1, Chapter 7)

m.e. of HCl neutralised by $NH_3 = 30 - 12 = 18.$

$\therefore$ m.e. of $NH_3 = 18$ (Eqn. 2, Chapter 7)

$$\text{Eq. of } NH_3 = \frac{18}{1000} = 0{\cdot}018. \quad \text{(Eqn. 3, Chapter 7)}$$

$$\therefore \quad \text{mole of } NH_3 = 0{\cdot}018. \quad \text{[Eqn. 6(ii), Chapter 7]}$$

Now mole of N in $NH_3 = 1 \times \text{mole of } NH_3$

$$= 1 \times 0{\cdot}018 = 0{\cdot}018.$$

Weight of nitrogen (N) $= 0{\cdot}018 \times 14 = 0{\cdot}252$ g.

$$\text{percentage of nitrogen} = \frac{0{\cdot}252}{1{\cdot}525} \times 100 = 16{\cdot}52\%.$$

Ex. 5. 0·42 g *of an organic compound containing* C, H, O *and* N *gave on combustion* 0·924 g *of* CO_2 *and* 0·243 g *of water.* 0·208 g *of the substance when distilled with* NaOH *gave* NH_3, *which required* 30 mL *of* (N/20) H_2SO_4 *solution for neutralization. Calculate the amount of each element in* 0·42 g *of the compound.*

***Solution* : Carbon**

Moles of C in $CO_2 = 1 \times \text{moles of } CO_2$

$$= 1 \times \frac{0{\cdot}924}{44} = 0{\cdot}021.$$

$$\text{Weight of C} = 0{\cdot}021 \times 12 = 0{\cdot}252 \text{ g.} \quad \text{(Rule 2, Chapter 1)}$$

Hydrogen

Moles of H in $H_2O = 2 \times \text{moles of } H_2O$

$$= 2 \times \frac{0{\cdot}243}{18} = 0{\cdot}027.$$

$$\text{Weight of H} = 0{\cdot}027 \times 1 = 0{\cdot}027 \text{ g.} \quad \text{(Rule 2, Chapter 1)}$$

Nitrogen

$$\text{m.e. of } H_2SO_4 = \frac{1}{20} \times 30 = 1{\cdot}5. \quad \text{(Eqn. 1, Chapter 7)}$$

$$\therefore \quad \text{m.e. of } NH_3 = 1{\cdot}5. \quad \text{(Eqn. 2, Chapter 7)}$$

$$\text{Eq. of } NH_3 = \frac{1{\cdot}5}{1000} = 0{\cdot}0015. \quad \text{(Eqn. 3, Chapter 7)}$$

$$\text{moles of } NH_3 = 0{\cdot}0015. \quad \text{[Eqn. 6(ii), Chapter 7]}$$

Now,

moles of N in $NH_3 = 1 \times \text{moles of } NH_3 = 0{\cdot}0015.$

Weight of N in 0·208 g of the compound $= 0{\cdot}0015 \times 14$

$$= 0{\cdot}021 \text{ g.}$$

$$\therefore \text{ weight of N in 0·42 g of the compound} = \frac{0{\cdot}021}{0{\cdot}208} \times 0{\cdot}42.$$

$$= 0{\cdot}042 \text{ g.}$$

Oxygen

$\therefore$ total weight of C, H and N $= (0{\cdot}252 + 0{\cdot}027 + 0{\cdot}042)$ g

$$= 0{\cdot}321 \text{ g.}$$

$$\therefore \quad \text{weight of O} = 0{\cdot}42 - 0{\cdot}321$$
$$= 0{\cdot}099 \text{ g.}$$

Halogen (From Carius Method)

Ex. 6. *In the estimation of chlorine in a given compound, it was found that* 0·0811 g *of a compound gave* 0·2368 g *of silver chloride. Calculate the percentage of chlorine in the given compound.* (Ag = 108, Cl = 35·5)

Solution : Moles of Cl in AgCl = 1 × mole of AgCl

$$= \frac{0{\cdot}2368}{143{\cdot}5}. \quad (\text{AgCl} = 143{\cdot}5)$$

$$\text{Weight of chlorine} = \frac{0{\cdot}2368}{143{\cdot}5} \times 35{\cdot}5$$
$$= 0{\cdot}0586 \text{ g.}$$

$$\text{Percentage of Cl in the compound} = \frac{0{\cdot}0586}{0{\cdot}0811} \times 100$$
$$= 72{\cdot}27\%.$$

Sulphur (From Carius Method)

Ex. 7. 0·36 g *of an organic compound containing sulphur produced* H_2SO_4 *by Carius method, which on treatment with* $BaCl_2$ *produced quantitatively* 0·2330 g *of* $BaSO_4$. *Calculate the percentage of S in the compound.*

(Ba = 137, S = 32, O = 16)

Solution : Moles of S in $BaSO_4$ = 1 × mole of $BaSO_4$

$$= \frac{0{\cdot}2330}{233} = 0{\cdot}001. \quad (BaSO_4 = 233)$$

$$\text{Weight of S} = 0{\cdot}001 \times 32 \text{ g}$$
$$= 0{\cdot}032 \text{ g.}$$

$$\% \text{ of S} = \frac{0{\cdot}032}{0{\cdot}36} \times 100 = 8{\cdot}89\%.$$

Phosphorus (From Carius Method)

Ex. 8. *In a gravimetric determination of phosphorus,* 0·248 g *of an organic compound was strongly heated in a Carius tube with concentrated nitric acid. Phosphoric acid so produced was precipitated as* $MgNH_4PO_4$ *which on ignition yielded* 0·444 g *of* $Mg_2P_2O_7$. *Find the percentage of phosphorus in the compound.*

(Mg = 24, P = 31, O = 16)

Solution : Moles of P in $Mg_2P_2O_7 = 2 \times$ moles of $Mg_2P_2O_7$

$$= 2 \times \frac{0{\cdot}444}{222} = 0{\cdot}004 \text{ g.}$$

$$\text{Weight of P} = 0{\cdot}004 \times 31 \text{ g}$$
$$= 0{\cdot}124 \text{ g.}$$

$$\text{percentage of P} = \frac{0{\cdot}124}{0{\cdot}248} \times 100 = 50\%.$$

PROBLEMS

(Answers bracketed with questions)

1. 0·3 g of a compound on combustion gave 0·54 g of water and 0·88 g of carbon dioxide. Find the percentages of carbon and hydrogen in the compound. (80%, 20%)

2. 0·2475 g of an organic compound on combustion gave 0·4950 g of CO_2 and 0·2025 g of H_2O. Calculate the percentage of oxygen in the compound. (36·37%)

3. 0·2060 g of a substance gave 18·8 mL of moist nitrogen at 17°C and 756 mm pressure. If the vapour tension at 17°C is 14·5 mm, find the percentage of nitrogen in the compound. (10·56%)

4. An organic compound of molecular formula C_3H_7N was analysed for nitrogen by Dumas method. Find the volume (in mL) of nitrogen evolved at NTP from 2 g of the substance. (393 mL)

5. 0·788 g of a substance after digestion with H_2SO_4 was distilled with an excess of NaOH. The liberated NH_3 was absorbed in 100 mL of N H_2SO_4 solution. The remaining acid required 73·7 mL of N NaOH solution for neutralization. Find the percentage of nitrogen in the compound. (46·7%)

6. Find the percentage of nitrogen in an organic compound analysed by Kjeldahl method. 1·61 g of the compound produced NH_3 which was absorbed in 250 mL of $\frac{N}{2}$ H_2SO_4 solution. The remaining acid was then diluted to one litre, 25 mL of which required 25·5 mL of N/10 NaOH for exact neutralization. (20%)

7. An organic compound contains C, H, N and O. 0·135 g of this compound on combustion produced 0·198 g of CO_2 and 0·108 g of H_2O while the same amount gave 16·8 mL of nitrogen at 0°C and 76 cm of pressure. Calculate the percentage of oxygen in the compound. (35·44%)

8. 0·1890 g of an organic compound gave 0·2870 g of silver chloride by Carius method. Find the percentage of chlorine in the compound. (37·57%)

9. 0·123 g of an organic compound produced 0·099 g of CO_2 and 0·0507 g of H_2O. 0·185 g of the same compound produced 0·319 g of AgBr. Find the percentages of carbon, hydrogen and bromine in the compound. (21·96%, 4·48%, 73·36%)

10. 0·2595 g of an organic compound yielded quantitatively 0·35 g of BaSO4. Find the percentage of sulphur in the compound. (Ba = 137·3, S = 32, O = 16) (18·5%)

11. 1·5 g of an organic compound in a quantitative determination of phosphorus gave 2·5090 g of $Mg_2P_2O_7$. Calculate the percentage of phosphorus in the compound. (46·71%)

12. For the estimation of N, 1.4 g of an organic compound was digested by Kjeldahl method and the evolved ammonia was absorbed in 60 mL of $\frac{M}{10}$ H_2SO_4. The unreacted acid required 20 mL of $\frac{M}{10}$ NaOH for complete neutralisation. The percentage of N in the compound is

(a) 6% (b) 10% (c) 3% (d) 5%

(b)

(IIT 2014 Main)

[**Hint:** For $H_2SO_4 : 0.1\ M = 0.2\ N$]

13. In Carius method of estimation of halogen, 250 mg of an organic compound gave 141 mg of AgBr. The percentage of bromine in the compound is (Ag = 108, Br = 80).

(a) 24 (b) 36 (c) 48 (d) 60

(a)

(IIT 2015 Main)

❑

CHAPTER TEN

EMPIRICAL, MOLECULAR AND STRUCTURAL FORMULAE

The utility of the mole concept is further illustrated by the problems of determining the empirical and molecular formulae of the compounds. Empirical formula represents the simplest set of whole numbers expressing the relative numbers of atoms in the compound and anything that can be said about relative numbers of atoms may be said about the relative numbers of moles of atoms. A calculation of the relative numbers of moles of each element in the compound will, therefore, lead us to the empirical formula of the compound. The empirical formula implies nothing about how many moles of atoms are actually in one mole of the compound. **In fact, the molecular formula expresses the actual numbers of moles of atoms of each element present in one mole of the compound.**

The molecular formula weight is the whole number multiple of the empirical formula weight for a given compound.

$$\frac{\text{Molecular formula weight}}{\text{Empirical formula weight}} = n \text{ (say)}$$

Thus if X represents the empirical formula of a compound, its molecular formula will be represented as $(X)_n$.

EXAMPLES

Ex. 1. *Find the empirical formula of chromium oxide containing* 68·4% *of chromium.* (Cr = 52, O = 16)

Solution : Let the weight of chromium oxide be 100 g.

∴ weight of chromium = 68·4 g

and weight of oxygen = 31·6 g.

$$\left.\begin{aligned} \therefore \quad \text{moles of chromium} &= \frac{68{\cdot}4}{52} = 1{\cdot}32 \\ \text{and} \quad \text{moles of oxygen} &= \frac{31{\cdot}6}{16} = 1{\cdot}98. \end{aligned}\right\} \quad \dots \text{(Eqn. 2)}$$

Relative numbers of moles of Cr and O atoms

$$= \frac{\text{moles of Cr}}{\text{moles of O}} = \frac{1{\cdot}32}{1{\cdot}98} = \frac{132}{198} = \frac{2 \times 66}{3 \times 66} = \frac{2}{3} \text{ (by inspection only).}$$

The empirical formula is Cr_2O_3.

Ex. 2. *Find the empirical formula of vanadium oxide if* 2·73 g *of the oxide contains* 1·53 g *of the metal.* (V = 51, O = 16)

Solution : Weight of vanadium = 1·53 g.

Weight of oxygen = 2·73 – 1·53 = 1·20 g.

$$\left.\begin{aligned}\text{Mole of V} &= \frac{1\cdot53}{51} = 0\cdot03.\\ \text{Mole of O} &= \frac{1\cdot20}{16} = 0\cdot075.\end{aligned}\right\}\quad \text{(Rule 2, Chap. 1)}$$

$$V : O = 0\cdot03 : 0\cdot075 = 30 : 75 = 2 \times 15 : 5 \times 15$$

$= 2 : 5$ (by inspection only).

∴ the empirical formula is V_2O_5.

Ex. 3. *Find the empirical formula of a substance whose composition includes hydrogen, carbon, oxygen and nitrogen in the mass ratio* 1 : 3 : 4 : 7.

Solution : Suppose the weights of each element in the compound are as follows:

$$\left.\begin{aligned}H &= 1\text{ g} = \frac{1}{1}\text{ mole}\\ C &= 3\text{ g} = \frac{3}{12}\text{ mole}\\ O &= 4\text{ g} = \frac{4}{16}\text{ mole}\\ N &= 7\text{ g} = \frac{7}{14}\text{ mole}\end{aligned}\right\}\quad \text{(Rule 2, Chap. 1)}$$

$$\therefore \qquad H : C : O : N = 1 : \frac{1}{4} : \frac{1}{4} : \frac{1}{2} = 4 : 1 : 1 : 2.$$

∴ the empirical formula is H_4CON_2.

Ex. 4. *A sample of metal chloride weighing* 0·22 g *required* 0·51 g *of* $AgNO_3$ *to precipitate the chloride completely. The specific heat of the metal is* 0·057. *Find out the molecular formula of the chloride, if the symbol of the metal is* 'M'. (Ag = 108, N = 14, O = 16, Cl = 35·5)

Solution : Let the valency of the metal M be x

$$MCl_x + AgNO_3 \rightarrow AgCl$$

As it is clear from the question that both Ag and Cl atoms are conserved, applying POAC for Ag atoms, we get

moles of Ag atoms in $AgNO_3$ = moles of Ag atoms in AgCl

$$1 \times \text{moles of } AgNO_3 = 1 \times \text{moles of } AgCl$$

$$\frac{0{\cdot}51}{170} = \text{moles of AgCl.} \quad \dots (1)$$

Again applying POAC for Cl atoms, we get

moles of Cl atoms in MCl_x = moles of Cl atoms in AgCl

$$x \times \text{moles of } MCl_x = 1 \times \text{moles of AgCl}$$

$$x \times \frac{0{\cdot}22}{\text{mol. wt. of } MCl_x} = \text{moles of AgCl.}$$

$$\text{Now, at. wt. of } M = \frac{6{\cdot}4}{0{\cdot}057} = 112{\cdot}3 \text{ (Dulong and Petit's law)}$$

$\therefore$ molecular weight of $MCl_x = (112{\cdot}3 + 35{\cdot}5x)$.

$$\therefore \quad x \times \frac{0{\cdot}22}{(112{\cdot}3 + 35{\cdot}5\,x)} = \text{moles of AgCl.} \quad \dots (2)$$

From equations (1) and (2),

$$x \approx 3$$

Since valency is a whole number

$\therefore$ $x = 3$ and the formula of the metal chloride is MCl_3.

Ex. 5. *Find the formula of the crystal hydrate of barium chloride knowing that 36·6 g of the salt when roasted, loses 5·4 g of mass.*

Solution : Let the formula be $BaCl_2 \cdot x\,H_2O$.

Since 5·4 g loss, on heating the salt, is due to the loss of H_2O molecules,

the weight of H_2O = 5·4 g, and

the weight of $BaCl_2$ = 36·6 − 5·4

= 31·2 g.

Now from the formula $BaCl_2 \cdot x\,H_2O$ we know,

$$x \times \text{moles of } BaCl_2 = 1 \times \text{moles of } H_2O \quad \text{(Rule 6, Chapter 2)}$$

$$x \times \frac{31{\cdot}2}{208{\cdot}3} = \frac{5{\cdot}4}{18} \quad (BaCl_2 = 208{\cdot}3, H_2O = 18)$$

$$x = 2.$$

The formula is $BaCl_2 \cdot 2H_2O$.

Ex. 6. *A hydrated sulphate of a metal contained 8·1% metal and 43·2% sulphate (SO_4^{2-}) by weight. Assuming the specific heat of the metal to be 0·24, determine the formula of the hydrated sulphate.*

Solution : Suppose the metal M has a valency x and number of moles of H_2O in one mole of the hydrated sulphate of the metal is y.

Hence the formula will be $M_2(SO_4)_x \cdot y\,H_2O$ from which we have (from Eqn. 6, Chapter 1) the following equations:

$$x \times \text{moles of } M = 2 \times \text{moles of } SO_4 \quad \dots (1)$$

and $$y \times \text{moles of } M_2(SO_4) = 1 \times \text{moles of } H_2O. \quad \dots (2)$$

$$\left.\begin{aligned} \text{Now wt. of M} &= 8{\cdot}1 \text{ g} \\ \text{Wt. of } SO_4 &= 43{\cdot}2 \text{ g} \\ \text{Wt. of } H_2O &= 48{\cdot}7 \text{ g} \end{aligned}\right\} \text{ in 100 g of } M_2(SO_4)_x \cdot yH_2O$$

Eqn. (1) becomes

$$x \times \frac{8{\cdot}1}{\text{at. wt. of M}} = 2 \times \frac{43{\cdot}2}{\text{ionic wt. of } SO_4^{2-}}$$

$$x \times \frac{8{\cdot}1}{27} = \frac{2 \times 43{\cdot}2}{96} \quad \left(\text{at. wt.} = \frac{6{\cdot}4}{0{\cdot}24} = 27\right)$$

(Dulong and Petit's law)

$$x = 3$$

From Eqn. (2), we have

$$y \times \frac{(8{\cdot}1 + 43{\cdot}2)}{342} = \frac{48{\cdot}7}{18} \quad \left\{\begin{aligned} M_2(SO_4)_3 &= 342 \\ H_2O &= 18 \end{aligned}\right\}$$

$$y = 18.$$

∴ formula is $Al_2(SO_4)_3 \cdot 18H_2O$.

Ex. 7. *Find the molecular formula of a compound of boron with hydrogen if the mass of* 1 *litre of this gas equals the mass of* 1 *litre of nitrogen under same condition and the boron content in the substance is* 78·2%.

(B = 11, N = 14, O = 16)

Solution : 100 g of the compound contains 78·2 g of B and 21·8 g of H.

$$\text{Moles of B} = \frac{78{\cdot}2}{11} = 7{\cdot}1.$$

$$\text{Moles of H} = \frac{21{\cdot}8}{1} = 21{\cdot}8.$$

$$\therefore \quad B : H = 1 : 3$$

Empirical formula is BH_3.

∴ empirical formula weight = 11 + 3 = 14.

Now, since equal volumes of two gases contain the same number of molecules or moles if their temperature and pressure are the same, the number of moles of the gaseous compound is equal to the number of moles of nitrogen, both the gases occupying 1 litre of volume.

Moles of the compound = moles of N_2.

$$\frac{\text{weight of the compound}}{\text{mol. wt. of the compound}} = \frac{\text{weight of } N_2}{\text{mol. wt. of } N_2}$$

$\because$ weight of compound = weight of N_2 (as given).

Molecular weight of the compound = molecular weight of $N_2 = 28$.

We know, $\frac{\text{molecular formula weight of the compound}}{\text{empirical formula wt. of the compound}} = \frac{28}{14} = 2.$

$\therefore$ molecular formula is $(BH_3)_2$, i.e., B_2H_6.

Ex. 8. *A mineral contained* $MgO = 31{\cdot}88\%$, $SiO_2 = 63{\cdot}37\%$ *and* $H_2O = 4{\cdot}75\%$. *Show that the simplest formula for the mineral is* $H_2Mg_3Si_4O_{12}$. $(H = 1,\ Mg = 24,\ Si = 28,\ O = 16)$

Solution : Suppose the weight of the mineral is 100 g. Then

weight of $MgO = 31{\cdot}88$ g,

weight of $SO_4 = 63{\cdot}37$ g,

weight of $H_2O = 4{\cdot}75$ g.

Moles of Mg in $MgO = 1 \times$ moles of MgO

$$= \frac{31{\cdot}88}{40} = 0{\cdot}797. \quad \text{... (Rule 1, Chapter 1)}$$

Moles of Si in $SiO_2 = 1 \times$ moles of SiO_2

$$= \frac{63{\cdot}37}{60} = 1{\cdot}0561. \quad \text{... (Rule 1, Chapter 1)}$$

Moles of H in $H_2O = 2 \times$ moles of H_2O

$$= \frac{2 \times 4{\cdot}75}{18} = 0{\cdot}5278. \quad \text{... (Rule 1, Chapter 1)}$$

Moles of O

$= $ moles of O in MgO + moles of O in SiO_2 + moles of O in H_2O

$= 1 \times$ moles of MgO + $2 \times$ moles of SiO_2 + $1 \times$ moles of H_2O

$$= \frac{31{\cdot}88}{40} + \frac{2 \times 63{\cdot}37}{60} + \frac{4{\cdot}75}{18} = 3{\cdot}172.$$

Moles of O = 3·172.

Now, by inspection, we have

moles of $H = 0{\cdot}5278 \quad = 0{\cdot}2639 \times 2$

moles of $Mg = 0{\cdot}797 \quad \approx 0{\cdot}2639 \times 3$

moles of $Si = 1{\cdot}0561 \quad \approx 0{\cdot}2639 \times 4$

moles of $O = 3{\cdot}172 \quad \approx 0{\cdot}2639 \times 12$

$\therefore \quad H : Mg : Si : O = 2 : 3 : 4 : 12.$

The formula is $H_2Mg_3Si_4O_{12}$.

Ex. 9. 0·596 g *of a gaseous compound containing only boron and hydrogen occupies* 484 mL *at* NTP. *When the compound was ignited in excess of oxygen, all its hydrogen was recovered as* 1·17 g *of* H_2O *and all the boron was present as* B_2O_3. *Find the molecular formula of the compound.* (B = 10·8, H = 1)

Solution : 484 mL of the gaseous compound at NTP weighs 0·596 g.

∴ molecular weight of the compound

= wt. of 1 mole of the compound

= wt. of 22400 mL at NTP

$$= \frac{0.596}{484} \times 22400 = 27.6.$$

Let the formula be B_xH_y.

The reaction may be written as, $B_xH_y + O_2 \rightarrow H_2O + B_2O_3$

$$\underset{0.596\text{ g}}{} \qquad\qquad \underset{1.17\text{ g}}{}$$

0·596 g (of B_xH_y) 1·17 g (of H_2O)

Applying POAC for H atoms,

$$y \times \text{moles of } B_xH_y = 2 \times \text{moles of } H_2O$$

$$y \times \frac{0.596}{27.6} = \frac{2 \times 1.17}{18}$$

$$y = 6.$$

Further, for the formula B_xH_y, mol. wt. = 27·6.

$$\therefore \quad 10.8\,x + 1y = 27.6$$

$$10.8x + 6 = 27.6$$

$$x = 2.$$

Hence the molecular formula is B_2H_6.

Ex. 10. *A hydrocarbon contains* 10·5 g *of carbon per gram of hydrogen.* 1 *litre of the hydrocarbon at* 127°C *and* 1 atm *pressure weighs* 2·8 g. *Find molecular formula.*

Solution : Volume at NTP $= \dfrac{p_1 \times V_1}{T_1} \times \dfrac{T_2}{P_2}$

$$= \frac{1 \times 1}{400} \times \frac{273}{1} = \frac{273}{400} \text{ litre.}$$

$\because \quad \dfrac{273}{400}$ litre weighs 2·8 g at NTP.

$\therefore \quad$ 22·4 litres (1 mole) weighs $= \dfrac{2.8 \times 400}{273} \times 22.4$ g.

$$= 91.89 \text{ g}$$

$$\approx 92 \text{ g.}$$

$\therefore$ molecular weight of the hydrocarbon = 92.

$$\therefore \begin{cases} C & 10{\cdot}5\text{ g} & 10{\cdot}5/12 = 0{\cdot}875 \\ H & 1\text{ g} & 1/1 = 1 \end{cases}$$

(columns: Weight, Mole)

or

$$\begin{aligned} C : H &= 0{\cdot}875 : 1 \\ &= 0{\cdot}875 \times 8 : 1 \times 8 \quad \text{(by inspection)} \\ &= 7 : 8. \end{aligned}$$

Thus empirical formula is C_7H_8.

Now empirical formula weight = 92 = molecular weight.

Hence molecular formula is also C_7H_8.

Ex. 11. *An organic compound contains* C, H *and* O. 0·30 g *of this compound on combustion yielded* 0·44 g *of* CO_2 *and* 0·18 g *of* H_2O. *If the weight of* 1 *mole of the compound is* 60, *show that the molecular formula is* $C_2H_4O_2$.

Solution : Moles of C in $CO_2 = 1 \times$ moles of $CO_2 = \dfrac{0{\cdot}44}{44} = 0{\cdot}01.$

... (Rule 1, Chapter 1)

Weight of $C = 0{\cdot}01 \times 12\text{ g} = 0{\cdot}12\text{ g}.$ ... (Rule 2, Chapter 1)

Moles of H in $H_2O = 2 \times$ moles of $H_2O = \dfrac{2 \times 0{\cdot}18}{18} = 0{\cdot}02.$

Weight of $H = 0{\cdot}02 \times 1 = 0{\cdot}02\text{ g}.$

$$\begin{aligned} \text{Weight of O} &= \text{weight of compound} - (\text{wt. of C} + \text{wt. of H}) \\ &= 0{\cdot}30 - (0{\cdot}12 + 0{\cdot}02)\text{ g} \\ &= 0{\cdot}16\text{ g}. \end{aligned}$$

$$\therefore \quad \text{mole of O} = \frac{0{\cdot}16}{16} = 0{\cdot}01.$$

$\therefore$ mole of C : H : O = 0·01 : 0·02 : 0·01 = 1 : 2 : 1.

$\therefore$ empirical formula = CH_2O.

Now, since the weight of 1 mole is the molecular weight in grams, the molecular weight of the compound is 60.

$$\therefore \quad \frac{\text{molecular formula weight}}{\text{empirical formula weight}} = \frac{60}{30} = 2.$$

Thus, the molecular formula is $(CH_2O)_2$, i.e., $C_2H_4O_2$.

Ex. 12. 0·2 g *of a monobasic organic acid containing* C, H *and* O *on combustion gave* 0·505 g *of* CO_2 *and* 0·0892 g *of* H_2O. 0·183 g *of this acid required* 15 cc *of* N/10 NaOH *for exact neutralisation. Find the molecular formula of the acid.*

Solution : Moles of $C = 1 \times$ moles of $CO_2 = \frac{0{\cdot}505}{44} = 0{\cdot}0115.$

$$\text{Wt. of } C = 0{\cdot}0115 \times 12 = 0{\cdot}1380 \text{ g.}$$

Moles of $H = 2 \times$ moles of $H_2O = 2 \times \frac{0{\cdot}0892}{18} = 0{\cdot}0099.$

Wt. of $H = 0{\cdot}0099 \times 1 = 0{\cdot}0099$ g.

Wt. of O = wt. of the compound – (wt. of C + wt. of H)

$$= 0{\cdot}2 - (0{\cdot}1380 + 0{\cdot}0099) = 0{\cdot}0521 \text{ g.}$$

Moles of $O = \frac{0{\cdot}0521}{16} = 0{\cdot}0033.$

$\therefore$ moles of $C : H : O = 0{\cdot}0115 : 0{\cdot}0099 : 0{\cdot}0033$

$$= 115 : 99 : 33$$

$$= 3{\cdot}5 : 3 : 1 = 7 : 6 : 2.$$

Empirical formula is $C_7H_6O_2$.

Now, since the acid is monobasic, mol. wt. = eq. wt. ...(Eqn. 6, Chapter 7)

$\therefore$ equivalent of the acid $= \frac{0{\cdot}183}{\text{eq. wt.}} = \frac{0{\cdot}183}{\text{mol. wt.}}$. ... (Eqn. 4, Chapter 7)

$\therefore$ m.e. of the acid $= \frac{0{\cdot}183}{\text{mol. wt.}} \times 1000$ ... (Eqn. 3, Chapter 7)

$$= \frac{183}{\text{mol. wt.}}.$$

m.e. of the base $= \frac{1}{10} \times 15 = 1{\cdot}5.$ ... (Eqn. 1, Chapter 7)

$\because$ m.e. of the acid = m.e. of the base ... (Eqn. 2, Chapter 7)

$\frac{183}{\text{mol. wt.}} = 1{\cdot}5.$ $\therefore$ mol. wt. = 122.

Since the empirical formula weight, i.e., $84 + 6 + 32 = 122$, is equal to the molecular formula weight, the molecular formula of the acid is the same as the empirical formula. Hence, the formula is $C_7H_6O_2$.

Ex. 13. *One volume of a gaseous compound of carbon, hydrogen and nitrogen on combustion produced* 2 *volumes of* CO_2, 3·5 *volumes of* H_2O *and* 0·5 *volume of nitrogen under identical conditions of temperature and pressure. Find the empirical and molecular formulae of the compound.*

Solution : Avogadro's principle states that equal volumes of gases contain the same number of molecules or moles.

Compound	→	CO_2	+	H_2O	+	N_2
1 vol.		2 vol.		3·5 vol.		0·5 vol.
1 mole		2 moles		3·5 moles		0·5 moles

Moles of C in $CO_2 = 1 \times$ moles of CO_2

$$= 1 \times 2 = 2.$$

Moles of H in $H_2O = 2 \times$ moles of H_2O

$$= 2 \times 3{\cdot}5 = 7.$$

Moles of N in $N_2 = 2 \times$ moles of $N_2 = 2 \times 0{\cdot}5 = 1$.

Hence the empirical formula is C_2H_7N.

Now we see that 1 mole of the compound contains 2 moles of CO_2 and 1 mole of CO_2 contains 1 mole of carbon atoms, so 1 mole of the compound should contain 2 moles of carbon atoms which is shown by the empirical formula. Hence, the molecular formula is the same as the empirical formula.

Ex. 14. *A gaseous compound which contains only* C, H *and* S *is burnt with oxygen under such conditions that individual volumes of the reactant and product can be measured at the same temperature and pressure. It is found that* 3 *volumes of the compound react with oxygen to yield* 3 *volumes of* CO_2, 3 *volumes of* SO_2 *and* 6 *volumes of water vapour. What volume of oxygen is required for the combustion? What is the formula of the compound? Is this an empirical formula or molecular formula?*

Solution :

	Compound	$+ O_2$	$\rightarrow$	CO_2	$+ SO_2$	$+ H_2O$
	3 vol.	v vol.(say)		3 vol.	3 vol.	6 vol.
or	3 moles	v moles		3 moles	3 moles	6 moles

Applying POAC for O atoms,

moles of O in O_2

= moles of O in CO_2 + moles of O in SO_2 + moles of O in H_2O

$2 \times$ moles of O_2

$= 2 \times$ moles of $CO_2 + 2 \times$ moles of $SO_2 + 1 \times$ moles of H_2O

$$2v = 2 \times 3 + 2 \times 3 + 1 \times 6 = 18\ ;\ v = 9.$$

$\therefore$ volume of O_2 required for combustion is 9.

Again, moles of C in $CO_2 = 1 \times$ moles of $CO_2 = 3$

moles of S in $SO_2 = 1 \times$ moles of $SO_2 = 3$

moles of H in $H_2O = 2 \times$ moles of $H_2O = 12$

$\therefore$ formula is $C_3S_3H_{12}$ or CSH_4.

Now we see 3 moles of the compound contain 3 moles of C atoms in $C_3S_3H_{12}$. Hence, 1 mole of the compound should contain 1 mole of C atoms and therefore, the molecular formula is CSH_4.

Ex. 15. 500 cc *of a hydrocarbon gas burnt in excess of oxygen yields* 2500 cc *of* CO_2 *and* 3 *litres of* H_2O *vapours, all the vapours being measured at the same temperature and pressure. What is the formula of the hydrocarbon gas?*

Solution : Hydrocarbon $\xrightarrow{\text{combustion}}$ CO_2 + H_2O

	0·5 lit.	2·5 lit.	3 lit.
	0·5 moles	2·5 moles	3 moles
∴	1 mole	5 moles	6 moles

Moles of C in CO_2 $= 1 \times$ moles of CO_2
$= 1 \times 5 = 5.$

Moles of H in $H_2O = 2 \times$ moles of H_2O
$= 2 \times 6 = 12.$

Since 1 mole of the compound contains 5 moles of C and 12 moles of H, the molecular formula of the hydrocarbon is C_5H_{12}.

Ex. 16. 0·45 g *of a dibasic organic acid upon combustion produced* 0·44 g *of* CO_2 *and* 0·09 g *of* H_2O. 0·76 g *of its silver salt when ignited gave* 0·54 g *of pure silver. What is the formula of the acid?* (Ag = 108)

Solution : Moles of C in $CO_2 = 1 \times$ moles of CO_2

$$= 1 \times \frac{0{\cdot}44}{44} = 0{\cdot}01.$$

Weight of C $= 0{\cdot}01 \times 12 = 0{\cdot}12$ g.

Moles of H in $H_2O = 2 \times$ moles of H_2O

$$= 2 \times \frac{0{\cdot}09}{18} = 0{\cdot}01.$$

Weight of H $= 0{\cdot}01 \times 1 = 0{\cdot}01$ g.

Weight of O = wt. of acid – (wt. of C + wt. of H)
$= 0{\cdot}45 - (0{\cdot}12 + 0{\cdot}01)$ g.
$= 0{\cdot}32$ g.

$$\text{Mole of O} = \frac{0{\cdot}32}{16} = 0{\cdot}02.$$

Mole of C : H : O = 0·01 : 0·01 : 0·02
= 1 : 1 : 2.

Empirical formula is CHO_2.

Now 0·76 g of Ag salt of the dibasic acid gives 0·54 g of pure silver, i.e.,

dibasic acid → Ag salt → Ag
0·76 g 0·54 g
(contains 2 Ag atoms)

We know that the Ag salt of the dibasic acid is formed by the replacement of 2 H atoms of the acid by 2 Ag atoms.

Applying POAC for the Ag atoms,

moles of Ag atom in Ag salt = moles of Ag in the product.

$$2 \times \text{moles of Ag salt} = \text{moles of Ag in the product}$$

$$2 \times \frac{0{\cdot}76}{\text{mol. wt. of Ag salt}} = \frac{0{\cdot}54}{108}.$$

Molecular weight of the salt = 304.

∴ molecular weight of the acid

$$= \text{mol. wt. of salt} - 2 \times \text{at. wt. of Ag} + 2 \times \text{at. wt. of H}$$

$$= 304 - 216 + 2 = 90.$$

Hence,

$$\frac{\text{molecular formula weight}}{\text{empirical formula weight}} = \frac{90}{45} = 2.$$

∴ molecular formula is $(CHO_2)_2$, i.e., $(COOH)_2$.

Ex. 17. 0·465 g *of an organic compound upon combustion produced* 1·32 g *of* CO_2 *and* 0·315 g *of* H_2O. 0·2325 g *of the same compound gave* 27·8 mL *of dry* N_2 *at* NTP. *If the compound contained* C, H *and* N, *find its formula.*

Solution : Moles of C in $CO_2 = 1 \times \text{moles of } CO_2 = \frac{1{\cdot}32}{44} = 0{\cdot}03.$

$$\text{Moles of H in } H_2O = 2 \times \text{moles of } H_2O = 2 \times \frac{0{\cdot}315}{18} = 0{\cdot}035.$$

$$\text{Moles of N in } N_2 = 2 \times \text{moles of } N_2 = 2 \times \frac{27{\cdot}8}{22400} = 0{\cdot}0025.$$

Now, 0·0025 mole of N is contained in 0·2325 g of the compound.

∴ moles of N in 0·465 g of the compound $= 0{\cdot}0025 \times \frac{0{\cdot}465}{0{\cdot}2325} = 0{\cdot}005.$

∴ moles of C : H : N = 0·030 : 0·035 : 0·005

= 30 : 35 : 5

= 6 : 7 : 1.

∴ empirical formula is C_6H_7N.

Ex. 18. 0·220 g *of a sample of a volatile compound, containing carbon, hydrogen and chlorine yielded on combustion in oxygen* 0·195 g *of* CO_2, 0·0804 g *of* H_2O. 0·120 g *of the compound occupied a volume of* 37·24 mL *at* 105°C *and* 768 mmHg *pressure. Calculate the molecular formula of the compound.*

Solution : Moles of C in $CO_2 = 1 \times \text{moles of } CO_2 = \frac{0{\cdot}195}{44} = 0{\cdot}00443.$

$$\text{Weight of C} = 0{\cdot}00443 \times 12 = 0{\cdot}05316 \text{ g.}$$

$$\text{Moles of H in } H_2O = 2 \times \text{moles of } H_2O = 2 \times \frac{0{\cdot}0804}{18} = 0{\cdot}00893.$$

$$\text{Weight of H} = 0{\cdot}00893 \times 1 = 0{\cdot}00893 \text{ g.}$$

$\therefore$ weight of $Cl = 0{\cdot}22 - (0{\cdot}05316 + 0{\cdot}00893)\text{ g} = 0{\cdot}15791$ g.

$$\text{Moles of Cl} = \frac{0{\cdot}15791}{35{\cdot}5} = 0{\cdot}00445.$$

$$\text{Moles of } C : H : Cl = 0{\cdot}00443 : 0{\cdot}00893 : 0{\cdot}00445$$
$$= 443 : 893 : 445 = 1 : 2 : 1$$

i.e., the empirical formula is CH_2Cl.

Now, volume of the vapour of 0·12 g of compound at NTP

$$= \frac{37{\cdot}24 \times 768}{378} \times \frac{273}{760} = 27{\cdot}18 \text{ mL.}$$

$$\text{Moles of the compound} = \frac{0{\cdot}12}{M} \qquad (M = \text{mol. wt.})$$

$$= \frac{27{\cdot}18}{22400}.$$ ... (Rule 3, Chapter 1)

$$\therefore \qquad M = 99.$$

$$\therefore \quad \frac{\text{molecular formula weight}}{\text{empirical formula weight}} = \frac{99}{49{\cdot}5} = 2.$$

Hence, the empirical formula is $(CH_2Cl)_2$, i.e., $C_2H_4Cl_2$.

Ex. 19. *A monoacid organic base gave the following data on analysis:*

(a) 0·2790 g *of the base gave* 0·7920 g *of* CO_2 *and* 0·1890 g *of* H_2O

(b) 0·1163 g *of the base gave* 14 mL *of dry nitrogen at* NTP

(c) 0·2980 g *of the platinichloride left* 0·0975 g *of* Pt (Pt = 195)

Calculate the molecular formula of the base.

Solution : Moles of $C = 1 \times \text{moles of } CO_2 = \dfrac{0{\cdot}792}{44} = 0{\cdot}018.$

$$\text{Moles of H} = 2 \times \text{moles of } H_2O = 2 \times \frac{0{\cdot}1890}{18} = 0{\cdot}021.$$

$$\text{Moles of N} = 2 \times \text{moles of } N_2 = 2 \times \frac{14}{22400} = 0{\cdot}00125.$$

$$\text{Moles of N in 0·2790 g of the base} = \frac{0{\cdot}00125}{0{\cdot}1163} \times 0{\cdot}2790$$
$$= 0{\cdot}003.$$

$$\therefore \quad \text{moles of } C : H : N = 0{\cdot}018 : 0{\cdot}021 : 0{\cdot}003$$
$$= 18 : 21 : 3$$
$$= 6 : 7 : 1.$$

Hence, the empirical formula is C_6H_7N or $C_6H_5NH_2$.

Suppose the monoacid base is B.

B	+ H_2PtCl_6	$\rightarrow$	$B_2H_2PtCl_6$	$\rightarrow$	Pt
Base (Monoacid)	Acid		Chloroplatinate (Platinichloride)		
			0·2980 g		0·0975 g

Applying POAC for Pt atoms:

1 × moles of $B_2H_2PtCl_6$ = moles of Pt in product

$$\frac{0{\cdot}2980}{\text{mol. wt. of } B_2H_2PtCl_6} = \frac{0{\cdot}0975}{195} \qquad (Pt = 195)$$

$\therefore$ mol. wt. of $B_2H_2PtCl_6 = 596$.

Molecular weight of B

$$= \frac{\text{mol. wt. of } B_2H_2PtCl_6 - \text{mol. wt. of } H_2PtCl_6}{2}$$

$$= \frac{596 - 410}{2} = 93.$$

Since empirical formula weight is also 93,

therefore, the molecular formula of the base is $C_6H_5NH_2$.

Ex. 20. *The analysis of an organic compound gave the following data:*

(a) 0·4020 g *gave* 0·6098 g *and* 0·2080 g *of* CO_2 *and* H_2O *respectively.*

(b) 1·01 g *by Kjeldahl method produced ammonia which was neutralised by* 23·2 mL *of* N/2HCl.

(c) 0·1033 g *of the compound gave* 0·2772 g *of* $BaSO_4$.

(d) 0·1015 g *when vaporised in Victor Meyer's apparatus displaced* 27·96 mL *of air at* 15°C *and* 766 *mm pressure.*

Calculate the molecular formula. (*Aqueous tension at* 15°C = 16 mm)

Solution : Moles of $C = 1 \times \text{moles of } CO_2 = 1 \times \frac{0{\cdot}6098}{44} = 0{\cdot}0138.$

$$\text{Moles of H} = 2 \times \text{moles of } H_2O = 2 \times \frac{0{\cdot}2080}{18} = 0{\cdot}0231$$

Calculation of mole of N:

$$\text{m.e. of HCl} = 23{\cdot}2 \times \frac{1}{2} = 11{\cdot}6.$$

$\therefore$ m.e. of $NH_3 = 11{\cdot}6$

mmol of $NH_3 = 11{\cdot}6$.

$$\text{Moles of } NH_3 = \frac{11{\cdot}6}{1000} = 0{\cdot}0116.$$

$\therefore$ moles of N = 1 × moles of NH_3 = 0·0116.

Moles of N in 0·4020 g of compound $= \dfrac{0{\cdot}0116}{1{\cdot}01} \times 0{\cdot}4020$

$= 0{\cdot}0046.$

Moles of S = 1 × moles of $BaSO_4$

$= \dfrac{0{\cdot}2772}{233{\cdot}3} = 0{\cdot}00119.$ ($BaSO_4$ = 233·3)

Moles of S in 0·4020 g of compound $= \dfrac{0{\cdot}00119}{0{\cdot}1033} \times 0{\cdot}4020$

$= 0{\cdot}0046.$

$\therefore$ mole of C : H : N : S = 0·0138 : 0·0231 : 0·0046 : 0·0046

= 3 : 5 : 1 : 1.

Hence the empirical formula is C_3H_5NS.

Calculation of molecular weight:

Pressure due to dry air only = 766 – 16 = 750 mm.

Volume of vapour of 0·1015 g of compound at NTP

$$= \frac{750 \times 27{\cdot}96 \times 273}{288 \times 760} = 26{\cdot}155 \text{ mL}.$$

$\therefore$ moles of vapour $= \dfrac{26{\cdot}155}{22400} = 0{\cdot}00116.$

Molecular weight $= \dfrac{\text{weight in grams}}{\text{no. of moles}}$

$$= \frac{0{\cdot}1015}{0{\cdot}00116} = 86{\cdot}97 \approx 87.$$

Since empirical formula weight is also equal to 87, the molecular formula is C_3H_5NS.

Ex. 21. *An organic compound* (A) *contains* C *and* H *only.* (C = 90%). (A) *produces another compound* (B) *on treatment with* HBr. (B) *contains* 79·2% *of* Br. *If the molecular weight of* (A) *is* 40, *find the formulae of* (A) *and* (B). (Br = 80)

Solution : Moles of C $= \dfrac{90}{12} = 7{\cdot}5.$

Moles of H $= \dfrac{10}{1} = 10.$

$\therefore$ moles of C : H = 7·5 : 10

= 3 : 4.

$\therefore$ empirical formula is C_3H_4.

As molecular weight is 40, the molecular formula, will be the same as the empirical formula, i.e., (A) is

$$CH_3 - C \equiv CH \text{ or, } CH_2 = C = CH_2$$

Since with HBr, (A) will produce $CH_3 - C(Br_2) - CH_3$ in which

$$\text{Br \%} = \frac{160}{202} \times 100 = 79{\cdot}2\%$$

which is the same as the given value, (B) will be

$$CH_3 - C(Br_2) - CH_3 .$$

Ex. 22. *A certain hydrocarbon* (A) *was found to contain* 85·7% *of* C *and* 14·3% *of* H. *This compound consumes* 1 *molar equivalent of hydrogen to give a saturated hydrocarbon* (B). 1 g *of hydrocarbon* (A) *just decolourised* 38·05 g *of a* 5% *solution (by weight) of* Br_2 *in* CCl_4. *Compound* (A), *on oxidation with concentrated* $KMnO_4$, *gave compound* (C) *(molecular formula* C_4H_8O*) and acetic acid. Compound* (C) *could easily be prepared by the action of acidic aqueous mercuric sulphate on 2-butyne. Determine the molecular formula of* (A) *and deduce the structures of* (A), (B) *and* (C).

Solution : Moles of C : H in (A) $= \frac{85{\cdot}7}{12} : \frac{14{\cdot}7}{1} = 7{\cdot}14 : 14{\cdot}3 = 1 : 2.$

$\therefore$ empirical formula of (A) is CH_2 and empirical formula weight is 14.

As (A) consumes 1 molar equivalent of hydrogen to give a saturated hydrocarbon (B), (A) should have one double bond. Reaction of (A) with Br_2 will also be an addition reaction in which 1 mole of (A) will combine with 1 mole of Br_2. Thus,

Moles of (A) = moles of Br_2

or $$\frac{1}{M} = \frac{(5 \times 38{\cdot}05)/100}{160}. \qquad \left\{\begin{array}{l} M = \text{mol. wt. of (A)} \\ Br_2 = 160 \end{array}\right\}$$

$\therefore$ $M = 84.$

$\therefore$ $$n = \frac{\text{molecular formula wt.}}{\text{empirical formula wt.}} = \frac{84}{14} = 6.$$

$\therefore$ molecular formula of (A) is C_6H_{12}.

Now from the given reaction sequence, we have,

$$\underset{\text{(A)}}{C_6H_{12}} \xrightarrow[\text{KMnO}_4]{\text{(O)}} \underset{\text{(C)}}{C_4H_8O} + CH_3COOH$$

I

$$\text{II} \uparrow H^+ / HgSO_4$$

$$CH_3 . C \equiv C . CH_3$$

2-butyne

We conclude that C_6H_{12} (A) must be represented as $C_4H_8 = CH . CH_3$ (step I) and C_4H_8O (C) must be a ketone

$$CH_3 \cdot \overset{\overset{C}{\|}}{C}\ CH_2 \cdot CH_3 \quad \text{(step II).}$$

Further, addition of H_2 to compound (A) will produce C_6H_{14} (B) (an alkane). The structure of (A), (B) and (C) may be represented as

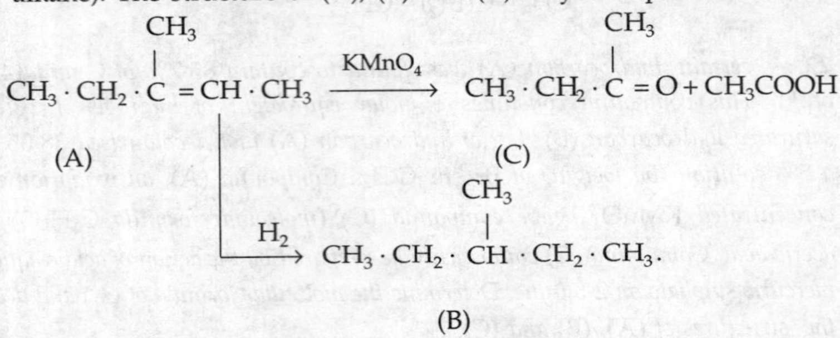

Ex. 23. *An organic compound* (X) *containing* C, H *and* O *has a vapour density* 37. 0·2750 g *of* (X) *produced* 0·6540 g *of* CO_2 *and* 0·3375 g *of* H_2O. *The compound* (X) *on dehydration gave a hydrocarbon* (Y) *containing* 85·71% *of* C. (Y), *on treatment with* HI *followed by hydrolysis, gave* (Z) *which was isomeric with* (X). *Give structural formulae of* (X), (Y) and (Z).

Solution : Moles of C in (X) = 1 × moles of CO_2

$$= 1 \times \frac{0{\cdot}6540}{44} = 0{\cdot}01486.$$

Moles of H in (X) = 2 × moles of H_2O

$$= 2 \times \frac{0{\cdot}3375}{18} = 0{\cdot}0375.$$

Wt. of oxygen in (X) = 0·2750 − (wt. of C + wt. of H)

$$= 0{\cdot}2750 - (0{\cdot}01486 \times 12 + 0{\cdot}0375 \times 1)$$

$$= 0{\cdot}0592 \text{ g.}$$

$$\therefore \quad \text{mole of O in (X)} = \frac{0{\cdot}0592}{16} = 0{\cdot}0037.$$

$$\therefore \quad \text{mole of C : H : O} = 0{\cdot}01486 : 0{\cdot}0375 : 0{\cdot}0037$$

$$= 4 : 10 : 1.$$

∴ empirical formula of (X) is $C_4H_{10}O$ (74).

As molecular weight of (X) = 2 × VD.

$$= 2 \times 37 = 74,$$

$\therefore$ molecular formula of X is $C_4H_{10}O$.

Now for the compound (Y):

$$\text{Moles of C : H} = \frac{85{\cdot}71}{12} : \frac{14{\cdot}29}{1}$$
$$= 7{\cdot}143 : 14{\cdot}29$$
$$= 1 : 2.$$

Empirical formula of (Y) is CH_2.

Since (Y) is produced from (X) by dehydration,

(Y) must be $C_4H_{10}O - H_2O$, i.e., C_4H_8.

Further, as C_4H_8 is an unsaturated hydrocarbon, (X), i.e., $C_4H_{10}O$ must be an alcohol.

$\therefore$ (X) may be $CH_3 \cdot CH_2 \cdot CH_2 \cdot CH_2 \cdot OH$ (primary)

or $CH_3 \cdot CH_2 \cdot CH(OH) \cdot CH_3$ (secondary)

and (Y) may be $CH_3 \cdot CH_2 \cdot CH = CH_2$

or $CH_3 \cdot CH = CH \cdot CH_3$.

But as given in the question,

$$\underset{(X)}{CH_3 \cdot CH_2 \cdot CH_2 \cdot CH_2OH} \xrightarrow[-H_2O]{HI} CH_3 \cdot CH_2 \cdot CHI \cdot CH_3 \xrightarrow{\text{hydrolysis}} \underset{(Z)}{CH_3 \cdot CH_2 \cdot CH(OH) \cdot CH_3}$$

[isomeric to $CH_3 \cdot CH_2 \cdot CH_2 \cdot CH_2OH$ (X)]

Thus,

(X) is $CH_3 \cdot CH_2 \cdot CH_2 \cdot CH_2OH$

(Y) is $CH_3 \cdot CH_2 \cdot CH = CH_2$

(Z) is $CH_3 \cdot CH_2 \cdot \underset{\displaystyle OH}{\underset{|}{CH}} \cdot CH_3$.

Ex. 24. 0·1588 g *of an organic compound* (A), *containing* C, H *and* O, *gave* 0·3615 g *of* CO_2 *and* 0·1479 g *of* H_2O. (A) *reduces Fehling's solution and on oxidation gave an acid* (B). *The silver salt of* (B) *contained* 59·6% *of* Ag. *If the vapour density of* (A) *is* 29, *find the structural formulae of* (A) *and* (B).

Solution : Moles of C in (A) = 1 × moles of CO_2

$$= \frac{0{\cdot}3615}{44} = 0{\cdot}0082.$$

Moles of H in (A) = 2 × moles of H_2O

$$= 2 \times \frac{0{\cdot}1479}{18} = 0{\cdot}0164.$$

$$\text{Wt. of O} = 0{\cdot}1588 - (\text{wt. of C} + \text{wt. of H})$$
$$= 0{\cdot}1588 - (0{\cdot}0082 \times 12 + 0{\cdot}0164 \times 1)$$
$$= 0{\cdot}044 \text{ g.}$$

$$\therefore \quad \text{mole of O in (A)} = \frac{0{\cdot}044}{16} = 0{\cdot}00275.$$

$$\therefore \quad \text{mole of C : H : O} = 0{\cdot}0082 : 0{\cdot}0164 : 0{\cdot}00275$$
$$= 3 : 6 : 1.$$

Empirical formula of (A) is C_3H_6O (58).

As the molecular wt. of (A) $= 2 \times 29 = 58$,

$\therefore$ molecular formula of (A) is C_3H_6O.

As (A) reduces Fehling's solution, it may be C_2H_5CHO, which on oxidation will produce C_2H_5COOH (B) and its silver salt, i.e., C_2H_5COOAg shall contain $\frac{108}{181} \times 100$, i.e., 59·6% of Ag which is the same as given. Thus,

(A) is C_2H_5CHO

and (B) is C_2H_5COOH.

Ex. 25. *An organic compound of empirical formula* $C_5H_{10}O$ *has a vapour density,* 43. *On treatment with sodium bisulphite, a crystalline addition product was formed. The compound did not reduce Fehling's solution. On oxidation, a mixture of* CH_3COOH *and* C_2H_5COOH *was formed. Find the structural formula of the organic compound.*

Solution : As the molecular weight of the organic compound is 2×43, i.e., 86, its molecular formula will be the same as the empirical formula ($C_5H_{10}O = 86$).

Since it gives a crystalline product and does not reduce Fehling's solution it must be a ketone which may be

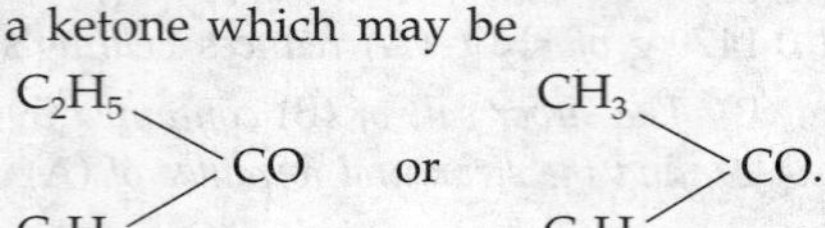

Further, since both the compounds, on oxidation, produce CH_3COOH and C_2H_5COOH, hence the organic compound may have any of the two structures given above.

Ex. 26. *An organic compound* (X), *on analysis, gives* 24·24% *of* C *and* 4·04% *of* H. *Further, sodium extract of* 1·0 g *of* (X) *gives* 2·90 g *of silver chloride with acidified silver nitrate solution. The compound* (X) *may be represented by two isomeric structures* (Y) *and* (Z). (Y), *on treatment with aqueous potassium hydroxide solution, gives a dihydroxy compound while* (Z), *on similar treatment, gives ethanal. Find out the molecular formula of* (X) *and give the structures of* (Y) *and* (Z). **(IIT 1989)**

Solution : Moles of Cl in (X) = moles of Cl in AgCl

$$= 1 \times \text{moles of AgCl} = \frac{2{\cdot}90}{143{\cdot}5} \quad (\text{AgCl} = 143{\cdot}5)$$

$$\text{Wt. of Cl} = \frac{2{\cdot}90}{143{\cdot}5} \times 35{\cdot}5 = 0{\cdot}7174 \text{ g.}$$

$$\therefore \qquad \% \text{ of Cl in (X)} = \frac{0{\cdot}7174}{1} \times 100 = 71{\cdot}74\%.$$

$$\therefore \qquad \text{moles of C : H : Cl} = \frac{24{\cdot}24}{12} : \frac{4{\cdot}04}{1} : \frac{71{\cdot}74}{35{\cdot}5}$$

$$= 2{\cdot}02 : 4{\cdot}04 : 2{\cdot}02 = 1 : 2 : 1.$$

∴ empirical formula of (X) is CH_2Cl.

One of the isomers of (X), on treatment with aq. KOH solution gives a dihydroxy compound which indicates that the compound (X) contains 2 Cl atoms. Hence (X) should be $C_2H_4Cl_2$, two isomers of which are

$$\begin{array}{ccc} CH_2ClCH_2Cl\ (Y) & \text{and} & CH_3 \cdot CH \cdot Cl_2\ (Z) \\ \downarrow \text{aq. KOH} & & \downarrow \text{aq. KOH} \\ \text{dihydroxy compound} & & CH_3CHO \text{ (ethanal)} \end{array}$$

Ex. 27. *An organic compound* (A) (*mol. wt.* = 59) *contains* C (40·68%), H (8·48%), N (23·72%) *and* O (27·12%). (A), *on distillation with* P_2O_5 *gave* (B) *which on hydrolysis gave* (C), *a monobasic acid of molecular weight* 60. (B), *on reduction gave* (D), *a monoacid organic base of molecular weight* 45. (A), *on being treated with* Br_2/KOH *produced* (E), *another monoacid base of molecular weight* 31. (A) *can also give* (C) *by boiling it with alkali. Assign structural formulae to* (A) *to* (E).

Solution : For the compound (A):

$$\text{Moles of C : H : N : O} = \frac{40{\cdot}68}{12} : \frac{8{\cdot}48}{1} : \frac{23{\cdot}72}{14} : \frac{27{\cdot}12}{16}$$

$$= 3{\cdot}39 : 8{\cdot}48 : 1{\cdot}69 : 1{\cdot}69$$

$$= 2 : 5 : 1 : 1.$$

Empirical formula of (A) is C_2H_5NO (59)

∴ molecular formula of (A) is also C_2H_5NO.

(A) may be an amide from the reaction sequence given.

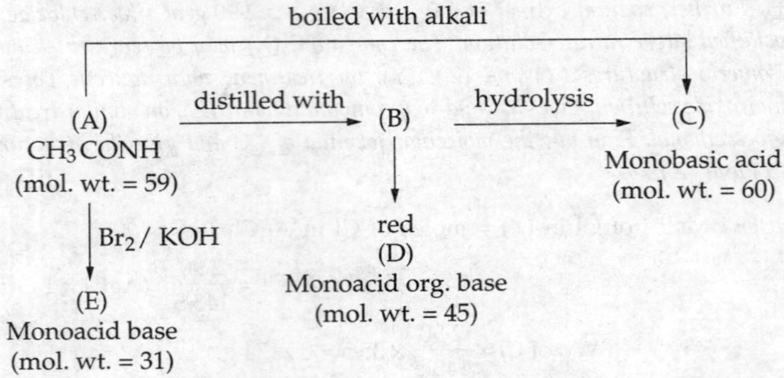

Thus, from the reaction sequence and mol. wt. values,

(A) is CH_3CONH_2

(B) is CH_3CN

(C) is CH_3COOH (Mol. wt. = 60)

(D) is $CH_3CH_2NH_2$ (Mol. wt. = 45)

and (E) is CH_3NH_2 (mol. wt. = 31).

Ex. 28. *An organic compound* (X) *(vapour density* = 37·5*) contains* C (32%), H (6·66%), N (18·67%) *and* O (42·67%). (X), *on reduction, gave a primary amine* (Y) *which, on treatment with* HNO_2, *gave* C_2H_5OH. (Y), *on warming with* $CHCl_3$ *and* KOH, *gave* (Z). (Z) *having an offensive odour, on reduction, produced* $C_2H_5NHCH_3$. *Assign structures of* (X), (Y) *and* (Z).

Solution : For the compound (X):

$$\text{Moles of C : H : N : O} = \frac{32}{12} : \frac{6{\cdot}66}{1} : \frac{18{\cdot}67}{14} : \frac{42{\cdot}67}{16}$$

$$= 2{\cdot}67 : 6{\cdot}66 : 1{\cdot}33 : 2{\cdot}67$$

$$= 2 : 5 : 1 : 2.$$

Empirical formula of (X) is $C_2H_5NO_2$ (75).

Mol. wt. of (X) = $2 \times 37{\cdot}5$ = 75 = empirical formula wt.

$\therefore$ the molecular formula of (X) is $C_2H_5NO_2$.

Now, the reaction sequence may be represented as

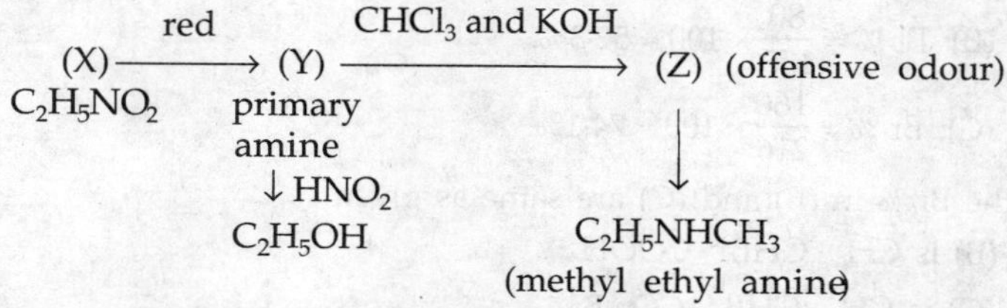

From the reaction sequence, we conclude that a primary amine (Y) is produced by the reduction of (X), (X) must be nitroethane ($C_2H_5NO_2$) and as (Y) on treatment with HNO_2 gave C_2H_5OH, (Y) must be ethylamine ($C_2H_5NH_2$). Further, $C_2H_5NH_2$, on treatment with $CHCl_3$ and KOH, gave (Z) (having an offensive odour) which on reduction gave $C_2H_5NHCH_3$, (Z) must be ethyl isocyanide ($C_2H_5{\cdot}NC$).

Ex. 29. *An organic compound* (A) (*mol. wt.* = 74) *contained* C (48·65%), H (8·11%) *and* O (43·24%). (A), *on treatment with bromine, gave* (B) *containing* 52·3% *of* Br, *while on treatment with phosphorus and bromine, gave* (C) *containing* 74·1% *of bromine. Both* (B) *and* (C), *on boiling with water, gave the same product* (D). (A), *on distillation with soda lime, gave ethane. Assign structural formulae to* (A), (B), (C) and (D).

Solution : For the compound (A):

$$\text{Moles of C : H : O} = \frac{48{\cdot}65}{12} : \frac{8{\cdot}11}{1} : \frac{43{\cdot}24}{16}$$
$$= 4{\cdot}054 : 8{\cdot}11 : 2{\cdot}70$$
$$= 1{\cdot}5 : 3 : 1$$
$$= 3 : 6 : 2.$$

∴ empirical formula of (A) is $C_3H_6O_2$ (74).

As the mol. wt. = emp. formula wt.,

∴ molecular formula of (A) is $C_3H_6O_2$.

As (A), on distillation with soda lime gave C_2H_6, (A) must be $CH_3 \cdot CH_2 \cdot COOH$ (propionic acid)

Now,

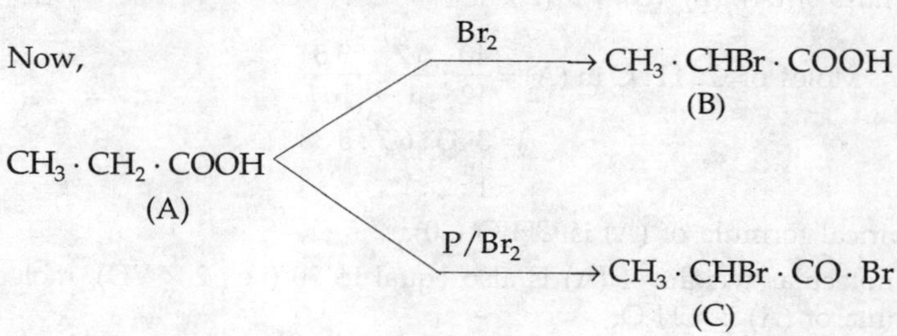

For (B): $Br\,\% = \frac{80}{153} \times 100 \approx 52{\cdot}3\%$.

For (C): $Br\,\% = \frac{160}{216} \times 100 \approx 74{\cdot}1\%$.

As the Br % in (B) and (C) are same as given

(B) is $CH_3 \cdot CHBr \cdot COOH$

and (C) is $CH_3 \cdot CHBr \cdot CO \cdot Br$.

Now since both (B) and (C) will give the same product $CH_3 \cdot CH(OH) \cdot COOH$ (α-hydroxy propionic acid), (D) must be $CH_3 \cdot CH(OH) \cdot COOH$.

Ex. 30. *An ester has a molecular weight of* 102. *On aqueous hydrolysis, it produces a monobasic acid and an alcohol. If* 0·185 g *of the acid produced completely neutralises* 25 mL *of* 0·1 N NaOH, *find out the structural formulae of the produced alcohol, acid and the ester.*

Solution : Let the equivalent weight of the acid formed be E.

m.e. of the acid = m.e. of NaOH

$$\frac{0{\cdot}185}{E} \times 1000 = 0{\cdot}1 \times 25$$

or $$E = 74.$$

As the acid is monobasic, its molecular weight is 74. Thus the reaction sequence may be represented as

$$\underset{\text{ethyl propionate}}{C_2H_5COOC_2H_5} \xrightarrow{\text{hydrolysis}} \underset{\substack{\text{(mol. wt. = 74)} \\ \text{propionic acid}}}{C_2H_5COOH} + \underset{\text{ethyl alcohol}}{C_2H_5OH}$$

Ex. 31. *An organic compound* (A) *contains* 40% *carbon,* 6·7% *hydrogen and rest, oxygen. Its vapour density is* 15. *On reacting with a concentrated solution of* KOH, *it gives two compounds,* (B) *and* (C). *When* (B) *is oxidised, the original compound* (A) *is obtained. When* (C) *is treated with concentrated* HCl, *it gives a compound* (D) *which reduces Fehling's solution as well as ammoniacal* $AgNO_3$ *solution and also gives effervescence with* $NaHCO_3$ *solution. Write the structures of* (A), (B), (C) *and* (D).

Solution : Moles of C : H : O in (A) $= \frac{40}{12} : \frac{6{\cdot}7}{1} : \frac{53{\cdot}3}{16}$

$$= 3{\cdot}33 : 6{\cdot}7 : 3{\cdot}33$$

$$= 1 : 2 : 1.$$

Empirical formula of (A) is CH_2O (30).

As molecular weight of (A) is also equal to 30 (i.e., 2 × VD), molecular formula of (A) is CH_2O.

Now from the following reaction sequence

$$\underset{(CH_2O)}{(A)} \xrightarrow{KOH} (B) + (C)$$

$$(B) \xrightarrow{(O)} \underset{(CH_2O)}{(A)} \qquad (C) \xrightarrow{HCl} (D)$$

(D) reduces Fehling's solution and amm. $AgNO_3$ and gives effervescence with $NaHCO_3$

It is clear that,

(A) is HCHO
(B) is CH_3OH
(C) is HCOOK
and (D) is HCOOH.

Ex. 32. *n-butane is produced by the monobromination of ethane followed by Wurtz reaction. Calculate the volume of ethane at* NTP *required to produce* 55 g *of n-butane, if the bromination takes place with* 90% *and the Wurtz reaction with* 85% *yield.* **(IIT 1989)**

Solution :

$$C_2H_6 \rightarrow C_2H_5Br$$

$$2C_2H_5Br + 2Na \xrightarrow[\text{reaction}]{\text{Wurtz}} C_4H_{10} + 2NaBr$$

1 mole of C_2H_6 gives 1 mole of C_2H_5Br and 2 moles of C_2H_5Br give 1 mole of C_4H_{10}.

Let the number of moles of C_2H_6 be n.

$\therefore$ n moles of C_2H_6 will give $0{\cdot}9\,n$ moles of C_2H_5Br (90%) and $0{\cdot}9\,n$ moles of C_2H_5Br will give $(0{\cdot}45\,n \times 0{\cdot}85)$ moles of n-butane (85%).

Thus, $\quad 0{\cdot}45\,n \times 0{\cdot}85 = \dfrac{55}{58} \quad (C_4H_{10} = 58)$

$$n = 2{\cdot}789 \text{ moles.}$$

$\therefore$ volume of C_4H_{10} at NTP $= 2{\cdot}789 \times 22{\cdot}4$

$$= 55{\cdot}53 \text{ litres.}$$

Ex. 33. *An aromatic hydrocarbon* (A) (*mol. wt.* = 78) *contains* 92·3% *of carbon.* (A), *on treatment with bromine in the dark, produced* (B) *which contains* 45·85% *of carbon,* 3·19% *of hydrogen and* 50·96% *of bromine.* (B), *on heating with* CH_3Br *and* Na *in etherical solution, gave* (C) *containing* 91·3% *of* C *and* 8·7% *of* H. (C), *on oxidation, produced a monobasic acid* (D). *The sodium salt of* (D), *on distillation with soda lime, gave* (A). *Determine the structures of* (A), (B), (C) *and* (D).

Solution : Moles of $C : H = \frac{92 \cdot 3}{12} : \frac{7 \cdot 7}{1}$

$= 7 \cdot 7 : 7 \cdot 7$

$= 1 : 1.$

Empirical formula of (A) is CH and empirical formula weight is 13. As the molecular weight of (A) is 78,

$$n = \frac{78}{13} = 6.$$

Molecular formula of (A) is C_6H_6.

In compound (B):

$$\text{moles of } C : H : Br = \frac{45 \cdot 85}{12} : \frac{3 \cdot 19}{1} : \frac{50 \cdot 96}{80}$$

$$= 3 \cdot 82 : 3 \cdot 19 : 0 \cdot 637$$

$$= 6 : 5 : 1.$$

As (B) must be C_6H_5Br, it has to be a substitution product and not the addition product.

Further, (C) is produced by heating C_6H_5Br with CH_3Br and Na (Fittig reaction), (C) must be $C_6H_5CH_3$. This is also supported by the following data, i.e., for (C) being C_7H_8:

Moles of $C : H = \frac{91 \cdot 3}{12} : \frac{8 \cdot 7}{1}$

$= 1 : 1 \cdot 143$

$= 7 : 8.$

(C) on oxidation, will give C_6H_5COOH (D), the sodium salt of which on distillation with soda lime will give C_6H_6 (A).

Ex. 34. *An organic compound contains* C (30·6%), H (1·7%) *and* Br (67·7%). 0·706 g *of this compound, when dissolved in* 10 g *of acetone, increased the b.p. of acetone by* 0·5°. *If* K_b *for acetone is* 1·67, *find the molecular formula of the compound* (Br = 80).

Solution : We have,

$$\Delta T_b = K_b . m \quad \text{... (Eqn. 8, Ch. 13)}$$

$$0 \cdot 5 = 1 \cdot 67 \times \frac{0 \cdot 706}{M} \times \frac{1000}{10}$$

or $M = 236.$ $\left\{ \begin{array}{l} M \text{ is the mol. wt. of the} \\ \text{organic compound.} \end{array} \right\}$

Now, in the given compound,

moles of $C : H : Br = \frac{30 \cdot 6}{12} : \frac{1 \cdot 7}{1} : \frac{67 \cdot 7}{80}$

$$= 2{\cdot}55 : 1{\cdot}7 : 0{\cdot}84$$
$$= 3 : 2 : 1.$$

$\therefore$ empirical formula is C_3H_2Br (118).

As the molecular weight is twice the empirical formula weight, molecular formula of the organic compound is $C_6H_4Br_2$.

Ex. 35. *An organic compound* (X) (mol. wt. = 94) *containing* C, H *and* O *gave on analysis* 76·6% C, 6·38% H *and* 17·02% O. *A solution of* (X) *in aqueous* NaOH *with* $FeCl_3$ *gave a violet colour while when heated with* CCl_4, *it produced an acid* (Y) *of molecular weight* 138. *Find* (X) *and* (Y).

Solution : Moles of C : H : O $= \dfrac{76{\cdot}60}{12} : \dfrac{6{\cdot}38}{1} : \dfrac{17{\cdot}02}{16}$

$$= 6{\cdot}38 : 6{\cdot}38 : 1{\cdot}05$$
$$= 6 : 6 : 1.$$

Empirical formula of (X) is C_6H_6O (94).

As mol. wt. of (X) is also 94,

molecular formula of (X) is C_6H_6O.

Further, since the solution of (X) in aqueous NaOH gives violet colour with $FeCl_3$, (X) must contain a phenolic group.

Hence (X) is C_6H_5OH.

From the next given information, i.e., alkaline solution of (X) with CCl_4, on heating, produces an acid of molecular weight 138, the acid (Y) must be $C_6H_4\begin{smallmatrix} \diagup OH \\ \diagdown COOH \end{smallmatrix}$

Ex. 36. *An organic compound* (A) *contains* 69·42% C, 5·78% H *and* 11·57% N. *Its vapour density is* 60·5. *It evolves ammonia when boiled with caustic potash. On heating with* P_2O_5, *it gives a compound* (B) *containing* 81·55% C, 4·85% H *and* 13·59% N. *On reduction with* Na *and alcohol,* (B) *forms a base which reacts with* HNO_3 *giving off nitrogen and yielding an alcohol* (C). *The alcohol can be oxidised to benzoic acid. Assign structural formulae to* (A), (B) *and* (C).

Solution : For compound (A):

Moles of C : H : N : O $= \dfrac{69{\cdot}42}{12} : \dfrac{5{\cdot}78}{1} : \dfrac{11{\cdot}57}{14} : \dfrac{13{\cdot}23}{16}$

$$= 5{\cdot}66 : 5{\cdot}78 : 0{\cdot}82 : 0{\cdot}82$$
$$= 7 : 7 : 1 : 1.$$

Empirical formula of (A) is C_7H_7NO (121).

$\therefore$ molecular formula of (A) is also C_7H_7NO.

$$\left(\begin{aligned} \text{mol. wt.} &= 2 \times \text{V. D.} \\ &= 2 \times 60{\cdot}5 = 121 \end{aligned}\right)$$

For compound (B):

$$\text{Moles of C : H : N} = \frac{81{\cdot}55}{12} : \frac{4{\cdot}85}{1} : \frac{13{\cdot}59}{14}$$

$$= 6{\cdot}8 : 4{\cdot}85 : 0{\cdot}97$$

$$= 7 : 5 : 1.$$

$\therefore$ empirical formula of (B) is C_7H_5N.

Molecular formula of (B) will also be C_7H_5N as it is derived from C_7H_7NO (A) by dehydration with P_2O_5.

As (A) evolves ammonia when boiled with caustic potash, it must be an amide ($C_6H_5 \cdot CO \cdot NH_2$). Thus, (B) and (C) can also be named from the given reaction sequence,

$$\underset{\substack{\text{(A)} \\ \text{Benzamide}}}{C_6H_5CONH_2} \xrightarrow[-H_2O]{P_2O_5} \underset{\text{(B)}}{C_6H_5CN} \xrightarrow[\text{Na/alc.}]{\text{Red.}} C_6H_5CH_2NH_2 \xrightarrow{HNO_2}$$

$$\underset{\text{Benzoic acid}}{C_6H_5COOH} \xleftarrow{(O)} \underset{\text{(C)}}{C_6H_5CH_2OH} + H_2O + N_2$$

The structures may be represented as

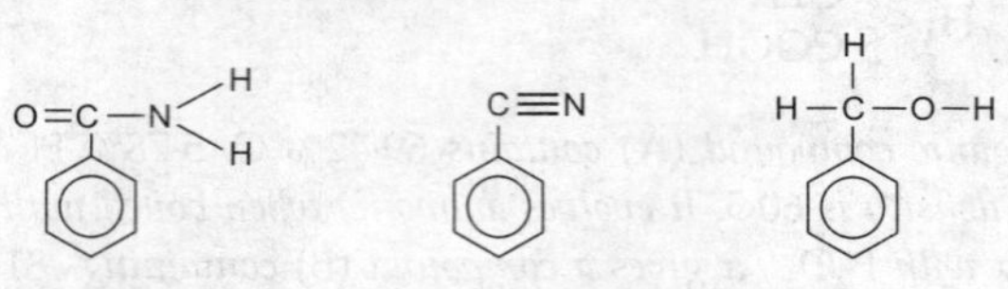

(A) (B) (C)

Ex. 37. *An aromatic compound* (A), *on treatment with a saturated solution of sodium bisulphite, produced a solid crystalline product.* (A) *gave a compound* (B) *and a sodium salt of an aromatic monobasic acid* (C) *on treatment with alkali.* (B) *contained* 77·8% C *and* 7·4% H *and rest, oxygen. The sodium salt on distillation with soda lime gave benzene. If* (B) *may be oxidised with* $KMnO_4$ *to* (C), *find the structural formulae of* (A), (B) *and* (C).

Solution : For the compound (B):

$$\text{Moles of C : H : O} = \frac{77{\cdot}8}{12} : \frac{7{\cdot}4}{1} : \frac{14{\cdot}8}{16}$$

$$= 6{\cdot}48 : 7{\cdot}4 : 0{\cdot}925$$

$$= 7 : 8 : 1.$$

$\therefore$ empirical formula of (B) is C_7H_8O.

The reaction sequence given may be represented as

$$(A) \xrightarrow{NaHCO_3} \text{solid crystalline product}$$

$\hookrightarrow$ (B) + sod. salt of aromatic monobasic acid (C)

(B) $\xrightarrow{(O)}$ (C) aromatic monobasic acid

sod. salt of (C) $\xrightarrow{\text{distl. with soda lime}}$ C_6H_6

(A) is an aromatic aldehyde as it forms crystalline product with $NaHSO_3$.

Sodium salt of (C), on distillation with soda lime gives C_6H_6, so (C) must be C_6H_5COOH. Further, since (C) is produced by the oxidation of (B), it must be a homologue of C_6H_6 or some compound containing an easily oxidisable side chain. From the formula of (B), it must be $C_6H_5 \cdot CH_2OH$ (benzyl alcohol). Again, since $C_6H_5 \cdot CH_2OH$ (B) and C_6H_5COOH (C) are produced from (A), on treatment with alkali, (A) must be C_6H_5CHO (benzaldehyde) (Cannizzaro reaction).

Ex. 38. *An organic compound $C_xH_{2y}O_y$ was burnt with twice the amount of oxygen needed for complete combustion to CO_2 and H_2O. The hot gases when cooled to 0°C and 1 atm pressure measured 2·24 litres. The water collected during cooling weighed 0·9 g. The vapour pressure of pure water at 20°C is 17·5 mmHg and is lowered by 0·104 mm when 50 g of the organic compound is dissolved in 1000 g of water. Give the molecular formula of the organic compound.*

Solution : (See Ex. 10, Chapter 3)

Ex. 39. *An aromatic compound contains 69·4% C and 5·8% H. A sample of 0·303 g of this compound was analysed for nitrogen by Kjeldahl method. The ammonia evolved was absorbed in 50 mL of 0·05 M H_2SO_4. The excess acid required 25 mL of 0·1 M NaOH for neutralisation. Determine the molecular formula of the compound if its molecular weight is 121. Draw two possible structures for this compound.*

Solution : Let us first calculate the percentage of N in the compound. As all the N is converted to NH_3, applying POAC for N atoms:

$$\text{Moles of N in the compound} = \text{moles of N in } NH_3$$
$$= 1 \times \text{moles of } NH_3$$
$$= \text{eq. of } NH_3$$

$$= \text{m.e. of } NH_3/1000$$
$$= \text{m.e. of } H_2SO_4/1000$$
$$= \frac{\text{total m.e. of } H_2SO_4 - \text{m.e. of excess } H_2SO_4}{1000}$$
$$= \frac{\text{total m.e. of } H_2SO_4 - \text{m.e. of NaOH}}{1000}$$
$$= \frac{50 \times 0{\cdot}1 - 25 \times 0{\cdot}1}{1000}$$
$$= 0{\cdot}0025.$$

$\therefore$ wt. of N $= 0{\cdot}0025 \times 14 = 0{\cdot}035$

and % of N $= \frac{0{\cdot}035}{0{\cdot}303} \times 100 = 11{\cdot}55\%$.

$\therefore$ % of O $= 100 - (69{\cdot}4 + 5{\cdot}8 + 11{\cdot}55)$
$= 13{\cdot}25\%$

$\therefore$ moles of C : H : N : O $= \frac{69{\cdot}4}{12} : \frac{5{\cdot}8}{1} : \frac{11{\cdot}55}{14} : \frac{13{\cdot}25}{16}$
$= 5{\cdot}8 : 5{\cdot}8 : 0{\cdot}825 : 0{\cdot}825$
$= 7 : 7 : 1 : 1.$

$\therefore$ empirical formula of the aromatic compound is C_7H_7NO (121).

As the molecular weight is also 121,
molecular formula is C_7H_7NO.

Since the compound is aromatic, it may be written as $C_6H_5CH{=}NOH$ (benzaldoxime) with the following isomeric structures:

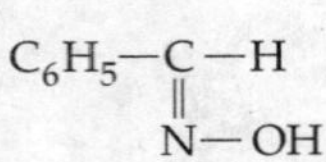

and

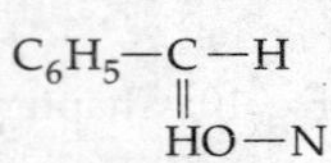

Ex. 40. *The f.p. constant of benzene is* 4·9 *and its m.p. is* 5·51°C. *A solution of* 0·816 g *of compound* (A) *when dissolved in* 7·5 g *of benzene freezes at* 1·59°C. *The compound* (A) *has* C, 70·58% *and* H, 5·88%. *Determine the molecular weight and molecular formula of* (A). *Compound* (A) *on heating with soda lime gives another compound* (B) *which on oxidation and subsequent acidification gives an acid* (C) *of equivalent weight* 122. (C) *on heating with soda lime gives benzene. Identify* (A), (B) *and* (C) *and explain the reactions involved.*

Solution : We have,

$$\Delta T_f = K_f \, . \, m \qquad \text{... (Eqn 7, Ch. 13)}$$

$$(5{\cdot}51 - 1{\cdot}59) = 4{\cdot}90 \times \left(\frac{0{\cdot}816}{M} \times \frac{1000}{7{\cdot}5}\right)$$

$$M = 136 \qquad [M \text{ is mol. wt. of compound (A)}]$$

$$\text{Moles of C : H : O} = \frac{70{\cdot}58}{12} : \frac{5{\cdot}88}{1} : \frac{23{\cdot}54}{16}$$

$$= 5{\cdot}88 : 5{\cdot}88 : 1{\cdot}471$$

$$= 4 : 4 : 1$$

$\therefore$ empirical formula of (A) = C_4H_4O (68).

As the molecular weight of (A) is 136,

molecular formula of (A) is $C_8H_8O_2$.

Now, from the given reaction sequence:

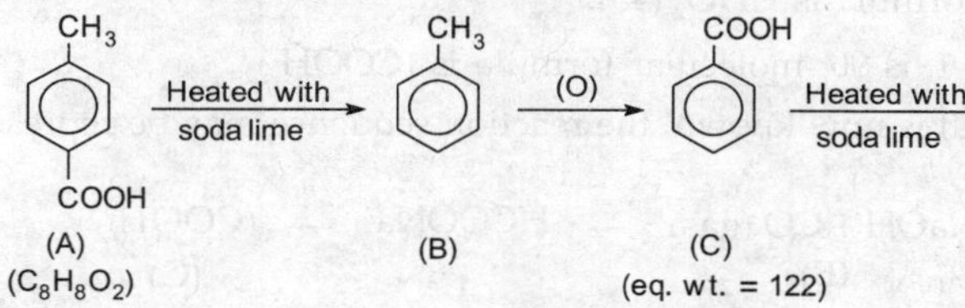

Ex. 41. *The sodium salt of carboxylic acid* (A) *was produced by passing a gas* (B) *into an aqueous solution of caustic alkali at an elevated temperature and pressure.* (A), *on heating in the presence of* NaOH, *followed by treatment with* H_2SO_4 *gave a dibasic acid* (C). *A sample of* 0·4 g *of* (C), *on combustion, gave* 0·08 g *of* H_2O *and* 0·39 g *of* CO_2. *The silver salt of the acid weighing* 1 g, *on ignition, yielded* 0·71 g *of* Ag *as residue. Identify* (A), (B) *and* (C). **(IIT 1990)**

Solution : Calculation of molecular weight of the acid (C):

As the acid is dibasic, the Ag salt of it will contain 2 atoms of Ag in one molecule of the acid (see Ag-salt method, Chapter 5).

Acid	$\rightarrow$	Salt containing	$\rightarrow$	Ag
(C)		2 Ag atoms		0·71 g
		1 g		

Applying POAC for Ag atom in the second step,

2 × moles of the salt = moles of Ag in the product

$$2 \times \frac{1}{M} = \frac{0{\cdot}71}{108} \cdot \qquad \left(\begin{matrix} M \equiv \text{mol. wt.} \\ \text{of salt} \end{matrix}\right)$$

$$\therefore \qquad M \approx 304 \cdot$$

Hence, mol. wt. of the acid = $(304 - 2 \times 108 + 2)$

$$= 90.$$

Calculation of empirical formula weight and molecular formula of the acid (C):

$$\text{Moles of C} = \frac{0{\cdot}39}{44} = 0{\cdot}0088.$$

$$\text{Wt. of C} = 0{\cdot}0088 \times 12 = 0{\cdot}1064 \text{ g}.$$

$$\text{Moles of H} = \frac{2 \times 0{\cdot}08}{18} = 0{\cdot}0088.$$

$$\text{Wt. of H} = 0{\cdot}0088 \times 1 = 0{\cdot}0088 \text{ g}.$$

$$\therefore \quad \text{wt. of O} = [0{\cdot}4 - (0{\cdot}1064 + 0{\cdot}0088)] = 0{\cdot}2848.$$

$$\text{Moles of O} = \frac{0{\cdot}2848}{16} = 0{\cdot}0178.$$

$\therefore$ moles of C : H : O = 0·0088 : 0·0088 : 0·0178 = 1 : 1 : 2.

$\therefore$ empirical formula is CHO_2 (45).

As the mol. wt. is 90, molecular formula is $(COOH)_2$.

As the acid (C) is now known, the reaction sequence may be represented as

$$\underset{(B)}{NaOH + CO \text{ (gas)}} \rightarrow \underset{\underset{\underset{(A)}{HCOOH}}{\uparrow}}{HCOONa} \rightarrow \underset{(C)}{(COOH)_2}$$

PROBLEMS

(Answers bracketed with questions)

1. A sulphide of Fe contains 46·5% of Fe by weight. Find the empirical formula of the sulphide. (FeS_2)

2. 1·60 g of an oxide of iron, on heating in a stream of hydrogen gas, completely converted to 1·12 g of iron. Find the empirical formula of the oxide. (Fe_2O_3)

3. Carbon combines with hydrogen to form three compounds, A, B and C. The percentages of hydrogen in A, B and C are 25%, 14·3% and 7·7% respectively. Find the empirical formula of the compound. (CH_4, CH_2, CH)

4. A compound of carbon, hydrogen and nitrogen contains three elements in the respective weight ratio of 9 : 1 : 3. Calculate its empirical formula. If its molecular weight is 108, what is its molecular formula? (C_3H_4N, $C_6H_8N_2$)

5. A hydrated salt of iron sulphate weighing 2 g, contains 0·9065 g of water of crystallisation. Find the formula of the hydrated salt. ($FeSO_4 \cdot 7H_2O$)

6. An iron compound yielded C = 64·4%, H = 5·5% and Fe = 29·9%. What is the simple formula of the compound? ($C_{10}H_{10}Fe$)

7. 1·763 g of hydrated $BaCl_2$ was heated to dryness. The anhydrous salt remained was 1·505 g. What is the formula of the hydrate? ($BaCl_2 \cdot 2H_2O$)

8. A boron–hydrogen compound weighing 0·0553 g created a pressure of 0·658 atm in a bulb of 40·7 mL volume at 100°C. Analysis showed it to be 85·7% boron. What is its molecular formula? (B_5H_9)

9. A mixture of one volume of gas and two volumes of oxygen, on explosion form two volumes of CO_2 and one volume of N_2 under the identical conditions of temperature and pressure. Find the formula of the gas. (C_2N_2)

10. 9 volumes of a gaseous mixture consisting of a gaseous organic compound A and just sufficient amount of oxygen required for complete combustion yielded on burning 4 volumes of CO_2, 6 volumes of water vapour and 2 volumes of N_2, all volumes measured at the same temperature and pressure. If the compound has only C, H and N,

(i) how many volumes of oxygen are required for complete combustion?

(ii) what is the molecular formula of the compound A? (7 vol., $C_2H_6N_2$)

[Hint: See Example 13]

11. A sample of gaseous hydrocarbon occupying 1·12 litres at NTP when completely burnt in air produced 2·2 g of CO_2 and 1·8 g of H_2O. Calculate the weight of the compound and the volume of the oxygen at NTP required for its burning. Find the molecular formula of the compound. (0·8 g, 2·24 litres, CH_4)

12. 0·21 g of an organic compound containing C, H, O and N gave upon combustion 0·462 g of CO_2 and 0·1215 g of H_2O. The ammonia produced on distillation of 0·104 g of this compound with NaOH, required 15 mL of N/20 H_2SO_4 for neutralisation. Find the empirical formula of the compound. ($C_7H_9NO_2$)

13. An organic compound on qualitative analysis was found to contain C, H, N and O. 1·0 g of it on oxidation with CuO and oxygen gave 1·239 g of CO_2 and 0·1269 g of H_2O. 2 g of the sample was digested with concentrated sulphuric acid and the residue was distilled after the addition of excess solution of sodium hydroxide. The ammonia evolved was absorbed in 50 mL of 1·0 N sulphuric acid. The resulting solution was diluted to 500 mL in a measuring flask. 25 mL of this solution required 21·8 mL of 0·05 N NaOH for complete neutralisation. Calculate the empirical formula of the compound. (C_2HNO_2)

14. 0·2614 g of an organic compound gave upon combustion 0·2324 g of CO_2 and 0·0950 g of H_2O. 0·1195 g of this compound gave 0·3470 g of AgCl. If the vapour density of the substance is 49·5, calculate its molecular formula. ($C_2H_4Cl_2$)

15. A dibasic anhydrous acid produced upon combustion 0·195 g of CO_2 and 0·04 g of H_2O. 0·5 g of the salt of this acid when ignited, converted to 0·335 g of pure silver. Find the molecular formula of the dibasic acid. [$(COOH)_2$]

16. A monoacid organic base gave the following results on analysis:

(i) 0·10 g of the base gave 0·2882 g of CO_2 and 0·0756 g of H_2O.

(ii) 0·20 g of the base produced 21·8 mL of N_2 at 15°C and 760 mm pressure.

(iii) 0·40 g of the platinichloride left 0·125 g of Pt.

Calculate the molecular formula of the base. (C_7H_9N)

17. An organic compound containing C, H, O and S gave the following data on analysis:

(a) The combustion of 0·1668 g of the compound resulted in the formation of 0·4540 g and 0·0663 g of CO_2 and H_2O respectively.

(b) 0·1254 g of the compound on heating with HNO_3 and $BaCl_2$ yielded 0·1292 g of $BaSO_4$.

Calculate the empirical formula of the organic compound. ($C_{14}H_{10}SO$)

18. An organic compound containing 92·3% of C, 7·7% of H had the molecular weight 26. When treated with bromine, it gave a product containing 92·5% of bromine but when treated with HBr, it gave a product containing 85·1% of bromine. What is the structural formula of the organic compound? ($CH \equiv CH$)

19. An organic aromatic compound (X) containing 52·2% of C, 3·7% of H and 44·1% of chlorine on oxidation with alkaline $KMnO_4$ gave a monobasic acid (Y), the sodium salt of which on distillation with soda lime gave benzene. What is the structural formula of (X) and (Y)? ($C_6H_5CH \cdot Cl_2$, C_6H_5COOH)

20. One mole of hydrocarbon (A) reacts with one mole of bromine giving a dibromocompound, $C_5H_{10}Br_2$. Substance (A) on treatment with cold, dilute alkaline $KMnO_4$ solution forms a compound $C_5H_{12}O_2$. On ozonolysis, (A) gives equimolar quantities of propanone and ethanal. Deduce the structural formula of (A).

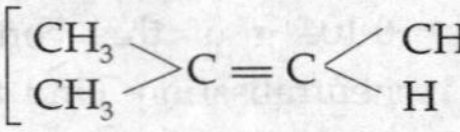

21. An alkene (A), on ozonolysis yields acetone and an aldehyde. The aldehyde is easily oxidised to an acid (B). When (B) is treated with bromine in presence of phosphorus, it yields compound (C) which on hydrolysis gives a hydroxy acid (D). The acid can also be obtained from acetone by the reaction with hydrogen cyanide followed by hydrolysis. Identify the compounds (A), (B), (C) and (D).

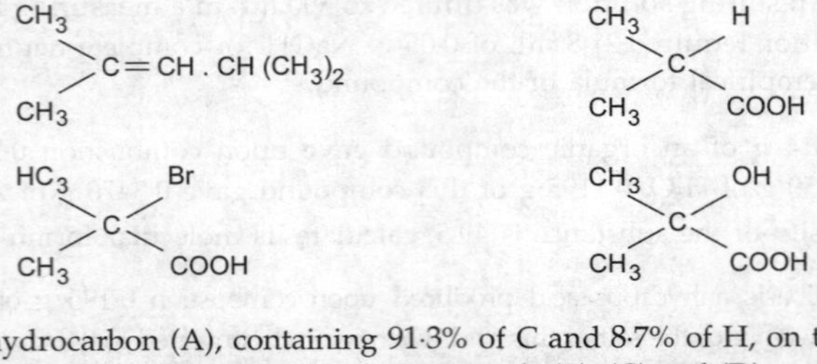

22. An aromatic hydrocarbon (A), containing 91·3% of C and 8·7% of H, on treatment with chlorine gave 3 isomeric monochlorocompounds (B), (C) and (D), each having 28% of chlorine. On oxidation with permanganate, all the three gave a monobasic acid. The acid from (B) on distillation with soda lime gave benzene while those from (C) and (D) gave monochlorobenzene on the same treatment. Assign formulae to (A), (B), (C) and (D).

[(A) $C_6H_5 \cdot CH_3$ (B) $C_6H_5CH_2Cl$
(C) $C_6H_4(CH_3)Cl$ (ortho) (D) (para)]

23. 0·15 g of an organic compound (A) gave 0·33 g of CO_2 and 0·18 g of H_2O. The molecular weight of (A) is 60. The compound (A) on dehydration gave a hydrocarbon (B) containing 85·7% of C. (B), on successive treatment with hydriodic acid and silver hydroxide gave a product (C), isomeric with (A). Find structural formulae of (A), (B) and (C).

[(A) $CH_3CH_2CH_2OH$
(B) $CH_3CH{=}CH_2$
(C) $CH_3CH(OH)CH_3$]

24. Compound (A) with molecular weight 108, contained 88·89% C and 11·11% H. It gave a white precipitate with ammoniacal silver nitrate. Complete hydrogenation atom of (A) gave another compound (B) with molecular weight 112. Oxidation of (A) gave an acid with equivalent weight 128. Decarboxylation of this acid gave cyclohexane. Give structures of (A) and (B).

[$C_6H_{11}C{\equiv}CH$ (A), $C_6H_{11}\,CH_2{-}CH_3$ (B)]

25. A compound (X) containing only C, H and O is unreactive towards sodium. It does not add bromine. It does not react with Schiff reagent. On refluxing with an excess of hydriodic acid, (X) yields only one organic product (Y). On hydrolysis, (Y) yields a new compound (Z) which can be converted to (Y) by reaction with red phosphorous and iodine. The compound (Z), on oxidation with $KMnO_4$ gives a carboxylic acid. The equivalent weight of this acid is 60. What are the compounds (X), (Y) and (Z). Write chemical equation leading to the conversion of (X) to (Y).

[$C_2H_5OC_2H_5$, C_2H_5I, C_2H_5OH
$C_2H_5OC_2H_5 + 2HI \rightarrow 2C_2H_5I + H_2O$]

26. 448 mL of a hydrocarbon (A) having C (87·80%), H (12·19%) weigh 1·64 g at NTP. On hydrogenation it gives 2 methyl pentane. Treatment of (A) with acidic $HgSO_4$ gives a new compound (B) of molecular weight $C_6H_{12}O$. Compound (A) does not react with ammoniacal $AgNO_3$. What is the structure of (A)?

[$CH_3{-}CH(CH_3){-}C{\equiv}C{-}CH_3$]

27. 0·369 g of a bromo derivative of a hydrocarbon (A) when vaporised occupied 67·2 mL at NTP. (A), on reaction with aqueous NaOH, gives (B). (B), when passed over alumina at 250°C, gives a neutral compound (C) while at 300°C, it gives a hydrocarbon (D). (D), when heated with HBr gives an isomer of (A). When (D) is treated with concentrated H_2SO_4 and the product is diluted with water and distilled, (E) is obtained. Identify (A) to (E).

[(A) $CH_3 \cdot CH_2CH_2Br$ (B) $CH_3 \cdot CH_2CH_2OH$
(C) $C_3H_7 \cdot O \cdot C_3H_7$ (D) $CH_3{-}CH{=}CH_2$
(E) $(CH_3)_2CH{-}OH$]

28. When 0·90 g of an organic compound $C_4H_{10}O_2$ (A) was treated with Na, 224 mL of H_2 were evolved at NTP, compound (A) could be separated into two fractions (B) and (C) by crystallisation, of which fraction (B) could be resolved into optical isomers, (D) and (E). Write down the structural formulae for (C), (D) and (E) with proper reasoning.

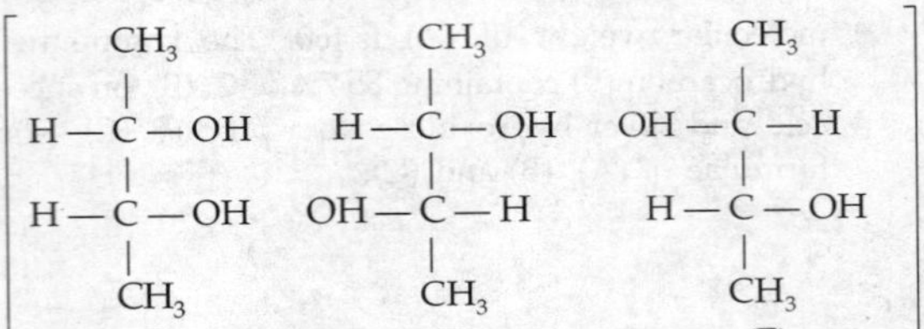

(C) (D) (E)

29. An organic compound (A) has 76·6% C and 6·38% H. Its vapour density is 47. It gives characteristic colour with aqueous $FeCl_3$ solution. (A), when heated with CO_2 and NaOH at 140°C under pressure, gives (B) which, on being acidified gives (C). (C) reacts with acetyl chloride to give (D) which is a well-known painkiller. Identify (A), (B), (C) and (D).

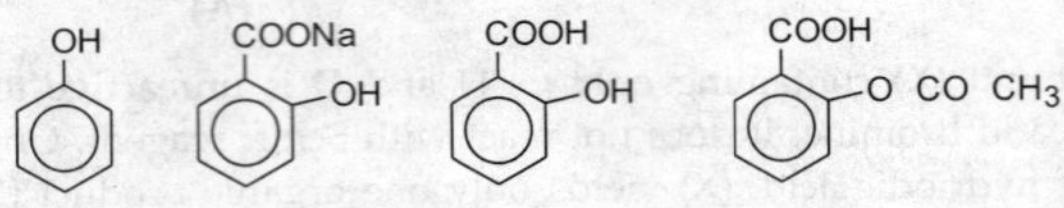

(A) (B) (C) (D)

30. An unknown compound of C, H and O contains 69·77% C and 11·63% H and has a molecular weight of 86. It does not reduce Fehling's solution but forms a bisulphite addition compound and gives a positive iodoform test. What are the possible structures?

$\left[\begin{matrix} CH_3 \\ C_3H_7 \end{matrix} > CO; \quad \begin{matrix} C_2H_5 \\ C_2H_5 \end{matrix} > CO\right]$

[Hint: See Example 25]

31. An organic compound (mol. wt. = 44) (X) contains 54·54% of C and 9·09% of H. With PCl_5, (X) gives a compound of molecular weight 99. On oxidation it gives an acid of molecular weight 60. What is (X)? (CH_3CHO)

32. A ketone (A) which undergoes haloform reaction gives compound (B) on reduction. (B) on heating with H_2SO_4 gives compound (C), which forms monozonide (D). (D) on hydrolysis in the presence of Zn dust gives only acetaldehyde. Identify (A), (B) and (C).

[(A) $CH_3COCH_2CH_3$ (B) $CH_3CH(OH)CH_2CH_3$ (C) $CH_3CH=CHCH_3$]

33. An organic compound (A) having molecular weight, 58, contained 62·06% of C and 10·35% of H and rest, oxygen. (A), on reduction gave (B) which gave iodoform test. (B), on dehydration, gave an unsaturated hydrocarbon (C) having molecular weight 42. Find (A), (B) and (C).

($CH_3 \cdot CO \cdot CH_3$; $CH_3CH(OH)CH_3$; $CH_3 \cdot CH{=}CH_2$)

34. An aromatic compound (X) contains C (79·25%) and H (5·66%). (X) on treatment with alkali, gave a neutral product (Y) containing C (77·78%) and H (7·41%) and the sodium salt of an aromatic organic acid (Z) which on distillation with soda lime gave benzene. Assign structural formulae to (X), (Y) and (Z).

(C_6H_5CHO, $C_6H_5CH_2OH$, C_6H_5COOH)

35. An aromatic compound (X) contained 58·5% of C, 4·1% of H and 11·4% of N. (X) may be obtained by the action of HNO_3 on a compound (Y). (X), on reduction, gives a monoacid organic base (Z). Give structural formulae of (X), (Y) and (Z).

($C_6H_5NO_2$, C_6H_6, $C_6H_5NH_2$)

36. An aliphatic amine (X) contains C (61·01%), H (15·25%) and N (23·74%). (X), on treatment with HNO_2, gave an alcohol (Y) having C(60%) and H (13·33%). (Y), on careful oxidation gave (Z) of vapour density 29. (Z) gave iodoform test. Assign structural formulae to (X), (Y) and (Z).

$$\left[\begin{matrix} CH_3 \\ CH_3 \end{matrix} \!\!>\! CH \cdot NH_2 ; \quad \begin{matrix} CH_3 \\ CH_3 \end{matrix} \!\!>\! CH \cdot OH ; \quad \begin{matrix} CH_3 \\ CH_3 \end{matrix} \!\!>\! CO \right]$$

37. An organic compound (A) contains C=32%, H=6·66% and N=18·67%. On reduction, it gives a primary amine (B) which gives ethyl alcohol with nitrous acid. (B) gives an offensive odour on warming with $CHCl_3$ and KOH and gives compound (C) which on reduction forms ethyl methyl amine. Assign the structures of (A), (B) and (C).

$$\left[\begin{matrix} (A)\ C_2H_5NO_2 \\ (B)\ C_2H_5NH_2 \\ (C)\ C_2H_5NC \end{matrix} \right]$$

[Hint: See Example 28]

38. Compound (A) gives positive Lucas test in 5 minutes. When 6 g of (A) is treated with Na metal, 1120 mL of H_2 is evolved at STP. Assuming (A) to contain one atom of oxygen per molecule, write the structural formula of (A).

[$CH_3 \cdot CH(OH) \cdot CH_3$]

39. An organic compound (A) of molecular weight 140·5 has 68·32% C, 6·4% H and 25·26% Cl. Hydrolysis of (A) with dilute acid gives compound (B), $C_8H_{10}O$. Compound (B) can be oxidised under mild condition to compound (C), C_8H_8O. Compound (C) forms a phenyl hydrazone (D) with $PhNHNH_2$ and gives a positive iodoform test. Give the structures of compounds (A) to (D).

$$\left[\begin{matrix} (A)\ C_6H_5\ CH \cdot Cl\,CH_3 & (B)\ \begin{matrix} C_6H_5 \\ CH_3 \end{matrix} \!\!>\! CH \cdot OH \\ (C)\ \begin{matrix} C_6H_5 \\ CH_3 \end{matrix} \!\!>\! CO & (D)\ C_6H_5{-}\underset{\displaystyle CH_3}{\underset{|}{C}}{=}N \cdot NH \cdot C_6H_5 \end{matrix} \right]$$

40. 10 g of a mixture of hexane and ethanol are reacted with Na to give 200 mL of hydrogen at 27°C and 760 mmHg pressure. What is the percentage of ethanol in the mixture.

(7·475%)

[Hint: $2C_2H_5OH + 2Na \rightarrow 2C_2H_5ONa + H_2$]

41. A mixture of 0·535 g of ethanol and acetaldehyde when heated with Fehling's solution gave 1·2 g of a red precipitate. What is the percentage of acetaldehyde in the mixture? (Cu = 63·5) (34·50%)

[**Hint:** Acetaldehyde reduces Fehling's solution to red cuprous oxide (Cu_2O) and $CH_3CHO + [O] \rightarrow CH_3COOH$]

42. An organic compound, (A) containing C, H, N and O, on analysis gives 49·32% carbon, 9·59% hydrogen and 19·18% nitrogen. (A) on boiling with NaOH gives off NH_3 and a salt which on acidification gives a monobasic nitrogen-free acid (B). The silver salt of (B) contains 59·67% silver. Deduce the structures of (A) and (B).

(IIT 1988) [(A) $C_2H_5CONH_2$ (B) C_2H_5COOH]

43. A certain compound was known to have a formula which would be represented as $[PdC_xH_yN_z](ClO_4)_2$. Analysis showed that the compound contained 30·15% C and 5·06% H. When converted to the corresponding thiocyanate, $[PdC_xH_yN_z](SCN)_2$, the analysis was 40·412% C and 5·94% H. Find x, y and z.

(14, 28, 4)

44. Gaseous hydrocarbon gives upon combustion 0.72 g of water and 3.08 g of CO_2. The empirical formula of the hydrocarbon is

(a) C_2H_4 (b) C_3H_8 (c) C_6H_5 (d) C_7H_8

(d)

(IIT 2013 Main)

45. The ratio of mass per cent of C and H of an organic compound ($C_xH_yO_z$) is 6 : 1. If one molecule of the above compound contains half as much oxygen as required to burn one molecule of the compound C_xH_y completely to CO_2 and H_2O. The empirical formula of the compound $C_xH_yO_z$ is

(a) $C_2H_4O_3$ (b) $C_3H_6O_3$ (c) C_2H_4O (d) $C_3H_4O_2$

(a)

[IIT 2018 Main]

[**Hint:** Mole ratio of C & H in 1 mole of $C_xH_yO_z = \dfrac{6/12}{1/1} = 1:2 = x:y$

As 2z mole of O (or z mole O_2) is used to combust one mole of C_xH_y

$$\underset{\text{1 mole}}{C_xH_y} + \underset{z\text{ mole}}{O_2} \rightarrow CO_2 + H_2O$$; applying POAC for C, H and O, we get,

$z = x + \dfrac{y}{4}$, if $x = 1$ and $y = 2$ $\therefore$ $z = 1.5$, i.e., $C_2H_4O_3$ should be the empirical formula.]

❑

CHAPTER ELEVEN

ATOMIC STRUCTURE AND RADIOACTIVITY

ATOMIC STRUCTURE

An atom consists of an extremely small and dense nucleus and an extranuclear space. The nucleus contains positively charged protons and neutral neutrons, and these particles are collectively called nucleons. In the extranuclear space, negatively charged electrons revolve around the nucleus. As the magnitude of the charge of an electron is the same as that of a proton, the number of electrons is equal to that of protons in an atom, the atom being neutral.

The number of protons present in the nucleus of an atom is termed as the atomic number of the element (Z). The sum of the number of protons and neutrons is called the mass number (A). The term 'nuclide' refers to a nucleus having a specific atomic number and specific mass number.

Since each proton and each neutron has a mass approximately equal to 1 (amu) on the atomic weight scale, the atomic weight of an element is approximately equal to the mass number. The electron has a negligible mass compared to the proton and neutron. The mass of an electron on the atomic weight scale is approximately equal to 1/1837 amu (1 amu = $1{\cdot}66 \times 10^{-24}$ g or $1{\cdot}66 \times 10^{-27}$ kg).

To calculate the radius (r) and energy (E) of a permissible orbit for one-electron species like H, He^+, Li^{2+}, etc., Bohr derived equations based on the following postulates.

Bohr's Postulates

1. The electrons revolve around the nucleus in certain orbits without losing energy because the energy in a fraction of a quantum can neither be lost nor gained.

2. Energy is absorbed or emitted only when an electron in an atom jumps from one orbit to another.

3. The electron is restricted to those orbits in which its angular momentum is an integral multiple of $h/2\pi$.

   Angular momentum $= mvr = n \times \dfrac{h}{2\pi}$ where $\boldsymbol{n}$ is an integer.

   The following equations were derived:

$$r = \frac{n^2 h^2}{4\pi^2 mZe^2} \qquad \dots (1)$$

$$E = -\frac{2\pi^2 Z^2 e^4 m}{n^2 h^2} \qquad \text{... (2)}$$

$$v = \frac{2\pi Ze^2}{nh} \qquad \text{... (3)}$$

In the cgs unit system, m is the mass in grams, e is the charge of an electron in esu, v is the electronic velocity in cm/s, r is the radius of the orbit in cm, h is the Planck constant in erg · s, E is the energy in ergs, Z is the atomic number and n is an integer like 1, 2, 3, etc., representing the first, second and third orbits respectively. Thus the radius of the first orbit ($n = 1$) of a hydrogen atom (or any one electron specie) is given by

$$r_1 = \frac{h^2}{4\pi^2 mZe^2}.$$

The radius of the nth orbit of a hydrogen atom (or any one-electron specie) is given by

$$r_n = r_1 n^2. \qquad \text{... (4)}$$

Again, the energy of the first orbit of a hydrogen atom (or any one-electron specie) is given by

$$E_1 = -\frac{2\pi^2 Z^2 e^4 m}{h^2}.$$

Thus the energy of the nth orbit is given by

$$E_n = \frac{E_1}{n^2}. \qquad \text{... (5)}$$

The negative sign of the energy indicates that the energy of an electron increases with the increase in n, that is to say the outer orbit has a greater energy than the inner one.

Now the velocity of an electron in the first orbit of a hydrogen atom (or any one-electron specie) may be given by

$$v_1 = \frac{2\pi Ze^2}{h}.$$

Thus the electronic velocity in the nth orbit will be given by

$$v_n = \frac{v_1}{n}. \qquad \text{... (6)}$$

When an electron drops from a higher orbit, say n_2, to a lower orbit, say n_1, radiations are emitted, the energy of which can be calculated using Planck's equation, $\Delta E = h\nu$, where ν is the frequency of the radiation emitted.

$$\Delta E = E_2 - E_1$$

$$\Delta E = -\frac{2\pi^2 Z^2 e^4 m}{n_2^2 h^2} - \left(-\frac{2\pi^2 Z^2 e^4 m}{n_1^2 h^2}\right)$$

or, $$h\nu = \frac{2\pi^2 Z^2 e^4 m}{h^2}\left(\frac{1}{n_1^2} - \frac{1}{n_2^2}\right).$$

But $c = \nu\,\lambda$, where c is the velocity of light and λ, the wavelength. So we have,

$$\text{wavenumber} = \frac{1}{\lambda} = \frac{2\pi^2 Z^2 e^4 m}{ch^3}\left(\frac{1}{n_1^2} - \frac{1}{n_2^2}\right) \quad \text{... (7)}$$

or, $$\frac{1}{\lambda} = RZ^2\left(\frac{1}{n_1^2} - \frac{1}{n_2^2}\right).$$

where $$R = \frac{2\pi^2 e^4 m}{ch^3}.$$

R is called Rydberg constant.

If the data in the problems are given in SI units, that is, m in kg, e in coulombs, v in metres/second, r in metres, h in joule · s and E in joules, the following formulae are to be used.

$$r = \frac{\varepsilon_0 n^2 h^2}{\pi m Z e^2} \quad \text{... (8)}$$

$$E = -\frac{Z^2 e^4 m}{8\varepsilon_0^2 n^2 h^2} \quad \text{... (9)}$$

$$v = \frac{Ze^2}{2nh\varepsilon_0} \quad \text{... (10)}$$

$$\left.\begin{aligned}\text{Wavenumber} = \frac{1}{\lambda} &= \frac{e^4 m Z^2}{8\varepsilon_0^2 ch^3}\left\{\frac{1}{n_1^2} - \frac{1}{n_2^2}\right\} \\ &= RZ^2\left\{\frac{1}{n_1^2} - \frac{1}{n_2^2}\right\}\end{aligned}\right], \quad \text{... (11)}$$

where $$R = \frac{e^4 m}{8\varepsilon_0^2 ch^3}$$

ε_0 (permittivity of free space)

$$= 8{\cdot}854 \times 10^{-12}\ \text{coulomb}^2/\text{newton} \cdot \text{metre}^2$$

Hydrogen Spectral Lines

Lines	n_1	n_2	Spectral Region
Lyman	1	2, 3, 4, ...	Ultraviolet (UV)
Balmer	2	3, 4, 5, ...	Visible
Paschen	3	4, 5, 6, ...	Infrared (IR)
Brackett	4	5, 6, 7, ...	Infrared (IR)
Pfund	5	6, 7, 8, ...	Infrared (IR)

The Bohr theory can be extended to one-electron species other than the hydrogen atom, for example, He^+, Li^{2+} and Be^{3+}. In these cases the energies are related to the quantum number, n, through the expression

$$E_n = -\frac{Z^2B}{n^2}\quad \left[\begin{array}{l} B = \dfrac{2\pi^2 e^4 m}{h^2} \text{ (cgs)} \\ B = \dfrac{e^4 m}{8\varepsilon_0^2 h^2} \text{ (SI)} \end{array}\right],$$

where Z is the atomic number of the species and $B = 2{\cdot}18 \times 10^{-11}$ erg [to calculate E in erg (cgs)] and $B = 2{\cdot}18 \times 10^{-18}$ J [to calculate E in J (SI)].

Nuclear Coulombic Barrier

The repulsive force between the nucleus and the approaching positively charged α-particle is called ***nuclear coulombic barrier.*** The potential energy of repulsion is given by

$$V = \frac{(Z_1e)\,(Z_2e)}{d} = \frac{Z_1\,Z_2\,e^2}{d}, \qquad \ldots (12)$$

where Z_1 and Z_2 are the atomic numbers of the nucleus and the α-particle respectively and d is the distance between them. Using the above equation, one can calculate V in ergs, substituting e in esu and d in cm (cgs units).

In SI system, use

$$V = \frac{Z_1\,Z_2\,e^2}{4\pi\varepsilon_0\,d}, \qquad \ldots (13)$$

where V is in J, e in coulombs, d in m and $\varepsilon_0 = 8{\cdot}854 \times 10^{-12}\ C^2/N \cdot m^2$

Failure of Bohr Theory

1. It does not explain the spectra of species having more than one electron.
2. It does not explain the fine spectral lines obtained under a spectroscope of strong resolution. However, it can be explained by Bohr–Sommerfeld theory of elliptical orbits.

3. It does not explain **Zeeman effect,** that is, splitting of spectral lines under magnetic field, and **Stark effect,** that is, splitting of spectral lines under electric field.

Bohr–Sommerfeld Theory

Bohr's assumption of circular orbits was modified by Sommerfeld, who assumed elliptical orbits. The angular momentum of an electron in elliptical orbit is also quantized and can have a limited number of values given by the factor $k \cdot \frac{h}{2\pi}$, where k is an integer, called the azimuthal quantum number. The principal quantum number, n, and the azimuthal quantum number, k, given by Bohr and Sommerfeld respectively are related by

$$\frac{n}{k} = \frac{\text{length of major axis}}{\text{length of minor axis}}.$$

For a given value of n, (except 1), k can have more than a single value. For example, for $n = 4$, k can have values 1, 2, 3 and 4. For $n = 1$, k will have only one value, 1, and the orbit will be circular.

As now the energy for electronic transition would depend on both n and k, the occurrence of fine spectral lines observed in the hydrogen spectrum could be explained.

It should be noted that the azimuthal quantum number, k, proposed by Sommerfeld is not the same as the azimuthal quantum number, l, suggested by quantum mechanics. For instance, say for quantum number, n,

$$k = 1, 2, 3, \ldots, n$$

whereas according to quantum mechanics

$$l = 0, 1, 2, 3, \ldots, (n-1).$$
$$s \;\; p \;\; d \;\; f$$

Thus although the Sommerfeld theory gives the correct total values of azimuthal quantum number of n but the actual values are incorrect. k and l are related as $l = k - 1$.

Theoretical Prediction of Wave Nature of Electron; de Broglie Equation

The French physicist de Broglie suggested that as light has the dual properties of wave and particle, the particles of matter such as electrons, protons etc., also have properties of waves. The wavelength of such a particle could then be derived. We know,

$$E = mc^2$$

$$\text{and } E = h\gamma$$

from which, we have

$$\lambda = \frac{h}{mc}$$

Here, c is the velocity of light. For the electron of mass m and velocity v,

$$\lambda = \frac{h}{mv} = \frac{h}{p}, \qquad \ldots (14)$$

where p is the momentum.

Quantization of Angular Momentum

Considering the wave characteristics of an electron moving in a circular orbit continuously in phase, the circumference of the circle must be an integral multiple of the wavelength.

$$2\pi r = n\lambda = n\frac{h}{mv}$$

or
$$mvr = n \cdot \frac{h}{2\pi} \quad \text{(angular momentum).}$$

Thus the wave mechanical picture of the de Broglie equation confirms Bohr's postulate of quantized angular momentum.

Verification of de Broglie Equation

Let an electron of charge e be accelerated by a potential V.

$$K \cdot E = V \cdot e = \frac{1}{2} mv^2$$

or
$$v = \sqrt{\frac{2V \cdot e}{m}} = \sqrt{\frac{2E}{m}} = \frac{h}{m\lambda}$$

or
$$\lambda = \frac{h}{\sqrt{2mVe}} = \frac{h}{\sqrt{2mE}}$$

or
$$\lambda = \frac{h}{\sqrt{2 \cdot V \cdot e \cdot m}} \qquad \ldots (15)$$

$$\lambda = \frac{12{\cdot}26 \times 10^{-10}}{\sqrt{V}} \text{ metre} = \frac{1.226}{\sqrt{V}} \text{ nm.}$$

Using the above equation, we can see that if the potential is varied between 10 to 10000 volts, λ varies between 0.387 nm and 0.123 nm. X-rays have the λ-value of this order which suggests electrons also behave like X-rays and have wave character. Later Davisson and Germer verified it experimentally.

Quantum Mechanical Picture of Atom

From the work of de Broglie, Davisson and Germer we understand effectively that the electron in an atom can be treated as a wave. Large objects as golf balls and moving automobiles obey laws of classical mechanics (Newton's laws) but very small particles such as electrons, atoms and molecules do not. A different kind of mechanism, called **wave mechanics** or **quantum mechanics,** which is based on the wave properties of matter, describes the behaviour of very small particles much better.

One of the underlying principles of quantum mechanics is the **Heisenberg Uncertainty Principle,** applicable to subatomic particles only and not to large objects.

Heisenberg stated that it is impossible to determine precisely both the position and the momentum (or mass × velocity) of a subatomic particle simultaneously. In other words, the position and the momentum of such particles can be determined only within certain limits of accuracy.

If Δx and Δp are the uncertainties in the position and momentum respectively, and ΔE and Δt are the uncertainties in energy and time respectively,

$$\left.\begin{aligned} \Delta x \cdot \Delta p &\geq \frac{h}{4\pi} \\ \Delta x \cdot \Delta v &\geq \frac{h}{4\pi m} \\ \Delta E \cdot \Delta t &\geq \frac{h}{4\pi} \end{aligned}\right] . \qquad \text{... (16)}$$

The Heisenberg uncertainty principle gives the concept of orbital, which is against the concept of Bohr's definite orbit.

The mathematical approach to quantum mechanics involves treating the electron in an atom as a standing wave or stationary wave. A standing wave is a wave that does not travel and therefore has at least one point at which it has zero amplitude. This point is called a **node.** Standing waves have limited frequencies and do not exhibit interference. A one-dimensional example of such a wave motion is a vibrating string fixed at both ends (nodes). In quantum mechanics, the electron is characterized by a 3-dimensional wave function, ψ. Schrödinger proposed the following equation based on the elements of Newton's classical mechanics and de Broglie's idea of particle-wave duality for one-electron species.

$$\nabla^2\psi + \frac{8\pi^2 m}{h^2}(E - V)\,\psi = 0, \qquad \text{... (17)}$$

where ∇^2 (del squared) is called the Laplacian operator.

$$\nabla^2 = \frac{\partial^2}{\partial x^2} + \frac{\partial^2}{\partial y^2} + \frac{\partial^2}{\partial z^2}$$

E is the total energy and V is the potential energy of the particle.

The Schrödinger equation is a second-order differential equation and therefore has a large number of solutions. For any physical situation, ψ **must be finite, single-valued and continuous.** In classical mechanics, ψ represents simply amplitude of a wave but in quantum mechanics, the physical significance of ψ is nebulous as ψ contains all the information we can hope to learn about the hydrogen atom, such as energies, momenta, spatial distribution, etc., by performing specific mathematical operations on ψ. However, ψ^2 in quantum mechanics can be interpreted as the probability of finding an electron at any point. An atomic **orbital** is a region in space in which the probability of finding an electron is high.

In classical mechanics, we have separate equations for particle motion and wave motion but in quantum mechanics we have only one equation, the Schrödinger equation in which the de Broglie relation, $\lambda = \frac{h}{mv}$, acts as a link between particle and wave properties.

One way of remembering the general behaviour of the wave function is to systematize their nodal properties. In a hydrogen-like wave function,

number of radial nodes $= n - l - 1$

number of angular nodes (nodal planes) $= l$.

$\therefore$ total number of nodes $= n - 1$.

The solution of the Schrödinger equation gives a set of numbers, called **quantum numbers,** that describe electronic configuration, energies of an electron in the atom and also give information about the shapes and orientations of the statistical probability distributions of electrons.

Principal Quantum Number (n) describes the main energy level, or shell, an electron occupies. It may be any positive integer.

$$n = 1, 2, 3, 4, \ldots .$$
$$K\ L\ M\ N$$

Azimuthal Quantum Number (l), or angular momentum quantum number, designates a sublevel or specific shape of atomic orbital that an electron occupies. Within a shell, that is, for a given value of n, different sublevels or subshells are possible, each with a characteristic shape. The number l can take integral values from 0 to $(n - 1)$.

$$^*l = 0, 1, 2, 3, \ldots, (n - 1)$$
$$s\ p\ d\ f$$

The orbital angular momentum (L) is given by

$$L = \sqrt{l(l+1)} \cdot \frac{h}{2\pi} = \sqrt{l(l+1)}\ \hbar,\ l = 0, 1, 2, \ldots, \qquad \ldots (18)$$

where $\hbar$ is called Dirac h and $\hbar = \frac{h}{2\pi}$.

Magnetic Quantum Number (m) designates the specific orbital within a subshell. Orbitals within a given subshell differ in their orientations in space, but not in their energies. Within each subshell m can take any integral values from $-l$ through 0 to $+l$.

$$m = (-l), \ldots 0, \ldots, (+l)$$

* The letters s, p, d and f come from the words **sharp, principal, diffuse** and **fundamental,** which were used to describe certain features of spectra before quantum mechanics was developed.

The maximum value of m depends on the value of l.

For $l = 1$ (p-subshell), $m = -1,\ 0,\ +1$

$p_x\ p_y\ p_z$

For $l = 2$ (d-subshell), $m = -2,\ -1,\ 0,\ +1,\ +2$

$d_{x^2-y^2}\ d_{z^2}\ d_{xy}\ d_{xz}\ d_{yz}$

$(d_1)\ (d_2)\ (d_3)\ (d_4)\ (d_5)$

Thus in the presence of a magnetic field, the energy levels have been split into $(2l + 1)$ levels (Zeeman effect).

When an atom has angular momentum, L, the atom acts like a small magnet. We say that it has a magnetic dipole moment, μ. The natural unit of magnetic dipole moment is expressed by **Bohr magneton** (μ_B) for electronic states.

$$\mu_B(BM) = \frac{eh}{4\pi m} = \frac{e\hbar}{2m}, \quad \text{... (19)}$$

where m is the mass of an electron.

$$\mu_B = 9.27 \times 10^{-24}\ \text{JT}^{-1}\ \text{(SI)}$$

$$= 9.27 \times 10^{-28}\ \text{JG}^{-1} \qquad (\because\ 1\text{T} = 10^4\ G)$$

$$= 9.27 \times 10^{-21}\ \text{erg G}^{-1}.$$

The SI unit of a magnetic field is the **tesla,** T, which is defined as

$$\text{T} = \text{NA}^{-1}\text{m}^{-1} = \text{kgs}^{-2}\text{A}^{-1}$$

where A is the ampere

$$\text{JT}^{-1} = \text{Am}^2.$$

The **Gauss, G,** is the unit for the magnetic field in the cgs Gaussian system of units.

Spin Quantum Number (s) refers to the spin of an electron and the orientation of the magnetic field produced by this spin. For every set of n, l and m values, s can take the value $+\frac{1}{2}$ or $-\frac{1}{2}$.

$$s = \pm\frac{1}{2}.$$

The values of n, l and m describe a particular atomic orbital. Each atomic orbital can accommodate no more than two electrons, one with $s = +1/2$ and another with $s = -1/2$.

Spin angular momentum is given by

$$S = \sqrt{s(s+1)} \cdot \frac{h}{2\pi} = \sqrt{s(s+1)}\ \hbar. \quad \text{... (20)}$$

Since $s = \frac{1}{2},\ S = \frac{\sqrt{3}}{2} \cdot \frac{h}{2\pi} = \frac{\sqrt{3}}{2}\hbar.$

Permissible Values of the Quantum Numbers Through $n = 4$					
n	l	m	s	Electron capacity of subshell $= 4l + 2$	Electron capacity of shell $= 2n^2$
1.	0 (1s)	0	$+\frac{1}{2}, -\frac{1}{2}$	2	2
2.	0 (2s)	0	$+\frac{1}{2}, -\frac{1}{2}$	2	8
	1 (2p)	−1, 0, +1	$\pm\frac{1}{2}$ for each value of m	6	
3.	0 (3s)	0	$+\frac{1}{2}, -\frac{1}{2}$	2	18
	1 (3p)	−1, 0, +1	$\pm\frac{1}{2}$ for each value of m	6	
	2 (3d)	−2, −1, 0, +1, +2	$\pm\frac{1}{2}$ for each value of m	10	
4.	0 (4s)	0	$+\frac{1}{2}, -\frac{1}{2}$	2	32
	1 (4p)	−1, 0, +1	$\pm\frac{1}{2}$ for each value of m	6	
	2 (4d)	−2, −1, 0, +1, +2	$\pm\frac{1}{2}$ for each value of m	10	
	3 (4f)	−3, −2, −1, 0, +1, +2, +3	$\pm\frac{1}{2}$ for each value of m	14	

The spin magnetic moment (μ) of the electron, excluding the orbital magnetic moment, is given by

$$\mu = \sqrt{n(n+2)}\ BM, \qquad \ldots (21)$$

where n is the number of unpaired electrons.

Quantum Numbers and Electronic Configuration

The following rules help us to predict electronic configuration of atoms with the help of quantum numbers

1. Electrons are added into orbitals in the way that gives the lowest total energy for the atom (**Aufbau Principle**).
2. Electrons are assigned to orbitals in order of increasing value of $(n + l)$. **See diagram below.**
3. For subshells with the same value of $(n + l)$, electrons are assigned first to the subshell with lower n.
4. No two electrons in an atom may have identical sets of four quantum numbers **(Pauli Exclusion Principle).**
5. Electrons occupy all the orbitals of a given subshell singly before pairing begins. These unpaired electrons have parallel spins. **(Hund Rule). Hund rule states that for degenerate orbitals, the lowest energy is attained when the number of electrons with the same spin is maximized.**

Order of Filling of Atomic Orbitals

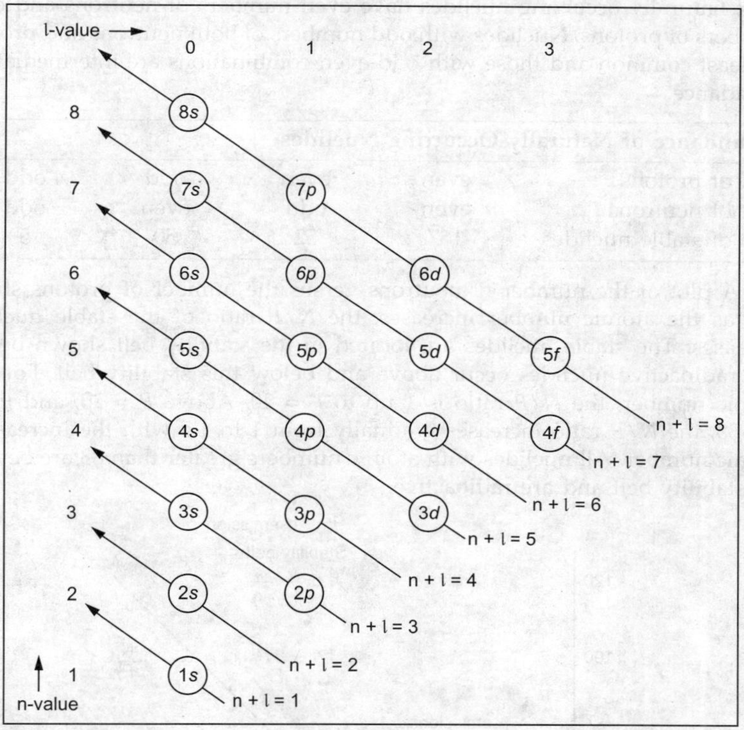

RADIOACTIVITY

The property of a nucleus emitting α, β and γ rays is known as **radioactivity** and the substance possessing this property is called a **radioactive substance.** The emission of these particles takes place because of the instability of the nucleus. α-particles are just the helium nucleus represented by ${}_{2}^{4}He$ (i.e., mass number = 4 and atomic number = 2) and hence 2+ charge. The β-particles are represented as electrons, ${}_{-1}^{0}e$ (mass = 0 and charge = –1) and carry a negative charge.

Nuclear Size

The radius, '*r*' of various nuclei can be approximately calculated in cm by using the relation

$$r = r_0 A^{1/3}, \qquad \dots (22)$$

where r_0 is a constant which is equal to 1.4×10^{-13} cm and A, the mass number of a given nucleus.

Neutron-Proton Ratio and Nuclear Stability

Most naturally occurring nuclides have even numbers of neutrons and even numbers of protons. Nuclides with odd numbers of both neutrons and protons are least common and those with odd-even combinations are intermediate in abundance.

Abundance of Naturally Occurring Nuclides				
No. of protons	even	even	odd	odd
No. of neutrons	even	odd	even	odd
No. of stable nuclides	157	52	50	5

A plot of the number of neutrons versus the number of protons shows that as the atomic number increases, the N/P ratio of the stable nuclides increases. The stable nuclides are located in the stability belt shown below. The radioactive nuclides occur above and below this stability belt. For low atomic number, the N/P ratio is 1 up to $P = 20$. Above $P = 20$, and up to $P = 83$, the N/P ratio increases gradually from 1 to 1.5 with the increase in atomic number. All nuclides with atomic numbers greater than 83 are beyond the stability belt and are radioactive.

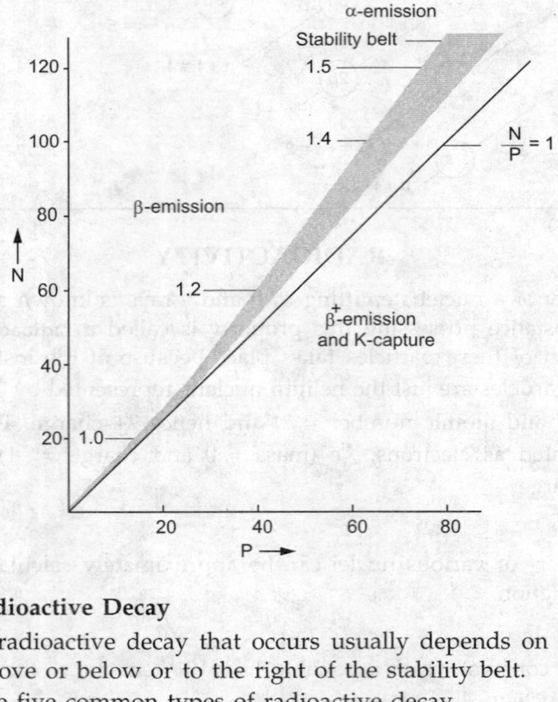

Types of Radioactive Decay

The type of radioactive decay that occurs usually depends on whether the nucleus is above or below or to the right of the stability belt.

There are five common types of radioactive decay.

1. Alpha Emission

All nuclides with atomic numbers greater than 83 are **beyond the band of stability** and are radioactive. Many of these decay by alpha emission.

$${}^{226}_{88}\text{Ra} \rightarrow {}^{222}_{86}\text{Rn} + {}^{4}_{2}\text{He (or } \alpha)$$

$${}^{210}_{84}\text{Po} \rightarrow {}^{206}_{82}\text{Pb} + {}^{4}_{2}\text{He (or } \alpha)$$

Some of the nuclides lying below the stability belt, especially the heavier ones, increase their $\frac{N}{P}$ ratio by undergoing alpha emission.

$${}^{204}_{82}\text{Pb} \rightarrow {}^{200}_{80}\text{Hg} + {}^{4}_{2}\text{He (or } \alpha)$$

The only stable nuclide with atomic number 83 is ${}^{209}_{83}\text{Bi}$.

2. Beta Emission

Nuclei **above the stability belt** have too high a ratio of neutrons and protons. Such nuclei undergo β-emission to decrease the ratio as β-emission is equivalent to the conversion of a neutron to a proton.

$${}^{1}_{0}\text{n} \rightarrow {}^{1}_{1}\text{p} + {}^{0}_{-1}\text{e (or } \beta)$$

Examples are

$${}^{228}_{88}\text{Ra} \rightarrow {}^{228}_{89}\text{Ac} + {}^{0}_{-1}\text{e (or } \beta)$$

$${}^{14}_{6}\text{C} \rightarrow {}^{14}_{7}\text{C} + {}^{0}_{-1}\text{e (or } \beta)$$

3. Positron Emission

Such emission occurs for the nuclei lying **below the stability belt.** Positron emission is most commonly encountered with artificially radioactive nuclei of the lighter elements. Positron (β^+) emission is equivalent to the conversion of a proton to a neutron to increase the N/P ratio.

$${}^{1}_{1}\text{p} \rightarrow {}^{1}_{0}\text{n} + {}^{0}_{1}\text{e (or } \beta^+)$$

Examples are

$${}^{38}_{19}\text{K} \rightarrow {}^{38}_{18}\text{Ar} + {}^{0}_{1}\text{e (or } \beta^+)$$

$${}^{15}_{8}\text{O} \rightarrow {}^{15}_{7}\text{N} + {}^{0}_{1}\text{e (or } \beta^+)$$

4. Electron Capture (K Capture)

Electron capture occurs with the nuclei **lying below the stability belt,** in which an electron from the K-shell is captured by the nucleus. In effect, a proton is changed to a neutron as in positron emission. Electron capture occurs most often with heavier elements.

$${}^{1}_{1}\text{p} + {}^{0}_{-1}\text{e} \rightarrow {}^{1}_{0}\text{n}$$

In electron capture, an X-ray photon is emitted. Examples are

$${}^{106}_{47}\text{Ag} + {}^{0}_{-1}\text{e} \rightarrow {}^{106}_{46}\text{Pd}$$

$${}^{40}_{19}\text{K} + {}^{0}_{-1}\text{e} \rightarrow {}^{40}_{18}\text{Ar}$$

5. Gamma Emission*

A metastable nucleus (an excited nucleus with life time < 10^{-9} s) decays by ${}^{0}_{0}\gamma$ emission. A metastable Technetium-99 is denoted by ${}^{99\,m}_{43}$Tc.

Example is

$$ {}^{99\,m}_{43}\text{Tc} \rightarrow {}^{99}_{43}\text{Tc} + {}^{0}_{0}\gamma. $$

There is no change in atomic number or mass number as the product is simply a lower energy state of the original nucleus.

Radioactive Disintegration

The conversion of an element to another due to emission of α and β particles is known as ***radioactive disintegration***. Radioactive disintegration is similar to a chemical reaction of first order and so we have the following equations for the radioactive disintegration, similar to that of a first-order reaction.

Initial number of atoms: N^0

$$ A \longrightarrow B $$

Number of atoms after time t: N

$$ \text{Rate of disintegration} = -\frac{d(N)}{dt} = \lambda\,(N) \qquad \text{... (23)} $$

(that is, number of disintegrations per unit time).

On integration, we get,

$$ \lambda = \frac{2{\cdot}303}{t} \log \frac{N^0}{N} \qquad \text{... (24)} $$

λ is known as disintegration constant or decay constant.

In contrast to the rate constant of a chemical reaction, the decay constant is completely independent of any external influence such as temperature and pressure.

This is to be noted that since $N°/N$ gives the ratio of number of atoms, it, therefore, gives the ratio of the number of moles of atoms.

Simultaneous Disintegration

Further, if an equimolar mixture of the two radioactive substances having decay constants λ_1 and λ_2, disintegrate simultaneously, the ratio of the nuclides at the end of time ***t***, that is, N_1/N_2, is given by,

$$ 2{\cdot}303 \log \frac{N_1}{N_2} = (\lambda_2 - \lambda_1)\,t. \qquad \text{... (25)} $$

Half-life Period

The time required for half of a radioactive substance to disintegrate is known as the ***half-life period,*** represented here by $t_{\frac{1}{2}}$.

---

* When a positron and an electron collide, both particles vanish with the emission of two γ-photons that carry away the energy. ${}^{0}_{1}\text{e} + {}^{0}_{-1}\text{e} \rightarrow 2\,{}^{0}_{0}\gamma$.

When $t = t_{\frac{1}{2}}$, $N = \frac{N^0}{2}$.

Equation (18) reduces to

$$t_{\frac{1}{2}} = \frac{0{\cdot}6932}{\lambda} \qquad \text{... (26)}$$

$t_{\frac{1}{2}}$ is independent of the initial number of atoms of the radioactive substance.

If N_0 and N are the initial number of nuclei and the number of nuclei undecayed at the end of the ***n***th half-life,

$$N = 2^{-n} \cdot N_0. \qquad \text{... (27)}$$

A similar expression holds for the mass of the undecayed isotope, m, i.e.,

$$m = 2^{-n} \cdot m_0, \qquad \text{... (28)}$$

where m_0 ***is the initial mass of the nuclei.***

Average-life Period (λ')

The reciprocal of the disintegration constant is known as the ***average-life period.***

$$\lambda' = \frac{1}{\lambda} = \frac{t_{\frac{1}{2}}}{0{\cdot}6932} = 1{\cdot}44\, t_{\frac{1}{2}}. \qquad \text{... (29)}$$

Units of Radioactivity

The standard unit of radioactivity (i.e., rate of disintegration) is 'curie' (Ci). A curie is a quantity of radioactive material decaying at the same rate as 1 g of Ra ($3{\cdot}7 \times 10^{10}$ dps) [1 millicurie (mCi) = $3{\cdot}7 \times 10^7$ dps, 1 microcurie (μCi) = $3{\cdot}7 \times 10^4$ dps]. Rutherford is a more recent unit (1 rutherford = 10^6 dps). The SI unit is becquerel (1 Bq = 1 dps).

Parallel Radioactive Disintegration

Quite often the radioactive substance can disintegrate to form different sets of products. For example, ^{227}Ac decays through two parallel paths, one leading to ^{227}Th and one leading to ^{223}Fr.

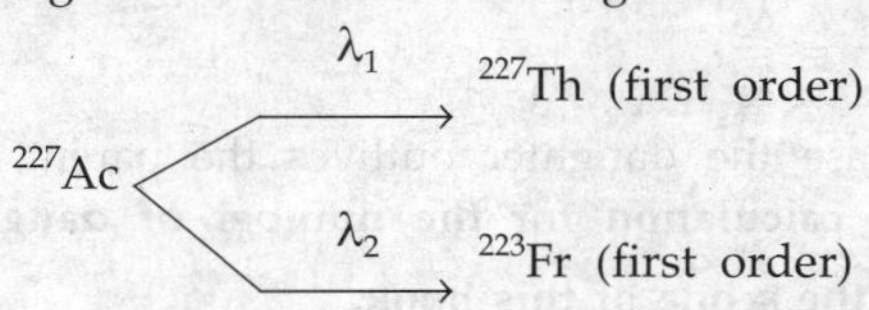

The overall decay constant for a set of parallel first-order reactions is equal to the sum of the separate decay constants, e.g.,

$$\lambda_{Ac} = \lambda_{Th} + \lambda_{Fr}$$

The fractional yield of either process is equal to the ratio of the decay constant for that process to the overall rate constant, e.g.,

$$\text{fractional yield of Th} = \frac{\lambda_{Th}}{\lambda_{Ac}}$$

and

$$\text{fractional yield of Fr} = \frac{\lambda_{Fr}}{\lambda_{Ac}}.$$

Successive Radioactive Disintegration

In this case the 'daughter nuclide' formed by the decay of the 'parent nuclide' also undergoes decay, e.g.,

initial number of nuclides: N

$$\underset{(t_{\frac{1}{2}})_1}{A} \xrightarrow{\lambda_1} \underset{(t_{\frac{1}{2}})_2}{B} \xrightarrow{\lambda_2} C$$

No. of nuclides after time t: N_1 N_2

We have,

$$N_1 = N\,e^{-\lambda_1 t}$$

$$N_2 = \frac{\lambda_1 N}{\lambda_2 - \lambda_1}(e^{-\lambda_1 t} - e^{-\lambda_2 t}). \qquad \text{... (30)}$$

Subscripts 1 and 2 stand for parent and daughter nuclides respectively. From the above equations, the following cases of equilibrium are often considered.

1. Transient equilibrium, in which the parent has a longer half-life than the daughter, i.e., $(t_{\frac{1}{2}})_1 > (t_{\frac{1}{2}})_2$ or $\lambda_1 < \lambda_2$ and

$$\frac{N_1}{N_2} = \frac{\lambda_2 - \lambda_1}{\lambda_1}. \qquad \text{... (31)}$$

2. Secular equilibrium, in which the activity of the parent does not decrease much during several half-lives of the daughter, i.e., $(t_{\frac{1}{2}})_1 \gg (t_{\frac{1}{2}})_2$ or $\lambda_1 \ll \lambda_2$ and

$$\frac{N_1}{N_2} = \frac{\lambda_2}{\lambda_1}. \qquad \text{... (32)}$$

3. Disequilibrium, in which case the daughter outlives the parent, i.e., $(t_{\frac{1}{2}})_1 < (t_{\frac{1}{2}})_2$ or $\lambda_1 > \lambda_2$ and **the calculation for the number of daughter nuclides in this case is beyond the scope of this book.**

Maximum Activity of Daughter Nuclide

The time required for the radioactive daughter to reach its maximum activity (i.e., the maximum concentration) is given by

$$t = \frac{2{\cdot}303\,(\log \lambda_1 - \log \lambda_2)}{\lambda_1 - \lambda_2} \qquad \text{... (33)}$$

Soddy–Fajans Group Displacement Law

In an α-particle emission, the resulting element has a mass number less by four units and atomic number less by two units and so lies two places to the left in the periodic table.

In a β-particle emission the resulting element has the same mass number but has an atomic number greater by one unit and so lies one place to the right in the periodic table.

Disintegration Series

The process of disintegration continues from one nucleus to another until a stable nuclide is formed. Disintegration series constitutes all such nuclei from the parent element to the final stable element. As the mass number is changed only by the emission of an α-particle (by four units), we have the following:

4*n* (Thorium Series) $^{232}_{90}Th \rightarrow ^{208}_{82}Pb$

4*n* + 1 (Neptunium Series) $^{237}_{93}Np \rightarrow ^{209}_{83}Bi$

4*n* + 2 (Uranium Series) $^{238}_{92}U \rightarrow ^{206}_{82}Pb$

4*n* + 3 (Actinium Series) $^{231}_{89}Ac \rightarrow ^{207}_{82}Pb$

Thus we can find out the series to which a nuclide belongs by dividing the mass number by 4.

In case,

remainder = 0; nuclide belongs to 4*n* series

remainder = 1; nuclide belongs to (4*n* + 1) series

remainder = 2; nuclide belongs to (4*n* + 2) series

remainder = 3; nuclide belongs to (4*n* + 3) series

Radioactive Isotopes and Isobars

A radioactive isotope is formed by the emission of one α- and two β-particles, while an isobar is formed by the emission of one β-particle.

	At. no. ($Z = P$)	**Mass no. ($A = P + N$)**	**Neutron (N)**
Isotopes	same	different	different
Isobars	different	same	different
Isotones	different	different	same
Isomers (nuclear)	same	same	same

Binding Energy of the Nucleus

It has been observed that the actual mass of an isotope of an element, obtained experimentally using mass spectrograph, is invariably less than the calculated mass of the isotope (i.e., the sum of masses of electrons, protons and neutrons). This mass difference is known as ***mass defect*** of the nucleus which is also expressed as the packing fraction of the nucleus.

$$\text{Packing fraction} = \frac{\text{isotopic mass} - \text{mass number}}{\text{mass number}} \times 10^4. \quad \ldots (34)$$

The loss of mass, i.e., the 'mass defect' is converted into energy that releases to stabilise the nucleus. This energy is known as binding energy which is calculated using Einstein's equation, $E = mc^2$, or just multiplying the mass defect with 931·5 MeV as 1 amu = 931·5 MeV (megaelectronvolt).

Plot of Binding Energy per nucleon vs Mass Number

The plot shows that the nuclides near the mass number 55 have the largest binding energies per nucleon. Thus **heavy nuclei** are expected to undergo **fission** to approach this mass number whereas light nuclei are expected to undergo **fusion**.

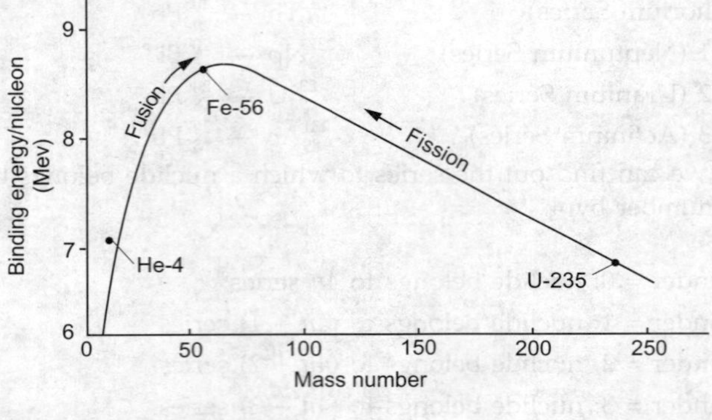

Equivalence of amu and Energy

Since 1 amu is exactly $\frac{1}{12}$ th of the mass of C^{12} atom,

$$\begin{aligned} 1 \text{ amu} &= \frac{1}{12} \times \text{mass of a } C^{12} \text{ atom} \\ &= \frac{1}{12} \times \frac{\text{mass of 1 mole of } C^{12} \text{ atom}}{\text{Avogadro constant}} \\ &= \frac{1}{12} \times \frac{12}{6{\cdot}022 \times 10^{23}} \text{ g} \\ &= 1{\cdot}66 \times 10^{-24} \text{ g.} \end{aligned}$$

According to the mass–energy equation of Einstein, we have the energy equivalent to 1 amu:

$$\begin{aligned} E &= mc^2 \\ &= 1{\cdot}66 \times 10^{-24} \times (2{\cdot}998 \times 10^{10})^2 \text{ ; } (c = 2{\cdot}998 \times 10^{10} \text{ cm/s}) \\ &= 1{\cdot}4925 \times 10^{-3} \text{ erg} \end{aligned}$$

$$E = \frac{1{\cdot}4925 \times 10^{-3}}{4{\cdot}184 \times 10^{7}} \text{ calorie} \quad (1 \text{ calorie} = 4{\cdot}184 \times 10^{7} \text{ erg})$$

$$= 0{\cdot}3567 \times 10^{-10} \text{ calorie}$$

$$= \frac{1{\cdot}4925 \times 10^{-3}}{10^{7}} \text{ joule} \quad (1 \text{ joule} = 10^{7} \text{ erg})$$

$$= 1{\cdot}4925 \times 10^{-10} \text{ joule}$$

$$= \frac{1{\cdot}4925 \times 10^{-10}}{1{\cdot}602 \times 10^{-19}} \text{ eV} \quad (1 \text{ eV} = 1{\cdot}602 \times 10^{-19} \text{ joule})$$

$$= 931{\cdot}64 \times 10^{6} \text{ eV}$$

$$= 931{\cdot}64 \text{ MeV} \quad (1 \text{ MeV} = 10^{6} \text{ eV})$$

$$\approx 931{\cdot}5 \text{ MeV}.$$

Electronvolt (eV)

1 eV is the energy necessary to move an electronic charge (e) through a potential of exactly 1 volt. In other words, if a particle being accelerated by a potential has a charge equal to the magnitude of the charge on an electron, the number of eV of energy is numerically equal to the potential in volts. Thus 1 eV = energy of 1 electron being accelerated by 1 volt.

$\therefore$ x eV = energy of 1 electron being accelerated by x volts.

Unit Conversions

$1 \text{ J} = 1 \text{ kg} \cdot \text{m}^2/\text{s}^2 = 10^7$ ergs; $1 \text{ J} = 1 \text{ V} \times 1 \text{ C}$;

$1 \text{ eV} = 1{\cdot}602 \times 10^{-19} \text{ J} = 1{\cdot}602 \times 10^{-12}$ erg.

EXAMPLES

Ex. 1. *Calculate the radius of the first Bohr orbit of a hydrogen atom. Given that* $h = 6{\cdot}63 \times 10^{-27}$ erg · s; $m = 9 \times 10^{-28}$ g *and charge of an electron* $= 4{\cdot}8 \times 10^{-10}$ esu.

Solution **:** We have for hydrogen, $Z = 1$, $n = 1$.

$$r_1 = \frac{h^2}{4\pi^2 me^2} \quad \text{... (Eqn. 1)}$$

$$= \frac{(6{\cdot}63 \times 10^{-27})^2}{4 \times (3{\cdot}14)^2 \times 9 \times 10^{-28} \times (4{\cdot}8 \times 10^{-10})^2}$$

$$= 0{\cdot}53 \times 10^{-8} \text{ cm} = 0{\cdot}53 \text{ Å}.$$

Ex. 2. *From the given data in* Ex. 1, *calculate the radius of the third orbit of a hydrogen atom.*

Solution : From Ex. 1, we have $r_1 = 0{\cdot}53 \times 10^{-8}$ cm.

Thus applying

$$r_n = n^2 r_1 \qquad \text{... (Eqn. 4)}$$

$$r_3 = 3^2 \times 0{\cdot}53 \times 10^{-8} \text{ cm}$$
$$= 4{\cdot}77 \times 10^{-8} \text{ cm} = 4{\cdot}77 \text{ Å}.$$

Ex. 3. *Calculate the energy of an electron in the first Bohr orbit of hydrogen.*

Solution : For hydrogen, $Z = 1$, $n = 1$; we have,

$$E_1 = -\frac{2\pi^2 e^4 m}{h^2} \qquad \text{... (Eqn. 2)}$$

$$= -\frac{2 \times (3{\cdot}14)^2 \times (4{\cdot}8 \times 10^{-10})^4 \times (9 \times 10^{-28})}{(6{\cdot}63 \times 10^{-27})^2}$$

$$= -2{\cdot}18 \times 10^{-11} \text{ erg.}$$

Ex. 4. *From the given data in* Ex. 3, *calculate the energy of an electron in the second Bohr orbit of an excited hydrogen atom.*

Solution : From Ex. 3 we have, $E_1 = -2{\cdot}18 \times 10^{-11}$ erg.

Thus applying

$$E_n = \frac{E_1}{n^2} \qquad \text{... (Eqn. 5)}$$

$$E_2 = -\frac{2{\cdot}18 \times 10^{-11}}{2^2} = -0{\cdot}545 \times 10^{-11} \text{ erg.}$$

Ex. 5. *Calculate the electronic velocity in the first Bohr orbit of hydrogen.*

Solution : For hydrogen; $Z = 1$, $n = 1$.

$$v_1 = \frac{2\pi e^2}{h} \qquad \text{... (Eqn. 3)}$$

$$= \frac{2 \times (3{\cdot}14) \times (4{\cdot}8 \times 10^{-10})^2}{6{\cdot}63 \times 10^{-27}}$$

$$= 2{\cdot}19 \times 10^8 \text{ cm/s.}$$

Ex. 6. *Calculate the velocity of an electron revolving in the second orbit of a hydrogen atom from the given data in* Ex. 5.

Solution : From Ex. 5, $v_1 = 2{\cdot}19 \times 10^8$ cm/s

We have, $v_n = \dfrac{v_1}{n}$. ... (Eqn. 6)

$$\therefore \qquad v_2 = \frac{2{\cdot}19 \times 10^8}{2}$$

$$= 1{\cdot}09 \times 10^8 \text{ cm/s.}$$

Ex. 7. *The electromagnetic radiation of wavelength* 242 nm *is just sufficient to ionise the sodium atom. Calculate the ionisation energy of sodium in kilojoule per mole.*

Solution **:** The energy required to ionise one atom is given by

$$E = \frac{hc}{\lambda} \left\{ \begin{array}{l} h = 6{\cdot}626 \times 10^{-34}\ \text{J} \cdot \text{s} \\ c = 3 \times 10^{8}\ \text{m s}^{-1} \\ \lambda = 242 \times 10^{-9}\ \text{metre} \end{array} \right\}.$$

Thus the energy required to ionise one mole (i.e., Av. const.) of Na atoms is

$$E = \frac{(6{\cdot}626 \times 10^{-34})\,(3 \times 10^{8})}{242 \times 10^{-9}} \times 6{\cdot}022 \times 10^{23}$$

$$= 4{\cdot}946 \times 10^{5}\ \text{J mole}^{-1}$$

$$= 4{\cdot}946 \times 10^{2}\ \text{kJ mole}^{-1}.$$

Ex. 8. *From the given data in* Ex. 6 *and* Ex. 1, *calculate the number of revolutions of an electron in the second Bohr orbit in one second.*

Solution **:** If r_2 is the radius of the second Bohr orbit, the distance travelled by an electron in one revolution will be $2\pi r_2$ (i.e., the circumference). From Ex. 6 we have calculated that an electron travels a distance of $1{\cdot}09 \times 10^{8}$ cm in one second in the second Bohr orbit. Hence,

$$\text{revolutions per second} = \frac{v_2}{2\pi r_2} = \frac{1{\cdot}09 \times 10^{8}}{2\pi r_2}.$$

Now,

$$r_2 = 2^2 \,.\, r_1 \qquad \text{... (Eqn. 4)}$$

$$= 2^2 \times {\cdot}53 \times 10^{-8}$$

$$= 2{\cdot}12 \times 10^{-8}$$

$$(n = 2,\ r_1 = 0{\cdot}53 \times 10^{-8}\ \text{from Ex. 1})$$

$$\therefore\ \text{revolutions per second} = \frac{1{\cdot}09 \times 10^{8}}{2 \times (3{\cdot}14) \times 2{\cdot}12 \times 10^{-8}}$$

$$= 8{\cdot}18 \times 10^{14}.$$

Ex. 9. *Calculate the ionisation energy in* eV *of a hydrogen atom in the ground state.*

Solution **:** Since ionisation energy is defined as the energy required to remove an electron from the outermost orbit of an atom, mathematically, we can say that since the hydrogen atom is in the ground state, ionisation energy will be the energy required to remove the electron from the first orbit to the infinite orbit (i.e., r_1 to r_∞).

Thus,

$$\text{ionisation energy} = E_\infty - E_1$$

$$= 0 - (-\,2{\cdot}18 \times 10^{-11})\ \text{erg}.$$

(from Ex. 3, $E_1 = -2{\cdot}18 \times 10^{-11}$ erg)

$\therefore$ ionisation energy $= 2{\cdot}18 \times 10^{-11}$ erg $= \dfrac{2{\cdot}18 \times 10^{-11}}{1{\cdot}602 \times 10^{-12}} = 13{\cdot}60$ eV.

(1 eV $= 1{\cdot}602 \times 10^{-12}$ erg)

Ex. 10. *What will be the wavelength of the emitted radiations when an electron jumps from a third orbit to a first orbit in a hydrogen atom? Rydberg constant,* $R = 109737\ \text{cm}^{-1}$.

Solution : We have for hydrogen ($Z = 1$)

$$\frac{1}{\lambda} = R\left(\frac{1}{n_1^2} - \frac{1}{n_2^2}\right). \qquad \text{... (Eqn. 7)}$$

Here $n_1 = 1$ and $n_2 = 3$.

$$\therefore \quad \frac{1}{\lambda} = 109737\left(\frac{1}{1^2} - \frac{1}{3^2}\right)$$

$$\lambda = 1{\cdot}025 \times 10^{-5}\ \text{cm/wave.}$$

Ex. 11. *Equations (1) to (7) have been applied to the above problems for hydrogen atom. Can these equations be applied to calculate r, E, v, etc., for* He^+ *and* Li^{2+} *ions?*

Solution : As the Bohr theory is applicable to a one-electron system, the said equations can be applied to He^+ and Li^{2+} ions as these species have only one electron each. The value of Z will be taken as 2 and 3 for He^+ and Li^{2+} respectively.

Ex. 12. *The ionisation energy of* He^+ *is* $19{\cdot}6 \times 10^{-18}$ J *per atom. Calculate the energy of the first stationary state of* Li^{2+}.

Solution : The ionisation energy of He^+ is $19{\cdot}6 \times 10^{-18}$ J per atom

$\therefore$ energy of the first orbit of He^+ $(Z = 2) = 19{\cdot}6 \times 10^{-18}$ J.

$\therefore$ energy of the first orbit of H^+ $(Z = 1) = \dfrac{19{\cdot}6 \times 10^{-18}}{4}$ J.

$\therefore$ energy of the first orbit of Li^{2+} $(Z = 3) = \dfrac{19{\cdot}6 \times 10^{-18}}{4} \times 9$

$= 4{\cdot}41 \times 10^{-17}$ J.

Ex. 13. *If the wavelength of light is* $5{\cdot}89 \times 10^{-5}$ cm, *what will be the energy in one photon in the given region of light?*

($h = 6{\cdot}63 \times 10^{-27}$ erg · s; $c = 3 \times 10^{10}$ cm/s)

Solution : We have,

$$E = h\nu$$

and $$c = \nu\lambda$$

from which we get,

$$E = \frac{hc}{\lambda}$$

$$= \frac{6{\cdot}63 \times 10^{-27} \times 3 \times 10^{10}}{5{\cdot}89 \times 10^{-5}}$$

$$= 3{\cdot}37 \times 10^{-12} \text{ erg.}$$

Ex. 14. *What electronic transition in the* He^+ *ion would emit the radiation of the same wavelength as that of the first Lyman transition of hydrogen (i.e., for an electron jumping from* $n = 2$ *to* $n = 1$*)? Neglect the reduced-mass effect. Also, calculate second ionisation potential of* He *and first Bohr orbit for* He^+.

($e = 1{\cdot}6 \times 10^{-19}$ coulomb, $m = 9{\cdot}1 \times 10^{-31}$ kg, $h = 6{\cdot}626 \times 10^{-34}$ J · s, $c = 2{\cdot}997 \times 10^8$ metre/s and $\epsilon_0 = 8{\cdot}854 \times 10^{-12}$ coulomb²/newton · metre²)

Solution : As He^+ has only one electron, Bohr equations may be applied and the Rydberg constant, R, for H and He^+ will be the same as the reduced-mass effect is neglected.

For first Lyman transition for hydrogen (i.e., when an electron jumps from $n = 2$ to $n = 1$), we have,

$$\frac{1}{\lambda} = RZ^2\left(\frac{1}{n_1^2} - \frac{1}{n_2^2}\right) = R\left(\frac{1}{1^2} - \frac{1}{2^2}\right) = \frac{3R}{4}.$$

Since wavelength of the radiation emitted in electronic transition in He^+ is the same as for hydrogen, therefore, for He^+ ($Z = 2$), we have,

$$\frac{1}{\lambda} = RZ^2\left(\frac{1}{n_1^2} - \frac{1}{n_2^2}\right) = 4R\left(\frac{1}{n_1^2} - \frac{1}{n_2^2}\right) = \frac{3R}{4}.$$

$\therefore$ n_1 and n_2 have to be 2 and 4 respectively as

$$\frac{1}{\lambda} = 4R\left(\frac{1}{2^2} - \frac{1}{4^2}\right) = \frac{3R}{4}.$$

As the second IP of He is the energy required to remove the second electron from first orbit to infinite orbit,

$$\therefore \quad \text{second IP of He} = \frac{Z^2 e^4 m}{8\varepsilon_0^2 h^2}\left(\frac{1}{1^2} - \frac{1}{\infty}\right) \quad \ldots \text{(Eqn. 9)}$$

$$\text{IP} = \frac{2^2 (1{\cdot}6 \times 10^{-19})^4 (9{\cdot}1 \times 10^{-31})}{8 (8{\cdot}854 \times 10^{-12})^2 (6{\cdot}626 \times 10^{-34})^2}$$

$$= 8{\cdot}67 \times 10^{-18} \text{ joules.}$$

Now to calculate r_1 for He (Z = 2), we have,

$$r_1 = \frac{\varepsilon_0 n^2 h^2}{\pi m Z e^2} \quad \text{... (Eqn. 8)}$$

$$= \frac{(8{\cdot}854 \times 10^{-12}) (1)^2 (6{\cdot}626 \times 10^{-34})^2}{(3{\cdot}14) (9{\cdot}1 \times 10^{-31}) (2) (1{\cdot}6 \times 10^{-19})^2}$$

$$= 2{\cdot}64 \times 10^{-11} \text{ metres.}$$

Ex. 15. *The energy of the electron in the second and third Bohr orbit of hydrogen atom is* $-5{\cdot}42 \times 10^{-12}$ *erg and* $-2{\cdot}41 \times 10^{-12}$ *erg respectively. Calculate the wavelength of the emitted radiation when the electron drops from the third to second orbit.* ($h = 6{\cdot}62 \times 10^{-27}$ erg · s, $c = 3 \times 10^{10}$ cm/s)

Solution **:** We have,

$$\Delta E = E_3 - E_2$$

$$= -2{\cdot}41 \times 10^{-12} - (-5{\cdot}42 \times 10^{-12}).$$

$$= 3{\cdot}01 \times 10^{-12} \text{ erg.}$$

We know, $\Delta E = h\nu$ and $c = \nu\lambda$.

Thus $$\Delta E = \frac{hc}{\lambda}.$$

$$\therefore \quad \frac{hc}{\lambda} = 3{\cdot}01 \times 10^{-12}.$$

$$\therefore \quad \lambda = \frac{hc}{3{\cdot}01 \times 10^{-12}}$$

$$= \frac{6{\cdot}62 \times 10^{-27} \times 3 \times 10^{10}}{3{\cdot}01 \times 10^{-12}}$$

$$= 6{\cdot}6 \times 10^{-5} \text{ cm.}$$

Ex. 16. *Calculate the wavelength in angstrom of the photon that is emitted when an electron in Bohr orbit* $n = 2$ *returns to the orbit* $n = 1$ *in the hydrogen atom. The ionisation potential of the ground state of hydrogen atom is* $2{\cdot}17 \times 10^{-11}$ *erg per atom.*

Solution **:** Since the hydrogen atom has only one orbit containing only one electron, the ionisation potential of the ground state of the hydrogen atom is the energy of the electron of the first orbit., i.e.,

$$E_1 = -2{\cdot}17 \times 10^{-11} \text{ erg.}$$

Thus, $E_2 = \dfrac{E}{n^2}$ ... (Eqn. 5)

$$= -\frac{2{\cdot}17 \times 10^{-11}}{2^2}. \qquad (n = 2)$$

$\therefore$ energy of the radiation emitted,

$$\Delta E = E_2 - E_1$$

$$= -\frac{2{\cdot}17 \times 10^{-11}}{2^2} - (-2{\cdot}17 \times 10^{-11})$$

$$= 1{\cdot}627 \times 10^{-11} \text{ erg.}$$

We know $\Delta E = h\nu = \dfrac{hc}{\lambda}$.

Thus $\dfrac{hc}{\lambda} = 1{\cdot}627 \times 10^{-11}$

$$\lambda = \frac{6{\cdot}62 \times 10^{-27} \times 3 \times 10^{10}}{1{\cdot}627 \times 10^{-11}} = 1{\cdot}22 \times 10^{-5} \text{ cm}$$

$$= 1220 \text{ Å}. \quad (1 \text{ Å} = 10^{-8} \text{ cm})$$

Ex. 17. *The electron energy in hydrogen atom is given by $E = (-21{\cdot}7 \times 10^{-12})/n^2$ erg. Calculate the energy required to remove an electron completely from the $n = 2$ orbit. What is the longest wavelength (in centimetres of light that can be used to cause this transition?*

Solution : The energy required to remove an electron from the second orbit is the same as the energy released (ΔE) when an electron will drop from the infinite orbit to the second orbit. Thus,

$$\Delta E = -\frac{21{\cdot}7 \times 10^{-12}}{\infty^2} - \left(-\frac{21{\cdot}7 \times 10^{-12}}{2^2}\right)$$

$$= 0 + \frac{21{\cdot}7 \times 10^{-12}}{2^2} = 5{\cdot}42 \times 10^{-12} \text{ erg.}$$

Now we know $\Delta E = h\nu = \dfrac{hc}{\lambda}$.

$\therefore$ $\dfrac{hc}{\lambda} = 5{\cdot}42 \times 10^{-12}$

$$\lambda = \frac{6{\cdot}62 \times 10^{-27} \times 3 \times 10^{10}}{5{\cdot}42 \times 10^{-12}} = 3{\cdot}66 \times 10^{-5} \text{ cm.}$$

Ex. 18. *According to Bohr theory, the electronic energy of a hydrogen atom in the nth Bohr atom is given by* $E_n = -\dfrac{21{\cdot}76\times10^{-19}}{n^2}$ J. *Calculate the longest wavelength of light that will be needed to remove an electron from the third Bohr orbit of the* He^+ *ion.*

($h = 6{\cdot}626\times10^{-34}$ J · s., $c = 3\times10^8$ m · s^{-1}) **(IIT 1990)**

Solution : Let us first calculate the energy required to remove an electron from the third orbit of the He^+ ion. This energy will be equal to the energy released (ΔE) when an electron will drop from ∞ orbit to third orbit. For He^+, Z = 2 and for H^+, Z = 1, Thus for He^+

$$\Delta E = -\frac{21{\cdot}76\times10^{-19}}{\infty^2}\times2^2 - \left(-\frac{21{\cdot}76\times10^{-19}}{3^2}\times2^2\right)$$

$$= 0 + 9{\cdot}67\times10^{-19}\text{ J.}$$

Now,

$$\Delta E = h\nu = \frac{hc}{\lambda}.$$

$$\therefore \qquad \lambda = \frac{hc}{\Delta E} = \frac{(6{\cdot}626\times10^{-34})(3\times10^8)}{9{\cdot}67\times10^{-19}}$$

$$= 2{\cdot}055\times10^{-7}\text{ metres.}$$

Ex. 19. *Assuming a spherical shape for the* F *nucleus, calculate the radius and the nuclear density of the* F *nucleus of mass number* 19.

Solution : We have,

$$r = (1{\cdot}4\times10^{-13})A^{1/3}$$

$$= 1{\cdot}4\times10^{-13}\times19^{1/3}$$

$$= 5{\cdot}07\times10^{-13}\text{ cm.} \qquad \text{... (Eqn. 8)}$$

$$\text{Volume of 1 F atom} = \frac{4}{3}\pi r^3$$

$$= \frac{4}{3}\times3{\cdot}14\times(5{\cdot}07\times10^{-13})^3$$

$$= 7{\cdot}17\times10^{-37}\text{ cm}^3.$$

$$\text{Mass of a single nucleus} = \frac{\text{mass of 1 mole of nucleus}}{\text{Av. no.}}$$

$$= \frac{19}{6{\cdot}022\times10^{23}}\text{ g.}$$

$$\text{Thus, density of the nucleus} = \frac{\text{mass of single nucleus}}{\text{volume of single nucleus}}$$

$$= \frac{19}{6.022 \times 10^{23}} \times \frac{1}{7.17 \times 10^{-37}}$$

$$= 0.44 \times 10^{14} \text{ g cm}^{-3}.$$

Ex. 20. *What is the relationship between the* eV *and the wavelength in metres of the energetically equivalent photon?*

Solution : The electronvolt is the energy imparted to an electron accelerated by a potential difference of 1 volt.

$$1 \text{ eV} = (\text{charge of an electron}) \times (\text{pot. diff.})$$
$$= 1.6022 \times 10^{-19} \text{ coulomb} \times 1 \text{ volt}$$
$$= 1.6022 \times 10^{-19} \text{ joule.}$$

[**Note:** 1 V = 1 J/C]

We know,

$$E = h\nu = \frac{hc}{\lambda}$$

or,

$$\lambda = \frac{hc}{E}.$$

$$= \frac{6.626 \times 10^{-34} \text{ (J} \cdot \text{s)} \times 2.988 \times 10^{8} \text{ (m} \cdot \text{s}^{-1})}{1.6022 \times 10^{-19} \text{ (J)}}$$

$$= 1.24 \times 10^{-6} \text{ metres.}$$

Thus wavelength equivalent to 1 eV is 1.24×10^{-6} metres.

Ex. 21. *The vapour of* Hg *absorbs some electrons accelerated by a potential difference of* 4·5 *volts as a result of which light is emitted. If the full energy of a single incident electron is supposed to be converted into light emitted by a single* Hg *atom, find the wavenumber* (1/λ) *of the light.*

Solution : From the previous example, we have,

$$4.5 \text{ eV} = \left(\begin{matrix}\text{charge of an electron} \\ \text{in coulomb}\end{matrix}\right)\left(\begin{matrix}\text{potential difference} \\ \text{in volt}\end{matrix}\right)$$

$$= (1.6022 \times 10^{-19})\,(4.5) \text{ joule.}$$

Now,

$$\text{wavenumber} = \frac{1}{\lambda} = \frac{E}{h.c}$$

$$= \frac{(1.6022 \times 10^{-19})\,(4.5)}{(6.626 \times 10^{-34})\,(2.998 \times 10^{8})}$$

$$= 3.63 \times 10^{6} \text{ metres}^{-1}.$$

Ex. 22. *The uncertainty in the momentum of a particle is* $2{\cdot}5 \times 10^{-16}$ g cm s^{-1}. *With what accuracy can its position be determined?* ($h = 6{\cdot}625 \times 10^{-27}$ erg · s, i.e., g cm^2 s^{-1})

Solution : We have,

$$\Delta x \cdot \Delta p = \frac{h}{4\pi} \qquad \text{... (Eqn. 16)}$$

or

$$\Delta x = \frac{h}{4\pi} \times \frac{1}{\Delta p}$$

$$= \frac{6{\cdot}625 \times 10^{-27}\ \text{g cm}^2\ \text{s}^{-1}}{4 \times 3{\cdot}14 \times (2{\cdot}5 \times 10^{-16}\ \text{g cm s}^{-1})}$$

$$= 2{\cdot}11 \times 10^{-12}\ \text{cm}.$$

Ex. 23. *What conclusion may be drawn from the following results of (a) and (b)?*
(a) If a 1×10^{-3}-kg *body is travelling along the x-axis at* 1 m/s *within* 0·01 m/s. *Calculate the theoretical uncertainty in its position.*
(b) If an electron is travelling at 100 m/s *within* 1 m/s, *calculate the theoretical uncertainty in its position.*
[$h = 6{\cdot}63 \times 10^{-34}$ J · s, *mass of electron* $= 9{\cdot}109 \times 10^{-31}$ kg]

Solution : (a) The velocity has an uncertainty of 0·02 m/s. (from 0·99 to 1·01 m/s)

$$\therefore\ \Delta x = \frac{h}{4\pi m \Delta v} = \frac{6{\cdot}63 \times 10^{-34}\ (\text{J} \cdot \text{s})}{(4 \times 3{\cdot}14)\ (1 \times 10^{-3}\ \text{kg})\ (0{\cdot}02\ \text{m/s})}$$

$$\approx 3 \times 10^{-30}\ \text{metre}.$$

(b) $$\Delta x = \frac{6{\cdot}63 \times 10^{-34}\ (\text{J} \cdot \text{s})}{(4 \times 3{\cdot}14)\ (9{\cdot}109 \times 10^{-31}\ \text{kg})\ (2\ \text{m/s})}$$

$$\approx 3 \times 10^{-5}\ \text{metres}.$$

[**Note:** From the values of Δx in (a) and (b), we see that the small mass of electron causes a very significant uncertainty in its position, whereas uncertainty in position of a macroscopic body is negligible compared to that of an electron. Hence it may be concluded that the uncertainty principle is not applicable to macroscopic bodies.]

Ex. 24. *Find the product of uncertainty in position and velocity for an electron of mass* $9{\cdot}10 \times 10^{-31}$ kg. ($h = 6{\cdot}62 \times 10^{-34}$ J · s., i.e., kg m^2 s^{-1})

Solution : We have,

$$\Delta x \,.\, \Delta p = \frac{h}{4\pi} \qquad \text{... (Eqn. 16)}$$

$$\Delta x \cdot (m \times \Delta v) = \frac{h}{4\pi}$$

or $$\Delta x \cdot \Delta v = \frac{h}{4\,\pi\, m}$$

$$= \frac{6{\cdot}62 \times 10^{-34}\ (\text{kg·m}^2\text{·s}^{-1})}{4 \times 3{\cdot}14 \times 9{\cdot}10 \times 10^{-31}\ (\text{kg})}$$

$$= 5{\cdot}8 \times 10^{-5}\ \text{m}^2\ \text{s}^{-1}.$$

Ex. 25. *Calculate the minimum uncertainty in velocity of a particle of mass* $1{\cdot}1 \times 10^{-27}$ kg *if uncertainty in its position is* 3×10^{-10} cm.
($h = 6{\cdot}62 \times 10^{-34}$ kg · m^2 s^{-1})

Solution : We have,

$$\Delta x \cdot \Delta p = \frac{h}{4\pi} \qquad \text{... (Eqn. 16)}$$

$$\Delta x \cdot (m\ \Delta v) = \frac{h}{4\pi}$$

or $$\Delta v = \frac{h}{4\pi} \cdot \frac{1}{m\ \Delta x}$$

$$= \frac{6{\cdot}62 \times 10^{-34}\ (\text{kg} \cdot \text{m}^2\ \text{s}^{-1})}{4 \times 3{\cdot}14 \times (1{\cdot}1 \times 10^{-27}\ \text{kg}) \times (3 \times 10^{-12}\ \text{m})}$$

$$= 1{\cdot}6 \times 10^4\ \text{m s}^{-1}.$$

Ex. 26. *The uncertainties in the position and velocity of a particle are* 10^{-10} m *and* $5{\cdot}27 \times 10^{-24}$ m s^{-1} *respectively. Calculate the mass of the particle.* ($h = 6{\cdot}625 \times 10^{-34}$ J · s)

Solution : We have, $\Delta x \cdot (m\ \Delta v) = \dfrac{h}{4\pi}$

or $$m = \frac{h}{4\pi} \times \frac{1}{\Delta x \cdot \Delta v}$$

$$= \frac{6{\cdot}625 \times 10^{-34}\ (\text{kg} \cdot \text{m}^2\ \text{s}^{-1})}{4 \times 3{\cdot}14 \times (10^{-10}\ \text{m})\ (5{\cdot}27 \times 10^{-24}\ \text{m s}^{-1})}$$

$$= 0{\cdot}10\ \text{kg}.$$

Ex. 27. *The first ionisation potential for* Li *is* 5·41 eV *and electron affinity of* Cl *is* 3·61 eV. *Calculate* ΔH *in* kJ/mol *for the reaction:*

$$\text{Li (g)} + \text{Cl (g)} \rightarrow \text{Li}^+ + \text{Cl}^-$$

Suppose that the resulting ions do not combine with each other.
(1 eV = $1{\cdot}60 \times 10^{-22}$ kJ, Av. const. = $6{\cdot}02 \times 10^{23}$)

Solution : Li (g) $\rightarrow$ $Li^+ + e$;

$$\Delta H_1 = +5{\cdot}41 \times 1{\cdot}60 \times 10^{-22} \text{ kJ / atom (IP)}$$
$$= +5{\cdot}41 \times 1{\cdot}60 \times 10^{-22} \times 6{\cdot}02 \times 10^{23} \text{ kJ/mole.}$$

$$Cl\,(g) + e \rightarrow Cl^-;\quad \Delta H_2 = -3{\cdot}61 \times 1{\cdot}60 \times 10^{-22} \text{ kJ/atom (EA)}$$
$$= -3{\cdot}61 \times 1{\cdot}60 \times 10^{-22} \times 6{\cdot}02 \times 10^{23} \text{ kJ/mole.}$$

On adding, we get,

$$Li\,(g) + Cl\,(g) \rightarrow Li^+ + Cl^-;$$
$$\Delta H = \Delta H_1 + \Delta H_2$$
$$= 173{\cdot}4 \text{ kJ/mole (on substituting values of } \Delta H_1 \text{ and } \Delta H_2).$$

Ex. 28. *Calculate the accelerating potential that must be imparted to a proton beam to give it an effective wavelength of* 0·005 nm.

Solution : Atomic weight of hydrogen is 1·008 amu.

$\therefore$ weight of 1 mole of proton = 1·008 g.

$$\therefore \text{ weight of 1 proton} = \frac{1{\cdot}008}{6{\cdot}022 \times 10^{23}} = 0{\cdot}167 \times 10^{-23} \text{ g}$$
$$= 0{\cdot}167 \times 10^{-26} \text{ kg.}$$

We have,

$$v = \frac{h}{m\lambda} = \frac{(6{\cdot}63 \times 10^{-34} \text{ J}\cdot\text{s})}{(0{\cdot}167 \times 10^{-26} \text{ kg})\,(0{\cdot}005 \times 10^{-9} \text{ m})} = 7{\cdot}94 \times 10^4 \text{ m/s.}$$

$$\text{Kinetic energy} = \frac{1}{2}mv^2 = \frac{1}{2}(0{\cdot}167 \times 10^{-26})\,(7{\cdot}94 \times 10^4)^2$$
$$= 5{\cdot}26 \times 10^{-18} \text{ J}$$
$$= \frac{5{\cdot}26 \times 10^{-18}}{1{\cdot}602 \times 10^{-19}} \text{ eV}$$
$$= 32{\cdot}8 \text{ eV.}$$

As the magnitude of charge of a proton is the same as that of an electron, the potential required is equal in magnitude to the number of eV, i.e., 32·8 volts.

Ex. 29. 13·6 eV *is needed for ionisation of a hydrogen atom. An electron in a hydrogen atom in its ground state absorbs* 1·50 *times as much energy as the minimum energy required for it to escape from the atom. What is the wavelength of the emitted electron?* ($m_e = 9{\cdot}109 \times 10^{-31}$ kg, $e = 1{\cdot}602 \times 10^{-19}$ *coulomb,* $h = 6{\cdot}63 \times 10^{-34}$ J · s)

Solution : 1·5 times of 13·6 eV, i.e., 20·4 eV, is absorbed by the hydrogen atom out of which 6·8 eV (20·4 – 13·6) is converted to kinetic energy.

$$KE = 6\cdot 8 \text{ eV} = 6\cdot 8\ (1\cdot 602 \times 10^{-19} \text{ coulomb})\ (1 \text{ volt}) = 1\cdot 09 \times 10^{-18} \text{ J}.$$

Now, $KE = \frac{1}{2} mv^2$

or $v = \sqrt{\frac{2KE}{m}} = \sqrt{\frac{2\ (1\cdot 09 \times 10^{-18} \text{ J})}{(9\cdot 109 \times 10^{-31} \text{ kg})}}$

$= 1\cdot 55 \times 10^6$ m/s.

$\therefore \quad \lambda = \frac{h}{mv} = \frac{(6\cdot 63 \times 10^{-34} \text{ J}\cdot\text{s})}{(9\cdot 109 \times 10^{-31} \text{ kg})\ (1\cdot 55 \times 10^6 \text{ m/s})}$

$= 4\cdot 70 \times 10^{-10}$ metres.

Ex. 30. *How many photons of light having a wavelength of* 4000 Å *are necessary to provide* 1 J *of energy?*

$(h = 6\cdot 63 \times 10^{-34} \text{ J}\cdot\text{s},\ c = 3 \times 10^8 \text{ m/s})$

Solution : Energy of a photon $= h\nu = \frac{hc}{\lambda}$

$= \frac{(6\cdot 63 \times 10^{-34} \text{ J.s})\ (3 \times 10^8 \text{ m/s})}{(4000 \times 10^{-10} \text{ m})}$

$= 4\cdot 97 \times 10^{-19}$ J.

$\therefore$ number of photons required $= \frac{1 \text{ J}}{4\cdot 97 \times 10^{-19} \text{ J}}$

$= 2\cdot 01 \times 10^{18}$.

Ex. 31. *With what velocity should an* α*-particle travel towards the nucleus of a* Cu *atom so as to arrive at a distance* 10^{-13} *metre from the nucleus of the* Cu *atom?* (Cu = 29, $e = 1\cdot 6 \times 10^{-19}$ C, $\varepsilon_0 = 8\cdot 85 \times 10^{-12}$ $\text{J}^{-1}\ \text{C}^2\ \text{m}^{-1}$, $m_e = 9\cdot 1 \times 10^{-31}$ kg)

(IIT 1997)

Solution : We have,

$$\frac{Z_1 Z_2 e^2}{4\pi\ \varepsilon_0\ d} = \frac{1}{2} m_e v^2. \qquad \ldots (13)$$

Substituting $Z_1 = 29$, $Z_2 = 2$ and the given values in the above equation, we get,

$v = 5\cdot 41 \times 10^8$ m/s.

Ex. 32. *The iodine molecule dissociates into atoms after absorbing light of* 4500 Å *if one quantum of radiation is absorbed by each molecule. Calculate the kinetic*

energy of iodine atoms. (Bond energy per molecule of I_2 = 240 kJ mol^{-1})

(IIT 1995)

Solution : Bond energy per molecule of $I_2 = \dfrac{240 \times 1000}{6{\cdot}022 \times 10^{23}}$ J

$= 3{\cdot}984 \times 10^{-19}$ J.

$$\text{Energy absorbed} = \frac{hc}{\lambda}$$

$$= \frac{6{\cdot}626 \times 10^{-34} \times 3 \times 10^{8}}{4500 \times 10^{10}} = 4{\cdot}417 \times 10^{-19} \text{ J.}$$

$\therefore$ KE of one I_2 molecule $= 4{\cdot}417 \times 10^{-19} - 3{\cdot}984 \times 10^{-19}$ J

$= 4{\cdot}33 \times 10^{-20}$ J.

$$\text{KE of one I atom} = \frac{4{\cdot}33 \times 10^{-20}}{2} = 2{\cdot}165 \times 10^{-20} \text{ J.}$$

Ex. 33. *Consider an* α*-particle just in contact with* $^{238}_{92}U$ *nucleus. Calculate the coulombic repulsion energy (i.e., the height of the coulombic barrier between* ^{238}U *and the* α*-particle) assuming that the distance between them is equal to the sum of their radii.* ($r_0 = 1{\cdot}4 \times 10^{-13}$, $e = 4{\cdot}8 \times 10^{-10}$ esu)

Solution : Distance between U^{238} and He^4 nuclei, *d*,

$= \text{radius of } U^{238} + \text{radius of } He^4$

$= 1{\cdot}4 \times 10^{-13} \times (238)^{\frac{1}{3}} + 1{\cdot}4 \times 10^{-13} \times (4)^{\frac{1}{3}}$ ... (Eqn. 14)

$= 1{\cdot}0899 \times 10^{-12}$ cm.

$\therefore$ coulombic barrier $= \dfrac{Z_1 Z_2 e^2}{d}$ ... (Eqn. 12)

$$= \frac{92 \times 2 \times (4{\cdot}8 \times 10^{-10})^2}{1{\cdot}0899 \times 10^{-12}}$$

$= 3{\cdot}899 \times 10^{-5}$ erg.

Ex. 34. *Calculate the electronegativity of fluorine from the following data:*

$E_{\text{H–H}}$ = 104·2 kcal mol^{-1}

$E_{\text{F–F}}$ = 36·6 kcal mol^{-1}

$E_{\text{H–F}}$ = 134·6 kcal mol^{-1}

Electronegativity of H = 2·05

Solution : Energy for 100% covalent bond (H–F) $= \sqrt{E_{\text{H–H}} \times E_{\text{F–F}}}$

$= \sqrt{104{\cdot}2 \times 36{\cdot}6}$

$= 61{\cdot}75$ kcal.

Resonance energy (Δ) = actual bond energy − energy for 100% covalent bond

$$= 134{\cdot}6 - 61{\cdot}75$$
$$= 72{\cdot}85 \text{ kcal.}$$

Electronegativity difference = $0{\cdot}18\sqrt{\Delta}$

Electronegativity of F − electronegativity of H = $0{\cdot}18\sqrt{\Delta}$

Electronegativity of F = $0{\cdot}18 \times \sqrt{72{\cdot}85} + 2{\cdot}05$

$$= 3{\cdot}586.$$

Note: Electronegativity difference = $0{\cdot}18\sqrt{\Delta}$; Δ in kcal/mole

$= 0{\cdot}088\sqrt{\Delta}$; Δ in kJ/mole

$= 0{\cdot}208\sqrt{\Delta}$; Δ in eV/mole.

Ex. 35. *Calculate the value of Bohr magneton. For a hydrogen-like atom with a 3d electron, what is the value of the orbital angular momentum in terms of* $\hbar$? ($e = 1{\cdot}6 \times 10^{-19}$c, $h = 6{\cdot}626 \times 10^{-34}$ J.s and $m_e = 9{\cdot}109 \times 10^{-31}$ kg)

Solution : We have,

$$\mu_B(BM) = \frac{eh}{4\pi m} \quad \dots (19)$$

$$= \frac{(1{\cdot}6 \times 10^{-19})(6{\cdot}626 \times 10^{-34})}{4 \times 3{\cdot}14 \times 9{\cdot}109 \times 10^{-31}}$$

$$= 9{\cdot}266 \times 10^{-24}\ \text{JT}^{-1}.$$

Further, we have,

$$L = \sqrt{l(l+1)}\,\frac{h}{2\pi} = \sqrt{l(l+1)}\,\hbar. \quad \dots (18)$$

For a 3*d* electron, $l = 2$

$$L = \sqrt{2(2+1)}\,\hbar$$
$$= \sqrt{6}\,\hbar.$$

Ex. 36. *A compound of vanadium has a magnetic moment of* 1.73 BM. *Work out the electronic configuration of the vanadium ion in the compound.* **(IIT 1997)**

Solution : See Q. 506 Ch. 21.

Ex. 37. *The wavelength associated with an electron moving in a potential difference of* 2000 V *is* 0.05 nm. *How slow must a* 0.01-kg *hockey ball travel to have the same de Broglie wavelength as a* 2000 V *electron?*

Solution : We have,

$$\lambda = \frac{h}{mv}$$

or $$v = \frac{h}{m\lambda}$$

$$= \frac{6{\cdot}626 \times 10^{-27}}{0{\cdot}01 \times 10^{-3} \times 0.05 \times 10^{-7}}$$

$$= 1.325 \times 10^{-19} \text{ cm s}^{-1}$$

Ex. 38. *Consider the hydrogen atom to be a proton embedded in a cavity of radius a_0 (Bohr radius) whose charge is neutralised by the addition of an electron to the cavity in vacuum, infinitely slowly. Estimate the average total energy of an electron in its ground state in a hydrogen atom as the work done in the above neutralisation process. Also, if the magnitude of the average kinetic energy is half the magnitude of the average potential energy, find the average potential energy.* **(IIT 1996)**

Solution : Work done in neutralisation $= \int_{a=\infty}^{a=a_0} (\text{force})\, da_0$.

Now potential energy of the electron, at a distance a_0 from the nucleus, is given by the work done in moving the electron from ∞ to that point.

$\therefore$ potential energy of an electron $= \int_{a=\infty}^{a=a_0} \frac{1}{4\pi\epsilon_0} \frac{e^2}{a_0^2}\, da_0$

$$= -\frac{e^2}{4\pi\epsilon_0 a_0}.$$

$\therefore$ average kinetic energy $= \frac{1}{2}\left(\frac{e^2}{4\pi\epsilon_0 a_0}\right)$ (given)

$$= \frac{e^2}{8\pi\epsilon_0 a_0}.$$

Total energy $= KE + PE = \frac{e^2}{8\pi\epsilon_0 a_0} + \left(-\frac{e^2}{4\pi\epsilon_0 a_0}\right)$

$$= -\frac{e^2}{8\pi\epsilon_0 a_0}.$$

Ex. 39. *What is the maximum number of electrons in an atom in which the last electron, filled, has the following quantum numbers?*

(*a*) $n = 3, l = 1$

(*b*) $n = 3, l = 2$

(*c*) $n = 3, l = 2$ *and* $m = -1$

(*d*) $n = 3, l = 1$ *and* $m = -1$

(*e*) $n = 3, l = 2, m = 0$ *and* $s = -\frac{1}{2}$

Solution :

Q. Nos.	*e* in last orbital	Electronic configuration	Max. No. of *e*
(a) $n=3, l=1$	$3p^6$	$1s^2, 2s^2, 2p^6, 3s^2, 3p^6$	18
(b) $n=3, l=2$	$3d^{10}$	$1s^2, 2s^2, 2p^6, 3s^2, 3p^6, 4s^2, 3d^{10}$	30
(c) $n=3, l=2, m=-1$	$3d^7$	$1s^2, 2s^2, 2p^6, 3s^2, 3p^6, 4s^2, 3d^7$	27
(d) $n=3, l=1, m=-1$	$3p^4$	$1s^2, 2s^2, 2p^6, 3s^2, 3p^4$	16
(e) $n=3, l=2, m=0, s=-\frac{1}{2}$	$3d^8$	$1s^2, 2s^2, 2p^6, 3s^2, 3p^6, 4s^2, 3d^8$	28

Ex. 40. *The following incorrect set of quantum numbers in the order n, l, m, s are written for paired electrons or for one electron in an orbital. Correct them, assuming n values are correct.*

(a) $1, 0, 0, +\frac{1}{2}, +\frac{1}{2}$ *(b)* $2, 2, 1, \pm\frac{1}{2}$

(c) $3, 2, 3, \pm\frac{1}{2}$ *(d)* $3, 1, 2, +\frac{1}{2}$

(e) $2, 1, -1, 0$ *(f)* $3, 0, -1, -\frac{1}{2}$

Solution : (a) $1, 0, 0, \pm\frac{1}{2}$ (b) $2, 1, 1, \pm\frac{1}{2}$

(c) $3, 2, 2, \pm\frac{1}{2}$ (d) $3, 1, 1, +\frac{1}{2}$

(e) $2, 1, -1, +\frac{1}{2}$ (f) $3, 0, 0, -\frac{1}{2}$

Ex. 41. *Plutonium-239 has been used as a power source for heart pacemakers. What is the energy obtained from the following decay of* 215 mg *of* Pu-239?

$$^{239}_{94}Pu \rightarrow \,^{4}_{2}He + \,^{235}_{92}U$$

Suppose the electric energy produced from this amount of Pu-239 *is* 25% *of this value, what is the minimum grams of* Zn *that would be needed for the standard voltaic cell*

$$Zn \mid Zn^{2+} \parallel Cu^{2+} \mid Cu$$

to obtain the same electric energy? Given that $E^0_{cell} = 1{\cdot}1$ V, He = 4.0015 amu, U = 234·9935 amu, Pu = 239·0006 amu.

Solution : $^{239}_{94}Pu \rightarrow \,^{4}_{2}He + \,^{235}_{92}U$

Mass defect = 239·0006 − 4·0015 − 234·9935 = 0·0056 amu.

From mass – energy equation of Einstein, 1 amu ≡ 931·5 MeV.

Energy released per Pu nuclide = 0·0056 × 931·5 = 5·2164 MeV.

Energy obtained from the decay of 215 mg of Pu-239

$$= 5{\cdot}2164 \times \frac{0{\cdot}215}{239} \times 6{\cdot}022 \times 10^{23} \text{ MeV}$$
$$= 2{\cdot}826 \times 10^{21} \text{ MeV}$$
$$= 2{\cdot}826 \times 10^{21} \times 1{\cdot}602 \times 10^{-19} \text{ J}$$
$$= 4{\cdot}527 \times 10^{2}. \text{ J}$$

$\therefore$ electric energy = 25% of total energy $= \dfrac{4{\cdot}527 \times 10^2}{4} = 1{\cdot}13 \times 10^2$ J.

Let the amount of Zn required to produce this energy from a standard galvanic cell be w g.

As 1 mole of Zn produces 2 moles of electrons,

moles of electrons produced by $\dfrac{w}{65}$ moles of Zn $= \dfrac{2w}{65}$.

Electric energy by a galvanic cell = nFE

$$= \frac{2w}{65} \times 96500 \times 1{\cdot}1$$
$$= 3266{\cdot}15\, w \text{ J}.$$

Thus, $3266{\cdot}15\, w = 1{\cdot}13 \times 10^2$

$$w = 3{\cdot}46 \times 10^{-2} \text{ g}.$$

[Note: Read the text in Chapter 18]

Ex. 42. *One of the nuclides in each of the following pairs is radioactive; the other is stable. Which one is radioactive and which is stable? Explain.*

(a) $^{208}_{84}Po$, $^{209}_{83}Bi$ *(b)* $^{39}_{19}K$, $^{40}_{19}K$ *(c)* $^{71}_{31}Ga$, $^{76}_{31}Ga$

Solution :

(a) Po has an atomic number greater than 83 and so it is radioactive. Bi-209 has 126 neutrons (a magic no.), so it is expected to be stable.

(b) $^{39}_{19}K$ has a magic no. of neutrons, (20), so it is expected to be stable. The isotope $^{40}_{19}K$ has an odd number of protons (19) and an odd number of neutrons (21). Since stable odd-odd nuclei are rare, $^{40}_{19}K$ is expected to be radioactive.

(c) Of the two isotopes $^{76}_{31}Ga$ lies farther from the centre of the stability belt, so it is expected to be radioactive and $^{71}_{31}Ga$ to be stable.

Note: Nuclei with a certain number of protons or neutrons called **Magic Number** appear to be very stable. For protons, the magic numbers are 2, 8, 20, 28, 50 and 82. Neutrons have the same magic numbers as well as the magic number 126. For protons, calculations show that 114 should also be a magic number.

Ex. 43. *The mass number and the atomic number of uranium are* 238 *and* 92 *respectively. If on nuclear disintegration* 6 α*- and* 4 β*-particles are emitted, find the mass number and atomic number of the atom formed.*

Solution : Applying group displacement law, emission of 6 α-particles will cause a decrease in mass number by 24 (6×4) units and in atomic number by 12 (6×2) units. Now with the emission of 4 β-particles, mass number will not change but atomic number will increase by 4 units. Thus the mass number and the atomic number of the atom formed will be 214 (i.e., $238 - 24$) and 84 (i.e., $92 - 12 + 4$) respectively.

Ex. 44. ${}^{234}_{90}\text{Th}$ *disintegrates to give* ${}^{206}_{82}\text{Pb}$ *as the final product. How many* α*- and* β*-particles are emitted during the process?* **(IIT 1986)**

Solution : Since with the emission of an α-particle, the mass number of the product decreases by four units. Thus,

$$\text{number of } \alpha\text{-particles emitted} = \frac{234 - 206}{4} = 7.$$

Now with the emission of 7 α-particles, the atomic number decreases by 14 units, i.e., the resulting atomic number will be $90 - 14 = 76$. But since Pb has an atomic number of 82, therefore, 6 β-particles will be emitted in order to increase the atomic number from 76 to 82.

Thus $${}^{234}_{90}\text{Th} \xrightarrow[6\beta]{7\alpha} {}^{206}_{82}\text{Pb}.$$

Ex. 45. ${}^{7}_{4}\text{Be}$ *captures a* K*-electron into its nucleus. What will be the mass number and atomic number of the nuclide formed?*

Solution : When a nucleus captures a K-electron, a proton is converted to a neutron. So the mass number does not change but the atomic number reduces by 1 unit. Thus the mass number and atomic number of the resulting nuclide will be 7 and 3 respectively.

Ex. 46. *To which radioactive series does* ${}^{227}_{89}\text{Ac}$ *belong?*

Solution : Dividing the mass number 227 by 4, the remainder is 3 and therefore, the radioactive series to which ${}^{227}_{89}\text{Ac}$ belongs to is ($4n + 3$), i.e., Actinium series.

Ex. 47. *A sample of radioactive* ${}^{133}\text{I}$ *gave with a Geiger counter* 3150 *counts per minute at a certain time and* 3055 *counts per minute exactly one hour later. Calculate the half-life period of* ${}^{133}\text{I}$.

Solution : $N^0 = 3150$; $N = 3055$, $t = 1$ hour.

We have, $\lambda = \dfrac{0.6932}{t_{1/2}}$.

Substituting these values in $\lambda = \dfrac{2.303}{t} \log \dfrac{N^0}{N}$

$$\frac{0.6932}{t_{\frac{1}{2}}} = \frac{2.303}{1} \log \frac{3150}{3055}$$

$$t_{\frac{1}{2}} = 22.63 \text{ hours.}$$

Ex. 48. *In a sample of pitchblende the ratio of* ^{206}Pb : ^{238}U = 0·2 : 1 *by weight. If the disintegration constant of* ^{238}U *is* 1.54×10^{-10} *per year and all* Pb *is supposed to be originated from uranium, what will be the age of the mineral?*

Solution : Initial moles: N^0

$$^{238}\text{U} \rightarrow {}^{206}\text{Pb}$$

Moles after time t: $(N^0 - x) \qquad x$

$$\frac{\text{wt. of } ^{206}\text{Pb}}{\text{wt. of } ^{238}\text{U}} = \frac{0.2}{1}.$$

$$\therefore \quad \frac{\text{mole of Pb}^{206}}{\text{mole of U}^{238}} = \frac{0.2/206}{1/238} = \frac{0.231}{1}$$

or $$\frac{x}{N^0 - x} = 0.231.$$

Adding 1 in both sides,

$$1 + \frac{x}{N^0 - x} = 0.231 + 1$$

$$\frac{N^0}{N^0 - x} = 1.231.$$

We have, $$\lambda = \frac{2.303}{t} \log \frac{N^0}{N} \qquad \text{... (Eqn. 24)}$$

or $$\lambda = \frac{2.303}{t} \log \frac{N^0}{N^0 - x}$$

$$1.54 \times 10^{-10} = \frac{2.303}{t} \log 1.231$$

$$t = 1.35 \times 10^9 \text{ years.}$$

Ex. 49. *For* ^{24}Na, $t_{\frac{1}{2}} = 14.8$ *hours. In what period of time will a sample of this substance lose* 90% *of its radioactive intensity?*

Solution : Let the initial radioactive intensity be 100 which corresponds to N^0. The radioactive intensity after a time period, say t hours, will be 10 (corresponding to N) as the substance has lost 90% of its radioactive intensity.

$$\lambda = \frac{0.6932}{t_{\frac{1}{2}}} = \frac{0.6932}{14.8} \qquad \text{... (Eqn. 26)}$$

We have,
$$\lambda = \frac{2.303}{t} \log \frac{N^0}{N}$$

$$\frac{0.6932}{14.8} = \frac{2.303}{t} \log \frac{100}{10}$$

$$t = 49.17 \text{ hours.}$$

Ex. 50. *Radioactivity is a first-order process. Radioactive carbon in wood sample decays with a half-life of* 5770 *years. What is the rate constant (in year*$^{-1}$*) for the decay? What fraction would remain after* 11540 *years?*

Solution : We have,
$$\lambda = \frac{0.6932}{t_{\frac{1}{2}}} \qquad \text{... (Eqn. 26)}$$

$$= \frac{0.6932}{5770} = 1.2 \times 10^{-4} \text{ per year.}$$

Further, as the duration of 11540 years is just the double of $t_{\frac{1}{2}}$ (i.e., 5770 years) we can find out the fraction remained after 11540 years without using equation (24) as follows:

After 5770 years, half of the substance would remain and after another 5770 years (i.e., a total of 11540 years) half of the half would remain, i.e., one fourth would remain after 11540 years.

Ex. 51. *Carbon-containing compounds in living organisms maintain an equilibrium with* ^{14}C *in the atmosphere. The activity associated with* C *in the equilibrium is* 15 *disintegrations/minute/gram of carbon* ($t_{1/2}$ *of* $^{14}C = 5730$ *years). An old tree, found in an burial, was subjected to radioactive dating. The decay rate associated with its* ^{14}C *content is* 10 *disintegrations/minute/gram of* C. *Find the time elapsed since the tree was cut down.*

Solution : $\lambda = \frac{0.6932}{5730} = 1.21 \times 10^{-4} \text{ yr}^{-1}$ ... (Eqn. 26)

Now,

rate of disintegration of ^{14}C (when equilibrium was destroyed, i.e., when the tree was cut down) $\propto N^0$.

And present rate of disintegration $\propto N$

i.e.,
$$\frac{N^0}{N} = \frac{15}{10}.$$

Using Eqn. (24),
$$\lambda = \frac{2.303}{t} \log \frac{N^0}{N}$$

or $$1{\cdot}21 \times 10^{-4} = \frac{2{\cdot}303}{t} \log \frac{15}{10}$$

or $$t = 3{\cdot}35 \times 10^3 \text{ years.}$$

Ex. 52. *The lowest level ^{14}C activity for experimental detection is* 0·03 dis./min/g *of C. What is the maximum age of an object that can be determined by the* C-14 *method? Use necessary data from previous example.*

Solution : From the previous example, 15 dis./min corresponds to N^0 and 0·03 dis./min/g of C corresponds to N.

$$\therefore \quad \frac{N^0}{N} = \frac{15}{0{\cdot}03}.$$

Let the maximum age of the object be t years.

Thus, $$\lambda = \frac{2{\cdot}303}{t} \log \frac{N^0}{N} \quad \text{... (Eqn. 24)}$$

$$\frac{0{\cdot}6932}{5730} = \frac{2{\cdot}303}{t} \log \frac{15}{0{\cdot}03}$$

or $$t = 5{\cdot}1 \times 10^4 \text{ years.}$$

Ex. 53. *What is the* α *activity in disintegrations per minute for* 0·001 g *sample of* $^{226}Ra\left(t_{\frac{1}{2}} = 1620 \text{ years}\right)$?

Solution : As the rate of disintegration is to be calculated as disintegrations per minute, $t_{\frac{1}{2}}$ has also to be taken in minutes.

$$t_{\frac{1}{2}} = 1620 \times 365 \times 24 \times 60 \text{ minutes}$$

$$\text{Mole of } ^{226}Ra = \frac{0{\cdot}001}{226}.$$

$$\therefore \quad \text{no. of nuclei of } ^{226}Ra = \text{mole} \times \text{Av. const.}$$

$$= \frac{0{\cdot}001}{226} \times 6{\cdot}022 \times 10^{23}.$$

Now, we have,

$$-\frac{d(N)}{dt} = \lambda(N) = \frac{0{\cdot}6932}{t_{\frac{1}{2}}} \times N \quad \text{... (23)}$$

$$= \frac{0{\cdot}6932}{1620 \times 365 \times 24 \times 60} \times \frac{0{\cdot}001 \times 6{\cdot}022 \times 10^{23}}{226}$$

$$= 2{\cdot}16 \times 10^{19} \text{ events/minute.}$$

Ex. 54. *What weight of* $^{14}C\left(t_{\frac{1}{2}} = 5760 \text{ years}\right)$ *will make one curie of it?* [*one curie gives* $3{\cdot}7 \times 10^{10}$ *disintegrations/second* (dps)].

Solution : Let the weight of ^{14}C be w gram

Number of ^{14}C nuclei = mole × Av. constant

$$= \frac{w}{14} \times 6{\cdot}022 \times 10^{23}$$

and $t_{\frac{1}{2}}$ in second $= 5760 \times 365 \times 24 \times 60 \times 60$

$$\text{Now, we have, } -\frac{d(N)}{dt} = \lambda(N) = \frac{0{\cdot}6932}{t_{\frac{1}{2}}} \times N \qquad \dots (23)$$

$$3{\cdot}7 \times 10^{10} = \frac{0{\cdot}6932}{5760 \times 365 \times 24 \times 60 \times 60} \times \frac{w}{14} \times 6{\cdot}022 \times 10^{23}.$$

$$w = 0{\cdot}225 \text{ g.}$$

Ex. 55. *A sample of radioactive substance shows an intensity of* 2·3 *millicurie at a time t and an intensity of* 1·62 *millicurie* 600 *seconds later. What is the half-life period of the radioactive material?*

Solution : Here $\frac{N^0}{N} = \frac{2{\cdot}30}{1{\cdot}62}$.

We have, $\lambda = \frac{2{\cdot}303}{600} \log \frac{2{\cdot}30}{1{\cdot}62} = 0{\cdot}000584.$

Now, $t_{\frac{1}{2}} = \frac{0{\cdot}6932}{\lambda} = \frac{0{\cdot}6932}{0{\cdot}000584} = 1187 \text{ seconds.}$

Ex. 56. *The radioisotope* $^{32}_{15}P$ *is used in biochemical studies. A sample containing this isotope has an activity* 1000 *times the detectable limit. How long could an experiment be run with the sample before the radioactivity could no longer be detected? Half-life of* $^{32}_{15}P$ *is* 14·2 *days.*

Solution : The minimum number of P atoms, the radioactivity of which could be detected, is one, and so the sample will initially contain 1000 atoms of P.

Thus, $\frac{N^0}{N} = \frac{1000}{1}$

Now, $\lambda = \frac{2{\cdot}303}{t} \log \frac{N^0}{N}$

$$\frac{0{\cdot}6932}{14{\cdot}2} = \frac{2{\cdot}303}{t} \log \frac{1000}{1}$$

or $t = 141{\cdot}5$ days.

Ex. 57. 1 *gram* ^{226}Ra *is placed in an evacuated tube whose volume is* 5 cc. *Assuming that each* Ra *nucleus yields four* He *atoms which are retained in the tube, what* *will be the partial pressures at* 27°C *of* He *produced at the end of a year?* $t_{\frac{1}{2}}$ *for* Ra *is* 1590 *years.*

Solution : Suppose x moles of Ra disintegrates into He after 1 year.

$$\begin{array}{lcc} \text{Initial moles:} & 1/226 & 0 \\ & \text{Ra} \rightarrow & 4\text{He} \\ \text{Moles after 1 yr:} & \left(\dfrac{1}{226} - x\right) & 4x \end{array}$$

(∵ 1 mole of Ra produces 4 moles of He)

We have,

$$\lambda = \frac{0{\cdot}6932}{t_{\frac{1}{2}}} = \frac{2{\cdot}303}{t} \log \frac{N^0}{N}$$

$$\frac{0{\cdot}6932}{1590} = \frac{2{\cdot}303}{1} \log \frac{1/226}{\left(\dfrac{1}{226} - x\right)}$$

$$= 2{\cdot}303 \log \frac{1}{(1 - 226\,x)}$$

On solving, we get, $x = \dfrac{0{\cdot}0005}{226}$.

$$\text{Total moles of He} = 4 \times x$$

$$= 4 \times \frac{0{\cdot}0005}{226} = \frac{0{\cdot}002}{226}.$$

Volume of the tube = 0·005 litre.

Temperature = (273 + 27) K

= 300 K.

Thus, partial pressure of He $= \dfrac{n}{V} RT$

$$p_{\text{He}} = \frac{0{\cdot}002}{226} \times \frac{0{\cdot}0821}{0{\cdot}005} \times 300$$

$$= 0{\cdot}04358 \text{ atm}$$

$$= 0{\cdot}04358 \times 760 \text{ mm}$$

$$= 33{\cdot}12 \text{ mm}.$$

Ex. 58. ^{90}Y *has a half-life of* 64 *hours and* ^{90}Sr, 28 *years.* ^{90}Sr *decays to* ^{90}Y *by* β*-emission. What will be the amount of* ^{90}Y *in equilibrium with* 1 g *of* ^{90}Sr?

Solution : Let the amount of ^{90}Y be w g.

Mole of $^{90}Y = \dfrac{w}{90}$.

$$\text{Mole of } {}^{90}\text{Sr} = \frac{1}{90}$$

$$(t_{1/2})_Y = 64 \text{ hours}$$

$$(t_{1/2})_{Sr} = 28 \times 365 \times 24 \text{ hours.}$$

Now, we have, for secular equilibrium,

$$\frac{(\text{mole})_Y}{(\text{mole})_{Sr}} = \frac{(t_{1/2})_Y}{(t_{1/2})_{Sr}} \quad \text{...(Eqn. 32)}$$

$$\frac{w/90}{1/90} = \frac{64}{28 \times 365 \times 24}.$$

Hence, $w = 3{\cdot}91 \times 10^{-4}$ g.

Ex. 59. *Calculate in* MeV *the energy equivalent to one gram-atomic weight of electrons at rest.*

Solution : The mass of 1 mole (1 gram-atomic weight) of electrons on atomic weight scale is $\frac{N}{1837}$ amu.

$\therefore$ energy equivalent to $\frac{N}{1837}$ amu of electrons

$$= \frac{6{\cdot}022 \times 10^{23}}{1837} \times 931{\cdot}5 = 3{\cdot}053 \times 10^{23} \text{ MeV.}$$

(1 amu = 931·5 MeV)

Ex. 60. *Calculate the binding energy per nucleon (in* MeV) *in* He *atom,* ${}^4_2$He *which has a mass of* 4·00260 amu. *Mass of one neutron* = 1·008665 amu *and mass of* 1 *hydrogen atom* = 1·007825 amu.

Solution : He atom contains 2 protons, 2 neutrons and 2 electrons.

Mass of (2 protons + 2 electrons) = 2 × mass of 1 H atom

$$= 2 \times 1{\cdot}007825 \text{ amu}$$

$$= 2{\cdot}01565 \text{ amu.}$$

$$\text{Mass of 2 neutrons} = 2 \times 1{\cdot}008665 \text{ amu}$$

$$= 2{\cdot}017330 \text{ amu.}$$

$$\therefore \quad \text{mass of He atom} = 2{\cdot}01565 + 2{\cdot}01733$$

$$= 4{\cdot}03298 \text{ amu.}$$

Actual mass of He atom = 4·00260 amu.

$$\therefore \quad \text{mass-defect} = 4{\cdot}03298 - 4{\cdot}00260$$

$$= 0{\cdot}03038 \text{ amu}$$

$$\therefore \quad \text{binding energy} = 0{\cdot}03038 \times 931{\cdot}5 \text{ MeV}$$

$$= 28{\cdot}298 \text{ MeV.}$$

$$(1 \text{ amu} = 931{\cdot}5 \text{ MeV})$$

$$\therefore \text{ binding energy per nucleon} = \frac{28{\cdot}298}{4}$$

$$= 7{\cdot}0745 \text{ MeV.}$$

Ex. 61. *A certain radioisotope* ${}_Z^AX$ $\left(t_{\frac{1}{2}} = 10 \text{ days}\right)$ *decays to give* ${}_{Z-2}^{A-4}Y$. *If 1 mole of* ${}_Z^AX$ *is kept in a sealed vessel, what volume of helium will accumulate in 20 days at* NTP?

Solution : ${}_Z^AX \rightarrow {}_{Z-2}^{A-4}Y + {}_2^4\text{He}$

Suppose that x moles of X give x moles of He after 20 days.

$\therefore$ moles of X after 20 days $= 1 - x$

Now, we have,

$$\lambda = \frac{2{\cdot}303}{t} \log \frac{N^0}{N} = \frac{0{\cdot}6932}{t_{\frac{1}{2}}}$$

or
$$\frac{2{\cdot}303}{20} \log \frac{1}{1-x} = \frac{0{\cdot}6932}{10}$$

$$x = 0{\cdot}75 \text{ mole.}$$

$\therefore$ volume of He at NTP $= 0{\cdot}75 \times 22{\cdot}4$ ... (Rule 3, Ch. 1)

$$= 16{\cdot}8 \text{ litres.}$$

Ex. 62. *You have* 0·1 *gram-atom of a radioactive isotope* ${}_Z^AX$ *(half-life* = 5 *days). How many number of atoms will decay during the eleventh day?*

Solution : Let the amount of X after 10 days be x moles

We have, $\lambda = \dfrac{2{\cdot}303}{t} \log \dfrac{N^0}{N} = \dfrac{0{\cdot}6932}{t_{\frac{1}{2}}}$

or
$$\frac{2{\cdot}303}{10} \log \frac{0{\cdot}1}{x} = \frac{0{\cdot}6932}{5}$$

or
$$x = 0{\cdot}025 \text{ mole.}$$

$\therefore$ number of atoms $= 0{\cdot}025 \times 6{\cdot}022 \times 10^{23}$ ... (Rule 4, Ch. 1)

$$= 1{\cdot}505 \times 10^{22}.$$

Thus the number of atoms on the eleventh day is $1{\cdot}505 \times 10^{22}$.

Now, no. of atoms decaying on the eleventh day

= no. of disintegration per day.

$$= -\frac{d(N)}{dt} = \lambda(N) = \frac{0{\cdot}6932}{t_{\frac{1}{2}}} \times N.$$

or $$-\frac{d(N)}{dt}=\frac{0\cdot6932}{5}\times1\cdot505\times10^{22}=2\cdot086\times10^{21}.$$

Ex. 63. *The limit of a particular detection system is* 0·002 dps *for* 1 g *sample. Find the maximum* $t_{\frac{1}{2}}$ *that this system could detect in a* 1-g *sample of a nuclide of mass number* 200.

Solution : We know, N = no. of atoms = moles × Av. const.

$$=\frac{1}{200}\times6\cdot022\times10^{23}.$$

$$-\frac{d(N)}{dt}=\lambda(N)$$

$$0\cdot002=\frac{0\cdot6932}{t_{\frac{1}{2}}}\times\frac{1}{200}\times6\cdot022\times10^{23}$$

$$\therefore \quad t_{\frac{1}{2}}=1\cdot0436\times10^{24}\ \text{seconds}$$

$$=3\cdot334\times10^{16}\ \text{years}.$$

Ex. 64. *An experiment requires minimum β-activity produced at the rate of* 346 *β-particles per minute. The* $t_{\frac{1}{2}}$ *of* $^{99}_{42}\text{Mo}$*, which is a β-emitter, is* 66·6 *hours. Find the minimum amount of* $^{99}_{42}\text{Mo}$ *required to carry out the experiment in* 6·909 *hours.* **(IIT 1989)**

Solution : $\beta\text{-activity}=-\frac{d(N)}{dt}=346\times60\ \beta\text{-particles/h}.$

$$\therefore \quad -\frac{d(N)}{dt}=\lambda(N)=\frac{0\cdot6932}{t_{\frac{1}{2}}}\times N$$

or $$346\times60=\frac{0\cdot6932}{66\cdot6}N$$

or $$N=2\times10^{6}.$$

Again,

$$\lambda=\frac{2\cdot303}{t}\log\frac{N^0}{N}$$

$$\therefore \quad \frac{0\cdot6932}{66\cdot6}=\frac{2\cdot303}{6\cdot909}\log\frac{N^0}{N}=\frac{1}{3}\log\frac{N^0}{N}.$$

Substituting the value of N in the above equation,

$$\log\frac{N^0}{N}=\frac{3\times0\cdot6932}{66\cdot6}=0\cdot0312$$

or $$\frac{N^0}{N} = 1{\cdot}074$$

or $$N^0 = 1{\cdot}074 \times N = 1{\cdot}074 \times 2 \times 10^6 = 2{\cdot}148 \times 10^6$$

$\therefore$ no. of moles of Mo $= \dfrac{2{\cdot}148 \times 10^6}{6{\cdot}022 \times 10^{23}} = 3{\cdot}567 \times 10^{-18}$

and wt. of Mo $= (3{\cdot}567 \times 10^{-18} \times 99)$ g (Mo = 99)

$= 3{\cdot}53 \times 10^{-16}$ g.

Ex. 65. *For the decay:*

$$^{228}\text{Ac} \xrightarrow{-\beta} {}^{228}\text{Th} \xrightarrow{-\alpha} {}^{224}\text{Ra}$$

where λ (Ac) $= 3{\cdot}14 \times 10^{-5}$ s^{-1} *and* λ (Th) $= 1{\cdot}148 \times 10^{-8}$ s^{-1}. *Determine the time for the radioactive daughter to reach its maximum activity.*

Solution : We have,

$$t = \frac{2{\cdot}303\,(\log \lambda_1 - \log \lambda_2)}{\lambda_1 - \lambda_2} \quad \text{... (Eqn. 33)}$$

$$= \frac{2{\cdot}303\,(\log 3{\cdot}14 \times 10^{-5} - \log 1{\cdot}148 \times 10^{-8})}{3{\cdot}14 \times 10^{-5} - 1{\cdot}148 \times 10^{-8}}$$

$$= 2{\cdot}52 \times 10^5 \text{ s.}$$

Ex. 66. ^{227}Ac *has a half-life of 21·8 years with respect to radioactive decay. The decay follows two parallel paths, one leading to* ^{227}Th *and the other leading to* ^{223}Fr. *The percentage yields of these two daughter nuclides are 1·2% and 98·8% respectively. What is the rate constant in* y^{-1}, *for each of the separate paths?*

Solution : We have,

$$\lambda_{\text{Ac}} = \frac{0{\cdot}6932}{t_{\frac{1}{2}}} = \frac{0{\cdot}6932}{21{\cdot}8} = 3{\cdot}18 \times 10^{-2} \text{ yr}^{-1}.$$

$$\lambda_{\text{Th}} = (\text{fractional yield of Th}) \times \lambda_{\text{Ac}}$$

$$= \frac{1{\cdot}2}{100} \times 3{\cdot}18 \times 10^{-2} = 3{\cdot}8 \times 10^{-4} \text{ yr}^{-1}.$$

Similarly,

$$\lambda_{\text{Fr}} = \frac{98{\cdot}8}{100} \times 3{\cdot}18 \times 10^{-2} = 3{\cdot}14 \times 10^{-2} \text{ yr}^{-1}.$$

[**Note:** λ_{Ac} is equal to the sum of λ_{Th} and λ_{Fr} .]

Ex. 67. $^{64}\text{Cu}(t_{1/2} = 12{\cdot}8 \text{ h})$ *decays by* β^-*-emission* (38%), β^+*-emission* (19%) *and electron capture* (43%). *Write decay products and calculate partial half-lives for each of the decay processes.* **(IIT 2002)**

Solution : $\lambda_{Cu} = \dfrac{0.6932}{12.8} = 0.054$

$$ {}^{64}_{29}Cu \xrightarrow{\lambda_1} {}^{64}_{30}Zn + {}^{0}_{-1}e \quad (38\%) $$

$$ {}^{64}_{29}Cu \xrightarrow{\lambda_2} {}^{64}_{28}Ni + {}^{0}_{+1}e \quad (19\%) $$

$$ {}^{64}_{29}Cu \xrightarrow[{}^{0}_{-1}e]{\lambda_3} {}^{64}_{28}Ni \quad (43\%) $$

$\therefore \quad \lambda_1 = \text{fractional yield of } Zn \times \lambda_{Cu} = \dfrac{38}{100} \times 0.054 = 0.0205.$

$\therefore \quad t_{1/2}$ for β^--emission $= \dfrac{0.6932}{0.0205} = 33.8$ h.

Similarly we can calculate, $t_{1/2}$ for β^+-emission $= 67.6$ h

and $t_{1/2}$ for electron capture $= 29.85$ h

Ex. 68. *A mixture of* ^{239}Pu *and* ^{240}Pu *has a specific activity of* 6.0×10^9 dis/s. *The half-lives of the isotopes are* 2.44×10^4 *and* 6.58×10^3 *years, respectively. Calculate the isotopic composition of this sample.*

Solution : Specific activity means activity per gram.

$$ \text{Sp. activity of } {}^{239}Pu = \lambda N = \left(\frac{0.6932}{2.44 \times 10^4}\right)\left(\frac{6.022 \times 10^{23}}{239}\right) $$
$$ = 7.15 \times 10^{16} \text{ /yr/g} $$
$$ = 2.27 \times 10^9 \text{ /s/g.} \quad (1 \text{ yr} = 3.15 \times 10^7 \text{ s}) $$

$$ \text{Sp. activity of } {}^{240}Pu = \left(\frac{0.6932}{6.58 \times 10^3}\right)\left(\frac{6.022 \times 10^{23}}{240}\right) $$
$$ = 2.64 \times 10^{17}\text{/yr/g} = 8.37 \times 10^9\text{/s/g.} $$

Thus,

$$ (2.27 \times 10^9)\,x + (8.37 \times 10^9)(1-x) = 6.0 \times 10^9 \quad (x \equiv \text{fraction of } {}^{239}Pu) $$
$$ x = 0.39 \text{ or } 39\%. $$

Ex. 69. *Consider*

$$ {}^{228}Th \xrightarrow{-\alpha} {}^{224}Ra \xrightarrow{-\alpha} {}^{220}Rn $$

where $t_{\frac{1}{2}}$ (^{228}Th) $= 1.913$ yr, $t_{\frac{1}{2}}$ (^{224}Ra) $= 3.64$ d. *Determine the N (Th)/N* (Ra) *ratio once transient equilibrium in which* $t_{\frac{1}{2}}$ *(parent)* > $t_{\frac{1}{2}}$ *(daughter) or* λ *(parent)* < λ *(daughter) has been established.*

Solution : We have,

$$\frac{N_1}{N_2} = \frac{\lambda_2 - \lambda_1}{\lambda_1}. \quad \text{... (Eqn. 31)}$$

Now,

$$\lambda_1 = \frac{0{\cdot}6932}{(t_{\frac{1}{2}})_1} = \frac{0{\cdot}6932}{1{\cdot}913 \times 365} = 9{\cdot}927 \times 10^{-4}\ \text{d}^{-1}$$

$$\lambda_2 = \frac{0{\cdot}6932}{(t_{\frac{1}{2}})_2} = \frac{0{\cdot}6932}{3{\cdot}64} = 0{\cdot}1900\ \text{d}^{-1}.$$

Thus,

$$\frac{N\,(\text{Th})}{N\,(\text{Ra})} = \frac{0{\cdot}1900 - 9{\cdot}927 \times 10^{-4}}{9{\cdot}927 \times 10^{-4}} = 190.$$

Ex. 70. *Consider*

$$^{224}\text{Ra} \xrightarrow{-\alpha} {}^{220}\text{Rn} \xrightarrow{-\alpha} {}^{216}\text{Po}$$

where $t_{\frac{1}{2}}$ *(Ra)* = 3·64 yr, $t_{\frac{1}{2}}$ *(Rn)* = 55 s. *Determine the N (Ra)/N (Rn) ratio once secular equilibrium in which* $t_{\frac{1}{2}}$ *(parent) >>* $t_{\frac{1}{2}}$ *(daughter) or* λ *(parent) <<* λ *(daughter) has been established.*

Solution : We have,

$$\frac{N_1}{N_2} = \frac{\lambda_2}{\lambda_1}. \quad \text{... (Eqn. 32)}$$

Now,

$$\lambda_1 = \frac{0{\cdot}6932}{(t_{\frac{1}{2}})_1} = \frac{0{\cdot}6932}{3{\cdot}64 \times 60 \times 60} = 2{\cdot}20 \times 10^{-6}\ \text{s}^{-1}$$

$$\lambda_2 = \frac{0{\cdot}6932}{(t_{\frac{1}{2}})_2} = \frac{0{\cdot}6932}{55} = 1{\cdot}26 \times 10^{-2}\ \text{s}^{-1}.$$

Thus,

$$\frac{N\,(\text{Ra})}{N\,(\text{Rn})} = \frac{1{\cdot}26 \times 10^{-2}}{2{\cdot}20 \times 10^{-6}} = 5727.$$

PROBLEMS

(Answers bracketed with questions)

1. Using Bohr theory, calculate the radii of the first and the tenth orbits in the hydrogen atom. ($r_1 = 0{\cdot}53 \times 10^{-8}$ cm, $r_{10} = 0{\cdot}53 \times 10^{-6}$ cm)

[**Hint:** See examples 1 and 2.]

2. Calculate the velocities of the electron in the first and the tenth orbits of the hydrogen atom. ($v_1 = 21{\cdot}9 \times 10^7$ cm/s)

3. Calculate the energy in calories required to produce, from neutral He atoms, 1 mole of (a) He^+ ions (b) 'He^{++} ions using Bohr's equations. [(b) 1,821000 cal]

4. Calculate the energy in eV required to ionise 1 mole of hydrogen.

[**Hint:** See Ex. 8.] ($8{\cdot}189 \times 10^{24}$ eV)

5. Calculate the frequency of the spectral line when an electron from the fifth orbit jumps to the second orbit in a hydrogen atom. (R = 109737 cm^{-1}) ($6{\cdot}91 \times 10^{14}$)

6. Find (i) the total number of neutrons (ii) the total mass of neutrons in 7 mg of ^{14}C. (Assume the mass of a neutron = mass of a hydrogen atom)

[(i) $24{\cdot}08 \times 10^{20}$ (ii) 4 mg]

7. Calculate the wave number and frequency of radiation having wavelength 5800 Å.

(172400 cm^{-1}, $5{\cdot}172 \times 10^{14}$ cycles s^{-1})

8. What total amount of energy in calories would be required to shift all the electrons from the first Bohr orbit to the sixth Bohr orbit in 1 mole of hydrogen?

Through what distance would each electron have to move?

What frequency of radiation would be emitted if the electrons returned to their initial state? (304800 calories)

9. The ionisation energy of hydrogen atom is 13·6 eV. What will be the ionisation energy of He^+ and Li^{2+} ions? (54·4 eV, 122·4 eV)

10. Calculate the frequency, energy and wavelength of the radiation corresponding to the spectral line of lowest frequency in Lyman series in the spectra of hydrogen atom. Also, calculate the energy for the corresponding line in the spectra of Li^{2+}.

($R_H = 1{\cdot}09678 \times 10^7\ m^{-1}$, $c = 3 \times 10^8$ m/s, $h = 6{\cdot}625 \times 10^{-34}$ J · s.)

($2{\cdot}176 \times 10^{-18}$ J, $1{\cdot}958 \times 10^{-17}$ J)

11. Considering the shape of a Ca nucleus like a sphere, calculate the density of the nucleus of Ca of mass number 40. ($1{\cdot}8 \times 10^{14}$ g cm^{-3})

12. For He^+ and Li^{2+}, the energies are related to the quantum number, n, through an expression

$$E_n = -\frac{Z^2 B}{n^2},$$

where Z is the atomic number of species and $B = 2{\cdot}179 \times 10^{-18}$ J.

(a) What is the energy of the lowest level (n = 1) of a He^+ ion?

(b) What is the energy of the level $n = 3$ of a Li^{2+} ion?

{(a) $8{\cdot}716 \times 10^{-18}$ J; (b) $2{\cdot}179 \times 10^{-18}$ J}

13. The line at 434 nm in the Balmer series of the hydrogen spectrum corresponds to a transition of an electron from the nth to second Bohr orbit. What is the value of n? (n = 5)

$$\left[\textbf{Hint: } \nu = 3\cdot2881 \times 10^{15}\ s^{-1}\left(\frac{1}{2^2} - \frac{1}{n^2}\right)\right]$$

14. What change in molar energy in joule would be associated with an atomic transition giving rise to radiation at 1 Hz? ($h = 6\cdot626 \times 10^{-34}$ J · s) ($3\cdot99 \times 10^{-10}$ J · mol^{-1})

[**Hint:** 1 Hz = s^{-1}, $\Delta E = Nh\nu$ and $\nu = 1$]

15. Calculate the uncertainty in the position of a particle when the uncertainty in the momentum is zero. (∞)

16. Calculate the uncertainty in the velocity of a wagon of mass 2000 kg, whose position is known to an accuracy of ± 10 metre. ($5\cdot25 \times 10^{-31}$ m s^{-1})

17. Calculate the uncertainty in velocity of an electron when the uncertainty in position is 0·1 nm. ($5\cdot786 \times 10^{5}$ m s^{-1})

18. For the gaseous reaction $K + F \rightarrow K^+ + F^-$, ΔH was calculated to be 19 kcal under conditions where the cations and anions were prevented by electrostatic separation from combining with each other. The ionisation potential of K is 4·3 eV. Find electron affinity of F. (3·47 eV)

[**Hint:** See Ex. 27]

19. The prominent yellow line in the spectrum of sodium vapour lamp has a wavelength of 590 nm. What minimum accelerating potential is needed to excite this line in an electron tube containing sodium vapour.
($h = 6\cdot63 \times 10^{-34}$ J · s, $c = 3 \times 10^8$ m/s) (2·11 V)

[**Hint:** $1\ eV = 1\cdot602 \times 10^{-19}$ J, $E = \frac{hc}{\lambda}$

and 1 eV = energy of 1 electron being accelerated by 1 volt.]

20. If the energy difference between the ground state of an atom and its excited state is $4\cdot4 \times 10^{-19}$ J, what is the wavelength of the photon required to produce this transition? ($4\cdot5 \times 10^{-7}$ m)

[**Hint:** $\lambda = \frac{hc}{\Delta E}$]

21. A body weighing $3\cdot0 \times 10^3$ kg is moving and its speed can be measured with an accuracy of ± 0·0025 mile per hour and its position with an accuracy of ± 0·01 mile. Is the uncertainty principle valid? (No)

22. An electron is accelerated by applying potential difference of 1000 eV. What is the de Broglie's wavelength associated with it? 1 eV = $1\cdot6 \times 10^{-12}$ erg.
($h = 6\cdot627 \times 10^{-27}$ erg s, $m_e = 9\cdot1 \times 10^{-28}$ g) ($3\cdot87 \times 10^{-9}$ cm)

23. A large object weighing $1\cdot0 \times 10^3$ kg is moving with a velocity of 50 km per hour. Does it have a wave motion? (λ is very low hence no)

24. Is it possible to locate the electron within 0·005 nm? (Use the standard values of h and m.) (not possible)

[**Hint:** Δv would be nearly as large as the velocity of light.]

25. Assume that 10^{-17} J of light energy is needed by the interior of the human eye to see an object. How many photons of green light ($\lambda = 495$ nm) are needed to generate this minimum energy? (25)

26. How are a 1s orbital and a 2s orbital in an atom similar? How do they differ? How are a $2p_x$ orbital and a $2p_y$ orbital in an atom similar? How do they differ? (read text)

27. What is the maximum number of electrons in an atom in which the last electron, filled, has the following quantum numbers?

(a) $n = 3$ (b) $n = 3$, and $l = 1$ (c) $n = 3, l = 1$ and $m = -1$

(d) $n = 3, l = 1, m = -1$ and $s = +\dfrac{1}{2}$ [(a) 30, (b) 18 (c) 16 (d) 13]

28. Which of the following equations describe(s) particle-like behaviour? Which describe(s) wavelike behaviour? Do any involve both types of behaviour?

(a) $C = \gamma\lambda$ (b) $E = mc^2$ (c) $\gamma = \dfrac{n^2 a_0}{Z}$ (d) $E = h\gamma$ (e) $\lambda = \dfrac{h}{mv}$

[Wavelike: (a) & (d), particle-like: (b) & (c), Both: (e)]

29. The quantum numbers listed below are for four different electrons in the same atom. Arrange them in order of increasing energy. Indicate whether any two have the same energy.

(a) $n = 4, l = 0, m = 0$ and $s = +\dfrac{1}{2}$

(b) $n = 3, l = 2, m = +1$ and $s = +\dfrac{1}{2}$

(c) $n = 3, l = 2, m = -2$ and $s = -\dfrac{1}{2}$

(d) $n = 3, l = 1, m = +1$ and $s = -\dfrac{1}{2}$ [(d) < (a) < (b) = (c)]

30. When compounds of barium are heated in a flame, green light of wavelength 554 nm is emitted. How much energy is lost when one mole of barium atoms each emit one photon of this wavelength? (216 kJ)

31. What is the energy in J evolved when 1 mole of He-4 nuclei is produced from protons and neutrons? How many litres of C_2H_6(g) at 25°C and 725 mmHg are needed to evolve the same quantity of energy when C_2H_6 is burnt in O_2? (neutron = 1·008867 amu, proton = 1·00728 amu, He = 4·0015 amu, $\Delta H_{comb.}(C_2H_6) = -1427{\cdot}81$ kJ/mol)

($2{\cdot}732 \times 10^{12}$ J, $4{\cdot}9 \times 10^{7}$ L)

32. A 1·0-mg sample of Tc–99 has an activity of $1{\cdot}7 \times 10^{-5}$ curie, decaying by β-emission. Calculate the decay constant. ($1{\cdot}0 \times 10^{-13}$ s^{-1})

33. Tritium, $^3_1$H is a radioactive nucleus of hydrogen. It is used in luminous watch dials. Tritium decays by β-emission with $t_{1/2} = 12.3$ years. What is the activity in curie of a sample containing 2·5 μg of tritium? The atomic mass of tritium is 3·02 amu. (0·024 Ci)

34. Predict the type of radioactive decay process that is likely for each of the following nuclides.

(a) $^{228}_{92}U$ (b) $^{8}_{5}B$ (c) $^{68}_{29}Cu$ [(a) α (b) $β^+$ or K–capture (c) β emission]

35. If 28·0% of a sample of Ag-112 decays in 1·52 h, what is the half-life of this isotope? (3·21 h)

36. $^{232}_{90}Th$ disintegrates to $^{x}_{y}Pb$ by emitting six α- and four β-particles. Find x and y. (208, 82)

37. The uranium (mass no. 238 and at. no. 92) emits an α-particle, the product has the mass no. and at. no.:

(a) 236 and 92 (b) 234 and 90 (c) 238 and 90 (d) 236 and 90 [b]

38. To which series will the following elements belong?

$^{257}_{103}Lr$, $^{254}_{99}Es$, $^{243}_{95}Am$ ($4n+1$, $4n+2$, $4n+3$)

39. A sample of carbon from an ancient frame gives 7 counts of ^{14}C per minute per gram of carbon. If freshly cut wood gives 15·3 counts of ^{14}C per minute per gram, what is the age of the frame? (Half-life period of ^{14}C = 5770 years)

[**Hint:** $N^0 = 15{\cdot}3$, $N = 7$] (6520 years)

40. Calculate the number of atoms disintegrating per minute in a mass of 0·001g of radium which is an α-emitter with a half-life period of 1620 years. ($2{\cdot}16 \times 10^9$)

41. In a sample of pitchblende the atomic ratio is $^{206}Pb : ^{238}U = 0{\cdot}23 : 1$. Calculate the age of the mineral, if half-life of uranium is taken as $4{\cdot}5 \times 10^9$ years. All lead originated from uranium. ($1{\cdot}34 \times 10^9$ years)

42. A sample of radon emitted initially 7×10^4 α-particles per second. After some time, the emission rate became $2{\cdot}1 \times 10^4$. If $t_{\frac{1}{2}}$ for radon is 3·8 days, find the age of the sample. (6·6 days)

43. ^{222}Rn has a half-life period of 3·83 days. What fraction of the sample will remain undecomposed at the end of 10 days? (0·164)

44. The number of α-particles emitted per second by 1 g of radium is $3{\cdot}608 \times 10^{10}$. Calculate decay constant and $t_{\frac{1}{2}}$.

$$\begin{Bmatrix} 1{\cdot}35 \times 10^{-11}\ s^{-1} \\ 5{\cdot}13 \times 10^{10}\ s \end{Bmatrix}$$

45. The rate of radioactive decomposition corresponding to $3{\cdot}7 \times 10^{10}$ disintegration per second is called a curie. What weight of ^{226}Ra, whose $t_{\frac{1}{2}} = 1620$ yr, will be required to yield 1 millicurie of radiation? (1×10^{-3} g)

46. Calculate the weight of $^{14}C\left(t_{\frac{1}{2}} = 5720\text{ yr}\right)$ atoms which will give $3{\cdot}70 \times 10^7$ disintegrations per second (dps). (0·2234 mg)

47. Calculate the number of disintegrations which 1 g of $^{226}Ra\left(t_{\frac{1}{2}} = 1600 \text{ yr}\right)$ would undergo per second. What quantity of $Na^{24}\left(t_{\frac{1}{2}} = 15 \text{ hours}\right)$ would undergo the same no. of disintegrations per second. ($3{\cdot}658 \times 10^{10}$ dps, $1{\cdot}37 \times 10^{-7}$ g)

48. A piece of wood was found to have $^{14}C/^{12}C$ ratio 0·7 times that in a living plant. Calculate the period (in years) when the plant died.
$\left(t_{\frac{1}{2}} \text{ for } C^{14} = 5760 \text{ yr}\right)$ (2964 yr)

49. 10·0 gram-atom of an α-active radioisotope is disintegrating in a sealed container. In one hour, the He gas collected at STP is 11·2 litres. Calculate the half-life of the isotope supposing each nucleus yielding one He atom. (13·49 h)

50. The disintegration rate for a sample containing $^{60}_{27}Co$ as the only radioactive nuclide, is found to be 240 atoms/minute. $t_{\frac{1}{2}}$ of Co is 5·2 years. Find the number of atoms of Co in the sample. How long must this radioactive sample be maintained before the rate falls to 100 disintegrations/minute. ($9{\cdot}6 \times 10^8$, 6·6 years)

51. Sample containing $^{234}_{88}Ra$, which decays by α-particle emission, is observed to disintegrate at the following rate expressed as counts per minute (cpm). Calculate half-life of this nuclide. $t = 0$, 1000 cpm; $t = 1$ h, 992 cpm; $t = 10$ h, 924 cpm; $t = 100$ h, 452 cpm; $t = 250$ h, 138 cpm. (3·63 days)

52. The thorium radioactive decay series produces one atom of ^{208}Pb as the final disintegration product of an atom of ^{232}Th. $t_{\frac{1}{2}}$ of ^{232}Th is $1{\cdot}39 \times 10^{10}$ years. A certain rock is found to have a mass ratio of ^{208}Pb and ^{232}Th as ·14 : 1. Determine the age of the rock. ($2{\cdot}97 \times 10^9$ years)

53. The ratio of the number of atoms of two radioactive elements *A* and *B*, in equilibrium with each other, is $3{\cdot}1 \times 10^9 : 1$. If $t_{\frac{1}{2}}$ of element *B* is 6·45 yrs, calculate that of element *A*. (2×10^{10})

54. Which nucleus has higher binding energy per nucleon : $^{58}_{28}Ni$ (57·941 amu) or $^{55}_{25}Mn$ (54·939 amu)? Mass of neutron is 1·00867 amu and that of proton is 1·00728 amu. (Ni)

55. For $^{238}_{92}U$ the binding energy per nucleon is 7·576 MeV. What is the atomic weight of this isotope? Use the mass of neutron and proton from Ex. 60.
[Hint: See Ex. 60] (239·93 amu)

56. The atomic masses of He and Ne are 4 and 20 respectively. The value of de Broglie wavelength of He gas at −73°C is M times that of the de Broglie wavelength of Ne at 727°C. M is ... . **(IIT 2013 Adv.)** (5)

[**Hint:** Equation 15, $\lambda = \frac{h}{\sqrt{2mE}}$

As $E \propto T$, $\lambda \propto \frac{h}{\sqrt{2mT}}$]

57. In an atom the total number of electrons having quantum numbers $n = 4$, $|m_e| = 1$ and $m_s = -\frac{1}{2}$ is ... . **(IIT 2014 Adv.)** (6)

58. The periodic table consists of 18 groups. An isotope of Cu, on bombardment with protons undergoes a nuclear reaction yielding element, X as shown below. To which group element X belongs to in the periodic table?

$^{63}_{29}Cu + ^{1}_{1}H \rightarrow 6^{1}_{0}n + ^{4}_{2}\alpha + 2\,^{1}_{1}H + X$ **(IIT 2012)** (8th)

[**Hint:** X is $^{52}_{26}X$, atomic no. = 26. Transition metal Fe($_4S^2\ _3d^6$)]

59. A closed vessel with rigid walls contains 1 mole of $^{238}_{92}U$ and 1 mole of air at 298 K. Considering complete decay of $^{238}_{92}U$ to $^{206}_{82}Pb$, the ratio of the final pressure to the initial pressure of the system at 298 K is ... . **(IIT 2015 Adv.)** (9)

[**Hint:** $^{238}_{92}U \rightarrow ^{206}_{82}P + 8\,^{4}_{2}He + 6\,^{0}_{-1}e$

$$\frac{P_{final}}{P_{initial}} = \frac{8(He) + 1\ (air)}{1\ (air)}$$]

Objective Problems

1. The frequency of a green light is 6×10^{14} Hz. Its wavelength is

(a) 500 nm (b) 5 nm

(c) 5000 nm (d) none of these

[**Hint:** 1 nm = 10^{-9} m]

2. The ratio of energy of a photon of 2000 Å wavelength radiation to that of 4000 Å radiation is

(a) 1/4 (b) 4 (c) 1/2 (d) 2

3. The values of charge on the oil droplets experimentally observed were $-1{\cdot}6 \times 10^{-19}$, $-2{\cdot}4 \times 10^{-19}$ and -4×10^{-19} coulomb. The value of the electronic charge, indicated by these results is

(a) $-1{\cdot}6 \times 10^{-19}$ (b) $-2{\cdot}4 \times 10^{-19}$ (c) -4×10^{-19} (d) $-0{\cdot}8 \times 10^{-19}$

[**Hint:** Find the highest common factor.]

4. Which of the following statements is/are correct?

(a) A photon is a positively charged nuclear particle.

(b) A photon is a particle of light energy.

(c) A photon is a quantum of light.

(d) A photon is a bundle of energy of definite magnitudes but not necessarily light energy.

5. For which of the following species, Bohr theory does not apply?
(a) H (b) H^+ (c) He^+ (d) Li^{2+}

6. The radius of the first Bohr orbit of hydrogen atom is r. The radius of the 3rd orbit would be
(a) $3r$ (b) $9r$ (c) $27r$ (d) none of these

7. The electronic velocity in the fourth Bohr orbit of hydrogen is v. The velocity of the electron in the first orbit would be
(a) $4v$ (b) $16v$ (c) $v/4$ (d) $v/16$

8. The ionisation potential of hydrogen atom is 13·6 eV. The energy required to remove an electron in the $n = 2$ state of hydrogen atom is
(a) 27·2 eV (b) 13·6 eV (c) 6·8 eV (d) 3·4 eV.

9. The energy of the second Bohr orbit in the hydrogen atom is – 3·41 eV. The energy of the second Bohr orbit of He^+ ion would be
(a) – 0·85 eV (b) – 13·64 eV (c) – 1·70 eV (d) – 6·82

10. The ratio of the radii of the first three Bohr orbits is
(a) 1 : 5 : 33 (b) 1 : 2 : 3 (c) 1 : 4 : 9 (d) 1 : 8 : 27

11. The ratio of the difference in energy between the first and second Bohr orbits to that between the second and third Bohr orbit is
(a) 1/2 (b) 1/3 (c) 4/9 (d) 27/5

12. For $l = 3$, which value of m is not possible?
(a) 4 (b) 0 (c) –3 (d) –1

13. The designation of an orbital with $n = 4$ and $l = 1$, is
(a) $4d$ (b) $4s$ (c) $4f$ (d) $4p$

14. Which of the following orbitals is meaningless?
(a) $6f$ (b) $2d$ (c) $7s$ (d) $3d$

15. Which of the following sets of quantum numbers is possible for an electron in a $4f$ orbital
(a) $n = 4, l = 2, m = 2, s = +\frac{1}{2}$ (b) $n = 4, l = 4, m = +4, s = +\frac{1}{2}$
(c) $n = 4, l = 3, m = +1, s = -\frac{1}{2}$ (d) $n = 4, l = 3, m = 4, s = +\frac{1}{2}$

16. The maximum number of electrons in an atom with quantum numbers $n = 3, l = 2$ is
(a) 2 (b) 6 (c) 10 (d) 30

17. The number of orbitals in $n = 3$ are
(a) 1 (b) 4 (c) 9 (d) 16

18. If the nitrogen atom had electronic configuration $1s^7$, it would have energy lower than that of the normal ground state configuration $1s^2, 2s^22p^3$, because the electrons would be closer to the nucleus. Yet, $1s^7$ is not observed because it violates
(a) Heisenberg uncertainty principle
(b) Hund rule
(c) Pauli exclusion principle
(d) Bohr postulate of stationary orbits **(IIT 2002)**

19. Rutherford's experiment, which establishes the nuclear model of the atom, used a beam of
(a) β-particles, which impinged on a metal foil and got absorbed
(b) γ-rays, which impinged on a metal foil and ejected electrons
(c) helium atoms, which impinged on a metal foil and got scattered
(d) helium nuclei, which impinged on a metal foil and got scattered **(IIT 2002)**

20. Total number of nodal planes are same in
(a) 3*s*, 4*d* (b) 4*s*, 3*p* (c) 5*s*, 4*d* (d) 4*s*, 4*p*

21. The number of waves made by an electron moving in an orbit having maximum quantum number (*m*) +3 is
(a) 3 (b) 4 (c) 5 (d) 6

[**Hint:** $l = 3$ and $n = 4$]

22. The electronic transition from $n = 2$ to $n = 1$ will produce shortest wavelength in
(a) Li^{2+} (b) He^{+} (c) H (d) H^{+}

23. The first emission line of Balmer series in He^{+}-spectrum has the wave no. in cm^{-1} equal to (*R*-Rydberg constant)
(a) $\frac{3R}{4}$ (b) $\frac{20R}{36}$ (c) $\frac{5R}{36}$ (d) $\frac{R}{6}$

24. If the radius of first Bohr orbit is *r*, the wavelength of an electron in the third orbit of a hydrogen atom is equal to
(a) $6\pi r$ (b) $2\pi r$ (c) $9\pi r$ (d) $3r$

25. If the shortest λ of hydrogen atom in Lyman series is *x*, the longest λ in Balmer series of He^{+} is
(a) $\frac{x}{4}$ (b) $\frac{5x}{9}$ (c) $\frac{6x}{5}$ (d) $\frac{9x}{5}$

26. Consider a large number of hydrogen atoms with electrons randomly distributed in the $n = 1, 2, 3$, and 4 orbits. How many different wavelengths of light are emitted by these atoms as the electrons fall into lower energy orbits.
(a) 1 (b) 3 (c) 6 (d) 9

27. How many times larger is a hydrogen atom than the radius of an H atom in its ground state if the H atom with an electron characterised by a quantum number of 106?
(a) 106 (b) 212 (c) 11236 (d) none of these

28. Which is larger, an He^{+} ion with an electron in an orbit with $n = 3$ or an Li^{2+} ion with an electron in an orbit with $n = 5$?
(a) He^{+} (b) Li^{2+} (c) both equal

29. Which of the following sets of quantum numbers is permissible for an electron in an atom?
(a) $n = 1, l = 1, m = 0, s = +\frac{1}{2}$ (b) $n = 3, l = 1, m = -2, s = -\frac{1}{2}$
(c) $n = 2, l = 1, m = 0, s = +\frac{1}{2}$ (d) $n = 2, l = 0, m = 0, s = 1$

30. Number of stable isotopes is least when the number of neutrons and that of protons in the isotopes are respectively
(a) odd and odd (b) even and odd
(c) odd and even (d) even and even

31. A radioactive nuclide generally disintegrates by α-emission when its N/P ratio is
(a) less than 1 (b) equal to one (c) equal to 1·2 (d) greater than 1·5

32. $^{x}_{8}X$ atom is isotone to $^{17}_{9}Y$ atom. The value of x is
(a) 8 (b) 16 (c) 9 (d) 17

33. If 5 g of a radioactive substance has a $t_{\frac{1}{2}} = 14$ h, 20 g of the same substance will have a $t_{\frac{1}{2}}$ equal to
(a) 56 h (b) 3·5 h (c) 14 h (d) 28 h

34. The designations of a proton and neutron are $^{1}_{1}H$ and $^{1}_{0}n$ respectively. Are the two particles of equal mass?
(a) Yes, both proton and neutron have the same mass, i.e., 1 amu.
(b) No, mass numbers reflect actual masses only to the nearest amu. The precise masses of proton and neutron are 1·0072765 and 1·008665 amu.

35. The nuclides $^{40}_{18}Ar$ and $^{41}_{19}K$ are
(a) isotopes (b) isobars (c) isotones (d) none of these

36. Which of the following nuclides are isobars?
(a) $^{40}_{21}Sc$ and $^{42}_{21}Sc$ (b) $^{40}_{18}Ar$ and $^{40}_{21}Sc$ (c) $^{40}_{18}Ar$ and $^{41}_{19}K$ (d) none of these

37. To what stable isotope would $^{257}_{103}Lr$ decay?
(a) $^{208}_{82}Pb$ (b) $^{209}_{83}Bi$ (c) $^{206}_{82}Pb$ (d) $^{207}_{82}Pb$

[Hint: $^{257}_{103}Lr$ belongs to $(4n+1)$ series]

38. The half-life of a radioactive isotope is 1·5 hours. The mass of it that remains undecayed after 6 hours is (if the initial mass of the isotope was 32 g)
(a) 32 g (b) 16 g (c) 4 g (d) 2 g

39. The radioactivity due to C-14 isotope $\left(t_{\frac{1}{2}} = 6000 \text{ yr}\right)$ of a sample of wood from an ancient tomb was found to be nearly half that of fresh wood; the tomb is, therefore, about
(a) 3000 years old (b) 6000 years old
(c) 9000 years old (d) 12000 years old

40. The half-life period of a radioactive nuclide is 3 hours. In 9 hours its activity will be reduced by a factor of
(a) 1/9 (b) 1/8 (c) 1/27 (d) 1/6

[Hint: Cal. $N/N°$]

41. If 3/4 quantity of a radioactive element disintegrates in two hours, its half-life would be
(a) 1 hour (b) 45 m (c) 30 m (d) 15 m

42. A sample of rock from moon contains equal number of atoms of uranium and lead $\left(t_{\frac{1}{2}} \text{ for U} = 4{\cdot}5 \times 10^9 \text{ years}\right)$. The age of the rock would be

(a) $4{\cdot}5 \times 10^9$ years (b) 9×10^9 years
(c) $13{\cdot}5 \times 10^9$ years (d) $2{\cdot}25 \times 10^9$ years

43. A radioactive isotope having a half-life of 3 days was received after 12 days. It was found that there were 3 g of the isotope in the container. The initial weight of the isotope when packed was

(a) 12 g (b) 24 g (c) 36 g (d) 48 g

44. If N^0 is the initial number of nuclei, number of nuclei remaining undecayed at the end of nth half-life is

(a) $2^{-n} N^0$ (b) $2^n N^0$ (c) $n^{-2} N^0$ (d) $n^2 N^0$

45. A radioactive substance is decaying with $t_{\frac{1}{2}} = 30$ days. On being separated into two fractions, one of the fractions, immediately after separation, decays with $t_{1/2} = 2$ days. The other fraction, immediately after separation, would show

(a) constant activity (b) increasing activity
(c) decay with $t_{1/2} = 30$ days (d) decay with $t_{1/2} = 28$ days

46. A radioactive substance has a constant activity of 2000 disintegrations/minute. The material is separated into two fractions, one of which has an initial activity of 1000 disintegrations per minute while the other fraction decays with $t_{\frac{1}{2}} = 24$ hours. The total activity in both samples after 48 hours of separation is

(a) 1500 (b) 1000 (c) 1250 (d) 2000

47. The orbital angular momentum of an electron in 2 s orbital is

(a) $+\frac{1}{2} \cdot \frac{h}{2\pi}$ (b) zero (c) $\frac{h}{2\pi}$ (d) $\sqrt{2}\,\frac{h}{2\pi}$

[Hint: $L = \sqrt{l\,(l+1)}\,\frac{h}{2\pi}$]

48. For a *d*-electron, the orbital angular momentum is

(a) $\sqrt{6}\,\hbar$ (b) $\sqrt{2}\,\hbar$ (c) $\hbar$ (d) $2\,\hbar$

[Hint: $L = \sqrt{l\,(l+1)}\,\hbar$; $l = 0, 1, 2, \ldots$; $\hbar = \frac{h}{2\pi}$; $\hbar$ is called Dirac h] **(IIT 1997)**

49. $^{27}_{13}$Al is a stable isotope. $^{29}_{13}$Al is expected to disintegrate by

(a) α-emission (b) β-emission
(c) positron emission (d) proton emission **(IIT 1996)**

50. Which of the following relates to light both as wave motion as well as particle?

(a) Diffraction and interference (b) Photoelectric effect
(c) $E = mc^2$ (d) $E = h\nu$

51. The kinetic energy of an electron in the second Bohr orbit of a hydrogen atom is (a_0 is Bohr radius)

(a) $\frac{h^2}{4\pi^2 m a_0^2}$ (b) $\frac{h^2}{16\pi^2 m a_0^2}$ (c) $\frac{h^2}{32\pi^2 m a_0^2}$ (d) $\frac{h^2}{64\pi^2 m a_0^2}$

[Hint: $mvr = \frac{nh}{2\pi}$, $mv = \frac{nh}{2\pi r}$ and

$$KE = \frac{1}{2}mv^2 = \frac{m^2v^2}{2m} \quad \text{or} \quad E = \frac{n^2h^2}{4\pi^2r^2} \times \frac{1}{2m}$$

use $r = n^2a_0$ $(a_0 = r_1)$] **(IIT 2012)**

52. Energy of an electron is given by

$$E = -2.178 \times 10^{-18}\,\text{J}\left(\frac{z^2}{n^2}\right).$$

The wavelength of light required to excite an electron in a hydrogen atom from level $n = 1$ to $n = 2$ will be $(h = 6.62 \times 10^{-34}$ Js, $C = 3.0 \times 10^8$ m s$^{-1})$

(a) 1.214×10^{-7} m (b) 2.816×10^{-7} m

(c) 6.560×10^{-7} m (d) 8.500×10^{-7} m **(IIT 2013 Main)**

53. The first ionisation potential of Na is 5.1 eV. The value of electron gain enthalpy of Na^+ will be

(a) –2.55 eV (b) –5.1 eV

(c) –10.2 eV (d) +2.55 eV **(IIT 2013 Main)**

54. The correct set of four quantum numbers for the valence electrons of rubidium atom ($Z = 37$) is

(a) 5, 0, 0, $+\frac{1}{2}$ (b) 5, 1, 0, $+\frac{1}{2}$

(c) 5, 1, 1, $+\frac{1}{2}$ (d) 5, 0, 1, $+\frac{1}{2}$ **(IIT 2015 Main)**

[Hint: Valence electron is 5 S′]

55. Which of the following is the energy of a possible excited state of hydrogen?

(a) +13.6 eV (b) –6.8 eV

(c) –3.4 eV (d) +6.8 eV **(IIT 2015 Main)**

56. A stream of electrons from a heated filament was passed between two charged plates kept at a potential difference V esu. If e and m are charge and mass of an electron respectively then the value of h/λ (where λ is wavelength associated with electron wave) is given by

(a) 2 m eV (b) $\sqrt{\text{m eV}}$

(c) $\sqrt{2\text{ m eV}}$ (d) m eV **(IIT 2016 Main)**

[Hint : See equation 15.]

57. Which of the following atoms has the highest first ionisation energy?

(a) Na (b) K

(c) Sc (d) Rb **(IIT 2016 Main)**

58. P is the probability of finding the 1 s electron of hydrogen atom in a spherical shell of infinitesimal thickness, dr at a distance, r from the nucleus. The volume of this shell is $4\pi r^2 dr$. The qualitative sketch of the dependence of P on r is

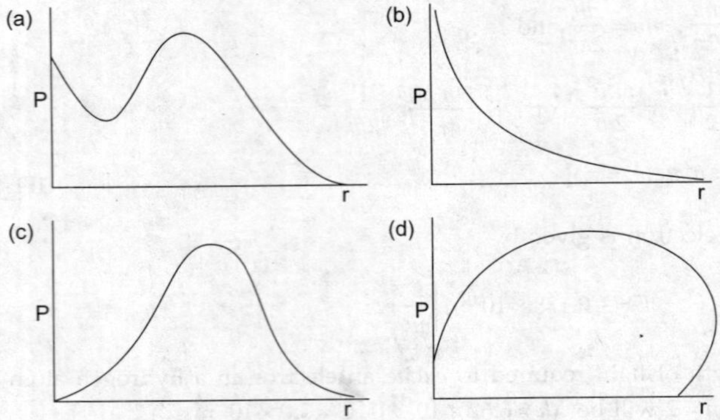

(IIT 2016 Adv.)

[**Hint:** See Mod. App. to Phy. Chemistry, Vol. II, Ch. 1]

59. The radius of second Bohr orbit for hydrogen atom is ($h = 6.6262 \times 10^{-34}$ Js, mass of electron $= 9.1091 \times 10^{-31}$ kg, charge $e = 1.60210 \times 10^{-19}$ C, permitivity of vacuum, $\in_0 = 8.854185 \times 10^{-12}$ kg^{-1} m^{-3} A^2)

(a) 1.65 Å (b) 4.76 Å
(c) 0.529 Å (d) 2.12 Å **(IIT 2017 Main)**

[**Hint :** $r = \dfrac{\in_0 n^2 h^2}{\pi m e^2 z}$, 1 Å $= 10^{-10}$ m]

60. The group having isoelectronic species is

(a) $O^{2-}, F^-, Na^+, Mg^{2+}$ (b) O^-, F^-, Na, Mg^{2+}
(c) O^{2-}, F^-, Na, Mg^{2+} (d) O^-, F^-, Na^+, Mg^{2+} **(IIT 2017 Main)**

61. The wave function ψ_{n}, l, m_e is a mathematical function whose value depends upon spherical polar coordinates (r, θ, ϕ) of the electron and characterised by the quantum number n, l and m_e. Here r is distance from nucleus, θ is colatitude and ϕ is azimuth. In mathematical functions given in the table, Z is atomic number and a_0 is Bohr radius.

Column 1	Column 2	Column 3
(1) 1 s orbital	(i) $\psi_{nlm_e} \propto \left(\dfrac{z}{a_0}\right)^{3/2} e^{-(zr/a_0)}$	(P) 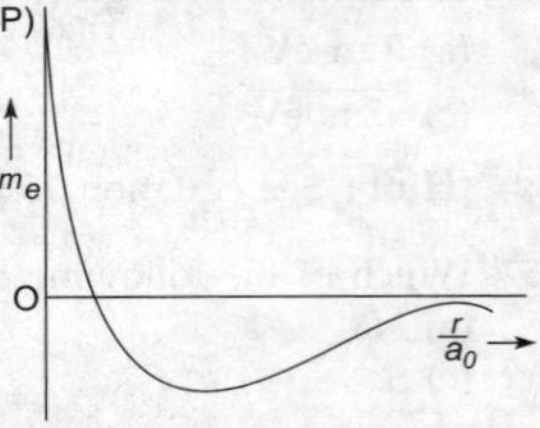
(2) 2 s orbital	(ii) One radial node	(Q) Probability density at nucleus $\propto \dfrac{1}{a_0^3}$

(3) 2 p_z orbital	(iii) $\psi_{nlm_e} \propto \left(\frac{z}{a_0}\right)^{5/2} r\, e^{-(zr/a_0)} \cos\theta$	(R) Probability density is max. at nucleus.
(4) $3d_x^2$ orbital	(iv) X–Y plane is a nodal plane	(S) Energy needed to excite electron from $n = 2$ state to $n = 4$ state is 27/32 times the energy needed to excite electron from $n = 2$ to $n = 6$.

For He^+ ion the only incorrect combination is

(a) (1) (i) (S) (b) (2) (ii) (Q)

(c) (1) (iii) (R) (d) (1) (i) (R) **(IIT 2017 Adv.)**

[**Hint:** For Q. 61, 62 and 63, read

Mod. App. to Physical Chemistry Vol. 2, Ch. 1]

62. Refer to Q. 61. For the given orbital in column 1, the only correct combination for any hydrogen like species is

(a) (2) (ii) (P) (b) (1) (ii) (S)

(c) (4) (iv) (R) (d) (3) (iii) (P) **(IIT 2017 Adv.)**

63. Refer to Q. 61. For hydrogen atom, the only correct combination is

(a) (1) (i) (P) (b) (1) (iv) (R)

(c) (2) (i) (Q) (d) (1) (i) (S) **(IIT 2017 Adv.)**

64. Bombardment of Al by α-particles leads to artificial disintegration in two ways (i) and (ii) as shown. Products are respectively X, Y and Z.

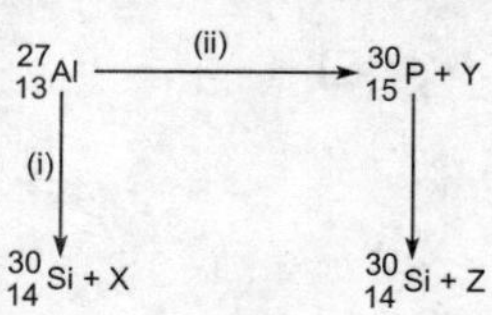

(a) proton, neutron, positron (b) neutron, positron, proton

(c) proton, positron, neutron (d) positron, proton, neutron **(IIT 2011)**

[**Hint :** neutron– ${}_0^1n$; proton– ${}_1^1p$; positron– ${}_1^0e$]

65. In the nuclear transmutation

$$ {}_4^9Be + X \rightarrow {}_4^8Be + Y $$

(X, Y) is (are)

(a) (γ, n) (b) (p, D)

(c) (n, D) (d) (γ, p) **(IIT 2013 Adv.)**

[**Hint :** ${}_4^9Be + {}_b^aX \rightarrow {}_4^8Be + {}_d^cY$

$4 + b = 4 + d$ gamma—${}_0^0\gamma$

$9 + a = 8 + C,$ deutorium– ${}_1^2D$

proton– ${}_1^1p$]

66. A plot of the number of neutrons(n) against the number of protons(p) of stable nuclei exhibits upward deviation from linearily for atomic number, $Z > 20$. For an unstable nucleus having n/p ratio less than 1, the possible mode(s) of decay is(are),

(a) β^--decay (β-emission) (b) orbital and K-electron capture

(c) neutron emission (d) β^+-decay (positron emission)

(IIT 2016 Adv.)

[**Hint:** See text.]

67. Ejection of the photoelectron from metal in the photoelectric effect experiment can be stopped by applying 0.5 V when the radiation of 250 nm is used. The work function of the metal is ... .

(a) 4.5 eV (b) 5 eV (c) 5.5 eV (d) 4 eV

[**Hint:** 1 eV = 1.602×10^{-19} J, energy to slop PEE = 0.5 eV

$$\frac{(hc/\lambda)}{1.602 \times 10^{-19}} \text{(eV)} = 0.5 \text{ eV} + \text{work function}]$$

Answers

1-a, 2-d, 3-d, 4-b,c, 5-b, 6-b, 7-a, 8-d, 9-b, 10-c, 11-d, 12-a 13-d, 14-b, 15-c, 16-d, 17-c, 18-c, 19-d, 20-d, 21-b, 22-a, 23-b, 24-a, 25-d, 26-c, 27-c, 28-b, 29-c, 30-a, 31-d, 32-b, 33-c, 34-b, 35-c, 36-b, 37-b, 38-d, 39-b, 40-b, 41-a, 42-a, 43-d, 44-a, 45-b, 46-d, 47-b, 48-a, 49-b, 50-d, 51-c, 52-a, 53-b, 54-a, 55-c, 56-c, 57-c, 58-c, 59-d, 60-a, 61-c, 62-a, 63-d, 64-a, 65-a, b, 66-b, d, 67-a.

❑

CHAPTER TWELVE

PROPERTIES OF GASES

The Ideal Gas Laws

1. Avogadro's Law

One mole of any substance contains the Avogadro constant ($6{\cdot}022 \times 10^{23}$) of molecules. It was Avogadro who discovered the law of nature for gases. This law is known as Avogadro's law, which states:

The volumes of the same number of moles of all gases measured at constant temperature and pressure are the same. That is, at the same temperature and pressure equal volumes of all gases contain equal number of moles or molecules.

Mathematically, at constant temperature and pressure:

Volume $\propto$ Number of moles

Volume $= K \times$ number of moles

Thus, for the same value of number of moles at constant temperature and pressure the proportionality constant, K will be a universal constant for all gases. When temperature and pressure are taken as 0°C and 1 atm respectively, K for 1 mole of the gas then represents the standard molar volume which is equal to 22·4 litres per mole or $22{\cdot}4 \times 10^{-3}$ metre^3 mole^{-1} (SI).

2. Boyle's Law

For a gas at constant temperature the pressure is inversely proportional to the volume, provided number of moles (*n*) of the gas does not change during the experiment.

$$p \propto \frac{1}{V}; \; pV = \text{constant} \qquad \dots (1)$$

$$p_1V_1 = p_2V_2 \quad \text{(T and } n \text{ are constant)}$$

Boyle's law can be expressed graphically in various ways:

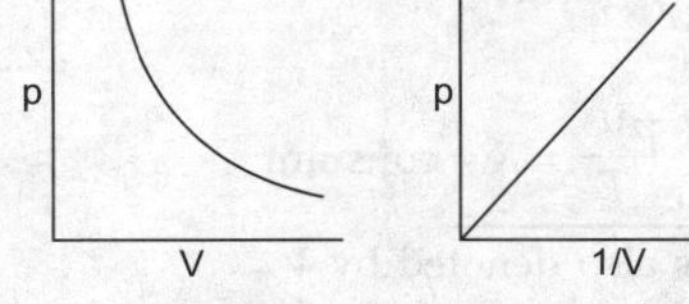

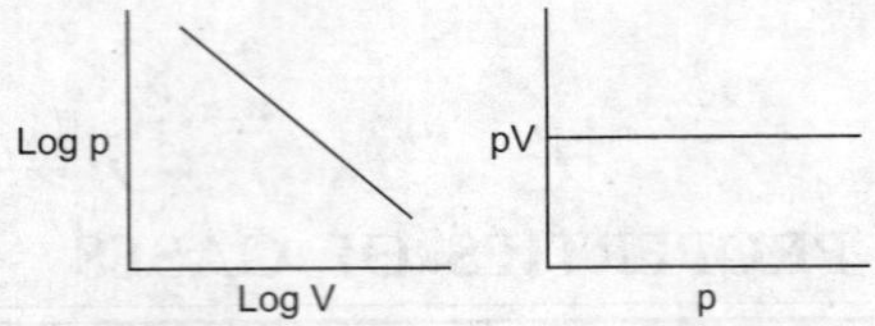

3. Charles's Law

The volume, for a given number of moles of a gas, is directly proportional to its absolute temperature if the pressure is held constant.

$$V \propto T$$

$$\frac{V}{T} = \text{constant} \qquad \ldots (2)$$

$$\frac{V_1}{T_1} = \frac{V_2}{T_2} \qquad (p \text{ and } n \text{ are constant})$$

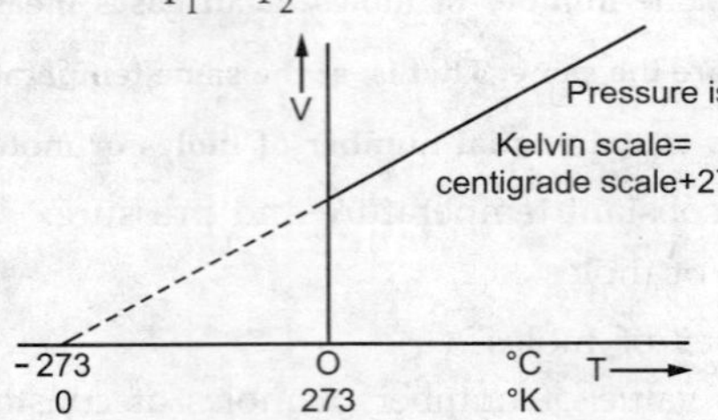

From the graph we see that the volume of the gas should be (hypothetically) zero at – 273°C or 0 K and this temperature is called absolute zero of temperature as any temperature lower than this will respond to a negative volume of the gas. At absolute zero, the kinetic energy is zero, the molecular motion is zero or the heat in the gas is zero. It is therefore impossible to cool the gas below absolute zero, since there is now no heat left to remove from the gas.

The volume of a gas increases or decreases by 1/273 of its original value per degree rise or fall in temperature respectively. If the temperature is raised from 0°C to 273°C, the increase in volume will, therefore, be 273/273 of its 0°C volume and so for a rise in temperature from 0°C to 273°C, the volume doubles.

4. Ideal Gas Equation of State

Combining the Boyle's and Charles's law, we get the ideal gas equation of state.

$$pV = nRT \qquad \ldots (3\text{ a})$$

For 1 mole, $n = 1$

$$\therefore \qquad \frac{pV^*}{T} = R = \text{constant}$$

* The molar volume, V, is also denoted by $\overline{V}$.

The constant R is known as universal gas constant.

$$\therefore \quad \frac{p_1V_1}{T_1}=\frac{p_2V_2}{T_1}. \qquad (n \text{ is constant}) \qquad \text{... (3 b)}$$

If R is divided by Avogadro constant, we get another constant, called Boltzmann constant, $k\left(=\frac{R}{N}\right)$, which is defined as the gas constant per molecule.

Further, if W, M and d represent weight in grams, molecular weight and density of a gas respectively, we get,

$$p=n\frac{RT}{V}=\frac{W}{M}\cdot\frac{RT}{V}=\frac{dRT}{M}\,^{*} \qquad \text{... (4)}$$

Physical Significance of R and its Values in Different Units

For 1 mole of an ideal gas:

$$R=\frac{pV}{T}=\frac{\text{pressure}\times\text{volume}}{\text{degrees}}$$

$$=\frac{(\text{force/area})\times(\text{area}\times\text{length})}{\text{degrees}}=\frac{\text{force}\times\text{length}}{\text{degrees}}$$

Thus R represents work done per degree per mole. The values of R may be calculated in different units.

Since 1 mole of any gas at NTP (273 K and 1 atm) occupies 22·4 litres,

$$R=\frac{1\times 22{\cdot}4}{273}=0{\cdot}0821 \text{ lit-atm per degree per mole.}$$

To calculate R in cgs unit,

$$n=1 \text{ mole}$$

$$p=1 \text{ atm}=76 \text{ cmHg}$$

$$=76\times 13{\cdot}6\times 981 \text{ dynes/cm}^2$$

$$V=22400 \text{ cc}$$

$$T=273 \text{ K}$$

$$R=\frac{76\times 13{\cdot}6\times 981\times 22400}{273}$$

$$=8{\cdot}314\times 10^7 \text{ ergs per degree per mole}$$

$$=8{\cdot}314 \text{ joules per degree per mole}$$

$$(1 \text{ joule}=1\times 10^7 \text{ ergs}) \text{ (SI)}$$

$$=1{\cdot}987\approx 2 \text{ calories per degree per mole.}$$

$$(1 \text{ cal}=4{\cdot}184\times 10^7 \text{ ergs})$$

---

* In SI units M should be in kg/mole.

5. Dalton's Law of Partial Pressure

The total pressure of a mixture of gases (not reacting with each other) is equal to the sum of partial pressures of the constituent gases at the same temperature, that is,

$$p = p_1 + p_2 + p_3 + \dots , \qquad \dots (5)$$

where p is the total pressure and p_s are the partial pressures.

The partial pressure of a constituent gas is defined as the pressure it would exert if it alone occupied the entire volume of the vessel at the same temperature. Partial pressure may be calculated using the formula:

partial pressure of a constituent gas

$$= \frac{\text{moles of constituent gas}}{\text{total moles}} \times \text{total pressure} \qquad \dots (6)$$

$$\text{partial pressure} = \text{mole fraction} \times \text{total pressure.}$$

If a gas is collected over a liquid, it must be borne in mind in calculations that its pressure is partial which is equal to the difference between the total pressure of the gas mixture and the partial pressure of the vapour of the liquid.

6. Amagat's Law of Partial Volume

The total volume of a gas mixture at a given temperature is the sum of partial volumes of individual gases of the mixture.

$$V = v_1 + v_2 + v_3 + \dots , \qquad \dots (7)$$

where V is the total volume and v_s are the partial volumes.

The partial volume of a constituent gas is defined as the volume which the particular constituent gas would occupy at the total pressure of the mixture, temperature remaining constant. Partial volume may be calculated using the formula:

Partial volume of a constituent gas

$$= \frac{\text{moles of constituent gas}}{\text{total moles}} \times \text{total volume.} \qquad \dots (8)$$

The concept of partial volume is a purely mathematical one and has least physical significance.

7. Graham's Law of Effusion or Diffusion

When a gas is made to pass through a fine hole made in the wall of the container under a difference of pressure, it is known as effusion. The rate of effusion depends on the density, the temperature of the gas and the pressure gradient.

Graham's law states that at a constant temperature and for constant pressure gradient the rates of effusion or diffusion of different gases are inversely proportional to the square root of their densities.

If r_1 and r_2 are the rates of diffusion of two gases having densities d_1 and d_2,

$$r_1 \propto \frac{1}{\sqrt{d_1}} \text{ and } r_2 \propto \frac{1}{\sqrt{d_2}}$$

or
$$\frac{r_1}{r_2} = \sqrt{\frac{d_2}{d_1}}\ .$$

Since the gas densities are directly proportional to their vapour densities and molecular weights, M_1 and M_2,

$$\frac{r_1}{r_2} = \sqrt{\frac{(\text{vapour density})_2}{(\text{vapour density})_1}} = \sqrt{\frac{M_2}{M_1}}\ .$$

Again if m_1 and m_2 are the number of moles of the two gases or t_1 and t_2 are the time of flow for equal volume of the gases, we have,

$$\frac{r_1}{r_2} = \frac{m_1}{m_2} = \frac{v_1}{v_2} = \frac{t_2}{t_1} = \sqrt{\frac{d_2}{d_1}} = \sqrt{\frac{M_2}{M_1}}\ . \qquad \ldots (9)$$

If two gases are allowed to diffuse under the identical conditions and p_1 and p_2 are their pressure-drops per unit time,

$$\frac{r_1}{r_2} = \frac{p_1}{p_2} = \frac{m_1}{m_2} \quad (\text{pressure} \propto \text{moles}). \qquad \ldots (10)$$

This law is very useful for calculating molecular weight, density, etc., of gases. However, it should be noted that the **law is true only for gases diffusing under low pressure gradient.**

8. Barometric Distribution Law

In the discussion of ideal gases it has been tacitly assumed that the pressure of the gas has the same value everywhere in the container and is not affected by the gravitational field. For gaseous systems of ordinary size the influence of gravity field is negligible but for those having high value of molecular weight, e.g., in polymers, the effect is quite pronounced, that is, the pressure will be different in different vertical positions in a container. The pressure of the gas decreases with increase in its height from the ground. The relationship between the pressure and height of the gas from the ground is given by,

$$p = p_0\, e^{-Mgh/RT},$$

where p_0 and p are the pressure of the gas at the ground level and at a height 'h' from the ground respectively, M, the molecular weight and 'g', the acceleration due to gravity.

Since this equation describes the distribution of pressure of the gas in the column, it is known as Barometric distribution law or gravitational distribution law. It is to be noted that this equation is valid only when the system is isothermal and at equilibrium.

As the density *'d'* is proportional to the pressure and number of molecules per cc *'n'* is proportional to the pressure, the above equation can be written in the other equivalent forms.

$$d = d_0\, e^{-Mgh/RT}$$

and $$n = n_0\, e^{-Mgh/RT}$$

In calculations, we generally use the logarithmic-form of the above equations:

$$\left.\begin{aligned} 2{\cdot}303 \log\left(\frac{p}{p_0}\right) &= -\left(\frac{Mgh}{RT}\right) \\ 2{\cdot}305 \log\left(\frac{d}{d_0}\right) &= -\left(\frac{Mgh}{RT}\right) \\ 2{\cdot}303 \log\left(\frac{n}{n_0}\right) &= -\left(\frac{Mgh}{RT}\right) \end{aligned}\right\} \qquad \ldots (11)$$

In applying all these equations, *R, M, g* and *h* are all expressed in cgs units.

Pressure Derived from Kinetic Molecular Theory of Ideal Gases

If *'n'* molecules of a gas, the mass of each being *'m'*, are enclosed in a container of volume *V*, the pressure exerted by them, derived from kinetic theory, is given by the equation,

$$pV = \tfrac{1}{3}\, mnC^2, \qquad \ldots (12)$$

where *C* is the root mean square (rms) speed.

The rms speed may be defined as the hypothetical speed which all the molecules of the gas would possess if the total kinetic energy is equally divided among them. For *n* molecules having speeds $C_1, C_2, C_3, \ldots, C_n$, the rms speed is given by the equation

$$C = \sqrt{\frac{C_1^2 + C_2^2 + C_3^2 + \ldots + C_n^2}{n}}.$$

The average speed may be given as follows:

$$\text{average speed} = \frac{C_1 + C_2 + \ldots + C_n}{n}.$$

There is a third type of molecular speed known as 'Most Probable Speed' which may be defined as the speed possessed by maximum number of molecules of a gas. These three types of speeds are closely related.

$$\text{rms speed} = \sqrt{3p/d} = \sqrt{3RT/M}$$

$$\text{Average speed} = \sqrt{(8/\pi)\,(p/d)} = \sqrt{8RT/\pi M}$$

$$\text{Most probable speed} = \sqrt{2p/d} = \sqrt{2RT/M}$$

Thus,
rms speed : average speed : most probable speed
= 1 : 0·9211 : 0·8165

Calculation of rms Speed

We have,

$$pV = \tfrac{1}{3} mnC^2$$

or
$$C = \sqrt{\frac{3\,pV}{mn}} \quad \dots (12\text{ a})$$

$$= \sqrt{\frac{3p}{mn/V}} \quad (mn = \text{total mass})$$

$$= \sqrt{\frac{3p}{d}}. \quad \dots (12\text{ b})$$

For 1 mole, $pV = RT$, $n = N$ (Av. const.)

$$C = \sqrt{\frac{3pV}{mN}}$$

$$= \sqrt{\frac{3pV}{M}} \quad (M^* \equiv \text{mol. wt., i.e., mass of 1 mole}) \quad \dots (12\text{ c})$$

$$= \sqrt{\frac{3RT}{M}}. \quad \dots (12\text{ d})$$

(Since Av. const. of molecules are contained by 1 mole)

Thus the heavier gas has smaller rms speed. Note that rms speed does not depend on p or V.

Kinetic Energy of Gaseous Molecules

We have, $pV = \frac{1}{3} mnC^2$.

For 1 mole of gases $pV = RT$, $n = N$ (Av. const.)

$$\therefore \quad RT = \frac{1}{3} mNC^2$$

$$RT = \frac{2}{3}\left(\frac{1}{2} mNC^2\right) \quad (mN \equiv \text{mass of 1 mole})$$

$$= \frac{2}{3} \text{KE of 1 mole of the gas}$$

$$\therefore \text{ KE of 1 mole of the gas} = \frac{3}{2} RT. \quad \dots (13\text{ a})$$

$$\text{Average KE of 1 molecule} = \frac{3}{2} \cdot \frac{R}{N} \cdot T$$

* In SI unit M should be taken in kg/mole, e.g., for O_2, $M = 0{\cdot}032$ kg/mole

$$= \frac{3}{2} kT, \quad \text{... (13 b)}$$

where k is the Boltzmann constant.

The equation 13 (a) is known as Maxwell generalisation which states that at a given temperature, the kinetic energy of a mole of every gas is the same which is equal to 3/2 *RT*.

KE of a gas depends on its mass, i.e., no. of moles and temperature but does not depend on volume or pressure.

For n moles of a gas,

$$\text{KE} = \frac{3}{2} n\, RT. \quad \text{... (13 c)}$$

Collision Frequency (n)

The number of collisions experienced by molecules per cc of a gas per second is known as ***collision frequency*** of the gas. From kinetic consideration,

Total no. of molecules colliding per cc per second is

$$Z = \sqrt{2}\, \pi\, v\, \sigma^2 N^2.$$

$$\text{Collision frequency, } n = \frac{Z}{2}$$

or
$$n = \frac{\pi\, v\, \sigma^2 N^2}{\sqrt{2}}, \quad \text{... (14)}$$

where v is the average velocity, σ is molecular diameter and N is the number of molecules per cc.

Mean Free Path (λ)

The mean free path is defined as the average distance travelled by a molecule between two successive collisions.

$$\lambda = \frac{\text{average distance travelled per second}}{\text{number of collisions by a molecule per second}}$$

$$= \frac{v}{\sqrt{2}\, \pi\, v\, \sigma^2 N}$$

or
$$\lambda = \frac{1}{\sqrt{2}\, \pi\, \sigma^2 N}.$$

Assuming that the collision diameter, σ, is independent of temperature, the temperature and pressure dependence of the mean free path may be obtained by substituting ideal gas law in the form: $N = \frac{p}{kT}$,

$$\therefore \quad \lambda = \frac{kT}{\sqrt{2} \pi \sigma^2 p}. \quad \text{... (15)}$$

Thus for bimolecular collisions

$$\lambda \propto \frac{1}{p} \quad \text{... at constant temperature}$$

$$\lambda \propto T \quad \text{... at constant pressure,}$$

Viscosity of Gases

Viscosity is the internal resistance that one part of a fluid offers to the flow of another part of the fluid. Like liquids, gases also show viscous behaviour. For gases we assume planes at a distance of mean free path. The coefficient of viscosity (η) is given by

$$\eta = \frac{5}{16\sigma^2}\left(\frac{mkT}{\pi}\right)^{1/2}, \qquad \text{... (16)}$$

where k is the Boltzmann constant equal to R/N, σ and m are diameter and mass of the molecule respectively. Reciprocal of η gives fluidity. The variation of σ with pressure and temperature gives the nature of intermolecular forces in the gas. The unit of viscosity is *poise* (dyne cm^{-2}. s) in cgs and Pascal second (Pa s) or newton second per square metre (Nsm^{-2}) in SI units.

Viscosity of gases, unlike for liquids, increases with increasing temperature and is independent of pressure.

Molar Heat Capacity of Ideal Gases

Specific heat capacity of a substance is defined as the amount of heat required to raise the temperature of 1 g of substance through 1°C, the unit of specific heat is calorie. 1 cal is defined as the amount of heat required to raise the temperature of 1 g of water through 1°C.

Molar heat capacity is defined as the amount of heat required to raise the temperature of 1 mole of a gas through 1°C. Thus:

Molar heat capacity = sp. heat × molecular wt. of the gas.

For gases there are two values of molar heats, viz., molar heat at constant pressure and molar heat at constant volume, respectively denoted by C_p and C_v. C_p is greater than C_v and $C_p - C_v = R \approx 2$ cal.

From the ratio of C_p and C_v, we get the idea of atomicity of a gas. For monoatomic gas, $C_p = 5$ cal and $C_v = 3$ cal.

$$\therefore \qquad \gamma = \frac{5}{3} = 1{\cdot}67.$$

For diatomic gas, $C_p = 7$ cal and $C_v = 5$ cal.

$$\therefore \qquad \gamma = \frac{7}{5} = 1{\cdot}40.$$

For triatomic gas, $C_p = 8$ cal and $C_v = 6$ cal.

$$\therefore \qquad \gamma = \frac{8}{6} = 1{\cdot}33.$$

For example, γ values of He, O_2 and CO_2 gases are 1·66, 1·42 and 1·33 respectively.

Real Gases: van der Waals Equation

To display the deviation of a gas from ideal behaviour clearly, the ratio of the observed molar volume, V, to the ideal molar volume $V_{id} = \left(\frac{RT}{p}\right)$ is plotted against pressure at constant temperature. This ratio is called *the compressibility factor*, Z.

$$Z = \frac{V}{V_{id}} = \frac{pV}{RT}$$

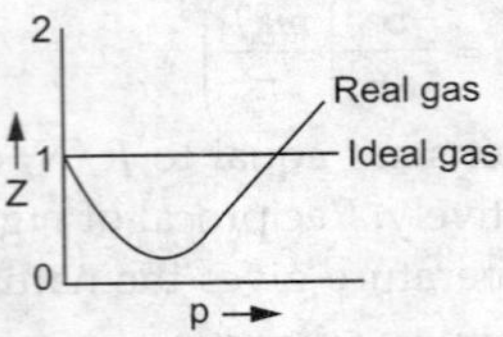

For an ideal gas, Z = 1 and is independent of p and t. For a real gas, $Z \neq 1$ and depends on p and t. If $Z < 1$ or $Z > 1$, the gas is more or less compressible compared to an ideal gas respectively. The curves for the real gases can be explained by van der Waals equation.

When the gases do not obey ideal gas laws, we call them real gases. The gases under high pressure and low temperature deviate from the ideal behaviour. The ideal gas equation $pV = nRT$ was modified by van der Waals for real gases and the new equation, known as van der Waals equation, was found to be quite valid under high pressure and low temperature. van der Waals equation for 'n' moles of the gas is,

$$\left(p + \frac{n^2a}{V^2}\right)(V - nb) = n\,RT, \quad \ldots (17)$$

where the factor $\frac{n^2a}{V^2}$ and nb are the pressure and volume correction factors respectively introduced in $pV = nRT$ due to intermolecular forces of attraction and incompressible volume of gas.

a and b are the van der Waals constants, their units being atm lit^2 mole^{-2} and litre mole^{-1} respectively.

Since 'b' is four times the actual volume of the gaseous molecules,

$$b = 4 \times \text{Av. const.} \times \frac{4}{3}\pi r^3, \quad \ldots (18)$$

where 'r' is the radius of the gaseous molecule.

Amagat observed experimentally that at ordinary temperature and high pressure, for all the gases, except H_2 and He, the compressibility factor, Z, first decreases and then increases with the increase in pressure. H_2 and He at ordinary temperature show only increase in Z with the increase in pressure. However, if the temperature is kept sufficiently low, H_2 and He may also behave like other gases. The abnormal behaviour of H_2 and He at ordinary temperature is because of their exceptionally small-sized molecules. It has also been observed that even at high pressure range, a real gas may behave ideally at a particular temperature. This temperature is called **Boyle Temperature, T_b.**

$$T_b = \frac{a}{Rb} \quad \ldots (19)$$

The Role of Intermolecular Distance in the Behaviour of Gases: Critical State

Under the condition of ordinary temperature and pressure, the gaseous molecules remain far away from each other and hence the intermolecular forces and the actual volume of the molecules compared to total volume of the gas were neglected. The gases behave ideally under this condition, that is, the gases obey ideal gas laws.

When the temperature of the gas is quite low and pressure is quite high, the molecules lie close to each other due to which the intermolecular forces as well as the actual volume of the molecules compared to the total volume of the gas could not be neglected. Under this condition, the gases do not behave ideally and the ideal gas equation was thus modified to van der Waals equation to explain the behaviour of nonideal or real gases.

Now if the temperature is sufficiently low and pressure is sufficiently high, the molecules may come so close to each other that the gaseous molecules may now condense into liquid. But it is an interesting observation that a gas does not liquefy above a certain temperature, however high the pressure may be. This characteristic temperature is called the **critical temperature (T_c).** The minimum pressure which is just sufficient to liquefy the gas at the critical temperature is called the **critical pressure (p_c).** The volume occupied by 1 mole of a gas at T_c and p_c is called the **critical volume (V_c).** T_c, p_c and V_c are called critical constants. The importance of critical temperature of a gas was first discovered by Andrews. His experiments on *p-V* relationship for CO_2 has been graphically represented below:

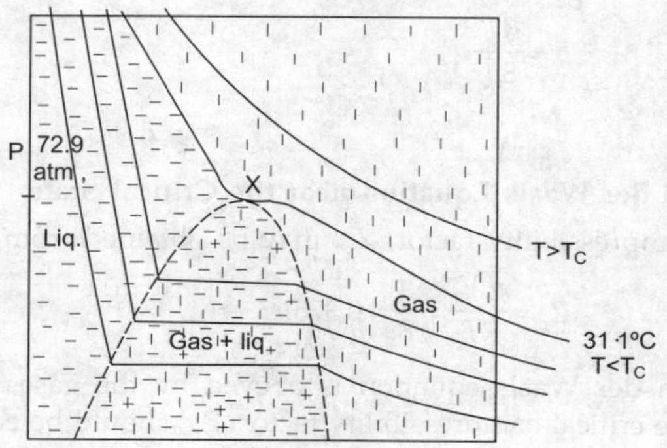

The horizontal portions of the curves signify condensation of the gas into liquid at different temperatures. The critical point 'X' represents critical temperature of 31·1°C and critical pressure of 72.9 atm for CO_2. Above this critical temperature the isotherm is continuous giving no evidence of liquefaction at all. Within the area of the boundary curve shown by dotted lines, both the liquid and gaseous states can coexist. But outside this area either liquid or gaseous state alone can exist and so the gas in this area may be condensed into a liquid or vice-versa **continuously**. The fact that it is not always possible to distinguish between a liquid and a gas is the principle of **continuity of state.**

The values of critical constants can be obtained from van der Waals equation in terms of the gas constants, 'a', 'b' and 'R'.

$$T_c = \frac{8a}{27Rb} \qquad \text{... (20)}$$

$$p_c = \frac{a}{27b^2} \qquad \text{... (21)}$$

$$V_c = 3b \qquad \text{... (22)}$$

from which a, b and R can be obtained.

$$a = 3p_c V_c^2 \qquad \text{... (23)}$$

$$b = \frac{V_c}{3} \qquad \text{... (24)}$$

$$R = \frac{8}{3}\frac{p_c V_c}{T_c}. \qquad \text{... (25)}$$

But since experimentally it is hard to determine V_c accurately, it would be better if a and b could be obtained from p_c and T_c only.

$$a = \frac{27}{64}\frac{R^2 T_c^2}{p_c} \qquad \text{... (26)}$$

$$b = \frac{RT_c}{8p_c}. \qquad \text{... (27)}$$

Failure of van der Waals Equation near the Critical State

The critical compressibility factor, Z_c, may be obtained from equation (25).

$$Z_c = \frac{p_c V_c}{RT_c} = \frac{3}{8} = 0.375. \qquad \text{... (25)}$$

If the van der Waals equation is obeyed by the gases at their critical points then the critical compressibility factor, Z_c, should be equal to 0.375 for every gas but the experimental values of Z_c generally disagree and the difference between the two values varies from gas to gas. It may be due to the fact that the van der Waals equation does not adequately describe the behaviour of the gas near the critical state.

The Law of Corresponding States

When the pressure, volume and temperature of a gas are expressed relative to their critical pressure, volume and temperature respectively, they are called reduced variables.

$$p_r = \frac{p}{p_c},\ V_r = \frac{V}{V_c},\ T_r = \frac{T}{T_c}.$$

Gases having the same values of their reduced variables deviate almost equally from ideality and thus are said to be in corresponding states. The law of corresponding states can also be expressed mathematically

$$\left(p_r + \frac{3}{V_r^2}\right)\left(V_r - \frac{1}{3}\right) = \frac{8}{3} T_r. \qquad \dots (28)$$

This equation is a more general form of van der Waals equation in terms of only reduced variables because it does not contain any gas constant like *a*, *b* and *R* and thus is capable of describing all gases.

Virial Equation

This is another equation for real gases. It expresses the compressibility factor, *Z* either as a function of $1/V$ or as a function of *p* as shown by the following equations.

$$Z = \frac{pV}{RT} = 1 + \frac{B}{V} + \frac{C}{V^2} + \dots \qquad \dots (29)$$

or $$Z = \frac{pV}{RT} = 1 + \frac{Bp}{(RT)} + \frac{Cp^2}{(RT)^2} + \dots \qquad \dots (30)$$

V is the molar volume, *B* and *C* are second and third virial coefficients respectively.

Since the contribution from the square cube and other higher power terms are necessarily smaller than the first two terms, the equation reduces to pV = constant, at a temperature at which $B = 0$. This temperature for any gas is known as **Boyle temperature (T_b)**. The virial coefficients *B* and *C* can be obtained as

$$B = b - \frac{a}{RT} \qquad \dots (31)$$

$$C = b^2. \qquad \dots (32)$$

Now when $B = 0$, $T = T_b$.

$$\therefore \quad T_b = \frac{a}{Rb} \qquad \dots (19)$$

Adsorption of Gases on Solids

The phenomenon of accumulation of a substance on the surface of a solid or liquid is called *adsorption*. Here a gas is considered as adsorbate and a solid, adsorbent. Forces holding the adsorbate and the adsorbent together may be either van der Waals forces or valence forces. The former is the case of physical adsorption and the latter, chemisorption. The process of adsorption is accompanied with evolution of heat, low in physical adsorption and high in chemisorption.

In case of a gas adsorbing on a solid, decrease in temperature and increase in pressure cause in the increase of adsorbed gas on a solid. If x grams of a gas is adsorbed per m gram of a solid, the adsorption pattern may be represented graphically.

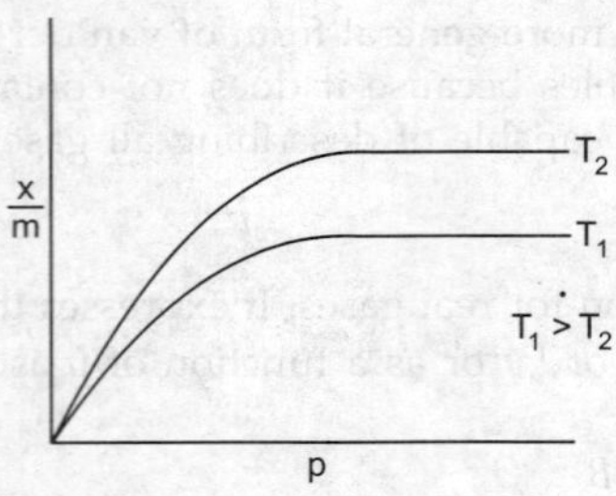

$$\frac{x}{m} \propto p \quad \text{... at low pressure}$$

$$\frac{x}{m} \propto p^{\circ} \quad \text{... at high pressure} \qquad \text{or,} \quad \frac{x}{m} = kp^{n}, \qquad \text{... (33)}$$

where $n = 1$ at low pressure and $n = 0$ at high pressure.

In other words, at lower pressure, adsorption of a gas on a solid increases with the increase in pressure but at higher pressure further adsorption stops as all the surface area gets covered by gas molecules.

Further, $\frac{x}{m} = kp^{n}$ or $\frac{x}{m} = kC^{n}$ (pressure $\propto$ concentration)

or $\log \frac{x}{m} = n \log p + \log k$

A plot of $\log \frac{x}{m}$ vs $\log p$ gives a straight line with a positive slope equal to n. The above pattern is called *Freundlich adsorption isotherm* which was further modified by Langmuir and Gibbs.

In this chapter, the questions are generally asked in mixed system of units. The students should, therefore, use the units carefully remembering particularly cgs and SI unit combinations.

Quantity	cgs	SI
Time	second	second
Temperature	kelvin	kelvin
Distance	cm	metre
Mass	g	kg
Velocity	cm/s	m/s
Volume	cm^3	$metre^3$
Density	g/cm^3	$kg/metre^3$
Pressure	$dynes/cm^2$	pascal ($newton/metre^2$)
Energy	erg	joule
Gas constant (R)	erg/K/mole	joule/K/mole

Moreover, some nonsystem units like litre for volume, atm for pressure, calorie for energy, etc., are still in use. The value of R should be used accordingly.

EXAMPLES

Ex. 1. *A large cylinder for storing compressed gas has a volume of* 1·5 ft^3. *If the gas is stored under a pressure of* 150 atm *at* 300 K, *how many moles of the gas are contained in the cylinder? What would be the weight of oxygen in such a cylinder?* (1 ft^3 = 28·32 litres)

Solution : $pV = n\,RT$

$$n = \frac{150 \times (1{\cdot}5 \times 28{\cdot}32)}{0{\cdot}0821 \times 300}$$

$$= 258{\cdot}70 \text{ moles.}$$

$$(R = 0{\cdot}0821 \text{ lit} \cdot \text{atm/K/mole})$$

Weight of oxygen = moles × mol. weight ... (Rule 1, Chapter 1)

$$= 258{\cdot}70 \times 32$$

$$= 8278{\cdot}40 \text{ g.}$$

Ex. 2. 5 g *of ethane are confined in a bulb of one-litre capacity. The bulb is so weak that it will burst if the pressure exceeds* 10 atm. *At what temperature will the pressure of the gas reach the bursting value?*

Solution : $pV = nRT$

$$pV = \frac{\text{wt. of } C_2H_6}{\text{mol. wt. of } C_2H_6} \cdot RT$$

$$\therefore \quad T = \frac{10 \times 1 \times 30}{0.0821 \times 5}$$
$$= 730.81 \text{ K}$$
$$= (730.81 - 273)°\text{C}$$
$$= 457.81°\text{C}.$$
$$(R = 0.0821 \text{ lit} \cdot \text{atm/K/mole})$$

Ex. 3. *A gas cylinder containing cooking gas can withstand a pressure of* 14·9 atm. *The pressure gauge of cylinder indicates* 12 atm *at* 27°C. *Due to sudden fire in the building, its temperature starts rising. At what temperature the cylinder will explode?*

Solution : Suppose the cylinder will burst at T_2 K.

When the pressure will increase from 12 atm to 14·9 atm we have,

$$\frac{p_1 V_1}{T_1} = \frac{p_2 V_2}{T_2} \qquad p_1 = 12 \text{ atm}, T_1 = (27 + 273); \; p_2 = 14.9 \text{ atm}, T_2 = ?$$

Here $V_1 = V_2$ as the volume does not change

$$\therefore \quad T_2 = \frac{p_2 T_1}{p_1} = \frac{14.9 \times 300}{12}$$
$$= 372.5 \text{ K}.$$

Ex. 4. *How high a column of air would be necessary to cause the barometer to read* 76 cmHg, *if the atmosphere were of uniform density* 1·2 kg/m³? *The density of* Hg = 13.6×10^3 kg/m³.

Solution : Let the height of the air column be h metres and acceleration due to gravity, g metre s^{-2}.

∴ pressure corresponding to

$$76 \text{ cm } (0.76 \text{ m}) \text{ of Hg} = 0.76 \times (13.6 \times 10^3) \times g$$

and pressure corresponding to 'h' metres of air = $h \times 1.2 \times g$.

Thus, $\quad h \times 1.2 \times g = 0.76 \times 13.6 \times 10^3 \times g$

or $\quad h = 8613$ metres.

Ex. 5. *The density of the vapour of a substance at* 373 K *and* 1.013×10^5 Pa *is* 2·55 g/lit. *Calculate its molecular weight.*

Solution : We have,

$$p = \frac{dRT}{M} \qquad \text{... (Eqn. 4)}$$

$$1.013 \times 10^5 = \frac{2.55 \times 8.314 \times 373}{M} \qquad \left[\begin{array}{l}\text{In SI units} \\ d = 2.55 \text{ g/L} = 2.55 \text{ kg/m}^3\end{array}\right]$$

$\therefore \qquad M = 0{\cdot}078 \text{ kg/mole}$

$= 78 \text{ g/mole.}$

$\therefore$ molecular weight = 78.

Ex. 6. *Oxygen is present in a one-litre flask at a pressure of* $7{\cdot}6 \times 10^{-10}$ mmHg. *Calculate the number of oxygen molecules in the flask at* 0°C.

Solution : $p = 7{\cdot}6 \times 10^{-10}$ mm

$= 0{\cdot}01 \times 10^{-10}$ atm (760 mm = 1 atm)

$T = (0 + 273) = 273$ K

We have,

$$pV = n\,RT$$

$$0{\cdot}01 \times 10^{-10} \times 1 = n \times 0{\cdot}0821 \times 273, \quad \therefore \; n = \frac{10^{-12}}{0{\cdot}0821 \times 273}.$$

$\therefore$ no. of oxygen molecules = mole × Av. const. ... (Rule 4, Chapter 1)

$$= \frac{10^{-12}}{0{\cdot}0821 \times 273} \times 6{\cdot}022 \times 10^{23}$$

$$= 2{\cdot}68 \times 10^{10}.$$

Ex. 7. *A* 10-cm *column of air is trapped by an* 8-cm-*long column of* Hg *in a capillary tube horizontally fixed as shown in the figure at* 1 atm *pressure. Calculate the length of air column when the tube is fixed*

(a) *vertically with open end up*

(b) *vertically with open end down*

(c) *at* 45° *from horizontal with the open end up*

Solution : (a) At the horizontal position of tube, air pressure is 1 atm, i.e., 76 cmHg. If the tube is now held in different given positions at the same temperature, the length of air column will vary. Applying Boyle's law, we have,

$$p_1V_1 = p_2V_2$$

or $\qquad p_1l_1\,a = p_2l_2\,a \quad \text{or} \quad p_1l_1 = p_2l_2.$

Now when the tube is held vertically with open end up,

air pressure $p_2 = 76$ cmHg + 8 cmHg

$= 84$ cmHg.

$$\therefore \qquad l_2 = \frac{p_1\,l_1}{p_2} = \frac{76 \times 10}{84} = 9{\cdot}04 \text{ cm.}$$

(b) When the tube is held vertically with open end down,

air pressure $p_2 = 76$ cmHg − 8 cmHg

$$= 68 \text{ cmHg}$$

$$\therefore \quad l_2 = \frac{p_1 l_1}{p_2} = \frac{76 \times 10}{68} = 11{\cdot}17 \text{ cm.}$$

(c) When the tube is held at 45° with the open end up, the wt. of Hg is borne partially by the gas and partially by the glass. Vertical height of Hg is the measure of the additional pressure on the gas. Thus vertical height of Hg = $8/\sqrt{2} = 5{\cdot}66$ cm.

$$\therefore \quad p_2 = 76 + 5{\cdot}66 = 81{\cdot}66 \text{ cmHg}$$

$$\therefore \quad l_2 = \frac{76 \times 10}{81{\cdot}66} = 9{\cdot}30 \text{ cmHg.}$$

Ex. 8. (a) *An open bulb containing air at 19°C was cooled to a certain temperature at which the number of moles of the gaseous molecules increased by 25%. What is the final temperature?*

(b) *An open vessel at 27°C is heated until three-fifths of the air in it has been expelled. Assuming the volume of the vessel remains constant, find the temperature to which the vessel has to be heated.*

Solution : (a) Suppose the volume of the bulb is V containing n moles at 19°C, i.e., 292 K.

Let the temperature be T K when n moles increases to $1{\cdot}25\, n$ (i.e., by 25%). Since $1{\cdot}25\, n$ moles at T K occupy a volume V

$\therefore$ n moles at T K should occupy $\dfrac{V}{1{\cdot}25}$.

Thus for n moles of the gas,

$T_1 = 292$ K $\qquad T_2 = T$ K

$V_1 = V \qquad V_2 = \dfrac{V}{1{\cdot}25}$

$p_1 = p \qquad p_2 = p$ ($p_1 = p_2$ as the bulb is open)

$$\therefore \quad \frac{pV}{292} = \frac{p \times V/1{\cdot}25}{T}$$

$$T = \frac{292}{1{\cdot}25} = 233{\cdot}6 \text{ K}$$

$$= -39{\cdot}4\text{°C.}$$

(b) Suppose the volume of the vessel at 27°C is V containing n moles of the gas.

Let the vessel be heated to T K when $2n/5$ moles remain (as three-fifth has been expelled).

Since $\frac{2n}{5}$ moles at T K occupy a volume V,

$\therefore$ n moles at T K should occupy $= \frac{5V}{2}$.

Thus for n moles of the gas,

$p_1 = p$ $\qquad$ $p_2 = p$ ($p_1 = p_2 = p$ as the vessel is open)

$V_1 = V$ $\qquad$ $V_2 = \frac{5V}{2}$

$T_1 = 300$ K $\qquad$ $T_2 = T$ K

$$\frac{pV}{300} = \frac{p \times (5V/2)}{T}$$

$$T = 750 \text{ K} = 477°\text{C}.$$

Ex. 9. *A stockroom supervisor measured the contents of a partially filled* 25·0 *gallon acetone drum on a day when the temperature was* 18°C *and atmospheric pressure was* 780 mmHg *and found that* 15·4 *gallon of the solvent remained. After tightly sealing the drum a student assistant dropped the drum while carrying it upstairs to the organic laboratory. The drum was dented and its internal volume was decreased to* 20·4 *gallon. What is the total pressure inside the drum after the accident? The vapour pressure of acetone at* 18°C *is* 400 mmHg.

Solution : At the time the drum was sealed, the pressure inside the drum, which is equal to the sum of the pressures of air and acetone, is equal to the atmospheric pressure. Thus,

$$p_{\text{air}} = 780 - 400 = 380 \text{ mm}$$

$$\text{Mole of air} = \frac{pV}{RT} = \frac{(380)\,(25 - 15{\cdot}4)}{RT}$$

After the accident, mole of air will remain the same.

$$p_{\text{air}} = \frac{nRT}{V}$$

$$= \frac{(380) \times 9{\cdot}6}{RT} \times \frac{RT}{(20{\cdot}4 - 15{\cdot}4)}$$

$$= 729{\cdot}6 \text{ mm.}$$

$\therefore$ total pressure $= p_{\text{air}} + p_{\text{acetone}}$

$$= 729{\cdot}6 + 400$$

$$= 1129{\cdot}6 \text{ mm.}$$

Ex. 10. *A mixture of* C_2H_6 *and* C_2H_4 *occupies* 40 *litres at* 1 atm *and at* 400 K. *The mixture reacts completely with* 130 g *of* O_2 *to produce* CO_2 *and* H_2O. *Assuming ideal gas behaviour, calculate the mole fractions of* C_2H_4 *and* C_2H_6 *in the mixture.* **(IIT 1995)**

Solution : Let the moles of C_2H_6 and C_2H_4 be n_1 and n_2 respectively. Applying ideal gas equation, $pV = n\,RT$,

$$1 \times 40 = (n_1 + n_2) \times 0{\cdot}0821 \times 400$$

or, $$(n_1 + n_2) = \frac{40}{0{\cdot}0821 \times 400} \quad \dots (1)$$

$$C_2H_6 + C_2H_4 + O_2 \rightarrow CO_2 + H_2O$$

Applying POAC for C,

$$2\,n_1 + 2\,n_2 = \text{moles of } CO_2 \quad \dots (2)$$

Applying POAC for H,

$$6\,n_1 + 4\,n_2 = 2 \times \text{moles of } H_2O \quad \dots (3)$$

Applying POAC for O,

$$2 \times \frac{130}{32} = 2 \times \text{moles of } CO_2 + \text{moles of } H_2O \quad \dots (4)$$

From equations (2), (3) and (4) we get,

$$7\,n_1 + 6\,n_2 = \frac{260}{32} \quad \dots (5)$$

Solving equations (1) and (5) we get,

$$n_1 = 0{\cdot}8168$$

$$n_2 = 0{\cdot}4012$$

$$\therefore \text{ mole fraction of } C_2H_6 = \frac{0{\cdot}8168}{0{\cdot}8168 + 0{\cdot}4012} = 0{\cdot}67.$$

$$\text{Mole fraction of } C_2H_4 = 1 - 0{\cdot}67 = 0{\cdot}33.$$

Ex. 11. *A bulb of unknown volume V contains an ideal gas at* 1 atm *pressure. This bulb was connected to another evacuated bulb of volume* 0·5 *litre through a stopcock. When the stopcock was opened the pressure at each bulb became* 530 mm *while the temperature remained constant. Find V in litres.*

Solution : Since, on opening the stopcock, the same number of moles of the gas occupy a volume of $(V + 0{\cdot}5)$ litre at the same temperature, we apply Boyle's law:

$$p_1V_1 = p_2V_2$$

$$1 \times V = \frac{530}{760} \times (V + 0{\cdot}5) \qquad \left(p_2 = \frac{530}{760} \text{ atm}\right)$$

$$V = 1{\cdot}152 \text{ litres.}$$

Ex. 12. *A bulb of unknown volume contained an ideal gas at* 650 mm *pressure. A certain amount of gas was withdrawn and found to occupy* 1·52 cc *at* 1 atm *pressure. The pressure of the gas remaining in the bulb was* 600 mm. *If all measurements were made at a constant temperature, find the volume of the bulb.*

Solution : When a certain amount of the gas is withdrawn, the pressure dropped from 650 mm to 600 mm. So the pressure difference (650 – 600) or 50 mm will be the pressure of the gas withdrawn which occupied the bulb of volume V_1 cc (suppose). Since the same amount of withdrawn gas occupied 1·52 cc at 1 atm (760 mm) pressure at the same temperature, we can apply Boyle's law:

$$p_1 V_1 = p_2 V_2$$
$$50 \times V_1 = 760 \times 1{\cdot}52$$
$$V_1 = 23{\cdot}1 \text{ cc.}$$

Ex. 13. *The stopcock, connecting the two bulbs of volumes* 5 *litres and* 10 *litres containing an ideal gas at* 9 atm *and* 6 atm *respectively, is opened. What is the final pressure in the two bulbs if the temperature remained the same?*

Solution : For the first bulb,

$$p_1 V_1 = n_1 RT.$$

For the second bulb,

$$p_2 V_2 = n_2 RT.$$

On adding: $$p_1 V_1 + p_2 V_2 = (n_1 + n_2)\, RT. \quad \ldots (1)$$

When the stopcock is opened, the total volume is $(V_1 + V_2)$ and total no. of moles is $(n_1 + n_2)$. Suppose the equilibrium pressure at each bulb is p atm. Then,

$$p(V_1 + V_2) = (n_1 + n_2)\, RT \quad \ldots (2)$$

From equations (1) and (2), we have,

$$p_1 V_1 + p_2 V_2 = p(V_1 + V_2)$$

Given that, $p_1 = 9$ atm, $V_1 = 5$ lit., $p_2 = 6$ atm, $V_2 = 10$ lit.

Substituting: $9 \times 5 + 6 \times 10 = p \times 15$

$$p = 7 \text{ atm.}$$

Ex. 14. *A large irregularly-shaped closed tank is first evacuated and then connected to a 50-litre cylinder containing compressed nitrogen gas. The gas pressure in the cylinder, originally at* 21·5 atm, *falls to* 1·55 atm *after it is connected to the evacuated tank. Calculate the volume of the tank.*

Solution : Let the volume of the tank be V litres.

As the no. of moles of N_2 before and after connecting it to the tank will be same,

$$\underset{\left(\begin{array}{c}\text{moles of } N_2 \text{ before}\\ \text{connection}\end{array}\right)}{\frac{21{\cdot}5 \times 50}{RT}} = \underset{\left(\begin{array}{c}\text{moles of } N_2 \text{ after}\\ \text{connection}\end{array}\right)}{\frac{1{\cdot}55 \times (50 + V)}{RT}} \qquad \left[n = \frac{pV}{RT}\right]$$

$$\therefore \quad V = 643{\cdot}5 \text{ litres.}$$

Ex. 15. *A spherical balloon of* 21 cm *diameter is to be filled up with hydrogen at* NTP *from a cylinder containing the gas at* 20 *atmosphere at* 27°C. *If the cylinder can hold* 2·82 *litres of water, calculate the number of balloons that can be filled up.* **(IIT 1987)**

Solution : Volume of the balloon $= \frac{4}{3}\pi r^3$

$$= \frac{4}{3} \times \frac{22}{7} \times (10{\cdot}5)^3 \text{ cc} \qquad \left(\text{radius} = \frac{21}{2} = 10{\cdot}5\right)$$

$$= 4851 \text{ cc.}$$

Balloon is at a temperature of 273 K and 1 atm pressure, i.e.,

$$\text{pressure} = 1 \times 76 \times 13{\cdot}6 \times 981 = 1{\cdot}014 \times 10^6 \text{ dynes/cm}^2.$$

$\therefore$ no. of moles of H_2, the balloon can contain at NTP $= \frac{pV}{RT}$

$$= \frac{1{\cdot}014 \times 10^6 \times 4851}{8{\cdot}314 \times 10^7 \times 273} = 0{\cdot}2167.$$

No. of moles in the cylinder $= \frac{(20 \times 76 \times 13{\cdot}6 \times 981) \times 2820}{8{\cdot}314 \times 10^7 \times 300}$

$$= 2{\cdot}2929.$$

While filling the last balloon, when the pressure of the cylinder will drop to 1 atm, gas cannot be withdrawn.

Now, no. of moles of H_2 remaining in the cylinder unused

$$= \frac{1{\cdot}014 \times 10^6 \times 2820}{8{\cdot}314 \times 10^7 \times 300} = 0{\cdot}1146$$

$\therefore$ no. of balloons that can be filled

$$= \frac{\text{no. of moles of } H_2 \text{ in the cylinder that can be used}}{\text{no. of moles of } H_2 \text{ one balloon can contain}}$$

$$= \frac{2{\cdot}2929 - 0{\cdot}1146}{0{\cdot}2167} = 10.$$

Ex. 16. *Two bulbs of equal volumes connected through a stopcock, contained* 0·7 *mole of* H_2 *gas at* 0·5 atm *pressure and* 27°C *(at the open position of the stopcock). If the first bulb was heated to* 127°C, *keeping the other at the same temperature, i.e.,* 27°C, *what will be the final pressure and moles in each bulb?*

Solution : Let the volume of each bulb be V litres.

For the two connected bulbs: $p = 0{\cdot}5$ atm, $n = 0{\cdot}7$ mole,

$$T = 27 + 273 = 300 \text{ K and vol.} = 2V.$$

We have, $pV = nRT$

$$0{\cdot}5\,(2V) = 0{\cdot}7 \times 0{\cdot}0821 \times 300$$

$$V = 17{\cdot}22 \text{ litres.}$$

When one of the bulbs is maintained at 127°C, i.e., 400 K and the other at 300 K, let the moles of H_2 in these bulbs be n_1 and n_2 respectively.

$\therefore \quad n_1 + n_2 = 0{\cdot}7 \quad \ldots (1)$

Since stopcock is open, the pressure in each bulb will be the same. Let it be p atm. Thus for the bulb at 400 K,

$$p \times V = n_1 \times 0{\cdot}0821 \times 400$$

or $\quad 17{\cdot}22\, p = 32{\cdot}8\, n_1$

or $\quad p = 1{\cdot}90\, n_1 \quad \ldots (2)$

And for the second bulb at 300 K,

$$p \times 17{\cdot}22 = n_2 \times 0{\cdot}0821 \times 300$$

$$p = 1{\cdot}42\, n_2. \quad \ldots (3)$$

From equations (1), (2) and (3), we get,

$p = 0{\cdot}571$ atm, $n_1 = 0{\cdot}3$ mole, $n_2 = 0{\cdot}4$ mole.

Ex. 17. *Two gases in adjoining vessels were brought into contact by opening a stopcock between them. One vessel measured* 0·25 *litre and contained* NO *at* 800 torr *and* 220 K; *the other measured* 0·1 *litre and contained* O_2 *at* 600 torr *and* 220 K. *The reaction to form* N_2O_4 (*solid*) *exhausts the limiting reactant completely.* (a) *Neglecting the vapour pressure of* N_2O_4 *what is the pressure of the gas remaining at* 220 K *after completion of the reaction?* (b) *What weight of* N_2O_4 *is formed?* (torr = mm)

Solution **:** Let us first calculate no. of moles of NO and O_2 before the reaction takes place. Let n_1 and n_2 be the no. of moles of NO and O_2 respectively in each vessel.

We have $pV = nRT$

For NO: $\dfrac{800}{760} \times 0{\cdot}25 = n_1 \times RT; \qquad n_1 = \dfrac{200}{760RT}.$

For O_2: $\dfrac{600}{760} \times 0{\cdot}10 = n_2 \times RT; \qquad n_2 = \dfrac{60}{760RT}.$

Since NO and O_2 react in 2 : 1 molar ratio

$$\left(\underset{(g)}{2NO} + \underset{(g)}{O_2} \rightarrow \underset{(s)}{N_2O_4}\right)$$

$\therefore \quad \dfrac{60}{760\,RT}$ mole of O_2 will react with $\dfrac{120}{760RT}$ mole of NO and so only $\left(\dfrac{200}{760RT} - \dfrac{120}{760RT}\right)$ mole, i.e., $\dfrac{80}{760RT}$ mole of NO shall remain after the completion of reaction. The pressure due to remaining NO can thus be calculated as:

$$pV = nRT$$

$$p(0{\cdot}25 + 0{\cdot}1) = \frac{80}{760RT} \times RT; \; p = 0{\cdot}30 \text{ atm or } 229 \text{ mm}.$$

Further, we know that $\frac{120}{760RT}$ mole of NO completely changed to N_2O_4.

$\therefore$ applying POAC for N atoms in NO $\xrightarrow{O_2} N_2O_4$

$$1 \times \text{moles of NO} = 2 \times \text{moles of } N_2O_4$$

$$\frac{120}{760RT} = 2 \times \frac{\text{wt. of } N_2O_4}{92}; \; \text{wt. of } N_2O_4 = {\cdot}402 \text{ g}.$$

$$\begin{Bmatrix} R = 0{\cdot}082 \text{ lit} \cdot \text{atm K}^{-1} \text{ mol}^{-1} \\ T = 220 \text{ K} \end{Bmatrix}.$$

Ex. 18. *The density of a mixture of* O_2 *and* N_2 *at* NTP *is* 1·3 g/L. *Calculate partial pressure of* O_2.

Solution : Mol. wt. of the mixture = density (g/L) × molar volume (L)

$$= 1{\cdot}3 \times 22{\cdot}4 = 29{\cdot}12.$$

Now let the no. of moles of O_2 and N_2 be n_1 and n_2 respectively.

$$\therefore \text{ mol. wt. of the mixture} = \frac{32n_1 + 28n_2}{n_1 + n_2}.$$

$$\therefore \quad \frac{32n_1 + 28n_2}{n_1 + n_2} = 29{\cdot}12$$

from which, we get mole fraction of $O_2 = \frac{n_1}{n_1 + n_2} = 0{\cdot}28.$

At NTP, $p = 1$ atm.

$$\therefore \text{ partial pressure of } O_2 = \frac{n_1}{n_1 + n_2} \times p = 0{\cdot}28 \times 1 = 0{\cdot}28 \text{ atm}.$$

Ex. 19. *Two gases A and B having molecular weights* 60 *and* 45 *respectively are enclosed in a vessel. The weight of A is* 0·5 g *and that of B is* 0·2 g. *The total pressure of the mixture is* 750 mm. *Calculate the partial pressure of the two gases.*

Solution : Mole of $A = \frac{0{\cdot}50}{60} = 0{\cdot}0083.$

Mole of $B = \frac{0{\cdot}20}{45} = 0{\cdot}0044.$

Total mole $= 0{\cdot}0127.$

Total pressure $= 750$ mm.

$$\text{Partial pressure of } A = \frac{\text{moles of } A}{\text{total moles}} \times \text{total pressure}$$

$$= \frac{0{\cdot}0083}{0{\cdot}0127} \times 750 = 490 \text{ mm.}$$

$$\text{Partial pressure of } B = \frac{0{\cdot}0044}{0{\cdot}0127} \times 750 = 260 \text{ mm.}$$

Ex. 20. *When* 2·0 g *of a gas A is introduced into an evacuated flask kept at* 25°C, *the pressure is found to be* 1 atm. *If* 3 g *of another gas B is then added to the same flask, the total pressure becomes* 1·5 atm. *Assuming ideal gas behaviour, calculate the ratio of molecular weights* $M_A : M_B$.

Solution : Total pressure = 1·5 atm.

$$\text{Moles of } A = \frac{2}{M_A}.$$

$$\text{Moles of } B = \frac{3}{M_B}.$$

$$\text{Total moles} = \frac{2}{M_A} + \frac{3}{M_B}.$$

$$\text{Partial pressure of } A = \frac{2/M_A}{2/M_A + 3/M_B} \times 1{\cdot}5$$

$$\text{and partial pressure of } B = \frac{3/M_B}{2/M_A + 3/M_B} \times 1{\cdot}5.$$

As given, partial pressure of $A = 1$ atm.

Partial pressure of $B = 1{\cdot}5 - 1 = 0{\cdot}5$ atm.

$$\text{Thus,} \quad \frac{\dfrac{2/M_A}{2/M_A + 3/M_B} \times 1{\cdot}5}{\dfrac{3/M_B}{2/M_A + 3/M_B} \times 1{\cdot}5} = \frac{1}{0{\cdot}5} = 2$$

$$\text{or} \quad \frac{2}{3} \times \frac{M_B}{M_A} = 2 \quad \text{or} \quad \frac{M_A}{M_B} = \frac{1}{3}.$$

Ex. 21. *A long rectangular box is filled with chlorine (at. wt.:* 35·45) *which is known to contain only* ^{35}Cl *and* ^{37}Cl. *If the box could be divided by a partition and the two types of chlorine molecules put in the two compartments respectively, calculate where should the partition be made if the pressure on both sides are to be equal. Is this pressure the same as the original pressure?*

Solution : Since 34·45 is the average atomic weight of ^{35}Cl (at. wt. = 35) and ^{37}Cl (at. wt. = 37) we have,

$$\frac{35n_1 + 37n_2}{n_1 + n_2} = 35{\cdot}45,$$

where n_1 and n_2 are the number of moles of ^{35}Cl and ^{37}Cl respectively.

$\therefore \quad \frac{n_1}{n_2} = 3{\cdot}44; n_1 : n_2 = 3{\cdot}44 : 1.$

$\therefore$ ratio of lengths is 3·44 : 1.

Since the pressures on both the sides of the partition is equal, the pressure before and after partition will be same (no. of moles per unit volume being same).

Ex. 22. *A mixture of* H_2O *vapour,* CO_2 *and* N_2 *was trapped in a glass apparatus with a volume of* 0·731 mL. *The pressure of the total mixture was* 1·74 mmHg *at* 27°C. *The sample was transferred to a bulb in contact with dry ice* (– 75°C) *so that* H_2O *vapour was frozen out. When the sample was returned to the measured volume, the pressure was* 1·32 mmHg. *The sample was then transferred to a bulb in contact with liquid nitrogen* (–195°C) *to freeze out the* CO_2 . *In the measured volume, the pressure was* 0·53 mmHg. *How many moles of each constituent are in the mixture?*

Solution : $p_{H_2O} + p_{CO_2} + p_{N_2} = 1{\cdot}74 \text{ mm}$

$$p_{CO_2} + p_{N_2} = 1{\cdot}32 \text{ mm}$$

$$p_{N_2} = 0{\cdot}53 \text{ mm}$$

$\therefore \qquad p_{CO_2} = 1{\cdot}32 - 0{\cdot}53 = 0{\cdot}79 \text{ mm}$

and $\qquad p_{H_2O} = 1{\cdot}74 - 1{\cdot}32 = 0{\cdot}42 \text{ mm}.$

Number of moles of each constituent is calculated using the equation

$$pV = nRT.$$

For H_2O: $\quad p_{H_2O} = \frac{0{\cdot}42}{760} \text{ atm}, V = \frac{0{\cdot}731}{1000} \text{ lit.}$

$$T = 27 + 273 = 300 \text{ K}; \; R = 0{\cdot}082 \text{ lit} \cdot \text{atm/K/mole}$$

$$\frac{0{\cdot}42}{760} \times \frac{0{\cdot}731}{1000} = n_{H_2O} \times 0{\cdot}082 \times 300$$

$$n_{H_2O} = 1{\cdot}64 \times 10^{-8}.$$

Similarly,

for CO_2: $\quad \frac{0{\cdot}79}{760} \times \frac{0{\cdot}731}{1000} = n_{CO_2} \times 0{\cdot}082 \times 300$

$$n_{CO_2} = 3{\cdot}08 \times 10^{-8}$$

and for N_2: $\quad \frac{0{\cdot}53}{760} \times \frac{0{\cdot}731}{1000} = n_{N_2} \times 0{\cdot}082 \times 300$

$$n_{N_2} = 2{\cdot}07 \times 10^{-8}.$$

Ex. 23. *A mixture of nitrogen and water vapours is admitted to a flask which contains a solid drying agent. Immediately after admission, the pressure of the flask is 760 mm. After some hours the pressure reached a steady value of 745 mm.*

(a) Calculate the composition, in mole per cent, of the original mixture.

(b) If the experiment is done at 20°C and the drying agent increases in weight by 0·15 g, what is the volume of the flask? (The volume occupied by the drying agent may be ignored.)

Solution : (a) $p_{N_2} + p_{H_2O} = 760$ mm

$$p_{N_2} = 745 \text{ mm}$$

$$p_{H_2O} = 760 - 745 = 15 \text{ mm}.$$

Mole % of N_2 = pressure % of N_2

$$= \frac{745}{760} \times 100 = 98{\cdot}03\%.$$

$\therefore$ mole % of $H_2O = 100 - 98{\cdot}03 = 1{\cdot}975\%$.

(b) Since the weight of drying agent increases by 0·15 g,

weight of $H_2O = 0{\cdot}15$ g.

Mol. wt. of $H_2O = 18$.

$$\therefore \quad \text{mole of } H_2O = \frac{0{\cdot}15}{18}$$

$$p_{H_2O} = 15 \text{ mm} = \frac{15}{760} \text{ atm.}$$

Applying $pV = nRT$ in the flask of volume V litres (say)

$$\frac{15}{760} \times V = \frac{0{\cdot}15}{18} \times 0{\cdot}082\,(273 + 20)$$

$$V = 10{\cdot}28 \text{ litres.}$$

Ex. 24. 2·69 g *of a sample of* PCl_5 *was placed in a 1-litre flask and completely vaporised to a temperature of* 250°C. *The pressure observed at this temperature was* 1 atm. *The possibility exists that some of* PCl_5 *may have dissociated according to the equation*

$$PCl_5\,(g) \rightleftharpoons PCl_3\,(g) + Cl_2\,(g).$$

What are partial pressures of PCl_5, PCl_3 *and* Cl_2 *under these experimental conditions?*

Solution : Let us first calculate the pressure supposing PCl_5 does not undergo dissociation.

$$pV = nRT$$

$$p \times 1 = \frac{2 \cdot 69}{208} \times 0 \cdot 082 \times 523.$$

[mol. wt. of $PCl_5 = 208$, $R = 0{\cdot}082$ lit · atm/K/mole, $T = (273 + 250)$ K]

$$p = 0 \cdot 553 \text{ atm.}$$

But PCl_5 undergoes dissociation in the following way,

Initial moles: a (say) 0 0

$$PCl_5 \rightleftharpoons PCl_3 + Cl_2 \quad (\alpha \equiv \text{deg. of dissociation})$$

Moles at eqb.: $a(1-\alpha)$ $a\alpha$ $a\alpha$

$\therefore$ total no. of moles $= a(1-\alpha) + a\alpha + a\alpha = a(1+\alpha)$

$\because$ pressure of a gas is proportional to no. of moles

$$\therefore \quad \frac{\text{moles before diss.}}{\text{moles after diss.}} = \frac{a}{a(1+\alpha)} = \frac{0 \cdot 553}{1}$$

or $\alpha = 0{\cdot}81.$

$$\text{Partial pressure of } PCl_5 = \frac{\text{moles of } PCl_5 \text{ at eqb.}}{\text{total moles}} \times \text{total pressure}$$

$$= \frac{a(1-\alpha)}{a(1+\alpha)} \times 1 = \frac{1 - 0 \cdot 81}{1 + 0 \cdot 81} = 0 \cdot 104 \text{ atm.}$$

Partial pressure of PCl_5 = partial pressure of Cl_2

$$= \frac{\text{moles of } PCl_3 \text{ or } Cl_2}{\text{total mole}} \times \text{total pressure}$$

$$= \frac{a\alpha}{a(1+\alpha)} \times 1 = \frac{0 \cdot 81}{1 \cdot 81} = 0 \cdot 447 \text{ atm.}$$

Ex. 25. *Helium is contained at* 30·2°C *in the system as shown in the figure. The levelling bulb (L) can be raised so as to fill the lower bulb with mercury and force the gas into the upper part of the device. The volume of bulb A to the mark 'a' is* 100·5 cm³ *and the volume of bulb B between the marks 'a' and 'b' is* 110 cm³. *The pressure exerted by the helium is measured by the difference between the mercury levels in the device and in the evacuated arm of the manometer. When mercury level is at 'b' the pressure is* 20·14 mmHg. *What is the mass of helium in the container?*

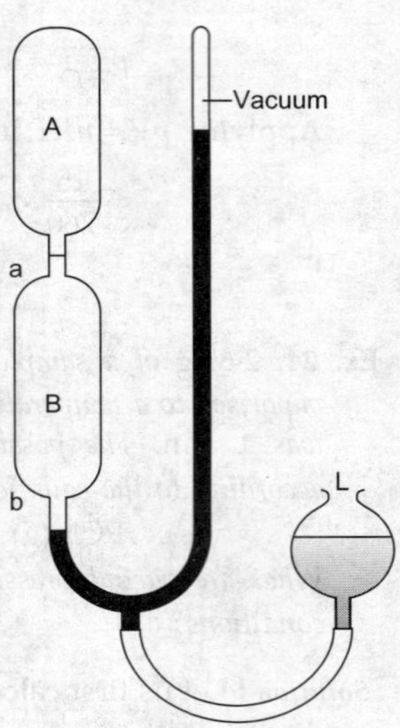

Solution : $pV = nRT$ or

$$pV = \frac{\text{wt. of He in g}}{\text{mol. wt. of He}} \times RT$$

$$\frac{20{\cdot}14}{760} \times \frac{(110 + 100{\cdot}5)}{1000}$$
(atm) (litre)

$$= \frac{\text{wt. of He in g}}{4} \times 0{\cdot}082 \times (273 + 30{\cdot}2)$$

$\therefore$ wt. of helium $= 0{\cdot}000897$ g.

Ex. 26. *A vertical cylinder closed at both ends is divided into two parts by a frictionless piston, each part containing one mole of air. At temperature* 300 K, *the volume of the upper part is four times than that of the lower part. At what temperature will the volume of the upper part be three times than that of the lower part?*

Solution : At 300 K, the position of the piston in the cylinder of volume, say V, is represented as:

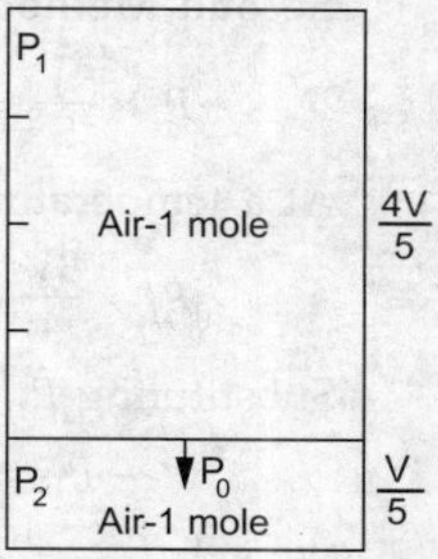

Let p_1 and p_2 be the pressure at the upper and lower part of the cylinder respectively. Let the pressure at the lower part due to the weight of the piston of the cylinder be p_0.

$$\therefore \quad p_2 = p_1 + p_0. \qquad \dots (1)$$

In the two parts of the cylinder, the no. of moles of air are same. As the temperature is also same (300 K),

we have, $$p_1 \times \frac{4V}{5} = p_2 \times \frac{V}{5};$$

or $$4p_1 = p_2 = p_1 + p_0 \,;\; p_1 = \frac{p_0}{3}. \qquad \dots (2)$$

Now let the temperature be T at which the volume of the upper part will be three times than that of the lower part.

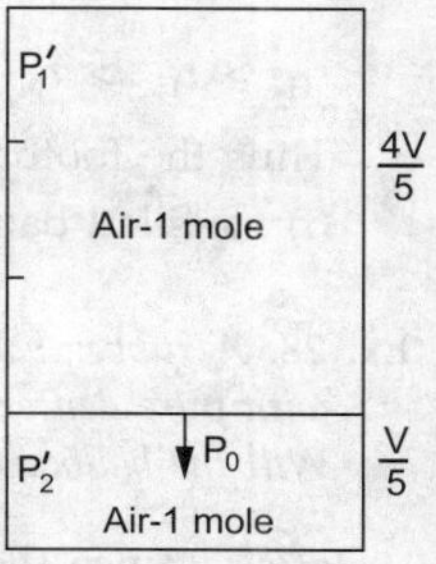

Let the pressures at the upper and lower parts be p'_1 and p'_2 respectively. Thus,

$$p'_2 = p'_1 + p_0. \qquad \dots (3)$$

Again in both the parts, temperature and no. of moles are the same, we have,

$$p'_1 \times \frac{3V}{4} = p'_2 \times \frac{V}{4},$$

or $$3p'_1 = p'_2 = p'_1 + p_0$$

or $$p'_1 = \frac{p_0}{2}. \qquad \dots (4)$$

From (2) and (4), we have, $p'_1 = \frac{3}{2} p_1$. $\qquad \dots (5)$

Now, for the upper part of the cylinder at temperature T,

we have, $p_1' \times \frac{3V}{4} = RT.$ $(n = 1)$

Substituting p_1' from (5), we have,

$$\frac{3}{2} p_1 \times \frac{3V}{4} = RT$$

$$\frac{9}{8} p_1 V = RT. \quad \dots (6)$$

Further,
for the upper part of the cylinder at 300 K,

we have, $p_1 \times \frac{4V}{5} = R\,(300)$ ... (7)

From (6) and (7), we get, $T = 421{\cdot}9$ K.

Second Method At 300 K,

$$P_1 \times \frac{4V}{5} = R \times 300 \text{ and } P_2 \times \frac{V}{5} = R \times 300$$

At a temperature T (say),

$$P_1' \times \frac{3V}{4} = RT \text{ and } P_2' \times \frac{V}{4} = RT$$

Substituting P_1, P_2, P_1' and P_2' from the above equations in

$$P_2 - P_1 = P_2' - P_1'$$

we get, $T = 421{\cdot}9$ K.

Ex. 27. *Three footballs are respectively filled with nitrogen, hydrogen and helium. In what order are these footballs to be reinflated?*

Solution : Since $r \propto \frac{1}{\sqrt{M}}$ ... (Eqn. 9)

$r_{H_2} > r_{He} > r_{N_2}$ as $M_{N_2} > M_{He} > M_{H_2}$

Thus the footballs are to be inflated in the following order:
(i) H_2-filled ball (ii) He-filled ball (iii) N_2-filled ball.

Ex. 28. *A rubber balloon, permeable to hydrogen in all its isotopic forms, is filled with pure deuterium* (D_2) *and then placed in a box containing pure hydrogen. Will the balloon expand or contract or remain as it is?*

Solution : Since the molecular weight of D_2 is greater than that of H_2, rate of diffusion of H_2 will be higher than that of D_2. Thus, the balloon will expand.

Ex. 29. 32 cc *of hydrogen diffuses through a fine hole in* 1 *minute. What volume of* CO_2 *will diffuse in* 1 *minute under the same conditions?*

Solution : We have,

$$\frac{V_1}{V_2} = \sqrt{\frac{M_2}{M_1}}$$

i.e.,
$$\frac{V_{H_2}}{V_{CO_2}} = \sqrt{\frac{M_{CO_2}}{M_{H_2}}} \quad \text{... (Eqn. 9)}$$

$$\frac{32}{V_{CO_2}} = \sqrt{\frac{44}{2}}$$

$$V_{CO_2} = 6{\cdot}82 \text{ cc.}$$

Ex. 30. *A tube with a porous wall allows* 0·53 *litre of* N_2 *to escape per minute from a pressure of* 1 atm *to an evacuated chamber. What will be the amount escaping under the same conditions for* He, CCl_4 *vapour and* UF_6*?*
(He = 4, N = 14, C = 12, Cl = 35·5, F = 19, U = 238)

Solution : Using Eqn. 9 for He,

$$\frac{V_{He}}{V_{N_2}} = \sqrt{\frac{M_{N_2}}{M_{He}}}\ ;\ \frac{V_{He}}{0{\cdot}53} = \sqrt{\frac{28}{4}} = \sqrt{7}$$

$$V_{He} = 0{\cdot}53 \times \sqrt{7} = 1{\cdot}40 \text{ litre per minute.}$$

Similarly, for CCl_4 vapour and UF_6,

$$\frac{V_{CCl_4}}{V_{N_2}} = \sqrt{\frac{M_{N_2}}{M_{CCl_4}}} = \sqrt{\frac{28}{152}}\ ;\ V_{CCl_4} = 0{\cdot}227 \text{ lit. per min.}$$

$$\frac{V_{UF_6}}{V_{N_2}} = \sqrt{\frac{M_{N_2}}{M_{UF_6}}} = \sqrt{\frac{28}{352}}\ ;\ V_{UF_6} = 0{\cdot}149 \text{ lit. per min.}$$

Ex. 31. *The reaction between gaseous* NH_3 *and* HBr *produces the white solid* NH_4Br. *Suppose that* NH_3 *and* HBr *are introduced simultaneously into the opposite ends of an open tube that is* 1 *metre long. Where would you expect the white solid to form?*

Solution : Suppose the two gases meet to form a white solid NH_4Br at a distance of r_1 cm from the NH_3 end. Thus from the HBr end, the distance will be $(100 - r_1)$ cm.

We have,

$$\frac{r_{NH_3}}{r_{HBr}} = \sqrt{\frac{M_{HBr}}{M_{NH_3}}} = \sqrt{\frac{81}{17}} \quad \text{... (Eqn. 9)}$$

Since the rate of diffusion is proportional to the distance, the molecules travel.

$$\therefore \qquad \frac{r_{NH_3}}{r_{HBr}} = \frac{r_1}{100 - r_1} = \sqrt{\frac{81}{17}} = 2{\cdot}18$$

$$\therefore \qquad r_1 = 65{\cdot}55 \text{ cm.}$$

Ex. 32. *At room temperature ammonia gas at* 1 atm *pressure and hydrogen chloride at p atmospheric pressure are allowed to diffuse through identical pin holes from opposite ends of a glass tube of* 1 *metre length and of uniform cross-section. Ammonium chloride is first formed at a distance of* 60 cm *from the end through which* HCl *gas is sent in. What is the value of p?*

Solution : In this problem the pressure under which the two gases are diffusing are different.

$$\frac{r_{NH_3}}{r_{HCl}} = \sqrt{\frac{M_{HCl}}{M_{NH_3}}} \times \frac{p_{NH_3}}{p_{HCl}} = \frac{40}{60}$$

or $$\sqrt{\frac{36{\cdot}5}{17}} \times \frac{1}{p_{HCl}} = \frac{40}{60}$$

$$p_{HCl} = 2{\cdot}198 \text{ atm.}$$

Ex. 33. *The composition of the equilibrium mixture for the equilibrium* $Cl_2 \rightleftharpoons 2Cl$ *at* 1470 K *may be determined by the rate of diffusion of the mixture through a pinhole. It is found that at* 1470 K, *the mixture diffuses* 1·16 *times as fast as krypton* (83·8) *diffuses under the same conditions. Find the degree of dissociation of* Cl_2 *at equilibrium.*

Solution : For the equilibrium,

	Cl_2	$\rightleftharpoons$	$2Cl$	
Initial moles :	1		0	
Moles at eqb.:	$1 - x$		$2x$	(x is the degree of dissociation)

Molecular weight of the mixture of Cl_2 and Cl at eqb., i.e., M_{mix} is calculated as

$$M_{mix} = \frac{(1-x)\,71 + (2x)\,35{\cdot}5}{(1-x) + 2x} = \frac{71}{1+x}.$$

Now we have,

$$\frac{r_{mix}}{r_{Kr}} = \sqrt{\frac{M_{Kr}}{M_{mix}}} = \sqrt{\frac{83{\cdot}8}{71/(1+x)}} = 1{\cdot}16. \qquad \text{... (Eqn. 9)}$$

$$\therefore \qquad x = 0{\cdot}14.$$

Ex. 34. 75 cc *of a gas was collected over mercury in a tube closed at the top by a porous plug. On standing in air for some time, and when the mercury level became constant again, the volume was found to be* 123 cc. *What is the molecular weight of the gas?* (1 *litre of air weighs* 1·293 g *at* NTP)

Solution : Molecular weight of air = weight of 1 mole of air

= wt. of 22·4 litres of air at NTP

$= 1{\cdot}293 \times 22{\cdot}4.$

$= 28{\cdot}96.$

From the given question it is clear that the time during which 75 cc of the gas diffuses out and the time during which 123 cc of air diffuses in are the same.

Thus
$$\frac{V_1}{V_2} = \sqrt{\frac{M_2}{M_1}} \qquad \text{... (Eqn. 9)}$$

$$\frac{75}{123} = \sqrt{\frac{28{\cdot}96}{M}}$$

$$M = 78.$$

Ex. 35. *A mixture of hydrogen and oxygen in the 2 : 1 volume ratio is allowed to diffuse through a porous partition. Calculate the composition of the gas coming out initially.*

Solution : We have,

$$\frac{r_{H_2}}{r_{O_2}} = \sqrt{\frac{M_{O_2}}{M_{H_2}}} = \sqrt{\frac{32}{2}} = 4.$$

Now since volume ratio of H_2 and O_2 in the initial stage is 2 : 1 (given) and the ratio of their rates is 4 : 1, the overall volume ratio of the gases diffusing through the porous partition at the initial stage will be 8 : 1.

Ex. 36. *A gaseous mixture of* O_2 *and* X *containing* 20% (mole %) *of* X, *diffused through a small hole in* 234 *seconds while pure* O_2 *takes* 224 *seconds to diffuse through the same hole. Find molecular weight of* X.

Solution : We have,

$$\frac{t_{mix}}{t_{O_2}} = \sqrt{\frac{M_{mix}}{M_{O_2}}} \qquad \text{... (Eqn. 9)}$$

$$\frac{234}{224} = \sqrt{\frac{M_{mix}}{32}};$$

$$\therefore \qquad M_{mix} = 34{\cdot}921.$$

As the mixture contains 20% (mole %) of X, the molar ratio of O_2 and X may be represented as $0{\cdot}8n : 0{\cdot}2n$, n being the total no. of moles.

$$\therefore \qquad M_{mix} = \frac{32 \times 0{\cdot}8n + M_x \times 0{\cdot}2n}{n} = 34{\cdot}921$$

$\therefore M_x$ (mol. wt. of X) = 46·6.

Ex. 37. *The pressure in a bulb dropped from* 2000 mm *to* 1500 mmHg *in* 47 *minutes when the contained oxygen leaked through a small hole. The bulb was then completely evacuated. A mixture of oxygen and another gas of molecular weight* 79 *in the molar ratio* 1 : 1 *at a total pressure of* 4000 mmHg *was introduced. Find the molar ratio of the two gases remaining in the bulb after a period of* 74 *minutes.*

Solution : Suppose p_{O_2} and p be the pressure-drops per minute for O_2 and an unknown gas x (say) respectively.

$$\therefore \quad p_{O_2} = \frac{2000 - 1500}{47} = 10{\cdot}64 \text{ mm/min.}$$

$$\therefore \quad \frac{r_{O_2}}{r_x} = \frac{p_{O_2}}{p_x} = \sqrt{\frac{M_x}{M_{O_2}}} \qquad \text{... (Eqn. 10)}$$

$$\frac{10{\cdot}64}{P_x} = \sqrt{\frac{79}{32}}\ ; \ p_x = 6{\cdot}77 \text{ mm/min.}$$

Since the bulb is now refilled with equal number of moles of O_2 and x, the partial pressures of each gas will be 2000 mm as the total pressure is 4000 mm. ... (Eqn. 6)

$\therefore$ pressure of O_2 after 75 min

= partial press. of O_2 – press. drop in 74 min

= 2000 – (10·64 × 74)

= 1212·64 mm

and pressure of x after 74 min

= partial press. of x – press. drop after 74 min

= 2000 – (6·77 × 74)

= 1499·02 mm.

We thus apply Eqn. 10.

$$\frac{\text{moles of } O_2 \text{ left after 74 min}}{\text{moles of } x \text{ left after 74 min}} = \frac{\text{pressure of } O_2 \text{ after 74 min}}{\text{pressure of } x \text{ after 74 min}}$$

$$= \frac{1212{\cdot}64}{1499{\cdot}02} = 0{\cdot}8089.$$

Hence, molar ratio of O_2 and x after 74 minutes is 0·8089 : 1.

Ex. 38. *What would the molecular weight of a gas have to be if the pressure of the gas is to fall to one-half of its value in a vertical distance of* 1 *metre?* (T = 25°C). *What type of molecules have molecular weight of this magnitude?*

Solution : We have,

$$2{\cdot}303 \log \frac{p}{p_0} = -\frac{Mgh}{RT} \qquad \text{... (Eqn. 11)}$$

$p = P$ (suppose)

$p_0 = 2P$

$g = 981\ \text{cm/s}^2$

$h = 100$ cm

$R = 8{\cdot}314 \times 10^7\ \text{erg} \cdot \text{K}^{-1}\ \text{mole}^{-1}$

$T = 25 + 273 = 298$ K.

$$\therefore \qquad 2{\cdot}303 \log \frac{P}{2P} = \frac{M \times 981 \times 100}{8{\cdot}314 \times 10^7 \times 298}$$

$$M = 175000.$$

Molecules with such a high value of molecular weight must be those of a polymer.

Ex. 39. *Calculate rms speed of oxygen molecules in the lungs at normal body temperature,* 37°C.

Solution **:** We have,

$$C = \sqrt{\frac{3RT}{M}} \qquad \text{... [Eqn. (12 d)]}$$

$$= \sqrt{\frac{3 \times 8{\cdot}314 \times 10^7 \times 310}{32}} = 4{\cdot}92 \times 10^4\ \text{cm/s}.$$

Ex. 40. *Calculate the temperature at which the root-mean-square speed of* SO_2 *molecules is the same as that of oxygen at* 27°C.

Solution : We have,

$$C = \sqrt{\frac{3RT_{SO_2}}{M_{SO_2}}} = \sqrt{\frac{3RT_{O_2}}{M_{O_2}}}$$

or
$$\sqrt{\frac{3RT_{SO_2}}{64}} = \sqrt{\frac{3R \cdot 300}{32}}$$

$$T_{SO_2} = 600\ \text{K}.$$

Ex. 41. *Calculate the rms speed of ozone kept in a closed vessel at* 20°C *and* 82 cmHg *pressure.* **(IIT 1985)**

Solution **:** Volume occupied by 1 mole of O_3 at 20°C and 82 cm pressure

$$= 22400 \times \frac{293}{273} \times \frac{76}{82} = 22282\ \text{cc} \qquad \text{... (Eqn. 3 b)}$$

$$p = 82 \times 13{\cdot}6 \times 981\ \text{dynes/cm}^2.$$

Now we have,

$$C=\sqrt{\frac{3pV}{M}} \qquad \text{... (Eqn. 12 c)}$$

$$=\sqrt{\frac{3\times 82\times 13{\cdot}6\times 981\times 22282}{48}}$$

$$=3{\cdot}9\times 10^4 \text{ cm/s.}$$

[**Note:** The other direct method is by the use of Eqn. 12 d].

Ex. 42. *A gas has a density of* 1·2504 g/L *at* 0°C *and a pressure of* 1 atm. *Calculate the rms, average and the most probable speeds of its molecules at* 0°C.

Solution : $p=76\times 13{\cdot}6\times 981$ dynes/cm^2

$d=1{\cdot}2504$ g/L $=0{\cdot}0012504$ g/cc

Now we have,

$$C=\sqrt{\frac{3P}{d}} \qquad \text{... (Eqn. 12 b)}$$

$$=\sqrt{\frac{3\times 76\times 13{\cdot}6\times 981}{0{\cdot}0012504}}=4{\cdot}93\times 10^4 \text{ cm/s.}$$

∴ average speed $=0{\cdot}9211\times 4{\cdot}93\times 10^4$

$=4{\cdot}59\times 10^4$ cm/s

and most probable speed $=0{\cdot}8165\times 4{\cdot}93\times 10^4$

$=4{\cdot}03\times 10^4$ cm/s.

[rms speed : av. speed : m.p. speed = 1 : 0·9211 : 0·8165]

Ex. 43. *Calculate the rms speed in* cm/s *at* 25°C *of a free electron and of a molecule of* UF_6. (H = 1, U = 238, F = 19)

Solution : Mass of electron on atomic wt. scale is 1/1837 amu

∴ mass of 1 mole of electron $=\frac{1}{1837}$ g.

$R=8{\cdot}314\times 10^7$ ergs/K/mole

$T=273+25=298$ K.

We have,

$$C=\sqrt{\frac{3RT}{M}}=\sqrt{\frac{3\times 8{\cdot}314\times 10^7\times 298}{1/1837}} \qquad \text{... (Eqn. 12 d)}$$

∴ rms speed of an electron = $1{\cdot}16\times 10^7$ cm/s.

To calculate rms spd. of UF_6, put $M=(238+6\times 19)$ in Eqn. 12 d.

Ex. 44. *The average speed at* T_1 K *and the most probable speed at* T_2 K *of* CO_2 *gas is* 9×10^4 cm s^{-1}. *Calculate the value of* T_1 *and* T_2. **(IIT 1990)**

Solution : We know, for 1 mole of an ideal gas,

$$\text{rms speed} = \sqrt{\frac{3RT}{M}} \qquad \text{... (Eqn. 12 d)}$$

and rms speed : average speed : most probable speed

$$= 1 : 0{\cdot}9211 : 0{\cdot}8165.$$

$$\therefore \quad \text{average speed at } T_1 \text{ K} = 0{\cdot}9211 \times \sqrt{\frac{3RT_1}{M}} = 9 \times 10^4 \qquad \text{... (1)}$$

$$\text{and most probable speed at } T_2 \text{ K} = 0{\cdot}8165 \times \sqrt{\frac{3RT_2}{M}} = 9 \times 10^4. \qquad \text{... (2)}$$

Substituting $R = 8{\cdot}314 \times 10^7$ ergs/K/mole and $M = 44$ in (1) and (2), we get,

$T_1 = 1684$ K and

$T_2 = 2143$ K.

Ex. 45. *The kinetic molecular theory attributes an average kinetic energy of* $\frac{3}{2}kT$ *to each particle. What rms speed would a mist particle of mass* 10^{-12} g *have at room temperature* (27°C) *according to the kinetic molecular theory?*

Solution : KE per molecule $= \frac{3}{2}kT$

$$= \frac{3}{2} \cdot \frac{R}{N} \cdot T.$$

If the mass of one molecule is m then KE of this molecule

$$= \frac{1}{2}mC^2, \text{ where C is the rms speed.}$$

$$\therefore \qquad \frac{1}{2}mC^2 = \frac{3}{2}\ \frac{R}{N} \cdot T$$

$$\text{or} \qquad C = \sqrt{3 \times \frac{R}{N} \times \frac{T}{m}}$$

$$= \sqrt{\frac{3 \times 8{\cdot}314 \times 10^7 \times 300}{6{\cdot}022 \times 10^{23} \times 10^{-12}}} = 0{\cdot}35 \text{ cm/s}.$$

Ex. 46. *Calculate rms speed of* N_2 *at* 298 K *in* metre/second.

Solution : In SI units,

$R = 8{\cdot}314$ J/K/mol, $T = 298$ K and $M = 0{\cdot}028$ kg/mol

$$\therefore \qquad C = \sqrt{\frac{3RT}{M}}$$

$$= \sqrt{\frac{3 \times 8{\cdot}314 \times 298}{0{\cdot}028}} = 515{\cdot}2 \text{ m/s.}$$

Ex. 47. *Calculate rms speed of* O_2 *at* 273 K *and* 1×10^5 Pa *(pascal) pressure. The density of* O_2 *under these conditions is* $1{\cdot}42 \text{ kg/m}^3$.

Solution : Data are given in SI units.

$$\therefore \qquad C = \sqrt{\frac{3p}{d}} = \sqrt{\frac{3 \times 10^5}{1{\cdot}42}} = 459{\cdot}63 \text{ m/s.}$$

Ex. 48. *Calculate the pressure in* atm, *exerted by* 10^{23} *gas particles each of mass* 10^{-22} g *in a container of volume* 1 *litre. The rms speed is* 10^5 cm/s. *What is the total kinetic energy (in cal) of these particles? What must be the temperature?*

Solution : We have,

$$pV = \frac{1}{3} mnC^2 \qquad \text{... (12)}$$

$$p \times 1000 = \frac{1}{3} \times 10^{-22} \times 10^{23} \times 10^5 \times 10^5$$

$$p = \frac{10^8}{3} \text{ dynes per cm}^2.$$

We know, $\quad 1 \text{ atm} = 76 \text{ cm} = (76 \times 13{\cdot}6 \times 981) \text{ dynes/cm}^2$

$$\therefore \qquad p = \frac{10^8}{3} \times \frac{1}{76 \times 13{\cdot}6 \times 981} = 33 \text{ atm.}$$

Total KE of molecules $= \frac{1}{2} \times (\text{total mass}) \times C^2$

$$= \frac{1}{2} \times 10^{-22} \times 10^{23} \times (10^5)^2 \text{ ergs}$$

$$= \frac{1}{2} \times \frac{10^{-22} \times 10^{23} \times 10^{10}}{4{\cdot}184 \times 10^7} = 1195 \text{ cal.}$$

$$(1 \text{ cal} = 4{\cdot}184 \times 10^7 \text{ ergs})$$

Now, number of moles $= \frac{10^{23}}{\text{Av. const.}} = \frac{10^{23}}{6{\cdot}022 \times 10^{23}}$

$$= \frac{1}{6{\cdot}022}. \qquad \text{... (Rule 4, Chapter 1)}$$

Again we have,

$$\text{KE} = \frac{3}{2} nRT \qquad \text{... (Eqn. 12 c)}$$

$$1195 = \frac{3}{2} \times \frac{1}{6{\cdot}022} \times 2 \times T$$

$$T = 2398{\cdot}7 \text{ K}.$$

Ex. 49. *Calculate total energy of* 1 mole *of an ideal monoatomic gas at* 27°C.

Solution : $R = 2$ cal/deg/mole

$T = 273 + 27 = 300$ K.

KE of 1 mole of a gas $= \frac{3}{2} RT$ ... (Eqn. 13 a)

$= \frac{3}{2} \times 2 \times 300 = 900$ calories.

Ex. 50. *Calculate the average kinetic energy in joules of the molecules in* 8 g *of methane at* 27°C.

Solution : No. of moles of CH_4

$$n = \frac{\text{wt. in g}}{\text{mol. wt.}} = \frac{8}{16} = 0{\cdot}5$$

$R = 8{\cdot}314$ joules/K/mole

$T = 273 + 27° = 300$ K

KE for n moles $= \frac{3}{2} n\,RT = \frac{3}{2} \times 0{\cdot}5 \times 8{\cdot}314 \times 300$ $(n = 0{\cdot}5)$

$= 1870{\cdot}65$ joules.

Ex. 51. $C_p - C_v$ *for an ideal gas is* (... ? ...)

Solution : $C_p - C_v = R \approx 2$ calories.

Ex. 52. 5·6 *litres of an unknown gas at* NTP *requires* 12·5 *calories to raise its temperature by* 10°C *at constant volume. Calculate* C_v *for the gas and its atomicity.*

Solution : No. of moles $= \frac{5{\cdot}6}{22{\cdot}4} = \frac{1}{4}$. ... (Rule 3, Chapter 1)

∴ heat in calories required to raise the temperature of 1 mole of the gas by 10°C is $12{\cdot}5 \times 4 = 50$.

As C_v is defined as the heat required to raise the temperature of 1 mole of the gas through 1°C at constant volume,

$$C_v = \frac{50}{10} = 5 \text{ cal.}$$

∴ $C_p = 5 + 2 = 7$ cal. $(\because C_p - C_v = 2)$

∴ $\frac{C_p}{C_v} = \frac{7}{5} = 1{\cdot}4.$

The gas is thus diatomic.

Ex. 53. *At 627°C and 1 atm pressure, SO_3 undergoes partial dissociation into SO_2 and O_2*

$$SO_3 \rightleftharpoons SO_2 + \frac{1}{2}O_2$$

If the observed density of the equilibrium mixture is 0·925 g/L, calculate degree of dissociation of SO_3.

Solution : Let the initial no. of moles of SO_3 be 1 and its degree of dissociation, x.

1		0		0	Initial no. of moles
SO_3	$\rightleftharpoons$	SO_2	+	$\frac{1}{2}O_2$	
$1-x$		x		$x/2$	Moles at equilibrium

$\therefore$ total no. of moles at eqb. $= 1 - x + x + \frac{x}{2} = 1 + \frac{x}{2}$.

Thus applying $pV = nRT$

$$1 \times V = \left(1 + \frac{x}{2}\right) \times 0{\cdot}0821 \times (627 + 273)$$

$$V = \left(1 + \frac{x}{2}\right) \times 73{\cdot}89 \text{ litres.}$$

Now wt. of 1 mole of SO_3 = 80 g and therefore,

from the law of conservation of mass, we have,

wt. of gases at eqb. = 80 g.

$$\therefore \text{ density} = \frac{\text{wt. in g}}{\text{vol. in litres}} = \frac{80}{\left(1 + \frac{x}{2}\right) \times 73{\cdot}89} = 0{\cdot}925 \text{ (given)}$$

or $\quad x = 0{\cdot}34.$

Ex. 54. *A sample of O_2 gas initially at* NTP *is transferred from a 1-litre container to a 2-litre container at a constant temperature. What effect does this change have on*

(a) the average kinetic energy of O_2 molecules

(b) the average speed of O_2 molecules

(c) the rms speed of O_2 molecules

(d) the total number of collisions of O_2 molecules with the container walls in a unit time

Solution : (a) No change in average kinetic energy as the temperature remains constant.

(b) & (c) No change in average speed and rms speed as the average kinetic energy remains unchanged.

(d) Number of collisions per unit time with the container walls decrease, as due to increase in volume, the molecules move longer distances between collisions.

Ex. 55. *Consider the arrangement of bulbs shown below:*

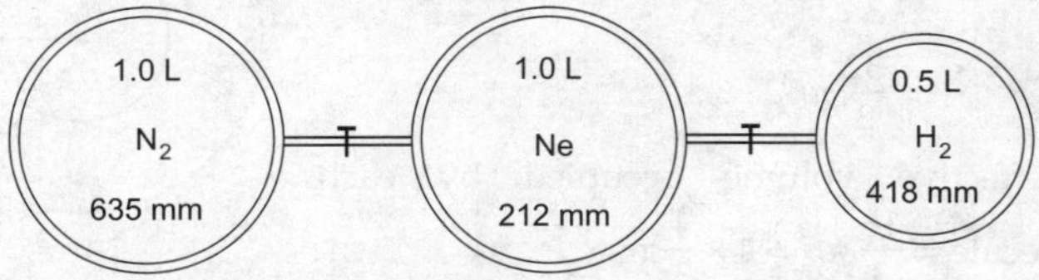

What is the pressure of the system when all the stopcocks are opened?

Solution : Total number of moles $= \frac{635}{760} \times \frac{1}{RT} + \frac{212}{760} \times \frac{1}{RT} + \frac{418}{760} \times \frac{\cdot 5}{RT}$.

Let the pressure of the gases be P atm when all the stopcocks are open. Applying ideal gas equation,

$$P(1 + 1 + \cdot 5) = \left(\frac{635}{760} \times \frac{1}{RT} + \frac{212}{760} \times \frac{1}{RT} + \frac{418}{760} \times \frac{\cdot 5}{RT}\right) RT; \; P = 0{\cdot}56 \text{ atm.}$$

Ex. 56. *A mixture of* CH_4 *and* C_2H_2 *occupies a certain volume at a total pressure of* 70·5 mmHg. *The sample is burnt, formed* CO_2 *and* H_2O. *The* H_2O *is removed and the remaining* CO_2 *is found to have a pressure of* 96·4 mmHg *at the same volume and temperature as the original mixture. What mole fraction of the gas was* C_2H_2*?*

Solution : Let the number of moles of CH_4 and C_2H_2 be x and y respectively.

$$CH_4 + C_2H_2 + O_2 \rightarrow CO_2 + H_2O$$

x moles $\;$ y moles

Applying POAC for C atoms,

$$1 \times \text{moles of } CH_2 + 2 \times \text{moles of } C_2H_2 = 1 \times \text{moles of } CO_2$$

$$x + 2y = \text{moles of } CO_2$$

As no. of moles α pressure at const. temperature and volume.

$$\frac{\text{no. of moles of } CH_4 \text{ and } C_2H_2}{\text{no. of moles of } CO_2} = \frac{70{\cdot}5}{96{\cdot}4}$$

or
$$\frac{x + y}{x + 2y} = \frac{70{\cdot}5}{96{\cdot}4}.$$

$$\therefore \quad \frac{y}{x + y} = \text{mole fraction of } C_2H_2 = 0{\cdot}368.$$

Ex. 57. *The radius of each molecule of a gas is r* cm. *What is the effective volume occupied by one molecule?*

Solution : Two molecules when touching from any direction will always have their centres separated by $2r$, where r is the radius of each molecule considering only bimolecular collisions,

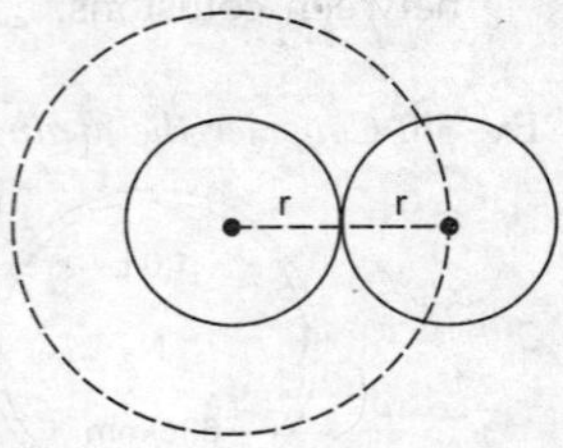

or, effective volume occupied by two molecules $= \frac{4}{3}\pi(2r)^3$

$= \frac{32}{3}\pi r^3.$

$\therefore$ effective volume occupied by each molecule $= \frac{16}{3}\pi r^3 = 4 \times \frac{4}{3}\pi r^3.$

Ex. 58. *The density of the vapour of a substance at* 1 atm *pressure and* 500 K *is* 0·36 kg m^{-3}. *The vapour effuses through a small hole at a rate of* 1·33 *times faster than oxygen under the same condition.*

(a) Determine (i) molecular weight (ii) molar volume (iii) compression factor (Z) of the vapour, and (iv) which forces among the gas molecules are dominating, the attractive or the repulsive?

(b) If the vapour behaves ideally at 1000 K, *determine the average transitional kinetic energy of a molecule.*

Solution : (a) (i) $\frac{\gamma_g}{\gamma_{O_2}} = \sqrt{\frac{M_{O_2}}{M_g}}$; $M_g = \frac{32}{1{\cdot}33^2} = 18{\cdot}09$ g/mole.

(ii) Molar volume $= \frac{\text{mol. wt. (g/mole)}}{\text{density (g/L)}}$

$= \frac{18{\cdot}09}{0{\cdot}36}$ $\quad (\text{kg m}^{-3} = \text{gL}^{-1})$

$= 50{\cdot}25$ L mole^{-1}.

(iii) $z = \frac{pV}{RT} = \frac{1 \times 50{\cdot}25}{{\cdot}0821 \times 500}$

$= 1{\cdot}224.$

(iv) As $z > 1$, repulsive force dominates.

(b) Av. KE per molecule $= \frac{3}{2}\frac{R}{N}T$

$= \frac{3}{2} \times \frac{8{\cdot}314}{6{\cdot}022 \times 10^{23}} \times 1000$

$= 2{\cdot}07 \times 10^{-20}$ J.

Ex. 59. *The compressibility factor for 1 mole of a van der Waals gas at 0°C and 100 atm pressure is found to be 0·5. Assuming that the volume of a gas molecule is negligible, calculate the van der Waals constant, a.* **(IIT 2001)**

Solution : For 1 mole of the gas,

$$Z = \frac{pV}{RT}$$

$$0{\cdot}5 = \frac{100 \times V}{{\cdot}0821 \times 273};\ V = 0{\cdot}112 \text{ L}$$

Neglecting b, van der Waals equation reduces to $\left(P + \frac{a}{V^2}\right)V = RT$

or $\quad pV + \frac{a}{V} = RT$

or $\quad 100 \times 0{\cdot}112 + \frac{a}{0{\cdot}112} = {\cdot}0821 \times 273$

$$a = 1{\cdot}25 \text{ L}^2\text{atm mol}^{-2}.$$

Ex. 60. *A vertical hollow cylinder of height* 1·52 m *is fitted with a movable piston of negligible mass and thickness. The lower half of the cylinder contains an ideal gas and the upper half is filled with mercury. The cylinder is initially at* 300 K. *When the temperature is raised, half of the mercury comes out of the cylinder. Find the temperature, assuming the thermal expansion of mercury to be negligible.*

Solution : Initially at 300 K, length of Hg column $= \frac{152}{2} = 76$ cm = 1 atm.

Let the volume of the cylinder be V litres

Applying ideal gas equation: $(1 + 1) \cdot \frac{V}{2} = nR{\cdot}300.$ ... (1)

Finally, say at a temp. T: length of Hg column $= \frac{76}{2}$ cm $= 0{\cdot}5$ atm.

$\therefore \quad (1 + 0{\cdot}5)\frac{3V}{4} = n\,RT.$ ... (2)

From equation (1) and (2), $T = 337{\cdot}5$ K.

Ex. 61. *What is the molar volume of* N_2 *at* 500 K *and* 600 atm *according to (a) ideal gas law (b) virial equation. The virial coefficient B of* $N_2(g)$ *at* 500 K = 0·0169 L mol^{-1}. *How do you interpret the results?*

Solution : (a) For 1 mole of an ideal gas

$$V = \frac{RT}{p} = \frac{0{\cdot}0821 \times 500}{600} = 6{\cdot}84 \times 10^{-2} \text{ L mol}^{-1}.$$

(b) Using virial equation, neglecting higher powers.

$$Z = \frac{pV}{RT} = 1 + \frac{B}{V} = 1 + \frac{BP}{RT}$$

$$Z = 1 + \frac{0{\cdot}0169 \times 600}{0{\cdot}0821 \times 500} = 1{\cdot}247.$$

For 1 mole of a real gas,

$$Z = \frac{pV}{RT}$$

or $$V = \frac{ZRT}{p} = \frac{1{\cdot}247 \times 0{\cdot}0821 \times 500}{600}$$

$$= 8{\cdot}53 \times 10^{-2} \text{ L mol}^{-1}.$$

The molar volume of the real gas is greater because of the finite volume of the molecules.

Ex. 62. *Calculate the compressibility factor,* $\frac{(p_{real})(V_{real})}{RT}$ *for a* 1·0*-mole sample of* NH_3*, under the following conditions: in a* 500 mL*-vessel at* –10·0°C *it exerts a pressure of* 30·0 atm. *What would be the ideal pressure for* 1·0*-mole of* NH_3 *at* –10·0°C *in a* 500-mL *vessel?*

Solution : Compressibility factor $(Z) = \frac{30 \times {\cdot}5}{{\cdot}0821 \times 263} = 0{\cdot}69.$

Applying ideal gas equation:

$$pV = nRT$$

$$p \times {\cdot}5 = 1 \times {\cdot}0821 \times 263$$

$$p = 43{\cdot}18 \text{ atm.}$$

Ex. 63. 2 *moles of ammonia occupied a volume of* 5 *litres at* 27°C. *Calculate the pressure if the gas obeyed van der Waals equation.*
($a = 4{\cdot}17$ atm lit^2 mole^{-2}, $b = 0{\cdot}0371$ lit mole^{-1})

Solution : $n = 2$

$V = 5$ litres

$T = 273 + 27 = 300$ K

$a = 4{\cdot}17$ atm lit^2 mole^{-2}

$b = 0{\cdot}0371$ lit mole^{-1}

$R = 0{\cdot}082$ lit atm deg^{-1} mole^{-1}

Applying van der Waals equation for n moles

$$\left(p + \frac{an^2}{V^2}\right)(V - n\,b) = n\,RT \qquad \text{... (Eqn. 17)}$$

$$\left(p+\frac{4\cdot17\times2^2}{5^2}\right)(5-2\times0\cdot0371)=2\times0\cdot082\times300$$

$$p=9\cdot33 \text{ atm.}$$

Ex. 64. *van der Waals constant 'b' for a real gas is* 0·02788 lit mole^{-1}. *Calculate the radius of the molecule of the gas.*

Solution : Volume of one molecule $=\frac{4}{3}\pi r^3$,

where r is the radius of the molecule.

$\therefore$ volume of 1 mole $=\frac{4}{3}\pi r^3\times$ Av. const.

(1 mole contains Av. const. of molecules)

We have, $$b=4\times\frac{4}{3}\pi r^3\times \text{Av. const.} \qquad \text{... (Eqn. 18)}$$

b = 0·02788 lit/mole = 27·88 mL/mole (given).

$$\therefore \quad 27\cdot88=4\times\frac{4}{3}\times\frac{22}{7}\times r^3\times6\cdot022\times10^{23}.$$

$$\therefore \quad r=1\cdot4\times10^{-8}\text{ cm.}$$

Ex. 65. *Calculate the weight of water in grams per litre of air at* 20°C *and* 45% *relative humidity. Vapour pressure of water at* 20°C *is* 17·5 mmHg.

Solution : We know that,

% relative humidity

$$=\frac{\text{partial pressure of } H_2O \text{ in air}}{\text{vapour pressure of } H_2O \text{ at the same temp.}}\times100.$$

$$\therefore \quad p_{H_2O}=\frac{45\times17\cdot5}{100}=7\cdot9\text{ mm}=0\cdot0104\text{ atm.}$$

Thus for 1 litre of air, mole of H_2O can be calculated as,

$$n=\frac{pV}{RT}=\frac{0\cdot0104\times1}{0\cdot0821\times293}=4\cdot3\times10^{-4}\text{ mole.}$$

$\therefore$ wt. of $H_2O=4\cdot3\times10^{-4}\times18=7\cdot74\times10^{-3}$ g/L of air.

Ex. 66. *One way of writing the equation of state for real gases is*

$$p\,\overline{V} = RT\left[1+\frac{B}{\overline{V}}+\ \ldots\right]$$

where B is constant. Derive an approximate expression for B in terms of the van der Waals constants, a and b.

Solution : The van der Waals equation for 1 mole of a gas is

$$\left(p+\frac{a}{\overline{V}^2}\right)(\overline{V}-b) = RT,$$

where $\overline{V}$ is the molar volume.

or $$\left(p+\frac{a}{\overline{V}^2}\right) = \frac{RT}{(\overline{V}-b)}$$

or $$\left(p\overline{V}+\frac{a}{\overline{V}}\right) = \frac{\overline{V}}{(\overline{V}-b)}RT$$

or $$\frac{p\overline{V}}{RT}+\frac{a}{\overline{V}RT} = \frac{\overline{V}}{\overline{V}-b}$$

$$\frac{p\overline{V}}{RT} = \frac{\overline{V}}{\overline{V}-b} - \frac{a}{RT\overline{V}}$$

$$= \frac{1}{\left(1-\frac{b}{\overline{V}}\right)} - \frac{a}{RT\overline{V}} = \left(1-\frac{b}{\overline{V}}\right)^{-1} - \frac{a}{RT\overline{V}}.$$

At low pressures, $\frac{b}{\overline{V}} \ll 1$ so that we can expand the first term using $(1-x)^{-1} = 1 + x + x^2 + \cdots$. This yields the virial equation in terms of volume:

$$\frac{p\overline{V}}{RT} = \left[1+\frac{b}{\overline{V}}+\left(\frac{b}{\overline{V}}\right)^2+\cdots\right]-\frac{a}{RT\overline{V}}$$

$$= 1 + \left(b-\frac{a}{RT}\right)\frac{1}{\overline{V}} + \left(\frac{b}{\overline{V}}\right)^2 + \cdots$$

or $$p\overline{V} = RT\left[1 + \left(b-\frac{a}{RT}\right)\frac{1}{\overline{V}} + \left(\frac{b}{\overline{V}}\right)^2 + \cdots\right].$$

Comparing with the given equation, we have

$$B = \left(b-\frac{a}{RT}\right).$$

Ex. 67. *Calculate the viscosity of molecular oxygen in* Pa s *at* 273 K *and* 1 bar. *Given that molecular diameter* $= 0{\cdot}36 \times 10^{-9}$ m, $k = 1{\cdot}38 \times 10^{-23}$ JK^{-1} *and Av. const.* $= 6{\cdot}022 \times 10^{23}$

Solution : Mass of one O_2 molecule, $m = \frac{32\times10^{-3}}{6{\cdot}022\times10^{23}}$

$$= 5{\cdot}314 \times 10^{-26} \text{ kg.}$$

We have,

$$\eta = \frac{5}{16\,\sigma^2}\left(\frac{mkT}{\pi}\right)^{1/2} \quad \text{... (Eqn 16)}$$

$$= \frac{5}{16 \times (0.36 \times 10^{-9})^2} \left(\frac{5.314 \times 10^{-26} \times 1.38 \times 10^{-23} \times 273}{22/7} \right)^{1/2}$$

$$= 1.926 \times 10^{-5} \text{ Pa s.}$$

PROBLEMS

(Answers bracketed with questions)

1. 1 g of helium gas is confined in a two-litre flask under a pressure of 2·05 atm. What is its temperature? (200 K)

2. The density of He is 0·1784 kg/m^3 at STP. If a given mass of He at STP is allowed to expand to 1·5 times its initial volume by changing the temperature and pressure, compute its resultant density. (0·1189 kg/m^3)

[**Hint:** 1 kg/m^3 = 1 g/L. Density is reduced by 1·5 times]

3. The density of an ideal gas A is 1·43 g/L at STP. Determine the density of A at 17°C and 700 torr (mm). (1·24 g/L)

4. A container has 3·2 g of a certain gas at NTP. What would be the mass of the same gas contained in the same vessel at 200°C and 16 atm pressure. (29·534 g)

[**Hint:** Mol. wt. $(M) = \frac{3.2}{V} \times 22.4$; V = vol. in lit. Apply $pV = \frac{w}{M} RT$;

w = wt. of the gas in g at 473 K and 16 atm]

5. Calculate the volume occupied by 5 g of acetylene gas at 50°C and 740 mm pressure. **(IIT 1991)** (5·2375 litres)

6. A bottle is heated with its mouth open from 15°C to 100°C. What fraction of air originally contained in the vessel is expelled? (23·5%)

7. An underwater bubble with a radius of 0·5 cm at the bottom of a tank, where the temperature is 5°C and the pressure is 3 atm., rises to the surface where the temperature is 25°C and pressure is 1 atm. What will be the radius of the bubble when it reaches the surface? (0·74 cm)

8. A good vacuum produced in common laboratory apparatus corresponds to 10^{-6} mm pressure at 25°C. Calculate number of molecules per cc at this pressure and temperature. (3.2×10^{10})

9. A sample of nitrogen gas is bubbled through liquid water at 25°C and then collected in a volume of 750 cc. The total pressure of the gas which is saturated with water vapour, is found to be 740 mm at 25°C. The vapour pressure of water at this temperature is 24 mm. How many moles of nitrogen are in the sample? (0·028)

10. A flask of volume 1 litre contains vapour of CH_3OH at a pressure of 1 atm and 25°C. The flask was then evacuated till the final pressure dropped to 10^{-4} mm. Find the number of molecules of methyl alcohol left in the flask. (3.2×10^{15})

11. Estimate the number of molecules left in a volume of the size of a pinhead about 1 cubic mm when the air is pumped out to give a vacuum of 10^{-6} mmHg at 25°C. ($3{\cdot}24 \times 10^7$)

12. A 500-cc bulb weighs 38·734 grams when evacuated and 39·3135 grams when filled with air at 1 atm pressure and 24°C. Assuming that air behaves as an ideal gas at this pressure, calculate effective mass of 1 mole of air. (28·2 g)

13. A desiccator of internal volume of 1 litre and containing nitrogen at 1 atm pressure is partially evacuated to a final pressure of 7·6 mmHg while the temperature remains constant. What is the volume of the gas at this stage? (1·00 litre)

14. A litre of air weighs 1·293 grams at NTP. At what temperature will a litre of air weigh 1 gram, the pressure being 72 cm? (61·3′c)

15. The vapour of a hydrocarbon is 2·47 times heavier than that of oxygen. What is its molecular weight? (79·04)

16. A gas cylinder contains 370 g of O_2 at 30 atm and 25°C. What mass of O_2 would escape if first the cylinder were heated to 75°C and then the valve were held open until the gas pressure was 1 atm, the temperature being maintained at 75°C? (359 g)

[**Hint:** First calculate the volume of the cylinder and then wt. of O_2 present at 1 atm and 75°C using $pV = nRT$]

17. A gaseous compound is composed of 85·7% by weight of C and 14·3% by weight of H. Its density is 2·28 g/L at 300 K and 1 atm pressure. Calculate the molecular formula of the compound. (C_4H_8)

18. A balloon filled with helium rises to a certain height at which it gets fully inflated to a volume of 1×10^5 litres. If at this altitude temperature and atmospheric pressure is 268 K and 2×10^{-3} atm respectively, what weight of helium will be required to fully inflate the balloon? (36·36 g)

19. Find the total pressure exerted by 1·6 g of methane and 2·2 g of CO_2 contained in a four-litre flask at 27°C. (0·9236 atm)

20. At 100°C and 1 atm pressure the density of water vapour is 0·0005970 g/cc.
(a) What is the molar volume and how does this compare with ideal gas value?
(b) What is the compressibility factor 'Z'?
[V (obs.) = 30·18 lit, and V (ideal) = 30·621 litres; Z = 0·986]

21. A box is divided by a thin partition into equal compartments and they are filled with an equal number of hydrogen and heavy hydrogen molecules respectively. If the pressure in the hydrogen compartment is 1 cmHg, what is the pressure in the other compartment? What will be the pressure if the partition is removed? (1 cmHg; 1 cmHg)

22. 1 g of N_2 and 1 g of O_2 are put in a two-litre flask at 27°C. Calculate partial pressure of each gas, the total pressure and the composition of the mixture in mole percentage. (0·44 atm.; 0·82 atm; 53·3%; 46·7%)

23. Into a gas bulb of 2·83 litres, are introduced 0·174 g of H_2 and 1·365 g of N_2 which can be assumed to behave ideally. The temperature is 0°C. What are the partial pressures of H_2 and N_2 and what is the total gas pressure? What are the mole fractions of each gas? What are pressure fractions? ($pH_2/p = nH_2/n = 0{\cdot}639$)

24. 100 cm^3 of NH_3 diffuses through a fine hole in 32·5 seconds. How much time will 60 cc of N_2 take to diffuse under the same conditions? (25 s)

25. A gas '*X*' diffuses five times as rapidly as another gas '*Y*'. Calculate the ratio of molecular weights of '*X*' and '*Y*'. (1/25)

26. The rate of diffusion of methane at a given temperature is twice that of a gas X. The molecular weight of X is

(A) 64 (B) 32 (C) 4 (D) 8 (A)

27. A mixture containing 1·12 litres of H_2 and 1·12 litres of D_2 at NTP is taken inside a bulb connected to another bulb by a stopcock with a small opening. The second bulb is fully evacuated, the stopcock opened for a certain time and then closed. The first bulb is now found to contain 0·05 g H_2. Determine the percentage composition by weight of the gases in the second bulb. (41·6; 58·33)

[**Hint:** H = 1, D = 2, Apply Eqn. 9]

28. A mixture consisting of 80 mole per cent hydrogen and 20 mole per cent deuterium at 25°C and a total pressure of 1 atm is permitted to effuse through a small orifice of area 0·20 mm^2. Calculate composition of the initial gas that passes through. (5·65 : 1)

[**Hint:** mole % = vol. %.]

29. A straight glass tube has two inlets '*X*' and '*Y*' at the two ends. The length of the tube is 200 cm. HCl gas through inlet *X* and NH_3 gas through inlet *Y* are allowed to enter the tube at the same time. White fumes first appear at a point *P* inside the tube. Find the distance of *P* from *X*. (81·1 cm)

30. A mixture of hydrogen and oxygen in 3 : 1 volume ratio is allowed to diffuse through a porous partition. What should be the composition of the initial gas diffusing out of the vessel? (12 : 1)

31. The time required by a certain volume of oxygen to diffuse through a small hole is 3600 seconds. Calculate the time required by the same volume of chlorine to diffuse through the same hole. (5360 s)

32. Calculate the relative rates of diffusion of 235 UF_6 and 238 UF_6 in the gaseous form. (1·0043 : 1)

33. The rate of diffusion of a sample of ozonised oxygen is 0·98 times more than that of pure oxygen. Find the percentage (by volume) of ozone in the ozonised sample. (8·25%)

34. The pressure in a vessel that contained pure oxygen dropped from 2000 mm to 1500 mm in 47 minutes as the oxygen leaked through a small hole into vacuum. When the same vessel was filled with another gas, the pressure dropped from 2000 mm to 1500 mm in 74 minutes. What is the molecular weight of the gas? (79)

35. A vessel of volume 100 mL contains 10% of oxygen and 90% of an unknown gas. The gases diffuse in 86 seconds through a small hole of the vessel. If pure oxygen, under the same conditions, diffuses in 75 seconds, find the molecular weight of the unknown gas. (43·2)

36. A balloon having a capacity of 10000 metre^3 is filled with helium at 20°C and 1 atm pressure. If the balloon is loaded with 80% of the load that it can lift at ground level, at what height will the balloon come to rest? Assume that the volume of the balloon is constant, the atmosphere is isothermal, 20°C, the molecular weight of air is 28·8 and the ground level pressure is 1 atm. The mass of the balloon is $1{\cdot}3 \times 10^6$ g.

[Hint: Use Eqn. 11.]

37. The density of hydrogen at 0°C and 760 mmHg pressure is 0·00009 g/cc. What is the rms speed of hydrogen molecules? ($183{\cdot}8 \times 10^3$ cm/s)

38. A one-litre gas bulb contains $1{\cdot}03 \times 10^{23}$ H_2 molecules. If the pressure exerted by these molecules is 760 mmHg, what must be average squared molecular speed? ($9{\cdot}43 \times 10^4$ cm/s)

39. At what temperature the rms speed of hydrogen is equal to escape velocity from the surface of the earth?

[Hint: Escape vel. = $\sqrt{2gR}$; R = earth's radius = $6{\cdot}37 \times 10^6$ cm.] (100·2 K)

40. At what temperature, hydrogen at 1 atm pressure has the same rms speed as that of oxygen at NTP? (–256°C)

41. Compute rms speed of (i) O_2 at 15°C and 77 cm pressure, (ii) NH_3 molecules at NTP, and (iii) average speed of CH_4 at 500°C.

($47{\cdot}3 \times 10^3$ cm/s, $6{\cdot}8 \times 10^4$ cm/s, $10{\cdot}9 \times 10^4$ cm/s)

42. Show that the ideal gas law can be written as $p = \frac{2}{3}E$, where E is the kinetic energy per unit volume.

43. At what temperature would the most probable speed of CO molecules be twice that at 0°C? (819°C)

44. At what temperature would N_2 molecules have the same average speed as He atoms at 300 K? (2100 K)

45. Calculate the kinetic energy of Avogadro constant of gaseous molecules at 0°C. ($3{\cdot}4 \times 10^{10}$ erg)

46. Calculate the average kinetic energy per mole of CO_2 gas at 27°C in different units. ($3{\cdot}74 \times 10^{10}$ ergs, 3742 joules, 894 calories)

47. A gas of molecular weight 40 has a specific heat 0·075 cal/g/deg at a constant volume. What is the C_v value for it and what is the atomicity of the gas?

(3 cal/mole, monoatomic)

48. At 627°C and 1 atm, SO_3 partially dissociates into SO_2 and O_2. One litre of the equilibrium mixture weighs 0·94 g under the above conditions. Calculate the partial pressures of the constituent gases in the mixture.

[Hint: See Example 53] (0·5986, 0·2676, 0·1338 atm)

49. The degree of dissociation of N_2O_4 according to the equation $N_2O_4 = 2NO_2$ at 70°C and atmospheric pressure is 65·6%. Calculate the apparent molecular weight of N_2O_4 under the above conditions. (55·56)

$$\textbf{[Hint: } \underset{1-x}{\overset{1}{N_2O_4}} \rightleftharpoons \underset{2x}{2NO_2}\text{; approx. mol. wt.} = \frac{92(1-x)+46(2x)}{1+x}\cdot\text{]}$$

50. 1 mole of a gas is changed from its initial state (15 lit ; 2 atm) to final state (4 lit, 10 atm) reversibly. If this change can be represented by a straight line in p–V curve, calculate maximum temperature, the gas attained. (698 K)

[Hint: Eqn. for the line is $\frac{p-2}{10-2} = \frac{V-15}{4-15}$; $11p + 8V = 142$

For $(pV)_{max}$, $11p = \frac{142}{2} = 71$ and $8V = \frac{142}{2} = 71$

or $p = \frac{71}{11}$ and $V = \frac{71}{8}$ then apply

$(pV)_{max} = nRT_{max}$ ·]

51. Calculate the volume occupied by 7 g of N_2 under a pressure of 100 atm at 27°C ($a = 1{\cdot}39$ atm lit^2 mole^{-2}, $b = 0{\cdot}391$ lit mole^{-1}) (58·8 mL)

52. The van der Waals constant b for a gas is $4{\cdot}2 \times 10^{-2}$ lit mole^{-1}. How close the nuclei of the two molecules come together ? ($3{\cdot}2 \times 10^{-8}$ cm)

53. Find the temperature at which 3 moles of SO_2 will occupy a volume of 10 litres at a pressure of 15 atms.

($a = 6{\cdot}71$ atm lit^2 mole^{-2}),

($b = 0{\cdot}0564$ lit mole^{-1}) (351°C)

54. A 2·55 g sample of NH_4NO_2 is heated in a test tube.

$$NH_4NO_2(s) \rightarrow N_2(g) + 2H_2O(g)$$

What volume of N_2 will be collected in the flask when the water and gas temperature is 26°C and barometric pressure is 745 mmHg? (1·013 L)

55. How is the rms speed of N_2 molecules in a gas sample changed by

(a) an increase in temperature?

(b) an increase in volume of sample?

(c) mixing with an Ar sample at the same temperature?

[(a) increases (b) no effect (c) no effect]

56. Nickel carbonyl, $Ni(CO)_4$, is one of the most toxic substances known. The present maximum allowable concentration in laboratory air during an 8-hr work day is 1 part in 10^9. Assume 24°C and 1 atm pressure what mass of $Ni(CO)_4$ is allowable in a laboratory that is 110 m^2 in area, with a ceiling height of 2·7 m? ($3{\cdot}53 \times 10^{-4}$ g)

57. A sample of gas has a molar volume of 10·1 L at a pressure of 745 mmHg and a temperature of −138°C. Is the gas behaving ideally? (No)

58. Cyanogen is 46·2% C and 53·8% N by mass. At a temperature of 25°C and a pressure of 750 mmHg, 1·0 g of cyanogen gas occupies 0·476 L. Find the formula of cyanogen. (C_2N_2)

59. Two containers of the same volume, one containing the gas *A* and the other containing the gas *B*, have the same number of molecules of each gas. The mass of the molecule *A* is twice the mass of the molecule *B*. The rms speed of *A* is also twice that of *B*. Calculate the pressure ratio of the gases *A* and *B*. (8 : 1)

60. Calculate the percentage of free volume available in 1 mole gaseous water at 1 atm pressure and 373 K. (99·94%)

61. To an evacuated vessel with movable piston under external pressure of 1 atm, 0.1 mole of He and 1.0 mole of an unknwon compound (V.P. = 0.68 atm at 0°C) are introduced. Considering the ideal gas behaviour, the total volume (in L) of the gases at 0°C is close to .... . **(IIT 2011)** (7 litres)

[**Hint:** As the piston is movable, the gas pressure is 1 atm.

$\therefore\ p_{He} = 1 - 0.68 = 0.32$ atm. Apply $pV = nRT$ for He.] (99·94%)

62. The diffusion coefficient of an ideal gas is proportional to its mean free path and mean speed. The absolute temperature of an ideal gas is increased 4 times and its pressure is increased 2 times. As a result the diffusion coefficient increases *x* times. The value of *x* is .... . **(IIT 2016 Adv.)** (4)

[**Hint:** D.C. $\propto \lambda = \dfrac{kT}{\sqrt{2}\pi\sigma^2 P}$.

Also D.C. $\propto$ Av. speed $= \sqrt{\dfrac{8RT}{\pi M}}$

$\therefore$ D.C. $\propto \dfrac{T^{3/2}}{P}$.]

63. A closed tank has two compartments A and B, both filled with oxygen (assumed to be ideal gas). The partition separating the two compartments is fixed and is a perfect heat insulator (figure 1). If the old partition is replaced by a new partition which can slide and conduct heat but does NOT allow the gas leak across (figure 2), the volume (in m^3) of the compartment A after the system attains equilibrium is .... . **(IIT 2018 Adv.)** (2.22 m^3)

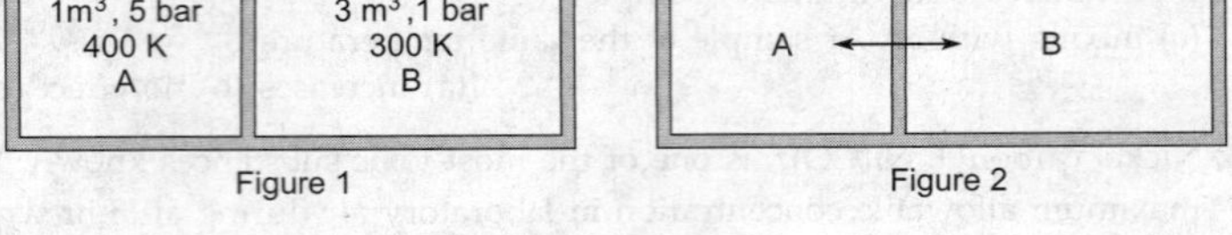

Figure 1 Figure 2

[**Hint:** With old partition,

$5 \times 1 = n_1 \times 400\,R$ ... for A,

$1 \times 3 = n_2 \times 300\,R$ ... for B

$\therefore \quad \dfrac{n_1}{n_2} = \dfrac{15}{12}$.

With new partition, pressure and temperature in A & B remain the same say p & T.

As the piston moves right, the volume of A & B at eqb. will be (1 + V) and (3 – V) respectively.

$$p(1 + V) = n_1RT$$
$$p(3 - V) = n_2RT$$

$$\therefore \quad \frac{1 - V}{3 - V} = \frac{n_1}{n_2} = \frac{15}{12};\ V{=}1.22$$

$\therefore$ Vol. of $A = 1 + V$.]

Objective Problems

1. Which of the following curves does not represent Boyle's law?

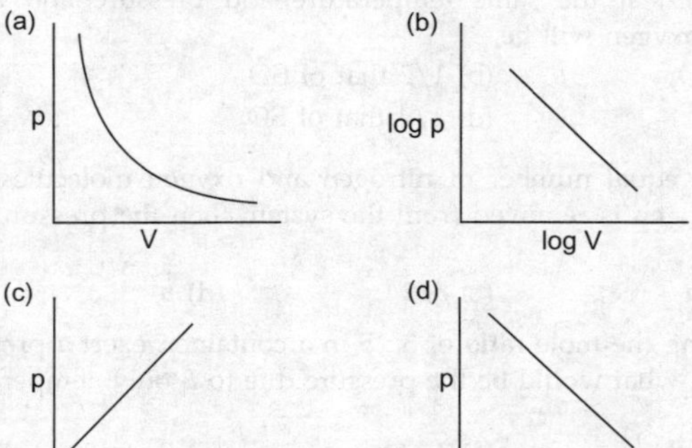

2. The temperature of 20 litres of nitrogen was increased from 100 K to 300 K at a constant pressure. Change in volume will be

(a) 80 litres (b) 60 litres (c) 40 litres (d) 20 litres

3. If the volume of a given mass of a gas at constant temperature becomes three times, the pressure will be

(a) $3p$ (b) $p/3$ (c) $9p$ (d) p

4. If the pressure of a given mass of a gas is reduced to half and temperature is doubled simultaneously, the volume will be

(a) same as before (b) twice as before

(c) four times as before (d) one-fourth as before

5. 10 g of a gas at NTP occupies a volume of 2 litres. At what temperature will the volume be double, pressure and amount of the gas remaining same?

(a) 273 K (b) 546 K (c) –273°C (d) 546°C

6. If 1 litre of N_2 at 27°C and 760 mm contains N molecules, 4 litres of O_2, under the same conditions of temperature and pressure, shall contain

(a) N molecules (b) $2N$ molecules (c) $\frac{N}{4}$ molecules (d) $4N$ molecules

7. Under identical conditions of temperature the density of a gas X is three times that of gas Y while molecular mass of gas Y is twice that of X. The ratio of pressures of X and Y will be

(a) 6 (b) 1/6 (c) 2/3 (d) 3/2

8. The molecules of a gas A travel four times faster than the molecules of gas B at the same temperature. The ratio of molecular weights (M_A/M_B) will be

(a) 1/16 (b) 4 (c) 1/4 (d) 16

9. The volume occupied by 22·4 g of a gas (vap. density = 11·2) at NTP is

(a) 22·4 litres (b) 11·2 litres (c) 44·8 litres (d) 1 litres

10. 32 g of oxygen and 3 g of hydrogen are mixed and kept in a vessel of 760 mm pressure and 0°C. The total volume occupied by the mixture will be nearly

(a) 22·4 litres (b) 33·6 litres (c) 56 litres (d) 44·8 litres

11. A pre-weighed vessel was filled with oxygen at NTP and weighed. It was then evacuated, filled with SO_2 at the same temperature and pressure and again weighed. The weight of oxygen will be

(a) the same as that of SO_2 (b) 1/2 that of SO_2
(c) twice that of SO_2 (d) 1/4 that of SO_2

12. A closed vessel contains equal number of nitrogen and oxygen molecules at a pressure of p mm. If nitrogen is removed from the system then the pressure will be

(a) p (b) $2p$ (c) $p/2$ (d) p^2

13. Two gases A and B, having the mole ratio of 3 : 5 in a container, exert a pressure of 8 atm. If A is removed, what would be the pressure due to B only, temperature remaining constant?

(a) 1 atm (b) 2 atm (c) 4 atm (d) 5 atm

14. Equal weight of methane and oxygen are mixed in an empty container at 25°C. The fraction of the total pressure exerted by oxygen is

(a) 1/3 (b) 1/2 (c) 2/3 (d) $\frac{1}{3} \times \frac{273}{298}$

15. In two separate bulbs containing ideal gases A and B respectively, the density of gas A is twice that of gas B while molecular weight of gas A is half that of gas B at the same temperature, pressure ratio P_A/P_B will be

(a) 1/4 (b) 1/2 (c) 4 (d) 1

16. If 1 litre of a gas A at 600 mm and 0·5 litre of gas B at 800 mm are taken in a 2-litre bulb, the resulting pressure is

(a) 1500 mm (b) 1000 mm (c) 2000 mm (d) 500 mm

17. If the ratio of the rates of diffusion of the two gases A and B is 4 : 1, the ratio of their densities is

(a) 1 : 16 (b) 1 : 4 (c) 1 : 8 (d) 1 : 2

18. Since the atomic weights of C, N and O are 12, 14 and 16 respectively, among the following pair of gases, the pair that will diffuse at the same rate is
(a) carbon dioxide and nitrous oxide
(b) carbon dioxide and nitrogen peroxide
(c) carbon dioxide and carbon monoxide
(d) nitrous oxide and nitrogen peroxide

19. Two grams of hydrogen diffuse from a container in 10 minutes. How many grams of oxygen would diffuse through the same container in the same time under similar conditions?
(a) 0·5 g (b) 4 g (c) 6 g (d) 8 g

20. Four rubber tubes are respectively filled with H_2, O_2, N_2 and He. The tube which will be required to be reinflated first is
(a) H_2-filled tube (b) O_2-filled tube (c) N_2-filled tube (d) He-filled tube

21. Equal weights of methane and hydrogen are mixed in an empty container at 25°C. The fraction of the total pressure exerted by hydrogen is
(a) 1/2 (b) 8/9 (c) 1/9 (d) 16/17

22. The specific gravity of CCl_4 vapour at 0°C and 76 cmHg in grams/litre is
(a) 11·2 (b) 77
(c) 6·88 (d) cannot be calculated

23. The rms speed of gas molecules at a temperature 27 K and pressure 1·5 bar is 1×10^4 cm/s. If both temperature and pressure are raised three times, the rms speed of the gas will be
(a) 9×10^4 cm/s (b) 3×10^4 cm/s
(c) $\sqrt{3} \times 10^4$ cm/s (d) 1×10^4 cm/s

24. The ratio of rms velocity to average velocity of gas molecules at a particular temperature is
(a) 1·086 : 1 (b) 1 : 1·086 (c) 2 : 1·086 (d) 1·086 : 2

25. The average velocity of an ideal gas molecule at 27°C is 0·3 m/s. The average velocity at 927°C will be
(a) 0·6 m/s (b) 0·3 m/s (c) 0·9 m/s (d) 3·0 m/s
(IIT 1986)

26. Kinetic energy per mole of an ideal gas
(a) is proportional to temperature (b) inversely proportional to temperature
(c) is independent of temperature (d) is zero at 0°C

27. The temperature of a sample of a gas is raised from 127°C to 527°C. The average kinetic energy of the gas
(a) does not change (b) is doubled
(c) is halved (d) cannot be calculated

28. The kinetic energy of N molecules of O_2 is x joule at –123°C. Another sample of O_2 at 27°C has a kinetic energy of $2x$ joules. The latter sample contains
(a) N molecules of O_2 (b) $2N$ molecules of O_2
(c) $N/2$ molecules of O_2 (d) none of these

29. A helium atom is two times heavier than a hydrogen molecule at 298 K, the average kinetic energy of helium is
(a) two times that of hydrogen molecules
(b) same as that of hydrogen molecules
(c) four times that of hydrogen molecules
(d) half that of hydrogen molecules

30. The kinetic energy of any gas molecule at 0°C is
(a) $5{\cdot}66 \times 10^{-21}$ J (b) 3408 J (c) 2 cal (d) 0

31. The ratio of the average molecular kinetic energy of UF_6 to that of H_2, both at 300 K, is
(a) 1 : 1 (b) 7 : 2 (c) 176 : 1 (d) 2 : 7

32. At what temperature will hydrogen molecules have the same kinetic energy as nitrogen molecules have at 35°C?
(a) $\left(\dfrac{28 \times 35}{2}\right)$°C (b) $\left(\dfrac{2 \times 35}{28}\right)$°C (c) $\left(\dfrac{2 \times 28}{35}\right)$°C (d) 35°C

33. A monoatomic gas, a diatomic gas and a triatomic gas are mixed, taking one mole of each. C_p/C_v for the mixture is
(a) 1·40 (b) 1·428 (c) 1·67 (d) 1·33

34. When 1 mole of a monoatomic ideal gas is heated to raise its temperature through 1°C, the fraction of the heat energy supplied which increases the kinetic energy of the gas is
(a) 2/5 (b) 3/5 (c) 3/7 (d) 5/7

35. The values of van der Waals constant '*a*' for the gases O_2, N_2, NH_3 and CH_4 are 1·36, 1·39, 4·17 and 2·253 L^2 atm $mole^{-2}$ respectively. The gas which can most easily be liquefied is
(a) O_2 (b) N_2 (c) NH_3 (d) CH_4
(IIT 1989)

36. One mole of an ideal monoatomic gas is mixed with 1 mole of an ideal diatomic gas. The molar specific heat of the mixture at constant volume is
(a) 3 cal (b) 4 cal (c) 8 cal (d) 5 cal

37. According to kinetic theory of gases, for a diatomic molecule,
(a) the pressure exerted by the gas is proportional to the mean velocity of the molecule
(b) the pressure exerted by the gas is proportional to the root mean square velocity of the molecule
(c) the root mean square velocity of the molecule is inversely proportional to the temperature
(d) the mean translational kinetic energy of the molecule is proportional to the absolute temperature
(IIT 1991)

38. The ratio between the rms velocity of H_2 at 50 K and that of O_2 at 800 K is
(a) 4 (b) 2 (c) 1 (d) 1/4
(IIT 1996)

39. The compressibility factor for an ideal gas is
(a) 1·5 (b) 1·0 (c) 2·0 (d) ∞
(IIT 1997)

40. *X* mL of H_2 gas effuses through a hole in a container in 5 seconds. The time taken for the effusion of the same volume of the gas specified below under identical conditions is
(a) 10 seconds : He (b) 20 seconds : O_2
(c) 25 seconds : CO (d) 55 seconds : CO_2
(IIT 1996)

41. One mole of N_2O_4 (g) at 300 K is kept in a closed container under one atmosphere. It is heated to 600 K when 20% by mass of N_2O_4 (g) decomposes to NO_2 (g). The resultant pressure is
(a) 1·2 atm (b) 2·4 atm (c) 2 atm (d) 1 atm
(IIT 1996)

42. A column of air 1 m^2 in cross section extending through the atmosphere has a mass of roughly 10,000 kg. The atmospheric pressure at the surface of the earth is
(a) 1×10^5 N (b) 1×10^2 k Pa (c) 1×10^5 k Pa (d) 1 atmosphere

43. If the product of the gas constant *R* i.e., ·0821 lit atm/K/mole and NTP temperature in kelvin equals 22·4, the compressibility factor of the gas at 1 atmospheric pressure is
(a) > 1 (b) < 1 (c) = 1 (d) = 0

44. The graph of the quantity d/p against pressure is extrapolated to zero pressure to obtain a limiting value. If this limiting d/p value for a certain nonideal gas is found to be 2·86 g/L-atm at 0°C, the molar mass of the gas will be
(a) 2·86 (b) 64·06 (c) 22·4 (d) none of these

45. The rms speed of N_2 molecules in a gas sample can be changed by
(a) an increase in volume of sample
(b) mixing with Ar sample
(c) an increase in pressure on the gas
(d) an increase in temperature

46. How high must a column of water be to exert a pressure equal to that of a column of Hg that is 760 mm high
(a) 760 mm (b) 55·9 mm (c) 74480 mm (d) 10336 mm

47. In the following graph in which volume is plotted versus temperature, the lines *A*, *B* and *C* represent the same mass of the same ideal gas at different pressures p_1, p_2 and p_3 respectively.
The correct relationship of pressures is
(a) $p_1 > p_2 > p_3$ (b) $p_1 = p_2 = p_3$
(c) $p_3 > p_2 > p_1$ (d) $p_1 > p_2 < p_3$

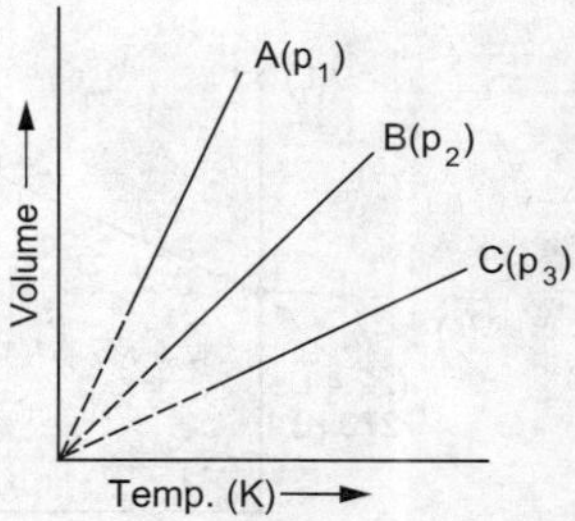

48. A 5·0-L reaction vessel contain hydrogen at a partial pressure of 0·588 atm and oxygen gas at a partial pressure of 0·302 atm. The limiting reactant in the following reaction

$$2H_2 + O_2 = 2H_2O$$

is

(a) H_2 (b) O_2 (c) none of these (d) both (a) and (b)

49. When a gas is passed through a small hole at a temperature greater than its critical temperature, Joule–Thomson effect will show

(a) cooling of the gas (b) warming of the gas
(c) no change in temperature (d) first cooling and then warming

50. Which of the following statements is wrong?

(a) Critical temperature is the highest temperature at which condensation of a gas is possible.
(b) Critical pressure is the highest pressure at which a liquid will boil when heated.
(c) Boyle temperature is the temperature at which the second virial coefficient is zero.
(d) Inversion temperature is the temperature above which the gas becomes cooler during Joule–Thomson expansion.

51. Which of the following choice (s) is (are) correct for a gas?

(a) $p_{ideal} > p_{real}$ (b) $V_{ideal} > V_{real}$
(c) $p_{ideal} < p_{real}$ (d) $V_{ideal} < V_{real}$

52. Which of the following volume (V) – temperature (T) plots represents the behaviour of one mole of an ideal gas at one atmospheric pressure?

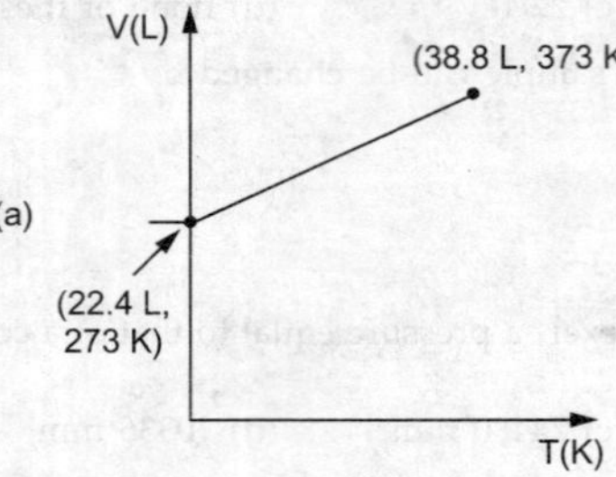

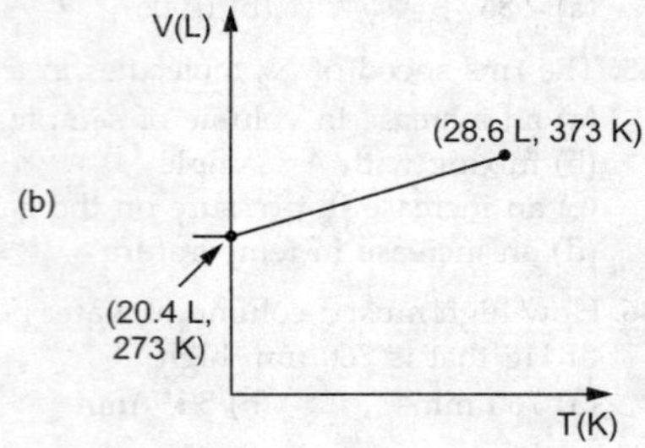

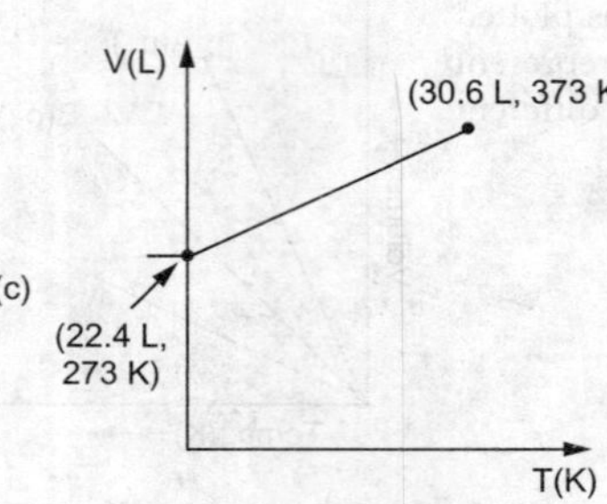

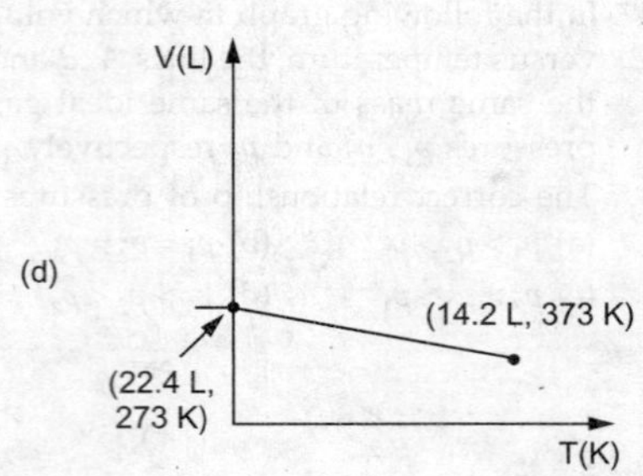

53. According to kinetic theory of gases

(a) collisions are always elastic.

(b) heavier molecules transfer more momentum to the walls of container.

(c) only a small number of molecules have very high velocity.

(d) between collisions, the molecules move in straight lines with constant velocity.

(IIT 2011)

54. For 1 mole of van der Waals gas when $b = 0$ and $T = 300$ K. The pV vs $1/V$ plot is shown below. The value of 'a' in atm L mol^{-2} is

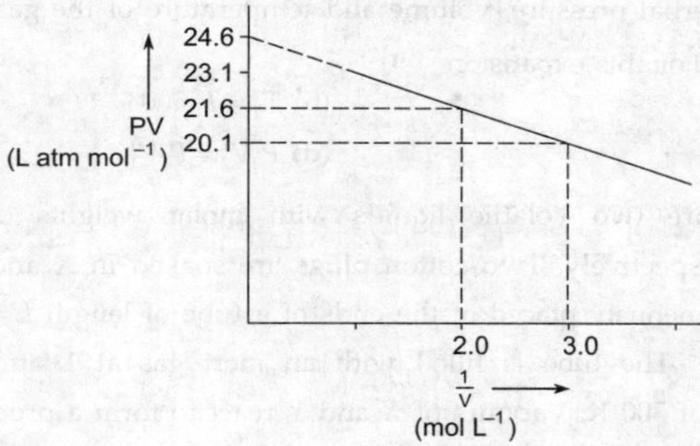

(a) 1.0 (b) 4.5 (c) 1.5 (d) 3.0 **(IIT 2012)**

[**Hint:** If $b = 0$, van der Waals eqn. is $pV + \frac{a}{V} = RT$. Calculate slope of the plot.]

55. For gaseous state if most probable speed is denoted by C^*, average speed by $\overline{C}$ and root square speed by C, then for a large number of molecules, the ratio of these speeds are

(a) $C^* : \overline{C} : C = 1.225 : 1.128 : 1$

(b) $C^* : \overline{C} : C = 1.128 : 1.225 : 1$

(c) $C^* : \overline{C} : C = 1 : 1.128 : 1.225$

(d) $C^* : \overline{C} : C = 1 : 1.225 : 1.128$

(IIT 2013 Main)

[**Hint:** See text.]

56. If Z is compressibility factor, van der Waals equation at low pressure can be written as

(a) $Z = 1 + \frac{RT}{pb}$

(b) $Z = 1 - \frac{a}{VRT}$

(c) $Z = 1 - \frac{pb}{RT}$

(d) $Z = 1 + \frac{pb}{RT}$ **(IIT 2014 Main)**

[**Hint:** Read Modern Approach to Physical Chemistry, Vol. I, Ch. 1.]

57. An ideal gas in thermally insulated vessel at internal pressure p_1, volume, V_1 and absolute temperature T_1 expands irreversibly against zero external pressure as shown in the diagram.

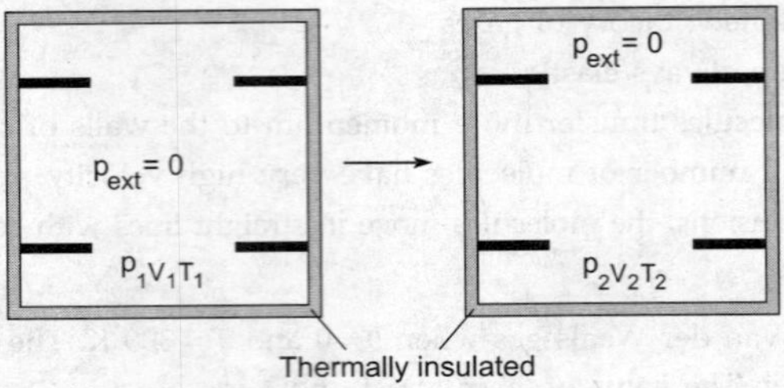

The final internal pressure, volume and temperature of the gas are p_2, V_2 and T_2 respectively. For this expansion

(a) $q = 0$ (b) $T_2 = T_1$

(c) $P_2V_2 = P_1V_1$ (d) $P_2V_2^{\gamma} = P_1V_1^{\gamma}$ **(IIT 2014 Adv.)**

58. X and Y are two volatile liquids with molar weights of 10 g mol^{-1} and 40 g mol^{-1} respectively. Two cotton plugs are soaked in X and other soaked in Y are simultaneously placed at the ends of a tube of length $L = 24$ cm, as shown in the figure. The tube is filled with an inert gas at 1 atm pressure and a temperature of 300 K. Vapours of X and Y react to form a product which is first observed at a distance d cm from the plug soaked in X. Take X and Y to have equal molecular diameters and assume ideal behaviour for the inert gas and the two vapours.

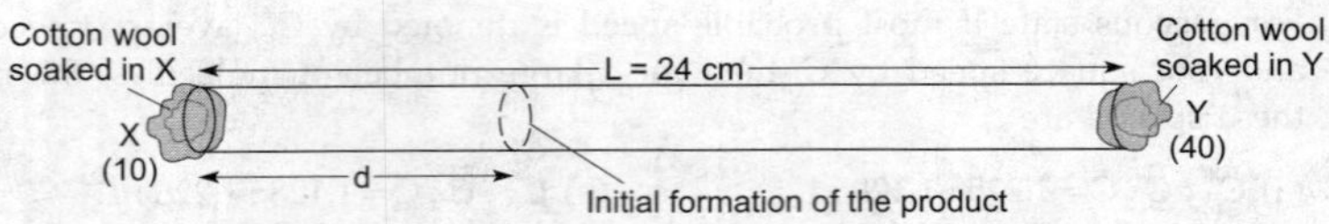

The value of d in cm as estimated from Graham's law is

(a) 8 (b) 12

(c) 16 (d) 20 **(IIT 2014 Adv.)**

59. Refer to Q. 58. The experimental value of d is found to be smaller than the estimate obtained using Graham's law. This is due to

(a) larger mean free path of X compared to that of Y.

(b) larger mean free path of Y compared to that of X.

(c) increased collision frequency of Y with the inert gas as compared to that of X with the inert gas.

(d) increased collision frequency of X with the inert gas as compared to that of Y with the inert gas. **(IIT 2014 Adv.)**

60. One mole of a monoatomic real gas satisfies the equation $p(V - b) = RT$, where b is a constant. The relationship of interatomic potential $V(r)$ and interatomic distance r for gas is given by

(a) 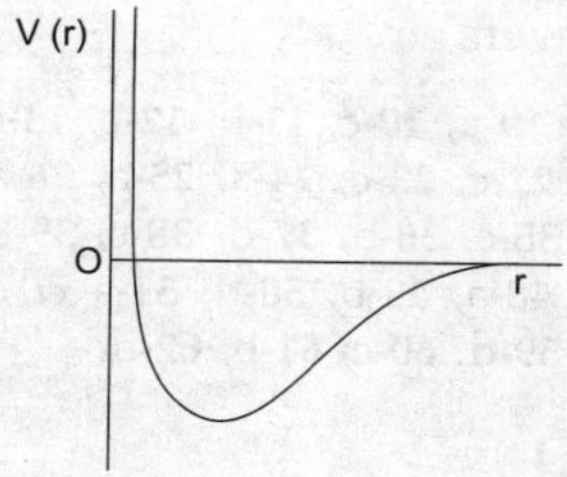

(b) 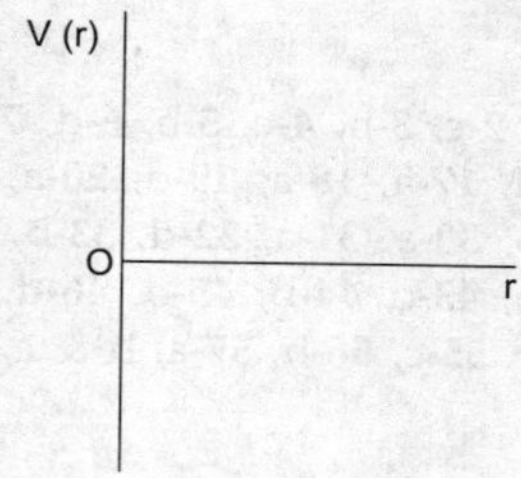

(c) 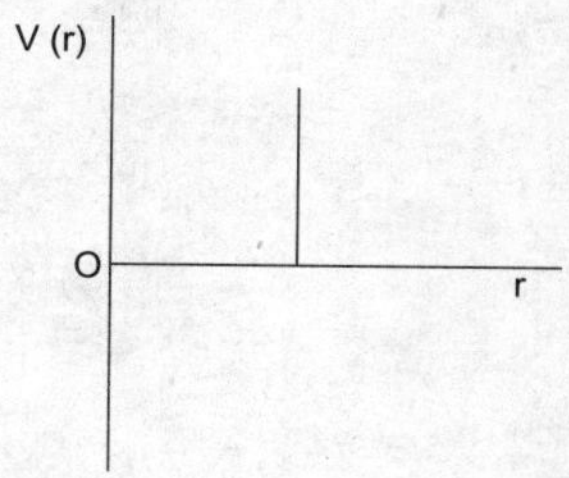

(d) 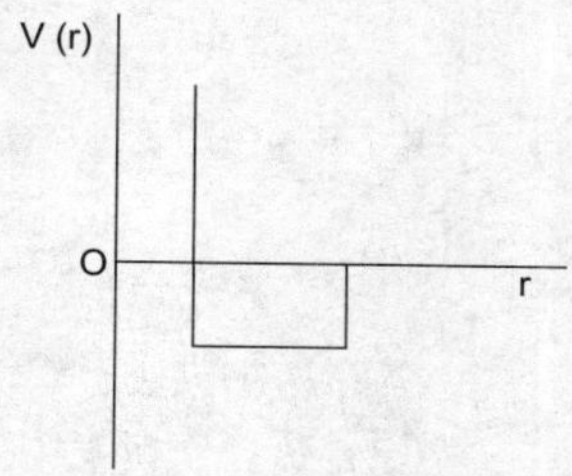

(IIT 2015 Adv.)

61. Two closed bulbs of equal volume (V) containing an ideal gas initially at pressure p_i and temperature T_1 are connected through a narrow pipe of negligible volume as shown in the figure below. The temperature of one of the bulbs is then raised to T_2. The final pressure p_f is

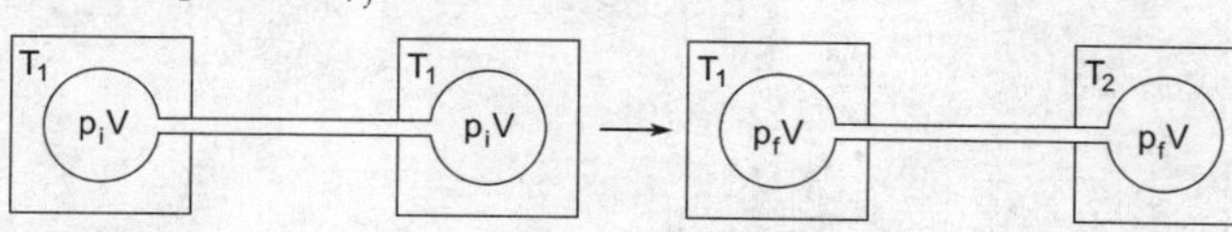

(a) $2p_i\left(\dfrac{T_1}{T_1+T_2}\right)$ (b) $2p_i\left(\dfrac{T_2}{T_1+T_2}\right)$

(c) $2p_i\left(\dfrac{T_1T_2}{T_1+T_2}\right)$ (d) $p_i\left(\dfrac{T_1T_2}{T_1+T_2}\right)$

(IIT 2016 Main)

[Hint: $p_i(2V) = nRT, p_fV = n_1RT_1$

$p_fV = n_2RT_2$ and $n = n_1 + n_2$]

62. If x gram of a gas is adsorbed by m gram of adsorbent at pressure p, the plot of $\log \dfrac{x}{m}$ versus $\log p$ is linear. The slope of the plot is

(n and k are constants and $n > 1$)

(a) $\log k$ (b) $\dfrac{1}{n}$

(c) $2k$ (d) n **(IIT 2018 Main)**

[Hint: Read Ch. 12, Adsorption of gases on solids]

Answers

1-d, **2**-c, **3**-b, **4**-c, **5**-b, **6**-d, **7**-a, **8**-a, **9**-a, **10**-c, **11**-b, **12**-c, **13**-d, **14**-a, **15**-c, **16**-d, **17**-a, **18**-a, **19**-d, **20**-a, **21**-b, **22**-c, **23**-c, **24**-a, **25**-a, **26**-a, **27**-b, **28**-a, **29**-b, **30**-a, **31**-a, **32**-d, **33**-b, **34**-b, **35**-c, **36**-b, **37**-d, **38**-c, **39**-b, **40**-b, **41**-b, **42**-b, **43**-c, **44**-b, **45**-d, **46**-d, **47**-c, **48**-a, **49**-b, **50**-d, **51**-a & d, **52**-c, **53**-a, **54**-c, **55**-c, **56**-b, **57**-a, b & c, **58**-c, **59**-d, **60**-c, **61**-b, **62**-d.

❑

CHAPTER THIRTEEN

DILUTE SOLUTION AND COLLIGATIVE PROPERTIES

A solution is a homogeneous mixture of chemical species. The solutions are of several types but we shall discuss here mainly the solutions containing a solid solute and a liquid solvent. The distinction between a solute and a solvent is an arbitrary one. However, the constituent present in the greater amount is generally known as the solvent, while those present in relatively smaller amounts are called the solutes.

Concentration Units

The concentrations of solutions may be expressed in the following common sets of units.

1. Molarity: The molarity of a solution is the number of moles of solute present in one litre of the solution. It is expressed by 'M'.

2. Demal unit: The concentrations are also expressed in Demal units. One demal unit represents one mole of solute present in one litre of the solution at 0°C. It is expressed by 'D'.

3. Molality: The molality is defined as the number of moles of the solute present in 1000 g of the solvent. It is symbolised by '*m*'.

4. Normality: The normality is the number of equivalents of solute present in one litre of the solution. It is expressed by 'N'.

5. Formality: The formality is the number of gram-formula weights of the solute per litre of the solution. It is expressed by 'F'.

6. Mole fraction: The ratio of the number of moles of the solute to the total number of moles of the solution is known as mole fraction of the solute. Similarly, the ratio of the number of moles of the solvent to the total number of moles of the solution is called mole fraction of the solvent.

If x_1 and x_2 represent the mole fraction of the solute and the solvent respectively, we have,

$$x_1 = \frac{n_1}{n_1 + n_2} \quad \text{and} \quad x_2 = \frac{n_2}{n_1 + n_2},$$

where n_1 and n_2 are the number of moles of the solute and the solvent respectively present in a solution.

7. Percentage: The concentration in percentage is generally expressed in three ways:

(i) weight of solute in grams per 100 mL of solution

(ii) weight of solute in grams per 100 g of solution
(iii) volume of solute in mL per 100 mL of solution

Dilute Solution or Ideal Solution

An ideal solution is that which is formed by dissolving such a small amount of the nonvolatile solute in the solvent that there is no absorption or evolution of heat. For such solutions,

vapour pressure of the solvent in the presence of solute ∝ mole fraction of the solvent

or $p_{\text{solution}} \propto x_{\text{solvent}}$

or $p_{\text{solution}} = p_{\text{solvent}} \cdot x_{\text{solvent}}$

p_{solvent} is the proportionality constant representing the vapour pressure of the pure solvent.

The above relation is known as **Raoult's law**. Any solution obeying this law is called an ideal solution.

Colligative Properties

Dilute solutions have the following related properties:

(i) vapour-pressure lowering
(ii) osmotic pressure
(iii) freezing-point depression
(iv) boiling-point elevation

All these properties have the common characteristic that they do not depend upon the nature of the solute present but only on the number of the solute particles. Since these properties are bound together through their common origin they are called colligative properties (In Latin: co–together, ligare–to bind). We shall first take up those solutes which are nonvolatile and do not undergo dissociation or association.

Vapour-Pressure Lowering: Raoult's Law

The vapour pressure of a liquid is the pressure (partial pressure) exerted by a vapour in equilibrium with its liquid. The addition of solute particles lowers vapour pressure of the solvent.

We know that,

lowering in vapour pressure of the solvent ∝ mole fraction of solute

$$p^0 - p \propto \frac{n_1}{n_1 + n_2}$$

$$p^0 - p = p^0 \frac{n_1}{n_1 + n_2}$$

or $$\frac{p^0 - p}{p^0} = \frac{n_1}{n_1 + n_2} \approx \frac{n_1}{n_2} \qquad \dots (1)$$

where p^0 and p are the vapour pressures of the pure solvent and that of the solvent in the presence of solute. $(p^0 - p)/p^0$ is known as relative lowering of vapour pressure. The above relation is known as Raoult's law.

Determination of Lowering of Vapour Pressure

(i) Differential tensimeter method (static method)

(ii) Isopiestic vapour-pressure method

(iii) Gas-solution method (dynamic method)

The dynamic method was suggested by **Ostwald** and **Walker.** Dry air is bubbled successively through the (i) solution (ii) the pure solvent, and (iii) a substance absorbing solvent vapour, e.g., $CaCl_2$ for water solvent. We have thus

$$\frac{p^0 - p}{p^0} = \frac{\text{loss in weight of solvent}}{\text{gain in weight of } CaCl_2}.$$

Raoult's Law in Binary Solution

Raoult's law can also be applied to an ideal binary solution in which both components are volatile. A binary solution is said to be ideal when there is no absorption or evolution of heat when the two volatile liquid components are mixed to form the solution. If x_A and x_B represent the mole fraction of the two components 'A' and 'B' in liquid respectively, we have, from Raoult's law, the following relations for each of the components.

$$\left.\begin{array}{l}\text{Partial vapour pressure of component } A;\ p_A = x_A \cdot p_A^0 \\ \text{and partial vapour pressure of component } B;\ p_B = x_B \cdot p_B^0\end{array}\right\} \quad \ldots (2)$$

where p_A^0 and p_B^0 are the vapour pressures of pure liquid components 'A' and 'B' respectively. The total vapour pressure of the solution, p will thus be,

$$p = x_A \cdot p_A^0 + x_B \cdot p_B^0. \quad \ldots (3)$$

Since $x_A + x_B = 1$

$$\therefore \quad p = x_A\, p_A^0 + (1 - x_A)\, p_B^0$$

$$= p_B^0 + x_A\,(p_A^0 - p_B^0).$$

Thus the addition of a solute may raise or lower the vapour pressure of the solvent, depending upon which one is more volatile.

Now it should be remembered that **molar ratio of the two components in the liquid and in its vapour would not be the same,** that is, the mole fraction of each component would therefore be different in the liquid and its vapour. It is a general fact that when an ideal binary solution is in equilibrium with its vapour, the vapour is always richer than the liquid in the more volatile component of the system.

If x'_A and x'_B represent the mole fraction of the components '*A*' and '*B*' in the vapour respectively, we have from **Dalton's law of partial pressure**

$$\left.\begin{aligned} x'_A &= \frac{p_A}{p} \\ \text{and} \qquad x'_B &= \frac{p_B}{p} \end{aligned}\right\} \quad \dots (4)$$

from which we can derive the following relation:

$$\frac{1}{p} = \frac{x'_A}{p_A^0} + \frac{x'_B}{p_B^0}. \quad \dots (5)$$

Raoult's law for an ideal solution of two volatile components, *A* and *B*, can be represented graphically. In the following graph, *B* is supposed to be more volatile than $A(p_B^0 > p_A^0)$.

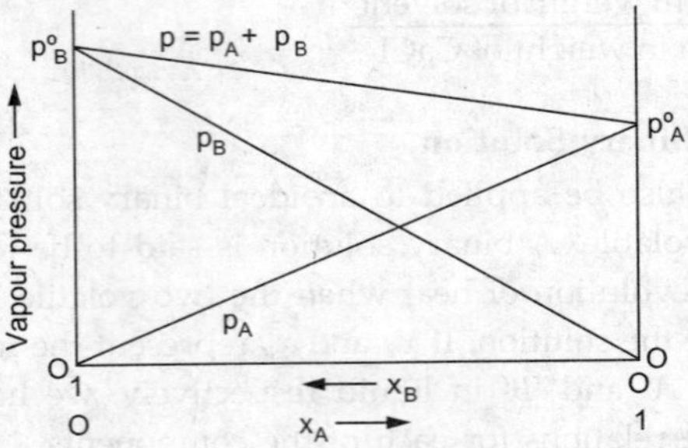

Nonideal or Real Binary Solutions

The nonideal binary solutions are those in which none of its components obeys Raoult's law of vapour pressure. Such solutions are of the following types:

Type I: The total pressure of the binary mixture is greater than the vapour pressures of either of the pure liquid components. Such solutions are said to show positive deviation from Raoult's law and boil at a lower temperature than either of the components. Examples are water–propyl alcohol, ethyl alcohol–chloroform, acetaldehyde–carbon disulphide mixtures, etc.

Type II: The total vapour pressure of the binary mixture is less than the vapour pressures of either of the pure liquid components. Such solutions are said to show negative deviations from Raoult's law and boil at a higher temperature than either of the components. Examples are water–HCl, water–H_2SO_4, water–HNO_3, acetone–chloroform, etc.

Azeotropic Mixture

An azeotropic mixture is a mixture of two real, miscible liquids which boil and distil at a constant temperature without undergoing any change in composition. In case of type I solutions, minimum-boiling-point azeotrope is formed while in case of type II solutions, maximum-boiling-point azeotrope is formed. An azeotropic mixture cannot be separated into pure components by fractional distillation.

A liquid mixture which distils with a change in composition is called a **zeotropic** mixture.

Osmotic Pressure

When a solution and its solvent are separated by a semipermeable membrane, the excess pressure applied to the solution to just stop osmosis is known as osmotic pressure of the solution.

Osmotic pressure ∝ molar concentration at a temperature T

or $$p = \frac{n}{V} RT = cRT \qquad \left(c = \frac{n}{V}\right), \qquad \dots (6)$$

where n is the number of moles of solute and V is the volume of solution in litres. n/V is thus the molar concentration.

Isotonic Solutions

Solutions having the same osmotic pressure at the same temperature are known to be isotonic. Isotonic solutions have the same molar concentrations.

Hypertonic and Hypotonic Solutions

For two solutions, the one having higher osmotic pressure is called a hypertonic solution, and the other with the lower value of osmotic pressure is called a hypotonic solution.

van't Hoff's Law of Osmotic Pressure

A substance in dilute solution behaves exactly like a gas and the osmotic pressure of a solution is equal to the pressure which the solute would exert if it were a gas at the same temperature and occupied the same volume as that occupied by the solution.

Determination of Osmotic Pressure

(i) Pfeffer's method

(ii) Morse and Frazer's method

(iii) Berkeley and Hartley's method

Freezing-Point Depression

Depression in f.p. ∝ molal concentration (m)

$$\Delta T_f \propto m \quad \text{or} \quad \Delta T_f = K_f \,.\, m. \qquad \dots (7)$$

If n is the number of moles of the solute and W, the weight in grams of the solvent, $m = \frac{n}{W} \times 1000$ (moles of solute per 1000 g of the solvent).

K_f is the molal depression constant or cryoscopic constant of the solvent, which is defined as the depression in f.p. produced when 1 mole of the solute is dissolved in 1000 g of the solvent.

K_f may also be calculated from the following thermodynamically derived equation

$$K_f = \frac{RT^2}{1000\, l_f},$$

where T is the freezing point of the solvent and l_f is the latent heat of fusion per gram of solvent.

Determination of ΔT_f

(i) Beckmann's method

(ii) Rast's method

Boiling-Point Elevation

Elevation in b.p. $\propto$ molal concentration

$$\Delta T_b \propto m$$

or $$\Delta T_b = K_b \,.\, m \qquad \text{... (8)}$$

K_b is known as the molal elevation constant or ebullioscopic constant of the solvent.

K_b may be calculated from the equation: $K_b = \dfrac{RT^2}{1000 l_v}$,

where T is the boiling point of the solvent and l_v is the latent heat of vaporization per gram of the solvent.

Determination of ΔT_b

(i) Landsberger's method

(ii) Cottrell's method

Molecular Weight of Nonvolatile Solute

As we see that all the equations (1, 5, 6 and 7) on colligative properties contain a term 'n', the number of moles of the solute, we can calculate molecular weight of the solute by substituting

$$\text{number of moles} = \frac{\text{weight in grams}}{\text{molecular weight}}$$

or $n = \dfrac{w}{M}$ in the said equations.

The following figure shows the vapour pressure versus temperature curves of a pure solvent and a solution of a nonvolatile solute, under a constant total pressure of 1 atm. The vapour pressure of the solid solvent is unaffected by the presence of solute if the solid freezes out without containing a significant concentration of solute, as is usually the case.

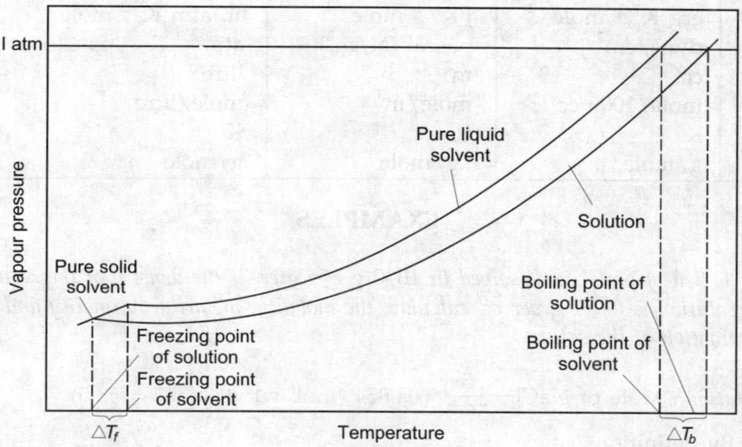

Abnormal Colligative Properties

van't Hoff Factor (*i*)

We have so far discussed ideal solutions containing the solute which neither undergoes dissociation nor association. In case where the solute undergoes dissociation or association, the number of solute particles change as a result of which abnormal results are obtained for colligative properties and molecular weights of solutes as the colligative properties depend upon the number of solute particles.

In order to account for such abnormal results, van't Hoff introduced a factor '*i*' known as van't Hoff factor, expressed as

$$i = \frac{\text{observed magnitude of any colligative property}}{\text{normal magnitude of the same colligative property}} \quad \dots (9)$$

$$= \frac{\text{normal molecular weight}}{\text{observed mol. weight}} \quad \left\{ \begin{array}{l} \text{as mol. wt. and colligative} \\ \text{properties are inversely} \\ \text{related} \end{array} \right\}$$

'*i*' can also be calculated as

$$i = \frac{\text{moles of solute particles after dissociation/association}}{\text{moles of solute particle originally taken}} \quad \dots (10)$$

$$= \frac{\text{observed molality}}{\text{calculated molality}}.$$

[**Note:** The following unit combinations must be remembered.]

	cgs	**SI**	**Nonsystem Units**
R	erg K^{-1} $mole^{-1}$	J K^{-1} $mole^{-1}$	lit. atm K^{-1} $mole^{-1}$
P	dyne/cm^2	pascal (N/m^2)	atm
V	cc	m^3	litre
c	mole/1000 cc	mole/m^3	mole/litre
T	K	K	K
M	g/mole	kg/mole	g/mole

EXAMPLES

Ex. 1. 5 g *of* NaCl *is dissolved in* 1000 g *of water. If the density of the resulting solution is* 0·997 g *per* cc, *calculate the molality, molarity, normality and mole fraction of the solute.*

Solution : Mole of NaCl $= \dfrac{5}{58 \cdot 5} = 0 \cdot 0854$ (mol. wt. of NaCl = 58·5)

By definition:

$$\text{Molality} = \frac{\text{moles}}{\text{wt. of solvent in grams}} \times 1000$$

$$= \frac{0 \cdot 0854}{1000} \times 1000 = 0 \cdot 0854 \text{ m.}$$

$$\text{Volume of the solution} = \frac{\text{wt. in grams}}{\text{density in grams/cc}} = \frac{1005}{0 \cdot 997} \text{ cc}$$

$$= 1008 \text{ mL} = 1 \cdot 008 \text{ litres.}$$

Again by definition

$$\text{molarity} = \frac{\text{moles}}{\text{volume of solution in litres}} = \frac{0 \cdot 0854}{1 \cdot 008} = 0 \cdot 085 \text{ M.}$$

$\therefore$ normality = 0·085 N (for NaCl, eq. wt. = mol. wt.)

Further, moles of $H_2O = \dfrac{1000}{18} = 55 \cdot 55$
(per 1000 g)

(Supposing 1 mL = 1 g for water having density 1 g/mL)

Total moles = moles of NaCl + moles of H_2O

$= 0 \cdot 0854 + 55 \cdot 55 = 55 \cdot 6409.$

$$\text{Mole fraction of NaCl} = \frac{\text{moles of NaCl}}{\text{total moles}} = \frac{0 \cdot 0854}{55 \cdot 6409}$$

$$= 1 \cdot 53 \times 10^{-3}.$$

Ex. 2. *A solution of ethanol in water is* 10% *by volume. If the solution and pure ethanol have densities of* 0·9866 g/cc *and* 0·785 g/cc *respectively, find the per cent by weight.*

Solution : Volume of ethanol = 10 mL; vol. of solution = 100 mL.

Weight of ethanol = volume × density

$$= 10 \times 0{\cdot}785 = 7{\cdot}85 \text{ g.}$$

Weight of solution $= 100 \times 0{\cdot}9866 = 98{\cdot}66$ g.

$$\therefore \quad \text{weight per cent} = \frac{7{\cdot}85}{98{\cdot}66} \times 100 = 7{\cdot}95\%.$$

Ex. 3. *Calculate the molarity of water.*

Solution : Supposing the density to be 1, one litre of water weighs 1000 g or $\frac{1000}{18}$ moles (H_2O = 18).

$$\therefore \quad \text{molarity} = \frac{1000}{18} = 55{\cdot}55 \text{ M.}$$

(moles/litre)

Ex. 4. *Calculate the molality and molarity of a solution of ethanol in water if the mole fraction of ethanol is* 0·05 *and the density of solution is* 0·997 g/cc.

Solution :

$$\text{Since mole fraction of } C_2H_5OH = \frac{\text{moles of } C_2H_5OH}{\text{moles of } C_2H_5OH + \text{moles of } H_2O}$$

$$= 0{\cdot}05 = \frac{5}{100}.$$

∴ 100 moles of solution contain 5 moles of C_2H_5OH.

Weight of C_2H_5OH = moles × mol. wt.

$$= 5 \times 46 = 230 \text{ g.}$$

Weight of $H_2O = 95 \times 18 = 1710$ g.

∴ weight of solution = 230 + 1710 = 1940 g.

$$\text{Volume of solution} = \frac{1940}{0{\cdot}997} \text{ cc} = 1945{\cdot}8 \text{ mL} = 1{\cdot}9458 \text{ litres.}$$

$$\therefore \quad \text{molality} = \frac{5}{1710} \times 1000 = 2{\cdot}92\ m.$$

$$\text{Molarity} = \frac{5}{1{\cdot}9458} = 2{\cdot}57 \text{ M.}$$

Ex. 5. *Calculate the molality and molarity of a solution made by mixing equal volumes of* 30% *by weight of* H_2SO_4 (density = 1·218 g/mL) *and* 70% *by weight of* H_2SO_4 (density = 1·610 g/mL).

Solution : Suppose that the solution contains 100 mL of each variety of H_2SO_4. Total volume is, therefore, 200 mL or 0·2 litre.

Wt. of 100 mL of H_2SO_4 solution (30%) $= 1{\cdot}218 \times 100$
$= 121{\cdot}8$ g

and wt. of 100 mL of H_2SO_4 solution (70%) $= 1{\cdot}610 \times 100$
$= 161$ g.

Wt. of H_2SO_4 (30%) $= 121{\cdot}8 \times \frac{30}{100} = 36{\cdot}54$ g.

Wt. of H_2SO_4 (70%) $= 161 \times \frac{70}{100} = 112{\cdot}7$ g.

Total wt. of H_2SO_4 (solute) $= 36{\cdot}54 + 112{\cdot}7 = 149{\cdot}24$ g.

$\therefore$ wt. of H_2O (solvent) = wt. of solution – wt. of solute
$= (121{\cdot}8 + 161) - 149{\cdot}24$
$= 133{\cdot}56$ g.

Moles of $H_2SO_4 = \frac{149{\cdot}24}{98} = 1{\cdot}5228$ (mol. wt. of $H_2SO_4 = 98$)

$$\text{Molality} = \frac{1{\cdot}5228}{133{\cdot}56} \times 1000 = 11{\cdot}4\ m.$$

$$\text{Molarity} = \frac{1{\cdot}5228}{0{\cdot}2} = 7{\cdot}6\ \text{M}.$$

Ex. 6. *The density of* 3 M *solution of sodium thiosulphate* ($Na_2S_2O_3$) *is* 1·25 g/mL. *Calculate*

(i) amount of sodium thiosulphate
(ii) mole fraction of sodium thiosulphate
(iii) molality of Na^+ *and* $S_2O_3^{2-}$ *ions*

Solution : (i) Let us consider one litre of sodium thiosulphate solution.

$\therefore$ wt. of the solution = density × volume (mL)
$= 1{\cdot}25 \times 1000 = 1250$ g.

Wt. of $Na_2S_2O_3$ present in 1 L of the solution
= molarity × mol. wt.
$= 3 \times 158 = 474$ g.

$$\text{Wt. \% of } Na_2S_2O_3 = \frac{474}{1250} \times 100 = 37{\cdot}92\%.$$

(ii) Wt. of solute ($Na_2S_2O_3$) = 474 g.

$$\text{Moles of solute} = \frac{474}{158} = 3.$$

Wt. of solvent (H_2O) $= 1250 - 474 = 776$ g.

$$\text{Moles of solvent} = \frac{776}{18} = 43{\cdot}11.$$

$$\therefore \text{mole fraction of } Na_2S_2O_3 = \frac{3}{3 + 43{\cdot}11} = 0{\cdot}063.$$

(iii) Molality of $Na_2S_2O_3 = \dfrac{\text{moles of } Na_2S_2O_3}{\text{wt. of solvent in grams}} \times 1000$

$$= \frac{3}{776} \times 1000 = 3{\cdot}865\ m.$$

$\because$ 1 mole of $Na_2S_2O_3$ contains 2 moles of Na^+ ions and 1 mole of $S_2O_3^{2-}$ ions.

$\therefore$ molality of $Na^+ = 2 \times 3{\cdot}865 = 7{\cdot}73\ m$.

Molality of $S_2O_3{}^{2-} = 3{\cdot}865\ m$.

Ex. 7. *One litre of milk weighs* 1·035 kg. *The butter fat it contains to the extent of* 4% *by volume has a density of* 875 kg/m^3. *Find the density of the fat-free 'skimmed' milk.* (1 m^3 = 10^3 litres)

Solution : Weight of 1 m^3 milk $= 1{\cdot}035 \times 10^3 = 1035$ kg.

Weight of the fat in 1 m^3 of milk $= 0{\cdot}04 \times 875$ kg.

(1 m^3 milk contains 0·04 m^3 of fat) = 35 kg.

$\therefore$ density of skimmed milk $= \dfrac{\text{wt. of skimmed milk}}{\text{vol. of skimmed milk}}$

$$= \frac{(1035 - 35)\ \text{kg}}{(1 - 0{\cdot}04)\ \text{m}^3} = 1042\ \text{kg/m}^3.$$

Ex. 8. *Calculate the molality of a 1-litre solution of* 93% H_2SO_4 *(wt./vol.). The density of the solution is* 1·84 g/mL. **(IIT 1990)**

Solution : The solution is 1000 mL, containing 930 g of H_2SO_4.

The weight of the solution will be 1840 g. The weight of the solvent (H_2O) will, therefore, be (1840 − 930), i.e., 910 g.

$$\therefore \text{ molality} = \frac{\text{mole of } H_2SO_4}{\text{wt. of } H_2O\ (g)} \times 1000$$

$$= \frac{930/98}{910} \times 1000 = 10{\cdot}428\ m.$$

Ex. 9. *Calculate the concentration of* NaOH *solution in* g/mL *which has the same normality as that of a solution of* HCl *of concentration* 0·04 g/mL.

Solution : Equivalents of HCl/mL $= \dfrac{0{\cdot}04}{36{\cdot}5}$ (eq. wt. of HCl = 36·5)

$$= 0{\cdot}001095 \text{ eq.}$$

$\therefore$ normality of HCl = eq. of HCl/1000 mL = 1·095 N.

$\therefore$ normality of NaOH = 1·095 N.

$\therefore$ g/1000 mL of NaOH = normality × eq. wt.

(eq. wt. of NaOH = 40)

$= 1{\cdot}095 \times 40 = 43{\cdot}8.$

$\therefore$ concentration of NaOH solution in g/mL = 0·0438 g/mL.

Ex. 10. *In which mode of expression, the concentration of solution remains independent of temperature?*
(a) Molarity *(b) Normality* *(c) Formality* *(d) Molality* **(IIT 1988)**

Solution : Since in the expression of molality, we take moles of solute per 1000 g of solvent and the mass is independent of temperature, the answer is (d).

Ex. 11. *The vapour pressure of benzene at* 25°C is 639·7 mmHg *and the vapour pressure of solution of a solute in benzene at the same temperature is* 631·9 mm *of mercury. Calculate the molality of the solution.*

Solution : We have,

$$\frac{p^0 - p}{p^0} = \frac{n_1}{n_2} \qquad \text{... (Eqn. 1)}$$

or $$\frac{n_1}{n_2} = \frac{\text{moles of solute}}{\text{moles of solvent}} = \frac{639{\cdot}7 - 631{\cdot}9}{639{\cdot}7} = 0{\cdot}0122.$$

Thus 1 mole of solvent (C_6H_6) contains 0·0122 mole of solute, or 78 g of solvent (C_6H_6) contains 0·0122 mole of solute ($C_6H_6 = 78$)

$$\therefore \text{ molality} = \frac{0{\cdot}0122}{78} \times 1000 = 0{\cdot}156\ m.$$

Ex. 12. *The vapour pressure of pure benzene at a certain temperature is* 640 mmHg. *A nonvolatile nonelectrolyte solid weighing* 2·175 g *is added to* 39 g *of benzene. The vapour pressure of the solution is* 600 mmHg. *What is the molecular weight of the solid substance?* **(IIT 1990)**

Solution : We know,

$$\frac{p^0 - p}{p^0} = \frac{n}{n + N} \qquad \text{... (Eqn. 1)}$$

Given that $p^0 = 640$ mm, $p = 600$ mm

$$n = \frac{2{\cdot}175}{M};\ N = \frac{39}{78} \qquad (C_6H_6 = 78)$$

(M is the mol. wt. of the solute)

$$\therefore \quad \frac{640 - 600}{640} = \frac{2{\cdot}175/M}{\frac{2{\cdot}175}{M} + 0{\cdot}5};\ M = 65{\cdot}25.$$

Ex. 13. *The vapour pressure of* H_2O *is* 23·756 mmHg *at* 25°C. *What is the vapour pressure of a solution of* 28·5 g *of sucrose,* $C_{12}H_{22}O_{11}$ (342) *in* 100 g *of* H_2O?

Solution : Mole of sucrose $= \frac{28 \cdot 5}{342} = 0 \cdot 0834.$

$$\text{Moles of } H_2O = \frac{100}{18} = 5 \cdot 55.$$

$$\therefore \text{ mole fraction of solute} = \frac{0 \cdot 0834}{0 \cdot 0834 + 5 \cdot 55} = 0 \cdot 0148.$$

We have,

$$\frac{p^0 - p}{p^0} = \text{mole fraction of solute} \qquad \text{... (Eqn. 1)}$$

$$\frac{23 \cdot 756 - p}{23 \cdot 756} = 0 \cdot 0148$$

$$p = 23 \cdot 404 \text{ mm.}$$

Ex. 14. *The vapour pressure of a dilute solution of glucose* ($C_6H_{12}O_6$) *is* 750 mmHg *at* 373 K. *Calculate (i) molality, and (ii) mole fraction of the solute.*

(IIT 1989)

Solution : Mole fraction of solute $= \frac{n}{n+N} = \frac{p^0 - p}{p^0}$ ... (Eqn. 1)

or $$\frac{n}{n+N} = \frac{760 - 750}{760} = 0 \cdot 0132.$$

$\therefore$ for $n = 0 \cdot 0132$, $(n + N)$ should be 1.

$\therefore$ mole of $H_2O = 1 - 0 \cdot 0132 = 0 \cdot 9868.$

Wt. of $H_2O = 0 \cdot 9868 \times 18 = 17 \cdot 763$ g.

$$\therefore \text{ molality} = \frac{0 \cdot 0132}{17 \cdot 763} \times 1000 = 0 \cdot 7404\ m.$$

Ex. 15. *Calculate vapour pressure of a* 5% *(by weight) solution of water in glycerol (mol. wt.* 92·1) *at* 100°C, *assuming Raoult's law to be valid and neglecting the vapour pressure of glycerol.*

Solution : Since pure water boils at 100°C, and b.p. is defined as that temperature at which the vapour pressure of the liquid is equal to atmospheric pressure, i.e., 760 mm, the vapour pressure of pure water at 100°C is 760 mm.

In this problem since glycerol (92·1) is the solute and water is the solvent and their weight per cents are 95% and 5% respectively, Raoult's law cannot be applied as the solution is not at all dilute. But according to the question, we apply Raoult's law

$$\frac{p^0 - p}{p^0} = \frac{n_1}{n_1 + n_2}. \qquad \text{... (Eqn. 1)}$$

Here $p^0 = 760$ mm, $n_1 = \text{mole of glycerol} = \frac{95}{92 \cdot 1}$, $n_2 = \text{mole of water} = \frac{5}{18}$ } supposing the weight of the solution as 100 g

Substituting in Eqn. (1), we get $p = 161{\cdot}2$ mm.

Ex. 16. *Calculate the molal lowering of vapour pressure for* H_2O *at* 100°C.

Solution : Molal lowering of vapour pressure is the lowering of vapour pressure of water when 1 mole of the solute is dissolved in 1000 g of the solvent (H_2O). Further, vapour pressure of pure water (p^0) at 100°C will be 760 mm as 100°C is its boiling point.

Now we have,

$$\text{lowering of v.p.} = p^0 - p = p^0\left(\frac{n}{n+N}\right)$$

$$= p^0\left(\frac{n}{N}\right)$$

$$= 760 \times \frac{1}{1000/18} \qquad \left(\begin{array}{l} n = 1 \\ N = \frac{1000}{18} \end{array}\right)$$

$$= 13{\cdot}68 \text{ mm.}$$

Ex. 17. *The vapour pressure of pure benzene at* 88°C *is* 957 mm *and that of toluene at the same temperature is* 379·5 mm. *Calculate the composition of a benzene–toluene mixture boiling at* 88°C.

Solution : Since the b.p. is that temperature at which the vapour pressure of the liquid is equal to the atmospheric pressure, i.e., 760 mm.

we have,

$$p = x_A \cdot p_A^0 + x_B \cdot p_B^0 \qquad \text{... (Eqn. 3)}$$

$$760 = x_{\text{benzene}}\, 957 + x_{\text{toluene}}\, 379{\cdot}5$$

As $x_{\text{benzene}} + x_{\text{toluene}} = 1$

$\therefore$ $760 = 957 \cdot x_{\text{benzene}} + 379{\cdot}5\,(1 - x_{\text{benzene}})$

$\therefore$ $x_{\text{benzene}} = 0{\cdot}6589.$

Ex. 18. *The vapour pressure in* mmHg *of a* $CH_3OH–C_2H_5OH$ *binary solution, p, at a certain temperature is represented by the equation*

$$p = 254 - 119\,x,$$

where x is the mole fraction of C_2H_5OH. *Find the vapour pressure of the pure components.*

Solution : We have,

$$p = x_A \cdot p_A^0 + x_B p_B^0 \qquad \text{... (Eqn. 3)}$$

or $$p = x_{CH_3OH} \cdot p^0{}_{CH_3OH} + x_{C_2H_5OH} \cdot p^0{}_{C_2H_5OH}$$

or $$p = (1 - x_{C_2H_5OH})\, p^0{}_{CH_3OH} + x_{C_2H_5OH} \cdot p^0{}_{C_2H_5OH}$$

or $$p = p^0{}_{CH_3OH} - (p^0{}_{CH_3OH} - p^0{}_{C_2H_5OH})\, x_{C_2H_5OH}.$$

Comparing this equation with the given equation, we get,

$$p^0{}_{CH_3OH} = 254 \text{ mm}$$

and $$p^0{}_{CH_3OH} - p^0{}_{C_2H_5OH} = 119 \text{ mm}$$

$\therefore$ $$p^0{}_{C_2H_5OH} = 254 - 119 = 135 \text{ mm.}$$

Ex. 19. *A solution is prepared by dissolving* 10 g *of nonvolatile solute in* 180 g *of* H_2O. *If the relative lowering of vapour pressure is* 0·005, *find the mol. wt. of the solute.*

Solution : Suppose the mol. wt. of the solute is M.

$$\text{Moles of solute} = \frac{10}{M}.$$

$$\text{Mole of solvent } (H_2O) = \frac{180}{18} = 10.$$

$$\text{Mole fraction of solute} = \frac{10/M}{\frac{10}{M} + 10} = \frac{1}{M+1}.$$

We know, relative lowering of vap. pressure

$$= \text{mole fraction of solute}$$

$$0{\cdot}005 = \frac{1}{M+1}$$

$$M = 199.$$

Ex. 20. *Benzene and toluene form an ideal solution. The vapour pressure of benzene and toluene are respectively* 75 mm *and* 22 mm *at* 20°C. *If the mole fractions of benzene and toluene in vapour are* 0·63 *and* 0·37 *respectively, calculate the vapour pressure of the ideal mixture.*

Solution : Since the mole fractions of the components given in the problem are of the vapour which is in equilibrium with the ideal mixture, we have,

$$\frac{1}{p}=\frac{x'_A}{p_A^0}+\frac{x'_B}{p_B^0} \quad \text{... (Eqn. 5)}$$

$$\frac{1}{p}=\frac{0\cdot63}{75}+\frac{0\cdot37}{22}$$

$$p=39\cdot65 \text{ mm.}$$

Ex. 21. *The vapour pressures of benzene and toluene at* 20°C *are* 75 mmHg *and* 22 mmHg *respectively.* 23·4 g *of benzene and* 64·4 g *of toluene are mixed. If the two form an ideal solution, calculate the mole fraction of benzene in the vapour phase if the vapours are in equilibrium with the liquid mixture at the same temperature.*

Solution : No. of moles of benzene in liquid $=\frac{23\cdot4}{78}=0\cdot3.$ $(C_6H_6=78)$

No. of moles of toluene in liquid $=\frac{64\cdot4}{92}=0\cdot7.$ $(C_6H_5CH_3=92)$

$\therefore$ mole fraction of benzene $=\frac{0\cdot3}{0\cdot3+0\cdot7}=0\cdot3.$

$\therefore$ mole fraction of toluene $=\frac{0\cdot7}{0\cdot3+0\cdot7}=0\cdot7.$

$$\left.\begin{array}{l}\text{Partial pressure of benzene} = 75\times0\cdot3=22\cdot5\\ \text{Partial pressure of toluene} = 22\times0\cdot7=15\cdot4\end{array}\right\} \quad \text{...(Eqn. 2)}$$

Total pressure $=22\cdot5+15\cdot4=37\cdot9$ mm.

Mole fraction of benzene in vapour phase $=\frac{\text{partial pressure}}{\text{total pressure}}$ ... (Eqn. 4)

$$=\frac{22\cdot5}{37\cdot9}=0\cdot59.$$

[**Note:** Mole fraction in vapour can also be calculated using Eqn. 5.]

Ex. 22. *Cyclohexane and ethanol at a particular temperature have vapour pressures of* 280 mm *and* 168 mm *respectively. If these two solutions having the mole fraction value of cyclohexane equal to* 32 *are mixed and the mixture has a total vapour pressure of* 376 mm, *will the mixture be an ideal solution?*

Solution : Mole fraction of cyclohexane $=0\cdot32.$

$\therefore$ mole fraction of ethanol $=1-0\cdot32=0\cdot68.$

Thus,

$$p=280\times0\cdot32+168\times0\cdot68 \quad \text{... (Eqn. 3)}$$

$$=203\cdot84 \text{ mm.}$$

For the solution to be an ideal one, the vapour pressure should be 203·84 but as the given value of vapour pressure is 376 mm, the solution is not ideal.

Ex. 23. *Two liquids A and B form an ideal solution. At* 300 K, *the vapour pressure of a solution containing* 1 *mole of A and* 3 *moles of B is* 550 mmHg. *At the same temperature, if one mole of B is added to this solution, the vapour pressure of the solution increases by* 10 mmHg. *Determine the vapour pressure of A and B in their pure states.*

Solution : Mole fraction of $A = \frac{1}{1+3} = \frac{1}{4}$.

Mole fraction of $B = \frac{3}{1+3} = \frac{3}{4}$.

Applying Eqn. 3 we get,

$$550 = \frac{1}{4} \times p_A^0 + \frac{3}{4} p_B^0. \quad \text{... (1)}$$

When one more mole of *B* was added to the solution:

Mole fraction of $A = \frac{1}{1+4} = \frac{1}{5}$.

Mole fraction of $B = \frac{4}{1+4} = \frac{4}{5}$.

Again applying Eqn. (3), we get,

$$550 + 10 = \frac{1}{5} p_A^0 + \frac{4}{5} p_B^0 \quad \text{... (2)}$$

From (1) and (2), we get,

$$p_A^0 = 400 \text{ mm}$$
$$p_B^0 = 600 \text{ mm.}$$

Ex. 24. *Benzene and toluene form nearly ideal solutions. If at* 300 K, $p^0_{\text{toluene}} = 32{\cdot}06$ mm *and* $p^0_{\text{benzene}} = 103{\cdot}01$ mm,

(a) calculate the vapour pressure of a solution containing 0·6 *molefraction of toluene*

(b) calculate the mole fraction of toluene in the vapour for this composition of liquid

Solution : (a) We have,

$$p = x_A \cdot p_A^0 + x_B \cdot p_B^0. \quad \text{... (Eqn. 3)}$$

Given that $x_{\text{toluene}} = 0{\cdot}6, \quad x_{\text{benzene}} = 1 - 0{\cdot}6 = 0{\cdot}4$

$$\therefore \quad p = 0{\cdot}6 \times 32{\cdot}06 + 0{\cdot}4 \times 103{\cdot}01$$
$$= 60{\cdot}44 \text{ mm.}$$

Again we have,

$$\frac{1}{p} = \frac{x'_A}{p_A^0} + \frac{x'_B}{p_B^0} \quad \text{... (Eqn. 5)}$$

$$\frac{1}{60{\cdot}44} = \frac{x'_{\text{toluene}}}{32{\cdot}06} + \frac{x'_{\text{benzene}}}{103{\cdot}01}$$

But $x'_{toluene} + x'_{benzene} = 1.$

$$\therefore \quad \frac{1}{60 \cdot 44} = \frac{x'_{toluene}}{32 \cdot 06} + \frac{1 - x'_{toluene}}{103 \cdot 01}$$

$$\therefore \quad x'_{toluene} = 0 \cdot 3182.$$

Ex. 25. *A very small amount of a nonvolatile solute (that does not dissociate) is dissolved in* 56·8 cm^3 *of benzene (density* 0·889 g cm^{-3}*). At room temperature, vapour pressure of this solution is* 98·88 mmHg *while that of benzene is* 100 mmHg. *Find the molality of this solution. If the freezing temperature of this solution is* 0·73 *degree lower than that of benzene, what is the value of molal freezing-point depression constant of benzene?* **(IIT 1997)**

Solution : We have,

$$\frac{p^0 - p}{p^0} = \frac{n}{N}$$

$$\frac{100 - 98 \cdot 88}{100} = \frac{n}{0 \cdot 6474} \qquad \left\{ N = \frac{56 \cdot 8 \times 0 \cdot 889}{78} = 0 \cdot 6474 \right\}$$

$$n = 0 \cdot 00725.$$

$$\text{Molality} = \frac{0 \cdot 00725 \times 1000}{56 \cdot 8 \times 0 \cdot 889} = 0 \cdot 1435 \; m.$$

Further,

$$K_f = \frac{\Delta T_f}{m} = \frac{0 \cdot 73}{0 \cdot 1435} = 5 \cdot 087.$$

Ex. 26. *The molar volume of liquid benzene (density* = 0·877 g/mL*) increases by a factor of* 2750 *as it vaporises at* 20°C *and that of liquid toluene (density* = 0·867 g/mL*) increases by a factor of* 7720 *at* 20°C. *A solution of benzene and toluene at* 20°C *has a vapour pressure of* 46 torr. *Find the mole fraction of benzene in the vapour above the solution.* **(IIT 1996)**

Solution : The volume of 1 mole of C_6H_6 (liquid) $= \frac{78}{0 \cdot 877}$ mL.

The volume of 1 mole of C_6H_6 (vapour) $= \frac{78}{0 \cdot 877} \times 2750$ mL

$= 244 \cdot 58$ litres

for 1 mole of C_6H_6 vapour at 20°C,

$$p_b^0 V = RT$$

$$p_b^0 = \frac{0 \cdot 0821 \times 293}{244 \cdot 58} \text{ atm} = 74 \cdot 75 \text{ mm}.$$

Similarly for 1 mole of $C_6H_5CH_3$ vapour at 20°C,

$$p_t^0 = \frac{0.0821 \times 293}{\left(\frac{92}{0.867} \times \frac{7720}{1000}\right)} \text{ atm} = 22.32 \text{ mm.}$$

Now,

$$\frac{1}{p} = \frac{x_b'}{p_b^0} + \frac{(1 - x_b')}{p_t^0}$$

$$\frac{1}{46} = \frac{x_b'}{74.75} + \frac{1 - x_b'}{22.32}$$

$$x_b' = 0.7336 .$$

Ex. 27. *Benzene and toluene form nearly ideal solutions. At* 300 K, $p^0_{\text{toluene}} = 32.06$ mm *and* $p^0_{\text{benzene}} = 103.01$ mm.

(a) A liquid mixture is composed of 3 moles of toluene and 2 moles of benzene. If the pressure over the mixture at 300 K is reduced, at what pressure does the first vapour form?

(b) What is the composition of the first trace of vapour formed?

(c) If the pressure is reduced further, at what pressure does the last trace of liquid disappear?

(d) What is the composition of the last trace of liquid?

Solution : (a) $p = \frac{3}{5} \times 32.06 + \frac{2}{5} \times 103.01$ ... (Eqn. 3)

$= 60.44$ mm.

(b) $\frac{1}{60.44} = \frac{x_t}{32.06} + \frac{(1 - x_t)}{103.01}$ ... (Eqn. 5)

$x_t = 0.3181.$

(c) $\frac{1}{p} = \frac{\frac{3}{5}}{32.06} + \frac{\frac{2}{5}}{103.01}$ ... (Eqn. 5)

$p = 44.25$ mm

(d) $44.25 = x_t \times 32.06 + (1 - x_t)\, 103.01$ ... (Eqn. 3)

$x_t = 0.8281.$

Ex. 28. *Calculate the osmotic pressure of a decinormal solution of cane sugar at* 0°C.

Solution : $\frac{n}{V}$ = molar concentration = 0·1 M

$R = 0.082$ lit · atm/K/ mole; $T = 273$ K

We have, $p = \frac{n}{V} RT$ ...(Eqn. 6)

$\therefore$ $p = 0{\cdot}1 \times 0{\cdot}082 \times 273$

$= 2{\cdot}24$ atm.

Ex. 29. *An aqueous solution contains* 10 g *of glucose* (*mol. wt.* = 180) *per* 0·5 L. *Assuming the solution to be ideal, calculate osmotic pressure at* 25°C.

Solution : Mole of glucose $= \frac{10}{180} = \frac{1}{18}$, $V = 0{\cdot}5$ L

$R = 0{\cdot}082$ lit · atm/K/mole $T = 273 + 25 = 298$ K.

We have,

$$p = \frac{n}{V} RT$$

$$= \frac{1/18}{0{\cdot}5} \times 0{\cdot}082 \times 298$$

$$= 2{\cdot}715 \text{ atm.}$$

Ex. 30. *A solution of* 1·73 g *of 'A' in* 100 cc *of water is found to be isotonic with a* 3·42% (*wt./vol.*) *solution of sucrose* ($C_{12}H_{22}O_{11}$). *Calculate molecular weight of A.* ($C_{12}H_{22}O_{11} = 342$)

Solution : We know that the isotonic solutions have the same molar concentration (i.e., moles/lit).

Let the molecular weight of A be M.

$$\text{Moles of } A = \frac{1{\cdot}73}{M}$$

$\therefore$ molar conc. of A (moles/litre) $= \frac{1{\cdot}73}{M} \times \frac{1000}{100} = \frac{17{\cdot}3}{M}$.

$$\text{Molar conc. of sucrose} = \frac{3{\cdot}42}{342} \times \frac{1000}{100} = 0{\cdot}1$$

$\therefore$ $\frac{17{\cdot}3}{M} = 0{\cdot}1;\ M = 173.$

Ex. 31. *Calculate the molecular weight of cellulose acetate if its* 0·2% (*wt./vol.*) *solution in acetone* (*sp. gr.* 0·8) *shows an osmotic rise of* 23·1 mm *against pure acetone at* 27°C.

Solution : 0·2 per cent solution means 0·2 g of cellulose acetate dissolved in 100 mL of solution.

Osmotic pressure = 2·31 cm of acetone

$$p = 2{\cdot}31 \times \frac{0{\cdot}80}{13{\cdot}6} \text{ cmHg} = 0{\cdot}136 \text{ cmHg}.$$

$$p = \frac{0{\cdot}136}{76} \text{ atm. (1 atm} = 76 \text{ cmHg)}$$

Suppose M is the molecular weight of cellulose acetate.

$$n = \frac{0{\cdot}2}{M},\ V = 100 \text{ mL} = 0{\cdot}1 \text{ litre};\ R = 0{\cdot}082 \text{ lit} \cdot \text{atm/K/mole}$$

and $T = 273 + 27 = 300$ K.

Now, $$p = \frac{n}{V} RT \qquad \text{... (Eqn. 6)}$$

$$\therefore \qquad \frac{0{\cdot}136}{76} = \frac{0{\cdot}2/M}{0{\cdot}1} \times 0{\cdot}082 \times 300$$

$$M = 27{,}500.$$

Ex. 32. *At* 10°C *the osmotic pressure of a urea solution is* 500 mm. *The solution is diluted and the temperature is raised to* 25°C, *when the osmotic pressure is found to be* 105·3 mm. *Determine the extent of dilution.*

Solution : Suppose V_1 litres of the solution contains n moles of the solute at 10°C which was diluted to V_2 litres at 25°C.

Thus we have,

$$\frac{500}{760} = \frac{n}{V_1} \times 0{\cdot}082 \times 283 \qquad \text{... (1)}$$

and $$\frac{105{\cdot}3}{760} = \frac{n}{V_2} \times 0{\cdot}082 \times 298 \qquad \text{... (2)}$$

... (Eqn. 6)

Dividing (1) by (2), we get,

$$\frac{V_2}{V_1} = 5.$$

Thus the solution was diluted 5 times.

Ex. 33. 2 g *of benzoic acid, dissolved in* 25 g *of benzene, produces a freezing-point depression of* 1·62°. *Calculate the molecular weight. Compare this with the molecular weight obtained from the formula for benzoic acid,* C_6H_5COOH. ($K_f = 4{\cdot}90$)

Solution : $$\text{Molality} = \frac{\Delta T_f}{K_f} \qquad \text{... (Eqn. 7)}$$

$$= \frac{1{\cdot}62}{4{\cdot}9}.$$

Suppose the molecular weight of benzoic acid is M.

$\therefore$ moles of benzoic acid $= \dfrac{2}{M}$.

$\therefore$ molality (moles/1000 g) $= \dfrac{2}{M} \times \dfrac{1000}{25} = \dfrac{80}{M}$.

Thus $\dfrac{80}{M} = \dfrac{1 \cdot 62}{4 \cdot 9}; M = 241 \cdot 97.$

The actual value of the molecular weight of benzoic acid, from its formula C_6H_5COOH, is 122. The observed (experimental) value, i.e., 241·97 is much greater than the normal value because of the association of benzoic acid in benzene.

Ex. 34. *Ethylene glycol,* $CH_2OH \cdot CH_2OH$, *the major component of permanent antifreeze, effectively depresses the freezing point of water in an automobile radiator. What minimum weight of ethylene glycol must be mixed with 6 gallons of water to protect it from freezing at −24°C.* (1 gallon = 3·785 lit, $K_f = 1 \cdot 86$)?

Solution : We have,

$$\Delta T_f = K_f \times \text{molality}.$$

$\therefore$ $$\text{molality} = \frac{\Delta T_f}{K_f} = \frac{24}{1 \cdot 86} = 12 \cdot 90. \qquad (\text{f.p. of } H_2O = 0°C)$$

Thus,

1 kg of H_2O should contain 12·90 moles of ethylene glycol.

$\therefore$ 22·7 kg (i.e., 6 gallons) of H_2O should contain $12 \cdot 90 \times 22 \cdot 7$ moles

$$= (12 \cdot 90 \times 22 \cdot 7 \times 62) \text{ g}$$
$$= 18155 \text{ g}$$
$$= 18 \cdot 155 \text{ kg of } CH_2OH \cdot CH_2OH.$$

$$(CH_2OH \cdot CH_2OH = 62)$$

Ex. 35. *What is the weight of* 1 *mole of a solute,* 0·132 g *of which in* 29·7 g *of benzene, gave a freezing-point depression of* 0·108°C?
(K_f *for benzene* = 5·12)

Solution : $$\text{Molality} = \frac{\Delta T_f}{K_f} = \frac{0 \cdot 108}{5 \cdot 12} \qquad \text{... (Eqn. 7)}$$

and also molality $= \dfrac{0 \cdot 32}{M} \times \dfrac{1000}{29 \cdot 7} m$ (moles per 1000 g)

(M = mol. wt. or wt. of 1 mole of solute)

Thus,

$$\frac{0 \cdot 132}{M} \times \frac{1000}{29 \cdot 7} = \frac{0 \cdot 108}{5 \cdot 12}$$

$$M = 211 \cdot 2 \text{ g mole}^{-1}.$$

Ex. 36. *K_f for water is $1 \cdot 86°C \cdot m^{-1}$. What is the molality of a solution which freezes at $-0 \cdot 192°C$? Assuming no change in the solute by raising the temperature, at what temperature will the solution boil? (K_b for $H_2O = 0 \cdot 515°C \cdot m^{-1}$)*

Solution : Molality $= \frac{\Delta T_f}{K_f}$

$$= \frac{0 \cdot 192}{1 \cdot 86} = 0 \cdot 103 \, m$$

Again $\Delta T_b = K_b \, . \, m$ ... (Eqn. 8)

$$= 0 \cdot 515 \times 0 \cdot 103 = 0 \cdot 0532°C.$$

Assuming the b.p. of pure water to be 100°C the b.p. of the solution will be 100·0532°C.

Ex. 37. *An aqueous solution of urea had a freezing point of $-0 \cdot 52°C$. Predict the osmotic pressure of the same solution at 37°C. Assume that the molar concentration and the molality are numerically equal.* ($K_f = 1 \cdot 86$)

Solution : Molality $= \frac{\Delta T_f}{K_f} = \frac{0 \cdot 52}{1 \cdot 86}$ = molarity (given)

Osmotic pressure $= c\, RT$

$$= \frac{0 \cdot 52}{1 \cdot 86} \times 0 \cdot 0821 \times (273 + 37)$$

$$= 7 \cdot 1 \text{ atm.}$$

Ex. 38. *The freezing temperature of pure benzene is 5·40°C. When 1·15 g of naphthalene is dissolved in 100 g of benzene, the resulting solution has a freezing point of 4·95°C. The molal f.p. depression constant for benzene is 5·12, what is the molecular weight of naphthalene?*

Solution : Molality $= \frac{\Delta T_f}{K_f} = \frac{5 \cdot 40 - 4 \cdot 95}{5 \cdot 12} = 0 \cdot 088$ mole per 1000 g.

$\because$ 1000 g of the solvent contains 11·5 g of naphthalene.

$\therefore$ mol. wt. of naphthalene $= \frac{\text{weight in grams/1000 g}}{\text{no. of moles/1000 g}}$

... (Rule 1, Chapter 1)

$$= \frac{11 \cdot 5}{0 \cdot 088} = 130.$$

Ex. 39. *An aqueous solution contains 5% and 10% of urea and glucose respectively (by weight). Find freezing point of the solution.* K_f *for water* = 1·86.

Solution : We have,

$$\Delta T_f = K_f \times \text{molality}$$

$$= K_f \times \text{ no. of moles of solute per 1000 g of solvent}$$

$$= 1{\cdot}86 \times \left(\frac{5}{60} + \frac{10}{180}\right) \times \frac{1000}{85} = 3{\cdot}03. \quad \left\{\begin{array}{r} \text{urea} = 60 \\ \text{glucose} = 180 \end{array}\right\}$$

Since pure water freezes at 0°C, the f.p. of the solution will be –3·03°C.

Ex. 40. *A mixture which contains* 0·550 g *of camphor and* 0·045 g *of an organic solute freezes at* 157°C. *The solute contains* 93·46% *of* C *and* 6·54% *of* H *by weight. What is the molecular formula of the compound? (Freezing point of camphor* = 178·4°C and K_f = 37·70)

Solution : Molality $= \dfrac{\Delta T_f}{K_f} = \dfrac{178{\cdot}4 - 157}{37{\cdot}70} = \dfrac{21{\cdot}4}{37{\cdot}70}$.

If the molecular wt. of solute is M then

$$\text{molality} = \frac{\text{moles of solute}}{\text{wt. of solvent in grams}} \times 1000$$

$$= \frac{0{\cdot}045/M}{0{\cdot}550} \times 1000 = \frac{4500}{55\,M}.$$

Thus, $\dfrac{4500}{55M} = \dfrac{21{\cdot}4}{37{\cdot}70}$; $M = 144{\cdot}14$.

Now, from the given weight % of C and H, we get,

$$\left.\begin{array}{l} \text{moles of C} = \dfrac{93{\cdot}46}{12} = 7{\cdot}79 \\ \text{moles of H} = \dfrac{6{\cdot}54}{1} = 6{\cdot}54 \end{array}\right\} \quad \text{(supposing 100 g of the solute)}$$

∴ molar ratio of C and H = 7·79 : 6·54, i.e., 1·2 : 1, i.e., 12 : 10.

∴ empirical formula is $C_{12}H_{10}$.

Empirical-formula weight = 12 × 12 + 1 × 10 = 154.

The empirical-formula weight is slightly greater than the molecular weight which might be due to some molecular change of the solute in the solvent. However, the molecular formula will be $C_{12}H_{10}$.

Ex. 41. 1000 g *of* 1 *m aqueous solution of sucrose is cooled and maintained at* –3·534°C. *Find how much ice will separate out.*

$K_f(H_2O) = 1{\cdot}86$ K· kg · mol^{-1}.

Solution : Let m' be the molality of the solution after the ice separates out at –3·534°C.

Now we have,

$$\Delta T_f = K_f \cdot m'$$

$$\therefore \qquad m' = \frac{\Delta T_f}{K_f} = \frac{3{\cdot}534}{1{\cdot}86} = 1{\cdot}9.$$

Let us now calculate the amount of ice separated.

Initially the molality is 1 m and wt. of solution is 1000 g.

1 mole of sucrose is dissolved in 1000 g of H_2O

or 342 g of sucrose is dissolved in 1000 g of H_2O.

$\therefore$ 1342 g of solution contained 342 g of sucrose.

$\therefore$ 1000 g of solution contained $\frac{342}{1342} \times 1000 = 254{\cdot}84$ g.

Amount of $H_2O = 1000 - 254{\cdot}84 = 745{\cdot}16$ g.

Now, when ice separates out, the molality is 1·9 and the weight of sucrose remains the same as before.

$\because$ $(1{\cdot}9 \times 342)$ g of sucrose is present in 1000 g of H_2O.

$\therefore$ 254·84 g of sucrose should be in $\frac{1000 \times 254{\cdot}84}{1{\cdot}9 \times 342}$

$= 392{\cdot}18$ g of H_2O.

Thus amount of ice separated $= 745{\cdot}16 - 392{\cdot}18$

$= 352{\cdot}98$ g.

Ex. 42. *If* 30 g *of a solute of molecular weight* 154 *is dissolved in* 250 g *of benzene, what will be the boiling point of the resulting solution under atmospheric pressure? The molal boiling-point elevation constant for benzene is* $2{\cdot}61°C \cdot m^{-1}$ *and the boiling point of pure benzene is* 80·1°C.

Solution : Suppose the elevation in b.p. is ΔT_b.

$$\text{Molality} = \frac{\Delta T_b}{K_b} = \frac{\Delta T_b}{2{\cdot}61}. \qquad \text{... (Eqn. 8)}$$

$$\text{Moles of solute} = \frac{30}{154}$$

$$\therefore \qquad \text{molality} = \frac{30}{154} \times \frac{1000}{250} \text{ moles/1000 g of solvent}$$

$$\text{Thus,} \qquad \frac{\Delta T_b}{2{\cdot}61} = \frac{30}{154} \times \frac{1000}{250}$$

$$\Delta T_b = 2°.$$

Thus the b.p. of the resulting solution is $80{\cdot}1 + 2 = 82{\cdot}1°C$.

Ex. 43. *A solution containing* 0·5216 g *of naphthalene (mol. wt.* = 128·16*) in* 50 mL *of* CCl_4 *shows boiling-point elevation of* 0·402° *while a solution of* 0·6216 g *of an unknown solute in the same weight of solvent gave a boiling-point elevation of* 0·647°. *Find the molecular mass of the unknown solute.*

Solution : Suppose the weight of 50 mL of solvent CCl_4 is W grams.

For the first solution, molality $= \dfrac{0{\cdot}5216/128{\cdot}16}{W} \times 1000$

$$= \frac{521{\cdot}6}{128{\cdot}16\,W}.$$

$$K_b = \frac{\Delta T_b}{m}$$

$$= \frac{0{\cdot}402}{521{\cdot}6/128{\cdot}16\,W}.$$

As the second solution is prepared in the same weight of solvent,

so for the second solution, molality $= \dfrac{0{\cdot}6216/M}{W} \times 1000$

$$= \frac{621{\cdot}6}{W\,M}$$

$\therefore$ $$K'_b = \frac{0{\cdot}647}{621{\cdot}6/W\,M}.$$

(M is the mol. wt. of the unknown solute.)

Since the solvent in both the solutions is the same,

$$K_b = K'_b.$$

Thus, $$\frac{0{\cdot}402}{521{\cdot}6/128{\cdot}16\,W} = \frac{0{\cdot}647}{621{\cdot}6/W\,M}$$

$$M = 94{\cdot}84.$$

Ex. 44. *A solution of urea in water has a boiling point of* 100·18°C. *Calculate the freezing point of the same solution. Molal constants for water* K_f *and* K_b *are* 1·86 and 0·512 *respectively.*

Solution : For a solution of molality m, we have,

$$m = \frac{\Delta T_f}{K_f} = \frac{\Delta T_b}{K_b}. \qquad \text{...(Eqns. 7 and 8)}$$

$\therefore$ $$\Delta T_f = \Delta T_b \frac{K_f}{K_b}; \qquad (\Delta T_b = 100{\cdot}18 - 100 = 0{\cdot}18°)$$

$$= 0{\cdot}18 \times \frac{1{\cdot}86}{0{\cdot}512} = 0{\cdot}654°.$$

As the f.p. of pure water is 0°C, the f.p. of the solution will be −0·654°C.

Ex. 45. *The b.p. of a solution of* 5 g *of sulphur in* 100 g *of carbon disulphide is* 0·476° *above that of pure solvent. Determine the molecular formula of sulphur in this solvent. The b.p. of pure carbon disulphide is* 46·30°C *and its heat of vaporisation is* 84·1 *calories per gram.*

Solution : We have, $K_b = \dfrac{RT^2}{1000\, l_v} = \dfrac{2 \times (46{\cdot}30 + 273)^2}{1000 \times 84{\cdot}1} = 2{\cdot}425.$

$\therefore \qquad m = \dfrac{\Delta T_b}{K_b} = \dfrac{0{\cdot}476}{2{\cdot}425}.$

And also, $\qquad m = \dfrac{\text{moles of solute}}{\text{wt. of solvent in grams}} \times 1000.$

$$= \frac{5}{M} \times \frac{1000}{100} = \frac{50}{M}$$

where M is the mol. wt. of sulphur.

Thus, $\qquad \dfrac{50}{M} = \dfrac{0{\cdot}476}{2{\cdot}425}$; $M = 255.$

Since the atomic wt. of sulphur is 32 and its mol. wt. is 255, therefore, number of sulphur atoms associated to form a single molecule is $\dfrac{255}{32} = 7{\cdot}96$, i.e., 8. Hence the formula is S_8.

Ex. 46. *The freezing point of an aqueous solution of* KCN *containing* 0·1892 mole/kg H_2O *was* – 0·704°C. *On adding* 0·095 *mole of* $Hg(CN)_2$, *the freezing point of the solution was* – 0·53°C. *Assuming that the complex is formed according to the equation*

$$Hg(CN)_2 + m\,CN^- \longrightarrow Hg(CN)^{m-}_{m+2}$$

and also $Hg(CN)_2$ *is the limiting reactant, calculate m.*

Solution : $K_f = \dfrac{\Delta T_f}{m} = \dfrac{0{\cdot}704}{2 \times 0{\cdot}1892}$ (KCN dissociates completely)

$= 1{\cdot}86$

0·095 mole		0·1892 mole		0 mole	... initially
$Hg(CN)_2$	+	$m\,CN^-$	$\longrightarrow$	$Hg(CN)^{m-}_{m+2}$	
0 mole		$(0{\cdot}1892 - 0{\cdot}095\,m)$		0·095 mole	... finally

Total molality after the addition of $Hg(CN)_2$

$= \text{molality of } K^+ + \text{molality of } CN^-$

$+ \text{molality of } Hg(CN)^{m-}_{m+2}$

$= 0{\cdot}1892 + (0{\cdot}1892 - 0{\cdot}095\,m) + 0{\cdot}095$

$= (0{\cdot}4734 - 0{\cdot}095\,m)$

Now, $K_f = \dfrac{\Delta T_f}{m}$

$$1{\cdot}86 = \frac{0{\cdot}53}{0{\cdot}4734 - 0{\cdot}095\,m}$$

$$m = 1{\cdot}984$$

or $m = 2.$

Ex. 47. *A* 0·01 *m aqueous solution of* $K_3[Fe(CN)_6]$ *freezes at* −0·062°C. *What is the apparent percentage of dissociation?* (K_f *for water* = 1·86)

Solution : We have, $\Delta T_f = m \times K_f$ ... (Eqn. 7)

$$= 0{\cdot}01 \times 1{\cdot}86$$

i.e., $(\Delta T_f)_{\text{normal}} = 0{\cdot}0186°$

and $(\Delta T_f)_{\text{observed}} = 0{\cdot}062°$ (given)

$$\therefore \quad i = \frac{\text{observed colligative property}}{\text{normal colligative property}} \quad \text{... (Eqn. 9)}$$

$$= \frac{0{\cdot}062}{0{\cdot}0186}.$$

Now suppose x is the degree of dissociation of $K_3[Fe(CN)_6]$

Thus,

moles before diss.: 1 mole 0 0

$$K_3[Fe(CN)_6] = 3K^+ + [Fe(CN)_6]^{3-}$$

moles after diss.: $(1 - x)$ $3x$ x

$$\therefore \quad i = \frac{(1-x) + 3x + x}{1} = \frac{0{\cdot}062}{0{\cdot}0186} \quad \text{... (Eqn. 10)}$$

$$x = 0{\cdot}78$$

∴ per cent dissociation = 78%

Ex. 48. *A* 0·2 *m aqueous solution of* KCl *freezes at* −0·68°C. *Calculate 'i' and the osmotic pressure at* 0°C. *Assume the volume of solution to be that of pure* H_2O *and* K_f *for* H_2O *is* 1·86.

Solution : $(\Delta T_f)_{\text{normal}} = K_f \times m$ ... (Eqn. 7)

$$= 1{\cdot}86 \times 0{\cdot}2$$

$$= 0{\cdot}372.$$

We have: $$i = \frac{\text{observed colligative property}}{\text{normal colligative property}} \quad \text{... (Eqn. 9)}$$

$$= \frac{0{\cdot}68}{0{\cdot}372} = 1{\cdot}83.$$

Again, $$i = \frac{\text{observed osmotic pressure}}{\text{normal osmotic pressure}}.$$

$\therefore$ observed osmotic pressure = $i \times$ normal osmotic pressure

$$= i \times c\,RT \qquad \text{... (Eqn. 6)}$$
$$= 1{\cdot}83 \times 0{\cdot}2 \times 0{\cdot}082 \times 273$$
$$= 8{\cdot}2 \text{ atm.}$$

Ex. 49. 1 g *of a monobasic acid when dissolved in* 100 g *of water lowers the freezing point by* 0·168°C. 0·2 g *of the same acid when dissolved and titrated required* 15·1 mL *of* N/10 *alkali. Calculate the degree of dissociation of the acid.* (K_f *for water is* 1·86)

Solution : Milliequivalent of alkali (m.e.) $= \frac{1}{10} \times 15{\cdot}1 = 1{\cdot}51$

(Eqn. 1, Chapter 7)

m.e. of the acid = 1·51. (Eqn. 2, Chapter 7)

$\therefore$ eq. of acid $= \frac{1{\cdot}51}{1000} = 0{\cdot}00151.$ (Eqn. 3, Chapter 7)

Now equivalent of acid $= \frac{\text{weight in grams}}{\text{equivalent weight}}.$

$$= \frac{0{\cdot}2}{\text{eq. wt.}}$$

$$= \frac{0{\cdot}2}{M} \text{ [for monobasic acid, eq. wt. = mol. wt. } (M)]$$

$$\therefore \quad \frac{0{\cdot}2}{M} = 0{\cdot}00151; \; M = 132{\cdot}45.$$

$\therefore$ molality of the acid (m) $= \frac{\text{moles of solute}}{\text{wt. of solvent in grams}} \times 1000$

$$= \frac{1}{132{\cdot}45} \times \frac{1000}{100} = 0{\cdot}076.$$

We have, $(\Delta T_f)_{\text{normal}} = K_f \,.\, m$

$$= 1{\cdot}86 \times 0{\cdot}076 = 0{\cdot}141$$

$$\therefore \quad i = \frac{(\Delta T_f)_{\text{observed}}}{(\Delta T_f)_{\text{normal}}} \qquad \text{... (Eqn. 9)}$$

$$= \frac{0{\cdot}168}{0{\cdot}141} = 1{\cdot}19.$$

The monobasic acid (say AH) ionises as

moles before diss.: 1 mole 0 0

$$AH \;=\; A^- + H^+$$

moles after diss.: $(1-x)$ x x (x = degree of dissociation)

$$i = \frac{1 - x + x + x}{1} = 1.19. \qquad \text{... (Eqn. 10)}$$

$$\therefore \qquad x = 0.19.$$

Ex. 50. 1·1 g *of* $CoCl_3 \cdot 6NH_3$ *(mol. wt. = 267) was dissolved in* 100 g *of* H_2O. *The freezing point of the solution was* –0·29°C. *How many moles of solute particles exist in solution for each mole of solute introduced?* K_f *for* $H_2O = 1.86°C \cdot m^{-1}$.

Solution : Molality (experimental) $= \frac{\Delta T_f}{K_f} = \frac{0.29}{1.86}$

$$= 0.156 \text{ mole/1000 g.}$$

$$\text{Molality (theoretical)} = \frac{\text{moles of solute}}{\text{wt. of solvent in g}} \times 1000$$

$$= \frac{1.1}{267} \times \frac{1000}{100} = 0.0412 \text{ mole/1000 g.}$$

Thus, number of moles of solute particles produced by 1 mole of solute

$$= \frac{0.156}{0.0412} \approx 4. \qquad \text{... (Eqn. 10)}$$

Ex. 51. 20·27 g *of benzene containing* 0·2965 g *of benzoic acid (mol. wt. =* 122*) freezes at* 0·317° *below the freezing point of pure benzene. If benzoic acid exists as dimer in benzene, find its degree of association.* K_f *for benzene is* $5.12°C \cdot m^{-1}$.

Solution : Molality (calculated) $= \frac{0.2965/122}{20.27} \times 1000$

$$= 0.12 \text{ mole/1000 g.}$$

$$\text{Molality (observed) } \frac{\Delta T_f}{K_f} = \frac{0.317}{5.12} = 0.0619 \text{ mole/1000 g.}$$

$$\therefore \quad i = \frac{\text{molality (observed)}}{\text{molality (calculated)}} = \frac{0.0619}{0.12} = 0.5158. \qquad \text{... (Eqn. 10)}$$

Since benzoic acid exists as dimer in benzene, we have

$$\begin{array}{lcc} \text{moles before association:} & 1 \text{ mole} & 0 \\ & 2C_6H_5COOH = & (C_6H_5COOH)_2 \\ \text{moles after association:} & (1 - x) & \frac{x}{2} \end{array} \quad (x = \text{degree of association})$$

$$\therefore \qquad i = \frac{1 - x + \frac{x}{2}}{1} = 0.5158 \qquad \text{... (Eqn. 10)}$$

$$x = 0.9684 \text{ or } 96.84\%.$$

Ex. 52. *A 0·025 M solution of a monobasic acid had a freezing point of –0·06°C. Calculate K_a for the acid. K_f (H_2O) = 1·86.*

Solution : ΔT_f (observed) = 0·06°C, (f.p. of H_2O = 0°C)

ΔT_f (calculated) = $0{\cdot}025 \times 1{\cdot}86 = 0{\cdot}0465$°C.

(for dilute aqueous solution, molarity = molality)

$$\therefore \quad i = \frac{\text{observed depression in f.p.}}{\text{calculated depression in f.p.}} = \frac{0{\cdot}06}{0{\cdot}0465} = 1{\cdot}29.$$

For the acid HA, we have

0·025		0		0	initial concentration
HA	=	H^+	+	A^-	
0·025(1 – x)		0·025x		0·025x	final concentration

(x is the degree of dissociation of HA)

$$i = \frac{1+x}{1} = 1{\cdot}29; \; x = 0{\cdot}29$$

$$\therefore \quad K_a = \frac{(0{\cdot}025\,x)\,(0{\cdot}025\,x)}{0{\cdot}025\,(1-x)} = 0{\cdot}025\,\frac{x^2}{1-x}$$

$$= \frac{0{\cdot}025 \times 0{\cdot}29^2}{1 - 0{\cdot}29} = 2{\cdot}96 \times 10^{-3}.$$

Ex. 53. *A decimolar solution of potassium ferrocyanide is* 50% *dissociated at* 300 K. *Calculate the osmotic pressure of the solution.*

($R = 8{\cdot}314\,\text{JK}^{-1}\,\text{mol}^{-1}$)

Solution :

1				
$K_4[Fe(CN)_6]$	$\rightarrow$	$4K^+$ +	$[Fe(CN)_6]^{4-}$	
1 – x		$4x$	x	($x \equiv$ deg. of diss.)

$$i = \frac{1 - x + 4x + x}{1} = 1 + 4x = 1 + 4 \times 0{\cdot}5 = 3$$

Observed osmotic pressure = $icRT$

$$= 3 \times (0{\cdot}1 \times 10^3) \times 8{\cdot}314 \times 300$$
$$= 7{\cdot}4826 \times 10^5 \text{ pascal } (\text{N/m}^2).$$

[**Note:** R is given in SI units. Hence unit of concentration 'c' would be in mole/m^3 (i.e., $c = 0{\cdot}1 \times 10^3$ mole/m^3) and that of pressure would be in pascal.]

Ex. 54. *The degree of dissociation of* $Ca(NO_3)_2$ *in a dilute aqueous solution containing* 7 g *of the salt per* 100 g *of water at* 100°C *is* 70 *per cent. If the vapour pressure of water at* 100°C *is* 760 mm, *calculate the vapour pressure of the solution.* **(IIT 1991)**

***Solution* :** 1

$$\underset{1-x}{Ca(NO_3)_2} \rightarrow \underset{x}{Ca^{2+}} + \underset{2x}{2NO_3^-}$$

van't Hoff factor, $i = \dfrac{1-x+x+2x}{1} = \dfrac{1+2x}{1} = 1 + 2 \times 0{\cdot}7 = 2{\cdot}4.$

Now, $$i = \frac{\text{observed lowering of vapour pressure}}{\text{normal lowering of vapour pressure}}.$$

$\therefore$ obs. lowering of vap. press. $= i \times$ nor. lowering of vap. press.

$$p^0 - p_{\text{obs.}} = i \times \left\{\frac{n}{n+N} \cdot p^0\right\}$$

$$760 - p_{\text{obs.}} = 2{\cdot}4 \times \frac{7/164}{\dfrac{7}{164} + \dfrac{100}{18}} \times 760 \quad \left\{\begin{matrix} Ca(NO_3)_2 = 164 \\ H_2O = 18 \end{matrix}\right\}$$

$$= 13{\cdot}90$$

$\therefore$ $$p_{\text{obs.}} = 760 - 13{\cdot}90 = 746{\cdot}10 \text{ mm.}$$

Ex. 55. *A sample of a drug* $C_{21}H_{23}O_5N$ *(mol. wt. = 369) mixed with lactose* $C_{12}H_{22}O_{11}$ *(mol. wt. = 342) was analysed by osmotic pressure measurement to determine the amount of sugar present. If* 100 mL *of solution containing* 1·0 g *of the drug-sugar mixture has an osmotic pressure of* 5·27 mmHg *at* 25°C, *what is the per cent sugar present?*

***Solution* :** Let the amount of lactose be x g.

$\therefore$ moles of the sugar-and-drug mixture $= \dfrac{x}{342} + \dfrac{(1-x)}{369}$

Now, osmotic pressure $= cRT$

$$\frac{527}{760} = \left[\left\{\frac{x}{342} + \frac{(1-x)}{369}\right\} \times \frac{1000}{100}\right] \times 0{\cdot}0821 \times 298$$

$$x = 0{\cdot}580 \text{ g.}$$

$\therefore$ percentage of sugar = 58·0%.

Ex. 56. *Dry air was passed through a solution containing* 20 g *of a substance in* 100·0 g *of water and then through pure water. The loss in mass of the solution was* 2·945 g *and that of pure water was* 0·059 g. *Calculate the molar mass of the substance.*

***Solution* :** We have,

$$\frac{\text{loss in mass of pure water}}{\text{loss in mass of solution}} = \frac{p^0 - p}{p}$$

or $$\frac{p^0 - p}{p} = \frac{0{\cdot}059}{2{\cdot}945} = 0{\cdot}02$$

$$\text{or} \qquad \frac{p^0}{p} - 1 = 0{\cdot}02$$

$$\text{or} \qquad \frac{p^0}{p} = 1{\cdot}02$$

$$\frac{p}{p^0} = \frac{1}{1{\cdot}02}$$

$$\frac{p^0 - p}{p^0} = 1 - \frac{1}{1{\cdot}02} = 0{\cdot}0196.$$

$$\therefore \qquad \frac{n}{N} = \frac{20/M}{100/18} = 0{\cdot}0196; \; M = 183{\cdot}6 \text{ g/moles.}$$

Ex. 57. *Two beakers are placed side by side in a sealed encloser. The first beaker contains* 100 mL *of pure water, the other, an equal volume of an aqueous solution of sugar. What would be the volumes of the liquids in the two beakers when all physical changes stop, that is, when the equilibrium is reached?*

Solution **:** The vapour pressure over water is higher than that over the sugar solution. The vapour pressure necessary to achieve equilibrium with pure water (p_1) is higher than that required with the sugar solution (p_2).

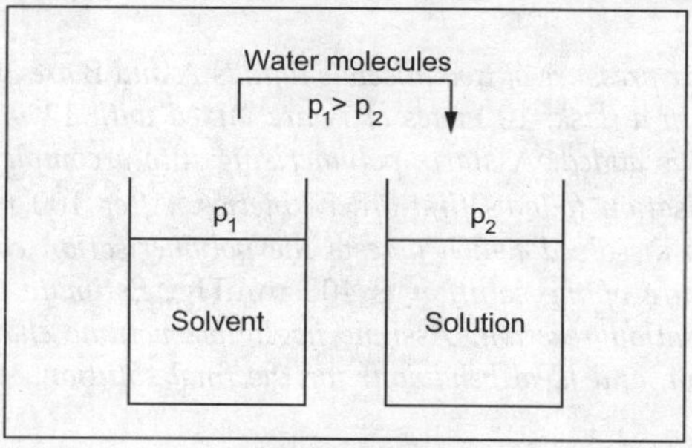

Consequently, as the pure water solvent in the first beaker seeks to reach equilibrium by forming water vapour, the sugar solution in the other beaker seeks to reach equilibrium by removing water molecules from the vapour phase. A net movement of water molecules from the pure water, in the first beaker, to the sugar solution, in the second beaker, results. The process continues until no free water in the first beaker remains. The other beaker shall then have 200 mL of sugar solution.

Ex. 58. *Two beakers, one containing* 20 mL *of a* 0·05 M *aqueous solution of a nonvolatile, nonelectrolyte and the other, the same volume of* 0·03 M *aqueous solution of* NaCl, *are placed side by side in a closed encloser. What are the volumes in the two beakers when equilibrium is attained.*

Solution : Moles of solute particles in the first beaker

$$= \frac{0.05}{1000} \times 20 = 0.001.$$

Moles of solute particles (Na^+ and Cl^-) in the other beaker

$$= \frac{2 \times 0.03}{1000} \times 20 = 0.0012.$$

As the solution in the first beaker is more dilute than that in the other beaker, the vapour pressure over the solution in the first beaker will be higher than that over the NaCl solution in the second beaker. The water molecules shall thus flow from the first beaker to the other till both the solutions have equal molarity. Let this volume of water molecules be v mL. Thus,

$$\frac{0.001}{20 - v} = \frac{0.0012}{20 + v}$$

or $v = 1.8$ mL.

$\therefore$ volume of the solution in the first beaker

$$= 20 - 1.8 = 18.2 \text{ mL}$$

and the volume of NaCl solution in the other beaker

$$= 20 + 1.8 = 21.8 \text{ mL}.$$

Ex. 59. *The vapour pressure of two miscible liquids* A *and* B *are* 300 *and* 500 mmHg *respectively. In a flask,* 10 *moles of* A *are mixed with* 12 *moles of* B. *However, as soon as* B *is added,* A *starts polymerising into a completely insoluble solid. The polymerisation follows first-order kinetics. After* 100 *minutes,* 0·525 *mole of a solute is dissolved which arrests the polymerisation completely. The final vapour pressure of the solution is* 400 mmHg. *Estimate the rate constant of the polymerisation reaction. Assume negligible volume change on mixing and polymerisation, and ideal behaviour for the final solution.*

(IIT 2001)

Solution : 10 mole

$$n\,A(l) \xrightarrow[\text{polymerisation}]{B(l)(12 \text{ mole})} An(s)$$

After 100 min : x moles (say).

Rate law : $\text{Rate} = -\frac{1}{n}\frac{d[A]}{dt} = k'[A]$

or $-\frac{1}{n}\frac{dx}{dt} = k'(x).$

On integration : $nk' = k = \frac{2.303}{t}\log\frac{a}{x}\cdot$ ... (1)

[In this problem n is not given. As n, i.e., no. of molecules of A polymerising to give one molecule, is a constant for this reaction and so the multiplication of n with the rate constant k' gives another constant k, which may also be called rate constant, though not according to its definition.]

Let us now calculate x to get the value of k from Equation (1).

When 0·525 mole of a solute is added, polymerisation stops and x moles of A remain. Just before the addition of the solute,

$$\text{moles of solvent } A = x$$

$$\text{moles of solvent } B = 12$$

$\therefore$ vapour pressure of the mixture of solvents $= x_A p_A^0 + x_B p_B^0$

$$p^0 = \frac{x}{x+12} \times 300 + \frac{12}{x+12} \times 500 \qquad \dots (2)$$

Vapour pressure of the solution, $p = 400$ ... (3)

(after the addition of 0·525 mole of a solute).

Applying Raoult's law,

$$\frac{p^0 - p}{p^0} = \text{mole fraction of the solute}$$

$$= \frac{0{\cdot}525}{0{\cdot}525 + (x+12)}. \qquad \dots (4)$$

Solving equations (2), (3) and (4), we get

$$x = 2{\cdot}84.$$

Substituting $x = 2{\cdot}84$, $t = 100$ m and $a = 10$ in Equation (1)

$$k = \frac{2{\cdot}303}{100} \log \frac{10}{2{\cdot}84} = 1{\cdot}2 \times 10^{-2} \text{ min}^{-1}.$$

PROBLEMS

(Answers bracketed with questions)

1. Calculate mole fractions of urea and water if 2·0 g of urea is dissolved in 31·4 g of aqueous solution. (0·02, 0·98)

2. Under what condition will the molality of a solute not be equal to its molarity? [**Hint:** Consider density of the solvent.]

3. A sugar syrup of weight 214·2 g contains 34·2 g of sugar ($C_{12}H_{22}O_{11}$). Calculate (i) molal concentration, and (ii) mole fraction of sugar in the syrup. **(IIT 1988)** (0·556 m, 0·0099)

4. Calculate mole fraction of solute in an aqueous 4 *m* solution assuming the density of the solution as 1·0 g/mL. (0·0674)

[**Hint:** See Example 6]

5. Nitric acid (70%) has a specific gravity of 1·42. Find the normality and molarity of the acid. (15·8 N, 15·8 M)

6. 1 g of NaCl is dissolved in 10 g of a solution, the density of which is 1·07 g/cc. Find the molality and molarity of NaCl. (1.899 *m*, 1.829 M)

[Hint: See Example 6]

7. Calculate mole fraction, molarity and molality of C_2H_5OH solution which is 50% by weight of C_2H_5OH in H_2O and has a density of ·9144 g per cc. (0·281, 9·93 M, 21·7 *m*)

8. The molarity and molality of a solution are *M* and *m* respectively. If the molecular weight of the solute is *M′*, calculate the density of the solution in terms of *M*, *m* and *M′*.

$$\left[D = M\left(\frac{1}{m} + \frac{M'}{1000}\right)\right]$$

9. When 400 g of a 20% solution was cooled, 50 g of the solute precipitated. What is the per cent concentration of the remaining solution? (8·6%)

10. In what mass of water must 25 g of $CuSO_4 \cdot 5H_2O$ be dissolved to obtain an 8% solution of $CuSO_4$? (174·775 g)

11. One litre of water was added to 500 mL of 32% HNO_3 of density 1·20 g/mL. What is the per cent concentration of HNO_3 in the solution obtained? (12·8%)

12. Determine the per cent concentration of a solution obtained by mixing 300 g of a 25% and 400 g of a 40% solution. (33·6%)

13. Calculate the per cent concentration of a 9·28 N NaOH solution of density 1·31 g/mL. (28·3%)

14. The pressure of the water vapour of a solution containing a nonvolatile solute is 2% below that of the vapour of pure water. Determine the molality of the solution. (1·134 *m*)

15. What is the vapour pressure at 100°C of a solution containing 15·6 g of water and 1·68 g of sucrose ($C_{12}H_{22}O_{11}$)? (75·57 cm)

16. The vapour pressure of an aqueous solution of cane sugar (mol. wt. 342) is 756 mm at 100°C. How many grams of sugar are present per 1000 g of water? (99·94 g)

17. At 20°C the vapour pressure of ether is 442 mmHg. When 6·1 g of a substance is dissolved in 50 g of ether (mol. wt. 74), the vapour pressure falls to 410 mm. What is the molecular weight of the substance? (115·4)

18. 0·5 g of a nonvolatile organic compound (mol. wt. 65) is dissolved in 100 mL of CCl_4. If the vapour pressure of pure CCl_4 is 143 mm, what would be the vapour pressure of the solution? (Density of CCl_4 solution is 1·58 g/cc). (141·93 mm)

19. Which of the following aqueous solutions has a higher vapour pressure if the density of water is 1 g/cc?

(a) Solution having mole fraction of cane sugar = 0·1

(b) Solution having molal concentration = 1 *m* (b)

20. An aqueous solution containing 20% by weight of liquid *X* (mol. wt. = 140) has a vapour pressure 160 mm at 60°C. Calculate the vapour pressure of pure liquid '*X*' if the vapour pressure of water is 150 mm at 60°C. (470·5 mm)

21. At 25°C, benzene and toluene have densities 0·879 and 0·867 g/cc respectively. Assuming that benzene–toluene solutions are ideal, establish the equation for the density of the solution.

$$d = \frac{1}{100}[0\cdot879\ V + 0\cdot867\ (100 - V)],$$

where V is the volume per cent of benzene.

22. Ethanol and methanol form a solution that is very nearly ideal. The vapour pressure of ethanol is 44·5 mm and that of methanol is 88·7 mm at 20°C.

(a) Calculate the mole fraction of methanol and ethanol in a solution obtained by mixing 60 g of ethanol with 40 g of methanol.

(b) Calculate the partial pressure and the total vapour pressure of this solution and the mole fraction of ethanol in the vapour. [(a) 0·49, 0·51 (b) 22·7, 43·5, 0·34]

23. The vapour pressure of pure benzene is 22 mm and that of pure toluene is 75 mm at 20°C. What is the composition of the solution of these two components that has a vapour pressure of 50 mm at this temperature? What is the composition of vapour in equilibrium with this solution?

[**Hint:** Use equations 3 and 5] (0·47 : 0·53 : 0·20 : 0·80)

24. A solution containing ethyl alcohol and propyl alcohol has a vapour pressure of 290 mm at 30°C. Find the vapour pressure of pure ethyl alcohol if its mole fraction in the solution is 0·65. The vapour pressure of propyl alcohol is 210 mm at the same temperature. (333·12 mm)

25. What are partial and total vapour pressures at 25°C above the solution having equal numbers of molecules of benzene and toluene? The vapour pressures of benzene and toluene at this temperature are 95·1 and 28·4 mmHg respectively. What is the composition of the vapour in equilibrium with benzene–toluene solution? [47·6, 14·2, 61·8 mmHg; 0·77 and 0·23]

[**Hint:** Mole fraction of each is 0·5.]

26. Benzene and toluene form nearly ideal solutions. If at 27°C the vapour pressures of pure toluene and pure benzene are 32·06 mm and 103·01 mm respectively,

(a) calculate the vapour pressure of a solution containing 0·60 mole fraction of toluene

(b) calculate the mole of fraction of toluene in vapour for this composition of the liquid [(a) 60·44 mm (b) 0·318]

27. At 50°C the vapour pressures of pure water and ethyl alcohol are, respectively 92·5 mm and 219·9 mmHg. If 6 g of nonvolatile solute of mol. wt. 120 is dissolved in 150 g of each of these solvents, what will be the relative vapour pressure lowerings in the two solvents? (0·006, 0·015)

28. Calculate the molecular weight of a substance, 10 g of which in 1 litre of solution exerts an osmotic pressure of 81 mmHg at 27 K. (207·99)

29. The water vapour pressure at 293 K is 2338 Pa and the vapour pressure of an aqueous solution is 2295·8 Pa. Determine its osmotic pressure at 313 K if the solution density at this temperature is 1010 kg/m^3. The molecular weight of the solute is 60. ($2{\cdot}56 \times 10^6$ Pa)

30. What is the osmotic pressure of a solution of 4·48 g of a substance of molecular weight 286 in 100 cm^3 water at 298 K?
(R = 82·1 cm^3 atm/deg/mole) (3·83 atm)

31. 10·1 g of a volatile liquid occupies a volume of 4 litres when vaporised at 100°C and 70 cm pressure. What would be the osmotic pressure of a 2% (grams per 100 cc) solution of this substance at 0°C? (5·34 atm)

32. The average osmotic pressure of human blood is 7·7 atm at 40°C.
(a) What should be the total concentration of various solutes in the blood?
(b) Assuming this concentration to be essentially the same as the molality, find the freezing point of blood. [K_f (H_2O) = 1·86] (0·29 mole/litre, –0·539°C)

33. A solution containing 10·2 g of glycerene per litre is found to be isotonic with a 2% solution of glucose. Calculate the molecular weight of glycerene (mol. wt. of glucose = 180). (91·8)

34. The osmotic pressure of an aqueous solution containing 45 g of sucrose (343) per litre of solution is 2·97 atm at 0°C. Find the value of the gas constant and compare the result with the accepted value. (0·0826 lit · atm/K/mole)

35. What is the freezing point of a solution containing 6·84 g of sucrose per 500 g of water? K_f for water is 1·86°C · m^{-1}. (–0·074°C)

36. What weight of glycerol would have to be added to 1000 g of water in order to lower its freezing point by 10°C? K_f = 1·86. (495 g)

37. An aqueous solution contains 10% by weight of urea (60·00) and 5% by weight of glucose (180·00). What will be its freezing point? K_f for water is 1·86. (–4·254°C)

38. If glycerene, $C_3H_5(OH)_3$, and methyl alcohol, CH_3OH, are sold at the same price per kg, which would be cheaper for preparing an antifreeze solution for the radiator of an automobile? (CH_3OH)

39. How much ice will separate if a solution containing 25 g of ethylene glycol [$C_2H_4(OH)_2$] in 100 g of water is cooled to –10°C ? K_f (H_2O) = 1·86. (25·05 g)

40. What approximate proportions by volume of water (d = 1 g/cc) and ethylene glycol $C_2H_6O_2$ (d = 1·12 g/cc) must be mixed to ensure protection of an automobile cooling system to –10°C? (10 : 3)

41. At 25°C a solution containing 0·2 g of polyisobutylene in 100 cc of benzene, developed a rise of 2·4 mm at osmotic equilibrium. Calculate the molecular weight of polyisobutylene if the density of the solution is 0·88 g/cc

[**Hint**: OP = $\dfrac{0{\cdot}24 \times 0{\cdot}88}{13{\cdot}6}$ cmHg] ($2{\cdot}4 \times 10^5$)

42. 0·1 g of an unknown substance was dissolved in 5 g of camphor and it was found that the melting point of camphor was depressed by 5·3°C. If K_f is 39·7, find the weight of 1 mole of the solute. (150)

43. 1·23 g of a substance dissolved in 10 g of water raised the boiling point of water to 100·39°C. Calculate the molecular weight of the substance. ($K_b = 0{\cdot}52°C\ m^{-1}$) (164)

44. Two elements 'A' and 'B' form compounds having formulae AB_2 and AB_4. When dissolved in 20 g of benzene, 1 g of AB_2 lowers the freezing point by 2·3°C whereas 1 g of AB_4 lowers the freezing point by 1·3°C. Calculate the atomic weights of A and B. (K_f for benzene = $5{\cdot}1°C\ m^{-1}$) (25·57, 42·65)

45. When 45 g of glucose was dissolved in 500 g of water, the solution has a freezing point of –0·93°C.

(a) What is the molecular weight of glucose?

(b) If the simplest formula is CH_2O, what is its molecular formula? (180; $C_6H_{12}O_6$)

46. Calculate K_b of water when 1 mole of the solute is dissolved in 1000 g of water. The latent heat of vaporisation of water is 539·9 calories per gram. (0·514)

47. Molal elevation constant of chloroform is 3·88. 0·3 g of camphor added to 25·2 g of chloroform raised the boiling point of the solvent by 0·299°C. Calculate the molecular weight of camphor. (154)

48. Calculate the molal depression constant of water. Latent heat of fusion of ice at 0° is 80 calories per gram. (1·84°)

49. K_b for CCl_4 is 5·02. The boiling point of pure CCl_4 is 76·8°C. Calculate the boiling point of a 1 molal solution of naphthalene ($C_{10}H_8$) in CCl_4. (81·8°C)

50. Calculate the K_b for chloroform from the following data:

(a) Boiling point of pure $CHCl_3$ = 61·3°C.

(b) The solution containing 5·02 g of naphthalene ($C_{10}H_8$) in 18 g of $CHCl_3$ boils at 69·5°C. ($3{\cdot}8°C\ m^{-1}$)

51. A solution containing 1·23 g of $Ca(NO_3)_2$ in 10 g of water boils at 100·975°C. Calculate the degree of ionisation of the nitrate ($K_b = 0{\cdot}52$). (75%)

52. A 0·5% aqueous solution of KCl was found to freeze at 272·76 K. Calculate the van't Hoff factor and the degree of dissociation of KCl at this concentration. $K_f(H_2O) = 1{\cdot}86°C\ m^{-1}$. (1·92, 0·92)

53. When 60·26 g of VCl_4 was added to 1000 g of solvent CCl_4, the freezing point of CCl_4 was depressed by 5·415°C. K_f for CCl_4 is 29·9. Compare the number of moles of particles with the number predicted by the formula. Calculate the number of dimers, V_2Cl_8 present.
[Hint: $2VCl_4 \rightarrow V_2Cl_8$**]**

$$\left(\begin{array}{l} V_2Cl_8 = 0{\cdot}134 \text{ mole} \\ VCl_4 = 0{\cdot}0473 \text{ mole} \end{array}\right)$$

54. At 25°C, a 0·1 m solution of CH_3COOH is 1·35% dissociated. Calculate the freezing point and osmotic pressure of the solution. Compare your results with those expected under conditions of no dissociation. K_f for water = 1·86°C m^{-1}.
[Hint: See Example 48] (–0·19°C, 2·47 atm)

55. The vapour pressure of a 0·01 m solution of a weak base BOH in water at 20°C is 17·536 mm. Calculate K_b for the base. Aq. tension at 20°C = 17·54 mm. ($9{\cdot}7 \times 10^{-4}$)

[Hint: $i = \dfrac{\text{obs. lowering in v.p.}}{\text{cal . lowering in v.p.}} = \dfrac{p - p_{\text{obs.}}}{p^0 \cdot \dfrac{n}{N}} = \dfrac{17{\cdot}540 - 17{\cdot}536}{17{\cdot}54 \times \dfrac{0{\cdot}01}{1000/18}}$.

Then apply $i = 1 + x$ and $K_b = \dfrac{0{\cdot}01x^2}{1 - x}$]

56. In an Ostwald–Walker experiment, dry air was first blown through a solution containing a certain amount of solute ($M = 278$) in 150 g of water, and then also through pure water. The loss in mass of water was found to be 0·0827 g while the mass of water absorbed in sulphuric acid was 3·317 g. Calculate the amount of the solute. (0·57 g)

57. Pheromones are compounds secreted by the females of many insect species to attract males. One of these compounds contains 80·78% C, 13·56% H, and 5·66% O. A solution of 1·0 g of this substance in 8·50 cc of benzene freezes at 3·37°C. What are the molecular weight and molecular formula of the compound?
(f.p. of C_6H_6 = 5·5°C, $K_f(C_6H_6)$ = 5·12°C/m) (282, $C_{19}H_{38}O$)

58. The vapour pressure of pure water at 25°C is 23·62 mmHg. What will be the vapour pressure of a solution of 1·5 g of urea in 50 g of water? (23·41 mmHg)

59. MX_2 dissociates into M^{2+} and X^- ions in an aqueous solution, with a degree of dissociation (α) of 0.5. The ratio of the observed depression of f.p. of the aqueous solution to the value of the depression of f.p. in the absence of ionic dissociation is ... . **(IIT 2014 Adv.)** (2)

[Hint: $\dfrac{(\Delta T_f)_{\text{obs}}}{(\Delta T_f)_{\text{cal}}} = i = 1 + 2\alpha$]

60. If the freezing point of 0.01 molal aqueous solution of a cobalt(III) chloride-ammonia complex (which behaves as strong electrolyte) is –0.0558°C, the number of chloride(s) in the coordination sphere of the complex is ... .
(k_f of water = 1.86 K kg mol^{-1}) **(IIT 2015 Adv.)** (1)

$$\left[\textbf{Hint: } \frac{(\Delta T_f)_{\text{obs}}}{(\Delta T_f)_{\text{cal}}} = i = \frac{0.0558}{1.86 \times 0.01} = 3 \text{ which shows} \quad [Co(NH_3)_5Cl]Cl_2 \rightarrow [Co(NH_3)_5Cl]^{2+} + 2Cl^- \right]$$

61. The mole fraction of a solute in a solution is 0.1. At 298 K molarity of this solution is the same as its molality. The density of this solution at 298 K is

2.0 g cm^{-3}. The ratio of the molecular weights of the solute and solvent, $m_{solute}/m_{solvent}$ is ... . **(IIT 2016 Adv.)** (9)

[Hint: $\dfrac{n_1}{n_1+n_2} = 0.1$ and $\dfrac{n_1 \times 1000}{(n_1M_1 + n_2M_2)d} = \dfrac{n_1 \times 1000}{n_2M_2}$ · Find M_1/M_2.]

62. Liquids A and B form ideal solution over the entire range of composition. At temperature, T, equimolar binary solution of liquids A and B has vapour pressure 45 torr. At the same temperature, a new solution of A and B having molefraction x_A and x_B, respectively, has vapour pressure of 22.5 torr. The value of x_A/x_B in the new solution is ... . **(IIT 2018 Adv.)** (19)

[Hint: Apply the equation $P = x_AP_A^0 + x_BP_B^0$ twice.]

63. The plot given below shows $P - T$ curves for two solvents X and Y and isomolal solution of NaCl in these solvents. NaCl completely dissociates in both the solvents

1. Solvent X
2. Solution of NaCl in solvent X
3. Solvent Y
4. Solution of NaCl in solvent Y

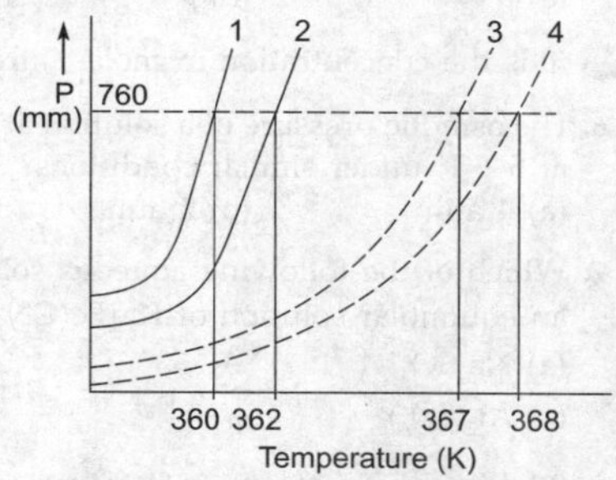

On addition of equal number of moles of a nonvolatile solute S in equal amount (in kg) of these solvents, the elevation in boiling point of solvent X is three times that of solvent Y. Solute S is known to undergo dimerisation in these solvents. If the degree of dimerisation is 0.7 in solvent Y, the degree of dimerisation in solvent X is ... . **(IIT 2018 Adv.)** (0.05)

[Hint: With NaCl undergoing 100% dissociation

$(\Delta Tb)_x = 2(Kb)_x m$, $(\Delta Tb)_y = 2(Kb)_y m$

From figure: $(\Delta Tb)_x = 362 - 360$

and $(\Delta Tb)_y = 368 - 367$.

With solute undergoing dimerisation,

$$(\Delta Tb)_x = \left(1 - \frac{x}{2}\right)(Kb)_x m$$

$$(\Delta Tb)_y = \left(1 - \frac{0.7}{2}\right)(Kb)_x m$$

Given that $(\Delta Tb)_x/(\Delta Tb)_y = 3$; find x.]

Objective Problems

1. A molal solution is one that contains one mole of a solute in

(a) 1000 g of the solvent (b) one litre of solvent

(c) one litre of solution (d) 22·4 litres of solution **(IIT 1986)**

2. Vapour pressure of a solvent is 17·5 mmHg while that of its dilute solution is 17·45. The mole fraction of the solvent is
(a) 0·997 (b) 0·075 (c) 17·48 (d) 1·05

3. Which of the following is not a colligative property?
(a) Vapour pressure (b) Depression in f.p.
(c) Elevation in b.p. (d) Osmotic pressure

4. When an ideal binary solution is in equilibrium with its vapour, molar ratio of the two components in the solution and in the vapour phases is
(a) same (b) different
(c) may or may not be same depending upon volatile nature of the two components

5. The osmotic pressure of a solution is given by the relation
(a) $p=\frac{RT}{c}$ (b) $p=\frac{cT}{R}$ (c) $p=\frac{Rc}{T}$ (d) $\frac{p}{c}=RT$
(c is the concentration in moles/litre)

6. The osmotic pressure of a solution at 0°C is 4 atm. What will be its osmotic pressure at 546 K under similar conditions?
(a) 4 atm (b) 2 atm (c) 8 atm (d) 1 atm

7. Which of the following aqueous solutions has osmotic pressure nearest to that of an equimolar solution of $K_4[Fe(CN)_6]$?
(a) Na_2SO_4 (b) $BaCl_2$
(c) $Al_2(SO_4)_3$ (d) $C_{12}H_{22}O_{11}$

8. 0·1 M solution of urea, at a given temperature, is isotonic with
(a) 0·1 M NaCl solution (b) 0·1 M glucose solution
(c) 0·02 M KCl solution (d) 0·1 M $BaCl_2$ solution

9. Which one of the following pairs of solution can we expect to be isotonic at the same temperature?
(a) 0·1 M urea and 0·1 M NaCl (b) 0·1 M urea and 0·2 M $MgCl_2$
(c) 0·1 M NaCl and 0·1 M Na_2SO_4 (d) 0·1 M $Ca(NO_3)_2$ and 0·1 M Na_2SO_4

10. Which statement is incorrect about osmotic pressure (p), volume (V) and temperature (T)?
(a) $p \propto 1/V$ if T is constant (b) $p \propto T$ if V is constant
(c) $p \propto V$ if T is constant (d) pV is constant if T is constant

11. Equal volume of 0·1 M urea and 0·1 M glucose are mixed. The mixture will have
(a) lower osmotic pressure (b) same osmotic pressure
(c) higher osmotic pressure (d) none of these

12. The factor $\Delta T_f/K_f$ represents
(a) molarity (b) formality
(c) normality (d) molality

13. The f.p. of 1% solution of $Ca(NO_3)_2$ in water will be
(a) below 0°C (b) 0° (c) 1°C (d) 2°C

14. Which has the highest f.p. at 1 atm?

(a) 0·1 M NaCl solution (b) 0·1 M $BaCl_2$ solution

(c) 0·1 M sugar solution (d) 0·1 M $FeCl_3$ solution

15. Which of the following 0·1 M aqueous solution will have the lowest f.p.?

(a) K_2SO_4 (b) NaCl

(c) Urea (d) Glucose **(IIT 1989)**

16. An aqueous solution contains 5% and 10% of urea and glucose respectively (by wt.). If K_f for water is 1·86, the f.p. of the solution is

(a) 3·03 K (b) 3·03°C

(c) –3·03°C (d) –3·03 K

17. When 1 mole of a solute is dissolved in 1 kg of H_2O, boiling point of solution was found to be 100·5°C. K_b for H_2O is

(a) 0·5 (b) 100

(c) 100·5 (d) 95·5

18. An aqueous solution of NaCl shall boil at

(a) 100°C (b) below 100°C

(c) above 100°C (d) 99·9°C

19. Which solution will have the highest b.p.?

(a) 1% solution of $C_6H_{12}O_6$ (b) 1% solution of NaCl

(c) 1% solution of $ZnSO_4$ (d) 1% solution of $(NH_2)_2CO$

20. Which solution will have the highest b.p.?

(a) 1 M $C_6H_{12}O_6$ solution (b) 1 M NaCl solution

(c) 1 M $BaCl_2$ solution (d) 1 M $(NH_2)_2CO$ solution

21. The temperature at which the vapour pressure of a liquid equals external pressure is called

(a) f.p. (b) b.p.

(c) m.p. (d) critical temp.

22. The ratio of the value of any colligative property for KCl solution to that for sugar solution is nearly

(a) 1·0 (b) 0·5 (c) 2·0 (d) 2·5

23. For an aqueous solution of NaCl, the observed molecular weight of NaCl will be

(a) less than 58·5 (b) more than 58·5

(c) 58·5 (d) $58{\cdot}5^2$

24. The weight of water in 1 litre of 2 M NaCl solution of density 1·117 g/mL is

(a) 1117 g (b) 1000 g

(c) 117 g (d) 883 g

25. Glucose is added to 1 litre of water to such an extent that $\Delta T_f / K_f$ becomes equal to 1/1000. The wt. of glucose added is

(a) 180 g (b) 18 g

(c) 1·8 g (d) 0·18 g

26. Which of the following colligative properties is associated with the concentration term 'molarity'?
(a) Lowering of vap. pressure (b) Osmotic pressure
(c) Depression in f.p. (d) Elevation in b.p.

27. Which of the following experimental methods is adopted to determine osmotic pressure?
(a) Berkley–Hartley's method (b) Beckmann's method
(c) Landsberger's method (d) Differential method

28. Which of the following solutes will produce the largest total molality of solute particles upon addition to 1 kg of water?
(a) 1·0 mole $Co(NO_3)_2$ (b) 2·0 mole KCl
(c) 3·0 mole C_2H_5OH (d) 3·0 mole of sugar

29. Consider separate solution of 0.5 M $C_2H_5OH(aq)$, 0.1 M $Mg_3(PO_4)_2(aq)$, 0.25 M KBr(aq) and 0.125 M $Na_3PO_4(aq)$ at 25°C. Which statement is true about these solutions, assuming all salts to be strong electrolytes?
(a) They all have the same osmotic pressure.
(b) 0.1 M $Mg_3(PO_4)_2(aq)$ has the highest osmotic pressure.
(c) 0.125 M $Na_3PO_4(aq)$ has the highest osmotic pressure.
(d) 0.5 M $C_2H_5OH(aq)$ has the highest osmotic pressure. **(IIT 2014 Main)**

30. The vapour pressure of acetone at 20°C is 185 torr. When 1.2 g of a nonvolatile substance was dissolved in 100 g of acetone at 20°C, its vapour pressure was 183 torr. The molar mass (g mol^{-1}) of the substance is
(a) 32 (b) 64 (c) 128 (d) 488
(IIT 2015 Main)

31. 18 g of glucose ($C_6H_{12}O_6$) is added to 178.2 g of water. The vapour pressure of water (in torr.) for this aqueous solution is
(a) 76.0 (b) 752.4 (c) 759.0 (d) 7.6
(IIT 2016 Main)

32. Mixture(s) showing positive deviation from Raoult's law of 35°C is(are)
(a) carbon tetrachloride + methanol (b) carbon disulphide + acetone
(c) benzene + toluene (d) phenol + aniline **(IIT 2016 Adv.)**

33. The freezing point of benzene decreases by 0.45°C when 0.2 g of acetic acid is added to 20 g of benzene. If acetic acid associates to form a dimer in benzene, percentage association of acetic acid in benzene will be (K_f for benzene = 5.12 K kg mol^{-1})
(a) 64.6% (b) 80.4% (c) 74.6% (d) 94.6%
(IIT 2017 Main)

34. For a solution formed by mixing liquids *L* and *M*, the vapour pressure of *L* plotted against the mole fraction of *M* in solution is shown in the following figure. Here

x_L and x_M represent mole fractions of L and M respectively in the solution. The correct statement(s) applicable to this system is (are)

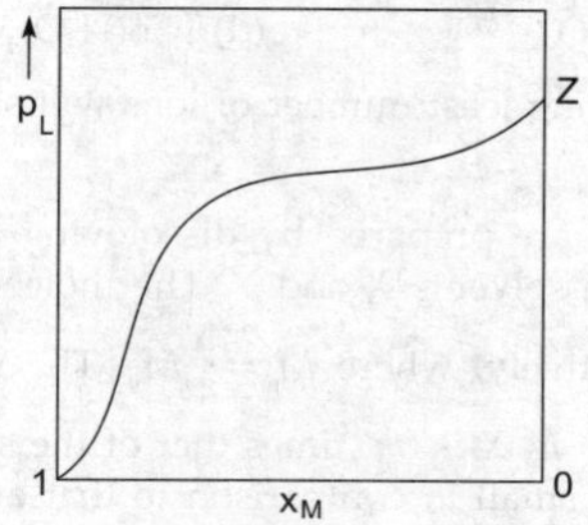

(a) The point Z represents vapour pressure of pure liquid M and Raoult's law is obeyed from $X_L = 0$ to $X_L = 1$.
(b) Attractive intermolecular interactions between $L - L$ in pure liquid L and $M - M$ in pure liquid M are stronger than those between $L - M$ when mixed in solution.
(c) The point Z represents vapour pressure of pure liquid M and Raoult's law is obeyed when $X_L \to 0$.
(d) The point Z represents vapour pressure of pure liquid L and Raoult's law is obeyed when $X_L \to 1$. **(IIT 2017 Adv.)**

35. Pure water freezes at 273 K and 1 bar. The addition of 34.5 g of ethanol to 500 g of water changes the freezing point of the solution. Use the *f. p.* depression constant of water as 2 K kg mol^{-1}. The figures shown below represents plot of vapour pressure (V.P.) versus temperature (T). (mol. wt. of ethanol = 46 g mol^{-1}). Among the following the option representing change in the *f. p.* is

(a)

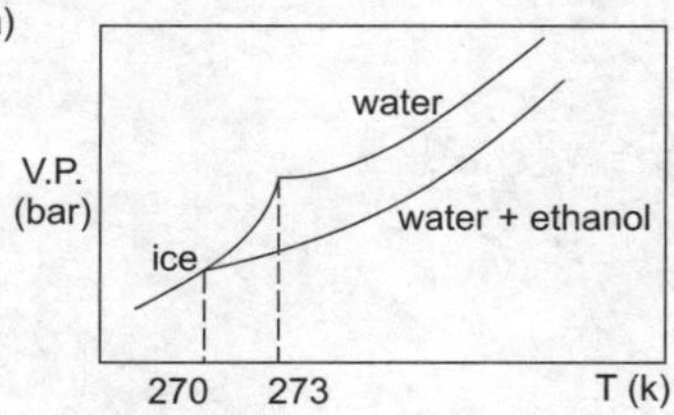

(b)

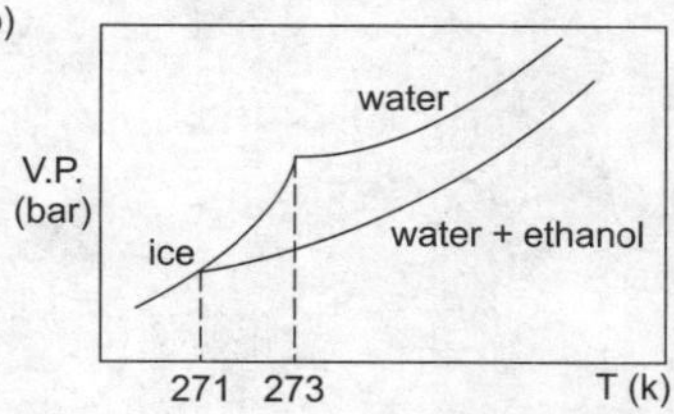

(c)

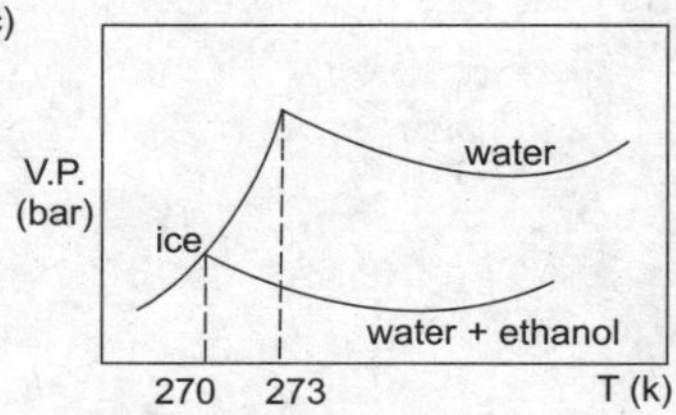

(d)

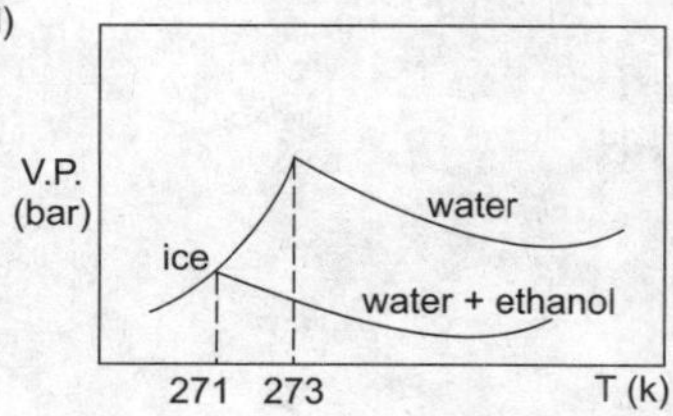

(IIT 2017 Adv.)

[Hint: Find ΔT_f; $\Delta T_f = K_f \cdot m$]

36. For 1 molal aqueous solution of the following compounds, which one will show the highest freezing point?

(a) $[Co(H_2O)_3Cl_3]3H_2O$ (b) $[Co(H_2O)_6]Cl_3$

(c) $[Co(H_2O)_5Cl]Cl_2 \cdot H_2O$ (d) $[Co(H_2O)_4Cl_2]Cl \cdot 2H_2O$

[**Hint:** The complex giving least number of ions will show highest freezing point.]

(IIT 2018 Main)

37. Two 5 molal solutions are prepared by dissoloving a nonelectrolyte nonvolatile solute separately in the solvents X and Y. The molecular weights of the solvents are M_x and N_y, respectively, where $M_x = \frac{3}{4} M_y$. The relative lowering of vapour pressure of the solution in X is 'm' times that of the solution in Y. Given that the moles of solute is very small in comparison to that of solvent, the value of 'm' is

(a) $\frac{4}{3}$ (b) $\frac{3}{4}$ (c) $\frac{1}{2}$ (d) $\frac{1}{4}$ **(IIT 2018 Main)**

[**Hint:** Relative lowering of V.P. $= \frac{n}{N} \cdot$]

Answers

1-a, **2**-a, **3**-a, **4**-c, **5**-d, **6**-c, **7**-c, **8**-b, **9**-d, **10**-c, **11**-b, **12**-d, **13**-a, **14**-c, **15**-a, **16**-c, **17**-a, **18**-c, **19**-b, **20**-c, **21**-b, **22**-c, **23**-a, **24**-b, **25**-d, **26**-b, **27**-a, **28**-b, **29**-a, **30**-b, **31**-b, **32**-a, b, **33**-d, **34**-b, d, **35**-a, **36**-a, **37**-b.

CHAPTER FOURTEEN

CHEMICAL THERMODYNAMICS

The various laws of thermodynamics are based on human experience about the behaviour of macroscopic systems (assemblage of a large number of molecules). Thermodynamics is concerned with the **equilibrium states** of the systems. An equilibrium state is one in which the macroscopic properties of the system, such as its temperature, density and chemical composition, are well defined and do not change with time. Thus, the subject of thermodynamics does not concern itself with the time element in any transformation and, therefore, has no valid application in the study of reaction kinetics.

The First Law of Thermodynamics

The first law of thermodynamics is a statement of the principle of conservation of energy. This law may be stated in various ways as given below:

(1) Energy can neither be created nor be destroyed. It can only be converted from one form of energy to another.

(2) Since energy and mass are interlinked by Einstein's equation $E = mc^2$, the total mass and energy of an isolated system remain unchanged.

(3) The conclusion of all the statements of the first law of thermodynamics is that it is impossible to construct a perpetual motion machine, i.e., a machine that can work without consuming energy.

Mathematical Formulation of the First Law

The amount of heat 'q' given to a closed system is used to increase the internal energy 'U' of the system and also to produce work 'W' in such a way that:

Heat absorbed = increase in internal energy + work done by the system.

$$q = \Delta U - W$$

or $$\Delta U = q + W \quad \ldots (1)$$

W is taken negative as the work is done by the system. The first law of thermodynamics states that the change in internal energy of a system, ΔU, equals q plus W.

The heat absorbed by a closed system in a process in which no work is done is equal to the internal energy of the system.

$$\Delta U = q$$

According to the old convention, W is taken positive when the work is done by the system. In such a case the first law is expressed as

$$q = \Delta U + W$$

or $\Delta U = q - W$.

Note that the use of any of the two sign conventions on W leads to the same result.

Internal Energy

Now, we need to understand as to what the internal energy of a system is. The internal energy is the sum of kinetic and potential energies of the particles making up the system. The kinetic energy includes the energy of the motion of the electrons and molecules. The potential energy results from the chemical bonding of atoms and from the attractions between molecules. Internal energy is the property of a system that depends only on its present state, which is completely determined by the variables such as temperature and pressure.

When there is a change in the physical state of a system (a liquid changing to a gas), the internal energy U changes even if the temperature remains constant. When the physical state of a system does not change (gas remains as gas), the internal energy will change only if the temperature changes. In other words, when a gaseous state changes to another gaseous state at constant temperature, the change in internal energy, ΔU, is equal to zero.

State Functions and Exact Differentials

The first law of thermodynamics is

$$\Delta U = q + W.$$

For infinitesimal changes

$$dU = dq + dW.$$

When a system changes from one state to another, q and W depend on the path of the process and ΔU does not. ΔU depends on the initial and final states of the system only. U is thus called a state function but q and W are not. The other state functions are enthalpy, H, entropy, S, and free energy, G.

State functions give exact differential. As U is a state function, dU is an exact differential. An exact differential can be integrated between limits. On the other hand, as q and W are not state functions, dq and dW are not exact differentials and hence dq and dW cannot be integrated between limits. If a system goes from state a to state b, we can write

$$\int_a^b dU = U_b - U_a = \Delta U.$$

And the work done is represented by

$$\int_a^b dW = W$$

but note that the result of the integration is not written as $W_b - W_a$, because W depends on the path that is followed between State a and State b. For example, when a gas is allowed to expand, the amount of work obtained may vary from zero (if the gas is allowed to expand into vacuum) to a maximum value that is obtained if the expansion is carried out reversibly. However, the sum of two inexact differentials can be an exact differential

$$dU = \delta q + \delta W$$

where d and δ stand for exact and inexact differentials respectively.

The mathematical test for exactness is given by **Euler's criterion** for exactness.

If $\phi = f(x, y)$

then $d\phi$ would be an exact differential if

$$\left[\frac{\partial}{\partial y}\left(\frac{\partial \phi}{\partial x}\right)_y\right]_x = \left[\frac{\partial}{\partial x}\left(\frac{\partial \phi}{\partial y}\right)_x\right]_y$$

or

$$\left(\frac{\partial M}{\partial y}\right)_x = \left(\frac{\partial N}{\partial x}\right)_y$$

is satisfied. ϕ can be considered as any thermodynamic function, U, H, G, etc., and x, y may be the thermodynamic variables, T, P, V, etc. T, P, V, etc., are also the state functions.

IUPAC Sign Convention on *W* and *q*

The IUPAC convention on W is that it is positive when work is done on the system (energy is added to the system) and negative when the system does work on the surroundings (energy is subtracted from the system). A similar convention is applied to heat, q, i.e., q is positive when heat is absorbed by the system from its surroundings. A negative sign for q means that the system gives up heat to its surroundings.

In the expansion of a gas, work is done by the system and hence W is taken negative. When a gas is compressed, work is done on the system by the surroundings and W is taken positive. When a system changes from one state to another without undergoing a change in volume, no work is done, i.e., $W = 0$.

Reversible and Irreversible Processes

The ordinary sense of a reversible process (say a reaction) is that of a reaction that can go both ways. But the thermodynamic sense of a reversible process is not that simple. Let us try to understand the thermodynamic sense of a reversible process.

When a process is carried out reversibly, the state functions of the system never differ by more than an infinitesimal amount from one moment to another. Also, the state functions of the system, like pressure and temperature, never differ from those of their surroundings by more than an infinitesimal amount. For example, in a reversible expansion or compression of a gas

$$p_{int} = p_{ext} \pm dp,$$

where p_{int} is the pressure of the system (gas), p_{ext} is the external pressure and dp is an infinitesimally small amount of pressure. '+' and '–' are for expansion and compression respectively.

In a reversible process, the system does not deviate from equilibrium by more than an infinitesimal amount and so **the direction of a reversible process can be reversed at any time just by making an infinitesimal change in the surroundings. A truly reversible process has thus to be carried out extremely slowly** in an infinite number of steps and would therefore require infinite time for its completion. A reversible process is thus an ideal process of conceptual importance only but has tremendous applications in thermodynamics.

On the other hand, an irreversible process occurs rapidly or spontaneously such that it does not remain in equilibrium during the transformation. **It cannot be reversed by an infinitesimal change in external conditions.** As all irreversible or spontaneous processes take a finite time for their completion, they are the real processes in actual practice.

Another of the important differences between reversible and irreversible processes is that the work done by the system in a reversible process (W_{rev}) is greater than the work done in the corresponding irreversible process (W_{irr}) between the same two states, say A and B. This can be illustrated as follows:

In case of a reversible expansion of an ideal gas from a volume V_1 to V_2, p_{int} is greater than p_{ext} by only an infinitesimally small pressure, dp.

$$p_{ext} = p_{int} - dp$$

or $$p_{ext} \approx p_{int}$$

$$\therefore \quad W_{rev} = -\int_{V_1}^{V_2} p_{ext}\, dV = -\int_{V_1}^{V_2} p_{int}\, dV$$

$$= -nRT \int_{V_1}^{V_2} \frac{dV}{V}$$

$$= -2{\cdot}303\, nRT \log \frac{V_2}{V_1}$$

whereas for an irreversible expansion,

$$p_{int} > p_{ext} \quad \text{or} \quad p_{int} \neq p_{ext}$$

$$W_{irr} = -\int_{V_1}^{V_2} p_{ext}\, dV$$

$$= -p_{ext}(V_2 - V_1)$$

Since $p_{int} > p_{ext}$, we conclude,

$$W_{rev} > W_{irr}.$$

The graphical representation of work done by the system (gas), W_{rev} and W_{irr} (given by the shaded area under p–V isotherm between V_1 and V_2), makes it clear that

$$W_{rev} > W_{irr}$$

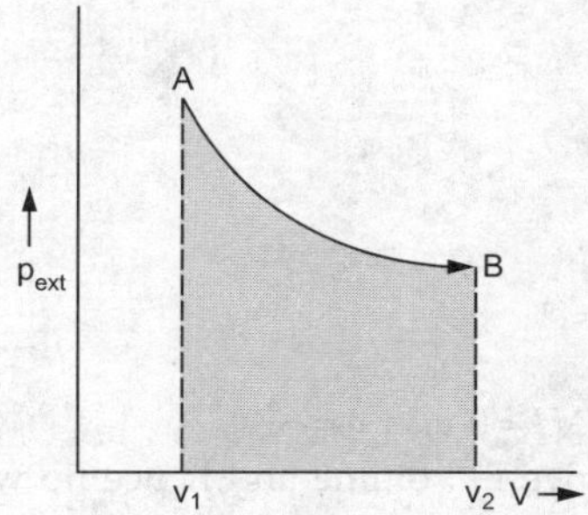

Reversible expansion of an ideal gas

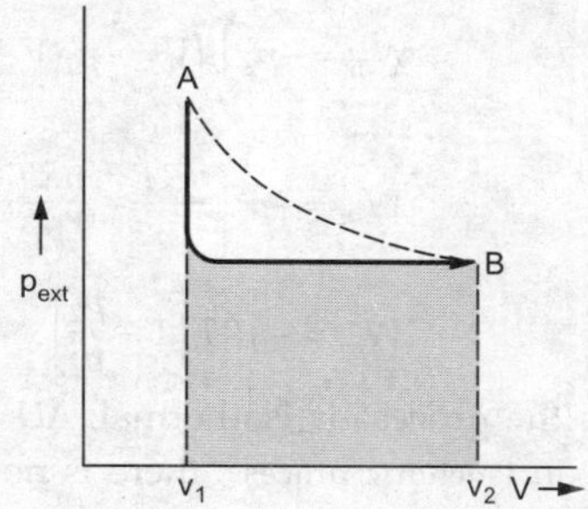

Irreversible expansion of an ideal gas

Work in Different Processes

If the volume of a system at a constant pressure 'p_{ext}' (isobaric process) increases by ΔV then the work done by the system, W, is given by

$$W = -p_{ext}\,\Delta V = -nR(T_2 - T_1);\ T_2 > T_1. \qquad \dots (2)$$

Work is often conventionally measured by the lifting or falling of masses. If h is the height through which the mass is lifted, the work, W, required to lift a mass m is given by

$$W = -mgh = -p\Delta V, \qquad \dots (3a)$$

where g is the acceleration due to gravity. If the mass is falling, work is done on the system and so the equation is given as

$$W = mgh = -p\Delta V. \qquad \dots (3b)$$

In the reversible **isothermal expansion** of a gas from volume V_1 to V_2 at a (constant) temperature T, pressure does not remain constant. In such a case, work is given by

$$W = -2{\cdot}303\, nRT \log \frac{p_1}{p_2} \qquad \dots (4a)$$

or

$$W = -2{\cdot}303\, nRT \log \frac{V_2}{V_1}. \qquad \dots (4b)$$

As temperature is constant, $\Delta U = 0$ and also $\Delta H = 0$.

$$\therefore \quad q = -W = 2{\cdot}303\, nRT \log \frac{p_1}{p_2}$$

$$= 2{\cdot}303\, nRT \log \frac{V_2}{V_1}.$$

In an **isothermal irreversible expansion** of a gas, suppose that the external pressure (p_{ext}) is released suddenly from initial pressure p_1 to final pressure $p_2 (p_2 < p_1)$ and then the gas expands against the pressure p_2. Work done by the gas is now against a constant external pressure p_2 and so W_{irr} is given by

$$W_{irr} = -p_2 \int_{V_1}^{V_2} dV = -p_2(V_2 - V_1)$$

or $$W_{irr} = -p_2 \left[\frac{nRT}{p_2} - \frac{nRT}{p_1}\right]$$

$$W_{irr} = -nRT\left[1 - \frac{p_2}{p_1}\right]. \qquad \ldots (5)$$

As the process is isothermal, $\Delta U = 0$, $\Delta H = 0$ and $q = -W_{irr}$.

In an **isochoric process,** there is no change in volume and hence no work is done. Thus $W = 0$ and $q = \Delta U$.

In an **adiabatic process,** the system is thermally isolated so that there is no exchange of heat with the surroundings ($q = 0$). Thus

$$W = \Delta U = nC_V(T_2 - T_1), \qquad \ldots (6)$$

where C_V is the molar heat capacity at constant volume.

The work done on a closed system in an adiabatic process is equal to increase in internal energy of the system. If a system does work on its surroundings, W is negative and furthermore, ΔU is negative (i.e., the internal energy of the system decreases) if the process is adiabatic.

The Heat Content, or Enthalpy of a System

It is often convenient, particularly in dealing with systems at constant pressure, to use a function H, called enthalpy, in place of internal energy U, the two being related by the equation,

$$H = U + pV$$

H represents the total energy stored in a system.

When the initial state of a system changes to a final state reversibly at constant pressure, the change in enthalpy (ΔH) and change in internal energy (ΔU) are related by

$$\Delta H = \Delta U + p\Delta V. \qquad \ldots (7)$$

The enthalpy change, ΔH, i.e., increase in enthalpy H is equal to heat absorbed at constant pressure and this explains the name, 'heat content' for the function H.

Heat of Reaction at Constant Volume and at Constant Pressure

The heat of a reaction at constant volume (q_v) is given by ΔU whereas at constant pressure, the heat of reaction (q_p) is given by ΔH.

ΔU and ΔH, i.e., q_v and q_p, are related by an equation,

$$\left.\begin{aligned} \Delta H &= \Delta U + \Delta n_g RT \\ q_p &= q_v + \Delta n_g RT \end{aligned}\right\} \quad \dots (8)$$

where Δn_g = moles of gaseous product – moles of gaseous reactant. For reactions involving only solids and liquids

$$\Delta H \cong \Delta U$$

Thermochemistry

The reactions are generally carried out in open vessels, i.e., the reactions are carried out at a constant pressure. The heat of reaction *ΔH, which is equal to q_p, is thus measured because of its greater practical importance. The chemistry dealing with heat changes in chemical reactions is known as *thermochemistry*.

Exothermic and Endothermic Reactions

A reaction is said to be exothermic when heat is evolved, i.e., heat or enthalpy of the product is less than that of the reactant. Heat of reaction of exothermic reactions is expressed in two ways.

$$\text{A} + \text{B} \rightarrow \text{Product} + Q \text{ cal} \quad \text{or} \quad \text{A} + \text{B} \rightarrow \text{Product}; \ \Delta H = -Q \text{ cal}$$

In endothermic reactions, heat is absorbed, i.e., heat of the product is greater than heat of the reactant. The heat of reaction of endothermic reactions may also be represented in two ways.

$$\text{A} + \text{B} \rightarrow \text{Product} - Q \text{ cal} \quad \text{or} \quad \text{A} + \text{B} \rightarrow \text{Product}; \ \Delta H = Q \text{ cal}$$

Heat of Formation

The heat of formation of a compound is defined as the heat change accompanying the formation of 1 mole of the compound from its elements under standard conditions. (1 atm pressure and 25°C)**. Heat of formation of CH_4 and CO_2 are given below:

$$C + 2H_2 \rightarrow CH_4; \ \Delta H_f = -18{\cdot}50 \text{ kcal}$$

$$C + O_2 \rightarrow CO_2; \ \Delta H_f = -96{\cdot}96 \text{ kcal}$$

The standard molar enthalpy of formation, ΔH_f^0, for any element in its standard state is zero by convention. ΔH_f^0 values are used to calculate ΔH^0 of the reactions.

* In general, the heats of reactions given are at their standard states and so ΔH may be denoted by ΔH^0.

** IUPAC recommends that the standard pressure be 1 bar (1×10^5 Pa). Thermodynamic tables are becoming available for 1 bar pressure and will probably replace those for 1 atm.

Heat of Combustion

The heat of combustion of a substance is defined as the heat evolved when 1 mole of that compound is burnt completely in oxygen. Heat of combustion of H_2 and S are given below:

$$H_2\,(g) + \frac{1}{2}O_2\,(g) \rightarrow H_2O\,(l);\ \Delta H = -\,68{\cdot}40\ \text{kcal}$$

$$S\,(g) + O_2\,(g) \rightarrow SO_2\,(g);\quad \Delta H = -\,71\ \text{kcal}$$

The heat of combustion is measured at constant volume generally in Berthelot's bomb calorimeter. So we get the value of ΔU for the heat of combustion which is then converted to ΔH using Eqn. (8).

Heat of Neutralisation

The heat of neutralisation of an acid or a base is defined as the heat evolved when 1 equivalent of the acid or base is neutralised by a strong base or strong acid respectively using dilute solutions. For neutralisation reaction,

ΔH (neutralisation) = ΔH (ionisation) + ΔH ($H^+ + OH^-$)

For strong acids, ΔH (ionisation) is zero and so

ΔH (neutralisation) = ΔH ($H^+ + OH^-$) = − 13·7 kcal per equivalent. For weak acids, ΔH (ionisation) > 0. Thus ΔH (neutralisation) for weak acids is always less than 13·7 kcal. In other words, the greater the value of the heat of neutralisation (negative value) of an acid, the more will be its strength.

Heat of Solution

The heat of solution is defined as the heat evolved or absorbed when 1 mole of the solute is dissolved completely in excess of solvent.

Heat of Hydration

The heat of hydration is defined as the heat change accompanying the formation of 1 mole of a specified hydrate from 1 mole of an anhydrous substance on combination with the required number of moles of water.

Laws of Thermochemistry

1. Lavoisier and Laplace law: The law states that the heat evolved in a chemical reaction is equal to the heat absorbed when the reaction is reversed.

2. Hess's law of Constant Heat Summation: The law states that at constant pressure the total enthalpy change (ΔH) accompanying a chemical reaction is the same whether the reaction takes place in one or more steps, e.g.,

for a reaction, $\quad A \rightarrow D \qquad \Delta H = q$

and for $\quad A \rightarrow B \rightarrow C \rightarrow D$

$$\Delta H = q_1 \quad \Delta H = q_2 \quad \Delta H = q_3$$

Thus, according to this law: $q = q_1 + q_2 + q_3$

Hess's law is a special case of the first law of thermodynamics.

Hess's law may be applied in calculating heats of reactions, keeping in mind the following guidelines for such thermochemical calculations:

1. Write down the balanced equation of the reaction for which the ΔH value is to be determined. Let us call it **Main Reaction.**
2. Try to come to the **main reaction** from the given thermochemical reactions by following the **Inspection Method** given below.

When the common reactant (or product) present in the main (or required) reaction and in one of the given reactions lies on the same side, the given thermochemical equation is to be added after equating the stoichiometric coefficient of the common reactant (or product) by multiplying the given reaction with a suitable number. And if the common reactant (or product) lies on the opposite sides in the main and in the given reaction, the given equation is to be subtracted after equating the stoichiometric coefficient of the common reactant (or product). This method can be more clearly followed by going through the solved examples.

ΔH^0 for any reaction can also be calculated from ΔH_f^0 values of the reactants and products.

ΔH^0= sum of ΔH_f^0 values of the products

– sum of ΔH_f^0 values of the reactants.

Kirchoff's Law: Dependence of ΔH and ΔU on Temperature

For a given reaction, ΔH and ΔU vary with temperature. If ΔH_1 and ΔH_2 are the enthalpies of the reaction at temperatures T_1 and T_2 respectively, and ΔC_p is the difference in the heat capacities of the products and the reactants at constant pressure, we have

$$\frac{\Delta H_2 - \Delta H_1}{T_2 - T_1} = \Delta C_p. \quad \text{... (9a)}$$

If the reaction takes place at constant volume then

$$\frac{\Delta U_2 - \Delta U_1}{T_2 - T_1} = \Delta C_v, \quad \text{... (9b)}$$

where ΔC_v is the difference in the heat capacities of the products and reactants at constant volume.

[**Note:** The heat capacities of all the substances taking part in the reaction must be included. The elements cannot be omitted as they remain in calculating enthalpy differences.]

Bond Energy

Bond energy may be defined as the average amount of energy required to break (or dissociate) 1 mole (i.e., Avogadro constant) of such bonds present in different compounds, the original molecules and the dissociation products being in the gaseous state. Since in this process heat is given to the system, the bond energy is always given a positive sign. The energy of formation of a bond is numerically equal to its corresponding bond energy but with a negative sign.

Calculation of Bond Energy

With the knowledge of heat of formation (or heat of reaction) of a substance and the heat of atomisation of the constituents, bond energy can be calculated.

Heat of formation (or heat of reaction) may be calculated by adding algebraically all the heats (evolved or absorbed) of the reactants and products. Calculation based on bond energy is satisfactory, provided there is no resonance in the molecules and in such cases a correction for resonance energy has to be made. **Resonance energy is always negative.**

Bond Energies in kcal (multiply by 4·1868 to get in kJ)

Bond	Energy	Bond	Energy
C − C	83·1	H − H	103·2
C = C	140	O − O	33
C ≡ C	198	O = O	118
C − H	99	N ≡ N	225
C − O	84	O − H	110·5
C = O	173	H − Cl	103
C − Cl	78·5	O − Cl	49
C − N	69·7		

Heats of Atomisation in kcal (multiply by 4·1868 to get in kJ)

Element	Atomisation	Heat of Atomisation
C	$C(s) \rightarrow C(g)$	170·9
H	$\frac{1}{2}H_2(g) \rightarrow H(g)$	52·1
O	$\frac{1}{2}O_2(g) \rightarrow O(g)$	59·6
N	$\frac{1}{2}N_2(g) \rightarrow N(g)$	113·0
Cl	$\frac{1}{2}Cl_2(g) \rightarrow Cl(g)$	28·9

[**Note:** If ΔH or W are given in joules (SI unit), R should be taken as 8·314 J/K/mol.]

The Second Law of Thermodynamics

The first law of conservation of energy is a fundamental law, but at the same time the law has its limitations. For example, it cannot explain firstly, why heat cannot flow from a cold body to a hot body, and secondly, why heat cannot be completely converted into work. This can be explained by another law of nature.

'All nonequilibrium states of a system tend to shift towards the equilibrium state on their own or in a natural way, or spontaneously or irreversibly, but the reverse cannot occur without external help to the system from the surroundings'.

The second law thus fixes the criterion of spontaneity or feasibility of physical and chemical processes.

To conclude how much heat can be converted to work in a heat engine, Carnot adopted a cyclic process of four steps, viz., **I. Isothermal expansion, II. Adiabatic expansion, III. Isothermal compression** and **IV. Adiabatic compression** of 1 mole of an ideal gas as shown on a *p*–*V* diagram. As this cycle produces only heat and work effects in the surroundings, it is called 'Heat Engine'.

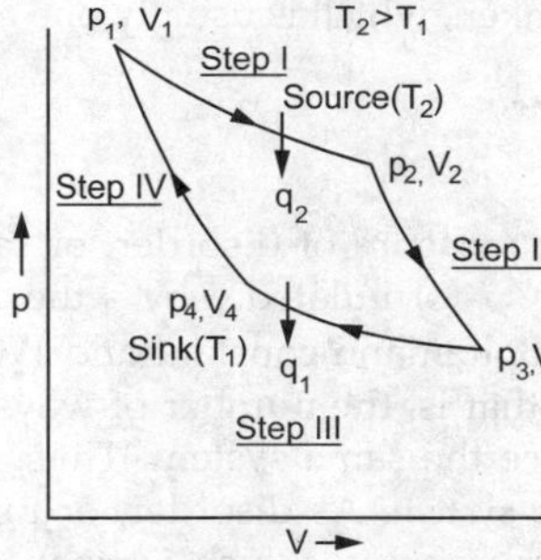

The relationship between heat and work in a process based on the **reversible** Carnot cycle is given by the equation

$$\eta = \frac{W}{q_2} = \frac{T_2 - T_1}{T_2} < 1; \qquad \text{... (10)}$$

$$(\text{where } W = q_2 - q_1 = q)$$

where η is the efficiency of the heat engine, q_2 is the amount of heat absorbed by the gas from the hot reservoir (source) at a temperature T_2(Step I), q_1 is the heat given up by the gas to the cold reservoir (sink) at a temperature T_1 (step III) and W is the net work done by the gas in all the four steps. q, that is, $(q_2 - q_1)$, is the net heat absorbed.

As W is always less than q_2, heat cannot be completely converted into work. In other words, the efficiency of the heat engine is always less than 1. The efficiency of all heat engines working reversibly between the same temperatures of the source and the sink is the same.

Entropy

From the Carnot cycle, we can derive

$$\frac{q_2}{T_2} \text{ (Step I)} = \frac{q_1}{T_1} \text{ (Step III)}$$

In general,

$$\frac{q_{rev}}{T} = \text{constant},$$

where q_{rev} is the quantity of heat exchanged in a process carried out reversibly at a temperature T, the quantity $\frac{q_{rev}}{T}$ represents a definite quantity, or state function, called the entropy change, ΔS.

$$\Delta S = \frac{q_{rev}}{T} \quad \text{or} \quad dS = \frac{dq_{rev}}{T}. \qquad \ldots (11)$$

Entropy is an extensive property. It is therefore necessary to mention the quantity of the substance taken, which is usually one mole. The unit of entropy is cal or joule per mole.

Significance of Entropy

The idea that entropy is a measure of disorder, or randomness, of a system, has been quantitatively formulated by the Boltzmann equation, $S = k \ln W$, where k is the Boltzmann constant and W is the thermodynamical probability of the system, that is, the number of ways in which the molecules can be arranged to produce the same system. Thus, the **increase in entropy is a measure of disorder of the system.** As disorder, or randomness, of molecules is due to the portion of the energy which is unavailable to the system for doing useful work, we can say that entropy is a measure of the extent to which the internal energy of the system has become unavailable for doing useful work.

Further, the entropy of a pure, perfectly crystalline substance (perfectly ordered) is zero at 0 K.

Criterion of Spontaneity of a Process by Entropy Change

The state functions U and H from the first law do not give the criterion of spontaneity of a process. The state function S (entropy) is very useful to know whether a physical or chemical process is feasible under the given conditions. The entropy change, both in the system and the surroundings, produces the following results.

For a spontaneous or irreversible process,

$$(\Delta S)_{universe} = (\Delta S)_{sys} + (\Delta S)_{surr} > 0 \text{ or positive}$$

and for a reversible process,

$$(\Delta S)_{universe} = (\Delta S)_{sys} + (\Delta S)_{surr} = 0 \qquad \ldots (12)$$

Since all natural processes are irreversible, the entropy of the universe goes on increasing, while the total energy of the universe is constant. No process that produces order in a system can proceed without producing an even larger disorder in its surroundings.

Entropy effects Associated with Melting and Freezing						
Change	Temperature	Sign of ΔS_{sys}	Sign of ΔS_{surr}	(Magnitude of ΔS_{sys}) Compared with (Magnitude of ΔS_{surr})	$\Delta S_{univ} = \Delta S_{sys} + \Delta S_{surr}$	Spontaneity
Melting (solid → liquid)	(a) > mp	+	−	>	> 0	Spontaneous
	(b) = mp	+	−	=	= 0	Equilibrium
	(c) < mp	+	−	<	< 0	Nonspontaneous
Freezing (liquid → solid)	(a) > mp	−	+	>	< 0	Nonspontaneous
	(b) = mp	−	+	=	= 0	Equilibrium
	(c) < mp	−	+	<	> 0	Spontaneous

Definition of the Second Law

From the previous discussion, the second law can be defined in various forms:

1. Heat cannot flow from a cold body to a hot body by itself.
2. It is not possible to convert heat into work completely. In other words, the efficiency of a heat engine is always less than 1.
3. A spontaneous, or natural, process is accompanied by an increase in the total entropy of the universe.

Combined Mathematical form of the First and Second Laws

We have from the first law,

$$dq = dU - dW$$

or

$$dq = dU + pdV$$

And from the second law,

$$dS = \frac{dq_{rev}}{T}$$

from which, we get the combined form of the first and second laws of thermodynamics.

$$dU = TdS - pdV \qquad \ldots (13)$$

This is one of the fundamental equations in thermodynamics.

Entropy Change of an Ideal Gas in Different Processes

Isothermal process ($\Delta T = 0$)

$$\Delta S_T = 2.303nR \log \frac{p_1}{p_2} = 2.303nR \log \frac{V_2}{V_1} \qquad \ldots (14)$$

Isobaric process ($\Delta p = 0$)

$$\Delta S_p = 2.303n\, C_p \log \frac{T_2}{T_1} \qquad \ldots (15)$$

Isochoric process ($\Delta V = 0$)

$$\Delta S_V = 2.303 nC_V \log \frac{T_2}{T_1} \quad \dots (16)$$

With temperature and pressure

$$\Delta S = 2.303 nC_p \log \frac{T_2}{T_1} - 2.303 nR \log \frac{p_2}{p_1} \quad \dots (17)$$

With temperature and volume

$$\Delta S = 2.303 nC_V \log \frac{T_2}{T_1} + 2.303 nR \log \frac{V_2}{V_1} \quad \dots (18)$$

Entropy Change accompanying Change of Phase

Solid changing to liquid: $\Delta S_f = \dfrac{\Delta H_f}{T_f}$ ... (19)

Liquid changing to vapour: $\Delta S_v = \dfrac{\Delta H_v}{T_b}$ ... (20)

ΔH_f and ΔH_v are the molar heat of fusion and the molar heat of vaporisation respectively. T_f and T_b are the freezing point and boiling point respectively. Both ΔH_f and ΔH_v are positive, since the process of fusion and vaporisation are both accompanied with an increase in entropy.

Gibbs Free Energy

The entropy changes in both the system and its surroundings give the criterion for the spontaneity of a process (reaction):

$(\Delta S)_{sys} + (\Delta S)_{surr} > 0$, or positive (irreversible process)

$(\Delta S)_s + (\Delta S)_{surr} = 0$ (reversible process)

But as it is not always very convenient to obtain $(\Delta S)_{surr}$, if the criterion of spontaneity were expressed in terms of properties of the system alone, it would be much easier to use. Thus, another thermodynamic function, free energy (G), was defined, as change in free energy (ΔG) of the system alone can give the criterion for spontaneity of a process (reaction):

For an irreversible process,

$(\Delta G)_{sys} < 0$, or negative

For a reversible process,

$(\Delta G)_{sys} = 0$ ... (21)

G is thus a better thermodynamic function than S in practice.

It is now necessary to first understand what actually this free energy function, G, is.

Theoretically, the spontaneous reactions can be used to obtain useful work. Useful work means energy that can be used directly to cause reaction.

In principle, if a reaction is carried out to obtain the maximum useful work, no entropy is produced. It can be shown that the maximum work W_{max} for a spontaneous reaction is ΔG.

$$W_{max} = \Delta G$$

The term 'free energy' comes from this result. The free energy change is the maximum energy available, or free, to do useful work. The concept of maximum work from a chemical reaction is an idealisation. In any real situation, less than this maximum work is obtained and some entropy is created.

When a system reaches an equilibrium state, it loses the capacity to do external work. Free energy of a system is the difference between the initial-state energy and the equilibrium-state energy. This amount of energy is available for doing external work. The equilibrium state energy of the system is regarded as nonavailable energy given by the product of T and S. Gibbs free energy (G) is the one left as a balance from enthalpy H of the system

$$G = H - TS \qquad \text{... at constant } T \text{ and } p.$$

As a reaction proceeds at a given temperature and pressure, reactants form products and the enthalpy H and entropy S change. These changes in H and S, denoted by ΔH and ΔS, result in a change in the free energy ΔG, given by the equation

$$\Delta G = \Delta H - T\Delta S. \qquad \text{... (22)}$$

When a reaction acquires the state of maximum entropy, or minimum free energy, it is said to have reached the equilibrium state.

This equation serves as a criterion for spontaneity of a reaction or process.

The Two Aspects of Spontaneity

Two factors affect the spontaneity of any physical or chemical change:

1. Spontaneity is favoured when heat is released ($\Delta H = -$ve).
2. Spontaneity is favoured when the change causes an increase in disorder ($\Delta S = +$ve).

The free energy change and spontaneity of a reaction depend on both enthalpy and entropy changes. Both ΔH and ΔS may be either positive or negative, so we can group reactions in four classes with respect to spontaneity. At constant temperature and pressure, chemical reactions proceed spontaneously (product-favoured) if $\Delta G < 0$, or negative.

$\Delta G = \Delta H - T\Delta S$			(Constant temperature and pressure)
Class 1.	$\Delta H = -$ (favourable)	$\Delta S = +$ (favourable)	Reactions are product-favoured at all temperatures
Class 2.	$\Delta H = -$ (favourable)	$\Delta S = -$ (unfavourable)	Reactions become product-favoured below a definite temperature
Class 3.	$\Delta H = +$ (unfavourable)	$\Delta S = +$ (favourable)	Reactions become product-favoured above a definite temperature
Class 4.	$\Delta H = +$ (unfavourable)	$\Delta S = -$ (unfavourable)	Reactions are reactant-favoured at all temperatures

A graphical representation of the dependence of ΔG and spontaneity on temperature for each of the four classes of reactions is shown below:

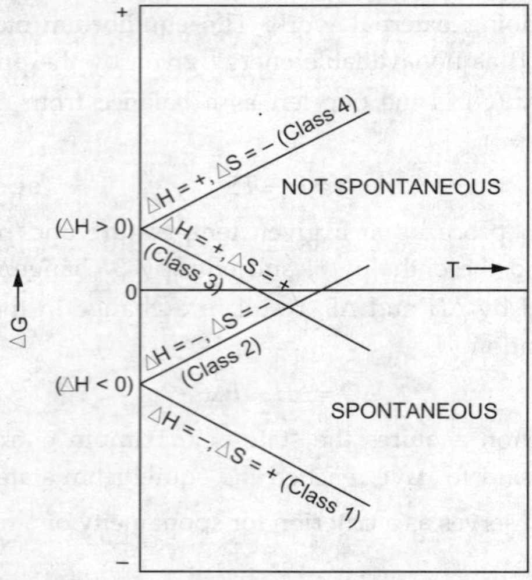

In Class 2 and 3 types of reactions, the spontaneity of the forward reaction depends on temperature. In Class 2, the reaction becomes product-favoured at a lower temperature, while in Class 3, the reaction becomes product-favoured at a higher temperature.

The equation $\Delta G = \Delta H - T\Delta S$ can also be used to estimate the temperature at which a process is in equilibrium. When a system is in equilibrium, $\Delta G = 0$.

Thus, $$T = \frac{\Delta H}{\Delta S}.$$

The values of ΔH and ΔS can be supposed to be equal to ΔH^0 and ΔS^0 respectively, **assuming that ΔH and ΔS do not change with temperature.** The values of ΔH^0 and ΔS^0 of the chemical reactions can be calculated from the ΔH_f^0 and S^0 values of different reactants and products. Thus, one can calculate the boiling point of a liquid, as at the b.p., $\Delta G = 0$ for the equilibrium: Liquid $\rightleftharpoons$ Vapour

At thermodynamic standard conditions,

$$\Delta G^0 = \Delta H^0 - T\Delta S^0. \qquad \dots (22)$$

The temperature at which $\Delta G^0 = 0$ is the temperature limit of spontaneity. The sign of ΔS^0 tells us whether the reaction is spontaneous below or above this limit.

Free Energy Change of an Ideal Gas in an Isothermal Process

One of the most important fundamental equations in thermodynamics is

$$dG = Vdp - SdT, \qquad \dots (23)$$

which gives the change in free energy when a system undergoes reversibly a change in pressure as well as a change in temperature.

If n moles of an ideal gas undergo a change of state at constant temperature,

$$dG = Vdp = nRT\frac{dp}{p}.$$

On integration, we get

$$\Delta G = 2.303nRT \log\frac{p_2}{p_1} = 2.303nRT \log\frac{V_1}{V_2}. \qquad \dots (24)$$

The initial and final states are designated by 1 and 2 respectively.

Unit Conversions

1 cal = 4·1868 J = $4{\cdot}1868 \times 10^7$ ergs

1 J = 10^7 ergs

1 eV = $1{\cdot}602 \times 10^{-19}$ J = $1{\cdot}602 \times 10^{-12}$ ergs

1 MeV = 10^6 eV

Remember: 1 cal > 1 J > 1 erg > 1 eV.

EXAMPLES

Ex. 1. *Calculate W and ΔU for the conversion of* 1 *mole of water into* 1 *mole of steam at a temperature of* 100°C *and at a pressure of* 1 *atmosphere. Latent heat of vaporisation of water is* 9720 cal/mole.

Solution : $p = 1$ atm

$= 76$ cm

$= 76 \times 13{\cdot}6 \times 981$ dynes/cm^2

$= 1{\cdot}013 \times 10^6$ dynes/cm^2.

V_1 = volume of 1 mole of water at 100°C = 18 mL.

V_2 = volume of 1 mole of steam at 100°C

$= \frac{373}{273} \times 22400 = 30605$ mL. (Charles's law)

Now we have,

$$W = -p(V_2 - V_1) \quad \text{... (Eqn. 2)}$$

$$= -1{\cdot}013 \times 10^6 \times (30605 - 18) \text{ ergs}$$

$$= \frac{-1{\cdot}013 \times 10^6 \times 30587}{4{\cdot}18 \times 10^7} \text{ calories}$$

$$= -741 \text{ calories.}$$

Again we have,

$$\Delta U = q + W \quad \text{... (Eqn. 1)}$$

$$= 9720 + (-741)$$

$$= 8979 \text{ calories.}$$

Ex. 2. *Calculate the maximum work done when pressure on* 10 g *of hydrogen is reduced from* 20 to 1 atm *at a constant temperature of* 273 K. *The gas behaves ideally. Will there be any change in internal energy? Also, calculate 'q'.*

Solution : We have,

$$W = -2{\cdot}303\, nRT \log \frac{p_1}{p_2} \quad \text{... (Eqn. 4a)}$$

n = number of moles of hydrogen $= \frac{\text{wt. in grams}}{\text{mol. wt.}} = \frac{10}{2} = 5$ moles.

Thus $W = -2{\cdot}303 \times 5 \times 2 \times 273 \times \log \frac{20}{1}$

$= -8180$ calories.

Further, the change in state of the system is from a gas to a gas and therefore, at constant temperature, internal energy will not change, i.e., $\Delta U = 0$.

Again, $q = \Delta U - W$

$= 0 - (-8180) = 8180$ calories.

Ex. 3. *Calculate the work performed when* 2 *moles of hydrogen expand isothermally and reversibly at* 25°C *from* 15 *to* 50 *litres.*

Solution : We have,

$$W = -2{\cdot}303\, nRT \log \frac{V_2}{V_1} \qquad \text{... (Eqn. 4b)}$$

$$= -2{\cdot}303 \times 2 \times 2 \times 298 \times \log \frac{50}{15}$$

$$= -1436 \text{ calories.}$$

Ex. 4. *For the combustion of 1 mole of liquid benzene at 25°C, the heat of reaction at constant pressure is given by,*

$$C_6H_6\,(l) + 7\tfrac{1}{2}\,O_2\,(g) \;\rightarrow\; 6CO_2\,(g) + 3H_2O\,(l);\; \Delta H = -\,780980 \text{ cal.}$$

What would be the heat of reaction at constant volume?

Solution : We have,

$$\Delta H = \Delta U + \Delta n_g\, RT \qquad \text{... (Eqn. 8)}$$

Here, $\Delta n_g = 6 - 7{\cdot}5 = -\,1{\cdot}5.$

Thus, $\Delta U = \Delta H - \Delta n_g\, RT$

$$= -\,780980 - (-\,1{\cdot}5) \times 2 \times 298$$

$$= -\,780090 \text{ calories.}$$

Ex. 5. *When* 1 *mole of ice melts at* 0°C *and at constant pressure of* 1 atm, 1440 *calories of heat are absorbed by the system. The molar volumes of ice and water are* 0·0196 *and* 0·0180 *litre respectively. Calculate* ΔH *and* ΔE.

Solution : Since $\Delta H = q_p$

= heat absorbed by the system at constant pressure

= 1440 calories.

In the equation $\Delta H = \Delta U + p\Delta V$

$$p\Delta V = 76 \times 13{\cdot}6 \times 981\,(18 - 19{\cdot}6) \text{ ergs}$$

$$= -\frac{76 \times 13{\cdot}6 \times 981 \times 1{\cdot}6}{4{\cdot}18 \times 10^7} \text{ calories}$$

$$= -0{\cdot}039 \text{ calories.}$$

($p = 1$ atm $= 76 \times 13{\cdot}6 \times 981$ dynes / cm^2, $V_2 = 18$ cc, $V_1 = 19{\cdot}6$ cc)

Since $p\Delta V$ is very small compared to ΔU, $p\Delta V$ can be neglected.

Thus, $\Delta H = \Delta U = 1440$ calories.

Ex. 6. 1 *mole of ice at* 0°C *and* 4·6 mmHg *pressure is converted to water vapour at a constant temperature and pressure. Find* ΔH and ΔU *if the latent heat of fusion of ice is* 80 cal/g *and latent heat of vaporisation of liquid water at* 0°C *is* 596 cal/g *and the volume of ice in comparison to that of water (vapour) is neglected.*

Solution : Latent heat of fusion of ice per mole = $80 \times 18 = 1440$ cal.

Latent heat of vaporisation of liquid water per mole = 596×18
= 10728 cal.

∴ total heat absorbed by 1 mole (18 grams) of ice in being converted to 1 mole of water vapour

$$= 1440 + 10728 = 12168 \text{ calories.}$$

Since the conversion took place at a constant pressure,

$$q_p = \Delta H = 12168 \text{ calories.}$$

Now we have,

$$\Delta U = \Delta H - p\Delta V = q_p - p\Delta V \quad \text{... (Eqn. 7)}$$

As the volume of ice is to be neglected, $V_1 = 0$

and V_2 = volume of 1 mole of water vapour at 0°C and 4·6 mmHg pressure

$$= 22400 \times \frac{760}{4{\cdot}6} = 3717000 \text{ mL.} \quad \text{(Charles's law)}$$

$$\therefore \quad \Delta V = (V_2 - V_1) = 3717000 \text{ mL}$$

$$p = 4{\cdot}6 \text{ mm} = 0{\cdot}46 \times 13{\cdot}6 \times 981$$

$$= 6137 \text{ dynes/cm}^2.$$

$$\therefore \quad p\Delta V = 6137 \times 3717000 \text{ ergs}$$

$$= \frac{6137 \times 3717000}{4{\cdot}18 \times 10^7} \text{ cal}$$

$$= 545 \text{ cal.}$$

$$\therefore \quad \Delta U = 12168 - 545 = 11623 \text{ calories. (from Eqn. 7)}$$

Ex. 7. *Two moles of a gas at* 1 *bar and* 298 K *are compressed at constant temperature by use of a constant pressure of* 5 *bar. How much work is done on the gas? If the compression is driven by a* 100-kg *mass, how far will it fall in the earth's gravitational field?* (g = 9·8 m s^{-2})

Solution : We have, for an irreversible isothermal process,

$$W = -p_2 (V_2 - V_1) \quad \text{... (Eqn. 2)}$$

$$= -p_2 \left(\frac{nRT}{p_2} - \frac{nRT}{p_1} \right)$$

$$= -nRT \left(1 - \frac{p_2}{p_1} \right)$$

$$= -2 \times 8{\cdot}314 \times 298 \left(1 - \frac{5}{1} \right)$$

$$= 19820 \text{ J.}$$

Further, $W = mgh$ ... (Eqn. 3b)

$$h = \frac{W}{mg} = \frac{19820}{100 \times 9{\cdot}8} \text{ m}$$

$$= 20{\cdot}22 \text{ m.}$$

The following problems are based on thermochemical calculations. In dealing with these problems, we apply the inspection method, as discussed previously.

Ex. 8. *The heats of combustion of ammonia and hydrogen are* 9·06 *and* 68·9 kcal *respectively. Calculate the heat of formation of ammonia.*

Solution : Given that,

(i) $NH_3\,(g) + \frac{3}{4} O_2\,(g) \rightarrow \frac{1}{2} N_2\,(g) + \frac{3}{2} H_2O;\ \Delta H = -\,9{\cdot}06$ kcal

(ii) $H_2\,(g) + \frac{1}{2} O_2\,(g) \rightarrow H_2O\,(g);\ \Delta H = -\,68{\cdot}9$ kcal

(The negative signs are taken as the combustion process is exothermic)

We have to calculate ΔH of the following reaction:

(iii) $\frac{1}{2} N_2\,(g) + \frac{3}{2} H_2\,(g) \rightarrow NH_3\,(g);\ \Delta H = ?$

Since H_2 in Equation (iii) and in Equation (ii) is on the same side and NH_3 in Equation (iii) and in Equation (i) is on opposite sides, we first multiply Equation (ii) by $\frac{3}{2}$ to equate number of moles of H_2 in equations (ii) and (iii). Then subtracting Equation (i) from equation (ii)

$\left\{ \text{i.e., } \frac{3}{2} \times \text{Equation (ii)} - \text{Equation (i)} \right\}$, we get,

$$\frac{3}{2} H_2\,(g) + \frac{3}{4} O_2\,(g) - NH_3\,(g) - \frac{3}{4} O_2\,(g) \rightarrow \frac{3}{2} H_2O\,(g) - \frac{1}{2} N_2\,(g) - \frac{3}{2} H_2O\,(g);$$

$$\Delta H = \frac{3}{2} \times (-\,68{\cdot}9) - (-\,9{\cdot}06)$$

or $\frac{1}{2} N_2\,(g) + \frac{3}{2} H_2\,(g) \rightarrow NH_3\,(g);\ \Delta H = -\,94{\cdot}29$ kcal.

Ex. 9. *Calculate enthalpy of formation of ethane at* 25°C *if the enthalpies of combustion of carbon, hydrogen and ethane are* 94·14, 68·47 *and* 373·3 kcal *respectively.*

Solution : Given that,

(i) $C\,(s) + O_2\,(g) \rightarrow CO_2\,(g);\ \Delta H = -\,94{\cdot}14$ kcal

(ii) $H_2 (g) + \frac{1}{2} O_2 (g) \rightarrow H_2O (l);$ $\Delta H = - 68{\cdot}47$ kcal

(iii) $C_2H_6 (g) + 3\frac{1}{2} O_2 (g) \rightarrow 2CO_2 (g) + 3H_2O (l);$

$\Delta H = - 373{\cdot}3$ kcal

We have to calculate ΔH of the equation,

(iv) $2C (s) + 3H_2 (g) \rightarrow C_2H_6 (g);$ $\Delta H = ?$

Carbon in Equation (i) and (iv), and hydrogen in equation (ii) and (iv) are on the same sides, but C_2H_6 in equations (iii) and (iv) is on opposite sides. Thus following [2 × Eqn. (i) + 3 × Eqn. (ii) – Eqn. (iii)], we get,

$$2C (s) + 2O_2 (g) + 3H_2 (g) + \frac{3}{2} O_2 (g) - C_2H_6 (g) - 3\frac{1}{2} O_2 (g) \rightarrow$$
$$2CO_2 (g) + 3H_2O (l) - 2CO_2 (g) - 3H_2O (l);$$
$$\Delta H = 2 \times (-94{\cdot}14) + 3 \times (-68{\cdot}47) - (-373{\cdot}3)$$

or $2C (s) + 3H_2 (g) \rightarrow C_2H_6 (g);$ $\Delta H = - 20{\cdot}39$ kcal.

Ex. 10. *The molar heats of combustion of* C_2H_2 (g), C(*graphite*) *and* H_2 (g) *are* 310·62 kcal, 94·05 kcal *and* 68·32 kcal *respectively. Calculate the standard heat of formation of* C_2H_2 (g).

Solution : Given that,

(i) $C (s) + O_2 (g) \rightarrow CO_2 (g);$ $\Delta H = - 94{\cdot}05$ kcal

(ii) $H_2 (g) + \frac{1}{2} O_2 (g) \rightarrow H_2O (g);$ $\Delta H = - 68{\cdot}32$ kcal

(iii) $C_2H_2 (g) + 2\frac{1}{2} O_2 (g) \rightarrow 2CO_2 (g) + H_2O (g);$

$\Delta H = - 310{\cdot}62$ kcal

We have to calculate ΔH of the equation

$$2C (s) + H_2 (g) \rightarrow C_2 H_2 (g);\ \Delta H = ?$$

Proceeding the same way as in Example 9, we apply,

[2 × Eqn. (i) + Eqn. (ii) – Eqn. (iii)] and we get,

$$2C (s) + 2O_2 (g) + H_2 (g) + \tfrac{1}{2} O_2 (g) - C_2H_6 (g) - 2\tfrac{1}{2} O_2 (g) \rightarrow$$
$$2CO_2 (g) + H_2O (g) - 2CO_2 (g) - H_2O (g);$$
$$\Delta H = 2 \times (-94{\cdot}05) + (-68{\cdot}32) - (-310{\cdot}62)$$

or $2C (s) + H_2 (g) \rightarrow C_2H_2 (g);$ $\Delta H = 54{\cdot}20$ kcal.

Ex. 11. *Given the following standard heats of reaction at constant pressure:*

(i) *Heat of formation of water* = – 68·3 kcal

(ii) *Heat of combustion of acetylene* = – 310·6 kcal

(iii) *Heat of combustion of ethylene* = – 337·2 kcal

Calculate the heat of reaction for the hydrogenation of acetylene at constant volume (25°C).

Solution : Given that,

(i) $H_2\ (g) + \frac{1}{2} O_2\ (g) \rightarrow H_2O\ (l);\ \Delta H = -\ 68{\cdot}3$ kcal

(ii) $C_2H_2\ (g) + 2\frac{1}{2} O_2\ (g) \rightarrow 2CO_2\ (g) + H_2O\ (l);\ \ \Delta H = -\ 310{\cdot}6$ kcal

(iii) $C_2H_4\ (g) + 3O_2\ (g) \rightarrow 2CO_2\ (g) + 2H_2O\ (l);\ \Delta H = -\ 337{\cdot}2$ kcal.

We have to calculate ΔU for the equation,

(iv) $C_2H_2\ (g) + H_2\ (g) \rightarrow C_2H_4\ (g);\ \Delta U = ?$

C_2H_2 in eqns. (ii) and (iv), and H_2 in eqns. (i) and (iv) are on the same sides, whereas C_2H_4 in eqns. (iii) and (iv) is on opposite sides. Thus applying [Eqn. (i) + Eqn. (ii) – Eqn. (iii)] we get

$$H_2\ (g) + \tfrac{1}{2} O_2\ (g) + C_2H_2\ (g) + 2\tfrac{1}{2} O_2\ (g) - C_2H_2\ (g) - H_2\ (g) \rightarrow$$
$$H_2O\ (l) + 2CO_2\ (g) + H_2O\ (l) - 2CO_2\ (g) - 2H_2O\ (l);$$
$$\Delta H = -\ 68{\cdot}3 + (-\ 310{\cdot}6) - (-\ 337{\cdot}2)$$

or $C_2H_2\ (g) + H_2\ (g) \rightarrow C_2H_4\ (g);\ \ \Delta H = -\ 41{\cdot}7$ kcal.

ΔH is the heat of reaction at constant pressure and to calculate heat at constant volume, we determine ΔU by using the equation,

$$\Delta H = \Delta U + \Delta n_g\, R\, T \qquad \text{... (Eqn. 8)}$$

Δn_g = moles of gaseous product – moles of gaseous reactant

$= 1 - 2 = -\ 1$

$R = 0{\cdot}002$ kcal/K/mole

$T = (273 + 25) = 298$ K

Thus we have,

$$\Delta U = -\ 41{\cdot}7 - (-\ 1 \times {\cdot}002 \times 298)$$
$$= -\ 41{\cdot}104 \text{ kcal.}$$

[**Note:** If the heat of reaction is given in kJ, R should be taken as $8{\cdot}314 \times 10^{-3}$ kJ/K/mole]

Ex. 12. *Calculate heat of the following reaction at constant pressure,*

$$F_2O\ (g) + H_2O\ (g) \rightarrow O_2\ (g) + 2HF\ (g)$$

The heats of formation of F_2O (g), H_2O (g) *and* HF (g) *are* 5·5 kcal, –57·8 kcal *and* –64·2 kcal *respectively.*

Solution : Given that,

(i) $F_2\,(g) + \frac{1}{2} O_2\,(g) \rightarrow F_2O\,(g);\ \Delta H = 5{\cdot}5$ kcal

(ii) $H_2\,(g) + \frac{1}{2} O_2\,(g) \rightarrow H_2O\,(g);\ \Delta H = -\,57{\cdot}8$ kcal

(iii) $\frac{1}{2} H_2\,(g) + \frac{1}{2} F_2\,(g) \rightarrow HF\,(g);\ \Delta H = -\,64{\cdot}2$ kcal

F_2O and H_2O in Eqns. (i) and (ii) and in the equation given in the problem are on the opposite sides, while HF in Eqn. (iii) and in the equation given in the problem is on the same sides.

Thus applying, [– Eqn. (i) – Eqn. (ii) + 2 × Eqn. (iii)], we get

$$-F_2\,(g) - \tfrac{1}{2} O_2\,(g) - H_2\,(g) - \tfrac{1}{2} O_2\,(g) + H_2 + F_2\,(g) \rightarrow$$

$$-F_2O\,(g) - H_2O\,(g) + 2HF\,(g);$$

$$\Delta H = -\,5{\cdot}5 - (-\,57{\cdot}8) + 2 \times (-\,64{\cdot}2)$$

or $F_2O\,(g) + H_2O\,(g) \rightarrow O_2\,(g) + 2HF\,(g);\ \Delta H = -\,76{\cdot}1$ kcal.

Ex. 13. *Calculate the heat of formation of anhydrous* Al_2Cl_6 *from the following data:*

(i) $2Al\,(s) + 6HCl\,(aq.) \rightarrow Al_2Cl_6\,(aq.) + 3H_2\,;\ \Delta H = -\,244$ kcal

(ii) $H_2\,(g) + Cl_2\,(g) \rightarrow 2HCl\,(g);\ \Delta H = -\,44$ kcal

(iii) $HCl\,(g) + aq. \rightarrow HCl\,(aq.);\ \Delta H = -\,17{\cdot}5$ kcal.

(iv) $Al_2Cl_6\,(s) + aq. \rightarrow Al_2Cl_6\,(aq.);\ \Delta H = -\,153{\cdot}7$ kcal

Solution : We have to calculate ΔH of the equation,

(v) $2Al\,(s) + 3Cl_2\,(g) \rightarrow Al_2Cl_6\,(s);\ \Delta H = ?$

In the given equation we see that Equation (iii) does not contain any species of Eqn. (v). Let us thus first consider Eqns. (i), (ii) and (iv) only and apply,

[Eqn. (i) + 3 × Eqn. (ii) – Eqn. (iv)], we get,

$$2Al\,(s) + 6HCl\,(aq.) + 3H_2\,(g) + 3Cl_2\,(g) - Al_2Cl_6\,(s) - aq. \rightarrow$$

$$Al_2Cl_6\,(aq.) + 3H_2\,(g) + 6\,HCl\,(g) - Al_2Cl_6\,(aq.);$$

$$\Delta H = -\,244 + 3 \times (-\,44) - (-\,153{\cdot}7)$$

or $2Al\,(s) + 3Cl_2\,(g) + 6HCl\,(aq) - aq = Al_2Cl_6\,(s) + 6HCl\,(g);$

$$\Delta H = -\,222{\cdot}3 \text{ kcal.}$$

Now multiplying Eqn. (iii) by 6 and adding it to the equation just above we get,

$$6HCl(g) + aq + 2Al\,(s) + 3Cl_2\,(g) + 6HCl\,(aq) - aq \rightarrow$$

$$6HCl\,(aq) + Al_2Cl_6\,(s) + 6HCl\,(g);$$

$$\Delta H = 6 \times (-\,17{\cdot}5) + (-\,222{\cdot}3)$$

or $2Al\,(s) + 3Cl_2\,(g) \rightarrow Al_2Cl_6\,(s);\ \Delta H = -\,327{\cdot}3$ kcal.

Ex. 14. *Heat of combustion of ethyl alcohol is* 325 kcal *and that of acetic acid is* 209·5 kcal. *Calculate the heat evolved in the following reaction:*

$$C_2H_5OH + O_2 \rightarrow CH_3COOH + H_2O.$$

Solution : Given that,

(i) $C_2H_5OH\,(l) + 3O_2\,(g) \rightarrow 2CO_2\,(g) + 3H_2O\,(g);\ \Delta H = -\,325$ kcal.

(ii) $CH_3COOH\,(l) + 2O_2\,(g) \rightarrow 2CO_2\,(g) + 2H_2O\,(g);$

$$\Delta H = -\,209{\cdot}5 \text{ kcal}$$

C_2H_5OH in Eqn. (i) and in the required equation is on the same sides, whereas CH_3COOH in Eqn. (ii) and in the required equation is on the opposite sides. Hence, applying

[Eqn. (i) – Eqn. (ii)], we get,

$$C_2H_5OH\,(l) + 3O_2\,(g) - CH_3COOH\,(l) - 2O_2\,(g) \rightarrow$$

$$2CO_2\,(g) + 3H_2O\,(l) - 2CO_2\,(g) - 2H_2O(1);$$

$$\Delta H = -\,325 - (-\,209{\cdot}5)$$

or $C_2H_5OH\,(l) + O_2\,(g) \rightarrow CH_3COOH\,(l) + H_2O\,(l);$

$$\Delta H = -\,115{\cdot}5 \text{ kcal}$$

Ex. 15. *The standard heats of formation at* 298 K *for* CCl_4 (g), H_2O (g), CO_2 (g) *and* HCl (g) *are* – 25·5, – 57·8, – 94·1 and – 22·1 kcal *per mole respectively. Calculate* ΔH^0 *at* 298 K *for the reaction,*

$$CCl_4\,(g) + 2H_2O\,(g) \rightarrow CO_2\,(g) + 4HCl\,(g).$$

Solution : Given that,

(i) $C\,(s) + 2Cl_2\,(g) \rightarrow CCl_4\,(g);\ \Delta H^0 = -\,25{\cdot}5$ kcal

(ii) $H_2\,(g) + \frac{1}{2}O_2\,(g) \rightarrow H_2O\,(g);\ \Delta H^0 = -\,57{\cdot}8$ kcal

(iii) $C\,(s) + O_2\,(g) \rightarrow CO_2\,(g);\ \Delta H^0 = -\,94{\cdot}1$ kcal

(iv) $\frac{1}{2}H_2\,(g) + \frac{1}{2}Cl_2\,(g) \rightarrow HCl\,(g);\ \Delta H^0 = -\,22{\cdot}1$ kcal

CO_2 and HCl in eqns. (iii) and (iv) respectively are on the same sides as in the required equation, while CCl_4 in Eqn. (i) and H_2O in Eqn. (ii) are on the opposite sides in the required equation. Hence applying,

[– Eqn. (i) – 2 × Eqn. (ii) + Eqn. (iii) + 4 × Eqn. (iv)], we get,

$$-\,C\,(s) - 2Cl_2\,(g) - 2H_2\,(g) - O_2\,(g) + C\,(s) + O_2\,(g) + 2H_2\,(g) + 2Cl_2\,(g)$$

$$\rightarrow -\,CCl_4\,(g) - 2H_2O\,(g) + CO_2\,(g) + 4HCl\,(g);$$

$$\Delta H^0 = -\,(-\,25{\cdot}5) - (2 \times -\,57{\cdot}8) + (-\,94{\cdot}1) + (4 \times -\,22{\cdot}1)$$

or $CCl_4 (g) + 2H_2O (g) \rightarrow CO_2 (g) + 4HCl (g); \Delta H^0 = -41.4$ kcal.

Ex. 16. *The standard molar heats of formation of ethane, carbon dioxide and liquid water are* – 21·1, – 94·1 *and* – 68·3 kcal *respectively. Calculate the standard molar heat of combustion of ethane.* **(IIT 1986)**

Solution : Given that,

(i) $2C (s) + 3H_2 (g) \rightarrow C_2H_6 (g); \Delta H^0 = -21.1$ kcal.

(ii) $C (s) + O_2 (g) \rightarrow CO_2 (g); \Delta H^0 = -94.1$ kcal

(iii) $H_2 (g) + \frac{1}{2} O_2 (g) \rightarrow H_2O (g); \Delta H^0 = -68.3$ kcal

We have to calculate ΔH^0 of the equation

$C_2H_6 (g) + 3\frac{1}{2} O_2 (g) \rightarrow 2CO_2 (g) + 3H_2O (g); \Delta H^0 = ?$

Applying the inspection method,

[– Eqn. (i) + 2 × Eqn. (ii) + 3 × Eqn. (iii)], we get,

$$-2C (s) - 3H_2 (g) + 2C (s) + 2O_2 (g) + 3H_2 (g) + 1\frac{1}{2} O_2 (g) \rightarrow$$

$$-C_2H_6 (g) + 2CO_2 (g) + 3H_2O (g);$$

$$\Delta H^0 = -(-21.1) + 2 \times (-94.1) + 3 \times (-68.3)$$

or $C_2H_6 (g) + 3\frac{1}{2} O_2 (g) \rightarrow 2CO_2 (g) + 3H_2O (g); \Delta H^0 = -372.0$ kcal.

Ex. 17. *The standard enthalpy of combustion at* 25°C *of hydrogen, cyclohexene* (C_6H_{10}) *and cyclohexane* (C_6H_{12}) *are* – 241, – 3800 *and* – 3920 kJ/mole *respectively. Calculate the heat of hydrogenation of cyclohexene.* **(IIT 1989)**

Solution : Given that,

(i) $H_2 (g) + \frac{1}{2} O_2 (g) \rightarrow H_2O (l); \Delta H = -241$ kJ

(ii) $C_6H_{10} (g) + 8\frac{1}{2} O_2 (g) \rightarrow 6CO_2 (g) + 5H_2O (l); \Delta H = -3800$ kJ

(iii) $C_6H_{12} (g) + 9O_2 (g) \rightarrow 6CO_2 (g) + 6H_2O (l); \Delta H = -3920$ kJ

We have to calculate ΔH of the reaction,

$$C_6H_{10} (g) + H_2 (g) \rightarrow C_6H_{12} (g); \Delta H = ?$$

Applying the inspection method, i.e.,

Eqn. (i) + Eqn. (ii) – Eqn. (iii), we get,

$$H_2 (g) + \frac{1}{2} O_2 (g) + C_6H_{10} (g) + 8\frac{1}{2} O_2 (g) - C_6H_{12} (g) - 9O_2 (g) \rightarrow$$

$$H_2O (l) + 6CO_2 (g) + 5H_2O (l) - 6CO_2 (g) - 6H_2O (l);$$

$$\Delta H = -241 + (-3800) - (-3920)$$

or $C_6H_{10} (g) + H_2 (g) \rightarrow C_6H_{12} (g); \Delta H = -121$ kJ.

Ex. 18. *From the following data of ΔH of the following reactions,*

$$C\,(s) + \frac{1}{2}O_2\,(g) \rightarrow CO\,(g);\ \Delta H = -110\text{ kJ}$$

and $$C\,(s) + H_2O\,(g) \rightarrow CO\,(g) + H_2\,(g);\ \Delta H = 132\text{ kJ},$$

calculate the mole composition of the mixture of steam and oxygen on being passed over coke at 1273 K, *keeping the reaction temperature constant.*

Solution : From the question we see that the first reaction is exothermic and the second one is endothermic. Thus, if a mixture of oxygen and steam (H_2O) is passed over coke and at the same time, the temperature does not change, the composition should be such that ΔH of both the reactions are numerically equal.

In the first reaction, consumption of 1/2 mole of O_2 evolves 110 kJ of energy, while in the second reaction, for 1 mole of steam (H_2O), 132 kJ of energy is absorbed.

$\therefore$ mole of O_2 needed to evolve 132 kJ $= \frac{0{\cdot}5}{110} \times 132 = 0{\cdot}6.$

$\therefore$ mole ratio of O_2 and steam (H_2O) $= 0{\cdot}6 : 1.$

Ex. 19. *A gas mixture of* 3·67 *litres of ethylene and methane on complete combustion at* 25°C *produces* 6·11 *litres of* CO_2. *Find out the amount of heat evolved on burning one litre of the gas mixture. The heats of combustion of ethylene and methane are* – 1423 *and* – 891 kJ mol^{-1} at 25°C. **(IIT 1991)**

Solution : 3.67 lit. (total of C_2H_4 and CH_4)

$$C_2H_4 \quad + \quad CH_4 \quad + \quad O_2 \quad \rightarrow \quad CO_2 \quad + \quad H_2O$$

x lit. (say) $\quad (3{\cdot}67 - x)$ lit. $\quad$ 6·11 lit.

or x moles $\quad (3{\cdot}67 - x)$ moles $\quad$ 6·11 moles

Applying POAC for C atoms,

$$2 \times x + 1 \times (3{\cdot}67 - x) = 1 \times 6{\cdot}11;\ x = 2{\cdot}44\text{ lit.}$$

Thus, the volume of $C_2H_4 = 2{\cdot}44$ lit., and volume of $CH_4 = 1{\cdot}23$ lit.

$\therefore$ volume of C_2H_4 in a 1-litre mixture $= \frac{2{\cdot}44}{3{\cdot}67} = 0{\cdot}665$ lit.

and volume of CH_4 in a 1-litre mixture $= 1 - 0{\cdot}665 = 0{\cdot}335$ lit.

Now, thermochemical reactions for C_2H_4 and CH_4 are

$$C_2H_4 + 3O_2 \rightarrow 2CO_2 + 2H_2O;\ \Delta H = -1423\text{ kJ}$$

$$CH_4 + 2O_2 \rightarrow CO_2 + 2H_2O;\ \Delta H = -891\text{ kJ}$$

As ΔH values given are at 25°C, i.e., 298 K, let us first calculate the volume occupied by one mole of any gas at 25°C (supposing pressure as 1 atm)

Volume per mole at 25°C $= \frac{298}{273} \times 22{\cdot}4 = 24{\cdot}45$ lit.

Thus, heat evolved in the combustion of 0·665 lit. of C_2H_4

$$= -\frac{1423}{24{\cdot}45} \times 0{\cdot}665 = -38{\cdot}70 \text{ kJ}$$

and heat evolved in the combustion of 0·335 lit. of CH_4

$$= -\frac{891}{24{\cdot}45} \times 0{\cdot}335 = -12{\cdot}20 \text{ kJ}.$$

∴ total heat evolved in the combustion of 1 litre of the mixture

$$= -38{\cdot}70 + (-12{\cdot}20)$$
$$= -50{\cdot}90 \text{ kJ}.$$

Ex. 20. *The heat of combustion of naphthalene (s) is* – 123·25 kcal. *If the heats of formation of* CO_2 (g) *and* H_2O (l) *are* – 97·0 *and* – 68·4 kcal *respectively, calculate the heat of formation of naphthalene.*

Solution : Given that,

(i) $C_{10}H_8 \text{ (s)} + 12O_2 \text{ (g)} \rightarrow 10CO_2 \text{ (g)} + 4H_2O\text{(l)};\ \Delta H = -123{\cdot}25 \text{ kcal}$

(ii) $C \text{ (s)} + O_2 \text{ (g)} \rightarrow CO_2 \text{ (g)};\ \Delta H = -97{\cdot}0 \text{ kcal}$

(iii) $H_2 \text{ (g)} + \frac{1}{2} O_2 \text{ (g)} \rightarrow H_2O \text{ (l)};\ \Delta H = -68{\cdot}4 \text{ kcal}$

We have to calculate ΔH of the equation,

$$10C \text{ (s)} + 4H_2 \text{ (g)} \rightarrow C_{10}H_8 \text{ (s)};\ \Delta H = ?$$

Applying the inspection method,

[– Eqn. (i) + 10 × Eqn. (ii) + 4 × Eqn. (iii)], we get,

$$-C_{10}H_8 \text{ (s)} - 12O_2 \text{ (g)} + 10C \text{ (s)} + 10O_2 \text{ (g)} + 4H_2 \text{ (g)} + 2O_2 \text{ (g)} \rightarrow$$
$$-10CO_2 \text{ (g)} - 4H_2O \text{ (l)} + 10CO_2 \text{ (g)} + 4H_2O \text{ (l)};$$
$$\Delta H = -(-123{\cdot}25) + 10 \times (-97{\cdot}0) + 4 \times (-68{\cdot}4)$$

or $10C \text{ (s)} + 4H_2 \text{ (g)} \rightarrow C_{10}H_8 \text{ (s)};\ \Delta H = -1120{\cdot}35 \text{ kcal}.$

Ex. 21. *Calculate the enthalpy change accompanying the conversion of* 10 g *of graphite into diamond if the heats of combustion of* C *(graphite) and* C *(diamond) are* – 94·05 *and* – 94·50 kcal *respectively.*

Solution : Given that,

(i) $C \text{ (graphite)} + O_2 \text{ (g)} \rightarrow CO_2 \text{ (g)};\ \Delta H = -94{\cdot}05 \text{ kcal}$

(ii) $C \text{ (diamond)} + O_2 \text{ (g)} \rightarrow CO_2 \text{ (g)};\ \Delta H = -94{\cdot}50 \text{ kcal}$

Thus, applying the inspection method,

[Eqn. (i) – Eqn. (ii)], we get,

$C \text{ (graphite)} + O_2 \text{ (g)} - C \text{ (diamond)} - O_2 \text{ (g)} \rightarrow CO_2 \text{ (g)} - CO_2 \text{ (g)};$

$$\Delta H = -94{\cdot}05 - (-94{\cdot}50)$$

or C (graphite) $\rightarrow$ C (diamond); $\Delta H = +0{\cdot}45$ kcal

Since this enthalpy change is only for conversion of 1 mole, i.e., 12 g of C (graphite) to C (diamond), therefore, for the conversion of 10 g of C (graphite) to C (diamond)

$$\Delta H = 0{\cdot}45 \times \frac{10}{12} = 0{\cdot}375 \text{ kcal.}$$

Ex. 22. *A person inhales* 640 g *of* O_2 *per day. If all the* O_2 *is used for converting sugar into* CO_2 *and* H_2O*, how much sucrose* ($C_{12}H_{22}O_{11}$) *is consumed in the body in one day and what is the heat evolved?* ΔH (*combustion of sucrose*) $= -5645$ kJ mole^{-1}.

Solution : Moles of O_2 inhaled by a person in one day $= \frac{640}{32} = 20.$

Given that,

$$C_{12}H_{22}O_{11} + 12O_2 \rightarrow 12CO_2 + 11H_2O; \ \Delta H = -5645 \text{ kJ.}$$

Thus, 12 moles of O_2 consume 1 mole of sucrose

or 12 moles of O_2 consume 342 g of sucrose

$\therefore$ 20 mole of O_2 consume $\frac{342}{12} \times 20$, i.e., 570 g of sucrose

Further,

342 g (1 mole) of sucrose liberates 5645 kJ

$\therefore$ 570 g of sucrose should liberate $\frac{5645}{342} \times 570 = 9408{\cdot}34$ kJ.

Ex. 23. *An athlete is given* 100 g *of glucose* ($C_6H_{12}O_6$) *of energy equivalent to* 1560 kJ. *He utilizes* 50% *of this gained energy in the event. In order to avoid storage of energy in the body, calculate the weight of water he would need to perspire. The enthalpy of evaporation of water is* 44 kJ mole^{-1}. **(IIT 1989)**

Solution : Energy remained in the body of the athlete after the event

$$= \frac{1560}{2} = 780 \text{ kJ.}$$

$\therefore$ weight of water to be evaporated by 780 kJ of energy

$$= \frac{18}{44} \times 780 = 319{\cdot}1 \text{ g.} \qquad (H_2O = 18)$$

Ex. 24. *Ethylene undergoes combustion according to the thermochemical equation:*

$$C_2H_4\,(g) + 3O_2\,(g) \rightarrow 2CO_2\,(g) + 2H_2O\,(l); \ \Delta H^0 = -337 \text{ kcal}$$

Assuming 70% *efficiency, how much water at* 20°C *can be converted into steam at* 100°C *by the combustion of* 10^3 *litres of* C_2H_4 *gas at* NTP?

Solution : Mole of $C_2H_4 = \dfrac{1000}{22{\cdot}4} = 44{\cdot}6.$

Total heat evolved $= 44{\cdot}6 \times 337 = 1{\cdot}5 \times 10^4$ kcal.

$\therefore$ useful heat $= 0{\cdot}7 \times 1{\cdot}5 \times 10^4 = 1{\cdot}05 \times 10^4$ kcal.

Now, water evaporation takes place in two stages:

$$H_2O\ (l, 20°C) \rightarrow H_2O\ (l, 100°C);\ \Delta H = 80\ \text{kcal/kg}.$$

and $$H_2O\ (l, 100°C) \rightarrow H_2O\ (g, 100°C);\ \Delta H = 540\ \text{kcal/kg}.$$

$$\therefore \quad H_2O\ (l, 20°C) \rightarrow H_2O\ (g, 100°C);\ \Delta H = 620\ \text{kcal/kg}.$$

$\therefore$ weight of water converted to steam

$$= \frac{\text{amount of heat available}}{\text{heat required per kg}}$$

$$= \frac{1{\cdot}05 \times 10^4}{620} = 16{\cdot}9\ \text{kg}.$$

Ex. 25. *An intimate mixture of ferric oxide* (Fe_2O_3) *and aluminium* (Al) *is used as solid rocket fuel. Calculate fuel value per gram and fuel value per cc of the mixture. Heats of formation and densities are as follows:*

$$\Delta H_f\,(Al_2O_3) = 399\ \text{kcal/mole}$$

$$\Delta H_f\,(Fe_2O_3) = 199\ \text{kcal/mole}$$

Density of $Fe_2O_3 = 5{\cdot}2$ g/cc

Density of Al $= 2{\cdot}7$ g/cc **(IIT 1988)**

Solution : Given that,

(i) $2Al + 3/2\ O_2 \rightarrow Al_2O_3;\ \Delta H = -\ 399$ kcal

(ii) $2Fe + 3/2\ O_2 \rightarrow Fe_2O_3;\ \Delta H = -\ 199$ kcal

Applying the inspection method, [Eqn. (i) – Eqn. (ii)], we get,

$$2Al + Fe_2O_3 \rightarrow Al_2O_3 + 2Fe;$$

$$\Delta H = -\ 399 - (-\ 199) = -\ 200\ \text{kcal}$$

Now, 2 moles of Al weigh (2×27), i.e., 54 g and 1 mole of Fe_2O_3 weighs 160 g.

Total weight of the mixture $= 54 + 160 = 214$ g.

$\therefore$ fuel value per gram of the mixture $= \dfrac{200}{214} = 0{\cdot}9345$ kcal/g.

Further, volume of 2 moles of Al $= \dfrac{54}{2{\cdot}7} = 20$ cc

and volume of 1 mole of $Fe_2O_3 = \dfrac{160}{5{\cdot}2} = 30{\cdot}76$ cc.

$\therefore$ total volume of the mixture $= 50{\cdot}76$ cc

$\therefore$ fuel value per cc of the mixture $= \dfrac{200}{50{\cdot}76}$

$= 3{\cdot}94$ kcal/cc.

Ex. 26. *Calculate the enthalpy of formation of sulphuric acid (l) from the following data:*

(i) $S\,(s) + O_2\,(g) \rightarrow SO_2\,(g);\ \Delta H = -\,71{\cdot}0$ kcal

(ii) $SO_2\,(g) + \frac{1}{2}O_2\,(g) \rightarrow SO_3\,(g);\ \Delta H = -\,23{\cdot}5$ kcal

(iii) $SO_3\,(g) + H_2O\,(l) \rightarrow H_2SO_4\,(l);\ \Delta H = -\,31{\cdot}2$ kcal

(iv) $H_2\,(g) + \frac{1}{2}O_2\,(g) \rightarrow H_2O\,(l);\ \Delta H = -\,68{\cdot}5$ kcal

Solution : From the above equations, we have to calculate ΔH for the equation

$$H_2\,(g) + S\,(s) + 2O_2\,(g) \rightarrow H_2SO_4\,(l);\ \Delta H = ?$$

Applying the inspection method,

[Eqn. (i) + Eqn. (ii) + Eqn. (iii) + Eqn. (iv)], we get,

$$S\,(s) + O_2\,(g) + SO_2\,(g) + \tfrac{1}{2}O_2\,(g) + SO_3\,(g) + H_2O\,(l) + H_2\,(g)$$
$$\rightarrow \tfrac{1}{2}O_2\,(g) \rightarrow SO_2\,(g) + SO_3\,(g) + H_2SO_4\,(l) + H_2O\,(l);$$
$$\Delta H = (-\,71{\cdot}0 - 23{\cdot}5 - 31{\cdot}2 - 68{\cdot}5)\ \text{kcal}$$

or $H_2\,(g) + S\,(s) + 2O_2\,(g) \rightarrow H_2SO_4\,(l);\ \Delta H = -\,194{\cdot}2$ kcal.

Ex. 27. *At* 18°C *the heat of solution of anhydrous* $CuSO_4$ *in a large volume of water is* –15·90 kcal *per mole while that of* $CuSO_4 \cdot 5H_2O$ *is* 2·75 kcal *per mole. What is the heat of hydration of* $CuSO_4$*?*

Solution : Given that,

(i) $CuSO_4\,(s) + aq \rightarrow CuSO_4\,(aq);\ \Delta H = -\,15{\cdot}9$ kcal

(ii) $CuSO_4 \cdot 5H_2O + aq \rightarrow CuSO_4\,(aq);\ \Delta H = +\,2{\cdot}75$ kcal

We have to calculate ΔH of the equation:

$$CuSO_4\,(s) + aq \rightarrow CuSO_4 \cdot 5H_2O;\ \Delta H = ?$$

Applying the inspection method, [Eqn (i) – Eqn (ii)],

we get, $CuSO_4\,(s) + aq \rightarrow CuSO_4 \cdot 5H_2O;$

$$\Delta H = -\,15{\cdot}9 - 2{\cdot}75 = -\,18{\cdot}65\ \text{kcal}.$$

Ex. 28. *The heats of neutralisation of* CH_3COOH, $HCOOH$, HCN *and* H_2S *are* – 13·2, – 13·4, – 2·9 *and* – 3·8 kcal *per equivalent respectively. Arrange these acids in increasing order of strength.*

Solution : Since we know that the greater the (negative) value of the heat of neutralisation, the more is the strength of the acid, the given acids may be arranged in increasing order of strength as

$$HCN < H_2S < CH_3COOH < HCOOH.$$

Ex. 29. *The heat of neutralisation of acetic acid and sodium hydroxide is* $-50{\cdot}6$ kJ eq^{-1}. *Find the heat of dissociation of* CH_3COOH *if the heat of neutralisation of a strong acid and a strong base is* $-55{\cdot}9$ kJ eq^{-1}.

Solution : We have,

ΔH (neutralisation)

$$= \Delta H \text{ (ionisation of } CH_3COOH) + \Delta H\,(H^+ + OH^-)$$

$$\therefore\ \Delta H \text{ (ionisation of } CH_3COOH) = -50{\cdot}6 - (-55{\cdot}9) = 5{\cdot}3 \text{ kJ mol}^{-1}$$

[ΔH (ionisation of NaOH) = 0 as NaOH is a strong base]

Ex. 30. *Only* N_2 *and* CO_2 *gases remain after* 15·5 g *of carbon is treated with* 25 *litres of air at* 25°C *and* 5·5 atm *pressure. Assume air of composition:* O_2 – 19%, N_2 – 80% *and* CO_2 – 1% *(by volume). Calculate the heat evolved under constant pressure. Given,*

$$C + O_2 \longrightarrow CO_2;\quad \Delta H = -94{\cdot}05 \text{ kcal/mole}$$

$$C + \frac{1}{2}O_2 \longrightarrow CO;\quad \Delta H = -26{\cdot}41 \text{ kcal/mole}$$

Solution : Moles of C $= \dfrac{15{\cdot}5}{12} = 1{\cdot}292.$

$$\text{Moles of } O_2 = \frac{pV}{RT} = \frac{5{\cdot}5 \times (0{\cdot}19 \times 25)}{0{\cdot}0821 \times 298} = 1{\cdot}068.$$

$$\underset{x \text{ moles (say)}}{C} + O_2 \longrightarrow \underset{x \text{ moles}}{CO_2}$$

$$\underset{(1{\cdot}292 - x) \text{ moles}}{C} + \tfrac{1}{2}O_2 \longrightarrow \underset{(1{\cdot}292 - x) \text{ moles}}{CO}$$

As O_2 is fully consumed,

number of moles of O before reaction

= number of moles of O after reaction

= moles of O in CO_2 + moles of O in CO

or $\quad 2 \times 1{\cdot}068 = 2x + (1{\cdot}292 - x)$

or $\quad x = 0{\cdot}844$

Mole of CO_2 = 0·844.

Mole of CO = 1·292 – 0·844 = 0·448.

Total heat evolved = 0·844 (–94·05) + 0·448 (–26·41) = –91.2 kcal.

Ex. 31. *Calculate ΔH at 358 K for the reaction*

$$Fe_2O_3\ (s) + 3H_2\ (g) \longrightarrow 2Fe\ (s) + 3H_2O\ (l)$$

Given that, $\Delta H_{298} = -33{\cdot}29$ kJ $mole^{-1}$ *and* C_p *for* Fe_2O_3 (s), Fe (s), H_2O (l) *and* H_2 (g) *are* 103·8, 25·1, 75·3 *and* 28·8 J/K mole.

Solution : $\Delta C_p = 2 \times 25{\cdot}1 + 3 \times 75{\cdot}3 - [103{\cdot}8 + 3 \times 28{\cdot}8]$

$= 85{\cdot}9$ J/K mole.

We have,

$$\frac{\Delta H_2 - \Delta H_1}{T_2 - T_1} = \Delta C_p \qquad \text{... (Eqn. 9a)}$$

$$\frac{\Delta H_{358} - (-33290)}{358 - 298} = 85{\cdot}9$$

$$\Delta H_{358} = -28136 \text{ J/mole}$$

$$= -28{\cdot}136 \text{ kJ/mole.}$$

Ex. 32. *Predict whether the entropy change of the system in each of the following processes is positive or negative.*

(a) $CaCO_3(s) \rightarrow CaO(s) + CO_2(g)$

(b) $N_2(g) + 3H_2(g) \rightarrow 2NH_3(g)$

(c) $N_2(g) + O_2(g) \rightarrow 2NO(g)$

(d) $HCl(g) + NH_3(g) \rightarrow NH_4Cl(s)$

(e) $2SO_2(g) + O_2(g) \rightarrow 2SO_3(g)$

(f) Cooling of N_2(g) *from* 20°C *to* −50°C

Solution : Gaseous substances generally possess more entropy than solids. So whenever the products contain more moles of a gas than the reactants, the entropy change is probably positive. And hence, Δ*S* is

(a) positive

(b) negative

(c) small, the sign of Δ*S* is impossible to predict

(d) negative

(e) negative

(f) negative

[**Note:** For a given substance at a given temperature, $S_{gas} > S_{liquid} > S_{solid}$]

Ex. 33. *One mole of an ideal gas undergoes reversible isothermal expansion from an initial volume* V_1 *to a final volume* $10V_1$, *and does* 10 kJ *of work. The initial pressure was* 1×10^7 Pa.

(a) Calculate V_1.

(b) If there were two moles of the gas, what had been its temperature?

Solution : For reversible isothermal expansion of an ideal gas,

(a) $W = -2{\cdot}303nRT \log \dfrac{V_2}{V_1} = -10{,}000$ (given)

$$\text{or} \quad 2{\cdot}303 \times 1 \times 8{\cdot}314 \times T \log \frac{10V_1}{V_1} = 10{,}000$$

$$T = 522{\cdot}27 \text{ K.}$$

We have,

$$pV = nRT$$

$$10^7 \times V_1 = 1 \times 8{\cdot}314 \times 522{\cdot}27$$

$$V_1 = 4{\cdot}342 \times 10^{-4} \text{ m}^3.$$

(b) Assuming pressure to remain constant, for 2 moles of the ideal gas.

$$T = \frac{522{\cdot}27}{2} = 261{\cdot}13 \text{ K.}$$

Ex. 34. *If the equation of state for* 1 *mole of a gas is*

$$\left(p + \frac{a}{V^2}\right)V = RT$$

prove that p is a state function and hence dp is an exact differential.

Solution : *dp* would be an exact differential if $\dfrac{\partial^2 p}{\partial V \partial T} = \dfrac{\partial^2 p}{\partial T \partial V}$ as $p = f(V, T)$.

Given that $\left(p + \dfrac{a}{V^2}\right)V = RT$

$$\text{or} \quad p = \frac{RT}{V} - \frac{a}{V^2}. \quad \dots (1)$$

Differentiating w.r.t. V at constant T, we get

$$\left(\frac{\partial p}{\partial V}\right)_T = -\frac{RT}{V^2} + \frac{2a}{V^3}.$$

Differentiating w.r.t. T at constant V, we get

$$\frac{\partial^2 p}{\partial T \partial V} = -\frac{R}{V^2}. \quad \dots (2)$$

Again differentiating Eqn. (1) first w.r.t. T at constant V and then w.r.t. V at constant T, we get,

$$\frac{\partial^2 p}{\partial V \partial T} = -\frac{R}{V^2}. \quad \dots (3)$$

From Eqns. (2) and (3) we have

$$\frac{\partial^2 p}{\partial T \partial V} = \frac{\partial^2 p}{\partial V \partial T}.$$

Thus dp is an exact differential and p is a state function.

Ex. 35. *K_a for* CH_3COOH *at* 25ºC *is* $1{\cdot}754 \times 10^{-5}$. *At* 50ºC, K_a *is* $1{\cdot}633 \times 10^{-5}$. *What are* ΔH^0 *and* ΔS^0 *for the ionisation of* CH_3COOH?

Solution : $(\Delta G^0)_{298} = -2{\cdot}303RT \log K$

$$= -2{\cdot}303 \times 8{\cdot}314 \times 298 \times \log(1{\cdot}754 \times 10^{-5})$$

$$= 27194 \text{ J}.$$

$$(\Delta G^0)_{323} = -2{\cdot}303 \times 8{\cdot}314 \times 323 \times \log(1{\cdot}633 \times 10^{-5})$$

$$= 29605 \text{ J}$$

$$\Delta G^0 = \Delta H^0 - T\Delta S^0$$

$$27194 = \Delta H^0 - 298\, \Delta S^0$$

$$29605 = \Delta H^0 - 323\, \Delta S^0$$

$$\therefore \quad \Delta H^0 = -1{\cdot}55 \text{ kJ/mol}$$

$$\Delta S^0 = -96{\cdot}44 \text{ J/mol·K}.$$

Ex. 36. *A certain reaction is nonspontaneous at* 298 K. *The entropy change during the reaction is* 121 J/K. *Is the reaction endothermic or exothermic? What is the minimum value of* ΔH *for the reaction?*

Solution : Given that $\Delta G = +ve$ for nonspontaneous process.

As, $\Delta G = \Delta H - T\Delta S$ and $\Delta S = +121$ J/K

ΔH has to be positive, that is, the reaction is endothermic.

To calculate the minimum value of ΔH, $\Delta G = 0$.

$\therefore \quad \Delta H = T\Delta S$

or $\quad \Delta H = 298 \times 121$ J

$\Delta H = 36{\cdot}06$ kJ.

Ex. 37. *Calculate* ΔG *at* 298 K *for the following reaction if the reaction mixture consists of* 1 atm *of* N_2, 3 atm *of* H_2 *and* 1 atm *of* NH_3.

$$N_2(g) + 3H_2(g) \rightleftharpoons 2NH_3(g); \ \Delta G^0 = -33{\cdot}32 \text{ kJ}$$

What is the effect of increasing pressure on the reaction?

Solution : $Q = \dfrac{p_{NH_3}^2}{p_{N_2} \cdot p_{H_2}^3} = \dfrac{1^2}{1 \times 3^3} = 3{\cdot}7 \times 10^{-2},$

where Q is the reaction quotient.

$\therefore \quad \Delta G = \Delta G^0 + 2{\cdot}303RT \log Q$

$$\Delta G = (-33{\cdot}32) + 2{\cdot}303 \times 8{\cdot}314 \times 10^{-3} \times 298 \log(3{\cdot}7 \times 10^{-2})$$
$$\Delta G = -41{\cdot}49 \text{ kJ}.$$

Thus, increasing the pressure of H_2 from 1 atm (std. condition) to 3 atm, free energy change, ΔG, becomes more negative, that is, from −33·32 kJ to −41·49 kJ. Hence increase in pressure favours the forward reaction.

Ex. 38. *How much heat is produced in burning a mole of* CH_4 *under standard conditions if reactants and products are brought to 298 K and* $H_2O(l)$ *is formed? What is the maximum amount of useful work that can be accomplished under standard conditions by this system?*

	$CH_4(g)$	$+\ 2O_2(g)$	$\rightarrow\ CO_2(g)$	$+\ 2H_2O(l)$
ΔHf^0 :	−74·8	0	−393·5	−285·85 kJ
ΔGf^0 :	−50·8	0	−394·4	−236·8 kJ

Solution : $\Delta H^0 = (-393{\cdot}5) + (2 \times -285{\cdot}85) - (-74{\cdot}8) - 2 \times 0$
$= -890{\cdot}4$ kJ/mole.

Now the free energy change for a process, ΔG, equals the maximum useful work that can be done by the system at constant temperature and pressure.

$$\therefore \quad W_{max} = \Delta G = (2 \times -236{\cdot}8) + (-394{\cdot}4) - (-50{\cdot}8) - 2 \times 0$$
$$= -817{\cdot}2 \text{ kJ/mole } CH_4.$$

Ex. 39. *Calculate the boiling point of bromine from the following data:* ΔH^0 *and* ΔS^0 *values of* $Br_2(l) \rightarrow Br_2(g)$ *are* 30·91 kJ/mole *and* 93·2 J/mol . K *respectively. Assume that* ΔH *and* ΔS *do not vary with temperature.*

Solution : Consider the process: $Br_2(l) \rightarrow Br_2(g)$.

The b.p. of a liquid is the temperature at which the liquid and the pure gas coexist at equilibrium at 1 atm.

$$\therefore \qquad \Delta G = 0.$$

As it is given that ΔH and ΔS do not change with temperature

$$\Delta H = \Delta H^0 = 30{\cdot}91 \text{ kJ}$$
$$\Delta S = \Delta S^0 = 93{\cdot}2 \text{ J/K} = 0{\cdot}0932 \text{ kJ/K}$$

We have,

$$\Delta G = \Delta H - T\Delta S = 0$$

$$\therefore \qquad T = \frac{\Delta H}{\Delta S} = \frac{30{\cdot}91}{0{\cdot}0932} = 331{\cdot}6 \text{ K}.$$

This is the temperature at which the system is in equilibrium, that is, the b.p. of bromine.

Ex. 40. *What is* ΔG^0 *at* 1000°C *for the following reaction?*

$$CaCO_3(s) \rightleftharpoons CaO(s) + CO_2(g)$$

Is this reaction spontaneous at 1000°C *and* 1 atm?

What is the value of K_p *at* 1000°C *for this reaction?*

What is partial pressure of CO_2?

Use the following data:

	$CaCO_3(s)$	$CaO(s)$	$CO_2(g)$
ΔH_f^0	−1206·9	−635·1	−393·5 kJ
S^o	92·9	38·2	213·7 J/K

Solution : For the given reaction,

$$\Delta H^0 = (-635{\cdot}1) + (-393{\cdot}5) - (-1206{\cdot}9) = 178{\cdot}3 \text{ kJ}$$

$$\Delta S^0 = (38{\cdot}2) + (213{\cdot}7) - (92{\cdot}9) = 159{\cdot}0 \text{ J/K}.$$

To determine whether the reaction is spontaneous in the forward direction, let us calculate ΔG^0.

$$\Delta G^0 = \Delta H^0 - T\Delta S^0$$

$$\Delta G^0 = 178{\cdot}3 - 1273 \times (159{\cdot}0 \times 10^{-3}) = -24{\cdot}1 \text{ kJ}.$$

The negative sign of ΔG^0 suggests that the reaction is spontaneous at 1273 K and 1 atm.

Further, $\log K = -\dfrac{\Delta G^0}{2{\cdot}303\, RT}$

$$\log K = -\frac{(-24{\cdot}1)}{2{\cdot}303 \times (8{\cdot}314 \times 10^{-3}) \times 1273}$$

$$= 0{\cdot}987$$

$$K = K_p = 9{\cdot}75.$$

$$\therefore \quad K_p = p_{CO_2} = 9{\cdot}75 \text{ atm}.$$

Ex. 41. *The heat of vaporisation of* CCl_4 *at* 298 K *is* 43·0 kJ/mole.

$$CCl_4(l) \rightarrow CCl_4(g);\ \Delta H = 43{\cdot}0 \text{ kJ}$$

If 1 *mole of liquid* CCl_4 *at* 298 K *has an entropy of* 214 J/K, *what is the entropy of* 1 *mole of the vapour in equilibrium with the liquid at this temperature?*

Solution : $\Delta S = \dfrac{\Delta H_v}{T} = \dfrac{43{\cdot}0 \times 10^3}{298} = 144{\cdot}3$ J/mol·K.

But $\Delta S = S_V - S_L$

or $S_V = \Delta S + S_L = 144{\cdot}3 + 214 = 358{\cdot}3$ J/mol·K.

Ex. 42. *Calculate the equilibrium constant K_{sp} for the reaction*

$$AgCl(s) \rightleftharpoons Ag^+(aq) + Cl^-(aq)$$

Using the data: $\Delta G_f^0(AgCl) = -109{\cdot}7$ kJ, $\Delta G_f^0(Ag^+) = 77{\cdot}1$ kJ *and* $\Delta G_f^0(Cl^-) = -131{\cdot}2$ kJ.

Solution : ΔG^0 for the reaction is calculated as

$$\Delta G^0 = 77{\cdot}1 + (-131{\cdot}2) - (-109{\cdot}7) = 55{\cdot}6 \text{ kJ}.$$

We have,

$$\Delta G^0 = -2{\cdot}303\, RT \log K$$

$$\log K = -\frac{\Delta G^0}{2{\cdot}303RT} = \frac{55{\cdot}6 \times 10^3}{2{\cdot}303 \times 8{\cdot}314 \times 298} = -9{\cdot}75$$

$$\therefore \quad K = K_{sp} = 1{\cdot}8 \times 10^{-10}.$$

Ex. 43. *For a certain process,* $\Delta H = 178$ kJ *and* $\Delta H = 160$ J/K. *What is the minimum temperature at which the process will be spontaneous? Assume that* ΔH *and* ΔS *do not vary with temperature.*

Solution : When the process is at equilibrium, $\Delta G = 0$.

$$\therefore \quad \Delta H = T\Delta S \qquad (\because\ \Delta G = \Delta H - T\Delta S)$$

$$\text{or} \quad T = \frac{\Delta H}{\Delta S} = \frac{178000}{160} = 1112{\cdot}5 \text{ K}.$$

Thus the process will become spontaneous above 1112·5 K.

Ex. 44. *Metallic mercury is obtained by roasting mercury (II) sulphide in a limited amount of air. Estimate the temperature range in which the standard reaction is product-favoured.*

$$HgS(s) + O_2(g) \rightarrow Hg(l) + SO_2(g)$$

$\Delta H^0 = -238{\cdot}6$ kJ/mole *and* $\Delta S^0 = +36{\cdot}7$ J/mole·K.

Solution : Assume that ΔH and ΔS values do not depend on temperature. As ΔH^0 is negative and ΔS^0 is positive, using the equation

$$\Delta G^0 = \Delta H^0 = T\Delta S^0$$

ΔG^0 will be negative at all temperatures and so the reaction is product-favoured at all temperatures. In this problem, both the factors ΔH^0 and ΔS^0 are favourable to spontaneity.

Ex. 45. *Estimate the temperature range for which the following standard reaction is product-favoured.*

$$SiO_2(s) + 2C(s) + 2Cl_2(g) \rightarrow SiCl_4(g) + 2CO(g)$$

$$\Delta H^0 = +32{\cdot}9 \text{ kJ/mole } \textit{and } \Delta S^0 = 226{\cdot}5 \text{ J/mole·K}$$

Solution : In this problem, the factor ΔS^0 being positive, is favourable to spontaneity, whereas the factor ΔH^0 being positive, is unfavourable. Thus the reaction becomes product-favoured above some temperature. We can set ΔG^0 equal to zero in the equation: $\Delta G^0 = \Delta H^0 - T\Delta S^0$ and solve for the temperature at which the reaction is at equilibrium and above which the reaction becomes product-favoured as then ΔG^0 becomes negative.

$$T = \frac{\Delta H^0}{\Delta S^0} = \frac{+32{\cdot}9}{+0{\cdot}2265} = 145{\cdot}25 \text{ K.}$$

Ex. 46. *The efficiency of the Carnot engine is* ⅙. *On decreasing the temperature of the sink by* 65 K, *the efficiency increases to* ⅓. *Find the temperatures of the source and the sink.*

Solution : We have,

$\eta = \dfrac{T_2 - T_1}{T_2}$; where T_1 and T_2 are the temperatures of sink and source respectively.

$$\therefore \quad \eta = \frac{T_2 - T_1}{T_2} = \frac{1}{6}. \qquad \text{... (1)}$$

Now the temperature of the sink is reduced by 65 K.

$\therefore$ temp. of the sink $= (T_1 - 65)$.

$$\therefore \quad \eta = \frac{T_2 - (T_1 - 65)}{T_2} = \frac{1}{3}. \qquad \text{... (2)}$$

On solving Eqns. (1) and (2), we get,

$$T_1 = 325 \text{ K}$$

$$T_2 = 390 \text{ K.}$$

Ex. 47. *A Carnot engine operates between temperatures* 600 K *and* 390 K. *It absorbs* 120 cal *of heat from the source. Calculate* η *and the heat rejected to the sink.*

Solution : We have, $\eta = \dfrac{T_2 - T_1}{T_2} = \dfrac{600 - 390}{600} = 0{\cdot}25$ or 25%.

Now, again we have,

$$\eta = \frac{W}{q_2} = \frac{q_2 - q_1}{q_2},$$

where q_2 is the heat absorbed by the system from the source and q_1 is the heat rejected to the sink.

$$\therefore \quad \frac{120 - q_1}{120} = 0{\cdot}25.$$

$$\therefore \quad q_1 = 90 \text{ cal.}$$

Ex. 48. *The state of a mole of an ideal gas changed from State A(2p, v) through four different processes and finally returns to initial State A reversibly as shown below.*

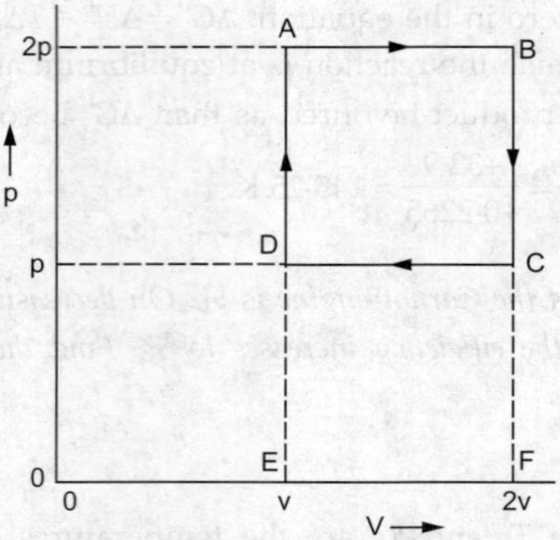

Calculate the total work done by the system and heat absorbed by the system in the cyclic process.

***Solution* : State A to State B** (Isobaric expansion)

Pressure is held constant at $2p$ and the gas is heated until the volume v becomes $2v$.

$\therefore \quad W_1 = -p\Delta V = -2p(2v - v) = -2pv$ (= area $ABFE$)

State B to State C (Isochoric process)

Volume is held constant at $2v$ and the gas is cooled until the pressure $2p$ reaches p.

$\therefore \quad W_2 = 0 \quad (\because \; \Delta V = 0)$

State C to State D (Isobaric compression)

Pressure is held constant at p and the gas is further cooled until the volume $2v$ becomes v.

$\therefore \quad W_3 = -p(v - 2v) = pv$(= area $CDEF$)

State D to State A (Isochoric process)

Volume is held constant at v and the gas is heated until the pressure p reaches $2p$.

$\therefore \quad W_4 = 0 \quad (\because \; \Delta V = 0)$

Total work done by the gas $= W = W_1 + W_2 + W_3 + W_4$

or $\quad W = -2pv + 0 + pv + 0 = -pv$ (= area $ABCD$)

As the process is cyclic, $\Delta U = 0$

$$q = \Delta U - W = -W = pv$$

where q is the heat absorbed in the cyclic process.

Ex. 49. *Two moles of a perfect gas undergo the following processes:*

(a) a reversible isobaric expansion from (1·0 atm, 20·0 L) *to* (1·0 atm, 40·0 L)

(b) a reversible isochoric change of state from (1·0 atm, 40·0 L) *to* (0·5 atm, 40·0 L)

(c) a reversible isothermal compression from (0·5 atm, 40·0 L) to (1·0 atm, 20·0 L)

(i) Sketch the labels of each of the processes on the same p-V diagram.

(ii) Calculate the total work (W) and the total heat change (q) involved in the above processes.

(iii) What will be the values of ΔU, ΔH and ΔS for the overall process?

(IIT 2002)

Solution **:** (i) State A : (1·0 atm, 20 L)

State B : (1·0 atm, 40 L)

State C : (0·5 atm, 40 L)

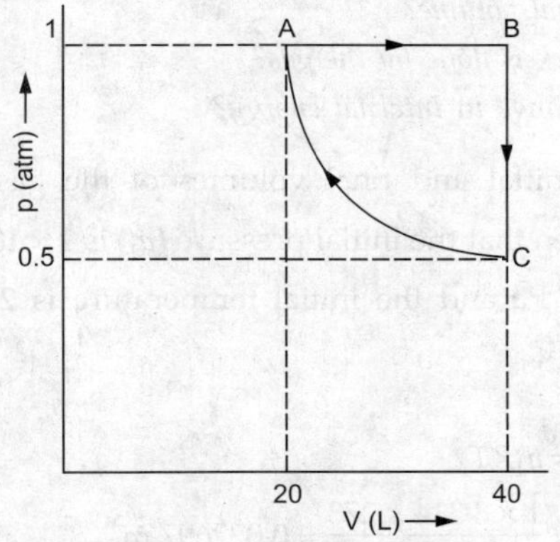

(ii) (a) **State A to State B** (Isobaric expansion)

Work done by the gas $= W_1 = -p(V_B - V_A)$

$= -1(40 - 20) = -20$ L · atm

(b) **State B to State C** (Isochoric process)

Work done by the gas $= W_2 = 0. \quad (\because \ \Delta V = 0)$

(c) **State C to State A** (Isothermal compression)

$$\text{Work done on the gas} = W_3 = -2{\cdot}303 nRT \log \frac{V_A}{V_C}$$

$$= -2{\cdot}303\, pV \log \frac{V_A}{V_C}$$

$$(pV = p_A V_A = p_B V_B = nRT = 20)$$

$$= -2{\cdot}303 \times 20 \log \frac{20}{40} = +13{\cdot}864 \text{ L} \cdot \text{atm}.$$

$\therefore$ total work done by the gas

$$= W_1 + W_2 + W_3 = -20 + 0 + 13{\cdot}864 = -6{\cdot}136 \text{ L} \cdot \text{atm}$$

$$= -\frac{6{\cdot}136 \times 8{\cdot}314}{0{\cdot}0821} \text{ J}$$

$$= -621{\cdot}37 \text{ J}.$$

And, $q = -W = 621{\cdot}37$ J. ($\because$ in cyclic process, $\Delta U = 0$)

(iii) The total process is cyclic and U, H and S are state functions.

$\therefore \quad \Delta U = 0, \ \Delta H = 0$ and $\Delta S = 0$.

Ex. 50. *A mole of a monoatomic ideal gas at* 1 atm *and* 273 K *is allowed to expand adiabatically against a constant pressure of* 0·395 *bar until equilibrium is reached.*

(a) What is the final temperature?

(b) What is the final volume?

(c) How much work is done by the gas?

(d) What is the change in internal energy?

Solution : Let the initial and final volumes of the gas be V_1 and V_2 m^3 respectively. Given that the initial pressure (p_1) is 1×10^5 Pa, final pressure (p_2) is $0{\cdot}395 \times 10^5$ Pa and the initial temperature is 273 K. Let the final temperature be T_2.

We have,

$$p_1V_1 = n_1RT_1$$

$$V_1 = \frac{1 \times 8{\cdot}314 \times 273}{1 \times 10^5} = 0{\cdot}022697 \text{ m}^3.$$

For an adiabatic expansion of 1 mole of monoatomic ideal gas against a constant external pressure (p_2), work done is given as

$$W = -p_2(V_2 - V_1) = C_V(T_2 - T_1) = \frac{3R}{2}(T_2 - T_1)$$

or $$-0{\cdot}395 \times 10^5(V_2 - 0{\cdot}022697) = \frac{3 \times 8{\cdot}314}{2}(T_2 - 273). \quad \ldots (1)$$

Again,

$$p_2V_2 = nRT_2$$

$$0{\cdot}395 \times 10^5 \times V_2 = 1 \times 8{\cdot}314 \times T_2. \quad \ldots (2)$$

Solving eqns. (1) and (2), we get,

(a) the final temperature, $T_2 = 207$ K

(b) the final volume, $V_2 = 0{\cdot}043578$ m^3

(c) the work done by the gas, $W = -p_{ext}(V_2 - V_1)$

$$= -0{\cdot}395 \times 10^5(0{\cdot}043578 - 0{\cdot}022697)$$
$$= -825 \text{ J/mole}$$

(d) as $q = 0$, and $q = \Delta U - W$

$\Delta U = W = -825$ J/mole.

Ex. 51. *(a) One mole of an ideal gas expands isothermally and* ***reversibly*** *at 25°C from a volume of 10 litres to a volume of 20 litres.*

(i) What is the change in entropy of the gas?

(ii) How much work is done by the gas?

(iii) What is q(surroundings)?

(iv) What is the change in the entropy of the surroundings?

(v) What is the change in the entropy of the system plus the surroundings?

(b) Also answer the questions (i) to (v) if the expansion of the gas occurs ***irreversibly*** *by simply opening a stopcock and allowing the gas to rush into an evacuated bulb of 10-L volume.*

Solution **:** (i) $\Delta S = 2{\cdot}303 nR \log \dfrac{V_2}{V_1} = 2{\cdot}303 \times 1 \times 8{\cdot}314 \times \log \dfrac{20}{10} = 5{\cdot}76 \text{ J/K}.$

(a) (ii) $W_{rev} = -2{\cdot}303 nRT \log \dfrac{V_2}{V_1}$

$$= -2{\cdot}303 \times 1 \times 8{\cdot}314 \times 298 \times \log \frac{20}{10} = -1718 \text{ J}.$$

(iii) For isothermal process, $\Delta U = 0$ and heat is absorbed by the gas,

$q_{rev} = \Delta U - W = 0 - (-1718) = 1718$ J

$\therefore \quad q_{surr} = 1718$ J. $\quad$ ($\because$ process is reversible)

(iv) $\Delta S_{surr} = -\dfrac{1718}{298} = -5{\cdot}76$ J/K.

As entropy of the system increases by 5·76 J, the entropy of the surroundings decreases by 5·76 J, since the process is carried out **reversibly.**

(v) $\Delta S_{sys} + \Delta S_{surr} = 0$ ... for reversible process.

(b) (i) $\Delta S = 5{\cdot}76$ J/K, which is the same as above because S is a state function.

(ii) $W = 0$. $\quad$ ($\because p_{ext} = 0$)

(iii) No heat is exchanged with the surroundings.

(iv) $\Delta S_{surr} = 0$.

(v) The entropy of the system plus surroundings increases by 5·76 J/K, as we expect entropy to increase in an **irreversible** process.

BOND ENERGY

Ex. 52. *The heat of formation of methane is* – 17·9 kcal. *If the heats of atomisation of carbon and hydrogen are* 170·9 *and* 52·1 kcal *per mole, calculate the* C—H *bond energy in methane.*

Solution : Given that,

$$C\,(s) + 2H_2\,(g) \rightarrow CH_4\,(g);\ \Delta H = -17{\cdot}9 \text{ kcal}$$

Energy change in reactants:

Heat of atomisation of 1 mole of C = 170·9 kcal

Heat of atomisation of 4 moles of H = 4 × 52·1 kcal

Energy change in product:

Heat of formation of 4 moles of C—H bonds = 4 × x kcal.

(where x is the energy of formation of C—H bonds in kcal/mole).

Since the algebraic sum of all the heat changes is equal to the heat of formation of the above given equation,

we have

$$170{\cdot}9 + 4 \times 52{\cdot}1 + 4x = -17{\cdot}9;\ x = -99{\cdot}3 \text{ kcal.}$$

Thus the bond energy = + 99·3 kcal/mole.

Ex. 53. *The heat of formation of ethylene is* 12·5 kcal. *Calculate* C=C *bond energy in ethylene from the following data. Heat of atomisation of* C = 170·9 kcal/mole, *Heat of atomisation of* H = 52·1 kcal/mole, *bond energy of* C—H = 99·3 kcal/mole.

Solution : Given that,

$$2C\,(s) + 2H_2\,(g) \rightarrow C_2H_4\,(g) \left\{ \begin{array}{c} \text{H}\quad\text{H} \\ |\quad\ \ | \\ \text{H—C=C—H} \end{array} \right\};\quad \Delta H = 12{\cdot}5 \text{ kcal}$$

For reactants:

Heat of atomisation of 2 moles of C = 2 × 170·9 kcal

Heat of atomisation of 4 moles of H = 4 × 52·1 kcal

For products:

Heat of formation of 4 moles of C—H bond = – 4 × 99·3 kcal

Heat of formation of 1 mole of C=C bond = 1 × x

[where x is the energy of formation of C=C bond in kcal/mole]

Adding up, we get the heat of formation of the reaction given above

i.e., $2 \times 170{\cdot}9 + 4 \times 52{\cdot}1 - 4 \times 99{\cdot}3 + x = 12{\cdot}5$

or $x = -140{\cdot}5$ kcal/mole.

Thus, the bond energy of C=C bond = + 140·5 kcal/mole.

Ex. 54. *The bond dissociation energy of gaseous* H_2 , Cl_2 *and* HCl *are* 104, 58 *and* 103 kcal/mole *respectively. Calculate the enthalpy of formation of* HCl *gas.*

(IIT 1985)

Solution : For the reaction $H_2(g) + Cl_2(g) \rightarrow 2HCl(g)$

For reactants:

Bond energy of 1 mole of H—H bond = 104 kcal

Bond energy of 1 mole of Cl—Cl bond = 58 kcal

For products:

Energy of formation of 2 moles of H—Cl bond = -2×103 kcal

Thus ΔH of the above reaction = $104 + 58 - 206 = -44$ kcal.

Since, for

$$H_2 + Cl_2 \rightarrow 2HCl; \ \Delta H = -44 \text{ kcal}$$

then for,

$$\frac{1}{2}H_2 + \frac{1}{2}Cl_2 \rightarrow HCl; \quad \Delta H = -22 \text{ kcal.}$$

Ex. 55. *The enthalpies for the following reactions* (ΔH^0) *at* 25°C *are given.*

$$\tfrac{1}{2}H_2(g) + \tfrac{1}{2}O_2(g) \rightarrow OH(g); \ \Delta H^0 = 10{\cdot}06 \text{ kcal}$$

$$H_2(g) \rightarrow 2H(g); \ \Delta H^0 = 104{\cdot}18 \text{ kcal}$$

$$O_2(g) \rightarrow 2O(g); \ \Delta H^0 = 118{\cdot}32 \text{ kcal}$$

Calculate O—H *bond energy in the hydroxyl radical.*

Solution : From the given data, we have,

$$H_2(g) + O_2(g) \rightarrow 2OH(g); \ \Delta H^0 = 20{\cdot}12 \text{ kcal}$$

For reactants:

Bond energy for 1 mole of H—H bonds = 104·18 kcal

Bond energy for 1 mole of O—O bonds = 118·32 kcal

and for product:

Energy of formation of 2 moles of O—H bond = $2 \times x$

(where x is the energy of formation of 1 mole of O—H bonds)

Adding up all the energy changes, we get heat of the above reaction.

(= 20·12 kcal)

i.e., $104{\cdot}18 + 118{\cdot}32 + 2 \times x = 20{\cdot}12$

$$x = -101{\cdot}19 \text{ kcal.}$$

Hence bond energy of O—H bond = + 101·19 kcal per mole.

Ex. 56. *Calculate the heat of formation of benzene from the following data, assuming no resonance. Bond energies:*

$$C—C = 83 \text{ kcal} \quad C{=}C = 140 \text{ kcal}$$
$$C—H = 99 \text{ kcal}$$

Heat of atomisation of C = 170·9 kcal

Heat of atomisation of H = 52·1 kcal

Solution : We have to calculate ΔH for the reaction

$$6C\ (s) + 3H_2\ (g) \rightarrow C_6H_6\ (g)$$

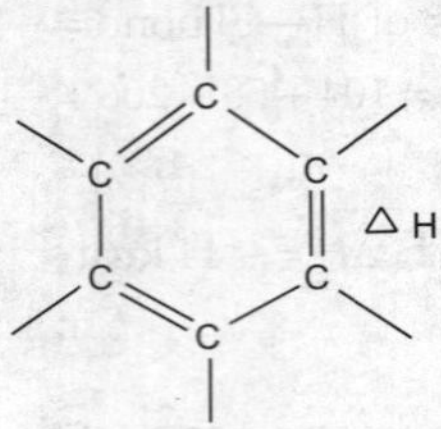

For reactants:

Heat of atomisation of 6 moles of C = 6 × 170·9 kcal

Heat of atomisation of 6 moles of H = 6 × 52·1 kcal

For products:

Heat of formation of 6 moles of C—H bonds = – 6 × 99

Heat of formation of 3 moles of C—C bonds = – 3 × 83

Heat of formation of 3 moles of C=C bonds = – 3 × 140

On adding, we get heat of formation of C_6H_6, i.e.,

$$\Delta H = 6 \times 170{\cdot}9 + 6 \times 52{\cdot}1 - 6 \times 99 - 3 \times 83 - 3 \times 140 = 75{\cdot}0 \text{ kcal.}$$

Ex. 57. *Calculate* ΔH *for the following homogeneous gaseous reaction*

$$CH_3COCH_3 + 2O_2 \rightarrow CH_3COOH + CO_2 + H_2O$$

from the following data:

Bond energies: C — H = 99 kcal

C — C = 83 kcal

C=O = 173 kcal

O=O = 118 kcal

C — O = 84 kcal

O — H = 110 kcal

Solution : We have,

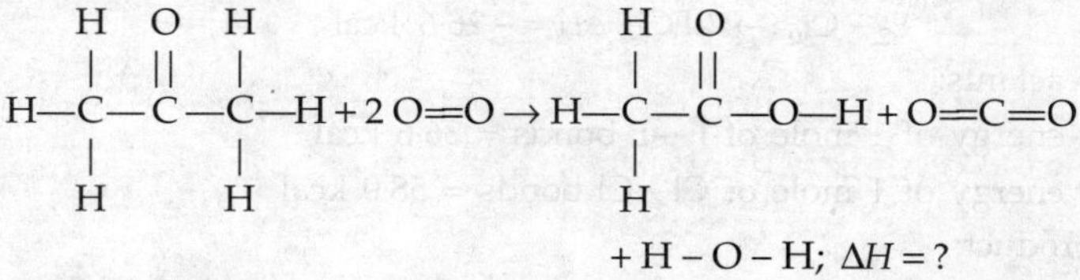

For reactants:

Bond energy of 6 moles of C — H bonds $= 6 \times 99$ kcal

Bond energy of 2 moles of C — C bonds $= 2 \times 83$ kcal

Bond energy of 1 mole of C$=$O bonds $= 1 \times 173$ kcal

Bond energy of 2 moles of O$=$O bonds $= 2 \times 118$ kcal

For products:

Energy of formation of 3 moles of C—H bonds $= -3 \times 99$ kcal

Energy of formation of 1 mole of C—C bonds $= -83$ kcal

Energy of formation of 1 mole of C$=$O bonds $= -173$ kcal

Energy of formation of 1 mole of C—O bonds $= -84$ kcal

Energy of formation of 1 mole of O—H bonds $= -110$ kcal

Energy of formation of 2 moles of C$=$O bonds $= -2 \times 173$ kcal

Energy of formation of 2 moles of O—H bonds $= -2 \times 110$ kcal

Adding up, we get ΔH of the required equation, i.e.,

$$\Delta H = -144 \text{ kcal.}$$

Ex. 58. *Calculate enthalpy of the reaction,*

$$F_2O\,(g) + H_2O\,(g) \rightarrow O_2\,(g) + 2HF\,(g)$$

if the bond energies of O—F, O—H, H—F *and* O$=$O *bonds are* 44, 111, 135 *and* 119 kcal *per mole respectively.*

Solution : We have to calculate ΔH of the reaction,

$$F_2O\,(g) + H_2O\,(g) \rightarrow O_2\,(g) + 2HF\,(g)$$

For reactants:

Bond energy of 2 moles of O—F bonds $= 2 \times 44$ kcal

Bond energy of 2 moles of O—H bonds $= 2 \times 111$ kcal

For products:

Bond formation energy of 1 mole of O$=$O bonds $= -119$ kcal

Bond formation energy of 2 moles of H—F bonds $= -2 \times 135$ kcal

Adding, we get ΔH of the required equation,

$$\Delta H = 2 \times 44 + 2 \times 111 - 119 - 2 \times 135 = -79 \text{ kcal.}$$

Ex. 59. *Bond energies of* F_2 *and* Cl_2 *are respectively* 36·6 *and* 58·0 kcal *per mole. If the heat liberated in the reaction* $F_2 + Cl_2 \rightarrow 2FCl$ *is* 26·6 kcal, *calculate the bond energy of* F—Cl *bond.*

Solution : Given that,

$$F_2 + Cl_2 \rightarrow 2FCl;\ \Delta H = -26{\cdot}6\ \text{kcal}$$

For reactants:

Bond energy of 1 mole of F—F bonds = 36·6 kcal

Bond energy of 1 mole of Cl—Cl bonds = 58·0 kcal

For product:

Energy of formation of 2 moles of F—Cl bond $= 2 \times x$

(where x is the bond formation energy of F—Cl bonds in kcal/mole)

Adding all the heat changes, we get ΔH of the given reaction, i.e.,

$$36{\cdot}6 + 58 + 2 \times x = -26{\cdot}6$$

$$x = -60{\cdot}6\ \text{kcal.}$$

Thus, the bond energy of F—Cl bonds is + 60·6 kcal per mole.

Ex. 60. *Calculate the bond energy of* C$\equiv$C *in* C_2H_2 *from the following data:*

(i) $C_2H_2\ (g) + 2\frac{1}{2} O_2\ (g) \rightarrow 2CO_2\ (g) + H_2O;\ \Delta H = -310\ \text{kcal}$

(ii) $C\ (s) + O_2\ (g) \rightarrow CO_2\ (g);\ \Delta H = -94\ \text{kcal}$

(iii) $H_2\ (g) + \frac{1}{2} O_2\ (g) \rightarrow H_2O\ (g);\ \Delta H = -68\ \text{kcal}$

Bond energy of C—H *bonds* = 99 kcal

Heat of atomisation of C = 171 kcal

Heat of atomisation of H = 52 kcal

Solution : Let us first calculate heat of formation of C_2H_2 from which we can calculate bond energy of C$\equiv$C bond,

$$2C\ (s) + H_2 \rightarrow C_2H_2\ (g);\ \Delta H = ?$$

Applying the inspection method, i.e.,

[2 × Eqn. (ii) + Eqn. (iii) – Eqn. (i)], we get,

$$2C\ (s) + 2O_2\ (g) + H_2\ (g) + \frac{1}{2} O_2\ (g) - C_2H_2\ (g) - 2\frac{1}{2} O_2 \rightarrow$$

$$2CO_2\ (g) + H_2O\ (g) - 2CO_2\ (g) - H_2O\ (g);$$

$$\Delta H = 2 \times (-94) + (-68) - (-310)\ \text{kcal}$$

or $2C\ (s) + H_2\ (g) \rightarrow C_2H_2\ (g);\ \Delta H = 54\ \text{kcal.}$

Now heat changes for reactants:

Heat of atomisation of 2 moles of C $= 2 \times 171$ kcal

Heat of atomisation of 2 moles of H $= 2 \times 52$ kcal

And heat change for the product (H—C$\equiv$C—H):

Heat of formation of 2 moles of C—H bonds $= -2 \times 99$ kcal

Heat of formation of 1 mole of C$\equiv$C bonds $= x$ (say)

Summing up, we get heat of formation of C_2H_2,

$$2 \times 171 + 2 \times 52 - 2 \times 99 + x = 54$$
$$x = -194 \text{ kcal.}$$

Hence bond energy of $C\equiv C$ bond in C_2H_2 is +194 kcal.

Ex. 61. *Using the data (all values are in* kcal *per mole at* 25°C) *given below, calculate the bond energy of* C—C *and* C—H *bonds.*

$$\Delta H^0_{\text{combustion}} \text{(ethane)} = -372$$
$$\Delta H^0_{\text{combustion}} \text{(propane)} = -530$$
$$\Delta H^0 \text{ for } C \text{ (graphite)} \rightarrow \text{C (g)} = 172$$
$$\text{Bond energy of H—H} = 104$$
$$\Delta H_f^0 \text{ of } H_2O \text{ (l)} = -68$$
$$\Delta H_f^0 \text{ of } CO_2 \text{ (g)} = -94$$

(IIT 1990)

Solution : The following are the given thermochemical reactions:

(i) $C_2H_6 \text{ (g)} + 3\frac{1}{2}O_2 \text{ (g)} \rightarrow 2CO_2 \text{ (g)} + 3H_2O \text{ (l)}; \quad \Delta H^0 = -372$

(ii) $C_3H_8 \text{ (g)} + 5O_2 \text{ (g)} \rightarrow 3CO_2 \text{ (g)} + 4H_2O \text{ (l)}; \quad \Delta H^0 = -530$

(iii) $H_2 \text{ (g)} + \frac{1}{2}O_2 \text{ (g)} \rightarrow H_2O \text{ (l)}; \quad \Delta H^0 = -68$

(iv) $C \text{ (s)} + O_2 \text{ (g)} \rightarrow CO_2 \text{ (g)}; \quad \Delta H^0 = -94$

Let us first calculate the heat of formation of C_2H_6 and C_3H_8 from which we shall calculate the bond energy of C—C and C—H bonds.

$$2C \text{ (s)} + 3H_2 \text{ (g)} \rightarrow C_2H_6 \text{ (g)}; \quad \Delta H^0 = ?$$

and $$3C \text{ (s)} + 4H_2 \text{ (g)} \rightarrow C_3H_8 \text{ (g)}; \quad \Delta H^0 = ?$$

Applying the inspection method, i.e.,

[– Eqn. (i) + 3 × Eqn. (iii) + 2 × Eqn. (iv)], we get,

$$2C \text{ (s)} + 3H_2 \text{ (g)} \rightarrow C_2H_6 \text{ (g)}; \quad \Delta H^0 = -20$$

$[\Delta H^0 = -(-372) + 3(-68) + 2(-94) = -20]$

Again applying the inspection method, i.e.,

[– Eqn. (ii) + 4 × Eqn. (iii) + 3 × Eqn. (iv)], we get,

$$3C \text{ (s)} + 4H_2 \text{ (g)} \rightarrow C_3H_8 \text{ (g)}; \quad \Delta H^0 = -24$$

$[\Delta H^0 = -(-530) + 4(-68) + 3(-94) = -24]$

Calculation of bond energy:

Now we have,

$$2C \text{ (s)} + 3H_2 \text{ (g)} \rightarrow C_2H_6 \text{ (g)}; \; \Delta H^0 = -20$$

For reactants:

Heat of atomisation of 2 moles of C atoms = 2 × 172

Heat of dissociation of 3 moles of H—H bonds = 3 × 104

For products:

Heat of formation of 1 mole of C—C bonds = x (say)

Heat of formation of 6 moles of C—H bonds = $6y$ (say)

On adding, we get ΔH^0 of formation of C_2H_6.

$$2 \times 172 + 3 \times 104 + x + 6y = -20$$

or $$x + 6y = -676 \qquad \dots (1)$$

Again, we have,

$$3C\ (s) + 4H_2\ (g) \rightarrow C_3H_8\ (g);\ \Delta H^0 = -24$$

For reactants:

Heat of atomisation of 3 moles of C atoms = 3×172

Heat of dissociation of 4 moles of H—H bonds = 4×104

For products:

Heat of formation of 2 moles of C—C bonds = $2x$

Heat of formation of 8 moles of C—H bonds = $8y$

On adding, we get,

$$3 \times 172 + 4 \times 104 + 2x + 8y = -24$$

or $$2x + 8y = -956 \qquad \dots (2)$$

From eqns. (1) and (2), we get, $x = -82$ and $y = -99$.

∴ C—C bond energy = 82 kcal

C—H bond energy = 99 kcal.

Ex. 62. *Calculate the resonance energy in* CH_3COOH *from the following data if the observed heat of formation of* CH_3COOH *is* – 439·7 kJ.

Bond energy (kJ)	*Heat of atomisation* (kJ)
C—H = 413	C = 716·7
C—C = 348	H = 218·0
C=O = 732	O = 249·1
C—O = 351	
O—H = 463	

Solution : Calculation of $\Delta H_f (CH_3COOH)$:

$$2C\ (s) + 2H_2\ (g) + O_2\ (g) \rightarrow CH_3COOH$$

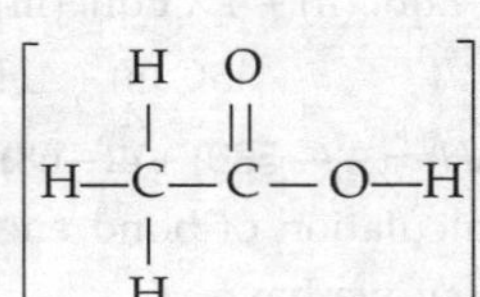

For reactants:

Heat of atomisation of 2 moles of C = $2 \times 716{\cdot}7 = 1433{\cdot}4$ kJ

Heat of atomisation of 4 moles of H = $4 \times 218{\cdot}0 = 872{\cdot}0$ kJ

Heat of atomisation of 2 moles of O = $2 \times 249{\cdot}1 = 498{\cdot}2$ kJ

For products:

Heat of formation of 3 moles of C—H = – (3 × 413) = – 1239 kJ

Heat of formation of 1 mole of C—C = – (1 × 348) = – 348 kJ

Heat of formation of 1 mole of C=O = – (1 × 732) = – 732 kJ

Heat of formation of 1 mole of C—O = – (1 × 351) = – 351 kJ

Heat of formation of 1 mole of O—H = – (1 × 463) = – 463 kJ

Resonance energy in CH_3COOH = x kJ (say)

Adding algebraically, we get ΔH_f of CH_3COOH.

$$-329{\cdot}4 + x = -439{\cdot}7 \text{ (given)}$$

$$\therefore \quad x = -110{\cdot}3 \text{ kJ mole}^{-1}.$$

Ex. 63. *From the following thermochemical equations*

$$C_2H_4 + H_2 \rightarrow C_2H_6 ; \quad \Delta H = -32{\cdot}7 \text{ kcal}$$

$$C_6H_6 + 3H_2 \rightarrow C_6H_{12} ; \quad \Delta H = -49{\cdot}2 \text{ kcal}$$

Calculate the resonance energy of benzene.

Solution : C_6H_6 has three isolated carbon-carbon double bonds. ΔH for hydrogenation should be close to three times the ΔH for hydrogenation of C_2H_4, with one double bond, but the given value is –49·2 kcal.

$$\therefore \quad \text{resonance energy} = 3 \times (-32{\cdot}7) - (-49{\cdot}2)$$

$$= -48{\cdot}9 \text{ kcal.}$$

Ex. 64. *The standard molar enthalpies of formation of cyclohexane (l) and benzene (l) at 25°C are* –156 *and* +49 kJ mol^{-1} *respectively. The standard enthalpy of hydrogenation of cyclohexene (l) at 25°C is* –119 kJ mol^{-1}. *Use these data to estimate the magnitude of the resonance energy of benzene.* **(IIT 1996)**

Solution : Enthalpy of formation of 3 carbon-carbon double bonds

= ΔH_f (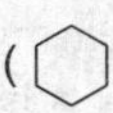) — ΔH_f (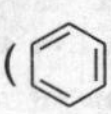)

$$= -156 - (+49) \text{ kJ}$$

$$= -205 \text{ kJ.}$$

Given that,

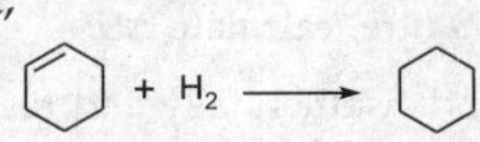

Δ ; H = – 119 kJ

Theoretical enthalpy of formation of 3 double bonds in benzene ring

$$= 3 \times (-119) \text{ kJ}$$

$$= -357 \text{ kJ.}$$

$$\therefore \quad \text{resonance energy of benzene} = -357 - (-205) \text{ kJ}$$

$$= -152 \text{ kJ mole}^{-1}.$$

Ex. 65. *Compute the heat of formation of liquid methyl alcohol in kilojoules per mole using the following data. The heat of vaporisation of liquid methyl alcohol* = 38 kJ/mole. *The heats of formation of gaseous atoms from the elements in their standard states:* H, 218 kJ/mole; C, 715 kJ/mole; O, 249 kJ/mole.
Average bond energies: C—H = 415 kJ/mole
C—O = 356 kJ/mole
O—H = 463 kJ/mole **(IIT 1997)**

Solution : $C + 2H_2 + \frac{1}{2}O_2 \rightarrow CH_3OH$; $\Delta H = ?$

For reactants:

Heat of atomisation of 1 mole of C = 715 kJ

Heat of atomisation of 4 moles of H = 4 × 218 kJ

Heat of atomisation of 1 mole of O = 249 kJ

For products:

Heat of formation of 3 moles of C—H bonds = −3 × 415 kJ

Heat of formation of 1 mole of C—O bonds = −356 kJ

Heat of formation of 1 mole of O—H bonds = −463 kJ

Heat of condensation of 1 mole of CH_3OH to liquid = −38 kJ

On adding, we get ΔH of formation of CH_3OH (l).

$\Delta H = -266$ kJ mole^{-1}.

PROBLEMS

(Answers bracketed with questions)

1. Find the work done when 1 mole of hydrogen expands isothermally from 15 to 50 litres against a constant pressure of 1 atm at 25°C. (−718 cal)

2. 100 g of argon is allowed to expand from a pressure of 10 atm to 0·1 atm at 100°C. Calculate the heat which is absorbed, assuming ideal behaviour. (8580 cal)

3. The molar volumes of ice and water are respectively 0·0196 and 0·0180 litres per mole at 273 K. If ΔH for the transition of ice to water is 1440 calories per mole at 1 atm pressure, find ΔU. (1440 cal)

4. For the reaction, $C_{graphite} + \frac{1}{2}O_2\,(g) = CO\,(g)$ at 298 K and 1 atm, $\Delta H = -26416$ cal. If the molar volume of graphite is 0·0053 litre, calculate ΔU. (−26712 cal)

5. One mole of an ideal gas at 300 K expands isothermally and reversibly from 5 to 20 litres. Calculate the work done and heat absorbed by the gas. (−832 kJ, 832 kJ)

6. Calculate the heat of formation of ethylene from the following data at 20°C:

$$H_2\,(g) + \frac{1}{2}O_2\,(g) \rightarrow H_2O\,(l);\ \Delta H = -65 \text{ kcal}$$

$$C\,(s) + O_2\,(g) \rightarrow CO_2\,(g);\ \Delta H = -97 \text{ kcal}$$

$$C_2H_4\,(g) + 3O_2\,(g) \rightarrow 2CO_2\,(g) + 2H_2O\,(l);\ \Delta H = -340 \text{ kcal}$$

(16 kcal)

7. Calculate the heat of formation of carbon monoxide from the following data:

$$C\ (s) + O_2\ (g) \rightarrow CO_2\ (g);\ \Delta H = -94\ \text{kcal}$$

$$2CO\ (g) + O_2\ (g) \rightarrow 2CO_2\ (g);\ \Delta H = -136\ \text{kcal}$$

(–26 kcal)

8. The heat of combustion of ethyl alcohol is 330 kcal. If the heats of formation of CO_2 (g) and H_2O (l) are 94·3 and 68·5 kcal respectively, calculate the heat of formation of ethyl alcohol. (–64·1 kcal)

9. The heats of formation of CO_2 (g) and H_2O (l) are – 94·05 kcal and –68·32 kcal respectively. The heat of combustion of methyl alcohol (l) is –173·65 kcal. Calculate the heat of formation of liquid methyl alcohol. (–57·04 kcal)

10. At constant volume at 27°C,

$$2C_6H_6\ (g) + 15O_2\ (g) \rightarrow 12CO_2\ (g) + 6H_2O\ (l);\ \Delta U = -1600\ \text{kcal}$$

$$2C_2H_2\ (g) + 5O_2\ (g) \rightarrow 4CO_2\ (g) + 2H_2O\ (l);\ \Delta U = -620\ \text{kcal}$$

Calculate the heat of polymerisation of acetylene to benzene at constant pressure. (–131·2 kcal)

11. Calculate the heat of formation of CH_3COOH (l) at 25°C from the following data:

$$CH_3COOH\ (l) + 2O_2\ (g) \rightarrow 2CO_2 + 2H_2O\ (l);\ \Delta H = -208{\cdot}34\ \text{kcal}$$

$$C\ (s) + O_2\ (g) \rightarrow CO_2\ (g);\ \Delta H = -94{\cdot}05\ \text{kcal}$$

$$H_2\ (g) + \frac{1}{2}O_2 \rightarrow H_2O\ (l)\ ;\ \Delta H = -68{\cdot}32\ \text{kcal}$$

(–116·4 kcal)

12. For the reaction at 25°C,

$$NH_3\ (g) \rightarrow \frac{1}{2}N_2\ (g) + \frac{3}{2}H_2\ (g);\ \Delta H^0 = 11{\cdot}04\ \text{kcal}$$

calculate ΔU^0 of the reaction at the given temperature. (10·44 kcal)

13. From the following equations, calculate the standard molar heat of formation of AgCl at 25°C.

$$Ag_2O\ (s) + 2HCl\ (g) \rightarrow 2AgCl\ (s) + H_2O(l);\ \Delta H^0 = -77{\cdot}61\ \text{kcal}$$

$$2Ag\ (s) + \frac{1}{2}O_2\ (g) \rightarrow Ag_2O\ (s);\ \Delta H^0 = -73{\cdot}1\ \text{kcal}$$

$$\frac{1}{2}H_2\ (g) + \frac{1}{2}Cl_2\ (g) \rightarrow HCl\ (g);\ \Delta H^0 = -22{\cdot}06\ \text{kcal}$$

$$H_2\ (g) + \frac{1}{2}O\ (g) \rightarrow H_2O\ (l);\ \Delta H^0 = -68{\cdot}32\ \text{kcal}$$

(–63·20 kcal)

14. Calculate the enthalpy of formation of HBr (g) from the following data:

$$SO_2\ (aq) + \frac{1}{2}O_2\ (g) \rightarrow SO_3\ (aq);\ \Delta H = -63{\cdot}7\ \text{kcal}$$

$$2Br\ (g) + SO_2\ (aq) + H_2O\ (l) \rightarrow 2HBr\ (aq) + SO_3\ (aq);\ \Delta H = -54\ \text{kcal}$$

$$H_2\ (g) + \frac{1}{2}O_2\ (g) \rightarrow H_2O\ (l);\ \Delta H = -68{\cdot}4\ \text{kcal}$$

$$HBr\ (g) + aq \rightarrow HBr\ (aq);\ \Delta H = -20\ \text{kcal}$$

(–9·4 kcal)

15. Calculate the enthalpy of formation of $Ca(OH)_2$ (s) from the following data:

$$H_2\ (g) + \frac{1}{2}O_2\ (g) \rightarrow H_2O\ (l);\ \Delta H = -68{\cdot}3\ \text{kcal}$$

$$CaO\,(s) + H_2O(l) \rightarrow Ca(OH)_2\,(s);\ \Delta H = -15{\cdot}3\text{ kcal}$$

$$Ca\,(s) + \frac{1}{2}O_2\,(g) \rightarrow CaO\,(s);\ \Delta H = -151{\cdot}8\text{ kcal} \qquad (-235{\cdot}4\text{ kcal})$$

16. Calculate the heat of formation of KOH from the following data:

$$K + H_2O + aq \rightarrow KOH\,(aq) + \frac{1}{2}H_2\,;\ \Delta H = -\,48{\cdot}4\text{ kcal}$$

$$H_2 + \frac{1}{2}O_2 \rightarrow H_2O;\ \Delta H = -\,68{\cdot}44\text{ kcal}$$

$$KOH + aq \rightarrow KOH(aq);\ \Delta H = -\,14{\cdot}01\text{ kcal} \qquad (-102{\cdot}46\text{ kcal})$$

17. Calculate the heat of formation of C_6H_6, given that the heats of combustion of benzene, carbon and hydrogen are 754, 94 and 68 kcal respectively. (–14 kcal)

18. Calculate the enthalpy change for the reaction

$$Fe_2O_3 + 3CO \rightarrow 2Fe + 3CO_2\,;\ \text{from the following data:}$$

$$2Fe + \frac{3}{2}O_2 \rightarrow Fe_2O_3\,;\ \Delta H = -\,177{\cdot}1\text{ kcal}$$

$$C + \frac{1}{2}O_2 \rightarrow CO;\ \Delta H = -\,32{\cdot}8\text{ kcal}$$

$$C + O_2 \rightarrow CO_2\,;\ \Delta H = -\,94{\cdot}3\text{ kcal} \qquad (-8{\cdot}4\text{ kcal})$$

19. The heats of formation of $Na_2B_4O_7$ (s) and $Na_2B_4O_7 \cdot 10H_2O$ (s) are –742 and –1460 kcal respectively. Calculate the heat of hydration of $Na_2B_4O_7$ in forming decahydrate. (–718 kcal)

20. The heats of combustion of graphite and diamond at 298 K are –393 and –395 kJ/mole respectively. The specific heats of these substances are 720 and 505 J kg^{-1} K^{-1} respectively. Calculate the heat of transformation of graphite into diamond at 273 K. (2·0645 kJ mole^{-1})

[**Hint:** $\Delta H = \{-393 - (395)\} + (0{\cdot}720 - 0{\cdot}505) \times \frac{12}{1000} \times (298 - 273)$]

21. The enthalpies of neutralisation for CH_3COOH with NaOH and NH_4OH with HCl are –50·6 and – 51·4 kJ eq^{-1} respectively. Calculate the enthalpy of neutralisation of CH_3COOH with NH_4OH. (–44·74 kJ eq.$^{-1}$)

22. Calculate the fuel efficiency in kJ/gram of C_2H_4 and C_4H_{10}. The heats of formation of C_2H_4, C_4H_{10}, CO_2 and H_2O are 52·3, –126·1, –393·5 and –285·8 kJ mole^{-1} respectively. (50·39 and 49·6 kJ g^{-1})

23. The heat of formation of Fe_2O_3 is – 821·32 kJ mole^{-1} at 298 K and 1 atm and that of Al_2O_3 is –1675·60 kJ mole^{-1} under the same condition. Calculate the heat of reaction of reduction of 1 mole of Fe_2O_3 with metallic aluminium. (–854·28 kJ)

24. A gas is enclosed in a cylinder with a piston. Weights are added to the piston, giving a total mass of 2·20 kg. As a result the gas is compressed and the weights are lowered 0·25 m. At the same time, 1·50 J of heat evolved from the system. What is the change in internal energy of the system? (3·89 J)

25. Calculate ΔG^0 for the reaction

$$CaF_2(s) \rightleftharpoons Ca^{2+}(aq) + 2F^-(aq)$$

Given that, $\Delta G_f^0\,(CaF_2(s)) = -1162$ kJ/mole.

$\Delta G_f^0\,(Ca^{2+}(aq)) = -553{\cdot}0$ and $\Delta G_f^0\,(F^-(aq)) = -276{\cdot}5$ kJ/mole.

Also calculate K_{sp} for this reaction at 25°C. (56 kJ, 2×10^{-10})

26. To what temperature must magnesium carbonate be heated to decompose it to MgO and CO_2 at 1 atm? Given:

	$MgCO_3(s)$	$MgO(s)$	$CO_2(g)$
ΔH_f^0	–1112	–601·2	–393·5 kJ
S^0	65·9	26·9	213·7 J/K

(671 K)

27. Predict the direction in which ΔG^0 for the equilibrium

$$N_2(g) + 3H_2(g) \rightleftharpoons 2NH_3(g);\ \Delta G_{298}^0 = -33{\cdot}32 \text{ kJ}$$

will change with increase in temperature. Calculate ΔG^0 at 500°C, assuming that ΔH and ΔS do not change with temperature, that is, $\Delta H_{298}^0 = \Delta H_{793}$ and $\Delta S_{298}^0 = \Delta S_{793}$.

$$\Delta H^0 = -92{\cdot}38 \text{ kJ and } \Delta S^0 = -198{\cdot}2 \text{ J/K}$$

(The eqn. shifts to left with increase in temp., +60·83 kJ)

28. A heat pump is used to draw water from a well. The temperature of the water is 15°C and that of the atmosphere is 40°C. If the amount of water drawn is 10 kg from the depth of 12 m, calculate the amount of heat supplied to the well. (14·72 kJ)

29. Given the following reactions with their enthalpy changes, at 25°C,

$$N_2(g) + 2O_2(g) \rightarrow 2NO_2(g);\ \Delta H = 16{\cdot}18 \text{ kcal}$$

$$N_2(g) + 2O_2(g) \rightarrow N_2O_4(g);\ \Delta H = 2{\cdot}31 \text{ kcal}$$

calculate the enthalpy of dimerisation of NO_2. Is N_2O_4 apt to be stable with respect to NO_2 at 25°C? (–13·87 kcal, N_2O_4 stable only at low temp.)

30. A system is changed from an initial state to a final state by a manner such that $\Delta H = q$. If the same change from the initial state to the final state were made by a different path, would ΔH be the same as that of the first path? Would q? (same, most probably different)

31. Two moles of an ideal gas are held by a piston under 10 atm pressure at 273 K. The pressure is suddenly released to 0·4 atm and the gas is allowed to expand isothermally. Calculate W, q, ΔU and ΔH. (–4358 J, 4358 J, 0, 0)

[**Hint:** The gas expands irreversibly.)

32. If the equation of state for 1 mole of a gas is

$$p(V - b) = RT,$$

prove that dp is an exact differential and p is a state function.

33. Show that the differential $d\overline{V}$ of the molar volume of an ideal gas is an exact differential and hence $\overline{V}$ is a state function.

[Hint: $\overline{V} = \frac{RT}{p}$; $\overline{V} = f(p, T)$]

34. Assume that the only change in volume is due to the production of hydrogen and calculate the work done in joules when 2·0 moles of Zn dissolve in hydrochloric acid, giving H_2 at 35°C and 1 atm.

$$Zn(s) + 2HCl(aq) \rightarrow ZnCl_2(aq) + H_2(g)$$ (–5·12 × 10³ J)

[Hint: $W = -p\Delta V = -\Delta n_{H_2}RT$]

35. Acetic acid forms a dimer in the gas phase

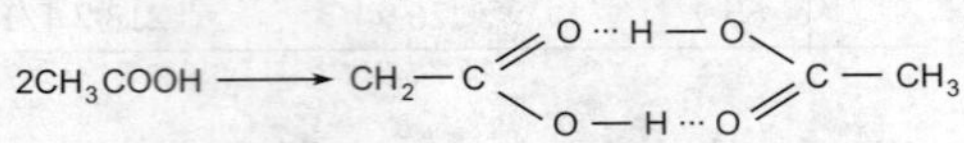

The dimer is held together by two hydrogen bonds with a total strength of 66·5 kJ per mole of dimer. At 25°C, the equilibrium constant for the dimerisation is $1{\cdot}3 \times 10^3$ (pressure in atm). What is ΔS^0 for the reaction? Assume that ΔH does not vary with temperature. (–0·11 kJ/mol· K)

36. A Carnot engine has an efficiency of 40%. If the temperature of the reservoir is 280 K, what is the temperature of the source? (466·6 K)

37. A Carnot engine whose temperature of the source is 400 K takes 200 cal of heat at this temperature and rejects 150 cal of heat to the sink. Calculate the temperature of the sink and the efficiency of the engine. (300 K, 25%)

38. One mole of an ideal gas at 22·4 litres is expanded isothermally and reversibly at 300 K to a volume of 224 litres at a constant pressure. Calculate $W, q, \Delta H, \Delta G$ and ΔS. (–5·74 kJ, 5·74 kJ, 0, –5·74 kJ, +19·1 J/K)

39. K_{sp} of AgCl at 25°C is $1{\cdot}782 \times 10^{-10}$. At 35°C, K_{sp} is $4{\cdot}159 \times 10^{10}$. What are ΔH^0 and ΔS^0 for the reaction: $AgCl(s) = Ag^+(aq) + Cl^-(aq)$? ($\Delta H^0 = 64{\cdot}655$ kJ, $\Delta S^0 = 30{\cdot}3$ J)

40. For the reaction

$$C(s) + O_2(g) \rightarrow CO_2(g)$$

$\Delta H^0 = -393{\cdot}51$ kJ/mole and $\Delta S^0 = 2{\cdot}86$ J/mole · K at 25°C. Does the reaction become more or less favourable as the temperature increases? (more favourable)

41. The $\Delta H^0, \Delta G^0$ and ΔS^0 values for the reaction

$$2H_2O_2(l) \rightarrow 2H_2O(l) + O_2(g)$$

at 25°C are –196·0 kJ/mole, –233·6 kJ/mole and +125·6 J/mole · K respectively. Is there any temperature at which $H_2O_2(l)$ is stable at 1 atm? Assume that ΔH and ΔS values do not change with temperature. (theoretically at –1586 K)

42. For each of the following processes, tell whether the entropy of the system increases, decreases or remains constant.

(a) Melting one mole of ice to water at 0°C

(b) Freezing one mole of water to ice at 0°C

(c) Freezing one mole of water to ice at –10°C

(d) Freezing one mole of water to ice at 0°C and then cooling it to –10°C

[(a) increases (b) decreases (c) decreases (d) decreases]

43. The heat of formation of ethane is –20·3 kcal. Calculate the bond energy of C—C bond in ethane if the heats of atomisation of carbon and hydrogen are respectively 170·9 and 52·1 kcal per mole and bond energy of C—H bond is 99·0 kcal.

(80·7 kcal)

44. The heat of reaction of $N_2 + 3H_2 \rightarrow 2NH_3$ is –20 kcal. If the bond energies of H—H and N—H bonds are 104 and 93 kcal/mole respectively, calculate the bond energy of N≡N bond. (226 kcal)

45. Calculate the heat of formation of acetone from the following data:

$$\frac{1}{2}H_2\,(g) \rightarrow H\,(g);\ \Delta H = 52{\cdot}1\ \text{kcal}$$

$$\frac{1}{2}O_2\,(g) \rightarrow O\,(g);\ \Delta H = 59{\cdot}16\ \text{kcal}$$

$$C\,(s) \rightarrow C\,(g);\ \Delta H = 171{\cdot}7\ \text{kcal}$$

Bond energies:

C—H = 99 kcal

C—C = 80 kcal

C=O = 81 kcal (51·86 kcal)

46. Calculate the heat of formation of methyl alcohol (liquid) from the following data:

Heat of atomisation of C = 170·9 kcal

Heat of atomisation of H = 52·1 kcal

Heat of atomisation of O = 59·6 kcal

Bond energies: C—H = 99 kcal

C—O = 84 kcal

O—H = 110·55 kcal

Heat of liquefaction of 1 mole of CH_3OH = –8·4 kcal. (–61·0 kcal)

47. Calculate the heat of the following gaseous reaction:

$$CH_4 + 4F_2 \rightarrow CF_4 + 4\,HF$$

The bond energies of C—H; F—F; C—F and H—F bonds are 99·3; 38; 116 and 135 kcal/mole respectively. (–454·8 kcal)

48. Estimate the heat of formation of gaseous isoprene

```
      CH3 H
       |  |
(H2C = C—C = CH2) from the following data:
```

Bond energies:

C—H = 98·8 kcal/mole

H—H = 104·0 kcal/mole

C—C = 83·1 kcal/mole

C=C = 147·0 kcal/mole

Heat of sublimation of carbon (s) = 171·7 kcal per mole. (23·9 kcal)

49. Using the required bond-energies data from the above problems, calculate the heat of hydrogenation of ethene to ethane. (–29·7 kcal)

50. Calculate the heat of the following homogeneous gaseous reaction

$$CH_3COCH_3 + 2O_2 \rightarrow CH_3COOH + CO_2 + H_2O$$

from the following data:

Bond energies (kJ):

C—H = 414·49	C—O = 967·13
C—C = 347·92	Resonance energy:
C=O = 724·32	— COOH = 117·23
O=O = 494·04	CO_2 = 138·16
O—H = 462·64	

(–554·33 kJ)

51. Calculate the resonance energy of N_2O from the following data:

$$\Delta H_f^0 \text{ of } N_2O = 82 \text{ kJ mole}^{-1}$$

Bond energies of $N \equiv N$, $N = N$, $O = O$ and $N = O$ bonds are 946, 418, 498 and 607 kJ mole^{-1} respectively. (–88 kJ)

Objective Problems

1. The heat of formation of HCl (g) from the reaction

$$H_2(g) + Cl_2(g) = 2HCl(g);\ \Delta H = -44 \text{ kcal is}$$

(a) +44 kcal (b) –44 kcal (c) +22 kcal (d) –22 kcal

2. Given $N_2(g) + 3H_2(g) = 2NH_3(g)$; $\Delta H^0 = -22$ kcal. The standard enthalpy of formation of NH_3 gas is

(a) –11 kcal/mole (b) 11 kcal/mole
(c) –22 kcal/mole (d) 22 kcal/mole

3. If for $H_2(g) = 2H(g)$; $\Delta H = 104$ kcal, heat of atomisation of hydrogen is

(a) 52 kcal (b) 104 kcal (c) 208 kcal (d) none of these

4. Heats of combustion of CH_4, C_2H_4, C_2H_6 are –890, –1411 and –1560 kJ/mole respectively. Which has the lowest fuel value in kJ/g?

(a) CH_4 (b) C_2H_4 (c) C_2H_6 (d) all same

5. 2·1 g of Fe combines with S evolving 3·77 kJ. The heat of formation of FeS in kJ/mole is

(a) –3·77 (b) –1·79 (c) –100·5 (d) none of these

6. When a certain amount of ethylene was combusted, 6226 kJ of heat was evolved. If the heat of combustion of ethylene is 1411 kJ, the volume of O_2 (at NTP) that entered into the reaction is

(a) 296·5 mL (b) 296·5 litres (c) 6226 × 22·4 litres (d) 22·4 litres

7. $\left[\begin{array}{l} H_2\,(g) + \frac{1}{2} O_2\,(g) = H_2O\,(g);\ \Delta H = -241{\cdot}8 \text{ kJ} \\ CO\,(g) + \frac{1}{2} O_2\,(g) = CO_2\,(g);\ \Delta H = -283 \text{ kJ} \end{array}\right]$

The heat evolved in the combustion of 112 litres of water gas (mixture of equal volumes of H_2 and CO) is
(a) 241·8 kJ (b) 283 kJ (c) 1312 kJ (d) 1586 kJ

8. $\left[\begin{array}{l} H_2\,(g) + \frac{1}{2} O_2\,(g) = H_2O\,(g);\ \Delta H = -241{\cdot}8 \text{ kJ} \\ C_2H_2\,(g) + 2\frac{1}{2} O_2\,(g) = 2CO_2\,(g) + H_2O\,(g);\ \Delta H = -1300 \text{ kJ} \end{array}\right]$

Equal volumes of C_2H_2 and H_2 are combusted under identical conditions. The ratio of heats evolved in the two cases is
(a) 5·37/1 (b) 1/5·37 (c) 1/1 (d) none of these

9. The heat of neutralisation of a strong dibasic acid in dilute solution by NaOH is nearly
(a) –27·4 kcal/eq (b) 13·7 kcal/mole
(c) –13·7 kcal/eq (d) –13·7 kcal/mole

10. The temperature of 5 mL of a strong acid increases by 5° when 5 mL of a strong base is added to it. If 10 mL of each is mixed, the temperature should increase by
(a) 5° (b) 10° (c) 15° (d) cannot be known

11. The heat of neutralisation of a strong acid by a strong base is equal to ΔH of:
(a) $H^+ + OH^- = H_2O$
(b) $H_2O + H^+ = H_3O^+$
(c) $2H_2 + O_2 = 2H_2O$
(d) $CH_3COOH + NaOH = CH_3COONa + H_2O$

12. In which case of mixing of a strong acid and a base, each being of 1 N concentration, the increase in temperature is the highest?
(a) 20 mL acid and 30 mL alkali (b) 10 mL acid and 40 mL alkali
(c) 25 mL acid and 25 mL alkali (d) 35 mL acid and 15 mL alkali

13. The heat of neutralisation of HCl by NaOH is – 55·9 kJ/mole. If the heat of neutralisation of HCN by NaOH is –12·1 kJ/mole, the energy of dissociation of HCN is
(a) –43·8 kJ (b) 43·8 kJ (c) 68 kJ (d) –68 kJ

14. The dissociation energy of CH_4 and C_2H_6 are respectively 360 and 620 kcal/mole. The bond energy of C—C is
(a) 260 kcal/mole (b) 180 kcal/mole
(c) 130 kcal/mole (d) 80 kcal/mole

15. In which of the following cases does entropy decrease?
(a) Solid changing to liquid (b) Expansion of a gas
(c) Crystals dissolve (d) Polymerisation

16. In which of the following reactions is ΔS positive?
(a) $H_2O\,(l) \rightarrow H_2O\,(s)$ (b) $3O_2\,(g) \rightarrow 2O_3\,(g)$

(c) $H_2O\ (l) \rightarrow H_2O\ (g)$ (d) $N_2\ (g) + 3H_2\ (g) \rightarrow 2NH_3\ (g)$

17. In the process of ice melting at –15°C

(a) $\Delta G < 0$ (b) $\Delta G > 0$ (c) $\Delta G = 0$ (d) none of these

18. In a reaction, ΔH and ΔS both are more than zero. In which of the following cases would the reaction be spontaneous?

(a) $\Delta H > T\,\Delta S$ (b) $T\,\Delta S > \Delta H$ (c) $\Delta H = T\,\Delta S$ (d) none of these

19. In which case is a reaction possible at any temperature?

(a) $\Delta H < 0,\ \Delta S > 0$ (b) $\Delta H < 0,\ \Delta S < 0$

(c) $\Delta H > 0,\ \Delta S > 0$ (d) in none of the cases

20. In which case is a reaction impossible at any temperature?

(a) $\Delta H > 0,\ \Delta S > 0$ (b) $\Delta H > 0,\ \Delta S < 0$

(c) $\Delta H < 0,\ \Delta S < 0$ (d) in all cases

21. The difference between the heats of reaction at constant pressure and constant volume for the reaction

$2C_6H_6\ (l) + 15\,O_2\ (g) \rightarrow 12CO_2\ (g) + 6H_2O\ (l)$ at 25° C in kJ is

(a) –7·43 (b) +3·72 (c) –3·72 (d) +7·43

22. Molar heat capacity of water in equilibrium with ice at constant pressure is

(a) zero (b) ∞

(c) 40·45 kJ K^{-1} mol^{-1} (d) 75·48 J K^{-1} mol^{-1}

23. Standard molar enthalpy of formation of CO_2 is equal to

(a) zero

(b) the standard molar enthalpy of combustion of gaseous carbon

(c) the sum of standard molar enthalpies of formation of CO and O_2

(d) the standard molar enthalpy of combustion of carbon (graphite)

24. When steam condenses to water at 90°C, the entropy of the system decreases. What must be true if the second law of thermodynamics is to be satisfied?

(a) Entropy of the universe also decreases.

(b) Entropy of the surroundings also decreases.

(c) Entropy of the surroundings increases to the same extent to which entropy of the system decreases.

(d) Increase in entropy in the surroundings is greater than decrease in entropy of the system.

[**Hint:** For a spontaneous process $\Delta S_{universe}$ must be positive.]

25. A certain reaction is spontaneous at 85°C. The reaction is endothermic by 34 kJ. The minimum value of ΔS for the reaction is

(a) 497·2 J/K (b) –497·2 J/K

(c) +2094 J/K (d) cannot be calculated

26. For the reaction,

$$\frac{1}{2}A_2 + \frac{1}{2}B_2 \rightarrow AB;\ \Delta H = -50 \text{ kcal.}$$

If the bond energies of A_2, B_2 and AB are respectively $x, \frac{x}{2}$ and x kcal, the value of x is

(a) 50 (b) 100 (c) 200 (d) 400

27. One mole of a nonideal gas undergoes a change of state (2·0 atm, 3·0 L, 95 K) $\rightarrow$ (4 atm, 5 L, 245 K) with a change in internal energy, $\Delta U = 30{\cdot}0$ L atm. The change in enthalpy, ΔH, of the process in L. atm is

(a) 40·0 (b) 42·3 (c) 44·0

(d) not defined, because pressure is not constant

[**Hint:** $\Delta H = \Delta U + (p_2V_2 - p_1V_1)$]

28. Which of the following thermodynamic variables is extensive?

(a) Pressure (b) Density (c) Temperature (d) Mass

29. Which of the following is an intensive property?

(a) Volume (b) Internal energy

(c) Entropy (d) Mass/volume

30. Which of the following is an intensive property?

(a) emf (b) Volume (c) Mass (d) Enthalpy

31. Which of the following is not a thermodynamic state function?

(a) Work (b) Internal energy

(c) Free energy (d) Temperature

32. Which of the following is not an exact differential?

(a) dq (b) dG (c) $d\overline{V}$ (d) dS

33. Which of the following is a state function and also an extensive property?

(a) Internal energy (b) Pressure

(c) Molar heat capacity (d) Temperature

34. Which of the following is not equal to zero in a cyclic process?

(a) ΔG (b) ΔW (c) ΔS (d) ΔH

35. The reversible expansion of an ideal gas under adiabatic and isothermal conditions is shown in the figure, which of the following statement(s) is(are) correct

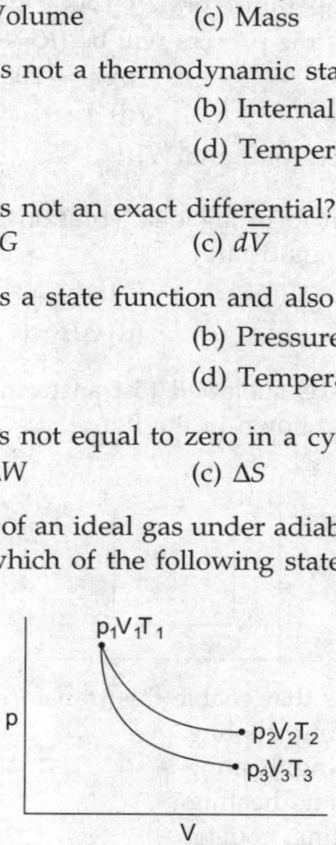

(a) $T_1 = T_2$ (b) $T_3 > T_1$

(c) $W_{\text{isothermal}} > W_{\text{adiabatic}}$ (d) $\Delta U_{\text{isothermal}} > \Delta U_{\text{adiabatic}}$

[**Hint:** Read Mod. App. to Phy. Chemistry, Vol. I

Ch. 2., Fig. 2.14, Upper curve is isotheramal] **(IIT 2012)**

36. For an ideal gas, consider only $p - V$ work in going from initial state X to the final state Z. The final state Z can be reached by either of the two paths shown in the figure.

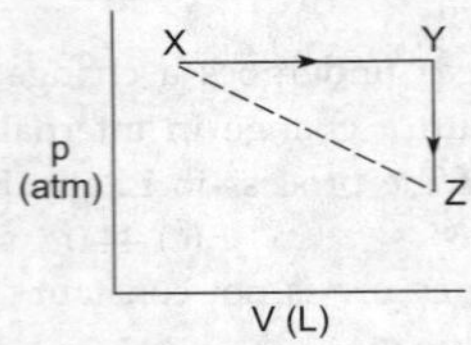

(take ΔS as change in entropy and W as work done)
Which of the following choice(s) is(are) correct?

(a) $\Delta S_{x \to z} = \Delta S_{x \to y} + \Delta S_{y \to z}$ (b) $W_{x \to z} = W_{x \to y} + W_{y \to z}$

(c) $W_{x \to y} = W_{x \to y}$ (d) $\Delta S_{x \to y \to z} = \Delta S_{x \to y}$ **(IIT 2012)**

[Hint: Work is shown by the area under the curve. S is a state function while W is not.]

37. A piston filled with 0.04 mole of an ideal gas expands reversibly from 50.0 mL to 375 mL at a constant temperature of 37°C. As it does so it absorbs 208 J of heat. The value of q and W for the process will be ($R = 8.314$ J/mol K, ln 7.5 = 2.01)

(a) $q = +208$ J, $W = -208$ J (b) $q = -208$ J, $W = -208$ J

(c) $q = -208$ J, $W = +208$ J (d) $q = +208$ J, $W = +208$ J

[Hint: See sign convension for q and W] **(IIT 2013 Main)**

38. Benzene and naphthalein form an ideal solution at room temperature. For this process, the true statement(s) is(are)

(a) ΔG is positive (b) ΔS_{sys} is positive

(c) $\Delta S_{sur} = 0$ (d) $\Delta H = 0$ **(IIT 2013 Adv.)**

39. A fixed mass m of a gas is subjected to transformation of states from K to L to M to N and back to K as shown in the figure.

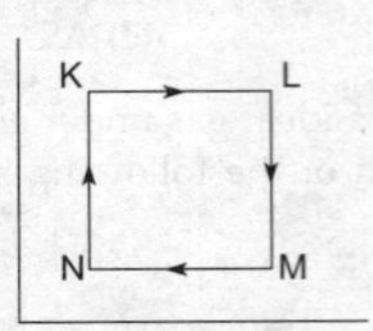

The successive operations that enable this transformation of states are

(a) heating, cooling, heating, cooling
(b) cooling, heating, cooling, heating
(c) heating, cooling, cooling, heating
(d) cooling, heating, heating, cooling **(IIT 2013 Adv.)**

[Hint: Follow heating at constant pressure,
cooling at constant volume
cooling at constant pressure
heating at constant volume]

40. Refer to Q. 39. The pair of isochoric processes among the transformation of states is

(a) K to L and L to M (b) L to M and N to K
(c) L to M and M to N (d) M to N and N to K **(IIT 2013 Adv.)**

[Hint : See hint of Q. 39]

41. For the complete combustion of ethanol,

$$C_2H_5OH(l) + 3O_2(g) \rightarrow 2CO_2(g) + 3H_2O(l)$$

the amount of heat produced as measured in bomb calorimeter is 1364.47 kJ mol^{-1} at 25°C. Assuming ideality, the enthalpy of combustion ΔH_{comb} for the reaction will be ($R = 8.314$ J K^{-1} mol^{-1})

(a) –1366.95 kJ mol^{-1} (b) –1361.95 kJ mol^{-1}
(c) –1460.50 kJ mol^{-1} (d) –1350.50 kJ mol^{-1} **(IIT 2014 Main)**

[Hint: $\Delta U = -1364.47$, $\Delta n_g = -1$, $R = \frac{8.314}{1000}$

Use $\Delta H = \Delta U + \Delta n_g RT$]

42. For the process

$$H_2O(l) \rightarrow H_2O(g)$$

at $T = 100°C$ and $p = 1$ atm, the correct choice is

(a) $\Delta S_{sys} > 0$ and $\Delta_{sur} > 0$ (b) $\Delta S_{sys} > 0$ and $\Delta S_{sur} < 0$
(c) $\Delta S_{sys} < 0$ and $\Delta_{sur} > 0$ (d) $\Delta S_{sys} < 0$ and $\Delta S_{sur} < 0$ **(IIT 2014 Adv.)**

43. 1 mole of an ideal gas at 300 K in thermal contact with surroundings expands isothermally from 1.0 L to 2.0 L against a constant pressure of 3.0 atm. In this process, the change in entropy of surroundings (ΔS_{sur}) in J K^{-1} is (1 L atm = 101.3 J)

(a) 5.763 (b) 1.013 (c) –1.013 (d) –5.763

(IIT 2016 Adv.)

[Hint: $\Delta U = q + w$. For isothermal process, $\Delta U = 0$

$\therefore \quad q = -w = -P\Delta V = -3(2-1) = -3$ L atm $= -303.9$ J

Now apply $\Delta S = \frac{q}{T}$]

44. ΔU is equal to

(a) isochoric work (b) isoboric work
(c) adiabatic work (d) isothermal work **(IIT 2017 Main)**

45. An ideal gas is expanded from $(p_1V_1T_1)$ to $(p_2V_2T_2)$ under different conditions. The correct statement(s) among the following is(are)

(a) The work done by the gas is less when it is expanded reversibly from V_1 to V_2 under adiabatic conditions as compared to that when expanded reversibly from V_1 to V_2 under isothermal conditions.

(b) The change in internal energy of the gas is (i) zero, if it is expanded reversibly with $T_1 = T_2$ and (ii) positive, if it is expanded reversibly under adiabatic conditions with $T_1 \neq T_2$.

(c) If the expansion is carried out freely, it is simultaneously both isothermal as well as adiabatic.

(d) The work done on the gas is maximum when it is compressed irreversibly from (p_2V_2) to (p_1V_1) against constant pressure p_1. **(IIT 2017 Adv.)**

46. For the reaction taking place in a container in equilibrium with its surroundings, the effect of temperature on its equilibrium constant K in terms of change in entropy is described by

(a) with increase in temperature, the value of K for endothermic reaction increases because unfavourable change in entropy of the surroundings decreases.

(b) with increase in temperature the value of K for exothermic reaction decreases because favourable change in entropy of the surroundings decreases.

(c) with increase in temperature, the value of K for endothermic reaction increases because change in entropy of the system is negative.

(d) with increase in temperature the value of K for exothermic reaction decreases because the entropy change of the system is positive. **(IIT 2017 Adv.)**

47. Using data provided, calculate the multiple bond energy (kJ mol^{-1}) of a C≡C bond in C_2H_2. The energy is (C—H bond energy = 350 kJ mol^{-1})

$$2C(s) + H_2(g) \rightarrow C_2H_2(g); \quad \Delta H = 225 \text{ kJ mol}^{-1}$$
$$2C(s) \rightarrow 2C(g); \quad \Delta H = 1410 \text{ kJ mol}^{-1}$$
$$H_2(g) \rightarrow 2H(g); \quad \Delta H = 330 \text{ kJ mol}^{-1}$$

(a) 1165 (b) 837 (c) 865 (d) 815 **(IIT 2012)**

[**Hint:** 1410 + 330 − (2 × 350) + x = 255
x is heat of formation of C≡C bond]

48. The standard enthalpy of formation of $CO_2(g)$, $H_2O(l)$ and glucose(s) at 25°C are −400.0 kJ mol^{-1}, −300.0 kJ mol^{-1} and −1300.0 kJ mol^{-1} respectively. The standard enthalpy of combustion per gram of glucose at 25°C is

(a) +2900 kJ (b) −2900 kJ (c) −16.11 kJ (d) +16.11 kJ **(IIT 2013 Adv.)**

49. The heats of combustion of C and CO are −393.5 and −283.5 kJ mol^{-1} respectively. The heat of formation (in kJ) of CO per mode is

(a) 676.5 (b) −676.5
(c) −110.5 (d) 110.5 **(IIT 2016 Main)**

50. Given

1. C(graphite) + $O_2(g) \rightarrow CO_2(g)$; $\Delta H° = -3935$ kJ mol^{-1}

2. $H_2(g) + \frac{1}{2}O_2(g) \rightarrow H_2O(l)$; $\Delta H° = -285$ kJ mol^{-1}

3. $CO_2(g) + 2H_2O(l) \rightarrow CH_4(g) + 2O_2(g)$; $\Delta H° = +890.3$ kJ mol^{-1}

Based on the above thermochemical equations the value of $\Delta_r H°$ at 298 K for the reaction

$$\text{C(graphite)} + 2\,H_2(g) \rightarrow CH_4(g)$$

will be

(a) +78.8 kJ mol^{-1} (b) +144.0 kJ mol^{-1}
(c) −74.8 kJ mol^{-1} (d) −1440 kJ mol^{-1} **(IIT 2017 Main)**

[**Hint:** Apply inspection method: eq. 1 + (2 × eq. 2) + eq. 3]

51. The standard state Gibbs energies of formation of C(graphite) and C(diamond) at $T = 298$ K are

$$\Delta_f G°[\text{C(graphite)}] = 0 \text{ kJ mol}^{-1}$$
$$\Delta_f G°[\text{C(diamond)}] = 2.9 \text{ kJ mol}^{-1}$$

The standard state means that the pressure should be 1 bar and substance should be pure at a given temperature. The conversion of graphite to diamond reduces its volume by 2×10^{-6} m^3 mol^{-1}. If C(graphite) is converted to C isothermally at $T = 298$ K, the pressure at which C(graphite) is in equilibrium with C(diamond) is ($1 \text{ J} = 1 \text{ kg m}^2\text{s}^{-2}$, $1 \text{ Pa} = 1 \text{ kg m}^{-1}\text{s}^{-2}$, $1 \text{ bar} = 10^5 \text{ Pa}$)

(a) 58001 bar (b) 1450 bar

(c) 14501 bar (d) 29001 bar **(IIT 2017 Adv.)**

[**Hint:** $\Delta G = V\Delta p - S\Delta T$, *As* $\Delta T = 0$, $\Delta G = V\Delta p$

$\therefore \quad G_{gr} - G^{\circ}_{gr} = V_{gr}\,\Delta P$ and $G_{dim} - G^{\circ}_{dim} = V_{dim}\Delta P$

$\therefore \quad (V_{dim} - V_{gr})\,\Delta P = (G_{dim} - G_{gr}) + (G^{\circ}_{gr} - G^{\circ}_{dim})$

at eqb. $G_{dim} = G_{gr}$

$\therefore (V_{dim} - V_{gr})\Delta P = G^{\circ}_{dim} - G^{\circ}_{gr} = 2.9 \times 10^3$

calculate Δp, i.e., $p - p_0$, then p i.e., ($p_0 = 1$).]

52. The combustion of benzene(l) gives CO_2(g) and H_2O(l). Given that heat of combustion of benzene at constant volume is –3263.9 kJ mol^{-1} at 25°C; heat of combustion in (kJ mol^{-1}) of benzene at constant pressure will be

(R = 8.314 JK^{-1} mol^{-1})

(a) –3267.6 (b) 4152.6 (c) –452.4 (d) 3260

[**Hint:** $\Delta H = \Delta U + \Delta n_g RT$] **(IIT 2018 Main)**

53. $\Delta_f G°$ at 500 K for substance '*S*' in liquid state and gaseous state are +100.7 kcal mol^{-1} and +103 kcal mol^{-1} respectively. Vapour pressure of liquid '*S*' at 500 K is approximately equal to (R = 2 cal K^{-1} mol^{-1})

(a) 10 atm (b) 0.1 atm (c) 1 atm (d) 100 atm

(IIT 2018 Main)

[**Hint:** $\Delta G_{\text{solid} \to \text{liquid}} = 2.303\, RT \log \frac{p_2}{p_1}$; $p_1 = 1$ atm, $p_2 = ?$]

54. Given

$2Fe_2O_3(s) \rightarrow 4Fe(s) + 3O_2(g)$; $\Delta G° = -1487$ kJ mol^{-1}

$2CO(g) + O_2(g) \rightarrow 2CO_2(g)$; $\Delta G° = -514.4$ kJ mol^{-1}

Free energy change $\Delta G°$ for the reaction $2Fe_2O_3(s) + 6CO(g) \rightarrow 4Fe(s) + 6CO_2(g)$ will be

(a) –168.2 kJ mol^{-1} (b) –208.0 kJ mol^{-1}

(c) –56.2 kJ mol^{-1} (d) –112.4 kJ mol^{-1} **(IIT 2018 Main)**

[**Hint:** Cal. $\Delta G°$ following {eqn. (i) + 3 eqn. (ii)}]

55. For which of the following reactions, ΔH is equal to ΔU?

(a) $2HI(g) \rightleftharpoons H_2(g) + I_2(g)$ (b) $2NO_2(g) \rightleftharpoons N_2O_4(g)$

(c) $N_2(g) + 3H_2(g) \rightleftharpoons 2NH_3(g)$ (d) $2SO_2(g) + O_2(g) \rightleftharpoons 2SO_3(g)$

(IIT 2018 Main)

[**Hint:** $\Delta n_g = 0$]

56. An ideal gas undergoes a cyclic process as shown in figure.

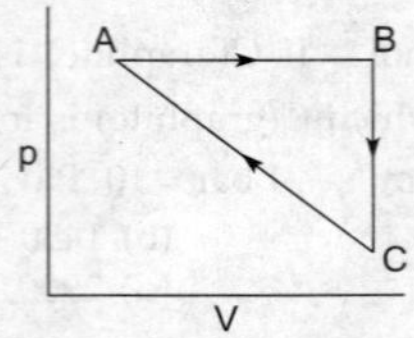

$$\Delta U_{BC} = -5 \text{ kJ mol}^{-1},\ q_{AB} = 2\text{kJ mol}^{-1}$$
$$W_{AB} = -5 \text{ kJ mol}^{-1},\ W_{CA} = 3\text{kJ mol}^{-1}$$

Heat absorbed by the system during process CA is,

(a) -5 kJ mol^{-1} (b) $+5$ kJ mol^{-1}

(c) -18 kJ mol^{-1} (d) $+18$ kJ mol^{-1} **(IIT 2018 Main)**

[**Hint:** For cyclic process,

$$\Delta U = \Delta U_{A\to B} + \Delta U_{B\to C} + \Delta U_{C\to A} = 0$$

Also use $\Delta U = q = w$]

57. A reversible cyclic process for an ideal gas is shown below. Here p, V and T are pressure, volume and temperature, respectively. The thermodynamic parameters, q, W, H and U are heat, work, enthalpy and internal energy, respectively

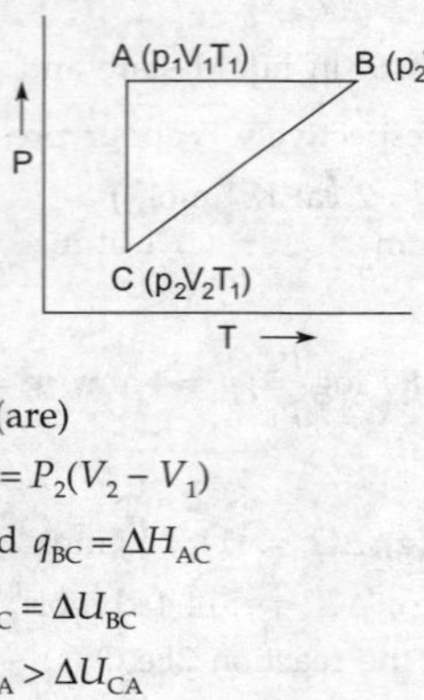

The correct option(s) is(are)

(a) $q_{AC} = \Delta U_{BC}$ and $W_{AB} = P_2(V_2 - V_1)$

(b) $W_{BC} = P_2(V_2 - V_1)$ and $q_{BC} = \Delta H_{AC}$

(c) $\Delta H_{CA} < \Delta U_{CA}$ and $q_{AC} = \Delta U_{BC}$

(d) $q_{BC} = \Delta H_{AC}$ and $\Delta H_{CA} > \Delta U_{CA}$ **(IIT 2018 Adv.)**

[**Hint:** AC – Isochoric: $W = \Delta U = nC_V(T_2 - T_1)$

$$q = 0$$

AB – Isothermal: $W = -2.303\, RT \log \dfrac{V_2}{V_1}$

$$q = -W$$

BC – Isobaric: $W = -p\Delta V = -nR(T_2 - T_1)$]

58. For a reaction, $A \rightleftharpoons P$, the plot of $[A]$ and $[P]$ with time at temperature T_1 and T_2 are given below:

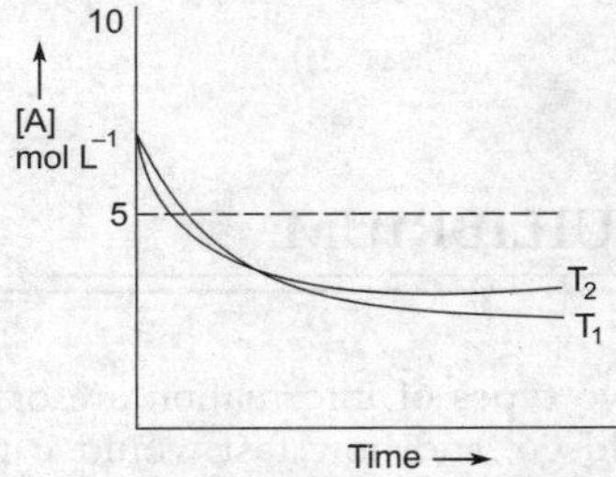

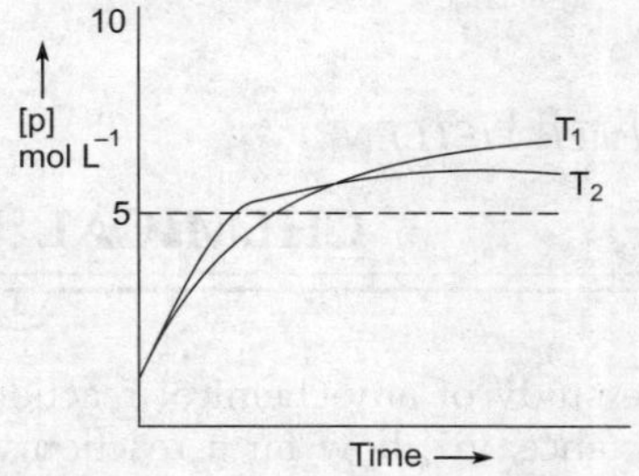

If $T_2 > T_1$, the correct statement(s) is(are)

(Assume ΔH° and ΔS° are independent of temperature and ratio of $\ln K$ at T_1 to $\ln K$ at T_2 is greater than T_2/T_1. Here H, S, G and K are enthalpy, entropy, Gibbs energy and equilibrium constant, respectively)

(a) $\Delta H^\circ < 0, \Delta S^\circ < 0$ (b) $\Delta G^\circ < 0, \Delta H^\circ > 0$

(c) $\Delta G^\circ < 0, \Delta S^\circ < 0$ (d) $\Delta G^\circ < 0, \Delta S^\circ > 0$ **(IIT 2018 Adv.)**

[**Hint:** As k_p increases with the increase in temperature, the reaction is exothermic, i.e., $\Delta H < 0$.

Further $T_1(\ln K)_1 > T_2(\ln K)_{T_2}$

or $-RT_1(\ln K)_1 < -RT_2(\ln K)_2$, i.e., $(\Delta G^\circ)_{T_1} < (\Delta G^\circ)_{T_2}$

or $\Delta H^\circ - T_1\Delta S^\circ < \Delta H^\circ - T_2\Delta S^\circ$, ΔS has to be < 0.]

Answers

1-d, **2**-a, **3**-a, **4**-b, **5**-c, **6**-b, **7**-c, **8**-a, **9**-c, **10**-a, **11**-a, **12**-c, **13**-b, **14**-d, **15**-d, **16**-c, **17**-b, **18**-b, **19**-a, **20**-b, **21**-a, **22**-b, **23**-d, **24**-d, **25**-a, **26**-c, **27**-c, **28**-d, **29**-d, **30**-a, **31**-a, **32**-a, **33**-a, **34**-b, **35**-a, c, d, **36**-a, c, **37**-a, **38**-b, c, d, **39**-c, **40**-b, **41**-a, **42**-b, **43**-d, **44**-c, **45**-a, c, d, **46**-a, b, **47**-d, **48**-c, **49**-c, **50**-c, **51**-c, **52**-a, **53**-a, **54**-c, **55**-a, **56**-b, **57**-b, c, **58**-a, c

CHAPTER FIFTEEN

CHEMICAL EQUILIBRIUM

In the study of any chemical reaction, two types of information are of vital importance, viz, how far a reaction would go, and how fast would it reach its goal. The answer to the first question forms the subject matter of this chapter.

Chemical Equilibrium and Equilibrium Constant

The reactions are generally reversible, that is, they can proceed both ways. A reaction is said to have attained equilibrium when the rate of the forward reaction equals that of the backward reaction.

Let us consider a general case of a reversible reaction,

$$aA + bB \rightleftharpoons mM + nN$$

Applying the law of mass action:

Rate of the forward reaction $\propto [A]^a\ [B]^b$

or rate of the forward reaction $= k_1\,[A]^a\ [B]^b$.

Rate of backward reaction $\propto [M]^m\ [N]^n$

or rate of backward reaction $= k_2\,[M]^m\ [N]^n$.

At equilibrium:

Rate of forward reaction = rate of backward reaction

$$k_1\,[A]^a\ [B]^b = k_2\,[M]^m\ [N]^n$$

$$\frac{k_1}{k_2} = \frac{[M]^m\ [N]^n}{[A]^a\ [B]^b}$$

or $$^*K_c = \frac{k_1}{k_2} = \frac{[M]^m\ [N]^n}{[A]^a\ [B]^b} \quad \ldots (1)$$

[ ] represents concentration in moles per litre. Here, all concentrations are at equilibrium. k_1 and k_2 are known as the rate constants of the forward and backward reactions respectively.

Chemical equilibrium is dynamic in the sense that individual molecules are continually reacting, even though the overall composition of the reaction mixture does not change. The other criteria of a chemical equilibrium are: the

---

* When there is no subscript on K, it is understood to be K_c. In thermodynamics, the equilibrium constant is defined in terms of activities rather than concentrations.

same equilibrium state can be attained from both sides of the reaction and a reaction which attains equilibrium is always incomplete.

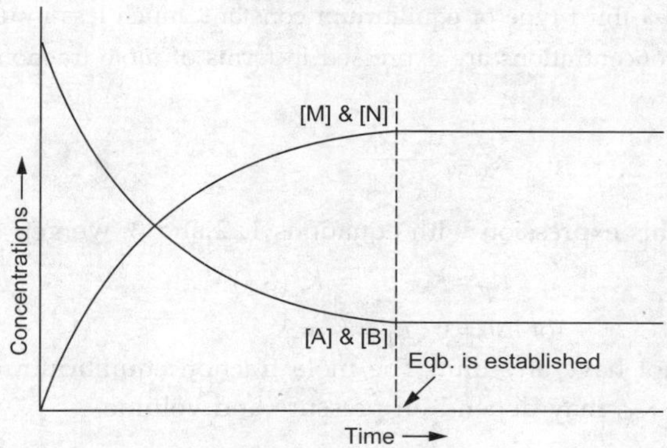

The rate constant of any reaction is defined as the rate of the reaction when the concentration of each reactant is unity. The rate constant of a reaction depends on the temperature and the catalyst.

K_c, the ratio of the rate constants, is known as the equilibrium constant. Now, if the above reaction is supposed to be a homogeneous gaseous one, the equilibrium constant may also be represented as

$$K_p = \frac{p_{\rm M}^m \cdot p_{\rm N}^n}{p_{\rm A}^a \cdot p_{\rm B}^b}, \qquad \dots (2)$$

where p's* represent the partial pressures at equilibrium.

The equilibrium constant K_p is known as pressure equilibrium constant. The magnitude of K_p and K_c is a measure of the extent to which the reaction occurs. The equilibrium constants depend only on temperature. For a given reaction, K_p and K_c may be equal and may also be different depending on the stoichiometry of the reaction. K_p and K_c are related as

$$K_p = K_c(RT)^{\Delta n}, \qquad \dots (3)$$

where, Δn = no. of moles of the gaseous product – no. of moles of the gaseous reactants (in the balanced equation and not in the reaction)

or $\Delta n = (m + n) - (a + b)$.

But both the equilibrium constants for a given reaction give the same information about the state of equilibrium when the reaction at equilibrium is subjected to a change in temperature, pressure or concentration as also expressed by Le Chatelier's principle. **The units of K_p and K_c are not fixed and**

* Partial pressure of any species of the reactant or product at equilibrium $= \dfrac{\text{moles of the species at eqb.}}{\text{total moles}} \times \text{total pressure.}$

depend on the stoichiometry of the reaction. In case the number of moles of the reactant and that of the product are same, K_p and K_c do not have any unit.

[**Note:** There is a third type of equilibrium constant, much less in use, called K_x, when concentrations are expressed in terms of mole fraction (x).]

Thus,

$$K_x = \frac{x_M^m \cdot x_N^n}{x_A^a \cdot x_B^b}.$$

Relating this expression with equations 1, 2 and 3, we get

$$K_p = K_x \cdot p^{\Delta n} = K_c (RT)^{\Delta n}$$

$$\text{for } \Delta n = 0 : K_p = K_c = K_x$$

K_x does not have any unit. The mole fraction equilibrium constant K_x, unlike K_p and K_c, may depend on pressure and volume.

Thermodynamic Definition of Equilibrium Constant

The thermodynamic definition of the equilibrium constant involves activities rather than concentration. The activity of a component of an ideal mixture is the ratio of its concentration or partial pressure to a standard concentration (1 M) or pressure (1 atm). Hence the activity of each species is a dimensionless quantity, that is, it has no unit. For any pure solid or pure liquid, the activity is taken as 1. Because of the use of activities, **the thermodynamic equilibrium constant has no unit.** The values of activities we put into K_c and K_p are numerically equal to molar concentrations or partial pressures but have no units.

Calculations involving equilibrium are, therefore, frequently carried out without units. But as in the present text the molar concentrations and partial pressures have been used instead of activities, K_p and K_c may be expressed with their units.

Calculation of K_p and K_c

From the relation $K_p = K_c (RT)^{\Delta n}$, we get two types of reactions, viz.,

(i) when $\Delta n = 0$, i.e., those reactions in which there is no change in the number of molecules, e.g., $H_2 + I_2 \rightleftharpoons 2HI$;

$$\Delta n = 2 - 2 = 0; \text{ for such a reaction, } K_p = K_c$$

(ii) when $\Delta n \neq 0$, i.e., those reactions in which there is a change in the number of molecules, e.g., $N_2 + 3H_2 \rightleftharpoons 2NH_3$; $\Delta n = 2 - 4 = -2$

For such a reaction, $K_p \neq K_c$.

The calculations of K_p and K_c for the above types of reactions depend mainly on the values of molar concentration (i.e., moles/litre) and partial pressure at equilibrium, as shown in Eqns. (1) and (2).

Let us now present the methods of calculating K_p and K_c by taking the following examples:

(1) Formation of HI

$$\begin{array}{cccccl} a & & b & & 0 & \text{Initial moles} \\ H_2\,(g) & + & I_2\,(g) & \rightleftharpoons & 2HI\,(g) & \\ (a-x) & & (b-x) & & 2x & \text{Moles at equilibrium} \\ \dfrac{(a-x)}{V} & & \dfrac{(b-x)}{V} & & \dfrac{2x}{V} & \text{Molar concentration at eqb.} \end{array}$$

where x is the number of moles of H_2 or I_2 converted to HI at equilibrium and V is the volume in litres of the container.

$$K_c = \frac{[HI]^2}{[H_2]\,[I_2]} = \frac{(2x/V)^2}{\dfrac{(a-x)}{V}\cdot\dfrac{(b-x)}{V}}$$

$$= \frac{4x^2}{(a-x)\,(b-x)}.$$

Further, total number of moles at equilibrium

$$= (a-x) + (b-x) + 2x$$
$$= a + b$$

Let the total pressure be p.

$$p_{H_2} = \frac{(a-x)}{(a+b)}\cdot p;\; p_{I_2} = \frac{(b-x)}{(a+b)}\cdot p \text{ and}$$

$$p_{HI} = \frac{2x}{(a+b)}\cdot p$$

$$K_p = \frac{p_{HI}^2}{p_{H_2}p_{I_2}} = \frac{4x^2}{(a-x)\,(b-x)}\cdot \quad \text{(substituting the value of } p\text{'s)}$$

(2) Formation of NH_3

$$\begin{array}{cccccl} a & & b & & 0 & \text{Initial moles} \\ N_2\,(g) & + & 3H_2\,(g) & \rightleftharpoons & 2NH_3 & \\ (a-x) & & (b-3x) & & 2x & \text{Moles at equilibrium} \\ \dfrac{(a-x)}{V} & & \dfrac{(b-3x)}{V} & & \dfrac{2x}{V} & \text{Molar concentration at equilibrium} \end{array}$$

where x is the number of moles of N_2 which converted to NH_3 at equilibrium.

$$\therefore \qquad K_c = \frac{[NH_3]^2}{[N_2]\,[H_2]^3} = \frac{\left(\dfrac{2x}{V}\right)^2}{\left(\dfrac{a-x}{V}\right)\left(\dfrac{b-3x}{V}\right)^3}.$$

Further, total number of moles at equilibrium

$$= a - x + b - 3x + 2x = (a + b - 2x)$$

$$K_p = \frac{p^2_{NH_3}}{p_{N_2} \cdot p^3_{H_2}} = \frac{\left\{\left(\frac{2x}{a+b-2x}\right) \cdot p\right\}^2}{\left\{\frac{a-x}{a+b-2x} \cdot p\right\}\left\{\frac{b-3x}{a+b-2x} \cdot p\right\}^3}.$$

(3) Dissociation of PCl_5

a	0	0	Initial moles
PCl_5 (g) $\rightleftharpoons$	PCl_3 (g) +	Cl_2 (g)	
$(a-x)$	x	x	Moles at equilibrium
$\frac{(a-x)}{V}$	$\frac{x}{V}$	$\frac{x}{V}$	Molar concentration at eqb.

Here, x is the number of moles of PCl_5 converted to the products at equilibrium. But if x represents degree of dissociation, i.e., fraction of the total number of molecules undergoing dissociation, the equilibrium moles and equilibrium molar concentration of each species will be as follows:

a	0	0	Initial moles
PCl_5 (g) $\rightleftharpoons$	PCl_3 (g) +	Cl_2 (g)	
$a(1-x)$	ax	ax	Moles at equilibrium
$\frac{a(1-x)}{V}$	$\frac{ax}{V}$	$\frac{ax}{V}$	Molar concentration at eqb.

From the values given above, equations for K_p and K_c can be derived.

(4) Dissociation of Halogen Molecules

a	0	Initial moles
X_2 (g) $\rightleftharpoons$	$2X$ (g);	{If x is the number of moles of X_2 converted to X at eqb.}
$(a-x)$	$2x$	Moles at equilibrium
$\left(\frac{a-x}{V}\right)$	$\left(\frac{2x}{V}\right)$	Molar concentration at equilibrium

But if x is the degree of dissociation:

X_2 (g) $\rightleftharpoons$	$2X$ (g)	
$a(1-x)$	$2ax$	Moles at equilibrium
$\left(\frac{a(1-x)}{V}\right)$	$\left(\frac{2ax}{V}\right)$	Molar concentration at eqb.

Derive K_p and K_c.

Degree of Dissociation from Vapour Density

In calculating K_p and K_c for reactions in which dissociation takes place, the value of the degree of dissociation may be needed. The degree of dissociation may be calculated from vapour density measurements.

If one molecule of the substance dissociates into n molecules, it can be shown that

$$\text{*degree of dissociation, } x = \frac{d_t - d_o}{(n-1)\, d_o} \quad \ldots (4)$$

where d_t = theoretical density (assuming no dissociation) and d_o = observed (or experimental, or actual) density, both under the same pressure. If all the d's are taken as vapour densities,

$$\left(d_t = \frac{\text{mol. weight}}{2}\right).$$

More About Equilibrium Constant

Let us now study important points concerning the use of equilibrium constants. These should be carefully noted.

(i) In equilibrium,

$$CaCO_3\,(s) \rightleftharpoons CaO\,(s) + CO_2\,(g)$$

$$\text{Equilibrium constant} = K' = \frac{[CaO]\,[CO_2]}{[CaCO_3]}$$

Since CaO and $CaCO_3$ are pure solids, their concentrations do not change, or simply remain constant.

If x_1 and x_2 represent the constant concentrations of CaO and $CaCO_3$ respectively

$$\frac{K' \,.\, x_2}{x_1} = [CO_2]$$

or new equilibrium constant $K = [CO_2]$.

The new equilibrium constant K includes the constant concentrations, x_1 and x_2, of pure solids. However, the active mass of solid is conventionally taken to be 1.

(ii) The equilibrium constant for the reaction

$$2H_2 + O_2 \rightleftharpoons 2H_2O, \text{ is given by,}$$

$$K_1 = \frac{[H_2O]^2}{[H_2]^2\,[O_2]}.$$

For the reverse reaction $2H_2O = 2H_2 + O_2$

$$\text{equilibrium constant, } K_2 = \frac{[H_2]^2\,[O_2]}{[H_2O]^2}$$

Comparing the above equations, we get,

* Eqn. 4 is not applicable to those reactions in which no. of moles of the reactants and those of products are same, e.g., in $2HI = H_2 + I_2$. In such cases the no. of molecules before and after dissociation remains the same.

$$K_2 = \frac{1}{K_1}. \qquad \text{... (5)}$$

Thus, the equilibrium constant of a forward reaction and that of its backward reaction are the reciprocal of each other.

(iii) Another important matter is illustrated by the relation between the two equilibrium constants, one for a simple reaction and the other when it is multiplied by a certain factor. For example,

$$NO + \tfrac{1}{2}O_2 \rightleftharpoons NO_2; \quad K_1 = \frac{[NO_2]}{[NO]\,[O_2]^{\frac{1}{2}}}.$$

And for the other one, (multiplying the above reaction by a factor, say 2)

$$2NO + O_2 \rightleftharpoons 2NO_2; \; K_2 = \frac{[NO_2]^2}{[NO]^2\,[O_2]}$$

Comparing the two equations,

$$K_2 = K_1^2. \qquad \text{... (6)}$$

In general, if a chemical equation is multiplied by a certain factor, its equilibrium constant must be raised to a power equal to that factor in order to obtain the equilibrium constant for the new reaction.

(iv) When a reaction takes place in more than one step, there can be a relationship among the equilibrium constants of the overall reaction and those of the step reactions.

The reaction,

$$H_2O_2 + 2H^+ + 2I^- \rightleftharpoons I_2 + 2H_2O, \text{ for which,}$$

$$K = \frac{[I_2]\,[H_2O]^2}{[H_2O_2]\,[H^+]^2\,[I^-]^2}$$

takes place in the following steps,

$$H_2O_2 + I^- \rightleftharpoons OH^- + HOI; \text{ for which } K_1 = \frac{[OH^-]\,[HOI]}{[H_2O_2]\,[I^-]}$$

$$H^+ + OH^- \rightleftharpoons H_2O; \text{ for which } K_2 = \frac{[H_2O]}{[H^+]\,[OH^-]} \text{ and}$$

$$HOI + H^+ + I^- \rightleftharpoons I_2 + H_2O; \text{ for which } K_3 = \frac{[I_2]\,[H_2O]}{[HOI]\,[H^+]\,[I^-]}$$

Multiplying equations for K_1, K_2 and K_3, we see that

$$K = K_1 \cdot K_2 \cdot K_3. \qquad \text{... (7)}$$

Thus, when a reaction takes place in more than one step, the equilibrium constant of the overall reaction is equal to the product of the equilibrium constants of each step.

Interpretation of Equilibrium Constant

The equilibrium constant for a reaction tells about the tendency of a reaction to proceed to products.

(i) If the concentration of each reactant and product in a general reaction

$$aA + bB \rightleftharpoons mM + nN$$

are such that,

$$\frac{[M]^m \cdot [N]^n}{[A]^a \cdot [B]^b} = K \text{ (equilibrium constant)} \qquad \ldots (8)$$

the reaction is at equilibrium.

(ii) If the concentrations are such that,

$$\frac{[M]^m \cdot [N]^n}{[A]^a \cdot [B]^b} < K \qquad \ldots (9)$$

the reaction will proceed from left to right in order to attain equilibrium, or in other words, the **concentration factor or the reaction quotient increases** and becomes equal to K.

(iii) And, if the concentrations are such that,

$$\frac{[M]^m \cdot [N]^n}{[A]^a \cdot [B]^b} > K \qquad \ldots (10)$$

the reaction will proceed from right to left in order to attain equilibrium, or in other words, the **concentration factor or the reaction quotient decreases** and becomes equal to K.

If the concentration factor, i.e.,

$$\frac{[M]^m \cdot [N]^n}{[A]^a \cdot [B]^b} \text{ (not necessarily eqb. concentrations)}$$

is expressed as reaction quotient, Q, we have,

1. $Q > K$: Reaction proceeds in backward direction until equilibrium is established

2. $Q = K$: Reaction is at equilibrium

3. $Q < K$: Reaction proceeds in forward direction until equilibrium is established

Le Chatelier's Principle and Equilibrium Constant

Le Chatelier's principle states that if a chemical reaction at equilibrium is subjected to a change in temperature, pressure or concentration, the equilibrium shifts to that direction by which the effect of the change is minimised.

One should note that the change in either pressure or concentration at equilibrium may shift the equilibrium, or more clearly, change the state of equilibrium, but it cannot change the equilibrium constant K_p or K_c which depends only on temperature.

However, when a reaction (except those for which $\Delta H = 0$) is subjected to a change in temperature, the equilibrium shifts to another equilibrium position at the new temperature and the equilibrium constant also changes.

In general, for an exothermic reaction (ΔH is negative), increase in temperature results in shifting the equilibrium from right to left, thereby decreasing the value of the equilibrium constant. For an endothermic reaction (ΔH is positive), the equilibrium constant increases as the temperature increases.

Effect of Temperature on Equilibrium Constant

The quantitative effect of temperature on the equilibrium constant can be determined by the following thermodynamically derived equation, known as van't Hoff equation,

$$\log K_{p_2} - \log K_{p_1} = \frac{\Delta H^\circ}{2{\cdot}303\,R}\left[\frac{T_2 - T_1}{T_1 \,.\, T_2}\right], \qquad \text{... (11)}$$

where K_{p_1} and K_{p_2} are the equilibrium constants at temperatures T_1 and T_2 respectively, ΔH° is standard heat of the reaction at constant pressure.

Here, it has been assumed that ΔH is constant in the temperature range between T_2 and T_1 ($T_2 > T_1$). We readily see that $K_{p_2} > K_{p_1}$, if ΔH^0 is positive, (endothermic) and $K_{p_2} < K_{p_1}$, if ΔH^0 is negative (exothermic).

Factors Influencing the Equilibrium State

The effects of various factors on the equilibrium state for the different types of reactions may be clearly understood by using the following expressions of K_p.

Type I reaction

$\Delta n = 0$: A (g) + B (g) = C (g) + D (g) ;

$$K_p = \frac{\text{moles of C} \times \text{moles of D}}{\text{moles of A} \times \text{moles of B}}$$

Type II reaction

Δn is positive: A (g) = C (g) + D (g) ;

$$K_p = \frac{\text{moles of C} \times \text{moles of D}}{\text{moles of A}} \times \frac{\text{total pressure}}{\text{total moles}}$$

Δn is negative: A (g) + B (g) = D (g) ;

$$K_p = \frac{\text{moles of D}}{\text{moles of A} \times \text{moles of B}} \times \frac{\text{total moles}}{\text{total pressure}}$$

1. Effect of Pressure (or volume)

Type of reaction	Effect of Pressure (or volume)
Type I: $\Delta n = 0$	Eqb. is not affected
Type II: $\Delta n = +ve$	Increase in p (or decrease in V) shifts the eqb. to the left
$\Delta n = -ve$	Increase in p (or decrease in V) shifts the eqb. to the right

2. Effect of Addition of Inert Gas

Type of reaction	Effect of addition of inert gas
Type I: $\Delta n = 0$	Eqb. is not affected either at constant pressure or at constant volume
Type II: $\Delta n = +ve$	At constant volume: eqb. is not affected At constant pressure: eqb. shifts to the right
$\Delta n = -ve$	At constant volume: eqb. is not affected At constant pressure: eqb. shifts to the left

3. Effect of Addition of Catalyst

Adding a catalyst to a reaction changes its rate but this cannot shift the equilibrium in favour of either products or reactants. Because a catalyst affects the activation energy of both forward and backward reactions equally, it changes both rate constants by the same factor, so their ratio, K_c, does not change. **Thus, adding a catalyst to a reaction at equilibrium has no effect; it changes neither the reaction quotient, Q, nor the equilibrium constant, K,** but only the time required to establish equilibrium is altered.

4. Effect of Addition or Removal of any of the Reactants or Products

If any of the reactants or products is added or removed from a system at equilibrium the equilibrium shifts in the direction that consumes or produces the added or removed substance respectively. **This is only true if the volume of the system is kept constant.** But under the constant pressure condition, shifting of equilibrium may be in the reverse direction depending upon the value and sign of Δn and number of molecules of the added or removed substance involved in the reaction.

Equilibrium Constant and Standard Free Energy Change

The equilibrium constant K is related to free energy change, ΔG, given by thermodynamics, as

$$\Delta G = \Delta G^0 + RT \ln Q,$$

where ΔG^0 is the standard free energy change for a reaction which would accompany complete conversion of all reactants initially present at standard conditions to all products at standard conditions, that is, at 1 M or 1 atm.

ΔG is the free energy change for any other concentrations or pressures. Q is the reaction quotient. At equilibrium,

$$\Delta G = 0 \quad \text{and} \quad Q = K$$

$$\therefore \quad \Delta G^0 = -RT \ln K = -2{\cdot}303\, RT \log K \qquad \dots (12)$$

when Equation (12) is used with

1. all gaseous reactants and products; K represents K_p
2. all solution reactants and products; K represents K_c
3. a mixture of solution and gaseous reactants; K represents the thermodynamic equilibrium constant and we do not make the distinction between K_p and K_c.

From Equation (12), we may conclude that for standard reactions, i.e., at 1 M or 1 atm,

when $\Delta G^0 = -\text{ve}$ or $K > 1$: forward reaction is feasible
$\Delta G^0 = +\text{ve}$ or $K < 1$: reverse reaction is feasible
$\Delta G^0 = 0$ or $K = 1$: reaction is at equilibrium (very rare)

Since ΔG^0 can be obtained from independent measurements, e.g., from thermal and spectroscopic data, it is possible to calculate K_p. The value of ΔG^0 for a spontaneous or feasible reaction is always negative. From Eqn. 12, ΔG^0 is obtained in the same energy unit in which the value of R is introduced, provided, the equilibrium constant K_p is just a pure number or it is calculated by taking partial pressures in atmosphere.

Calculation of ΔH^0 and ΔS^0 from K

Further, we have from thermodynamics

$$\Delta G^0 = \Delta H^0 - T\Delta S^0. \qquad \dots (13)$$

From equations (12) and (13), we get

$$\log K = -\frac{\Delta H^0}{2{\cdot}303\, R} \cdot \frac{1}{T} + \frac{\Delta S^0}{2{\cdot}303\, R}.$$

This equation is important because the curve between $\log K$ and $\frac{1}{T}$ gives a straight line with negative slope from which ΔH^0 can be calculated by measuring the slope and multiplying it by $2{\cdot}303R$.

The intercept on the $\log K$ axis gives the value of $\Delta S^0/2{\cdot}303R$. Thus, by measuring the intercept and multiplying it by $2{\cdot}303R$, standard entropy change ΔS^0 for the reaction can be calculated.

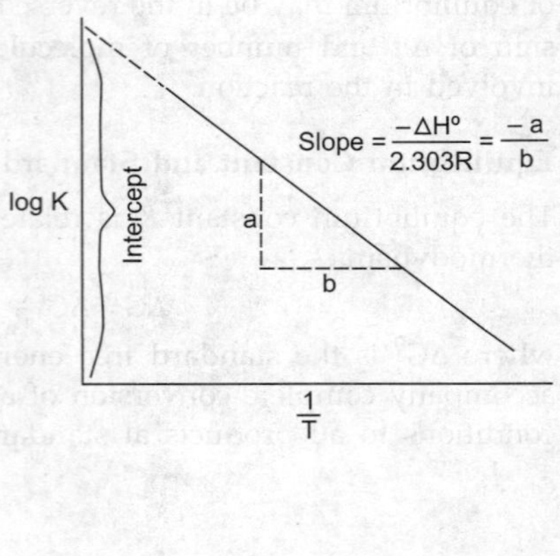

EXAMPLES

Ex. 1. (a) *For which of the following reactions, K_p is equal to K_c?*

(i) $H_2 + I_2 \rightleftharpoons 2HI$ (ii) $N_2 + 3H_2 \rightleftharpoons 2NH_3$

(iii) $PCl_5 \rightleftharpoons PCl_3 + Cl_2$

Solution : For reaction $H_2 + I_2 \rightleftharpoons 2HI$ only

$$\Delta n = 2 - 2 = 0,$$

i.e., Eqn. (3) reduces to $K_p = K_c$.

(b) *For which of the following cases does the reaction go farthest to completion: $K = 1$, $K = 10^{10}$, $K = 10^{-10}$?*

Solution : The reaction having $K = 10^{10}$ will go farthest to completion because the ratio [product]/[reactant] is maximum in this case [Eqn. (1)].

Ex. 2. *Both the metals* Mg *and* Fe *can reduce the metal copper from a solution having copper ions* (Cu^{2+})*, according to the equilibria:*

$$Mg\ (s) + Cu^{2+} \rightleftharpoons Mg^{2+} + Cu\ (s);\ K_1 = 6 \times 10^{90}$$

$$Fe\ (s) + Cu^{2+} \rightleftharpoons Fe^{2+} + Cu\ (s);\ K_2 = 3 \times 10^{26}$$

Which metal will remove cupric ions from the solution to a greater extent?

Solution : Since $K_1 > K_2$, the product in the first reaction is much more favoured than in the second one. Mg thus removes more Cu^{2+} from solution than does Fe.

(K_1 and K_2 include the constant concentration of the solid species.)

Ex. 3. *The equilibrium constant of the reaction,*

$$H_2\ (g) + I_2\ (g) \rightleftharpoons 2HI\ (g)$$

at 426°C *is* 55·3, *what will be the value of the equilibrium constant*

(a) if the reaction is reversed, and

(b) if the given reaction is represented as

$$3H_2 + 3I_2 \rightleftharpoons 6HI?$$

Solution : (a) The reverse reaction of the given reaction is

$$2HI \rightleftharpoons H_2 + I_2$$

$\therefore$ equilibrium constant $= \dfrac{1}{55 \cdot 3}$. ... (Eqn. 5)

(b) The reaction $3H_2 + 3I_2 \rightleftharpoons 6HI$ has been obtained by multiplying the reaction

$H_2 + I_2 \rightleftharpoons 2HI$ by 3. Hence, $K = (55 \cdot 3)^3$. ... (Eqn. 6)

Ex. 4. *What will be the effect on the equilibrium constant for the reaction* $N_2 + 3H_2 \rightleftharpoons 2NH_3$; $\Delta H = -22{\cdot}4$ kcal, *when (a) pressure is increased (b) concentration of* N_2 *is increased, and (c) temperature is raised at equilibrium?*

Solution : (a) No effect. (b) No effect. } As K does not depend on pressure and concentration

(c) Equilibrium constant will decrease as the temperature is increased. ... (Eqn. 11)

Ex. 5. K_c *for the reaction* $SO_2 + \frac{1}{2}O_2 \rightleftharpoons SO_3$ *at* 600°C *is* 61·7. *Calculate* K_p. *What is the unit of* K_p *for the above equilibrium?* ($R = 0{\cdot}0821$ lit · atm per deg. per mole)

Solution : Δn = moles of product – moles of reactant

$$= 1 - \left(1 + \frac{1}{2}\right) = -\frac{1}{2}$$

$$K_p = K_c\,(RT)^{\Delta n}$$

$$K_p = 61{\cdot}7\{0{\cdot}0821 \times (600 + 273)\}^{-\frac{1}{2}} = 7{\cdot}29.$$

$$K_p = \frac{p_{SO_3}}{p_{SO_2} \cdot p_{O_2}^{\frac{1}{2}}} = \frac{\text{atm}}{\text{atm. atm}^{\frac{1}{2}}}$$

Unit of K_p is atmosphere$^{-\frac{1}{2}}$.

Ex. 6. *One mole of nitrogen is mixed with 3 moles of hydrogen in a 4-litre container. If 0·25% of nitrogen is converted to ammonia by the following reaction,* $N_2\,(g) + 3H_2\,(g) \rightleftharpoons 2NH_3\,(g)$, *calculate the equilibrium constant* (K_c) *in concentration units. What will be the value of* K_c *for the following equilibrium?*

$$\frac{1}{2}N_2\,(g) + \frac{3}{2}H_2\,(g) \rightleftharpoons NH_3\,(g)$$

Solution :

N_2	+ $3H_2$	$\rightleftharpoons$ $2NH_3$	
1	3	0	Initial moles
$(1 - 0{\cdot}0025)$	$(3 - 3 \times 0{\cdot}0025)$	$(2 \times 0{\cdot}0025)$	Moles at equilibrium
$\frac{0{\cdot}9975}{4}$	$\frac{2{\cdot}9925}{4}$	$\frac{0{\cdot}005}{4}$	Molar concentration at eqb.

$$K_c = \frac{[NH_3]^2}{[N_2]\,[H_2]^3} = \frac{\left(\frac{0{\cdot}005}{4}\right)^2}{\left(\frac{0{\cdot}9975}{4}\right)\left(\frac{2{\cdot}9925}{4}\right)^3} = 1{\cdot}481 \times 10^{-5}\,(\text{mol/L})^{-2}.$$

Since the reaction $\frac{1}{2}N_2 + \frac{3}{2}H_2 \rightleftharpoons NH_3$ has been obtained by multiplying the equation $N_2 + 3H_2 \rightleftharpoons 2NH_3$ by $\frac{1}{2}$, the new equilibrium constant

$= (1{\cdot}481 \times 10^{-5})^{\frac{1}{2}}$ ... (Eqn. 6)

$K_c = 3{\cdot}84 \times 10^{-3}$ (mole/litre)$^{-1}$.

Ex. 7. *From the given data of equilibrium constants of the following reactions,*

$$CoO\,(s) + H_2\,(g) \rightleftharpoons Co\,(s) + H_2O\,(g); \quad K = 67$$
$$CoO\,(s) + CO\,(g) \rightleftharpoons Co\,(s) + CO_2\,(g); \quad K = 490$$

Calculate the equilibrium constant of the reaction,

$$CO_2\,(g) + H_2\,(g) \rightleftharpoons CO\,(g) + H_2O\,(g)$$

Solution : $CoO\,(s) + H_2\,(g) \rightleftharpoons Co\,(s) + H_2O\,(g) \quad K_1 = 67$ ... (1)

Now reversing the second reaction,

$$Co\,(s) + CO_2\,(g) \rightleftharpoons CoO\,(s) + CO\,(g) \quad K_2 = \frac{1}{490}$$ ... (2) (Eqn. 5)

Adding the two reactions, we get,

$$CO_2\,(g) + H_2\,(g) \rightleftharpoons CO\,(g) + H_2O\,(g)$$

for which $K = K_1 \,.\, K_2 = 67 \times \frac{1}{490} = 0{\cdot}137$. ... (Eqn. 7)

Ex. 8. *Given that at* 1000 K

$2SO_2\,(g) + O_2\,(g) \rightleftharpoons 2SO_3\,(g); \quad K = 261$

Calculate K for the following equations:

(i) $2SO_3\,(g) \rightleftharpoons 2SO_2\,(g) + O_2\,(g)$

(ii) $SO_3\,(g) \rightleftharpoons SO_2\,(g) + \frac{1}{2}O_2\,(g)$

(iii) $SO_2\,(g) + \frac{1}{2}O_2\,(g) \rightleftharpoons SO_3\,(g)$

Solution : Equation (i) is the reverse of the given equation

$\therefore K = \frac{1}{261} = 0{\cdot}0038$ (for (i)). ... (Eqn. 5)

As the equation (ii) has been obtained by dividing the equation (i) by 2, K for the equation (ii) will be $(0{\cdot}0038)^{\frac{1}{2}} = 0{\cdot}0619$. ... (Eqn. 6)

Reversing the equation (ii) we get the equation (iii), K for which will be

$$\frac{1}{0{\cdot}0619} = 16{\cdot}155.$$ ... (Eqn. 5)

Ex. 9. *Calculate the volume per cent of chlorine at equilibrium in* PCl_5 *under a total pressure of* 1·5 atm ($K_p = 0{\cdot}202$).

Solution :

$$\begin{array}{cccl} 1 & & 0 \quad 0 & \text{Initial moles} \\ PCl_5 & \rightleftharpoons & PCl_3 + Cl_2 & \\ (1-x) & & x \quad x & \text{Moles at equilibrium} \end{array}$$

Total moles at equilibrium $= 1 - x + x + x = (1 + x)$

$$K_p = \frac{p_{PCl_3} \cdot p_{Cl_2}}{p_{PCl_5}} = \frac{\left(\frac{x}{1+x} \cdot p\right)\left(\frac{x}{1+x} \cdot p\right)}{\left(\frac{1-x}{1+x} \cdot p\right)}.$$

$$\left\{\text{partial pressure of a species} = \frac{\text{moles of species}}{\text{total moles}} \times \text{total pressure}\right\}$$

$$K_p = \frac{x^2}{1-x^2} \cdot p.$$

Substituting the values of K_p and p

$$0{\cdot}202 = \frac{x^2}{1-x^2} \times 1{\cdot}5; \; x = 0{\cdot}343.$$

$\therefore$ moles of Cl_2 at equilibrium $= 0{\cdot}343$

and total moles at equilibrium $= 1 + x$

$= 1 + 0{\cdot}343$

$= 1{\cdot}343.$

$\therefore$ volume percentage of chlorine = mole percent of chlorine

$$= \frac{\text{moles of } Cl_2}{\text{total moles}} \times 100$$

$$= \frac{0{\cdot}343}{1{\cdot}343} \times 100$$

$$= 25{\cdot}5\%.$$

Ex. 10. *The concentrations of* A, B *and* C *in a reaction of the type* $3A + B \rightleftharpoons 2C + D$ *are respectively* 0·03, 0·01 *and* 0·008 mole/L. *Calculate the initial concentrations of* A *and* B.

Solution : Let the initial concentrations of A and B be a and b mole/L respectively.

$$\begin{array}{ccccl} a & b & & 0 \quad 0 & \text{Initial concentration} \\ 3A \; + & B & \rightleftharpoons & 2C + D & \\ (a-3x) & (b-x) & & 2x \quad x & \text{Eqb. concentration} \end{array}$$

Thus, $[A] = a - 3x = 0{\cdot}03$

$[B] = b - x = 0{\cdot}01$

and $[C] = 2x = 0{\cdot}008.$

from which, we get, $a = 0{\cdot}042$ mole/L

$b = 0{\cdot}014$ mole/L.

Ex. 11. *The equilibrium concentrations of* A, B *and* C *for the reaction* $A \rightleftharpoons B + C$ *are* 4·6, 2·3 *and* 2·3 moles/L *respectively at* 25°C. *If* 2 *moles per litre of* A *are removed, calculate the equilibrium concentrations of* A, B *and* C *at the same temperature.*

Solution : For equilibrium: $A \rightleftharpoons B + C$

4·6 2·3 2·3 concentration at equilibrium

$$K = \frac{2\cdot3 \times 2\cdot3}{4\cdot6} = 1\cdot15.$$

Now suppose the initial concentration is '*a*' moles/litre, and *x* moles/litre of A changed to the products at equilibrium

$\therefore$ concn. of A at eqb. $= (a - x) = 4\cdot6$

and concn. of B or C at eqb. $= x = 2\cdot3.$

$\therefore$ initial concn. of $A = a = (a - x) + x = 4\cdot6 + 2\cdot3$

$= 6\cdot9$ moles/litre.

Since 2 moles per litre of A are removed, the initial concentration of A will be (6·9 – 2) i.e., 4·9 moles per litre.

Suppose x' moles/litre of A will change to the product when the new equilibrium is attained.

4·9		0		0	Initial concentration
A	$\rightleftharpoons$	B	+	C	
$(4\cdot9 - x')$		x'		x'	Concentration at equilibrium

$\therefore K = \dfrac{x' \cdot x'}{4\cdot9 - x'} = 1\cdot15$ (*K* remains the same as the temperature is not change)

On solving, we get,

$$x' = 1\cdot86.$$

$\therefore$ at equilibrium,

$[A] = 4\cdot9 - 1\cdot86 = 3\cdot04$ moles/litre,

$[B] = 1\cdot86$ moles/litre,

$[C] = 1\cdot86$ moles/litre.

Ex. 12. K_c *for* $PCl_5\,(g) \rightleftharpoons PCl_3\,(g) + Cl_2\,(g)$ *is* 0·04 *at* 250°C. *How many mole of* PCl_5 *must be added to a 3-litre flask to obtain a* Cl_2 *concentration of* 0·15 M?

Solution : At equilibrium, number of moles of Cl_2 in 3 litres is $0\cdot15 \times 3$, i.e., 0·45.

Thus,

x (say)		0		0	Initial moles
PCl_5	$\rightleftharpoons$	PCl_3	+	Cl_2	
$(x - 0\cdot45)$		0·45		0·45	Moles at equilibrium
$\dfrac{x - 0\cdot45}{3}$		0·15		0·15	Equilibrium concentration

Now,

$$K_c = \frac{0{\cdot}15 \times 0{\cdot}15}{\left(\dfrac{x - 0{\cdot}45}{3}\right)} = 0{\cdot}04 \quad \text{(given)}$$

$\therefore$ $x = 2{\cdot}1$ moles.

Ex. 13. *At 27°C and 1 atm, N_2O_4 is 20% dissociated into NO_2. Find (a) K_p (b) the per cent dissociation at 27°C and at a total pressure of 0·1 atm.*

Solution : (a)

$$\begin{array}{lll} 1 & 0 & \text{Initial moles} \\ N_2O_4 & \rightleftharpoons 2NO_2 & \\ (1 - 0{\cdot}2) & 0{\cdot}4 & \text{Moles at equilibrium} \end{array}$$

$$\therefore \quad p_{NO_2} = \frac{0{\cdot}4}{1 - 0{\cdot}2 + 0{\cdot}4} \times 1 = \frac{1}{3};$$

$$p_{N_2O_4} = \frac{0{\cdot}8}{1 - 0{\cdot}2 + 0{\cdot}4} \times 1 = \frac{2}{3}.$$

$$\therefore \quad K_p = \frac{(p_{NO_2})^2}{p_{N_2O_4}} = \frac{(1/3)^2}{2/3} = 0{\cdot}17.$$

(b) Let the degree of dissociation of N_2O_4 at 27°C and 0·1 atm be x.

Thus, for,

$$\begin{array}{ll} N_2O_4 & \rightleftharpoons 2NO_2 \\ 1 - x & 2x \end{array}$$

$$p_{NO_2} = \frac{2x}{1 + x} \times 0{\cdot}1; \; p_{N_2O_4} = \frac{1 - x}{1 + x} \times 0{\cdot}1$$

$$\therefore \quad K_p = \frac{\left(\dfrac{2x}{1 + x} \times 0{\cdot}1\right)^2}{\left(\dfrac{1 - x}{1 + x} \times 0{\cdot}1\right)} = 0{\cdot}17 \text{ [as temperature is same as in (a)]}$$

$\therefore$ $x = 0{\cdot}55$ or 55%.

Ex. 14. *The equilibrium constant of the reaction A_2 (g) + B_2 (g) $\rightleftharpoons$ 2AB (g) at 100°C is 50. If a one-litre flask containing one mole of A_2 is connected to a two-litre flask containing two moles of B_2, how many moles of AB will be formed at 373 K?* **(IIT 1985)**

Solution :

$$\begin{array}{llll} 1 & 2 & 0 & \text{Initial moles} \\ A_2\,(g) + & B_2\,(g) \rightleftharpoons & 2AB\,(g) & \\ (1 - x) & (2 - x) & 2x & \text{Moles at equilibrium} \\ \dfrac{(1 - x)}{3} & \dfrac{(2 - x)}{3} & \dfrac{2x}{3} & \text{Molar concentration at eqb.} \end{array}$$

where x is moles of A_2 converted to AB at equilibrium and the total volume of the container = 1 + 2 = 3 litres.

Hence,

$$K_c = \frac{[AB]^2}{[A_2]\,[B_2]} = \frac{\left(\frac{2x}{3}\right)^2}{\frac{(1-x)}{3}\cdot\frac{(2-x)}{3}}$$

$$\frac{4x^2}{(1-x)\,(2-x)} = 50 \text{ (given that } K_p = 50 \text{ at 373 K)}$$

$$x = 0{\cdot}93.$$

(the other value of x may be neglected as it is greater than 1)

Moles of AB = 2 × 0·93 = 1·86.

Ex. 15. 25 mL *of hydrogen and* 18 mL *of iodine when heated in a closed container, produced* 30·8 mL *of* HI *at equilibrium. Calculate the degree of dissociation of* HI *at the same temperature.*

Solution : In this problem, volume of the species is proportional to their concentration. Thus,

	25		18		0	Initial volume
	H_2	+	I_2	$\rightleftharpoons$	2HI	
	$(25 - x)$		$(18 - x)$		$2x$ (= 30·8)	Volume at equilibrium
or	(25 – 15·4)		(18 – 15·4)		30·8	$\left(x = \frac{30{\cdot}8}{2} = 15{\cdot}4\right)$
	= 9·6		= 2·6			

$$K_c = \frac{[HI]^2}{[H_2]\,[I_2]} = \frac{30{\cdot}8^2}{9{\cdot}6 \times 2{\cdot}6} = 38{\cdot}01.$$

Now, if the dissociation of HI is carried out at the same temperature then for the reaction having the degree of dissociation, x', we have,

1		0		0	Initial moles
2HI	$\rightleftharpoons$	H_2	+	I_2	
$(1 - x')$		$\frac{x'}{2}$		$\frac{x'}{2}$	Moles at equilibrium

Equilibrium constant $K'_c = \frac{1}{K_c} = \frac{1}{38{\cdot}01}$. ... (Eqn. 5)

$$\therefore \qquad K'_c = \frac{\left(\frac{x'}{2}\right)\left(\frac{x'}{2}\right)}{(1-x')^2} = \frac{1}{38{\cdot}01}$$

$$x' = 0{\cdot}245.$$

Ex. 16. *At 427°C, for the gaseous reaction*

$$H_2 + I_2 = 2HI, \quad K = 55$$

(a) What weight of HI *will be formed at equilibrium, if* 1 *mole of* H_2 *and* 1 *mole of iodine are placed in a* 1*-litre vessel at* 427°C?

(b) Will there be any reaction at 427°C *in a mixture consisting of* 0·07 atm *of* HI *and* 0·02 atm *of each of* H_2 *and* I_2*? If so, will* HI *be consumed or formed?* (I = 127, H = 1).

Solution : (a)

1	1	0	Initial moles
H_2 +	I_2 ⇌	2HI	
$(1-x)$	$(1-x)$	$2x$	Moles at equilibrium
$\frac{(1-x)}{1}$	$\frac{(1-x)}{1}$	$\frac{2x}{1}$	Molar concentration at eqb.

x is the number of moles of H_2 or I_2 converted into HI at equilibrium.

$$K_c = \frac{[HI]^2}{[H_2]\,[I_2]} = \frac{(2x)^2}{(1-x)\,(1-x)} = \frac{4x^2}{(1-x)^2} = 55$$

or $\quad x = 0{\cdot}8.$

∴ moles of HI at equilibrium $= 2x = 2 \times 0{\cdot}8 = 1{\cdot}6.$

Weight of HI = moles × molecular weight

$$= 1{\cdot}6 \times 128 = 204{\cdot}8 \text{ grams.}$$

(b) In this part the given concentrations of H_2, I_2 and HI are not at equilibrium.

Now, for $H_2 + I_2 \rightleftharpoons 2HI$, the reaction quotient is given by

$$\frac{[HI]^2}{[H_2]\,[I_2]} = \frac{(0{\cdot}07)^2}{(0{\cdot}02)\,(0{\cdot}02)} = 12{\cdot}25.$$

{The relative molar concentrations of H_2, I_2 and HI will be 0·02, 0·02 and 0·07 respectively.}

Now, from Equation (9) we see that

$$\frac{[HI]^2}{[H_2]\,[I_2]} < K. \qquad (\because 12{\cdot}25 < 55)$$

Hence, the equilibrium will proceed from left to right, or HI will be formed.

Ex. 17. *If* 1 : 3 *(molar ratio) mixture of* N_2 *and* H_2 *yields* 17·8% *(by volume) of* NH_3 *at* 30 atm, *calculate* K_p *for the equilibrium represented by* $N_2 + 3H_2 \rightleftharpoons 2NH_3$.

Solution :

1	3	0	Initial moles (say)
N_2 +	$3H_2$ ⇌	$2NH_3$	
$1-x$	$3-3x$	$2x$	Moles at eqb.

x is the moles of N_2 converted into the product at equilibrium.

Now,

$\because$ mole % of NH_3 = volume % of NH_3

$$\therefore \quad \frac{\text{moles of } NH_3}{\text{total moles}} \times 100 = 17{\cdot}8$$

$$\frac{2x}{4-2x} \times 100 = 17{\cdot}8$$

(total moles = $1 - x + 3 - 3x + 2x = 4 - 2x$)

$$\therefore \quad x = 0{\cdot}3$$

$$K_p = \frac{p^2_{NH_3}}{p_{N_2} \times p^3_{H_2}} = \frac{\left(\frac{2x}{4-2x} \times p\right)^2}{\left(\frac{1-x}{4-2x} \times p\right)\left(\frac{3-3x}{4-2x} \times p\right)^3}.$$

Substituting $x = 0{\cdot}3$ and $p = 30$ atm.

we get, $K_p = 0{\cdot}00071$.

Ex. 18. *When alcohol (C_2H_5OH) and acetic acid are mixed together in equimolecular proportions at 25°C, 66·5% is converted into ester. Calculate how much ester will be formed if 1 mole of acetic acid is treated with 0·5 mole of alcohol at the same temperature.*

***Solution* :**

1	1	0	0	Initial moles
C_2H_5OH (l) +	CH_3COOH (l) $\rightleftharpoons$	$CH_3COOC_2H_5$ (l) +	H_2O (l)	
(1 − 0·665)	(1 − 0·665)	(0·665)	(0·665)	Moles at eqb.
$\frac{0{\cdot}335}{V}$	$\frac{0{\cdot}335}{V}$	$\frac{0{\cdot}665}{V}$	$\frac{0{\cdot}665}{V}$	Molar concn. at eqb.

where V is the volume in litres of the reaction mixture.

$$\therefore \quad K_c = \frac{[CH_3COOC_2H_5]\,.\,[H_2O]}{[C_2H_5OH]\,.\,[CH_3COOH]} = \frac{\frac{0{\cdot}665}{V} \times \frac{0{\cdot}665}{V}}{\frac{0{\cdot}335}{V} \times \frac{0{\cdot}335}{V}} = 4.$$

Now, for the second reaction

0·5	1	0	0	Initial moles
C_2H_5OH +	CH_3COOH $\rightleftharpoons$	$CH_3COOC_2H_5$ +	H_2O	
(0·5 − x)	(1 − x)	x	x	Moles at eqb.
$\frac{(0{\cdot}5-x)}{V}$	$\frac{(1-x)}{V}$	$\frac{x}{V}$	$\frac{x}{V}$	Molar concn. at eqb.

x is the moles of ester produced at equilibrium.

$$\therefore \qquad K_c = \frac{\frac{x}{V} \cdot \frac{x}{V}}{\frac{0.5 - x}{V} \cdot \frac{1 - x}{V}}$$

$$= \frac{x^2}{(0.5 - x)(1 - x)} = 4$$

(K_c has the same value as temperature remains the same)

$x = 0.423$ mole.

Ex. 19. *A mixture of SO_2, SO_3 and O_2 gases is maintained in a 10-litre flask at a temperature at which K_c = 100 for the reaction*

$$2SO_2 + O_2 \rightleftharpoons 2SO_3$$

(a) If the numbers of moles of SO_2 and SO_3 in the flask are equal, how much O_2 is present?

(b) If the number of moles of SO_3 in the flask is twice the number of moles of SO_2, how much O_2 is present?

Solution : (a) We know, at equilibrium,

$$K_c = \frac{[SO_3]^2}{[SO_2]^2\,[O_2]} = 100.$$

As moles of SO_2 is equal to that of SO_3

$$[SO_2] = [SO_3].$$

$$\therefore \qquad [O_2] = \frac{1}{100}.$$

$$\frac{\text{Mole of } O_2}{10} = \frac{1}{100} \qquad \text{(volume = 10 L)}$$

$$\therefore \qquad \text{mole of } O_2 = 0.1.$$

(b) Further, if moles of SO_3 are twice that of SO_2

$$\frac{\text{moles of } SO_3}{\text{moles of } SO_2} = 2 \qquad \ldots (1)$$

We again have,

$$K_c = \frac{[SO_3]^2}{[SO_2]^2\,[O_2]} = 100$$

$$\text{or} \qquad \frac{\left(\frac{\text{moles of } SO_3}{10}\right)^2}{\left(\frac{\text{moles of } SO_2}{10}\right)^2 \left(\frac{\text{moles of } O_2}{10}\right)} = 100$$

$$\text{or} \qquad \left(\frac{\text{moles of } SO_3}{\text{moles of } SO_2}\right)^2 \left(\frac{10}{\text{moles of } O_2}\right) = 100. \qquad \ldots (2)$$

From Eqns. (1) and (2), we get,

$$2^2 \times \frac{10}{\text{moles of } O_2} = 100$$

or $\text{mole of } O_2 = 0{\cdot}4.$

Ex. 20. *Prove that the pressure necessary to obtain* 50% *dissociation of* PCl_5 *at* 250°C *is numerically three times of* K_p.

Solution :

1		0	0	Initial moles (say)
PCl_5	$\rightleftharpoons$	PCl_3 +	Cl_2	
$(1 - 0{\cdot}5)$		$0{\cdot}5$	$0{\cdot}5$	Moles at equilibrium

Total moles at equilibrium $= 0{\cdot}5 + 0{\cdot}5 + 0{\cdot}5 = 1{\cdot}5$

$$K_p = \frac{p_{PCl_3} \cdot p_{Cl_2}}{p_{PCl_5}} = \frac{\left(\frac{0{\cdot}5}{1{\cdot}5}p\right)\left(\frac{0{\cdot}5}{1{\cdot}5}p\right)}{\left(\frac{0{\cdot}5}{1{\cdot}5}p\right)} \quad (p \equiv \text{total pressure})$$

or $K_p = \frac{1}{3} \cdot p$

or $p = 3K_p$.

Ex. 21. 1 *mole of* H_2, 2 *moles of* I_2 *and* 3 *moles of* HI *were taken in a* 1*-litre flask. If the value of* K_c *for the equation* $H_2(g) + I_2(g) \rightleftharpoons 2HI(g)$ *is* 50 *at* 440°C, *what will be the concentration of each species at equilibrium?*

Solution : Suppose that x moles each of H_2 and I_2 are converted to HI at equilibrium.

1	2		3	Initial moles
H_2 +	I_2	$\rightleftharpoons$	$2HI$	
$(1-x)$	$(2-x)$		$(3+2x)$	Moles at equilibrium
$\frac{(1-x)}{1}$	$\frac{(2-x)}{1}$		$\frac{(3+2x)}{1}$	Molar concentration at eqb.

$$K_c = \frac{[HI]^2}{[H_2][I_2]} = \frac{(3+2x)^2}{(1-x)(2-x)} = 50.$$

$\therefore$ $x = 0{\cdot}7$ (approx.).

$\therefore$ $[H_2] = (1-x) = 1 - 0{\cdot}7 = 0{\cdot}3$

$[I_2] = 2 - x = 2 - 0{\cdot}7 = 1{\cdot}3$

$[HI] = 3 + 2x = 3 + 1{\cdot}4 = 4{\cdot}4.$

Ex. 22. *At a certain temperature the equilibrium constant* (K_c) *is* 16 *for the reaction*

$$SO_2(g) + NO_2(g) \rightleftharpoons SO_3(g) + NO(g)$$

If we take one mole of each of the four gases in a one-litre container, what would be the equilibrium concentration of NO *and* NO_2? **(IIT 1987)**

Solution : We have,

1 mole		1 mole		1 mole		1 mole	Initial moles
SO_2	+	NO_2	$\rightleftharpoons$	SO_3	+	NO	
$1-x$		$1-x$		$1+x$		$1+x$	Moles at eqb. or concentration at eqb.

where x is the number of moles of each reactant changed to the products at equilibrium.

$$K=\frac{(1+x)(1+x)}{(1-x)(1-x)}=\frac{(1+x)^2}{(1-x)^2}=16 \text{ (given)}$$

or $$\frac{1+x}{1-x}=4\ ;\ x=0{\cdot}6$$

$$[NO]=1+x=1+0{\cdot}6=1{\cdot}6 \text{ moles/litre}$$

$$[NO_2]=1-x=1-0{\cdot}6=0{\cdot}4 \text{ moles/litre}$$

Ex. 23. *An equilibrium mixture at* 300 K *contains* N_2O_4 *and* NO_2 *at* 0·28 *and* 1·1 atm *pressures respectively. If the volume of the container is doubled, calculate the new equilibrium pressure of the two gases.*

Solution : $N_2O_4 \rightleftharpoons 2NO_2$

0·28 atm   1·1 atm   ... at equilibrium

$$\therefore \quad K_p=\frac{(1{\cdot}1)^2}{0{\cdot}28}=4{\cdot}32.$$

Further, when the volume of the container is doubled, the pressures of N_2O_4, and NO_2 shall be temporarily halved till some new equilibrium is attained.

0·14 atm		0·55 atm	... when vol. is halved
N_2O_4	$\rightleftharpoons$	$2NO_2$	
$(0{\cdot}14-x)$ atm		$(0{\cdot}55+2x)$ atm	... at the new eqb.

$$K_p=\frac{(0{\cdot}55+2x)^2}{0{\cdot}14-x}=4{\cdot}32 \text{ (temp. remaining same)}$$

$$x=0{\cdot}045.$$

$$\therefore \quad p_{N_2O_4}=0{\cdot}14-x=0{\cdot}14-0{\cdot}045=0{\cdot}095 \text{ atm}$$

and $$p_{NO_2}=0{\cdot}55+2x=0{\cdot}55+2\times 0{\cdot}045=0{\cdot}64 \text{ atm.}$$

Ex. 24. K_c *for* N_2O_4 (g) $\rightleftharpoons$ $2NO_2$ (g) *is* 0·00466 *at* 298 K. *If a one-litre container initially contained* 0·8 *mole of* N_2O_4, *what would be the concentrations of* N_2O_4 *and* NO_2 *at equilibrium? Also, calculate equilibrium concentration of* N_2O_4 *and* NO_2 *if the volume is halved at the same temperature.*

Solution : Suppose x moles of N_2O_4 change to NO_2 at equilibrium.

$$N_2O_4 \rightleftharpoons 2NO_2$$

$$0{\cdot}8 - x \qquad 2x \qquad \ldots \text{ mole/L at eqb.}$$

$$\therefore \quad K_c = \frac{(2x)^2}{(0{\cdot}8 - x)} = 0{\cdot}00466; \quad x = 0{\cdot}03 \text{ M.}$$

$$\therefore \text{ at equilibrium, } [N_2O_4] = 0{\cdot}8 - x = 0{\cdot}8 - 0{\cdot}03 = 0{\cdot}77 \text{ M}$$

$$[NO_2] = 2x = 2 \times 0{\cdot}03 = 0{\cdot}06 \text{ M.}$$

Now when the volume is halved, pressure will increase. From Le Chatelier's principle we know that the equilibrium will shift to the left-hand side with the increase in pressure (or decrease in volume). Further, when volume is halved, concentration will be doubled.

Concentration of $N_2O_4 = 0{\cdot}77 \times 2 = 1{\cdot}54$ M

and concentration of $NO_2 = 0{\cdot}06 \times 2 = 0{\cdot}12$ M.

The equilibrium concentration now will be

$$N_2O_4 \rightleftharpoons 2NO_2$$

$$(1{\cdot}54 + y) \qquad (0{\cdot}12 - 2y)$$

$$K_c = \frac{(0{\cdot}12 - 2y)^2}{(1{\cdot}54 + y)} = 0{\cdot}00466; \quad y = 0{\cdot}104 \text{ and } 0{\cdot}017$$

$y = 0{\cdot}104$ is unacceptable as in this case $2y > 0{\cdot}12$. $\therefore\ y = 0{\cdot}017$.

$\therefore$ at new equilibrium

$$[N_2O_4\,] = 1{\cdot}54 + 0{\cdot}017 = 1{\cdot}557 \text{ M}$$

$$[NO_2\,] = 0{\cdot}12 - 2 \times 0{\cdot}017 = 0{\cdot}086 \text{ M.}$$

Ex. 25. *At* 800 K *a reaction mixture contained* 0·5 *mole of* SO_2, 0·12 *mole of* O_2 *and* 5 *moles of* SO_3 *at equilibrium.* K_c *for the equilibrium* $2SO_2 + O_2 \rightleftharpoons 2SO_3$ *is* 833 lit/mol. *If the volume of the container is* 1 *litre, calculate how much* O_2 *is to be added at this equilibrium in order to get* 5·2 *moles of* SO_3 *at the same temperature.*

Solution : Suppose x moles of O_2 is added by which equilibrium shifts to the right-hand side and y mole of O_2 changes to SO_3. The new equilibrium concentration may be represented as

	$2SO_2$ +	O_2 $\rightleftharpoons$	$2SO_3$
First eqb.:	0·5 mole	0·12 mole	5 moles
Second eqb.:	$(0{\cdot}5 - 2y)$	$(0{\cdot}12 + x - y)$	$(5 + 2y)$

but $5 + 2y = 5{\cdot}2$ (given) $\therefore\ y = 0{\cdot}1$.

$$\text{Now, } K_c = \frac{[SO_3]^2}{[SO_2]^2\,[O_2]} = \frac{(5 + 2y)^2}{(0{\cdot}5 - 2y)^2\,(0{\cdot}12 + x - y)} = 833 \quad \text{(vol. = 1 L.)}$$

Substituting $y = 0{\cdot}1$, we get,

$$x = 0{\cdot}34 \text{ mole.}$$

Ex. 26. *The equilibrium constant K_p of the reaction is* 900 atm *at* 800 K. *A mixture containing*

$$2SO_2(g) + O_2(g) \rightleftharpoons 2SO_3(g)$$

SO_3 *and* O_2 *having initial partial pressure of* 1 *and* 2 atm *respectively is heated at constant volume to equilibrate. Calculate the partial pressure of each gas at* 800 K. **(IIT 1989)**

Solution : From the unit of K_p it is obvious that $K_p = 900$ is for the equilibrium

1		0		2	Initial pressure
$2SO_3$	$\rightleftharpoons$	$2SO_2$	+	O_2	
$(1-2x)$		$2x$		$(2 + x)$	Eqb. pressure

where $2x$ is the partial pressure of SO_2 at equilibrium.

Thus, $$K_p = \frac{(p_{SO_2})^2 (p_{O_2})}{(p_{SO_3})^2} = \frac{(2x)^2(2+x)}{(1-2x)^2} = 900.$$

On solving, we get, $x = 0{\cdot}475$.

$\therefore$ $$p_{SO_3} = 1 - 2x = 1 - 2 \times 0{\cdot}475 = 0{\cdot}05 \text{ atm}$$

$$p_{SO_2} = 2x = 2 \times 0{\cdot}475 = 0{\cdot}95 \text{ atm}$$

and $$p_{O_2} = 2 + x = 2 + 0{\cdot}475 = 2{\cdot}475 \text{ atm.}$$

Ex. 27. *A flask containing* 0·50 atm *of ammonia contains some solid* NH_4HS *which undergoes dissociation according to*

$NH_4HS(s) \rightleftharpoons NH_3(g) + H_2S(g)$. *Calculate the pressure of* NH_3 *and* H_2S *at equilibrium* ($K_p = 0{\cdot}11$). *Also, calculate the total pressure.*

Solution : Since 1 mole of solid NH_4HS, on dissociation, produces 1 mole of NH_3 and 1 mole of H_2S, the partial pressure of NH_3 should be equal to that of H_2S if the flask contains no other substance. Let this be x atm. But as the flask contains 0·50 atm of NH_3 besides solid NH_4HS, the partial pressure of NH_3 will be $(0{\cdot}50 + x)$ atm.

Now for the equilibrium,

$$NH_4HS(s) \rightleftharpoons NH_3(g) + H_2S(g)$$

$K_p = p_{NH_3} \times p_{H_2S}$ As NH_4HS is solid, K_p includes the constant active mass of NH_4HS

$$0{\cdot}11 = (0{\cdot}5 + x) \times x$$

$$x = 0{\cdot}17.$$

$\therefore$ $$p_{NH_3} = 0{\cdot}50 + 0{\cdot}17 = 0{\cdot}67 \text{ atm.}$$

$$p_{H_2S} = 0{\cdot}17 \text{ atm.}$$

Total pressure $= p_{NH_3} + p_{H_2S} = 0{\cdot}84$ atm.

Ex. 28. *The dissociation of solid* NH_4HS *in a closed container produces a pressure of* 1·12 atm *at* 105°C. *Calculate* K_p *for the equilibrium,*

$$NH_4HS\ (s) \rightleftharpoons NH_3\ (g) + H_2S\ (g)$$

Solution :

$$\begin{array}{llll} 1 & 0 & 0 & \text{Initial moles (suppose)} \\ NH_4HS\ (s) \rightleftharpoons & NH_3\ (g) + & H_2S\ (g) & \\ 1-x & x & x & \text{Moles at equilibrium} \end{array}$$

x is the degree of dissociation of NH_4HS

Total moles of gases $= x + x = 2x$.

Now we have,

$$K_p = p_{NH_3} \times p_{H_2S} = \left(\frac{x}{2x} \times p\right)\left(\frac{x}{2x} \times p\right) = \left(\frac{p}{2}\right)^2 \quad (p \text{ is the total pressure})$$

$$K_p = \left(\frac{1\cdot 12}{2}\right)^2 = 0\cdot 3136.$$

Ex. 29. *Would* 1% CO_2 *in the air be sufficient to prevent any loss in weight when* Ag_2CO_3 *is dried at* 120°C? $Ag_2CO_3\ (s) \rightleftharpoons Ag_2O\ (s) + CO_2\ (g)$; $k_p = 0\cdot 0095$ atm *at* 120°C. *How low would the partial pressure of* CO_2 *have to be to promote this reaction at* 120°C?

Solution :

$$K_p = \frac{p_{Ag_2O} \times p_{CO_2}}{p_{Ag_2CO_3}} = p_{CO_2}$$

$$\therefore \quad p_{CO_2} = 0\cdot 0095 \text{ atm} = \text{constant at } 120°C.$$

Thus, if Ag_2CO_3 is taken in a closed container, a small amount of it would decompose to give CO_2 gas until the partial pressure of CO_2 reaches 0·0095 atm. As this is the equilibrium pressure of CO_2, the decomposition would then stop.

Now since partial pressure of CO_2 in air is 0·01 atm ($\because$ CO_2 is 1% in air) which is much greater than 0·0095 atm, the equilibrium would practically shift to the left-hand side completely, or in other words, there would be no loss in weight of Ag_2CO_3 (by decomposition) if placed in air containing 1% CO_2.

Further, if the partial pressure of CO_2 in air is less than the equilibrium pressure of 0·0095 atm, the decomposition of Ag_2CO_3 would continue till the CO_2 pressure around Ag_2CO_3 becomes 0·0095 atm.

Ex. 30. 0·0755 g *of selenium vapour occupying a volume of* 114·2 mL *at* 700°C *exerts a pressure of* 185 mm. *The selenium is in a state of equilibrium according to the reaction* $Se_6\ (g) \rightleftharpoons 3Se_2\ (g)$. *Calculate (i) degree of dissociation of selenium, (ii)* K_p*, and (iii)* K_c (Se = 79).

Solution : (i) Volume at NTP $= \dfrac{185}{760} \times \dfrac{114{\cdot}2}{973} \times 273$ mL;

$$= 7{\cdot}799. \qquad \left(\frac{p_1V_1}{T_1} = \frac{p_2V_2}{T_2}\right)$$

$\therefore$ moles at NTP $= \dfrac{7{\cdot}799}{22400}$. ... (Rule 3, Chapter 1)

Let the observed molecular weight of selenium be M.

$\therefore$ no. of moles $= \dfrac{0{\cdot}0755}{M}$. ... (Rule 1, Chapter 1)

Hence, $\dfrac{0{\cdot}0755}{M} = \dfrac{7{\cdot}799}{22400}$

$$M = 217{\cdot}6.$$

$\therefore$ observed vapour density $= \dfrac{217{\cdot}6}{2} = 108{\cdot}8.$

Theoretical vapour density $= \dfrac{79 \times 6}{2} = 237.$ (mol. wt. of $Se_6 = 79 \times 6$)

(i.e., when there is no diss.)

Since 1 molecule of selenium produces 3 molecules

$\therefore$ $n = 3.$

Substituting the above values in Eqn. (4), we get

degree of dissociation of $Se_6 = \dfrac{(237 - 108{\cdot}8)}{(3-1) \times 108{\cdot}8}$

$$= 0{\cdot}59.$$

(ii) 1 0 Initial moles (suppose)

$Se_6\,(g) \rightleftharpoons 3Se_2\,(g)$

$(1 - x)$ $3x$ Moles at eqb. (x = deg. of diss.)

Total moles at equilibrium $= 1 - x + 3x = 1 + 2x$

$$K_p = \frac{p^3_{Se_2}}{p_{Se_6}} = \frac{\left(\dfrac{3x}{1+2x} \times p\right)^3}{\left(\dfrac{1-x}{1+2x} \times p\right)} = \frac{27x^3}{(1+2x)^2\,(1-x)} \times p^2.$$

Substituting $x = 0{\cdot}59$ and $p = \dfrac{185}{760}$ atm,

$$K_p = 0{\cdot}1687.$$

(iii) $K_p = K_c\,(RT)^{\Delta n}$... (Eqn. 3)

$0{\cdot}1687 = K_c\,(0{\cdot}0821 \times 973)^2$ $(\Delta n = 3 - 1 = 2)$

$$K_c = \frac{0{\cdot}1687}{(0{\cdot}0821 \times 973)^2} = 0{\cdot}2645 \times 10^{-4}.$$

Ex. 31. *For the reaction* $CO(g) + 2H_2(g) \rightleftharpoons CH_3OH(g)$, *hydrogen gas is introduced into a five-litre flask at 327°C, containing 0·2 mole of CO (g) and a catalyst, until the pressure is 4·92 atm. At this point 0·1 mole of CH_3OH is formed. Calculate the equilibrium constants K_p and K_c.* **(IIT 1990)**

Solution : Suppose x moles of H_2 are introduced into the flask

0·2	x	0	Initial moles
CO +	$2H_2$ ⇌	CH_3OH	
(0·2 − 0·1)	$(x - 0{\cdot}2)$	0·1	Moles at eqb.
$\left(\frac{0{\cdot}1}{5} = 0{\cdot}02\right)$	$\left(\frac{x - 0{\cdot}2}{5}\right)$	$\left(\frac{0{\cdot}1}{5} = 0{\cdot}02\right)$	Concentration at eqb.

Total moles $= 0{\cdot}1 + x - 0{\cdot}2 + 0{\cdot}1 = x$

$$\therefore \quad K_p = \frac{p_{CH_3OH}}{p_{CO} \times p_{H_2}^2} = \frac{\left(\frac{0{\cdot}1}{x} \times 4{\cdot}92\right)}{\left(\frac{0{\cdot}1}{x} \times 4{\cdot}92\right)\left(\frac{x - 0{\cdot}2}{x} \times 4{\cdot}9\right)^2}$$

$$\text{or} \quad K_p = \left[\frac{x}{(x - 0{\cdot}2)\, 4{\cdot}9}\right]^2. \qquad \ldots (1)$$

Further,

$$K_c = \frac{[CH_3OH]}{[CO]\,[H_2]^2} = \frac{0{\cdot}02}{0{\cdot}02 \times \left(\frac{x - 0{\cdot}2}{5}\right)^2} = \left(\frac{5}{x - 0{\cdot}2}\right)^2. \qquad \ldots (2)$$

Applying $K_p = K_c\,(RT)^{\Delta n}$

$$\frac{K_p}{K_c} = (RT)^{\Delta n} = (0{\cdot}0821 \times 600)^{-2}. \qquad \ldots (3)$$

From eqns. (1) and (2), we have,

$$\frac{K_p}{K_c} = \left(\frac{x}{(x - 0{\cdot}2)\, 4{\cdot}92}\right)^2 \left(\frac{x - 0{\cdot}2}{5}\right)^2 = \left(\frac{x}{4{\cdot}92 \times 5}\right)^2 \qquad \ldots (4)$$

From eqns. (3) and (4), we have,

$$\left(\frac{x}{4{\cdot}92 \times 5}\right)^2 = (0{\cdot}0821 \times 600)^{-2}; \; x = 0{\cdot}5.$$

Substituting x in eqns. (1) and (2), we get,

$$K_p = 0{\cdot}1147 \text{ atm}^{-2}$$

$$K_c = 277{\cdot}78 \text{ (moles/litre)}^{-2}.$$

Ex. 32. *The pressure of iodine gas at* 1273 K *is found to be* 0·112 atm, *whereas the expected pressure is* 0·074 atm. *The increased pressure is due to dissociation* $I_2 \rightleftharpoons 2I$. *Calculate* K_p.

Solution : 1 0 Initial moles (suppose)

$$I_2 \rightleftharpoons 2I$$

$1 - x$ $2x$ Moles at equilibrium

x is the degree of dissociation.

Total moles at equilibrium $= 1 - x + 2x = 1 + x$.

Since pressure is proportional to number of moles,

$$\frac{\text{experimental value of pressure}}{\text{expected value of pressure}} = \frac{1+x}{1}$$

or $$\frac{0{\cdot}112}{0{\cdot}074} = 1 + x \;\; ; \;\; x = 0{\cdot}51$$

$$K_p = \frac{p_I^2}{p_{I_2}} = \frac{\left(\frac{2x}{1+x} \times p\right)^2}{\left(\frac{1-x}{1+x} \times p\right)}$$

or $$K_p = \frac{4x^2}{1-x^2} \times p.$$

Putting $x = 0{\cdot}51$ and $p = 0{\cdot}112$ atm, we get,

$$K_p = 0{\cdot}1575.$$

Ex. 33. *For the reaction* $Br_2 \rightleftharpoons 2Br$, *the equilibrium constants at* 327°C *and* 527°C *are respectively* $6{\cdot}1 \times 10^{-12}$ *and* $1{\cdot}0 \times 10^{-7}$. *What is the nature of the reaction?*

Solution : We have,

$$\log K_{p_2} - \log K_{p_1} = \frac{\Delta H^0}{2{\cdot}303R} \times \frac{T_2 - T_1}{T_2 T_1}. \qquad \text{... (Eqn. 11)}$$

As we know from Eqn. (11) that if on increasing temperature, K_p increases, ΔH^0 becomes positive, i.e., the reaction is endothermic. Thus from the given data, we see that the reaction is endothermic.

Ex. 34. *From the following data*

(i) $H_2\,(g) + CO_2\,(g) \rightleftharpoons H_2O\,(g) + CO\,(g)$; $K_{2000\,K} = 4{\cdot}40$

(ii) $2H_2O\,(g) \rightleftharpoons 2H_2\,(g) + O_2\,(g)$; $K^{I}_{2000\,K} = 5{\cdot}31 \times 10^{-10}$

(iii) $2CO\,(g) + O_2\,(g) \rightleftharpoons 2CO_2\,(g)$; $K_{1000\,K} = 2{\cdot}24 \times 10^{22}$

show whether the reaction (iii) is exothermic or endothermic.

Solution : Let us consider reactions (i) and (ii).

Multiplying Equation (i) by 2, we get,

(iv) $2H_2\,(g) + 2CO_2\,(g) \rightleftharpoons 2H_2O\,(g) + 2CO\,(g)$

$$K^{II}_{2000\text{ K}} = K^2_{2000\text{ K}} = 4{\cdot}40^2 \qquad \text{... (Eqn. 6)}$$

Adding (iv) and (ii), we have,

$$2CO_2\,(g) \rightleftharpoons 2CO + O_2$$

$$K^{III}_{2000\text{ K}} = K^{II}_{2000\text{ K}} \times K^{I}_{2000\text{K}} \qquad \text{... (Eqn. 7)}$$

$$= 4{\cdot}40^2 \times 5{\cdot}31 \times 10^{-10}$$

$\therefore$ by reversing the reaction, we get the reaction (iii) but at 2000 K

(v) $2CO\,(g) + O_2\,(g) \rightleftharpoons 2CO_2\,(g)$; for which,

$$K^{IV}_{2000\text{ K}} = \frac{1}{K^{III}_{2000\text{ K}}} \qquad \text{... (Eqn. 5)}$$

$$= \frac{1}{4{\cdot}40^2 \times 5{\cdot}31 \times 10^{-10}}$$

$$= 9{\cdot}72 \times 10^7.$$

From (iii) and (v) we see that as the temperature increases, the equilibrium constant decreases ($9{\cdot}72 \times 10^7 < 2{\cdot}24 \times 10^{22}$). The reaction (iii) is thus exothermic because ΔH^0 will be negative. (Eqn. 11)

Ex. 35. *The partial pressure of* CO_2 *in the reaction*

$$CaCO_3\,(s) \rightleftharpoons CaO\,(s) + CO_2\,(g)$$

is 0·773 mm *at* 500°C. *Calculate* K_p at 600°C *for the above reaction.* ΔH^0 *of the reaction is* 43·2 kcal *per mole and does not change in the given range of temperature.*

Solution : $CaCO_3\,(s) \rightleftharpoons CaO\,(s) + CO_2\,(g)$

$$K_p = p_{CO_2}$$

Here, K_p includes the constant active masses of $CaCO_3$ and CaO which are solids.

Thus,

$$K_{p_1} = 0{\cdot}773 \text{ mm at } 500°\text{C}$$

$$K_{p_2} = p \text{ mm (say) at } 600°\text{C}$$

$$\Delta H^0 = 43{\cdot}2 \text{ kcal per mole}$$

$$= 43200 \text{ cal.}$$

Now we have,

$$\log \frac{K_{p_2}}{K_{p_1}} = \frac{\Delta H^0}{2{\cdot}303\,R}\left(\frac{T_2 - T_1}{T_1\,T_2}\right) \qquad \text{... (Eqn. 11)}$$

$$\log \frac{p}{0{\cdot}773} = \frac{43200}{2{\cdot}303 \times 1{\cdot}98}\left(\frac{873 - 773}{873 \times 773}\right) \qquad (R = 1{\cdot}98 \text{ cal deg}^{-1}\text{ mol}^{-1})$$

$$p = 19{\cdot}6 \text{ mm.}$$

$\therefore$ K_p at 600°C is 19·6 mm.

Ex. 36. *In the preparation of quicklime from limestone, the reaction is*

$$CaCO_3\,(s) \rightleftharpoons CaO\,(s) + CO_2\,(g).$$

Experiments carried out between 850°C and 950°C led to a set of K_p *values fitting an empirical equation*

$$\log K_p = 7{\cdot}282 - \frac{8500}{T},$$

where T is the absolute temperature. If the reaction is carried out in quiet air, what temperature would be predicted from this equation for complete decomposition of the limestone?

Solution : For the equilibrium,

$$CaCO_3\,(s) \rightleftharpoons CaO\,(s) + CO_2\,(g)$$

$$K_p = p_{CO_2}.$$

In quiet air, the decomposition of $CaCO_3$ would continue till the pressure developed due to CO_2 equals 1 atm (atmospheric pressure).

$\therefore$ when the decomposition is complete

$$K_p = 1 \text{ atm}.$$

Substituting K_p in the given empirical equation,

$$\log 1 = 7{\cdot}282 - \frac{8500}{T} = 0.$$

$$\therefore \quad T = 1167\text{ K}$$

$$= 894°C.$$

Ex. 37. *The value of* K_p *at 298 K for the reaction* $\frac{1}{2}N_2 + \frac{3}{2}H_2 \rightleftharpoons 2NH_3$ *is found to be 826·0, partial pressures being measured in atmospheric units. Calculate* ΔG^0 *at 298 K.*

Solution : $\Delta G^0 = -2{\cdot}303\,RT \log K_p$ ... (Eqn. 12)

$$= -2{\cdot}303 \times 1{\cdot}98 \times 298 \times \log 826$$

$$= -3980 \text{ calories.}$$

Ex. 38. *In a reaction at equilibrium, x moles of the reactant A decompose to give 1 mole of each of B and C. It has been found that the fraction of A decomposed at equilibrium is independent of the initial concentration of A. Calculate x.*

Solution : According to the problem, the equilibrium reaction may be represented as

$$x\,A \rightleftharpoons B + C.$$

Let *'a'* be the initial concentration and α be the degree of dissociation of A, i.e., the fraction of A decomposed at equilibrium. Thus we have,

$$\begin{array}{cccccl} a & & 0 & & 0 & \text{Initial concentration} \\ x\,\text{A} & \rightleftharpoons & \text{B} & + & \text{C} & \\ a(1-\alpha) & & a\dfrac{\alpha}{x} & & a\dfrac{\alpha}{x} & \text{Concentration at eqb.} \end{array}$$

$$\therefore \qquad K_c = \frac{[\text{B}]\,[\text{C}]}{[\text{A}]^x} = \frac{\dfrac{a\,\alpha}{x} \times \dfrac{a\,\alpha}{x}}{\{a(1-\alpha)\}^x}$$

or
$$K_c = \frac{\alpha^2}{x^2\,(1-\alpha)^x} \times a^{2-x}.$$

Now we see that only for $x = 2$ the above equation will reduce to

$$K_c = \frac{\alpha^2}{x^2\,(1-\alpha)^x}$$

which does not contain the term *'a'* and so K_c being constant, α will not depend on a.

Ex. 39. *A saturated solution of iodine in water contains* 0·33 g *of* I_2 *per litre of solution. More than this can dissolve in* KI *solution because of the following equilibrium,*

$$I_2\ (\text{aq.}) + I^- \rightleftharpoons I_3^-$$

A 0·1 M KI *solution actually dissolves* 12·5 g *of* I_2*/litre, most of which is converted to* I_3^-*. Assuming that the concentration of* I_2 *in all saturated solutions is the same, calculate the equilibrium constant for the above reaction.*

Solution :

$$\begin{array}{ccccll} \dfrac{12{\cdot}5}{254}\,\text{M} & & 0{\cdot}1\,\text{M} & & 0\,\text{M} & \text{... initially} \\ I_2 & + & I^- & \rightleftharpoons & I_3^- & \\ \dfrac{0{\cdot}33}{254}\,\text{M} & & \left(0{\cdot}1 - \dfrac{12{\cdot}17}{254}\right)\text{M} & & \left(\dfrac{12{\cdot}5}{254} - \dfrac{0{\cdot}33}{254}\right)\text{M} & \text{... at eqb.} \\ & & = \dfrac{13{\cdot}23}{254}\,\text{M} & & = \dfrac{12{\cdot}17}{254}\,\text{M} & \end{array}$$

Thus,

$$K = \frac{[I_3^-]}{[I_2]\,[I^-]} = \frac{\dfrac{12{\cdot}17}{254}}{\dfrac{0{\cdot}33}{254} \times \dfrac{13{\cdot}23}{254}} = 708.$$

Ex. 40. *A sample of air consisting of* N_2 *and* O_2 *was heated to* 2500 K *until the equilibrium,*

$$N_2\,(g) + O_2\,(g) \rightleftharpoons 2\,NO\,(g)$$

was established with an equilibrium constant $K_c = 2{\cdot}1 \times 10^{-3}$*. At equilibrium,* *the mole % of* NO *was* 1·8. *Estimate the initial composition of air in mole fraction of* N_2 *and* O_2. **(IIT 1997)**

Solution : Let the total number of moles of N_2 and O_2 be 100 containing '*a*' mole of N_2 initially.

$$\begin{array}{cccccl} a & & (100-a) & & 0 & \text{Initial moles} \\ N_2 & + & O_2 & \rightleftharpoons & 2\,NO & \\ (a-x) & & (100-a-x) & & 2x & \text{Moles at eqb.} \end{array}$$

Mole % of NO at equilibrium

$$= \frac{2x}{(a-x)+(100-a-x)+2x} \times 100 = 1{\cdot}8.$$

$\therefore \quad x = 0{\cdot}9.$

Thus, at equilibrium,

moles of $N_2 = (a - 0{\cdot}9)$

moles of $O_2 = (100 - a - 0{\cdot}9) = (99{\cdot}1 - a)$

moles of NO $= 2x = 2 \times 0{\cdot}9 = 1{\cdot}8$

$$K_c = \frac{(2x/V)^2}{\left(\frac{a-0{\cdot}9}{V}\right)\left(\frac{99{\cdot}1-a}{V}\right)} = \frac{(2x)^2}{(a-0{\cdot}9)\,(99{\cdot}1-a)}$$

$$2{\cdot}1 \times 10^{-3} = \frac{(1{\cdot}8)^2}{(a-0{\cdot}9)\,(99{\cdot}1-a)}$$

$$a = 79{\cdot}46\%$$

(we know that percentage of N_2 in the air is more than that of O_2)

mole fraction of $N_2 = 0{\cdot}7946$

mole fraction of $O_2 = 0{\cdot}2054$.

Ex. 41. *If* K_c *is not numerically equal to* K_p, *how can both of the following equations be valid?*

$$\Delta G^0 = -\,2{\cdot}303\,RT\log K_c\,;\ \Delta G^0 = -\,2{\cdot}303\,RT\log K_p$$

Solution : ΔG^0 will have two different values but the first equation refers to the standard state in which all reactants and products are 1 M, and the second equation refers to the standard state in which all reactants and products are at 1 atm.

Ex. 42. *The rate of the elementary reaction at* 25°C

$$H_2 + I_2 \rightarrow 2HI$$

is given by

$$rate = 1{\cdot}7 \times 10^{-18}[H_2][I_2].$$

The rate of decomposition of gaseous HI *to* H_2 *and* I_2 *at* 25°C *is given by*

$$rate = 2{\cdot}4 \times 10^{-21}[HI]^2.$$

What is the equilibrium constant for the formation of HI *from* H_2 *and* I_2 *at* 25°C?

Solution : $H_2 + I_2 \underset{k_2}{\overset{k_1}{\rightleftharpoons}} 2HI$

for which, $k_1 = 1{\cdot}7 \times 10^{-18}$

and $k_2 = 2{\cdot}4 \times 10^{-21}$.

$$\therefore \quad K = \frac{k_1}{k_2} = \frac{1{\cdot}7 \times 10^{-18}}{2{\cdot}4 \times 10^{-21}} = 7{\cdot}08 \times 10^2.$$

Ex. 43. *At* 25°C *and* 1 atm, *the partial pressures in an equilibrium mixture of* N_2O_4(g) *and* NO_2(g) *are* 0·70 atm *and* 0·30 atm *respectively. Calculate their partial pressures when they are at equilibrium at* 9·0 atm *and* 25°C.

Solution : For the equilibrium,

$$\underset{0{\cdot}3\text{ atm}}{2NO_2} \rightleftharpoons \underset{0{\cdot}7\text{ atm}}{N_2O_4}$$

$$K_p = \frac{p_{N_2O_4}}{p^2_{NO_2}} = \frac{0{\cdot}7}{(0{\cdot}3)^2} = 7{\cdot}77.$$

Let the equilibrium pressures of NO_2 and N_2O_4 be p_1 and p_2 atm at 9·0 atm at the same temperature.

$$K_p = \frac{p_2}{p_1^2} = 7{\cdot}77 \qquad \dots (1)$$

and $$p_1 + p_2 = 9{\cdot}0. \qquad \dots (2)$$

Solving eqns. (1) and (2)

$$p_1 = 1{\cdot}01 \text{ atm}$$

$$p_2 = 7{\cdot}99 \text{ atm}$$

Ex. 44. *If* 50% *of* CO_2 *converts to* CO *at the following equilibrium:*

$$C(s) + CO_2(g) \rightleftharpoons 2CO(g)$$

and the equilibrium pressure is 12 atm, *calculate* K_p.

Solution : Initial pressure: p atm

$$C(s) + CO_2(g) \rightleftharpoons 2CO(g)$$

Equilibrium pressure: $(p - {\cdot}5p)$ $\quad p$

As given: $p - {\cdot}5p + p = 12;\ p = 8$ atm.

$\therefore$ at equilibrium: $p_{CO} = 8$ atm

$p_{CO_2} = 4$ atm.

$$K_p = \frac{p^2_{CO}}{p_{CO_2}} = \frac{8^2}{4} = 16 \text{ atm.}$$

Ex. 45. *The density of an equilibrium mixture of* N_2O_4 *and* NO_2 *at* 1 atm *and* 348 K *is* 1·84 g/L. *Calculate* K_c *for the equilibrium*

$$N_2O_4(g) \rightleftharpoons 2NO_2(g)$$

Solution :

$$\begin{array}{lcl} 1\text{ M} & 0 & \text{... initial} \\ N_2O_4 & \rightleftharpoons 2NO_2 & \\ (1-x) & 2x & \text{... at eqb.} \end{array}$$

$$\text{Mol. wt. of mix.} = \frac{92(1-x) + 46 \times 2x}{1+x} = \frac{92}{1+x}.$$

We have,
$$p = \frac{dRT}{M}$$

$$1 = \frac{1{\cdot}84 \times 0{\cdot}0821 \times 348}{92/(1+x)}$$

$$x = 0{\cdot}75.$$

$$\therefore \qquad K_c = \frac{(2x)^2}{1-x} = \frac{4 \times 0{\cdot}75^2}{1 - 0{\cdot}75} = 9{\cdot}0.$$

Ex. 46. *A* 10·0-*litre vessel contains* 2 *moles of* B *at* 100 K. *How many moles of* C *should be added to drive the following backward reaction practically to completion?*

$$A(s) \rightleftharpoons B(s) + 2C(g); \; K_p = 16 \text{ atm}^2$$

Solution : Let us first calculate the number of moles of gaseous C at equilibrium.

$$K_p = p_c^2 = 16; \; p_c = 4 \text{ atm}$$

Applying ideal gas law, for C,

$$pV = nRT$$

$$4 \times 10 = n_c \times 0{\cdot}0821 \times 100$$

$$n_c = 4{\cdot}872 \text{ moles.}$$

From the stoichiometry of the given equation, 2 moles of B shall combine with 4 moles of C to give back A. Thus total moles of C required to drive the backward reaction practically to completion will be 4·872 + 4·0 = **8·872** so that the value of K_p remains constant.

Ex. 47. *Two solids* A *and* C *dissociate into gas products as follows.*

$$A(s) \rightleftharpoons B(g) + D(g); \; K_{p_1} = 400$$

$$C(s) \rightleftharpoons E(g) + D(g); \; K_{p_2} = 900$$

At 25°C, the pressure over excess solid A only is 40 atm, and that over solid C only is 60 atm. Find the pressure over the solid mixture.

Solution : When the two solids are taken together in a closed container, both decompose to give gases B, D and E. As D is the common gas, the dissociation of both the solids A and C shall be suppressed. Suppose that the partial pressures of B and D due to dissociation of only A are p_1 atm each, and the partial pressures of E and D due to dissociation of only C are p_2 atm each.

$$\underset{}{A(s)} \rightleftharpoons \underset{p_1}{B(g)} + \underset{p_1}{D(g)};\ Kp_1 = 400$$

$$C(s) \rightleftharpoons \underset{p_2}{E(g)} + \underset{p_2}{D(g)};\ Kp_2 = 900$$

$\therefore \quad Kp_1 = p_B \cdot p_D = p_1(p_1 + p_2) = 400 \quad \ldots (1)$

and, $Kp_2 = p_E \cdot p_D = p_2(p_1 + p_2) = 900. \quad \ldots (2)$

Solving eqns. (1) and (2),

$$p_1 = 11{\cdot}097 \text{ and } p_2 = 24{\cdot}96 \text{ atm.}$$

$$\therefore \quad \text{pressure over solid mixture} = 2(p_1 + p_2) = 2(11{\cdot}097 + 24{\cdot}96) = 72{\cdot}114 \text{ atm.}$$

Ex. 48. *Hot copper turnings can be used as an 'oxygen getter' for inert gas supplies by slowly passing the gas over the turnings at 600 K.*

$$2Cu(s) + \frac{1}{2}O_2(g) \rightarrow Cu_2O(s);\ \Delta G^{\circ} = -124{\cdot}944 \text{ kJ}$$

How many molecules of O_2 are left in 1·0 litre of a gas supply after equilibrium has been reached?

Solution : We have, $\Delta G^0 = -2{\cdot}303\, RT \log K_p$.

$$\therefore \quad \log K_p = \frac{(-124{\cdot}944 \times 10^3)}{-2{\cdot}303 \times 8{\cdot}314 \times 600} = 10{\cdot}8757$$

$$K_p = 7{\cdot}5 \times 10^{10}.$$

Now for the equilibrium,

$$2Cu(s) + \frac{1}{2}O_2(g) \rightleftharpoons Cu_2O(s)$$

$$K_p = \frac{1}{(p_{O_2})^{1/2}}.$$

$$\therefore \quad p_{O_2} = \left(\frac{1}{K_p}\right)^2 = \left(\frac{1}{7{\cdot}5 \times 10^{10}}\right)^2$$

$$p_{O_2} = 1{\cdot}77 \times 10^{-22} \text{ atm.}$$

Note that ΔG^0 and R are in the same energy unit, provided K_p is just a pure number or it is calculated by taking partial pressures in atm.

Again we have, $pV = nRT$

$$n = \frac{pV}{RT}$$

$$= \frac{1{\cdot}77 \times 10^{-22} \times 1}{0{\cdot}0821 \times 600}$$

$$= 3{\cdot}593 \times 10^{-24} \text{ mole.}$$

$\therefore$ number of O_2 molecules left $= (3{\cdot}593 \times 10^{-24})(6{\cdot}022 \times 10^{23})$

$= 2{\cdot}163$

$\approx$ 2molecules per litre.

Ex. 49. *A mixture of air at* 1·0 *bar and* 2000 K *was passed through an electric arc to produce nitric oxide*

$$N_2(g) + O_2(g) \rightleftharpoons 2NO(g);\ K = 4{\cdot}0 \times 10^{-4}$$

What are partial pressures of gases at equilibrium?

What fraction of O_2 *is converted? Assume that the composition of air is* N_2 *and* O_2 *in a* 4 : 1 *molar ratio.*

Solution : 0·8 bar 0·2 bar

$$\underbrace{N_2(g) + O_2(g)}_{\text{air}} \rightleftharpoons 2NO(g)$$

$(0{\cdot}8-p) \quad (0{\cdot}2-p) \quad 2p$ ... at equilibrium

$$\therefore \quad K_p = \frac{(2p)^2}{(0{\cdot}8-p)(0{\cdot}2-p)} = 4 \times 10^{-4};\ p = 4 \times 10^{-3}.$$

$\therefore \quad p_{N_2} = 0{\cdot}8 - 4 \times 10^{-3} \approx 0{\cdot}8$ bar

$p_{O_2} = 0{\cdot}2 - 4 \times 10^{-3} \approx 0{\cdot}2$ bar

$p_{NO} = 2 \times 4 \times 10^{-3} = 8 \times 10^{-3}$ bar

Further,

$$\text{fraction of } O_2 \text{ converted} = \frac{p}{0{\cdot}2}$$

$$= \frac{4 \times 10^{-3}}{0{\cdot}2}$$

$$= 0{\cdot}02.$$

Ex. 50. *The average molar mass of an equilibrium mixture of* $F_2(g)$ *and* F(g) *at* 1000 K *was* 36·739 g/mole.

Calculate K_x for the reaction

$$F_2(g) \rightleftharpoons 2F(g)$$

in terms of mole fraction. Write the unit of K_x.

Solution : Let the mole fractions of F_2 and F in the equilibrium mixture be x_{F_2} and x_F respectively.

$\therefore \quad x_{F_2} + x_F = 1$

and, $19x_{F_2} + 38x_F = 36{\cdot}739$.

Solving these equations, we get, $x_{F_2} = 0{\cdot}066$

$$x_F = 0{\cdot}934$$

$$K_x = \frac{x_F^2}{x_{F_2}} = \frac{(0{\cdot}934)^2}{0{\cdot}066} = 13{\cdot}2.$$

K_x has no unit because mole fraction has no unit.

Ex. 51. *Two equilibria are simultaneously existing in a vessel at* 25°C

$$NO(g) + NO_2(g) \rightleftharpoons N_2O_3(g);\ K_{p_1}\text{(say)}$$

$$2NO_2(g) \rightleftharpoons N_2O_4(g);\ K_{p_2} = 8\ \text{atm}^{-1}$$

If initially only NO *and* NO_2 *are present in a* 3 : 5 *mole ratio and the total pressure at equilibrium is* 5·5 atm *with the pressure of* NO_2 *at* 0·5 atm, *calculate* K_{p_1}.

Solution : As mole ratio represents pressure ratio in gases, for the following equilibria, we have,

Initial pressure:	$3p$ atm (say)	$5p$ atm (say)		0 atm
	NO(g) +	$NO_2(g)$	$\rightleftharpoons$	$N_2O_3(g)$; Kp_1
Equilibrium pressure:	p_1 (say)	0·5 atm		p_2 (say)

And, $\quad 2NO_2(g) \rightleftharpoons N_2O_4(g); \quad Kp_2 = 8$

Equilibrium pressure: $\quad$ 0·5 atm $\quad$ 2 atm

$$\because \quad Kp_2 = \frac{p_{N_2O_4}}{p_{NO_2}^2} = \frac{p_{N_2O_4}}{0{\cdot}5^2} = 8;\ p_{N_2O_4} = 2\ \text{atm}.$$

Now, out of $3p$ atm of NO, p_2 atm of it converts to p_2 atm of N_2O_3,

$\therefore \quad 3p - p_2 = p_1. \qquad \ldots (1)$

As NO_2 converts to both N_2O_3 and N_2O_4, out of $5p$ atm of NO_2, p_2 atm converted to p_2 atm of N_2O_3 and 4 atm of it gave 2 atm of N_2O_4 and 0·5 atm of NO_2 remained at equilibrium.

$\therefore \quad 5p - p_2 - 4 = 0{\cdot}5. \qquad \ldots (2)$

Further, at equilibrium, the total pressure is given by

$$p_{NO} + p_{NO_2} + p_{N_2O_3} + p_{N_2O_4} = 5 \cdot 5$$

or $\quad p_1 + 0 \cdot 5 + p_2 + 2 = 5 \cdot 5.$ ... (3)

Solving eqns. (1), (2) and (3),

$$p_1 = 2 \cdot 5 \quad \text{and} \quad p_2 = 0 \cdot 5.$$

$$\therefore \quad Kp_1 = \frac{p_{N_2O_3}}{p_{NO} \times p_{NO_2}} = \frac{p_2}{p_1 \times 0 \cdot 5} = \frac{0 \cdot 5}{2 \cdot 5 \times 0 \cdot 5} = 0 \cdot 40 \text{ atm}^{-1}.$$

Ex. 52. *For the equilibrium:*

$$CaCO_3(s) \rightleftharpoons CaO(s) + CO_2(g)$$

$\Delta H_f^0(CaCO_3) = -1207 \cdot 1$ kJ/mole, $\Delta H_f^0(CaO) = -635 \cdot 5$ kJ/mole

$\Delta H_f^0(CO_2) = -393 \cdot 5$ kJ/mole

(a) How would K_p depend on temperature?

(b) The equilibrium constant for this reaction is much less than 1. Why, then, does heating $CaCO_3(s)$ in an open container lead to a complete conversion to the products?

Solution : (a) $\Delta H^0 = [\Delta H_f^0(CaO) + \Delta H_f^0(CO_2)] - \Delta H_f^0(CaCO_3)$

$$= [-635 \cdot 5 + (-393 \cdot 5)] - (-1207 \cdot 1)$$

$$= +178 \cdot 1 \text{ kJ}.$$

Because the given reaction is endothermic, the equilibrium constant will increase with increasing temperature.

(b) On heating $CaCO_3$, K_p increases which favours the dissociation of $CaCO_3$. Also, in an open container, CO_2 gas escapes and more $CaCO_3$ dissociates to replace it until no more $CaCO_3$ remains.

Ex. 53. *Calculate K_p at 298 K for the gas-phase reaction*

$$CO + H_2O \rightleftharpoons CO_2 + H_2$$

from the following data.

	CO	H_2O	CO_2	H_2
ΔH_f^0 (kJ/mole)	–110·5	–241·8	–393·5	0
S^0 (J/mole. K)	197·6	188·7	213·6	130·6

Solution : For the given reaction,

$$\Delta H^0 = [\Delta H_f^0(CO_2) + \Delta H_f^0(H_2)] - [\Delta H_f^0(CO) + \Delta H_f^0(H_2O)]$$

$$= (-393 \cdot 5 + 0) - (-110 \cdot 5 - 241 \cdot 8)$$

$$= -41 \cdot 2 \text{ kJ} = -41200 \text{ J/mole}.$$

$$\Delta S^0 = [S^0(CO_2) + S^0(H_2)] - [S^0(CO) + S^0(H_2O)]$$
$$= (213{\cdot}6 + 130{\cdot}6) - (197{\cdot}6 + 188{\cdot}7)$$
$$= -42{\cdot}1 \text{ J/mole.}$$

Applying the thermodynamic equation,

$$\Delta G^0 = \Delta H^0 - T\Delta S^0$$
$$= -41200 - 298 \times (-42{\cdot}1)$$
$$= -28655 \text{ J.}$$

Now, we have,

$$\Delta G^0 = -2{\cdot}303\, RT \log K_p$$
$$-28655 = -2{\cdot}303 \times 8{\cdot}314 \times 298 \times \log K_p$$
$$\log K_p = 5{\cdot}022.$$
$$\therefore \quad K_p = 1{\cdot}052 \times 10^5.$$

Ex. 54. *Calculate K_p for the following reaction at 25°C*

$$2N_2O(g) \rightleftharpoons 2N_2(g) + O_2(g)$$

$\Delta G_f^0(N_2O) = 104{\cdot}2$ kJ/mole, $\Delta G_f^0(N_2) = 0$ *and* $\Delta G_f^0(O_2) = 0$

Solution : $\Delta G^0 = [2 \times \Delta G_f^0(N_2) + \Delta G_f^0(O_2)] - 2 \times \Delta G_f^0(N_2O)$

$$= (2 \times 0 + 0) - 2 \times 104{\cdot}2$$
$$= -208{\cdot}4 \text{ kJ/mole}$$
$$= -2{\cdot}084 \times 10^5 \text{ J/mole.}$$

This is a gas-phase reaction, so ΔG^0 is related to K_p by

$$\Delta G^0 = -2{\cdot}303\, RT \log K_p$$
$$-2{\cdot}084 \times 10^5 = -2{\cdot}303 \times 8{\cdot}314 \times 298 \log K_p$$
$$K_p = 3{\cdot}3 \times 10^{36}.$$

Ex. 55. *The equilibrium constant for the reaction*

$$H_2(g) + Br_2(l) \rightleftharpoons 2HBr(g)$$

is $K_p = 4{\cdot}5 \times 10^{18}$ atm *at* 298 K. *The vapour pressure of liquid* Br_2 *at this temperature is* 0·28 atm.

(a) Find K_p at 298 K *for the reaction*

$$H_2(g) + Br_2(g) \rightleftharpoons 2HBr(g)$$

(b) How will the equilibrium in part (a) be shifted by an increase in the volume of the container if (i) liquid Br_2 *is absent, and (ii) liquid* Br_2 *is present.*

Solution : For the eqb: $H_2(g) + Br_2(g) \rightleftharpoons 2HBr(g)$

(a) $K_p = \dfrac{p_{HBr}^2}{p_{H_2} \cdot p_{Br_2}} = \dfrac{p_{HBr}^2}{p_{H_2}} \times \dfrac{1}{p_{Br_2}}$

$$= \frac{4{\cdot}5 \times 10^{18}}{0{\cdot}28} = 1{\cdot}6 \times 10^{19}.$$

(b) (i) In part (a) $\Delta n = 0$, so change in volume has no effect on equilibrium.
(ii) If liquid Br_2 is present, $\Delta n = +1$, so increase in volume will shift the equilibrium to the right.

Ex. 56. *At 25°C the equilibrium constant for the reaction*

$$NH_4SH(s) \rightleftharpoons NH_3(g) + H_2S(g)$$

is $K_p = 12$. *A* 10·0*-litre flask is charged with* 1 atm *of pure* $NH_3(g)$ *at* 25°C. *What is the minimum mass in grams of* NH_4HS *that must be added to the flask in order to achieve equilibrium?*

Solution : Suppose that NH_4SH dissociates to give p atm each of NH_3 and H_2S at equilibrium in the presence of 1 atm of NH_3.

$$\therefore \quad K_p = p_{NH_3} \cdot p_{H_2S}$$

$$12 = (p+1) \times p; \quad p = 3 \text{ atm.}$$

Applying the ideal gas equation for H_2S

$$p_{H_2S} \times V = n_{H_2S} \times R \times T$$

$$3 \times 10 = n_{H_2S} \times 0{\cdot}0821 \times 298$$

$$n_{H_2S} = 1{\cdot}2262.$$

From the stoichiometry of the reaction given,
moles of NH_4SH to give 1·2262 moles of $H_2S = 1{\cdot}2262$.

$$\therefore \quad \text{wt. of } NH_4SH = 1{\cdot}2262 \times 51 = 62{\cdot}54 \text{ g.}$$

Ex. 57. *When 1-pentyne* (A) *is treated with* 4 N *alcoholic* KOH *at* 175°C, *it is converted slowly into an equilibrium mixture of* 1·3% *1-pentyne* (A), 95·2% *2-pontyne* (B) *and* 3.5% *of* 1, 2*-pentadiene* (C). *The equilibrium was maintained at* 175°C. *Calculate* ΔG^0 *for the following equilibria:*

$$B \rightleftharpoons A \qquad \Delta G_1^0 = ?$$

$$B \rightleftharpoons C \qquad \Delta G_2^0 = ?$$

From the calculated values of ΔG_1^0 *and* ΔG_2^0 *indicate the stability of* A, B *and* C.

Solution : For the equilibrium: $B \rightleftharpoons A$,

$$K = \frac{1{\cdot}3}{95{\cdot}2} = 0{\cdot}0136$$

$$\Delta G_1^0 = -2{\cdot}303\, RT \log K$$

$$= -2{\cdot}303 \times 8{\cdot}314 \times 448 \times \log 0{\cdot}0136$$

$$= +16010 \text{ J}$$

$$= +16{\cdot}01 \text{ kJ.}$$

For the equilibrium: B $\rightleftharpoons$ C,

$$K = \frac{3{\cdot}5}{95{\cdot}2} = 0{\cdot}0367$$

$$\Delta G_2^0 = -2{\cdot}303 \times 8{\cdot}314 \times 448 \times \log 0{\cdot}0367$$
$$= +12312 \text{ J}$$
$$= +12{\cdot}312 \text{ kJ}.$$

Further, negative value of ΔG^0 favours the forward reaction

∴ for, B $\rightleftharpoons$ A; $\Delta G^0 = +16{\cdot}01$ kJ; stability order is: B > A.

And for, B $\rightleftharpoons$ C; $\Delta G^0 = +12{\cdot}312$ kJ; stability order is: B > C.

To determine the order of stability between A and C,

A $\rightleftharpoons$ B; $\Delta G^0 = -16{\cdot}01$ kJ

B $\rightleftharpoons$ C; $\Delta G^0 = +12{\cdot}312$ kJ

Add: A $\rightleftharpoons$ C; $\Delta G^0 = -3{\cdot}698$ kJ; stability order is: C > A.

Thus, the decreasing order of stability is B > C > A.

Ex. 58. *The value of* K_p *is* 1×10^{-3} atm^{-1} *at* 25°C *for the reaction:* $2NO + Cl_2 \rightleftharpoons 2NOCl$. *A flask contains* NO *at* 0·02 atm *and at* 25°C. *Calculate the moles of* CO_2 *that must be added if* 1% *of* NO *is to be converted to* NOCl *at equilibrium. The volume of the flask is such that* 0·2 *mole of gas produces* 1 atm *pressure at* 25°C. *(Ignore probable association of* NO *to* N_2O_2.)

Solution **:** If 0·2 mole of a gas produces 1 atm pressure at 25°C then *n* moles (say) of Cl_2 under similar conditions shall produce 5*n* atm pressure.

	2NO	+	Cl_2	$\rightleftharpoons$	2NOCl
Initial pressure:	0·02 atm		5*n* atm		0
Eqb. pressure:	(0·02 − 0·0002)		(5*n* − 0·0001)		0·0002

$$K_p = \frac{p_{NOCl}^2}{p_{NO}^2 \cdot p_{Cl_2}} = 1 \times 10^{-3}$$

or
$$\frac{(0{\cdot}0002)^2}{(0{\cdot}0198)^2(5n - 0{\cdot}0001)} = 1 \times 10^{-3}$$

$$n = 0{\cdot}0204 \text{ mole of } Cl_2.$$

Ex. 59. *Calculate K for the following reaction at* 300 K *and constant pressure.*

$$A(g) + B(g) \rightleftharpoons C(g) + D(g) + E(g)$$

using the ideal gas law and the following data:

$$\Delta E^0 = -90{\cdot}0 \text{ kcal}, \; \Delta S^0 = 100 \text{ cal/K}$$

Solution : $\Delta H^0 = \Delta E^0 + \Delta n_g RT$

$$= (-90{,}000) + (-1) \times 2 \times 300$$
$$= -89400 \text{ cal.}$$
$$\Delta G^0 = \Delta H^0 - T\Delta S^0$$
$$= (-89400) - 300 \times 100$$
$$= -119400 \text{ cal.}$$

And then,

$$\Delta G^0 = -2{\cdot}303\, RT \log K$$
$$-119400 = -2{\cdot}303 \times 2 \times 300 \times \log K$$
$$\log K = 86{\cdot}4090$$
$$K = 2{\cdot}564 \times 10^{86}.$$

Ex. 60. *Assuming that the heat capacities of* $H_2(g)$, $N_2(g)$ *and* $NH_3(g)$ *do not vary with temperature, and further assuming that* ΔS *is independent of temperature for the reaction*

$$3H_2(g) + N_2(g) \rightleftharpoons 2NH_3(g)$$

estimate the minimum temperature at which this reaction will occur spontaneously with all reactants at unit concentration. Given: $\Delta H = -92{\cdot}38$ kJ *and* $\Delta S = -197{\cdot}7$ J/K.

Solution : At the minimum temperature at which the reaction occurs spontaneously, $\Delta G = \Delta H - T\Delta S = 0$. (Any higher temperature would make ΔG negative; any lower temperature would make ΔG positive.)

$$\therefore \quad \Delta H = T\Delta S$$
$$T = \frac{\Delta H}{\Delta S} = \frac{(-92{\cdot}38)}{(-0{\cdot}1977)} = 467{\cdot}27 \text{ K.}$$

Ex. 61. *The value of K for the reaction*

$$O_3(g) + OH(g) \rightleftharpoons H(g) + 2O_2(g)$$

changed from 0·096 *at* 298 K *to* 1·4 *at* 373 K. *Above what temperature will the reaction become thermodynamically spontaneous in the forward direction assuming that* ΔH^0 *and* ΔS^0 *values for the reaction do not change with change in temperature? Given that,* $\Delta S^0_{298} = 102{\cdot}96\ JK^{-1}$.

Solution : We have,

$$\log \frac{K_2}{K_1} = \frac{\Delta H^0}{2{\cdot}303\, R} \; \frac{(T_2 - T_1)}{T_1 T_2}$$
$$\log \frac{1{\cdot}4}{0{\cdot}096} = \frac{\Delta H^0}{2{\cdot}303 \times 8{\cdot}314} \left(\frac{373 - 298}{373 \times 298} \right)$$
$$\Delta H^0 = 33025 \text{ J.}$$

Now the temperature above which the forward reaction will be spontaneous is actually the temperature at which the reaction attains equilibrium, that is, when $K = 1$ or $\log K = 0$.

$$\therefore \quad \Delta G^0 = -2{\cdot}303\, RT \log K$$
$$= -2{\cdot}303\, RT \log 1{\cdot}0$$
$$= 0.$$

From thermodynamics, we have

$$\Delta G^0 = \Delta H^0 - T\Delta S^0$$
$$0 = 33025 - T \times 102{\cdot}96$$
$$\text{or} \quad T = 320{\cdot}75 \text{ K}.$$

PROBLEMS

(Answers bracketed with questions)

1. K_c for the equilibrium $N_2O_4\,(g) \rightleftharpoons 2NO_2\,(g)$ at 298° C is $5{\cdot}7 \times 10^{-9}$. Which species has a higher concentration at equilibrium? (N_2O_4)

2. The equilibrium constant of an equilibrium represented by A (g) $\rightleftharpoons$ B (g) is 1·1. Which of the two gases A and B has a molar concentration greater than 1 at equilibrium? (B)

3. At a particular temperature, the equilibrium constant for

$$2NO_2\,(g) \rightleftharpoons N_2O_4\,(g)$$

is one. The same reaction is carried out in a container of volume just half of the former. Will the value of $[N_2O_4]/[NO_2]^2$ be equal to 1 if no reaction occurred? Will the equilibrium constant change by this change in volume? (No. No)

4. Would you expect the equilibrium constant for the reaction $I_2\,(g) \rightleftharpoons 2I\,(g)$ to increase or decrease as temperature increases? Why?

[**Hint:** The forward reaction is endothermic as energy is required to break I_2 into I.]

5. For the reaction $NOBr\,(g) \rightleftharpoons NO\,(g) + \frac{1}{2} Br_2\,(g)$; $K_p = 0{\cdot}15$ atm at 90°C. If 0·50 atm of NOBr, 0·40 atm of NO, and 0·20 atm of Br_2 are mixed at this temperature, will Br_2 be consumed or formed?

[**Hint:** Use eqns. 8, 9 and 10.] (Consumed)

6. The ammonia in equilibrium with a 1 : 3 N_2–H_2 mixture at 20 atm and 427°C amounts to 16%. Calculate K_p and K_c for

$$\frac{1}{2} N_2 + \frac{3}{2} H_2 \rightleftharpoons NH_3 \qquad (3{\cdot}49 \times 10^{-2};\ 2)$$

7. State whether the following statement is false or true. If the equilibrium constant for $A_2 + B_2 \rightleftharpoons 2AB$ is K, the equilibrium constant for $AB \rightleftharpoons \frac{1}{2} A_2 + \frac{1}{2} B_2$ is $1/K$.

(False)

8. One mole of N_2 and 3 moles of PCl_5 are placed in a 100-litre vessel heated to 227°C. The equilibrium pressure is 2·05 atm. Assuming ideal behaviour, calculate the degree of dissociation for PCl_5 and K_p for the reaction:

$$PCl_5\,(g) \rightleftharpoons PCl_3\,(g) + Cl_2\,(g) \qquad (0{\cdot}33;\ 0{\cdot}205\ \text{atm})$$

[Hint: While calculating partial pressure, include moles of N_2 in the total moles]

9. Twenty grams of HI is heated at 327°C in a bulb of 1-litre capacity. Calculate the volume percentage of H_2, I_2 and HI at equilibrium. Given that the mass law constant for the equation $2HI \rightleftharpoons H_2 + I_2$ is 0·0559 at 327°C when concentrations are expressed in moles/litre. (HI = 67·9%, $H_2 = I_2 = 16{\cdot}05\%$)

10. 25 mL of H_2 and 18 mL of I_2 vapours were heated in a sealed tube at 456°C, when at equilibrium 30·8 mL of HI was formed. Calculate the degree of dissociation of pure HI at 456°C. (24·5%)

11. Bodenstein found that at 443°C the dissociation of HI according to the equation $2\,HI \rightleftharpoons H_2 + I_2$ was 21·98%. For experiments at the same temperature, starting with varying amounts of H_2 and I_2, the amount of HI present when equilibrium was attained is given by the following numbers:

I_2 (mL) (vap.) initial	H_2 (mL) (vap.) initial	HI (mL) (vap.) at equib.
2·94	8·10	5·65
9·27	8·07	13·46
33·10	7·89	15·41

Verify whether the amounts of HI obtained are according to the law of mass action.

12. HI is introduced into three identical 500-mL bulbs at 350°C. Each bulb is opened at different time intervals and analysed for I_2 by titrating with 0·015 M hypo solution.

$$(I_2 + 2Na_2S_2O_3 \rightarrow Na_2S_4O_6 + 2NaI)$$

Bulb number	Initial mass of HI	Time of opening bulb	Vol. of hypo required
1	0·3 g	2	20·96 mL
2	0·406 g	20	41·50 mL
3	0·28 g	40	28·68 mL

Calculate K_C for $2HI \rightleftharpoons H_2 + I_2$ at 350°C. $(1{\cdot}49 \times 10^{-2})$

[Hint: Moles of $I_2 = \frac{1}{2}$(moles of $Na_2S_2O_3$)

$$= \frac{1}{2}(\text{molarity} \times \text{vol. in L})]$$

13. Five grams of PCl_5 (molecular wt. 208·5) was completely vapourised to an equilibrium state at 250°C in a vessel of 1·9-litre capacity. The equilibrium mixture exerted a pressure of 1 atm. Calculate degree of dissociation, K_c and K_p for the reaction. (84·6%, 0·0587, 2·53)

14. Ammonia under a pressure of 15 atm at 27°C is heated to 347°C in a closed vessel in the presence of a catalyst. Under these conditions NH_3 is partially decomposed according to the equation

$$2NH_3 \rightleftharpoons N_2 + 3H_2.$$

The vessel is such that the volume remains effectively constant, whereas pressure increases to 50 atm. Calculate the percentage of NH_3 actually decomposed. Pressure of NH_3 at 27°C or 300 K = 15 atm. (61·3%)

15. In a gaseous reaction of the type $A + 2B \rightleftharpoons 2C + D$, the initial concentration of B was 1·5 times that of A. At equilibrium the equilibrium concentrations of A and D were equal. Calculate the equilibrium constant. (4)

16. A mixture of SO_2 and O_2 at 1 atm in the mole ratio of 2 : 1 is passed through a catalyst at 1170°C at a rate sufficient for attainment of equilibrium. The existing gas, suddenly chilled and analysed, is found to contain 87% SO_3 by volume. Calculate K_p for the reaction: $SO_2 + \frac{1}{2}O_2 \rightleftharpoons SO_3$ (57·76)

17. For the reaction:

$$2Fe^{3+}(aq) + Hg_2^{2+}(aq) \rightleftharpoons 2Fe^{2+}(aq) + 2Hg^{2+}(aq)$$

$K_c = 9{\cdot}14 \times 10^{-6}$ at 25°C. If the initial concentrations of the ions are: $[Fe^{3+}] = 0{\cdot}5$ M, $[Hg_2^{2+}] = 0{\cdot}5$ M, $[Fe^{2+}] = 0{\cdot}03$ M and $[Hg^{2+}] = 0{\cdot}03$ M, what will be the ionic concentrations at equilibrium?

$$\left[\begin{array}{l} [Fe^{3+}] = 0{\cdot}497\text{ M},\ [Hg_2^{2+}] = 0{\cdot}499\text{ M} \\ [Fe^{2+}] = 3{\cdot}27 \times 10^{-2}\text{ M},\ [Hg^{2+}] = 3{\cdot}27 \times 10^{-2}\text{ M} \end{array}\right]$$

[**Hint:** Calculate reaction quotient and compare with K_c to establish the direction *of the reaction*.]

18. In a 1-litre vessel at 1000 K are introduced 0·1 mole each of NO and Br_2 and 0·01 mole of NOBr

$$2NO(g) + Br_2(g) \rightleftharpoons 2NOBr(g); \qquad K_c = 1{\cdot}32 \times 10^{-2}$$

Determine the direction of the net reaction and calculate the partial pressure of NOBr in the vessel at equilibrium. (Reverse, 0·30 atm)

19. An air sample containing 21 : 79 of O_2 and N_2 (mole ratio) is heated to 2400°C. If the mole per cent of NO at equilibrium is 1·8%, calculate K_p for the reaction $N_2 + O_2 \rightleftharpoons 2NO$. $(2{\cdot}1 \times 10^{-3})$

20. In phosgene gas reaction at 400°C, the initial pressures are $p_{CO} = 342$ mm and $p_{Cl_2} = 352$ mm and the total pressure at equilibrium is 440 mm.

$$CO + Cl_2 \rightleftharpoons COCl_2$$

Calculate percentage dissociation of phosgene at 400°C at 1 atm. (20·6%)

21. If the heat of reaction at constant volume exceeds that at constant pressure for an endothermic reaction at 25°C by 1190 cal, what is the value of the ratio $\frac{K_p}{K_c}$?

[**Hint:** Apply $q_p = q_v + \Delta nRT$ and $K_p = K_c \cdot (RT)^{\Delta n}$

We get, $\frac{K_p}{K_c} = (RT)^{-2} = (0{\cdot}082 \times 298)^{-2}$ in atm and litre units]

22. At 400°C for the gas-phase reaction:

$$2H_2O\,(g) + 2Cl_2\,(g) \rightleftharpoons 4HCl\,(g) + O_2\,(g)$$

the K_p is 0·035 when partial pressures are measured in atmospheric units. Calculate K_c value for it, concentration being measured in mole per litre units. State the unit. ($6{\cdot}35 \times 10^{-4}$ mole per litre)

23. One mole each of acetic acid and ethyl alcohol are mixed at 25°C. When the mixture attains equilibrium it is found that 12 g of water is formed. Find the value of K_c. What weight of ethyl acetate will be formed when two moles of ethyl alcohol are further added and the equilibrium is attained? (4, 79·2 g)

24. The equilibrium constant of the ester formation of propionic acid with ethyl alcohol is 7·36 at 50°C. Calculate the weight of ethyl propionate, in grams, existing in an equilibrium mixture when 0·5 mole of propionic acid is heated with 0·5 mole of ethyl alcohol at 50°C. (37·29 g)

25. 0·1 mole of each of ethyl alcohol and acetic acid are allowed to react and at equilibrium the acid was exactly neutralised by 100 mL of 0·85 N NaOH. If no hydrolysis of ester is supposed to have undergone, find the equilibrium constant. (0·031)

26. The K_p value for the equilibrium $H_2\,(g) + I_2\,(s) \rightleftharpoons 2HI\,(g)$ is 871 at 25°C. If the vapour pressure of iodine is 4×10^{-4} atm, calculate the equilibrium constant in terms of partial pressures at the same temperature for the reaction; $H_2\,(g) + I_2\,(s) \rightleftharpoons 2HI\,(g)$. (0·3484 atm)

27. In the reaction $CuSO_4 \cdot 3H_2O \rightleftharpoons CuSO_4 \cdot H_2O + 2H_2O$ (vap.), the dissociation pressure is 7×10^{-3} atm at 25°C and $\Delta H^0 = 2700$ cal. What will be the dissociation pressure at 127°C? ($1{\cdot}247 \times 10^{-2}$ atm)

28. Under what pressure conditions will $CuSO_4 \cdot 5H_2O$ be efflorescent at 25°C? How good a drying agent is $CuSO_4 \cdot 3H_2O$ at the same temperature? For the reaction

$$CuSO_4 \cdot 5H_2O\,(s) \rightleftharpoons CuSO_4 \cdot 3H_2O\,(s) + 2H_2O\,(g)$$

$K_p = 1{\cdot}086 \times 10^{-4}$ atm^2 at 25°C. Vapour pressure of water at 25°C is 23·8 mm (Hg).

[Efflorescence occurs when partial press. of H_2O vap. in air is less than 7·92 mm. $CuSO_4 \cdot 3H_2O$ can reduce moisture when partial press. of H_2O vap. in air is 7·92 mm.]

[Hint: An efflorescent salt is one that loses water to the atmosphere.

$$K_p = p_{H_2O}^2 = 1.086 \times 10^{-4};$$

$$p_{H_2O} = 1.042 \times 10^{-2} \text{ atm} = 7.92 \text{ mm.}]$$

29. Vapour density of N_2O_4 which dissociated according to the equation $N_2O_4\,(g) \rightleftharpoons 2NO_2\,(g)$ is 25·67 at 100°C and a pressure of 1 atm. Calculate the degree of dissociation and K_p for the reaction. (0·792, 6·7)

30. Equilibrium constant (K_p) for the reaction

$$2H_2S\,(g) \rightleftharpoons 2H_2\,(g) + S_2\,(g)$$

is 0·0118 at 1065°C and heat of dissociation is 42·4 kcal. Find equilibrium constant at 1132°C. (0·025)

31. In the gaseous reaction $2A + B \rightleftharpoons A_2B$, $\Delta G^0 = 1200$ cal at 227°C. What total pressure would be necessary to produce 60% conversion of B into A_2B when 2 : 1 mixture is used?

[Hint: Use Equation 12] (5·02 atm)

32. Equilibrium constants (K_p) for the reaction $\frac{3}{2}H_2\,(g) + \frac{1}{2}N_2\,(g) \rightleftharpoons NH_3\,(g)$ are 0·0266 and 0·0129 at 350°C and 400°C respectively. Calculate the heat of formation of gaseous ammonia. (12140 cal)

33. Equilibrium constant K_c for the equilibrium

$$A\,(g) \rightleftharpoons B\,(g) + C\,(g) \text{ is } 0.45 \text{ at } 200°C.$$

One litre of a container contains 0·2 mole of A, 0·3 mole of B and 0·3 mole of C at equilibrium. Calculate the new equilibrium concentrations of A, B and C if the volume of the container is (a) doubled (b) halved at 200°C.

[(a) [A] = 0·07 M, [B] = [C] = 0·18 M
(b) [A] = 0·52 M, [B] = [C] = 0·48 M]

34. A 2-litre vessel contains 0·48 mole of CO_2, 0·48 mole of H_2, 0·96 mole of H_2O and 0·96 mole of CO at equilibrium.

$$\{CO_2\,(g) + H_2\,(g) \rightleftharpoons H_2O\,(g) + CO\,(g)\}$$

(a) How many moles and how many grams of H_2 must be added to bring the concentration of CO to 0·60 M?

(b) How many moles and how many grams of CO_2 must be added to bring the CO concentration to 0·60 M?

(c) How many moles of H_2O must be removed to bring the CO concentration to 0·60 M?

[(a) H_2 – 1·2 mole, 2·4 g
(b) CO_2 – 1·2 mole, 53 g
(c) H_2O – 1 mole]

35. For the reaction $F_2 \rightleftharpoons 2F$, calculate the degree of dissociation and density of fluorine at 4 atm and 1000 K, when $K_p = 1.4 \times 10^{-2}$ atm.

If $K_p\,(760°) = 2 \times 10^{-5}$ atm and $K_p\,(960°) = 4 \times 10^{-3}$ atm, calculate ΔH^0 for the dissociation of fluorine. (0·03, 18·46)

36. From the following data at 1000 K

$$COCl_2\,(g) \rightleftharpoons CO\,(g) + Cl_2\,(g); \qquad K_1 = 0{\cdot}329$$

$$2CO\,(g) + O_2\,(g) \rightleftharpoons 2CO_2\,(g); \qquad K_2 = 2{\cdot}24 \times 10^{22}$$

Calculate equilibrium constant at 1000 K for

$$2COCl_2\,(g) + O_2\,(g) \rightleftharpoons 2CO_2\,(g) + 2Cl_2\,(g)$$ (2·43 × 10^{21})

37. For the equilibrium $COCl_2\,(g) \rightleftharpoons CO\,(g) + Cl_2\,(g)$, $K_{1000\,K}$ is 0·329. Suppose that x moles of $COCl_2$ is allowed to reach equilibrium in a one-litre container. What will be the value of x that must be used in order that half the chlorine atoms remain as $COCl_2$? (0·658 mole)

38. N_2O_4 is 25% dissociated at 37 °C and 1 atm pressure. Calculate (i) K_p, and (ii) the percentage dissociation at 0·1 atm and 37 °C.
[**Hint:** See Example 14] **(IIT 1988)** (0·267, 63·27%)

39. Sulphide ion in alkaline solution reacts with solid sulphur to form polysulphide ions having formulae S_2^{2-}, S_3^{2-}, S_4^{2-}, and so on. The equilibrium constant for the formation of S_2^{2-} is 12 and for the formation of S_3^{2-} is 130, both from S and S^{2-}. Find the equilibrium constant for the formation of S_3^{2-} from S_2^{2-} and S. (10·83)

[**Hint:** $S\,(s) + S^{2-} = S_2^{2-}; \quad K_1 = [S_2^{2-}]/[S^{2-}]$

$2S\,(s) + S^{2-} = S_3^{2-}; \quad K_2 = [S_3^{2-}]/[S^{2-}]$

$S\,(s) + S_2^{2-} = S_3^{2-}; \quad K_3 = [S_3^{2-}]/[S_2^{2-}] = \dfrac{K_2}{K_1}$]

40. The equilibrium constant K_p at 80°C is 1·57 for the reaction,

$$PH_3BCl_3(s) \rightleftharpoons PH_3(g) + BCl_3(g)$$

(a) Calculate the equilibrium pressures of $PH_3(g)$ and $BCl_3(g)$ if a sample of PH_3BCl_3 is placed in a closed vessel at 80°C and allowed to decompose until equilibrium is attained.
(b) What is the minimum amount of PH_3BCl_3 that must be placed in a 0·5-litre vessel at 80°C if equilibrium is to be attained? (1·253 atm; 4·568 g)

41. A mixture of 3·0 moles of SO_2, 4·0 moles NO_2, 1·0 mole of SO_3 and 4·0 moles of NO is placed in a 2·0-litre vessel.

$$SO_2(g) + NO_2(g) \rightleftharpoons SO_3(g) + NO(g)$$

At equilibrium, the vessel is found to contain 1·0 mole of SO_2.
(a) Calculate the equilibrium concentrations of SO_2, NO_2, SO_3 and NO.
(b) Calculate the value of K_c. (0·5 M, 1·0 M, 1·5 M, 3·0 M; 9·0)

42. When 20·0 g of $CaCO_3$ in a 10·0-litre flask is heated to 800°C, 35% of it did not dissociate, calculate K_p for the equilibrium

$$CaCO_3(s) \rightleftharpoons CaO(s) + CO_2(g)$$ (1·145 atm)

43. At 298 K, 550 g of D_2O (20 g/mole, density 1·10 g/mL) and 498·5 g of H_2O (18 g/mole, density 0·997 g/mL) are mixed. The volumes are additive. 47·0% of the H_2O reacts to form HDO. Calculate K_c at 298 K for the reaction

$$H_2O + D_2O \rightleftharpoons 2HDO$$

(3·18)

44. At 300 K, the equilibrium constant for the reaction

$$2SO_2(g) + O_2(g) \rightleftharpoons 2SO_3(g)$$

is $6{\cdot}98 \times 10^{24}$. Calculate ΔG^0 of the reaction and $\Delta G_f^0(SO_3)$. Given that $\Delta G_f^0(SO_2) = -300{\cdot}12$ kJ/mole. (142·7 kJ/mole, 371·5 kJ/mole)

45. What kind of equilibrium constant can be calculated from a ΔG^0 value for a reaction involving only gases? (K_p)

46. Calculate K_p for the reaction

$$2SO_2(g) + O_2(g) \rightleftharpoons 2SO_3(g)$$

if at a particular temperature and a total pressure of 112·0 atm, the equilibrium mixture consists of 56·6 mole per cent SO_2, 10·6 mole per cent O_2, and 32·8 mole per cent SO_3. (0·0283)

47. The standard Gibbs free energy change for the reaction:

$$2AB \rightleftharpoons A_2 + B_2$$

is 11·8 kJ at 230 K and 1 atm. Calculate the degree of dissociation of HI at 230 K. (0·084)

48. Determine the equilibrium concentrations that result from the reaction of a mixture of 0·10 mole of H_2 and 0·050 mole of F_2 in a 1·0-litre flask according to the equation

$H_2(g) + F_2(g) \rightleftharpoons 2HF(g)$; K = 115 $([H_2] = 0{\cdot}052, [F_2] = 0{\cdot}002, [HF] = 0{\cdot}097$ M)

49. A stream of gas containing H_2 at an initial partial pressure of 0·20 atm is passed through a tube in which CuO is kept at 500 K. The reaction

$$CuO(s) + H_2(g) \rightleftharpoons Cu(s) + H_2O(g)$$

comes to equilibrium. For this reaction, $K_p = 1{\cdot}6 \times 10^9$. What is the partial pressure of H_2 in the gas leaving the tube? Assume that the total pressure of the stream is unchanged. (negligibly small)

50. At 973 K, K_p is 1·50 for the reaction

$$C(s) + CO_2(g) \rightleftharpoons 2CO(g)$$

Suppose the total gas pressure at equilibrium is 1·0 atm. What are the partial pressures of CO and CO_2? (CO–0·686 atm, CO_2–0·314 atm)

51. A flask contains $NH_4Cl(s)$ in equilibrium with its decomposition products.

$$NH_4Cl(s) \rightleftharpoons NH_3(g) + HCl(g)$$

For this reaction, $\Delta H = 176$ kJ/mole. How is the mass of NH_3 in the flask affected by each of the following disturbances?

(a) The temperature is decreased

(b) NH_3 is added

(c) HCl is added

(d) NH_4Cl is added with no appreciable change in volume

(e) A large amount of NH_4Cl is added decreasing the volume available to the gases. [(a) Increase (b) Increase (c) Decrease (d) No effect (e) Decrease]

52. Chlorine molecules are 1·0% dissociated at 975 K at a pressure of 1·0 atm (1·0% of the pressure is due to Cl atoms).

$$Cl_2(g) \rightleftharpoons 2Cl$$

Calculate K_p and K_c. $(1{\cdot}01 \times 10^{-4}, 1{\cdot}26 \times 10^{-6})$

53. Show that: $2{\cdot}303 \log K = -\dfrac{\Delta H^0}{RT} + \dfrac{\Delta S^0}{R}$

[Hint: Use equations 12 and 22 (Ch. 14)]

54. Calculate the equilibrium ratio of C to A if 2·0 moles each of A and B were allowed to come to equilibrium at 300 K

$$A + B \rightleftharpoons C + D;\ \Delta G^0 = 460 \text{ cal}$$ (0·679)

55. Calculate ΔE^0 for the reaction:

$$2A(g) + B(g) \rightleftharpoons A_2B(g)$$

for which $\Delta S^0 = 5{\cdot}0$ J/K, $K = 1{\cdot}0 \times 10^{-10}$ and $T = 300$ K

[Hint: Apply $\Delta G^0 = -2{\cdot}303\, RT \log K$, $\Delta H^0 = \Delta G^0 + T\Delta S^0$ and then $\Delta U^0 = \Delta H^0 - \Delta n_g\, RT$] (63·8 kJ)

56. The surface of copper gets tarnished by the formation of copper oxide. N_2 gas was passed to prevent the oxide formation during heating of copper at 1250 K. However, the N_2 gas contains 1 mole per cent of water vapour as impurity. The water vapour oxidises copper as par the reaction given below:

$$2Cu(s) + H_2O(g) \rightarrow Cu_2O(s) + H_2(g)$$

p_{H_2} is the minimum partical pressure of H_2 (1 bar) needed to prevent the oxidation at 1250 K. The value of $\ln(p_{H_2})$ is ... .

[Given: Total pressure = 1 bar,

R (universal gas constant) = 8 J K^{-1} mol^{-1}, ln (10) = 2.3, Cu(s) and $Cu_2O(s)$ are mutually immiscible.

At 1250 K : $2Cu(s) + \frac{1}{2}O_2(g) \rightarrow Cu_2O(s)$, $\Delta G° = -78{,}000$ J mol^{-1}

$$H_2(g) + \frac{1}{2}O_2(g) \rightarrow H_2O(g);\ \Delta G° = -1{,}78{,}000 \text{ J mol}^{-1}]$$

(IIT 2018 Adv.) (–14.6)

[Hint: From the given equations,

$2Cu(s) + H_2O(g) \rightarrow Cu_2O(s) + H_2(g)$, $\Delta G° = +178000 - 78000s = 100000$.

Now, $\Delta G° = -RT \ln \dfrac{p_{H_2}}{p_{H_2O}}$

$100000 = -8 \times 1250 (\ln p_{H_2} - \ln 0.01)$

Calculate $\ln p_{H_2}$ (ln 0.01 = 2.3 log 0.01)].

57. Consider the following reversible reaction:

$$A(g) + B(g) \rightleftharpoons AB(g).$$

The activation energy of the backward reaction exceeds that of the forward reaction by 2 RT(in J mol^{-1}). If the pre-exponential factor of the forward reaction is four

times that of the reverse reaction, the absolute value of ΔG° (in J mol^{-1}) for the reaction at 300 K is ... .
(Given: ln 2 = 0.7, $RT = 2500$ J mol^{-1} at 300 K and G is the Gibbs energy)

(IIT 2018 Adv.) (–8500 J mol^{-1})

[Hint: $K = \frac{k_f}{k_b} = \frac{A_f \cdot e^{-E_f/RT}}{A_b \cdot e^{-E_b/RT}}$.

Substitute $\frac{A_f}{A_b} = 4$ and $E_b - E_f = 2RT$, calculate K, then apply $\Delta G^\circ = -RT \ln K$]

Objective Problems

1. 120 g of urea is dissolved in 5 litres. The active mass of urea is
(a) 0·08 (b) 0·4 (c) 120/5 (d) 5/120

2. For the equilibrium $A + B \rightleftharpoons C + D$, equilibrium constant may be equal to
(a) $\frac{[C] \cdot [A]}{[D] \cdot [B]}$ (b) $\frac{[A] \cdot [B]}{[C] \cdot [D]}$ (c) $\frac{[C] + [D]}{[A] + [B]}$ (d) all wrong

3. For which of the following reactions, K_p may be equal to 0·5 atm?
(a) $2HI \rightleftharpoons H_2 + I_2$ (b) $PCl_5 \rightleftharpoons PCl_3 + Cl_2$
(c) $N_2 + 3H_2 \rightleftharpoons 2NH_3$ (d) $2NO_2 \rightleftharpoons N_2O_4$

4. Which of the following is correct?
(a) K_p will always have some unit
(b) K_c will always have some unit
(c) K_x will never have any unit
(d) When $\Delta n = 0$, $K_p = K_c = K_x$ then all the three Ks have the same unit

5. K_p/K_c for the reaction $CO + \frac{1}{2}O_2 \rightleftharpoons CO_2$ is
(a) RT (b) $1/\sqrt{RT}$ (c) $\sqrt{RT}$ (d) 1

6. If the initial number of moles/L of N_2, H_2 and NH_3 are 1, 2 and 3 respectively, their concentrations at equilibrium will be

	N_2 +	$3H_2$ $\rightleftharpoons$	$2NH_3$
(a)	$(1-x)$	$(2-3x)$	$2x$
(b)	$(1-x/3)$	$(2-x)$	$2x/3$
(c)	$(1-x)$	$(2-x)$	$(3+x)$
(d)	$(1-x)$	$(2-3x)$	$(3+2x)$

7. A 1-litre container contains 2 moles of PCl_5 initially. If at equilibrium, K_c is found to be 1, the degree of dissociation of PCl_5 is
(a) 1 (b) –1 (c) $\frac{1}{2}$ (d) 50

8. The vapour density of undecomposed N_2O_4 is 46. When heated, the vapour density decreases to 24·5 due to its dissociation to NO_2. The percentage dissociation of N_2O_4 at the final temperature is
(a) 87 (b) 60 (c) 40 (d) 70

9. If the equilibrium constants of the following equilibrium

$$SO_2 + 1/2O_2 \rightleftharpoons SO_3 \quad \text{and} \quad 2SO_3 \rightleftharpoons 2SO_2 + O_2$$

are given by K_1 and K_2 respectively, which of the following relations is correct?

(a) $K_2 = \left(\frac{1}{K_1}\right)^2$ (b) $K_1 = \left(\frac{1}{K_2}\right)^3$ (c) $K_2 = \frac{1}{K_1}$ (d) $K_2 = (K_1)^2$

10. For the reactions

$$A \rightleftharpoons B \qquad K_c = 2$$
$$B \rightleftharpoons C \qquad K_c = 4$$
$$C \rightleftharpoons D \qquad K_c = 6$$

K_c for the reaction $A \rightleftharpoons D$ is

(a) $(2+4+6)$ (b) $\frac{2\times 4}{6}$ (c) $\frac{4\times 6}{2}$ (d) $2\times 4\times 6$

11. For the equilibrium $CaCO_3(s) \rightleftharpoons CaO(s) + CO_2(g)$, which of the following expressions is correct?

(a) $K_p = \frac{[CaO].[CO_2]}{[CaCO_3]}$ (b) $K_p = \frac{p_{CaO}\times p_{CO_2}}{p_{CaCO_3}}$

(c) $K_p = p_{CO_2}$ (d) $K_p = \frac{p_{CaO}+p_{CO_2}}{p_{CaCO_3}}$

12. For the reaction $C(s) + CO_2(g) \rightleftharpoons 2CO(g)$, the partial pressures of CO_2 and CO are respectively 4 and 8 atm. K_p for the reaction is

(a) 16 (b) 2 (c) 0·5 (d) 4

13. K_c for $A + B \rightleftharpoons 3C$ is 20 at 25°C. If a 2-litre vessel contains 1, 2 and 4 moles of A, B and C respectively, the reaction at 25°C shall

(a) proceed from left to right (b) proceed from right to left
(c) be at equilibrium (d) not occur

14. K_c for $A + B \rightleftharpoons C + D$ is 10 at 25°C. If a container contains 1, 2, 3 and 4 moles of A, B, C and D respectively at 25°C, the reaction shall

(a) proceed from left to right (b) proceed right to left
(c) be at equilibrium (d) not occur

15. A 1-litre vessel contains 2 moles each of gases A, B, C and D at equilibrium. If 1 mole each of A and B are removed, K_c for $A + B \rightleftharpoons C + D$ will be

(a) 4 (b) 1 (c) 1/4 (d) 2

16. Two samples of HI each of 5 g were taken separately in two vessels of volume 5 and 10 litres respectively at 27°C. The extent of dissociation of HI will be

(a) more in the 5-litre vessel (b) more in the 10-litre vessel
(c) equal in both vessels (d) nil at both

17. For a reversible reaction if the concentrations of the reactants are doubled, the equilibrium constant will be

(a) doubled (b) halved
(c) one-fourth (d) the same

18. If one-third of HI decomposes at a particular temperature, K_c for $2HI \rightleftharpoons H_2 + I_2$ is
(a) 1/16 (b) 1/4 (c) 1/6 (d) 1/2

19. 28 g of N_2 and 6 g of H_2 were mixed. At equilibrium 17 g NH_3 was produced. The weights of N_2 and H_2 at equilibrium are respectively
(a) 11 g, 0 g (b) 1 g, 3 g
(c) 14 g, 3 g (d) 11 g, 3 g

20. For the reaction: $H_2(g) + I_2(g) \rightleftharpoons 2HI(g)$, the equilibrium constant K_p changes with
(a) total pressure (b) catalyst
(c) the amount of H_2 and I_2 present (d) temperature

21. The oxidation of SO_2 by O_2 to SO_3 is an exothermic reaction. The yield of SO_3 will be maximum if
(a) temperature is increased and pressure is kept constant
(b) temperature is reduced and pressure is increased
(c) both temperature and pressure are increased
(d) both temperature and pressure are reduced

22. For the gas phase reaction $C_2H_4 + H_2 \rightleftharpoons C_2H_6$ ($\Delta H = -32{\cdot}7$ kcal) carried out in a vessel, the equilibrium concentration of C_2H_4 can be increased by
(a) increasing the temperature (b) decreasing the pressure
(c) removing some H_2 (d) adding some C_2H_6

23. Pure ammonia is placed in a vessel at a temperature where its dissociation constant is appreciable. At equilibrium,
(a) K_p does not change significantly with pressure
(b) α does not change with pressure
(c) concentration of NH_3 does not change with pressure
(d) concentration of H_2 is less than that of N_2

24. An example of a reversible reaction is
(a) $Pb(NO_3)_2(aq) + 2NaI(aq) = PbI_2(s) + 2NaNO_3(aq)$
(b) $AgNO_3(aq) + HCl(aq) = AgCl(s) + HNO_3(aq)$
(c) $2Na(s) + H_2O(l) = 2NaOH(aq) + H_2(g)$
(d) $KNO_3(aq) + NaOH(aq) = KOH(aq) + NaNO_3(aq)$ **(IIT 1985)**

25. When $NaNO_3$ is heated in a closed vessel, O_2 is liberated and $NaNO_2$ is left behind. At equilibrium
(a) addition of $NaNO_2$ favours reverse reaction
(b) addition of $NaNO_3$ favours forward reaction
(c) increasing temperature favours forward reaction
(d) increasing pressure favours forward reaction **(IIT 1986)**

26. The equilibrium $SO_2Cl_2(g) \rightleftharpoons SO_2(g) + Cl_2(g)$ is attained at 25°C in a closed container and an inert gas, helium, is introduced. Which of the following statements are correct?
(a) Concentrations of SO_2, Cl_2 and SO_2Cl_2 change
(b) More Cl_2 is formed

(c) Concentration of SO_2 is reduced

(d) More SO_2Cl_2 is formed **(IIT 1989)**

27. The decomposition of N_2O_4 into NO_2 is carried out at 280 K in chloroform. When equilibrium has been established, 0·2 mole of N_2O_4 and 2×10^{-3} mole of NO_2 are present in 2 litres of solution. The equilibrium constant for the reaction $N_2O_4 \rightleftharpoons 2NO_2$ is

(a) 1×10^{-2} (b) 2×10^{-3}

(c) 1×10^{-5} (d) 2×10^{-5}

28. The equilibrium constants for the reaction $Br_2 \rightleftharpoons 2Br$ at 500 K and 700 K are 1×10^{-10} and 1×10^{-5} respectively. The reaction is

(a) endothermic (b) exothermic

(c) fast (d) slow

29. Which oxide of nitrogen is most stable at 273 K?

(a) $2N_2O_5 \rightleftharpoons 2N_2 + 5O_2$; $K_c = 1 \times 10^{34}$

(b) $2N_2O \rightleftharpoons 2N_2 + O_2$; $K_c = 1 \times 10^{32}$

(c) $2NO \rightleftharpoons N_2 + O_2$; $K_c = 1 \times 10^{30}$

(d) $2NO_2 \rightleftharpoons N_2 + 2O_2$; $K_c = 1 \times 10^{15}$

30. If a chemical reaction is at equilibrium, which of the following is not correct?

(a) $\Delta G^0 = 0$ (b) $K_p = 1$

(c) $K_c = 1$ (d) $\Delta G^0 = 1$

31. If pressure is applied to the equilibrium system of solid $\rightleftharpoons$ liquid, the melting point of the solid

(a) will not change

(b) may increase or decrease depending upon its nature

(c) will always increase

(d) will always decrease

32. If pressure is applied to the following equilibrium of liquid $\rightleftharpoons$ vapour, the boiling point of the liquid

(a) will decrease (b) will increase

(c) may increase or decrease (d) will not change

33. For the equilibrium $PCl_5 \rightleftharpoons PCl_3 + Cl_2$, $K_p = \dfrac{\alpha^2}{(1-\alpha)V}$; temperature remaining constant,

(a) K_p will increase with the increase in volume

(b) K_p will increase with the decrease in volume

(c) K_p will not change with the change in volume

(d) K_p may increase or decrease with the change in volume

34. When $CaCO_3$ is heated at a constant temperature in a closed container, the pressure due to CO_2 produced will

(a) change with the amount of $CaCO_3$ taken
(b) change with the size of the container
(c) remain constant so long as temperature is constant
(d) remain constant even if temperature is changed

35. For the reaction $PCl_5\,(g) \rightleftharpoons PCl_3\,(g) + Cl_2\,(g)$, the forward reaction at constant temperature is favoured by
(a) introducing an inert gas at constant volume
(b) introducing chlorine gas at constant volume
(c) introducing an inert gas at constant pressure
(d) increasing the volume of the container
(e) introducing PCl_5 at constant volume

36. Which of the following curves represents a very rare standard reaction at equilibrium?

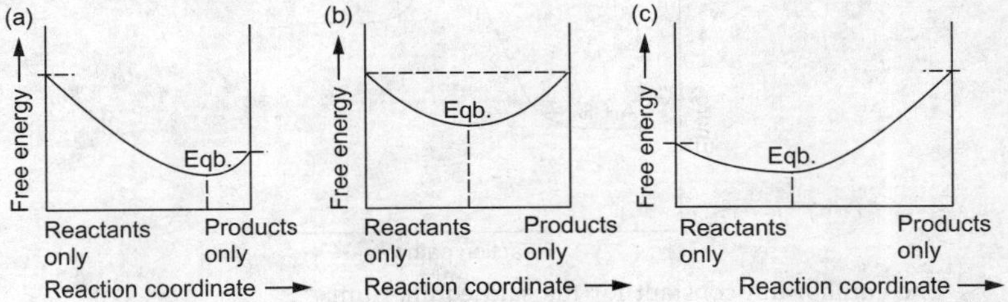

37. Which of the curves given in Q. 36 represents a standard reaction spontaneous in forward direction?

38. Which of the curves given in Q. 36 represents a standard reaction for which $\Delta G^0 > 0$?

39. Which of the curves given in Q. 36 represents a standard reaction with $K = 1$?

40. Each of the mixtures listed below was placed in a closed container and allowed to stand. Which of the following mixtures is not capable of attaining the equilibrium: $CaCO_3(s) \rightleftharpoons CaO(s) + CO_2(g)$?
(a) Pure $CaCO_3$
(b) CaO and a pressure of CO_2 greater than K_p
(c) Some $CaCO_3$ and a pressure of CO_2 greater than K_p
(d) $CaCO_3$ and CaO

41. Consider the following equilibrium in a closed container:

$$N_2O_4(g) \rightleftharpoons 2NO_2(g)$$

At a fixed temperature, the volume of the reaction container is halved. For this change which of the following statements holds true regarding the equilibrium constant (K_p) and degree of dissociation (α)?
(a) Neither K_p nor α changes
(b) Both K_p and α change

(c) K_p changes, but α does not change

(d) K_p does not change but α changes **(IIT 2002)**

42. The value of the reaction quotient before any reaction occurs is

(a) ∞ (b) -1 (c) 0 (d) 1

43. What is the minimum mass of $CaCO_3$, below which it decomposes completely, required to establish equilibrium in a 6·50-litre container for the reaction: $CaCO_3(s) \rightleftharpoons CaO(s) + CO_2(g)$? ($K_c = 0{\cdot}05$ mole/litre)

(a) 32·5 g (b) 24·6 g (c) 40·9 g (d) 8·0 g

44. The energy profile of the reaction:

$$A + B \rightleftharpoons C$$

is shown as,

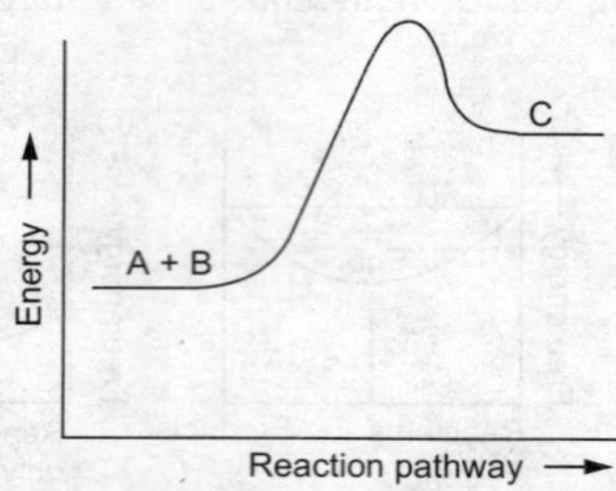

The equilibrium constant for the said equilibrium

(a) increases with the increase in temperature

(b) decreases with the increase in temperature

(c) does not change with the change in temperature

(d) is equal to the rate constant of the forward reaction

45. NiO is to be reduced to Ni in an industrial process by the use of the reaction

$$NiO(s) + CO(g) \rightleftharpoons Ni(s) + CO_2(g)$$

At 1600 K, the equilibrium constant for the reaction is 600. If a CO pressure of 150 mmHg is to be employed in the furnace and the total pressure never exceeds 760 mmHg, will the reduction occur?

(a) Yes (b) No

46. Which of the following curves between log K and $\frac{1}{T}$ is correct?

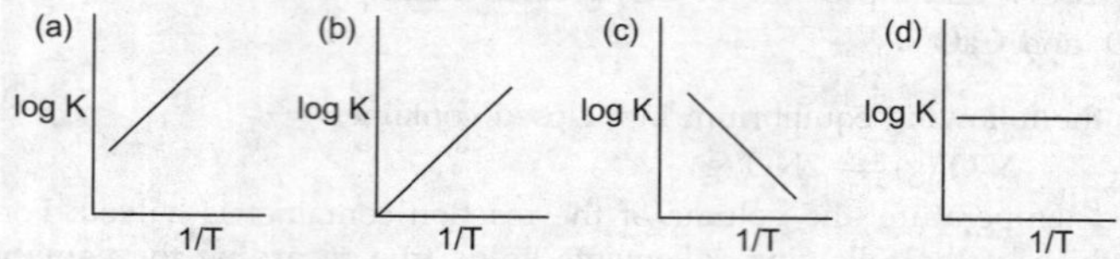

47. The thermal dissociation of equilibrium of $CaCO_3(s)$ is studied under different conditions

$$CaCO_3(s) \rightleftharpoons CaO(s) + CO_2(g).$$

For this equilibrium, the correct statement(s) is(are)

(a) ΔH is dependent on temperature
(b) K is independent of initial amount of $CaCO_3$
(c) K is dependent on the pressure of CO_2 at a given temperature
(d) ΔH is independent of the catalyst, if any. **(IIT 2013 Adv.)**

48. For the reaction

$$SO_2(g) + \frac{1}{2}O_2(g) \rightleftharpoons SO_3(g).$$

If $K_p = K_c(RT)^x$, the value of x is (assuming ideality)

(a) –1 (b) $-\frac{1}{2}$ (c) $\frac{1}{2}$ (d) 1 **(IIT 2014 Main)**

49. The following reaction is performed at 298 K

$$2NO(g) + O_2(g) \rightleftharpoons 2NO_2(g)$$

The standard free energy of formation of NO(g) is 86.6 kJ mol^{-1} at 298 K. What is the standard free energy of formation of $NO_2(g)$ at 298 K? ($K_p = 1.6 \times 10^{12}$)

(a) $R(298)\ln(1.6 \times 10^{12}) - 86600$ (b) $86600 = R(298)\ln(1.6 \times 10^{12})$

(c) $86600 - \dfrac{\ln(1.6 \times 10^{12})}{R(298)}$ (d) $0.5[2 \times 86600 - R(298)\ln(1.6 \times 10^{12})]$

[Hint: Use eqn. 12] **(IIT 2015 Main)**

50. The standard Gibbs energy change at 300 K for the reaction $2A \rightleftharpoons B + C$ is 2494.2 J. At a given time, the composition of the reaction mixture is $[A] = \frac{1}{2}$, $[B] = 2$ and $[C] = \frac{1}{2}$. The reaction proceeds in the (R = 8.314 J/K mol, e = 2.718)

(a) forward direction because $Q > K_c$ (b) reverse direction because $Q > K_c$
(c) forward direction because $Q < K_c$ (d) reverse direction because $Q < K_c$

(IIT 2015 Main)

[Hint: $\Delta G° = -RT\ln K$ and $Q = \dfrac{\left(\frac{1}{2}\right)2}{(1/2)^2}$]

51. The percentage yield of ammonia is a function of time in the reaction

$$N_2(g) + 3H_2(g) \rightleftharpoons 2NH_3(g);\ \Delta H < 0$$

at (p_1T_1) is given below:

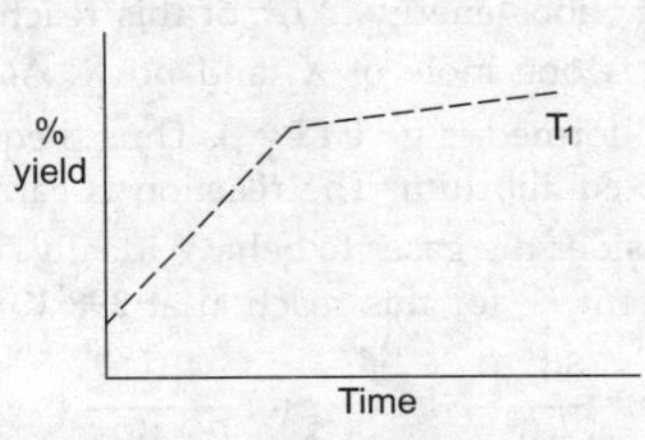

If this reaction is conducted (p_1T_1) with $T_2 > T_1$, the percentage yield of ammonia as a function of time is represented by

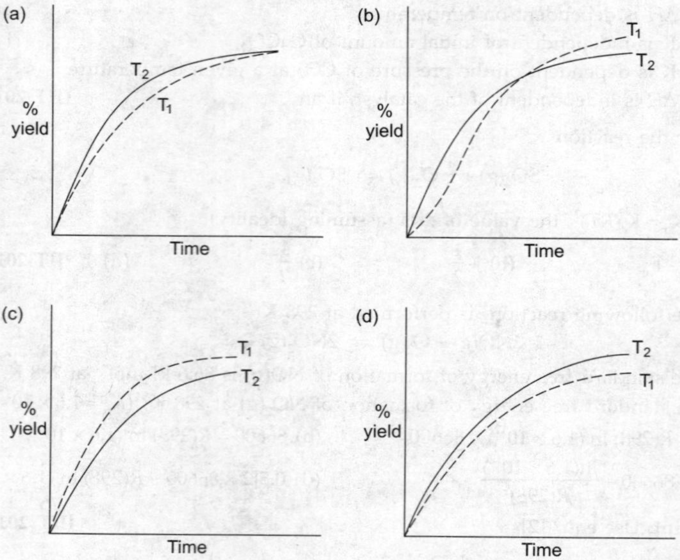

[IIT 2015 Adv.]

[Hint: As the reaction is exothermic, increase in temperature shifts the equilibrium to left, i.e., $[NH_3]$ should be less. But an increase in temperature increases the rate and so NH_3 will be formed at a faster rate initially at high temperature.]

52. The equilibrium constant at 298 K for a reaction

$$A + B \rightleftharpoons C + D$$

is 100. If the initial concentration of all the four species were 1 M each, then equilibrium concentration of D(mol L^{-1}) will be

(a) 0.818 (b) 1.818 (c) 1.182 (d) 0.182

(IIT 2016 Main)

53. Thermal decomposition of gaseous X_2 to X at 298 K takes place according to the following equation

$$X_2(g) \rightleftharpoons 2X(g).$$

The standard reaction Gibbs energy, $\Delta_r G^\circ$ of this reaction is positive. At the start of the reaction, there is one mole of X_2 and no X. As the reaction proceeds the number of moles of X formed is given by β. Thus β equilibrium is the number of moles of X formed at equilibrium. The reaction is carried out at a constant total pressure of 2 bar. Consider the gases to behave ideally. ($R = 0.083$ L bar K^{-1} mol^{-1}). The equilibrium constant K_p for this reaction at 298 K in terms of β_{eqb} is

(a) $\dfrac{8\beta_{eqb}^2}{2-\beta_{eqb}}$ (b) $\dfrac{8\beta_{eqb}^2}{4-\beta_{eqb}^2}$ (c) $\dfrac{4\beta_{eqb}^2}{2-\beta_{eqb}}$ (d) $\dfrac{4\beta_{eqb}^2}{4-\beta_{eqb}^2}$

(IIT 2016 Adv.)

[Hint: $\underset{1-\beta}{X_2(g)} \rightleftharpoons \underset{2\beta}{2X(g)}$, $\beta_{eqb} = 2\beta$. Find K_p.]

54. Refer to Q. 53. The incorrect statement among the following for this reaction is
(a) decrease in total pressure will result in the formation of more moles of gaseous X.
(b) At the start of the reaction, dissociation of gaseous X_2 takes place spontaneouly.
(c) $\beta_{eqb.} = 0.7$
(d) $K_C < 1$ **(IIT 2016 Adv.)**

55. Which of the following lines correctly show the temperature dependence of equilibrium constant K, for an exothermic reaction?

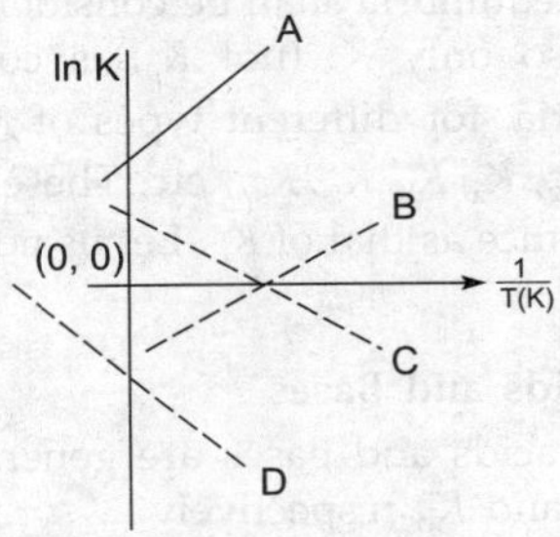

(a) A and D (b) A and B (c) B and C (d) C and D
(IIT 2018 Main)

[Hint: See eqn. 13. ln K vs $1/T$ curve has a negative slope. But as ΔH is –ve (exo), curve will show positive slope.]

56. At a certain temperature in a 5-L vessel, 2 moels of CO and 3 moles of chlorine were allowed to reach equilibrium according to the reaction

$$CO + Cl_2 \rightleftharpoons COCl_2.$$

At equilibrium, if one mole of CO is present, then equilibrium constant (K_c) for the reaction is
(a) 2 (b) 2.5 (c) 4 (d) 3 **(IIT 2018 Main)**

57. In which of the following reactions, an increase in volume of the container will favour the formation of products?
(a) $2NO_2(g) \rightleftharpoons 2NO(g) + O_2(g)$
(b) $4NH_3(g) + 5O_2(g) \rightleftharpoons 4NO(g) + 6H_2O(l)$
(c) $3O_2(g) \rightleftharpoons 2O_3(g)$
(d) $H_2(g) + I_2(g) \rightleftharpoons 2HI(g)$
(IIT 2018 Main)

[Hint: Apply Le Chatelier's principle]

Answers

1-b, **2**-b, **3**-b, **4**-c, **5**-b, **6**-d, **7**-c, **8**-a, **9**-a, **10**-d, **11**-c, **12**-a, **13**-a, **14**-a, **15**-b, **16**-c, **17**-d, **18**-a, **19**-c, **20**-d, **21**-b, **22**-All correct, **23**-a, **24**-d, **25**-c, **26**-All wrong, **27**-c, **28**-a, **29**-d, **30**-d, **31**-b, **32**-b, **33**-c, **34**-c, **35**-c, d and e, **36**-b, **37**-a, **38**-c, **39**-b, **40**-c, **41**-d, **42**-c, **43**-a, **44**-a, **45**-a, **46**-c, **47**-All correct, **48**-b, **49**-d, **50**-b, **51**-b, **52**-b, **53**-b, **54**-c, **55**-b, **56**-b, **57**-a.

❑

CHAPTER SIXTEEN

IONIC EQUILIBRIUM IN AQUEOUS SOLUTIONS

This chapter is an extension to the previous chapter 'Chemical Equilibrium'. Here, only such chemical equilibria shall be considered which involve ions in aqueous solutions, and so only K_c (not K_p) is considered as equilibrium constant. In ionic equilibria, for different types of reactions, K_c is expressed by different notations, viz., K_a, K_b, K_w, K_{sp}, etc. These constants will thus have the same physical significance as that of K_c. Let us now discuss different types of ionic equilibria.

Relative Strengths of Acids and Bases

The relative strengths of acids and bases are generally determined by their dissociation constants K_a and K_b respectively.

For an acid of the type HA

$$\begin{array}{cccccl} c & & 0 & & 0 & \text{Initial concentrations} \\ HA & \rightleftharpoons & H^+ & + & A^- & \\ c(1-x) & & cx & & cx & \text{Concentration at equilibrium} \end{array}$$

$$K_a = \frac{[H^+][A^-]}{[HA]} = \frac{cx \cdot cx}{c(1-x)} = \frac{cx^2}{1-x}$$

$= cx^2$ (if x, i.e., degree of dissociation, is very small).

Similarly, for a base of the type of BOH

$$\begin{array}{ccccc} c & & 0 & & 0 \\ BOH & \rightleftharpoons & B^+ & + & OH^- \\ c(1-x) & & cx & & cx \end{array}$$

$$K_b = \frac{[B^+][OH^-]}{[BOH]} = \frac{cx \cdot cx}{c(1-x)} = \frac{cx^2}{(1-x)} = cx^2 \qquad \text{(if } x \text{ is very small).}$$

K_a and K_b are just the equilibrium constants and hence depend only on temperature. The greater the value of dissociation constant of the acid (K_a), the more is the strength of the acid and similarly, the greater the value of dissociation constant (K_b), the more is the strength of the base.

The strengths of acids and bases depend on the roles of the solvents used. The greater the tendency of the solvent to accept protons, the more will be the strength of the acid giving protons. Acids like HCl, H_2SO_4, etc., dissociate completely in water (solvent) as water accepts all their protons. Thus, these acids in water have equal acid strength **(levelling effect).** But the same acids, when taken in acetic acid (solvent), dissociate partially and act as weak acids.

Again, acetic acid in water is a weak acid, whereas in the solvent ammonia, it is a strong acid.

Modern Method of Expressing the Strengths of Acids and Bases (p*K*)

The modern method is to convert K_a as a power of 10 and express acid strength by the power of 10 with a sign change and call this new unit pK_a. Thus if K_a for an acid is equal to 10^{-4}, $pK_a = 4$, so a higher pK_a value means lower acid strength. Expressing algebraically,

$$pK_a = -\log K_a.$$

This is exactly similar to the method of expressing H^+ concentration by pH.

For a conjugate acid and base pair $HX - X^-$ in aqueous solution,

$$HX \rightleftharpoons H^+ + X^-;\ K_a$$

$$X^- + H_2O \rightleftharpoons HX + OH^-;\ K_b$$

On adding, $H_2O \rightleftharpoons H^+ + OH^-;\ K_w = K_a \cdot K_b$

or $pK_a + pK_b = pK_w = 14$ at 25°C.

Hence, the strength of a base is expressed in terms of pK_a of its conjugate acid, e.g., if K_b value of ammonia is 10^{-5}, its pK_b is 5. In modern practice, we say that pK_a of ammonia is 9 (i.e., $14 - 5 = 9$).

It should be noted that the higher the pK_a value, the lower the acid strength and therefore, higher the base strength.

Relative Acid Strength

The strength of an acid depends on how easily the proton H^+ is lost or removed from H – X bond in the acid species. Two factors are important in determining relative acid strengths.

1. **Polarity of the bond** to which the H atom is attached. The more polarised the bond is in this direction, the more easily the proton is removed and the greater the acid strength. The polarity of the bond develops due to electronegativity difference of the two atoms, H and X, forming the bond.
2. **Bond strength,** that is, how tightly the proton is held. This in turn depends on the size of the atom X in H – X. The larger the atom X, the weaker is the bond and the greater is the acid strength.

   The acid strengths increase in the following order:

   (i) $HF < HCl < HBr < HI$

   (ii) $H_2O < HF$

   (iii) $HIO < HBrO < HClO$

   (iv) $\underset{+1}{HClO} < \underset{+3}{HClO_2} < \underset{+5}{HClO_3} < \underset{+7}{HClO_4}$

   (v) $HSO_4^- < H_2SO_4$

The following list gives the overall relative acid strength in increasing order.

$NH_3 < CH_3OH < H_2O < HCO_3^- < HCN < NH_4^+ < H_2S < H_2CO_3$
$< CH_3COOH < HF < HNO_2 < H_3PO_4 < H_2SO_3 < HSO_4^- < H_3O^+$
$< HNO_3 < HCl < HBr < HI < H_2SO_4 < HClO_4$

Ionic Product of Water

The dissociation constant K of H_2O, dissociating very slightly into H^+ and OH^- ($H_2O \rightleftharpoons H^+ + OH^-$)* is given by

$$K = \frac{[H^+][OH^-]}{[H_2O]}.$$

The concentration of an undissociated molecule of H_2O (i.e., $[H_2O]$) is practically equal to the initial molar concentration of water as water dissociates feebly. $[H_2O]$ may thus be taken as constant, the value of which is equal to 1000/18, i.e., 55·54 moles per litre supposing the density of water as 1 g per mL.

$$K \times [H_2O] = [H^+][OH^-]$$

or $$K_w = [H^+][OH^-],$$

K_w is known as ionic product of water which increases with the increase in temperature.

At 25°C, $K_w = 1 \times 10^{-14}$ (mole/L)^2.

At the body temperature (37°C), $K_w = 2{\cdot}5 \times 10^{-14}$ (mole/L)^2

For pure water (neutral) $[H^+] = [OH^-]$

$$\therefore \quad [OH^+] = [OH^-] = \sqrt{K_w} = \sqrt{10^{-14}} = 10^{-7} \text{ mole/L}.$$

pH

Since K_w at a given temperature is constant, the addition of H^+ ions to an aqueous solution will reduce OH^- ions and vice versa. Sorensen introduced the pH scale to determine the concentration of H^+ in the solution. pH is defined as the negative logarithm of the concentration of H^+, in moles per litre.

$$pH = -\log [H^+] = \log \frac{1}{[H^+]} \quad \text{... (1)}$$

For a neutral solution at 25°C

$$pH = -\log 10^{-7} = 7$$

Thus, if the solution is acidic $[H^+] > 10^{-7}$, i.e., $pH < 7$.

---

* The equilibrium is also expressed as

$$H_2O + H_2O \rightleftharpoons H_3O^+ + OH^-$$

and, if the solution is basic $[H^+] < 10^{-7}$, i.e., $pH > 7$.

Similarly, we can also write $pOH = -\log [OH^-]$

$$= \log \frac{1}{[OH^-]} \quad \text{... (2)}$$

Adding eqns. (1) and (2) we get,

$$pH + pOH = pK_w = -\log 10^{-14} = 14 \text{ at } 25°C \quad \text{... (3)}$$

[**Note:** Solutions may have a pH value of zero and even negative at concentrations greater than 1 molar, but in such calculations, concentrations must be replaced by the corresponding activities of hydrogen ions.]

pH of Buffer Solutions

Buffer solutions are generally prepared either by mixing a weak acid and its salt (e.g., CH_3COOH and CH_3COONa) or by mixing a weak base and its salt (e.g., NH_4OH and NH_4Cl). In other words, buffers contain either a weak acid and its conjugate base (e.g., CH_3COOH and CH_3COO^-) or a weak base and its conjugate acid (e.g., NH_4OH and NH_4^+).

For a buffer mixture of a weak acid and its salt, we have,

$$[H^+] = K_a \frac{[\text{acid}]}{[\text{salt}]} = K_a \cdot \frac{\text{moles or millimoles of acid}}{\text{moles or millimoles of salt}} \quad \text{... (4a)}$$

the logarithm equation of which is given by

$$pH = pK_a + \log \frac{[\text{salt}]}{[\text{acid}]}$$

$$= pK_a + \log \frac{\text{moles (or millimoles) of salt}}{\text{moles (or millimoles) of acid}} \quad \text{... (4b)}$$

And for a buffer mixture of a weak base and its salt, we have,

$$[OH^-] = K_b \frac{[\text{base}]}{[\text{salt}]}$$

$$= K_b \times \frac{\text{moles or millimoles of base}}{\text{moles or millimoles of salt}} \quad \text{... (5a)}$$

the logarithm equation of which is given by

$$pOH = pK_b + \log \frac{[\text{salt}]}{[\text{base}]}$$

$$= pK_b + \log \frac{\text{moles or millimoles of salt}}{\text{moles or millimoles of base}} \quad \text{... (5b)}$$

$(pK_a = -\log K_a \text{ and } pK_b = -\log K_b)$

K_a and K_b are the dissociation constants of the acid and base respectively.

Equations (4) and (5) are known as Henderson–Hasselbalch equations.

The condition for maximum buffer capacity is that the ratio [S]/[A] or [S]/[B] should be unity. In other words the buffer solution must be prepared from a conjugate acid–base pair in which $K_a \approx [H^+]$ (or $pK_a \approx pH$) or $K_b \approx [OH^-]$ (or $pK_b \approx POH$) as the case may be. A buffer solution has generally lost its usefulness when one component of the buffer pair is less than about 10% of the other. Weak acids and their salts are better as buffers for pH < 7; weak bases and their salts are better as buffers for pH > 7.

Buffer Capacity

It may be defined as the number of moles of the acid or base which when added to one litre of a buffer solution, changes its pH by one unit.

Mathematically,

$$\text{buffer capacity} = \frac{\text{no. of moles of acid/alkali added per litre}}{\text{change in pH}}$$

Hydrolysis of Salts

The phenomenon of interaction of cations of a salt with OH^- ions and anions with H^+ ions in an aqueous salt solution is known as hydrolysis of the salt. As a result of hydrolysis, the resulting solution may be acidic, basic or even neutral.

Hydrolysis is the reverse process of neutralisation and therefore, is an endothermic process.

The salts of strong acids and strong bases (e.g., NaCl) do not undergo hydrolysis as the said interactions are not possible.

Salts of Weak Acids and Strong Bases

In the hydrolysis of salts of this category (e.g., CH_3COONa), the anions of the salt (e.g., CH_3COO^-) combine with H^+ forming a feebly dissociated acid, whereas the cations of the salt (e.g., Na^+) do not combine with OH^- (since NaOH is a strong electrolyte) as a result of which OH^- concentration increases and H^+ concentration decreases. The solution thus becomes alkaline.

Let us consider a uniunivalent salt, say CH_3COONa, undergoing hydrolysis:

$$\underbrace{CH_3COO^- + Na^+}_{\text{salt}} + H_2O \rightleftharpoons CH_3COOH + Na^+ + OH^-$$

The equilibrium constant K may be given by

$$K = \frac{[CH_3COOH]\,[Na^+]\,[OH^-]}{[CH_3COO^-]\,[Na^+]\,[H_2O]} = \frac{[CH_3COOH]\,[OH^-]}{[CH_3COO^-]\,[H_2O]}$$

or $K \times [H_2O] = \dfrac{[CH_3COOH]\,[OH^-]}{[CH_3COO^-]}$.

Since $[H_2O]$ will be practically constant

$$\therefore \qquad K_h = \frac{[CH_3COOH]\,[OH^-]}{[CH_3COO^-]}. \qquad \text{... (6)}$$

K_h is known as the hydrolysis constant.

If we consider the dissociation of CH_3COOH and H_2O which may undergo very slight dissociation, we shall have,

for, $CH_3COOH \rightleftharpoons CH_3COO^- + H^+$

$$K_a = \frac{[H^+]\,[CH_3COO^-]}{[CH_3COOH]} \qquad \text{... (7)}$$

and for, $H_2O \rightleftharpoons H^+ + OH^-$

$$K_w = [H^+]\,[OH^-] \qquad \text{... (8)}$$

From eqns. (6), (7) and (8) we get,

$$K_h = \frac{K_w}{K_a} \qquad \text{... (9)}$$

Let us now determine the degree of hydrolysis of the salts, which is the ratio of the moles of the salt hydrolysed and the initial moles of the salt. Consider again the hydrolysis equilibrium,

$$\begin{array}{lcccl} a & & 0 & 0 & \text{Initial concentration} \\ CH_3COO^- + H_2O & \rightleftharpoons & CH_3COOH & + \; OH^- & \\ a(1-x) & & ax & ax & \text{Equilibrium concentration} \end{array}$$

Here x is the degree of hydrolysis.

$$\therefore \qquad K_h = \frac{ax \cdot ax}{a(1-x)} = \frac{ax^2}{1-x} \qquad \text{... (10a)}$$

$$\text{or} \qquad K_h = \frac{K_w}{K_a} = ax^2 \quad \text{(if } x \text{ is very small)} \qquad \text{... (10b)}$$

[**Note:** The value of x should be neglected only when it is less than 0·05, i.e., 5%.]

$$\therefore \qquad x = \sqrt{\frac{K_h}{a}}$$

$$\text{or} \qquad x = \sqrt{\frac{K_w}{K_a \times a}} \quad \text{(if } x \text{ is very small)} \qquad \text{... (11a)}$$

$$\text{Further,} \quad [OH^-] = ax = a \times \sqrt{\frac{K_w}{K_a \cdot a}}$$

or $$[\mathrm{H}^+] = \frac{K_\mathrm{w}}{[\mathrm{OH}^-]} = \sqrt{\frac{K_\mathrm{w} \cdot K_\mathrm{a}}{a}}.$$

Taking negative of the log on both sides, we get,

$$\mathrm{pH} = \frac{1}{2}(\mathrm{p}K_\mathrm{w} + \mathrm{p}K_\mathrm{a} + \log a) \quad \text{... (11b)}$$

Salts of Weak Bases and Strong Acids

In the hydrolysis of these salts (e.g., NH_4Cl), cations of the salt (e.g., NH_4^+) combine with OH^-, forming a feebly dissociated base whereas anions of the salt (e.g., Cl^-) do not combine with H^+ (HCl being a strong electrolyte) as a result of which the solution becomes acidic.

Thus for the hydrolysis equilibrium,

$$\underbrace{\mathrm{NH_4^+} + \mathrm{Cl^-}}_{\text{salt}} + \mathrm{H_2O} \rightleftharpoons \mathrm{NH_4OH} + \mathrm{H^+} + \mathrm{Cl^-}$$

we may have the following equations:

$$K_\mathrm{h} = \frac{K_\mathrm{w}}{K_\mathrm{b}} = \frac{ax^2}{1-x} \quad \text{... (12)}$$

(K_b is the dissociation constant of NH_4OH)

$$= ax^2 \quad \text{(if } x \text{ is very small)}$$

$$x = \sqrt{\frac{K_\mathrm{h}}{a}} = \sqrt{\frac{K_\mathrm{w}}{K_\mathrm{b} \times a}} \quad \text{... (13a)}$$

$$\text{and } \mathrm{pH} = \frac{1}{2}(\mathrm{p}K_\mathrm{w} - \mathrm{p}K_\mathrm{b} - \log a) \quad \text{... (13b)}$$

Salts of Weak Acids and Weak Bases

In this case the cations of the salt combine with OH^- and anions with H^+, forming a feebly dissociated base and acid respectively and therefore the resultant solution may be neutral, acidic or alkaline, depending on the degree of dissociation of the acid and the base formed.

For a salt of this category (e.g., CH_3COONH_4) we have the following equilibrium,

a	a		0	0	Initial concn.
CH_3COO^- +	NH_4^+ + H_2O	$\rightleftharpoons$	CH_3COOH +	NH_4OH	
$a(1-x)$	$a(1-x)$		ax	ax	Concn. at eqb.

$$\text{for which we have, } K_\mathrm{h} = \frac{ax \cdot ax}{a(1-x) \cdot a(1-x)} = \frac{x^2}{(1-x)^2}$$

$$\therefore \quad K_h = \frac{K_w}{K_a \times K_b} = \frac{x^2}{(1-x)^2} \quad \text{... (14)}$$

$$= x^2 \quad \text{(if } x \text{ is very small)}$$

$$x = \sqrt{K_h} = \sqrt{\frac{K_w}{K_a \times K_b}} \quad \text{... (15a)}$$

and

$$\text{pH} = \frac{1}{2}(\text{p}K_w + \text{p}K_a - \text{p}K_b) \quad \text{... (15b)}$$

When, $K_a = K_b$; pH = 7; the resulting solution is neutral,

$K_a > K_b$; pH < 7; the resulting solution is acidic,

and $K_a < K_b$; pH > 7; the resulting solution is alkaline.

K_a and K_b are the dissociation constants of the acid and the base respectively formed during hydrolysis.

[**Note:** (i) Equations (11), (13) and (15) should be used only when the degree of hydrolysis 'x' is negligible.

(ii) Equations 9 to 15 have been derived for uniunivalent salts as considered above. For other types of salts (say unibivalent or bibivalent salts, etc.) we have to derive the similar equations by following the same method. The term 'a' in the above equations is the concentration of that ion of the salt which undergoes hydrolysis.]

Solubility Product

In a saturated solution, there exists an equilibrium between the undissociated solute and the dissociated ions furnished by the dissolved part of the solute. Let us consider a saturated solution of a sparingly soluble salt say, Ag_2SO_4 containing some undissociated salt.

$$\underset{\text{(solid)}}{Ag_2SO_4} \rightleftharpoons \underset{\text{(in solution)}}{2Ag^+ + SO_4^{2-}}$$

Eqb. constant, $K = \dfrac{[Ag^+]^2\,[SO_4^{2-}]}{[Ag_2SO_4]}$

As Ag_2SO_4 is a solid, $[Ag_2SO_4] = 1$

$$K_{sp} = [Ag^+]^2\,[SO_4^{2-}] \quad \text{... (16)}$$

K_{sp} is known as the solubility product of a salt. For soluble salts the concentration terms in the solubility product equation are replaced by activity of the ions. For dilute solutions, activity is equal to concentration. **The value of K_{sp} for a given salt is constant at a given temperature irrespective of the source from which ions originate.**

Thus, if at a constant temperature the common ions are added to the saturated solution of a salt, its ionic product (i.e., reaction quotient) exceeds the K_{sp} value, as a result of which the solution gets precipitated to keep the solubility product value constant.

If ionic product > K_{sp}; the solution gets precipitated

ionic product = K_{sp}; the solution is just saturated

ionic product < K_{sp}; the solution is not saturated.

Applications of Solubility Product Principle

1. Separation of Group II and Group IV cations:

The solubility product of sulphides of group IV cations are comparatively higher than those of group II cations. S^{2-} concentration is controlled by HCl (common-ion effect) in such a way that the ionic product value may exceed the K_{sp} value of group II sulphides but not the K_{sp} of group IV sulphides, as a result of which only group II cations are precipitated as sulphides. In the group II itself CdS has a relatively higher value of K_{sp} than the sulphides of its other cations. The solution is thus diluted to reduce the common-ion effect in order to increase the S^{2-} ion concentration so that CdS may be precipitated.

Just as acids are used to lower the concentration of anions in solution, complexing agents are used in some cases to lower the concentration of cations, e.g., NH_3 is added to a solution containing Ag^+, to convert it to a complex ion, $Ag(NH_3)_2^+$, to lower $[Ag^+]$.

2. Precipitation of Group III cations:

In the group III analysis, OH^- ions (from NH_4OH) concentration is controlled by the addition of NH_4Cl so that the cations of only group III get precipitated as hydroxides.

[**Note:** (i) When the problems involve a precipitation (or complexation) reaction, the reaction should be assumed to go to completion and then the equilibrium dissolving process should be considered.

(ii) While calculating the solubility of MnS, the problem differs from analogous problems dealing with chromates, oxalates, halides, sulphates and iodates. The difference lies in the extensive hydrolysis of the sulphide ions.

(iii) All relevant equilibria must be satisfied when more than one equilibrium is required to describe a solution.]

Ostwald Theory of Indicators

An indicator, generally organic weak acids or weak bases, is a substance which is used to determine the end point in a titration. They change their colour within a certain pH range. Generally, the colour change is due to shifting of the indicator equilibrium, e.g., phenolphthalein, a weak acid, has the following equilibrium.

$$\underset{\text{(colourless)}}{HPh} \rightleftharpoons H^+ + \underset{\text{(pink)}}{Ph^-}$$

And shifting of this equilibrium from left to right produces a pink colour.

$$\text{Indicator constant} = K_{In} = \frac{[H^+][Ph^-]}{[HPh]}.$$

When $[Ph^-] = [HPh]$, i.e., for neutral colour

$$K_{In} = [H^+] \text{ or } pK_{In} = pH$$

Indicator	**pH range**	**Colour change**	**pK_{In} (Neutral colour pH)**
Methyl orange	3.2–4.5	Pink to yellow	3.7
Methyl red	4.4–6.5	Red to yellow	5.1
Litmus	5.5–7.5	Red to blue	7.0
Phenol red	6.8–8.4	Yellow to red	7.8
Phenolphthalein	8.3–10.5	Colourless to pink	9.6

EXAMPLES

Ex. 1. *The degree of dissociation of acetic acid in a* 0·1 M *solution is* $1{\cdot}32 \times 10^{-2}$. *Find out the dissociation constant of the acid and its* pK_a *value.*

Solution :

$$\begin{array}{cccl} 0{\cdot}1 & 0 & 0 & \text{Initial concentration} \\ CH_3COOH \rightleftharpoons & CH_3COO^- + & H^+ & \\ 0{\cdot}1(1-0{\cdot}0132) & 0{\cdot}1\times 0{\cdot}0132 & 0{\cdot}1\times 0{\cdot}0132 & \text{Concentration at eqb.} \end{array}$$

$$K_a = \frac{[CH_3COO^-][H^+]}{[CH_3COOH]} = \frac{0{\cdot}1\times 0{\cdot}0132\times 0{\cdot}1\times{\cdot}0132}{0{\cdot}1\,(1-0{\cdot}0132)}$$

$$= 1{\cdot}76\times 10^{-5}.$$

Further,

$$pK_a = -\log K_a = -\log(1{\cdot}76\times 10^{-5}) = 4{\cdot}75.$$

Ex. 2. *How much water must be added to* 300 mL *of a* 0·2 M *solution of* CH_3COOH *for the degree of dissociation of the acid to double?*

$$K_a\,(CH_3COOH) = 1{\cdot}8\times 10^{-5}$$

Solution : Let the degree of dissociation of CH_3COOH be x. Thus

$$\begin{array}{cccl} 0{\cdot}2 & 0 & 0 & \text{Initial concentration} \\ CH_3COOH \rightleftharpoons & CH_3COO^- + & H^+ & \\ 0{\cdot}2(1-x) & 0{\cdot}2x & 0{\cdot}2x & \text{Concentration at eqb.} \end{array}$$

$$K_a = \frac{0{\cdot}2x \times 0{\cdot}2x}{0{\cdot}2\,(1-x)} = 1{\cdot}8 \times 10^{-5}\,;\; x = 9{\cdot}5 \times 10^{-3}.$$

Now, let the solution be diluted to c mole per litre in order to double the degree of dissociation, x. Since for the degree of dissociation '$2x$' the dissociation constant will be the same, therefore, we have,

$$K_a = \frac{(c \times 2x)\,(c \times 2x)}{c\,(1-2x)} = \frac{c\,(2x)^2}{(1-2x)}$$

$$1{\cdot}8 \times 10^{-5} = \frac{c \times (2 \times 9{\cdot}5 \times 10^{-3})^2}{(1 - 2 \times 9{\cdot}5 \times 10^{-3})}\,;\; c = 0{\cdot}05\text{ M}.$$

Now suppose V mL of water is to be added to dilute 300 mL of CH_3COOH solution to change its concentration from 0·2 M to 0·05 M. As millimoles of CH_3COOH before and after dilution will be same,

$$0{\cdot}2 \times 300 = 0{\cdot}05 \times (300 + V) \qquad \text{(Rule 5, Chap. 1)}$$

$$V = 900\text{ mL}.$$

Ex. 3. *The degree of dissociation of pure water at* 18°C *is found to be* $1{\cdot}8 \times 10^{-9}$. *Find ionic product of water and its dissociation constant at* 18°C.

Solution : Concentration of water $= \frac{1000}{18} = 55{\cdot}56$ moles/litre

(supposing density of water = 1 g/mL).

Thus,

$$\begin{array}{ccccccl} 55{\cdot}56 & & 0 & & 0 & & \text{Initial concentration} \\ H_2O & \rightleftharpoons & H^+ & + & OH^- & & \end{array}$$

$$55{\cdot}56\,(1 - 1{\cdot}8 \times 10^{-9})\;(55{\cdot}56 \times 1{\cdot}8 \times 10^{-9})\;(55{\cdot}56 \times 1{\cdot}8 \times 10^{-9})$$

Eqb. concentration

$$\therefore \quad K_w = [H^+]\,[OH^-] = (55{\cdot}56 \times 1{\cdot}8 \times 10^{-9})\,(55{\cdot}56 \times 1{\cdot}8 \times 10^{-9})$$

$$= 1 \times 10^{-14}\text{ (mole/litre)}^2.$$

Further,

$$K = \frac{[H^+]\,[OH^-]}{[H_2O]} \qquad (K \equiv \text{dissociation const. of } H_2O)$$

or
$$K = \frac{K_w}{[H_2O]} = \frac{1 \times 10^{-14}}{55{\cdot}56} = 1{\cdot}8 \times 10^{-16}\text{ mole/litre}.$$

Ex. 4. *For the reaction* $Ag(CN)_2^- = Ag^+ + 2CN^-$, *the equilibrium constant at* 25°C *is* 4×10^{-19}. *Calculate the* Ag^+ *concentration in a solution which was originally* 0·1 molar *in* KCN *and* 0·03 molar *in* $AgNO_3$.

Solution :

0·1 M	0·03 M	0	Initial concentration
$2KCN$ +	$AgNO_3$ →	$Ag(CN)_2^-$	
$(0{\cdot}1 - 0{\cdot}03 \times 2) = 0{\cdot}04$	$(0{\cdot}03 - 0{\cdot}03) = 0$	0·03	Concentration at the end of reaction

In this reaction $AgNO_3$ shall be fully consumed. The concentration of $Ag(CN)_2^-$ will be 0·03 M. The solution will also contain 0·04 M of free CN^- ions as the concentration of unreacted KCN is 0·04 M. The concentration of CN^- produced by $Ag(CN_2)^-$ may be neglected because of the very low value of dissociation constant of the equilibrium.

Applying law of chemical equilibrium to the given equilibrium

$$Ag(CN)_2^- \rightleftharpoons Ag^+ + 2CN^-$$

we have,

$$K = \frac{[Ag^+]\,[CN^-]^2}{[Ag(CN)_2^-]} = 4 \times 10^{-19}$$

or

$$[Ag^+] = \frac{[Ag(CN)_2^-]}{[CN^-]^2} \times 4 \times 10^{-19}$$

$$= \frac{0{\cdot}03 \times 4 \times 10^{-19}}{(0{\cdot}04)^2} = 7{\cdot}5 \times 10^{-18} \text{ mole/litre.}$$

Ex. 5. *The dissociation constant of a weak acid* HA *is* $4{\cdot}9 \times 10^{-8}$. *After making the necessary approximations, calculate (i) percentage ionisation (ii)* pH *(iii)* OH^- *concentration in a decimolar solution of the acid. Water has a* pH *of* 7.

Solution :

0·1M		0	0	Initial concentration
HA	⇌	H^+ +	A^-	
$0{\cdot}1\,(1-x)$		$0{\cdot}1x$	$0{\cdot}1x$	Concentration at eqb.

(i) $$K_a = \frac{[H^+]\,[A^-]}{[HA]} = \frac{0{\cdot}1x \times 0{\cdot}1x}{0{\cdot}1\,(1-x)} = 4{\cdot}9 \times 10^{-8}$$

$$x = 7 \times 10^{-4}$$

∴ percentage ionisation $= 7 \times 10^{-2}$ %.

(ii) $[H^+] = 0{\cdot}1 \times x = 0{\cdot}1 \times 7 \times 10^{-4} = 7 \times 10^{-5}$ mole/litre.

∴ $\quad pH = -\log(7 \times 10^{-5}) = 4{\cdot}15$

(iii) We know,

$$K_w = [H^+]\,[OH^-]$$

$$1 \times 10^{-14} = (7 \times 10^{-5})\,[OH^-]$$

∴ $\quad [OH^-] = 1{\cdot}43 \times 10^{-10}$ mole/litre.

Ex. 6. *Calculate the* pH *of a* 10^{-3} M *solution of* $Ba(OH)_2$ *if it undergoes complete ionisation.* $K_W = 1 \times 10^{-14}$.

Solution : Since 1 mole of $Ba(OH)_2$ gives 2 moles of OH^-,

concentration of $OH^- = 2 \times$ concentration of $Ba(OH)_2 = 2 \times 10^{-3}$.

$$\therefore \quad [H^+] = \frac{K_W}{[OH^-]} = \frac{1 \times 10^{-14}}{2 \times 10^{-3}} = 0{\cdot}5 \times 10^{-11}.$$

$$\therefore \quad pH = -\log [H^+] = -\log (0{\cdot}5 \times 10^{-11}) = 11{\cdot}30.$$

Ex. 7. *Compute the* pH *of the following solutions assuming complete dissociation of* HCl:

(i) 0·01 N HCl (ii) 1·0 N HCl (iii) 1·50 N HCl

Solution : (i) $[H^+] = [HCl] = 0{\cdot}01 = 10^{-2}$ M

(for HCl ; normality = molarity)

$\therefore$ pH $= -\log [H^+] = -\log 10^{-2} = 2$.

(ii) $[H^+] = [HCl] = 1$ M

$\therefore$ pH $= -\log [H^+] = -\log 1 = 0$.

(iii) $[H^+] = [HCl] = 1{\cdot}5$ M

$\therefore$ pH $= -\log [H^+] = -\log 1{\cdot}5 = -0{\cdot}17$.

[**Note:** Solutions may have negative pH value but at a concentration greater than about 1·0 M, concentration should be replaced by activity (i.e., effective concentration.)]

Ex. 8. 0·01 *mole of sodium hydroxide is added to* 10 *litres of water. How will the* pH *of water change?*

Solution : Concentration of NaOH $= \frac{0{\cdot}01}{10}$ mole per litre $= 10^{-3}$ M

$\therefore$ $[OH^-] = 10^{-3}$ M

$$[H^+] = \frac{K_W}{[OH^-]} = \frac{1 \times 10^{-14}}{10^{-3}} = 10^{-11}$$

$\therefore$ pH $= -\log [H^+] = -\log 10^{-11} = 11$.

Since pH of water is 7, its pH increases by 4, i.e., (11 – 7).

Ex. 9. *How much should the* H^+ *concentration of a solution be changed to raise the* pH *by unity?*

Solution : We know,

$$pH = -\log [H^+]$$

or

$$[H^+] = 10^{-pH}.$$

Hence to raise the pH by unity, the concentration of H^+ should be ten times reduced.

Ex. 10. *Calculate the number of* H^+ *present in* 1 mL *of a solution whose* pH *is* 13.

Solution : $\because$ pH = 13 $\therefore$ $[H^+] = 10^{-13}$ mole per litre.

Thus,

$$\text{mole of } H^+ \text{ present in 1 mL} = \frac{10^{-13}}{10^3} = 10^{-16}.$$

$\therefore$ no. of H^+ in 1 mL = mole in 1 mL × Av. const. ... (Rule 4, Chapter 1)

$$= 10^{-16} \times 6{\cdot}022 \times 10^{23}$$
$$= 6{\cdot}022 \times 10^7.$$

Ex. 11. *The* pH *of an aqueous solution of ammonia is* 11·5. *Find the molarity of the solution.* $K_b(NH_4OH) = 1{\cdot}8 \times 10^{-5}$.

Solution : pH = 11·5, i.e., $-\log[H^+] = 11{\cdot}5$

$\therefore$ $[H^+] = 3{\cdot}162 \times 10^{-12}$

$$\therefore \quad [OH^-] = \frac{K_w}{[H^+]} = \frac{10^{-14}}{3{\cdot}162 \times 10^{-12}} = 0{\cdot}0032\text{ M.}$$

Let the initial concentration of NH_4OH be x moles/litre

x		0	0	Initial concentration
NH_4OH	$\rightleftharpoons$	NH_4 +	OH^-	
$(x - 0{\cdot}0032)$		0·0032	0·0032	Concentration at eqb.

$$\therefore \quad K_b = \frac{0{\cdot}0032 \times 0{\cdot}0032}{x - 0{\cdot}0032} = 1{\cdot}8 \times 10^{-5} \quad \text{(given)}$$

$$\therefore \quad x = 0{\cdot}57\text{ M.}$$

Ex. 12. *Find the* pH *of a solution prepared by mixing* 25 mL *of a* 0·5 M *solution of* HCl, 10 mL *of a* 0·5 M *solution of* NaOH *and* 15 mL *of water.*

Solution : We know that for HCl and NaOH, m.e. = m.m.

$$\therefore \quad \left\{\begin{array}{l}\text{m.e. of HCl} \quad = 0{\cdot}5 \times 25 = 12{\cdot}5 \\ \text{m.e. of NaOH} = 0{\cdot}5 \times 10 = 5{\cdot}0\end{array}\right\} \quad \text{... (Eqn. 1, Chapter 7)}$$

m.e. of HCl in the resultant mixture = 12·5 − 5·0 = 7·5.

Total volume = (25 + 10 + 15) mL = 50 mL.

$$\therefore \text{ normality of HCl} = \frac{\text{m.e.}}{\text{vol (mL)}} = \frac{7{\cdot}50}{50}. \quad \text{... (Eqn. 1, Chapter 7)}$$

$$\therefore \quad \text{molarity} = \frac{7{\cdot}5}{50}.$$

$$\therefore \qquad [H^+] = [HCl] = \frac{7 \cdot 5}{50}.$$

$$\therefore \qquad pH = -\log \frac{7 \cdot 5}{50} = 0 \cdot 8239.$$

Ex. 13. *Calculate the* pH *at which an acid indicator with* $K_a = 1 \times 10^{-5}$ *changes colour when the indicator is* 1×10^{-3} M.

Solution : When the indicator (HIn) is added to a solution, there exists the equilibrium,

$$\underset{\text{colour A}}{HIn} \rightleftharpoons H^+ + \underset{\text{colour B}}{In^-}$$

$$K_{In} = \frac{[H^+][In^-]}{[HIn]}.$$

The midpoint of the colour range of an indicator HIn is the point at which $[In^-] = [HIn]$

$$\therefore \qquad K_{In} = [H^+] = 1 \times 10^{-5}; \;\; pH = 5.$$

Thus at pH = 5 of the solution, the indicator will change its colour.

Ex. 14. *An acid-type indicator,* HIn, *differs in colour from its conjugate base* (In^-). *The human eye is sensitive to colour differences only when the ratio* $[In^-]/[HIn]$ *is greater than* 10 *or smaller than* 0·1. *What should be the minimum change in the* pH *of the solution to observe a complete colour change?* ($K_a = 1 \cdot 0 \times 10^{-5}$) **(IIT 1997)**

Solution : $HIn \rightleftharpoons H^+ + In^-$

$$K_a = \frac{[H^+][In^-]}{[HIn]}$$

When $\frac{[In^-]}{[HIn]} = 10$; $\quad K_a = [H^+] \times 10$

$[H^+] = 1 \times 10^{-5}/10 = 10^{-6}$

pH = 6

When $\frac{[In^-]}{[HIn]} = \frac{1}{10}$; $\quad [H^+] = 1 \times 10^{-5} \times 10 = 10^{-4}$

pH = 4

$\therefore$ minimum change in pH = 6 − 4 = 2.

Ex. 15. *An acid–base indicator has* K_a *of* 3×10^{-5}. *The acid form of the indicator is red and the basic form is blue. By how much must the* pH *change in order to change the indicator from* 75% *red to* 75% *blue?*

Solution : $HIn \rightleftharpoons H^+ + In^-$

$$K_{In} = K_a = \frac{[H^+][In^-]}{[HIn]} = \frac{[H^+][base]}{[acid]}$$

$$[H^+] = K_a \frac{[acid]}{[base]}$$

For 75% red: $[H^+] = \frac{(3 \times 10^{-5})(75)}{25} = 9 \times 10^{-5}$

$\therefore$ $pH = 4{\cdot}05$

For 75% blue: $[H^+] = \frac{(3 \times 10^{-5})(25)}{75} = 1 \times 10^{-5}$

$pH = 5$

The change in pH = 5 − 4·05 = 0·95.

Ex. 16. *The ionisation constant of* NH_4^+ *in water is* $5{\cdot}6 \times 10^{-10}$ *at 25°C. The rate constant for the reaction of* NH_4^+ *and* OH^- *to form* NH_3 *and* H_2O *at 25°C is* $3{\cdot}4 \times 10^{10}$ $L\ mol^{-1}\ s^{-1}$. *Calculate the rate constant for proton transfer from water to* NH_3. **(IIT 1996)**

Solution : $NH_4^+ \rightleftharpoons NH_3 + H^+$ $\qquad K_1 = 5{\cdot}6 \times 10^{-10}$

$H_2O \rightleftharpoons H^+ + OH^-$ $\qquad K_2 = 1 \times 10^{-14}$

From which, we have,

$$NH_4^+ + OH^- \underset{k_2}{\overset{k_1}{\rightleftharpoons}} NH_3 + H_2O \qquad K = K_1/K_2$$

$$K = \frac{5{\cdot}6 \times 10^{-10}}{1 \times 10^{-14}}$$

$$= 5{\cdot}6 \times 10^4$$

Further, at equilibrium,

$$K = \frac{k_1}{k_2}$$

$$k_2 = \frac{k_1}{K} = \frac{3{\cdot}4 \times 10^{10}}{5{\cdot}6 \times 10^4} = 6{\cdot}07 \times 10^5.$$

Ex. 17. *Calculate the* pH *of a* 10^{-5} M HCl *solution if* 1 mL *of it is diluted to* 1000 mL. $K_w = 1 \times 10^{-14}$.

Solution : When 1 ml of 10^{-5} M HCl is diluted to 1000 mL, the concentration of HCl will be 10^{-8} M and accordingly the pH of the diluted solution of

HCl should be 8 which comes in the basic range of pH. But it is not true. It should be noted that in this case while calculating pH we have to consider the H^+ ions produced by the dissociation of water in addition to H^+ ions given by HCl, i.e., 10^{-8} mole/litre, because H^+ concentration from acid and water are comparable.

To calculate the H^+ concentration from water, let us consider the following equilibrium in the presence of HCl; the dissociation of H_2O is suppressed by HCl by the common-ion effect

$$\underset{}{H_2O} \rightleftharpoons \underset{x}{H^+} + \underset{x}{OH^-}$$

where x is $[H^+]$ or $[OH^-]$.

$\therefore$ total H^+ concentration $= (x + 10^{-8})$ $\quad([H^+] = [HCl] = 10^{-8})$

Now, $\quad K_w = [H^+][OH^-]$

$$1 \times 10^{-14} = (x + 10^{-8})x; \quad x = 9.5 \times 10^{-8}$$

$\therefore$ total $\quad [H^+] = (9.5 \times 10^{-8} + 10^{-8})$.

Thus,

$$pH = -\log[H^+] = -\log(9.5 \times 10^{-8} + 10^{-8})$$
$$= -\log(1.05 \times 10^{-7}) = 6.98.$$

Ex. 18. *Calculate the concentration of* H^+ *and* OH^- *in a* 0·05 M HNO_3 *solution at* 25°C.

Solution : As HNO_3 is a strong acid, 0·05 M HNO_3 will ionise to give 0·05 M H^+ ions. Further, let us suppose that H_2O in the presence of 0·05 M H^+ dissociates to give x mole/litre of each of H^+ and OH^-, i.e.,

$$H_2O \rightleftharpoons \underset{x}{H^+} + \underset{x}{OH^-}$$

$\therefore \quad K_w = [H^+][OH^-] = 10^{-14}$ (at 25°C)

$$(0.05 + x)(x) = 10^{-14}$$
$$0.05x = 10^{-14} \qquad (0.05 + x \approx 0.05)$$
$$x = 2 \times 10^{-13}$$

Thus, $\quad [H^+] = 0.05 + x \approx 0.05$ M

and $\quad [OH^-] = x = 2 \times 10^{-13}$ M.

Ex. 19. *What is the pH of a 1 M solution of acetic acid? To what volume must one litre of this solution be diluted so that the pH of the resulting solution will be twice the original value? Given:* $K_a = 1.8 \times 10^{-5}$. **(IIT 1990)**

Solution :

$$\begin{array}{ccc} 1 & & \\ CH_3COOH & \rightleftharpoons & CH_3COO^- + H^+ \\ 1-x & & x \qquad x \end{array}$$

$$K_a = \frac{x^2}{1-x} \approx x^2 = 1{\cdot}8 \times 10^{-5}\ ;\ x = 4{\cdot}2 \times 10^{-3} = [H^+]$$

$\therefore$ $pH = -\log [H^+] = -\log (4{\cdot}2 \times 10^{-3}) = 2{\cdot}37.$

Now, let 1 litre of 1 M CH_3COOH be diluted to V litres so that the pH of the solution doubles. Let the concentration of the diluted solution be c moles/litre. Thus,

$$\begin{array}{ccc} c & & \\ CH_3COOH & \rightleftharpoons & CH_3COO^- + H^+ \\ (c-x') & & x' \qquad x' \end{array}$$

$$K_a = \frac{x' \cdot x'}{(c-x')} = 1{\cdot}8 \times 10^{-5} \qquad \dots (1)$$

Further, $pH = -\log x' = 2 \times 2{\cdot}37 = 4{\cdot}74$ (pH doubles on dilution)

or $\log x' = -4{\cdot}74 = \bar{5}{\cdot}26$

$$x' = 1{\cdot}8 \times 10^{-5}.$$

Substituting x' in (1), we get, $c = 3{\cdot}6 \times 10^{-5}$ M.

As the number of moles of CH_3COOH before and after dilution will be the same

$\therefore$ moles of CH_3COOH = molarity × volume in litres.

$\therefore$ $3{\cdot}6 \times 10^{-5} \times V = 1 \times 1$ $\begin{pmatrix}\text{Initial molarity} = 1\\ \text{Initial volume} = 1\end{pmatrix}$

$$V = 2{\cdot}78 \times 10^4 \text{ litres.}$$

Ex. 20. *Find the concentration of* H^+, HCO_3^- *and* CO_3^{2-} *in a* 0·01 M *solution of carbonic acid if the* pH *of this is* 4·18. $K_a(H_2CO_3) = 4{\cdot}45 \times 10^{-7}$, $K_a(HCO_3^-) = 4{\cdot}69 \times 10^{-11}$.

Solution : We have,

$$pH = -\log [H^+]$$

$$4{\cdot}18 = -\log [H^+]$$

or $\log [H^+] = -4{\cdot}18 = \bar{5}{\cdot}82$

Taking antilog: $[H^+] = 6{\cdot}61 \times 10^{-5}$ mole/litre.

Now, for equilibrium,

$$H_2CO_3 \rightleftharpoons HCO_3^- + H^+$$

$$K_a = \frac{[HCO_3^-][H^+]}{[H_2CO_3]}$$

or $$4 \cdot 45 \times 10^{-7} = \frac{[HCO_3^-](6 \cdot 61 \times 10^{-5})}{0 \cdot 01}$$

$$\therefore \quad [HCO_3^-] = \frac{4 \cdot 45 \times 10^{-7} \times \cdot 01}{6 \cdot 61 \times 10^{-5}}$$

$$= 6 \cdot 73 \times 10^{-5} \text{ mole/litre.}$$

Again for the dissociation of HCO_3^-, we have,

$$HCO_3^- \rightleftharpoons CO_3^{2-} + H^+$$

$$K_a = \frac{[CO_3^{2-}][H^+]}{[HCO_3^-]}$$

$$\therefore \quad [CO_3^{2-}] = \frac{K_a \times [HCO_3^-]}{[H^+]} = \frac{4 \cdot 69 \times 10^{-11} \times 6 \cdot 73 \times 10^{-5}}{6 \cdot 61 \times 10^{-5}}$$

$$= 4 \cdot 8 \times 10^{-11} \text{ mole/litre.}$$

Ex. 21. *Calculate the concentrations of all species of significant concentration present in* 0·1M H_3PO_4 *solution.* $K_1 = 7 \cdot 5 \times 10^{-3}$, $K_2 = 6 \cdot 2 \times 10^{-8}$, $K_3 = 3 \cdot 6 \times 10^{-13}$.

Solution : As the value of K_1 is much larger than K_2 and K_3, it may be assumed that H^+ formed in the second and third steps of dissociation of H_3PO_4 is insignificant. Further, as H_2O dissociates feebly, H^+ ions produced by the dissociation of H_2O may also be neglected. The species of significant concentration are H_3PO_4, H^+ and $H_2PO_4^-$.

Now, first dissociation of H_3PO_4 is

$$\underset{(0 \cdot 1 - x)}{\overset{0 \cdot 1}{H_3PO_4}} \rightleftharpoons \underset{x}{H_2PO_4^-} + \underset{x}{H^+} \quad \text{Concentration at eqb.}$$

$$\therefore \quad K_1 = \frac{x \cdot x}{0 \cdot 1 - x} = 7 \cdot 5 \times 10^{-3}; \quad x = 0 \cdot 024$$

Thus, from the first step of dissociations

$$[H^+] = [H_2PO_4^-] = 0 \cdot 024 \text{ M}$$

and $$[H_3PO_4] = 0 \cdot 1 - 0 \cdot 024 = 0 \cdot 076 \text{ M}$$

For the second dissociation, initial concentration of $H_2PO_4^-$ is 0·024 M

$$\underset{(0 \cdot 024 - y)}{H_2PO_4^-} \rightleftharpoons \underset{y}{HPO_4^{2-}} + \underset{y + 0 \cdot 024}{H^+} \quad \text{Concentration at eqb.}$$

└→ from first step

$$\therefore \qquad K_2 = \frac{y\,(y + 0{\cdot}024)}{0{\cdot}024 - y} = 6{\cdot}2 \times 10^{-8}$$

As $0{\cdot}024 >> y$, $\dfrac{0{\cdot}024y}{0{\cdot}024} = 6{\cdot}2 \times 10^{-8}$; $y = 6{\cdot}2 \times 10^{-8}$

or $\qquad [HPO_4^{2-}] = K_2$

Thus, 0·1 M H_3PO_4 contains $[H_3PO_4] = 0{\cdot}076$ M

$$[H^+] = y + 0{\cdot}024 \approx 0{\cdot}024 \text{ M}$$

$$[H_2PO_4^-] = 0{\cdot}024 - y \approx 0{\cdot}024 \text{ M.}$$

[**Note:** For weak polyprotic acids containing no other electrolyte, the concentration of anion produced in the second step of dissociation is always equal to K_2 if the concentration is reasonable.]

Ex. 22. *Assuming that the buffer in blood is* $CO_2 - HCO_3^-$, *calculate the ratio of conjugate base to acid necessary to maintain blood at its proper* pH, 7·4. K_1 $(H_2CO_3) = 4{\cdot}5 \times 10^{-7}$.

Solution : CO_2 with H_2O forms H_2CO_3.

$$CO_2 + H_2O \rightleftharpoons H^+ + HCO_3^-$$

$$K_1 = \frac{[H^+]\,[HCO_3^-]}{[CO_2]} = 4{\cdot}5 \times 10^{-7}$$

Now, $pH = -\log\,[H^+] = 7{\cdot}4$; $\qquad [H^+] = 4{\cdot}0 \times 10^{-8}$

Thus, $\dfrac{[HCO_3^-]}{[CO_2]} = \dfrac{4{\cdot}5 \times 10^{-7}}{4 \times 10^{-8}} = 11.$

Ex. 23. *A solution of a monobasic acid was titrated with* NaOH *solution and the end point came when* 36·12 mL *of* 0·1 N NaOH *was added. Further* 18·06 mL *of* 0·1 N HCl *was added and the* pH *was then found to be* 4·92. *Find the* K_a *of the acid.*

Solution : At the end point,

m.e. (or millimoles) of the salt produced

= m.e. of NaOH
= 0·1 × 36·12
= 3·612

{ Since salt formed will be uniunivalent and so for such salts m.m. = m.e.

m.e. (or millimole) of HCl added = 0·1 × 18·06 = 1·806.

The addition of 1·806 m.e. of HCl will produce the same number of m.e. of the unknown acid and reduce the amount of the salt by 1·806 m.e.

∴ m.e. (or millimole) of the unknown acid = 1·806

and $\qquad$ m.e. (or millimole) of the salt = 3·612 − 1·806 = 1·806.

Using,

$$\text{pH} = \text{p}K_a + \log \frac{\text{millimole of salt}}{\text{millimole of acid}} \quad \text{... (Eqn. 4b)}$$

$$4{\cdot}92 = \text{p}K_a + \log \frac{1{\cdot}806}{1{\cdot}806}$$

$$\text{p}K_a = 4{\cdot}92$$

$$-\log K_a = 4{\cdot}92$$

$$\log K_a = -4{\cdot}92 = \bar{5}{\cdot}08, \; K_a = 1{\cdot}2 \times 10^{-5}.$$

Ex. 24. 0·1 *mole of sodium formate was added to one litre of* 0·2 M *solution of formic acid* ($K_a = 1{\cdot}8 \times 10^{-4}$). *How will the* H^+ *concentration of the acid diminish assuming the salt to be completely dissociated?*

Solution : Let us first find $[H^+]$ of HCOOH before adding HCOONa.

For the equilibrium,

$$HCOOH \rightleftharpoons H^+ + HCOO^-$$

$$K_a = \frac{[H^+][HCOO^-]}{[HCOOH]} = \frac{[H^+]^2}{[HCOOH]} \qquad (\because [H^+] = [HCOO^-])$$

$$\therefore \quad [H^+] = \sqrt{K_a \cdot [HCOOH]}$$

$$= \sqrt{1{\cdot}8 \times 10^{-4} \times 0{\cdot}2}$$

$$= 6 \times 10^{-3} \text{ mole/litre.}$$

Now on the addition of sodium formate in the acid, we have,

$$[H^+] = K_a \cdot \frac{[\text{acid}]}{[\text{salt}]} \quad \text{... (Eqn. 4a)}$$

$$= 1{\cdot}8 \times 10^{-4} \times \frac{0{\cdot}2}{0{\cdot}1}$$

$$= 3{\cdot}6 \times 10^{-4}.$$

Thus, the addition of the salt caused the $[H^+]$ to diminish to $\frac{3{\cdot}6 \times 10^{-4}}{6 \times 10^{-3}}$ i.e., 1/16·6 of its original value.

Ex. 25. *A certain buffer solution contains equal concentrations of* X^- *and* HX. *The* K_b *for* X^- *is* 10^{-10}. *The* pH *of the buffer is*

(a) 4 (b) 7 (c) 10 (d) 14

Solution : $\text{p}K_b = -\log K_b = -\log 10^{-10}$

$$= 10.$$

$\therefore$ pK_a for HX which is conjugate acid to the base X^- will be 14 – 10, i.e., 4 $\quad (\because pK_a + pK_b = 14)$

Now for a buffer solution containing the acid HX and the salt X^- (e.g., NaX) we have,

$$pH = pK_a + \log \times \frac{[\text{salt}]}{[\text{acid}]} \cdot \quad \text{... (Eqn. 4b)}$$

Since $\qquad [X^-] = [HX]$

$\therefore \qquad pH = pK_a = 4.$

Hence the answer is (a).

Ex. 26. 4 g *of* NaOH *was dissolved in one litre of a solution containing one mole of acetic acid and one mole of sodium acetate. Find the* pH *of the resulting solution. The dissociation constant for acetic acid is* $1{\cdot}8 \times 10^{-5}$.

Solution : Concentration of NaOH $= \frac{4}{40}$ mole per litre

$= 0{\cdot}1$ M. $\quad$ (NaOH = 40)

Since 0·1 mole of NaOH is added to one litre of the buffer solution containing 1 mole each of CH_3COOH and CH_3COONa, 0·1 mole of NaOH will neutralise 0·1 mole of CH_3COOH, reducing the concentration of CH_3COOH from 1 M to 0·9 M and increasing the concentration of CH_3COONa from 1 M to 1·1 M,

i.e., $\qquad [CH_3COOH] = 1 - 0{\cdot}1 = 0{\cdot}9$ M

$\qquad [CH_3COONa] = 1 + 0{\cdot}1 = 1{\cdot}1$ M

Now we have,

$$pH = pK_a + \log \frac{[\text{salt}]}{[\text{acid}]} \quad \text{... (Eqn. 4b)}$$

$$= -\log (1{\cdot}8 \times 10^{-5}) + \log \frac{1{\cdot}1}{0{\cdot}9} \cdot$$

On calculating, we get,

$$pH = 4{\cdot}83.$$

Ex. 27. 20 mL *of* 0·2 M *sodium hydroxide is added to* 50 mL *of* 0·2 M *acetic acid to give* 70 mL *of the solution. What is the* pH *of this solution? Calculate the additional volume of* 0·2 M NaOH *required to make the* pH *of the solution* 4·74. *The ionisation constant of acetic acid is* $1{\cdot}8 \times 10^{-5}$.

Solution : We know,

number of millimoles = molarity × volume (mL) ... (Rule 5, Chapter 1)

$\therefore$ millimoles of NaOH $= 0{\cdot}2 \times 20 = 4$.

Millimoles of $CH_3COOH = 0{\cdot}2 \times 50 = 10.$

Millimoles of CH_3COONa produced = 4.

$\therefore$ millimoles of CH_3COOH remained = 10 − 4 = 6.

Now,

$$pH = pK_a + \log \frac{\text{millimoles of salt}}{\text{millimoles of acid}} \qquad \text{... (Eqn. 4b)}$$

$$= -\log (1{\cdot}8 \times 10^{-5}) + \log \frac{4}{6}$$

$$= 4{\cdot}56.$$

Let the additional volume of NaOH be v mL to make the pH 4·74.

Millimoles of added NaOH = $0{\cdot}2v$.

Since $0{\cdot}2v$ m.m. of NaOH will neutralise $0{\cdot}2v$ m.m. of CH_3COOH and form $0{\cdot}2v$ m.m. of CH_3COONa;

millimoles of $CH_3COOH = (6 - 0{\cdot}2v)$,

millimoles of $CH_3COONa = (4 + 0{\cdot}2v)$.

Again, using Equation 4 (b), we get

$$pH = pK_a + \log \frac{4 + 0{\cdot}2v}{6 - 0{\cdot}2v}$$

$$4{\cdot}74 = -\log (1{\cdot}86 \times 10^{-5}) + \log \frac{4 + 0{\cdot}2v}{6 - 0{\cdot}2v}$$

$$4{\cdot}74 = 4{\cdot}7447 + \log \frac{4 + 0{\cdot}2v}{6 - 0{\cdot}2v}$$

$$-0{\cdot}0047 = \log \frac{4 + 0{\cdot}2v}{6 - 0{\cdot}2v}$$

$$\log \frac{6 - 0{\cdot}2v}{4 + 0{\cdot}2v} = 0{\cdot}0047.$$

Taking antilog, $\dfrac{6 - 0{\cdot}2v}{4 + 0{\cdot}2v} = 1{\cdot}011, \quad v = 4{\cdot}86$ mL.

Ex. 28. *How many moles of sodium propionate should be added to one litre of an aqueous solution containing* 0·02 *mole of propionic acid to obtain a buffer solution of* pH 4·75? *What will be the* pH *if* 0·01 *mole of hydrogen chloride is dissolved in the above buffer solution? Compare the last* pH *value with the* pH *of a* 0·01 *molar* HCl *solution. Dissociation constant of propionic acid,* K_a, *at* 25°C *is* $1{\cdot}34 \times 10^{-5}$.

Solution : Suppose x moles of sodium propionate is added to one litre of an aqueous solution containing 0·02 mole of propionic acid.

Using, $$pH = pK + \log \frac{[\text{salt}]}{[\text{acid}]} \qquad \text{... (Eqn. 4b)}$$

$$4{\cdot}75 = -\log(1{\cdot}34 \times 10^{-5}) + \log \frac{x}{0{\cdot}02}$$

$$\therefore \quad x = 1{\cdot}5 \times 10^{-2}$$

$$= 0{\cdot}015 \text{ M.}$$

Further, addition of 0·01 mole of HCl will produce an extra 0·01 mole of propionic acid and remove 0·01 mole of sodium propionate in a one-litre solution.

$$[C_2H_5COOH] = 0{\cdot}02 + 0{\cdot}01 = 0{\cdot}03 \text{ M}$$

$$[C_2H_5COONa] = 0{\cdot}015 - 0{\cdot}01 = 0{\cdot}005 \text{ M}$$

$$\therefore \quad pH = -\log(1{\cdot}34 \times 10^{-5}) + \log \frac{0{\cdot}005}{0{\cdot}03} = 4{\cdot}06.$$

Now, since the pH of 0·01 M, i.e., 10^{-2} M HCl is 2, therefore, the pH of the buffer solution is about two times greater than that of 0·01 M HCl.

Ex. 29. *The base imidazole has a K_b of $9{\cdot}8 \times 10^{-8}$ at 25°C.*

(a) In what amounts should 0·02 M HCl *and* 0·02 M *imidazole be mixed to make* 100 mL *of a buffer at* pH 7?

(b) If the resulting buffer is diluted to one litre, find pH *of the diluted buffer.*

Solution : (a) As pH = 7, pOH = 14 − 7 = 7 (at 25°C)

$$pK_b = -\log K_b = -\log(9{\cdot}8 \times 10^{-8}) = 7{\cdot}0088.$$

Applying

$$pOH = pK_b + \log \frac{[\text{salt}]}{[\text{base}]},$$

$$7 = 7{\cdot}0088 + \log \frac{[\text{salt}]}{[\text{base}]}$$

$$\log \frac{[\text{salt}]}{[\text{base}]} = -0{\cdot}0088$$

$$\log \frac{[\text{base}]}{[\text{salt}]} = 0{\cdot}0088.$$

Taking antilog,
$$\frac{[\text{base}]}{[\text{salt}]} = 1{\cdot}019$$

or
$$\frac{\text{m.m. of base}}{\text{m.m. of salt}} = 1{\cdot}019. \qquad \dots (1)$$

Suppose v_1 mL of HCl is mixed with v_2 mL of imidazole (base) to make the buffer.

$$\text{m.m. of HCl} = 0{\cdot}02\, v_1$$

$$\text{m.m. of imidazole} = 0{\cdot}02\, v_2.$$

As the buffer is of the base and its salt, 0·02 v_1 m.m. of HCl will combine with 0·02 v_1 m.m. of base to give 0·02 v_1 m.m. of salt.

$\therefore$ m.m. of salt = m.m. of HCl = 0·02 v_1

and m.m. of base left = $0{\cdot}02\, v_2 - 0{\cdot}02\, v_1$

$= 0{\cdot}02\,(v_2 - v_1)$

$\therefore$ from (1), we get,

$$\frac{0{\cdot}02\,(v_2 - v_1)}{0{\cdot}02\, v_1} = 1{\cdot}019$$

or $$\frac{v_2 - v_1}{v_1} = 1{\cdot}019. \qquad \ldots (2)$$

Given that $v_1 + v_2 = 100$. ... (3)

From (2) and (3) we get, $v_1 = 33$ mL

$v_2 = 67$ mL.

(b) pH shall remain the same on dilution as both K_b and [salt]/[base] will not change.

Ex. 30. *A* 40 mL *solution of a weak base,* BOH, *is titrated with* 0·1 N HCl *solution. The* pH *of the solution is found to be* 10·04 *and* 9·14 *after the addition of* 5 mL and 20 mL *of the acid respectively. Find out the dissociation constant of the base.* **(IIT 1991)**

Solution : Let the normality of BOH be n.

m.e. of BOH = 40 n.

On the addition of 5 mL of 0·1N HCl into BOH solution,

m.e. of HCl = 0·1 × 5 = 0·5

m.e. of salt formed = 0·5

m.e. of BOH used = 0·5

$\therefore$ m.e. of remaining BOH = (40n − 0·5).

Applying Henderson's equation

$$\text{pOH} = \text{p}K_b + \log\frac{[\text{salt}]}{[\text{base}]}$$

$$(14 - 10{\cdot}04) = \text{p}K_b + \log\left(\frac{0{\cdot}5}{40n - 0{\cdot}5}\right). \qquad \ldots (1)$$

Similarly, on the addition of 20 mL of HCl,

$$(14 - 9{\cdot}14) = \text{p}K_b + \log\left(\frac{2}{40n - 2}\right). \qquad \ldots (2)$$

Subtracting (1) from (2), we get,

$n = 0{\cdot}088$.

Substituting n in (1), we get,

$$pK_b = 4{\cdot}7412$$

$\therefore$ $\quad -\log K_b = 4{\cdot}7412$

or $\quad \log K_b = -4{\cdot}7412 = \bar{5}{\cdot}2588$

Taking antilog, $\quad K_b = 1{\cdot}82 \times 10^{-5}$.

Ex. 31. *What is the* pH *of the solution when* 0·2 *mole of* HCl *is added to one litre of a solution containing:*
(i) 1 M *each of* CH_3COOH *and acetate ion*
(ii) 0·1 M *each of* CH_3COOH *and acetate ion*
Assume that the total volume is 1 *litre.*

$$K_a \text{ of } CH_3COOH = 1{\cdot}8 \times 10^{-5}.$$

(IIT 1987)

Solution : (i) Addition of 0·2 mole of HCl increases the concentration of CH_3COOH by 0·2 and reduces the concentration of CH_3COO^- by 0·2 mole/L.

$\therefore$ $[CH_3COOH] = 1 + 0{\cdot}2 = 1{\cdot}2$ M

$[CH_3COO^-] = 1 - 0{\cdot}2 = 0{\cdot}8 \text{ M} = [CH_3COONa]$

Using $\quad pH = pK_a + \log \dfrac{[\text{salt}]}{[\text{acid}]}$ ... (Eqn. 4b)

$$= -\log(1{\cdot}8 \times 10^{-5}) + \log \frac{0{\cdot}8}{1{\cdot}2}$$

$$= 4{\cdot}57.$$

(ii) In this case 0·1 M of CH_3COO^- will combine with 0·1 M of HCl forming 0·1 M CH_3COOH.
Total $[CH_3COOH] = 0{\cdot}1 + 0{\cdot}1 = 0{\cdot}2$ M.
But in the presence of the remaining 0·1 M HCl, the dissociation of 0·2 M CH_3COOH will be further suppressed by the common-ion effect. Hence neglecting $[H^+]$ produced by CH_3COOH, the pH may be calculated for just 0·1 M HCl.

$$pH = -\log[H^+] = -\log(0{\cdot}1) = 1{\cdot}0.$$

Ex. 32. *When* 0·05 *mole of* NaOH *was added to one litre of a buffer solution, its* pH *changed from* 5·70 *to* 5·85. *Find the buffer capacity.*

Solution : Buffer capacity $= \dfrac{\text{mole of alkali added per litre}}{\text{change in pH}}$

$$= \frac{0{\cdot}05}{5{\cdot}85 - 5{\cdot}70} = 0{\cdot}3334.$$

Ex. 33. *The* pH *of a* 0·1N *solution of* NH_4Cl *is* 5·4. *Find the hydrolysis constant supposing degree of hydrolysis as very small.*

Solution : NH_4Cl is a salt of a strong acid and a weak base.

If x is the degree of hydrolysis,

$$\begin{array}{cccl} 0{\cdot}1 & & 0 \quad 0 & \text{Initial concentration.} \\ NH_4^+ \; + \; H_2O & \rightleftharpoons & NH_4OH + H^+ & \\ 0{\cdot}1(1-x) & & 0{\cdot}1x \quad 0{\cdot}1x & \text{Concentration at eqb.} \end{array}$$

$$pH = -\log [H^+] = 5{\cdot}4; \quad [H^+] = 3{\cdot}981 \times 10^{-6}$$

$$\because \quad [H^+] = 0{\cdot}1x \quad \therefore \quad x = \frac{[H^+]}{0{\cdot}1} = 3{\cdot}981 \times 10^{-5}.$$

In the above equilibrium,

$$K_h = \frac{[NH_4OH]\,[H^+]}{[NH_4^+]} = \frac{(0{\cdot}1x)\,(0{\cdot}1x)}{0{\cdot}1(1-x)} = \frac{0{\cdot}1x^2}{1-x} \qquad \text{... (Eqn. 13)}$$

or $$K_h = 0{\cdot}1x^2 \text{ (}x\text{ is very small)}$$

$$= 0{\cdot}1 \times (3{\cdot}981 \times 10^{-5})^2$$

$$= 1{\cdot}58 \times 10^{-10}.$$

Ex. 34. *Calcium lactate is a salt of a weak organic acid and is represented as* $Ca(Lac)_2$. *A saturated solution of* $Ca(Lac)_2$ *contains* 0·13 *mole of this salt in a* 0·50-*litre solution. The* pOH *of this solution is* 5·60. *Assuming a complete dissociation of the salt, calculate* K_a *of lactic acid.*

Solution : We have, for the salts of weak acids and strong bases,

$$pH = \tfrac{1}{2}(pK_w + pK_a + \log a), \qquad \text{... (Eqn. 11b)}$$

where a is the concentration of lactate ions, i.e., $(2 \times 0{\cdot}26)$ mole per litre.

$$(14 - 5{\cdot}6) = \tfrac{1}{2}(14 - \log K_a + \log 0{\cdot}52)$$

$$\log K_a = -3{\cdot}084 = \bar{4}{\cdot}916$$

Taking antilog,

$$K_a = 8{\cdot}24 \times 10^{-4}.$$

Ex. 35. *Calculate the degree of hydrolysis of a salt of aniline and acetic acid in* 0·1 M *solution.* $K_a\,(CH_3COOH) = 1{\cdot}8 \times 10^{-5}$, $K_b\,(C_6H_5NH_2) = 4{\cdot}6 \times 10^{-10}$.

Solution : Since aniline and acetic acid both are weak electrolytes, we have,

$$K_h = \frac{K_w}{K_a \cdot K_b} = \frac{1 \times 10^{-14}}{1{\cdot}8 \times 10^{-5} \times 4{\cdot}6 \times 10^{-10}} = 1{\cdot}2 \qquad \text{... (Eqn. 15)}$$

$$\therefore \quad x = \sqrt{K_h} = \sqrt{1{\cdot}2} = 1{\cdot}09.$$

This value of x being greater than 1 is insignificant. So Eqn. (14) has to be considered. The above equation may now be written as

$$K_h = \frac{x^2}{(1-x)^2} \quad \text{... (Eqn. 14)}$$

$$1{\cdot}2 = \frac{x^2}{(1-x)^2}$$

$$\frac{x}{1-x} = \sqrt{1{\cdot}2} = 1{\cdot}096$$

$$\therefore \quad x = 0{\cdot}523.$$

Ex. 36. *Calculate the hydrolysis constant of* KF. *Determine the degree of hydrolysis of the salt in a* 0·01 M *solution and the* pH *of the solution.* K_a (HF) $= 6{\cdot}6 \times 10^{-4}$.

Solution **:** KF is a salt of a weak acid and a strong base,

$$K^+ + F^- + H_2O \rightleftharpoons K^+ + OH^- + HF$$

0·01   0   0   Initial concentration

or $\quad F^- + H_2O \rightleftharpoons OH^- + HF$

$0{\cdot}01\,(1-x)$ $0{\cdot}01x$ $0{\cdot}01x$ Concentration at eqb.

We have,

$$K_h = \frac{K_w}{K_a} \quad \text{... (Eqn. 9)}$$

$$= \frac{1 \times 10^{-14}}{6{\cdot}6 \times 10^{-4}} = 1{\cdot}5 \times 10^{-11}.$$

Further,

$$x = \sqrt{K_h / a} \quad \text{... (Eqn. 11)}$$

$$= \sqrt{\frac{1{\cdot}5 \times 10^{-11}}{0{\cdot}01}} = 3{\cdot}87 \times 10^{-5}.$$

Now, we see from the above equilibrium that

$$[OH^-] = 0{\cdot}01 \times x = 0{\cdot}01 \times 3{\cdot}87 \times 10^{-5} = 3{\cdot}87 \times 10^{-7}.$$

$$\therefore \quad [H^+] = \frac{K_w}{[OH^-]} = \frac{1 \times 10^{-14}}{3{\cdot}87 \times 10^{-7}} = 2{\cdot}583 \times 10^{-8}.$$

$$\therefore \quad pH = -\log(2{\cdot}583 \times 10^{-8}) = 7{\cdot}59.$$

Ex. 37. *A certain weak acid has a dissociation constant of* 1×10^{-4}. *The equilibrium constant for its reaction with a strong base is*

(a) 1×10^{-1} (b) 1×10^{-10} (c) 1×10^{10} (d) 1×10^{14}

Solution **:** When a weak acid combines with a strong base, their salt is formed which undergoes hydrolysis. Since hydrolysis may be considered as the reverse of neutralisation, the equilibrium constant will thus be equal to $1/K_h$. Therefore,

$$\frac{1}{K_h} = \frac{1}{K_w / K_a} = \frac{K_a}{K_w} = \frac{1 \times 10^{-4}}{1 \times 10^{-14}} = 1 \times 10^{10}.$$

Hence the answer is (c).

Ex. 38. *Calculate the approximate* pH *of a* 0·1 M H_2S (*aq*) *solution. The first and second dissociation constants of* H_2S *are* 1×10^{-7} *and* $1{\cdot}3 \times 10^{-14}$ *respectively.*

Solution **:** Most of the H^+ results from primary dissociation, i.e.,

$$H_2S \rightleftharpoons H^+ + HS^-$$

Let x be the concentration of $[H^+]$ or $[HS^-]$ at equilibrium.

$\therefore$ $[H_2S] = 0{\cdot}1 - x \approx 0{\cdot}1$ (x is very small)

$$K_1 = \frac{[H^+]\,[HS^-]}{[H_2S]}$$

or $1 \times 10^{-7} = \dfrac{x \cdot x}{0{\cdot}1}$

or $x = 1 \times 10^{-4}$

or $[H^+] = 1 \times 10^{-4}$

$\therefore$ $pH = -\log [H^+] = -\log (10^{-4}) = 4.$

Ex. 39. *Calculate the* pH *of a* 0·1 M K_3PO_4 *solution. The third dissociation constant of orthophosphoric acid is* $1{\cdot}3 \times 10^{-12}$. *Assume that the hydrolysis proceeds only in the first step.*

Solution **:** $K_3PO_4 + H_2O \rightleftharpoons K_2HPO_4 + KOH$

0·1	0	0	Initial concentration

or $PO_4^{3-} + H_2O \rightleftharpoons HPO_4^{2-} + OH^-$

$0{\cdot}1(1-x)$	$0{\cdot}1x$	$0{\cdot}1x$	Concentration at eqb.

Since K_h is determined by the dissociation constant of the weak acid HPO_4^{2-}, i.e., by the third dissociation constant of H_3PO_4,

$$K_h = \frac{K_w}{K_a} = \frac{1 \times 10^{-14}}{1{\cdot}3 \times 10^{-12}} = 7{\cdot}7 \times 10^{-3}$$

$$x = \sqrt{\frac{K_h}{a}} \quad \text{... (Eqn. 11)}$$

$$= \sqrt{\frac{7{\cdot}7 \times 10^{-3}}{0{\cdot}1}} = 0{\cdot}28$$

$$[OH^-] = 0{\cdot}1 \times x = 0{\cdot}1 \times 0{\cdot}28 = 2{\cdot}8 \times 10^{-2}.$$

$$\therefore \qquad [H^+] = \frac{K_w}{[OH^-]} = \frac{10^{-14}}{2{\cdot}8 \times 10^{-2}}$$

$$= 3{\cdot}57 \times 10^{-13}$$

$$pH = -\log [H^+] = -\log (3{\cdot}57 \times 10^{-13})$$

$$= 12{\cdot}45.$$

Ex. 40. *Calculate* H^+ *concentration in the following solutions:*

(a) a mixture of 5 mL *of* $\frac{N}{10}$ CH_3COOH *and* 5 mL *of* $\frac{N}{10}$ NaOH

(b) a mixture of 5 mL *of* $\frac{N}{10}$ *ammonia and* 5 mL *of* $\frac{N}{10}$ HCl.

Assume that K_a *(acetic acid)* = K_b *(ammonia)* $= 1 \times 10^{-5}$

Solution **:** (a) m.e. (or millimole) of $CH_3COOH = \frac{1}{10} \times 5 = 0{\cdot}5.$

(Rule 5, Chapter 1)

m.e. (or millimole) of $NaOH = \frac{1}{10} \times 5 = 0{\cdot}5.$

$\therefore$ m.e. (or millimole) of CH_3COONa produced = 0·5.

$\because$ total volume = 5 + 5 = 10 mL.

$\therefore$ molarity of $CH_3COONa = \frac{\text{millimoles}}{\text{volume (mL)}} = \frac{0{\cdot}5}{10} = 0{\cdot}05$ M.

As CH_3COONa undergoes hydrolysis, we have,

$$\begin{array}{llll} 0{\cdot}05 & 0 & 0 & \text{Initial concentration.} \\ CH_3COO^- + H_2O \rightleftharpoons & CH_3COOH + & OH^- & \\ 0{\cdot}05\,(1-x) & 0{\cdot}05\,x & 0{\cdot}05\,x & \text{Concentration at eqb.} \end{array}$$

$$K_h = \frac{K_w}{K_a} = \frac{1 \times 10^{-14}}{1 \times 10^{-5}} = 10^{-9}$$

$$x = \sqrt{\frac{K_h}{a}} = \sqrt{\frac{10^{-9}}{0{\cdot}05}} = 1{\cdot}41 \times 10^{-4}$$

$$[OH^-] = 0{\cdot}05 \times x = 0{\cdot}05 \times 1{\cdot}41 \times 10^{-4} = 7{\cdot}05 \times 10^{-6}.$$

$$\therefore \qquad [H^+] = \frac{K_w}{[OH^-]} = \frac{1 \times 10^{-14}}{7{\cdot}05 \times 10^{-6}} = 1{\cdot}42 \times 10^{-9} \text{ mole/litre.}$$

(b) m.e. (or millimole) of ammonia $= \frac{1}{10} \times 5 = 0{\cdot}5$

m.e. (or millimole) of HCl $= \frac{1}{10} \times 5 = 0{\cdot}5.$

m.e. (or millimole) of NH_4Cl produced = 0·5.

$\because$ total volume = 5 + 5 = 10 mL

$\therefore$ molarity of $NH_4Cl = \frac{\text{millimole}}{\text{volume (mL)}} = \frac{0{\cdot}5}{10} = 0{\cdot}05$ M.

As NH_4Cl undergoes hydrolysis according to

$$\begin{array}{lcccl} 0{\cdot}05 & & 0 & 0 & \text{Initial concentration} \\ NH_4^+ + H_2O & \rightleftharpoons & NH_4OH & + \ H^+ & \\ 0{\cdot}05\,(1-x) & & 0{\cdot}05x & 0{\cdot}05x & \text{Concentration at eqb.} \end{array}$$

$$\therefore \quad K_h = \frac{K_w}{K_b} = \frac{1 \times 10^{-14}}{1 \times 10^{-5}} = 10^{-9}$$

$$x = \sqrt{\frac{K_h}{a}} = \sqrt{\frac{10^{-9}}{0{\cdot}05}} = 1{\cdot}41 \times 10^{-4}.$$

$$\therefore \quad [H^+] = 0{\cdot}05x = 0{\cdot}05 \times 1{\cdot}41 \times 10^{-4} = 7{\cdot}05 \times 10^{-6} \text{ mole/litre.}$$

Ex. 41. *The solubility product of chalk is* $9{\cdot}3 \times 10^{-8}$. *Calculate its solubility in grams per litre.*

Solution : $CaCO_3 \rightleftharpoons Ca^{2+} + CO_3^{2-}$

Let the solubility of $CaCO_3$ be S moles per litre.

$$\therefore \quad K_{sp} = [Ca^{2+}][CO_3^{2-}] = S.S$$

$$\therefore \quad S = \sqrt{K_{sp}} = \sqrt{9{\cdot}3 \times 10^{-8}} = 0{\cdot}000304 \text{ mole/litre.}$$

Solubility in g/L = mole/litre × mol. wt. of $CaCO_3$

$= 0{\cdot}000304 \times 100$

$= 0{\cdot}0304$ gram/litre.

Ex. 42. *The solubility of* CaF_2 *in water at 18°C is* $2{\cdot}05 \times 10^{-4}$ *mole per litre. Calculate its solubility product.*

Solution : $CaF_2 \rightleftharpoons Ca^{2+} + 2F^-$

Since the solubility of CaF_2 in water is $2{\cdot}05 \times 10^{-4}$

$$\therefore \quad [Ca^{2+}] = 2{\cdot}05 \times 10^{-4}$$

$$[F^-] = 2 \times 2{\cdot}05 \times 10^{-4}$$

$$K_{sp} = [Ca^{2+}][F^-]^2$$
$$= (2{\cdot}05 \times 10^{-4})(2 \times 2{\cdot}05 \times 10^{-4})^2$$
$$= 3{\cdot}45 \times 10^{-11}.$$

Ex. 43. *Calculate the solubility of* PbI_2 *in water at 25°C which is 90% dissociated.* $K_{sp}(PbI_2) = 1{\cdot}39 \times 10^{-8}$ *at* 25°C.

Solution : Let the solubility of PbI_2 in water be S moles per litre.

$$PbI_2 \rightleftharpoons Pb^{2+} + 2I^-$$

Since PbI_2 is 90% dissociated

$\therefore \quad [Pb^{2+}] = 0{\cdot}9 \times S$ and $[I^-] = 2 \times 0{\cdot}9 \times S = 1{\cdot}8\ S$

$$K_{sp} = [Pb^{2+}][I^-]^2 = (0{\cdot}9\ S)(1{\cdot}8\ S)^2$$
$$1{\cdot}39 \times 10^{-8} = (0{\cdot}9\ S)(1{\cdot}8\ S)^2$$
$$S = 1{\cdot}68 \times 10^{-3} \text{ mole/litre.}$$

Ex. 44. *The solubilities of* AgCl *in water,* 0·01 M $CaCl_2$, 0·01 M NaCl *and* 0·05 M $AgNO_3$ *are denoted by* S_1, S_2, S_3 *and* S_4 *respectively. Which of the following relationships is correct?*

(a) $S_1 > S_2 > S_3 > S_4$ (b) $S_1 > S_3 > S_2 > S_4$

(c) $S_1 > S_2 = S_3 > S_4$ (d) $S_1 > S_3 > S_4 > S_2$

K_{sp} (AgCl) $= 1{\cdot}8 \times 10^{-10}$

Solution : Solubility of AgCl in water:

$$AgCl = Ag^+ + Cl^-$$
$$K_{sp} = [Ag^+][Cl^-] = (S_1)(S_1) = S_1^2$$
$$\therefore \quad S_1 = \sqrt{K_{sp}} = \sqrt{1{\cdot}8 \times 10^{-10}} = 1{\cdot}34 \times 10^{-5} \text{ mole/litre.}$$

Solubility of AgCl in 0·01 M $CaCl_2$:

$$[Ag^+] = S_2\ ;\ [Cl^-] = (2 \times 0{\cdot}01 + S_2)$$
$$K_{sp} = [Ag^+][Cl^-]$$
$$1{\cdot}8 \times 10^{-10} = S_2(0{\cdot}02 + S_2)$$
$$\therefore \quad S_2 = 9 \times 10^{-9} \text{ mole/litre.}$$

Solubility of AgCl in 0·01 M NaCl:

$$[Ag^+] = S_3\ ;\ [Cl^-] = (0{\cdot}01 + S_3)$$
$$\therefore \quad 1{\cdot}8 \times 10^{-10} = S_3(0{\cdot}01 + S_3)$$
$$\therefore \quad S_3 = 1{\cdot}8 \times 10^{-8} \text{ mole/litre.}$$

Solubility of AgCl in 0·05 M $AgNO_3$:

$$[Ag^+] = (0\cdot05 + S_4);\ [Cl^-] = S_4$$

$$\therefore \quad 1\cdot8 \times 10^{-10} = (0\cdot05 + S_4)\, S_4$$

$$\therefore \quad S_4 = 3\cdot6 \times 10^{-9} \text{ mole/litre.}$$

From the values of solubility we get $S_1 > S_3 > S_2 > S_4$.

Hence answer (b) is correct.

Ex. 45. *The precipitate of* CaF_2 ($K_{sp} = 1\cdot7 \times 10^{-10}$) *is obtained by mixing equal volumes of which of the following?*

(a) 10^{-4} M Ca^{2+} + 10^{-4} M F^-

(b) 10^{-2} M Ca^{2+} + 10^{-3} M F^-

(c) 10^{-5} M Ca^{2+} + 10^{-3} M F^-

(d) 10^{-3} M Ca^{2+} + 10^{-5} M F^-

Solution : We know that precipitation occurs only when ionic product exceeds the value of solubility product.

When equal volumes of solutions containing Ca^{2+} and F^- are mixed, $[Ca^{2+}]$ and $[F^-]$ will be halved. Let us now calculate the ionic product in each case.

(a) Ionic product = $[Ca^{2+}]\,[F^-]^2$

$$= \left(\frac{1}{2} \times 10^{-4}\right)\left(\frac{1}{2} \times 10^{-4}\right)^2 = \frac{1}{8} \times 10^{-12} < K_{sp}; \text{ no precipitation.}$$

$(K_{sp} = 1\cdot7 \times 10^{-10})$

$$\text{(b) Ionic product} = \left(\frac{1}{2} \times 10^{-2}\right)\left(\frac{1}{2} \times 10^{-3}\right)^2$$

$$= \frac{1}{8} \times 10^{-8} > K_{sp}\,;\ \text{precipitation.}$$

$$\text{(c) Ionic product} = \left(\frac{1}{2} \times 10^{-5}\right)\left(\frac{1}{2} \times 10^{-3}\right)^2$$

$$= \frac{1}{8} \times 10^{-11} < K_{sp}\,;\ \text{no precipitation.}$$

$$\text{(d) Ionic product} = \left(\frac{1}{2} \times 10^{-3}\right)\left(\frac{1}{2} \times 10^{-5}\right)^2$$

$$= \frac{1}{8} \times 10^{-13} < K_{sp}\,;\ \text{no precipitation.}$$

Hence answer (b) is correct.

Ex. 46. *Assuming the complete dissociation of* HCl *and the lead salt, calculate how much* HCl *is to be added to a* 0·001 M *lead salt solution to just prevent precipitation when saturated with* H_2S. *The concentration of* H_2S *in its saturated solution is* 0·1 M.

$$K_a\,(H_2S) = 1{\cdot}1 \times 10^{-23};\ K_{sp}\,(PbS) = 3{\cdot}4 \times 10^{-28}$$

Solution : We know,

$$K_{sp}\,(PbS) = [Pb^{2+}]\,[S^{2-}].$$

Since the lead salt is completely dissociated, $[Pb^{2+}]$ is equal to the concentration of the lead salt, i.e., $[Pb^{2+}] = 0{\cdot}001$ M. If $[S^{2-}]$ is the concentration of S^{2-} required to just start precipitation of PbS,

$$[S^{2-}] = \frac{3{\cdot}4 \times 10^{-28}}{0{\cdot}001} = 3{\cdot}4 \times 10^{-25}.$$

Now, the addition of HCl will suppress the dissociation of H_2S to that extent that $[S^{2-}] = 3{\cdot}4 \times 10^{-25}$ M.

Since HCl is completely ionised, $\therefore\ [H^+] = [HCl]$

Let [HCl] be x', therefore, $[H^+] = x'$

$$H_2S \rightleftharpoons 2H^+ + S^{2-}$$

At eqb. $[H_2S] = 0{\cdot}1 - 3{\cdot}4 \times 10^{-25} \approx 0{\cdot}1$

$$[H^+] = 2 \times 3{\cdot}4 \times 10^{-25} + x' \approx x'$$

$$[S^{2-}] = 3{\cdot}4 \times 10^{-25}$$

$$\therefore \qquad K_a = \frac{[H^+]^2\,[S^{2-}]}{[H_2S]}.$$

$$1{\cdot}1 \times 10^{-23} = \frac{x'^2\,(3{\cdot}4 \times 10^{-25})}{0{\cdot}1}$$

$$x'^2 = \frac{0{\cdot}1 \times 1{\cdot}1 \times 10^{-23}}{3{\cdot}4 \times 10^{-25}}$$

$$x'^2 = 3{\cdot}2356$$

$$x' = 1{\cdot}80.$$

$$\therefore \qquad [HCl] = 1{\cdot}80\text{ M}.$$

Thus, any concentration of HCl greater than 1·80 M will just prevent precipitation of PbS.

Ex. 47. *What* $[H^+]$ *must be maintained in a saturated* H_2S (0·1 M) *to precipitate* CdS *but not* ZnS, *if* $[Cd^{2+}] = [Zn^{2+}] = 0{\cdot}1$ M *initially?*

$$K_{sp}\,(CdS) = 8 \times 10^{-27}$$
$$K_{sp}\,(ZnS) = 1 \times 10^{-21}$$
$$K_a\,(H_2S) = 1{\cdot}1 \times 10^{-21}$$

Solution : In order to prevent precipitation of ZnS,

$$[Zn^{2+}]\,[S^{2-}] < K_{sp}\,(ZnS) = 1 \times 10^{-21}$$
(ionic product)

or $(0{\cdot}1)\,[S^{2-}] < 1 \times 10^{-21}$

or $[S^{2-}] < 1 \times 10^{-20}$.

This is the maximum value of $[S^{2-}]$ before ZnS will precipitate.

Let $[H^+]$ to maintain this $[S^{2-}]$ be x.

Thus for

$$H_2S \rightleftharpoons 2H^+ + S^{2-},$$

$$K_a = \frac{[H^+]^2\,[S^{2-}]}{[H_2S]} = \frac{x^2\,(1 \times 10^{-20})}{0{\cdot}1} = 1{\cdot}1 \times 10^{-21}$$

or $x = [H^+] = 0{\cdot}1$ M.

$\therefore$ no ZnS will precipitate at a concentration of H^+ greater than 0·1 M.

Ex. 48. *The solubility of* CaF_2 *in water at* 18°C *is* $2{\cdot}04 \times 10^{-4}$ *mole/litre. Calculate: (a)* K_{sp}, *and (b) solubility in* 0·01 M NaF *solution.*

Solution : (a) For $CaF_2 \rightleftharpoons Ca^{2+} + 2F^-$
$\qquad\qquad S \qquad 2S$

$$K_{sp}\,(CaF_2) = [Ca^{2+}]\,[F^-]^2$$

If S is the solubility,

$$K_{sp} = S\,(2S)^2 = 4S^3$$
$$= 4 \times (2{\cdot}04 \times 10^{-4})^3$$
$$= 3{\cdot}4 \times 10^{-11}.$$

(b) Let the solubility of CaF_2 in NaF be S' moles/litre.

Assuming NaF to be completely ionised,

$$[F^-] = (2S' + 0{\cdot}01)$$

and $[Ca^{2+}] = S'$

$\therefore \quad K_{sp} = S'\,(2S' + 0{\cdot}01)^2 = 3{\cdot}4 \times 10^{-11}.$

Neglecting S'^2 and S'^3 terms, (S' being very small), we get,

$$S' = 3{\cdot}4 \times 10^{-7} \text{ mole/litre.}$$

Ex. 49. *Solid* $AgNO_3$ *is added to a solution which is* 0·1 M *in* Cl^- *and* 0·1 M *in* CrO_4^{2-}. *If* K_{sp} *values for* AgCl *and* Ag_2CrO_4 *are* $1{\cdot}7 \times 10^{-10}$ *and* $1{\cdot}9 \times 10^{-12}$ *respectively, which one will precipitate first? Find the concentration of* Cl^- *when* Ag_2CrO_4 *starts precipitating.*

Solution : $[Ag^+]$ required to precipitate AgCl and Ag_2CrO_4 may be calculated as follows:

$$K_{sp}\,(AgCl) = [Ag^+]\,[Cl^-]$$

$$\therefore \quad [Ag^+] = \frac{1{\cdot}7 \times 10^{-10}}{0{\cdot}1} = 1{\cdot}7 \times 10^{-9}\ M.$$

(to precipitate AgCl).

$$K_{sp}\,(Ag_2CrO_4) = [Ag^+]^2\,[CrO_4^{2-}]$$

$$\therefore \quad [Ag^+] = \sqrt{\frac{1{\cdot}9 \times 10^{-12}}{0{\cdot}1}} = 4{\cdot}34 \times 10^{-6}\ M.$$

(to precipitate Ag_2CrO_4)

Since $[Ag^+]$ required to precipitate AgCl is less than that required to precipitate Ag_2CrO_4, AgCl will precipitate first. Now since $[Ag^+] = 4{\cdot}34 \times 10^{-6}$ when Ag_2CrO_4 starts precipitating, $[Cl^-]$ at this stage may be calculated as follows:

$$K_{sp}\,[AgCl] = [Ag^+]\,[Cl^-]$$

or $$1{\cdot}7 \times 10^{-10} = (4{\cdot}34 \times 10^{-6})\,[Cl^-]$$

or $$[Cl^-] = 3{\cdot}9 \times 10^{-5}\ M.$$

Ex. 50. *Solid* $AgNO_3$ *is slowly added to a solution that is* 0·001 M *each in* NaCl, NaBr *and* NaI.

(a) Calculate the concentration of Ag^+ *required to initiate the precipitation of each of* AgCl, AgBr *and* AgI.

(b) Calculate the percentage of I^- *precipitated before* AgBr *precipitates.*

(c) Calculate the percentage of I^- *precipitated before* Cl^- *precipitates.*

$K_{sp}\,(AgI) = 1{\cdot}5 \times 10^{-16}$, $K_{sp}\,(AgBr) = 3{\cdot}3 \times 10^{-13}$, $K_{sp}\,(AgCl) = 1{\cdot}8 \times 10^{-10}$.

Solution : (a) NaCl, NaBr and NaI are completely ionised.

$$\therefore \quad [Cl^-] = [Br^-] = [I^-] = 0{\cdot}001\ M$$

$$K_{sp}\,(AgI) = [Ag^+]\,[I^-] = 1{\cdot}5 \times 10^{-16}$$

$$\therefore\ [Ag^+] \text{ to start precipitation of AgI} = \frac{1{\cdot}5 \times 10^{-16}}{0{\cdot}001} = 1{\cdot}5 \times 10^{-13}\ M.$$

Similarly, $[Ag^+]$ to initiate precipitation of AgBr and AgCl may be calculated.

$$\therefore [Ag^+] \text{ to start precipitation of AgBr} = \frac{K_{sp}(AgBr)}{[Br^-]} = \frac{3{\cdot}3 \times 10^{-13}}{0{\cdot}001}$$

$$= 3{\cdot}3 \times 10^{-10} \text{ M.}$$

And,

$$[Ag^+] \text{ to start precipitation of AgCl} = \frac{K_{sp}(AgCl)}{[Cl^-]} = \frac{1{\cdot}8 \times 10^{-10}}{0{\cdot}001}$$

$$= 1{\cdot}8 \times 10^{-7} \text{ M.}$$

(b) When AgBr just starts precipitating, $[Ag^+] = 3{\cdot}3 \times 10^{-10}$ M

At this stage $[I^-]$ may be calculated as

$$K_{sp}(AgI) = [Ag^+][I^-]$$

or

$$[I^-] = \frac{K_{sp}(AgI)}{[Ag^+]} = \frac{1{\cdot}5 \times 10^{-16}}{3{\cdot}3 \times 10^{-10}}$$

$$= 4{\cdot}5 \times 10^{-7} \text{ M.}$$

$$\therefore \text{ percentage of } I^- \text{ not precipitated} = \frac{4{\cdot}5 \times 10^{-7}}{0{\cdot}001} \times 100 = 0{\cdot}045\%.$$

$\therefore$ percentage of I^- precipitated before AgBr begins to precipitate

$$= 99{\cdot}955\%.$$

(c) Similarly, $[I^-]$ may be calculated when AgCl begins to precipitate.

Here $$[Ag^+] = 1{\cdot}8 \times 10^{-7}$$

$$K_{sp}(AgI) = [Ag^+][I^-]$$

$$[I^-] = \frac{K_{sp}(AgI)}{[Ag^+]} = \frac{1{\cdot}5 \times 10^{-16}}{1{\cdot}8 \times 10^{-7}} = 8{\cdot}3 \times 10^{-10} \text{ M.}$$

$$\therefore \text{ percentage of } I^- \text{ not precipitated} = \frac{8{\cdot}3 \times 10^{-10}}{0{\cdot}001} \times 100 = 0{\cdot}000083\%.$$

$\therefore$ percentage of I^- precipitated before AgCl begins to precipitate

$$= 99{\cdot}999917\%.$$

Ex. 51. *A solution contains a mixture of* Ag^+ (0·1 M) *and* Hg_2^{2+} (0·1 M) *which are to be separated by selective precipitation. Calculate the maximum concentration of iodide ion at which one of them gets precipitated almost completely. What percentage of that metal ion is precipitated?*

$$\{K_{sp}(AgI) = 8{\cdot}5 \times 10^{-17}\}$$
$$\{K_{sp}(Hg_2I_2) = 2{\cdot}5 \times 10^{-26}\}$$

Solution : Let us first calculate $[I^-]$ to precipitate AgI and Hg_2I_2:

$$K_{sp}\,[AgI] = [Ag^+]\,[I^-]$$

$$8.5 \times 10^{-17} = (0.1)\,[I^-]$$

$$\therefore \quad [I^-] \text{ to precipitate AgI} = \frac{8.5 \times 10^{-17}}{0.1} = 8.5 \times 10^{-16}\text{ M}$$

$$K_{sp}\,(Hg_2I_2) = [Hg_2^{2+}]\,[I^-]^2$$

$$2.5 \times 10^{-26} = (0.1)\,[I^-]^2$$

$\therefore$ $[I^-]$ to precipitate $Hg_2I_2 = 5.0 \times 10^{-13}$ M.

$[I^-]$ to precipitate AgI is smaller, therefore, AgI will start precipitating first. On further addition of I^- more AgI will precipitate and when $[I^-] \geq 5.0 \times 10^{-13}\text{ M}, Hg_2I_2$ will start precipitating. The maximum concentration of Ag^+ at this stage will thus be calculated as

$$K_{sp}\,(AgI) = [Ag^+]\,[I^-]$$

$$8.5 \times 10^{-17} = [Ag^+]\,(5.0 \times 10^{-13})$$

$$[Ag^+] = \frac{8.5 \times 10^{-17}}{5.0 \times 10^{-13}} = 1.7 \times 10^{-4}\text{ M.}$$

$$\therefore \text{ percentage of } Ag^+ \text{ remained unprecipitated} = \frac{1.7 \times 10^{-4}}{0.1} \times 100 = 0.17\%.$$

Thus, percentage of Ag^+ precipitated = 99·83%.

Ex. 52. $Pb(IO_3)_2$, *a sparingly soluble salt in water, has a* K_{sp} *value of* 2.6×10^{-13}. *To* 35 mL *of a* 0·15 M $Pb(NO_3)_2$ *solution,* 15 mL *of a* 0·8 M KIO_3 *solution is added and a precipitate of* $Pb(IO_3)_2$ *formed. Find the concentrations of* Pb^{2+} *and* IO_3^- *left in the solution.*

Solution : m.m. of $Pb(NO_3)_2$ solution = 0·15 × 35 = 5·25. (Rule 5, Chapter 1)

m.m. of KIO_3 solution = 0·8 × 15 = 12.

According to reaction

$$Pb(NO_3)_2 + 2KIO_3 \rightarrow Pb(IO_3)_2 + 2KNO_3$$

5·25 m.m. of $Pb(NO_3)_2$ will combine with 10·50 m.m. (i.e., 2 × 5·25) of KIO_3 to give 5·25 m.m. of $Pb(IO_3)_2$.

$\therefore$ m.m. of KIO_3 remained unreacted = 12 − 10·5 = 1·5.

$$\therefore \text{ molarity of } KIO_3 = \frac{\text{m.m.}}{\text{volume (mL)}} = \frac{1.5}{35 + 15} = 0.03\text{ M.}$$

$\therefore \qquad [IO_3^-] = 0.03$ M

Now, for the equilibrium,

$$Pb(IO_3)_2 \rightleftharpoons Pb^{2+} + 2IO_3^-$$

$$K_{sp} = [Pb^{2+}][IO_3^-]^2$$

$$2.6 \times 10^{-13} = [Pb^{2+}](0.03)^2$$

or $$[Pb^{2+}] = \frac{2.6 \times 10^{-13}}{9 \times 10^{-4}} = 2.9 \times 10^{-9} \text{ mole/litre.}$$

Ex. 53. *Calculate* $[Pb^{2+}]$ *in a* 1 M HCl *solution that is saturated in* PbS. $K_a(H_2S) = 1.1 \times 10^{-21}$, $K_{sp}(PbS) = 8 \times 10^{-28}$.

Solution : We have,

$$PbS(s) \rightleftharpoons Pb^{2+} + S^{2-}; \qquad K_{sp} = 8 \times 10^{-27}$$

and $$\underset{\text{(from HCl)}}{2H^+} + \underset{\text{(from PbS)}}{S^{2-}} = H_2S; \qquad K' = \frac{1}{K_a(H_2S)} = \frac{1}{1.1 \times 10^{-21}}$$

On adding we get,

$$\underset{}{PbS(s)} + \underset{1-2x}{\overset{1}{2H^+}} \rightleftharpoons \underset{x}{Pb^{2+}} + \underset{x}{H_2S}$$

for which, applying Eqn. 7, (chapter on 'Chemical Equilibrium') we get

$$K = \frac{[Pb^{2+}][H_2S]}{[H^+]^2} = K_{sp} \times K' = \frac{8 \times 10^{-28}}{1.1 \times 10^{-21}} = 7.3 \times 10^{-7}$$

or $$\frac{x \cdot x}{(1-2x)^2} = 7.3 \times 10^{-7}.$$

Supposing x to be very small,

we get, $$x = [Pb^{2+}] = 8.55 \times 10^{-4} \text{ M.}$$

Ex. 54. *The solubility of* $Mg(OH)_2$ *in pure water is* 9.57×10^{-3} g/L. *Calculate its solubility (in gram per litre) in* 0.02 M $Mg(NO_3)_2$ *solution.* **(IIT 1986)**

Solution : Solubility of $Mg(OH)_2$ in pure water in moles/litre

$$= \frac{9.57 \times 10^{-3}}{58.31} \qquad [Mg(OH)_2 = 58.31]$$

$$= 1.64 \times 10^{-4}$$

$\therefore \qquad [Mg^{2+}] = 1.64 \times 10^{-4}$

and $\qquad [OH^-] = 2 \times 1.64 \times 10^{-4}$.

$$\therefore \quad K_{sp}[\text{Mg(OH)}_2] = [\text{Mg}^{2+}][\text{OH}^-]^2$$
$$= (1{\cdot}64 \times 10^{-4})(2 \times 1{\cdot}64 \times 10^{-4})^2 = 1{\cdot}764 \times 10^{-11}.$$

Let the solubility of $Mg(OH)_2$ in 0·02 M $Mg(NO_3)_2$ solution be S moles per litre. In this case,

$$[\text{Mg}^{2+}] = S + 0{\cdot}02$$
$$[\text{OH}^-] = 2S$$
$$\therefore \quad K_{sp} = (S + 0{\cdot}02)(2S)^2 = 1{\cdot}764 \times 10^{-11}$$
$$S = 1{\cdot}48 \times 10^{-5} \text{ mole per litre}$$
$$= 1{\cdot}48 \times 10^{-5} \times 58{\cdot}31 \text{ g per litre}$$
$$= 8{\cdot}63 \times 10^{-4} \text{ g per litre.}$$

Ex. 55. *What is the solubility of* AgCl *in* 0·1 M $AgNO_3$ *solution? What is its solubility in a* 0·1 M NaCl *solution?* $K_{sp}(\text{AgCl}) = 1{\cdot}8 \times 10^{-10}$.

Solution : Let the solubility of AgCl in 0·1 M $AgNO_3$ solution be S moles/litre.

$$\therefore \quad [\text{Ag}^+] = (S + 0{\cdot}1) \text{ and } [\text{Cl}^-] = S$$
$$K_{sp} = [\text{Ag}^+][\text{Cl}^-] = (S + 0{\cdot}1)\,S$$
$$\therefore \quad (S + 0{\cdot}1)\,S = 1{\cdot}8 \times 10^{-10}$$
$$S = 1{\cdot}8 \times 10^{-9} \quad \text{mole per litre.}$$

Let the solubility of AgCl in 0·1 M NaCl solution be S' moles/litre.

$$\therefore \quad [\text{Ag}^+] = S' \text{ and } [\text{Cl}^-] = (S' + 0{\cdot}1)$$
$$\therefore \quad S'(S' + 0{\cdot}1) = 1{\cdot}8 \times 10^{-10}$$
$$\text{or} \quad S' = 1{\cdot}8 \times 10^{-9} \text{ mole/litre.}$$

Ex. 56. *Will* MnS *be precipitated from a decimolar solution of* Mn^{2+} *containing* 0·01 M H^+ *ions by passing* H_2S? *Neglect hydrolysis of* S^{2-} *ions.*

$$K_{sp}(\text{MnS}) = 1{\cdot}4 \times 10^{-15}, \; K_{\text{H}_2\text{S}} = 1{\cdot}1 \times 10^{-21}.$$

Solution : The dissociation of H_2S is suppressed by the presence of H^+ ions. Therefore, $[S^{2-}]$ in presence of H^+ may be calculated as follows:

$$\text{H}_2\text{S} \rightleftharpoons 2\text{H}^+ + \text{S}^{2-}$$
$$K(\text{H}_2\text{S}) = \frac{[\text{H}^+]^2[\text{S}^{2-}]}{[\text{H}_2\text{S}]}$$
$$1{\cdot}1 \times 10^{-21} = \frac{(0{\cdot}01)^2[\text{S}^{2-}]}{0{\cdot}1} \qquad ([\text{H}_2\text{S}] = 0{\cdot}1 \text{ M})$$
$$[\text{S}^{2-}] = 1{\cdot}1 \times 10^{-18}$$

Ionic product of MnS $= [Mn^{2+}][S^{2-}] = 0{\cdot}1 \times 1{\cdot}1 \times 10^{-18}$

$= 1{\cdot}1 \times 10^{-19} < 1{\cdot}4 \times 10^{-15} \, (K_{sp})$.

Since ionic product of MnS is less than its solubility product value, there will be no precipitation of MnS.

[**Note:** Concentration of H_2S is taken as 0·1 M if not given.]

Ex. 57. *Will* $Fe(OH)_3$ (s) *precipitate from a buffer solution prepared by mixing* 0·5 M CH_3COOH *and* 0·15 M CH_3COONa *at* 25°C, *if the solution contains* 0·25 M Fe^{3+}?

$$K_{sp}(Fe(OH)_3) = 4 \times 10^{-38}, \; K_a(CH_3COOH) = 1{\cdot}74 \times 10^{-5}.$$

Solution : We have,

$$[H^+] = K_a \frac{[\text{acid}]}{[\text{salt}]}$$

$$= 1{\cdot}74 \times 10^{-5} \times \frac{0{\cdot}5}{0{\cdot}15} = 5{\cdot}8 \times 10^{-5}$$

$$\therefore \quad [OH^-] = \frac{K_w}{[H^+]} = \frac{1 \times 10^{-14}}{5{\cdot}8 \times 10^{-5}} = 1{\cdot}7 \times 10^{-10}.$$

Now, ionic product of $Fe(OH)_3 = [Fe^{3+}][OH^-]^3$

$= (0{\cdot}25)(1{\cdot}7 \times 10^{-10})^3 = 1{\cdot}23 \times 10^{-30}$.

As this value is greater than the given $K_{sp}(Fe(OH)_3)$, i.e., 4×10^{-38}, $Fe(OH)_3$ shall precipitate.

Ex. 58. *Freshly precipitated aluminium and magnesium hydroxides are stirred vigorously in a buffer solution containing* 0·25 *mole/ litre of ammonium chloride and* 0·05 *mole/litre of ammonium hydroxide. Calculate the concentration of aluminium and magnesium ions in the solution.*

$$K_b(NH_4OH) = 1{\cdot}8 \times 10^{-5}; \; K_{sp}(Mg(OH)_2) = 6 \times 10^{-10};$$

$$K_{sp}(Al(OH)_3) = 6 \times 10^{-32}.$$

(IIT 1989)

Solution : $[OH^-]$ of the buffer solution of NH_4OH and NH_4Cl may be calculated as

$$[OH^-] = K_b \cdot \frac{[\text{base}]}{[\text{salt}]} = 1{\cdot}8 \times 10^{-5} \times \frac{0{\cdot}05}{0{\cdot}25} = 3{\cdot}6 \times 10^{-6}. \quad \text{... (Eqn. 5a)}$$

The addition of $Al(OH)_3$ and $Mg(OH)_2$ does not change the OH^- concentration of the buffer solution, hence ignoring OH^- ions produced by the dissociation of $Al(OH)_3$ and $Mg(OH)_2$, we calculate $[Al^{3+}]$ and $[Mg^{2+}]$ as follows:

$$[Al^{3+}] = \frac{K_{sp}\,(Al(OH)_3)}{[OH^-]^3} = \frac{6 \times 10^{-32}}{(3{\cdot}6 \times 10^{-6})^3} = 1{\cdot}286 \times 10^{-15}\ M$$

and $$[Mg^{2+}] = \frac{K_{sp}\,(Mg(OH)_2)}{[OH^-]^2} = \frac{6 \times 10^{-10}}{(3{\cdot}6 \times 10^{-6})^2} = 46{\cdot}29\ M.$$

Ex. 59. *Find the simultaneous solubility of* AgSCN *and* AgBr *if their solubility products are* $1{\cdot}0 \times 10^{-12}$ *and* 5×10^{-13} *respectively.*

Solution : $$\frac{K_{sp}\,(AgSCN)}{K_{sp}\,(AgBr)} = \frac{[Ag^+]\,[SCN^-]}{[Ag^+]\,[Br^-]} = \frac{1 \times 10^{-12}}{5 \times 10^{-13}} = 2$$

or $$\frac{[SCN^-]}{[Br^-]} = 2.$$

Since the concentration of SCN^- ions is twice that of Br^- ion concentration in the solution, the solubility of AgSCN is double that of AgBr.

Let the solubility of AgBr be S moles per litre.

$\therefore$ the solubility of AgSCN = $2S$

Now, for $$\underset{}{AgBr} \rightleftharpoons \underset{S}{Ag^+} + \underset{S}{Br^-}$$

and $$AgSCN \rightleftharpoons \underset{2S}{Ag^+} + \underset{2S}{SCN^-}$$

$$[Ag^+] \text{ in solution} = S + 2S = 3S$$

$$[Br^-] = S$$

$\therefore$ $$K_{sp}\,(AgBr) = [Ag^+]\,[Br^-]$$

$$5 \times 10^{-13} = 3S \times S$$

$\therefore$ $$S = 4{\cdot}083 \times 10^{-7} \text{ mole/litre}$$

$\therefore$ solubility of AgBr $= 4{\cdot}083 \times 10^{-7}$ mole/litre

and solubility of AgSCN $= 8{\cdot}166 \times 10^{-7}$ mole/litre.

Ex. 60. *Calculate* $[F^-]$ *in a solution saturated with respect to both* MgF_2 *and* SrF_2 *assuming no hydrolysis of* F^-. $K_{sp}\,(MgF_2) = 6{\cdot}5 \times 10^{-9}$, $K_{sp}\,(SrF_2) = 2{\cdot}9 \times 10^{-9}$.

Solution : $$\frac{K_{sp}\,(MgF_2)}{K_{sp}\,(SrF_2)} = \frac{[Mg^{2+}]\,[F^-]^2}{[Sr^{2+}]\,[F^-]^2} = \frac{6{\cdot}5 \times 10^{-9}}{2{\cdot}9 \times 10^{-9}}\,;\quad \frac{[Mg^{2+}]}{[Sr^{2+}]} = 2{\cdot}24\,.$$

Thus, the solubility of MgF_2 is 2·24 times that of SrF_2.

Let the solubility of SrF_2 be S moles/litre.

$$\underset{}{SrF_2} \rightleftharpoons \underset{S}{Sr^{2+}} + \underset{2S}{2F^-}$$

$$\underset{}{MgF_2} \rightleftharpoons \underset{2\cdot24\,S}{Mg^{2+}} + \underset{4\cdot48\,S}{2F^-}$$

Now, $K_{sp}\,(SrF_2) = [Sr^{2+}]\,[F^-]^2 = (S)\,(2S + 4\cdot48S)^2 = 2\cdot9 \times 10^{-9}$

$\therefore$ $S = 0\cdot41 \times 10^{-3}$.

Thus, $[F^-] = 2S + 4\cdot48\,S = 6\cdot48\,S = 6\cdot48 \times 0\cdot41 \times 10^{-3}$

$= 2\cdot65 \times 10^{-3}$.

Ex. 61. Ag_2SO_4 *and* $SrSO_4$ *are shaken up with pure water. Calculate the concentration of* Ag^+ *and* Sr^{2+} *in mole/litre in the resulting saturated solution.* $K_{sp}\,(Ag_2SO_4) = 1\cdot5 \times 10^{-5}$, $K_{sp}\,(SrSO_4) = 2\cdot8 \times 10^{-7}$.

Solution : From the K_{sp} data, it is clear that SO_4^{2-} are mainly produced by Ag_2SO_4 because $K_{sp}\,(Ag_2SO_4)$ is much greater than $K_{sp}\,(SrSO_4)$ and $K_{sp}\,(Ag_2SO_4)$, is the product of three ion concentrations.

Let $[SO_4^{2-}] = S$ for Ag_2SO_4

$\therefore$ $K_{sp}\,(Ag_2SO_4) = (2S)^2\,S = 1\cdot5 \times 10^{-5}$

$S = 1\cdot55 \times 10^{-2}$

$\therefore$ $[Ag^+] = 2S = 3\cdot1 \times 10^{-2}$.

And for $SrSO_4$, $K_{sp} = [Sr^{2+}]\,[SO_4^{2-}]$

$2\cdot8 \times 10^{-7} = [Sr^{2+}]\,(1\cdot55 \times 10^{-2})$

$\therefore$ $[Sr^{2+}] = 1\cdot8 \times 10^{-5}$.

Ex. 62. *The solubility product of* $Ag_2C_2O_4$ *at 25°C is* $1\cdot29 \times 10^{-11}$ mole^3 L^{-3}. *A solution of* $K_2C_2O_4$ *containing* 0·1520 *moles in* 500 mL *water is shaken at* 25°C *with excess* Ag_2CO_3 *till the following equilibrium is reached:*

$$Ag_2CO_3 + K_2C_2O_4 \rightleftharpoons Ag_2C_2O_4 + K_2CO_3$$

At equilibrium, the solution contains 0·0358 *mole of* K_2CO_3. *Assuming the degree of dissociation of* $K_2C_2O_4$ *and* K_2CO_3 *to be equal, calculate the solubility product of* Ag_2CO_3.

Solution : In the solution, we have the equilibrium,

(excess)	0·304 M		0	0	Initial concn.
Ag_2CO_3 +	$K_2C_2O_4$	$\rightleftharpoons$	$Ag_2C_2O_4$ +	K_2CO_3	
(solid)	(0·304 – 0·0716) = 0·2324		(ppt.) 0·0716 M	0·0716 M	Eqb. concn.

$[Ag^+]$ in the solution may be calculated from the K_{sp} value of $Ag_2C_2O_4$ (supposing $K_2C_2O_4$ to be completely ionised).

$$K_{sp}(Ag_2C_2O_4) = [Ag^+]^2[C_2O_4^{2-}]$$
$$1{\cdot}29 \times 10^{-11} = [Ag^+]^2 (0{\cdot}2324)$$
$$[Ag^+] = 7{\cdot}45 \times 10^{-6} \text{ M}$$

($C_2O_4^{2-}$ concn. from $Ag_2C_2O_4$ is negligible)

Supposing K_2CO_3 to be completely ionised,

$$[CO_3^{2-}] = [K_2CO_3]$$
$$= 0{\cdot}0716 \text{ M}$$

$\therefore \quad K_{sp}(Ag_2CO_3) = [Ag^+]^2[CO_3^{2-}] \quad ([CO_3^{2-}]$ from Ag_2CO_3 is negligible)

$$= (7{\cdot}45 \times 10^{-6})^2 (0{\cdot}0716)$$
$$= 3{\cdot}97 \times 10^{-12} \text{ (mole/litre)}^3.$$

Ex. 63. *What is the* $[CO_3^{2-}]$ *in a* 0·001 M Na_2CO_3 *solution after the hydrolysis reactions have come to an equilibrium?* $K_a(HCO_3^-) = 4{\cdot}7 \times 10^{-11}$

Solution : The first step of hydrolysis of CO_3^{2-} ions is predominant.

$$\underset{(0{\cdot}001 - x)}{\overset{0{\cdot}001}{CO_3^{2-}}} + H_2O \rightleftharpoons \underset{x}{HCO_3^-} + \underset{x}{OH^-} \qquad \text{... At eqb.}$$

$$K_h = \frac{x^2}{(0{\cdot}001 - x)} = \frac{K_w}{K_2} = \frac{1 \times 10^{-14}}{4{\cdot}7 \times 10^{-11}}$$

or $\quad x = 3{\cdot}7 \times 10^{-4}$

$\therefore \quad [CO_3^{2-}] = 0{\cdot}001 - 0{\cdot}00037 = 0{\cdot}00063$ M.

Ex. 64. *A sample of* AgCl *was treated with* 5·0 mL *of* 1·5 M Na_2CO_3 *solution to give* Ag_2CO_3. *The remaining solution contained* 0·0026 g *of* Cl^- *per litre. Calculate the solubility product of* AgCl. $[K_{sp}(Ag_2CO_3) = 8{\cdot}2 \times 10^{-12}]$

(IIT 1997)

Solution : Initial $[Na_2CO_3] = [CO_3^{2-}] = 1{\cdot}5$ M

$$\text{Equilibrium } [Cl^-] = [NaCl] = \frac{0{\cdot}0026}{35{\cdot}5} = 0{\cdot}0000732 \text{ M}$$

$$2AgCl\,(s) + \underset{\substack{(1{\cdot}5 - 0{\cdot}0000366)\\ = 1{\cdot}5\text{ M}}}{\overset{1{\cdot}5\text{ M}}{Na_2CO_3}} \rightleftharpoons Ag_2(CO)_3\,(s) + \underset{0{\cdot}0000732\text{ M}}{\overset{0}{2NaCl}} \quad \begin{matrix}\text{– initial concn.}\\ \text{– eqb. concn.}\end{matrix}$$

$$[Ag^+] = \sqrt{\frac{K_{sp}(Ag_2CO_3)}{[CO_3^{2-}]}} = \sqrt{\frac{8\cdot2\times10^{-12}}{1\cdot5}} = 2\cdot338\times10^{-6}\text{ M}$$

$$\therefore \quad K_{sp} = [Ag^+][Cl^-] = (2\cdot338\times10^{-6})(0\cdot0000732) = 1\cdot71\times10^{-10}.$$

Ex. 65. *Calculate the minimum mass of* NaCl *required to dissolve* 0·01 *mole of* AgCl *in* 100 *litres of solution.* K_{sp} (AgCl) = 1×10^{-10}, K_d ($AgCl_2^-$) = $3\cdot3\times10^{-6}$.

Solution : Molar concentration of AgCl = $\frac{0\cdot01}{100} = 10^{-4}$ M.

As AgCl dissolves completely, $[Ag^+] = 10^{-4}$ M. In complexation, we assume that 10^{-4} M Ag^+ first combines with Cl^- completely to produce 10^{-4} M $AgCl_2^-$ and then we consider the equilibrium for the decomposition of $AgCl_2^-$ in the presence of minimum concentration, say, x molar NaCl (i.e., $[Cl^-] = x$) solution.

$$AgCl_2^- \longrightarrow Ag^+ + 2Cl^-$$

$$K_d = \frac{[Ag^+][Cl^-]^2}{[AgCl_2^-]} = \frac{K_{sp}(AgCl)\times[Cl^-]}{10^{-4}} = 3\cdot3\times10^{-6}$$

or $$\frac{10^{-10}\times x}{10^{-4}} = 3\cdot3\times10^{-6}$$

or x = 3·3 moles/litre

= 3·3 × 58·5 grams/litre

= 193·05 grams/litre.

∴ amount of NaCl per 100 litre = 19305 g = 19·3 kg.

Ex. 66. *Equal volumes of* 0·02 M $AgNO_3$ *and* 0·02 M HCN *were mixed. Calculate* $[Ag^+]$ *at equilibrium assuming no cyano-complex formation.* K_{sp} (AgCN) = $2\cdot2\times10^{-16}$, K_a (HCN) = $6\cdot2\times10^{-10}$.

Solution : In precipitation reaction, we first assume complete precipitation.

$$Ag^+ + HCN \longrightarrow AgCN + H^+$$

Since equal volumes of $AgNO_3$ and HCN are mixed, $[H^+]$ = 0·01 M. Now for the equilibria,

$$AgCN \rightleftharpoons Ag^+ + CN^-; \quad K_{sp} = [Ag^+][CN^-] = 2\cdot2\times10^{-16}$$

$$HCN \rightleftharpoons H^+ + CN^-; \quad K_a = \frac{[H^+]\cdot[CN^-]}{[HCN]} = 6\cdot2\times10^{-10}$$

Since every CN^- which hydrolyses produces one HCN

$$\therefore \quad [Ag^+] = [CN^-] + [HCN]$$

$$\frac{2.2 \times 10^{-16}}{[CN^-]} = [CN^-] + \frac{[H^+][CN^-]}{6.2 \times 10^{-10}}$$

$$= [CN^-] + \frac{(0.01)[CN^-]}{6.2 \times 10^{-10}}$$

$$[CN^-] = 3.7 \times 10^{-12}$$

$$\therefore \quad [Ag^+] = \frac{K_{sp}(AgCN)}{[CN^-]} = \frac{2.2 \times 10^{-16}}{3.7 \times 10^{-12}} = 5.9 \times 10^{-5} \text{ M.}$$

Ex. 67. *The solubility product of* AgCl *at* 25°C *is* 1×10^{-10}. *A solution of* Ag^+ *at a concentration of* 4×10^{-3} M *just fails to yield a precipitate of* AgCl *with concentration of* 1×10^{-3} M *of* Cl^- *when the concentration of* NH_3 *in the solution is* 2×10^{-2} M. *Calculate the equilibrium constant for*

$$[Ag(NH_3)_2]^+ \rightleftharpoons Ag^+ + 2NH_3.$$

Solution : $K_{sp} = [Ag^+][Cl^-]$

$$10^{-10} = [Ag^+](10^{-3}); \quad [Ag^+] = 10^{-7}$$

Now,

$$\begin{array}{cccccl} 4 \times 10^{-3}\text{M} & & 10^{-7}\text{M} & & 2 \times 10^{-2}\text{M} & \text{... Initially} \\ Ag(NH_3)_2^+ & \rightleftharpoons & Ag^+ & + & 2NH_3 & \\ (4 \times 10^{-3} - x)\text{ M} & & (x + 10^{-7})\text{ M} & & (2x + 2 \times 10^{-2})\text{M} & \text{... At eqb.} \\ = 4 \times 10^{-3} & & = 10^{-7} & & = 2 \times 10^{-2} & \end{array}$$

Thus,

$$K = \frac{[Ag^+][NH_3]^2}{[Ag(NH_3)_2^+]} = \frac{10^{-7} \times (2 \times 10^{-2})^2}{4 \times 10^{-3}} = 10^{-8}.$$

Ex. 68. *What is the concentration of* Ag^+ *ions in* 0·01 M $AgNO_3$ *that is also* 1·0 M NH_3? *Will* AgCl *precipitate from a solution that is* 0·01 M $AgNO_3$, 0·01 M NaCl *and* 1 M NH_3? $K_d(Ag[NH_3]_2^+) = 5.88 \times 10^{-8}$, K_{sp} (AgCl) $= 1.8 \times 10^{-10}$.

Solution : Let us first assume that 0·01 M $AgNO_3$ shall combine with 0·02 M NH_3 to form 0·01 M $Ag(NH_3)_2^+$ and then consider its dissociation.

$$\begin{array}{ccccc} 0.01\text{ M} & & 1\text{ M} & & 0 \\ AgNO_3 & + & 2NH_3 & \longrightarrow & Ag(NH_3)_2^+ \\ 0 & & (1 - 0.02) = 0.98\text{ M} & & 0.01\text{ M} \end{array}$$

$$\begin{array}{cccccl} 0.01\text{ M} & & 0 & & 0.98\text{ M} & \text{... Initial concn.} \\ Ag(NH_3)_2^+ & \rightleftharpoons & Ag^+ & + & 2NH_3 & \\ (0.01 - x) & & x & & (0.98 + 2x) & \text{... Eqb. concn.} \\ = 0.01\text{ M} & & & & = 0.98\text{ M} & \end{array}$$

$$K_d = \frac{[Ag^+][NH_3]^2}{[Ag(NH_3)_2^+]} = 5{\cdot}88 \times 10^{-8}$$

$$\therefore \quad [Ag^+] = \frac{5{\cdot}88 \times 10^{-8} \times 0{\cdot}01}{(0{\cdot}98)^2} = 6{\cdot}12 \times 10^{-10} \text{ M.}$$

Further, ionic product of AgCl $= [Ag^+][Cl^-]$

$$= (6{\cdot}12 \times 10^{-10})(0{\cdot}01)$$

$$= 6{\cdot}12 \times 10^{-12}.$$

Because the ionic product is smaller than $K_{sp} = 1{\cdot}8 \times 10^{-10}$, no precipitate should form.

Ex. 69. *Calculate the solubility of* AgCN *in a buffer solution of* pH 3·00 *assuming no complex formation.* K_a (HCN) $= 6{\cdot}2 \times 10^{-10}$ *and* K_{sp}(AgCN) $= 2{\cdot}2 \times 10^{-16}$.

Solution : Let the solubility of AgCN be x moles/litre. On dissolution silver remains as Ag^+ but CN^- ions are converted mostly to HCN due to the fixed acidity of the buffer.

$$K_a = \frac{[H^+][CN^-]}{[HCN]}$$

$$\text{or} \quad \frac{[HCN]}{[CN^-]} = \frac{[H^+]}{K_a} = \frac{1 \times 10^{-3}}{6{\cdot}2 \times 10^{-10}} = 1{\cdot}6 \times 10^6. \qquad \dots (1)$$

As each CN^- ion hydrolyses to yield one HCN

$$x = [Ag^+] = [CN^-] + [HCN]$$

from the ratio $[HCN]/[CN^-]$, we see $[CN^-] << [HCN]$.

$$\therefore \quad x = [Ag^+] = [HCN]$$

Thus from the equation (1) we have,

$$[CN^-] = \frac{x}{1{\cdot}6 \times 10^6}$$

Now, $K_{sp} = [Ag^+][CN^-]$

$$2{\cdot}2 \times 10^{-16} = x\left(\frac{x}{1{\cdot}6 \times 10^6}\right)$$

$$x = 1{\cdot}9 \times 10^{-5} \text{ M.}$$

Ex. 70. *Calculate the solubility of* MnS *in pure water,* K_{sp} (MnS) $= 2{\cdot}5 \times 10^{-10}$. *Assume hydrolysis of* S^{2-} *ions.* K_1 *and* K_2 *for* H_2S *are* 1×10^{-7} *and* 1×10^{-14} *respectively.*

Solution : Let the solubility of MnS be x mole/litre. On dissolution $[Mn^{2+}] = x$ but $[S^{2-}] \neq x$ because S^{2-} undergoes extensive hydrolysis. Let us consider the two stages of hydrolysis.

$$S^{2-} + H_2O \rightleftharpoons HS^- + OH^- ; K_h' = \frac{K_w}{K_2} = \frac{1 \times 10^{-14}}{1 \times 10^{-14}} = 1{\cdot}0$$

$$HS^- + H_2O \rightleftharpoons H_2S + OH^- ; K_h'' = \frac{K_w}{K_1} = \frac{1 \times 10^{-14}}{1 \times 10^{-7}} = 1 \times 10^{-7}$$

As $K_h' >> K_h''$, first stage of hydrolysis is almost complete.

$\therefore \quad x = [Mn^{2+}] = [HS^-] = [OH^-]$

Consider first stage of hydrolysis,

$$K_h' = \frac{[HS^-][OH^-]}{[S^{2-}]}$$

or $$[S^{2-}] = \frac{[HS^-][OH^-]}{K_h'} = \frac{x^2}{1{\cdot}0}.$$

At equilibrium,

$$[Mn^{2+}][S^{2-}] = K_{sp} = 2{\cdot}5 \times 10^{-10}$$

$$x = \frac{2{\cdot}5 \times 10^{-10}}{x^2}$$

$$x = 6{\cdot}3 \times 10^{-4} \text{ M.}$$

Ex. 71. *What is the nature of* 0·01 M $NaHCO_3$ *solution? Calculate its* pH. K_1 *and* K_2 *for* H_2CO_3 *are* $4{\cdot}5 \times 10^{-7}$ *and* $4{\cdot}7 \times 10^{-11}$ *respectively.*

Solution : Na^+ ions do not hydrolyse. HCO_3^- may undergo hydrolysis as well as ionisation.

$$HCO_3^- + H_2O \rightleftharpoons OH^- + H_2CO_3$$

$$K_h = \frac{[OH^-][H_2CO_3]}{[HCO_3^-]} = \frac{K_w}{K_1} = \frac{1 \times 10^{-14}}{4{\cdot}5 \times 10^{-7}} = 2{\cdot}2 \times 10^{-8}. \quad \text{... (1)}$$

And,

$$HCO_3^- \rightleftharpoons H^+ + CO_3^{2-}$$

$$K_2 = \frac{[H^+][CO_3^{2-}]}{[HCO_3^-]} = 4{\cdot}7 \times 10^{-11}. \quad \text{... (2)}$$

As $K_h > K_2$, i.e., $[OH^-] > [H^+]$, the solution will be alkaline. In this solution, the electrical neutrality can be preserved by maintaining anionic charge as the cationic charge remains at 0·01 M (= $[Na^+]$) For every negative charge removed by converting HCO_3^- to H_2CO_3, another negative charge is created by converting HCO_3^- to CO_3^{2-}. Thus,

$$[H_2CO_3] = [CO_3^{2-}] = x \text{ (say)}$$

$$\therefore \quad [HCO_3^-] = 0{\cdot}01 - 2x \approx 0{\cdot}01 \text{ M.}$$

Multiplying eqns. (1) and (2),

$$K_h \times K_2 = \frac{[H^+][OH^-][H_2CO_3][CO_3^{2-}]}{[HCO_3^-]^2} = (2{\cdot}2 \times 10^{-8})(4{\cdot}7 \times 10^{-11})$$

Substituting $[H^+][OH^-] = K_w = 1 \times 10^{-14}$

$$x = 1{\cdot}02 \times 10^{-4}.$$

Now using the eqn. (2) again, we get

$$K_2 = \frac{[H^+](1{\cdot}02 \times 10^{-4})}{0{\cdot}01} = 4{\cdot}7 \times 10^{-11}$$

or $\quad [H^+] = 4{\cdot}6 \times 10^{-9}$

$pH = -\log(4{\cdot}6 \times 10^{-9}) = 8{\cdot}34.$

Ex. 72. *The basic ionisation constant for hydrazine,* N_2H_4, *is* $9{\cdot}6 \times 10^{-7}$. *What would be the per cent hydrolysis of* $0{\cdot}10$ M N_2H_5Cl, *a salt containing the acid ion conjugate to hydrazine base?*

Solution : $N_2H_5^+ + H_2O \rightleftharpoons N_2H_4 + H_3O^+$

$$K_h = \frac{K_w}{K_b} = \frac{1 \times 10^{-14}}{9{\cdot}6 \times 10^{-7}} = 1{\cdot}04 \times 10^{-8}$$

$$h = \sqrt{\frac{K_h}{c}} = \sqrt{\frac{1{\cdot}04 \times 10^{-8}}{0{\cdot}10}} = 3{\cdot}225 \times 10^{-4}.$$

$\therefore$ per cent hydrolysis = 0·03225%.

Ex. 73. *What is the pH of a* $0{\cdot}10$ M *solution of* Na_2SO_4? $K_2(H_2SO_4) = 1 \times 10^{-2}$. *Assume hydrolysis of* SO_4^{2-} *to* HSO_4^-.

Solution : As Na_2SO_4 is a salt of NaOH and H_2SO_4, it appears that it will not undergo hydrolysis being a salt of a strong acid and a strong base and the pH should, therefore, be 7. But SO_4^{2-} ions may undergo hydrolysis as HSO_4^- is a weak acid with $K_a = 1 \times 10^{-2}$. The hydrolysis equilibrium is

$$SO_4^{2-} + H_2O \rightleftharpoons HSO_4^- + OH^-$$

As Na_2SO_4 is a strong electrolyte, $[SO_4^{2-}] = [Na_2SO_4] = 0{\cdot}10$.

$$\therefore \quad pH = \frac{1}{2}\{pK_w + pK_a + \log a\} \qquad \text{... (11.b)}$$

$$= \frac{1}{2}\{-\log 10^{-14} + (-\log 10^{-2}) + \log(0{\cdot}10)\}$$

$$= 7{\cdot}5.$$

Ex. 74. *A* 50·0-mL *aliquot of a* 0·01 M *solution of* HCOOH *was titrated with* 0·10 M NaOH. *Predict the* pH *of the solution (a) at the half-equivalence point*

(b) at the equivalence point, and (c) after an excess 10 mL *of the base is added.* K_a (HCOOH) $= 1{\cdot}772 \times 10^{-4}$

Solution : (a) m.m. of HCOOH taken $= 0{\cdot}01 \times 50 = 0{\cdot}5$.

Because NaOH and HCOOH combine in 1 : 1 molar ratio, m.m. of NaOH added at the half-equivalence point $= 0{\cdot}25$

$\therefore$ in the mixture,

m.m. of HCOONa formed $= 0{\cdot}25$

m.m. of HCOOH left unreacted $= 0{\cdot}5 - 0{\cdot}25 = 0{\cdot}25$.

Thus, we have for buffer solution of HCOOH – HCOONa,

$$pH = pK_a + \log \frac{\text{m.m. of salt}}{\text{m.m. of acid}}$$

$$pH = -\log(1{\cdot}772 \times 10^{-4}) + \log \frac{0{\cdot}25}{0{\cdot}25} = 3{\cdot}7515.$$

(b) At the equivalence point 0·5 m.m. of HCOOH combines with 0·5 m.m. of NaOH to form 0·5 m.m. of HCOONa.

$$\text{Volume of NaOH required at the equivalence point} = \frac{0{\cdot}01 \times 50}{0{\cdot}1} = 5 \text{ mL}.$$

Total volume at the equivalence point $= 50 + 5 = 55$ mL.

$$\therefore \quad [\text{HCOONa}] = \frac{0{\cdot}5}{55} = [\text{HCOO}^-]$$

As $HCOO^-$ ions undergo hydrolysis,

$$pH = \frac{1}{2}\{pK_w + \log K_a + \log a\}$$

$$pH = \frac{1}{2}\left\{14 + 3{\cdot}7515 + \log \frac{0{\cdot}5}{55}\right\} = 7{\cdot}85.$$

(c) m.m. of the excess $NaOH = 0{\cdot}1 \times 10 = 1$ total volume $= 55 + 10 = 65$ mL.

$$\therefore \quad \text{molarity of NaOH} = \frac{1}{65} \text{ M}.$$

In the presence of the strong electrolyte NaOH, OH^- ions produced due to hydrolysis of HCOONa may be neglected.

$$\therefore \quad pOH = -\log[OH^-] = -\log\left(\frac{1}{65}\right) = 0{\cdot}8129.$$

$$\therefore \quad pH = pK_w - pOH = 14 - 0{\cdot}8129 = 13{\cdot}1871.$$

Ex. 75. 0·01 *mole of* $AgNO_3$ *is added to one litre of a solution which is* 0·1 M *in* Na_2CrO_4 *and* 0·005 M *in* $NaIO_3$. *Calculate the mole of precipitate*

formed at equilibrium and the concentrations of Ag^+, IO_3^- *and* CrO_4^{2-}.
$K_{sp}(Ag_2CrO_4) = 10^{-8}$, $K_{sp}(AgIO_3) = 10^{-13}$.

Solution : Let x and y be the molar concentrations of Ag^+ required to precipitate of Ag_2CrO_4 and $AgIO_3$ respectively.

$$\therefore \quad x = \sqrt{\frac{K_{sp}(Ag_2CrO_4)}{[CrO_4^{2-}]}} = \sqrt{\frac{10^{-8}}{0\cdot1}} = 3\cdot16 \times 10^{-4}$$

$$y = \frac{K_{sp}(AgIO_3)}{[IO_3^-]} = \frac{10^{-13}}{0\cdot005} = 2\cdot0 \times 10^{-11}$$

As $y << x$, $AgIO_3$ will precipitate first

$$\begin{array}{lcccl} 0\cdot01 \text{ mole} & & 0\cdot005 \text{ mole} & & \\ AgNO_3 & + & NaIO_3 & = & AgIO_3(s) + NaNO_3 \\ 0\cdot01 - \cdot005 & & 0 & & 0\cdot005 \text{ mole} \\ = 0\cdot005 \text{ mole} & & & & \end{array}$$

And then, for remaining 0·005 mole of $AgNO_3$,

$$\begin{array}{lcccl} 0\cdot005 \text{ mole} & & 0\cdot01 \text{ mole} & & \\ 2AgNO_3 & + & Na_2CrO_4 & \rightarrow & Ag_2CrO_4(s) + 2NaNO_3 \\ 0 & & 0\cdot1 - 0\cdot0025 & & 0\cdot0025 \text{ mole} \\ & & = 0\cdot0975 \text{ mole} & & \end{array}$$

$\therefore$ mole of $AgIO_3$ ppt. = 0·005

and mole Ag_2CrO_4 ppt. = 0·0025

Now, for the equilibrium

$$Ag_2CrO_4 \rightleftharpoons 2Ag^+ + CrO_4^{2-}$$

$$K_{sp} = [Ag^+]^2[CrO_4^{2-}]$$

$$10^{-8} = [Ag^+]^2(0\cdot0975) \qquad \because \ [CrO_4^{2-}] = 0\cdot0975 \text{ M}$$

$$[Ag^+] = \sqrt{\frac{10^{-8}}{0\cdot0975}} = 3\cdot2 \times 10^{-4} \text{ mole/L.}$$

Further, for the equilibrium

$$AgIO_3 \rightleftharpoons Ag^+ + IO_3^-$$

$$K_{sp} = [Ag^+][IO_3^-]$$

$$10^{-13} = (3\cdot2 \times 10^{-4})[IO_3^-]$$

$$[IO_3^-] = 3\cdot125 \times 10^{-10} \text{ mole/L}$$

Ex. 76. 500 mL *of* 0·2 M *aqueous solution of acetic acid is mixed with* 500 mL *of* 0·2 M HCl *at* 25°C

(i) Calculate the degree of dissociation of acetic acid in the resulting solution and pH *of the solution.*

(ii) If 6 g *of* NaOH *is added to the above solution, determine the final* pH. *(Assume there is no change in volume on mixing;*
$K_a(CH_3COOH) = 1\cdot75 \times 10^{-5}$ mol/L*)* **(IIT 2002)**

Solution : (i) As equal volumes of CH_3COOH and HCl are mixed, molarity of each will be 0·1 mole/L or 0·1 M.

$$\begin{array}{llccc} & 0\cdot1\text{ M} & & & \\ & CH_3COOH & \rightleftharpoons & CH_3COO^- & + \quad H^+ \\ \text{At. eqb.:} & 0\cdot1(1-x) & & 0\cdot1x & (0\cdot1x+0\cdot1) \\ & & & & = 0\cdot1 \quad \uparrow\text{(due to HCl)} \end{array}$$

where x is the degree of dissociation of CH_3COOH in the presence of HCl.

$$\therefore \quad K_a = \frac{0\cdot1x \times 0\cdot1}{0\cdot1(1-x)} = 0\cdot1x \quad (x \text{ is very small})$$

$$x = \frac{1\cdot75 \times 10^{-5}}{0\cdot1} = 1\cdot75 \times 10^{-4}$$

$pH = -\log[H^+] = -\log(0\cdot1) = 1.$ (H^+ ions from CH_3COOH neglected)

(ii) Molarity of NaOH $= \frac{6}{40}$ mole/L = 0·15 M.

0·15 mole of NaOH, when added to this acid mixture, will neutralise 0·1 mole of HCl and 0·05 mole of CH_3COOH. The resulting solution will be 0·05 M in CH_3COOH and 0·05 M in CH_3COONa which is a buffer solution.

$$\therefore \quad pH = pK_a + \log\frac{[\text{salt}]}{[\text{acid}]}$$

$$pH = -\log(1\cdot75 \times 10^{-5}) + \log\frac{0\cdot05}{0\cdot05}$$

$$pH = 4\cdot757.$$

Ex. 77. pH *of the half-equivalence point of the titration of valine,* $(CH_3)_2CH\,CHNH_2COOH$, *is* 2·286 *with* HCl *and* 9·719 *with* NaOH. *Calculate the isoelectric point of valine, that is, the* pH *at which the dipole ion does not migrate in an electric field.*

Solution : The equilibrium of the zwitterion can be given as

$$\begin{array}{c}NH_3^+\\|\\HCH(CH_3)_2\\|\\COOH\end{array} \xrightleftharpoons{(pH)_1 = 2\cdot286} \begin{array}{c}NH_3^+\\|\\HCH(CH_3)_2\\|\\COO^-\end{array} \xrightleftharpoons{(pH)_2 = 9\cdot719} \begin{array}{c}NH_2\\|\\HCH(CH_3)_2\\|\\COO^-\end{array}$$

$$\text{where, } pH = \frac{(pH)_1 + (pH)_2}{2} = \frac{2\cdot286 + 9\cdot719}{2} = 6\cdot003.$$

PROBLEMS

(Answers bracketed with questions)

1. Calculate the degree of ionisation of 0·4 M acetic acid in water. Dissociation constant of acetic acid is $1{\cdot}8 \times 10^{-5}$. ($6{\cdot}71 \times 10^{-3}$)

2. Calculate the degree of dissociation of 0·2 N of a monobasic acid at 25°C. The dissociation constant of acetic acid at this temperature is $1{\cdot}8 \times 10^{-5}$. What will be the H^+ concentration? ($9{\cdot}48 \times 10^{-3}$, $1{\cdot}89 \times 10^{-3}$ M)

[**Hint:** $K = \dfrac{0{\cdot}2x^2}{(1-x)} = 1{\cdot}8 \times 10^{-5}$; $[H^+] = 0{\cdot}2x$, x is the degree of dissociation]

3. Calculate $[OH^-]$ for a solution whose pH is 6·2. ($1{\cdot}6 \times 10^{-8}$ M)

4. The dissociation constant of HCN is $4{\cdot}8 \times 10^{-10}$. What is the concentration of H_3O^+, OH^- and HCN in a solution prepared by dissolving 0·16 mole of NaCN in 450 mL of water at 25°C? ($[OH^-] = [HCN] = 2{\cdot}72 \times 10^{-3}$ M)

[**Hint:** Apply Equation 11(a).]

5. At what concentration of the solution will the degree of dissociation of nitrous acid be 0·2? K_a for HNO_2 is 4×10^{-4}. (0·008 M)

6. The degree of dissociation of acetic acid in a 0·1 N solution is $1{\cdot}32 \times 10^{-2}$. At what concentration of nitrous acid will its degree of dissociation be the same as that of acetic acid? $\{K_a(HNO_2) = 4 \times 10^{-4}\}$ (2·3 mole/litre)

7. Calculate the pH of the following aqueous solutions:

(i) 5×10^{-8} M HCl (ii) 5×10^{-10} M HCl

(iii) 10^{-8} M NaOH (iv) 10^{-10} M NaOH (6·89, 7, 7·02, 7)

8. Compute the pH of a solution at 25°C which is twice as alkaline as pure water. (7·3)

9. How many times is the H^+ concentration in the blood (pH = 7·36) greater than in the spinal fluid (pH = 7·53)? (1·5 times)

10. Calculate the pH of an NaOH solution, the concentration of which is 0·1 g/L. Assume the dissociation of NaOH to be complete. (11·40)

11. Find the pH of a 0·01 M solution of acetic acid, dissociating to the extent of 4·2%. (3·38)

12. Determine the pH value of a solution obtained by mixing 25 mL of 0·2 M HCl and 50 mL of 0·25 N NaOH solutions. (13)

13. Calculate how many H^+ ions are present in one millionth part of 1 mL of pure water. The ionic product of water is $1 \times 10^{-14}\ (\text{mol/L})^2$ (60·3 million)

14. Assuming the first step of dissociation to be complete, find the concentrations of all species in a 0·1 M H_2SO_4 solution. $K_2 = 1{\cdot}2 \times 10^{-2}$.

$([H_2SO_4] = 0\text{ M},\ [HSO_4^-] = 0{\cdot}09\text{ M},\ [H^+] = 0{\cdot}11\text{ M})$

15. Calculate the concentrations of various species in a 0·1 M H_2S saturated solution. $K_1 = 1 \times 10^{-7}$ and $K_2 = 1{\cdot}3 \times 10^{-13}$.

$$\left[\begin{array}{r}[H_2S] \approx 0{\cdot}1\text{ M} \\ [S^{2-}] = 1{\cdot}3 \times 10^{-13}\text{ M} \\ [HS^-] = 1 \times 10^{-4}\text{ M} \\ [H^+] = 1 \times 10^{-4}\text{ M}\end{array}\right]$$

16. A weak base BOH of concentration 0·02 mole/litre has a pH value of 10·45. If 100 mL of this base is mixed with 10 mL of 0·1 M HCl, what will be the pH of the mixture? (8·59)

17. How will the pH increase if 0·05 mole of sodium acetate is added to 1 litre of a 0·005 M acetic acid solution? $K_a\ (CH_3COOH) = 1{\cdot}8 \times 10^{-5}$. (pH increases by 2·21)

18. Calculate the pH of 0·1 M acetic acid solution if its dissociation constant is $1{\cdot}8 \times 10^{-5}$. If 1 litre of this solution is mixed with 0·05 mole of HCl, what will be pH of the mixture? (2·87, 1·3)

19. 2·05 g of sodium acetate was added to 100 mL of 0·1 M HCl solution. Find the H^+ ion concentration of the resulting solution. If 6 mL of 1 M HCl is further added to it, what will be the new H^+ concentration? $(1{\cdot}23 \times 10^{-5}\text{ M},\ 1{\cdot}36 \times 10^{-5}\text{ M})$

20. Calculate the pH of a buffer solution prepared by dissolving 30 g of Na_2CO_3 in 500 mL of an aqueous solution containing 150 mL of 1 M HCl. $K_a(H_2CO_3) = 4{\cdot}2 \times 10^{-7}$; $K_a(HCO_3^-) = 4{\cdot}8 \times 10^{-11}$. (10·3)

21. The pH of a buffer solution containing 0·1 M CH_3COOH and 0·1 M CH_3COONa is 4·74. What will be the pH if 0·05 mole of HCl is added to one litre of this buffer solution? $K_a\ (CH_3COOH) = 1{\cdot}8 \times 10^{-5}$. (4·27)

22. The concentration of hydrogen ion in a 0·2 M solution of formic acid is $6{\cdot}4 \times 10^{-3}$ mole per litre. To this solution sodium formate is added so as to adjust the concentration of sodium formate to one mole per litre. What will be the pH of this solution? The dissociation constant of formic acid is $2{\cdot}4 \times 10^{-4}$ and the degree of dissociation of sodium formate is 0·75. **(IIT 1985)** (4·19)

23. When 0·20 M acetic acid is neutralised with 0·20 M NaOH in 0·50 litre of solution, the resulting solution is slightly alkaline. Calculate the pH of the resulting solution. $K_a\ (CH_3COOH) = 1{\cdot}8 \times 10^{-5}$. (9·02)

24. The pH of a 0·1 M solution of NH_4Cl is 5·13. Find the dissociation constant of NH_4OH. $(1{\cdot}8 \times 10^{-5})$

25. A buffer solution is prepared by dissolving 0·2 mole of sodium formate and 0·25 mole of formic acid in approximately 200 (± 50) mL of water. What will be the concentration of H^+ and OH^-? K_a (HCOOH) $= 1{\cdot}8 \times 10^{-4}$ ($[OH^-] = 4{\cdot}4 \times 10^{-11}$)

26. A buffer solution was prepared by dissolving 0·05 mole formic acid and 0·06 mole sodium formate in enough water to make 1 litre of solution. K_a(HCOOH) $= 1{\cdot}8 \times 10^{-4}$.

(a) Calculate the pH of the solution.

(b) If this solution were diluted to 10 times its volume, what would be the pH? (3·83, 3·83)

27. How many gram moles of HCl will be required to prepare one litre of a buffer solution (containing NaCN and HCl) of pH 8·5 using 0·01 gram-formula weight of NaCN? $K_{dissociation}$ (HCN) $= 4{\cdot}1 \times 10^{-10}$ (0·0088 mole)

28. The pK_a of acetyl salicylic acid (aspirin) is 3·5. The pH of gastric juice in the human stomach is about 2–3 and the pH in the small intestine is about 8. Aspirin will be

(a) un-ionised in the small intestine and in the stomach

(b) completely ionised in the small intestine and in the stomach

(c) ionised in the stomach and almost un-ionised in the small intestine

(d) ionised in the small intestine and almost un-ionised in the stomach (d)

29. Calculate the degree of hydrolysis of an N/10 KCN solution at 25°C. K_a (HCN) $= 7{\cdot}2 \times 10^{-10}$; $K_w = 1 \times 10^{-14}$. (1·18%)

30. Calculate the degree of hydrolysis of CH_3COOK in 0·1 M and the pH of the solution. $K_a(CH_3COOH) = 1{\cdot}8 \times 10^{-5}$ ($7{\cdot}5 \times 10^{-5}$; 8·88)

31. Calculate the hydrolysis constant of NH_4Cl; determine the degree of hydrolysis of this salt in 0·01 M solution and the pH of the solution. $K_b(NH_4OH) = 1{\cdot}8 \times 10^{-5}$ ($5{\cdot}6 \times 10^{-10}$, $2{\cdot}4 \times 10^{-4}$, 5·63)

32. A 0·02 M solution of CH_3COONa in water at 25°C is found to have an H^+ concentration of 3×10^{-9} g ionic weight per litre. What is the hydrolytic constant of the salt? $K_w = 1{\cdot}01 \times 10^{-14}$, K_a (CH_3COOH) $= 1{\cdot}75 \times 10^{-5}$. ($5{\cdot}68 \times 10^{-10}$)

33. Calculate the hydrolysis constant of the reaction

$$HCO_2^- + H_2O \rightleftharpoons HCO_2H + OH^-$$

and find the concentrations of H_3O^+, OH^-, HCO_2^- and HCO_2H in a solution of 0·15 M HCO_2Na. K_a (HCOOH) $\rightleftharpoons 1{\cdot}8 \times 10^{-4}$.

[**Hint:** See equations 10 and 11.] ($5{\cdot}56 \times 10^{-11}$)

34. Calculate the pH of each of the following solutions:

(a) 100 mL 0·1 M CH_3COOH mixed with 100 mL of 0·1 M NaOH

(b) 100 mL of 0·1 M CH_3COOH mixed with 50 mL of 0·1 M NaOH

(c) 50 mL of 0·1 M CH_3COOH mixed with 100 mL of 0·1 M NaOH

(K_a (CH_3COOH) = $1{\cdot}8 \times 10^{-5}$, $K_w = 1 \times 10^{-14}$) [(a) 8·72 (b) 4·75 (c) 12·52]

[**Hint:** (a) Calculate pH due to hydrolysis of CH_3COONa produced.

(b) Calculate pH of the buffer solutions of CH_3COOH and CH_3COONa produced.

(c) Calculate pH only due to NaOH remained.]

35. Determine the solubility of AgCl (in mole/litre) in water.
K_{sp} (AgCl) = $1{\cdot}8 \times 10^{-10}$. ($1{\cdot}3 \times 10^{-5}$ mole/litre)

36. What is the solubility product of Ag_2CrO_4 if 0·0166 g of the salt dissolves in 500 mL of water at 18°C? [K_{sp} (Ag_2CrO_4) = 4×10^{-12}]

37. The solubility of lead sulphate in water is $1{\cdot}03 \times 10^{-4}$. Calculate its solubility in a centinormal solution of $H_2SO_4 \cdot K_{sp}$ ($PbSO_4$) = $1{\cdot}6 \times 10^{-8}$. ($2{\cdot}1 \times 10^{-6}$)

38. The solubility of bismuth sulphide in water at 20°C is $1{\cdot}7 \times 10^{-15}$ mol/L. Calculate the value of K_{sp}. ($1{\cdot}5 \times 10^{-72}$)

39. Calculate the solubility of $Mg(OH)_2$ in 0·05 M NaOH.
K_{sp} ($Mg(OH)_2$) = $8{\cdot}9 \times 10^{-12}$. ($3{\cdot}6 \times 10^{-9}$ mole/litre)

40. Equal volumes of 0·02 N solutions of $CaCl_2$ and Na_2SO_4 are mixed. Will there be a formation of $CaSO_4$ precipitate? K_{sp} ($CaSO_4$) = $1{\cdot}3 \times 10^{-4}$. (No)

41. 450 mL of 0·001 N solution of $AgNO_3$ is added to 50 mL of 0·001 N solution of HCl. Will there be a formation of precipitate of AgCl?
K_{sp} (AgCl) = $1{\cdot}8 \times 10^{-10}$. (Yes)

42. The solubility of CaF_2 in water at 18°C is $2{\cdot}04 \times 10^{-4}$ mole/litre. Calculate K_{sp} of CaF_2 and its solubility in 0·01 molar NaF solution.
($3{\cdot}4 \times 10^{-11}$; $3{\cdot}4 \times 10^{-7}$ mole/litre)

43. Will a precipitate of silver sulphate form if equal volumes of 1 N H_2SO_4 and 0·02 M $AgNO_3$ solutions are mixed? K_{sp} (Ag_2SO_4) = 2×10^{-5}. (Yes)

44. Will a precipitate of $CaSO_4$ form if

(i) equal volumes of 0·02 M $CaCl_2$ and 0·0004 M Na_2SO_4 solutions are mixed?

(ii) equal volumes of 0·08 M $CaCl_2$ and 0·02 M Na_2SO_4 are mixed?

K_{sp} ($CaSO_4$) = $2{\cdot}4 \times 10^{-5}$. {(i) No (ii) Yes}

45. A solution containing 0·01 mole/litre of $CaCl_2$ and 0·01 mole/litre of $SrCl_2$ is slowly added to a 0·01 N solution of H_2SO_4. Which substance begins to precipitate earlier? (a) $SrSO_4$ (b) $CaSO_4 \cdot K_{sp}$ ($SrSO_4$) = $3{\cdot}2 \times 10^{-7}$; K_{sp} ($CaSO_4$) = $1{\cdot}3 \times 10^{-4}$. ($SrSO_4$)

46. If the solubility product of silver oxalate is 1×10^{-11}, what will be the weight of $Ag_2C_2O_4$ in 2·5 litres of a saturated solution? (0·103 g)

47. Find the solubility of CaF_2 in 0·05 M solution of $CaCl_2$ and water. How many times is the solubility in the second case greater than in the first? K_{sp} $(CaF_2) = 4 \times 10^{-11}$. $(1{\cdot}4 \times 10^{-5},\ 2{\cdot}15 \times 10^{-4}$ mole/litre; 15·4 times)

48. How will the concentration of Ag^+ in a saturated solution of AgCl diminish if such an amount of HCl is added to it that the concentration of the Cl^- in the solution becomes equal to 0·03 mole/litre? K_{sp} $(AgCl) = 1{\cdot}8 \times 10^{-10}$.

$\left(\frac{1}{2230}\text{ of its initial value}\right)$

49. How does the solubility of CaC_2O_4 in a 0·1 M solution of $(NH_4)_2C_2O_4$ decrease in comparison with its solubility in water? Assume that the ionisation of $(NH_4)_2C_2O_4$ is complete. $K_{sp}(CaC_2O_4) = 2 \times 10^{-9}$.

$\left(\frac{1}{2200}\text{ of its solubility in water}\right)$

50. Solid $AgNO_3$ is gradually added to a solution containing Cl^- and I^-. If K_{sp} values of AgCl and AgI are respectively $1{\cdot}7 \times 10^{-10}$ and $1{\cdot}5 \times 10^{-16}$, which one will precipitate first? Also, find the relative concentration of $[I^-]$ to $[Cl^-]$ just before the precipitation of AgCl.

$\left(\text{AgI},\ \frac{[I^-]}{[Cl^-]} = 10^{-6}\right)$

51. Given that 2×10^{-4} mole each of Mn^{2+} and Cu^{2+} was contained in one litre of a 0·003 M $HClO_4$ solution, and this solution was saturated with H_2S. Determine whether or not each of these ions, Mn^{2+} and Cu^{2+}, will precipitate as sulphide. The solubility of H_2S, 0·1 mole per litre, is assumed to be independent of the presence of other materials in the solution.

$$K_{sp}\ (MnS) = 3 \times 10^{-14},\ K_{sp}\ (CuS) = 8 \times 10^{-37}.$$

K_1 and K_2 for H_2S are 1×10^{-7} and $1{\cdot}1 \times 10^{-14}$ respectively. Also, calculate the percentage of Cu remaining unprecipitated. Will MnS precipitate if the above solution is made neutral by lowering the $[H^+]$ to 10^{-7} M?

(CuS precipitates; $3{\cdot}27 \times 10^{-14}$ %; MnS precipitates)

52. What pH must be maintained in a solution saturated in H_2S (0·1 M) and 10^{-3} M in Zn^{2+} to prevent ZnS from precipitating?
$K_{sp}\ (ZnS) = 1 \times 10^{-21}$, $K_a\ (H_2S) = 1{\cdot}1 \times 10^{-21}$. (pH < 2)

53. Should FeS precipitate from a solution that is saturated in H_2S (0·1 M), 0·002 M in Fe^{2+} and at a pH = 3·5?
$K_{sp}\ (FeS) = 6{\cdot}3 \times 10^{-18}$, $K_a\ (H_2S) = 1{\cdot}1 \times 10^{-21}$. (No)

54. A buffer solution is 0·25 M CH_3COOH – 0·15 M CH_3COONa, saturated in H_2S (0·1 M) and has $[Mn^{2+}] = 0{\cdot}015$ M. $K_a\ (CH_3COOH) = 1{\cdot}74 \times 10^{-5}$,

$K_a\ (H_2S) = 1{\cdot}1 \times 10^{-21}$ and $K_{sp}\ (MnS) = 2{\cdot}5 \times 10^{-13}$.

(a) Will MnS precipitate?

(b) Which buffer component should be increased in concentration and to which minimum value to just start precipitation of MnS?

[(a) No (b) $[CH_3COO^-] = 1{\cdot}7$ M]

55. When equal volumes of the following solutions are mixed, precipitation of AgCl ($K_{sp} = 1{\cdot}8 \times 10^{-10}$) will occur only with

(a) 10^{-4} M (Ag^+) and 10^{-4} M (Cl^-)

(b) 10^{-5} M (Ag^+) and 10^{-5} M (Cl^-)

(c) 10^{-6} M (Ag^+) and 10^{-6} M (Cl^-)

(d) 10^{-10} M (Ag^+) and 10^{-10} M (Cl^-) **(IIT 1988)** (a)

56. How much NH_3 must be added to a 0·004 M Ag^+ solution to prevent the precipitation of AgCl when $[Cl^-]$ reaches 0·001 M? K_{sp} (AgCl) $= 1{\cdot}8 \times 10^{-10}$. Dissociation constant for $Ag(NH_3)_2^+ = 6{\cdot}0 \times 10^{-8}$. (0·044 mole/litre)

57. Calculate the simultaneous solubility of CaF_2 and SrF_2.

$K_{sp}\ (CaF_2) = 3{\cdot}9 \times 10^{-11}$.

$K_{sp}\ (SrF_2) = 2{\cdot}9 \times 10^{-9}$.

$\begin{bmatrix} 1{\cdot}2 \times 10^{-5} \text{ mole/litre} \\ 9 \times 10^{-4} \text{ mole/litre} \end{bmatrix}$

58. Aniline is a weak organic base in aqueous solutions. Suggest a solvent in which aniline would become a strong base. (Acetic acid)

59. Distinguish between acid strength and acid concentration. (Read text)

60. Liquid NH_3, like water, is an amphiprotic solvent. Write the equation for the auto-ionisation of NH_3. ($2NH_3 \rightleftharpoons NH_4^+ + NH_2^-$)

61. Calculate the sulphate ion concentration in 0·15 M H_2SO_4. $K_2 = 1{\cdot}02 \times 10^{-2}$

(**Hint:** First ionisation of H_2SO_4 is 100%) ($8{\cdot}9 \times 10^{-3}$)

62. A 50·0 mL sample of a 0·01 M solution of HCOOH was titrated with 0·10 M NaOH. Calculate the pH of the solution when 10 mL of NaOH was added.

K_a for $HCOOH = 1{\cdot}772 \times 10^{-4}$ (11·92)

63. What is the pH of a 0·10 M solution of ethylenediaminetetraacetic acid (edta).

$K_{a_1} = 1 \times 10^{-2}, K_{a_2} = 2{\cdot}1 \times 10^{-3}, K_{a_3} = 6{\cdot}9 \times 10^{-7}$ and $K_{a_4} = 5{\cdot}5 \times 10^{-11}$. (1·54)

[**Hint:** The only two major contributions to the concentration of H^+ are from the first two ionisation steps]

64. A concentrated strong acid is added to a solid mixture of 0·015-mole samples of $Fe(OH)_2$ and $Ca(OH)_2$ placed in one litre of water. At what value of pH will the dissolution of each hydroxide be complete? (Assume negligible volume change)

$K_{sp}[Fe(OH)_2] = 7{\cdot}9 \times 10^{-15}$ and $K_{sp}[Cu(OH)_2] = 1{\cdot}6 \times 10^{-19}$ (7·86)

65. In 1 L saturated solution of AgCl ($K_{sp} = 1.6 \times 10^{-10}$), 0.1 mole of CuCl ($K_{sp} = 1.0 \times 10^{-6}$) is added. The resultant concentration of Ag^+ is 1.6×10^{-x}. The value of x is .... . (7) **(IIT 2011)**

[Hint: See Ex. 60 and Ex. 61]

66. The solubility of a salt of weak acid (AB) at pH 3 is $y \times 10^{-3}$ mol L^{-1}. The value of y is .... .
(Given that the volume of solubility product of $AB(K_{sp}) = 2 \times 10^{-10}$ and the value of ionisation constant of $HB(K_a) = 1 \times 10^{-8}$) **(IIT 2018 Adv.)** (4.47)

[Hint: $\underset{}{AB} \rightleftharpoons \underset{x}{A^+} + \underset{x-y}{B^-}$; $K_{sp} = x \cdot (x-y) = 2 \times 10^{-10}$

$\underset{x-y}{B^-} + \underset{10^{-3}}{H^+} \rightleftharpoons \underset{y}{BH}$ $\quad \dfrac{1}{K_a} = \dfrac{y}{(x-y) \cdot 10^{-3}} = 10^8$; cal. x]

Objective Problems

1. Dissociation constant of H_2O at 25°C is

(a) 1×10^{-14} (b) 1×10^{14} (c) 14 (d) $1{\cdot}8 \times 10^{-16}$

2. K_a value for the acid HA is 1×10^{-6}. The value of K for

$$A^- + H_3O^+ = HA + H_2O \text{ is}$$

(a) 1×10^{-6} (b) 1×10^{12} (c) 1×10^{-12} (d) 1×10^{6}

3. pK_a of a base ($K_b = 1 \times 10^{-5}$) is

(a) 5 (b) – 9 (c) – 5 (d) 9

4. What molar concentration of NH_3 provides a $[OH^-]$ of $1{\cdot}5 \times 10^{-3}$?
($K_b = 1{\cdot}8 \times 10^{-5}$.)

(a) 0·125 (b) $(0{\cdot}125 + 1{\cdot}5 \times 10^{-3})$

(c) $(0{\cdot}125 - 1{\cdot}5 \times 10^{-3})$ (d) $(1{\cdot}5 \times 10^{-3})$

[Hint: $NH_3 + H_2O \rightleftharpoons NH_4^+ + OH^-$]

5. In which of the following cases is the acid strength highest?

(a) $K_a = 10^{-6}$ (b) $pK_a = 5$ (c) $pK_b = 10$ (d) $K_b = 10^{-11}$

6. The values of K_w in 0·1 M NaOH and 0·1 M NaCl are

(a) same (b) different (c) same only at 25°C

7. At 90°C, pure water has $[H_3O^+] = 10^{-6}$ mole/litre. The value of K_w at 90°C is

(a) 10^{-6} (b) 10^{-8} (c) 10^{-12} (d) 10^{-14} **(IIT 1984)**

8. The pH of a 0·01 N monobasic acid is 4. The acid must be

(a) strong (b) weak

9. 10^{-2} mole of KOH is dissolved in 10 litres of water. The pH of the solution is

(a) 12 (b) 2 (c) 3 (d) 11

10. The pH of the solution containing 0·1 N HCl and 0·1 N CH_3COOH is
(a) 1 (b) 0·7 (c) 2 (d) 1·3

11. The pH of a 10^{-8} M HCl solution is
(a) 8 (b) 7·02 (c) 7 (d) 6·98

12. pH of 10^{-11} M HCl is
(a) 11 (b) 3 (c) 6·8 (d) 7

13. If pH of a 0·01 N monobasic acid is 2·0, the acid must be
(a) strong (b) weak

14. If the temperature of water is increased from 25°C to 45°C, the pH of water at 45°C will be
(a) 7 (b) slightly greater than 7
(c) < 7 (d) 8

15. pH of a 10^{-3} M NaCl solution (aq) at 25°C is
(a) 7 (b) 11 (c) 3 (d) all wrong

16. pH of an aq. NaCl solution at 85°C should be
(a) 7 (b) > 7 (c) < 7 (d) 0

17. The pH of 7×10^{-8} M CH_3COOH is ($K_w = 1 \times 10^{-14}$)
(a) 8·1 (b) 7·9 (c) 7·1 (d) 6·85

18. The dissociation constant of an acid HA is 1×10^{-5}. The pH of 0·1 molar solution of the acid will be approximately
(a) 3 (b) 5 (c) 1 (d) 6

19. 1 cc of 0·1 N HCl is added to 999 cc solution of NaCl. The pH of the resulting solution will be
(a) 7 (b) 4 (c) 2 (d) 1

20. If a solution has a pOH value of 14 at 25°C, H^+ concentration should be
(a) 0 (b) 10 (c) 1 (d) none of these

21. What will be the hydrogen ion concentration in mole/litre of a solution of pH = 0?
(a) 0 (b) 10^{-7} (c) 10° (d) pH cannot be zero

22. Which of the following would decrease the pH of 25 cm^3 of a 0·01 M solution of HCl?
(a) The addition of Mg
(b) The addition of 25 cm^3 of 0·02 M HCl
(c) The addition of 25 cm^3 of 0·005 M HCl
(d) None of these

23. The pH of a 0·1 M NH_3 solution ($K_b = 1.8 \times 10^{-5}$) is
(a) 11·13 (b) 1 (c) 13 (d) none of these

[**Hint:** $NH_3 + H_2O \rightleftharpoons NH_4^+ + OH^-,\ K_b = \dfrac{[NH_4^+][OH^-]}{[NH_3]}$]

24. The pH of a buffer solution of 0·1 M NH_4OH and 0·1 M NH_4Cl is ($pK_b = 4·0$)
(a) 1 (b) 4 (c) 10 (d) 13

25. In a buffer solution of a weak acid and its salt, if the ratio of the concentration of the salt to that of the acid is increased ten times, the pH of the buffer
(a) increases by 1 (b) increases 10 times
(c) decreases 10 times (d) decreases by 1

26. The process of hydrolysis is
(a) always exothermic (b) always endothermic
(c) either exothermic or endothermic (d) neither exothermic nor endothermic

27. When a salt of a weak acid and a weak base is dissolved in water at 25°C, the pH of the resulting solution will always
(a) be 7 (b) be greater than 7
(c) be less than 7 (d) depend upon K_a and K_b values

28. The degree of hydrolysis of a salt of a weak acid and a weak base in its 0·1 M solution is found to be 50%. If the molarity of the solution is 0·2 M, the percentage hydrolysis of the salt should be
(a) 100% (b) 50% (c) 25% (d) none of these

29. If a salt of a strong acid and a weak base hydrolyses appreciably, which of the following formulae is to be used to calculate degree of hydrolysis 'X'?
(a) $X = \sqrt{\dfrac{K_w}{K_a \cdot a}}$ (b) $X = \sqrt{\dfrac{K_w}{K_b \cdot a}}$ (c) $X = \sqrt{\dfrac{K_w}{K_a \cdot K_b}}$ (d) none of these

30. K_{sp} for AgCl in water at 25°C is $1·8 \times 10^{-10}$. If 10^{-5} mole of Ag^+ ions are added to this solution, K_{sp} will be
(a) $1·8 \times 10^{-15}$ (b) $1·8 \times 10^{-10}$ (c) $1·8 \times 10^{-5}$ (d) none of these

31. In which of the following cases is the solution of AgCl unsaturated?
(a) $[Ag^+][Cl^-] < K_{sp}$ (b) $[Ag^+][Cl^-] > K_{sp}$
(c) $[Ag^+][Cl^-] = K_{sp}$

32. If the solubility of $Al(OH)_3$ is S moles/litre, the solubility product is
(a) S^3 (b) $27S^4$ (c) S^2 (d) $4S^3$

33. The volume of water needed to dissolve 1g of $BaSO_4$ ($K_{sp} = 1·1 \times 10^{-10}$) at 25°C is
(a) 820 litres (b) 450 litres (c) 205 litres (d) none of these

34. The solubility of $BaSO_4$ in water is 0·00233 g per litre at 30°C. The solubility of $BaSO_4$ in 0·1 M $(NH_4)_2SO_4$ solution at the same temperature is
(a) 10^{-5} mole/litre (b) 10^{-6} mole/litre (c) 10^{-8} mole/litre (d) 10^{-9} mole/litre

35. When equal volumes of the following solutions are mixed, precipitation of AgCl ($K_{sp} = 1·8 \times 10^{-10}$) will occur only with
(a) 10^{-4} M (Ag^+) and 10^{-4} M (Cl^-) (b) 10^{-5} M (Ag^+) and 10^{-5} M (Cl^-)
(c) 10^{-6} M (Ag^+) and 10^{-6} M (Cl^-) (d) 10^{-10} M (Ag^+) and 10^{-10} M (Cl^-)
(IIT 1988)

36. If the salts M_2X, QY_2 and PZ_3 have the same solubilities $(< \frac{4}{27})$, their K_{sp} values are related as

(a) $K_{sp}(M_2X) = K_{sp}(QY_2) > K_{sp}(PZ_3)$ (b) $K_{sp}(M_2X) > K_{sp}(QY_2) = K_{sp}(PZ_3)$

(c) $K_{sp}(M_2X) = K_{sp}(QY_2) = K_{sp}(PZ_3)$ (d) $K_{sp}(M_2X) > K_{sp}(QY_2) > K_{sp}(PZ_3)$

37. If pK_b for fluoride ion at 25°C is 10·83, the ionisation constant of hydrofluoric acid in water at this temperature is

(a) $1{\cdot}74 \times 10^{-5}$ (b) $3{\cdot}52 \times 10^{-3}$ (c) $6{\cdot}75 \times 10^{-4}$ (d) $5{\cdot}38 \times 10^{-2}$

(IIT 1997)

38. The solubility of A_2X_3 is y mol/dm^3. Its solubility product is

(a) $6y^4$ (b) $64y^4$ (c) $36y^5$ (d) $108y^5$ **(IIT 1997)**

39. Which of the following statements about buffer solutions is wrong?

(a) Weak acids and their salts are better as buffers for pH < 7.

(b) Weak bases and their salts are better as buffers for pH > 7.

(c) A buffer solution has generally lost its usefulness when one component of the buffer pair is less than about 10% of the other.

(d) For most effective buffering in the acid range or basic range, the two components of the buffer should have almost the same mass per unit volume.

40. 10 mL of 0·1 M HCl is titrated with 0·1 M NaOH. When the volume of NaOH added from the burette is from 9·99 mL to 10·01 mL, the pH jumps approximately from

(a) 4 to 10 (b) 6 to 8 (c) 6·9 to 7·1 (d) 1 to 14

41. When one drop of a concentrated HCl solution is added to one litre of pure water at 25°C, the pH drops suddenly from 7 to about 4. When the second drop of the same acid is added, the pH of the solution further drops to about

(a) 3·7 (b) 2·0 (c) 1·0 (d) 0

42. In which of the following aqueous solutions is the degree of dissociation of water maximum?

(a) NH_4Cl solution (b) CH_3COONa solution

(c) CH_3COONH_4 solution (d) NaCl solution

43. pH of an aqueous 1×10^{-8} M NaOH solution is

(a) 8 (b) 7·02

(c) 7 (d) 6

44. The pH of 1×10^{-3} M H_2O_2 solution ($K_a = 2{\cdot}2 \times 10^{-12}$) is

(a) ≈ 3 (b) slightly less than 7

(c) slightly greater than 7 (d) = 7

45. The pH of an aqueous solution of 0·01 M CH_3COONH_4 at 25°C is

$$(K_a(CH_3COOH) = K_b(NH_4OH) = 1{\cdot}8 \times 10^{-5})$$

(a) > 7 (b) < 7 (c) 7 (d) 8

46. The equilibrium: $2Cu^{I} \rightleftharpoons Cu^{0} + Cu^{II}$ in aqueous medium at 25°C shifts towards the left in the presence of

(a) NO_3^- (b) Cl^- (c) SCN^- (d) CN^-

[**Hint:** Cl^-, SCN^- and CN^- form precipitate with Cu^I.] **(IIT 2011)**

47. How many litres of water must be added to 1 L of an aqueous solution of HCl with a pH of 1 to create an aqueous solution with pH of 2?
(a) 0.1 L (b) 0.9 L (c) 2.0 L (d) 9.0 L
(IIT 2013 Main)

48. The K_{sp} of Ag_2CrO_4 is 1.1×10^{-12} at 298 K. The solubility (in mol L^{-1}) of Ag_2CrO_4 in a 0.1 M $AgNO_3$ solution is
(a) 1.1×10^{-11} (b) 1.1×10^{-10} (c) 1.1×10^{-12} (d) 1.1×10^{-9}
(IIT 2013 Adv.)

49. The reagent(s) that can selectively precipitate S^{2-} from a mixture of S^{2-} and SO_4^{2-} in aqueous solution is(are)
(a) $CuCl_2$ (b) $BaCl_2$
(c) $Pb(OOCCH_3)_2$ (d) $Na_2[Fe(CN)_5NO]$ **(IIT 2016 Adv.)**

50. pK_a of a weak acid (HA) and pK_b of a weak base (BOH) are 3.2 and 3.4 respectively. The pH of their salt (AB) solution is
(a) 7.2 (b) 6.9 (c) 7.0 (d) 1.0
(IIT 2017 Main)

[**Hint:** $pH = \frac{1}{2}\{pK_w + pK_a - pK_b\}$]

51. In the following reactions
(i) $ZnO + Na_2O \rightarrow Na_2ZnO_2$
(ii) $ZnO + CO_2 \rightarrow ZnCO_3$
ZnO is respectively acting as a(an)
(a) base and acid (b) base and base
(c) acid and acid (d) acid and base **(IIT 2017 Main)**

52. An alkali is titrated with methyl orange as indicator, which of the following is a correct combination?

	Base	Acid	End Point
(a)	strong	strong	pink to colourless
(b)	weak	strong	colourless to pink
(c)	strong	strong	pinkish red to yellow
(d)	weak	strong	yellow to pinkish red

[**Hint:** Read text] **(IIT 2018 Main)**

53. An aqueous solution contains an unknown concentration of Ba^{2+}, when 50 mL of 1 M solution of Na_2SO_4 is added $BaSO_4$ just begins to precipitate. The final volume is 500 mL. The solubility product of $BaSO_4$ is 1×10^{-10}. What is the original concentration of Ba^{2+}?
(a) 1.0×10^{-10} M (b) 5×10^{-9} M
(c) 2×10^{-9} M (d) 1.1×10^{-9} M **(IIT 2018 Main)**

[**Hint:** Volume of original Ba^{2+} solution = 500 − 50 = 450 mL

$$\text{mole of } SO_4^{2-} = \text{mole of } Na_2SO_4 = \frac{\text{mm of } Na_2SO_4}{1000} = \frac{M \times V(\text{mL})}{1000} = \frac{1 \times 50}{1000} = \frac{1}{20}$$

concentration of $SO_4^{2-} = \frac{1}{20} \times \frac{1000}{500} = 0.1$ M (mole/L)

$\therefore\ K_{sp} = [Ba^{2+}][SO_4^{2-}]$ or $1 \times 10^{-9} = [Ba^{2+}](0.1)$; $[Ba^{2+}] = 10^{-9}$ M

Let the molarity of original Ba^{2+} solution be x

$\therefore\ x \times 450 = 10^{-9} \times 500.]$

54. Which of the following are Lewis acids?

(a) BCl_3 and $AlCl_3$ (b) PH_3 and BCl_3

(c) $AlCl_3$ and $SiCl_4$ (d) pH_3 and $SiCl_4$ **(IIT 2018 Main)**

55. An aqueous solution contains 0.10 M H_2S and 0.20 M HCl. If the equilibrium constants for the formation of HS^- from H_2S is 1.0×10^{-7} and that of S^{2-} from HS^- ions is 1.2×10^{-13} then the concentration of S^{2-} ions in aqueous solution is

(a) 5×10^{-19} (b) 5×10^{-8} (c) 3×10^{-20} (d) 6×10^{-21}

(IIT 2018 Main)

[Hint: $H_2S \rightleftharpoons 2H^+ + S^{2-}$; $K = K_1 \cdot K_2$

$(0.1 - x) \approx 0.1 \quad (2x + 0.2) \approx 0.2 \quad x \quad \therefore\ K = \frac{0.2^2 x}{0.1} = (1 \times 10^{-7})\ (1.2 \times 10^{-13})$]

56. Following four solutions are prepared by mixing different volumes of NaOH and HCl of different concentrations, pH of which of them will be equal to 1?

(a) 55 mL $\frac{M}{10}$ HCl + 45 mL $\frac{M}{10}$ NaOH (b) 100 mL $\frac{M}{10}$ HCl + 100 mL $\frac{M}{10}$ NaOH

(c) 60 mL $\frac{M}{10}$ HCl + 40 mL $\frac{M}{10}$ NaOH (d) 75 mL $\frac{M}{5}$ HCl + 25 mL $\frac{M}{5}$ NaOH

(IIT 2018 Main)

[Hint: mm = M × V(mL) and Molarity = $\frac{\text{mm}}{\text{Volume(mL)}}$]

57. The minimum volume of water required to dissolve 0.1 g lead(II) chloride to get a saturated solution (K_{sp} of $PbCl_2 = 3.2 \times 10^{-8}$, Pb = 207 u) is

(a) 17.98 L (b) 0.18 L

(c) 1.798 L (d) 0.36 L **(IIT 2018 Main)**

[Hint: First calculate solubility of $PbCl_2$]

58. Dilution processes of different aqueous solutions, with water, are given in LIST–I. The effects of dilution of the solution on $[H^+]$ are given in LIST–II.

[**Note:** degree of dissociation (α) of weak acid and weak base is << 1, degree of hydrolysis of salt << 1, $[H^+]$ represents the concentration of H^+ (ions)]

LIST–I	LIST–II
(P) (10 mL of 0.1 M NaOH + 20 mL of 0.1 M acetic acid) diluted to 60 mL	(1) The value of $[H^+]$ does not change on dilution.
(Q) (20 mL of 0.1 M NaOH + 20 mL of 0.1 M Acetic acid) diluted to 80 mL	(2) The value of $[H^+]$ changes to half of its initial value on dilution.
(R) (20 mL of 0.1 M HCl + 20 mL of 0.1 M ammonia solution) diluted to 80 mL	(3) The value of $[H^+]$ changes to two times of its initial value on dilution.

(S) 10 mL saturated solution of $Ni(OH)_2$ in equilibrium with excess solid $Ni(OH)_2$ is diluted to 20 mL (solid $Ni(OH)_2$ is still present after dilution).

(4) The value of $[H^+]$ changes to $1/\sqrt{2}$ times of its initial value on dilution.

(5) The value of $[H^+]$ changes to $\sqrt{2}$ times of its initial value on dilution.

Match each process given in LIST–I with one or more effect(s) in LIST–II. The correct option is

(a) $P \to 4;\ Q \to 2;\ R \to 3;\ S \to 1$
(b) $P \to 4;\ Q \to 3;\ R \to 2;\ S \to 3$
(c) $P \to 1;\ Q \to 4;\ R \to 5;\ S \to 3$
(d) $P \to 1;\ Q \to 5;\ R \to 4;\ S \to 1$

(IIT 2018 Adv.)

[Hint:

(P) Buffer of $CH_3COOH + CH_3COONa$ is formed. Hence no change in $[H^+]$ on dilution $(P \to 1)$

(Q) Hydrolysis of salt CH_3COONa for which $[H^+] = \sqrt{\dfrac{K_w \cdot K_a}{c}}$ or $[H^+] \propto \dfrac{1}{\sqrt{c}}$. $[H^+]$ changes $\sqrt{2}$ times on diluting 2 times $(Q \to 5)$

(R) Hydrolysis of salt NH_4Cl for which $[H^+] = \sqrt{\dfrac{K_w \cdot c}{K_b}}$ or $[H^+] \propto \sqrt{c}$. $[H^+]$ changes $\dfrac{1}{\sqrt{2}}$ times on diluting 2 times $(R \to 4)$

(S) $K_{sp}(Ni(OH)_2)$ is constant at a given temperature. Hence no change in $[H^+]$ on dilution $(S \to 1)$]

Answers

1-d, **2**-d, **3**-d, **4**-b, **5**-d, **6**-a, **7**-c, **8**-b, **9**-d, **10**-a, **11**-d, **12**-d, **13**-a, **14**-c, **15**-a, **16**-c, **17**-d, **18**-a, **19**-b, **20**-c, **21**-c, **22**-b, **23**-a, **24**-c, **25**-a, **26**-b, **27**-d, **28**-b, **29**-d, **30**-b, **31**-a, **32**-b, **33**-b, **34**-d, **35**-a, **36**-a, **37**-c, **38**-d, **39**-d, **40**-a, **41**-a, **42**-c, **43**-b, **44**-b, **45**-c, **46**-b, c, d **47**-d, **48**-b, **49**-a, **50**-b, **51**-d, **52**-d, **53**-d, **54**-a, c, **55**-c, **56**-d, **57**-b, **58**-d.

❑

CHAPTER SEVENTEEN

CHEMICAL KINETICS

In the chapter 'Chemical Equilibrium' we have discussed how far a reaction would proceed to attain equilibrium but nothing has been said about how fast that equilibrium would be attained. This will be the subject matter of the present chapter.

For a reaction at equilibrium the net reaction rate is zero, i.e., the forward rate is equal to the reverse rate. In this section we shall limit ourselves mainly to a forward reaction which is yet to reach equilibrium.

Qualitative Measurement of the Rate of a Reaction; Differential Rate Law and Rate Constant

The rate of a reaction is defined as the change in concentration of any of the reactants or products per unit time. For a reaction

$$A + B \rightarrow C + D$$

the rate of the reaction is equal to the decrease in concentration of either A or B, or increase in concentration of either C or D per unit time.

Let us consider the general reaction

$$aA + bB \rightarrow cC + dD.$$

The stoichiometric coefficients *a, b, c* and *d* signify that for the disappearance of *a* moles of A and *b* moles of B at any instant, *c* moles of C and *d* moles of D will appear. **The rate may, therefore, be more accurately defined as the rate of disappearance of A or B per mole, which in turn, is equal to the rate of appearance of C or D per mole.**

Thus, the rate of the general reaction

= rate of disappearance of A or B per mole

= rate of appearance of C or D per mole

$$\text{or} \qquad \text{rate} = -\frac{1}{a}\frac{d[A]}{dt} = -\frac{1}{b}\frac{d[B]}{dt}$$

$$= +\frac{1}{c}\frac{d[C]}{dt} = +\frac{1}{d}\frac{d[D]}{dt} \qquad \dots (1)$$

[ ] represents the concentration in mole per litre, whereas '*d*' represents an infinitesimally small change in concentration. The negative sign shows that the concentrations of the reactants A and B are decreasing, whereas the positive sign indicates the increase in concentration of the products C and D with the progress of the reaction.

Note that the rate of disappearance or appearance of different reactants and products may be same or different depending on the stoichiometry of the equation, but the rate of disappearance or appearance per mole of any reactant or product will always be the same as it represents the rate of the reaction.

As the rate of a reaction decreases gradually with the progress of the reaction, $-\frac{1}{a}\frac{d[A]}{dt}$, etc., represent **instantaneous rate** of the reaction at time t, because these involve concentration changes over an infinitesimally short time interval, dt, but if it is written as $-\frac{1}{a}\frac{\Delta[A]}{\Delta t}$, it represents **average rate** over a finite time interval, Δt.

The reaction rate is also proportional to the product of concentrations of the reactants, each raised to some power.

Accordingly,

$$\text{rate} \propto [\text{A}]^m [\text{B}]^n$$

or

$$\text{rate} = k\,[\text{A}]^m [\text{B}]^n. \quad \text{... (2)}$$

Combining the expressions (1) and (2), the differential rate law may have the form

$$-\frac{1}{a}\cdot\frac{d[\text{A}]}{dt} = -\frac{1}{b}\cdot\frac{d[\text{B}]}{dt} = +\frac{1}{c}\cdot\frac{d[\text{C}]}{dt} = +\frac{1}{d}\cdot\frac{d[\text{D}]}{dt}$$

$$= k\,[\text{A}]^m [\text{B}]^n. \quad \text{... (3)}$$

Thus, the **differential rate law** is that mathematical expression which shows how the rate of a reaction depends on concentration.

The exponents *m* and *n* are known as the order of the reaction with respect to A and B respectively. $(m+n)$ gives the overall order of the reaction. *m* and *n* are experimental values.

One should note that ***m* or *n* may and may not be equal to stoichiometric coefficients *a* and *b* respectively in the net reaction**, e.g., for the reaction

$$H_2 + I_2 \rightarrow 2HI$$

the differential rate law obtained experimentally is

$$-\frac{d[H_2]}{dt} = -\frac{d[I_2]}{dt} = +\frac{1}{2}\times\frac{d[HI]}{dt} = k\,[H_2]\,[I_2],$$

which shows that the order of the reaction w.r.t. H_2 is 1 and that w.r.t. I_2 is also 1, and so, the overall order is 2. In this case

$$a = m = 1 \text{ and } b = n = 1.$$

But for the reaction $H_2 + Br_2 \rightarrow 2HBr$, the experiment shows that the rate law is

$$-\frac{d[H_2]}{dt} = -\frac{d[Br_2]}{dt} = +\frac{1}{2}\times\frac{d[HBr]}{dt} = k'\,[H_2]\,[Br_2]^{\frac{1}{2}}$$

in which, order w.r.t. H_2 is 1 while w.r.t. Br_2 is $\frac{1}{2}$, and so, overall order is $1\frac{1}{2}$. In this case $a = m = 1$ but $b \neq n$, i.e., $b = 1$ and $n = \frac{1}{2}$.

The constant k, appearing in the expression (2), is known as rate constant. It is defined as the rate of a reaction when the concentration of each reactant is equal to unity. Each reaction is characterised by its own rate constant. From the numerical value of rate constant, the rate of a reaction under a particular concentration condition can be calculated.

It is important to remember that **the value of k is for a specific reaction, represented by a balanced equation.** The units of k depend on the overall order of the reaction. The value of k does not depend on concentrations of either reactants or products and also on time. It depends on temperature and the catalyst used in the reaction.

Concept of Molecularity and Order in Elementary and Complex Reactions*

1. Molecularity of an elementary reaction is the total number of molecules or atoms taking part in the reaction. In case of complex reactions, molecularity as such has no significance, rather each elementary reaction involved in the complex reaction has its own value.

The order of a simple reaction is defined as the number of molecules or atoms whose concentrations alter as a result of chemical change at a given temperature. Mathematically, order is defined as the sum of the powers to which the concentration terms are raised in order to determine the rate of a reaction.

Accordingly, for a reaction like A + B + C $\rightarrow$ Product

if experiments show that

Rate $= k[A]^0$ or $k[B]^0$ or $k[C]^0$; the reaction is of zero order

Rate $= k[A]$ or $k[B]$ or $k[C]$; the reaction is of first order

Rate $= k[A]^2$ or $k[B]^2$ or $k[C]^2$ or $k[A][C]$ or $k[B][C]$

or $k[A][B]$; the reaction is of second order

Rate $= k[A]^3$ or $k[B]^3$ or $k[C]^3$ or $k[A][B][C]$ or $k[A]^2[B]$

or $k[A][B]^2$, etc; the reaction is of third order

In complex reactions the order of the slowest elementary reaction gives the order of the complex reaction.

---

* In chemical kinetics, reactions which occur in a single step are termed as simple or elementary reactions and those which take place in two or more steps are termed as complex reactions.

For elementary reactions, the law of mass action and the rate law expressions are generally same.

2. Kinetic order is an experimental property derived from the experimental rate law, whereas, molecularity is a theoretical property associated with a chemical equation of an elementary reaction.

3. The order can change with the conditions such as pressure, temperature, etc., whereas molecularity is invariant for a chemical equation.

The decomposition of NH_3 on tungsten is a zero order reaction whereas on hot quartz, it is a first order reaction.

4. As discussed a little earlier, order may or may not be equal to the molecularity but in an elementary reaction they are generally the same.

5. Molecularity must always be an integer and never be zero, whereas order may be an integer, a fractional quantity, zero or even negative.

6. Most of the reactions are of first or second order and very few reactions are of third order and hardly any one of higher order. Similarly, there are no known single steps of higher molecularity than termolecular.

Integrated Rate Law

Differential rate law equations involve infinitesimally small quantities like d[A], d[B], dt, etc., which cannot be practically measured. Hence the differential rate law equations are integrated to obtain equations involving measurable quantities so that the rate of a reaction or the rate constant can be determined. But we shall not give here the process of integration but simply present the integrated rate law equations for reactions of various orders. Note that the rate law expression relates rate and the concentration while the integrated rate equation relates time and concentration.

Zero-Order Reaction

In a zero-order reaction, rate is independent of the concentration of the reactants. Mathematically, a reaction is said to be of zero-order when the rate is proportional to zero power of the reactant concentration.

Suppose a zero-order reaction is of the form

a	0	Initial concentration
A $\rightarrow$	Product	
$(a-x)$	x	Concentration after time t

Differential rate law of this reaction will be

$$-\frac{d\,[\mathrm{A}]}{dt} = \frac{d\,[\text{product}]}{dt} = k_0\,[\mathrm{A}]^0$$

or

$$-\frac{d(a-x)}{dt} = +\frac{dx}{dt}$$

$$= k_0\,(a-x)^0 = k_0.$$

On integrating, we get the integrated rate law as

$$k_0 = \frac{x}{t} \quad \text{or,} \quad x = k_0\, t \qquad \dots (4)$$

or

$$k_0 t = a - (a - x),$$

where k_0 is the rate constant of a zero-order reaction, the unit of which is concentration per time. In zero-order reactions, **the rate constant is equal to the rate of the reaction at all concentrations.**

First-Order Reaction

Let a first-order reaction be represented as:

$$\begin{array}{cccl} a & & 0 & \text{Initial concentration} \\ \mathrm{A} & \rightarrow & \text{Product} & \\ (a-x) & & x & \text{Concentration after time } t \end{array}$$

Differential rate law:

$$-\frac{d(a-x)}{dt} = +\frac{dx}{dt} = k_1\,(a-x).$$

On integration, we get,

integrated rate law:

$$k_1 = \frac{2{\cdot}303}{t} \log\left(\frac{a}{a-x}\right), \qquad \dots (5)$$

where k_1 is the rate constant of a first-order reaction, the unit of which is per time.

Second-Order Reaction

Type I:

$$\begin{array}{cccl} a & & 0 & \text{Initial concentration} \\ \mathrm{A} & \rightarrow & \text{Products} & \\ (a-x) & & x & \text{Concentration after time } t \end{array}$$

Differential rate law:

$$-\frac{d(a-x)}{dt} = +\frac{dx}{dt} = k_2(a-x)^2$$

Integrated rate law:

$$k_2 = \frac{1}{t} \cdot \frac{x}{a(a-x)} \qquad \dots (6)$$

Type II: (i) Reactants A and B have the same initial concentration

$$\begin{array}{cccccl} a & & a & & 0 & \text{Initial concentration} \\ \mathrm{A} & + & \mathrm{B} & \rightarrow & \text{Products} & \\ (a-x) & & (a-x) & & x & \text{Concentration after time, } t \end{array}$$

Differential Rate Law:

$$-\frac{d(a-x)}{dt} = +\frac{dx}{dt} = k_2(a-x)(a-x) = k_2(a-x)^2$$

Integrated Rate Law:

$$k_2 = \frac{1}{t} \cdot \frac{x}{a(a-x)} \quad \text{(same as for A} \rightarrow \text{products)}$$

(ii) Reactants A and B have different initial concentrations.

a		b		0	Initial concentration
A	+	B	$\rightarrow$	Products	
$(a-x)$		$(b-x)$		x	Concentration after time t

Differential Rate Law:

$$-\frac{d(a-x)}{dt} = -\frac{d(b-x)}{dt} = +\frac{dx}{dt} = k_2(a-x)(b-x)$$

Integrated Rate Law:

When $a > b : k_2 = \frac{2 \cdot 303}{t(a-b)} \log \frac{b(a-x)}{a(b-x)}$ ... (7)

When $b > a : k_2 = \frac{2 \cdot 303}{t(b-a)} \log \frac{a(b-x)}{b(a-x)}$ ... (8)

The unit of the rate constant, k_2, is concentration^{-1} time^{-1}.

Third-Order reaction

We choose here the simplest type of third-order reactions

a		0	Initial concentration
A	$\rightarrow$	Products	
$(a-x)$		x	Concentration after time t

Differential Rate Law:

$$-\frac{d(a-x)}{dt} = +\frac{dx}{dt} = k_3(a-x)^3$$

Integrated Rate Law:

$$k_3 = \frac{1}{2t}\left\{\frac{x(2a-x)}{a^2(a-x)^2}\right\}. \quad \text{... (9)}$$

The unit of the rate constant, k_3, is concentration^{-2} time^{-1}.

Half-life Period

The time required for half the reaction to change is known as half-life period.

Let us see how the half-life period depends on the initial reactant concentration *'a'* in reactions of various orders.

Zero-Order Reaction

In a zero-order reaction we know,

$$k_0 = \frac{x}{t} \quad \text{or} \quad t = \frac{x}{k_0}.$$

Let $t_{\frac{1}{2}}$ be the half-life period,

i.e., when $t = t_{\frac{1}{2}}, \; x = \frac{a}{2}$

$$\therefore \qquad t_{\frac{1}{2}} = \frac{a}{2k_0} \quad \text{or} \quad t_{\frac{1}{2}} \propto a \left(\frac{1}{2k_0} \text{ is constant}\right).$$

Thus $t_{\frac{1}{2}}$ for zero order reaction is proportional to a.

First-Order Reaction

For first-order reactions we know

$$k_1 = \frac{2{\cdot}303}{t} \log \frac{a}{a-x}$$

for t to be $t_{\frac{1}{2}}$; $\quad x = \frac{a}{2}$

$$\therefore \qquad k_1 = \frac{2{\cdot}303}{t_{\frac{1}{2}}} \log \frac{a}{a - \frac{a}{2}} = \frac{2{\cdot}303}{t_{\frac{1}{2}}} \log 2$$

$$k_1 = \frac{0{\cdot}6932}{t_{\frac{1}{2}}} \quad \text{or} \quad t_{\frac{1}{2}} = \frac{0{\cdot}6932}{k_1}. \qquad \ldots (10)$$

As the above expression does not contain a, $t_{\frac{1}{2}}$ for first order reaction is independent of a. It should be noted that not only the half-life period, but the time required to complete any definite fraction $\left(\text{say } \frac{1}{n}\right)$ of the reaction is independent of initial concentration in first order kinetics.

$$t_{\frac{1}{n}} = \frac{2{\cdot}303}{k_1} \log \frac{a}{a - \frac{a}{n}} = \frac{2{\cdot}303}{k_1} \log \frac{1}{1 - \frac{1}{n}}$$

As the **radioactive disintegration** follows the first order rate equation, the equations (5) and (10) are also applicable to the radioactive process. In radioactivity, k_1 is termed as disintegration constant (λ) and the reciprocal of k_1 is known as average life of the radioactive substance.

Second-Order Reaction

For the second-order reaction of the type A $\rightarrow$ Product or A + B $\rightarrow$ Product, in which A and B have the same initial concentrations, we have

$$k_2 = \frac{1}{t} \times \frac{x}{a(a-x)}$$

For t to be $t_{\frac{1}{2}}$,
$$x = \frac{a}{2}$$

$$t_{\frac{1}{2}} = \frac{1}{k_2} \times \frac{1}{a}$$

or
$$t_{\frac{1}{2}} \propto \frac{1}{a}.$$

Thus $t_{\frac{1}{2}}$ **for a second-order reaction is inversely proportional to '*a*'.**

Note that in reactions with more than one reactant, $t_{1/2}$ of the reaction is calculated from the limiting reactant.

Third-Order Reaction

For a third-order reaction, we can similarly obtain

$$t_{\frac{1}{2}} \propto \frac{1}{a^2}.$$

Thus for nth order of reaction

$$t_{\frac{1}{2}} \propto \frac{1}{a^{n-1}} \qquad \dots (11)$$

Half-life of an nth-order reaction is given by

$$t_{\frac{1}{2}} = \frac{2^{n-1} - 1}{k_n (n-1)\, a^{n-1}} \quad \text{where} \quad n \geq 2.$$

[**Note:** For a zero, second or a third-order reaction, $t_{\frac{1}{2}}$ is dependent on the initial concentration in contrast to the situation for a first-order reaction. Therefore, the concept of half-life is much less useful for zero, second and third-order reactions.]

For a reaction of general type: $nA \rightarrow$ products, to be of

Zero Order

$$k_0 = \frac{x}{nt} \quad \text{and} \quad t_{1/2} = \frac{a}{2nk_0}$$

First Order

$$k_1 = \frac{2{\cdot}303}{nt} \log \frac{a}{(a-x)} \quad \text{and} \quad t_{1/2} = \frac{0{\cdot}6932}{nk_1}$$

Second Order

$$k_2 = \frac{1}{nt} \left\{ \frac{x}{a(a-x)} \right\} \quad \text{and} \quad t_{1/2} = \frac{1}{nak_2}$$

Third Order

$$k_3 = \frac{1}{2nt} \left\{ \frac{x(2a-x)}{a^2(a-x)^2} \right\} \quad \text{and} \quad t_{1/2} = \frac{3}{2na^2k_3}$$

Effect of Temperature on Reaction Rate
Energy of Activation

The speed of the reactions generally increases between 2–3 times for a rise of temperature of 10°C. The ratio of the rate constants of a reaction at two different temperatures differing by 10°C is known as temperature coefficient.

$$\text{Temperature coefficient} = \frac{k_{35°}}{k_{25°}} = 2 \text{ to } 3.$$

Variation of reaction speed with temperature has been represented by Arrhenius's equation.

$$k = Ae^{-E/RT},$$

where k is the rate constant representing the rate of reaction, E, the energy of activation, A is known as frequency factor and R and T have their usual notations. If k_1 and k_2 are the values of rate constants at temperatures T_1 and T_2 respectively, one can derive from Arrhenius's equation

$$\log \frac{k_2}{k_1} = \frac{E}{2{\cdot}303\, R}\left[\frac{T_2 - T_1}{T_1\, T_2}\right]. \qquad \ldots (12)$$

Determination of Order

1. Graphical Methods

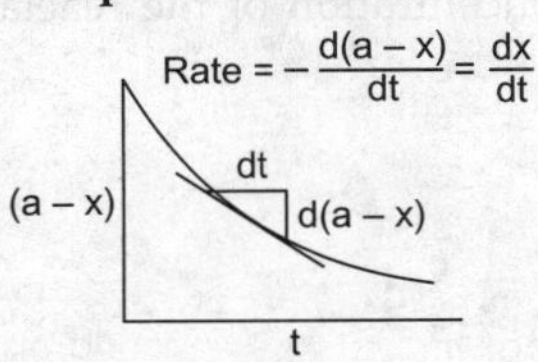

Fig. 1

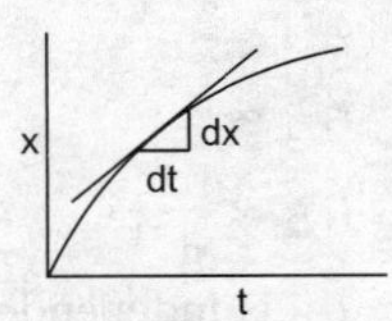

Fig. 2

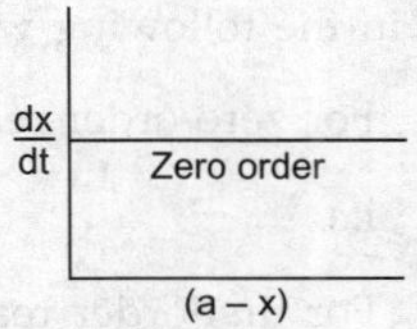

Fig. 3

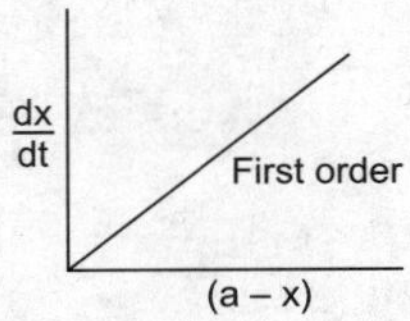

Fig. 4

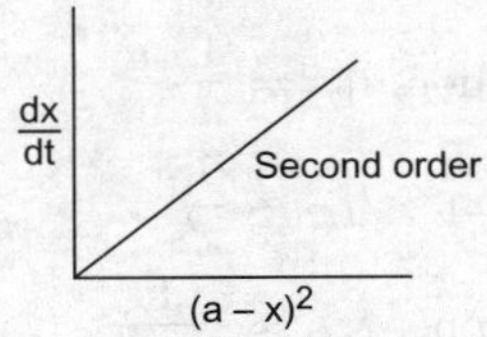

Fig. 5

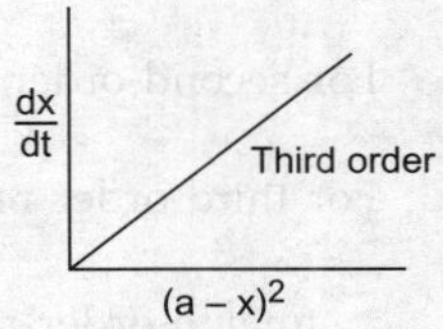

Fig. 6

(a) A graph is plotted between time t and the concentration of reactant $(a - x)$ or product (x) from which a tangent is set up for the curve for a particular instant of time, the slope of which gives the rate of reaction $\left(\frac{dx}{dt}\right)$ for the selected time instant. (Figures 1 and 2)

The various values of the rate $\left(\frac{dx}{dt}\right)$ are now plotted against the corresponding concentration $(a-x)$ or $(a-x)^2$ or $(a-x)^3$ from which we draw the following graphical conclusions:

(b) Alternatively, the order (n) can also be determined from the slope of the curve plotted between log $\left(\frac{dx}{dt}\right)$ and log $(a-x)$:

$$\frac{dx}{dt} = k(a-x)^n$$

$$\log\left(\frac{dx}{dt}\right) = \log k + n\log(a-x)$$

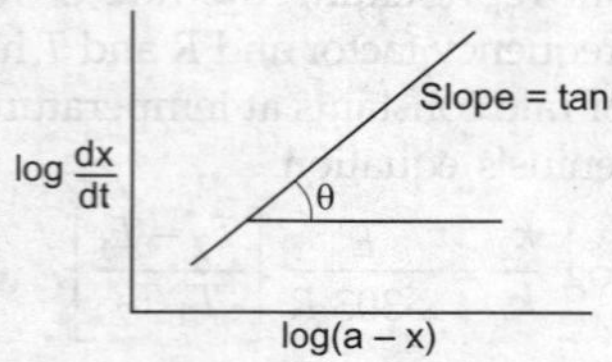

Fig. 7

(c) We know that the time required to complete a definite fraction (say one half) of the reaction depends on the initial concentration of the reactant (a) in the following way:

For zero-order reaction, $t_{1/2} \propto a$

i.e., $t_{1/2} \propto \frac{1}{a^{-1}}$.

For first-order reaction, $t_{1/2}$ is independent of a

i.e., $t_{1/2} \propto \frac{1}{a^0}$

For second-order reaction, $t_{1/2} \propto \frac{1}{a}$.

For third-order reaction, $t_{1/2} \propto \frac{1}{a^2}$

$\therefore$ for nth-order reaction, $t_{1/2} \propto \frac{1}{a^{(n-1)}}$.

Thus we get the following plots for reactions of various orders:

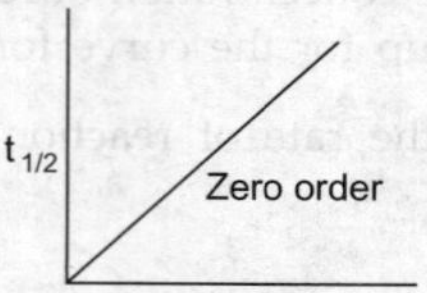

Fig. 8

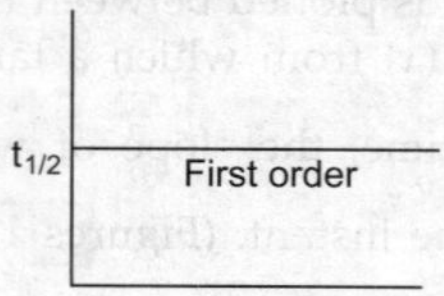

Fig. 9

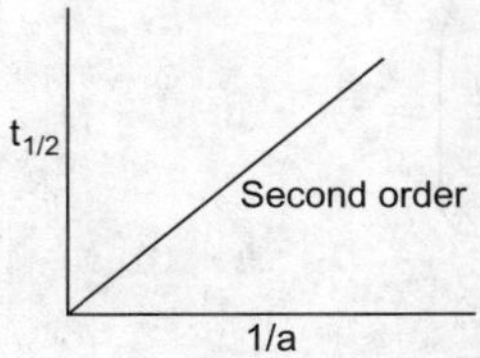

Fig. 10

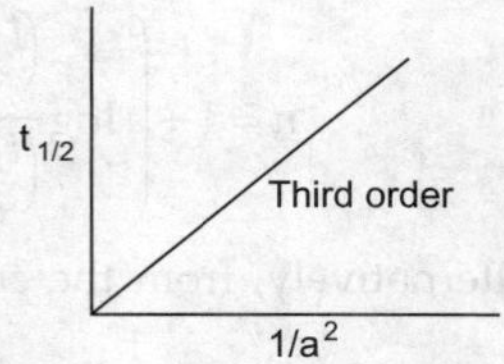

Fig. 11

(d) From integrated rate law equations (4, 5, 6 and 9) of various reactions of different orders, we may have the following plots:

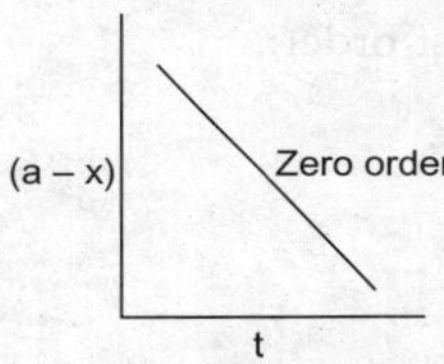

Fig. 12

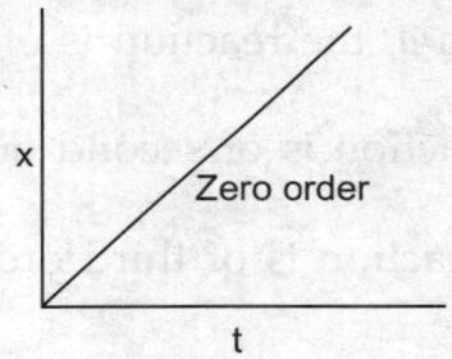

Fig. 13

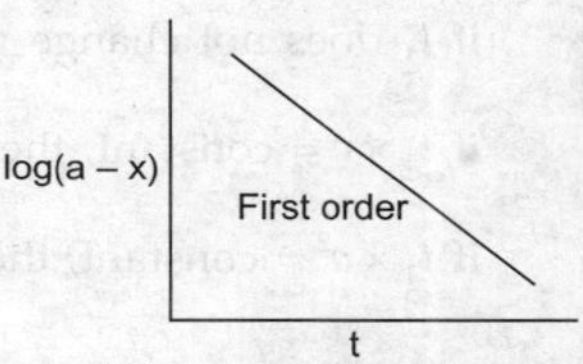

Fig. 14

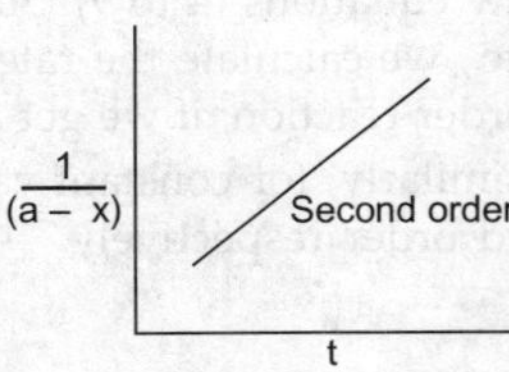

Fig. 15

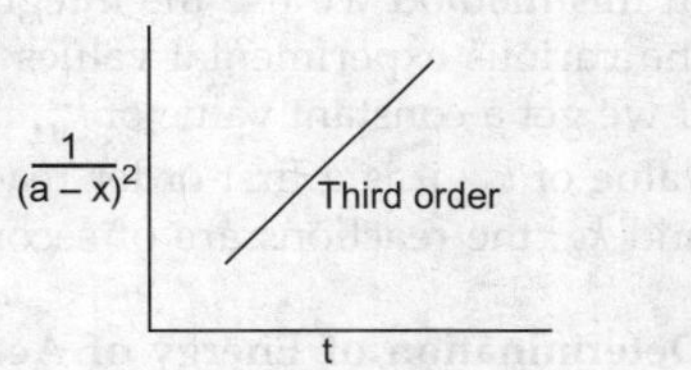

Fig. 16

In case of gaseous reactions, concentration may be replaced by pressure.

2. Method of Equifractional Parts: Half-life Method

(a) As we know that

$$t_{\frac{1}{2}} \propto \left(\frac{1}{a}\right)^{n-1}, \qquad \text{where } n \text{ is the order}$$

we may have the following relation

$$\frac{\left(t_{\frac{1}{2}}\right)_1}{\left(t_{\frac{1}{2}}\right)_2} = \left\{\frac{a_2}{a_1}\right\}^{n-1}, \qquad \ldots (13)$$

where, $\left(t_{\frac{1}{2}}\right)_1$ and $\left(t_{\frac{1}{2}}\right)_2$ are the half-life periods, or time for a definite fractional change of a reaction when the respective initial concentrations of the reactants are a_1 and a_2.

Taking log we have,

$$n = 1 + \left(\log \frac{\left(t_{\frac{1}{2}}\right)_1}{\left(t_{\frac{1}{2}}\right)_2} \Bigg/ \log \frac{a_2}{a_1} \right)$$

(b) Alternatively, from the equation

$$t_{\frac{1}{2}} \propto \frac{1}{a^{n-1}}, \quad \text{we conclude that}$$

if $t_{\frac{1}{2}} \times a^{-1}$ = constant, the reaction is of zero order;

if $t_{\frac{1}{2}}$ does not change with a, the reaction is of first order;

if $t_{\frac{1}{2}} \times a$ = constant, the reaction is of second order;

if $t_{\frac{1}{2}} \times a^2$ = constant, the reaction is of third order.

3. Use of Integrated Rate Law

In this method we use the integrated rate law equations (4 to 9). Substituting the various experimental values of a, x, t, etc., we calculate the rate constant. If we get a constant value of k_0, it is a zero order reaction; if we get a constant value of k_1, it is a first-order reaction; and similarly, for constant values of k_2 and k_3, the reactions are of second and third order respectively.

Determination of Energy of Activation

(a) Graphical Method

Arrhenius's equation may be represented as

$$\log k = -\frac{E}{2{\cdot}303\,R} \cdot \frac{1}{T} + \log A.$$

Now plotting $\log k$ against $1/T$, slope of the straight line is equal to $-\dfrac{E}{2{\cdot}303\,R}$ from which E can be calculated.

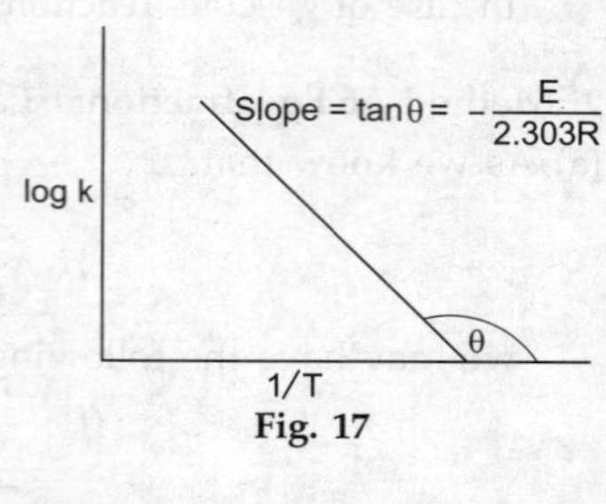

Fig. 17

(b) Use of Equation (12)

Energy of activation E can be calculated by using Eqn. (12) if for a given reaction, rate constants k_1 and k_2 are known at two different temperatures T_1 and T_2 respectively.

Kinetics of Simultaneous Reactions

The kinetics of simultaneous reactions, viz., parallel reactions and successive or consecutive reactions, considering all reactions to be of first order (as is true in radioactive disintegrations), has been discussed in Chapter 11. Thus, the equations from 23 to 26 (Chapter 11) are applicable just by replacing the decay constant (λ) by the rate constant (k).

Effect of Catalyst on the Reaction Rate

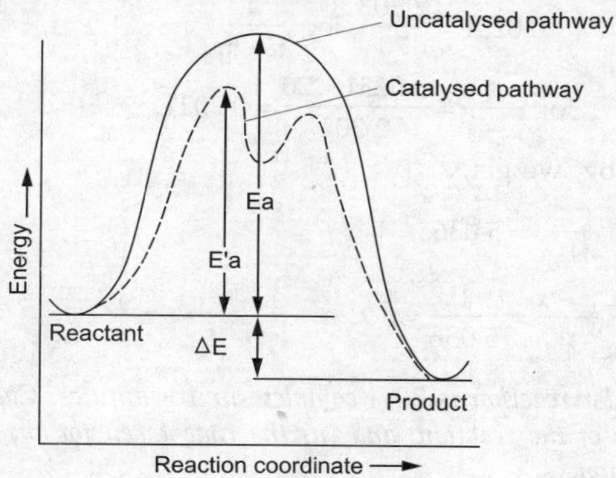

The catalyst provides a different path of the reaction corresponding to a lower value of energy of activation for the formation of the products. This enhances the rate of the reaction. A catalysed reaction typically occurs in several steps, as shown above, with its own value of energy of activation, but the overall energy of activation is lower than that for the uncatalysed reaction. Thus more molecules possess the minimum kinetic energy necessary for reaction. This is analogous to allowing more students to pass a course by lowering the requirements. Though the energy of activation for the catalysed reaction ($E'a$) and that for the uncatalysed reaction (Ea) are different, the energy released or absorbed in the course of reaction is the same for either path.

EXAMPLES

Ex. 1. *The half-life period of a first order reaction is 30 minutes. Calculate the specific reaction rate of the reaction. What fraction of the reactant remains after 70 minutes?*

Solution : $k_1 = \dfrac{0{\cdot}6932}{t_{\frac{1}{2}}} = \dfrac{0{\cdot}6932}{30} = 0{\cdot}0231 \text{ min}^{-1}$. ... (Eqn. 10)

Let the reaction be

a		0	Initial concentration
A	$\rightarrow$	Product	
$(a - x)$		x	Concentration after 70 minutes

$\therefore$ fraction of the reactant remained unreacted $= \dfrac{(a-x)}{a}$.

Now,

$$k_1 = \frac{2{\cdot}303}{t} \log \frac{a}{a-x} \qquad \text{... (Eqn. 5)}$$

or

$$0{\cdot}0231 = \frac{2{\cdot}303}{70} \log \frac{a}{a-x}$$

$$\log \frac{a}{a-x} = \frac{0{\cdot}0231 \times 70}{2{\cdot}303} = 0{\cdot}7021.$$

Taking antilog, we get,

$$\frac{a}{a-x} = 5{\cdot}036.$$

$$\therefore \quad \frac{a-x}{a} = \frac{1}{5{\cdot}036} \approx 0{\cdot}2.$$

Ex. 2. *A first-order reaction is 20% complete in 10 minutes. Calculate (i) specific rate constant of the reaction, and (ii) the time taken for the reaction to go to 75% completion.*

Solution : $k = \frac{2{\cdot}303}{t} \log \frac{a}{a-x}$ ... (Eqn. 5)

$$= \frac{2{\cdot}303}{10} \log \frac{a}{0{\cdot}8a} \cdot \left\{ x = \frac{20}{100} \times a = 0{\cdot}2a;\ (a-x) = a - 0{\cdot}2a = 0{\cdot}8a \right\}$$

$$\therefore \quad k = 0{\cdot}0223 \text{ min}^{-1}.$$

Let the time taken for 75% completion of the reaction be t.

$$k_1 = \frac{2{\cdot}303}{t} \log \frac{a}{(a-x)}$$

$$\because \quad (a-x) = a - \frac{75}{100} a = 0{\cdot}25\, a$$

$$\therefore \quad 0{\cdot}0223 = \frac{2{\cdot}303}{t} \log \frac{a}{0{\cdot}25a}.$$

$$\therefore \quad t = 62{\cdot}18 \text{ minutes.}$$

Ex. 3. *The specific reaction rate of a first-order reaction is* $0{\cdot}02\ \text{s}^{-1}$. *The initial concentration of the reactant is 2 moles/litre. Calculate (a) initial rate, and (b) rate after 60 seconds.*

Solution : We know that for a first-order reaction:

Rate of a reaction = $k \times$ molar concentration of the reactant.

(a) $\therefore$ initial rate = $k \times$ initial concentration

= $0{\cdot}02 \times 2 = 0{\cdot}04$ mole/litre/second.

(b) Now, to calculate rate after 60 seconds, let us first calculate concentration of the reactant after 60 seconds.

$$k = \frac{2\cdot303}{60} \log \frac{2}{\text{concentration after 60 seconds}} = 0\cdot02. \qquad \text{... (Eqn. 5)}$$

∴ concentration of the reactant after 60 seconds = 0·60 M.

∴ rate after 60 seconds = $k \times$ concentration of the reactant after 60 seconds.

$$= 0\cdot02 \times 0\cdot60$$
$$= 0\cdot012 \text{ mole/litre/second.}$$

Ex. 4. *The rate law for the reaction* A + B → Product is

$$rate = -\frac{d\,[A]}{dt} = +\frac{d\,[\text{Product}]}{dt} = k\,[A]^{\frac{3}{2}}\,[B]^{-\frac{1}{2}}.$$ *Find the order.*

Solution : The order of the reaction by definition is equal to,

$$\frac{3}{2} + \left(-\frac{1}{2}\right) = 1.$$

Ex. 5. *In a reaction* 2A → Product, *the concentration of* A *decreases from* 0·5 *mole/litre to* 0·4 *mole/litre in* 10 *minutes. Calculate the rate during this interval.*

Solution : Rate = rate of disappearance of A per mole

$$= -\frac{1}{2}\frac{\Delta\,[A]}{\Delta t} = -\frac{1}{2}\frac{(0\cdot5 - 0\cdot4)}{10} = -0\cdot005.$$

The negative sign simply indicates the fall in concentration of A. Thus ignoring the negative sign, the rate of the reaction is 0·005 mole/litre/minute.

Ex. 6. *The rate constant is numerically the same for three reactions of first, second and third order respectively, the unit of concentration being in moles per litre. Which reaction should be the fastest and is this true for all ranges of concentrations?*

Solution : Suppose R_1, R_2 and R_3 are the rates of three reactions of first, second and third-order respectively and k is the rate constant, which is the same for the three reactions,

$$\therefore \quad R_1 = k\,[A]$$
$$R_2 = k\,[A]^2$$
$$R_3 = k\,[A]^3$$

[A] being the concentration of the reactant A in moles per litre.

Now if, $[A] = 1$, $R_1 = R_2 = R_3$;

$[A] < 1$, $R_1 > R_2 > R_3$;

and $[A] > 1$, $R_3 > R_2 > R_1$.

Ex. 7. *Radioactive decay is a first-order process. Radioactive carbon in wood sample decays with a half-life of 5770 years. What is the rate constant (in year^{-1}) for the decay? What fraction would remain after 11540 years?*

Solution : Disintegration constant, $k = \dfrac{0.6932}{t_{\frac{1}{2}}}$

(or rate constant)

$$k = \frac{0.6932}{5770} = 0.00012\ \text{yr}^{-1}. \qquad \text{... (Eqn. 10)}$$

Again, for the radioactive disintegration process say,

$$\begin{array}{lll} a & & \text{Initial concentration} \\ \text{A} \rightarrow & \text{Product} & \\ (a-x) & x & \text{Concentration after 11540 years} \end{array}$$

We have,
$$k = \frac{2.303}{t} \log \frac{a}{(a-x)}$$

$$0.00012 = \frac{2.303}{11540} \log \frac{a}{(a-x)}$$

$$\log \frac{a}{(a-x)} = \frac{0.00012 \times 11540}{2.303} = 0.6013$$

Taking antilog,

$$\frac{a}{(a-x)} = 3.993$$

$$\therefore \text{ fraction of A remained} = \frac{(a-x)}{a} = \frac{1}{3.993} \approx \frac{1}{4}.$$

Ex. 8. *A first-order gas reaction has $k = 1.5 \times 10^{-6}$ per second at 200°C. If the reaction is allowed to run for 10 hours, what percentage of the initial concentration would have changed in the product? What is the half-life of this reaction?* **(IIT 1987)**

Solution : We know from first-order reaction,

$$k = \frac{2.303}{t} \log \frac{a}{a-x}$$

$$1.5 \times 10^{-6} = \frac{2.303}{10 \times 60 \times 60} \log \frac{a}{a-x}. \qquad (t = 10 \times 60 \times 60\ \text{s})$$

$$\therefore \qquad \log \frac{a}{a-x} = \frac{1.5 \times 10^{-6} \times 10 \times 60 \times 60}{2.303}$$

$$= 0.02344$$

$$\text{or} \qquad \log \frac{a-x}{a} = -0.02344 = \bar{1}.9765.$$

$$\therefore \quad \frac{a-x}{a} = 0{\cdot}9473.$$

$$\text{Percentage of reactant remained} = \frac{a-x}{a} \times 100$$

$$= 0{\cdot}9473 \times 100 = 94{\cdot}73\%.$$

$\therefore$ percentage of the initial concentration changed to product

$$= 100 - 94{\cdot}73 = 5{\cdot}27\%.$$

$$\text{Further,} \quad t_{\frac{1}{2}} = \frac{0{\cdot}6932}{k} = \frac{0{\cdot}6932}{1{\cdot}5 \times 10^{-6}} = 0{\cdot}462 \times 10^{6} \text{ s.}$$

Ex. 9. *For a reaction, the concentration of the reactant C varies with time t as shown by the following graph:*

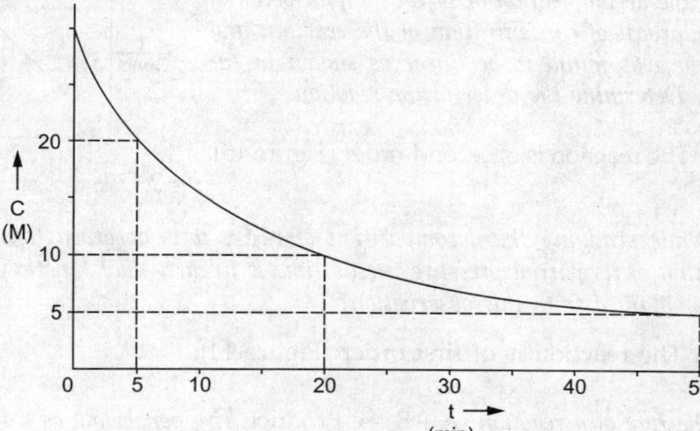

what do you conclude about the order of the reaction?

Solution **:** The change in concentration of the reactant from 20 M to 10 M takes 15 minutes and when it changes from 10 M to 5 M, it takes 30 minutes. So, for the given reaction, $t_{1/2}$ is inversely proportional to the initial concentration of the reactant suggesting that the reaction is of second order.

Ex. 10. *For a reaction, a graph was plotted between reactant concentration C and time t as shown in the figure.*

Identify the order of the reaction with respect to the reactant. Can the concentration of the reactant be theoretically zero after infinite time?

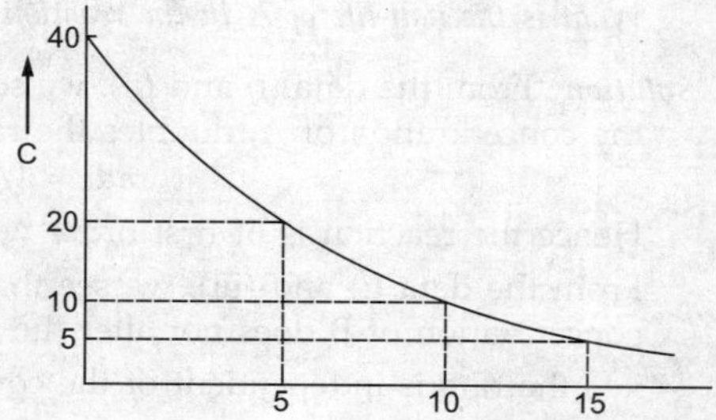

Solution : From the graph we see that the concentration reduces from 40 to 20 in 5 minutes, reduces from 20 to 10 in 5 minutes, and reduces from 10 to 5 in again 5 minutes, which shows that half-life period is independent of concentration which is the characteristic of a first order reaction (Figure 9). In a first order reaction the concentration of the reactant can never be theoretically zero.

Ex. 11. *In the decomposition of* N_2O *into* N_2 *and* O_2, *the half-life period was found to be inversely proportional to the initial pressure of* N_2O. *What is the order of the reaction?*

Solution : Reaction is of second-order (Figure 10).

Ex. 12. *In the decomposition of* N_2O_5 *the plot between the reciprocal of concentration of the reactant and the time was found to be linear as shown in the figure. Determine the order of the reaction.*

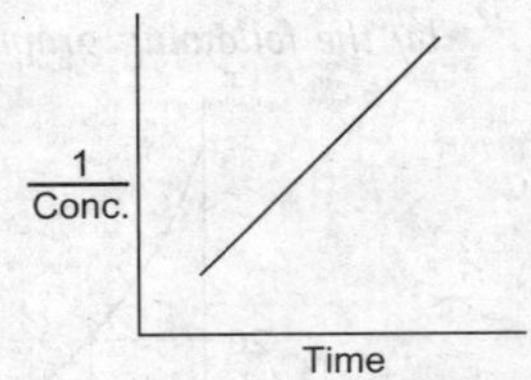

Solution : The reaction is of second-order (Figure 15).

Ex. 13. *While studying the decomposition of* N_2O_5 *it is observed that a plot of logarithm of its partial pressure versus time is linear. What kinetic parameters can be obtained from this observation?* **(IIT 1985)**

Solution : The reaction is of first order (Figure 14).

Ex. 14. *The rate of a reaction,* A + B → Product, *is given below as a function of different initial concentrations of* A *and* B.

	[A] (moles/litre)	[B] (moles/litre)	*Initial Rate* (moles/litre/min)
(i)	0·01	0·01	0·005
(ii)	0·02	0·01	0·010
(iii)	0·01	0·02	0·005

Determine the order of the reaction with respect to A *and with respect to* B. *What is the half-life of* A *in the reaction?*

Solution : From the data (i) and (ii), we see that [B] is constant and doubling the concentration of A, doubles the rate, i.e.,

$$\text{rate} \propto [A].$$

Hence the reaction is of first order with respect to A.

From the data (i) and (iii), we see that [A] is constant and changing the concentration of B does not alter the rate,

i.e., the rate is independent of the concentration of B.

Hence the reaction is of zero-order with respect to B.

Again, $\text{rate} = k\,[A]^m\,[B]^n$. ... (Eqn. 2)

We have calculated above that $m = 1$ and $n = 0$.

Considering the data (i),

$$0{\cdot}005 = k\,(0{\cdot}01)^1\,(0{\cdot}01)^0$$

$$\therefore \quad k = \frac{0{\cdot}005}{0{\cdot}01} = 0{\cdot}5 \text{ minute}^{-1}$$

Now, $t_{\frac{1}{2}}$ is given by

$$t_{\frac{1}{2}} = \frac{0{\cdot}6932}{k} = \frac{0{\cdot}6932}{0{\cdot}5} = 1{\cdot}39 \text{ min.} \qquad \text{... (Eqn. 10)}$$

Ex. 15. *For the reaction* $2NO + Cl_2 \rightarrow 2NOCl$, *it is found that doubling the concentration of both reactants increases the rate by a factor of 8, but doubling the* Cl_2 *concentration alone, only doubles the rate. What is the order of the reaction with respect to* NO *and* Cl_2*?*

Solution : $\text{Rate} = k\,[NO]^m\,[Cl_2]^n$ ... (Eqn. 2)

Let the concentrations of NO and Cl_2 be x and y respectively.

According to the question,

$$R_1 = k\,x^m y^n$$

and,

$$R_2 = k\,(2x)^m\,(2y)^n$$

$$= k \cdot x^m\,y^n \cdot 2^{m+n}.$$

$$\therefore \quad \frac{R_2}{R_1} = 2^{m+n} = 8 = 2^3. \qquad \text{(given)}$$

$$\therefore \quad m + n = 3.$$

Again,

$$R_3 = k\,(x)^m\,(2y)^n$$

$$= k\,x^m\,y^n \cdot 2^n.$$

$$\therefore \quad \frac{R_3}{R_1} = 2^n = 2. \qquad \text{(given)}$$

$$\therefore \quad n = 1$$

$$m = 3 - 1 = 2.$$

Ex. 16. *The rate of a reaction was found to be equal to its rate constant at any concentration of the reactant. What is the order of the reaction?*

Solution : Suppose for a reaction A $\rightarrow$ Product

$\text{rate} = k[A]^n$ when n is order.

Rate will be equal to k at any concentration of A only when $n = 0$.
Thus, the reaction should be of zero order.

Ex. 17. *In a reaction* $2N_2O_5 \rightarrow 4NO_2 + O_2$, *the rate can be expressed as*

(i) $-\dfrac{d\,[N_2O_5]}{dt} = k_1\,[N_2O_5]$

(ii) $\dfrac{d\,[NO_2]}{dt} = k_2\,[N_2O_5]$

(iii) $\dfrac{d\,[O_2]}{dt} = k_3\,[N_2O_5]$

How are k_1, k_2 *and* k_3 *related?*

Solution : The rate law of the given reaction is

$$\text{rate} = -\frac{1}{2} \times \frac{d\,[N_2O_5]}{dt} = \frac{1}{4} \times \frac{d\,[NO_2]}{dt} = \frac{d\,[O_2]}{dt} = k\,[N_2O_5]$$

$$\therefore \quad -\frac{d\,[N_2O_5]}{dt} = 2k\,[N_2O_5] = k_1\,[N_2O_5]$$

$$\frac{d\,[NO_2]}{dt} = 4k\,[N_2O_5] = k_2\,[N_2O_5]$$

$$\frac{d\,[O_2]}{dt} = k\,[N_2O_5] = k_3\,[N_2O_5]$$

$$\therefore \quad \begin{cases} k_1 = 2k \\ k_2 = 4k \\ k_3 = k \end{cases}$$

or $\quad k = \dfrac{k_1}{2} = \dfrac{k_2}{4} = k_3$

or $\quad 2k_1 = k_2 = 4k_3$.

Ex. 18. *The reaction,* $2N_2O_5 \rightarrow 4NO_2 + O_2$, *is forming* NO_2 *at the rate of* 0·0072 mole/L/s *at some time.*

(a) What is the rate of change of $[O_2]$ *at this time?*

(b) What is the rate of change of $[N_2O_5]$ *at this time?*

(c) What is the rate of reaction at this time?

Solution : The rate of the reaction is expressed as

$$\text{rate} = -\frac{1}{2}\frac{d[N_2O_5]}{dt} = +\frac{1}{4}\frac{d[NO_2]}{dt} = +\frac{d[O_2]}{dt}$$

And given that $\dfrac{d[NO_2]}{dt} = 0{\cdot}0072$ mole/L/s.

(a) Rate of appearance of $O_2 = \frac{1}{4} \times$ rate of appearance of NO_2

$$\frac{d[O_2]}{dt} = \frac{1}{4} \times \frac{d[NO_2]}{dt}$$

$$= \frac{1}{4} \times 0{\cdot}0072$$

$$= 0{\cdot}0018 \text{ mole/L/s.}$$

(b) Rate of disappearance of $N_2O_5 = \frac{1}{2} \times$ rate of appearance of NO_2

$$-\frac{d[N_2O_5]}{dt} = \frac{1}{2} \times \frac{d[NO_2]}{dt}$$

$$\frac{d[N_2O_5]}{dt} = -\frac{1}{2} \times 0{\cdot}0072$$

$$= -0{\cdot}0036 \text{ mole/L/s.}$$

(c) Rate of reaction $= \frac{1}{4} \times \frac{d[NO_2]}{dt}$

$$= \frac{1}{4} \times 0{\cdot}0072$$

$$= 0{\cdot}0018 \text{ mole/L/s.}$$

Ex. 19. *At some temperature, the rate constant for the decomposition of* HI *on a gold surface is* $0{\cdot}08 \text{ M·s}^{-1}$.

$$2HI(g) \rightarrow H_2(g) + I_2(g)$$

What is the order of the reaction? How long will it take for the concentration of HI to drop from 1·50 M *to* 0·30 M?

Solution : From the unit of k given, the reaction is of zero order.

As this reaction is of the type $2A \rightarrow$ Products, we apply,

$$k_0 = \frac{x}{nt}, \text{ where } n = 2$$

or $$k_0 = \frac{a - (a - x)}{nt}$$

or $$0{\cdot}08 = \frac{1{\cdot}50 - 0{\cdot}30}{2t}$$

$$t = 7{\cdot}5 \text{ s.}$$

Ex. 20. *The reaction* $2N_2O_5(g) \rightarrow 2N_2O_4(g) + O_2(g)$ *obeys the rate law,*

$$\text{rate} = k[N_2O_5] \text{ where } k = 0{\cdot}0084 \text{ s}^{-1}.$$

If 2·50 *moles of* N_2O_5 *were taken in a* 5·0-*litre flask, how many moles of* N_2O_4 *would remain after* 60 *seconds?*

Solution : The given first-order reaction is of the type

$$2A \rightarrow \text{Products}$$

we thus apply,

$$k = \frac{2 \cdot 303}{2t} \log \frac{2 \cdot 50/5}{x/5},$$

where x is supposed to be the number of moles of N_2O_5 remaining after 60 seconds.

$$0 \cdot 0084 = \frac{2 \cdot 303}{2 \times 60} \log \frac{2 \cdot 50}{x}$$

$$x = 0 \cdot 9125 \text{ mole.}$$

Ex. 21. *The rate of a first order reaction is* 0·04 mole/L/s *at* 10 *minutes and* 0·03 mole/L/s *at* 20 *minutes after initiation. Find the half-life of the reaction.* **(IIT 2001)**

Solution : Let the concentrations of the reactant after 10 min and 20 min be C_1 and C_2 respectively.

$\therefore$ rate after 10 min $= k \cdot C_1 = \cdot 04 \times 60$

and rate after 20 min $= k \cdot C_2 = \cdot 03 \times 60$.

$$\therefore \quad \frac{c_1}{c_2} = \frac{4}{3}.$$

Supposing the reaction starting after 10 minutes

$$k = \frac{2 \cdot 303}{10} \log \frac{c_1}{c_2} = \frac{2 \cdot 303}{10} \log \frac{4}{3} = \cdot 02878.$$

$$\therefore \quad t_{1/2} = \frac{0 \cdot 6932}{k} = \frac{0 \cdot 6932}{0 \cdot 02878} = 24 \cdot 086 \text{ min.}$$

Ex. 22. *The rate law of a chemical reaction*

$$2NO\,(g) + O_2\,(g) = 2NO_2\,(g)$$

is given as rate $= k\,[NO]^2\,[O_2]$. *How will the rate of the reaction change if the volume of the reaction vessel is reduced to one-fourth of its initial volume?*

Solution : Since the volume is reduced to one-fourth, concentration of each reactant will increase four times.

$\therefore$ rate $= k\,(4\,[NO])^2\,(4\,[O_2]) = 64\,(k\,[NO]^2\,[O_2])$.

Comparing this equation with the given one, we see that the rate grows sixty-four times.

Ex. 23. *In a reaction* $H_2 + I_2 \rightarrow 2HI$ *the rate of disappearance of* I_2 *is found to be* 10^{-6} *mole per litre per second. What would be the corresponding rate of appearance of HI?*

Solution : In the given reaction

rate = rate of disappearance of H_2 per mole

= rate of disappearance of I_2 per mole

= rate of appearance of HI per mole

$$\therefore \quad -\frac{d[H_2]}{dt} = -\frac{d[I_2]}{dt} = +\frac{1}{2} \times \frac{d[HI]}{dt} = 10^{-6}$$

$$\therefore \quad \frac{d[HI]}{dt} = 2 \times 10^{-6} \text{ mole L}^{-1}\text{ s}^{-1}.$$

Ex. 24. *Show that in the case of a first-order reaction, the time required for 99·9% of the reaction to take place is almost about ten times that required for half the reaction.*

Solution : For a first-order reaction,

$$t_{1/2} = \frac{0·6932}{k}. \quad \text{... (1)}$$

Now, suppose a first-order reaction, represented as

a	0	Initial concentration
A $\rightarrow$	Product	
$(a - 0·999\,a)$	$0·999\,a$	Concentration after time t
or $0·001\,a$	$0·999\,a$	

$$\therefore \quad t = \frac{2·303}{k} \log \frac{a}{0·001\,a}$$

$$= \frac{2·303}{k} \log 1000 = \frac{2·303}{k} \log 10^3$$

$$= \frac{2·303 \times 3}{k} = \frac{6·909}{k}. \quad \text{... (2)}$$

From (1) and (2), we see that t is about ten times of $t_{1/2}$.

Ex. 25. *For a reaction at 800°C,* $2NO + 2H_2 \rightarrow N_2 + 2H_2O$, *the following data were obtained:*

$[NO] \times 10^{-4}$ mole/litre	$[H_2] \times 10^{-3}$ mole/litre	$d[NO]/dt \times 10^{-4}$ mole litre^{-1} min^{-1}
(i) 1·5	4·0	4·4
(ii) 1·5	2·0	2·2
(iii) 0·5	2·0	0·24

What is the order of this reaction with respect to NO and H_2*?*

Solution : From the data (i) and (ii), we see that when the concentration of H_2 is halved, the rate is also halved at constant concentration of NO. Hence the reaction is of first order with respect to H_2. Let us now consider the data (ii) and (iii) to determine the order with respect to NO as $[H_2]$ is constant.

The rate law of the above reaction is

$$\text{rate} = -\frac{1}{2} \times \frac{d\,[\text{NO}]}{dt} = k\,[\text{NO}]^m\,[\text{H}_2]^1,$$

where m is the order with respect to NO ... (Eqn. 3)

or $$-\frac{d\,[\text{NO}]}{dt} = 2k\,[\text{NO}]^m\,[\text{H}_2]$$

Substituting data (ii) and (iii), we get

$$2{\cdot}2 \times 10^{-4} = 2k\,(1{\cdot}5 \times 10^{-4})^m \cdot (2 \times 10^{-3}) \quad \text{... (1)}$$

$$0{\cdot}24 \times 10^{-4} = 2k\,(0{\cdot}5 \times 10^{-4})^m \cdot (2 \times 10^{-3}) \quad \text{... (2)}$$

Dividing (1) by (2),

$$\frac{2{\cdot}2}{0{\cdot}24} = \frac{(1{\cdot}5 \times 10^{-4})^m}{(0{\cdot}5 \times 10^{-4})^m} = 3^m$$

or $$\frac{220}{24} = 3^m.$$

Taking log, $$\log 220 - \log 24 = m \log 3$$

$$2{\cdot}3424 - 1{\cdot}3802 = m \times 0{\cdot}4771$$

or $$0{\cdot}9622 = 0{\cdot}4771\,m$$

or $$m = \frac{0{\cdot}9622}{0{\cdot}4771} \approx 2.$$

Hence the reaction is of second and first-order with respect to NO and H_2 respectively.

Ex. 26. *The following data were obtained for a gaseous reaction:*

$$A_2 + 2B \rightarrow 2AB$$

$[A_2]$ mole L^{-1}	$[B]$ mole L^{-1}	$-\frac{d\,[A_2]}{dt}$ mole/lit./min
(i) 0·10	0·01	0·072
(ii) 0·10	0·04	0·288
(iii) 0·20	0·01	0·144

Calculate the rate constant and rate of formation of AB, *when* $[A_2] = 0{\cdot}01$ *and* $[B] = 0{\cdot}02$.

Solution : From (i) and (ii) we see that $[A_2]$ is constant and when [B] increases 4 times, the rate also increases 4 times, i.e., rate = k[B].

Again from (i) and (iii) we see [B] is constant and doubling the concentration of A_2 increases the rate by two times showing again,

$$\text{rate} = k\,[A_2].$$

Thus the rate law will be

$$-\frac{d\,[A_2]}{dt} = k\,[A_2]\,[B]$$

Substituting the data from (i),

$$0{\cdot}072 = k\,(0{\cdot}1)\,(0{\cdot}01)$$

$$k = 72\ \text{L mole}^{-1}\ \text{min}^{-1}$$

Rate of formation of AB (per mole) $= \dfrac{1}{2}\dfrac{d\,[AB]}{dt} = k\,[A_2]\,[B]$

$\therefore$ rate of formation of AB $= \dfrac{d\,[AB]}{dt} = 2 \times 72 \times 0{\cdot}01 \times 0{\cdot}02$

$$= 0{\cdot}0288\ \text{mole L}^{-1}\ \text{min}^{-1}.$$

Ex. 27. *Calculate the order of reaction* A + B → Product, *with respect to* A *and* B *from the following data:*

[A] (mole/L)	[B] (mole/L)	Rate × 10^{-5} (mole/L/s)
(i) 0·01	0·010	5·0
(ii) 0·01	0·015	5·0
(iii) 0·02	0·010	14·1

Solution : From data (i) and (iii), [B] is constant,

$\therefore$ for data (i) rate $= k\,[A]^n$

(n is the order with respect to A)

$$5 \times 10^{-5} = k\,(0{\cdot}01)^n. \quad \ldots (1)$$

For data (iii), $14{\cdot}1 \times 10^{-5} = k\,(0{\cdot}02)^n.$ ... (2)

Dividing (1) by (2), $\dfrac{5 \times 10^{-5}}{14{\cdot}1 \times 10^{-5}} = \dfrac{k\,(0{\cdot}01)^n}{k\,(0{\cdot}02)^n} = \left(\dfrac{1}{2}\right)^n$

$$n = 1{\cdot}5$$

Thus, order of the reaction with respect to A is 1·5.

Again we see from data (i) and (ii) that at constant concentration of A, change in concentration in B does not affect the rate. Hence the order is zero with respect to B.

Ex. 28. *What will be the order of reaction of the type* A → Product, *for which the following data were obtained?*

	[A]	$\frac{d\,[A]}{dt}$
(i)	1	1
(ii)	2	8
(iii)	3	27

Solution : Since the specific reaction rate k is the rate when the reactant concentration is unity, we see from data (i)

$$k = 1 \text{ as}$$
$$d\,[A]/dt = 1 \text{ at } [A] = 1.$$

From data (ii) or (iii) we thus have,

$$\frac{d\,[A]}{dt} = k\,[A]^m; \quad (m \text{ is the order})$$

$$8 = 1 \cdot (2)^m$$

or $$2^3 = 2^m \quad \text{or} \quad m = 3.$$

Again, $$27 = 1 \cdot (3)^m$$

$$3^3 = 3^m$$

or $$m = 3.$$

Hence the order is three.

Ex. 29. *Calculate the order of the reaction,* A → Product, *from the following data:*

[A] moles/L	d[product]/dt (moles/L/s)
0·0025	$10{\cdot}0 \times 10^{-5}$
0·005	$5{\cdot}0 \times 10^{-5}$
0·01	$2{\cdot}5 \times 10^{-5}$

Solution : Let the order be n.

Applying rate law for the first and second set of readings, we have,

$$\text{rate} = 10 \times 10^{-5} = k\,(0{\cdot}0025)^n \qquad \ldots(1)$$

and, $$\text{rate} = 5 \times 10^{-5} = k\,(0{\cdot}005)^n \qquad \ldots(2)$$

Dividing (1) by (2), we get,

$$2 = \left(\frac{1}{2}\right)^n$$

or $$n = -1.$$

Thus the order of the reaction is –1.

Ex. 30. *For a chemical reaction* A + B → Product, *the order is* 1 *with respect to each of* A *and* B. *Find x and y from the given data.*

Rate (moles/L/s)	[A]	[B]
0·10	0·20 M	0·05 M
0·40	x	0·05 M
0·80	0·40 M	y

Solution : The rate law may be written as

$$\text{rate} = k\,[A]\,[B]$$

Substituting the first set of data in the rate law, we get,

$$0{\cdot}10 = k \times 0{\cdot}20 \times 0{\cdot}05$$
$$k = 10.$$

Now substituting the second and third sets of data, we get,

$$0{\cdot}40 = 10 \times x \times 0{\cdot}05$$
$$x = 0{\cdot}80 \text{ M}.$$

And,
$$0{\cdot}80 = 10 \times 0{\cdot}40 \times y$$
$$y = 0{\cdot}20 \text{ M}.$$

Ex. 31. *The following data were obtained at a certain temperature for the decomposition of ammonia in contact with tungsten:*

p (mm)	50	100	200
Relative $t_{\frac{1}{2}}$	3·64	1·82	0·91

Find the order of the reaction.

Solution : We know,

$$\frac{\left(t_{\frac{1}{2}}\right)_1}{\left(t_{\frac{1}{2}}\right)_2} = \left(\frac{a_2}{a_1}\right)^{n-1} \qquad \text{... (Eqn. 13)}$$

$$\frac{3{\cdot}64}{1{\cdot}82} = \left(\frac{100}{50}\right)^{n-1}$$

$$2 = 2^{n-1}.$$

$$\therefore \qquad n - 1 = 1$$
$$n = 2.$$

Ex. 32. *In the decomposition of* H_2O_2 *at* 300 K, *the energy of activation was found to be* 18 kcal/mole, *while it decreased to* 6 kcal/mole *when the decomposition was carried out in the presence of a catalyst at* 300 K. *How many times is the catalysed reaction faster than the uncatalysed one?*

Solution : Suppose E_1 and E_2 are the energies of activation when the reaction is carried out in the absence and presence of a catalyst respectively.

Thus, $k_1 = Ae^{-E_1/RT}$

$k_2 = Ae^{-E_2/RT}$.

Taking log, $\ln k_1 = \ln A - \frac{E_1}{RT}$

$\ln k_2 = \ln A - \frac{E_2}{RT}$

$$\therefore \quad \ln k_2 - \ln k_1 = -\frac{E_2}{RT} + \frac{E_1}{RT}$$

$$\text{or} \quad \ln\frac{k_2}{k_1} = \frac{E_1}{RT} - \frac{E_2}{RT} = \frac{1}{RT}(E_1 - E_2)$$

$$\text{or} \quad 2{\cdot}303 \log\frac{k_2}{k_1} = \frac{1}{0{\cdot}002 \times 300}(18 - 6) = \frac{12}{0{\cdot}002 \times 300}$$

$$2{\cdot}303 \log\frac{k_2}{k_1} = 20$$

$$\log\frac{k_2}{k_1} = \frac{20}{2{\cdot}303}$$

$$= 8{\cdot}6843$$

Taking antilog, $\frac{k_2}{k_1} = 4{\cdot}8 \times 10^8$.

The ratio of rate constants shows that the reaction will be faster by $4{\cdot}8 \times 10^8$ times due to the addition of the catalyst.

Ex. 33. *The energy of activation and specific rate constant for a first-order reaction at* 25°C

$$\underset{\text{(in } CCl_4)}{2N_2O_5} \rightarrow \underset{\text{(in } CCl_4)}{2N_2O_4} + O_2$$

are 100 kJ/mole *and* $3{\cdot}46 \times 10^{-5}\ s^{-1}$ *respectively. Determine the temperature at which the half-life of the reaction is* 2 *hours.*

Solution : Let us calculate the rate constant (say k_2) at a temperature (say T_2) at which $t_{1/2}$ is given to be $2 \times 60 \times 60$ seconds.

The rate constant at a temperature 298 K (say T_1) is given as $3{\cdot}46 \times 10^{-5}\ s^{-1}$ (say k_1).

$$k_2 = \frac{0{\cdot}6932}{2 \times 60 \times 60} = 9{\cdot}62 \times 10^{-5}\ s^{-1}.$$

Thus, $$\log\frac{k_2}{k_1} = \frac{E}{2{\cdot}303\ R}\left(\frac{T_2 - T_1}{T_1 T_2}\right) \quad \text{... (Eqn. 12)}$$

$$\log \frac{9.62 \times 10^{-5}}{3.46 \times 10^{-5}} = \frac{100000}{2.303 \times 8.314}\left(\frac{T_2 - 298}{298\, T_2}\right)$$

or $$T_2 = 310 \text{ K}.$$

Ex. 34. *In Arrhenius's equation for a certain reaction, the value of A and E (activation energy) are* $4 \times 10^{13}\ \text{s}^{-1}$ *and* 98.6 kJ mol^{-1} *respectively. If the reaction is of first order, at what temperature will its half-life period be ten minutes?* **(IIT 1990)**

Solution **:** We have,

$$k = Ae^{-E/RT}$$

$$\ln k = \ln A - \frac{E}{RT}$$

$$2.303 \log k = 2.303 \log A - \frac{E}{RT}$$

or $$\log k = \log A - \frac{E}{2.303\, RT}. \qquad \dots (1)$$

Given that $$A = 4 \times 10^{13}\ \text{s}^{-1},\ E = 98.6 \text{ kJ mol}^{-1}.$$

$$t_{\frac{1}{2}} = 10 \times 60 \text{ s}.$$

For first-order reaction $k = \dfrac{0.6932}{t_{\frac{1}{2}}} = \dfrac{0.6932}{600}\ \text{s}^{-1}.$

Thus (1) becomes,

$$\log \frac{0.6932}{600} = \log (4 \times 10^{13}) - \frac{98.6}{2.303 \times 8.314 \times 10^{-3} \times T}.$$

$$[R = 8.314 \times 10^{-3} \text{ kJ/K/mol}]$$

$\therefore$ $$T = 311.2 \text{ K}.$$

Ex. 35. *Two reactions proceed at 25°C at the same rate. The temperature coefficient of the rate of the first reaction is 2 and that of the second is 2·5. Find the ratio of the rates of these reactions at 95°C.*

Solution **:** At 25°C the rates of the two reactions are same (say R). Since the temperature coefficient gives the number of times the rate of a reaction increases by a rise in temperature of 10°, we have the following results:

Temperature	Rate of	
	1st reaction	2nd reaction
25°	R	R
35°	$2\ R$	$2{\cdot}5\ R$
45°	$2^2\ R$	$(2{\cdot}5)^2\ R$
55°	$2^3\ R$	$(2{\cdot}5)^3\ R$
65°	$2^4\ R$	$(2{\cdot}5)^4\ R$
75°	$2^5\ R$	$(2{\cdot}5)^5\ R$
85°	$2^6\ R$	$(2{\cdot}5)^6\ R$
95°	$2^7\ R$	$(2{\cdot}5)^7\ R$

Thus, at 95°,

$$\frac{\text{rate of second reaction}}{\text{rate of first reaction}} = \frac{2{\cdot}5^7\ R}{2^7\ R} = 4{\cdot}768.$$

Ex. 36. *What is the energy of activation of a reaction if its rate doubles when the temperature is raised from* 290 K *to* 300 K?

Solution : We have,

$$\log\frac{k_2}{k_1} = \frac{E}{2{\cdot}303\ R}\left(\frac{T_2 - T_1}{T_1 T_2}\right) \qquad \text{... (Eqn. 12)}$$

$$\log 2 = \frac{E}{2{\cdot}303 \times 2}\left(\frac{300 - 290}{290 \times 300}\right) \qquad (R = 2\ \text{cal/K/mole})$$

$$E = 12062\ \text{calories}$$
$$\approx 12\ \text{kcal.}$$

Ex. 37. *For the reaction* $A + B \rightarrow C + D$; $\Delta H = +20$ kJ/mole, *the activation energy of the forward reaction is 85 kJ/mole. Calculate the activation energy of the reverse reaction.*

Solution : Energy of activation of the reverse reaction (say E_2) can be determined from ΔH and energy of activation of the forward reaction graphically as follows:

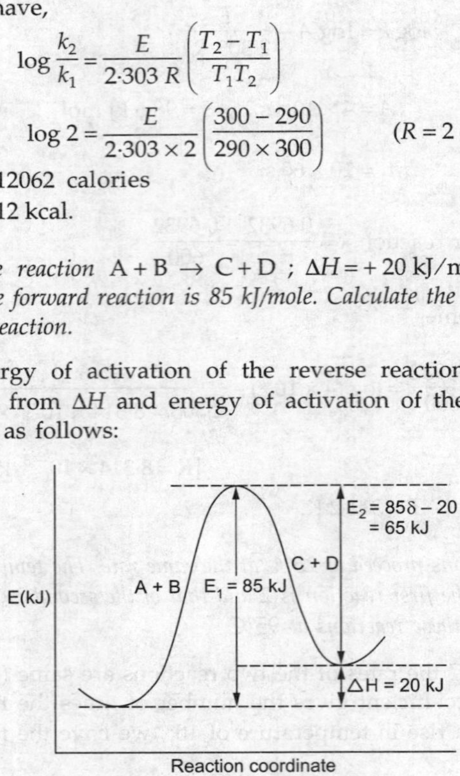

Ex. 38. *The activation energy for the reaction,*

$$O_3(g) + NO(g) \rightarrow NO_2(g) + O_2(g)$$

is 9·6 kJ/mole. *Prepare an activation energy plot if* ΔH° *for this reaction is* −200 kJ/mole. *What is the energy of activation for the reverse reaction?*

Solution : Energy of activation for reverse reaction

$$= 9{\cdot}6 + 200 \text{ kJ}$$
$$= 209{\cdot}6 \text{ kJ}.$$

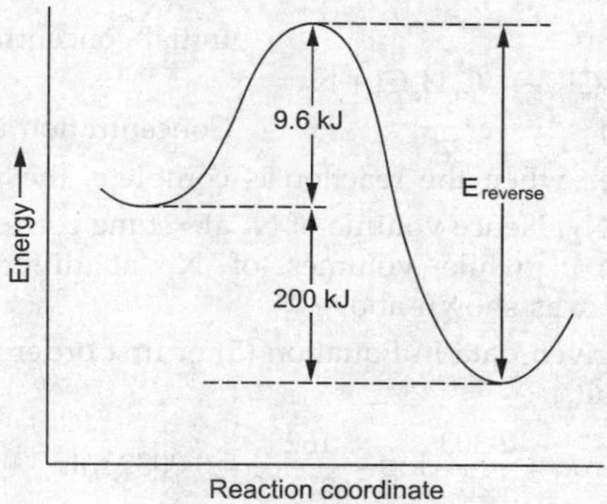

Ex. 39. *The decomposition of* H_2O_2 *in aqueous solution is a reaction of first order. How can it be justified from the following data:*

Time (min)	0	10	20	30
x':	25	20	15·7	12·5

where x' is the volume (mL) *of* $KMnO_4$ *required to titrate a definite volume of* H_2O_2?

Solution : a Initial concentration

$$H_2O_2 \rightarrow H_2O + \tfrac{1}{2}O_2$$

$(a - x)$ Concentration after time t

Since $(a - x)$ corresponds to the undecomposed H_2O_2, at different time intervals at zero time, it will correspond to initial concentration '*a*'.

Now substituting the given data in Equation (5) of a first order reaction we get,

for $t = 10$ min; $k_1 = \dfrac{2{\cdot}303}{10} \log \dfrac{25}{20} = 0{\cdot}022 \text{ min}^{-1}$

$t = 20$ min; $k_1 = \dfrac{2{\cdot}303}{20} \log \dfrac{25}{15{\cdot}7} = 0{\cdot}023 \text{ min}^{-1}$

$t = 30$ min; $k_1 = \dfrac{2{\cdot}303}{30} \log \dfrac{25}{12{\cdot}5} = 0{\cdot}023 \text{ min}^{-1}$

The constant value of k_1 shows that the given reaction is of first order.

Ex. 40. *From the following data for the decomposition of diazobenzene chloride, show that the reaction is of first order:*

Time: (min)	20	50	70	∞
Vol. of N_2: (mL)	10	25	33	162

Solution :

$$\begin{array}{lcl} a & & \text{Initial concentration} \\ C_6H_5N_2Cl \rightarrow C_6H_5Cl + N_2 & & \\ (a-x) & x & \text{Concentration after time } t \end{array}$$

At ∞ time, i.e., when the reaction is complete, the whole of $C_6H_5N_2Cl$ converts into N_2. Hence volume of N_2 at ∞ time corresponds to the initial concentration *'a'* while volumes of N_2 at different time intervals correspond to x as shown above.

Inserting the given data in Equation (5) of first order reaction, we get the following results.

For $t = 20$ min, $k_1 = \dfrac{2{\cdot}303}{20} \log \dfrac{162}{162-10} = 0{\cdot}0032 \text{ min}^{-1}$

For $t = 50$ min, $k_1 = \dfrac{2{\cdot}303}{50} \log \dfrac{162}{162-25} = 0{\cdot}0033 \text{ min}^{-1}$

For $t = 70$ min, $k_1 = \dfrac{2{\cdot}303}{70} \log \dfrac{162}{162-33} = 0{\cdot}0032 \text{ min}^{-1}$

The constancy of k_1 shows that the decomposition of $C_6H_5N_2Cl$ is a first order reaction.

Ex. 41. 1 mL *of methyl acetate was added to* 20 mL *of* 0·5 N HCl *and* 2 mL *of the mixture was withdrawn from time to time during the progress of hydrolysis of the ester and titrated with a solution of alkali. The amount of alkali needed for titration at various intervals is given below:*

Time: (min)	0	20	119	∞
Alkali used: (mL)	19·24	20·73	26·6	42·03

Establish that the reaction is of first order.

Solution :

$$\begin{array}{lcl} a & & \text{Initial concentration} \\ CH_3COOCH_3 + H_2O \xrightarrow[{[H^+]}]{HCl} CH_3COOH + CH_3OH & & \\ (a-x) & x & \text{Concentration after time } t \end{array}$$

HCl acts as a catalyst. Since in every titration the amount of HCl is the same, the alkali used against HCl is subtracted from the total alkali used (given in the data) to get the volume of alkali used only against CH_3COOH. At zero time no CH_3COOH is formed, so alkali used at zero time is only for HCl.

∴ we thus have,

Time (min):	0	20	119	∞
Vol. of alkali (mL) used against CH_3COOH	: 19·24 − 19·24 = 0	20·73 − 19·24 = 1·49 (x)	26·6 − 19·24 = 7·36 (x)	42·03 − 19·24 = 22·79 (a)

Now following the same way as in Example 40, we get the following results:

For $t = 20$; $k_1 = \frac{2{\cdot}303}{20} \log \frac{22{\cdot}79}{22{\cdot}79 - 1{\cdot}49} = 0{\cdot}0033 \text{ min}^{-1}$

For $t = 119$; $k_1 = \frac{2{\cdot}303}{119} \log \frac{22{\cdot}79}{22{\cdot}79 - 7{\cdot}36} = 0{\cdot}0032 \text{ min}^{-1}$

The constancy of k_1 shows that the reaction is of first order.

Ex. 42. *The following data were obtained for the saponification of ethyl acetate using equal concentration of ester and alkali.*

Time: (min)	0	4	10	20
Vol. of acid used: (mL)	8·04	5·30	3·50	2·22

Show that the reaction is of second order.

Solution :

$$\underset{(a-x)}{\overset{a}{CH_3COOC_2H_5}} + \underset{(a-x)}{\overset{a}{NaOH}} \rightarrow \underset{x}{CH_3COONa} + \underset{x}{C_2H_5OH}$$

Initial concn. (top row); Concentration after time t (bottom row)

Volume of the acid used in the titration corresponds to the amount of unused NaOH. Therefore, at zero time, volume of acid corresponds to initial concentration 'a' and volume after time t corresponds to $(a - x)$.

Thus, inserting the data in Equation (6) of a second order reaction, we get

$$k_2 = \frac{1}{t} \cdot \frac{x}{a\,(a-x)} \qquad \text{... (Eqn. 6)}$$

$a = 8{\cdot}04$; $x = a - (a - x)$.

For, $t = 4$ min, $a = 8{\cdot}04$.

$$x = a - (a - x) = 8{\cdot}04 - 5{\cdot}30 = 2{\cdot}74$$

$$\therefore \quad k_2 = \frac{1}{4} \times \frac{2{\cdot}74}{8{\cdot}04 \times 5{\cdot}30} = 0{\cdot}016 \text{ (mole/litre)}^{-1} \text{ (min)}^{-1}.$$

For, $t = 10$ min, $a = 8{\cdot}04$; $x = 8{\cdot}04 - 3{\cdot}50 = 4{\cdot}54$

$$k_2 = \frac{1}{10} \times \frac{4{\cdot}54}{8{\cdot}04 \times 3{\cdot}50} = 0{\cdot}016 \text{ (mole/litre)}^{-1} \text{ (min)}^{-1}.$$

Similarly, for t = 20 minutes we can calculate another value of k_2. Since k_2 comes out to be constant, the reaction is of second order.

Ex. 43. *The optical rotation of cane sugar in* 0·5 N *lactic acid at* 25°C *at various time intervals are given below:*

Time: (min)	0	1435	11360	∞
Rotation (°):	34·50°	31·10°	13·98°	–10·77°

Show that the reaction is of first order.

Solution :
$$\underset{\substack{\text{Sucrose}\\\text{Dextrorotatory}}}{C_{12}H_{22}O_{11}} + \underset{\text{(excess)}}{H_2O} \xrightarrow{\text{Lactic acid}} \underbrace{\underset{\substack{\text{Glucose}\\\text{Dextro}}}{C_6H_{12}O_6} + \underset{\substack{\text{Fructose}\\\text{Laevo}}}{C_6H_{12}O_6}}_{\text{Laevorotatory}}$$

Since in this reaction dextro form changes to laevo form, the optical rotation decreases with the progress of the reaction. Thus change in rotation is proportional to the amount of sugar remained after different time intervals. We now have,

Time (min)	0	1435	11360	∞
Change in rotation (°)	34·50 – (–10·77) = 45·27 (a)	31·10 – (–10·77) = 41·87 $(a - x)$	13·98 – (–10·77) = 24·75 $(a - x)$	– 10·77 – (–10·77) = 0

Substituting the data in Equation (5),

for $t = 1435$ min

$$k_1 = \frac{2{\cdot}303}{1435} \log \frac{45{\cdot}27}{41{\cdot}87} = 5{\cdot}442 \times 10^{-5}$$

and, for $t = 11360$ min

$$k_1 = \frac{2{\cdot}303}{11360} \log \frac{45{\cdot}27}{24{\cdot}75} = 5{\cdot}311 \times 10^{-5}.$$

The values of k_1 are fairly constant and so the reaction is of first order.

Ex. 44. *In the course of the reaction*

$$CH_3COOCH_3\,(g) \rightarrow C_2H_4\,(g) + H_2\,(g) + CO(g)$$

the initial pressure was found to be 0·42 atm *while after* 6·5 *minutes, it was* 0·54 atm. *If the reaction follows first order kinetics, find the rate constant.*

Solution **:** Let the initial number of moles of CH_3COOCH_3 be a.

Initial moles: a 0   0   0

$$CH_3COOCH_3 \rightarrow C_2H_4 + H_2 + CO$$

Moles after time t: $(a-x)$ x x x

According to the given data,

$$a = 0{\cdot}42 \text{ (initial moles corresponds to 0·42 atm)}$$

and $\{(a-x)+x+x+x\} = 0{\cdot}54$

(Mole after 6·5 min corresponds to 0·54 atm)

or $\quad a + 2x = 0{\cdot}54$

or $\quad x = \dfrac{0{\cdot}54 - 0{\cdot}42}{2} = 0{\cdot}06.$

Applying first-order rate equation,

$$k = \frac{2{\cdot}303}{6{\cdot}5}\log\frac{0{\cdot}42}{0{\cdot}42 - 0{\cdot}06} = 0{\cdot}0237 \text{ min}^{-1}.$$

Ex. 45. *The second order rate constant for* $I^- + ClO^- \rightarrow IO^- + Cl^-$ *is* 0·06 $(\text{mole/litre})^{-1}\,s^{-1}$. *If a solution initially contains* 1 M I^- *and* $1{\cdot}5 \times 10^{-4}$ M ClO^- *can you calculate the concentration of* ClO^- *after* 5 *minutes. If no, what additional information is required?*

Solution **:** No, rate law is required to calculate $[ClO^-]$ after 5 minutes because the rate law could be any of the following forms:

$$\text{Rate} = k\,[I^-]\,[ClO^-]$$

$$\text{Rate} = k\,[I^-]^2$$

$$\text{Rate} = k\,[ClO^-]^2, \text{ etc.}$$

Ex. 46. *If the rate law has the form: Rate* $= k\,[A]\,[B]^{\frac{3}{2}}$, *can the reaction be an elementary process?*

Solution **:** No, an elementary process would generally have a rate law with orders equal to its molecularity, which are necessarily integers.

Ex. 47. *The decomposition of* N_2O_5 *according to the equation*

$$2N_2O_5\,(g) \rightarrow 4NO_2\,(g) + O_2\,(g)$$

is a first-order reaction. After 30 *minutes from the start of the decomposition in a closed vessel, the total pressure developed is found to be* 284·5 mmHg *and on complete decomposition the total pressure is* 584·5 mmHg. *Calculate the rate constant of the reaction.* **(IIT 1991)**

***Solution* :**

$$\begin{array}{llll} & a & & \\ & 2N_2O_5 & \rightarrow 4NO_2 & + \; O_2 \\ \text{After 30 min:} & a-x & 2x & x/2 \end{array}$$

$$\therefore \quad (a-x)+2x+\frac{x}{2}=284{\cdot}5$$

$$\text{or} \quad a+\frac{3x}{2}=284{\cdot}5. \quad \ldots (1)$$

After complete decomposition of N_2O_5:

$$\begin{array}{lll} 2N_2O_5 & \rightarrow 4NO_2 & + \; O_2 \\ 0 & 2a & a/2 \end{array}$$

$$\therefore \quad 2a+\frac{a}{2}=584{\cdot}5$$

$$\text{or} \quad \frac{5a}{2}=584{\cdot}5 \quad \ldots (2)$$

From (1) and (2), we get,

$$a=233{\cdot}5,\ x=34$$

Thus, for a first-order reaction of the type 2A → products

$$k=\frac{1}{2t}\log\frac{a}{a-x}$$

$$k=\frac{2{\cdot}303}{2\times 30}\log\frac{233{\cdot}5}{233{\cdot}5-34}$$

$$=2{\cdot}625\times 10^{-3}\ \text{min}^{-1}.$$

Ex. 48. *Which reaction will have the greater temperature dependence for the rate constant—one with a small value E or one with large value of E?*

***Solution* :** We have, $k = A\,e^{-E/RT}$.

The temperature dependence of k can be found by taking the derivative of the above equation with respect to T. Thus,

$$\frac{dk}{dT} = A\,e^{-E/RT}\cdot\frac{E}{RT^2} = \frac{kE}{RT^2}.$$

Thus the temperature dependence will be greater for reactions with large value of E.

Ex. 49. *What is the value of the rate constant, predicted by the Arrhenius's equation if $T \rightarrow \infty$? Is this value physically reasonable?*

***Solution* :** We have, $k = A\,e^{-E/RT}$

If $T \longrightarrow \infty$, $k = A$ $\qquad (e^x = 1 + x + \frac{x^2}{2!} + \frac{x^3}{3!} + \cdots)$

If $k = A$, $e^{-E/RT} = 1$

$$\therefore \quad \frac{-E}{RT} = 0 \text{ or } E = 0.$$

Since activation energy cannot be zero, the result is not physically reasonable.

Ex. 50. *Bicyclohexane was found to undergo two parallel first order rearrangements. At 730 K, the first order rate constant for the formation of cyclohexane was measured as* $1{\cdot}26 \times 10^{-4}$ s^{-1}*, and for the formation of methyl cyclopentene the rate constant was* $3{\cdot}8 \times 10^{-5}$ s^{-1}*. What is the percentage distribution of the rearrangement products?*

Solution : Bicyclohexane $\xrightarrow{k_1}$ Cyclohexene

Bicyclohexane $\xrightarrow{k_2}$ Methylcyclopentene

$$\text{Percentage of cyclohexene} = \frac{k_1}{k_1 + k_2} \times 100$$

$$= \frac{1{\cdot}26 \times 10^{-4}}{1{\cdot}26 \times 10^{-4} + 3{\cdot}8 \times 10^{-5}} \times 100$$

$$= 77\%.$$

$\therefore$ percentage of methylcyclopentene = 23%.

Ex. 51. *The decomposition of* N_2O *into* N_2 *and O in the presence of argon follows second-order kinetics with* $k = (5{\cdot}0 \times 10^{11})\, e^{-29000\text{ K/T}}$. *Calculate the energy of activation.*

Solution : Comparing the given equation with Arrhenius's equation, we have,

$$\frac{E}{RT} = \frac{29000 \text{ K}}{T}$$

$$E = 29000 \times 8{\cdot}314 = 241 \text{ kJ/mole.}$$

Ex. 52. *For the displacement reaction*

$$[Co(NH_3)_5Cl]^{2+} + H_2O \longrightarrow [Co(NH_3)_5(H_2O)]^{3+} + Cl^-$$

the rate constant is given by

$$\ln [k/(\text{min}^{-1})] = -\frac{11067 \text{ K}}{T} + 31{\cdot}33.$$

Evaluate k, E and A for the chemical reaction at 25°C.

Solution : Substituting T = 298 K in the given equation, we get,

$$k = 3{\cdot}06 \times 10^{-3}\ \text{min}^{-1} = 5{\cdot}10 \times 10^{-5}\ \text{s}^{-1}.$$

Further, we have,

$$k = A\,e^{-E/RT}$$

$$\text{or}\quad \ln k = \ln A - \frac{E}{RT}$$

Comparing this equation with the given equation, we get,

$$\ln A = 31{\cdot}33$$

$$\text{or}\quad 2{\cdot}303 \log A = 31{\cdot}33\,; \qquad A = 4{\cdot}04 \times 10^{13}\ \text{min}^{-1}$$

$$= 6{\cdot}73 \times 10^{11}\ \text{s}^{-1}$$

$$\text{and,}\quad -\frac{E}{RT} = -\frac{11067\ \text{K}}{T}$$

$$E = (11067K)\,(8{\cdot}314\ \text{J K}^{-1}\ \text{mol}^{-1})$$

$$= 92011\ \text{J/mole} = 92{\cdot}011\ \text{kJ/mole}.$$

Ex. 53. *The complexation of* Fe^{2+} *with the chelating agent dipyridyl has been studied kinetically in both forward and reverse directions.*

$$Fe^{2+} + 3\ \text{dipy} \longrightarrow Fe(\text{dipy})_3^{2+}$$

Rate (forward) $= (1{\cdot}45 \times 10^{13})\,[Fe^{2+}]\,[\text{dipy}]^3$

and rate (reverse) $= (1{\cdot}22 \times 10^{-4})\,[Fe\,(\text{dipy})_3^{2+}]$.

Find the stability constant for the complex.

Solution : At dynamic equilibrium,

rate of formation of complex = rate of its decomposition

$$(1{\cdot}45 \times 10^{13})\,[Fe^{2+}]\,[\text{dipy}]^3 = (1{\cdot}22 \times 10^{-4})\,[Fe(\text{dipy})_3^{2+}]$$

$$K_s = \frac{[Fe(\text{dipy})_3^{2+}]}{[Fe^{2+}]\,[\text{dipy}]^3} = \frac{1{\cdot}45 \times 10^{13}}{1{\cdot}22 \times 10^{-4}} = 1{\cdot}19 \times 10^{17}.$$

Ex. 54. *The approach to the following equilibrium was observed kinetically from both directions.* $PtCl_4^{2-} + H_2O \rightleftharpoons Pt(H_2O)Cl_3^- + Cl^-$ *at 25°C. It was found that*

$$-\frac{d\,[PtCl_4^{2-}]}{dt} = (3{\cdot}9 \times 10^{-5})\,[PtCl_4^{2-}] - (2{\cdot}1 \times 10^{-3})\,[Pt(H_2O)Cl_3^-]\,[Cl^-]$$

Calculate the equilibrium constant for the complexation of the fourth Cl^- *by* Pt(II).

Solution : At equilibrium, $\dfrac{d\,[PtCl_4^{2-}]}{dt} = 0$.

Hence, $3{\cdot}9 \times 10^{-5}\,[PtCl_4^{2-}] = 2{\cdot}1 \times 10^{-3}\,[Pt(H_2O)Cl_3^-]\,[Cl^-]$

$$\text{or} \qquad K = \frac{[PtCl_4^{2-}]}{[Pt(H_2O)Cl_3^-]\,[Cl^-]} = \frac{2.1\times10^{-3}}{3.9\times10^{-5}} = 53.85.$$

Ex. 55. *For the reaction* $2\,H_2 + 2\,NO \longrightarrow N_2 + 2\,H_2O$, *the following mechanism has been suggested:*

$$2\,NO \rightleftharpoons N_2O_2 \qquad \text{... Eqb. constant} = K_1$$

$$N_2O_2 + H_2 \xrightarrow{k_2} N_2O + H_2O \qquad \text{... Rate-determining step}$$

$$N_2O + H_2 \longrightarrow N_2 + H_2O \qquad \text{... Fast}$$

Establish the rate law for this reaction.

Solution : $K_1 = \dfrac{[N_2O_2]}{[NO]^2}$

$$[N_2O_2] = K_1[NO]^2$$

Thus, the rate law is

$$\text{rate} = k_2[N_2O_2]\,[H_2]$$

$$= k_2\,(K_1[NO]^2)\,[H_2]$$

$$= k\,[NO]^2\,[H_2].$$

Ex. 56. *The possible mechanism for the reaction*

$$2\,NO + Br_2 \longrightarrow 2\,NOBr$$

is $NO + Br_2 \longrightarrow NOBr_2$ *(fast);*

$NOBr_2 + NO \longrightarrow 2\,NOBr$ *(slow).*

Establish the rate law.

Solution : $K_1 = \dfrac{[NOBr_2]}{[NO]\,[Br_2]}$

$$[NOBr_2] = K_1\,[NO]\,[Br_2]$$

Thus, the rate law is

$$\text{rate} = k_2\,[NOBr_2]\,[NO]$$

$$= k_2\,(K_1\,[NO]\,[Br_2])\,[NO]$$

$$= k\,[NO]^2\,[Br_2].$$

Ex. 57. *For the reaction*

$$[Cr(H_2O)_4Cl_2]^+\,(aq) \xrightarrow{k_1} [Cr(H_2O)_5Cl]^{2+}\,(aq) \xrightarrow{k_2} [Cr(H_2O)_6]^{3+}(aq)$$

$k_1 = 1.78\times10^{-3}\ s^{-1}$ *and* $k_2 = 5.8\times10^{-5}\ s^{-1}$ *for the initial concentration of*

$[Cr(H_2O)_4Cl_2]^+$ *is* 0·0174 *mole/litre at* 0°C. *Calculate the value of t at which the concentration of* $[Cr(H_2O)_5Cl]^{2+}$ *is maximum.*

Solution **:** We have,

$$t = \frac{2{\cdot}303\,(\log k_1 - \log k_2)}{k_1 - k_2} \qquad \text{... (Eqn. 33 Ch. 11)}$$

$$= \frac{2{\cdot}303\,(\log 1{\cdot}78 \times 10^{-3} - \log 5{\cdot}8 \times 10^{-5})}{1{\cdot}78 \times 10^{-3} - 5{\cdot}8 \times 10^{-5}}$$

= 1990 seconds.

Ex. 58. *For the reaction:* $A + 2B \rightarrow P$ *the rate law is given by*

$$rate = k[A][B]^2.$$

From the following data, calculate x and y.

	[A] (moles/L)	[B] (moles/L)	$t_{1/2}$ (min)
1.	2	10	50
2.	4	10	x
3.	10	4	100
4.	10	8	y

Solution **:** In sets 1 and 2, A is the limiting reactant. The half-life of the reaction is thus the time during which half of A converts to P. From the rate law, the reaction is first order w.r.t. A for which $t_{1/2}$ is independent of its concentration and so increasing the concentration of A from 2 moles/L to 4 mole/L, $t_{1/2}$ shall remain the same.

$\therefore \quad x = 50$ minutes.

In sets 3 and 4, B is the limiting reagent. The half-life of the reaction is thus the time during which half of B converts to P. As the reaction is of second order w.r.t. B for which $t_{1/2}$ is inversely proportional to [B] and so increasing the concentration of B twice from 4 mole/L to 8 mole/L, $t_{1/2}$ will be halved.

$$\therefore \quad y = \frac{100}{2} = 50 \text{ minutes.}$$

Ex. 59. *Some* $PH_3(g)$ *is introduced into a flask at* 600°C *containing an inert gas.* PH_3 *proceeds to decompose into* $P_4(g)$ *and* $H_2(g)$ *and the reaction goes to completion. The total pressure is given below as a function of time. Find the order of the reaction and calculate the rate constant for the reaction:*

$$4PH_3 \rightarrow P_4 + 6H_2$$

Time (s):	0	60	120	∞
p mmHg:	262·40	272·90	275·51	276·40

Solution : Let the initial partial pressures of PH_3 and the inert gas be p and p_i mm respectively and p' mm of PH_3 decomposes at different time intervals.

Initial partial pressure: p

$$4PH_3 \rightarrow P_4 + 6H_2$$

Partial pressure at different times: $p - p'$ $\quad p'/4 \quad 6p'/4$

As given, at $t = 60$ seconds:

$$p + p_i = 262{\cdot}40 \qquad \ldots (1)$$

and,
$$p - p' + \frac{p'}{4} + \frac{6p'}{4} + p_i = 272{\cdot}90 \qquad \ldots (2)$$

At $t = \infty$:
$$\frac{p}{4} + \frac{6p}{4} + p_i = 276{\cdot}40. \qquad \ldots (3)$$

Solving eqns. (1), (2) and (3), we get,

$$p = 18{\cdot}67 \text{ and } p' = 14.$$

Similarly, at $t = 120$ seconds

$$p - p' + \frac{p'}{4} + \frac{6p'}{4} + p_i = 275{\cdot}51. \qquad \ldots (4)$$

Solving eqns. (1), (4) and (3), we get,

$$p = 18{\cdot}67 \text{ and } p' = 17{\cdot}48.$$

As the given reaction is of the type $nA \rightarrow$ products, where $n = 4$, we have the following equation for first-order kinetics

$$k_1 = \frac{2{\cdot}303}{4t} \log \frac{a}{a-x} = \frac{2{\cdot}303}{4t} \log \frac{p}{p-p'}.$$

Thus, at $t = 60s$; $k_1 = \dfrac{2{\cdot}303}{4 \times 60} \log \dfrac{18{\cdot}67}{18{\cdot}67 - 14} = 5{\cdot}8 \times 10^{-3}\ s^{-1}$

$$t = 120\ s;\ k_1 = \frac{2{\cdot}303}{4 \times 120} \log \frac{18{\cdot}67}{18{\cdot}67 - 17{\cdot}48} = 5{\cdot}8 \times 10^{-3}\ s^{-1}.$$

As the values of k_1 are constant, the given reaction follows the first order kinetics.

Ex. 60. *The vapour pressures of two miscible liquids* A *and* B *are* 300 *and* 500 mmHg *respectively. In a flask* 10 *moles of* A *is mixed with* 12 *moles of* B. *However, as soon as* B *is added,* A *starts polymerising into a completely insoluble solid. The polymerisation follows first-order kinetics. After* 100 *minutes,* 0·525 *mole of a solute is dissolved which arrests the polymerisation completely. The final vapour pressure of the solution is* 400 mmHg *Estimate the rate constant of the polymerisation reaction. Assume negligible volume change on mixing and polymerisation, and ideal behaviour for the final solution.* **(IIT 2001)**

(See solved ex. 59. Ch. 13)

Ex. 61. *From the following reaction scheme, write the rate law for the disappearance of* A, B *and* C.

1. $A + B \xrightarrow{k_1} C + D$

2. $C + D \xrightarrow{k_2} A + B$

3. $B + C \xrightarrow{k_3} E + D$

Solution : The reactant A is removed in Step 1 and produced in Step 2.

$$\therefore \quad -\frac{d[A]}{dt} = k_1[A][B] - k_2[C][D].$$

Similarly,

$$-\frac{d[B]}{dt} = k_1[A][B] + k_3[B[C] - k_2[C][D]$$

and,
$$-\frac{d[C]}{dt} = k_2[C][D] + k_3[B[C] - k_1[A][B].$$

PROBLEMS

(*Answers bracketed with questions*)

1. A first order reaction takes 69·3 minutes for 50% completion. How much time will be needed for 80% completion? (160·9 minutes)

2. In a certain first order reaction, half the reaction was decomposed in 500 seconds. How long will it be until one-tenth is left? (1661 seconds)

3. Find the value of the rate constant for the reaction $A + B \rightarrow AB$, if the rate of the reaction is 5×10^{-5} (mol/L) min^{-1} and [A] and [B] are respectively 0·05 and 0·01 mol/L. [0·1 (mol/L)$^{-1}$ (min)$^{-1}$]

4. For three reactions of first, second and third order, $k_1 = k_2 = k_3$, when concentration is expressed in mol/L. What will be relation among k_1, k_2 and k_3 if the concentration is expressed in mol/mL? ($k_1 = k_2 \times 10^{-3} = k_3 \times 10^{-6}$)

5. How many times will the rate of the reaction 2A + B → A_2B change if the concentration of substance A is doubled and that of substance B is halved? (increase twofold)

6. The rate law for the reaction

$$RCl + NaOH\,(aq) \rightarrow ROH + NaCl$$

is given by

rate = k_1 [RCl], the rate of reaction will be

(a) doubled on doubling the concentration of NaOH

(b) halved on reducing the concentration of RCl to one-half

(c) increased on increasing the temperature of the reaction

(d) unaffected by increasing the temperature of the reaction

[Hint: See Example 14] **(IIT 1988)** (b)

7. How many times must the concentration of substance B_2 in the system $2A_2\,(g) + B_2\,(g) \rightarrow 2A_2B\,(g)$ be increased for the rate of the forward reaction to remain unchanged when the concentration of substance A_2 is lowered to one-fourth of its initial value? (16 times)

8. The order of the reaction $2A + B + C \rightarrow$ Product, is found to be 1, 2 and 0 w.r.t. A, B and C respectively. If the concentration of each reactant is increased by two times, what will be the effect on the rate of the reaction?

[Hint: See solved example 15] (8 times)

9. In the thermal decomposition of C_2H_5Br to C_2H_4 and HBr, the pressure changed from an initial value of 200 mmHg to 390 mmHg at the end of the reaction. What fraction remained unchanged when the pressure of the mixture was 300 mmHg? (0·47)

10. Catalytic decomposition of nitrous oxide by gold at 900°C at an initial pressure of 200 mm was 50% in 53 minutes and 73% in 100 minutes. Find the order of the reaction. How much will it decompose in 100 minutes at the same temperature but at an initial pressure of 600 mm? (First, 73%)

11. A certain reaction is of first order. After 540 seconds, 32·5% of the reactant remains.

(a) Calculate the rate constant.

(b) How long would it require for 25% of the reactant to be decomposed?

($2{\cdot}08 \times 10^{-3}\ s^{-1}$; 139 s)

12. The specific reaction rate constant for a first order reaction is $1 \times 10^{-2}\ s^{-1}$. If the initial concentration is 1 M, find the initial rate. What is the reaction rate after two minutes? ($1 \times 10^{-2}, 3 \times 10^{-3}$)

13. For the nonequilibrium process $A + B \rightarrow P$, the reaction is of first order with respect to A and of second order with respect to B. If 1 mole each of A and B are introduced into a 1-litre flask, and the initial rates were 1×10^{-2} mole/litre · s, calculate the rate when half the reactants have converted to the product.

($1{\cdot}2 \times 10^{-3}$ mole $L^{-1}\ s^{-1}$)

14. The rate law for the reaction,

$$CH_3COOC_2H_5 + H_2O \xrightarrow{[H^+]} CH_3COOH + C_2H_5OH$$

is $$\frac{dx}{dt} = k\,[CH_3COOC_2H_5]\,[H_2O]^{\circ}.$$

What will be the effect on the rate if

(i) the concentration of the ester is doubled?

(ii) the concentration of H^+ is tripled? [(i) Rate doubles (ii) No effect]

15. Prove that the half-life period of a reaction of *n*th order in a reaction of the type $A \rightarrow$ Product, is inversely proportional to the $(n-1)$th power of the initial concentration.

16. A living plant acquires a definite fraction of $^{14}_{6}C$ nuclei in carbon content. If a freshly cut piece of wood gives 16·1 counts per minute per gram and an old wooden bowl gives 9·6 counts per minute per gram of carbon, calculate the age of the wooden bowl. The half-life of $^{14}_{6}C$ is 5770 years. (4304 years)

17. The following data gives pressure of a gaseous N_2O_5 as a function of time at 45°C. Plot them first in $1/p$ vs t and then as p vs t. Determine the order and rate constant.

t (s)	p (mm)	t (s)	p (mm)
0	348	3600	58
600	247	4800	33
1200	185	6000	18
2400	105	7200	10

(First order, $k_1 = 5 \times 10^{-4}\ s^{-1}$)

18. A substance decomposes according to second order rate law. If the rate constant is $6{\cdot}8 \times 10^{-4}$ L mole^{-1} s^{-1}, calculate half-life of the substance, if the initial concentration is (i) 0·05 mole/L and (ii) 0·01 mole/L. ($2{\cdot}94 \times 10^4$ s; $1{\cdot}47 \times 10^5$ s)

19. For a certain reaction, it takes 5 minutes for the initial concentration of 0·5 mole/L to become 0·25 mole/L and another 5 minutes to become 0·125 mole/L. What is the order and specific rate constant of the reaction? (First; 0·138 minute^{-1})

20. The half-life period of a gaseous substance undergoing thermal decomposition was measured for various initial pressures (p) with the following results:

p (mm)	250	300	400	450
$t_{\frac{1}{2}}$ (min)	136	112·5	85	75·5

Calculate the order of the reaction. (Second)

21. The kinetics of decomposition of N_2O_5 in CCl_4 solution is studied by measuring the evolved oxygen. If 24 mL of the gas was evolved in one hour while 35 mL of the gas was evolved when no more oxygen was coming out, calculate the fraction of N_2O_5 decomposed in one hour. (0·686)

22. The following rate data were obtained at 30°C for the decomposition of N_2O_5 in CCl_4 solution:

$[N_2O_5]$ (mole/litre)	$d[N_2O_5]/dt$ (mole/litre/hour)
0·34	0·10
0·68	0·20
1·36	0·40

Calculate the order of the reaction and the rate constant at 30°C. (First; 0·29 h^{-1})

23. From the following data calculate the order with respect to each reactant A, B and C:

[A] (mole/L)	[B] (mole/L)	[C] (mole/L)	$d[B]/dt \times 10^{-5}$ (mole/L/s)
0·010	0·005	0·010	5·0
0·015	0·005	0·010	5·0
0·010	0·010	0·010	2·5
0·010	0·005	0·020	14·1

(0; −1; 3/2)

24. For a given reaction $A + B \rightarrow P$, the orders w.r.t. A and B are 1 and 2 respectively. Fill in the blanks from the following data:

Rate ($M\ s^{-1}$)	[A]	[B]
0·10	1·0 M	0·20 M
...	2·0 M	0·20 M
...	2·0 M	0·40 M

[Hint: See Example 30] ($R_2 = 0{\cdot}20$, $R_3 = 0{\cdot}80$)

25. The reaction $A + B \rightarrow C + D$; $\Delta H = 25$ kJ/mole should have an activation energy

(a) –25 kJ/mole (b) < +25 kJ/mole

(c) > + 25 kJ/mole

(d) either answer (b) or (c) depending upon experiment (c)

[Hint: See Example 37]

26. The half-life period for the reaction, $N_2O_5 \rightarrow 2NO_2 + \frac{1}{2}O_2$ is 2·4 hours at 30°C. What time would be required to reduce 5×10^{10} molecules of N_2O_5 to 10^8 molecules? (21·5 hours)

27. One mole of a gas A and two moles of a gas B are introduced into one vessel and 2 moles of A and 1 mole of B into a second vessel having the same capacity. The temperature is the same in both vessels. Will the rate of reaction between A and B in these vessels differ if it is expressed by the equation

(i) rate = k [A] [B] and (ii) rate = $k\,[A]^2\,[B]$? [(i) No (ii) Yes]

28. How many times will the rate of the reaction $2A + B \rightarrow A_2B$ change if the concentration of the substance A is tripled and that of the substance B is halved? (Increases 4·5 times)

29. What is the temperature coefficient of the reaction if the rate grows 15·6 times when the temperature is increased by 30 K? (2·5)

[Hint: See solved example 35]

30. The temperature coefficient of the rate of a reaction is 2·3. How many times will the rate of the reaction increase if the temperature is raised by 25 K? (8·02 times)

31. In a second order reaction when the concentration of both the reactants are equal, the reaction is 20% completed in 500 seconds. How long will it take the reaction to go to 60% completion? (3000 s)

[Hint: Use Equation 6]

32. For the reaction $A + B \rightarrow C$, the following data were obtained: In the first experiment when the initial concentration of both A and B is 0·1 M, the observed initial rate of formations of C is 1×10^{-4} mole per minute. In the second experiment, when the initial concentrations of A and B are 0·1 M and 0·3 M respectively, the initial rate is 9×10^{-4} mole per minute. In the third experiment with the initial concentration of both A and B, 0·3 M, the initial rate is 2.7×10^{-3} mole per minute. Write the rate law and calculate the rate constant for the reaction.

[Rate = $0.1 \times [A]^1 [B]^2$]

[**Hint:** Calculate m and n in rate = $k[A]^m [B]^n$; (See solved example 14)]

33. For the reaction $A \rightarrow B + C$ the following data were obtained:

t (s)	0	900	1800
[A]	50·8	19·7	7·62

Prove that the reaction is of first order. (k_1 is found to be constant)

34. Nitrosyl chloride, NOCl decomposes to NO and Cl_2.

$$2NOCl(g) \rightarrow 2NO(g) + Cl_2(g)$$

From the following data, determine the rate law, the rate constant and the overall order for this reaction:

[NOCl]:	0·10	0·20	0·30*ii*
Rate (mol/L/s)	8×10^{-10}	3.2×10^{-9}	7.2×10^{-9}

(Rate = $k[NOCl]^2$, $k = 8 \times 10^{-8}$ L/mol/s, two)

35. What is the half-life for the decomposition of NOCl when [NOCl] = 0·15 M? Given that for $2NOCl \rightarrow 2NO + Cl_2$: $-\dfrac{d[NOCl]}{dt}$ = (8.0×10^{-8} L/mol/s) $[NOCl]^2$

(8.34×10^7s)

36. The rate constant for the first order decomposition of N_2O_5 dissolved in chloroform at 45°C is $3.1 \times 10^{-4} min^{-1}$.

$$2N_2O_5 \rightarrow 4NO_2 + O_2$$

(a) What is the rate of decomposition of N_2O_5 when $[N_2O_5] = 0.40$ M?

(b) What are the rates of formation of NO_2 and of O_2 when $[N_2O_5] = 0.40$ M?

(c) What is the rate of this reaction.

[(a) 2.48×10^{-4} mole/L/min (b) 4.96×10^{-4} and 1.24×10^{-4} mol/L/min (c) 1.24×10^{-4} mol/L/min]

[**Hint:** First-order reaction is of the type: $2A \rightarrow$ Products]

37. For the reaction: $SO_2CL_2(g) \rightarrow SO_2(g) + Cl_2(g)$, it is found that a plot of $\ln[SO_2Cl_2]$ versus time is linear, and that in 240 seconds the $[SO_2Cl_2]$ decreases from 0·4 M to 0·28 M. What is the rate constant? (1.49×10^{-3} s^{-1})

38. A certain physiologically important first-order reaction has an activation energy equal to 45·0 kJ/mol at normal body temperature (37°C). Without a catalyst, the rate constant for the reaction is $5.0 \times 10^{-4} s^{-1}$. To be effective in the human body, where the reaction is catalysed by an enzyme, the rate constant must be at least $2{\cdot}0 \times 10^{-2} s^{-1}$. If the activation energy is the only factor affected by the presence of the enzyme, by how much must the enzyme lower the activation energy of the reaction to achieve the desired rate? (≈ 10 kJ/mol)

39. A drop (0·05 mL) of a solution contains $3{\cdot}0 \times 10^{-6}$ mole of H^+ ions. If the rate constant of disappearance of H^+ is $1{\cdot}0 \times 10^7$ mol/L/s, how long would it take for H^+ ions in the drop to disappear? $(6{\cdot}0 \times 10^{-9} s)$

40. The gas-phase decomposition of NOBr is second order in [NOBr], with $k = 0{\cdot}81\ M^{-1}{\cdot}s^{-1}$ at 10°C. Initial concentration of NOBr in the flask at 10°C $= 4{\cdot}00 \times 10^{-3}$ M. In how many seconds does it take up $1{\cdot}50 \times 10^{-3}$ M of this NOBr?

$$2NOBr \rightarrow 2NO + Br_2$$

(92·6 s)

[Hint: $k_2 = \frac{1}{2t}\left\{\frac{x}{a(a-x)}\right\}$]

41. The reaction, $A \rightarrow C + D$ was found to be second order in A. The rate constant for the reaction was determined to be 2·42 L/mol/s. If the initial concentration is 0·5 mole/L, what is the value of $t_{1/2}$. (0·8264 s)

[Hint: $k_2 = \frac{1}{t}\left\{\frac{x}{a(a-x)}\right\}$]

42. Both Technetium-99 and Thallium-201 are used to image heart muscle in patients who may have heart problems. The half-lives are 6 hours and 73 hours respectively. What per cent of radioactivity would remain for each of the isotopes after 2 days? (0·39%, 63%)

43. Two reactions have identical values for energy of activation. Does this ensure that they will have the same rate constant if run at the same temperature? (No, $k \propto A$)

44. The rate of the haemoglobin (Hb)– carbon monoxide reaction,

$$4Hb + 3CO \rightarrow Hb_4(CO)_3$$

has been studied at 20°C. Concentrations are expressed in μ mole/L.

[Hb] (μ mole/L)	[CO] (μ mole/L)	Rate of disappearance of Hb (μ mole/L/s)
3·36	1·00	0·941
6·72	1·00	1·88
6·72	3·00	5·64

(a) Calculate the rate constant for the reaction

(b) Calculate the rate of the reaction at the instant when

[Hb] = 1·50 and [CO] = 0·60 μ mole/L.

((a) 7×10^{-2} L/μ mole/s (b) $6{\cdot}3 \times 10^{-2}$ μ mole/L/s)

[Hint: Rate of reaction $= -\frac{1}{4}\frac{d[Hb]}{dt} = k[Hb][CO]$]

45. A first order reaction is 50% complete in 30 min at 27°C and in 10 min at 47°C. Calculate the reaction rate constant at 27°C and the energy of activation of the reaction in kJ/mole. **(IIT 1988)** (0·0231 min^{-1}, 43·84 kJ)

[**Hint:** $k(27°) = \dfrac{0{\cdot}6932}{(t_{1/2})_{27}}$

The reaction at 47°C is 3 times faster than that at 27°C. Use Equation 12, put $R = 8{\cdot}314 \times 10^{-3}$ kJ.]

46. The decomposition of arsine (AsH_3) into arsenic and hydrogen is a first order reaction. The decomposition was studied at constant volume and at constant temperature. The pressures at different times are as follows:

t (h):	0	5·5	6·5	8
p (atm):	0·9654	1·06	1·076	1·1

Calculate the velocity constant. $(4 \times 10^{-2}\,h^{-1})$

[**Hint:** For $AsH_3\,(g) \rightarrow As\,(s) + \frac{3}{2}H_2\,(g)$; $p_0 \propto a$ and $p_t \propto \left\{(a-x) + \frac{3x}{2}\right\}$

$\therefore\ p_t - p_0 \propto \frac{x}{2}$]

47. The rate constant of the first order reaction, that is, decomposition of ethylene oxide into CH_4 and CO, may be described by the following equation

$$\log k\,(s^{-1}) = 14{\cdot}34 - \frac{1{\cdot}25 \times 10^4}{T}\,K.$$

Find (a) energy of activation, and (b) rate constant at 397°C.

$\left[\text{(a) } 239{\cdot}34 \text{ kJ (b) } 4{\cdot}8 \times 10^{-5}\,s^{-1}\right]$

[**Hint:** Compare the given equation with $\log k = \log A - \dfrac{E}{2{\cdot}303\,RT}$]

48. For a homogeneous gaseous reaction $A \rightarrow B + C + D$ the initial pressure was p_0 while pressure after time t was p. Derive an expression for rate constant k in terms of p_0, p and t.

$\left[k = \dfrac{2{\cdot}303}{t}\log\dfrac{2p_0}{3p_0 - p}\right]$

[**Hint:** See solved example 44]

Objective Problems

1. The rate law for the single-step reaction $2A + B \rightarrow 2C$ is given by

(a) rate = $k\,[A] \cdot [B]$ (b) rate = $k\,[A]^2 \cdot [B]$

(c) rate = $k\,[2A] \cdot [B]$ (d) rate = $k\,[A]^2 \cdot [B]^0$

2. Which of the following rate laws has an overall order of 0·5 for the reaction $A + B + C \rightarrow$ Product?

(a) $R = k\,[A] \cdot [B] \cdot [C]$ (b) $R = k\,[A]^{.5}\,[B]^{.5}\,[C]^{.5}$

(c) $R = k\,[A]^{1.5}\,[B]^{-1}\,[C]^0$ (d) $R = k\,[A]\,[B]^0\,[C]^{.5}$

3. The rate law of the reaction $A + 2B \rightarrow$ Product is given by $\dfrac{d\,(\text{product})}{dt} = k\,[A]^2 \cdot [B]$. If A is taken in large excess, the order of the reaction will be

(a) 0 (b) 1 (c) 2 (d) 3

4. Which of the following statements is not correct?

(a) Law of mass action and rate law expressions are same for single-step reactions.

(b) Order of the slowest elementary reaction of a complex reaction gives the order of the complex reaction.

(c) Both order and molecularity have normally a maximum value of 3.

(d) Molecularity of a complex reaction $A + 2B \rightarrow C$ is 3.

5. If the volume of a closed vessel in which the equilibrium $2SO_2 + O_2 \rightleftharpoons 2SO_3$ is set in is halved, the rate of

(a) forward reaction will remain same as that of backward reaction

(b) forward reaction will become double that of backward reaction

(c) forward reaction will be halved that of backward reaction

(d) all are wrong

6. The rate of the simple reaction $2NO + O_2 \rightarrow 2NO_2$, when the volume of the reaction vessel is doubled,

(a) will grow eight times of its initial rate

(b) reduce to one-eighth of its initial rate

(c) will grow four times of its initial rate

(d) reduce to one-fourth of its initial rate

7. Rate of which reactions increase with temperature?

(a) Any

(b) Exothermic reactions

(c) Endothermic reactions

(d) None of the above

8. The specific rate constant of a first order reaction depends on the

(a) concentration of the reactant

(b) concentration of the product

(c) time

(d) temperature

9. If the rate constant k of a reaction is $1{\cdot}6 \times 10^{-3}$ (mol/L) (min^{-1}), the order of the reaction is

(a) 0 (b) 1 (c) 2 (d) cannot be known

10. If for any reaction, the rate constant is equal to the rate of the reaction at all concentrations, the order is

(a) 0 (b) 2 (c) 1 (d) 3

11. Which of the following procedures will lead to a change in the rate constant 'k' of a reaction?

(a) A change in the pressure

(b) Change in temperature

(c) Change in the volume of the reaction vessel

(d) Introduction of a catalyst

12. If a reaction with $t_{1/2} = 69{\cdot}3$ seconds, has a rate constant value of 10^{-2} per second, the order is

(a) 0 (b) 1 (c) 2 (d) 3

13. The specific reaction rate constant for a first order reaction is $1 \times 10^{-3}\ s^{-1}$. If the initial concentration of the reactant is 0·1 mole per litre, the rate is

(a) 10^{-4} (b) 10^{-3} (c) 10^{-2} (d) 10^{-1}

14. k for a zero order reaction is $2 \times 10^{-2}\ mol \cdot L^{-1} s^{-1}$. If the concentration of the reactant after 25 s is 0·5 M, the initial concentration must have been

(a) 0·5 M (b) 1·25 M (c) 12·5 M (d) 1·0 M

[**Hint:** Rate $= \dfrac{a - 0{\cdot}5}{25} = 2 \times 10^{-2}$]

15. A first order reaction is carried out with an initial concentration of 10 moles per litre and 80% of the reactant changed into the product. Now if the same reaction is carried out with an initial concentration of 5 moles per litre, the percentage of the reactant changing to the product is

(a) 40 (b) 80

(c) 160 (d) cannot be calculated

16. Which of the following curves represents a first order reaction?

(a)

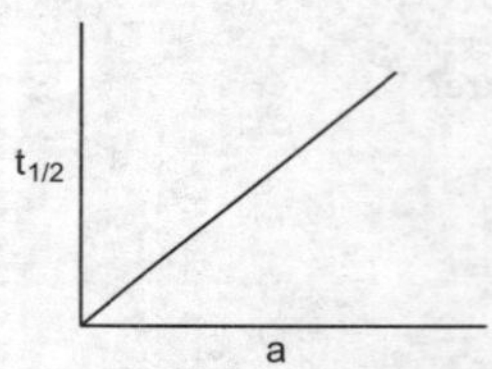

(b)

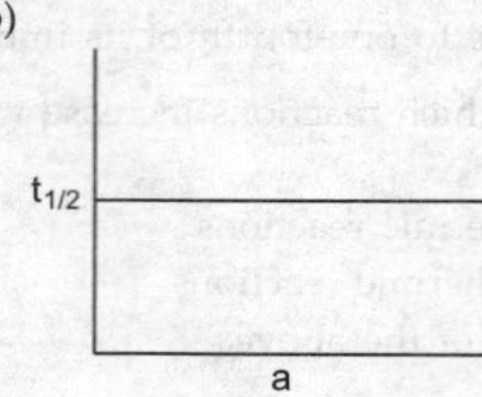

(c)

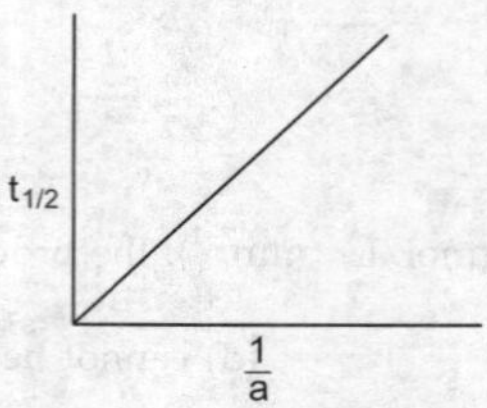

(d)

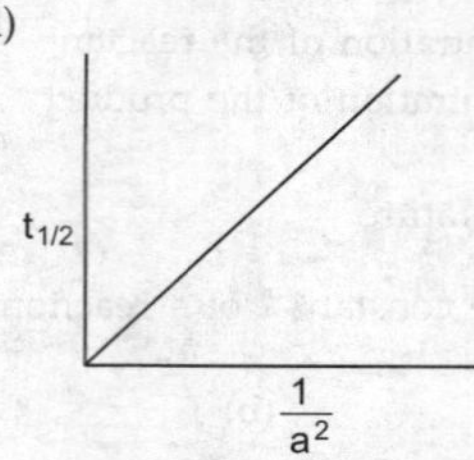

17. Which of the following curves represents a zero order reaction?

[$(a - x)$ = reactant concn.]

(a)

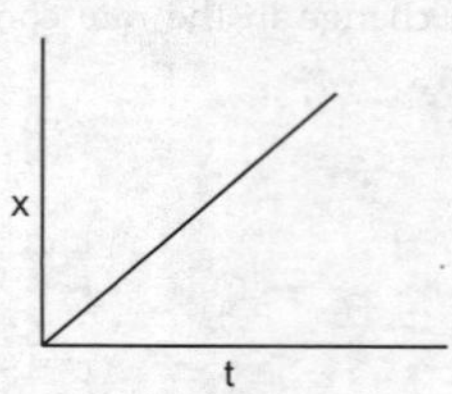

(b)

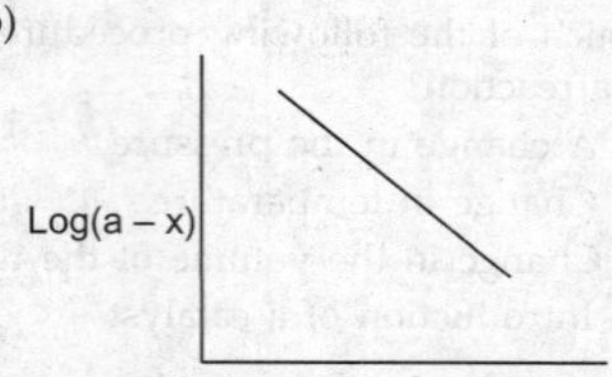

(c) 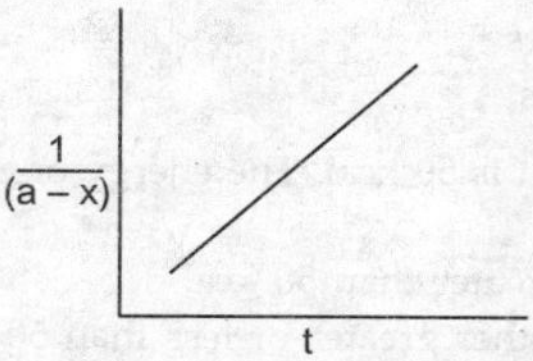

(d) 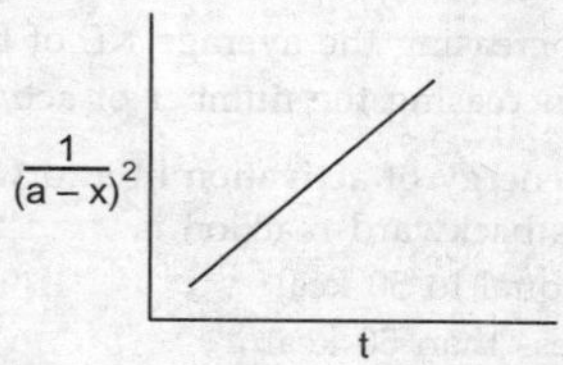

18. 75% of a first order reaction was completed in 32 min. When was 50% of the reaction completed?
(a) 24 min (b) 16 min (c) 8 min (d) 4 min

19. If doubling the initial concentration of a reactant doubles $t_{\frac{1}{2}}$ of the reaction, the order of the reaction is
(a) 3 (b) 2 (c) 1 (d) 0

20. For a given reaction the logarithm of the concentration of the reactant plotted against time gave a straight line with negative slope. The order of the reaction is
(a) 3 (b) 2 (c) 1 (d) 0

21. The concept of $t_{\frac{1}{2}}$ is useful for the reactions of
(a) zero order (b) first order (c) second order (d) all orders

22. The half-life for a given reaction was halved as the initial concentration of a reactant was doubled. The order for this component is
(a) 0 (b) 1 (c) 2 (d) 3

23. The rate constant for a second order reaction is $8 \times 10^{-5}\ M^{-1}\ min^{-1}$. How long will it take a 1 M solution to be reduced to 0·5 M in reactant?
(a) $8{\cdot}665 \times 10^3$ min (b) 8×10^{-5} min (c) $1{\cdot}25 \times 10^4$ min (d) 4×10^{-5} min

24. The rate for a first-order reaction is $0{\cdot}6932 \times 10^{-2}$ mol · L^{-1} min^{-1} and the initial concentration of the reactant is 0·1 M. $t_{\frac{1}{2}}$ is equal to
(a) $0{\cdot}6932 \times 10^{-2}$ min (b) $0{\cdot}6932 \times 10^{-3}$ min
(c) 10 min (d) 6·932 min

25. What fraction of a reactant remains after 40 min if $t_{\frac{1}{2}}$ is 20 min?
(a) 1/4 (b) 1/2 (c) 1/8 (d) 1/6

26. For a given reaction, the half-life period was found to be directly proportional to the initial concentration of the reactant. The order is
(a) 0 (b) 1 (c) 2 (d) 3

27. From different sets of data of $t_{1/2}$ at different initial concentration say *'a'* for a given reaction, the product $(t_{1/2} \times a)$ is found to be constant. The order of the reaction is
(a) 0 (b) 1 (c) 2 (d) 3

28. A catalyst increases the rate of a chemical reaction by
(a) increasing the activation energy

(b) decreasing the activation energy
(c) increasing the average KE of the molecules
(d) increasing the number of active molecules

29. The energy of activation of a forward reaction is 50 kcal. The energy of activation of its backward reaction is
(a) equal to 50 kcal (b) greater than 50 kcal
(c) less than 50 kcal (d) either greater or less than 50 kcal

30. The rate constant, the activation energy and the Arrhenius parameter of a chemical reaction at 25°C are $3 \times 10^{-4}\ s^{-1}$, $104{\cdot}4\ kJ\ mol^{-1}$ and $6 \times 10^{14}\ s^{-1}$ respectively. The value of the rate constant as $T \to \infty$ is
(a) $2 \times 10^{18}\ s^{-1}$ (b) $6 \times 10^{14}\ s^{-1}$ (c) ∞ (d) $3{\cdot}6 \times 10^{30}\ s^{-1}$
(IIT 1996)

31. Which of the following statements is wrong about reactions?
(a) There can be only three values of molecularity, that is, 1, 2, and 3.
(b) There can be only four values of order, that is, 0, 1, 2, and 3.
(c) There can be infinite number of values for order.
(d) The order involves rate while molecularity does not.

32. The temperature coefficient of a reaction is 2. The rate of this reaction, on raising the temperature by 25°, shall increase by
(a) 4 times (b) 8 times (c) 6 times (d) 5·65 times

33. The rate constant of a reaction, $2A \rightarrow$ Products, with initial reactant concentration a mole/L, is k mole/L/min. The $t_{1/2}$ for the reaction is equal to
(a) $\frac{0{\cdot}6932}{k}$ min (b) $\frac{1}{ak}$ min (c) $\frac{a}{2k}$ min (d) $\frac{a}{4k}$ min

34. The rate constant of a reaction, $A \rightarrow$ Product, with initial reactant concentration a mole/L, is k L/mole/min. The $t_{1/2}$ for the reaction is equal to
(a) $\frac{0{\cdot}6932}{k}$ (b) $\frac{1}{ak}$ (c) $\frac{1}{2ak}$ (d) $\frac{a}{2k}$

35. The rate of reaction doubles when the concentration of the reactant is increased four times. The order is
(a) ½ (b) 2 (c) 1 (d) 4

36. When the concentration of a reactant, A, in a reaction: $A \rightarrow$ Products, is doubled the rate of the reaction increases seven times, the order of the reaction is between
(a) 0 & 1 (b) 1 & 2 (c) 2 & 3 (d) 3 & 4

37. When the concentration of a reactant in a reaction, $A \rightarrow$ Products, is doubled, the increase in the rate of reaction cannot be more than
(a) 2 times (b) 4 times (c) 6 times (d) 8 times

38. For a second-order reaction of the type:

$$\underset{a-x}{\overset{a}{2A}} \rightarrow \underset{x/2}{\text{Products}},$$

the rate law is given by

$$\text{rate} = -\frac{1}{2}\frac{d(a-x)}{dt} = +\frac{d(x/2)}{dt} = k_2(a-x)^2.$$

The integrated rate law will be

(a) $k = \frac{1}{t}\left\{\frac{x}{a(a-x)}\right\}$ (b) $k = \frac{1}{2t}\left\{\frac{x}{a(a-x)}\right\}$ (c) $k = \frac{1}{t}\left\{\frac{a}{x(a-x)}\right\}$ (d) $k = \frac{1}{2t}\left\{\frac{a}{x(a-x)}\right\}$

39. 50% of a zero order reaction completes in 10 minutes. 100% of the same reaction shall complete in

(a) 5 min (b) 10 min (c) 20 min (d) ∞ time

40. The rate constant, k, of a second-order reaction, A $\rightarrow$ Products is given by

$$k = \frac{1}{t}\left\{\frac{x}{a(a-x)}\right\}.$$

The ratio $t_{3/4}/t_{1/2}$ is equal to

(a) 1/5 (b) 2/1 (c) 3/1 (d) 5/1

41. In which of the reactions of the following orders the molecularity and order can never be same?

(a) Zero order (b) First order (c) Second order (d) Third order

42. Which of the following curves represent(s) a zero-order reaction?

(a)

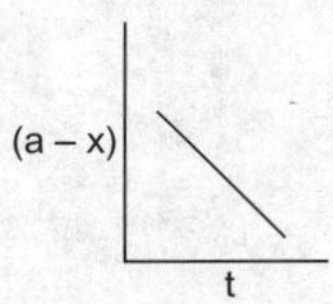

(b)

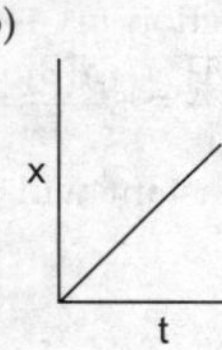

(c)

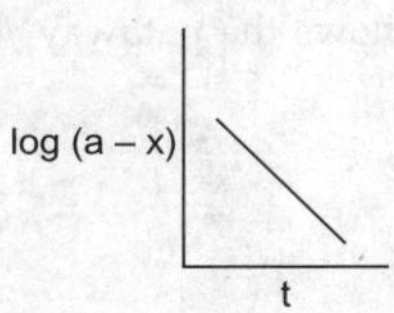

(d)

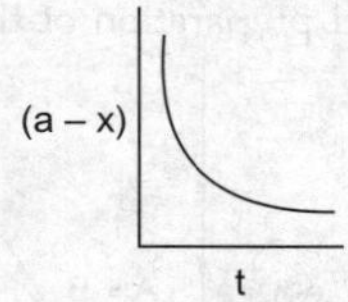

43. A plot of reactant concentration versus time for a reaction is a straight line with a negative slope giving the rate constant, and the intercept, giving the initial concentration of the reactant. The order of the reaction is

(a) 0 (b) 1 (c) 2 (d) none of these

44. Consider the chemical reaction

$$N_2(g) + 3H_2(g) \rightarrow 2NH_3(g)$$

The rate of this reaction can be expressed in terms of time derivative of concentration of $N_2(g)$, $H_2(g)$ or $NH_3(g)$. Identify the correct relationship amongst the rate expressions.

(a) Rate $= -\frac{d[N_2]}{dt} = -\frac{1}{3}\frac{d[H_2]}{dt} = \frac{1}{2}\frac{d[NH_3]}{dt}$

(b) Rate $= -\frac{d[N_2]}{dt} = 3\frac{d[H_2]}{dt} = 2\frac{d[NH_3]}{dt}$

(c) Rate $= \frac{d[N_2]}{dt} = \frac{1}{3}\frac{d[H_2]}{dt} = \frac{1}{2}\frac{d[NH_3]}{dt}$

(d) Rate $= -\frac{d[N_2]}{dt} = -\frac{d[H_2]}{dt} = \frac{d[NH_3]}{dt}$ **(IIT 2002)**

45. If the rate constant for a reaction represented by $2HI \rightarrow H_2 + I_2$ is denoted by k then for the same reaction if represented by $HI \rightarrow \frac{1}{2}H_2 + \frac{1}{2}I_2$, the rate constant shall be equal to

(a) k (b) $2k$ (c) $k/2$ (d) $(k+2)$

46. For the chemical reaction of the type

$$2A \rightarrow \frac{1}{2}B$$

the correct relationship amongst the rate expressions is

(a) $-2\frac{d[A]}{dt} = +\frac{1}{2}\frac{d[B]}{dt}$ (b) $+2\frac{d[A]}{dt} = -\frac{1}{2}\frac{d[R]}{dt}$

(c) $-\frac{1}{2}\frac{d[A]}{dt} = +2\frac{d[B]}{dt}$ (d) $+\frac{1}{2}\frac{d[A]}{dt} = -2\frac{d[B]}{dt}$

47. For what type of the following reactions is the law of mass action, never obeyed?

(a) Zero order (b) First order (c) Second order (d) Third order

48. If the rate law of a reaction $nA \rightarrow B$ is expressed as

$$\underset{(mol/L/s)}{Rate} = -\frac{1}{n}\frac{d[A]}{dt} = +\frac{d[B]}{dt} = k[A]^x$$

The unit of the rate constant will be

(a) $mol^x/L^x/s$ (b) $L^x/mol^x/s$

(c) $mol^{(1-x)} \cdot L^{(x-1)} \cdot s^{-1}$ (d) $mol^{(x-1)} \cdot L^{(1-x)} \cdot s^{-1}$

49. The exothermic reaction between substances A and B is presented in the plot below. Catalyst-induced preparation of the substance AB follows the pathway

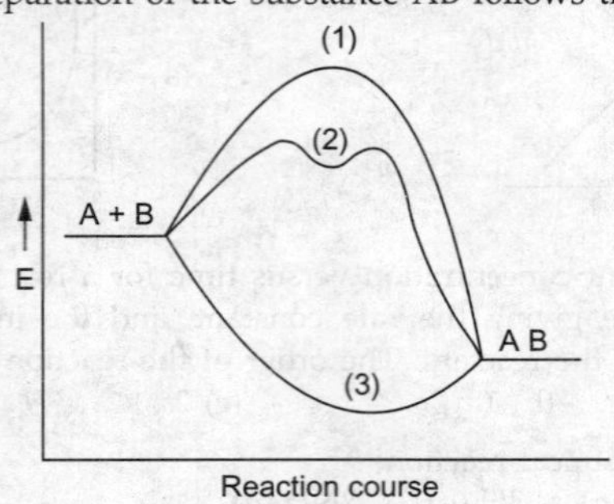

(a) 1 (b) 2 (c) 3

50. Which curve corresponds to the temperature dependence of the rate R of a simple one-step reaction?

(a) (b) (c) (d)

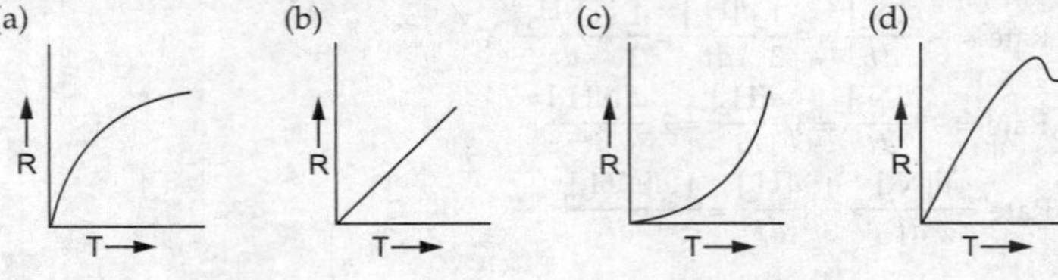

51. For the first-order reaction

$$2N_2O_5(g) \rightarrow 4NO_2(g) + O_2(g),$$

(a) the concentration of the reactant decreases exponentially with time.
(b) the half-life of the reaction decreases with increasing temperature.
(c) the half-life of the reaction depends on the initial concentration of the reactant.
(d) the reaction proceed to 99.6% completion eight half-life duration. **(IIT 2011)**

52. The rate of reaction doubles when its temperature changes from 300 K to 310 K. Activation energy of such a reaction will be ($R = 8.314$ J/mol K and log 2 = 0.301).

(a) 53.6 kJ mol^{-1} (b) 48.6 kJ mol^{-1} (c) 58.5 kJ mol^{-1} (d) 60.5 kJ mol^{-1}

(IIT 2013 Main)

53. In the reaction, $P + Q \rightarrow R + S$, the time taken for 75% reaction of P is twice the time taken for 50% reaction of P. The concentration of Q varies with reaction time as shown in the figure. The overall order of the reaction is

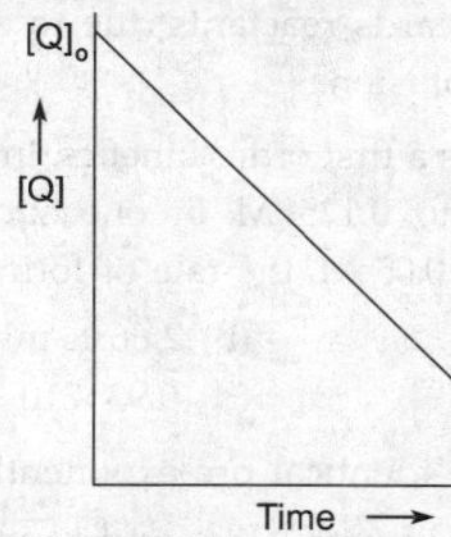

(a) 2 (b) 3 (c) 0 (d) 1 **(IIT 2013 Adv.)**

[**Hint:** Find k values using $k = \frac{2.303}{t} \log \frac{a}{a-x}$]

54. The initial rate of hydrolysis of methyl acetate (1 M) by a weak acid (HA, 1 M) is $\left(\frac{1}{100}\right)$th of that of a strong acid (HX, 1 M) at 25°C. The K_a(HA) value is

(a) 1×10^{-4} (b) 1×10^{-5} (c) 1×10^{-6} (d) 1×10^{-3}

(IIT 2013 Adv.)

[**Hint:** $R_{HA} = k[H^+]_{HA}$ and $R_{HX} = k[H^+]_{HX}$ $\therefore \frac{[H^+]_{HX}}{[H^+]_{HA}} = 100$

or $[H^+]_{HA} = \frac{1}{100}$; for Ha : $K_a = \frac{x^2}{1-x}$]

55. For the nonstoichiometric reaction,

$$2A + B \rightarrow C + D,$$

the following kinetic data were obtained in the three separate experiments, all at 298 K.

$[A]_{\text{initial}}$	$[B]_{\text{initial}}$	Initial rate of formation of C ($mol\ L^{-1}s^{-1}$)
1. 0.1 M	0.1 M	1.2×10^{-3}
2. 0.1 M	0.2 M	1.2×10^{-3}
3. 0.2 M	0.1 M	2.4×10^{-3}

The rate law for the formation of C is

(a) $\frac{dC}{dt} = k[A][B]$ (b) $\frac{dc}{dt} = k[A]^2[B]$

(c) $\frac{dC}{dt} = k[A][B]^2$ (d) $\frac{dC}{dt} = k[A]$ **(IIT 2014 Main)**

[Hint: Find orders w.r.t. A and B]

56. For the elementary reaction $M \rightarrow N$ the rate of disappearance of M increases by a factor of 8 upon doubling the concentration of M. The order of the reaction with respect to M is

(a) 4 (b) 3 (c) 2 (d) 1 **(IIT 2014 Adv.)**

57. Higher order (> 3) are rare due to

(a) low probability of simultaneous collisions of all the reacting species.

(b) increase in entropy and activation energy as more molecules are involved.

(c) shifting of equilibrium towards reactants due to elastic collisions.

(d) loss of active species on collision. **(IIT 2015 Main)**

58. Decomposition of H_2O_2 follows a first order kinetics. In 50 minutes, the concentration of H_2O_2 decreases from 0.5 to 0.125 M in one such decomposition. When the concentration of H_2O_2 reaches 0.05 M, the rate of formation of O_2 will be

(a) 6.93×10^{-4} mol min^{-1} (b) 2.66 L min^{-1} at STP

(c) 1.34×10^{-2} mol min^{-1} (d) 6.93×10^{-2} mol min^{-1} **(IIT 2016 Main)**

59. Two reactions R_1 and R_2 have identical pre-exponential factors. Activation energy of R_1 exceeds that of R_2 by 10 kJ mol^{-1}. If k_1 and k_2 are rate constants for reactions R_1 and R_2 respectively at 300 K, then $\ln\left(\frac{k_2}{k_1}\right)$ is equal to (R = 8.314 J mol^{-1} K^{-1})

(a) 8 (b) 12 (c) 6 (d) 4 **(IIT 2017 Main)**

60. In a bimolecular reaction, the steric factor P was experimentally determined to be 4.5. The correct option(s) among the following is(are)

(a) The activation energy of reaction is unaffected by the value of the steric factor.

(b) Experimentally determined value of frequency factor is higher than that predicted by Arrhenius's equation.

(c) The value of frequency factor predicted by Arrhenius's equation is higher than that determined experimentally.

(d) Since $P = 4.5$, the reaction will not proceed unless an effective catalyst is used. **(IIT 2017 Adv.)**

[Hint: Read Modern Approach to Physical Chemistry, Vol. II, Ch. 4]

61. At 518°C the rate of decomposition of a sample of gaseous acetaldehyde at a pressure of 363 torr, was 1.00 torr s^{-1} when 5% had reacted and 0.5 torr s^{-1} when 33% had reacted. The order of the reaction is

(a) 0 (b) 2 (c) 3 (d) 1 **(IIT 2018 Main)**

[Hint: Rate = k(pressure of $CH_3CHO)^n$,

$$\therefore \quad 1 = k\left(363 - \frac{5}{100} \times 363\right)^n \quad \text{and} \quad 0.5 = k\left(363 - \frac{33}{100} \times 363\right)^n.]$$

62. For a first-order reaction $A \rightarrow P$, $t_{1/2}$ is 10 days. The time required for $\frac{1}{4}$th conversion of A (in days) is: (ln 2 = 0.692, ln 3 = 1.1)
(a) 3.2 (b) 4.1 (c) 5 (d) 2.5 **(IIT 2018 Main)**

63. N_2O_5 decomposes to NO_2 and O_2 and follows first order kinetics. After 50 minutes the pressure inside the vessel increases from 50 mmHg to 87.5 mmHg. The pressure of the gaseous mixture after 100 minutes at constant temperature will be
(a) 116.25 mmHg (b) 106.25 mmHg
(c) 136.25 mmHg (d) 175.0 mmHg **(IIT 2018 Main)**

[Hint: $\overset{50}{N_2O_5} \rightarrow 2NO_2 + \frac{1}{2}O_2;\ 50 - p + 2p + \frac{p}{2} = 87.5$

after 50 minutes : $50 - p \quad 2p \quad p/2 \qquad p = 25$

after 100 minutes : p'

$$k = \frac{2.303}{50}\log\frac{50}{25} = \frac{2.303}{100}\log\frac{50}{p'};\ p' = 12.5.$$

Pressure after 100 minutes : $= 12.5 + 2 \times 37.5 + \frac{37.5}{2}$]

64. For a first-order reaction, $A(g) \rightarrow 2B(g) + C(g)$ at constant volume and 300 K, the total pressure at the beginning $(t = 0)$ and at time t are p_0 and p_t, respectively. Initially, only A is present with concentration $[A]_0$ and $t_{1/2}$ is the time required for the partial pressure of A to reach $\frac{1}{3}$rd of its initial value. The correct option(s) is(are) (Assume that all these gases behave as ideal gases).

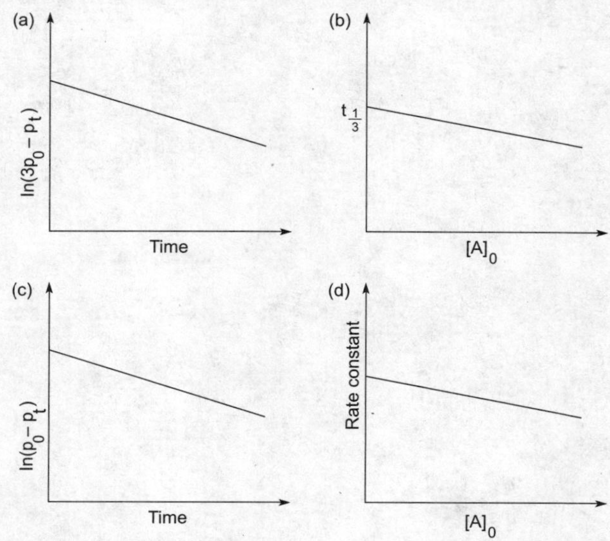

[Hint: $k = \frac{1}{t}\ln\frac{2p_0}{3p_0 - p_t}$ $\quad\therefore$ (a) is correct] **(IIT 2018 Adv.)**

Answers

1-b, **2**-c, **3**-b, **4**-d, **5**-b, **6**-b, **7**-a, **8**-d, **9**-a, **10**-a, **11**-b & d, **12**-b, **13**-a, **14**-d, **15**-b, **16**-b, **17**-a, **18**-b, **19**-d, **20**-c, **21**-b, **22**-c, **23**-c, **24**-c, **25**-a, **26**-a, **27**-c, **28**-b & d, **29**-d, **30**-b, **31**-b, **32**-d, **33**-d, **34**-b, **35**-a, **36**-c, **37**-d, **38**-b, **39**-c, **40**-c, **41**-a, **42**-a & b, **43**-a, **44**-a, **45**-b, **46**-c, **47**-a, **48**-c, **49**-b, **50**-c, **51**-a, b, d, **52**-a, **53**-d, **54**-a, **55**-d, **56**-b, **57**-a, **58**-a, **59**-d, **60**-a, **61**-b, **62**-b, **63**-b, **64**-a.

❑

CHAPTER EIGHTEEN

ELECTROMOTIVE FORCE

Electrochemical cells are of two types: 'electrolytic cells' and 'galvanic cells'. In electrolytic cells, the electrical energy is absorbed producing thereby chemical reactions. This process can also be reversed in which the chemical energy is converted to electrical energy. Any device which accomplishes this is called a galvanic cell or voltaic cell.

In an electrolytic cell electrons are fed into the cell from the external circuit. The cathode receives the electrons and becomes the negative electrode, and therefore, the anode becomes the positive electrode. Oxidation and reduction occur at the anode and cathode respectively. In a galvanic cell, oxidation occurs at the anode; the electrons so released by the anode are sent to the external circuit by it. The anode is thus regarded as the negative electrode and the cathode, where reduction occurs, thus becomes the positive electrode. Thus **the signs of cathode and anode in the electrolytic cell and in the galvanic cell are just the reverse.** Galvanic cells are of two types: chemical cell and concentration cell. In a chemical cell there is an overall cell reaction, whereas in a concentration cell, there is no overall cell reaction and the emf arises due to a concentration difference in the two half-cells. Let us consider a Daniell cell, an example of a **chemical cell** (galvanic cell)

$$\mathrm{Zn} \left| \begin{matrix} \mathrm{ZnSO_4} \\ \mathrm{(aq)} \end{matrix} \right\| \begin{matrix} \mathrm{CuSO_4} \\ \mathrm{(aq)} \end{matrix} \Big| \mathrm{Cu}$$

or, $\mathrm{Zn} \mid \mathrm{Zn^{2+}(aq)} \parallel \mathrm{Cu^{2+}(aq)} \mid \mathrm{Cu}$

Anode (–) (oxidation) Cathode (+) (reduction)

The cell reactions are

$$\mathrm{Zn} \rightleftharpoons \mathrm{Zn^{2+}} + 2e; \quad \text{(Oxidation): at anode}$$

$$\mathrm{Cu^{2+}} + 2e \rightleftharpoons \mathrm{Cu}; \quad \text{(Reduction): at cathode}$$

The convention of representing such cells are:

(i) The electrode at which reduction takes place is written on the right-hand side and the electrode at which oxidation takes place is written on the left-hand side.

(ii) The single vertical line generally represents the separation of solid phase from the electrolytic solution. The double vertical lines represent the separation of two half cells having indirect electrical contact by a salt bridge. The salt bridge also eliminates the liquid-junction potential which develops at the junction of the two liquids of different concentrations, due to different velocities of cations and anions crossing the junction. The salt bridge is made of agar-agar and KCl (or KNO_3 or NH_4NO_3).

Work Done by a Galvanic Cell

The electrical work expended in moving a charge through a conductor is

Electrical work	=	charge	×	potential difference
(joules)		(coulombs)		(volts)

The Faraday constant, *F*, is the magnitude of the charge on one mole of electrons. The faraday is a unit of charge equal to 96500 coulombs. In moving one faraday of charge from one electrode to another, the numerical value of the work done by a voltaic cell is equal to *F* × potential difference between the electrodes.

$$W = F \times \text{potential difference}$$

The maximum potential difference between the electrodes of a voltaic cell, that is, the limiting value of potential difference measured as the current goes to zero in the external circuit, is known as electromotive force (emf) of the cell, E_{cell} or the reversible cell potential. The maximum work (according to IUPAC) obtainable from a voltaic cell is expressed by

$$W_{max} = -nFE_{cell},$$

where *n* equals the number of electrons in either half-cell reaction.

Electrical work is usually expressed in energy units of watts times time. The watt is a unit of electrical power, that is, the rate of energy expenditure.

$$1 \text{ watt} = \frac{1\text{J}}{\text{S}}$$

$$\therefore \quad 1 \text{ J} = 1 \text{ W} \cdot \text{S}.$$

The unit employed by electric utilities is the kilowatt-hour (kWh).

$$1 \text{ kWh} = 1000 \times (1 \times 60 \times 60) = 3{\cdot}6 \times 10^6 \text{ J}$$

Reversible and Irreversible Cells

An electrochemical cell may be reversible or irreversible. In a **reversible cell** the following conditions should be fulfilled.

1. An infinitesimally small current is drawn from it.
2. When connected to an external source of emf (driving force), exactly equal to that of the cell, no cell reaction occurs (equilibrium) and no current flows.
3. If the driving emf is increased or decreased by an infinitesimally small amount, the same small current should flow in the opposite direction.

When these conditions are not satisfied, the cell is said to be **irreversible.** The potentiometric measurement in which no current is drawn from the cell is ideally suited for the measurement of **reversible** emf. The emf of an **irreversible** cell, in which a larger current is drawn, is measured by a voltmeter.

A familiar example of a reversible cell is the Daniell cell in which the net cell reaction is

$$Zn + Cu^{2+} \rightleftharpoons Zn^{2+} + Cu.$$

When the two electrodes are connected to an external source of emf, infinitesimally smaller than the cell emf, the small current flows from Zn-electrode to Cu-electrode.

$$Zn + Cu^{2+} \rightarrow Zn^{2+} + Cu.$$

And, if the external emf is infinitesimally greater than the cell emf, the same small current flows in the opposite direction and the cell reaction is reversed.

$$Cu + Zn^{2+} \rightarrow Cu^{2+} + Zn.$$

An example of an irreversible cell is Zn and Cu electrodes dipped in sulphuric acid solution. When the external source of emf is slightly smaller than the cell emf, the reaction occurring is

$$Zn + H_2SO_4 \rightarrow ZnSO_4 + H_2.$$

If the external source of emf is slightly greater than the cell emf, the cell reaction is

$$Cu + H_2SO_4 \rightarrow CuSO_4 + H_2.$$

As the two reactions are different (not reversible), that is, they have different values of cell potential, a comparatively larger current flows on changing the direction.

Reversible Electrodes

The two electrodes (half cells) of a reversible cell are called reversible electrodes. A reversible electrode can act both as cathode as well as anode. For example, when a reversible electrode, M/M^{n+}, acts as cathode, the half-cell reaction is

$$M^{n+} + ne = M. \qquad \text{(reduction)}$$

And, when the same electrode acts as anode, the half-cell reaction is

$$M = M^{n+} + ne. \qquad \text{(oxidation)}$$

A reversible electrode is reversible with respect to the ion whose concentration decreases and increases when the electrode acts as a cathode and anode respectively. For example, the electrode M/M^{n+} is reversible with respect to M^{n+} ions.

Half-Cell Potential (Single-Electrode Potential)

When a metal is dipped into a solution containing its own ions, a half cell or a single electrode is formed. In a half cell there are two opposing tendencies. Firstly, the metal, say M, may dissolve in the solution or rather may go into the solution in the form of ions ($M \rightleftharpoons M^{n+} + ne$; oxd.) and secondly, the ions, M^{n+}, from the solution may deposit on the electrode ($M^{n+} + ne \rightleftharpoons M$; red.). When one of these two tendencies dominates over the other, there develops a half-cell potential or electrode potential.

The tendency to lose electrons, i.e., to get oxidised, is called **oxidation potential**, and similarly, the tendency to gain electrons, i.e., to get reduced, is called **reduction potential.**

Since any half-cell reaction can be written as a reversible process, e.g., $Cu^{2+} + 2e \rightleftharpoons Cu$, the reduction potential and oxidation potential for a single electrode are equal in magnitude but opposite in sign. For the electrode $Cu/CuSO_4$ (1 M), the reduction potential, $E^0_{Cu^{2+}, Cu} = +0{\cdot}34$ V and so its oxidation potential, $E^0_{Cu, Cu^{2+}} = -0.34$ V at 25°C. **The half-cell potentials cannot be directly determined as there is no way of isolating a single half-cell reaction.** The electrode potential can be determined by coupling it with a standard hydrogen electrode (i.e., by forming a cell). As the electrode potential of a standard hydrogen electrode has been arbitrarily fixed as zero volt at 25°C, the emf of such a cell gives the single electrode potential or emf of half cell. A standard hydrogen electrode is represented as

$$\text{Pt, } H_2 \text{ (1 atm), HCl } ([H^+] = 1 \text{ M}).$$

Sign Convention (IUPAC)

The reduction potential of a half cell is given a positive sign when the half-cell reaction involves reduction, when coupled with a standard hydrogen electrode; and a negative sign when the half cell reaction involves oxidation, when connected with a standard hydrogen electrode.

Standard Half-Cell Potential (Standard Electrode Potential) and Electrochemical Series

It will be discussed a little later that the half-cell potential at a given temperature depends upon the concentration of ions of the dissolved material. If for the half cell $M \mid M^{n+}$ (aq), $[M^{n+}] = 1$ M* at 25°C, its potential is termed *standard half-cell potential* or *standard electrode potential* represented by E^0.

Standard half-cell potential, like half-cell potential is also measured on standard hydrogen electrode scale.

The E^0 values of various electrodes may be arranged in any of the following orders:

(i) In decreasing order of E^0 (red.) values (IUPAC)

(ii) In increasing order of E^0 (red.) values (European convention)

(iii) In decreasing order of E^0 (oxd.) values (American convention)

Such a list of E^0 values of various half cells arranged in any of the given orders is known as an electrochemical series.

---

* The concentration should not be, in fact, 1 molar (i.e., 1 M) but 1 molal (i.e., 1 *m*).

Some of the half cells or electrodes arranged in decreasing order of their E^0 (reduction) values are as follows:

The standard half-cell potential, E^0, listed below is for reduction reaction, i.e., for the reaction: oxidant + $ne \rightleftharpoons$ reductant, to proceed from left to right. Thus higher the E^0 (red.) for an electrode, more will be its tendency to get reduced (i.e., greater tendency to accept electrons). In the given electrochemical series, Ag^+ has a greater tendency to undergo reduction than Li^+, or Li is a better reducing agent compared to Ag. In other words, Li can reduce or displace Ag^+ ions and Fe can displace Cu^{2+} ions but not Zn^{2+} ions from their solutions. F_2 (g) and Li (s) are the strongest oxidising and reducing agents respectively. The equilibrium constant at 25°C for the reaction

$$\frac{1}{2}F_2\ (g) + Li\ (s) = F^-\ (aq) + Li^+\ (aq)$$

is about 10^{100}.

Table

Electrochemical series: Standard reduction potential at 25°C

Electrode	E^0(red.) volt	Electrode	E^0(red.) volt	Electrode	E^0(red.) volt
F_2 (g), $2F^-$	2·87	O_2, OH^-	0·401	Ni^{2+}, Ni	− 0·24
Co^{3+}, Co^{2+}	1·82	Cu^{2+}, Cu	0·34	Cd^{2+}, Cd	− 0·40
Pb^{4+}, Pb^{2+}	1·69	Hg_2Cl_2 (s), Cl^-; Hg	0·28	Fe^{2+}, Fe	− 0·44
Cl_2 (g), $2Cl^-$	1·36	AgCl (s), Cl^-; Ag	0·22	Cr^{3+}, Cr	− 0·74
Br_2 (l), $2Br^-$	1·06	Sn^{4+}, Sn^{2+}	0·15	Zn^{2+}, Zn	− 0·76
$2Hg^{2+}$, Hg_2^{2+}	0·92	Cu^{2+}, Cu^+	0·15	H_2O, OH^-	− 0·83
Hg^{2+}, Hg	0·85	Hg_2Br_2(s), Br^-; Hg	0·13	H_2, $2H^-$	− 2·25
Ag^+, Ag	0·80	AgBr (s), Br^-; Ag	0·07	Mg^{2+}, Mg	− 2·36
Hg_2^{2+}, 2Hg	0·79	$2H^+$, H_2 (g); Pt	± 0·00	Na^+, Na	− 2·71
Fe^{3+}, Fe^{2+}	0·77	Fe^{3+}, Fe	− 0·04	Ca^{2+}, Ca	− 2·87
Hg_2SO_4, SO_4^{2-}; Hg	0·61	Pb^{2+}, Pb	− 0·13	K^+, K	− 2·92
I_2 (s), $2I^-$	0·53	Sn^{2+}, Sn	− 0·14	Li^+, Li	− 3·04
Cu^+, Cu	0·52	AgI (s), I^-; Ag	− 0·15		

While constructing a galvanic cell by coupling any of the two half cells given in the electrochemical series, it is to be remembered that the half cell having higher value of E^0 undergoes reduction and so acts as a cathode (positive) and the other having lower value of E^0 undergoes oxidation and so acts as an anode (negative) as already represented in the Daniell cell.

[**Note:** (i) The standard half-cell potential, E^0, is an intensive property, like temperature or molar volume, and so E^0 shall be the same for half-cell reaction whether it is represented as $2X^+ + 2e \rightleftharpoons X_2$ or $X^+ + e \rightleftharpoons \frac{1}{2}X_2$. But unlike E^0, $\Delta G°$ (standard free energy change) is an extensive property which depends upon the mass, that is to say, if a half-cell reaction, $2X^+ + 2e \rightleftharpoons X_2$ is represented as $X^+ + e \rightleftharpoons \frac{1}{2}X_2$, $\Delta G°$ of the latter would be half that of the former.

(ii) If the direction of a half-cell (cell) reaction is reversed, its potential has the same magnitude but opposite sign, e.g.,

if, for $Cu^{2+} + 2e \rightleftharpoons Cu$; $E^0 = +0{\cdot}34$ V

then, for $Cu \rightleftharpoons Cu^{2+} + 2e$; $E^0 = -0{\cdot}34$ V.

The same is also true for free energy change.

(iii) Potentials are not thermodynamic functions and may not be added but the potential may be calculated from the free energy change ΔG, using $\Delta G° = -nFE^0$. For example,

$$Fe^{3+} + e \rightarrow Fe^{2+};\ E^0 = +0{\cdot}77\text{ V};\ \Delta G° = -1(+0{\cdot}77)\,F = -0{\cdot}77F$$

$$Fe^{2+} + 2e \rightarrow Fe;\ \ E^0 = -0{\cdot}44\text{ V};\ \Delta G° = -2(-0{\cdot}44)\,F = +0{\cdot}88F$$

$$Fe^{3+} + 3e \rightarrow Fe \qquad \text{On adding : } \Delta G° = +0{\cdot}11F$$

$$\therefore \quad E^0 \text{ for } (Fe^{3+} + 3e \rightarrow Fe) = \frac{\Delta G°}{-nF} = \frac{0{\cdot}11F}{-3F} = -0{\cdot}04\text{ V}$$

but not $(0{\cdot}77 - 0{\cdot}44) = 0{\cdot}33$ V.]

The emf of Cells

The galvanic cells are formed by coupling two half cells. The emf of such cells develops due to the potential difference between the two half cells and thus E_{cell} can be calculated as,

E_{cell} = red. pot. of right-hand electrode − red. pot. of left-hand electrode

or E_{cell} = oxd. pot. of left-hand electrode + red. pot. of right-hand electrode

or E_{cell} = oxd. pot. of left-hand electrode − oxd. pot. of right-hand electrode

Let us illustrate it with an example as to how to calculate emf of galvanic cells, E_{cell}, by following the above methods. For the Daniell cell,

$Zn \mid Zn^{2+}(1\text{ M}) \mid\mid Cu^{2+}(1\text{ M}) \mid Cu$, given that

$E_{Zn^{2+},\,Zn} = -0{\cdot}76$ V, $E_{Cu^{2+},\,Cu} = 0.34\ V$ (red. potentials)

$\therefore$ $E_{Zn,\,Zn^{2+}} = +0{\cdot}76$ V, $E_{Cu,\,Cu^{2+}} = -0{\cdot}34$ (oxd. potentials)

Substituting the above values, we get,

$$E_{cell} = E_{Cu^{2+},Cu} \text{ (red.)} - E_{Zn^{2+},Zn} \text{ (red.)}$$

$$= 0{\cdot}34 - (-0{\cdot}76) = 1{\cdot}1 \text{ V}$$

or $$E_{cell} = E_{Zn,Zn^{2+}} \text{ (oxd.)} + E_{Cu^{2+},Cu} \text{ (reduction)}$$

$$= 0{\cdot}76 + 0{\cdot}34 = 1{\cdot}1 \text{ V}$$

or $$E_{cell} = E_{Zn,Zn^{2+}} \text{ (oxd.)} - E_{Cu,Cu^{2+}} \text{ (oxidation)}$$

$$= 0{\cdot}76 - (-0{\cdot}34) = 1{\cdot}1 \text{ V.}$$

[**Note:** (i) In order to avoid confusion we shall adopt here only the first method to calculate the emf of a cell which is also in accordance with the latest convention. All half-cell potentials will thus be taken in reduction terms.

(ii) For a given half cell, the potential may be in negative or positive sign, but the emf of the cell calculated from half-cell potentials should always be positive.]

Concentration Cell

If in a galvanic cell, there is an overall chemical reaction, the cell is referred to as a chemical cell. For example in a Daniell cell, the overall (cell) reaction is $Cu^{2+} + Zn = Cu + Zn^{2+}$. In some galvanic cells, however, there is no resultant chemical reaction but the emf arises due to the transfer of matter from one half cell to another because of concentration difference. Such cells are called *concentration cells*. For example,

$$(Pt)H_2 \text{ (1 atm)} \mid HCl\ (C_1) \parallel HCl\ (C_2) \mid H_2 \text{ (1 atm)(Pt)}$$

where $C_2 > C_1$.

Half-cell reactions are

$$\frac{1}{2}H_2 = H^+(C_1) + e \qquad \text{... at anode}$$

$$H^+(C_2) + e = \frac{1}{2}H_2 \qquad \text{... at cathode}$$

and, the overall cell process is

$$H^+(C_2) = H^+(C_1).$$

Though oxidation and reduction take place at the anode and cathode respectively, there is no resultant chemical reaction. The net cell process is the flow of H^+ from right half cell to left half cell. E_{cell} of the concentration cells are derived in the same way as for chemical cells.

$$E_{cell} = \frac{2{\cdot}303\,RT}{nF}\log\frac{C_2}{C_1}.$$

Note that there is no E^0_{cell} term in the above equation as the E^0 of the two half cells are same and get cancelled.

Concentration Dependence of emf

The Nernst Equation

As in a half cell the potential depends on concentration of ions, the emf of a galvanic cell also depends on the concentration of the species involved in the cell reaction. Nernst equation relates the emf of the cell (or half cell) and the concentrations of the species involved.

For the general cell reaction

$$aA + bB \rightleftharpoons cC + dD$$

the Nernst equation is given by

$$E_{cell} = E^0 - \frac{2{\cdot}303\,RT}{nF}\log\frac{[C]^c\cdot[D]^d}{[A]^a\cdot[B]^b}, \quad \ldots (1)$$

where $\frac{[C]^c[D]^d}{[A]^a[B]^b} = Q$ (reaction quotient)

$E°$ is the standard emf of the cell (or half cell) which may be defined as the emf of the cell when concentration of each species of the cell reaction is unity, n is the number of electrons involved in the reaction.

If $T = 25°$ C, i.e., 298 K; $R = 8{\cdot}314\ J\ K^{-1}\ mol^{-1}$; $F = 96500$ coulombs $\Big\}$ $\frac{2{\cdot}303\,RT}{F} = 0{\cdot}0591$

$$\therefore\ E_{cell} = E^0 - \frac{0{\cdot}0591}{n}\log\frac{[C]^c\cdot[D]^d}{[A]^a\cdot[B]^b} \quad \ldots \text{ at } 25°C.$$

The Nernst equation is valid for the complete cell (galvanic cell) as well as for the half-cell reaction. For half-cell reaction, E_{cell} is replaced by emf of the half cell and E^0 would represent standard emf of the half cell.

Nernst equation may be derived from the thermodynamic equation

$$\Delta G = \Delta G° + RT\ln Q$$

where $Q = \frac{[C]^c\,[D]^d}{[A]^a\,[B]^b}$.

Equilibrium Constant and Standard Free Energy Change

The cell reaction taking place in a galvanic cell is a reversible one. When the reaction attains equilibrium, emf of the cell, E_{cell}, becomes zero; and $Q = K$.

$$\therefore\quad E^0 = \frac{2{\cdot}303\,RT}{nF}\log K \quad \ldots (2)$$

K, being equilibrium constant of the cell reaction. Since $\Delta G^0 = -\,n\,F\,E^0$ under standard conditions, $\Delta G°$ being the standard free energy change

$$\therefore\quad -\Delta G° = 2{\cdot}303\,RT\log K. \quad \ldots (3)$$

Thus knowing E^0, i.e., standard emf of the cell, one can calculate K and ΔG° and thereby obtain information regarding the feasibility of the cell reaction.

Feasibility of Cell Reactions

From thermodynamics we know that $\Delta G = -nFE$ and $\Delta G^{\circ} = -nFE^0$, where ΔG is the free energy change; E, the emf of the cell and ΔG° and E^0 are the standard free energy change and standard emf of the cell respectively. The negative sign emphasises the fact that a galvanic cell with a net negative free energy change ($\Delta G < 0$) will produce positive emf. Thus if E_{cell} is positive or ΔG is negative, the cell reactions are spontaneous or feasible and for negative value of E_{cell} or positive value of ΔG, the cell reactions are not feasible and in such a case two half cells forming the galvanic cell should be interchanged.

Further we see that E_{cell} is proportional to $\left(-\frac{\Delta G}{n}\right)$, the decrease in free energy of the cell reaction per electron transferred. The cell emf is, therefore, an **intensive** property of the system; it does not depend on the size of the cell or on stoichiometric coefficients chosen to balance the chemical equation for the cell reaction.

Spontaneity of forward reaction	ΔG°	K	E^0_{cell}
Spontaneous	–	>1	+
At equilibrium	0	1	0 (Steady-state condition)
Nonspontaneous	+	<1	–

It should be carefully noted that E^0_{cell} tells us only whether the cell reaction is feasible or not under standard conditions*, e.g., E^0_{cell} of cell reaction Cu^{2+} (aq) + Zn (s) $\rightleftharpoons$ Cu (s) + Zn^{2+} (aq) is + 1·1 V which tells us only that if both Cu^{2+} and Zn^{2+} are present at 1 M concentrations, the given cell reaction is feasible. However, if Cu^{2+} and Zn^{2+} are not taken at their standard concentration conditions then the direction of the spontaneous reaction may be reversed. (See Q. 14)

[**Note:** In the problems, the potentials may be given both in terms of oxidation and reduction, but the students are advised to compute emf only from reduction potential values to avoid confusion. The potential if given in oxidation term, should first be converted to reduction potential by changing the sign.]

* The standard condition is that in which all reagents have the concentration of 1 M or 1 atm pressure.

Determination of ΔG, ΔS and ΔH for the Cell Reaction

The following equations can be derived from Gibbs–Helmholtz equation and $\Delta G = -nFE$.

$$\Delta G = \Delta H - nFT\left(\frac{\partial E}{\partial T}\right)_P$$

$$\Delta S = nF\left(\frac{\partial E}{\partial T}\right)_P$$

$$\Delta H = nF\left[T\left(\frac{\partial E}{\partial T}\right)_P - E\right] = T\Delta S + \Delta G$$

ΔG, ΔS and ΔH can be calculated using the above equations by measuring the emf of the cell, E, and its **temperature coefficient,** $\left(\frac{\partial E}{\partial T}\right)_P$.

If $\left(\frac{\partial E}{\partial T}\right)_P = 0$: No heating or cooling of the cell occurs during its working.

If $\left(\frac{\partial E}{\partial T}\right)_P = +ve$: Temperature of the cell falls during its working.

If $\left(\frac{\partial E}{\partial T}\right)_P = -ve$: Temperature of the cell increases during its working.

EXAMPLES

Ex. 1. *Derive expressions to calculate emf (reduction) of the following half cells at 25°C:*

(i) $H^+ \mid H_2$ (Pt) (1 atm)

(ii) Cl_2 (g) $\mid 2Cl^-$ (Pt) (1 atm)

(iii) $Fe^{3+} \mid Fe^{2+}$ (Pt)

(iv) AgCl (s), $Cl^- \mid$ Ag

(v) Hg_2Cl_2 (s), $Cl^- \mid$ Hg

(vi) $Cu^{2+} \mid$ Cu (s)

Solution : We have the Nernst equation for $aA + bB \rightleftharpoons -cC + dD$

$$E = E^0 - \frac{2{\cdot}303\,RT}{nF}\log\left\{\frac{[C]^c\,[D]^d}{[A]^a\,[B]^b}\right\}. \qquad \text{... (Eqn. 1)}$$

Substituting:

$R = 8{\cdot}314\ JK^{-1}\ mol^{-1}$

$F = 96500$ coulombs and

$T = 25 + 273 = 298$ K, we get,

$$E = E^0 - \frac{0{\cdot}0591}{n} \log \left\{ \frac{[C]^c [D]^d}{[A]^a [B]^b} \right\}$$

(i) For the electrode $H^+ \mid H_2$ (Pt), the half-cell reaction is

$$H^+ + e \rightleftharpoons \underset{(1 \text{ atm})}{\frac{1}{2} H_2 (g)} \qquad \text{(reduction)}$$

$$\therefore \qquad E_{H^+, H_2} = E^0_{H^+, H_2} - \frac{0{\cdot}0591}{1} \log \frac{[H_2]^{\frac{1}{2}}}{[H^+]}$$

$$= 0 - 0{\cdot}0591 \log \frac{1}{[H^+]} \quad (E^0_{H^+, H_2} = 0 \text{ volt})$$

$$= 0{\cdot}0591 \log [H^+].$$

(ii) $\frac{1}{2} Cl_2 (g) + e \rightleftharpoons Cl^-$ (reduction)
(1 atm)

$$E_{Cl_2, Cl^-} = E^0_{Cl_2, Cl^-} - \frac{0{\cdot}0591}{1} \log \frac{[Cl^-]}{[Cl_2]^{\frac{1}{2}}}$$

$$= E^0_{Cl_2, Cl^-} - 0{\cdot}0591 \log [Cl^-]$$

(iii) $Fe^{3+} + e \rightleftharpoons Fe^{2+}$ (reduction)

$$E_{Fe^{3+}, Fe^{2+}} = E^0_{Fe^{3+}, Fe^{2+}} - \frac{0{\cdot}0591}{1} \log \frac{[Fe^{2+}]}{[Fe^{3+}]}$$

$$E_{Fe^{3+}, Fe^{2+}} = E^0_{Fe^{3+}, Fe^{2+}} + 0{\cdot}0591 \log \frac{[Fe^{3+}]}{[Fe^{2+}]}$$

(iv) $AgCl (s) + e^- \rightleftharpoons Ag (s) + Cl^-$ (reduction)

$$E_{AgCl, Cl^-} = E^0_{AgCl, Cl^-} - \frac{0{\cdot}0591}{1} \log \frac{[Ag][Cl^-]}{[AgCl]}$$

$$= E^0_{AgCl, Cl^-} - 0{\cdot}0591 \log [Cl^-]$$

For Ag and AgCl, both being solids, $[Ag] = [AgCl] = 1$.

(v) $\frac{1}{2} Hg_2Cl_2 (s) + e \rightleftharpoons Hg (l) + Cl^-$ (reduction)

$$E_{Hg_2Cl_2, Cl^-} = E^0_{Hg_2Cl_2, Cl^-} - \frac{0{\cdot}0591}{1} \log \frac{[Hg][Cl^-]}{[Hg_2Cl_2]}$$

$$= E^0_{Hg_2Cl_2, Cl^-} - 0{\cdot}0591 \log [Cl^-]$$

(vi) $Cu^{2+} + 2e \rightleftharpoons Cu (s)$ (reduction)

$$E_{Cu^{2+}, Cu} = E^0_{Cu^{2+}, Cu} - \frac{0 \cdot 0591}{2} \log \frac{[Cu]}{[Cu^{2+}]}$$

$$= E^0_{Cu^{2+}, Cu} + \frac{0 \cdot 0591}{2} \log [Cu^{2+}]$$

Ex. 2. *Calculate the reduction potential for the following half cells at 25°C.*

(i) $Mg \mid Mg^{2+} (1 \times 10^{-4} M)$; $E^0_{Mg, Mg^{2+}} = +2 \cdot 36$ V

(ii) $Cl_2 \mid Cl^- (2 \times 10^{-5} M)$; $E^0_{Cl_2, Cl^-} = +1 \cdot 36$ V

(iii) $Pt \mid Fe^{2+} (0 \cdot 1 M) - Fe^{3+} (0 \cdot 01 M)$; $E^0_{Fe^{3+}, Fe^{2+}} = +0 \cdot 77$ V

Solution : (i) $Mg^{2+} + 2e \rightleftharpoons Mg$ (reduction)

$$E_{Mg^{2+}, Mg} = E^0_{Mg^{2+}, Mg} + \frac{0 \cdot 0591}{2} \log (1 \times 10^{-4}) \text{ ... [See Example 1 (vi)]}$$

$$= -2 \cdot 36 - \frac{4 \times 0 \cdot 0591}{2} = -2 \cdot 4782 \text{ V.}$$

(ii) $\frac{1}{2} Cl_2 + e \rightleftharpoons Cl^-$ (reduction)

$$E_{Cl_2, Cl^-} = E^0_{Cl_2, Cl^-} - \frac{0 \cdot 0591}{n} \log [Cl^-] \quad \text{... [See Example 1 (ii)]}$$

$$= 1 \cdot 36 - \frac{0 \cdot 0591}{1} \log (2 \times 10^{-5}) = 1 \cdot 6377 \text{ V}$$

(iii) $Fe^{3+} + e \rightleftharpoons Fe^{2+}$ (reduction)

$$E_{Fe^{3+}, Fe^{2+}} = E^0_{Fe^{3+}, Fe^{2+}} - \frac{0 \cdot 0591}{n} \log \frac{[Fe^{2+}]}{[Fe^{3+}]}$$

$$= 0 \cdot 77 - \frac{0 \cdot 0591}{1} \log \frac{0 \cdot 1}{0 \cdot 01} = 0 \cdot 7109 \text{ V.}$$

Ex. 3. *The standard electrode potential of* Cu, Cu^{2+} *is* $-0 \cdot 34$ *volt. At what concentration of* Cu^{2+} *ions will this electrode potential be zero?*

Solution : For the electrode reaction (reduction)

$$Cu^{2+} + 2e \rightleftharpoons Cu, \; E^0 = +0 \cdot 34 \text{ volt,}$$

$$E_{Cu^{2+}, Cu} = E^0_{Cu^{2+}, Cu} - \frac{0 \cdot 0591}{2} \log \frac{1}{[Cu^{2+}]}$$

$$= E^0 + \frac{0 \cdot 0591}{2} \log [Cu^{2+}].$$

$$\therefore \quad 0 = 0 \cdot 34 + \frac{0 \cdot 0591}{2} \log [Cu^{2+}]$$

or $\log [Cu^{2+}] = -11{\cdot}5059 = \overline{12}{\cdot}4941.$

Taking antilog,

$$[Cu^{2+}] = 3{\cdot}12 \times 10^{-12}\ M.$$

Ex. 4. *Will* Ag *metal reduce* Sn^{2+}*?*

$$E^0_{Ag^+, Ag} = +0{\cdot}8 \text{ volt}; \ E^0_{Sn^{2+}, Sn} = -0{\cdot}14 \text{ volt}.$$

Solution : $Ag = Ag^+ + e;\ E^0 = -0{\cdot}8$ volt;

$$\Delta G° = -1\,(-0{\cdot}8)\,F = 0{\cdot}8F$$

or $2Ag = 2Ag^+ + 2e;\ E^0 = -0{\cdot}8$ volt; $\Delta G° = 1{\cdot}6\ F$ ... (1)

$Sn^{2+} + 2e = Sn;\ E^0 = -0{\cdot}14$ volt; $\Delta G° = -2\,(-0{\cdot}14)\,F = 0{\cdot}28\ F$ ... (2)

On adding (1) and (2),

$$2Ag + Sn^{2+} = 2Ag^+ + Sn;\ \Delta G° = +1{\cdot}6\ F + 0{\cdot}28\ F = +1{\cdot}88\ F$$

$$\therefore \quad E^0 = \frac{\Delta G°}{-nF} = \frac{1{\cdot}88\ F}{-2\ F} = -0{\cdot}94 \text{ volt.}$$

The negative value of E^0 indicates that the above reaction is not feasible, i.e., Ag will not reduce Sn^{2+} to Sn.

Ex. 5. *Construct galvanic cells from the following pairs of half cells and calculate their emf at* 25°C:

(i) Fe^{3+} (0·1 M), Fe^{2+} (1 M) (Pt) $\quad E^0_{Fe^{3+}, Fe^{2+}} = 0{\cdot}77$ volt

and AgCl (s), Cl^- (0·001 M) | Ag $\quad E^0_{AgCl, Cl^-} = 0{\cdot}22$ volt

(ii) Cd^{2+} | Cd $\quad E^0_{Cd^{2+}, Cd} = -0{\cdot}40$ volt
(1 M)

and Hg_2Cl_2 (s), Cl^- (0·1 M) | Hg $\quad E^0_{Hg_2Cl_2, Cl^-} = 0{\cdot}28$ volt

(iii) Sn, $SnCl_2$ (1 M) $\quad E^0_{Sn^{2+}, Sn} = -0{\cdot}14$ volt

and Fe | $FeSO_4$ (1 M) $\quad E^0_{Fe^{2+}, Fe} = -0{\cdot}44$ volt

Solution : (i) For the half-cell reaction

$$Fe^{3+} + e \rightleftharpoons Fe^{2+},$$

$$E_{Fe^{3+}, Fe^{2+}} = E^0_{Fe^{3+}, Fe^{2+}} - \frac{0{\cdot}0591}{1} \log \frac{[Fe^{2+}]}{[Fe^{3+}]}$$

$$= 0{\cdot}77 - \frac{0{\cdot}0591}{1} \log \frac{1}{0{\cdot}1} = 0{\cdot}71 \text{ volt.}$$

And, for the half-cell reaction

$$AgCl\,(s) + e = Ag\,(s) + Cl^-,$$

$$E_{AgCl, Cl^-} = E^0_{AgCl, Cl^-} - \frac{0.0591}{1} \log [Cl^-] = 0.22 - \frac{0.0591}{1} \log 0.001$$

$$= 0.3973 \text{ volt.}$$

On coupling the two half cells the first half cell shall undergo reduction while the other, oxidation, as the reduction potential of the former is more positive.

∴ the cell may be represented as

$$\underset{\text{(oxidation)}}{Ag \mid Cl^-, AgCl\,(s)} \mid\mid \underset{\text{(reduction)}}{Fe^{3+}, Fe^{2+}\,(Pt)}$$

E_{cell} = red. pot. of RHS electrode – red. pot. of LHS electrode

$= 0.71 - 0.3973 = 0.3127$ volt.

(ii) For the half-cell reaction

$$Cd^{2+} + 2e \rightleftharpoons Cd,$$

$$E_{Cd^{2+}, Cd} = E^0_{Cd^{2+}, Cd} - \frac{0.0591}{2} \log \frac{1}{[Cd^{2+}]}$$

$$= -0.40 - \frac{0.0591}{2} \log \frac{1}{1} = -0.40 \text{ volt.}$$

And, for the half-cell reaction

$$Hg_2Cl_2 + 2e \rightleftharpoons 2Hg\,(l) + 2Cl^-,$$

$$E_{Hg_2Cl_2, Cl^-} = E^0_{Hg_2Cl_2, Cl^-} - \frac{0.0591}{2} \log [Cl^-]^2$$

$$= 0.28 - \frac{0.0591}{2} \times 2 \log 0.1$$

$$= 0.28 + 0.0591 = 0.3391 \text{ volt.}$$

In this case, the first half cell undergoes oxidation while the other, reduction, as the latter has a higher value of electrode potential. The cell may be represented as

$Cd \mid Cd^{2+} \mid\mid Hg_2Cl_2\,(s), Cl^- \mid Hg$

$E_{cell} = 0.3391 - (-0.40) = 0.7391$ volt

(iii) For the first half-cell reaction

$$Sn^{2+} + 2e \rightleftharpoons Sn,$$

$$E_{Sn^{2+}, Sn} = E^0_{Sn^{2+}, Sn} - \frac{0.0591}{2} \log \frac{1}{[Sn^{2+}]}$$

$$= -0.14 - \frac{0.0591}{2} \log \frac{1}{1} = -0.14 \text{ volt.}$$

And, for the half-cell reaction

$$Fe^{2+} + 2e \rightleftharpoons Fe,$$

$$E_{Fe^{2+}, Fe} = E^0_{Fe^{2+}, Fe} - \frac{0.0591}{2} \log \frac{1}{[Fe^{2+}]}$$

$$= -0{\cdot}44 - \frac{0{\cdot}0591}{2}\log\frac{1}{1}$$

$$= -0{\cdot}44 \text{ volt.}$$

As $E_{Sn^{2+}, Sn}$ is higher than $E_{Fe^{2+}, Fe}$ the cell will be represented as

$$Fe \mid Fe^{2+} \mid\mid Sn^{2+} \mid Sn$$

$$E_{cell} = -0{\cdot}14 - (-0{\cdot}44)$$

$$= 0{\cdot}30 \text{ volt.}$$

Ex. 6. *The* E^0_{cell} *for the reaction* $Fe + Zn^{2+} \rightleftharpoons Zn + Fe^{2+}$ *is* $-0{\cdot}32$ volt *at* 25°C. *What will be the equilibrium concentration of* Fe^{2+} *when a piece of iron is placed in a* 1 M Zn^{2+} *solution?*

Solution : We have the Nernst equation at equilibrium at 25°C

$$E^0 = \frac{0{\cdot}0591}{n}\log K. \qquad \text{... (Eqn. 2)}$$

Since E^0_{cell} for the given reaction is negative, therefore, the reverse reaction is feasible for which E^0_{cell} will be $+0{\cdot}32$ V, Thus, for,

$$Zn + Fe^{2+} \rightleftharpoons Fe + Zn^{2+}; \qquad E^0_{cell} = +0{\cdot}32 \text{ V}$$

Applying Equation (2), we get

$$E^0 = \frac{0{\cdot}0591}{n}\log\frac{[Zn^{2+}]}{[Fe^{2+}]}$$

or

$$0{\cdot}32 = \frac{0{\cdot}0591}{2}\log\frac{1}{[Fe^{2+}]}$$

$$\log[Fe^{2+}] = -10{\cdot}829 = \overline{11}{\cdot}171.$$

Taking antilog,

$$[Fe^{2+}] = 1{\cdot}483 \times 10^{-11} \text{ M.}$$

Ex. 7. *A galvanic cell is constructed with* $Ag^+ \mid Ag$ *and* $Fe^{3+} \mid Fe^{2+}$ *electrodes. Find the concentration of* Ag^+ *at which the emf of the cell is zero at equimolar concentrations of* Fe^{2+} *and* Fe^{3+}. ($E^0_{Ag^+, Ag} = 0{\cdot}80$ V *and* $E^0_{Fe^{3+}, Fe^{2+}} = 0{\cdot}77$ V)

Solution : As E^0 for silver electrode is more positive, the cell may be represented as

$$Fe^{2+}, Fe^{3+} \mid\mid Ag^+, Ag$$

E_{cell} = red. pot. of RHS electrode – red. pot. of LHS electrode

$$E = \left\{E^0_{Ag^+, Ag} + \frac{0{\cdot}0591}{n}\log[Ag^+]\right\} - \left\{E^0_{Fe^{3+}, Fe^{2+}} + \frac{0{\cdot}0591}{n}\log\frac{[Fe^{3+}]}{[Fe^{2+}]}\right\}$$

$$0 = \left\{0.80 + \frac{0.0591}{1}\log[Ag^+]\right\} - \left\{0.77 + \frac{0.0591}{1}\log 1\right\} \quad ([Fe^{2+}] = [Fe^{3+}])$$

$$0 = 0.03 + 0.0591 \log [Ag^+]$$

or $\log [Ag^+] = -\dfrac{0.03}{0.0591} = -0.5085 = \bar{1}.4915$

$\therefore \quad [Ag^+] = 0.3121$ M.

Ex. 8. *The standard reduction potential of* $Cu^{2+}|Cu$ *and* $Ag^+|Ag$ *electrodes are* 0·337 *and* 0·799 volt *respectively. Construct a galvanic cell using these electrodes so that its standard emf is positive. For what concentration of* Ag^+ *will the emf of the cell at* 25°C *be zero if the concentration of* Cu^{2+} *is* 0·01 M? **(IIT 1990)**

Solution : As $E^0_{Ag^+, Ag}$ is more positive than $E^0_{Cu^{2+}, Cu}$, the former will act as cathode and the latter, anode. Thus the cell may be represented as

$$Cu, Cu^{2+} \;||\; Ag^+, Ag.$$

Further, given that $[Cu^{2+}] = 0.01$ M and $E_{cell} = 0$, we have,

$$E_{cell} = E_{Ag^+, Ag} - E_{Cu^{2+}, Cu} = 0$$

or $\left\{E^0_{Ag^+, Ag} + \dfrac{0.0591}{1}\log[Ag^+]\right\} - \left\{E^0_{Cu^{2+}, Cu} + \dfrac{0.0591}{2}\log[Cu^{2+}]\right\} = 0$

or $\left\{0.799 + \dfrac{0.0591}{1}\log[Ag^+]\right\} - \left\{0.337 + \dfrac{0.0591}{2}\log(0.01)\right\} = 0$

or $\log [Ag^+] = -8.8172 = \bar{9}.1828.$

Taking antilog,

$$[Ag^+] = 1.52 \times 10^{-9} \text{ M}.$$

Ex. 9. *Zinc granules are added in excess to* 500 mL *of* 1 M *nickel nitrate solution at* 25°C *until the equilibrium is reached. If the standard reduction potentials of* $Zn^{2+} | Zn$ *and* $Ni^{2+} | Ni$ *are* −0·75 V *and* −0·24 V *respectively, find out the concentration of* Ni^{2+} *in solution at equilibrium.* **(IIT 1991)**

Solution : The cell reaction for the cell,

$$Zn, Zn^{2+} \;||\; Ni^{2+}, Ni \quad \text{is}$$

$$Zn + Ni^{2+} \rightleftharpoons Zn^{2+} + Ni \quad \text{for which,}$$

$$E_{cell} = E^0_{cell} - \frac{0.0591}{2}\log\frac{[Zn^{2+}]}{[Ni^{2+}]}$$

At equilibrium, $E_{cell} = 0$.

$$\therefore \qquad E^0_{\text{cell}} = \frac{0.0591}{2} \log \frac{[Zn^{2+}]}{[Ni^{2+}]}$$

$$E^0_{Ni^{2+}, Ni} - E^0_{Zn^{2+}, Zn} = \frac{0.0591}{2} \log \frac{[Zn^{2+}]}{[Ni^{2+}]}$$

$$-0.24 - (-0.75) = \frac{0.0591}{2} \log \frac{[Zn^{2+}]}{[Ni^{2+}]}$$

$$\log \frac{[Zn^{2+}]}{[Ni^{2+}]} = 17.259.$$

Taking antilog,

$$\frac{[Zn^{2+}]}{[Ni^{2+}]} = 1.816 \times 10^{17}.$$

This concentration ratio shows that almost whole of the Ni^{2+} ions are reduced to Ni and therefore the concentration of Zn^{2+} produced from Zn would be nearly 1 M [$\because$ $Ni(NO_3)_2 = 1$ M]. Thus,

$$\frac{1}{[Ni^{2+}]} = 1.816 \times 10^{17}$$

or $$[Ni^{2+}] = 5.5 \times 10^{-18} \text{ M}.$$

Ex. 10. *The reduction potential diagram (Latimer diagram) for* Cu *in acid solution is*

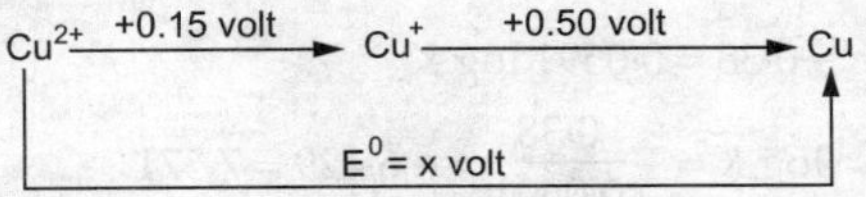

Calculate x. Does Cu^+ *disproportionate in the solution?*

Solution :

$Cu^{2+} + e \rightarrow Cu^+$; $E^0 = 0.15$; $\Delta G° = -nFE^0 = -1 \times 0.15\, F = -0.15\, F$

$Cu^+ + e \rightarrow Cu$: $E^0 = 0.50$; $\Delta G° = -1 \times 0.50\, F = -0.50\, F$

On adding,

$$Cu^2 + 2e \rightarrow Cu; \ \Delta G° = -0.65\, F$$

$$E^0_{Cu^{2+}, Cu} = \frac{\Delta G°}{-nF} = \frac{-0.65\, F}{-2\, F} = 0.325 \text{ volt}$$

$$x = 0.325 \text{ volt}.$$

Further,

$Cu^+ \rightarrow Cu^{2+} + e$; $\qquad \Delta G° = -nFE^0 = -1 \times (-0.15)\, F = 0.15\, F$

$Cu^+ + e \rightarrow Cu$; $\qquad \Delta G° = -nFE^0 = -1 \times 0.50 \times F = -0.50\, F$

On adding,

$$2Cu^+ \rightarrow Cu^{2+} + Cu; \quad \Delta G° = 0{\cdot}15\,F + (-\,0{\cdot}50\,F) = -\,0{\cdot}35\,F$$

$$\therefore \quad E^0 = -\frac{\Delta G°}{nF} = \frac{0{\cdot}35\,F}{F} = +\,0{\cdot}35\text{ V.} \qquad (n = 1)$$

Since E^0 of the above reaction is positive, the reaction is feasible. In other words, Cu^+ disproportionates because it can change into both Cu^{2+} and Cu simultaneously. Cu^+ thus behaves both as a self-reducing and a self-oxidising agent.

Ex. 11. *From the following standard electrode potentials at* 25°C

$Cu^{2+} + 2e = Cu; \quad E^0 = +\,0{\cdot}34$ V

$Cu^{2+} + e = Cu^+; \quad E^0 = +\,0{\cdot}15$ V

calculate equilibrium constant of the reaction $Cu + Cu^{2+} = 2Cu^+$.

Solution : $Cu = Cu^{2+} + 2e; \quad E^0 = -\,0{\cdot}34; \quad \Delta G° = -\,2\,(-\,0{\cdot}34)\,F = 0{\cdot}68\,F$

$2Cu^{2+} + 2e = 2\,Cu^+; \quad E^0 = +\,0{\cdot}15; \quad \Delta G° = -\,2\,(0{\cdot}15)\,F = -\,0{\cdot}30\,F$

On adding,

$$Cu + Cu^{2+} = 2Cu^+; \quad \Delta G° = +\,0{\cdot}38\,F$$

$$-\,\Delta G° = 2{\cdot}303\ RT\ \log K$$

$$-\,0{\cdot}38\,F = 2{\cdot}303\,RT \log\ K$$

or
$$-\,0{\cdot}38 = \frac{2{\cdot}303\,RT}{F} \log K$$

$$-\,0{\cdot}38 = 0{\cdot}0591 \log K$$

or
$$\log K = -\frac{0{\cdot}38}{0{\cdot}0591} = -\,6{\cdot}429 = \bar{7}{\cdot}571$$

$$\therefore \quad K = 3{\cdot}72 \times 10^{-7}.$$

Ex. 12. *What is* K_c *for the following reaction at* 25°C?

$$Cu^{2+}\,(aq) + Sn^{2+}\,(aq) \rightarrow Sn^{4+}\,(aq) + Cu\,(s)$$

$E^0_{Cu^{2+}, Cu} = 0{\cdot}34$ volt, $\quad E^0_{Sn^{4+}, Sn^{2+}} = 0{\cdot}15$ volt

Solution : The cell may be represented as

$$Sn^{2+} \mid Sn^{4+} \parallel Cu^{2+},\ Cu.$$

To calculate K_c, the cell reaction should be at equilibrium, i.e., $E_{cell} = 0$

$$E_{cell} = E_{Cu^{2+}, Cu} - E_{Sn^{4+}, Sn^{2+}} = 0$$

or
$$\left\{E^0_{Cu^{2+}, Cu} + \frac{0{\cdot}0591}{2} \log\,[Cu^{2+}]\right\} - \left\{E^0_{Sn^{4+}, Sn^{2+}} + \frac{0{\cdot}0591}{2} \log \frac{[Sn^{4+}]}{[Sn^{2+}]}\right\} = 0$$

or $$\left\{E^0_{Cu^{2+}, Cu} - E^0_{Sn^{4+}, Sn^{2+}}\right\} + \frac{0.0591}{2} \log \frac{[Cu^{2+}][Sn^{2+}]}{[Sn^{4+}]} = 0$$

or $$(+0.34 - 0.15) - \frac{0.0591}{2} \log \frac{[Sn^{4+}]}{[Cu^{2+}][Sn^{2+}]} = 0$$

or $$0.19 - \frac{0.0591}{2} \log K_c = 0$$

or $$\log K_c = 6.42$$

or $$K_c = 2.63 \times 10^6.$$

Ex. 13. *Show that the potentials are additive for the process in which half-reactions are added to yield an overall reaction but they are not additive when added to yield a third half-reaction.*

Solution : When half-reactions are added, ΔG of the overall reaction is nearly the sum of changes in free energy of the two half-reactions.

$$\Delta G_3 = \Delta G_1 + \Delta G_2$$

or $$-n_3 FE_3 = -n_1 FE_1 - n_2 FE_2 = -F(n_1 E_1 + n_2 E_2)$$

or $$E_3 = \frac{n_1 E_1 + n_2 E_2}{n_3}.$$

But, when the two half-reactions are added to yield an overall reaction, the number of moles of electrons involved in each half-reaction and the overall reaction are necessarily the same, i.e., $n_1 = n_2 = n_3 = n$

$\therefore$ $$E_3 = \frac{n E_1 + n E_2}{n} = E_1 + E_2.$$

However, when two half-reactions are added to yield a third half-reaction, the numbers of moles of electrons in all the three half-reactions cannot be the same.

Ex. 14. *Determine the emf of the galvanic cell.*

Ag | $AgNO_3$ (0·001 M) || $AgNO_3$ (0·1 M) | Ag

In what direction will electrons travel in the external circuit when the cell operates, $E^0_{Ag^+, Ag} = 0.80$ volt?

Solution : Let E_1 and E_2 be the reduction electrode potentials of left and right electrodes respectively.

$$E_{cell} = E_2 - E_1$$

$$= \left\{E^0_{Ag^+, Ag} + \frac{0.0591}{n} \log [Ag^+]_{right}\right\} - \left\{E^0_{Ag^+, Ag} + \frac{0.0591}{n} \log [Ag^+]_{left}\right\}$$

$$= +\frac{0.0591}{n} \log [Ag^+]_{right} - \frac{0.0591}{n} \log [Ag^+]_{left}$$

$$= \frac{0.0591}{n} \log \left\{ \frac{[Ag^+]_{right}}{[Ag^+]_{left}} \right\}$$

$$= \frac{0.0591}{1} \log \frac{0.1}{0.001} = 0.0591 \log 100$$

$$= 0.1182 \text{ volt.}$$

Since E_{cell} is positive, the right electrode will act as cathode (where reduction occurs) and the left electrode will act as anode (where oxidation occurs). The electrons will thus travel from left to right in external circuit.

Ex. 15. *What is the standard electrode potential for the electrode, $MnO_4^- \mid MnO_2$ in an acid solution?*

$$E^0_{MnO_4^-, Mn^{2+}} = 1.51 \text{ V}, \quad E^0_{MnO_2, Mn^{2+}} = 1.23 \text{ V}$$

Solution : $MnO_4^- + 8H^+ + 5e \rightarrow Mn^{2+} + 4H_2O \; ; \; E^0 = 1.51$ volts

$$\therefore \Delta G^\circ = -5(1.51)\,F = -7.55\,F$$

$$MnO_2 + 4H^+ + 2e \rightarrow Mn^{2+} + 2H_2O \; ; \; E^0 = 1.23 \text{ volts}$$

$$\therefore \Delta G^\circ = -2(1.23)\,F = -2.46\,F$$

On subtracting,

$$MnO_4^- - MnO_2 + 4H^+ + 3e \rightarrow 2H_2O;$$

$$\Delta G^\circ = -7.55\,F - (-2.46\,F) = -5.09\,F$$

or $MnO_4^- + 4H^+ + 3e \rightarrow MnO_2 + 2H_2O \; ; \; \Delta G^\circ = -5.09\,F$

$$\therefore \quad E^0_{MnO_4^-, MnO_2} = \frac{\Delta G^\circ}{-nF} = \frac{-5.09\,F}{-3F} = 1.70 \text{ volts.}$$

Ex. 16. *Two electrochemical cells are assembled in which the following reactions occur.*

$$V^{2+} + VO^{2+} + 2H^+ \rightarrow 2V^{3+} + H_2O \; ; \; E^0_{cell} = 0.616 \text{ V}$$

$$V^{3+} + Ag^+ + H_2O \rightarrow VO^{2+} + 2H^+ + Ag\,(s) \; ; \; E^0_{cell} = 0.439 \text{ V}$$

Calculate E^0 for the half-cell reaction

$$V^{3+} + e \rightarrow V^{2+} \qquad (E^0_{Ag^+, Ag} = 0.799 \text{ volt})$$

Solution : Let us find the number of electrons involved in each reaction

... (1)

$$\underset{+2}{V^{2+}} + \underset{+4}{VO^{2+}} + 2H^+ \rightarrow \underset{+3}{V^{3+}} + \underset{+3}{V^{3+}} + H_2O$$

Change in ON = 1

Change in ON = 1

$$V^{2+} + Ag^{+} + H_2O \rightarrow Ag(s) + VO^{2+} + 2H^{+} \quad \ldots (2)$$

Oxidation numbers: V^{2+}: +3; Ag^{+}: +1; Ag(s): 0; VO^{2+}: +4

Change in ON = 1 (Ag, +1 → 0)

Change in ON = 1 (V, +3 → +4)

Adding (1) and (2), we get,

$$V^{2+} + Ag^{+} \rightarrow V^{3+} + Ag\,(s)\ ;\ \Delta G^{\circ} = -(1 \times F \times 0{\cdot}616 + 1 \times F \times 0{\cdot}439)$$
$$= -1{\cdot}055\,F$$

$$\therefore \quad E^0_{cell} = -\frac{\Delta G^{\circ}}{nF} = \frac{1{\cdot}055\,F}{F} = 1{\cdot}055 \text{ V}.$$

Further, for the above cell,

$$E^0_{cell} = E^0_{Ag^+, Ag} - E^0_{V^{3+}, V^{2+}}$$

or $$1{\cdot}055 = 0{\cdot}799 - E^0_{V^{3+}, V^{2+}}$$

or $$E^0_{V^{3+}, V^{2+}} = 0{\cdot}799 - 1{\cdot}055$$
$$= -0{\cdot}256 \text{ volt.}$$

Ex. 17. *How much is the oxidising power of the* MnO_4^- (1 M) | Mn^{2+} (1 M) *couple decreased if the* H^+ *concentration is decreased from* 1 M to 10^{-4} M *at* 25°C?

Solution : The half-cell reaction is $MnO_4^- + 8H^+ + 5e = Mn^{2+} + 4H_2O$. Let us suppose that only the H^+ concentration deviates from 1 mole/litre.

$$E_{MnO_4^-, Mn^{2+}} = E^0_{MnO_4^-, Mn^{2+}} - \frac{0{\cdot}0591}{5} \log \frac{[Mn^{2+}]}{[MnO_4^-]\,[H^+]^8}$$

$$E - E^0 = -0{\cdot}0118 \log \frac{(1)}{(1)\,(10^{-4})^8} = -0{\cdot}38 \text{ volt.}$$

The couple MnO_4^- | Mn^{2+} has thus moved down by 0·38 volt from its standard value, i.e., to a position of less oxidising power.

Ex. 18. *Consider the cell,*

$$Zn \mid Zn^{2+} (aq)\ (1{\cdot}0\ M) \mid\mid Cu^{2+} (aq)\ (1{\cdot}0\ M) \mid Cu$$

The standard reduction potentials are + 0·35 V *for*

$2e + Cu^{2+} (aq) \rightarrow Cu$ and $-0{\cdot}763$ V for $2e + Zn^{2+} (aq) \rightarrow Zn$

(i) *Write down the cell reaction.*

(ii) *Calculate the emf of the cell.*

(iii) *Is the cell reaction spontaneous or not?*

Solution : (i) Cell reactions are as follows:

RHS electrode: $Cu^{2+} + 2e = Cu$ (reduction)

LHS electrode: $Zn = Zn^{2+} + 2e$ (oxidation)

(ii) $E_{cell} = E_{right\ electrode} - E_{left\ electrode}$

$$= \left\{E^0_{Cu^{2+}, Cu} + \frac{0.0591}{2}\log[Cu^{2+}]\right\} - \left\{E^0_{Zn^{2+}, Zn} + \frac{0.0591}{2}\log[Zn^{2+}]\right\}$$

$$= \left\{+0.35 + \frac{0.0591}{2}\log(1)\right\} - \left\{(-0.763) + \frac{0.0591}{2}\log(1)\right\}$$

$$= 0.35 + 0.763$$

$$= 1.13 \text{ volt.}$$

(iii) Since E_{cell} is positive the cell reactions as mentioned above are spontaneous.

Ex. 19. *The reversible reduction potential of pure water is* -0.414 V *under* 1 atm H_2 *pressure. If the reduction is considered to be* $2H^+ + 2e \rightarrow H_2$, *calculate the* $[H^+]$ *of pure water.*

Solution : $2H^+ + 2e \rightarrow H_2;\ E^0 = 0$ volt

$$E_{H^+, H_2} = E^0_{H^+, H_2} - \frac{0.0591}{2}\log\frac{1}{[H^+]^2}$$

$$-0.414 = 0 + 0.0591\log[H^+]$$

$$\log[H^+] = -6.99;\quad [H^+] = 1.02\times10^{-7}\ \text{M.}$$

Ex. 20. *The Edison storage cell is symbolised,*

$$Fe \mid FeO \mid KOH \mid\mid Ni_2O_3 \mid NiO \mid Ni$$

The half-cell reactions are:

$Ni_2O_3\ (s) + H_2O\ (l) + 2e = 2NiO\ (s) + 2OH^-$; $E^0_1 = 0.40$ V

$FeO\ (s) + H_2O\ (l) + 2e = Fe + 2OH^-$; $E^0_2 = -0.87$ V

(i) *What is the cell reaction?*

(ii) *How does the cell potential depend on the concentration of* KOH?

Solution : (i) Since E^0_1 (reduction) is higher than E^0_2 (reduction), first half cell will undergo reduction while the other, oxidation.

Thus,

at cathode: $Ni_2O_3\ (s) + H_2O\ (l) + 2e = 2NiO\ (s) + 2OH^-$ (reduction)

at anode: $Fe + 2OH^- = FeO\ (s) + H_2O\ (l) + 2e$ (oxidation)

Combining, we have the cell reaction,

$$Ni_2O_3\,(s) + H_2O\,(l) + 2e + Fe + 2OH^-$$

$$= 2NiO\,(s) + 2OH^- + FeO\,(s) + H_2O\,(l) + 2e$$

or $\quad Ni_2O_3\,(s) + Fe = 2NiO(s) + FeO\,(s)$

(ii) Further, $E_1 = E_1^0 - \dfrac{0{\cdot}0591}{2} \log\,[OH^-]^2$

$$E_2 = E_2^0 - \frac{0{\cdot}0591}{2} \log\,[OH^-]^2$$

$\therefore \qquad E_{cell} = E_2 - E_1 = E_2^0 - E_1^0.$

Since the above equation does not contain the term $[OH^-]$, i.e., OH^- concentration, E_{cell} does not depend on OH^- concentration.

Ex. 21. *Calculate the energy obtainable from a lead storage battery in which* 0·1 *mole of* Pb *is consumed. Assume constant concentration of* 10·0 M H_2SO_4. *Cell reaction with* E^0 *value is given below.*

$$PbO_2 + Pb + 4H^+ + 2SO_4^{2-} = 2PbSO_4 + 2H_2O\,; \quad E^0 = 2{\cdot}01V$$

Solution : $E = E^0 - \dfrac{0{\cdot}0591}{2} \log \left\{\dfrac{1}{[H^+]^4\,[SO_4^{2-}]^2}\right\}$

$$= 2{\cdot}01 - \frac{0{\cdot}0591}{2} \log \left\{\frac{1}{(20)^4\,(10)^2}\right\}$$

$$= 2{\cdot}20 \text{ V.}$$

As 0·1 mole of Pb is consumed,

number of faraday = no. of eq. of Pb consumed

= 0·2 F.

$\therefore \qquad$ charge $= 0{\cdot}2 \times 96500$ coulombs.

Energy $= (0{\cdot}2 \times 96500) \times 2{\cdot}20 = 42460$ J

$= 42{\cdot}46$ kJ.

Ex. 22. *Consider the reaction,* $2Ag^+ + Cd \rightarrow 2Ag + Cd^{2+}$;

$E^0_{Ag^+, Ag} = 0{\cdot}80$ volt; $E^0_{Cd^{2+}, Cd} = -0{\cdot}40$ volt.

(i) *What is the standard potential* $E°$ *for this reaction?*

(ii) *For the electrochemical cell in which this reaction takes place, which electrode is negative electrode?*

(iii) *Will the total emf of the reaction be more positive or more negative if the concentration of* Cd^{2+} *is* 0·10 M *rather than* 1 M?

Solution : (i) Since $E^0_{Ag^+, Ag}$ is higher than $E^0_{Cd^{2+}, Cd}$ the former undergoes reduction and the latter undergoes oxidation.

$$\therefore \quad E^0_{cell} = E^0_{Ag^+; Ag} - E^0_{Cd^{2+}, Cd}$$
$$= 0{\cdot}80 - (-0{\cdot}40)$$
$$= 1{\cdot}20 \text{ volts.}$$

(ii) In a galvanic cell, the anode is regarded as the negative electrode. The electrode Cd | Cd^{2+} is negative, being the anode where oxidation occurs.

(iii) Further, $E_{cell} = E_{Ag^+, Ag} - E_{Cd^{2+}, Cd}$

$$= \left\{E^0_{Ag^+, Ag} + \frac{0{\cdot}0591}{1}\log[Ag^+]\right\} - \left\{E^0_{Cd^{2+}, Cd} + \frac{0{\cdot}0591}{2}\log[Cd^{2+}]\right\}$$

$$= \left[E^0_{Ag^+, Ag} - E^0_{Cd^{2+}, Cd}\right] + \frac{0{\cdot}0591}{1}\log\frac{[Ag^+]}{[Cd^{2+}]^{\frac{1}{2}}}$$

$$= E^0_{cell} + 0{\cdot}0591 \log\frac{[Ag^+]}{[Cd^{2+}]^{\frac{1}{2}}}.$$

Since $[Cd^{2+}]$ is in the denominator, decrease in $[Cd^{2+}]$ will make E_{cell} more positive.

Ex. 23. *The following galvanic cell*

Zn	$Zn(NO_3)_2$ (aq)	||	$Cu(NO_3)_2$ (aq)	Cu
Anode	(100 mL, 1 M)		(100 mL, 1 M)	*Cathode*

was operated as an electrolysis cell as Cu *as the anode and* Zn *as the cathode. A current of* 0·48 *ampere was passed for* 10 *hours and then the cell was allowed to function as galvanic cell. What would be the emf of the cell at* 25°C. *Assume that the only electrode reactions occurring were those involving* Cu/Cu^{2+} *and* Zn/Zn^{2+}.

$(E^0_{Cu^{2+}, Cu} = +0{\cdot}34 \text{ V}, \ E^0_{Zn^{2+}, Zn} = -0{\cdot}76 \text{ V})$

Solution : When the cell acts as electrolysis cell, Cu being anode and Zn being cathode, the concentration of Cu^{2+} will increase due to dissolution ($Cu \rightarrow Cu^{2+} + 2e$) and Zn^{2+} concentration will decrease due to deposition ($Zn^{2+} + 2e \rightarrow Zn$).

Eq. of Zn deposited = no. of faraday of electricity passed
(at cathode)

$$= \frac{\text{no. of coulombs}}{96500}$$

$$= \frac{0{\cdot}48 \times (10 \times 60 \times 60)}{96500}$$

$$= 0{\cdot}18.$$

$\therefore$ eq. of Cu dissolved = 0·18
(at anode)

Thus, mole of Zn^{2+} removed from the cathodic compartment

$$= \frac{0 \cdot 18}{2} = 0 \cdot 09$$

and mole of Cu^{2+} gone to anodic compartment = 0·09

$$\therefore \begin{cases} \text{mole of } Zn^{2+} \text{ initially present} = 1 \times 0 \cdot 1 = 0 \cdot 1 \\ \text{and mole of } Cu^{2+} \text{ initially present} = 1 \times 0 \cdot 1 = 0 \cdot 1 \end{cases}$$

$$\therefore \begin{cases} \text{mole of } Zn^{2+} \text{ present after electrolysis} = 0 \cdot 1 - 0 \cdot 09 = 0 \cdot 01 \\ \text{and mole of } Cu^{2+} \text{ present after electrolysis} = 0 \cdot 1 + 0 \cdot 09 = 0 \cdot 19 \end{cases}$$

$\therefore$ $[Zn^{2+}] = 0 \cdot 1$ M and $[Cu^{2+}] = 1 \cdot 9$ M

As the electrolytic cell is now allowed to act as a galvanic cell as represented below.

$$Zn \mid Zn^{2+}\,(0 \cdot 1\,M) \mid\mid Cu^{2+}\,(1 \cdot 9\,M) \mid Cu,$$

$$E_{cell} = E_{Cu^{2+}, Cu} - E_{Zn^{2+}, Zn}$$

$$= \left\{E^0_{Cu^{2+}, Cu} + \frac{0 \cdot 0591}{2} \log [Cu^{2+}]\right\} - \left\{E^0_{Zn^{2+}, Zn} + \frac{0 \cdot 0591}{2} \log [Zn^{2+}]\right\}$$

$$= \left\{+ 0 \cdot 34 + \frac{0 \cdot 0591}{2} \log 1 \cdot 9\right\} - \left\{- 0 \cdot 76 + \frac{0 \cdot 0591}{2} \log 0 \cdot 1\right\}$$

$$= 1 \cdot 137 \text{ volts.}$$

Ex. 24. *For the reaction,*

H_2 (g) + 2 AgCl (s) = 2Ag (s) + $2H^+$ (aq) (0·1 M) + $2Cl^-$ (aq) (0·1 M)
(1 atm)
$\Delta G° = - 42927$ *joules at* 25°C. *Calculate the emf of the cell in which the given reaction takes place.*

Solution : We know,

$$\Delta G° = - nF E^0$$

$$\therefore \quad E^0 = - \frac{\Delta G°}{nF} = \frac{42927}{2 \times 96500} = 0 \cdot 2224 \text{ volt.}$$

Now, for the above cell reaction

$$E_{cell} = E^0 - \frac{0 \cdot 0591}{2} \log \frac{[H^+]^2\,[Cl^-]^2}{[H_2]}$$

$$= 0 \cdot 2224 - \frac{0 \cdot 0591}{2} \log \frac{(0 \cdot 1)^2\,(0 \cdot 1)^2}{(1)}$$

$$= 0 \cdot 3406 \text{ volt.}$$

Ex. 25. *The emf of the cell,* H_2 (g) | *Buffer* || *Normal Calomel Electrode is* 0·6885 *volt at* 40°C *when barometric pressure is* 760 mm. *What is the pH of the buffer solution?* $E_{calomel} = 0{\cdot}28$ *volt* .

Solution : Let us first calculate the electrode potential of hydrogen electrode. The pressure of H_2 is 1 atm (i.e., 760 mm)

$$E_{H^+, H_2} = E^0_{H^+, H_2} - \frac{2{\cdot}303\, RT}{nF} \log \frac{1}{[H^+]}$$

Substituting $R = 8{\cdot}314$ JK^{-1} mol^{-1}

$T = 273 + 40 = 313$ K

$F = 96500$ coulombs

$n = 1$

and, $E^0_{H^+, H_2} = 0$ volt.

We get,

$$E_{H^+, H_2} = -0{\cdot}0621 \log \frac{1}{[H^+]} = -0{\cdot}0621 \text{ pH}$$

Now, $E_{cell} = E_{calomel} - E_{hydrogen}$

$0{\cdot}6885 = 0{\cdot}28 - (-0{\cdot}0621 \text{ pH})$

∴ pH = 6·57.

Ex. 26. *Calculate the emf of the cell at* 25°C.

(Pt) H_2	|	HCl	||	HCl	|	H_2 (Pt)
(1 atm)		(*pH* 2·95)		(*pH* 1·47)		(1 atm)

Solution : $E_{cell} = E_{hydrogen}$ (RHS) $- E_{hydrogen}$ (LHS)

$= -0{\cdot}0591$ pH (RHS) $-$ [$-$ 0·0591 pH(LHS)] at 25°C

(See Example 25)

$= -0{\cdot}0591 \times 1{\cdot}47 + 00{\cdot}0591 \times 2{\cdot}95$

$= 0{\cdot}0591\,(2{\cdot}95 - 1{\cdot}47)$

$= 0{\cdot}0591 \times 1{\cdot}48$

$= 0{\cdot}0874$ volt.

Ex. 27. *The standard emf and the standard free energy change of a cell in which the reaction is*

$$\frac{1}{2}\text{Cd (s)} + \frac{1}{2}\text{Br}_2\text{ (l)} \rightarrow \frac{1}{2}\text{Cd}^{2+}\,(c = 1) + \text{Br}^-\,(c = 1)$$

are 1·47 *volt and* $-$ 141·5 kJ *respectively. What difference would there be in these values if the reaction is written as*

$$\text{Cd (s)} + \text{Br}_2\text{ (l)} \rightarrow \text{Cd}^{2+}\,(c = 1) + 2\text{Br}^-\,(c = 1)?$$

Solution : The standard emf for the reaction

$$Cd\,(s) + Br_2\,(l) \rightarrow Cd^{2+} + 2Br^- \quad \ldots (1)$$

will be the same (i.e., 1·47 volt) as that of

$$\frac{1}{2}Cd\,(s) + \frac{1}{2}Br_2\,(l) \rightarrow \frac{1}{2}Cd^{2+} + Br^- \quad \ldots (2)$$

because, E^0 *is an intensive property.*

But, $\Delta G°$ is an extensive property that depends on mass. Thus, $\Delta G°$ for the reaction (1) will be twice that of reaction (2)
[i.e., $2 \times (-141{\cdot}5)$ kJ] as the amounts have been doubled.

Ex. 28. *Mercury was added to a* 0·001 M *solution of* Fe^{3+} *at* 25°C. *On attainment of equilibrium,* 95·4 % *of* Fe^{3+} *changed into* Fe^{2+}. *Find standard emf of* $Hg_2^{2+} \mid Hg$ *if the cell reaction is*

$$2Hg + 2Fe^{3+} = Hg_2^{2+} + 2Fe^{2+}$$
$$E^0_{Fe^{3+}, Fe^{2+}} = 0{\cdot}77 \text{ V}.$$

Solution :

	$2Fe^{3+}$	Hg_2^{2+}	$2Fe^{2+}$	
$2Hg +$	0·001	0	0	Initial concentration
	$4{\cdot}6 \times 10^{-5}$	$47{\cdot}7 \times 10^{-5}$	$95{\cdot}4 \times 10^{-5}$	Concn. at eqb.

$$\left[\begin{array}{l} [Fe^{3+}] = 0{\cdot}001 - \frac{95{\cdot}4}{100} \times 0{\cdot}001 = 4{\cdot}6 \times 10^{-5}\text{ M} \\ [Fe^{2+}] = \frac{95{\cdot}4}{100} \times 0{\cdot}001 = 95{\cdot}4 \times 10^{-5}\text{ M} \\ [Hg_2^{2+}] = \frac{1}{2} \times 95{\cdot}4 \times 10^{-5} = 47{\cdot}7 \times 10^{-5}\text{ M} \end{array}\right]$$

$$E_{cell} = E_{Fe^{3+}, Fe^{2+}} - E_{Hg_2^{2+}, Hg}$$

At equilibrium, $E_{cell} = 0$

$$\therefore \quad E_{Fe^{3+}, Fe^{2+}} = E_{Hg_2^{2+}, Hg}$$

or $$E^0_{Fe^{3+}, Fe^{2+}} + \frac{0{\cdot}0591}{1} \log \frac{[Fe^{3+}]}{[Fe^{2+}]} = E^0_{Hg_2^{2+}, Hg} + \frac{0{\cdot}0591}{2} \log [Hg_2^{2+}]$$

$$0{\cdot}77 - E^0_{Hg_2^{2+}, Hg} = 0{\cdot}0591 \log \frac{[Hg_2^{2+}]^{\frac{1}{2}}[Fe^{2+}]}{[Fe^{3+}]}$$

$$= 0{\cdot}0591 \log \frac{(47{\cdot}7 \times 10^{-5})^{\frac{1}{2}}(95{\cdot}4 \times 10^{-5})}{(4{\cdot}6 \times 10^{-5})}$$

$$= 0{\cdot}0591 \log (45 \times 10^{-3})$$

$$= 0{\cdot}0591\,(2{\cdot}6532 - 3) = -0{\cdot}021$$

$\therefore \qquad E^0_{Hg_2^{2+}, Hg} = 0.77 + 00.021 = 0.791$ volt.

Ex. 29. *The standard reduction potential at* 25°C *of the reaction* $2H_2O + 2e^- \rightleftharpoons H_2 + 2OH^-$ *is* – 0·8277 *volt. Calculate the equilibrium constant for the reaction* $2H_2O = H_3O^+ + OH^-$ *at* 25°C. **[IIT 1989]**

Solution : Let the equilibrium constants for the following equations be as follows:

$$H_2O + H_2O \rightleftharpoons H_3O^+ + OH^- \quad \text{——} K_1 \qquad \ldots (1)$$

$$H_3O^+ + e \rightleftharpoons \frac{1}{2} H_2 + H_2O \quad \text{——} K_2 \qquad \ldots (2)$$

Adding them, we get

$$H_2O + e \rightleftharpoons \frac{1}{2} H_2 + OH^- \quad \text{——} K_3 \qquad \ldots (3)$$

We have now to calculate K_1.

Now for eqn. (2), $\quad E^0_{H_3O^+, H_2} = 0$

$$\therefore \qquad K_2 = 1 \left(\because E^0 = \frac{0.0591}{1} \log K \right) \qquad \ldots \text{(Eqn. 2)}$$

For eqn. (3), $E^0 = -0.8277$ volt (given)

$$\therefore \qquad E^0 = \frac{0.0591}{1} \log K_3 = -0.8277$$

$$\log K_3 = -\frac{0.8277}{0.0591}; \quad K_3 = 1 \times 10^{-14}$$

Further, as eqn. (3) is the sum of eqns. (1) and (2), we have

$$K_3 = K_1 \cdot K_2 = K_1$$

$$\therefore \qquad K_1 = 1 \times 10^{-14}.$$

Ex. 30. *Distinguish clearly between the meanings of* $E_{cell} = 0$ *and* $E^0_{cell} = 0$.

Solution : $E_{cell} = 0$ means that the concentration ratio in the Nernst equation for the cell reduces the E^0_{cell} value to zero while $E^0_{cell} = 0$ indicates a concentration cell.

Ex. 31. *Calculate the number of coulombs delivered by a Daniell cell, initially containing* 1 *litre each of* 1 M Cu^{2+} *and* 1 M Zn^{2+} *ion which is operated until its potential drops to* 1V.

$E^0_{Cu^{2+}, Cu} = 0.34$ V, $E^0_{Zn^{2+}, Zn} = -0.76$ V.

Solution : For the Daniell cell: Zn, Zn^{2+} || Cu^{2+}, Cu

$E_{cell} = E_{RHS}$ (red.) – E_{LHS} (red.)

$$E_{cell} = \left\{E^0_{Cu^{2+}, Cu} - \frac{0{\cdot}0591}{2}\log\frac{1}{[Cu^{2+}]}\right\} - \left\{E^0_{Zn^{2+}, Zn} - \frac{0{\cdot}0591}{2}\log\frac{1}{[Zn^{2+}]}\right\}$$

$$E_{cell} = \left(E^0_{Cu^{2+}, Cu} - E^0_{Zn^{2+}, Zn}\right) - \frac{0{\cdot}0591}{2}\log\frac{[Zn^{2+}]}{[Cu^{2+}]}$$

$$1 = \{0{\cdot}34 - (-0{\cdot}76)\} - \frac{0{\cdot}0591}{2}\log\frac{[Zn^{2+}]}{[Cu^{2+}]}$$

$$\therefore \quad \frac{[Zn^{2+}]}{[Cu^{2+}]} = 2390.$$

As in Daniell cell, Zn dissolves, the maximum concentration of Zn^{2+} ions in the solution may be 2M. Taking the ratio as

$$\frac{[Zn^{2+}]}{[Cu^{2+}]} = \frac{2}{2/2390} = \frac{2}{0{\cdot}00084},$$

we see that when the emf of the cell drops to 1 V, for the maximum concentration of Zn^{2+} (equal to 2M), $[Cu^{2+}]$ should be 0·00084 M. It is, therefore, clear that essentially all the copper has been consumed and so 2 moles of electrons or $2F$ of electricity or 2×96500, i.e., 193000 coulombs has been delivered. (for Cu : 1 mole $= 2$ eq. $\equiv 2F$)

Ex. 32. *Calculate the potential of a Daniell cell, initially containing* 1 *litre each of* 1M Cu^{2+} *and* 1M Zn^{2+}, *after a passage of* 1×10^5 *coulombs of charge.* $E^0_{Cu^{2+}, Cu} = +0{\cdot}34$ V, $E^0_{Zn^{2+}, Zn} = -0{\cdot}76$ V.

Solution : Moles of electricity passed $= \dfrac{10^5}{96500} = 1{\cdot}04F$

$\therefore$ eq. of Cu^{2+} removed $= 1{\cdot}04$

or mole of Cu^{2+} removed $= 0{\cdot}52$

and mole of Zn^{2+} produced $= 0{\cdot}52$

Thus, $[Zn^{2+}] = 1{\cdot}52$ M and $[Cu^{2+}] = 0{\cdot}48$ M

For the Daniell cell, $Cu^{2+} + Zn = Cu + Zn^{2+}$

$$E = 1{\cdot}10 - \frac{0{\cdot}0591}{2}\log\frac{10{\cdot}52}{0{\cdot}48} = 1{\cdot}09 \text{ V}.$$

Ex. 33. *The emf of a particular voltaic cell with the cell reaction*

$$Hg_2^{2+} + H_2 \rightleftharpoons 2\,Hg + 2\,H^+$$

is 0·65 V. *Calculate the maximum electrical work of this cell when* 0·5 g H_2 *is consumed.*

Solution : We have, $W_{max} = -nFE$.

To get n, we split the cell reaction into two half-reactions.

$$Hg_2^{2+} + 2e \rightleftharpoons 2\,Hg$$
$$H_2 \rightleftharpoons 2\,H^+ + 2e$$

We see, $n = 2$

$$W_{max} = -2 \times 96500 \times 0{\cdot}65 = -1{\cdot}25 \times 10^5 \text{ J.}$$

From the stoichiometry of the cell reaction, it is clear that for 1 mole of H_2 consumed, the work produced is $-1{\cdot}25 \times 10^5$ J. Thus,

work produced when $\frac{1}{4}$ mole (0·5 g) of H_2 reacts

$$= \frac{1}{4}(-1{\cdot}25 \times 10^5) \text{ J}$$
$$= -3{\cdot}12 \times 10^4 \text{ J.}$$

Ex. 34. *The standard reduction potential for* Cu^{2+}/Cu *is* +0·34 V. *Calculate the reduction potential at pH = 14 for the above couple.*
K_{sp} $(Cu[OH]_2) = 1 \times 10^{-19}$. **(IIT 1996)**

Solution **:** Given that pH = 14, i.e., $[H^+] = 10^{-14}$ M and $[OH^-] = 1$ M

$$K_{sp} = [Cu^{2+}][OH^-]^2 = 10^{-19}$$

$$[Cu^{2+}] = \frac{10^{-19}}{[OH^-]^2} = \frac{10^{-19}}{1} = 10^{-19}.$$

For the half-cell reaction,

$$Cu^{2+} + 2e \rightleftharpoons Cu$$

$$E = E^0 - \frac{0{\cdot}0591}{2}\log\frac{1}{[Cu^{2+}]}$$
$$= 0{\cdot}34 - \frac{0{\cdot}0591}{2}\log\frac{1}{10^{-19}}$$
$$= -0{\cdot}22 \text{ V.}$$

Ex. 35. *Calculate the equilibrium constant for the equilibrium,*

$$Cd^{2+} + 4NH_3 \rightleftharpoons Cd(NH_3)_4^{2+}$$

from the given $E^0_{\text{electrode}}$ *data though the reaction is not a redox reaction.*

$$E^0_{Cd^{2+}, Cd} = -0{\cdot}40\text{V}, \; E^0_{Cd(NH_3)_4^{2+}, Cd} = -0{\cdot}61\text{V}$$

Solution **:** The reaction cannot be split into two half-cell reactions as the reaction is not a redox reaction. We, therefore, arbitrarily introduce Cd, the reduced species for both Cd^{2+} and $Cd(NH_3)_4^{2+}$. Then the half-cell reactions are

$$Cd^{2+} + 2e \rightleftharpoons Cd; \qquad E^0 = -0{\cdot}40\text{V}$$

$$Cd(NH_3)_4^{2+} + 2e \rightleftharpoons Cd + 4NH_3\,; \qquad E^0 = -0{\cdot}61V$$

From which we get,

$$Cd^{2+} + 4NH_3 \rightleftharpoons Cd(NH_3)_4^{2+}\,; \qquad E^0 = -0{\cdot}40 - (-0{\cdot}61) = +0{\cdot}21V$$

We have,

$$E^0 = \frac{0{\cdot}0591}{n}\log K$$

$$\log K = \frac{2 \times 0{\cdot}21}{0{\cdot}0591} = 7{\cdot}1065$$

or $K = 1{\cdot}27 \times 10^7$.

Ex. 36. *When* Cl_2 *dissolves in water, it disproportionates, producing* Cl^- *ion and* HClO. *Find at what* H_3O^+ *concentration the potential for the disproportionation changes from a negative value to a positive value, assuming* 1·0 atm *of pressure and concentration of* 1·0 M *for all species except hydronium ions. Given:* $E^0_{Cl_2, Cl^-} = 1{\cdot}36V$ *and* $E^0_{HClO, Cl_2} = 1{\cdot}63V$ *at* 25°C.

Solution : $Cl_2 + 2e = 2Cl^-$ $\qquad E^0 = 1{\cdot}36V$

$$2HOCl + 2H^+ + 2e = Cl_2 + 2H_2O; \qquad E^0 = 1{\cdot}63V$$

The latter electrode has a higher value of standard reduction potential and so shall act as cathode. The cell reaction will, therefore, be

$$2HOCl + 2H^+ + 2Cl^- = 2H_2O + 2Cl_2$$

which is just the reverse of disproportionation reaction of Cl_2. As only $[H^+]$ is variable, there will be a value of $[H^+]$ at which negative $E_{cell-value}$ of disproportionation reaction of Cl_2 will just change to a positive value of its reverse reaction, that is, when $E_{cell} = 0$.

Thus,

$$E_{cell} = E^0 = \frac{{\cdot}0591}{2}\log\frac{1}{[H^+]^2[Cl^-]^2} = 0$$

$$E_{cell} = (1{\cdot}63 - 1{\cdot}36) - \frac{{\cdot}0591}{2}\log\frac{1}{[H^+]^2} = 0$$

or $[H^+] = 2{\cdot}70 \times 10^{-5}$ M.

Ex. 37. *The electrolysis of an aqueous solution of* $CuCl_2$ *using* Pt *electrodes produces* Cu(s) *and* Cl_2(g) *at cathode and anode respectively. What is the minimum emf required for this process under standard conditions?*
$E^0_{Cu^{2+}, Cu} = +0{\cdot}34V$ *and* $E^0_{Cl_2, Cl^-} = +1{\cdot}36V$

Solution : The reverse of an electrolytic cell is a galvanic cell. Thus, if the given electrolytic or electrolysis cell acts as a galvanic cell in which Cu^{2+} is reduced to Cu at cathode and Cl^- is oxidised to Cl_2 at anode under standard conditions, the emf can be calculated as

$$E^0_{cell} = E^0_{Cl_2, Cl^-} - E^0_{Cu^{2+}, Cu}$$

$$E^0_{cell} = 1{\cdot}36 - 0{\cdot}34 = 1{\cdot}02\text{V}.$$

Thus, a minimum emf of 1·02V has to be applied in the electrolytic cell to produce Cu and Cl_2 at cathode and anode respectively.

Ex. 38. *Calculate the standard reduction potential for the reaction:* $H_2O + e \rightarrow \frac{1}{2} H_2 + OH^-$, *using the Nernst equation and the fact that the standard reduction potential for the reaction:* $H^+ + e \rightarrow \frac{1}{2} H_2$ *is by definition equal to* 0·00 V *at* 25°C.

Solution : Given that,

$$H_2O + e \rightarrow \frac{1}{2} H_2 + OH^- ; \qquad E^0 = x\text{V (say)}$$

$$\frac{1}{2} H_2 \rightarrow H^+ + e; \qquad E^0 = 0\text{V (given)}$$

∴ for the reaction,

$$H_2O \rightarrow H^+ + OH^-; \qquad E^0 = x\text{V}$$

The Nernst equation is

$$E_{cell} = E^0 - \frac{0{\cdot}0591}{1} \log [H^+][OH^-].$$

At equilibrium, $E_{cell} = 0$ and $[H^+][OH^-] = K_w = 1 \times 10^{-14}$

$$\therefore \; 0 = x - \frac{{\cdot}0591}{1} \log 10^{-14}$$

or $x = -0{\cdot}8274\text{V}$.

Ex. 39. *Given the following* E^0 *values at* 25°C. *Calculate* K_{sp} *for silver bromide,* AgBr.

$$Ag^+(aq) + e = Ag(s); \qquad E^0_1 = 0{\cdot}80\text{V}$$

$$AgBr(s) + e = Ag(s) + Br^-(aq); \qquad E^0_2 = 0{\cdot}07\text{V}$$

Also, calculate ΔG^o *at* 25°C *for the process*

$$AgBr(s) \rightleftharpoons Ag^+(aq) + Br^-(aq)$$

Solution : Reduction potential $E^0_1 > E^0_2$, so cell reaction is

$$Ag^+(aq) + Br^-(aq) = AgBr(s)$$

for which,

$$E^0_{cell} = E^0_1 - E^0_2 = 0{\cdot}80 - 0{\cdot}07 = 0{\cdot}73\text{V}$$

We have,

$$E^0 = \frac{2{\cdot}303RT}{nF} \log K$$

$$0{\cdot}73 = \frac{2{\cdot}303 \times 8{\cdot}314 \times 298}{1 \times 96500} \log K$$

$$\log K = 12{\cdot}3515$$

or $$K = 2{\cdot}246 \times 10^{12}.$$

Thus, for the eqb: $AgBr(s) \rightleftharpoons Ag^+(aq) + Br^-(aq)$

$$K_{sp} = \frac{1}{K} = \frac{1}{2{\cdot}246 \times 10^{12}} = 4{\cdot}45 \times 10^{-13}$$

Further, for the reaction $Ag^+(aq) + Br^-(aq) \rightleftharpoons AgBr(s)$

$$\begin{aligned}\Delta G^o &= -2{\cdot}303RT \log K \\ &= -2{\cdot}303 \times 8{\cdot}314 \times 298 \times 12{\cdot}3515 \\ &= -70475{\cdot}7 \text{ J/mole} \\ &= -70{\cdot}47 \text{ kJ/mole.}\end{aligned}$$

$\therefore$ for $AgBr(s) \rightleftharpoons Ag^+(aq) + Br^-(aq)$

$$\Delta G^o = +70{\cdot}47 \text{ kJ/mole.}$$

Ex. 40. *Calculate the emf of the following cell at* 25°C,

$$H_2(g)(1\text{ atm})/H^+(aq) \,||\, OH^-(aq)/O_2(g)(1\text{ atm})$$

from the following data:

(i) $H_2 + \frac{1}{2}O_2 = H_2O;\ \Delta G^o = -226{\cdot}8$ kJ/mole

(ii) $H_2O = H^+ + OH^-;\ \Delta G^o = 76{\cdot}2$ kJ/mole

Solution : The electrodic processes are,

$\frac{1}{2}O_2 + H_2O + 2e = 2OH^-$ ... at cathode (RHS)

$H_2 = 2H^+ + 2e$ ... at anode (LHS)

The cell reaction is

$$\frac{1}{2}O_2 + H_2O + H_2 = 2H^+ + 2OH^-$$

for which,

$$E_{cell} = E^0_{cell} - \frac{{\cdot}0591}{2} \log [H^+]^2[OH^-]^2$$

$$E_{cell} = E^0_{cell} - \frac{{\cdot}0591}{2} \log (K_w)^2$$

$$E_{cell} = E^0_{cell} + 0{\cdot}8274 \ (K_w = 10^{-14}). \quad \ldots$$

(1)

Now multiply the given eqn. (ii) by 2 and add eqn. (i). We get the cell reaction.

$$H_2 + \frac{1}{2}O_2 + H_2O = 2H^+ + 2OH^-; \ \Delta G^\circ = -226{\cdot}8 + 2 \times 76{\cdot}2$$

$$= -74{\cdot}4 \text{ kJ}$$

$$= -74400 \text{ J}.$$

Again we have,

$$\Delta G^\circ = -2{\cdot}303RT \log K$$

$$-74400 = -2{\cdot}303 \times 8{\cdot}314 \times 298 \times \log K$$

$$\log K = 13{\cdot}04$$

$$\therefore \quad E^0 = \frac{2{\cdot}303RT}{nF} \log K$$

$$= \frac{{\cdot}0591}{2} \times 13{\cdot}04$$

$$= 0{\cdot}3853.$$

Substituting the E^0 value in eqn. (1)

$$E_{cell} = 0{\cdot}3853 + 0{\cdot}8274$$

$$= 1{\cdot}2128\text{V}.$$

Ex. 41. *Under standard conditions for all concentrations, the following reaction is spontaneous at* 25°C.

$$O_2(g) + 4H^+(aq) + 4Br^-(aq) \rightarrow 2H_2O(l) + 2Br_2(l); E^0 = 0{\cdot}16\text{V}$$

If $[H^+]$ *is adjusted by adding a buffer of* 0·10 M NaOCN *and* 0·10 M HOCN ($K_a = 3{\cdot}5 \times 10^{-4}$), *what value will* E_{cell} *have and will the reaction be spontaneous at this* $[H^+]$?

***Solution* :** For a buffer solution of HOCN and NaOCN, we have

$$[H^+] = K_a \times \frac{[HOCN]}{[NaOCN]} = 3{\cdot}5 \times 10^{-4} \times \frac{0{\cdot}1}{0{\cdot}1} = 3{\cdot}5 \times 10^{-4}$$

Thus,

$$E_{cell} = E^0_{cell} - \frac{0{\cdot}0591}{4} \log \frac{1}{[H^+]^4[Br^-]^4}$$

$$= 0{\cdot}16 - \frac{0{\cdot}0591}{4} \log \frac{1}{(3{\cdot}5 \times 10^{-4})^4(1{\cdot}0)^4}$$

$$= -0{\cdot}044\text{V}.$$

Negative value of E_{cell} suggests that the given cell reaction is nonspontaneous when $[H^+]$ is adjusted by a buffer solution.

Ex. 42. *The emf of a standard* Cd– *cell is* 1·018V *at* 25°C. *The temperature coefficient of the cell is* $-5{\cdot}0 \times 10^{-5}$VK^{-1}. *Calculate* ΔG, ΔH *and* ΔS *for the cell reaction. Will the cell temperature change during operation?*

Solution : $\Delta G = -nFE$

$$= -2 \times 96500 \times 1{\cdot}018 \text{ J mole}^{-1} \qquad (Cd^{2+} + 2e = Cd)$$

$$= -196{\cdot}474 \text{ kJ mole}^{-1}.$$

$$\Delta S = nF\left(\frac{\partial E}{\partial T}\right)_P$$

$$= 2 \times 96500 \times (-5{\cdot}0 \times 10^{-5}) \text{ Jk}^{-1}\text{mole}^{-1}$$

$$= -9{\cdot}65 \text{ JK}^{-1}\text{mole}^{-1}.$$

$$\Delta H = T\Delta S + \Delta G$$

$$= 298 \times (-9{\cdot}65) + (-196474) \text{ J mole}^{-1}$$

$$= -199349 \text{ J mole}^{-1}$$

$$= -199{\cdot}349 \text{ kJ mole}^{-1}.$$

The negative value of ΔH suggests that the cell temperature will increase during operation.

Ex. 43. *The standard potential of the following cell is* 0·23V *at* 15°C *and* 0·21V *at* 35°C

$$(Pt)\ H_2(g)\ |\ HCl(aq)\ |\ AgCl(s)\ |\ Ag(s)$$

(i) Write the cell reaction.

(ii) Calculate ΔH° *and* ΔS° *for the cell reaction by assuming that these quantities remain unchanged in the range* 15°C *to* 35°C.

(iii) Calculate the solubility of AgCl *in water at* 25°C. *Given:*

$$E^0_{Ag^+, Ag} = 0{\cdot}80\text{V at } 25°\text{C}$$ **(IIT 2001)**

Solution : (i) Cell reaction : $AgCl + \frac{1}{2}H_2 = Ag + Cl^- + H^+$

(ii) $(\Delta G°)_{288} = -nFE^0 = -1 \times 96500 \times 0{\cdot}23 = -22195 \text{ J} = -22{\cdot}195 \text{ kJ}$

$(\Delta G°)_{308} = -1 \times 96500 \times 0{\cdot}21 = -20265 \text{ J} = -20{\cdot}265 \text{ kJ}$

Now we have,

$$\Delta G° = \Delta H° - T\Delta S°$$

$\therefore \quad -22{\cdot}195 = \Delta H° - 288\ \Delta S°$

and $-20{\cdot}265 = \Delta H° - 308{\cdot}\Delta S°$

from these equations, we get

$$\Delta H° = 49{\cdot}98 \text{ kJ}$$

$$\Delta S° = -96{\cdot}50 \text{ kJ}$$

(iii) In the given cell

$$E^0_{cell} = E^0_{AgCl, Cl^-} - E^0_{H^+, H_2}$$

$$0{\cdot}22 = E_{AgCl, Cl^-} - 0 \quad \text{... (Average value of } E^0_{cell} \text{ is considered)}$$

or $E_{AgCl, Cl^-} = 0{\cdot}22.$

Let us now consider the following cell to calculate K_{sp} (AgCl).

$$Ag = Ag^+ + e; \qquad E^0 = -0{\cdot}8 \text{ (given)}$$

$$AgCl + e = Ag + Cl^-; \qquad E^0 = 0{\cdot}22 \text{ (calculated)}$$

∴ for the cell reaction: $AgCl = Ag^+ + Cl^-$

$$E_{cell} = (E^0_{AgCl, Cl^-} - E^0_{Ag^+, Ag}) - \frac{0{\cdot}0591}{1} \log [Ag^+][Cl^-]$$

At equilibrium, $E_{cell} = 0$ and $R.Q. = K_{sp}$.

$$\therefore \; E^0_{AgCl, Cl^-} - E^0_{Ag^+, Ag} = 0{\cdot}0591 \log K_{sp}$$

$$0{\cdot}22 - 0{\cdot}8 = 0{\cdot}0591 \log K_{sp}$$

$$\therefore \; K_{sp} = 1{\cdot}535 \times 10^{-10}$$

$$\therefore \; \text{solubility of AgCl} = \sqrt{K_{sp}} = \sqrt{1{\cdot}535 \times 10^{-10}}$$

$$= 1{\cdot}239 \times 10^{-5} \text{ mole/L.}$$

Ex. 44. *A cell utilises the following reaction:*

$$2Co^{3+}(aq) + Zn(s) \rightarrow 2Co^{2+}(aq) + Zn^{2+}(aq)$$

What is the effect on cell emf of each of the following changes?

(a) Co (II) *nitrate is dissolved in the cathode compartment*

(b) Co (III) *nitrate is dissolved in the cathode compartment*

(c) The size or area of Zn (s) *electrode is doubled*

(d) Additional water is added to the anode compartment

Solution : For the given cell, we have

$$E_{cell} = E^0_{cell} - \frac{0{\cdot}0591}{2} \log \frac{[Co^{2+}]^2[Zn^{2+}]}{[Co^{3+}]^2}$$

(a) Using the above equation, we know that if $[Co^{2+}]$ is increased E_{cell} will decrease.

(b) Similarly, if $[Co^{3+}]$ is increased, E_{cell} will increase.

(c) The emf is an intensive property and so the size of the electrode shall not affect the emf of the cell.

(d) On diluting the solution in anodic compartment, $[Zn^{2+}]$ shall decrease and so E_{cell} will increase.

Ex. 45. *A solution containing* 4·5 mmol *of* $Cr_2O_7^{2-}$ *and* 15 mmol *of* Cr^{3+} *shows a* pH *of* 2·0. *Calculate the potential of the half-reaction.*

$$Cr_2O_7^{2-} \rightarrow Cr^{3+}; \qquad E^0 = 1{\cdot}33V$$

Solution : The complete half-cell reaction is

$$Cr_2O_7^{2-} + 14H^+ + 6e \rightarrow 2Cr^{3+} + 7H_2O; \quad E^0 = 1{\cdot}33V$$

$$\text{Given: } [Cr_2O_7^{2-}] = \frac{4{\cdot}5}{1000} M, [Cr^{3+}] = \frac{15}{1000} M \text{ and } [H^+] = 10^{-2} M$$

$$E = E^0 - \frac{0{\cdot}0591}{6} \log \frac{[Cr^{3+}]^2}{[Cr_2O_7^{2-}][H^+]^{14}}$$

$$E = 1{\cdot}33 - \frac{0{\cdot}0591}{6} \log \frac{(0{\cdot}015)^2}{(0{\cdot}0045)(10^{-2})^{14}}$$

$$E = 1{\cdot}067V.$$

PROBLEMS

(*Answers bracketed with questions*)

[**Note:** See E^0 values from the table (p. 668), if not given.]

1. Calculate the reduction potentials for the following half cells:

(i) Ag | Ag^+ (10^{-5} M); $E^0_{Ag^+, Ag} = 0{\cdot}80$ V

(ii) Cu | Cu^{2+} (0·2 M); $E^0_{Cu^{2+}, Cu} = 0{\cdot}34$ V (0·50 V, 0·32 V)

2. Construct galvanic cells from the following pairs of half cells and calculate their emf.

(i) (Pt) H_2 (1 atm) | HCl (1 M) and Pt (Cl_2) (1 atm) | HCl (1 M)

(ii) Cu | Cu^{2+} (1 M) and Cl^- (1 M) | Cl_2 (Pt) (1 atm)

Use the emf series.

$$\left\{ \begin{array}{l} H_2 \mid H^+ \parallel Cl_2 \mid Cl^-;\ E = 1{\cdot}36\text{ V} \\ Cu \mid Cu^{2+} \parallel Cl_2 \mid Cl^-;\ E = 1{\cdot}02\text{ V} \end{array} \right\}$$

3. Can F^- be oxidised to F_2 by any substance listed in the electrochemical series? (No, but it can be oxidised electrolytically)

4. A Cu rod is dipped in 0·1 M $CuSO_4$ solution. Calculate the potential of this half cell if $CuSO_4$ undergoes 90% dissociation at this dilution at 25°C. $E^0_{Cu, Cu^{2+}} = -0{\cdot}34$ V. (0·31 V)

5. If excess metallic iron is added to an N-$CuSO_4$ solution, calculate the approximate concentration of Cu^{2+} when equilibrium is established. (3×10^{-27} M)

6. Will Mg reduce $CuSO_4$? $E^0_{Mg, Mg^{2+}} = +2{\cdot}36$ V and $E^0_{Cu, Cu^{2+}} = -0{\cdot}34$ V? (Yes)

7. Calculate emf of the following cells at 25°C in which the following reactions are taking place: use E^0 values from table.

(i) $Mg + Cl_2 \text{ (1 atm)} \rightleftharpoons Mg^{2+} (10^{-2} \text{ M}) + 2Cl^- (2 \times 10^{-2} \text{ M})$

(ii) $Zn + Fe^{2+} (10^{-3} \text{ M}) \rightleftharpoons Zn^{2+} (10^{-4} \text{ M}) + Fe$ (3·879 V, 0·2595 V)

8. E^0_{cell} for $Zn (s) + Pb^{2+} (1 M) \rightarrow Zn^{2+} (1 M) + Pb (s)$ is + 0·66 volt. E_{cell} for the reaction

$$Zn (s) + Pb^{2+} (0{\cdot}1 M) \rightarrow Zn^{2+} (0{\cdot}1 M) + Pb (s) \quad \text{is}$$

(a) + 0·63 volt (b) + 0·66 volt (c) + 0·69 volt (d) + 0·72 volt (b)

9. E^0_{cell} for the reaction $Cu^{2+} + 2Cl^- \rightarrow Cu (s) + Cl_2 (g)$ is – 1·02 V. This reaction
(a) can be made to produce electricity in a voltaic cell
(b) occurs whenever Cu^{2+} and Cl^- are brought together in an aqueous solution
(c) can be made to occur in an electrolytic cell
(d) can occur in an acidic solution but not in a basic solution (c)

10. E^0_{cell} for the reaction $Co (s) + Ni^{2+} \rightarrow Co^{2+} + Ni (s)$ is + 0·03 volt. If cobalt metal is added to an aqueous solution having $[Ni^{2+}] = 1$ M,
(a) the reaction will not proceed in the forward direction at all
(b) the displacement of Ni^{2+} from solution by Co will go to completion
(c) the displacement of Ni^{2+} from solution by Co will proceed to a considerable extent, but the reaction will stop before the Ni^{2+} is completely displaced
(d) only the reverse reaction will occur (c)

11. From the electrochemical series given in the text, determine the approximate value of E^0 for $X^{2+} (aq) + 2e \rightarrow X (s)$.
(a) The metal X dissolves in nitric acid but not in hydrochloric acid. It can displace Ag^+ but not Cu^{2+}.
(b) The metal X dissolves in hydrochloric acid producing H_2 but does not displace either Zn^{2+} or Fe^{2+}.

$$\begin{bmatrix} \text{(a)} & 0{\cdot}34 \text{ V} < E^0 < 0{\cdot}80 \text{ V} \\ \text{(b)} & -0{\cdot}44 \text{ V} < E^0 < 0{\cdot}00 \text{ V} \end{bmatrix}$$

12. The standard reduction potential of a calomel half cell is 0·28 V at 25°C. Calculate half-cell potential when 0·1 N KCl solution is used.
[**Hint:** $Hg_2Cl_2 (s) + 2e^- \rightarrow 2Hg + 2Cl^- (0{\cdot}1 \text{ N KCl})$] (0·339 V)

13. What is the potential of a cell containing two hydrogen electrodes, the negative one in contact with 10^{-8} molar H^+ and the positive one in contact with 0·025 molar H^+? (0·379 volt)

14. Copper can reduce zinc ions if the resultant copper ions can be kept at a sufficiently low concentration by the formation of an insoluble salt. What is the maximum concentration of Cu^{2+} in solution if this reaction is to occur, when Zn^{2+} is 1 molar?
[**Hint:** for Cu, $Cu^{2+} \| Zn^{2+}, Zn$; $E_{cell} = E_{red} (Zn) - E_{red} (Cu) = 0$] ($6 \times 10^{-38}$ M)

15. A cell contains two hydrogen electrodes. The negative electrode is in contact with a solution of 10^{-6} M hydrogen ions. The emf of the cell is 0·118 V at 25°C. Calculate the concentration of hydrogen ions at the positive electrode.
[**Hint:** Positive electrode is cathode. See Example 26] **(IIT 1988)** (10^{-4} M)

16. The standard potentials, E^0 for the half-reactions are as

$$Zn = Zn^{2+} + 2e^-;\quad E^0 = 0{\cdot}76\ V$$

$$Fe = Fe^{2+} + 2e^-;\quad E^0 = 0{\cdot}41\ V$$

The emf for the cell reaction $Fe^{2+} + Zn \rightarrow Zn^{2+} + Fe$ is

(a) – 0·35 V (b) + 0·35 V (c) + 1·17 V (d) – 1·17 V

(IIT 1988) (b)

17. Given that $Fe^{3+} + e \rightleftharpoons Fe^{2+};\ E^0 = 0{\cdot}77\ V$

$$Fe^{2+} + 2e \rightleftharpoons Fe;\ E^0 = -0{\cdot}44\ V$$

What will be the E^0 value for the following half cell?

$$Fe^{3+} + 3e \rightleftharpoons Fe$$ (– 0·04 V)

18. From the following values of E^0 drawn from the emf series, calculate standard emf and the equilibrium constant for the reaction,

$$Hg^{2+} + Hg \rightleftharpoons Hg_2^{2+}$$

$E^0_{Hg_2^{2+},\ Hg} = 0{\cdot}788\ V;\ E^0_{Hg^{2+},\ Hg_2^{2+}} = 0{\cdot}92\ V.$ (0·132 V; $1{\cdot}72 \times 10^2$)

19. Will Fe (s) be oxidised to Fe^{2+} by reacting with 1 M HCl?

$$E^0_{Fe,\ Fe^{2+}} = +0{\cdot}44\ V$$ (Yes)

20. A galvanic cell is composed of a standard Zn electrode and a chromium electrode immersed in a solution containing Cr^{3+}. At what concentration of Cr^{3+} will the emf of the cell be zero? (0·1 M)

21. The standard electrode potential corresponding to the reduction $Cr^{3+} + e \rightarrow Cr^{2+}$ is $E^0 = -0{\cdot}407$ volt. If excess Fe (s) is added to a solution in which $[Cr^{3+}] = 1$ M, what will be $[Fe^{2+}]$ when equilibrium is established at 25°C?

$$\{Fe\,(s) + 2Cr^{3+} \rightleftharpoons Fe^{2+} + 2Cr^{2+}\}$$ $\left([Fe^{2+}] = 0{\cdot}42\ M\right)$

22. The emf of a cell consisting of a copper and a lead electrode immersed in 1 M solution of salts of these metals is 0·47 V. Will the emf change if 0·001 M solutions are taken? (No)

23. What is the potential of a hydrogen electrode at pH = 10? (– 0·59 V)

24. We have an oxidation–reduction system:

$$[Fe\,(CN)_6]^{3-} + e \rightleftharpoons [Fe(CN)_6]^{4-};\ E^0 = +0{\cdot}36\ V$$

At what ratio of the concentrations of the oxidised and reduced forms will the potential of the system be 0·28 V? (0·044)

25. Calculate the emf of the following cell at 25°C,

$$\underset{}{Fe}\ |\ \underset{(0{\cdot}1\ M)}{FeSO_4}\ ||\ \underset{(0{\cdot}01\ M)}{CuSO_4}\ |\ Cu$$

Given that E^0 (oxd.) of Fe and Cu are 0·44 V and – 0·34 V respectively. (0·75 V)

26. Calculate the emf of the following cells, find their cell reactions using E^0 values from the table.

(i) Ag | Ag^+ (0·01 M) || Zn^{2+} (0·1 M) | Zn

(ii) Pt | Fe^{2+} (1 M), Fe^{3+} (0·1 M) || Cl^- (0·001 M) | AgCl | Ag

(iii) Zn | ZnO_2^{2-} (0·1 M), OH^- (1 M) | HgO | Hg

In each case, is the reaction as written spontaneous or not?

$$\left\{\begin{array}{l}\text{(i) } Zn^{2+}(0\cdot1) + 2Ag = Zn + 2Ag^+(0\cdot01);\ -1\cdot473 \text{ volt} \\ \text{(ii) } AgCl + Fe^{2+}(1\cdot0) = Ag + Cl^-(0\cdot001) + Fe^{3+}(0\cdot1);\ -0\cdot313 \text{ volt} \\ \text{(iii) } HgO + Zn + 2OH^-(1\cdot0) = Hg + ZnO_2^{2-}(0\cdot1) + H_2O;\ 1\cdot343 \text{ volts}\end{array}\right\}$$

[cell (iii) is spontaneous, (i) and (ii) are not]

27. Neglecting the liquid-junction potential, calculate the emf of the following cell at 25°C.

H_2 (1 atm) | 0·5 M HCOOH || 1 M CH_3 COOH | H_2 (1 atm)

The dissociation constants of HCOOH and CH_3 COOH are $1\cdot77 \times 10^{-4}$ and $1\cdot8 \times 10^{-5}$ respectively. (– 0·0246 volt)

28. The emf of the cell,

(Pt) H_2 | H^+ (c = unknown) || KCl soln. | Hg_2Cl_2 | Hg is 0·4783 V.

If the electrode potential of the calomel chloride is +0·2420 volt (reduction), calculate pH of the solution. (3·994)

29. Calculate the cell potential for the following galvanic cell:

The first electrode consists of Fe^{3+} | Fe^{2+} couple in which $[Fe^{3+}] = 1$ M and $[Fe^{2+}] = [0\cdot1$ M]

The second electrode consists of MnO_4^- | Mn^{2+} couple in acidic solution in which $[MnO_4^-] = 1 \times 10^{-2}$ M, $[Mn^{2+}] = 1 \times 10^{-4}$ M and $[H^+] = 1 \times 10^{-3}$ M.

($E^0_{Fe^{3+}, Fe^{2+}} = 0\cdot771$ V, $E^0_{MnO_4^-, Mn^{2+}} = 1\cdot51$ V)

[**Hint:** Cell reactions are: $Fe^{2+} = Fe^{3+} + e$ at anode

$MnO_4^- + 8H^+ + 5e = Mn^{2+} + 4H_2O$ at cathode] (0·42 V)

30. Find the equilibrium constant at 25°C for the reaction,

$$Hg_2(NO_3)_2 + 2Fe(NO_3)_2 = 2Hg + 2Fe(NO_3)_3$$

$$E^0_{Hg_2^{2+}, Hg} = 0\cdot79 \text{ volt};\ E^0_{Fe^{3+}, Fe^{2+}} = 0\cdot77 \text{ volt}$$

(4·76)

31. Calculate the potential of a silver electrode in a saturated solution of AgBr ($K_{sp} = 6 \times 10^{-13}$) containing, in addition, 0·1 mole per litre KBr.

$$E^0_{Ag^+, Ag} = 0\cdot80 \text{ volt}$$

(0·14 volt)

32. Calculate the potential of a cell in which hydrogen electrode is immersed in pure water, in a solution with a pH of 3·5 and in a solution with a pH of 10·7.

(0·425 volt)

33. A galvanic cell is constructed of two hydrogen electrodes, one immersed in a solution with H^+ at 1 M and the other in 1 M KOH. Calculate E_{cell}. If 1 M KOH solution is replaced by 1 M NH_3, will E_{cell} be higher or lower than in 1 M KOH? (0·83 V, lower)

34. Using the electrochemical series table, explain why Cu (I) sulphate does not exist in aqueous solution.

[**Hint:** Positive E^0 value of $2Cu^+ \rightarrow Cu^{2+} + Cu$ shows that Cu (I) is not stable.]

35. For the cell,

(Pt) H_2 (1 atm) | H^+ (pH unknown) | | H^+ (pH = 1) | H_2 (1 atm). The measured cell potential at 25°C is 0·16 volt. Calculate the unknown pH. (3·7)

36. Write the cell reaction and calculate the potential of the cell,

$$Cl_2\ (p = 0{\cdot}9\ \text{atm})\ |\ \text{NaCl solution}\ |\ Cl_2\ (p = 0{\cdot}1\ \text{atm}).$$

Will the cell reaction be spontaneous as written? (– 0·0282 volt)

37. The emf of a galvanic cell composed of two hydrogen electrodes is 272 mV. What is the pH of the solution in which the anode is immersed if the cathode is in a solution with a pH of 3? (7·6)

38. Calculate the standard emf, standard free energy change and equilibrium constant of a cell in which the following reaction takes place at 25°C.

$$\frac{1}{2}\text{Cu (s)} + \frac{1}{2}Cl_2\ (g) = \frac{1}{2}Cu^{2+} + Cl^-$$

$E^0_{Cl_2, Cl^-} = +\ 1{\cdot}36$ volt; $E^0_{Cu^{2+}, Cu} = +\ 0{\cdot}34$ volt (1·02 volt, – 98·43 kJ, 2×10^{17})

39. What must be the pressure of fluorine gas to produce a reduction potential of 2·75V in a solution that contains 0·38 M F^-? $E^0_{F_2, F^-} = 2{\cdot}87$V. ($1{\cdot}25 \times 10^{-5}$ atm)

40. Show by calculation that $E^0 = -1{\cdot}662$V for the reduction of Al^{3+} to Al(s), regardless of whether the equation for the reaction is written as

(a) $\frac{1}{3}Al^{3+} + e \rightarrow \frac{1}{3}Al(s);\ \Delta G^\circ = 160{\cdot}4\ \text{kJ/mole}$

or (b) $Al^{3+} + 3e \rightarrow Al(s);\ \Delta G^\circ = 481{\cdot}2\ \text{kJ/mole}$

41. If in a galvanic cell, say, Daniell cell, an inert platinum is used instead of a salt bridge, will the cell still produce a potential. (No)

42. Does the physical size of a galvanic cell govern the potential that it will deliver? What does the size affect? (No)

43. Consider the electrochemical cell represented by

$$Mg\ |\ Mg^{2+}\ |\ |\ Fe^{3+}\ |\ Fe^{2+}$$

If 150 mA is to be drawn from this cell for a period of 20 minutes, what is the minimum mass for the magnesium electrode? (0·0224 g)

44. Given the following E^0 values at 25°C, calculate K_{sp} for CdS.

$$Cd^{2+}(aq) + 2e \rightarrow Cd(s); \quad E^0 = -0{\cdot}403V$$
$$CdS(s) + 2e \rightarrow Cd(s) + S^{2-}(aq); \quad E^0 = -1{\cdot}21V$$

Also evaluate ΔG° at 25°C for the process

$$CdS(s) \rightleftharpoons Cd^{2+}(aq) + S^{2-}(aq)$$ ($4{\cdot}90 \times 10^{-28}$, +155·8 kJ /mole)

45. Under standard conditions what is the maximum electrical work, in joules, that a cell employing the cell reaction

$$Cd(s) + 2H^{+}(aq) \rightarrow Cd^{2+}(aq) + H_2(g)$$

can accomplish if 0·780 mole of Cd is consumed?

$$E^0_{Cd^{2+}, Cd} = -0{\cdot}40V, \quad E^0_{H^+, H_2} = 0{\cdot}0 \text{ V}$$

If the work comes with a negative sign, what does it indicate

(–60·2 kJ, negative sign means work is done by the cell)

46. A cell contains 0·04 M Cr^{3+} in one compartment and 1·0 M Cr^{3+} in the other with Cr electrodes in both. Which is the anode compartment? (Dilute solution side)

47. Under standard conditions for all concentrations, the following reaction is spontaneous at 25°C

$$O_2(g) + 4H^{+}(aq) + 4Br^{-}(aq) \rightarrow 2H_2O(l) + 2Br_2(l)$$

If $[H^+]$ is decreased so that the pH = 3·6, what value will E_{cell} have, and will the reaction be spontaneous at this $[H^+]$? Given:

$$O_2 + 4H^{+} + 4e = 2H_2O; \quad E^0 = 1{\cdot}23 \text{ V}$$
$$Br_2 + 2e = 2Br^{-}; \quad E^0 = 1{\cdot}07 \text{ V}$$ (–0·05V; No)

48. An electrode is prepared by dipping a silver strip into a solution saturated with silver thiocyanate, AgSCN, and containing 0·10 M SCN^-. The emf of the voltaic cell constructed by connecting this electrode as the cathode to the standard hydrogen half cell as the anode is 0·45 V. What is the solubility product of AgSCN? (1×10^{-7})

49. Calculate equilibrium constant for the equilibrium,

$$2MnO_4^{-} + 6H^{+} + 5H_2C_2O_4 \rightleftharpoons 2Mn^{2+} + 8H_2O + 10CO_2$$

Given that, $E^0_{MnO_4^-, Mn^{2+}} = 1{\cdot}51V$ and $E^0_{CO_2, C_2O_4^{2-}} = -0{\cdot}49$ V (10^{338})

50. All the energy released from the reaction; $\Delta_r G^0 = -193$ kJ mol^{-1} is used for oxidising M^+ as

$$M^{+} \rightarrow M^{3+} + 2e; \quad E^0 = -0.25 \text{ V}$$

Under standard conditions, the number of moles of M^+ oxidised when one mole of X is converted to y is .... .

(F = 96500 c mol^{-1}) **(IIT 2015 Adv.)** (4)

[**Hint:** x mole M^+ releases $2x$ mole of electrons. Apply $\Delta G° = -nFE^0$, when $n = 2x$.]

51. For the electrochemical cell,

$$Mg(s) \mid Mg^{2+}(aq, 1 M) \parallel Cu^{2+}(aq, 1 M) \mid Cu(s)$$

the standard emf of the cell is 2.70 V at 300 K. When the concentration is changed to x M, the cell potential changes to 2.67 V at 300 K. The value of x is ... .

(Given: $\frac{F}{R} = 11500 \text{ K V}^{-1}$, $\ln(10) = 2.30$) **(IIT 2018 Adv.)** (10)

[**Hint:** $E = E^0 - \frac{RT}{2F} \ln \frac{[Mg^{2+}]}{[Cu^{2+}]}$; $E^0 = 2.70$ V

$[Cu^{2+}] = 1$, Calculate $[Mg^{2+}]$]

52. Consider an electrochemical cell,

$$A(s) \mid A^{n+}(aq, 2M) \parallel B^{2n+}(aq, 1\,M) \mid B(s).$$

The value of ΔH^0 for the cell reaction is twice that of ΔG^0 at 300 K. If the emf of the cell is zero, the ΔS^0(in J K^{-1} mol^{-1}) of the cell reaction per mole of B formed at 300 K is ... .

(Given: ln (2) = 0.7, R = 8.3 J K^{-1} mol^{-1}, H, S and G are enthalpy, entropy and Gibbs energy, respectively.) **(IIT 2018 Adv.)** (–11.62 J K^{-1} mol^{-1})

[**Hint:** $2A(s) + B^{2n+}(aq) \rightarrow 2A^{n+}(aq) + B(s)$

$$E = E^0 - \frac{RT}{2nF} \ln \frac{[A^{n+}]^2}{[B^{2n+}]} = 0$$

$$E^0 = \frac{RT}{2nF} \ln \frac{2^2}{1} = \frac{RT}{2nF} \ln 4$$

$\therefore$ $$\Delta G^\circ = -2nFE^0 = -RT \ln 4$$

$$\Delta G^\circ = \Delta H^\circ - T\,\Delta S^\circ$$

or $$\Delta G^\circ = 2\Delta G^\circ - T\,\Delta S^\circ$$

$$\Delta S^\circ = \frac{\Delta G^\circ}{T} = -\frac{RT \ln 4}{T}.]$$

Objective Problems

1. Which of the following statements is wrong about galvanic cells?
(a) Cathode is the positive electrode.
(b) Cathode is the negative electrode.
(c) Electrons flow from anode to cathode in the external circuit.
(d) Reduction occurs at cathode.

2. In a galvanic cell
(a) chemical reaction produces electrical energy
(b) electrical energy produces chemical reaction
(c) reduction occurs at anode
(d) oxidation occurs at cathode

3. In the galvanic cell Cu | Cu^{2+} (1 M) | | Ag^{+} (1 M) | Ag, the electrons will travel in the external circuit
(a) from Ag to Cu (b) from Cu to Ag
(c) electrons do not travel in the external circuit (d) none of these

4. Is 1 M H^+ solution under hydrogen gas at 1 atm capable of oxidising Ag metal in the presence of 1 M Ag^+?

(a) Yes (b) No

5. The potential of hydrogen electrode is –118 mV. The concentration of H^+ in the solution is

(a) 0·01 M (b) 2 M (c) 10^{-4} M (d) 1 M

6. $E°$ for the half cell $Zn^{2+} \mid Zn$ is – 0·76 V. emf of the cell

$Zn \mid Zn^{2+}(1\ M) \mid\mid 2H^+(1\ M) \mid H_2(1\ atm)$ is

(a) – 0·76 V (b) + 0·76 V (c) – 0·38 V (d) + 0·38 V

7. The standard reduction potentials at 298 K for the following half-reactions are given against each.

$$Zn^{2+}(aq) + 2e = Zn(s);\ -0{\cdot}762\ V$$
$$Cr^{3+}(aq) + 3e = Cr(s);\ -0{\cdot}74\ V$$
$$2H^+ + 2e = H_2(g);\ \pm 0{\cdot}0\ V$$
$$Fe^{3+}(aq) + e = Fe^{2+}(aq);\ +0{\cdot}77\ V$$

Which is the strongest reducing agent?

(a) Zn (s) (b) Cr (s) (c) H_2 (g) (d) Fe^{3+} (aq)

8. The standard reduction potentials, E^0, for the half-reactions are as

$$Zn = Zn^{2+} + 2e;\ E^0 = +0{\cdot}76\ V$$
$$Fe = Fe^{2+} + 2e;\ E^0 = +0{\cdot}41\ V$$

the emf for the cell reaction,

$$Fe^{2+} + Zn = Zn^{2+} + Fe \text{ is}$$

(a) – 0·35 V (b) + 0·35 V (c) + 1·17 V (d) – 1·17 V

(IIT 1988)

9. From the following E^0 values of half cells,

(i) $A + e \rightarrow A^-$; $E^0 = -0{\cdot}24$ V

(ii) $B^- + e \rightarrow B^{2-}$; $E^0 = +1{\cdot}25$ V

(iii) $C^- + 2e \rightarrow C^{3-}$; $E^0 = -1{\cdot}25$ V

(iv) $D + 2e \rightarrow D^{2-}$; $E^0 = +0{\cdot}68$ V

What combination of two half cells would result in a cell with the largest potential?

(a) (ii) and (iii) (b) (ii) and (iv) (c) (i) and (iii) (d) (i) and (iv)

10. From the following E^0 values of half cells

(i) $A \rightarrow A^+ + e$; $E^0 = +1{\cdot}2$ V (ii) $B^- \rightarrow B + e$; $E^0 = -2{\cdot}1$ V

(iii) $C \rightarrow C^{2+} + 2e$; $E^0 = -0.{\cdot}38$ V (iv) $D^{-2} \rightarrow D^- + e$; $E^0 = -0{\cdot}59$ V

What combination of two half cells would result in a cell with the largest potential?

(a) (i) and (iv) (b) (ii) and (iii) (c) (iii) and (iv) (d) (i) and (ii)

11. From the following E^0 values of half cells

(i) $A^{3-} \rightarrow A^{2-} + e$; $E^0 = 1{\cdot}5$ V

(ii) $B^+ + e \rightarrow B$; $E^0 = -0{\cdot}5$ V

(iii) $C^{2+} + e \rightarrow C^+$; $E^0 = +0.5$ V

(iv) $D \rightarrow D^{2+} + 2e$; $E^0 = -1.5$ V

What combination of two half cells would result in a cell with the largest potential?

(a) (i) and (iii) (b) (i) and (iv) (c) (ii) and (iv) (d) (iii) and (iv)

12. If the following half cells have the E^0 values as

$Fe^{3+} + e \rightarrow Fe^{2+}$; $E^0 = +0.77$ V and $Fe^{2+} + 2e \rightarrow Fe$; $E^0 = -0.44$ V; the E^0 of the half cell $Fe^{3+} + 3e \rightarrow Fe$ will be

(a) 0·33 V (b) 1·21 V (c) –0·04 V (d) 0·605 V

13. E^0 (red.) values of the half cells Mg^{2+}/Mg and Cl_2/Cl^- are respectively –2·36 V and + 1·36 V. The E^0 value of the cell $Mg \mid Mg^{2+} \parallel Cl_2 \mid Cl^-$ is

(a) 3·72 V (b) 1 V (c) 0·18 V (d) 2·64 V

14. For the cell reaction $Zn\ (s) + Mg^{2+}\ (1\ M) = Zn^{2+}\ (1\ M) + Mg$, the emf has been found to be 1·60 V. E^0 of the cell is

(a) –1·60 V (b) 1·60 V (c) 0·0 V (d) 0·16 V

15. E^0 for $F_2 + 2e = 2F^-$ is 2·8 V, E^0 for $\frac{1}{2}F_2 + e = F^-$ is

(a) 2·8 V (b) 1·4 V (c) – 2·8 V (d) – 1·4 V

16. $\Delta G°$ of the cell reaction $AgCl\ (s) + \frac{1}{2}H_2\ (g) = Ag\ (s) + H^+ + Cl^-$ is – 21·52 kJ. $\Delta G°$ of $2AgCl\ (s) + H_2\ (g) = 2Ag\ (s) + 2H^+ + 2Cl^-$ is

(a) –21·52 kJ (b) – 10·76 kJ (c) – 43·04 kJ (d) 43·04 kJ

17. The value of equilibrium constant for a feasible cell reaction is

(a) < 1 (b) 0 (c) $= 1$ (d) > 1

18. E^0 for the reaction $Fe + Zn^{2+} = Zn + Fe^{2+}$ is – 0·35 V. The given cell reaction is

(a) feasible (b) not feasible (c) in equilibrium (d) none of these

19. A galvanic cell is composed of two hydrogen electrodes, one of which is a standard one. In which of the following solutions should the other electrode be immersed to get maximum emf?

(a) 0·1 M HCl (b) 0·1 M CH_3COOH

(c) 0·1 M H_3PO_4 (d) 0·1 M H_2SO_4

20. $\frac{1}{2}H_2\ (g) + AgCl\ (s) = H^+\ (aq) + Cl^-\ (aq) + Ag\ (s)$ occurs in the galvanic cell:

(a) $Ag/AgCl\ (s) \mid KCl\ (sol) \mid \mid AgNO_3\ (sol) \mid Ag$

(b) $Pt/H_2\ (g) \mid HCl\ (sol) \mid \mid AgNO_3\ (sol) \mid Ag$

(c) $Pt/H_2\ (g) \mid HCl\ (sol) \mid \mid AgCl\ (s) \mid Ag$

(d) $Pt/H_2\ (g) \mid KCl\ (sol) \mid \mid AgCl\ (s) \mid Ag$ **(IIT 1985)**

21. For the cell $Tl \mid Tl^+\ (0.001\ M) \mid \mid Cu^{2+}\ (0.1\ M) \mid Cu$. E_{cell} at 25°C is 0·83 V which can be increased

(a) by increasing $[Cu^{2+}]$ (b) by increasing $[Tl^+]$

(c) by decreasing $[Cu^{2+}]$ (d) by decreasing $[Tl^+]$

22. How much will the potential of Zn/Zn^{2+} change if the solution of Zn^{2+} is diluted 10 times?
(a) Increase by 0·03 V (b) Decrease by 0·03 V
(c) Increase by 0·059 V (d) Decrease by 0·059 V

23. The half-cell potential of a hydrogen electrode at pH = 10 will be
(a) 0·59 V (b) – 0·59 V (c) 0·059 V (d) – 0·059 V

24. How much will the potential of a hydrogen electrode change when its solution initially at pH = 0 is neutralised to pH = 7?
(a) Increase by 0·059 V (b) Decrease by 0·059 V
(c) Increase by 0·41 V (d) Decrease by 0·41 V

25. A dilute aqueous solution of Na_2SO_4 is electrolysed using platinum electrodes. The products at the anode and cathode are
(a) O_2, H_2 (b) $S_2O_8^{2-}$, Na (c) O_2, Na (d) $S_2O_8^{2-}$, H_2
(IIT 1996)

26. The standard reduction potential of Cu^{2+}/Cu and Cu^{2+}/Cu^{+} are 0·337 and 0·153 respectively. The standard electrode potential of Cu^{+}/Cu half cell is
(a) 0·184 V (b) 0·827 V (c) 0·521 V (d) 0·490 V **(IIT 1997)**

27. Which of the following facts about the chemical cell and concentration cell is correct?
(a) Chemical cell is an electrolytic cell whereas concentration cell is a galvanic cell.
(b) Chemical cell has an overall cell reaction whereas concentration cell has no overall reaction.
(c) Two half cells of both the chemical and concentration cells are chemically different.
(d) E_{cell} equations (Nernst equation) of both the cells have the term E^0_{cell}.

28. The temperature coefficient of a galvanic cell is $+5{\cdot}0 \times 10^{-5}\,VK^{-1}$. During the discharge of the cell, the cell temperature
(a) increases (b) decreases
(c) does not change (d) first increases and then decreases

29. Standard electrode potential data are useful for understanding the suitability of an oxidant in a redox titration. Some half-cell reactions and their standard potentials are given below:

$MnO_4^-(aq) + 8H^+(aq) + 5e \rightarrow Mn^{2+}(aq) + 4H_2O(l)$; $E^0 = 1{\cdot}51$ V
$Cr_2O_7^{2-}(aq) + 14H^+(aq) + 6e \rightarrow 2Cr^{3+}(aq) + 7H_2O(l)$; $E^0 = 1{\cdot}38$ V
$Fe^{3+}(aq) + e \rightarrow Fe^{2+}(aq)$; $E^0 = 0{\cdot}77$ V
$Cl_2(g) + 2e \rightarrow 2Cl^-(aq)$; $E^0 = 1{\cdot}40$ V

Identify the only incorrect statement regarding the quantitative estimation of aqueous $Fe(NO_3)_2$.
(a) MnO_4^- can be used in aqueous HCl
(b) $Cr_2O_7^{2-}$ can be used in aqueous HCl
(c) MnO_4^- can be used in aqueous H_2SO_4
(d) $Cr_2O_7^{2-}$ can be used in aqueous H_2SO_4

30. Using the following Latimer diagram for bromine,

$$\text{pH} = 0;\ BrO_4 \xrightarrow{1.82\ V} BrO_3^- \xrightarrow{1.50\ V} HBrO \xrightarrow{1.595\ V} Br_2 \xrightarrow{1.0652\ V} Br^-$$

the species undergoing disproportionation is

(a) BrO_4 (b) BrO_3^- (c) HBrO (d) Br_2

[**Hint:** If the potential to the left of a given chemical species is less than that to the right, the species will undergo disproportionation]

31. Consider the following cell reaction

$$2Fe(s) + O_2(g) + 4H^+(aq) \rightarrow 2Fe^{2+}(aq) + 2H_2O(l)$$

At $[Fe^{2+}] = 10^{-3}$ M, $p(O_2) = 0.1$ atm and pH = 3, the cell potential at 25°C is

(a) 1.47 V (b) 1.77 V (c) 1.87 V (d) 1.57 V **(IIT 2011)**

32. $AgNO_3$(aq) was added to an aqueous KCl solution gradually and the conductivity of the solution was measured. The plot of conductance (*G*) versus the volume of $AgNO_3$ is

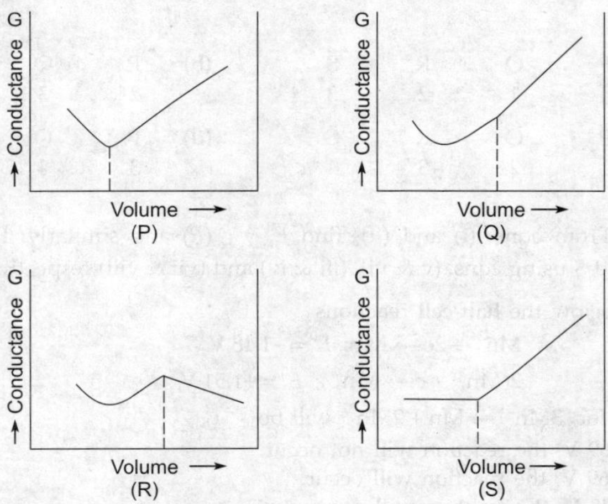

(a) *P* (b) *Q* (c) *R* (d) *S* **(IIT 2011)**

33. Given: $E^0_{Cr^{3+},Cr} = -0.74$ V

$E^0_{MnO_4^-, Mn^{2+}} = 1.51$ V

$E^0_{Cr_2O_7^{2-}, Cr^{3+}} = 1.33$ V

$E^0_{Cl, Cl^-} = 1.36$ V

Based on the data given above strongest oxidising agent will be

(a) Cl (b) Cr^{3+} (c) Mn^{2+} (d) MnO_4^-

(IIT 2013 Main)

34. The standard reduction potential data at 25°C is given below

(i) $E^0(Fe^{3+}, Fe^{2+}) = +0.77$ V

(ii) $E^0(Fe^{2+}, Fe) = -0.44$ V

(iii) $E^0(Cu^{2+}, Cu) = +0.34$ V
(iv) $E^0(Cu^+, Cu) = +0.52$ V
(v) $E^0[O_2(g) + H^+ + 4e \rightarrow 4OH^-] = +1.23$ V
(vi) $E^0[O_2(g) + 2H_2O + 4e \rightarrow 4OH^-] = +0.40$ V
(vii) $E^0(Cr^{3+}, Cr) = -0.74$ V
(viii) $E^0(Cr^{2+}, Cr) = +0.91$ V

Match E^0 of the redox pair in List I with the values given in List II and select the correct answer using the code given below the lists.

	List I		List II
P	$E^0(Fe^{3+}, Fe)$	1	– 0.18 V
Q	$E^0(4H_2O \rightleftharpoons 4H^+ + 4OH^-)$	2	– 0.4 V
R	$E^0(Cu^{2+}, Cu \rightarrow 2Cu^+)$	3	– 0.04 V
S	$E^0(Cr^{3+}, Cr^{2+})$	4	– 0.83 V

Codes

	P	Q	R	S		P	Q	R	S
(a)	4	1	2	3	(b)	2	3	4	1
(c)	1	2	3	4	(d)	3	4	1	2

(IIT 2013 Adv.)

[**Hint:** From eqns. (i) and (ii), find $E^0_{Fe^{3+}, Fe}$ (*P*) and similarly. Find E^0 values of *Q*, *R* and *S* using eqns. (v & vi), (iii & iv) and (vii & viii) respectively. Then match.]

35. Given below the half-cell reactions

$$Mn^{2+} + 2e \rightarrow Mn;\ E^0 = -1.18 \text{ V}$$
$$2(Mn^{3+} + e \rightarrow Mn^{2+});\ E^0 = +1.51 \text{ V}$$

The E^0 for, $3Mn^{2+} \rightarrow Mn + 2Mn^{3+}$ will be
(a) – 2.69 V; the reaction will not occur.
(b) – 2.69 V; the reaction will occur.
(c) – 0.33 V; the reaction will not occur.
(d) – 0.33 V; the reaction will occur. **(IIT 2014 Main)**

36. In a galvanic cell, the salt bridge
(a) does not participate chemically in the reaction.
(b) stops the diffusion of ions from one electrode to other.
(c) is necessary for the occurence of the cell reaction.
(d) ensures mixing of two electrolytic solutions. **(IIT 2014 Adv.)**

37. Galvanisation is applying a coating of
(a) Cr (b) Cu (c) Zn (d) Pb **(IIT 2016 Main)**

38. For the following electrochemical cell at 298 K

$$Pt(s) \mid H_2(g)\ (1\ bar) \mid H^+(aq)\ 1\ M \parallel M^{4+}(aq), M^{2+}(aq) \mid Pt(s)$$

$$E_{cell} = 0.092 \text{ V when } \frac{[M^{2+}]}{[M^{4+}]} = 10^x.$$

Given: $E^0_{M^{4+},M^{2+}} = 0.151$ V, $2.303\dfrac{RT}{F} = 0.059$.

The value of x is

(a) –2 (b) –1 (c) 1 (d) 2 **(IIT 2016 Adv.)**

39. Given: $E^0_{Cl_2,Cl^-} = 1.36$ V, $E^0_{Cr^{3+},Cr} = -0.74$ V

$E^0_{Cr_2O_7^{2-},Cr^{3+}} = 1.33$ V, $E^0_{MnO_4^-,Mn^{2+}} = 1.51$ V

Among the following, the strongest reducing agent is

(a) Cr (b) Mn^{2+}

(c) Cr^{3+} (d) Cl^- **(IIT 2017 Main)**

40. For the following cell

$$Zn(s) \mid ZnSO_4(aq) \mid\mid CuSO_4(aq) \mid Cu(s)$$

when $[Zn^{2+}]$ is ten times of $[Cu^{2+}]$, the expression ΔG(in J mol^{-1}) is

(a) 2.303 RT + 1.1 F (b) 1.1 F

(c) 2.303 RT – 2.2 F (d) 2.2 F **(IIT 2017 Adv.)**

[**Hint:** $\Delta G = \Delta G° + 2.303RT \log \dfrac{[Zn^{2+}]}{[Cu^{2+}]}$

and $\Delta G° = -nFE^0 = -2 \times 1.1\, F$; find ΔG ($E^0_{cell} = 1.1$ V)]

Answers

1-b, **2**-a, **3**-b, **4**-b, **5**-a, **6**-b, **7**-a, **8**-b, **9**-a, **10**-d, **11**-b, **12**-c, **13**-a, **14**-b, **15**-a, **16**-c, **17**-d, **18**-b, **19**-b, **20**-c, **21**-a, d, **22**-b, **23**-b, **24**-d, **25**-a, **26**-c, **27**-b, **28**-b, **29**-a, **30**-c, **31**-d, **32**-d, **33**-d, **34**-d, **35**-a, **36**-a, **37**-c, **38**-d, **39**-a, **40**-c.

❑

CHAPTER NINETEEN

OXIDATION NUMBER AND BALANCING OF REDOX REACTIONS

THE OXIDATION-NUMBER CONCEPT

The oxidation number, or oxidation state, of an atom in a substance is defined as the actual charge of the atom if it exists as a monatomic ion, or a hypothetical charge assigned to the atom in the substance according to some accepted rules.

1. Electrons shared between two unlike atoms are counted to be with the more electronegative atom. Electrons between two like atoms are divided equally between the two atoms. For example, in HCl, the oxidation numbers of H and Cl atoms are +1 and –1 respectively and the oxidation numbers of H in H_2, Cl in Cl_2, etc., are all zero.

2. The oxidation number of an atom in an element in its uncombined state is always zero.

3. The oxidation number of H in a compound is assigned as +1 except in metallic hydrides, where it is –1.

4. The oxidation number of O in compounds is assigned as –2 except in peroxides (H_2O_2, Na_2O_2, BaO_2, etc.), where it is –1; in OF_2, where it is +2 and in superoxide (KO_2), where it is $-\frac{1}{2}$.

5. Fluorine has an oxidation number of –1 in all of its compounds.

6. The oxidation number of halogens in halides is –1; sulphur in sulphides is –2; alkali metals, +1; alkaline earth metals, +2.

7. The algebraic sum of oxidation numbers of all atoms in a compound or ion is equal to the charge on it.

8. In coordination compounds, the ligands, for example, NH_3, CO, NO and H_2O, are neutral. Hence, their oxidation numbers are taken as zero. The oxidation numbers of CH_3, C_6H_5 are +1 and CN, OH, Cl are all –1.

The oxidation-number concept is of great utility to chemistry. In redox reactions, the oxidation number changes. The decrease in oxidation number is called reduction and increase in oxidation number, oxidation. An *oxidising agent* shows a decrease in its oxidation number, and a *reducing agent* shows an increase in its oxidation number.

Oxidation number is not the same as valency. Valency represents the charge in whole numbers with no plus or minus sign, while oxidation number may or may not be equal to valency. It may be positive, negative, zero, a whole number or fractional.

The rules for assigning oxidation numbers are consistent with the requirement of charge balance or the conservation of charge, while the oxidation numbers themselves do not denote actual charges on atoms but rather 'formal charges' which are arbitrarily assigned to atoms to help in balancing chemical equations with respect to stoichiometry and charge.

Calculations for ON of the underlined atoms

$K_2\underline{Cr_2}O_7$ (x) : $2 \times 1 + 2 \times x + 7(-2) = 0; x = +6$

$H_2\underline{S}O_4$ (x) : $2 \times 1 + x + 4(-2) = 0; x = +6$

$\underline{Mn}O_4^-$ (x) : $x + 4(-2) = -1; x = +7$

$\underline{C_6}H_{12}O_6$ (x) : $6 \times x + 12 \times 1 + 6(-2) = 0; x = 0$

$Na_2\underline{B_4}O_7$ (x) : $2 \times 1 + 4 \times x + 7(-2) = 0; x = +3$

$C_6H_5\underline{N}O_2$ (x) : $1 + x + 2(-2) = 0; x = +3$

$\underline{N_3}H$ (x) : $3 \times x + 1 = 0; x = -\frac{1}{3}$

$H_2\underline{S}O_5$ (x) : Two oxygen atoms form peroxide linkage
$2 \times 1 + x + 3(-2) + 2(-1) = 0; x = +6$

$H_2\underline{O}_2$ (x) : $2 \times 1 + 2 \times x = 0; x = -1$

$H_2\underline{S}_2O_8$ (x) : Two oxygen atoms form peroxide linkage
$2 \times 1 + 2 \times x + 6(-2) + 2(-1) = 0; x = +6$

$(\underline{N}H_4)\underline{N}O_3$ (x) (y) : In $\underline{N}H_4^+$: $x + 4 \times 1 = +1; x = -3$
In $\underline{N}O_3^-$: $y + 3(-2) = -1; y = +5$

$(\underline{N}O)\underline{Cl}O_4$ (x) (y) : In $\underline{N}O^+$: $x + 1(-2) = +1; x = +3$
In ClO_4^- : $y + 4(-2) = -1; y = +7$

$\underline{Cr}O_5$ (x) : Four oxygen atoms form peroxide linkage
$x + 1(-2) + 4(-1) = 0; x = +6$

$Ca(O\underline{Cl})\underline{Cl}$ (x) (y) : In $O\underline{Cl}^-$: $1(-2) + x = -1; x = +1$
In $Ca(OCl)\underline{Cl}$: $2 - 1 + y = 0; y = -1$

$\underset{(x)}{\overline{[Fe}}(H_2O)_5(NO)]SO_4$: $x + 5 \times 0 + 1 \times 0 + (-2) = 0; x = +2$

$\underset{(x)}{\overline{[Co}}(NH_3)_5Br]SO_4$: $x + 5 \times 0 + (-1) + (-2) = 0; x = +3$

$\underset{(x)}{\overline{[Sn}}(C_2O_4)_3]^{4-}$: $x + 3(-2) = -4; x = +2$

$\underset{(x)}{\overline{[Fe}}(CN)_6]^{4-}$: $x + 6(-1) = -4; x = +2$

$Na_2\underset{(x)}{\overline{S}}_2$: $2 \times 1 + 2x = 0; x = -1$

$(CH_3)_2\underset{(x)}{\overline{S}}O$: $2 \times 1 + x - 2 = 0; x = 0$

$\underset{(x)}{\overline{Ni}}(CO)_4$: $x + 4 \times 0 = 0; x = 0$

$(\underset{(x)}{\overline{N}}_2H_5)_2SO_4$: $2(2x + 5 \times 1) + (-2) = 0; x = -2$

$Ba(H_2\underset{(x)}{\overline{P}}O_2)_2$: $2 + 2(2 \times 1 + x + 2(-2)) = 0; x = +1$

$\underset{(x)}{\overline{N}}H_2 \cdot \underset{(x)}{\overline{N}}H_2$: $x + 2 + x + 2 = 0; x = -2$

Formula-Unit Equation and Net Ionic Equation

The equation to be balanced may be given in the form of an ionic equation as well as a formula-unit equation. One should thus know how to convert a formula-unit equation to an ionic equation and vice versa.

Consider a formula-unit equation in aqueous medium:

$$10FeSO_4 + 2KMnO_4 + 8H_2SO_4 = 5Fe_2(SO_4)_3 + 2MnSO_4 + K_2SO_4 + 8H_2O$$

Writing in total ionic form,

$$10(Fe^{2+} + SO_4^{2-}) + 2(K^+ + MnO_4^-) + 8(2H^+ + SO_4^{2-})$$
$$= 5(2Fe^{3+} + 3SO_4^{2-}) + 2(Mn^{2+} + SO_4^{2-}) + (2K^+ + SO_4^{2-}) + 8H_2O$$

Note: Insoluble and nonelectrolytic substances are written in formula-unit form.

Removing the cationic **'spectator'** ion, K^+, and the anionic **'spectator'** ion, SO_4^{2-}, from both sides of the equation, we get the net ionic equation.

$$10Fe^{2+} + 2MnO_4^- + 16H^+ = 10Fe^{3+} + 2Mn^{2+} + 8H_2O$$

Note: The ions which do not take part in the reaction are called **'spectator'** ions.

Consider another example:

Formula-unit equation: $2AgNO_3 + Cu = Cu(NO_3)_2 + 2Ag$

Total ionic equation: $2(Ag^+ + NO_3^-) + Cu = (Cu^{2+} + 2NO_3^-) + 2Ag$

Net ionic equation: $2Ag^+ + Cu = Cu^{2+} + 2Ag$

Here, NO_3^- ions are 'spectator' ions and hence, are cancelled from both sides of the equation.

Balancing of Redox Reactions

The rules for assigning oxidation numbers are constructed so that in all redox reactions,

'the total increase in oxidation numbers must equal the total decrease in oxidation numbers.'

This equivalence provides the basis for balancing redox reactions. All balanced equations must satisfy two criteria.

1. **Mass balance,** that is, there should be the same number of atoms of each kind in reactants and products.

2. **Charge balance,** that is, the sums of actual charges on both sides of the equation must equal each other.

Although there is no single 'best method' for balancing all redox reactions, the following methods are particularly useful.

The Half-Reaction Method

In this method the two equations describing oxidation and reduction in the redox reaction are separated and completely balanced. The number of electrons gained and lost in each half-reaction are equalised and finally the half-reactions are added to give the overall balanced equation. This method is generally applied in electrochemistry and hence is also known as the **half-cell method.** The use of half-reaction permits us to balance equations using only the principles of atom and charge conservation.

The following examples illustrate the half-reaction method:

Ex. 1. *Balance:* $I_2 + S_2O_3^{2-} = I^- + S_4O_6^{2-}$

First, separate oxidation and reduction processes

$$S_2O_3^{2-} \rightarrow S_4O_6^{2-} \text{ (oxd.)}$$

$$I_2 \rightarrow I^- \text{ (red.)}$$

Now, balance the number of each atom like I and S, undergoing reduction and oxidation and then add electrons to the electrically positive side of the equation to balance the charge on both sides of the equation.

$$2S_2O_3^{2-} \rightarrow S_4O_6^{2-} + 2e$$

$$I_2 + 2e \rightarrow 2I^-$$

Each balanced half-reaction involves two electrons. Add them. Electrons get cancelled.

$$I_2 + 2S_2O_3^{2-} = S_4O_6^{2-} + 2I^-$$

The formula-unit equation may now be written as

$$I_2 + 2Na_2S_2O_3 = Na_2S_4O_6 + 2NaI$$

Here, Na^+ ions are 'spectator' ions.

Note: In this example, oxygen atoms are automatically balanced. In case they are not, the rules given in the 'Oxidation-Number-Change Method' have to be followed to balance atoms other than those undergoing a change in oxidation number.

Ex. 2. *Balance:* $C_6H_5CHO + Cr_2O_7^{2-} \rightarrow C_6H_5COOH + Cr^{3+}$

First, separate oxidation and reduction processes,

$$C_6H_5CHO \rightarrow C_6H_5COOH \quad \text{(oxidation)}$$

$$Cr_2O_7^{2-} \rightarrow Cr^{3+} \quad \text{(reduction)}$$

Now, balance atoms (see Rule 5 below)

$$C_6H_5CHO + H_2O = C_6H_5COOH + 2H^+$$

$$14H^+ + Cr_2O_7^{2-} = 2Cr^{3+} + 7H_2O$$

Then add electrons to the electrically positive side to balance the charge.

$$C_6H_5CHO + H_2O = C_6H_5COOH + 2H^+ + 2e^-$$

$$6e^- + 14H^+ + Cr_2O_7^{2-} = 2Cr^{3+} + 7H_2O$$

Multiply the first equation by 3 to make the number of electrons involved in both the half-reactions equal and then add them.

$$3C_6H_5CHO + Cr_2O_7^{2-} + 8H^+ = 3C_6H_5COOH + 2Cr^{3+} + 4H_2O$$

Oxidation-Number-Change Method

This method is based on the fact that in a balanced redox reaction, the total increase in oxidation number is equal to the total decrease in oxidation number. The following **rules** are followed.

1. Assign oxidation numbers to elements that undergo changes in oxidation numbers.
2. First, balance only those atoms of elements that undergo oxidation-number changes and write the oxidation number for all the atoms balanced
3. Connect the atoms of the element undergoing oxidation by drawing a bracket and do the same for the reduction process. Show increase in oxidation number for oxidation process and decrease in oxidation number for reduction process.
4. Insert coefficients into the equation to make the total increase and decrease in oxidation numbers equal.

5. Now balance the atoms of other elements except H and O by inspection. H and O atoms in aqueous solutions are balanced in the following way:

(a) **Acidic Medium:** For one excess oxygen atom, add one H_2O on the other side and two H^+ on the same side.

(b) **Alkaline Medium:** For one excess oxygen atom, add one H_2O on the same side and two OH^- on the other side. In case H-atoms are still unbalanced, for one excess H-atom, add one OH^- on the same side and one H_2O, on the other side.

(c) In some cases both H and O atoms are balanced together by adding H_2O only.

In addition to the above rules, some tricks are applied in some typical reactions to balance them.

Ex. 3. *Balance the ionic equation:* $Fe^{2+} + MnO_4^- \rightarrow Fe^{3+} + Mn^{2+}$ *(acidic medium) and convert it to a balanced formula-unit equation.*

Step I: Apply rules 1 to 3.

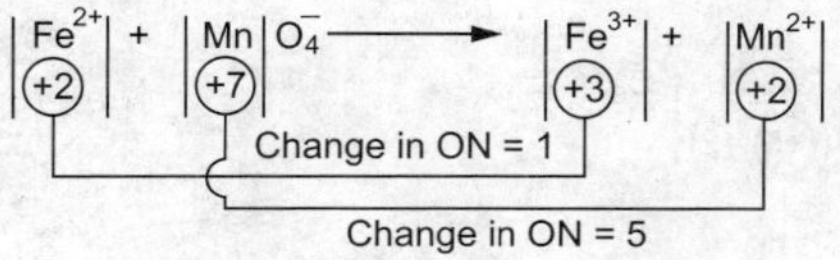

Step II: Apply Rule 4.

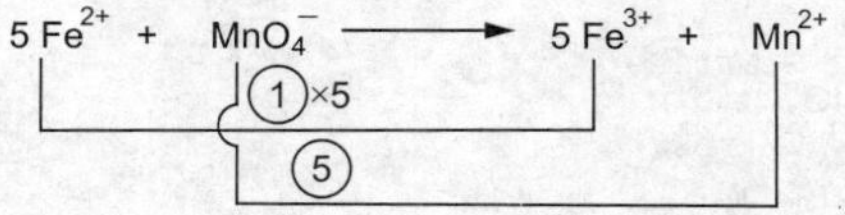

Step III: Apply Rule 5 (a).

$$5Fe^{2+} + MnO_4^- + 8H^+ \rightarrow 5Fe^{3+} + Mn^{2+} + 4H_2O$$

To change the ionic equation to formula-unit equation, add 'spectator' ions, that is, SO_4^{2-} ions and K^+ ions, in both sides to neutralise the charge.

$$5FeSO_4 + KMnO_4 + 4H_2SO_4 \rightarrow \frac{5}{2} Fe_2(SO_4)_3 + MnSO_4 + 4H_2O$$

or, $10FeSO_4 + 2KMnO_4 + 8H_2SO_4 \rightarrow 5Fe_2(SO_4)_3 + 2MnSO_4 + 8H_2O + K_2SO_4$

Ex. 4. *Balance:* $Cu_2S + SO_4^{2-} \rightarrow Cu^{2+} + SO_3^{2-}$ *(acidic medium)*

Step I: As both Cu_2S and SO_4^{2-} oxidise and reduce respectively to SO_3^{2-}, add one more SO_3^{2-} on the right side and apply rules 1 to 3.

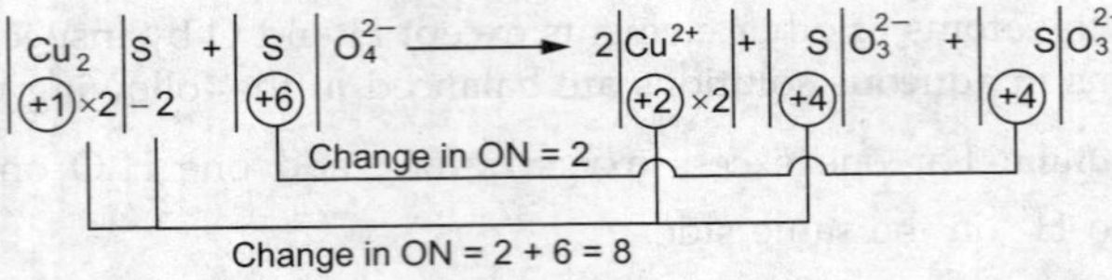

Step II: Apply Rule 4.

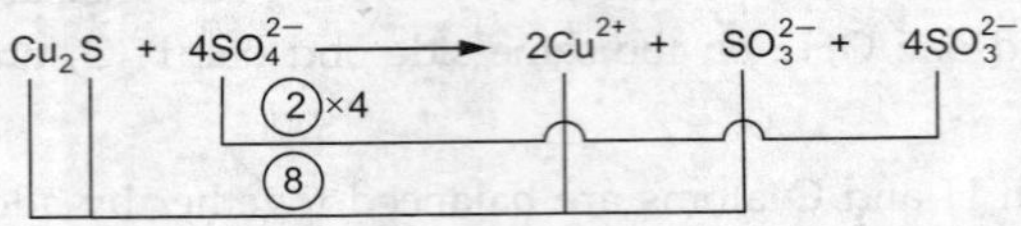

Step III: Apply Rule 5 (a).

$$Cu_2S + 4SO_4^{2-} + 2H^+ \rightarrow 2Cu^{2+} + 5SO_3^{2-} + H_2O$$

Ex. 5. *Balance:* $I_2 \rightarrow I^- + IO_3^-$ *(alkaline medium)*

Step I: As I_2 oxidises to IO_3^- and reduces to I^-, add one more I_2 on the left side of the equation to separate oxidation and reduction processes. Now apply rules 1 to 3.

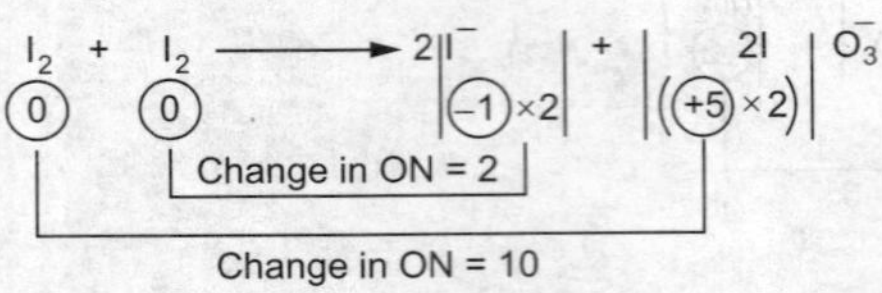

Step II: Apply Rule 4.

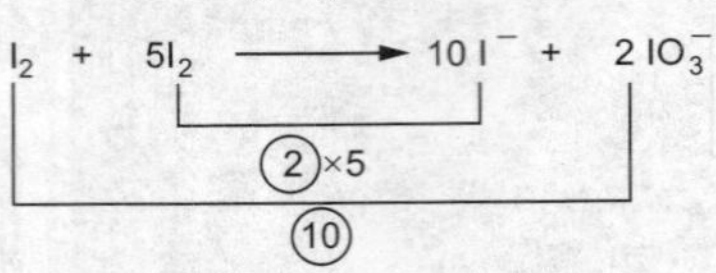

Step III: Apply Rule 5 (b).

$$6I_2 + 12OH^- \rightarrow 10I^- + 2IO_3^- + 6H_2O$$

Ex. 6. *Balance:* $P_4 \rightarrow H_2PO_2^- + PH_3$ *(alkaline medium)*

Step I: As P_4 oxidises as well as reduces, write one more P_4 on the left side of the equation. Now apply rules 1 to 3.

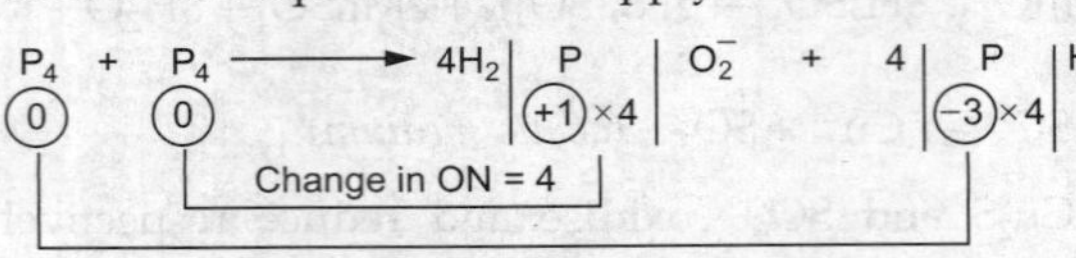

Step II: Apply Rule 4.

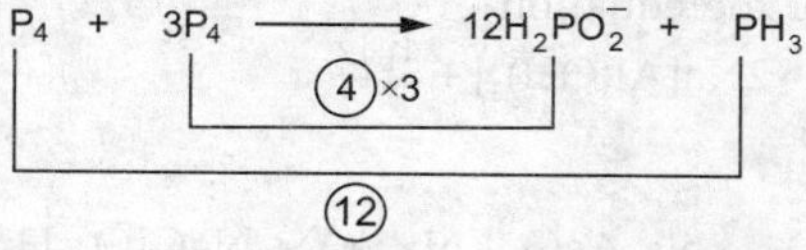

Step III: Apply Rule 5(b)

$$4P_4 + 12OH^- + 12H_2O \rightarrow 12H_2PO_2^- + 4PH_3$$

Ex. 7. *Balance:* $Cu + HNO_3 \rightarrow Cu(NO_3)_2 + NO + H_2O$

Step I: Apply rules 1 to 3.

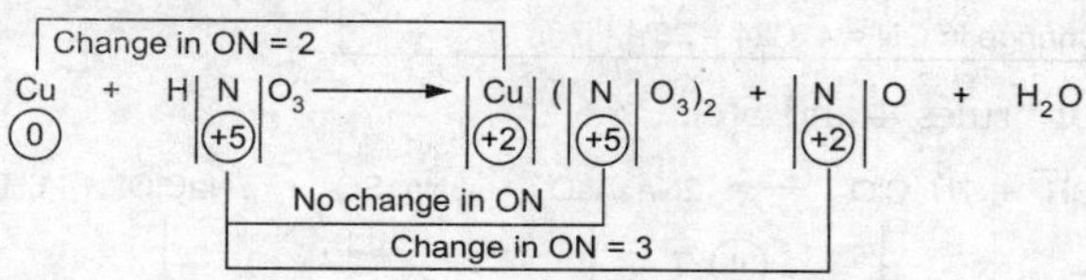

Step II: Apply Rule 4.

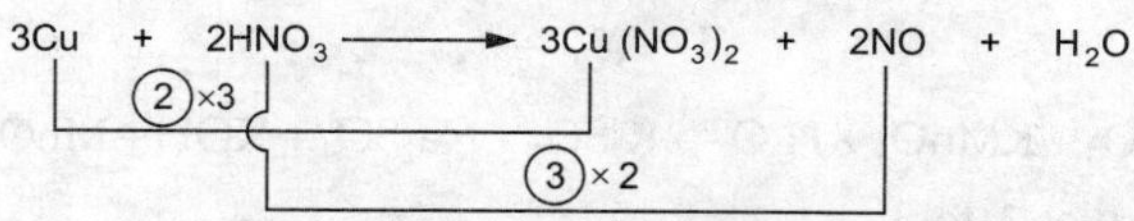

As all HNO_3 molecules do not reduce to NO, a part of it changes to $Cu(NO_3)_2$. Thus to balance 6 excess NO_3^- ions on the right side of the equation, add $6HNO_3$ on the left side.

$$3Cu + 8HNO_3 \rightarrow 3Cu(NO_3)_2 + 2NO + H_2O$$

Step III: Apply Rule 5(c).

$$3Cu + 8HNO_3 \rightarrow 3Cu(NO_3)_2 + 2NO + 4H_2O$$

Note: Out of $8NO_3^-$ ions, 6 are 'spectator' ions

Ex. 8. *Balance:* $NaOH + Al + H_2O \rightarrow Na[Al(OH)_4] + H_2$

Step I: Apply rules 1 to 3.

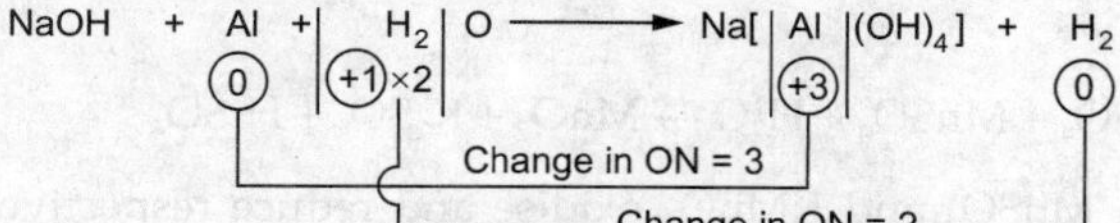

In this reaction H_2O is reduced to H_2 and Al is oxidised.

Step II: Apply Rule 4.

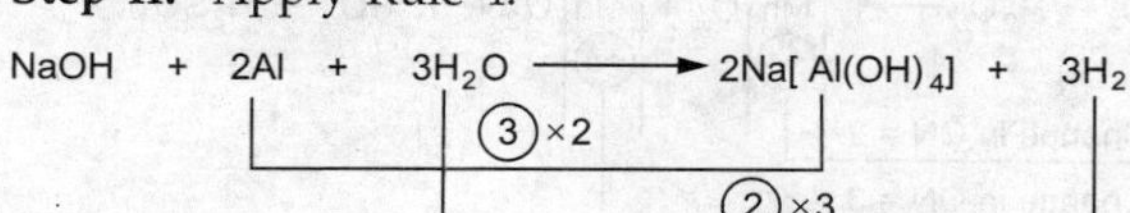

Step III: Apply Rule 5. First balance Na atoms. Then to balance H and O, add $3H_2O$ on the left side of the equation

$$2NaOH + 2Al + 6H_2O \rightarrow 2Na[Al(OH)_4] + 3H_2$$

Ex. 9. *Balance:*

$$As_2S_3 + NaOH + NaClO_3 \rightarrow Na_3AsO_4 + Na_2SO_4 + NaClO + H_2O$$

Step I: Apply rules 1 to 3.

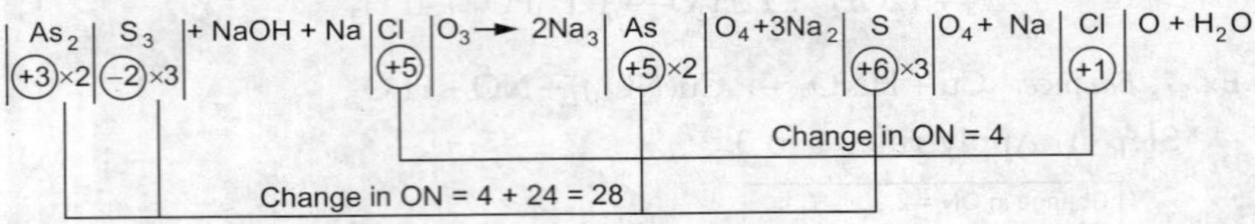

Step II: Apply rules 4 and 5(c).

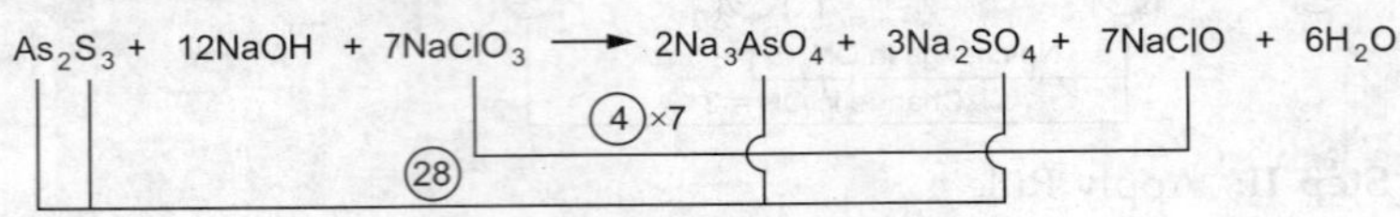

Ex. 10. *Balance:*

$$Na_2S_2O_3 + KMnO_4 + H_2O \rightarrow K_2SO_4 + Na_2SO_4 + KOH + MnO_2$$

Step I: Apply rules 1 to 3.

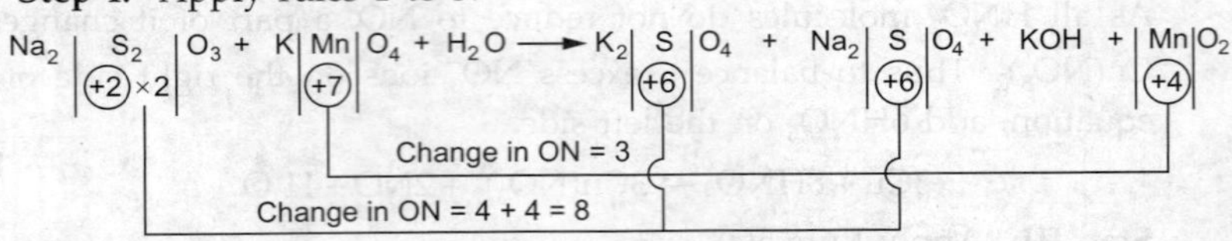

Step II: Apply rules 4 and 5(c) after balancing K.

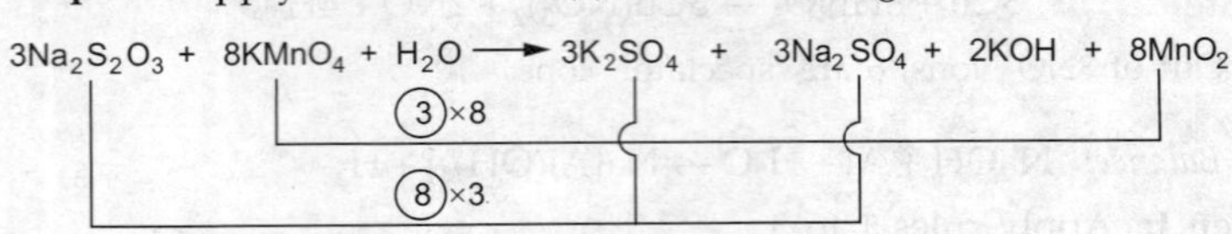

Ex. 11. *Balance:*

$$KMnO_4 + MnSO_4 + H_2O \rightarrow MnO_2 + K_2SO_4 + H_2SO_4$$

Step I: As both $MnSO_4$ and $KMnO_4$ oxidise and reduce respectively to MnO_2, add one more MnO_2 to the right side. Now apply rules 1 to 3.

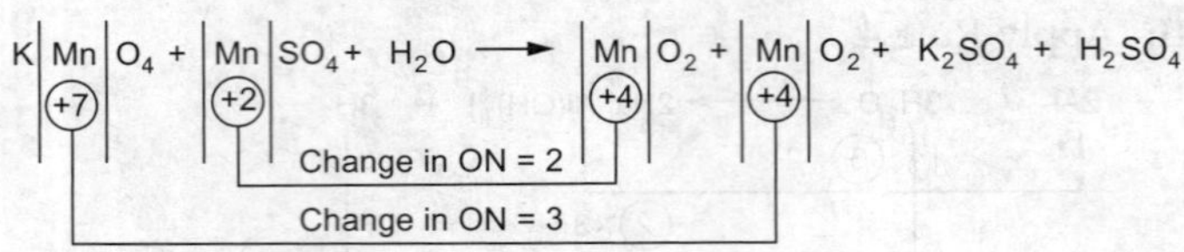

Step II: Apply Rule 4.

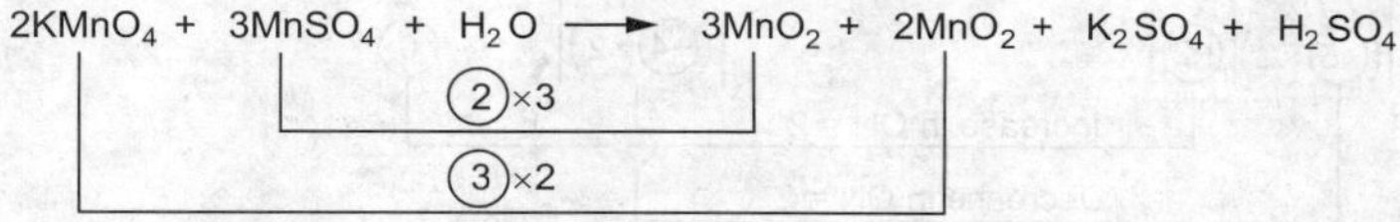

Step III: K atoms are balanced. Now balance SO_4 as a whole and then apply Rule 5(c) to assign the coefficient to H_2O.

$$2KMnO_4 + 3MnSO_4 + 2H_2O \rightarrow 5MnO_2 + K_2SO_4 + 2H_2SO_4$$

Ex. 12. *Balance:*

$$KClO_3 + H_2SO_4 \rightarrow KHSO_4 + HClO_4 + ClO_2 + H_2O$$

Step I: As $KClO_3$ disproportionates to $HClO_4$ and ClO_2, add one more $KClO_3$ on the left side and then apply rules 1 to 3.

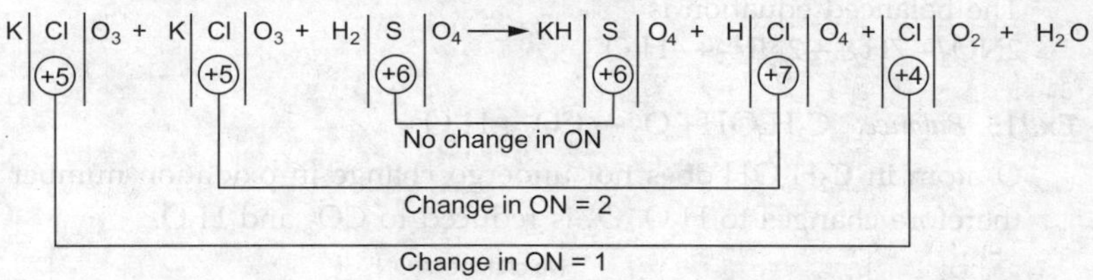

Step II: Apply Rule 4.

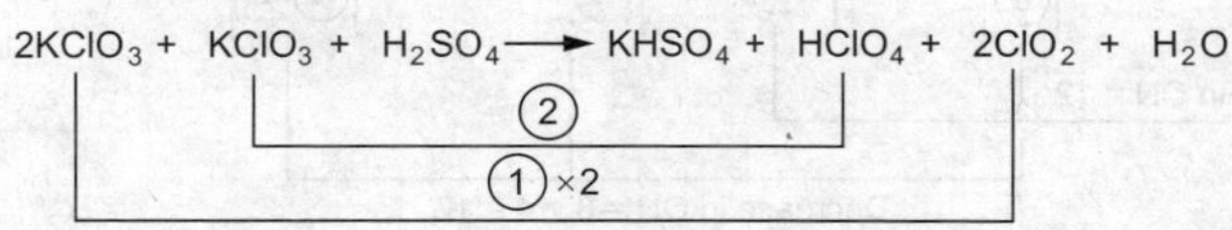

Step III: First balance K and then SO_4 as a whole. Now apply Rule 5(c) to assign the coefficient to H_2O.

$$3KClO_3 + 3H_2SO_4 \rightarrow 3KHSO_4 + HClO_4 + 2ClO_2 + H_2O$$

The said rules 1 to 5 are very useful in balancing the majority of redox equations. However, there are some problems where the application of the said rule sometimes becomes tedious. In such cases, the atoms of all the elements are balanced by inspection only i.e., by checking that the increase in oxidation number equals the decrease in oxidation number (Rule 4). The following examples illustrate it.

Ex. 13. *Balance:* $Cu(NO_3)_2 \rightarrow CuO + NO_2 + O_2$

Cu does not undergo change in oxidation number. Only N and O (but not all O atoms) undergo change in oxidation number. Thus balance N and O atoms by inspection so that increase in oxidation number is equal to decrease in oxidation number, both being equal to 2 as shown in the following figure.

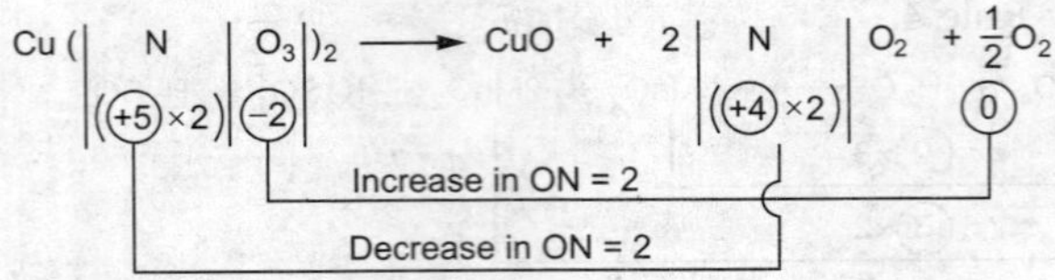

The balanced equation is

$$2Cu(NO_3)_2 \rightarrow 2CuO + 4NO_2 + O_2$$

Ex. 14. *Balance:* $NO_2 + H_2 \rightarrow NH_3 + H_2O$

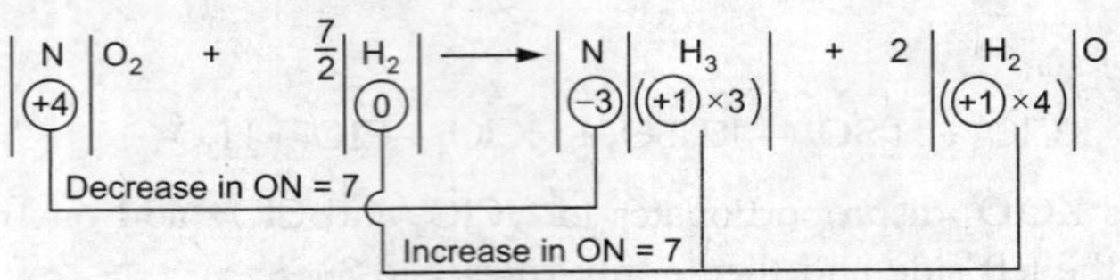

The balanced equation is

$2NO_2 + 7H_2 \rightarrow 2NH_3 + 4H_2O$

Ex. 15. *Balance:* $C_2H_5OH + O_2 \rightarrow CO_2 + H_2O$

O atom in C_2H_5OH does not undergo change in oxidation number and therefore changes to H_2O. O_2 is reduced to CO_2 and H_2O.

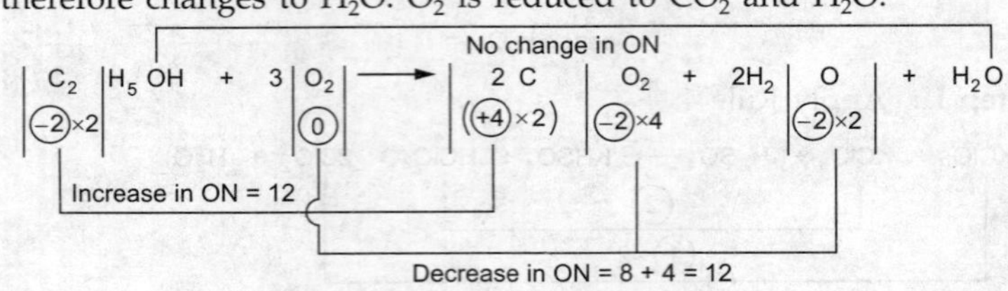

Ex. 16. *Balance:* $FeS_2 + O_2 \rightarrow Fe_2O_3 + SO_2$

Step I: As all the three elements, viz., Fe, S and O, undergo change in oxidation number, balance them in the order of Fe, S and then O. Now, assign oxidation numbers to all the atoms.

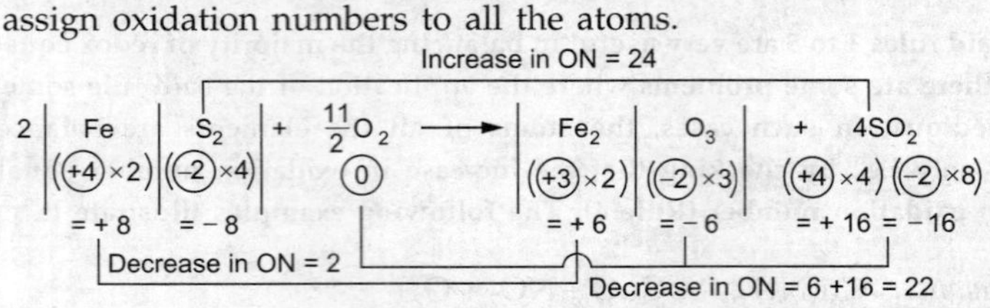

The equation is now balanced which can also be verified by seeing that increase in oxidation number is equal to decrease in oxidation number, both being equal to 24.

Thus, the balanced equation is

$$4FeS_2 + 11O_2 \rightarrow 2Fe_2O_3 + 8SO_2$$

PROBLEMS

Assign ON to atoms of only those elements which undergo ON change in the following redox reactions, and then balance the equations.

1. $C_2H_5OH + Cr_2O_7^{2-} + H^+ = Cr^{3+} + C_2H_4O + H_2O$
2. $Sn(OH)_3^- + Bi(OH)_3 + OH^- = Sn(OH)_6^{2-} + Bi$
3. $IO_3^- + N_2H_4 + HCl = N_2 + ICl_2^- + H_2O$
4. $NO_2 + OH^- = NO_3^- + NO_2^- + H_2O$
5. $Hg_2Cl_2 + NH_3 = Hg + HgNH_2Cl + NH_4Cl$
6. $Zn + NO_3^- + H^+ = Zn^{2+} + NH_4^+ + H_2O$
7. $I_2 + NO_3^- + H^+ = IO_3^- + NO_2 + H_2O$
8. $MnO_4^- + SO_2^{2-} + H_2O = MnO_2 + SO_4^{2-} + OH^-$
9. $H_2O_2 + ClO_2 + OH^- = ClO_2^- + O_2 + H_2O$
10. $ClO^- + CrO_2^- + OH^- = Cl^- + CrO_4^{2-} + H_2O$
11. $I_2 + Cl_2 + H_2O = HIO_3 + HCl$
12. $Cl_2 + KOH = KOCl + KCl + H_2O$
13. $Cl_2 + KOH = KClO_3 + KCl + H_2O$
14. $H_2O_2 + I_2 = HIO_3 + H_2O$
[**Hint:** Change in ON = 10]
15. $H_2O_2 + KMnO_4 = MnO_2 + KOH + O_2 + H_2O$
16. $HNO_2 + KMnO_4 + H_2SO_4 = HNO_3 + KMnO_4 + K_2SO_4 + H_2O$
17. $NaNO_2 + NaI + H_2SO_4 = NO + I_2 + Na_2SO_4 + H_2O$
18. $N_2H_4 + AgNO_3 + KOH = N_2 + Ag + KNO_3 + H_2O$
19. $N_2H_4 + Zn + KOH + H_2O = NH_3 + K_2[Zn(OH)_4]$
20. $Fe + N_2H_4 + H_2O = Fe(OH)_2 + NH_3$
21. $H_2S + HNO_3 = NO + S + H_2O$
22. $P + HNO_3 = HPO_3 + NO + H_2O$
23. $K_2Cr_2O_7 + HCl = KCl + CrCl_3 + H_2O + Cl_2$
24. $MnO_4^- + C_2O_4^{2-} + H^+ = CO_2 + Mn^{2+} + H_2O$
25. $Cr_2O_7^{2-} + C_2O_4^{2-} + H^+ = Cr^{3+} + CO_2 + H_2O$
26. $Cr(OH)_3 + IO_3 + OH^- = I^- + CrO_4^{2-} + H_2O$

27. $KMnO_4 + H_2S + H_2SO_4 = KHSO_4 + MnSO_4 + S + H_2O$

28. $NO_3^- + Cl^- + H^+ = NO + Cl_2 + H_2O$

29. $H_2O_2 + I^- + H^+ = I_2 + H_2O$

30. $Cr_2O_7^{2-} + NO_2^- + H^+ = Cr^{3+} + NO_3^- + H_2O$

31. $N_2O_4 + BrO_3^- + H_2O = NO_3^- + Br^- + H^+$

32. $S_2O_3^{2-} + Sb_2O_5 + H^+ + H_2O = SbO + H_2SO_3$

33. $Fe_2(SO_4)_3 + Fe = FeSO_4$

34. $Cu(NH_3)_4Cl_2 + KCN + H_2O = K_2Cu(CN)_3 + NH_3 + KCNO + NH_4Cl + KCl$

35. $Ag + KCN + H_2O + O_2 = KAg(CN)_2 + KOH$

36. $Zn + NO_3^- + H^+ = Zn^{2+} + NO_2 + H_2O$

37. $MnO_4^- + CN^- + H_2O = MnO_2 + CNO^- + OH^-$

38. $AsO_3^{3-} + IO_3^- = AsO_4^{3-} + I^-$

39. $Fe_3O_4 + MnO_4^- + H_2O = Fe_2O_3 + MnO_2 + OH^-$

40. $H_2S + Cr_2O_7^{2-} + H^+ = Cr^{3+} + S_8 + H_2O$

41. $ZnS + O_2 = ZnO + SO_2$

42. $KNO_3 + FeSO_4 + H_2SO_4 = KHSO_4 + Fe_2(SO_4)_3 + NO + H_2O$

43. $H_2S + K_2Cr_2O_7 + H_2SO_4 = KHSO_4 + Cr_2(SO_4)_3 + S + H_2O$

44. $KI + H_2SO_4 = KHSO_4 + SO_2 + I_2 + H_2O$

45. $C_2H_5OH + MnO_4^- + OH^- = C_2H_3O^- + MnO_2 + H_2O$

46. $Al + KMnO_4 + H_2SO_4 = KHSO_4 + Al_2(SO_4)_3 + MnSO_4 + H_2O$

47. $K_2Cr_2O_7 + H_2O + S = SO_2 + KOH + Cr_2O_3$

48. $MnO_2 + HCl = MnCl_2 + Cl_2 + H_2O$

49. $MnO_4^- + SO_3^{2-} + H_2O = MnO_2 + SO_4^{2-} + OH^-$

50. $Cr_2O_7^{2-} + SO_3^{2-} + H^+ = Cr^{3+} + SO_4^{2-} + H_2O$

51. $I_2 + SO_2 + H_2O = SO_4^{2-} + I^- + H^+$

52. $Sn + NO_3^- + H^+ = SnO_2 + NO_2 + H_2O$

53. $MnO_4^- + SO_2 + H_2O = Mn^{2+} + SO_4^{2-} + H^+$

54. $MnO_4^- + SO_3^{2-} + OH^- = MnO_4^{2-} + SO_4^{2-} + H_2O$

55. $ClO^- + Br^- = BrO_3^- + Cl^-$

56. $Zn + NO_3^- + H^+ = Zn^{2+} + NH_4^+ + H_2O$

57. $KMnO_4 + HCl = Cl_2 + KCl + MnCl_2 + H_2O$

58. $BaCrO_4 + KI + HCl = BaCl_2 + I_2 + KCl + CrCl_3 + H_2O$

59. $ClO_3^- + SO_2 + H^+ = ClO_2 + HSO_4^-$

60. $Mn^{2+} + S_2O_8^{2-} + H_2O = MnO_4^- + HSO_4^- + H^+$

61. $Cl_2 + IO_3^- + OH^- = Cl^- + IO_4^- + H_2O$

62. $H_2SO_3 + Cr_2O_7^{2-} + H^+ = HSO_4^- + Cr^{3+} + H_2O$

63. $ClO_2 + SbO_2^- + OH^- = ClO_2^- + Sb(OH)_6^- + H_2O$

64. $Zn + NO_3^- + OH^- = NH_3 + ZnO_2^{2-} + H_2O$

65. $AsO_3^{3-} + MnO_4^- = AsO_4^{3-} + MnO_2 + H_2O$

66. $KMnO_4 + H_2O_2 + H_2SO_4 = MnSO_4 + K_2SO_4 + O_2 + H_2O$

67. $H_2O_2 + PbS = PbSO_4 + H_2O$

68. $MnO_4^- + C_2O_4^{2-} + H^+ = Mn^{2+} + CO_2 + H_2O$

69. $Fe^{2+} + Cr_2O_7^{2-} + H^+ = Fe^{3+} + Cr^{3+} + H_2O$

70. $AsO_3^{3-} + I_2 + H_2O = AsO_4^{3-} + H^+ + I^-$

71. $S_2O_3^{2-} + I_2 = S_4O_6^{2-} + I^-$

72. $HNO_3 + I_2 = HIO_3 + NO_2 + H_2O$

73. $K_2Cr_2O_7 + H_2SO_4 + KI = I_2 + K_2SO_4 + Cr_2(SO_4)_3 + H_2O$

74. $CuO + NH_3 = Cu + N_2 + H_2O$

75. $As_2S_5 + HNO_3 = H_3AsO_4 + H_2SO_4 + NO_2 + H_2O$

76. $PbS + O_3 = PbSO_4 + O_2$

77. $Cl_2 + SeO_3^{2-} + H_2O = SeO_4^{2-} + Cl^- + H^+$

78. $Cu_3P + H^+ + Cr_2O_7^{2-} = Cu^{2+} + H_3PO_4 + Cr^{3+} + H_2O$

79. $Na_2SnO_2 + Bi(OH)_3 = Bi + Na_2SnO_3 + H_2O$

80. $H_2O + SbCl_3 = SbOCl + 2HCl$

81. $K_4Fe(CN)_6 + Ce(NO_3)_4 + KOH = Ce(OH)_3 + Fe(OH)_3 + H_2O + K_2CO_3 + KNO_3$

82. $S + OH^- = S^{2-} + S_2O_3^{2-} + H_2O$

83. $IO_4^- + I^- + H^+ = I_2 + H_2O$

84. $KMnO_4 = K_2MnO_4 + MnO_2 + O_2$
[**Hint:** $KMnO_4$ reduces to both K_2MnO_4 and MnO_2 and change in ON = 4]

85. $(NH_4)_2Cr_2O_7 = N_2 + H_2O + Cr_2O_3$

86. $MnO_4^{2-} + H^+ = MnO_4^- + MnO_2 + H_2O$

87. $H_3PO_3 = H_3PO_4 + PH_3$

88. $Zn + HNO_3 = Zn(NO_3)_2 + N_2O + H_2O$

89. $CuS + NO_3^- + H^+ = Cu^{2+} + S_8 + NO + H_2O$

90. $CuSO_4 + KI = Cu_2I_2 + I_2 + K_2SO_4$

91. $FeSO_4 = Fe_2O_3 + SO_2 + SO_3$

92. $NaOH + Cl_2 = NaCl + NaClO + H_2O$

93. $NH_3 + Hg_2Cl_2 = Hg + Hg(NH_2)Cl + HCl$
[**Hint:** Change in ON = 1]

94. $Br_2 + OH^- = BrO^- + Br^- + H_2O$

95. $As_2S_5 + HNO_3 = H_2SO_4 + NO_2 + H_3AsO_4 + H_2O$

96. $Cr_2O_7^{2-} + SO_2 + H^+ = Cr^{3+} + HSO_4^- + H_2O$

97. $HCl + WO_3 + SnCl_2 = H_2SnCl_6 + W_3O_8 + H_2O$

98. $HCl + FeCl_3 + V(OH)_4Cl = VOCl_2 + H_2O + FeCl_3$

99. $Au + KCN + H_2O + O_2 = KAu(CN)_4 + KOH$

100. $KOH + KMnO_4 = K_2MnO_4 + O_2 + H_2O$

Answers

1. $3C_2H_5OH + Cr_2O_7^{2-} + 8H^+ = 2Cr^{3+} + 3C_2H_4O + 7H_2O$
2. $Sn(OH)_3^- + 2Bi(OH)_3 + 3OH^- = 3Sn(OH)_6^{2-} + 2Bi$
3. $IO_3^- + N_2H_4 + 2HCl = N_2 + ICl_2^- + 3H_2O$
4. $2NO_2 + 2OH^- = NO_3^- + NO_2^- + H_2O$
5. $2Hg_2Cl_2 + 4NH_3 = 2Hg + 2HgNH_2Cl + 2NH_4Cl$
6. $4Zn + NO_3^- + 10H^+ = 4Zn^{2+} + NH_4^+ + 3H_2O$
7. $I_2 + 10NO_3^- + 8H^+ = 2IO_3^- + 10NO_2 + 4H_2O$
8. $2MnO_4^- + 3SO_3^{2-} + H_2O = 2MnO_2 + 3SO_4^{2-} + 2OH^-$
9. $H_2O_2 + 2ClO_2 + 2OH^- = 2ClO_2^- + O_2 + 2H_2O$
10. $3ClO^- + 2CrO_2^- + 2OH^- = 3Cl^- + 2CrO_4^{2-} + H_2O$

11. $I_2 + 5Cl_2 + 6H_2O = 2HIO_3 + 10HCl$

12. $2Cl_2 + 4KOH = 2KOCl + 2KCl + 2H_2O$

13. $3Cl_2 + 6KOH = KClO_3 + 5KCl + 3H_2O$

14. $5H_2O_2 + I_2 = 2HIO_3 + 4H_2O$

15. $3H_2O_2 + 2KMnO_4 = 2MnO_2 + 2KOH + 3O_2 + 2H_2O$

16. $5HNO_2 + 2KMnO_4 + 3H_2SO_4 = 5HNO_3 + 2KMnO_4 + K_2SO_4 + 3H_2O$

17. $2NaNO_2 + 2NaI + 2H_2SO_4 = 2NO + I_2 + 2Na_2SO_4 + 2H_2O$

18. $N_2H_4 + 4AgNO_3 + 4KOH = N_2 + 4Ag + 4KNO_3 + 4H_2O$

19. $N_2H_4 + Zn + 2KOH + 2H_2O = 2NH_3 + K_2[Zn(OH)_4]$

20. $Fe + N_2H_4 + 2H_2O = Fe(OH)_2 + 2NH_3$

21. $3H_2S + 2HNO_3 = 2NO + 3S + 4H_2O$

22. $3P + 5HNO_3 = 3HPO_3 + 5NO + H_2O$

23. $K_2Cr_2O_7 + 8HCl = 2KCl + 2CrCl_3 + 7H_2O + 3Cl_2$

24. $2MnO_4^- + 5C_2O_4^{2-} + 16H^+ = 10CO_2 + 2Mn^{2+} + 8H_2O$

25. $Cr_2O_7^{2-} + 3C_2O_4^{2-} + 14H^+ = 2Cr^{3+} + 6CO_2 + 7H_2O$

26. $2Cr(OH)_3 + IO_3 + 4OH^- = I^- + 2CrO_4^{2-} + 5H_2O$

27. $2KMnO_4 + 5H_2S + 4H_2SO_4 = 2KHSO_4 + 2MnSO_4 + 5S + 8H_2O$

28. $2NO_3^- + 6Cl^- + 8H^+ = 2NO + 3Cl_2 + 4H_2O$

29. $H_2O_2 + 2I^- + 2H^+ = I_2 + 2H_2O$

30. $Cr_2O_7^{2-} + 3NO_2^- + 8H^+ = 2Cr^{3+} + 3NO_3^- + 4H_2O$

31. $3N_2O_4 + BrO_3^- + 3H_2O = 6NO_3^- + Br^- + 6H^+$

32. $S_2O_3^{2-} + 2Sb_2O_5 + 6H^+ + 3H_2O = 4SbO + 6H_2SO_3$

33. $Fe_2(SO_4)_3 + Fe = 3FeSO_4$

34. $2Cu(NH_3)_4Cl_2 + 7KCN + H_2O = K_2Cu(CN)_3 + 6NH_3 + KCNO + 2NH_4Cl + 2KCl$

35. $4Ag + 8KCN + 2H_2O + O_2 = 4KAg(CN)_2 + 4KOH$

36. $Zn + 2NO_3^- + 4H^+ = Zn^{2+} + 2NO_2 + 2H_2O$

37. $2MnO_4^- + 3CN^- + H_2O = 2MnO_2 + 3CNO^- + 2OH^-$

38. $3AsO_3^{3-} + IO_3^- = 3AsO_4^{3-} + I^-$

39. $6Fe_3O_4 + 2MnO_4^- + H_2O = 9Fe_2O_3 + 2MnO_2 + 2OH^-$

40. $24H_2S + 8Cr_2O_7^{2-} + 64H^+ = 16Cr^{3+} + 3S_8 + 56H_2O$

41. $2ZnS + 3O_2 = 2ZnO + 2SO_2$

42. $2KNO_3 + 6FeSO_4 + 5H_2SO_4 = 2KHSO_4 + 3Fe_2(SO_4)_3 + 2NO + 4H_2O$

43. $3H_2S + K_2Cr_2O_7 + 5H_2SO_4 = 2KHSO_4 + Cr_2(SO_4)_3 + 3S + 7H_2O$

44. $2KI + 3H_2SO_4 = 2KHSO_4 + SO_2 + I_2 + 2H_2O$

45. $3C_2H_5OH + 2MnO_4^- + OH^- = 3C_2H_3O^- + 2MnO_2 + 5H_2O$

46. $10Al + 6KMnO_4 + 27H_2SO_4 = 6KHSO_4 + 5Al_2(SO_4)_3 + 6MnSO_4 + 24H_2O$

47. $2K_2Cr_2O_7 + 2H_2O + 3S = 3SO_2 + 4KOH + 2Cr_2O_3$

48. $MnO_2 + 4HCl = MnCl_2 + Cl_2 + 2H_2O$

49. $2MnO_4^- + 3SO_3^{2-} + H_2O = 2MnO_2 + 3SO_4^{2-} + 2OH^-$

50. $Cr_2O_7^{2-} + 3SO_3^{2-} + 8H^+ = 2Cr^{3+} + 3SO_4^{2-} + 4H_2O$

51. $I_2 + SO_2 + 2H_2O = SO_4^{2-} + 2I^- + 4H^+$

52. $Sn + 4NO_3^- + 8H^+ = SnO_2 + 4NO_2 + 4H_2O$

53. $2MnO_4^- + 5SO_2 + 2H_2O = 2Mn^{2+} + 5SO_4^{2-} + 4H^+$

54. $2MnO_4^- + SO_3^{2-} + 2OH^- = 2MnO_4^{2-} + SO_4^{2-} + H_2O$

55. $3ClO^- + Br^- = BrO_3^- + 3Cl^-$

56. $4Zn + NO_3^- + 10H^+ = 4Zn^{2+} + NH_4^+ + 3H_2O$

57. $2KMnO_4 + 16HCl = 5Cl_2 + 2KCl + 2MnCl_2 + 8H_2O$

58. $2BaCrO_4 + 6KI + 16HCl = 2BaCl_2 + 3I_2 + 6KCl + 2CrCl_3 + 8H_2O$

59. $2ClO_3^- + SO_2 + H^+ = 2ClO_2 + HSO_4^-$

60. $2Mn^{2+} + 5S_2O_8^{2-} + 8H_2O = 2MnO_4^- + 10HSO_4^- + 6H^+$

61. $Cl_2 + IO_3^- + 2OH^- = 2Cl^- + IO_4^- + H_2O$

62. $3H_2SO_3 + Cr_2O_7^{2-} + 5H^+ = 3HSO_4^- + 2Cr^{3+} + 4H_2O$

63. $2ClO_2 + SbO_2^- + 2OH^- = 2ClO_2^- + Sb(OH)_6^- + 2H_2O$

64. $4Zn + NO_3^- + 7OH^- = NH_3 + 4ZnO_2^{2-} + 2H_2O$

65. $3AsO_3^{3-} + 2MnO_4^- = 3AsO_4^{3-} + 2MnO_2 + 2H_2O$

66. $2KMnO_4 + 5H_2O_2 + 3H_2SO_4 = 2MnSO_4 + K_2SO_4 + 5O_2 + 8H_2O$

67. $4H_2O_2 + PbS = PbSO_4 + 4H_2O$

68. $2MnO_4^- + 5C_2O_4^{2-} + 16H^+ = 2Mn^{2+} + 10CO_2 + 8H_2O$

69. $6Fe^{2+} + Cr_2O_7^{2-} + 14H^+ = 6Fe^{3+} + 2Cr^{3+} + 7H_2O$

70. $AsO_3^{3-} + I_2 + H_2O = AsO_4^{3-} + 2H^+ + 2I^-$

71. $2S_2O_3^{2-} + I_2 = S_4O_6^{2-} + 2I^-$

72. $10HNO_3 + I_2 = 2HIO_3 + 10NO_2 + 4H_2O$

73. $K_2Cr_2O_7 + 7H_2SO_4 + 6KI = 3I_2 + 4K_2SO_4 + Cr_2(SO_4)_3 + 7H_2O$

74. $3CuO + 2NH_3 = 3Cu + N_2 + 3H_2O$

75. $As_2S_5 + 40HNO_3 = 2H_3AsO_4 + 5H_2SO_4 + 40NO_2 + 12H_2O$

76. $PbS + 4O_3 = PbSO_4 + 4O_2$

77. $Cl_2 + SeO_3^{2-} + H_2O = SeO_4^{2-} + 2Cl^- + 2H^+$

78. $6Cu_3P + 124H^+ + 11Cr_2O_7^{2-} = 18Cu^{2+} + 6H_3PO_4 + 22Cr^{3+} + 53H_2O$

79. $3Na_2SnO_2 + 2Bi(OH)_3 = 2Bi + 3Na_2SnO_3 + 3H_2O$

80. $H_2O + SbCl_3 = SbOCl + 2HCl$

81. $K_4Fe(CN)_6 + 61Ce(NO_3)_4 + 258KOH = 61Ce(OH)_3 + Fe(OH)_3$
$+ 36H_2O + 6K_2CO_3 + 250KNO_3$

82. $4S + 6OH^- = 2S^{2-} + S_2O_3^{2-} + 3H_2O$

83. $IO_4^- + 7I^- + 8H^+ = 4I_2 + 4H_2O$

84. $2KMnO_4 = K_2MnO_4 + MnO_2 + O_2$

85. $(NH_4)_2Cr_2O_7 = N_2 + 4H_2O + Cr_2O_3$

86. $3MnO_4^{2-} + 4H^+ = 2MnO_4^- + MnO_2 + H_2O$

87. $4H_3PO_3 = 3H_3PO_4 + PH_3$

88. $4Zn + 10HNO_3 = 4Zn(NO_3)_2 + N_2O + 5H_2O$

89. $24CuS + 16NO_3^- + 64H^+ = 24Cu^{2+} + 3S_8 + 16NO + 32H_2O$

90. $2CuSO_4 + 4KI = Cu_2I_2 + I_2 + 2K_2SO_4$

91. $2FeSO_4 = Fe_2O_3 + SO_2 + SO_3$ **[Hint:** Change in ON = 2]

92. $12NaOH + 6Cl_2 = 10NaCl + 2NaClO + 6H_2O$

93. $NH_3 + Hg_2Cl_2 = Hg + Hg(NH_2)Cl + HCl$

94. $Br_2 + 2OH^- = BrO^- + Br^- + H_2O$

95. $As_2S_5 + 40HNO_3 = 5H_2SO_4 + 40NO_2 + 2H_3AsO_4 + 12H_2O$

96. $Cr_2O_7^{2-} + 3SO_2 + 5H^+ = 2Cr^{3+} + 3HSO_4^- + H_2O$

97. $4HCl + 3WO_3 + SnCl_2 = H_2SnCl_6 + W_3O_8 + H_2O$

98. $2HCl + FeCl_2 + V(OH)_4Cl = VOCl_2 + 3H_2O + FeCl_3$

99. $4Au + 16KCN + 6H_2O + 3O_2 = 4KAu(CN)_4 + 12KOH$

100. $4KOH + 4KMnO_4 = 4K_2MnO_4 + O_2 + H_2O$

Objective Problems

1. Consider the following reaction:

$$xMnO_4^- + yC_2O_4^{2-} + zH^+ \rightarrow xMn^{2+} + 2yCO_2 + \frac{z}{2}H_2O$$

The values of x, y and z in the reaction are respectively,

(a) 5, 2, 16
(b) 2, 5, 8
(c) 2, 5, 16
(d) 5, 2, 8 **(IIT 2013 Main)**

2. From the following statements regarding H_2O_2, choose the incorrect statement.

(a) It can act only as an oxidising agent
(b) It decomposes on exposure to light.
(c) It has to be stored in plastic or wax-lined glass bottle in dark.
(d) It has to be kept away from dust. **(IIT 2013 Main)**

3. In which of the following reactions, H_2O_2 acts as reducing agent?

I. $H_2O_2 + 2H^+ + 2e \rightarrow 2H_2O$

II. $H_2O_2 - 2e \rightarrow O_2 + 2H^+$

III. $H_2O_2 + 2e \rightarrow 2OH^-$

IV. $H_2O_2 + 2OH^- - 2e \rightarrow O_2 + 2H_2O$

(a) I and II
(b) III and IV
(c) I and III
(d) II and IV **(IIT 2014 Main)**

4. For the reaction

$$I^- + ClO_3^- + H_2SO_4 \rightarrow Cl^- + HSO_4^- + I_2$$

The correct statement(s) in the balanced equation is(are)

(a) Stoichiometric coefficient of HSO_4^- is 6.
(b) Iodide is oxidised.
(c) Sulphur is reduced.
(d) H_2O is one of the products. **(IIT 2014 Adv.)**

5. The correct statement(s) about Cr^{2+} and Mn^{3+} is(are)
(Cr = 24, Mn = 25)

(a) Cr^{2+} is reducing agent.
(b) Mn^{3+} is an oxidising agent.
(c) Both Cr^{2+} and Mn^{3+} exhibit d^4 electronic configuration.
(d) When Cr^{2+} is used as reducing agent, the chormium ion attains d^5 electronic configuration. **(IIT 2015 Adv.)**

6. The pair in which phosphorus atoms have a formal oxidation state of +3 is

(a) pyrophosphorous and hypophosphoric acids
(b) orthophosphorous and hypophosphoric acids

(c) pyrophosphorous and pyrophosphoric acids

(d) orthophosphorous and pyrophosphorous acids **(IIT 2016 Main)**

[Hint: Find oxidation states from the formula of acids.]

7. Which of the following reactions is an example of redox reactions?

(a) $XeF_4 + O_2F_2 \rightarrow XeF_6 + O_2$ (b) $XeF_2 + PF_5 \rightarrow [XeF]^+PF_6^-$

(c) $XeF_6 + H_2O \rightarrow XeOF_4 + 2HF$ (d) $XeF_6 + 2H_2O \rightarrow XeO_2F_2 + 4HF$

(IIT 2017 Main)

8. The order of the oxidation state of the phosphorus atom in H_3PO_2, H_3PO_4, H_3PO_3 and $H_4P_2O_6$ is

(a) $H_3PO_4 > H_3PO_2 > H_3PO_3 > H_4P_2O_6$

(b) $H_3PO_4 > H_4P_2O_6 > H_3PO_3 > H_3PO_2$

(c) $H_3PO_2 > H_3PO_3 > H_4P_2O_6 > H_3PO_4$

(d) $H_3PO_3 > H_3PO_2 > H_3PO_4 > H_4P_2O_6$ **(IIT 2017 Adv.)**

9. Hydrogen peroxidise oxidises $[Fe(CN)_6]^{-4}$ to $[Fe(CN)_6]^{-3}$ in acidic medium but reduces $[Fe(CN)_6]^{-3}$ to $[Fe(CN)_6]^{-4}$ in alkaline medium. The other products formed are, respectively.

(a) H_2O and $(H_2O + OH^-)$ (b) $(H_2O + O_2)$ and H_2O

(c) $(H_2O + O_2)$ and $(H_2O + OH^-)$ (d) H_2O and $(H_2O + O_2)$ **(IIT 2018 Main)**

10. The oxidation states of Cr in

$$[Cr(H_2O)_6]Cl_3,\ [Cr(C_6H_6)_2] \text{ and } K_2[Cr(CN)_2(O)_2(O)_2(NH_3)]$$

respectively are

(a) +3, 0 and +4 (b) +3, +4 and +6

(c) +3, +2 and +4 (d) +3, 0 and +6 **(IIT 2018 Main)**

[Hint: Oxidation state of H_2O, C_6H_6 and NH_3 are taken as zero.]

Answers

1-c, **2**-a, **3**-d, **4**-a, b, d, **5**-a, b, c, **6**-d, **7**-a, **8**-b, **9**-d, **10**-d.

❑

CHAPTER TWENTY

SOLID AND LIQUID STATES

THE SOLID STATE

Solid substances are frequently classified as either crystalline or amorphous. Crystalline solids are characterised by a regular, ordered arrangement of particles. However, a small class of noncrystalline solids, known as amorphous solids, has no well-defined ordered structure. Examples are rubber, some kinds of plastics, amorphous sulphur, etc. Glass is sometimes called an amorphous solid and sometimes called an undercooled or supercooled liquid of high viscosity.

Crystalline Solid	Amorphous Solid
1. Definite and regular geometry with flat faces and sharp edges	1. No definite geometrical shape
2. It breaks up into smaller crystals of the same geometrical shape	2. Broken pieces are not generally flat
3. Sharp melting point	3. No sharp melting point
4. Anisotropic, i.e., physical properties are different in different directions	4. Isotropic, i.e., physical properties are same in all directions

Types of Crystalline Solids

	Ionic	Metallic	Covalent	Molecular
Particles occupying lattice points	Anions, cations	Metal ions in electron cloud	Atoms	Molecules (or atoms)
Binding force	Electrostatic attraction	Metallic bonds	Covalent bonds	Van der waals dipole-dipole
Properties	Hard, brittle, poor thermal and electrical conductors	Soft to very hard, good thermal and electrical conductors	Very hard, poor thermal and electrical conductors	Soft, poor thermal and electrical conductors
Examples	NaCl, $CaBr_2$, KNO_3, etc.	Li, K, Ca, Cu, Na, etc.	C (diamond), SiO_2 (quartz), etc.	H_2O, H_2, CO_2, Ar, etc.

Types of Symmetry in Cubic Crystals

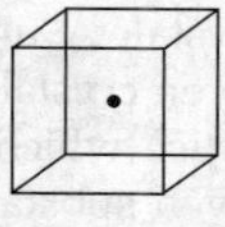

Centre of symmetry

Centre of Symmetry: It is an imaginary point within the crystal such that any line drawn through it intersects the surface of the crystal at equal distances in both directions. There is only one centre of symmetry in a cubic crystal.

Plane of Symmetry: It is an imaginary plane which passes through the centre of symmetry of a crystal and divides it into two equal portions such that one part is exactly the mirror image of the other.

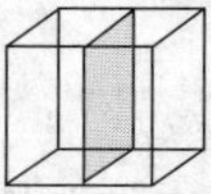

Rectangular plane of symmetry (3)

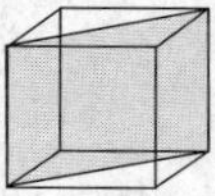

Diagonal plane of symmetry (6)

A cubic crystal has in all $3 + 6 = 9$ planes of symmetry.

Axis of Symmetry: It is an imaginary straight line about which if the crystal is rotated, it will present the same appearance more than once during a complete revolution. If the same appearance of a crystal is repeated through an angle of $360/n$, about an imaginary axis, the axis is called an n-fold axis.

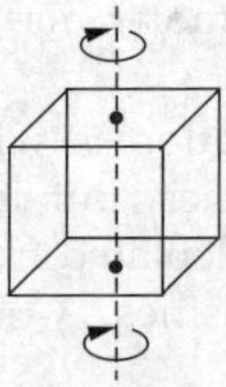

Axis of 4-fold symmetry (3)

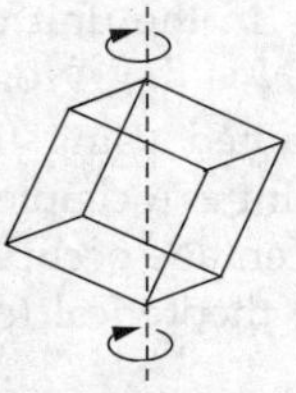

Axis of 3-fold symmetry (4)

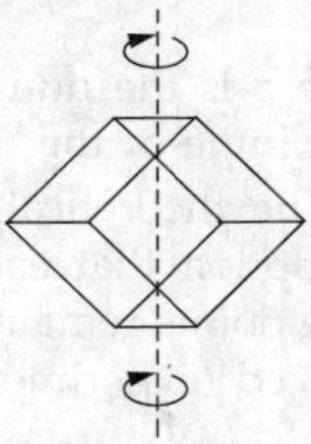

Axis of 2-fold symmetry (6)

A cubic crystal has in all $3 + 4 + 6 = 13$ axes of symmetry.

The total number of centres of symmetry, planes of symmetry and axes of symmetry possessed by a crystal is termed **elements of symmetry** of the crystal. **A cubic crystal possesses 23 elements of symmetry.**

Centre of symmetry $= 1$

Planes of symmetry $= 3 + 6 = 9$

Axes of symmetry $= 3 + 4 + 6 = 13$

Elements of symmetry $= 1 + 9 + 13 = 23$

Space Lattice and Unit Cell

A space lattice is an array (pattern) of points showing how molecules, atoms or ions in a crystal are arranged in an ordered, repetitive fashion in three-dimensional space.

A unit cell is the smallest sample that represents the picture of the entire crystal. A unit cell of a crystal possesses all the structural properties of the given crystal. In other words, a unit cell is the smallest repeating unit in space lattice which when repeated over and over again results in a crystal of the given substance.

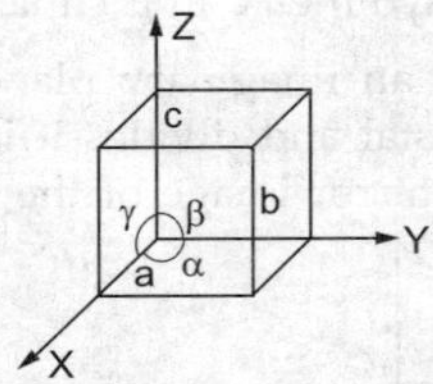

For a cubic crystal, $a = b = c$ and $\alpha = \beta = \gamma = 90°$

In order to describe a unit cell, one should know (i) the distances *a*, *b* and *c* which give the lengths of the edges of the unit cell, and (ii) the angles α, β and γ which give the angles between the three imaginary axes *OX*, *OY* and *OZ*.

Theoretical Density: Knowing the unit-cell dimensions, the theoretical density (ρ) of a cubic crystal can be calculated from the equation

$$\rho = \frac{zM}{NV}, \quad \ldots (1)$$

where *z* is the number of atoms in the unit cell, *M* is the molar mass, *V* is the volume of the unit cell and *N* is the Avogadro constant.

The theoretical density obtained using the above equation is with the assumption that each lattice point is occupied by the species. But if some lattice points remain vacant, percentage occupancy can be calculated from the observed (experimental) and the theoretical (calculated) densities.

$$\text{Percentage occupancy} = \frac{\rho_{exp}}{\rho_{cal}} \times 100$$

$$V = abc(1 - \cos^2\alpha - \cos^2\beta - \cos^2\gamma + 2\cos\alpha\cos\beta\cos\gamma)^{\frac{1}{2}}$$

If $\alpha = \beta = \gamma = 90°$

$$V = abc.$$

Separation of Atoms

The distance *l* between two atoms in a unit cell can be calculated from their coordinates (x_1, y_1, z_1) and (x_2, y_2, z_2) as

$$l = \left[a^2 (x_2 - x_1)^2 + b^2 (y_2 - y_1)^2 + c^2 (z_2 - z_1)^2\right]^{\frac{1}{2}}.$$

If $a = b = c$, as for a cubic unit cell,

$$l = a\left[(x_2 - x_1)^2 + (y_2 - y_1)^2 + (z_2 - z_1)^2\right]^{\frac{1}{2}}. \quad \ldots (2)$$

Crystal Systems: Seven types

It can be shown from geometrical considerations that theoretically, there can be 32 different combinations of elements of symmetry of a crystal. These are called **32 point groups or 32 systems.** However, some of the systems have been grouped together so that there are only seven basic crystal systems, viz., cubic, tetragonal, orthorhombic, monoclinic, hexagonal, rhombohedral (or trigonal) and triclinic.

Crystal System	Edge Length	Angles
Cubic	$a = b = c$	$\alpha = \beta = \gamma = 90°$
Tetragonal	$a = b \neq c,$	$\alpha = \beta = \gamma = 90°$
Orthorhombic	$a \neq b \neq c$	$\alpha = \beta = \gamma = 90°$
Monoclinic	$a \neq b \neq c$	$\alpha = \beta = 90°, \gamma \neq 90°$
Hexagonal	$a = b \neq c$	$\alpha = \beta = 90°, \gamma = 120°$
Rhombohedral	$a = b = c$	$\alpha = \beta = \gamma \neq 90°$
Triclinic	$a \neq b \neq c$	$\alpha \neq \beta \neq \gamma \neq 90°$

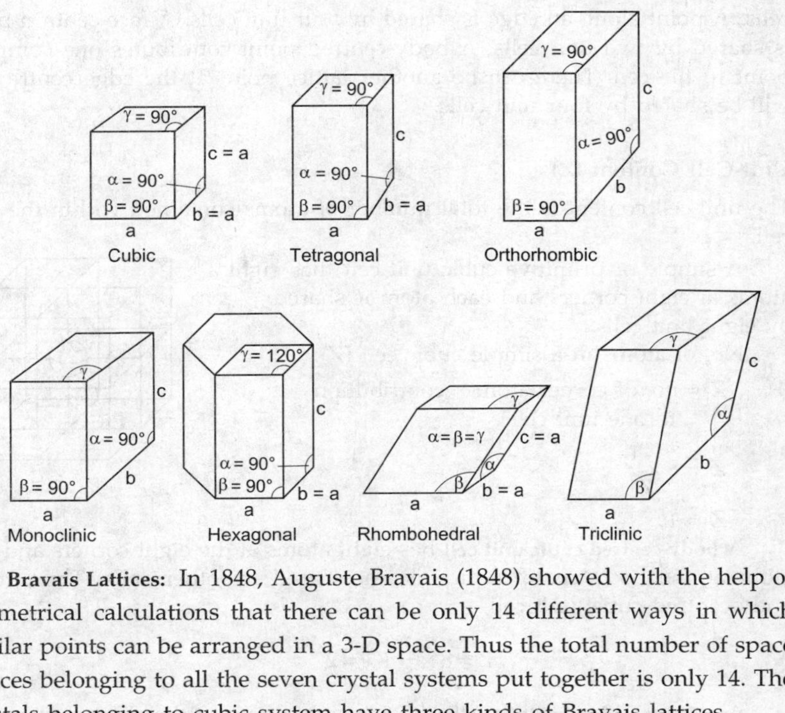

Bravais Lattices: In 1848, Auguste Bravais (1848) showed with the help of geometrical calculations that there can be only 14 different ways in which similar points can be arranged in a 3-D space. Thus the total number of space lattices belonging to all the seven crystal systems put together is only 14. The crystals belonging to cubic system have three kinds of Bravais lattices.

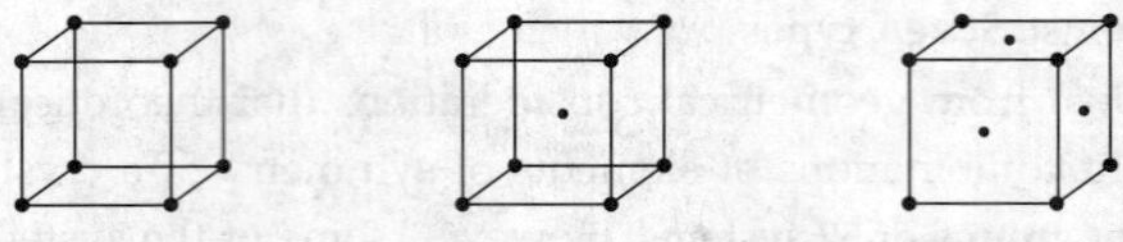

Crystal systems	No. of Bravais lattices	
1. Cubic	3	
2. Tetragonal	2	
3. Orthorhombic	4	
4. Monoclinic	2	Total = 14
5. Hexagonal	1	
6. Rhombohedral	1	
7. Triclinic	1	

Lattice Points: There are mainly three kinds of lattice points, that is, at the corners, at the face centres and within the unit cell.

A point that lies at the corner of a unit cell is shared among eight unit cells. A point along an edge is shared by four unit cells. A face-centred point is shared by two unit cells. A body-centred point contributes one complete point to the cell. There can be another lattice point at the edge centre that will be shared by four unit cells.

Unit-Cell Content (Z)

The unit-cell content is the total number of atoms contained within the unit cell.

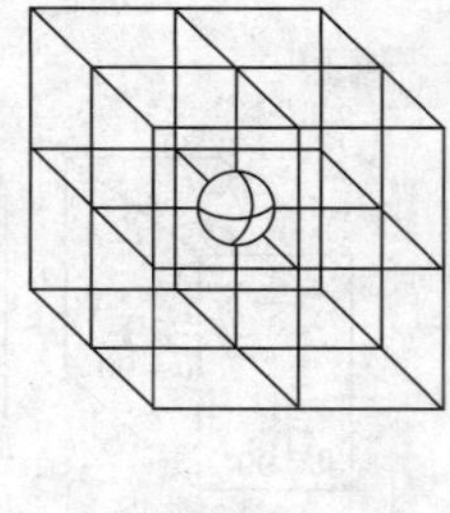

A simple or primitive cubic unit cell has eight atoms at eight corners and each atom is shared by eight unit cells.

No. of atoms in a simple cubic cell (Z):

Z = no. of corner atom × contribution to one unit cell

$$Z = 8 \times \frac{1}{8}$$

$$Z = 1$$

A **body-centred cubic unit cell** has eight atoms at the eight corners and one atom at the body centre which is not shared by other cells. Thus, for a body-centred cubic unit cell,

$$Z = 8 \times \frac{1}{8} + 1 = 2.$$

A **face-centred cubic unit cell** contains eight atoms at the eight corners and six atoms at the six faces. The corner atoms are shared by eight unit cells, while the atom at the face is shared by two unit cells. Thus, for a face-centred cubic unit cell,

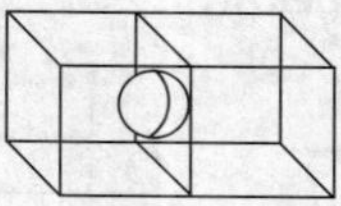

$$Z = 8 \times \frac{1}{8} + 6 \times \frac{1}{2} = 4.$$

Designation of Planes by Miller Indices

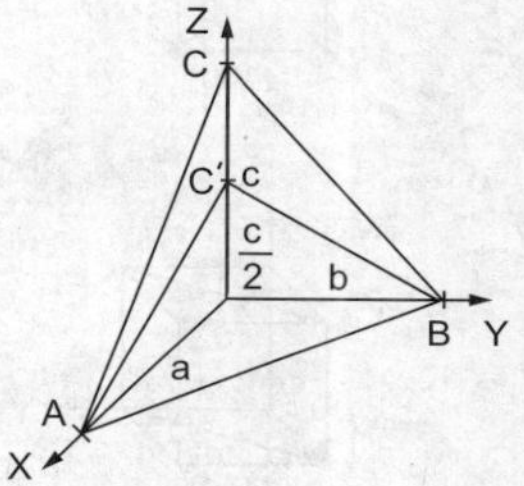

The faces of a crystal are described in terms of a plane, called the unit plane or standard plane. Plane *ABC* which cuts the axes *X*, *Y* and *Z* at *A*, *B*, and *C* respectively with intercepts *a*, *b* and *c* is called the unit plane.

W H Miller in 1839 represented the planes of a crystal by a set of three integers *h*, *k* and *l*, called the Miller indices. To obtain the Miller indices for any crystal plane, the ratios of the intercepts made by the unit plane to those made by the given plane (or its parallel plane) are converted to clear fractions.

The **procedure** for determining the Miller indices for a plane is as follows:

1. Prepare a three-column table with the unit cell axis at the tops of the columns.
2. In each column enter the intercept (expressed as a multiple of *a*, *b*, or *c*) of the plane with that axis.
3. Invert all numbers.
4. Clear fractions to obtain *h*, *k* and *l*.

For the plane *ABC′*, the Miller indices can be obtained as follows.

The plane *ABC′* intersects the axes at $x = a$, $y = b$ and $z = \frac{c}{2}$ · Let us prepare the table as described below.

a	*b*	*c*	
1	1	1/2	Intercepts
1	1	2	Reciprocals
1	1	2	Clear fractions

The plane *ABC′* is then said to be (1 1 2). The unit plane is labelled as (1 1 1). Miller indices are enclosed within parentheses and the negative sign in Miller indices is indicated by placing a bar on the integer. In practice, an index greater than 5 never occurs.

For a plane running parallel to the *XY* plane, the intercepts on the three axes are ∞, ∞ and 1. The Miller indices of such a plane are $\frac{1}{\infty}, \frac{1}{\infty}, \frac{1}{1}$ or (0 0 1). Similarly, the faces parallel to the *YZ* and *ZX* planes will have the Miller indices (1 0 0) and (0 1 0) respectively. **Two parallel planes in a crystal**

have the same Miller indices. Miller indices of some shaded planes are given below.

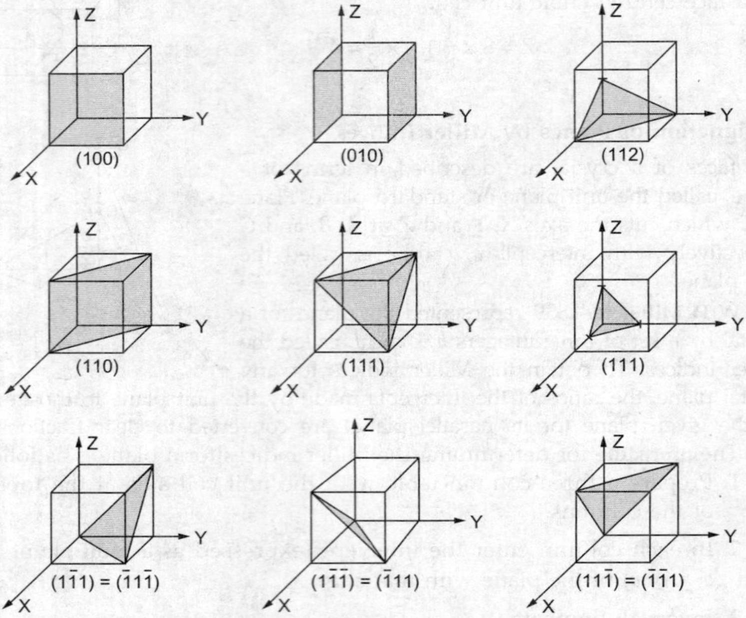

d-Spacings

The distance between two parallel planes in a cubic crystal (*d*-spacing) is given by

$$d = \frac{a}{\sqrt{h^2 + k^2 + l^2}}, \qquad \dots (3)$$

where a is the length of the side of the cube and h, k and l are Miller indices of the parallel planes.

For the planes (1 0 0), (1 1 0) and (1 1 1) in a cubic unit cell,

$$d_{100} = \frac{a}{\sqrt{1^2 + 0^2 + 0^2}} = a$$

$$d_{110} = \frac{a}{\sqrt{1^2 + 1^2 + 0^2}} = \frac{a}{\sqrt{2}}$$

$$d_{111} = \frac{a}{\sqrt{1^2 + 1^2 + 1^2}} = \frac{a}{\sqrt{3}}$$

Atomic Radius (r)

In a **simple or primitive cubic cell,** the two adjacent corner atoms are supposed to touch each other. In a **body-centred cubic cell,** the atom at the centre of the cube is supposed to touch the corner atoms. In a **face-centred cubic cell,** the atom at the face centre is supposed to touch its adjacent corner atoms. These are clearly shown by the figures. The atomic radius may thus be calculated by applying simple geometry.

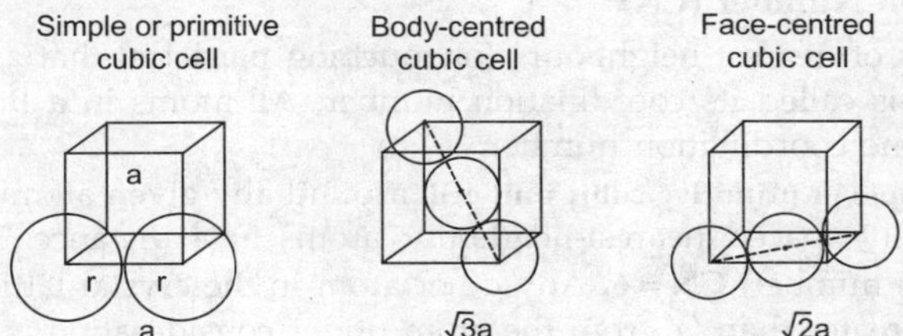

$$r = \frac{a}{2} \qquad r = \frac{\sqrt{3}a}{4} \qquad r = \frac{\sqrt{2}a}{4} \quad \dots (4)$$

Packing Fraction and Empty Space in the Closest Packing

In the closest packing, spherical balls must have some vacant space in the crystal. The fraction of the total volume of the unit cell occupied by the atom(s) is known as packing fraction. The percentage of vacant space can be calculated from the percentage packing fraction.

In a **simple or primitive cubic cell,** there is only one atom per unit cell.

$$\text{Volume of the atom} = \frac{4}{3}\pi r^3 = \frac{4}{3}\pi\left(\frac{a}{2}\right)^3 = \frac{\pi a^3}{6}.$$

$$\text{Volume of one unit cell} = a^3$$

$$\therefore \text{ packing fraction} = \frac{\pi a^3/6}{a^3} = \frac{\pi}{6} = 0{\cdot}5236 = 52{\cdot}36\%. \quad \dots (5)$$

Percentage of vacant space $= (100 - 52{\cdot}36)\% = 47{\cdot}64\%$.

In a **body-centred cubic cell,** there are two atoms per unit cell.

$$\text{Volume of two atoms} = 2 \times \left(\frac{4}{3}\pi r^3\right)$$

$$= 2 \times \frac{4}{3} \times \pi \times \left(\frac{\sqrt{3}\,a}{4}\right)^3 = \frac{\sqrt{3}\pi a^3}{8}$$

$$\therefore \text{ packing fraction} = \frac{\sqrt{3}\pi a^3/8}{a^3} = \frac{\sqrt{3}\pi}{8} = 0.6802 = 68{\cdot}02\%. \quad \dots (6)$$

Percentage of vacant space $= (100 - 68{\cdot}02)\% = 31{\cdot}98\%$.

In a **face-centred cubic cell,** there are four atoms per unit cell.

$$\text{Volume of four atoms} = 4 \times \left(\frac{4}{3}\pi r^3\right)$$

$$= 4 \times \frac{4}{3} \times \pi \times \left(\frac{\sqrt{2}\,a}{4}\right)^3 = \frac{\sqrt{2}\pi a^3}{6}.$$

$$\therefore \text{ packing fraction} = \frac{\sqrt{2}\pi a^3/6}{a^3} = \frac{\sqrt{2}\pi}{6} = 0{\cdot}7406 = 74{\cdot}06\%. \quad \ldots (7)$$

Percentage of vacant space = (100 − 74·06)% = 25·94%.

Coordination Number (CN)*

The number of nearest neighbours (or touching particles) that a particle has in a crystal is called its coordination number. All atoms in a Bravais lattice have the same coordination number.

In a **simple or primitive cubic unit cell,** around any given atom there would be six equally spaced nearest-neighbour atoms at a distance *'a'*. Thus the coordination number, **CN = 6.** Any other atom in the crystal lattice will be at a distance greater than *'a'* from the atom under consideration.

In a body-centred cubic unit cell, the atom at the centre of the unit cell is surrounded by eight nearest-neighbour atoms at the corners of the cube. Thus the coordination number, **CN = 8;** there are four nearest neighbours within the most closely packed layer, two in the layer above and two in that below.

The two arrangements of closest-packed layers are **hexagonal close-packed structure** (h.c.p.) and **cubic close-packed** (c.c.p.) or **face-centred cubic structure** (f.c.c.). These unit cells are the hexagonal body-centred cell (**Z** = 2) and the cubic face-centred cell (**Z** = 4) respectively. In each of these structures, every sphere (atom) is in contact with twelve others; six in its own layer, three in the layer above and three in the layer below. The coordination number, **CN = 12.** The high coordination number (12) in these structures results in a crystal of comparatively high density.

The following diagrams show the expanded structures to clarify the difference between hexagonal close-packed (h.c.p.) and cubic close-packed (c.c.p.) or face-centred cubic (f.c.c.) structures.

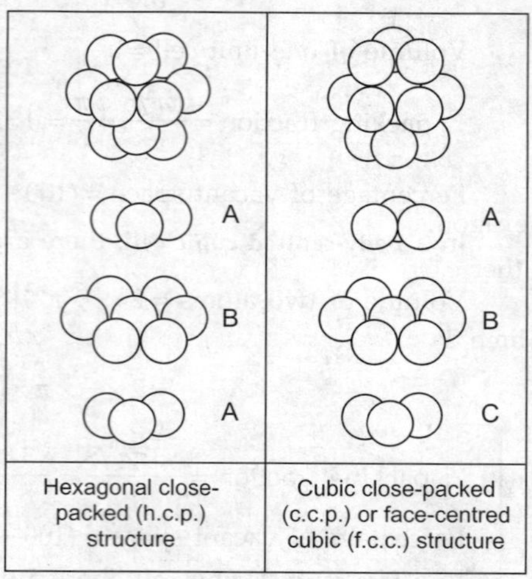

Hexagonal close-packed (h.c.p.) structure	Cubic close-packed (c.c.p.) or face-centred cubic (f.c.c.) structure

In an h.c.p. structure, the triangular layers are oriented in the same direction while in a c.c.p. or

* The term coordination number is used in crystallography in a somewhat different sense than it is in coordination compounds.

f.c.c. structure, the triangular layers are oriented in opposite directions. The extended patterns of arrangement of planes in h.c.p. and c.c.p. are of the type *AB AB AB* ... and *ABC ABC ABC* ... respectively, where *A*, *B* and *C* represent different planes. The vacant space in both h.c.p. or c.c.p. arrangement is the same (25·94%) as calculated previously for f.c.c.

In a c.c.p. arrangement, there is always a sphere at each face centre of a cube and c.c.p. is therefore called f.c.c. arrangement.

As in c.c.p. arrangement there is always a sphere at each face centre of a cube, the c.c.p. lattice is identical to the lattice having f.c.c. unit cell. To see this, take portions of four layers from the c.c.p. array. When these are placed together, these layers form an f.c.c. unit cell.

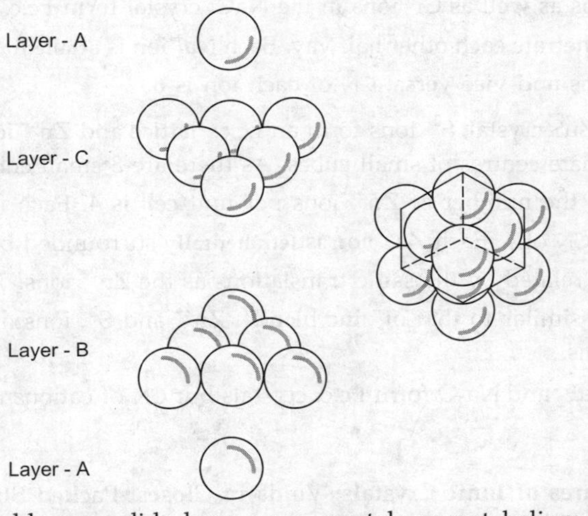

All the noble-gas solids have c.c.p. crystals except helium, which is a hexagonal close-packed h.c.p. crystal. Many of the metallic elements do have either c.c.p. or h.c.p. crystals, e.g., Au, Cu, Ag, Ni, Pt, etc.

Some Types of Cubic Ionic Solids

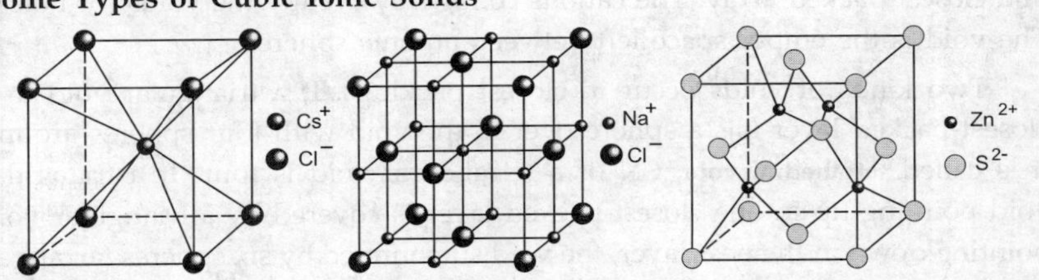

Caesium chloride CsCl — Sodium chloride NaCl — Zinc blende ZnS

Structure	Coordination number	Cell content (no. of units per cell)
CsCl	$Cs^+ - 8$, $Cl^- - 8$	1
NaCl	$Na^+ - 6$, $Cl^- - 6$	4
ZnS	$Zn^{2+} - 4$, $S^{2-} - 4$	4
CaF_2	$Ca^{2+} - 8$, $F^- - 4$	4
Na_2O	$Na^+ - 4$, $O^{2-} - 8$	4

CsCl is a simple cubic crystal, the Cs^+ ion being at the body centre of a cube of Cl^- ions, and the Cl^- ion at the body centre of the cube of Cs^+ ions. Coordination number of each ion is 8.

Na^+ ions as well as Cl^- ions in the NaCl crystal form f.c.c. lattice and the two interpenetrate each other halfway. Each Na^+ ion is **octahedrally** surrounded by 6 Cl^- ions and vice versa. CN of each ion is 6.

In the ZnS crystal, S^{2-} ions form an f.c.c. lattice and Zn^{2+} ions are present at the alternate centres of small cubes. As there are 8 small cubes in one unit cell of ZnS, the number of Zn^{2+} ions per unit cell is 4. Each ion in ZnS has therefore a CN of 4. Each Zn^{2+} ion is **tetrahedrally** surrounded by S^{2-} ions. The S^{2-} ions are related by the same translations as the Zn^{2+} ions. The unit cell of diamond is similar to that of zinc blende. Zn^{2+} and S^{2-} ions are replaced by carbon atoms.

Both CaF_2 and Na_2O form f.c.c. crystals but CN of cations and anions are different.

The Structures of Ionic Crystals: Voids in Closest-Packed Structures

Ionic crystals consist of cations and anions of different sizes. **The packing of these ions into a crystal structure is more complex than packing of the metal atoms in metallic crystals which are of the same size.**

In simple ionic crystals, anions are normally larger than cations arranged in a closest-packed array. The cations commonly occupy the voids (or holes). The void is the empty space left between anionic spheres.

Two kinds of voids occur in closest packings. If a triangular void in a closest-packed layer has a sphere over it, the void with four spheres around it is called **tetrahedral void.** CN of a tetrahedral void is four. If a triangular void pointing up in one closest-packed layer is covered by a triangular void pointing down in the next layer, the void surrounded by six spheres (arranged on the corners of an octahedron) is called an **octahedral void.** CN of an octahedral void is six.

The number of octahedral voids per sphere (anion)

$$= \frac{\text{no. of octahedral voids around a sphere (anion)}}{\text{no. of spheres around a void}} = \frac{6}{6} = 1.$$

In NaCl, Na^+ ions occupy the octahedral voids while Cl^- ions form the f.c.c. structure.

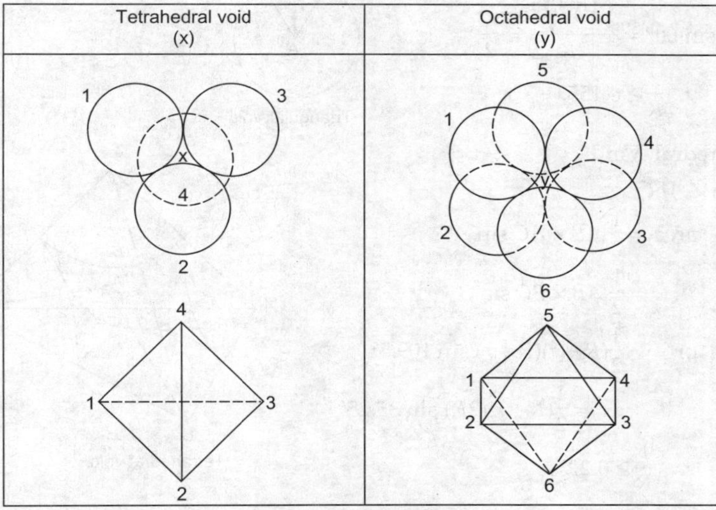

The number of tetrahedral voids per sphere (anion)

$$= \frac{\text{no. of tetrahedral voids around a sphere (anion)}}{\text{no. of spheres around a void}} = \frac{8}{4} = 2.$$

In ZnS, Zn^{2+} ions occupy one-half of the tetrahedral voids while S^{2-} ions form the f.c.c. structure.

Thus, in a closest-packed arrangement, the number of octahedral voids is equal to the number of spheres (anions), while the number of tetrahedral voids is double the number of spheres (anions). Note that the voids are generally occupied by cations.

If all the octahedral voids are occupied by cations, the number of cations is equal to the number of anions. And if all the tetrahedral voids are occupied by cations, the number of cations is twice the number of anions.

If r is the radius of the spherical voids and R is the radius of the spheres forming the voids, the critical r/R values for trigonal, tetrahedral, octahedral and cubic voids are found to be equal to 0.155, 0.225, 0.414 and 0.732 respectively. The calculations of these values have been shown below.

The decreasing order of the size of the void is

cubic > octahedral > tetrahedral > trigonal.

Calculations of r/R or $r+/r-$: The critical radius ratio of the void (cation) and sphere (anion), that is, r/R or $r+/r-$ can be calculated by solid geometry as follows.

Triangular Void

In $\triangle ABC$,

$$\sin C = \frac{AB}{AC}$$

$$\sin 60^\circ = \frac{\frac{3}{2}(R+r)}{2R} = \frac{\sqrt{3}}{2}$$

$$\therefore \ \frac{r}{R} \geq 0.155.$$

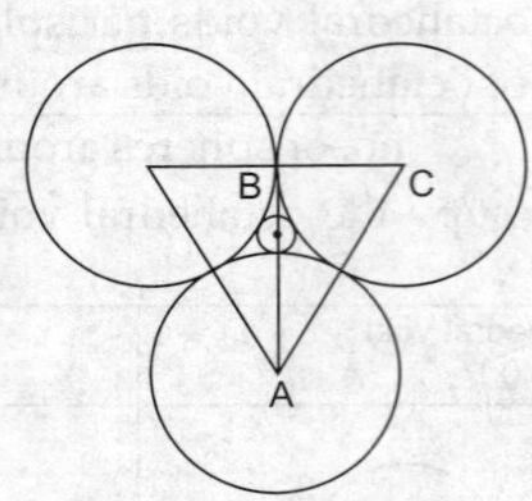

Triangular void

Tetrahedral Void

In $\triangle ABC$,

$$\text{area} = \frac{1}{2} AB \times AC \sin A$$

$$= \frac{1}{2} AB \times BC \sin B$$

$$\text{or} \quad \frac{1}{2}(R+r)(R+r) \sin 109.5^\circ$$

$$= \frac{1}{2}(R+r)(2R) \sin 35.25^\circ$$

$$\therefore \quad \frac{r}{R} \geq 0.225.$$

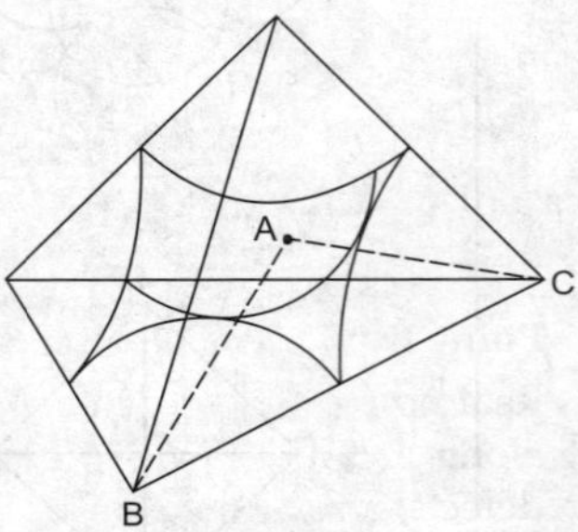

Tetrahedral void

Octahedral Void

In $\triangle ABC$,

$$\sin ACB = \frac{AB}{AC}$$

$$\sin 45^\circ = \frac{2R}{2(R+r)} = \frac{1}{\sqrt{2}}$$

$$\therefore \ \frac{r}{R} \geq 0.414.$$

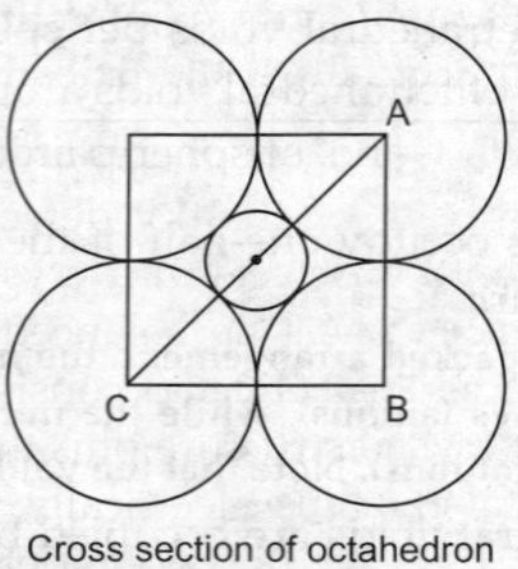

Cross section of octahedron

Cubic Void

In $\triangle ABC$,

$$AB^2 + BC^2 = AC^2$$

$$(2\sqrt{2}R)^2 + (2R)^2 = [2(R+r)]^2$$

$$\therefore \ \frac{r}{R} \geq 0.732.$$

Thus relatively small cations occupy tetrahedral holes and larger cations occupy octahedral holes. If cations are too large to fit into the octahedral holes, the anions may occupy the larger cubic holes made possible by the more open spacing.

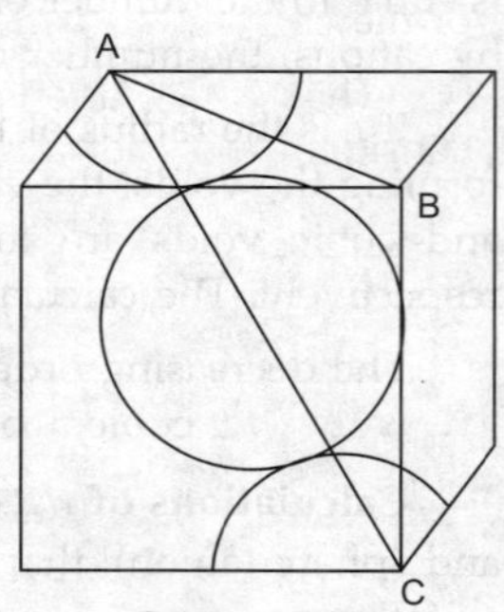

Cubic void

Radius Ratio and Coordination Number (CN)

In ionic solids, the radius ratio is the ratio of the cation-radius to the anion-radius. As cations get surrounded by anions, therefore, larger the radius ratio, higher is the coordination number of the cation from the radius ratio and hence the geometry of the ionic crystals.

Range of r_+/r_-	CN	Geometry
–0.155	2	Linear
0.155–0.225	3	Triangular
0.225–0.414	4	Tetrahedral
0.414–0.732	6	Octahedral
0.732–1.0	8	Cubic
1.0–	12	Closest-sphere packing

Point Defects

Real crystals have defects or imperfections. These defects may be at a point, along a line or over a plane. Hence, they are respectively known as point defects, line defects and plane defects.

A point defect may result from (a) the absence of an atom (vacancy), (b) displacement of an atom to an interstitial site (self-interstitial), (c) presence of an impurity atom at an interstice (interstitial impurity atom), or (d) presence of an impurity atom at a lattice site (substitutional impurity atom).

The **Schottky defect** consists of a pair of vacancies of opposite signs. In NaCl, there is roughly one Schottky defect for 10^6 ions. The presence of a large number of Schottky defects in a crystal lowers its density. In an ionic crystal, crystal vacancies must be balanced so that the crystal as a whole is electrically neutral. The **Frenkel defect** consists of an ion that has moved to an interstitial site in the crystal structure plus the vacancy left behind. In a Frenkel defect too, the crystal remains electrically neutral, but the density does not change significantly. Both the said point defects do not alter the stoichiometry of the crystal.

The concentration of point defects can be calculated from simple statistical considerations. For a crystal with N lattice sites and n vacancies

$$\frac{n}{N-n} = e^{-E_v/kT},$$

where $e^{-E_v/kT}$ is the Boltzmann factor, k is the Boltzmann constant and E_v is the energy required to make a defect.

As $n \ll N$ at ordinary temperature,

$$\frac{n}{N} = e^{-E_v/kT}.$$

The crystal defects may have special electrical properties that are useful. Semiconductors, transistors, insulators, etc., are made from synthetic crystals that have carefully controlled defects.

Bragg's Equation

The Bragg's equation

$$n\lambda = 2d \sin \theta \qquad \dots (8)$$

is very useful in the study of crystals. d is the distance between two successive planes, λ is the wavelength of the incident X-rays used and n is a whole number, 1, 2, 3, ... which denotes reflections of the first, second, third ... orders respectively. The angles at which the observed intensity of the diffracted beams is maximum, are the 'θ's of the Bragg's equation. The smallest angle for maximum intensity corresponds to $n = 1$ (first order); similarly, the higher angles correspond to $n = 2$ (second order), $n = 3$ (third order), etc.

THE LIQUID STATE

Viscosity of liquids

Viscosity is the resistance to flow of a liquid. Glycerine has a high viscosity at room temperature and freely flowing gasoline has a low viscosity.

The resistance that one part of a liquid, flowing with a certain velocity, offers to another part of the liquid, flowing with a different velocity, is known as the viscosity. If A is the area of each of the two successive layers of a liquid, separated by a distance dx, and du, their velocity difference, then the tangential force (F), required to maintain a constant velocity difference du is given by

$$F = \eta \cdot A \cdot \frac{du}{dx},$$

where η is called the coefficient of viscosity and $\frac{du}{dx}$, the velocity gradient.

Thus, **the coefficient of viscosity may be defined as the tangential force per unit area required to maintain a unit velocity gradient.**

The unit of viscosity is poise (dyne cm^{-2} s) in cgs and pascal second (N m^{-2} s) in SI.

Determination of Viscosity

The rate of laminar flow of a liquid through a capillary tube at a constant pressure is related to the viscosity of the liquid, given by the Poiseuille equation

$$\eta = \frac{\pi p r^4 t}{8Vl},$$

where V is the volume of the liquid of viscosity η which flows in time t through a capillary tube of radius r and length l under a driving pressure p.

The viscosity of liquids is measured with the help of the **Ostwald viscometer.** Equal volumes of two liquids are allowed to flow through the capillary tube of the same viscometer under similar conditions. Then using the Poiseuille equation, we get,

$\frac{\eta_1}{\eta_2} = \frac{t_1 d_1}{t_2 d_2}$, where d is the density of the liquid and t is the time of flow.

Knowing t_1 and t_2 experimentally and d_1, d_2 and η_2 from the text, we can calculate η_1. The viscosity of a **highly viscous liquid** is determined by using a **falling-sphere viscometer.** By measuring the velocity u of the spherical ball of known radius r and density d through a vertical column of liquid of density d_0, the coefficient of viscosity, η, can be calculated using the following equation known as **Stokes' law.**

$$\eta = \frac{2r^2(d - d_0)g}{9u}$$

A few more terms of viscosity

If the viscosity of a solution is denoted by η and that of its solvent by η_0, η/η_0 is called the relative viscosity.

$$\eta_r = \frac{\eta}{\eta_0}$$

Specific viscosity, η_{sp}, is given by

$$\eta_{sp} = \frac{\eta - \eta_0}{\eta_0}$$

And intrinsic viscosity, $[\eta]$, is given by

$$[\eta] = \lim_{C \to 0}(\eta_{sp}/C).$$

Factors Influencing Viscosity

The intermolecular forces between the molecules of a liquid, the size and shape of the molecules and the temperature determine how easily a liquid flows. In general, the stronger the intermolecular forces of attraction, the more viscous the liquid is. Increasing the size and surface of molecules generally results in increased viscosity, due to the increased dispersion forces. As temperature increases and the molecules move more rapidly, their kinetic energies are better able to overcome the forces that hold them together. Thus, **viscosity of liquids decreases with increasing temperature,** as long as no changes in composition occur. However, the coefficient of viscosity of gases increases with the increasing temperature.

The equation proposed by **Arrhenius and Guzman** relates viscosity of a liquid with temperature.

$$\eta = Ae^{E/RT}$$

A is a constant and E is the activation energy for the viscous flow. If at two temperatures T_1 and T_2, viscosities are η_1 and η_2 respectively then,

$$\log\frac{\eta_1}{\eta_2} = \frac{E}{2.303R}\left(\frac{T_2 - T_1}{T_1 T_2}\right).$$

Surface Tension

The molecules below the surface of a liquid are influenced by intermolecular attraction forces from all directions uniformly. Those on the surface are attracted towards the interior; these attractions pull the surface layer towards the centre. Alternatively, we may say that the liquid surface is under tension due to unbalanced forces.

Surface tension of a liquid may be defined as **the force per unit length acting perpendicular to the tangential line on the surface of the liquid.** Surface tension may also be defined as the **work done in expanding the surface of a liquid by unit area.** The units of surface tension are **J m^{-2}** or **N m^{-1} (SI)** and **dyne cm^{-1} (cgs).**

All forces holding a liquid together are called **cohesive** forces. The forces of attraction between a liquid and another surface are called **adhesive** forces. Capillary action occurs when one end of a capillary tube is immersed in a liquid. If adhesive forces exceed cohesive forces, the liquid creeps up the sides of the tube until a balance is reached between adhesive forces and the weight of the liquid. The smaller the diameter of the capillary tube, the higher the liquid climbs. In case of water its miniscus has a concave shape as water adheres to glass surface. On the other hand mercury does not wet glass because its cohesive forces are much stronger than its attraction to glass. Thus its miniscus is convex. The angle between the tangent to the liquid surface at the point of contact and the solid surface inside the liquid is known as the angle of contact, θ. For liquids which wet the solid surface, $\theta < 90°$ and for those which do not, $\theta > 90°$.

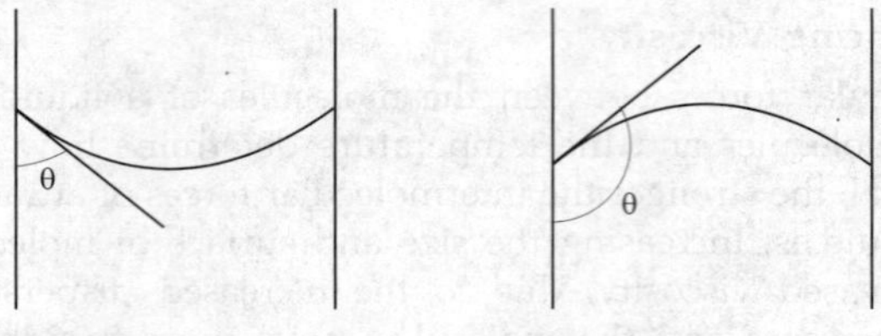

Determination of Surface Tension

Capillary tube method: In this method, the liquid, the surface tension (γ) of which is to be determined, is allowed to rise in a capillary tube of radius r. The height h to which the liquid rises in the tube is measured and then γ is calculated using the equation

$$\gamma = \frac{hdrg}{2\cos\theta},$$

where g is the gravitational constant, d is the density of the liquid and θ is the angle of contact.

If the angle of contact, $\theta = 0$ then $\cos\theta = 1$.

$$\gamma = \frac{hdrg}{2} = \frac{mg}{2\pi r},$$

where m is the mass of the liquid column.

Falling-Drop Method

The liquids are filled in a stalagmometer, one by one, and the falling drops are counted for the same volume of the two liquids. If the two liquids have surface tension γ_1 (unknown) and γ_2 (known) and their drop weights are m_1 and m_2 respectively then

$$\frac{\gamma_1}{\gamma_2} = \frac{m_1}{m_2}.$$

If n_1 and n_2 are the numbers of drops for the same volume of the two liquids whose densities are d_1 and d_2 respectively

$$\frac{\gamma_1}{\gamma_2} = \frac{d_1 n_2}{d_2 n_1}.$$

Thus, γ_1 can be calculated using this equation.

Effect of Temperature on Surface Tension

Surface tension of a liquid decreases with the increase in temperature and vanishes at the critical temperature as the intermolecular forces decrease with the rise in temperature.

The temperature dependence of the surface tension of a liquid is given by the Sugden equation

$$\gamma = \gamma_0 \left(1 - \frac{T}{T_C}\right)^n,$$

where γ_0 is a constant for a given liquid, T_C is the critical temperature, and n is a constant (≈ 1.2 for liquids that are not highly associated).

EXAMPLES

Ex. 1. *A metal crystallizes in a simple cubic unit cell. The length of the edge of the unit cell, a, is* 6.22 Å. *Find the radius of each atom of the metal.*

Solution : We have, for simple cubic structure,

$$\text{radius} = \frac{a}{2} = \frac{6.22}{2} = 3.11 \text{ Å}.$$

Ex. 2. *Copper metal has a face-centred cubic structure with the unit-cell length equal to* 0.361 nm. *Picturing copper ions in contact along the face diagonal, what is the apparent radius of a copper ion?*

Solution : For a face-centred cube, we have,

$$\text{radius} = \frac{\sqrt{2}\,a}{4} = \frac{\sqrt{2} \times 0.361}{4}\ \text{nm} = 0.128\ \text{nm}.$$

Ex. 3. *Find the distance between the body-centred atom and one corner atom in sodium.* $a = 0.424$ nm.

Solution : The atom at the centre of the cube is supposed to touch the two nearest corner atoms. The longest diagonal at which all these three atoms lie will be of length $4r$.

$\therefore$ distance between the centre atom and the corner atom $= 2r$

$$= 2 \times \frac{\sqrt{3}\,a}{4} = \frac{2 \times \sqrt{3} \times 0.424}{4}\ \text{nm} = 0.367\ \text{nm}.$$

Ex. 4. *Potassium crystallizes in a body-centred cubic unit cell. What is the unit-cell content (Z) for potassium?*

Solution : One-eighth of each corner atom and the entire body-centred atom are contained within the unit cell of K.

Thus, $Z = 8\left(\frac{1}{8}\right) + 1(1) = 2.$

Ex. 5. *A compound alloy of gold and copper crystallizes in a cube lattice in which the gold atoms occupy the lattice points at the corners of a cube and the copper atoms occupy the centres of each of the cube faces. Determine the formula of this compound.*

Solution : One-eighth of each corner atom (Au) and one-half of each face-centred atom (Cu) are contained within the unit cell of the compound.

Thus, number of Au atoms per unit cell $= 8 \times \frac{1}{8} = 1$ and number of Cu atoms per unit cell $= 6 \times \frac{1}{2} = 3$. The formula of the compound is $AuCu_3$.

Ex. 6. *What is the simplest formula of a solid whose cubic unit cell has the atom* A *at each corner, the atom* B *at each face centre and a* C *atom at the body centre?*

Solution : An atom at the corner of a cube is shared among 8 unit cells. As there are 8 corners in a cube,

number of corner atom (A) per unit cell $= 8 \times \frac{1}{8} = 1.$

A face-centred atom in a cube is shared by two unit cells. As there are 6 faces in a cube,

number of face-centred atoms (B) per unit cell $= 6 \times \frac{1}{2} = 3.$

An atom in the body of the cube is not shared by other cells.

$\therefore$ number of atoms (C) at the body centre per unit cell = 1.

Hence the formula of the solid is AB_3C.

Ex. 7. *Calculate the packing fraction for the* Ca *unit cell, given that* Ca *crystallizes in a face-centred cubic unit cell.*

Solution : One-eighth of each corner atom and one-half of each face-centred atom are contained within the unit cell of Ca giving

$$Z = 8\left(\frac{1}{8}\right) + 6\left(\frac{1}{2}\right) = 4.$$

Further, atomic radius, $r = \frac{\sqrt{2}\,a}{4}$ ... (Eqn. 4)

$$\text{Volume of 4 atoms} = 4 \times \frac{4}{3}\pi r^3 = 4 \times \frac{4}{3} \times \pi\left(\frac{\sqrt{2}\,a}{4}\right)^3 = \frac{\sqrt{2}\,\pi\,a^3}{6}.$$

Packing fraction $= \frac{\sqrt{2}\pi\,a^3}{6} \Big/ a^3 = \frac{\sqrt{2}\,\pi}{6} = 0.74.$ ... (Eqn. 7)

Ex. 8. *Metallic gold crystallizes in the face-centred cubic lattice. The edge length of the cubic unit cell,* $a = 4.070$ Å. *Calculate the closest distance between gold atoms and the density of gold. Atomic mass of* Au = 197 amu.

Solution : In a face-centred cubic cell,

radius $= \frac{\sqrt{2}a}{4}$ ... (Eqn. 3)

$\therefore$ the closest distance between two atoms = diameter $= 2 \times \frac{\sqrt{2}a}{4} = \frac{a}{\sqrt{2}}$

$$= \frac{4.070}{\sqrt{2}}\ \text{Å} = 2.878\ \text{Å}.$$

Number of atoms in a face-centred unit cell $= 8\left(\frac{1}{8}\right) + 6\left(\frac{1}{2}\right)$

$= 4.$

Mass of 4 atoms per unit cell $= 4 \times 197$ amu

$= 4 \times 197 \times (1.66 \times 10^{-24})$ g

$= 1.308 \times 10^{-21}$ g.

Volume of the unit cell $= a^3$

$= (4.07 \times 10^{-8})^3$ cc.

$$\therefore \text{ density of gold} = \frac{1.308 \times 10^{-21}}{(4.07 \times 10^{-8})^3} = 19.40 \text{ g/cc.}$$

[**Note:** This problem can be solved directly using Eqn (1)]

Ex. 9. *Calculate the value of the Avogadro constant from the internuclear distance of adjacent ions in* NaCl, 0.282 nm, *and the density of solid* NaCl, 2.17×10^3 kg/m³. *A unit cell contains* 4 NaCl *formula units.*

Solution : To calculate the Avogadro constant, that is, number of NaCl formula units per mole, let us take 1 mole of NaCl.

$$\text{Volume of 1 mole of NaCl} = \frac{\text{mass of 1 mole}}{\text{density}}$$

$$= \frac{58.5 \times 10^{-3}}{2.17 \times 10^3} = 2.7 \times 10^{-5} \text{ m}^3.$$

$$\text{Volume of 1 unit cell} = a^3 = (0.564 \times 10^{-9})^3 = 1.79 \times 10^{-28} \text{ m}^3.$$

$$\therefore \text{ number of unit cells per mole} = \frac{2.7 \times 10^{-5}}{1.79 \times 10^{-28}} = 1.51 \times 10^{23}.$$

Since 1 unit cell of NaCl has 4 NaCl formula units,

$$\text{Avogadro constant (formula units per mole)} = 4 \times (1.51 \times 10^{23})$$
$$= 6.04 \times 10^{23}.$$

Ex. 10. *The density of a particular crystal of* LiF *is* 2.65 g/cc. *X-ray analysis shows that* Li^+ *and* F^- *ions are arranged in a cubic array at a spacing of* 2.01 Å. *From these data calculate the apparent Avogadro constant.*

Solution : See Example 29, Chapter 1.

Ex. 11. *Gold has a close-packed structure which can be viewed as spheres occupying* 0.74 *of the total volume. If the density of gold is* 19.3 g/cc, *calculate the apparent radius of a gold ion in the solid.* (Au = 197 amu)

Solution : Gold has a close-packed structure with a packing fraction value of 0.74. This shows that it has a face-centred cubic cell. The number of ions in a face-centred unit cell is 4.

$$\text{Now, density} = \frac{\text{mass of unit cell}}{\text{volume of unit cell}}$$

$$\text{or} \qquad 19.3 = \frac{4 \times (197) \times 1.66 \times 10^{-24}}{a^3}; \; a = 4.07 \times 10^{-8} \text{ cm.}$$

In a face-centred cubic cell,

$$\text{radius} = \frac{\sqrt{2}a}{4} = \frac{\sqrt{2} \times 4.07 \times 10^{-8}}{4} = 1.439 \times 10^{-8} \text{ cm.}$$

Ex. 12. *An element exists in the body-centred cubic structure whose cell edge is* 2.88 Å. *The density of the element is* 7.20 g/cc. *Calculate the number of atoms in* 104 g *of the element.*

Solution : Volume of unit cell $= (2.88 \times 10^{-8})^3$ cc $= 2.39 \times 10^{-23}$ cc.

Volume of the element weighing 104 g $= \dfrac{\text{mass}}{\text{density}}$

$$= \frac{104}{7.20} = 14.44 \text{ cc.}$$

$\therefore$ number of unit cells present in 104 g of the element

$$= \frac{14.44}{2.39 \times 10^{-23}} = 6.04 \times 10^{23}.$$

Since each body-centred cubic cell contains 2 atoms,

number of atoms $= 2 \times 6.04 \times 10^{23} = 1.208 \times 10^{24}$.

Ex. 13. *Calculate the distance between* (1 1 1) *planes in a crystal of* Ca. *Repeat the calculation for the* (2 2 2) *planes. Which planes are closer?* ($a = 0.556$ nm)

Solution : We have,

$$d = \frac{a}{\sqrt{h^2 + k^2 + l^2}} \qquad \text{... (Eqn. 3)}$$

$$d_{111} = \frac{0{\cdot}556}{\sqrt{1^2 + 1^2 + 1^2}} = 0{\cdot}321 \text{ nm}$$

and

$$d_{222} = \frac{0{\cdot}556}{\sqrt{2^2 + 2^2 + 2^2}} = 0{\cdot}161 \text{ nm}$$

The separation of the (1 1 1) planes is twice as great as that of the (2 2 2) planes.

Ex. 14. *Calculate the ratio of separation between successive* (1 0 0), (1 1 0) *and* (1 1 1) *lattice planes in a cubic cell.*

Solution : We have,

$$d = \frac{a}{\sqrt{h^2 + k^2 + l^2}}.$$

Thus,

$$d_{100} = \frac{a}{\sqrt{1^2 + 0^2 + 0^2}} = a$$

$$d_{110} = \frac{a}{\sqrt{1^2 + 1^2 + 0^2}} = \frac{a}{\sqrt{2}}$$

and $$d_{111} = \frac{a}{\sqrt{1^2 + 1^2 + 1^2}} = \frac{a}{\sqrt{3}}$$

Hence, $d_{100} : d_{110} : d_{111} = a : \frac{a}{\sqrt{2}} : \frac{a}{\sqrt{3}} = 1 : \frac{1}{\sqrt{2}} : \frac{1}{\sqrt{3}}$.

Ex. 15. *The coordinates of the three corners of a shaded face on a cubic unit cell are* $\left(\frac{1}{2}, \frac{1}{2}, 1\right)$, $\left(0, 1, \frac{1}{2}\right)$ *and* $\left(1, 1, \frac{1}{2}\right)$ *as shown below in the figure. Determine the Miller indices of the plane.*

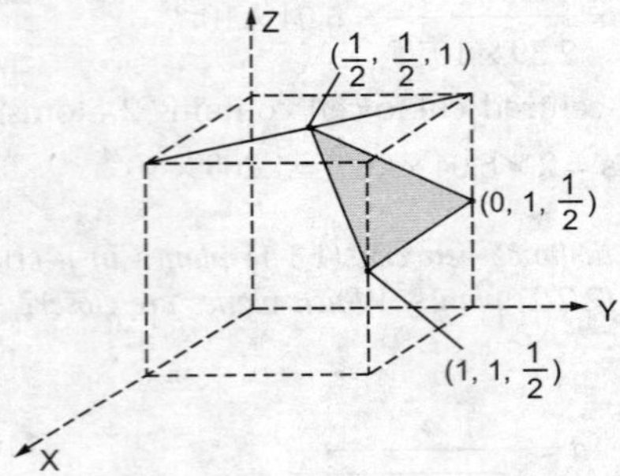

Solution : The intercepts of the shaded plane with the axes are ∞, 2*b* and 2*c* respectively.

a	*b*	*c*	
∞	2	2	intercepts
0	1/2	1/2	reciprocals
0	1	1	clear fractions

Thus, the Miller indices are (0 1 1).

Ex. 16. *Determine the Miller indices of the shaded plane. Coordinates of the corners of the plane are shown in the figure below.*

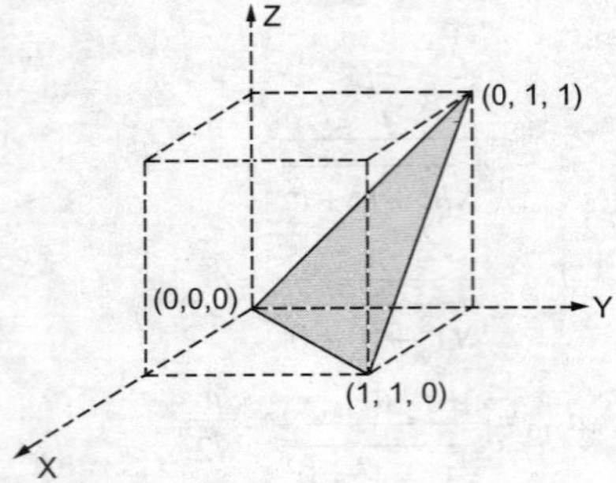

Solution : A plane parallel to the plane between (0, 0, 0), (1, 1, 0) and (0, 1, 1) will intersect the axes at $x = a, y = -b$ and $z = c$.

a	b	c	
1	− 1	1	intercepts
1	− 1	1	reciprocals
1	− 1	1	clear fractions

Thus the Miller indices are $(1\,\bar{1}\,1)$ or $(\bar{1}\,1\,\bar{1})$.

Ex. 17. *The first-order reflection of a beam of X-rays of wavelength* 1.54 Å *from the* (1 0 0) *plane of a crystal of the simple cubic type occurs at an angle of* 11.29°. *Calculate the length of the unit cell.* (sin 11.29° = 0.1991)

Solution : We have,

$$n\lambda = 2d \sin \theta \quad \text{... (Eqn. 6)}$$

$$1 \times (1.54 \times 10^{-8} \text{ cm}) = 2 \times d \times \sin 11.29$$

$$d = 3.68 \times 10^{-8} \text{ cm}$$

Further,
$$d = \frac{a}{\sqrt{h^2 + k^2 + l^2}} \quad \text{...(Eqn. 2)}$$

or
$$3.68 \times 10^{-8} = \frac{a}{\sqrt{1^2 + 0^2 + 0^2}}$$

or
$$a = 3.68 \times 10^{-8} \text{ cm.}$$

Ex. 18. *Find the distance between two* Po *atoms that lie along a body diagonal, given that* a = 0·336 nm.

Solution : The coordinates of the two atoms are (0, 0, 0) and (1, 1, 1) respectively. Thus,

$$l = 0{\cdot}336\,[(1-0)^2 + (1-0)^2 + (1-0)^2]^{\frac{1}{2}} \quad \text{... (Eqn. 2)}$$

$$= 0{\cdot}582 \text{ nm.}$$

Ex. 19. *Determine the number of formula units of* NaCl *in the unit cell.* NaCl *is a face-centred cubic crystal. See the diagram of the* NaCl *unit cell in the text.*

Solution : Either of Na^+ or Cl^- ions may be chosen as the lattice points. Choosing Cl^- ions (as shown in the text), the cube contains a Cl^- ion at the centre of each face as well as at the corners of the unit cell. There are 12 Na^+ ions at the centres of 12 edges and 1 Na^+ ion at the centre of the unit cell.

$\therefore$ at the 8 corners: $8\times\frac{1}{8}=1\ Cl^-$ ion

At the 6 faces: $6\times\frac{1}{2}=3\ Cl^-$ ion

Along the 12 edges: $12\times\frac{1}{4}=3Na^+$ ions

At the centre: 1 Na^+ ion

Hence, the unit cell contains 4 NaCl units.

Ex. 20. *The ionic radii of* Cs^+ *and* Cl^- *ions are* 1.69 Å *and* 1.81 Å *respectively. Predict the coordination number of* Cs^+.

Solution : $\frac{r_+}{r_-}=\frac{1.69}{1.81}=0.934.$

As the radius ratio lies between 0.732 and 1.0, the coordination number of Cs^+ is 8 (see table in the text).

Ex. 21. *The coordination number of* Ba^{2+} *in* BaF_2 *is* 8. *What must be the coordination number of* F^-?

Solution : The coordination numbers and the charges of ions always balance out to give neutrality. The CN of Ba^{2+} indicates that each Ba^{2+} ion is surrounded by 8 F^- ions. In order to balance 8 negative charges of F^- ions, 4 Ba^{2+} ions are needed. Hence the CN of F^- is 4.

Ex. 22. *How many unit cells are there in a*
(a) 1.0 g *cubic crystal of* NaCl? *and*
(b) along each edge of the crystal?
The cell content of NaCl *is* 4 *and molar mass of* NaCl *is* 58.5 g.

Solution : Mass of 1 unit cell $=4\times$ mass of 1 NaCl unit

$$=4\times\frac{58.5}{6.022\times10^{23}}\text{ g}=3.885\times10^{-22}\text{ g.}$$

$$\therefore \text{ no. of unit cells per 1.0 g of NaCl}=\frac{1.0}{3.885\times10^{-22}}=2.57\times10^{21}.$$

$$\text{No. of unit cells per edge}=\sqrt[3]{2.57\times10^{21}}=1.37\times10^{7}.$$

Ex. 23. *Zinc sulphide forms a cubic unit cell.* Zn^{2+} *ions form an f.c.c. lattice and* S^{2-} *ions occupy the centre of the alternate small cubes. Find the number of each ion in a unit cell.*

Solution **:** Effective no. of Zn^{2+} ions in unit cell $= 8 \times \frac{1}{8} + 6 \times \frac{1}{2} = 4.$

As there are 8 small cubes in one unit cell of zinc sulphide and S^{2-} ions are present in alternate cubes, effective no. of S^{2-} in unit cell $= \frac{8}{2} = 4.$

Hence, the molecular formula is ZnS and 4 ZnS formula units are present in one unit cell.

Ex. 24. *Zinc sulphide crystallizes with zinc ions occupying one-half of the tetrahedral holes in a closest-packed array of sulphide ions. What is the formula of zinc sulphide?*

Solution **:** Because there are two tetrahedral holes per anion (sulphide ion) and one-half of these holes are occupied by zinc ions, there must be $\frac{1}{2} \times 2 = 1$ zinc ion per sulphide ion. Thus, the formula is ZnS.

Ex. 25. *Aluminium oxide crystallizes with aluminium ion in two-thirds of the octahedral holes in a closest-packed array of oxide ions. What is the formula of aluminium oxide?*

Solution **:** Because there is one octahedral hole per anion (oxide ion) and only two-thirds of these holes are occupied, the ratio of Al and O should be $\frac{2}{3} : 1$, that is, 2 : 3. Thus the formula is Al_2O_3.

Ex. 26. *What minimum value of r_+/r_- is needed to prevent anion-anion contact in* NaCl *crystal?*

Solution **:** CN of each Na^+ and Cl^- is 6. Na^+ ion occupies the void surrounded by 6 Cl^- ions arranged on the corners of an octahedron. The minimum (limiting) value of r_+/r_- required to prevent anion-anion contact in NaCl may be calculated as follows.

In ΔABC

$$\sin ABC = \frac{AC}{AB}$$

$$\sin 45° = \frac{2r_-}{2(r_+ + r_-)} = \frac{1}{\sqrt{2}}$$

$$\therefore \quad \frac{r_+}{r_-} = 0.414.$$

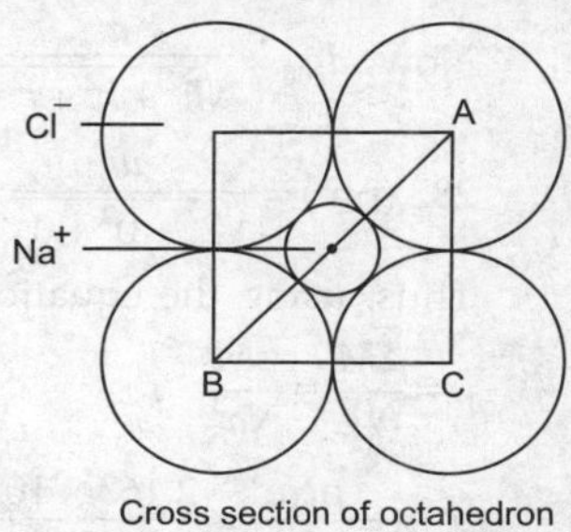

Cross section of octahedron

If r_+/r_- value is greater than 0.414, the two Cl^- ions shall not touch each other.

Ex. 27. *Why does* ZnS ($r_+/r_- = 0.402$) *not crystallize in the* NaCl *structure?*

Solution : The r_+/r_- ratio is 0.402 which is too low to avoid anion-anion contact in the NaCl structure. For anion-anion contact, the minimum r_+/r_- value should be 0.414.

Ex. 28. *For a primitive cubic crystal with* $a = 3 \times 10^{10}$ m, *what is the smallest diffraction angle* θ, *for* (1 1 0) *plane for* $\lambda = 1.50 \times 10^{-10}$ m?

Solution : We have,

$$n\lambda = 2d\sin\theta$$

and $d = \dfrac{a}{\sqrt{h^2+k^2+l^2}}$

$$\therefore\ n\lambda = \frac{2a}{\sqrt{h^2+k^2+l^2}}\sin\theta.$$

For first-degree reflection, $n = 1$

$$1 \times 1.50 \times 10^{-10} = \frac{2 \times 3 \times 10^{10}}{\sqrt{1^2+1^2+0^2}}\sin\theta$$

$$\sin\theta = \frac{\sqrt{2} \times 1.50 \times 10^{-10}}{6 \times 10^{10}}$$

$$\therefore\ \theta = 20.70°.$$

Ex. 29. *The density of sodium chloride at* 25°C *is* $2.163 \times 10^3\ \text{kg m}^{-3}$. *When X-rays from a palladium target having a wavelength of* 58.1 pm *are used, the* (2 0 0) *reflection of sodium chloride occurs at an angle of* 5.90°. *How many* Na^+ *and* Cl^- *ions are there in a unit cell?* (NaCl = 58.443)

Solution : We have, $n\lambda = 2d\sin\theta$.

For $n = 1$, $d_{200} = \dfrac{\lambda}{2\sin\theta} = \dfrac{58.1\ (\text{pm})}{2\sin 5.9°} = 282$ pm.

Now, $d_{200} = \dfrac{a}{\sqrt{h^2+k^2+l^2}}$

$$282\ (\text{pm}) = \frac{a}{\sqrt{2^2+0^2+0^2}};\ a = 564\ \text{pm}$$

Thus, using the equation

$$\rho = \frac{zM}{NV} = \frac{zM}{Na^3}$$

$$\therefore\ z = \frac{\rho Na^3}{M} = \frac{2.163 \times 10^3 \times (6.022 \times 10^{23}) \times (564 \times 10^{-12})^3}{58.443 \times 10^{-3}} = 3.999 \approx 4.$$

Thus, a unit cell contains 4 Na^+ and 4 Cl^- ions as expected.

Ex. 30. *What fraction (n/N) of the lattice sites are vacant at* 298 K *for a crystal for which the energy required to make a defect is* 1 eV. (1 eV = 1.602×10^{-19} J)

Solution : We have,

$$\frac{n}{N}=e^{-\left(\frac{E_v}{kT}\right)}$$

$$\ln\frac{n}{N}=-\frac{E_v}{kT}$$

$$\text{or}\ \log\frac{n}{N}=-\frac{E_v}{2.303\,kT}=-\frac{1.602\times 10^{-19}}{2.303\times(8.314/6.022\times 10^{23})\times 298}.$$

On solving, $\frac{n}{N}=1.24\times 10^{-17}$.

Ex. 31. *Lithium forms a b.c.c. lattice. If the lattice constant is* 3.50×10^{-10} m *and the experimental density is* 5.30×10^{2} kg m^{-3}, *calculate the percentage occupancy of* Li *metal.* (Li = 7)

Solution : We have

$$\text{theoretical density}=\frac{zM}{NV}=\frac{zM}{N(a^3)}$$

For a b.c.c. lattice: $z=2$ and given that $a=3.50\times 10^{-10}$ m and $M=7\times 10^{-3}$ kg/mole

$$d_{\text{cal}}=\frac{2\times(7\times 10^{-3})}{6.022\times 10^{23}\times(3.50\times 10^{-10})^3}$$

$$=5.42\times 10^{2}\ \text{kg m}^{-3}.$$

$$\therefore\ \text{percentage occupancy}=\frac{\rho_{\text{exp}}}{\rho_{\text{cal}}}\times 100$$

$$=\frac{5.30\times 10^{2}}{5.42\times 10^{2}}\times 100=97.78\%.$$

Ex. 32. *Find the radius of the smallest atom that can fit in the octahedral hole.*

Solution : See text: Calculation of r/R.

Ex. 33. *In a measurement of surface tension by the falling-drop method,* 5 *drops of a liquid of density* 0.797 g/mL *weighed* 0.220 g. *Calculate the surface tension of the liquid.*

Solution : Mass of the average drop $=\frac{0.22}{5}=0.044$ g

$$\text{volume of the drop}=\frac{\text{mass}}{\text{density}}$$

$$= \frac{0.044}{0.797} = 0.0552 \text{ mL.}$$

Assuming the drop to be spherical in shape

$$\text{volume of the drop} = \frac{4}{3}\pi r^3 = 0.0552$$

$$r = \sqrt[3]{\frac{3 \times 0.0552}{4 \times 3.14}} = 0.236 \text{ cm}$$

$$\gamma = \frac{mg}{2\pi r} = \frac{0.044 \times 981}{2 \times 3.14 \times 0.236}$$

$$= 29.12 \text{ dyne cm}^{-1}$$

$$= 0.02912 \text{ N m}^{-1}.$$

Ex. 34. *A film of pyridine filled a rectangular wire loop in which one side could be moved. Given that the wire loop is* 8.53 cm *wide and that a force of* 6.48×10^{-3} N *is needed to move the side, determine the value of the surface tension. What is the work necessary to stretch the film a distance of* 0.1 cm?

Solution : The force is related to the surface tension γ and the width l by

$$F = 2l\gamma$$

$$\therefore \quad \gamma = \frac{F}{2l} = \frac{6.48 \times 10^{-3}}{2 \times 8.53 \times 10^{-2}}$$

$$= 3.79 \times 10^{-2} \text{ N m}^{-1}.$$

As $\quad \gamma = \dfrac{\text{work done}(w)}{\text{area}(A)};$

$$w = \gamma \times A = 2(3.79 \times 10^{-2}) \times (8.53 \times 10^{-2}) \times (0.1 \times 10^{-2});$$

$$[A = 2 \times (8.53 \times 10^{-2}) \times (0.1 \times 10^{-2}) \text{ m}^2]$$

$$= 6.5 \times 10^{-6} \text{ J.}$$

Ex. 35. *A steel ball of density* 8.0×10^3 kg/m^3 *and radius* 2 mm *is observed to fall with a terminal velocity* 1.0×10^{-2} m/s *in a liquid of density* 1.8×10^3 kg/m^3. *Calculate the viscosity of the liquid.*

Solution : We have,

$$\eta = \frac{2r^2(d - d_0)g}{9u}$$

$$= \frac{2(2 \times 10^{-3})^2(8.0 \times 10^3 - 1.8 \times 10^3) \times 9.8}{9 \times 1.0 \times 10^{-2}}$$

$$= 5.4 \text{ N m}^{-2} \text{ s.}$$

Ex. 36. *Water rises in a capillary tube to a height of* 4.8 cm *at* 25°C. *The density of water at* 25°C *is* 0.9984 g/cc. *The same capillary, when filled with mercury, contained* 40.5 g/cm *of the capillary. Calculate the surface tension of water. Density of* Hg = 13.6 g/cc *and* $g = 981$ cm s^{-2}.

Solution : Let us first calculate the radius of the capillary tube (r).

Wt. of Hg in the capillary = mass of Hg × g

$$w_{Hg} = \text{volume} \times \text{density} \times g$$

$$w_{Hg} = \pi r^2 h \times d \times g$$

or
$$\frac{w_{Hg}}{h} = \pi r^2 dg = 40.5 \text{ g/cm}.$$

∴
$$r = \sqrt{\frac{40.5}{(22/7) \times 13.6 \times 981}} = 0.031 \text{ cm}$$

We have,

$$\gamma_{H_2O} = \frac{hdrg}{2}$$

$$= \frac{4.8 \times 0.9984 \times 0.031 \times 981}{2}$$

$$= 72.86 \text{ dyne cm}^{-1}.$$

Ex. 37. *The surface tension of ethanol at* 30°C *is* 2.189×10^{-2} N m^{-1} *and its density* = 0.780 g/cc. *To what height will this liquid rise in a capillary tube of radius* 0.002 cm? *What pressure is needed to push the miniscus level back with the surrounding liquid?*

***Solution* :** We have, $h = \dfrac{2\gamma}{rdg}$

$$d = 0.78 \text{ g/cc} = 0.78 \text{ kg/L} = 0.78 \times 10^3 \text{ kg/m}^3$$

$$r = 0.002 \text{ cm} = 2.0 \times 10^{-4} \text{ m}, g = 9.8 \text{ m/s}^2$$

∴
$$h = \frac{2 \times 2.189 \times 10^{-2}}{2.0 \times 10^{-4} \times 0.78 \times 10^3 \times 9.8} = 0.0286 \text{ m}.$$

Further, the pressure (p) required to push the liquid back with the surrounding liquid is given by

$$p = h \cdot d \cdot g$$

$$= 0.0286 \times 0.78 \times 10^3 \times 9.8$$

$$= 218.6 \text{ Pa (or N m}^{-2}).$$

Ex. 38. *The viscosity of molten sodium is* 4.5×10^{-4} N m^{-2} s *at* 473 K *and* 2.12×10^{-4} N m^{-2} s *at* 873 K. *Calculate the activation energy for the viscous flow. Also, calculate the viscosity at* 673 K.

Solution : We have,

$$\log \frac{\eta_1}{\eta_2} = \frac{E}{2.303R}\left(\frac{T_2 - T_1}{T_1\,T_2}\right)$$

$$\log \frac{4.5 \times 10^{-4}}{2.12 \times 10^{-4}} = \frac{E}{2.303 \times 8.314}\left(\frac{873 - 473}{473 \times 873}\right)$$

$$E = 6461\ \text{J/mole}$$
$$= 6.461\ \text{kJ/mole.}$$

Further, let the viscosity at 673 K be η_2.

Again, we have, $T_1 = 473\,\text{K};\ \ \eta_1 = 4.5 \times 10^{-4}$

$$T_2 = 673\ \text{K};\ \ \eta_2 = ?$$

$$\log \frac{4.5 \times 10^{-4}}{\eta_2} = \frac{6461}{2.303 \times 8.314}\left(\frac{673 - 473}{473 \times 673}\right)$$

$$\eta_2 = 2.76 \times 10^{-4}\ \text{N m}^{-2}\ \text{s.}$$

Ex. 39. *The surface tension of water at* $21°\text{C}$ *is* $72.75 \times 10^{-3}\ \text{N m}^{-1}$. *A* 33.24% (*vol./vol.*) *solution of ethanol has* $\gamma = 33.24 \times 10^{-3}\ \text{N m}^{-1}$ *at the same temperature. Given density (solution)* $= 0.9614 \times 10^3\ \text{kg m}^{-3}$ *and density (water)* $= 0.9982 \times 10^3\ \text{kg m}^{-3}$. *How much less will the alcohol solution rise in the same capillary? Angle of contact,* $\theta = 0°$.

Solution : We have,

$$\gamma = \frac{hdrg}{2}.$$

As the same capillary is used for both ethanol and water, r and g are the same in both.

$$\therefore \qquad \frac{h_1}{h_2} = \frac{\gamma_1 \cdot d_2}{\gamma_2 \cdot d_1}.$$

'1' and '2' stand for ethanol and water respectively.

$$\frac{h_1}{h_2} = \frac{(33.24 \times 10^{-3}) \times 0.9982 \times 10^3}{(72.75 \times 10^{-3}) \times 0.9614 \times 10^3} = 0.474.$$

Thus, ethanol will rise only 47.4% as far as pure water.

Ex. 40. *A liquid rises to* 1.0 cm *in a capillary tube of radius* r_1. *How much will it rise if the cross-sectional area of the capillary tube is doubled?*

Solution : Let the cross-sectional areas of the tubes be a and $2a$.

$$\therefore \qquad a = \pi r_1^{\,2};\ r_1 = \sqrt{\frac{a}{\pi}}$$

$$\text{and,} \qquad 2a = \pi r_2^{\,2};\ r_2 = \sqrt{\frac{2a}{\pi}}.$$

We have, $\gamma = \frac{hdrg}{2}$ and $h_1 = 1$ cm.

$\therefore$ for the same liquid, using the above-given equations, we get

$$\frac{h_2}{h_1} = \frac{r_1}{r_2} = \frac{1}{\sqrt{2}}; \quad h_2 = \frac{1}{\sqrt{2}} \times 1 = 0.707 \text{ cm.}$$

Ex. 41. *In an Ostwald viscometer, water takes* 25 s *to flow between the lower and upper marks, while the liquid* X *takes* 38 s. *Given* $d(H_2O) = 0.9983$ kg/dm^3, $d(X) = 0.7894$ kg/dm^3, $\eta(H_2O) = 1.005$ *centipoise. Calculate* η *of liquid* X.

Solution : $\frac{\eta_1}{\eta_2} = \frac{t_1 \cdot d_1}{t_2 \cdot d_2}$

Suppose 1 and 2 stand for X and water respectively.

$$\frac{\eta_1}{1.005} = \frac{38 \times 0.7894}{25 \times 0.9983}$$

$\eta_1 = 1.208$ centipoise.

Ex. 42. *Calculate the rate of flow (volume per second) of a liquid through a capillary tube of diameter* 0.2×10^{-3} m *and length* 1 m; $\eta = 3.00 \times 10^{-3}$ N m^{-2} s *and pressure gradient* = 10 atm.

Solution : $r = \frac{0.2 \times 10^{-3}}{2} = 0.1 \times 10^{-3}$ m

$= 10^{-4}$ m,

$p = 10$ atm

$= 10 \times 1.01 \times 10^5$ Pa

$= 1.01 \times 10^6$ Pa.

We have,

$$\eta = \frac{\pi p r^4 t}{8Vl}.$$

$\therefore$ rate of flow $= \frac{V}{t} = \frac{\pi p r^4}{8\eta l}$

$$= \frac{(22/7) \times (1.01 \times 10^6)(10^{-4})^4}{8 \times (3.0 \times 10^{-3}) \times 1}$$

$= 1.32 \times 10^{-8}$ m^3 s^{-1}.

Ex. 43. *How much work is required to break up* 1 *mole of water at* 20°C *into spherical droplets of radius* 1 cm? *Given that* $\gamma(H_2O) = 72.75 \times 10^{-3}$ N m^{-1}, *density* $(H_2O) = 0.998$ g cm^{-3}, $H_2O = 18.015$ g/mole.

Solution : Supposing 1 mole of water to be spherical,

$$\text{volume of 1 mole of water} = \frac{18.015}{0.998 \times 10^6}$$

$$= 1.8047 \times 10^{-5}\ \text{m}^3.$$

$$\text{Volume}(V) = \frac{4}{3}\pi r^3$$

$$\therefore \quad r = \sqrt[3]{\frac{3 \times V}{4\pi}} = \left(\frac{3 \times 1.8047 \times 10^{-5}}{4 \times 3.14}\right)^{1/3} = 1.6272 \times 10^{-2}\ \text{m}.$$

$$\text{Surface area} = 4\pi r^2 = 4 \times 3.14 \times (1.6272 \times 10^{-2})^2$$

$$= 3.3273 \times 10^{-3}\ \text{m}^2.$$

$$\text{Volume of each droplet} = \frac{4}{3} \times \pi \times (1 \times 10^{-2})^3 \approx 4 \times 10^{-6}\ \text{m}^3.$$

$$\text{Number of droplets} = \frac{1.8047 \times 10^{-5}}{4 \times 10^{-6}} \approx 5.$$

$$\text{Surface area of 5 droplets} = 5 \times 4\pi r^2$$

$$= 5 \times 4 \times 3.14 \times (1 \times 10^{-2})^2$$

$$= 6 \times 10^{-3}\ \text{m}^2.$$

When a drop of 1 mole of water breaks into 5 droplets,

$$\text{increase in surface area} = (6 \times 10^{-3} - 3.3273 \times 10^{-3})\ \text{m}^2$$

$$= 2.6727 \times 10^{-3}\ \text{m}^2.$$

Work done to break up 1 mole of water

= surface tension × increase in area

$$= 72.75 \times 10^{-3} \times (2.6727 \times 10^{-3})\ \text{J}$$

$$= 1.94 \times 10^{-4}\ \text{J}. \qquad (1\ \text{J} = 1\ \text{N m})$$

Ex. 44. *The densities of water and isopropyl alcohol are* 0.9982 *and* 0.7887 g/cc. *At* 20°C, *isopropyl alcohol flowed through a viscometer in* 624 s *and an equal volume of water flowed through the same viscometer in* 200 s. *If* η *(water)* $= 1.009 \times 10^{-3}$ N m^{-2} s, *calculate* η *(isopropyl alcohol) at* 20°C.

Solution : We have,

$$\frac{\eta_1}{\eta_2} = \frac{t_1 d_1}{t_2 d_2}.$$

Suppose that '1' and '2' stand for alcohol and water respectively,

$$\therefore \eta_1 = \frac{624 \times 0.7887}{200 \times 0.9982} \times (1.009 \times 10^{-3})$$

$$= 2.487 \times 10^{-3}\ \text{N m}^{-2}\ \text{s}.$$

PROBLEMS

(Answers bracketed with questions)

1. Potassium metal crystallizes in a face-centred arrangement of atoms where the edge of the unit cell is 0.574 nm. What is the shortest separation of any two potassium nuclei?

[Hint: Calculate diameter of K] (0.406 nm)

2. A simple cubic lattice consists of eight identical spheres of radius R in contact, placed at the corners of a cube. What is the volume of the cubical box that will just enclose these eight spheres and what fraction of this volume is actually occupied by the spheres? ($64R^3$, 52.36%)

3. Copper has a face-centred cubic structure with a unit-cell edge length of 3.61 Å. What is the size of the largest atom which could fit into the interstices of the copper lattice without distorting it?

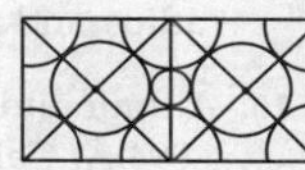

[Hint: Calculate the radius of the smallest circle in the figure.] (0.53 Å)

4. Calculate the packing fraction for the K unit cell. K crystallizes in a body-centred cubic unit cell. (0.68)

5. Calculate the percentage of vacant space in a Si unit cubic cell. The unit-cell content for Si is 8 and $r = \frac{\sqrt{3}\,a}{8}$. (See hint of Q. 7) (66%)

6. Calculate the packing factor for spheres occupying (a) a body-centred cubic structure, and (b) a simple cubic structure, where closest neighbours in both cases are in contact. [(a) 0.68 (b) 0.524]

7. Silicon crystallizes in a unit cell to that of diamond. Find the unit-cell content for Si. (8)

[Hint: Diamond has a face-centred cubic unit cell containing a tetrahedron of atoms.]

8. The intermetallic compound LiAg crystallizes in a cubic lattice in which both Li and Ag atoms have coordination numbers of 8. To what crystal class does the unit cell belong? (Cubic structure)

9. A compound formed by elements A and B crystallizes in the cubic structure where A atoms are at the corners of a cube and B atoms are at the face centre. Determine the simple formula of the compound. (AB_3)

10. A cubic solid is made of two elements A and B. Atoms of B are at the corners of the cube and of A, at the body centre. Determine the formula of the compound. (AB)

11. A mineral having the formula AB_2 crystallizes in the cubic close-packed lattice, with the A atoms occupying the lattice points. What are the coordination numbers of the A and B atoms? (8, 4)

12. Calculate the Avogadro constant from the following data:
Density of solid NaCl = 2.165 g/cc.

Distance between centres of adjacent Na^+ and Cl^- = 0.2819 nm.
Also, calculate the edge length of a cube containing 1 mole of NaCl and the number of ions ($Na^+ + Cl^-$) along one edge of the cube. (6.02×10^{23}, 3.0 cm, 1.064×10^8)

13. Metallic rhodium crystallizes in a face-centred cubic lattice with a unit-cell edge length of 3.803 Å. Calculate the molar volume of rhodium including the empty spaces. (8.28 cc)

14. The atomic radius of palladium is 1.375 Å. The unit cell of palladium is a face-centred cube. Calculate the density of palladium. (12.01 g/cc)

15. The unit cell of tungsten is a face-centred cube having a volume of 31.699 $Å^3$. The atom at the centre of each face just touches the atoms at the corners. Calculate the radius and atomic volume of tungsten. (1.1189 Å, 5.8676 $Å^3$)

16. Aluminium crystallizes in a face-centred cubic unit cell with an edge length of 4.094 Å. Calculate the approximate Avogadro constant. (5.83×10^{23})

17. An unknown metal is found to have a specific gravity of 10.2 at 25°C. It is found to crystallize in a body-centred cubic lattice with a unit cell edge length of 3.147 Å. Calculate the atomic weight. (95.7)

18. Zinc selenide, ZnSe, crystallizes in a face-centred cubic unit cell and has a density of 5.267 g/cc. Calculate the edge length of the unit cell. (5.667 Å)

19. A face-centred cubic solid of an element (atomic mass 60) has a cube edge of 4.0 Å. Calculate its density. (6.23 g/cc)

20. Polonium crystallizes in a simple cubic unit cell. Its atomic mass is 209 and density is 91.5 kg m^{-3}. What is the edge length of its unit cell? (1.56×10^{-7} cm)

21. A metallic element has cubic lattice. Each edge of the unit cell is 3.0 Å. The density of the metal is 8.5 g/cc. How many unit cells will be present in 50 g of the metal? (2.178×10^{23})

22. The d_{111} spacing for crystalline K is 0.3079 nm. Calculate the length of the cubic unit cell. (0.5333 nm)

23. Calculate the Miller indices of crystal planes which cut through the crystal axes at (a, b, c), $(2a, b, c)$ and $(2a, -3b, -3c)$. [(1 1 1), (1 2 2), $(3\,\bar{2}\,\bar{2})$]

24. For a primitive cubic crystal with $a = 3 \times 10^{10}$ m, what are the smallest diffraction angles θ for (a) (1 0 0) and (b) (1 1 1) planes for $\lambda = 1.50 \times 10^{-10}$ m? [(a) 15.48°, (b) 25.66°]

25. Potassium crystallizes with b.c.c. lattice and has a density of 0.856×10^3 kg m^{-3}. What is the length of the side of the unit cell, a, and what is the distance between (2 0 0), (1 1 0), and (2 2 2) planes? What is the closest distance between atoms, and what is the potassium atom radius, r?
(533.3 pm; 266.7, 377.1 and 154.0 pm; 462.0, 231.0 pm)

26. The ionic radii of Na^+ and Cl^- ions are 0.98×10^{-10} m and 1.81×10^{-10} m respectively. Find the coordination number of each ion. (6, 6)

27. What is the critical radius ratio for the CsCl structure? (0.732)

[Hint: See text for cubic void]

28. Calculate the fractional void volume in the c.c.p. and h.c.p. structures of hard spheres. (0.2594, 0.2594)

29. Ice crystallizes in a hexagonal lattice. At the low temperature at which the structure was determined, the lattice constants were $a = 4.53$ Å and $c = 7.41$ Å. How many H_2O molecules are contained in a unit cell? The density of ice is 0.92 g/cc at 0°C. A unit cell of H_2O is shown below: (4)

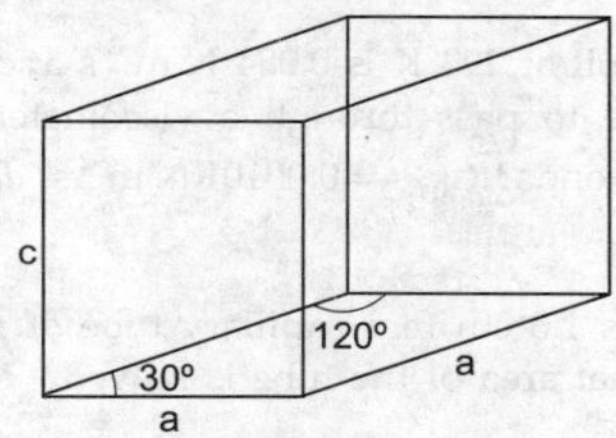

30. The surface tension of glacial acetic acid was determined using the 'bubble pressure' method in which the pressure needed to dislodge bubbles of air from the end of a capillary tube immersed in the liquid is measured. Given that the radius of the tube (r) is 1.1 mm, the depth of the tube in the liquid (h) is 3.56 cm, the pressure is 420 Pa and density (CH_3COOH) = 1.0492 g·cm^{-3}. Determine γ. (2.9×10^{-2} N m^{-1})

[Hint: Laplace equation: $\gamma = \frac{r}{2}(p - hdg)$]

31. The surface tension of toluene at 298 K is 0.0284 Nm^{-1} and its density is 0.866 g/cc. What is the largest radius of the capillary that will permit the liquid to rise 2×10^{-4} m? Assume $\theta = 0$. (3.34×10^{-4} m)

32. Calculate the capillary depression of Hg in a tube of diameter 1.0 mm. Assume that the contact angle is zero. The density of Hg is 13.6×10^3 kg m^{-3} and the surface tension of Hg is 0.460 Nm^{-1}. (1.38×10^{-2} m)

33. The time taken by a metal ball to drop through a liquid A of height h is 5.0 s, whereas that in liquid B is 7.5 s. If the densities of the metal ball, liquids A and B are 7.8×10^3, 1.5×10^3 and 4.6×10^3 kgm^{-3} respectively, calculate the viscosity of liquid A. η(B) = 2.5 cP. (3.28×10^{-2} cP)

34. A certain liquid has a viscosity of 1.0×10^4 poise and a density of 3.2 g/mL. How long will it take for a platinum ball with a 2.5-mm-radius to fall 1.0 cm through the liquid? The density of platinum is 21.4 g/cc. (40.5 s)

35. Two capillary tubes of radius 0.2 and 0.1 mm were placed into a sample of liquid H_2O_2. The difference between the heights of the liquid in the tubes is 5.50 cm. Given that density (H_2O_2) = 1.41 g cm^{-3}, determine γ. (7.61×10^{-2} N m^{-1})

36. 50 drops each of water and ether, weigh 3.64 g and 0.852 g respectively. Determine the surface tension of ether if the surface tension of water is 72.75 dyne cm^{-1}.

(17.03 dyne cm^{-1})

37. The number of drops of water, counted in falling-drop method using stalagmometer is 100, whereas the number of drops of an organic liquid is 280. Calculate the surface tension of the organic liquid if the surface tension of water is 0.07275 N m^{-1} and the densities of water and the organic liquid are 0.998×10^3 kg m^{-3} and 0.755×10^3 kg m^{-3} respectively. (0.01966 N m^{-1})

38. The viscosity of olive oil at 293 K is 0.084 N m^{-2} s and density is 1.1×10^3 kg m^{-3}. How long will it take to pass through a viscometer if water under the same conditions takes 30 seconds? ($\eta_{H_2O} = 0.00101$ N m^{-2} s, $d_{H_2O} = 0.998 \times 10^3$ kg m^{-3})

(37 min 43.7 s)

39. An organic liquid rises 1.0 cm in a capillary tube of radius r. How much will it rise if the cross-sectional area of the tube is halved? (1.414 cm)

40. A crystalline solid of pure substance has a face-centred cubic structure with a cell edge of 400 pm. If the density of the substance in the crystal is 8 g cm^{-3} then the number of atoms present in 256 g of the crystal is $N \times 10^{24}$. The value of N is ... .

(IIT 2017 Adv.) (2)

41. Consider an ionic solid MX with NaCl structure. Construct a new structure (Z) whose unit cell is constructed from the unit cell of MX following the sequential instructions given below. Neglect the charge balance.

(i) Remove all the anions except the central one.

(ii) Replace all the face-centered cations (M) by anions (X).

(iii) Remove all the corner cations (M).

(iv) Replace the central anion (X) with cation (M).

The value of $\left(\frac{\text{no. of anions}}{\text{no. of cations}}\right)$ in Z is ... . **(IIT 2018 Adv.)** (3)

[**Hint:** Both M^+ and X^- form f.c.c. lattice, $Z = 4$

M^+	X^-	
4	4 ...	Initial
4	1 ...	1st step
1	4 ...	2nd step
0	4 ...	3rd step
1	3 ...	4th step]

Objective Problems

1. Which of the following is an amorphous solid?

(a) Diamond (b) Graphite
(c) Glass (d) Common salt

2. Which of the following is not a property of crystalline solids?

(a) Isotropic (b) Sharp melting point
(c) Definite geometry (d) High intermolecular forces

3. A crystal may have one or more planes of symmetry as well as one or more axes of symmetry but it has
(a) two centres of symmetry (b) no centre of symmetry
(c) one centre of symmetry (d) four centres of symmetry

4. The number of basic crystal systems is
(a) 7 (b) 8 (c) 6 (d) 4

5. The total number of elements of symmetry in a cubic crystal is
(a) 9 (b) 23 (c) 9 (d) none of these

6. The number of Bravais lattices in a cubic crystal is
(a) 3 (b) 1 (c) 4 (d) 14

7. In a sodium chloride crystal, each chloride ion is surrounded by
(a) 6 sodium ions (b) 6 chloride ions
(c) 8 sodium ions (d) 4 sodium ions

8. The structure of CsCl crystal is
(a) body-centred cubic lattice (b) face-centred cubic lattice
(c) octahedral (d) none of (a), (b) and (c)

9. The structure of NaCl crystal is
(a) body centred (b) face-centred cube
(c) octahedral (d) square plane

10. The arrangement of Cl^- ions in CsCl structure is
(a) h.c.p. (b) simple cubic (c) f.c.c. (d) b.c.c.

11. The number of atoms per unit cell in a face-centred cube is
(a) 2 (b) 3 (c) 4 (d) 14

12. The coordination number of a body-centred atom in cubic structure is
(a) 4 (b) 6 (c) 8 (d) 12

13. Close packing is maximum in the crystal which is a
(a) simple cube (b) face-centred cube
(c) body-centred cube (d) primitive cube

14. In a body-centred cubic arrangement the ion A occupies the centre while the ions of B occupy the corners of a cube. The formula of the crystal is
(a) AB (b) A_2B (c) AB_2 (d) AB_3

15. The number of atoms per unit cell in a simple cube, face-centred cube and body-centred cube are respectively
(a) 1, 4, 2 (b) 1, 2, 4 (c) 8, 14, 9 (d) 8, 4, 2

16. The radius of an ion in a body-centred cube of edge a is
(a) $\frac{a}{2}$ (b) $\frac{\sqrt{2}a}{4}$ (c) $\frac{\sqrt{3}a}{4}$ (d) a

17. The volume occupied by an atom in a simple cubic unit cell is
(a) a^3 (b) $\frac{4\pi a^3}{3}$ (c) $\frac{\pi a^3}{6}$ (d) $\frac{\sqrt{3}\pi}{8}$

18. In a body-centred cubic cell, an atom at the body centre is shared by
(a) 1 unit cell (b) 4 unit cells
(c) 2 unit cells (d) 8 unit cells

19. In a face-centred cubic cell, an atom at the face centre is shared by
(a) 6 unit cells (b) 1 unit cell (c) 4 unit cells (d) 2 unit cells

20. An atom at the corner of a simple cubic cell is shared by
(a) 2 unit cells (b) 4 unit cells (c) 8 unit cells (d) 1 unit cell

21. The atomic radius in a face-centred cubic cell is
(a) $\frac{a}{2}$ (b) $\frac{\sqrt{2}a}{4}$ (c) $\frac{\sqrt{3}a}{4}$ (d) $\frac{a}{4}$

22. The fraction of the total volume occupied by atoms in a simple cube is
(a) $\frac{\pi}{2}$ (b) $\frac{\sqrt{3}\pi}{8}$ (c) $\frac{\sqrt{2}\pi}{6}$ (d) $\frac{\pi}{6}$

23. The Miller indices of two parallel planes in a crystal are
(a) same (b) different

24. The Miller indices of the shaded plane shown in the figure below are
(a) (0 0 1) (b) (0 1 0)
(c) (0 1 1) (d) (1 0 0)

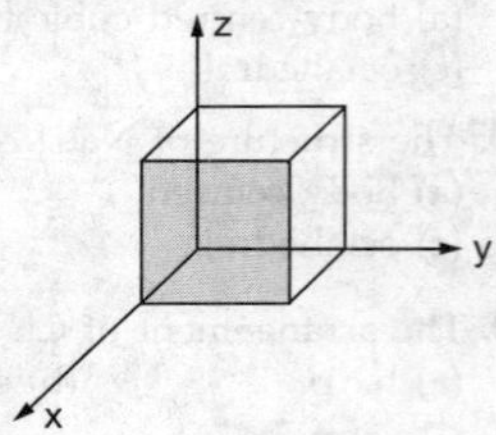

25. The Miller indices of the shaded plane shown in the figure below are
(a) (1 0 0) (b) $(1\,\bar{1}\,0)$
(c) (1 1 1) (d) (0 0 1)

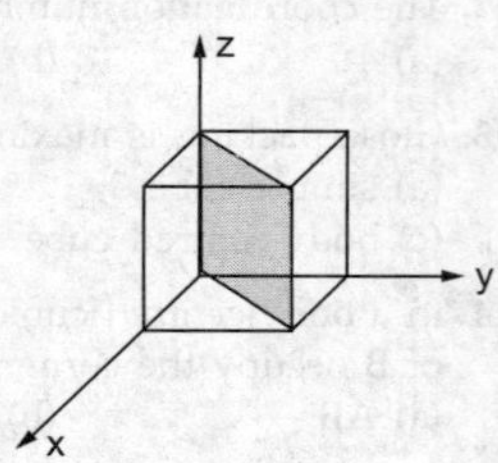

26. Bragg's law is given by the equation
(a) $n\lambda = 2\sin\theta$ (b) $n\lambda = 2d\sin\theta$ (c) $2d = n\lambda\sin\theta$ (d) $n\lambda = d\sin\theta$

27. A mineral having the formula AB_2 crystallizes in the c.c.p. lattice, with A atoms occupying the lattice points. The CN of A is 8 and that of B is 4. What percentage of the tetrahedral sites is occupied by B atoms?
(a) 25% (b) 50% (c) 75% (d) 100%

28. The number of octahedral sites per sphere in a c.c.p. (f.c.c.) structure is
(a) 0 (b) 1 (c) 2 (d) 4

29. The density of crystalline CsCl is 3.988 g/cc. The volume effectively occupied by a single CsCl ion pair in the crystal is (CsCl = 168.4)
(a) 7.014×10^{-23} cc (b) 2.81×10^{-22} cc (c) 6.022×10^{23} cc (d) 3.004×10^{-23} cc

30. The CsCl structure is observed in alkali halides only when the radius of the cation is sufficiently large to keep its eight nearest-neighbour anions from touching. What minimum value of r_+/r_- is needed to prevent this contact?
(a) 0.155 (b) 0.225 (c) 0.414 (d) 0.732

31. A substance A_xB_y crystallizes in an f.c.c. lattice in which atoms of 'A' occupy each corner of the cube and atoms of 'B' occupy the centres of each face of the cube. Identify the correct composition of the substance A_xB_y.
(a) AB_3 (b) A_4B_3
(c) A_3B (d) Composition cannot be specified

32. In a solid AB of NaCl structure, A atoms occupy the corners of the cubic unit cell. If all the corner atoms are removed then the formula of the unit cell will be
(a) A_4B_4 (b) B (c) A_3B_4 (d) AB

33. In a crystal AB, which of the following crystal systems will have parameters, $a \neq b \neq c$ and $\alpha = \beta = \gamma = 90°$?
(a) Cubic (b) Orthorhombic (c) Monoclinic (d) Triclinic

34. In the NaCl crystal, which of the following facts is not true?
(a) Na^+ ions form f.c.c. lattice (b) Cl^- ions form f.c.c. lattice
(c) Na^+Cl^- units form f.c.c. structure (d) CN of each Na^+ and Cl^- is 6

35. The decreasing order of the size of the void is
(a) cubic > octahedral > tetrahedral > trigonal
(b) trigonal > tetrahedral > octahedral > cubic
(c) trigonal > octahedral > tetrahedral > cubic
(d) cubic > tetrahedral > octahedral > trigonal

36. The greater the value of r_+/r_-,
(a) the lower will be the CN
(b) the higher the value of CN
(c) the higher will be the number of cations
(d) the lower will be the number of anions

37. An organic liquid rises 2.0 cm in a capillary tube. How much will it rise if the cross-sectional area of the tube is doubled?
(a) 2.0 cm (b) 1.0 cm (c) 4.0 cm (d) 1.4 cm

38. When the temperature is increased, surface tension of water
(a) increases (b) decreases
(c) remains constant (d) shows irregular behaviour

39. The units of surface tension in cgs and SI units are respectively,
(a) dyne cm^{-1}, N m^{-2} (b) dyne cm^{-2}, N m^{-2}
(c) dyne cm^{-1}, J m^{-2} (d) dyne cm^{-2}, N m^{-1}

40. The units of viscosity in cgs and SI units are respectively,
(a) dyne cm^{-2} s, N m^{-2} s (b) poise, Pa s^2
(c) poise, N m^{-2} s^{-1} (d) dyne · s, N s

41. The rise of a liquid in a capillary tube is due to
(a) osmosis (b) surface tension (c) viscosity (d) diffusion

42. The coefficient of viscosity of a solution and its solvent are respectively η and η_0. The specific viscosity η_{sp} may be expressed as

(a) $\frac{\eta}{\eta_0}$ (b) $\frac{\eta - \eta_0}{\eta_0}$ (c) $\frac{\eta - \eta_0}{\eta}$ (d) $\frac{\eta_0}{\eta}$

43. Which of the following properties of liquids increases with the increase in temperature?

(a) Vapour pressure (b) Surface tension
(c) Viscosity (d) None of these

44. The surface tension of several alcohols at 20°C is $\gamma(CH_3OH) = 22.61$ dyne·cm^{-1}, $\gamma(C_2H_5OH) = 2.275 \times 10^{-2}$ N m^{-1} and $\gamma(n-C_3H_7OH) = 23.78$ mJ m^{-2}. The alcohol having the highest surface tension is

(a) CH_3OH (b) C_2H_5OH
(c) $n-C_3H_7OH$ (d) All same

[**Hint:** 1 dyne $= 10^{-5}$ N and 1 J = 1 N m]

45. A compound M_pX_q has cubic-close packing (c.c.p.) arrangement of X. Its unit-cell structure is shown below. The empirical formula of the compound is

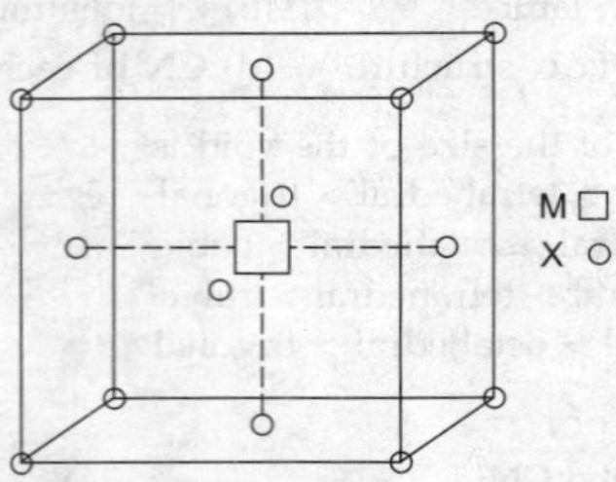

(a) MX (b) MX_2 (c) M_2X (d) M_5X_{14} **(IIT 2012)**

[**Hint:** Z = 2 (b.c.c.) and Z = 4 (f.c.c.)]

46. The packing fraction of the two dimensional square unit cell shown below is (all the circles are of the same radius)

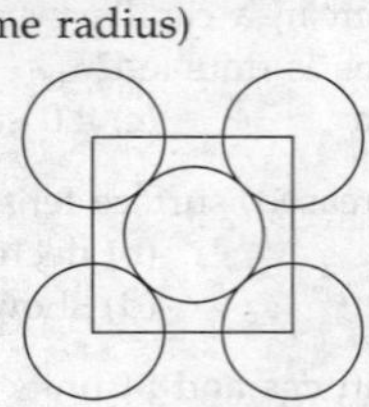

(a) 39.27% (b) 68.02% (c) 74.05% (d) 78.54% **(IIT 2010)**

[**Hint:** Number of atoms in 2-D square unit cell

= Number of corner atoms × contribution to one unit cell

$= 4 \times \frac{1}{4} = 1$

As the central atom is not shared by other cells.

Total number of atoms per square unit cell = 1 + 1 = 2

$\therefore$ Packing fraction

$$= \frac{\text{Area of atoms of one sq. unit cell}}{\text{Area of one sq. unit cell}}$$

$$= \frac{2\pi r^2}{l^2},$$

where $4r = \sqrt{2}l$ (figure)]

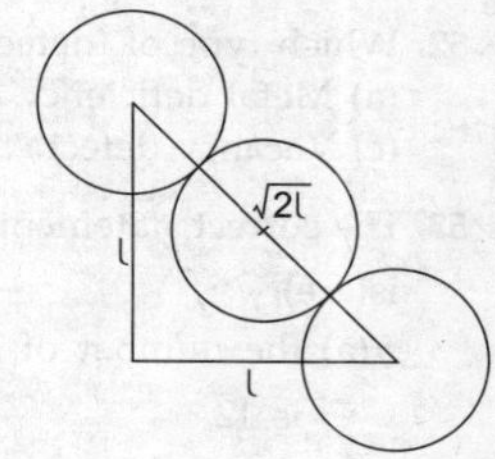

47. Which of the following exists as covalent crystals in solid state?
(a) Iodine (b) Silicon (c) Sulphur (d) Phosphorus
(IIT 2013 Main)

48. The arrangement of X^- ions around A^+ ion in solid AX is given in the figure (not drawn to scale). If the radius of X^- is 250 pm, the radius of A^+ is

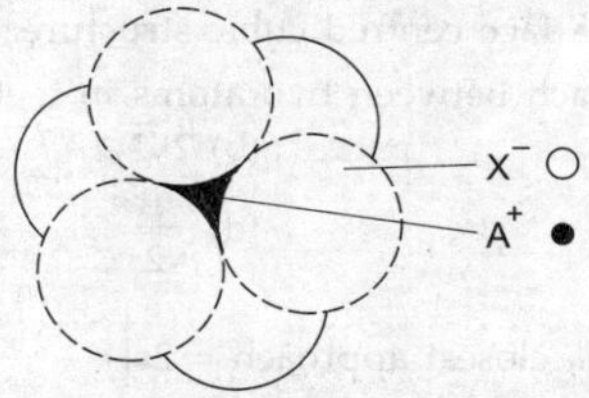

(a) 104 pm (b) 125 pm (c) 183 pm (d) 57 pm
(IIT 2013 Adv.)

[**Hint:** See text, octahedral void and radius ratio.]

49. CsCl crystallises in body centred cubic lattice. If 'a' is its edge length, then which of the following expressions is correct?

(a) $r_{Cs^+} + r_{Cl^-} = 3a$ (b) $r_{Cs^+} + r_{Cl^-} = \frac{3a}{2}$

(c) $r_{Cs^+} + r_{Cl^-} = \frac{\sqrt{3}a}{2}$ (d) $r_{Cs^+} + r_{Cl^-} = \sqrt{3}a$ **(IIT 2014 Main)**

50. Sodium metal crystallises in a body centred cubic lattice with a unit cell edge of 4.29 Å. The radius of Na atom is approximately
(a) 1.86 Å (b) 3.22 Å
(c) 5.72 Å (d) 0.93 Å **(IIT 2015 Main)**

51. If the unit cell of a mineral has cubic packed (c.c.p.) array of oxygen atoms with m fraction of octahedral holes occupied by aluminium ions and n fraction of tetrahedral holes occupied by magnesium ions, m and n respectively are

(a) $\frac{1}{2}, \frac{1}{8}$ (b) $1, \frac{1}{4}$

(c) $\frac{1}{2}, \frac{1}{2}$ (d) $\frac{1}{4}, \frac{1}{8}$ **(IIT 2015 Adv.)**

[**Hint:** c.c.p. is f.c.c. arrangement ($z = 4$) so number of O^{2-} ions are 4. Octahedral and tetrahedral holes are 4 and 8 respectively. As 8 negative changes can be neutralised by 2 Al^{3+} and 1 Zn^{2+} ion, m and n are $\frac{2}{4} = \frac{1}{2}$ and $\frac{1}{8}$ respectively.]

52. Which type of 'defect' has the presence of cations in the interstitial sites?
(a) Metal deficiency defect
(b) Schottky defect
(c) Vacancy defect
(d) Frenkel defect
(IIT 2018 Main)

53. The correct statement(s) for cubic-closed packed (c.c.p.) three dimensional structure is(are)

(a) the number of nearest neighbours of an atom present in the topmost layer is 12.

(b) the packing efficiency of atom is 74%.

(c) the number of octahedral and tetrahedral voids per atom are 1 and 2 respectively.

(d) the unit cell edge length is $2\sqrt{2}$ times the radius of the atom.

(IIT 2016 Adv.)

[Hint: Read text.]

54. A metal crystallises in a face centred cubic structure. If the edge length of its cell is 'a', the closest approach between two atoms in metallic crystal will be
(a) $2a$
(b) $2\sqrt{2}\,a$
(c) $\sqrt{2}\,a$
(d) $\dfrac{a}{\sqrt{2}}$
(IIT 2017 Main)

[Hint: for f.c.c.: $r = \frac{\sqrt{2}}{4}a$, closest approach = $2r$]

55. The qualitative sketches I, II and III given below show the variation of surface tension with molar concentration of three different aqueous solutions of KCl, CH_3OH and $CH_3(CH_2)_{11}OSO_3^-Na^+$ at room temperature.

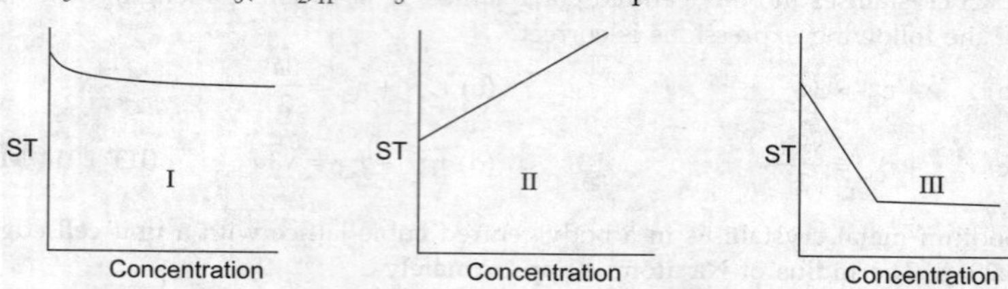

The correct assignment of the sketches is

(a) I KCl
II CH_3OH
III $CH_3(CH_2)_{11}OSO_3^-Na^+$

(b) I $CH_3(CH_2)_{11}OSO_3^-Na^+$
II CH_3OH
III KCl

(c) I KCl
II $CH_3(CH_2)_{11}OSO_3^-Na^+$
III CH_3OH

(d) I CH_3OH
II KCl
III $CH_3(CH_2)_{11}OSO_3^-Na^+$

(IIT 2016 Adv.)

56. All of the following share the same crystal structure except
(a) RbCl
(b) LiCl
(c) CsCl
(d) NaCl
(IIT 2018 Main)

Answers

1-c, **2**-a, **3**-c, **4**-a, **5**-b, **6**-a, **7**-a, **8**-a, **9**-b, **10**-b, **11**-c, **12**-c, **13**-b, **14**-a, **15**-a, **16**-c, **17**-c, **18**-a, **19**-d, **20**-c, **21**-b, **22**-d, **23**-a, **24**-d, **25**-b, **26**-b, **27**-d, **28**-b, **29**-a, **30**-d, **31**-a, **32**-c, **33**-b, **34**-c, **35**-a, **36**-b, **37**-d, **38**-b, **39**-c, **40**-a, **41**-b, **42**-b, **43**-a, **44**-c, **45**-b, **46**-d, **47**-c, **48**-a, **49**-c, **50**-a, **51**-a, **52**-d, **53**-b, c, d, **54**-d, **55**-d, **56**-b.

❑

CHAPTER TWENTY-ONE

MISCELLANEOUS PROBLEMS FOR REVISION

1. A sample of a pure compound contains 2.04 grams of sodium, 2.65×10^{22} atoms of carbon and 0.132 mole of oxygen atoms. Find the empirical formula of the compound. (Na_2CO_3)

2. The total number of molecules of hydrogen and oxygen that may be obtained from a given amount of H_2O is 6000. Find the amount of H_2O. (1.195×10^{-19} g)

3. The total population of the world is now believed to be about 4.2×10^9. How many moles of people is this? If you had one sulphur atom for each person, what would be the weight of the sulphur sample? (6.9×10^{-15} mole, 2.2×10^{-13} g)

4. Does 1 g of all elements contain nucleons equal to the Avogadro constant? Explain. (Yes)

5. A solution contains 0.18 g/mL of a substance X, whose molecular weight is approximately 68000. It is found that 0.27 mL of oxygen at 760 mmHg and 30°C will combine with the amount of X contained in 1.0 mL of the solution. How many molecules of oxygen will combine with one molecule of X? (4)

6. Find the simplest formula of a solid whose cubic unit cell has an '*x*' atom at each corner, a '*y*' atom at each face centre, and a '*z*' atom at the body centre. (xy_3z)

7. Calculate the work of an isobaric reversible expansion of three moles of an ideal gas while it is heated from 298 K to 400 K.
[**Hint:** $W = -nR(T_2 - T_1)$] (–2.54 kJ)

8. Calculate the work of an isothermal reversible expansion of three moles of water vapour from 5.0×10^4 to 2.0×10^4 Pa at 330 K.
[**Hint:** $W = -2.303\, nRT \log \frac{p_1}{p_2}$] (–7.54 kJ)

9. The heats of solution of one mole of Na and that of Na_2O in water under standard conditions are –183.79 kJ/mol and –237.94 kJ/mol respectively, water being taken in large excess in both the cases. Calculate the standard heat of formation of sodium oxide if the standard heat of formation of water is –285.84 kJ/mol.

$$\left[\begin{array}{ll}\textbf{Hint:} & Na + H_2O = NaOH + \frac{1}{2}H_2 \\ & Na_2O + H_2O = 2NaOH\end{array}\right]$$

(–415.48 kJ/mol)

10. The pressure of the water vapour of a solution containing a nonvolatile solute is 2% below that of the vapour of pure water. Calculate the molality of the solution. (1.134)

11. An aqueous solution freezes at 271.5 K. Determine its boiling point and vapour pressure at 298 K. The cryoscopic constant of water is 1.86°, its ebullioscopic

constant is 0.516° and the water vapour pressure at 298 K is 3168 Pa.
(373.42, 3124 Pa)

12. The water vapour pressure at 293 K is 2338.5 Pa, and the vapour pressure of an aqueous solution is 2295.8 Pa. Determine the osmotic pressure of this solution at 313 K if the solution density at this temperature is 1010 kg/m^3. The molecular weight of the solute is 60. (2.56×10^6 Pa)

13. If the vapour pressure of pure liquids A and B are 300 mm and 800 mmHg at 75°C, calculate the composition of the mixture such that it boils at 75°C. Find the composition of the vapour phase. (0.08, 0.92; 0.0316, 0.9684)

14. For the cells

$$2\,Ag + Pt^{2+} \rightarrow 2\,Ag^+ + Pt\,;\; E^0 = 0.4\text{ V}$$
$$2\,Ag + F_2 \rightarrow 2\,Ag^+ + 2\,F^-;\; E^0 = 2.07\text{ V}$$

if the potential for the reaction $Pt \rightarrow Pt^{2+} + 2e$ is assigned as zero, determine the potential for the following electrodes.

(i) $Ag \rightarrow Ag^+ + e$

(ii) $F^- \rightarrow \frac{1}{2} F_2 + e$

(–0.4 V, –2.47 V)

15. The standard reduction potential of the $Ag^+|Ag$ electrode at 298 K is 0.799 V. Given that for AgI, $K_{sp} = 8.7 \times 10^{-17}$, evaluate the potential of the $Ag^+|Ag$ electrode in a saturated solution of AgI. Also, calculate the standard reduction potential of the $I^-|AgI|Ag$ electrode. **(IIT 1994)**

(0.3244 V, –0.15 V)

[Hint: $[Ag^+] = \sqrt{K_{sp}\,(AgI)} = \sqrt{8.7 \times 10^{-17}}$

$$E_{Ag^+, Ag} = E^0_{Ag^+, Ag} - \frac{0.0591}{1} \log \frac{1}{[Ag^+]}$$

For standard $I^-|AgI|Ag$ electrode,

$$[I^-] = 1, \quad \therefore \quad [Ag^+] = \frac{K_{sp}(AgI)}{1} = 8.7 \times 10^{-17}.$$

The standard potential for this electrode, if the Nernst equation is used for $Ag^+|Ag$ electrode, may be calculated considering $[Ag^+] = 8.7 \times 10^{-17}$.]

16. A blown-up balloon has a volume of 500 mL at 5°C. The balloon is distended to 7/8 of its maximum stretching capacity. Will the balloon burst at 30°C? Determine the minimum temperature above which it will burst. (No, 44.7°C)

17. On the surface of the earth at 1 atm pressure, a balloon filled with H_2 gas occupies 500 mL. This volume is 5/6 of its maximum stretching capacity. The balloon is left in air. It starts rising. Calculate the height above which the balloon will burst if the temperature of the atmosphere remains constant and the pressure decreases 1 mm for every 100-cm rise in height. (126.67 m)

18. A spherical balloon of radius 30 cm weighs 100 g. Find the minimum amount of hydrogen the balloon should contain just to rise from the ground. The density of air is 1.29 g/L. (45.95 g)

19. A compound exists in the gaseous state both as a monomer (A) and a dimer (A_2). The molecular weight of the monomer is 48. In an experiment, 96 g of the compound was confined in a vessel of volume 33.6 litres and heated to 273°C. Calculate the pressure developed if the compound exists as a dimer to the extent of 50% by weight under these conditions. (2 atm)

20. The equilibrium concentrations of HI, H_2 and I_2 were found to be 0.49, 0.08 and 0.06 mole per litre respectively, when the reaction $H_2 + I_2 \rightleftharpoons 2HI$ was initially started with some amounts of H_2 and I_2. Calculate the new equilibrium concentrations of each gas if an additional 0.3 mole per litre of HI was added. (0.724 M, 0.113 M, 0.093 M)

21. Calculate K_c for the reaction: $A(g) + B(g) \rightleftharpoons 2C(g)$, if 1 mole of A, 1.4 moles of B and 0.50 mole of C are placed in a one-litre vessel and allowed to reach equilibrium. The equilibrium concentration of C is 0.75 mole per litre. (0.50)

22. At 700 K, CO_2 and H_2 react to form CO and H_2O. For this process K is 0.11. A mixture of 0.45 mole of CO_2 and 0.45 mole of H_2 is heated to 700 K.

(i) Find the amount of each gas at equilibrium.

(ii) After the equilibrium is reached, another 0.34 mole of CO_2 and 0.34 mole of H_2 are added to the reaction mixture. Find the composition of the new equilibrium state. [(i) 0.34, 0.11 (ii) 0.594, 0.196]

23. K_p for the reaction: $N_2O_4(g) \rightleftharpoons 2NO_2(g)$ is 0.66 atm at 320 K. Calculate the degree of dissociation of N_2O_4 at 320 K and 380 mm. Also calculate the partial pressures of N_2O_4 and NO_2 at equilibrium. (0.497; 0.332 atm, 0.168 atm)

24. 100 mL of 0.6 N $CuSO_4$ solution is electrolysed between two Pt electrodes till the concentration in the residual liquid is 0.1 N when a steady current of 5.0 amp is used. How long should the current be passed to get the said change? (965 s)

25. Electrolysis of an acetate solution produces ethane according to the reaction:

$$2CH_3COO^- \rightarrow C_2H_6(g) + 2CO_2(g) + 2e$$

The efficiency of the reaction is 82%. What volume of ethane and CO_2 would be produced at 27°C and 740 mmHg if a current of 0.5 amp is passed through the solution for 420 minutes? (1.354 L, 2.708 L)

26. An aqueous solution of NaCl on electrolysis gives $H_2(g)$, $Cl_2(g)$ and NaOH according to the reaction:

$$2Cl^-(aq) + 2H_2O = 2OH^-(aq) + H_2(g) + Cl_2(g)$$

A direct current of 25 amp with a current efficiency of 62% is passed through 20 litres of NaCl solution (20% by weight). Write down the reactions taking place at the anode and at the cathode. How long will it take to produce 1 kg of Cl_2? What will be the molarity of the solution with respect to hydroxide ions? (Assume no loss due to evaporation.) **(IIT 1992)**

[Hint: See Example 15, Chapter 8.] (49 h, 1.408 M)

27. For the reaction: $2A + B_2 + C \rightarrow A_2B + BC$, the rate law expression has been

determined experimentally to be $R = k[A]^2[C]$ with
$k = 3.0 \times 10^{-4}\ M^{-2}\ min^{-1}$.

(i) Determine the initial rate of the reaction, started with concentrations:

$$[A] = 0.1\ M,\ [B_2] = 0.35\ M \text{ and } [C] = 0.25\ M.$$

(ii) Determine the rate after 0.04 mole per litre of A has reacted.

$(7.5 \times 10^{-7}\ M\ min^{-1}, 2.5 \times 10^{-7}\ M\ min^{-1})$

28. The decomposition of AB_2 to AB and B is first-order with $k = 2.8 \times 10^{-7}\ s^{-1}$ at 1000°C.

$$AB_2 \rightarrow AB + B$$

Atomic weights of A and B are 12 and 32 respectively.

(i) Find the half-life of this reaction at 1000°C.

(ii) How many days would pass before 1 g of AB_2 had decomposed to the extent that 0.60 g of AB_2 remained?

(iii) With reference to (ii), how many grams of AB would be present after this length of time?

(iv) How much of a 1-g sample of AB_2 would remain after 35 days?

(28.58 days, 20.84 days, 0.23 g, 0.43 g)

29. Nicotinic acid ($K_a = 1.4 \times 10^{-5}$) is represented by the formula HNiC. Calculate its per cent dissociation in a solution which contains 0.10 mol of nicotinic acid per 2.0 L of solution. (1.67%)

30. The solution containing dissolved CO_2 is 0.035 M. If the dissociation constants for the reactions

$$CO_2 + H_2O = H^+ + HCO_3^-$$

and

$$HCO_3^- = H^+ + CO_3^{2-}$$

are 4.16×10^{-7} and 4.84×10^{-11} respectively, calculate $[H^+]$ in the solution.

$(1.2 \times 10^{-4}\ M)$

31. Hydrozoic acid HN_3 is a weak acid which hydrolyses in water according to

$$HN_3 + H_2O = H_3O^+ + N_3^-$$

$$pK_a\ (HN_3) = 4.72$$

(i) Calculate $[H_3O^+]$, $[HN_3]$, $[N_3^-]$ and $[OH^-]$ in 0.1 M acid solution.

(ii) Calculate pH of the acid.

$\left[\begin{array}{l}\text{(i) } 1.4 \times 10^{-3}\ M, 0.1\ M, 1.4 \times 10^{-3}\ M, 7.2 \times 10^{-12}\ M \\ \text{(ii) } 2.86\end{array}\right]$

32. How many moles of NaOH can be added to one litre of a solution of 0.1 M in NH_3 and 0.1 M in NH_4Cl without changing pOH more than one unit? Assume no change in volume, $pK_b = 4.75$. (0.082 mole)

33. A 4 : 1 molar mixture of He and CH_4 is contained in a vessel at 20 bar pressure. Due to a hole in the vessel the gas mixture leaks out. What is the composition of the mixture effusing out initially? **(IIT 1994)** (8 : 1)

[Hint: See Example 35, Chapter 12]

34. An LPG (liquefied petroleum gas) cylinder weighs 14.8 kg when empty. When full, it weighs 29 kg and shows a pressure of 2.5 atm. In the course of use at 27°C, the weight of the full cylinder reduced to 23.2 kg. Find out the volume of the gas used up in cubic metres at the normal usage conditions, and the final pressure inside the cylinder. Assume LPG to be *n*-butane with normal boiling point of 0°C.

(IIT 1994) (2.463 m^3, 1.4788 atm)

[**Hint:** $1 \times V_{\text{used}}$ (27°C and 1 atm) $= \dfrac{5800}{58} \times 0.0821 \times 300$]

35. A balloon of diameter 20 m weighs 100 kg. Calculate its payload if it is filled with He at 1.0 atm and 27°C. Density of air is 1.2 kg m^{-3}. (4249.5 kg)

36. An organic compound containing C, H and O exists in two isomeric forms (A) and (B). An amount of 0.108 g of one of the isomers gives on combustion 0.308 g of CO_2 and 0.072 g of H_2O. (A) is insoluble in NaOH and $NaHCO_3$ while (B) is soluble in NaOH. (A) reacts with conc. HI to give compounds (C) and (D). (C) can be separated from (D) by the ethanolic $AgNO_3$ solution and (D) is soluble in NaOH. (B) reacts readily with bromine water to give compound (E) of molecular formula $C_7H_5OBr_3$.

Identify (A), (B), (C), (D) and (E). **(IIT 1991)**

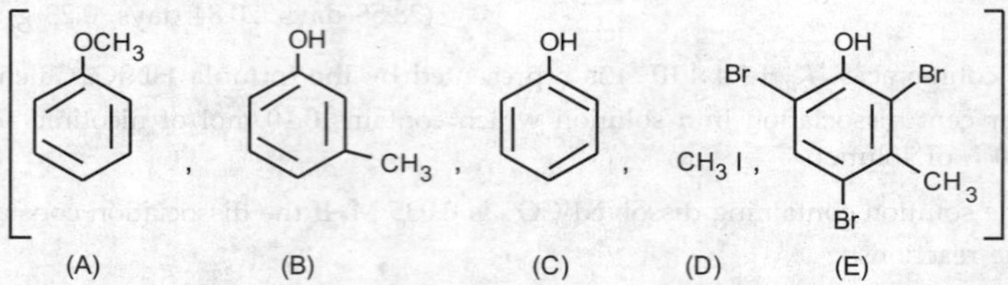

37. In the Kjeldahl method the gas evolved from a 1.325-g sample of a fertilizer is passed into 50.0 mL of 0.2030 N H_2SO_4. 25.32 mL of 0.1980 N NaOH is required for the titration of the unused acid. Calculate the percentage of nitrogen in the fertilizer. (5.43%)

38. When 0.0088 g of a compound (A) was dissolved in 0.50 g of camphor, the m.p. of camphor was lowered by 8°C. Analysis of (A) gave 68.18% C and 13.16% H. Compound (A) showed the following reactions: (i) It reacted with acetyl chloride and evolved hydrogen with sodium. (ii) When reacted with HCl + $ZnCl_2$, a dense oily layer separated out immediately. Compound (A) was passed over Al_2O_3 at 350°C to give compound (B). (B) on ozonolysis followed by hydrolysis gave two neutral compounds (C) and (D), which gave a positive test with carbonyl reagents but only (C) gave a positive test with Fehling's solution and resinous substance with NaOH. Identify (A), (B), (C), and (D). K_f for camphor = 40 K mol^{-1} kg.

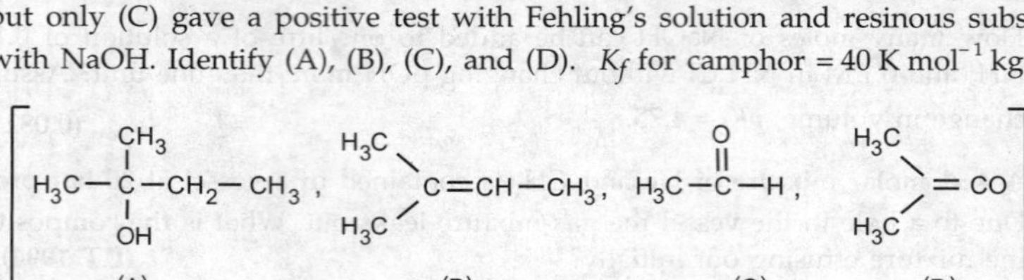

39. An excess of liquid mercury was added to a 10^{-3} M acidic solution of Fe^{3+}. It was found that only 4.6% of the iron remained as Fe^{3+} at equilibrium at 25°C. Calculate $E^0_{Hg_2^{2+}, Hg}$. Assume that the only reaction taking place is

$$2\,Hg + 2\,Fe^{3+} = Hg_2^{2+} + 2\,Fe^{2+}$$

Given that $E^0_{Fe^{3+}, Fe^{2+}} = 0.771$ volt. (0.791 V)

40. At what relative concentration of Zn^{2+} ions and Fe^{2+} ions will Zn(s) and Fe(s) have equal oxidation potential?

$$E^0_{Zn, Zn^{2+}} = 0.76\text{ V}, E^0_{Fe, Fe^{2+}} = 0.44\text{ V}$$ (7.031×10^{10})

41. The standard reduction potential for the half cell

$$NO_3^-(aq) + 2H^+(aq) + e \rightarrow NO_2(g) + H_2O$$

is 0.78 volt.

(i) Calculate the reduction potential in 8 M H^+.

(ii) What will be the reduction potential of the half cell in a neutral solution? Assume all the other species to be at unit concentration. **(IIT 1993)**

$\left[\textbf{Hint: } \text{Use } E = E^0 - \dfrac{0.0591}{1} \log \dfrac{1}{[H^+]^2}\right]$ (0.8867 V, – 0.0474 V)

42. The dipole moment of KCl is 3.336×10^{-29} coulomb·metre which indicates that it is a highly polar molecule. The interatomic distance between K^+ and Cl^- in this molecule is 2.6×10^{-10} m. Calculate the dipole moment of KCl molecule if there were opposite charges of one fundamental unit located at each nucleus. Calculate the percentage ionic character of KCl. ($e = 1.602 \times 10^{-19}$ coulomb).

(IIT 1993) (80.09%)

[**Hint:** Dipole moment $= e \times r$ coulomb·metre,

$$\% \text{ ionic character} = \frac{3.336 \times 10^{-29}}{1.602 \times 10^{-19} \times 2.6 \times 10^{-10}} \times 100]$$

43. Sodium metal crystallizes in a body-centred cubic lattice with the cell edge $a = 4.29$ Å. What is the radius of the sodium atom? **(IIT 1994)** (1.8576 Å)

[**Hint:**

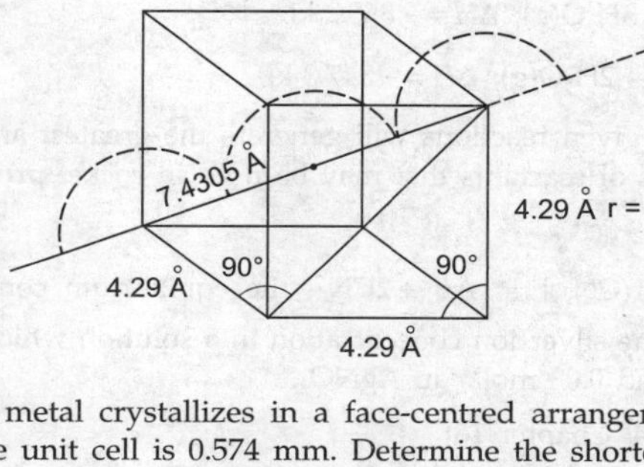

4.29 Å $r = \dfrac{7.4305}{4}$]

44. Potassium metal crystallizes in a face-centred arrangement of atoms where the edge of the unit cell is 0.574 mm. Determine the shortest separation of any two potassium nuclei. (0.406 mm)

[**Hint:** Calculate the diameter.]

45. 5.0 g of a polymer of molecular weight 50.0 kg mol^{-1} is dissolved in 1 dm^3 of water. If the density of this solution is 0.96 kg dm^{-3}, calculate the height of water that will represent this pressure. (26.52 mm)

46. The two liquids A and B have the same molecular weight and form an ideal solution. The solution has the vapour pressure 700 mmHg at 80°C. The above solution is distilled without reflux till 3/4 of the solution is collected as condensate. The composition of the condensate is $x'_A = 0.75$ and that of the residue is $x_A = 0.3$. If the vapour pressure of the residue at 80°C is 600 mmHg, calculate x_A, P^0_A and P^0_B. (0.635, 809.39 mm, 509.69 mm)

47. A solution containing compound X in water and a solution containing urea in water were put in a closed system. By doing this some water vapour was removed from one solution and got condensed in the other. It is found that when both the solutions were at equilibrium vapour pressure, one solution contains 2% of X and the other 5% by weight. Find the molecular weight of X. (23.26)

[**Hint:** At equilibrium, the relative lowering of vapour pressure of the two solutions is equal.]

48. The heat of Fe_2O_3 formation from simple substances is –821.32 kJ/mole at 298 K and standard pressure, and that of Al_2O_3 formation is –1675.60 kJ/mole under the same conditions. Calculate the heat of reaction of reduction of 1 mole Fe_2O_3 with metallic aluminium. (–854.28 kJ)

49. The heat of combustion of graphite at 298 K is –393.795 kJ/mole, while that of diamond's combustion at the same temperature is –395.692 kJ/mole. The specific heats for these substances are 720.83 and 505.58 J kg^{-1} K^{-1} respectively. Calculate the heat of graphite's transformation into diamond at 273 K. (1.962 kJ/mole)

[**Hint:** Use Eqn. 9a, Chapter 14.]

50. Which oxidizing agent, O_2, O_3 or H_2O_2, will generate the greatest amount of energy for 1 mole of $H_2(g)$?

(i) $2H_2(g) + O_2(g) \rightarrow 2H_2O(g);\ \Delta H = -483.6$ kJ

(ii) $3H_2(g) + O_3(g) \rightarrow 3H_2O(g);\ \Delta H = -868.2$ kJ

(iii) $H_2(g) + H_2O_2(g) \rightarrow 2H_2O(g);\ \Delta H = -347.3$ kJ

Which of the above given reactions will generate the greatest amount of energy on a total mass basis of reactants that may be used in rocket propulsion?

[(iii), (ii)]

51. For the reaction $[Ag(CN)_2] \rightleftharpoons Ag^+ + 2CN^-$, the equilibrium constant at 25°C is 4×10^{-19}. Calculate the silver ion concentration in a solution which was originally 0.1 molar in KCN and 0.03 molar in $AgNO_3$. **(IIT 1994)**

(**Hint:** See Example 4, Chapter 16) (7.5×10^{-18} M)

52. The Edison storage cell is represented as

$$Fe(s) \mid FeO(s) \mid KOH(aq) \mid Ni_2O_3(s) \mid Ni(s).$$

The half cell reactions are

$$Ni_2O_3(s) + H_2O(l) + 2e = 2NiO(s) + 2OH^-$$
$$E^0 = +0.40 \text{ V}$$
$$FeO(s) + H_2O(l) + 2e = Fe(s) + 2OH^-$$
$$E^0 = -0.87 \text{ V}$$

(i) What is the cell reaction?
(ii) What is the cell emf? How does it depend on the concentration of KOH?
(iii) What is the maximum amount of electrical energy that can be obtained from one mole of Ni_2O_3? **(IIT 1994)**

[(i) $Ni_2O_3(s) + Fe(s) \rightarrow 2NiO(s) + FeO(s)$
(ii) 1.27 V; independent of [KOH]
(iii) 245.11 kJ/mole of Ni_2O_3]

[**Hint:** See Example 20, Chapter 18 and for (iii) Electrical energy = nFE]

53. An aqueous solution of NaCl containing 5.85 g/L of NaCl was electrolysed by using platinum electrodes. Hydrogen and chlorine gases evolved at the cathode and anode respectively. Calculate the pH of the solution after electrolysis assuming complete electrolysis of NaCl. (13)

54. Anodic oxidation of ammonium hydrogen sulphate (NH_4HSO_4) produces ammonium persulphate.

$$NH_4HSO_4 = NH_4SO_4^- + H^+$$
$$2NH_4SO_4^- = (NH_4)_2S_2O_8 + 2e \quad \text{(at anode)}$$
$$2H^+ + 2e = H_2 \quad \text{(at cathode)}$$

$(NH_4)_2S_2O_8$ hydrolyses according to

$$(NH_4)_2S_2O_8 + 2H_2O = 2NH_4HSO_4 + H_2O_2.$$

Current efficiency in electrolytic process is 60%. Calculate the amount of current required to produce 85 g of H_2O_2 per hour if hydrolysis reaction has 100% efficiency. (223.38 amp)

55. Calculate the weight of dilute sulphuric acid solution (sp. gr. = 1) which was electrolysed to give H_2 at 300 K and 1 atm to fill a balloon of capacity 680 mL. (0.49 g)

56. The reaction $Cl_2(g) + S_2O_3^{2-} \rightarrow SO_4^{2-} + Cl^-$, is to be carried out in alkaline solution. Starting with 0.15 mole of Cl_2, 0.01 mole of $S_2O_3^{2-}$ and 0.30 mole of OH^-. How many mole of OH^- will be left in the solution after the reaction is complete? Assume that no other reaction takes place. (0.2 mole)

[**Hint:** $4Cl_2 + S_2O_3^{2-} + 10OH^- \rightarrow 2SO_4^{2-} + 8Cl^- + 5H_2O$]

57. 12.5 g of a sample of arsenious oxide was dissolved in water containing 7.5 g of sodium bicarbonate and the resulting solution was diluted to 250 mL. 25 mL of this solution was completely oxidized by 22.4 mL of a solution of iodine, 25 mL of which reacted with 24.5 mL of a solution of $Na_2S_2O_3 \cdot 5H_2O$ containing 27.7 g of this salt in 1 litre. Calculate the percentage of arsenious oxide in the sample.

Reactions involved are

$$As_2O_3 + 6NaHCO_3 \rightarrow 2Na_3AsO_3 + 3H_2O + 6CO_2$$

$$Na_3AsO_3 + 2I_2 + 2H_2O \rightarrow 2Na_3AsO_4 + 4HI$$

(9.74%)

58. 2.10 g of mixture of $NaHCO_3$ and $KClO_3$ requires 100 mL of 0.1 N HCl for complete reaction. Calculate the amount of residue that would be obtained on heating 2.20 g of the same mixture strongly. (1.358 g)

59. A compound was dissolved in water at 27°C. It is found that the vapour pressure lowering at 27°C is 0.72 mm. If the vapour pressure of water at 27°C is 26.74 mm, calculate the osmotic pressure of the solution. (0.028 atm)

60. The vapour pressure of water is 3167.2 Pa at 25°C. What would be the vapour pressure of a solution of sucrose (with mole fraction of sucrose = 0.1) and of a solution of levulose (with mole fraction of levulose = 0.1)?

[**Hint:** $p_{\text{soln}} = p_{\text{solvent}}(1 - x_{\text{solute}})$] (2850 Pa)

61. A beaker containing 0.01 mole of $C_{12}H_{22}O_{11}$ in 100 g of H_2O and a beaker containing 0.02 mole of $C_{12}H_{22}O_{11}$ in 100 g of H_2O are placed in a chamber and allowed to equilibrate. What is the concentration (mole fraction) of $C_{12}H_{22}O_{11}$ in the resulting solutions? (0.00269)

[**Hint:** Water vapour will be transferred from the more dilute solution to the more concentrated solution until both solutions have the same concentration.]

62. What mass of a solute ($M = 345$) is needed to decrease the vapour pressure of 100 g of H_2O at 25°C by 1 mmHg? Vapour pressure of water at 25°C is 23.756 mm. (84.2 g)

63. Azomethane, $(CH_3)_2N_2$ decomposes with a first-order rate according to the equation

$$(CH_3)_2N_2(g) \rightarrow N_2(g) + C_2H_6(g)$$

In the beginning the initial pressure was 36.2 mm and after 15 min. the total pressure was 42.4 mm. Calculate the rate constant. ($1.25 \times 10^{-2}\ \text{min}^{-1}$)

64. The rate of decomposition of a gas at a certain temperature is 5.14 and 7.25 in some units for 20% and 5% decomposition respectively. Calculate the order of the reaction. (2)

65. The order of the reaction: $A + B \rightarrow C$, is 1 with respect to each of the reactants. Find the approximate concentration of A remaining after 100 seconds if its initial concentration is 0.1 M and that of B is 6 M. Rate constant of the reaction is $5 \times 10^{-3}\ M^{-1}\ s^{-1}$. ($4.96 \times 10^{-3}$ M)

66. The decomposition of Cl_2O_7 at 400 K in the gas phase to Cl_2 and O_2 is a first-order reaction.

(i) After 55 seconds at 400 K, the pressure of Cl_2O_7 falls from 0.062 to 0.044 atm. Calculate the rate constant.

(ii) Calculate the pressure of Cl_2O_7 after 100 seconds of decomposition at this temperature. ($6.23 \times 10^{-3}\ s^{-1}$, 0.033 atm)

67. When a solution of formic acid was titrated with KOH solution, the pH of the solution was 3.65 when half the acid was neutralized. Calculate K_a(HCOOH).

(2.24×10^{-4})

68. Calculate $[OH^-]$ in a 1 M solution of NaOCN. K_a(HOCN) $= 3.3 \times 10^{-4}$. (5.5×10^{-6})

69. 0.1 M NH_4OH is 1% ionized. Find the extent of hydrolysis of 0.1 M NH_4Cl.

(10^{-4})

70. Calculate the change in pH of one litre of buffer solution containing 0.1 mole each of NH_3 and NH_4Cl upon addition of

(i) 0.02 mole of dissolved gaseous HCl

(ii) 0.02 mole of dissolved NaOH

Assume no change in volume, K_b for $NH_3 = 1.8 \times 10^{-5}$ (0.1761)

[**Hint:** See Example 28, Chapter 16]

71. The pH of the bloodstream is maintained by a proper balance of H_2CO_3 and $NaHCO_3$ concentrations. What volume of 5 M $NaHCO_3$ solution should be mixed with a 10 mL sample of blood which is 2 M in H_2CO_3, in order to maintain a pH of 7.4? K_a for H_2CO_3 in blood is 7.8×10^{-7}. **(IIT 1993)** (78.36 mL)

[**Hint:** mm of $NaHCO_3 = 5 \times x$

mm of $H_2CO_3 = 2 \times 10$

Apply Henderson's equation and calculate x.]

72. Chromium metal can be plated out from an acidic solution containing CrO_3 according to the following equation

$$CrO_3(aq) + 6H^+(aq) + 6e \rightarrow Cr(s) + 3H_2O$$

Calculate (i) how many grams of chromium will be plated out by 24000 coulombs, and (ii) how long will it take to plate out 1.5 g of chromium by using 12.5 amp current? **(IIT 1993)** (2.1554 g, 1336 s)

[**Hint:** Eq. of Cr deposited = faraday of electricity passed

$$= \frac{24000}{96500}$$

and eq. wt. of chromium $= \frac{52}{6}$]

73. In starting a car, the battery delivers roughly 50 amperes. (i) During the 5 seconds that it might take to start a car, totally how many grams of Pb and PbO_2 are consumed in the battery? (ii) If the car were run strictly from batteries, totally how many grams of Pb and PbO_2 would be consumed per mile if 50 amperes made it go at 5 mph? The cell reaction is

$$Pb + PbO_2 + 2H_2SO_4 \rightarrow 2PbSO_4 + 2H_2O$$ (0.5782 g)

[**Hint:** The reaction involves 2 moles of electrons]

74. The total pressure at equilibrium of a mixture of SO_2 and O_2 in the molar ratio 2 : 1 when kept over a platinum catalyst at 723 K is 10 atm. If 60% of SO_2 is converted to SO_3, calculate K_p for the reaction:

$$2SO_2(g) + O_2(g) \rightleftharpoons 2SO_3(g)$$ $(1.35\ atm^{-1})$

75. What would be the partial pressure of oxygen gas to get equal moles of SO_2 and SO_3?

$$2SO_2(g) + O_2(g) \rightleftharpoons 2SO_3(g)$$

Given that partial pressures of SO_2, O_2 and SO_3 are 0.662 atm, 0.101 atm and 0.331 atm respectively. (0.404 atm)

76. 8.0 moles of SO_2 and 4.0 moles of O_2 are mixed in a closed vessel. The reaction proceeds at constant temperature. By the moment equilibrium sets in, 80% of the initial amount of SO_2 enters the reaction. Determine the pressure of the gas mixture in equilibrium if the initial pressure was 2.96 atm. (2.17 atm)

77. The value of K_p for the reaction: $PCl_5 \rightleftharpoons PCl_3 + Cl_2$ is 1.78 at 250°C. Calculate the fraction of dissociation at equilibrium when 0.40 mole of PCl_5 is vaporized in a vessel containing 0.20 mole of Cl_2 gas (i) when a constant pressure of 2 atm is maintained, and (ii) when the volume is kept constant at 4 litres. (0.247, 0.332)

78. The density of a gas at 27°C and 760 mm pressure is 3.0 g/L. If the pressure remains constant, find out the temperature at which the density will be 2.4 g/L. (102°C)

79. A volume of a gas weighing 8 g was allowed to expand at constant temperature until the pressure of the gas reduced to one-half of its former value. It was found that 500 mL of the rarefied gas weighed 1.25 g.
(i) What was the original volume of the gas?
(ii) Determine the density of the gas in g/L. (1.6 L, 5.0 g/L)

80. A certain quantity of a gas occupied 50 mL, when collected over water at 15°C and 750 mm pressure. If the dry gas occupies 45.95 mL at NTP, calculate the aqueous tension at 15°C. (13.3 mm)

81. Assume that the centre of the sun consists of gases whose average molecular weight is 2. The density and pressure of the gases are 1.3 g/mL and 1.12×10^9 atm respectively. Find the temperature. (2.1×10^7 K)

82. Calculate the equilibrium constant at 25°C for the reaction:

$$2Fe^{3+} + 2I^- = 2Fe^{2+} + I_2$$

Given that $E^0_{Fe^{3+}, Fe^{2+}} = 0.77$ V, $E^0_{I_2, I^-} = 0.536$ V (8.29×10^7)

83. If the concentrations of Fe^{2+} and Fe^{3+} are equal, what should be the concentration of Ag^+ ions to have zero voltage for a galvanic cell made up of $Ag^+ | Ag$ and $Fe^{3+} | Fe^{2+}$ electrodes. Also, calculate the equilibrium constant at 25°C for the following cell reaction: $Fe^{2+} + Ag^+ = Fe^{3+} + Ag$.
Given that $E^0_{Ag^+, Ag} = 0.799$ V, $E^0_{Fe^{3+}, Fe^{2+}} = 0.771$ V (0.33 M, 3.0)

84. For the galvanic cell

$$Ag | AgCl(s), KCl(0.2\ M) || KBr(0.001\ M), AgBr(s) | Ag,$$

calculate the emf generated and assign correct polarity to each electrode for a spontaneous process after taking into account the cell reaction at 25°C.

$$K_{sp}(AgCl) = 1.8 \times 10^{-10}, K_{sp}(AgBr) = 3.3 \times 10^{-13}$$ (0.037 V)

[Hint: Calculate $[Ag^+]$ from K_{sp} values for both the half cells and then calculate E_{cell} for $Ag|Ag^+(c_1)|Ag^+(c_2)|Ag$.]

85. 30 mL of methanol (density 0.7980 g/mL) on mixing with 70 mL of water (density 0.9984 g/mL) at 298 K gave a solution of density 0.9575 g/mL. Calculate (i) mole fraction, (ii) molality, (iii) molarity, and (iv) f.p. of the solution. $K_f(H_2O) = 1.86$.

(0.1615, 10.7043, 7.6337, −19.91°C)

86. At 25°C the vapour pressure of methyl alcohol is 96 torr. What is the mole fraction of CH_3OH in a solution in which the (partial) vapour pressure of CH_3OH is 23 torr at 25°C?

(0.24)

87. The vapour pressure of a 5% solution of a nonvolatile organic substance in water at 373 K is 0.9935×10^5 N m^{-2}. Calculate the molecular mass of the solute (1 atm $= 1.0132 \times 10^5$ N m^{-2})

(0.0458 kg/mol)

88. How many grams of sugar, $C_6H_{12}O_6$, should be dissolved in 0.5 kg of water at 25°C to reduce the vapour pressure of water by 1%?

(50.4 g)

89. Find the molality of a solution (containing nonvolatile solute) if its vapour pressure is 2% below the vapour pressure of pure water.

(1.134 *m*)

90. Gaseous ozone is bubbled through water–ice mixture at 0°C. As the $O_3(g)$ decomposes to form $O_2(g)$, the enthalpy of reaction is absorbed by the resulting ice. Given that the heat of fusion of ice is 6.0095 kJ mol^{-1}, calculate the mass of ice that melts for each gram of O_3 that decomposes.

$$2O_3(g) \rightarrow 3O_2(g);\ \Delta H = -285.4 \text{ kJ}$$

(8.91 g)

91. The 'calorie' used in nutrition is actually a kilocalorie. Assume that a human requires '2500 cal' of energy each day for metabolic activity. What mass of ethanol is needed to provide this energy?

$$C_2H_5OH(l) + 3O_2(g) \rightarrow 2CO_2(g) + 3H_2O(l);\ \Delta H = -1371 \text{ kJ}$$

If the human body is considered to be a closed system, what would be the temperature increase resulting from this energy intake? Assume a mass of 75 kg and a specific heat of 4 J K^{-1}g^{-1}. If the body temperature is to be maintained at a constant value by the evaporation of water, what mass of water must evaporate? Assume that the heat of vaporization of water is 44 kJ mol^{-1}.

(352 g, 35 K day^{-1}, 4292 g)

92. One gram of commercial $AgNO_3$ is dissolved in 50 mL of water. It is treated with 50 mL of a KI solution. The silver iodide thus precipitated is filtered off. Excess of KI in the filtrate is titrated with (M/10) KIO_3 solution in the presence of 6 M HCl till all I^- ions are converted into ICl. It requires 50 mL of (M/10) KIO_3 solution. 20 mL of the same stock solution of KI requires 30 mL of (M/10) KIO_3 under similar conditions. Calculate the percentage of $AgNO_3$ in the sample.

(IIT 1992) (85%)

[Hint: Reaction: $KIO_3 + 2KI + 6HCl = 3ICl + 3KCl + 3H_2O$]

93. A mixture of pure $K_2Cr_2O_7$ and pure $KMnO_4$ weighing 0.561 g was treated with excess of KI in acid medium. Iodine liberated required 100 mL of 0.15 N hypo solution for exact oxidation. What is the percentage of each in the mixture? Reactions involved are

$$2KMnO_4 + 8H_2SO_4 + 10KI \rightarrow 6K_2SO_4 + 2MnSO_4 + 8H_2O + 5I_2$$
$$K_2Cr_2O_7 + 6KI + 7H_2SO_4 \rightarrow 4K_2SO_4 + Cr_2(SO_4)_3 + 7H_2O + 3I_2$$
$$I_2 + 2Na_2S_2O_3 \rightarrow 2NaI + Na_2S_4O_6$$

(39.3%, 60.7%)

94. If PCl_5 is heated to 250°C and allowed to reach equilibrium, 50% of PCl_5 is dissociated. Calculate how many moles of Cl_2 must be mixed with one mole of PCl_5 to reduce the dissociation to 40%, volume remaining constant. (0.35 mole)

95. Given that $K_c = 13.7$ at 546 K for $PCl_5(g) \rightleftharpoons PCl_3(g) + Cl_2(g)$.
Calculate what pressure will develop in a 10-litre box at equilibrium at 546 K when 1 mole of PCl_5 is introduced into the empty box. (8.92 atm)

96. Under what pressure must an equimolar mixture of Cl_2 and PCl_3 be placed at 250°C in order to obtain 80% conversion of PCl_3 into PCl_5? K_p for $PCl_3(g) + Cl_2(g) \rightleftharpoons PCl_5(g)$ is 1.78. (13.48 atm)

97. If the degree of dissociation of PCl_5 at a certain temperature and 1 atm is 0.2, calculate the pressure at which the substance will be half-dissociated at the same temperature. (0.125 atm)

98. In the dissociation of HI, it is found that 20% of the acid is dissociated when equilibrium is reached. Calculate the pressure equilibrium constant. (1/64)

99. 0.45 g of an organic compound (A) on ignition gives 0.905 g CO_2 and 0.185 g H_2O. 0.35 g (A) on boiling with HNO_3 and adding $AgNO_3$ solution gives 0.574 g of AgCl. The vapour density of (A) is 87.5. (A) on hydrolysis with $Ca(OH)_2$ yields (B) which on mild reduction gives an optically active compound (C). On heating (C) with I_2 and NaOH, iodoform is produced along with (D). With HCl, (D) gives a solid which is markedly more soluble in hot water than in cold. Identify (A), (B), (C) and (D).

[$C_6H_5-C(Cl)_2-CH_3$ (A), $C_6H_5-\overset{O}{\overset{\|}{C}}-CH_3$ (B), $C_6H_5CHOHCH_3$ (C), C_6H_5COONa (D)]

100. Two organic compounds containing C = 62.1%, H = 10.34%, O = 27.6% with KCN and H_2SO_4 gave compounds which on hydrolysis gave two isomeric monobasic acids with molecular mass 104. Name the compounds and also the third isomer.

[$(CH_3)_2C=O$ (acetone), C_2H_5CHO (propionaldehyde), $C_2H=CHCH_2OH$ (ally alcohol)]

101. Compound (X) of molecular formula C_5H_8O does not react appreciably with Lucas reagent at room temperature but gives a precipitate with amm. silver nitrate. With excess of MeMgBr, 0.42 g of (X) gives 224 mL of CH_4 at STP. Treatment of (X) with

H_2 in the presence of Pt catalyst followed by boiling with excess HI, gives *n*-pentane. Suggest the structure for (X). **(IIT 1992)**

($HC \equiv C\,CH_2\,CH_2\,CH_2\,OH$)

102. 0.037 g of an alcohol, ROH, was added to $CH_3\,MgI$ and the gas evolved measured 11.2 cm^3 at STP. What is the molecular weight of ROH? On dehydration, ROH gives an alkene which on ozonolysis gives acetone as one of the products. ROH on oxidation easily gives an acid containing the same number of carbon atoms. Give structures of ROH and the acid. (C_4H_9OH, C_3H_7COOH)

103. PCl_5 dissociates into PCl_3 and Cl_2. If the total pressure of the system in equilibrium is p at a density, d, and temperature, T, show that

$$\alpha = \frac{pM}{dRT} - 1,$$

where α is the degree of dissociation, M is the relative molar mass of PCl_5. If the vapour density has the value 62 when the temperature is 230°C, what is the value of p/d? (0.333 L atm g^{-1})

[**Hint:** $pV = n(1+\alpha)RT$; n is the initial no. of moles of PCl_5]

104. An unspecified quantity of an ideal gas has an initial pressure of 5 atm and temperature of 30°C. The gas is expanded at 30°C until the volume has increased by 60% of the initial value. Next, the quantity of the gas in the vessel is increased by 20% of the initial value while the volume is maintained constant. Finally the temperature is adjusted at constant volume until the gas pressure is again 5 atm. What is the final temperature? (404 K)

105. In a gaseous reaction $A \rightarrow B + C$, the pressure of A falls from 0.2 atm to 0.15 atm in one hour. Calculate the rate constant if it is a first-order reaction. What will be the pressure of A after 1.5 hours? (0.2878, 0.13 atm)

106. The decomposition of PH_3 at 950 K according to

$$4PH_3(g) \rightarrow P_4(g) + 6H_2(g)$$

is a first-order reaction. The following measurements were made on a system containing only PH_3 initially.

Time (min):	0	40	80
p (total mmHg):	100	150	166.7

Calculate the rate constant. (2.7×10^{-2} min^{-1})

107. The gas phase decomposition of dimethyl ether follows first-order kinetics:

$$CH_3 - O - CH_3(g) \rightarrow CH_4(g) + H_2(g) + CO\,(g)$$

The reaction is carried out in a constant-volume container at 500°C and has a half-life of 14.5 min. Initially only dimethyl ether is present at a pressure of 0.4 atm. What is the total pressure of the system after 12 minutes? Assume ideal gas behaviour. **(IIT 1993)** (0.75 atm)

108. The standard reduction potential of the electrode $OH^- \mid H_2(Pt)$ (1 atm) is $E^0 = 0.828$ volt. Calculate K_w at 298 K if Nernst equation takes the form

$$E = E^0_{H_2O,\,OH^-} - 0.059 \log [OH^-]$$

[**Hint:** Compare equation $E = E^0_{H_2O, OH^-} - 0.0591 \log [OH^-]$

and $E = E^0_{H^+, H_2} - 0.0591 \log \frac{1}{[H^+]}$

$= E^0_{H^+, H_2} - 0.0591 \log \frac{[OH^-]}{K_w}$] (1.03×10^{-14})

109. The standard reduction potential of the electrode $Cl^- \mid CuCl$ is $E^0 = 0.137$ volt. Calculate K_{sp} of CuCl at 298 K. ($E^0_{Cu^+, Cu} = 0.521$ volt). $[3.19 \times 10^{-7} \text{ (mol/ L)}^2]$

[**Hint:** $CuCl + e \rightleftharpoons Cu + Cl^-$; $E^0 = 0.137$ V

$Cu^+ + e \rightleftharpoons Cu$; $E^0 = 0.521$ V

for which $E_{cell} = E^0_{cell} + 0.0591 \log K_{sp} \text{ (CuCl)} = 0$]

110. Calculate the emf of the cell

$$Pb \mid \underset{\text{sat. soln.}}{PbSO_4} \mid\mid \underset{\text{sat. soln.}}{PbI_2} \mid Pb$$

at 25°C. $K_{sp}(PbSO_4) = 1.6 \times 10^{-8}$ and $K_{sp}(PbI_2) = 8 \times 10^{-9}$. (0.029 V)

111. The half-life of a substance in a first-order reaction is 100 minutes at 323.2 K and 15 minutes at 353.2 K. Calculate the temperature coefficient of the rate constant of this reaction. (1.88)

112. A metal object is to be coated with a nickel layer 0.3 mm thick. The surface area of the object is 100 cm^2. The density of nickel is 9.0 g/cm^3. How long will it take to pass a 3-amp current if the current yield is 90%? (9 h 8 min 25 s)

113. Electrolysis of a 20% potassium acetate solution at 290 K yields, among other things, ethane evolved on the anode according to the equation

$$2CH_3COO^- = C_2H_6 + 2CO_2 + 2e$$

The ratio between the molecules of ethane evolved on the anode and those of hydrogen evolved on the cathode is 0.8. Find current yield of ethane. (80%)

114. During an electrochemical experiment, 0.2773 g of Ag was transferred from one electrode to the other electrode in a coulometer. What electric charge did pass through the circuit? (248.1 C)

115. A vessel at 1000 K contains CO_2 with a pressure of 0.5 atm. Some of the CO_2 is converted to CO on addition of graphite. Calculate the value of K, if the total pressure at equilibrium is 0.8 atm. (1.8 atm)

[**Hint:** For the eqb. $\underset{(0.5-p)}{\overset{0.5 \text{ atm}}{CO_2 (g)}} + C(s) \rightleftharpoons \underset{2p}{2CO (g)}$

$0.5 - p + 2p = 0.8$ (given)]

116. PCl_5 was found to dissociate to the extent of 42% at 227°C and 1 atm. Find the equilibrium constant at (i) constant pressure, and (ii) constant volume. (0.214, 0.005)

117. The Deacon reaction is the oxidation of HCl by O_2:

$$HCl(g) + \frac{1}{4}O_2(g) \rightleftharpoons \frac{1}{2}Cl_2(g) + \frac{1}{2}H_2O\ (g)$$

At a pressure of 730 mm and with an initial mixture containing 8% HCl and 92% O_2, the degree of decomposition of the HCl is 0.80. What is the equilibrium partial pressure of oxygen? (660 mm)

118. What is the vapour density of PCl_5 at 250°C when dissociated to the extent of 80%? (57.9)

119. The density of a 0.33 M solution of $MgBr_2$ in water at 373 K is 1.055 g/mL. Calculate the vapour pressure of water above this solution. Assume ideal behaviour. (746.5 mm)

120. The vapour pressure of a 0.01 molal solution of a weak monobasic acid in water is 17.536 mm at 25°C. Calculate the degree of dissociation of the acid. Aqueous tension of water at 25°C is 17.54 mm. (0.266)

121. What volume of 98% sulphuric acid should be mixed with water to obtain 200 mL of 15% solution of sulphuric acid by weight? The density of water, sulphuric acid (98%) and sulphuric acid (15%) are 1 g/mL, 1.88 g/mL and 1.12 g/mL respectively. (18.2 mL)

122. What weight of the nonvolatile solute, urea ($NH_2-CO-NH_2$) needs to be dissolved in 100 g of water in order to decrease the vapour pressure of water by 25%? What will be the molality of the solution? **(IIT 1993)**
(111.12 g, 18.52 m)

123. In an ore the only oxidizable material is Sn^{2+}. This ore is titrated with a dichromate solution containing 2.5 g of $K_2Cr_2O_7$ in 0.50 litre. A 0.40 g sample of the ore required 10.0 cm^3 of titrant to reach equivalence point. Calculate the percentage of tin in the ore.

[Hint: $3Sn^{2+} + Cr_2O_7^{2-} + 14H^+ \rightarrow 3Sn^{4+} + 2Cr^{3+} + 7H_2O$] (15%)

124. Upon mixing 45.0 mL of 0.25 M lead nitrate solution with 25.0 mL of 0.10 M chromic sulphate solution, precipitation of lead sulphate takes place. How many moles of lead sulphate are formed? Also calculate the molar concentrations of species left behind in the final solution. Assume that lead sulphate is completely insoluble. **(IIT 1993)**

$$\begin{pmatrix} PbSO_4 = 0.0075\text{ mole} \\ [Pb^{2+}] = 0.05357\text{ M}, [NO_3^-] = 0.3214\text{ M} \\ [Cr^{3+}] = 0.0714\text{ M} \end{pmatrix}$$

[Hint: Apply the concept of limiting reagent]

125. During the operation of a cell with the cell reaction

$$Mn\ (OH)_2 + H_2O_2 \rightarrow MnO_2 + 2H_2O$$

for 45 seconds, 0.136 g of MnO_2 was produced. Calculate the average electric current produced by the cell. (6.7 amp)

126. What amount of work is needed to move an electron against a potential difference of 1.0 volt? What amount of work is needed for 1 mole of electrons and what does this value represent? (1.602×10^{-19} J, 96470 J mole^{-1}, Faraday constant)

127. The cathodic reaction of a $Zn - MnO_2$ dry cell is represented as

$$2MnO_2(s) + Zn^{2+} + 2e = Zn\,Mn_2O_4(s).$$

If there is 8.0 g MnO_2 present in the cathodic chamber, how long will the cell function to supply 4×10^{-3} ampere of current? (25.67 days)

$$\left[\textbf{Hint:}\ \frac{4 \times 10^{-3} \times t\,(s)}{96500} = \frac{8}{87};\ E_{MnO_2} = 87\right]$$

128. Calculate the total number of coulombs carried by Cl^- ions weighing 20.1 g. (5.46×10^4 C)

129. The dissociation constant of an acid HA at 25°C is 1.34×10^{-5}. How many moles of sodium salt of this acid should be added to one litre of an aqueous solution containing 0.02 mole of this acid to obtain a buffer solution of pH 4.75? What will be the pH if 0.01 mole of HCl is dissolved in the buffer solution? **(IIT 1993)** (0.015 mole)

[**Hint:** See Example 28, Chapter 16.]

130. An aqueous solution of a metal bromide MBr_2 (0.05 M) is saturated with H_2S. What is the minimum pH at which MS will precipitate?

$K_{sp}(MS) = 6.0 \times 10^{-21}$; concentration of saturated $H_2S = 0.1$ M;

$K_1 = 1 \times 10^{-7}$ and $K_2 = 1.3 \times 10^{-13}$ for H_2S. **(IIT 1993)** (0.983)

[**Hint:** See Example 46, Chapter 16.]

131. What change would be observed in sulphide ion concentration of 0.05 M H_2S solution if 0.1 mole of HCl is added to one litre of it? $K_a(H_2S) = 1 \times 10^{-22}$.

(1.075×10^{-8} M to 5.0×10^{-22} M)

132. The pH of 0.05 M aqueous solution of diethylamine is 12.0. Calculate its K_b. (2.5×10^{-3})

[**Hint:** Hydrolysis reaction is $(C_2H_5)_2NH + H_2O = (C_2H_5)_2NH_2^+ + OH^-$]

133. Given that

(i) $\frac{1}{2}Cl_2(g) + O_2(g) \rightarrow ClO_2(g)$; $\Delta H = 102.5$ kJ

(ii) $\frac{1}{2}Cl_2(g) + \frac{3}{2}O_2(g) \rightarrow ClO_3(g)$; $\Delta H = 155$ kJ

(iii) $Cl_2(g) + \frac{7}{2}O_2(g) \rightarrow Cl_2O_7(g)$; $\Delta H = 272$ kJ

Calculate ΔH for the reaction

$$3ClO_3(g) \rightarrow Cl_2O_7(g) + ClO_2(g)$$ (–90.5 kJ)

134. Given that

(i) $Ag(s) \rightarrow Ag^+(aq) + e$ $\Delta H = 105.58$ kJ

(ii) $\frac{1}{2}Cl_2(g) + e \rightarrow Cl^-(aq)$ $\Delta H = -167.16$ kJ

(iii) $Br_2(l) + 2e \rightarrow 2Br^-(aq)$ $\Delta H = -243.10$ kJ

(iv) $AgCl(s) \rightarrow Ag^+(aq) + Cl^-(aq)$ $\Delta H = 65.49$ kJ

(v) $AgBr(s) \rightarrow Ag^+(aq) + Br^-(aq)$ $\Delta H = 84.40$ kJ

Determine which of the thermochemical equations

$$Ag(s) + \frac{1}{2}Cl_2(g) \rightarrow AgCl(s)$$

$$Ag(s) + \frac{1}{2}Br_2(l) \rightarrow AgBr(s)$$

is more exothermic. (Silver–chlorine reaction)

135. Use the following thermochemical equations to determine the average bond enthalpy for C – H bonds.

$CH_4(g) \rightarrow CH_3(g) + H(g)$; $\Delta H = 438.47$ kJ

$CH_3(g) \rightarrow CH_2(g) + H(g)$; $\Delta H = 462.65$ kJ

$CH_2(g) \rightarrow CH(g) + H(g)$; $\Delta H = 423.40$ kJ

$CH(g) \rightarrow C(g) + H(g)$; $\Delta H = 338.85$ kJ (415.84 kJ mole^{-1})

[**Hint:** First calculate ΔH for $CH_4(g) \rightarrow C(g) + 4H(g)$]

136. Hot carbon reacts with steam to produce an equimolar mixture of CO(g) and $H_2(g)$ known as water gas. What is the energy released as water gas is used as fuel?

$$CO(g) + \frac{1}{2}O_2(g) \rightarrow CO_2(g);\ \Delta H = -282.984 \text{ kJ}$$

$$H_2(g) + \frac{1}{2}O_2(g) \rightarrow H_2O(g);\ \Delta H = -241.818 \text{ kJ}$$

(– 524.802 kJ)

137. Ethanol can undergo decomposition to form two sets of products:

$$C_2H_5OH(g) \rightarrow \begin{Bmatrix} C_2H_4(g) + H_2O(g); & \Delta H = 45.54 \text{ kJ} \\ CH_3CHO\ (g) + H_2(g); & \Delta H = 68.91 \text{ kJ} \end{Bmatrix}$$

If the molar ratio of C_2H_4 to CH_3CHO is 8:1 in a set of product gases, determine the energy involved in the decomposition process. (48.14 kJ)

[**Hint:** $\Delta H = \left\{\frac{8}{9}(45.54) + \frac{1}{9}(68.91)\right\}$ kcal]

138. Given that

(i) $2ClO_2(g) + O_3(g) \rightarrow Cl_2O_7(g)$; $\Delta H = -75.7$ kJ

(ii) $O_3(g) \rightarrow O_2(g) + O(g)$; $\Delta H = 106.7$ kJ

(iii) $2ClO_3(g) + O(g) \rightarrow Cl_2O_7(g)$; $\Delta H = -287$ kJ

(iv) $O_2(g) \rightarrow 2O(g)$; $\Delta H = 498.34$ kJ

Calculate enthalpy of the reaction

$$ClO_2(g) + O(g) \rightarrow ClO_3(g)$$

(–196 kJ)

139. A solution of nitric acid of specific gravity 1.46 contains 60% nitric acid. What weight of this solution is theoretically required to dissolve 5 g of cupric oxide? (19.28 g)

140. A flash bulb used for taking photographs in poor light contains 30 mL of O_2 at

780 mm pressure at 27°C. Supposing that the metal wire flashed is pure Al which is oxidized to Al_2O_3 in the process of flashing, calculate the minimum weight of Al wire that is to be used for maximum efficiency. (0.045 g)

141. How many years would it take to spend the Avogadro constant of rupees at the rate of 10 lakh rupees per second? (1.91×10^{10} yrs)

142. To a sample of an element X (at. wt. 70) another element Y (at. wt. 120) is to be added as an impurity. The ratio of the atoms in the mixture is to be $1 : 10^{-7}$. How many grams of Y will be required for 35 g of X? (6.0×10^{-6} g)

143. How much calcium is there in the amount of $Ca(NO_3)_2$ that contains 20 g of nitrogen? (28.60 g)

144. At temperature *T*, a compound AB_2(g) dissociates according to the reaction

$$2AB_2(g) \rightleftharpoons 2AB(g) + B_2(g)$$

with a degree of dissociation, *x*, which is small compared to unity. Deduce the expression for *x* in terms of the equilibrium constant, K_p, and the total pressure, *p*.

(IIT 1994) $\left[x = \left(\frac{2K_p}{p}\right)^{1/3}\right]$

145. 8.0575×10^{-2} kg of Glauber's salt is dissolved in water to obtain 1 dm^3 of a solution of density 1077.2 kg m^{-3}. Calculate the molarity, molality and mole fraction of Na_2SO_4 in the solution. **(IIT 1994)**

[**Hint:** 1 dm^3 = 1 L, kg m^{-3} = gL^{-1}] (0.5674 M, 0.5693 *m*, 0.01)

146. The freezing point of a solution containing 0.2 g of acetic acid in 20.0 g of benzene is lowered by 0.45°C. Calculate the degree of association of acetic acid in benzene. $K_f(C_6H_6) = 5.12$ K mol^{-1} kg. (0.9453)

[**Hint:** Acetic acid exists as dimer in benzene.]

147. Calculate the mole fraction of a benzene-toluene liquid solution that is in equilibrium with a vapour phase that contains 62 mole % C_6H_6. Vapour pressures of benzene and toluene at 25°C are 95.1 mm and 28.4 mm respectively. (0.328 for C_6H_6)

148. Calculate the electrode potential for

$$\underset{(1\text{ atm})}{(Pt)\ H_2} \mid H^+ (c = 0.1)$$

(– 0.0591 V)

149. Calculate the emf of the cell of 25°C:

$$\underset{(1\text{ atm})}{(Pt)\ H_2} \mid \underset{c\,=\,0.1\text{ M}}{CH_3COOH} \parallel \underset{c\,=\,0.01\text{ M}}{NH_4OH} \mid \underset{(1\text{ atm})}{H_2\ (Pt)}$$

$K_a(CH_3COOH) = 1.8 \times 10^{-5}$ and $K_b(NH_4OH) = 1.8 \times 10^{-5}$ (0.458 V)

150. A solution contains 0.06 M of Cu^{2+} ions and Ag^+ of unknown concentration. Find the concentration of Ag^+ ions so that both the metals can be codeposited. $E^0_{Cu^{2+}Cu} = +0.337$ V and $E^0_{Ag^+, Ag} = +0.7991$ V. (3.7×10^{-9} M)

151. Calculate the ratio of the oxidized to the reduced form at half-cell potential of 0.1 volt for the half cell $Fe^{3+}, Fe^{2+} | Pt$. $E^0_{Fe^{3+}, Fe^{2+}} = 0.7591$. (1.41×10^{11})

152. Cadmium amalgam is prepared by electrolysis of a solution of $CdCl_2$ using Hg cathode. For how long should electrolysis be carried out in order to prepare 12% by weight of cadmium amalgam using 22.0 g of Hg as cathode and a current strength of 5 ampere? (Cd = 112) (17 min 14 s)

153. Electrolytic oxidation of H_2SO_4 gives rise to the formation of persulphuric acid, $H_2S_2O_8$.

$$2H_2SO_4 \rightarrow H_2S_2O_8 + 2H^+ + 2e$$

If 4 litres of O_2 and 11.2 litres of H_2 were produced at NTP, determine the mass of $H_2S_2O_8$ produced. (27.72 g)

[**Hint:** $2H_2SO_4 \rightarrow H_2S_2O_8 + H_2$

$2H_2O \rightarrow 2H_2 + O_2$

Eq. of $H_2S_2O_8$ = eq. of H_2 − eq. of O_2]

154. An alloy of Pb(II) and Tl(I) containing 70% Pb and 30% Tl by weight can be electroplated on to a cathode from a perchloric acid solution. How much time would be required to deposit 5.0 g of this alloy at a current of 1.1 amp? (Pb = 207.19, Tl = 204.37) (1 hour)

155. The tungsten used in filaments for light bulbs can be prepared from tungsten (VI) oxide by reduction with hydrogen at 1200°C

$$WO_3(s) + 3H_2(g) = W(s) + 3H_2O(g); \quad \Delta H = 114.9 \text{ kJ}$$

ΔH_f for $H_2O(g) = -241.8$ kJ $mole^{-1}$. Calculate ΔH_f for $WO_3(s)$. (− 840.3 kJ)

156. Water gas is produced by the action of superheated steam on red hot coke.

$$C(s) + H_2O(g) + 131.4 \text{ kcal} = \underbrace{CO(g) + H_2(g)}_{\text{water gas}}$$

ΔH_f for $H_2O(g) = -241.8$ kJ $mole^{-1}$. Calculate ΔH_f for CO(g). (−110.4 kJ)

157. The thermochemical equation for the dissociation of hydrogen gas into atoms may be written as:

$$H_2 \rightarrow 2H; \quad \Delta H = 436 \text{ kJ}.$$

What is the ratio of the energy yield on combustion of hydrogen atoms to steam to the yield on combustion of an equal mass of hydrogen molecule to steam? Heat of formation of $H_2O(g)$ is − 241.81 kJ $mole^{-1}$. (2.8)

158. Calculate ΔH°_f for chloride ion from the following data.

$$\frac{1}{2}H_2(g) + \frac{1}{2}Cl_2(g) \rightarrow HCl(g); \; \Delta H^\circ_f = -92.4 \text{ kJ}$$

$$HCl(g) + xH_2O \rightarrow H^+(aq) + Cl^-(aq); \; \Delta H_{298} = -74.8 \text{ kJ}$$

$$\Delta H^\circ_f [H^+(aq)] = 0.0 \text{ kJ}$$

(−167.2 kJ)

159. The data below after the reaction of NO and Cl_2 to form NOCl at 295 K:

$[Cl_2]$	[NO]	Initial rate ($mol\ L^{-1}s^{-1}$)
0.05 M	0.05 M	1×10^{-3}
0.15 M	0.05 M	3×10^{-3}
0.05 M	0.15 M	9×10^{-3}

(i) What is the order with respect to NO and Cl_2?
(ii) Write the rate expression.
(iii) Calculate the rate constant.
(iv) Determine the reaction rate when the concentrations of Cl_2 and NO are 0.2 M and 0.4 M respectively. [(i) 2, 1 (iii) 8.0 $L^2\ mol^{-2}\ s^{-1}$ (iv) 0.256 $mol\ L^{-1}\ s^{-1}$]

[Hint: See Example 25, Chapter 17]

160. The order of the reaction: $A + B \rightarrow C$, is one with respect to each of the reactants. Fill in the blanks in the following table.

Initial [A]	Initial [B]	Initial rate
0.2 M	0.05 M	$0.1\ M\ s^{-1}$
?	0.05 M	$0.4\ M\ s^{-1}$
0.4 M	?	$0.8\ M\ s^{-1}$

(0.8 M, 0.2 M)

161. An unspecified quantity of an ideal gas has a volume of 30 litres at 20°C. The gas is first compressed at 20°C until the pressure has doubled and then the temperature is raised to 100°C, while the pressure is kept constant. Determine the final volume of the gas. (19.10 litres)

162. A glass tube *AD* of uniform cross section of length 100 cm sealed at both ends contains two columns of ideal gas *AB* and *CD* separated by a column of mercury of length 20 cm. When the tube is held horizontally, $AB = 20$ cm and $CD = 60$ cm. When the tube is held vertically with the end *A* up, the mercury column moves down 10 cm. What will be the length of gas column *AB* when the tube is held vertically with the end *D* up? (13.88 cm)

163. A vertical cylinder of height 100 cm contains air at a constant temperature. The top is closed by a frictionless light piston. The atmospheric pressure is equal to 75 cmHg. Mercury is slowly poured over the piston. Find the maximum height of the mercury column that can be put on the piston.

[Hint: Boyle's law: $75 \times 100 = p \times (100 - l)$; $p = (75 + l)$; calculate l] (25 cm)

164. A glass tube sealed at both ends is 100 cm long. It lies horizontally with the middle 10 cm containing Hg. The two ends of the tube containing air at 27°C and at a pressure 76 cmHg. The air column on one side is maintained at 0°C and on the other side at 127°C. Calculate the length of the air column on the cooler side. Neglect the changes in the volume of mercury and of the glass. (36.5 cm)

165. Mercury diffusion pumps may be used in the laboratory to produce a high vacuum. Cold traps are generally placed between the pump and the system to be evacuated. These cause the condensation of Hg vapour, and prevent mercury from diffusing back into the system. The maximum pressure of mercury that can exist in the system is the vapour pressure of mercury at the temperature of the cold trap. Calculate the number of mercury-vapour molecules per cc in a cold trap maintained at −120°C. The vapour pressure of mercury at this temperature is 10^{-6} mm. (6×10^{10})

166. Saccharin ($K_a = 2 \times 10^{-12}$) is a weak acid represented by the formula HSac. 4×10^{-4} mole saccharin is dissolved in 200 cm^3 aqueous solution of pH 3. Assuming no change in volume, calculate the concenration of Sac^{-1} ions in the resulting solution at equilibrium. (4.0×10^{-12})

167. K_a for butyric acid is 2.0×10^{-5}. Calculate pH and hydroxyl ion concentration of 0.2 M aqueous solution of sodium butyrate. (9.45)

[**Hint:** Use Equation 11 (b), Chapter 16]

168. A solution has 0.05 M Mg^{2+} and 0.05 M NH_3. Calculate the concentration of NH_4Cl required to prevent the formation of $Mg(OH)_2$ in the solution. $K_{sp}(Mg(OH)_2) = 9.0 \times 10^{-12}$ and $K_b(NH_3) = 1.8 \times 10^{-5}$. (0.067 M)

169. The solubility product of $Mg(OH)_2$ at 25°C is 3.4×10^{-11} and that of $Fe(OH)_3$ is 1.1×10^{-35}.

(i) How many grams per litre of Mg^{2+} and Fe^{3+} can remain dissolved in 100 mL of 0.1 M NH_4OH solution? $K_b(NH_4OH) = 1.01 \times 10^{-5}$.

(ii) How many gram per litre of Mg^{2+} and Fe^{3+} can remain in 100 mL of 0.1 M NH_4OH containing sufficient amount of NH_4Cl to make $[NH_4^+] = 2.0$ M?

[(i) 4.8×10^{-4} g/L, 2.8×10^{-26} g/L (ii) 1.14×10^{-3} g/L, 1.02×10^{-16} g/L]

170. The equilibrium constant for the following reaction:

$$CO(g) + H_2O(g) \rightleftharpoons CO_2(g) + H_2(g)$$

at 986°C is 0.63. A mixture of 1.0 mole of water vapour and 3.0 mole of CO is allowed to come to equilibrium. The equilibrium pressure is 2.0 atm.

(i) How many moles of H_2 are present at equilibrium?

(ii) Calculate the partial pressures of gases in the equilibrium mixture.

(0.68 mole; 0.34 atm, 0.34 atm, 1.16 atm, 0.16 atm)

171. 0.15 mole of CO taken in a 2.5-litre flask is maintained at 750 K along with a catalyst so that the following reaction can take place:

$$CO(g) + 2H_2(g) \rightleftharpoons CH_3OH(g)$$

Hydrogen is introduced until the total pressure of the system is 8.5 atm at equilibrium and 0.08 mole of methanol is formed. Calculate (i) K_p and K_c, and (ii) the final pressure if the same amount of CO and H_2 as before are used, but with no catalyst so that the reaction does not take place.

[(i) 4.94×10^{-2} atm^{-2}, 1.8684×10^2 $(mol/L)^{-2}$ (ii) 12.43 atm]

172. The degree of dissociation of N_2O_4 is 0.633 under a pressure of 3 atm. What must be the pressure if dissociation is to be 80%? (0.31)

173. Calculate the energy emitted when electrons of 1.0 g of hydrogen atoms undergoes transition giving the spectral line of lowest energy in the visible region of its atomic spectrum.
$R_H = 1.1 \times 10^7\ m^{-1}$, $c = 3 \times 10^8\ m\ s^{-1}$, $h = 6.62 \times 10^{-34}$ Js. (182.656 kJ)

174. What transition in the hydrogen spectrum would have the same wavelength as the Balmer transition ($n = 4$ to $n = 2$ of He^+ spectrum)? **(IIT 1993)**
[**Hint:** See Example 14, Chapter 11] ($n = 2$ to $n = 1$)

175. Electromagnetic radiation of wavelength 242 nm is just sufficient to ionize sodium atom. Calculate the ionization energy of sodium atom. $c = 3 \times 10^8\ m\ s^{-1}$, $h = 6.626 \times 10^{-34}$ Js. (8.203×10^{-22} kJ)
[**Hint:** See Example 7, Chapter 11]

176. Copper metal has a face-centred cubic structure with unit cell length equal to 0.361 nm. Picturing copper ions in contact along the face diagonal, find the apparent radius of a copper ion. (0.128 nm)

177. How many moles of electrons are involved in balancing the following redox equations?

$$H_2S + NO_3^- \rightarrow S + NO$$

and $$Mn(OH)_2 + H_2O_2 \rightarrow MnO_2$$ (6, 2)

178. Write the half reactions and number of moles of electrons involved in the overall cell reaction for the electrochemical cell designated by
Pt|Ag(s)|AgCl(s)|Cl^-($c = 1$)|Cl_2($c = 1$)|C (graphite)|Pt.

$$\left[\begin{array}{r} Ag(s) + Cl^-(aq) \rightarrow AgCl(s) + e, \\ Cl_2 + 2e \rightarrow 2Cl^-, \\ n = 2 \end{array}\right]$$

179. Given $E^0 = 0$ V for the H^+/H_2 couple and -0.8281 V for $H_2O/H_2, OH^-$ couple. Determine K_w at 25°C. (1.01×10^{-14})

180. Given $E^0 = -0.268$ V for the $PbCl_2/Pb$ couple and -0.126 V for Pb^{2+}/Pb couple. Determine K_{sp} for $PbCl_2$ at 25°C. (1.6×10^{-5})

$$\left[\begin{array}{l} \textbf{Hint: } PbCl_2 + 2e \rightarrow Pb + 2Cl^- \\ \qquad\qquad Pb \rightarrow Pb^{2+} + 2e \end{array}\right]$$

181. The vapour pressure of heptane and toluene at 50°C are 141 mm and 93 mm respectively. Find the composition of the solution (containing toluene and heptane) at which the partial pressure of toluene is equal to that of heptane. (0.602, 0.398)

182. A liquid mixture of A and B is placed in a cylinder-and-piston arrangement. The piston is slowly pulled out isothermally so that the volume of the liquid decreases and that of the vapour increases. At the instant when the quantity of the liquid still remaining is negligibly small, the mole fraction of A in the vapour is 0.4. $p_A^0 = 0.4$ atm, $p_B^0 = 1.2$ atm at the temperature in question. Calculate the total pressure at which the liquid has almost evaporated. Assume ideal behaviour. (0.667 atm)

183. Air was drawn through a solution containing 38 g of solute in 100 g of water, and then through water. The loss of weight of water was 0.0551 g and the total weight of water absorbed in sulphuric acid tube was 2.2117 g. Find the molecular weight of the dissolved substance. (267.7)

[**Hint:** Raoult's law: $\frac{\text{lowering of VP}}{\text{VP of solvent}} = \text{mole fraction of solute}$

or $\frac{0.0551}{2.2117} = \frac{38/M}{\frac{38}{M} + \frac{100}{18}}$]

184. Calculate the mole per cent of ammonia formed at 350°C and equilibrium pressure of 10 atm when nitrogen and hydrogen taken are in 1 : 3 molar ratio. $K_p = 7.08 \times 10^{-4}$. (9.47%)

185. A mixture of nitrogen and hydrogen in a proportion of 1 : 3 by volume was subjected to a pressure of 30 atm and a temperature of 723 K. After equilibrium was established, and cooled, the analysis indicated that the mixture contained 6% of ammonia by volume. Calculate K_p. (4.85×10^{-5} atm^{-2})

186. The equilibrium constant for the reaction

$$H_2(g) + I_2(g) \rightleftharpoons 2HI(g)$$

at 440°C is 50. How many moles of hydrogen per mole of iodine are required to convert 90% of iodine into hydrogen iodide? (1.548 moles)

187. At 21.5°C and a total pressure of 0.0787 atm, N_2O_4 is 48.3% dissociated into NO_2. Calculate K_c for the reaction: $N_2O_4(g) \rightleftharpoons 2NO_2(g)$. At what total pressure will the per cent dissociation be 10%? (0.00396, 2.371 atm)

188. Calculate the electron affinity of chlorine from the given data:

$Na(g) = Na^+(g) + e$ $\quad \Delta H° = 499.8$ kJ

$\frac{1}{2}Cl_2(g) = Cl(g)$ $\quad \Delta H° = 120.9$ kJ

$Na(s) = Na(g)$ $\quad \Delta H° = 108.3$ kJ

$Na(s) + \frac{1}{2}Cl_2(g) = NaCl(s)$ $\quad \Delta H° = -411.3$ kJ

$Na^+(g) + Cl^-(g) = NaCl(s)$ $\quad \Delta H° = -775.4$ kJ (−364.9 kJ mole^{-1})

189. A gas expands from a volume of 3.0 dm^3 to 5.0 dm^3 against a constant external pressure of 3.0 atm. The work done during expansion is used to heat 10 moles of water at a temperature 290 K. Calculate the final temperature of water. Specific heat of water = 4.184 J g^{-1} K^{-1}.

[**Hint:** $W = -p \, . \, \Delta V$; 1 atm = 101325 Pa, 1 dm^3 = 10^{-3} m^3] (290.807 K)

190. Determine the enthalpy of the reaction

$$C_3H_8(g) + H_2(g) \rightarrow C_2H_6(g) + CH_4(g) \text{ at } 25°C,$$

using the given heat of combustion values under standard conditions.

Compound:	$H_2(g)$	$CH_4(g)$	$C_2H_6(g)$	C (graphite)
$\Delta H°$ (kJ/mole):	–285.8	–890.0	–1560.0	–393.5

The standard heat of formation of $C_3H_8(g)$ is – 103.8 kJ/mole. (–55.7 kJ)

191. A cylindrical tube of length 30 cm is partitioned by a tight-fitting separator. The separator is very thin, very weakly conducting and can freely slide along the tube. Ideal gases are filled in the two parts of the vessel. In the beginning, the temperature in the parts *A* and *B* are 400 K and 100 K respectively. The separator slides to a momentary equilibrium position at which the length of the tube at part *A* is 20 cm. Find the final equilibrium position of the separator, reached after a long time. (10 cm along part A)

[**Hint:** Apply $\frac{p_1V_1}{T_1} = \frac{p_2V_2}{T_2}$ for part *A* under the given two conditions and again for part *B*. Remember that at both the equilibrium positions, both sides will have the same pressure.]

192. By how many folds the temperature of a gas would increase when the rms velocity of the gas molecules in a container of fixed volume is increased from 5×10^4 cm/s to 15×10^4 cm/s? (9 times)

193. The order of a reaction, $A + 2B \rightarrow C$, is 1 and 2 with respect to A and B respectively. Fill in the blanks in the following table:

[A]	[B]	Rate
1.0 M	0.2 M	$0.1\,M\,s^{-1}$
2.0 M	0.2 M	?
2.0 M	?	$0.8\,M\,s^{-1}$

(0.4 M, $0.2\,M\,s^{-1}$)

194. The activation energy for a first-order reaction is 104.5 kJ/mole and the factor *A* in the Arrhenius equation is $5 \times 10^{13}\,s^{-1}$. Find the temperature at which the half-life period of the reaction is 1 minute. (75°C)

195. The activation energy of a certain uncatalysed reaction at 300 K is 76 kJ per mole. The activation energy is lowered to 57 kJ per mole by the use of a catalyst. By what factor is the rate of the catalysed reaction increased? (2031 times)

196. Two reactions of the same order have equal pre-exponential factors, but their activation energies differ by 24.9 kJ per mole. Calculate the ratio between the rate constants of these reactions at 27°C. (2.2×10^{-4})

197. Two bulbs *A* and *B* of equal capacity are filled with He and SO_2 respectively at the same temperature.

(i) If the pressure in the two bulbs is the same, what will be the ratio of velocities of the molecules of the two gases?

(ii) At what temperature will the velocity of SO_2 molecules become half the velocity of He molecules at 27°C?

(iii) How will the velocities be affected if the volume of *B* becomes 4 times that of *A*?

(iv) How will the velocities be affected if half the molecules of SO_2 are removed from *B*? [(i) 4 (ii) 927°C (iii) (iv)—no change]

198. A reaction mixture for the combustion of SO_2 was prepared by opening a stopcock connecting two separate chambers, one having a volume of 2.125 litres filled with SO_2 at 0.75 atm and the other having 1.5 litres volume filled with oxygen at 0.50 atm; both gases were at 80°C.

(i) What were the mole fractions of SO_2 and O_2 in the mixture, and the total pressure?

(ii) If the mixture was passed over a catalyst that promoted the formation of SO_3 and was then returned to the original two connected vessels, what were the mole fractions in the final mixture and what was the final total pressure? Assume that the conversion of SO_2 is complete to the extent of the availability of O_2.

$$\left[\begin{array}{l}\text{(i) 0.68, 0.32; 0.64 atm}\\ \text{(ii) 0.06, 0.94; 0.44 atm}\end{array}\right]$$

199. The standard electrode potential of a standard hydrogen half cell is 0 volt. If the standard state is considered when $[OH^-] = 10^{-7}$ or $[H^+] = 10^{-7}$, calculate E^0 under these conditions. (–0.4137 V, –0.4137 V)

200. The potential for the reaction

$$O_2\,(g) + 4H^+ + 4e = 2H_2O\,(l)$$

is 1.23 V in 0.1 N acid solution. Calculate the potential of this couple in aqueous solutions having (a) pH = 10 (b) pH = 14.

[**Hint:** First calculate E^0 for the given half cell and then the potential at the given pH.] (0.6981 V, 0.462 V)

201. Calculate the solubility product of the reaction

$$Fe(OH)_3 = Fe^{3+} + 3OH^-$$

Given that,

$$Fe(OH)_3(s) + 3e = Fe(s) + 3OH^-\ ;\quad E^0 = -0.77\ V$$

$$Fe^{3+} + 3e = Fe(s)\ ;\quad E^0 = -0.036\ V$$

(8.699×10^{-38})

202. Calculate the potential of the cell

Cu | Mn (s) | $MnCl_2$(0.001 M), HCl (0.01 M) | O_2(0.25 bar) | Pt | Cu.

Given that $E^0 = -1.185$ V for Mn^{2+} | Mn couple and 1.229 V for the O_2 | H_2O, H^+ couple. (2.452 V)

203. Estimate the difference in energy between first and second Bohr orbit for a hydrogen atom. At what minimum atomic number, a transition from $n = 2$ to $n = 1$ energy level would result in the emission of X-rays with $\lambda = 3.0 \times 10^{-8}$ m? Which hydrogen-atom like species does this atomic number correspond to? ($R = 109677\ cm^{-1}$, $c = 3 \times 10^{10}\ cm\ s^{-1}$) ($1.63 \times 10^{-11}$ erg, Z = 2, He^+)

[**Hint:** Apply $\Delta E = RchZ^2\left(\frac{1}{n_1^2} - \frac{1}{n_2^2}\right)$]

204. The nuclide ratio, ${}_1^3H$ to ${}_1^1H$ in a sample of water is 8.0×10^{-18} : 1. Tritium undergoes decay with a half-life period of 12.3 years. How many tritium atoms would 10.0 g of such a sample contain 40 years after the original sample is collected?

(5.62×10^5)

[**Hint:** Given that, $\dfrac{\text{mol of } T_2O \text{ molecules}}{\text{mol of } H_2O \text{ molecules}} = \dfrac{8 \times 10^{-18}}{1}$

$\therefore$ 10 g of sample contains $(2 \times 8 \times 10^{-18}) \times \dfrac{10}{18}$ mole tritium atoms]

205. A solution of Na_2CO_3 that is 2.0 M in CO_3^{2-} ions was boiled with excess of CaF_2. Very small amounts of $CaCO_3$ and F^- were formed. If the solubility product of $CaCO_3$ is x and molar solubility of CaF_2 is y, find the molar concentration of F^- in the resulting solution after equilibrium is attained.

$\left(\sqrt{\dfrac{8y^3}{x}}\right)$

206. The solubility product of AgCl is 1.0×10^{-10}. The equilibrium constant of the reaction

$$AgCl(s) + Br^- \rightleftharpoons AgBr(s) + Cl^-$$

is 2×10^2 and that of the reaction

$$2AgBr(s) + S^{2-} \rightleftharpoons Ag_2S(s) + 2Br^-$$

is 1.6×10^{24}. Calculate K_{sp} of Ag_2S.

(1.56×10^{-49})

207. What amount of $BaSO_4$ will dissolve in 500 mL of aqueous solution? K_{sp} $(BaSO_4) = 1 \times 10^{-10}$.

(0.0011 g)

208. Calculate the solubility product of Ag_2CrO_4 at 25°C if the concentration of Ag^+ ions is 1.5×10^{-4} mole/litre in a saturated solution of Ag_2CrO_4 at 25°C.

(1.69×10^{-12})

209. A sample of hard water contains 0.005 mole of $CaCl_2$ per litre. What is the minimum concentration of Na_2SO_4 which must be exceeded for removing the calcium ions from the water sample? K_{sp} $(CaSO_4) = 2.4 \times 10^{-5}$ at 25°C. (0.0048 mole/L)

210. In a 3-litre vessel the following equilibrium partial pressures are measured: $N_2 = 190$ mm, $H_2 = 317$ mm, $NH_3 = 1000$ mm. Hydrogen is removed from the vessel until the pressure of nitrogen at equilibrium is equal to 250 mm. Calculate the pressure of other substances under the new condition. $(p_{H_2} = 265.8 \text{ mm})$

211. Pure phosphine originally present at 2.5 atm and 300 K decomposes slowly according to the equation:

$$4PH_3(g) \rightleftharpoons P_4(g) + 6H_2(g)$$

What is the vapour density of phosphine if it dissociates to the extent of 40%?

(13.08)

212. For equilibrium $AB(g) \rightleftharpoons A(g) + B(g)$, K_p is equal to four times the total pressure. Calculate the number of moles of B formed. ($2/\sqrt{5}$ times initial mol of AB)

213. When equimolar quantities of acetic acid and ethanol are at 25°C, 1/3 of acetic acid remains unreacted as shown by titration with standard alkali. Calculate the concentration equilibrium constant. (4.0)

214. At room temperature the following reactions proceed nearly to completion:

$$2NO + O_2 \rightarrow 2NO_2 \rightarrow N_2O_4$$

The dimer, N_2O_4 solidifies at 262 K. A 250-mL flask and a 100-mL flask are separated by a stopcock. At 300 K the nitric oxide in the larger flask exerts a pressure of 1.053 atm and the smaller one contains oxygen at 0.789 atm. The gases are mixed by opening the stopcock and after the end of the reaction the flasks are cooled to 220 K. Neglecting the vapour pressure of the dimer, find out the pressure and composition of the gas remaining at 220 K. Assume the gases to behave ideally. **(IIT 1992)**

[**Hint:** See Example 17, Chapter 12.] (0.221 atm, NO– 0.0043 mol)

215. At 27°C, hydrogen leaks through a tiny hole in a vessel for 20 minutes. Another unknown gas at the same temperature and pressure as that of H_2 leaks through the same hole for 20 minutes. After the effusion of the gases, the mixture exerts a pressure of 6 atm. The hydrogen content of the mixture is 0.7 mole. If the volume of the container is 3 litres, what is the molecular weight of the unknown gas? **(IIT 1992)** (1032)

[**Hint:** $\frac{r_X}{r_{H_2}} = \sqrt{\frac{2}{M}} = \frac{n_X}{n_{H_2}}; pV = (n_X + n_{H_2})RT$]

216. The emf of the cell

Ag | AgCl, KCl (0.05 M) || $AgNO_3$ (0.05 M) | Ag

is 0.788 volt. Find K_{sp} of AgCl. (1.16×10^{-16})

[**Hint:** For the cell reaction: $Ag^+(0.05) = Ag^+(c)$,

where c is the Ag^+ concentration in LHS half cell,

$$E_{cell} = \frac{0.0591}{1}\log\frac{0.05}{c};\ K_{sp}(AgCl) = c \times 0.05]$$

217. Calculate the solubility product of AgCl from the two half reactions and standard electrode potentials at 25°C

$Ag^+ + e \rightarrow Ag(s)$ $E^0 = 0.799$ V

$AgCl + e \rightarrow Ag(s) + Cl^-$ $E^0 = 0.222$ V (1.66×10^{-10})

$$\left[\begin{array}{l}\textbf{Hint: } AgCl \rightarrow Ag^+ + Cl^-;\ E^0 = 0.222 - 0.799 \\ E^0 = 0.0591 \log[Ag^+][Cl^-] = 0.0591 \log K_{sp}\end{array}\right]$$

218. Calculate emf of Ag | AgCl electrode immersed in 1 M KCl at 25°C. $K_{sp}(AgCl) = 1.8 \times 10^{-10}$, $E^0_{Ag^+, Ag} = 0.799$ volt. (0.223 V)

219. Determine the temperature at which the half-life for the decomposition of N_2O_5 is two hours. At 298 K, the rate constant is $3.46 \times 10^{-5}\ s^{-1}$ and $E_a = 106\ kJ\ mol^{-1}$.

(305 K)

220. Two reactions (i) $A \rightarrow P$ (ii) $B \rightarrow P$ follow first-order kinetics. The rate of the reaction (i) is doubled when the temperature is raised from 300 K to 310 K. The half-life for this reaction at 310 K is 30 minutes. At the same temperature B decomposes twice as fast as A. If the energy of activation for the reaction (ii) is half that of reaction (i), calculate the rate constant of the reaction (ii) at 300 K.

($3.23 \times 10^{-2}\ min^{-1}$)

[Hint:

Reactions	E_a	Rate constants	
		300 K	310 K
(i) $A \rightarrow P$	E_1	k_1	$2k_1$ (given)
(ii) $B \rightarrow P$	E_2	k_2 (?)	k'_2

Given that $2k_1 = \frac{0.6930}{30}$; $\frac{2k_1}{k'_2} = \frac{1}{2}$ and $\frac{E_1}{E_2} = 2$.]

221. A gas bulb of 1-litre capacity contains 2×10^{21} molecules of nitrogen exerting a pressure of $7.57 \times 10^3\ N\ m^{-2}$. Calculate the root-mean-square (rms) speed and the temperature of the gas molecules. If the ratio of the most probable speed to the root-mean-square speed is 0.82, calculate the most probable speed for these molecules at this temperature. **[IIT 1993]**

($494.18\ m\ s^{-1}$, 274.15 K, $405.22\ m\ s^{-1}$)

$$\left[\textbf{Hint: } \text{Use } C = \sqrt{\frac{3pV}{mn}};\ V = 10^{-3}\ m^3,\ m = \frac{28 \times 10^{-3}}{\text{Av. const.}}\right]$$

222. A mixture of 0.5 mole of CO and 0.5 mole of CO_2 is taken in a pot and allowed to effuse out through a pinhole into another vessel which is vacuous. If a total of Z moles has effused out in time t, show that

$$M_1 Z + M_2(1 - Z) = 36,$$

where M_1 and M_2 are the mean molar masses of the mixture that has effused out and the mixture still remaining in the pot respectively.

223. In order to get maximum calorific output, a burner should have an optimum fuel-to-oxygen ratio which corresponds to 3 times as much oxygen as is required theoretically for complete combustion of the fuel. A burner which has been adjusted for methane as fuel (with x litre/hour of CH_4 and $6x$ litre/hour of O_2) is to be readjusted for butane, C_4H_{10}. In order to get the same calorific output, what should be the rate of supply of butane and oxygen? Assume that losses due to incomplete combustion, etc., are the same for both fuels, and that the gases behave ideally. Heat of combustion:

$CH_4 = 809$ kJ/mole, $C_4H_{10} = 2878$ kJ/mole.

$$\left[\begin{array}{l}\textbf{Hint: } \text{Vol. of } C_4H_{10} \text{ required per h} = \dfrac{804}{2878}\,\text{L} \\[2ex] \text{Vol. of } O_2 \text{ required per h} \left(\dfrac{804x}{2878}\right) \times \dfrac{13}{2} \times 3\,\text{L}\end{array}\right] \quad \left[\begin{array}{r}C_4H_{10} = 0.281x\ \text{L/h} \\ O_2 = 5.481x\ \text{L/h}\end{array}\right]$$

224. 0.16 g of methane was subjected to combustion at 27°C in a bomb calorimeter. The temperature of the calorimeter system (including water) was found to rise by 0.5°C. Calculate the heat of combustion of methane at (i) constant volume; (ii) constant pressure. The thermal capacity of the calorimeter system is 17.7 kJ K^{-1}. ($R = 8.31\ JK^{-1}\ mol^{-1}$)

$\left[\textbf{Hint: } \Delta E = \dfrac{17.7 \times 0.5}{0.16/16}\ \text{kJ mol}^{-1}\right]$ (–885 kJ, –889.986 kJ)

225. The rates of effusion of two gases A and B under the same conditions of temperature and pressure are in the ratio $\gamma_A : \gamma_B = 2 : 1$. What would be the ratio of the rms speeds of molecules of A and B if $T_A : T_B = 2 : 1$? ($2\sqrt{2} : 1$)

226. There were 201 rows of spectators sitting in a hall. A magician releases laughing gas (N_2O) from the front, and tear gas (mol. wt. 176) from the rear of the hall simultaneously. Which row of spectators from the front will have a tendency to smile and weep simultaneously? (134)

227. At what temperature does the average translational kinetic energy of a molecule in a gas become equal to the kinetic energy of an electron accelerated from rest through a potential difference of 1 volt?

1 eV = 1.602×10^{-12} erg. (7.73×10^3 K)

228. The K_p value for the reaction equilibrium:

$$H_2(g) + I_2(s) \rightleftharpoons 2HI(g)$$

is 871 at 25°C. If the vapour pressure of iodine is 4×10^{-4} atm, calculate the equilibrium constant in terms of partial pressures at the same temperature.

(0.3484 atm)

229. Solid NH_4HS is taken in an evacuated vessel and allowed to dissociate at a certain temperature until the total gas pressure is 0.66 atm. What would be the value of K_p for the following reaction?

$$NH_4HS(s) \rightleftharpoons NH_3(g) + H_2S(g)$$

What would be the partial pressure of H_2S if additional NH_3 is introduced into the equilibrium mixture at the same temperature until the partial pressure of NH_3 is 0.921 atm? (0.1089 atm)

230. Solid ammonium carbamate, NH_2COONH_4, dissociates on heating to NH_3 and CO_2:

$$NH_2COONH_4(s) \rightleftharpoons 2NH_3(g) + CO_2(g)$$

When pure carbamate is put into a closed container and allowed to come to equilibrium with the gaseous products at a constant temperature, 35°C, the total pressure is found to be 0.30 atm. Determine K_p for this reaction at 35°C.

(4.0×10^{-3} atm^3)

231. Water gas is produced by the reaction

$$C + H_2O(g) \rightarrow H_2 + CO.$$

The heat required for this endothermic reaction may be supplied by adding a limited amount of air and burning some carbon to carbon dioxide. Calculate the amount of carbon to be burnt to CO_2 to provide enough heat for the water gas ($H_2 + CO$) conversion of 100 g of carbon. $\Delta H°_f$ for CO = – 110.53 kJ, $\Delta H°_f$ for H_2O = – 241.81 kJ and $\Delta H_{combustion}$ for C = – 393.51 kJ. (33.4 g)

232. A first-order reaction A → B, requires activation energy of 70 kJ mole^{-1}. When a 20% solution of A was kept at 25°C for 20 minutes, 25% decomposition took place. What will be the per cent decomposition in the same time in a 30% solution maintained at 40°C? Assume that activation energy remains constant in this range of temperature. **(IIT 1993)** (67.17%)

[**Hint:** First calculate k_{25}, then k_{40}. Again find out per cent decomposition at 40°C.]

233. In Arrhenius's equation for a certain reaction, the values of A and E_a are 4×10^{13} s^{-1} and 98.6 kJ mole^{-1} respectively. If the reaction is of the first-order, at what temperature will its half-life period be ten minutes? (311.2 K)

234. The energy of activation of the reaction: A + B → C is 24.6 kcal/mole. If the rate of formation of C is 0.133 M min^{-1} at 40°C, what would it be at 80°C?

(11.41 M min^{-1})

235. A reaction proceeds with the energy of activation of 55.3 kcal/mole. If ΔH of the reaction is 1 kcal, what would be the energy of activation of the reverse reaction?

(54.3 kcal)

236. Isopropyl alcohol and *n*-propyl alcohol, both have the same molecular formula C_3H_8O. A solution of the two, that is, 25% isopropyl alcohol by mass, has a total vapour pressure of 0.09 atm at a given temperature. A solution of the two, that is, 75% isopropyl alcohol by mass, has a total vapour pressure of 0.12 atm at the same temperature. Find the vapour pressures of the pure alcohols at this temperature.

(0.135 atm, 0.075 atm)

237. The immiscible liquid system aniline–water boils at 98°C under a presssure of 760 mm. At this temperature the vapour pressure of water is 707 mm. If this system is distilled in steam, what fraction of total weight of the distillate will be aniline?

(28%)

238. A current of dry air was passed through a solution containing 2.5 g of B in 100 g of solution, and through water alone. The loss of weight of solution was 1.25 g and that of water was 0.05 g. Determine the molecular weight of B. (117)

239. In nature a decay chain series starts with $^{232}_{90}Th$ and finally terminates at $^{208}_{82}Pb$. A thorium ore sample was found to contain 8×10^{-5} mL of helium at STP and 5×10^{-7} g of ^{232}Th. Find the age of the ore sample assuming the source of helium to be only the decay of ^{232}Th. Also, assume complete retention of helium within the ore. Half-life of $^{232}\text{Th} = 1.39 \times 10^{10}$ years.

[**Hint:** $^{232}_{90}\text{Th} \rightarrow {}^{208}_{82}\text{Pb} + 6\,{}^{4}_{2}\text{He} + 4\,{}^{0}_{-1}\text{e}$] ($4.89 \times 10^9$ years)

240. A 0.5-g sample containing MnO_2 is treated with HCl, liberating Cl_2. The Cl_2 is passed into a solution of KI and 30.0 cm^3 of 0.1 M $Na_2S_2O_3$ are required to titrate the liberated iodine. Calculate the percentage of MnO_2 in the sample. (26.10%)

241. A 2.0-g sample of a mixture containing sodium carbonate, sodium bicarbonate and sodium sulphate is gently heated till the evolution of CO_2 ceases. The volume of CO_2 at 750 mmHg pressure and at 298 K is measured to be 123.9 mL. 1.5 g of the same sample requires 150 mL of (M/10) HCl for complete neutralization. Calculate the percentage composition of the components of the mixture. **(IIT 1992)** (26.5%, 42.0%, 31.5%)

242. 5.6 g of a steel sample containing sulphur impurity was burnt in oxygen. SO_2, so produced, was then oxidised to sulphate by H_2O_2 solution to which 30 mL of 0.004 M NaOH solution had been added. 22.48 mL of 0.024 M HCl was required to neutralize the base remaining after oxidation reaction. Calculate percentage of S in the given sample of steel. (0.1886%)

[**Hint:** $SO_2 + H_2O_2 + 2OH^- = SO_4^{2-} + 2H_2O$

$$\text{mmol of S} = \text{mmol of } SO_2 = \frac{1}{2} \times \text{mmol of } OH^- = (30 \times 0.04 - 22.48 \times 0.024)]$$

243. Calculate $[NH_4^+]$ (derived from NH_4Cl) to prevent $Mg(OH)_2$ from precipitating in a one-litre solution, containing 0.01 M NH_3 and 0.001 M Mg^{2+} ions.
$K_{sp}[Mg(OH)_2] = 1.2 \times 10^{-11}$, $K_b(NH_4OH) = 1.8 \times 10^{-5}$. ($1.6 \times 10^{-3}$)

244. What is the maximum pH of a solution of 0.1 M in Mg^{2+} from which $Mg(OH)_2$ will not precipitate? $K_{sp}\,[Mg(OH)_2] = 1.2 \times 10^{-11}$. (9.04)

245. Calculate the maximum possible concentration of Ni^{2+} ions in a solution which is also 0.15 M in HCl and 0.1 M in H_2S.
$K_{sp}(NiS) = 2 \times 10^{-21}$, $K(H_2S) = 1 \times 10^{-21}$. (0.45 M)

246. Calculate the pH at which $Mg(OH)_2$ begins to precipitate from a solution containing 0.1 M Mg^{2+} ions, K_{sp} for $Mg(OH)_2 = 1 \times 10^{-11}$. (9.0)

247. A particular water sample has 131 ppm $CaSO_4$. What fraction of the water must be evaporated in a container before solid $CaSO_4$ begins to deposit?
$K_{sp}(CaSO_4) = 9.0 \times 10^{-6}$. (68%)

248. In the reaction

$$CH_3COCH_3(g) \rightarrow C_2H_4(g) + H_2(g) + CO(g)$$

the total pressure changes as follows:

Time (min):	0	6.5	13.0	19.9
Pressure (Pa):	41489.6	54386.6	65050.4	74914.6

Show that the reaction is of first-order.

249. Two reactions of the same order have equal pre-exponential factors but their activation energies differ by 41.9 kJ/mole. Calculate the ratio between the rate constants of these reactions at 600 K. (0.0002)

250. The dissociation of a substance is a first-order reaction with an activation energy of 231 kJ/mole. At 300 K, this substance dissociates at a rate of 95% within an hour. Calculate the temperature at which the dissociation rate of the substance is 0.1% per minute. (350 K)

251. A diazonium salt dissociates according to the equation

$$CH_3C_6H_4N_2Cl + H_2O \rightarrow CH_3C_6H_4OH + N_2 + HCl.$$

The dissociation process is a first-order reaction whose rate constants at 297.9 K and 303.2 K are 9×10^{-3} and 13×10^{-3} min^{-1} respectively. Calculate the rate constant at 308.2 K and the time for 99% dissociation of the salt at 308.2 K.

(16.37×10^{-3} min^{-1}, 282 min)

252. A thermally insulated container initially holds N_0 molecules of an ideal monoatomic gas at an absolute temperature T_0. Molecules escape from the container through a small hole in the wall, and it can be shown that in such a process at a temperature T, the average kinetic energy of the escaping molecules is 2 kT. How many molecules remain in the container when the temperature has fallen to $T_0/2$?

($N_0/8$)

[Hint: No. of molecules at T_0 is N_0. Let at any time the temp. of the container be T and the no. of molecules present be N. Both T and N are variable. Let during the time dt, the no. of molecules escaped be dN. Then

$$\frac{3}{2}kTN - 2\,kT\,dN = \frac{3}{2}k(T - dT)\,(N - dN)$$

or, $$\frac{3}{2}\int_{T_0}^{T_0/2}\frac{dT}{T} = \frac{1}{2}\int_{N_0}^{N}\frac{dN}{N} \quad \text{(neglecting } dTdN \text{ factor)}$$

or, $$N = \frac{N_0}{8}.\,]$$

253. K_p for the equilibrium: $CaCO_3(s) \rightleftharpoons CaO(s) + CO_2(g)$ is 1.16 atm at 1073 K. If 20 g of $CaCO_3$ was heated to 1073 K in a 10-litre vessel, find the percentage of $CaCO_3$ remaining unreacted at equilibrium. (34.20%)

254. Solid ammonium carbamate dissociates according to the reaction

$$NH_2COONH_4(s) \rightleftharpoons 2NH_3(g) + CO_2(g).$$

At 298 K, the total pressure of the gases in equilibrium with the solid is 0.116 atm. Calculate K_p.

(2.31×10^{-4} atm^3)

255. Carbon was heated with 1.0 g of hydrogen in a 5-litre tank to 1000°C. At equilibrium 0.22 g of methane was found in the tank. Calculate the equilibrium constant for the equilibrium

$$C(s) + 2H_2(g) \rightleftharpoons CH_4(g). \quad (0.31)$$

256. If the reaction between steam and iron proceeds as

$$3\,Fe(s) + 4\,H_2O(g) \rightleftharpoons Fe_3O_4(s) + 4H_2(g)$$

and partial pressures of steam and hydrogen are 50 mm and 940 mm respectively at 250°C, calculate the partial pressure of steam at equilibrium when partial pressure of hydrogen is 1800 mm. (95.7 mm)

257. The equilibrium constant of the reaction $H_2 + I_2 \rightleftharpoons 2HI$ at 693 K is 50.25 (K_c). Calculate the yield of HI if 0.846×10^{-3} kg of I_2 and 0.0212×10^{-3} kg of H_2 have been placed in a vessel having a capacity of 10^{-3} m^3. (0.821×10^{-3} kg)

258. At 823 K and 1.0133×10^5 Pa, the degree of dissociation of phosgene ($COCl_2$) into CO and Cl_2 is 77%. Find K_p and K_c. (1.456, 0.0215)

259. Express the equilibrium constant K_p of the reaction $H_2 + 0.5O_2 = H_2O(v)$ in terms of total pressure p and degree of dissociation α of water vapour.

$$\left[K_p = \frac{(1-\alpha)\,(\alpha+2)^{1/2}}{p^{1/2}\,\alpha^{3/2}}\right]$$

260. The equilibrium constant of the reaction

$$CO + H_2O \rightleftharpoons CO_2 + H_2$$

at 800 K is 4.12. A mixture of 20% CO and 80% H_2O is heated to 800 K. Determine the composition of the mixture in the state of equilibrium and the yield of hydrogen if 1 kg of water vapour is taken.

$$\begin{bmatrix} CO - 1.447\%, & H_2O - 59.858\% \\ CO_2 - 37.013\%, & H_2 - 1.68\% \end{bmatrix}$$

261. At 945°C and 1 atm, 1.7 g of H_2S occupies a volume of 5.384 litres. Calculate the degree of dissociation of hydrogen sulphide if the reaction proceeds according to the equation

$$H_2S = H_2 + 0.5\,S_2(v). \qquad (0.156)$$

262. The degree of dissociation of N_2O_4 according to the equation $N_2O_4 \rightleftharpoons 2NO_2$ at 70°C and 1 atm is 65.6%. Calculate the apparent mol. wt. of N_2O_4 under the given conditions. (55.56)

263. A bulb contains 2 moles of H_2 at a pressure of 0.8 atm and temperature T K. 0.6 mole of O_2 is added to this bulb and the temperature of the bulb is lowered by 15 K to keep the same pressure. Calculate the volume of the bulb and its temperature T. Also, calculate the partial pressure of each gas.

(13.34 dm^3, 65 K; $p_{H_2} = 0.615$ atm, $p_{O_2} = 0.185$ atm)

264. The proportion of O_2, SO_2 and SO_3 present in the mixture of gases is 0.5 : 0.3 : 0.2. They are allowed through a pinhole at 27°C. Calculate the composition of the mixture leaving initially. (3.953 : 1.677 : 1.0)

265. The polymerization of ethylene to linear polyethylene is represented by the reaction

$$n\,CH_2 = CH_2 \rightarrow (-CH_2 - CH_2-)_n,$$

where n has large integral value. Given that the average enthalpies of bond dissociation for C = C and C − C at 298 K are +590 and +331 kJ mole^{-1} respectively,

calculate the enthalpy of polymerization per mole of ethylene at 298 K.
[**Hint:** $\Delta H = 1 \times 590 - 2 \times 331$] **(IIT 1994)** (– 72 kJ)

266. At 337 K the vapour pressure of ethanol is 0.526 atm and the vapour pressure of water is 0.236 atm. A solution is prepared from equimolar amounts of water and ethanol at this temperature. The vapour above the solution is removed and condensed. The condensed solution is heated to 337 K and the vapour above the solution is removed and condensed. Determine the mole fraction of the condensed solution. (0.91, 0.09)

267. A current of dry air was bubbled through a bulb containing 26.66 g of an organic substance in 200 g of water, then through a bulb at the same temperature containing pure water, and finally through a tube containing fused calcium chloride. The loss in weight of water bulb was 0.087 g and the gain in weight of the calcium chloride tube was 2.036 g. Determine the molecular weight of the organic substance. (56.1)

268. A hydrocarbon (A) [C – 90.56%, vapour density – 53] was subjected to vigorous oxidation to give a dibasic acid (B). 0.1 g of (B) required 24.1 mL of 0.05 N NaOH for complete neutralization. Nitration of (B) gave a single mononitro derivative. When (B) was heated strongly with soda lime, it gave benzene. Identify (A) and (B).

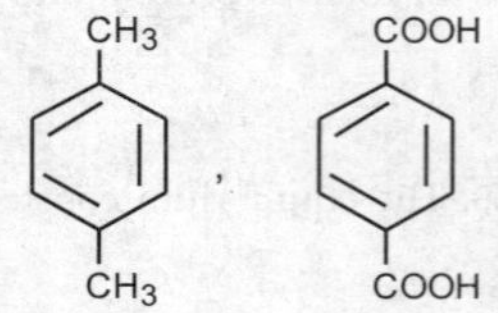

(A) (B)

269. A chloro compound (A) showed the following properties:
(i) Decolorized bromine in CCl_4
(ii) Absorbed hydrogen catalytically
(iii) Gave a precipitate with amm. cuprous chloride
(iv) When vaporized, 1.49 g of (A) gave 448 mL of vapour at STP.
Identify (A). ($ClCH_2C \equiv CH$)

270. A basic, volatile nitrogen compound gave a foul-smelling gas when treated with chloroform and alcoholic potash. A 0.295-g sample of the substance dissolved in aq. HCl and treated with $NaNO_2$ solution at 0°C, liberated a colourless, odourless gas whose volume corresponded to 112 mL at STP. After the evolution of the gas was complete, the aqueous solution was distilled to give an organic liquid which did not contain nitrogen, and which on warming with an alkali and iodine, gave a yellow precipitate. Identify the original substance. Assume that it contains one N atom per molecule. **(IIT 1993)** [$(CH_3)_2CHNH_2$ (isopropylamine)]

271. One mole of an organic amide (A) upon alkaline hydrolysis gives one mole of NH_3 and one mole of monobasic acid of equivalent weight 74. What is the molecular formula of (A)? ($C_2H_5CONH_2$)

272. How many grams of CaF_2 must be dissolved in 100 mL of water at 25°C to make the solution saturated? $K_{sp}(CaF_2) = 3.9 \times 10^{-11}$. ($1.66 \times 10^{-3}$ g)

273. A solution which is 0.1 M in NaI and also 0.1 M in $Na_2(SO_4)$ is treated with solid $Pb(NO_3)_2$. Which compound, PbI_2 or $PbSO_4$, will precipitate first? What is the concentration of anions of the least soluble compound when the more soluble one starts precipitating?

$K_{sp}(PbI_2) = 8.7 \times 10^{-9}$, $K_{sp}(PbSO_4) = 1.8 \times 10^{-8}$ ($PbSO_4$, 0.021 M)

274. The solubility product of $Ca(OH)_2$ at 25°C is 4.42×10^{-5}. 500 mL of saturated solution of $Ca(OH)_2$ is mixed with equal volume of 0.4 M NaOH. How much $Ca(OH)_2$ in mg is precipitated? (747 mg)

275. Calculate the amount of $(NH_4)_2SO_4$ in grams which must be added to 500 mL of 0.2 M NH_3 to yield a solution with pH = 9.35. $K_b(NH_3) = 1.78 \times 10^{-5}$. (10.256 g)

276. The ClO radicals decay by second-order reaction. If the initial concentration is 2.5×10^{-5} mole dm^{-3}, calculate its (a) first half-life, (b) second half-life, and (c) concentration of ClO radical after 4 min. The rate constant of the process is 2.25×10^7 dm^3 mol^{-1} s^{-1}. [(a) 1.778 milliseconds (b) 3.556 milliseconds (c) 7.69×10^{-6} mol dm^{-3}]

277. The data of a chemical reaction is plotted as $1/C$ vs time and the plot is a straight line. If intercept is 2×10^{-3} mol^{-1} dm^3 and slope, 2×10^{-2} mol^{-1} dm^3 s^{-1}, calculate the half-life period of the reaction. (0.1 s)

278. A reaction is 50% complete in 10 minutes. It is allowed to proceed another 5 minutes. How much of the reaction would be complete at the end of these 15 minutes if the reaction follows zero-order kinetics? (75%)

279. For the gaseous reaction, say, A $\rightarrow$ product, the rate is often described in terms of $d(P_A)/dt$ instead of $d[A]/dt$ or $d(n_A)/dt$. What is the relation among these three expressions? $\left[\frac{d(P_A)}{dt} = \frac{RT}{V}\frac{d(n_A)}{dt} = RT\frac{d[A]}{dt}\right]$

280. A certain mass of a substance in 100 g of C_6H_6 lowers the f.p. by 1.28°C. The same mass of the substance in 100 g of water lowers the f.p. by 1.395°C. If the substance has a normal molecular weight in C_6H_6 and is completely dissociated in water, calculate the number of moles of ions produced by the dissociation of 1 mole of the substance in water. K_f for H_2O and C_6H_6 are 1.86 and 5.00 respectively. (3)

281. A solution of 3.795 g sulphur in 100 g of CS_2 (b.p. = 46.3°C) boils at 46.66°C. Determine the formula of sulphur molecule in the solution. $K_b(CS_2) = 2.42$. (S_8)

282. An aqueous solution of cane sugar (mol. wt. = 342) has an osmotic pressure of 1.5 atm at 18°C. If 100 g of this solution is cooled to −3.0°C, what mass of ice will separate out? (94.10 g)

283. An aqueous solution of mannitol in water has a vapour pressure of 17.504 mm at 20°C, at which temperature, the vapour pressure of pure water is 17.535 mm. What is the f.p. depression for this solution? $K_f(H_2O) = 1.86$. (0.183°C)

284. Potassium sulphate is a strong electrolyte which dissociates completely in aqueous solution. Urea is a nonelectrolyte substance. A 0.01 molar solution (aq) of potassium sulphate depressed the f.p. of water by 0.0558°C. What will be the depression caused by a 0.01 molar solution of urea? (0.0186°C)

285. The rate law of the reaction $H_2 + Br_2 \rightarrow 2HBr$ is

$$\frac{d[\text{HBr}]}{dt} = k\,[\text{H}_2]\,[\text{Br}_2]^{½}.$$

What is the order of the reaction? Is this an elementary reaction? $\left(\frac{3}{2}, \text{No}\right)$

286. Find the rate law of the reaction $2H + Ar \rightarrow H_2 + Ar$ which is elementary. $(R = k[H]^2\,[Ar])$

287. A chemical reaction is known to be of zero-order with $k = 5 \times 10^{-8}$ mole lit^{-1} s^{-1}.

(a) How long does it take for reactant concentration to decrease from 4×10^{-4} mole lit^{-1} to 2×10^{-4} mole lit^{-1}?

(b) Will the time to decrease the reactant concentration from 2×10^{-2} mol lit^{-1} to 1×10^{-2} mole lit^{-1} be the same as in (a)? $(4 \times 10^3$ s, No)

288. What is the ratio of $t_{1/2}$ to $t_{1/3}$ for a first-order reaction? (1·71)

289. Calculate the average life for a reaction undergoing a first-order reaction. $\left(\frac{1}{k}\right)$

290. Calculate K_p and K_c of the reaction: $I_2 \rightleftharpoons 2I$ when 1.513×10^{-3} mole of iodine is heated to 1073 K if its vapour occupies a volume of 249.3×10^{-6} m^3 at 5.81×10^4 Pa. (0.01113, 0.1264)

291. At 525 K, the equilibrium constant of the reaction $PCl_5 \rightleftharpoons PCl_3 + Cl_2$ is 1.78 atm (K_p). At what pressure should an equimolar mixture of Cl_2 and PCl_3 be taken for the pressure of PCl_5 to be 5×10^4 Pa at equilibrium, volume remaining constant? $(28.99 \times 10^4$ Pa)

292. The pressure dependence of the equilibrium conditions for the equation

$$\frac{1}{2}N_2(g) + \frac{3}{2}H_2(g) \rightleftharpoons NH_3(g)$$

was studied by measuring the mole fraction of NH_3 produced at various pressures for 1 : 3 mixtures of N_2 and H_2 at 500°C. K_p was found to increase with the increase in pressure. Why is K_p pressure dependent?

293. A 0.1-mole sample of NO_2 was placed in a 10-litre container and heated to 750 K. The total pressure of the equilibrium mixture as a result of the decomposition $2NO_2(g) \rightarrow 2NO(g) + O_2(g)$ was 0.827 bar. What is the value of K_p at this temperature? What amount of NO_2 must be placed in this container to obtain an equilibrium concentration of NO_2 of 0.1 mole per litre? (0.704, 1.61 mole)

294. Calculate the mass of NH_4Cl that must be completely dissolved in 1 litre of aqueous solution to attain an osmotic pressure of 5 atm at 298 K. Assume ideal behaviour. (5.46 g)

295. 10 g of a substance was dissolved in water and the solution was made up to 250 cm^3. The osmotic pressure of the solution was found to be 8×10^5 N m^{-2} at 288 K. Find the molecular weight of the solute. (119.72)

[Hint: 1 cc = 10^{-6} m^3]

296. A solution of crab hemocyanin, a pigmented protein extracted from crabs, was prepared by dissolving 0.75 g in 125 cc of an aqueous medium. At 4°C, an osmotic pressure rise of 2.6 mm of the solution was observed. The solution had a density of 1 g/cc. Determine the molecular weight of the protein. Solve this problem in cgs units. (5.4×10^5 g/mole)

297. 4.0 g of a substance, A, dissolved in 100 g of water depressed the f.p. of water by 0.1°C, while 4.0 g of another substance, B, depressed the f.p. by 0.2°C. Which of the two substances has the higher molecular weight? (A)

298. Calculate the pressure in pascal (Pa) by a 760-mmHg column. (101328 Pa)

299. Dibutyl phthalate is commonly used as a liquid in manometers. What pressure in N m^{-2} is equivalent to a centimetre of this liquid? The density of the liquid is 1.0465 g/cc. (102.63 N m^{-2})

300. What is the approximate mass of the atmosphere of the earth? Assume the radius of the earth to be 6370 km. (5.27×10^{18} kg)

[Hint: Surface area of the earth = $4\pi r^2$ and atmospheric pressure = 1 atm = 101325 Pa]

301. A commercial gas cylinder contains 75 litres of helium at 15 bar (gauge pressure). Assuming ideal gas behaviour for the isothermal expansion, how many 3-litre balloons at a pressure of 1.1 bar can be filled by the gas in this cylinder? (315 balloons)

302. A diver at a depth of 45 m exhales a bubble of air that is 1 cm in radius. Assuming ideal gas behaviour, what will be the radius of this bubble as it breaks at the surface of the water? (1.8 cm)

[Hint: $p_1 = \rho gh$ + atm. pressure; p_2 = atm. pressure
Apply $p_1V_1 = p_2V_2$, calculate V_2 and then the radius]

303. 0.001 mole each of Fe^{2+} and Cd^{2+} is present in one litre of 0.02 M HCl, saturated with H_2S. Find whether each of these ions shall precipitate as sulphides. Calculate [Cd^{2+}] in the solution at equilibrium.
$K_a(H_2S) = 1 \times 10^{-21}$, $K_{sp}(CdS) = 8 \times 10^{-27}$, $K_{sp}(FeS) = 3.7 \times 10^{-19}$.

(CdS precipitates, [Cd^{2+}] = 3.86×10^{-8} M)

304. A solution contains both Ag^+ (0.30 M) and Ba^{2+} (0.05 M).
(i) If solid Na_2SO_4 is added very slowly to this solution, which will precipitate first, Ag_2SO_4 or $BaSO_4$?
(ii) The addition of Na_2SO_4 is continued until the second cation just starts

precipitating as sulphate. What is the concentration of the first cation at this point? $K_{sp}(Ag_2SO_4) = 1.2 \times 10^{-5}$, $K_{sp}(BaSO_4) = 1.5 \times 10^{-9}$. ($BaSO_4$, 1.15×10^{-5} M)

305. At a pressure of 1.0 bar, an equilibrium exists at 2000 K between 0.25 mole of $Br_2(g)$, 0.75 mole of $F_2(g)$ and 0.497 mole of $BrF_3(g)$. What will be the amounts of each gas after the pressure on the system has been increased to 2.0 bar and the equilibrium at 2000 K re-established?
$Br_2(g) + 3F_2(g) \rightarrow 2BrF_3(g)$ (0.189 mole, 0.567 mole, 0.619 mole)

306. For the equation $CaCO_3(s) \rightarrow CaO(s) + CO_2(g)$, $K_{1000} = 0.059$. Exactly 10 g of $CaCO_3$ is placed in a 10-litre container at 1000 K. After equilibrium is reached, what mass of $CaCO_3$ remains? (9.29 g)

307. Hot copper turnings can be used as an 'oxygen getter' for inert gas supplies by slowly passing the gas over the turnings at 600 K:

$$2Cu(s) + \frac{1}{2}O_2(g) \rightleftharpoons Cu_2O(s); \; K_p = 7.5 \times 10^{10}$$

How many molecules of O_2 are left in one litre of a gas supply after equilibrium has been reached? ($2.17 \approx 2$)

308. For $O_3(g) + OH(g) \rightleftharpoons H(g) + 2O_2(g)$, $K = 0.096$ at 298 K and $K = 1.4$ at 373 K. Above what temperature will the reaction become thermodynamically spontaneous? ($T > 361$ K)

[**Hint:** For spontaneous process $K_p > 1$. Use Equation 11, Chapter 15]

309. For the reaction $A + 2B \rightarrow$ product, the reaction rate was halved as the concentration of A was doubled. What is the order of reaction with respect to A? (– 1)

310. The total pressure of the system at 279.0°C for the equation $SO_2Cl_2(g) \rightarrow SO_2(g) + Cl_2(g)$ was observed as a function of time. Find the order of the reaction.

Time (s):	204	3270	7500	8400	∞
p (mm):	325	365	415	425	594.2

(First)

311. $t_{1/2}$ for a given reaction was doubled as the initial concentration of a reactant was doubled. What is the order of the reaction for this component? (Zero)

312. Which reaction will have the greater temperature dependence for the rate constant—one with a small value of energy of activation (E) or one with a large value of E? (Large value of E)

$$\left[\textbf{Hint: } \frac{dk}{dt} = Ae^{-E/RT} \frac{E}{RT^2} = k\frac{E}{RT^2} \right]$$

313. The half-time of the first-order decomposition of nitramide is 2.1 hour at 15°C,

$$NH_2NO_2(aq) \rightarrow N_2O(g) + H_2O(l).$$

If 6.2 g of NH_2NO_2 is allowed to decompose, calculate

(i) the time taken for 99% decomposition, and

(ii) the volume of dry N_2O produced at this point measured at STP.

[(i) 21 hours (ii) 2.2176 litres]

314. A toy balloon originally held 1.0 g of helium gas and had a radius of 10.0 cm. During the night, 0.25 g of the gas effused from the balloon. Assuming ideal gas behaviour under these constant pressure and temperature conditions, what was the radius of the balloon the next morning? (9.0 cm)

315. Assuming ideal gas behaviour, how many atoms of Ar are contained in a typical human breath of 0.5 litre at 1.0 bar and 37°C? Air consists of 1% Ar atoms. Assuming that the argon atoms from the last breath of Plato have been distributed randomly throughout the atmosphere (5×10^{18} m^3), how long would it take to breathe one of these atoms? A typical adult breath rate is 10 min^{-1}.

(1×10^{20} $breath^{-1}$, 10 min)

316. The total pressure of a mixture of H_2 and O_2 is 1.0 bar. The mixture is allowed to react to form water, which is completely removed to leave only pure H_2 at a pressure of 0.35 bar. Assuming ideal gas behaviour, and that all pressure measurements were made under the same temperature and volume conditions, calculate the composition of the original mixture. (0.78, 0.22)

317. A mixture of He and CO_2 has a volume of 63.5 mL at 1.0 bar and 28°C. The system containing the mixture is cooled in liquid nitrogen, and the remaining gas is evacuated. The system is restored to 1.0 bar and 28°C, and the volume is 40.5 mL. Find the composition of the original mixture. (0.638, 0.362)

318. The average molar mass of the vapour above solid NH_4Cl is nearly 26.5 g $mole^{-1}$. Find the composition of the vapour. (0.5, 0.5)

319. The escape velocity is given by

$$v = \sqrt{2\,gr}$$

where $r = 6.37 \times 10^6$ m for the earth. At what temperature will the rms velocity of an H_2 molecule attain the escape velocity? (1.02×10^4 K)

320. An He atom at 25°C is released from the surface of the earth to travel upwards. Assuming that it undergoes no collisions with other molecules, how high will it travel before coming to rest? (9.47×10^4 m)

[**Hint:** Use $\frac{3}{2} kT = mgh$]

321. Addition of 0.643 g of a compound to 50 mL of benzene (density = 0.879 g/mL) lowers the f.p. from 5.51°C to 5.03°C. If K_f for benzene is 5.12, calculate the molecular weight of the compound. **(IIT 1992)** (156.06)

322. A solution of cane sugar at 27°C develops an osmotic pressure of 4.93 atm. Calculate the f.p. of this solution (molecular depression constant for 100 g of water is 18.6). (0.372°C)

323. What relative proportions of ethylene glycol $C_2H_6O_2$ and water (by weight) should be mixed to form an antifreeze solution that will not start to freeze until the temperature reaches – 37°C? K_f for $H_2O = 1.86$. (1.23 : 1)

324. In a cold climate water gets frozen causing damage to the radiator of a car. Ethylene glycol is used as antifreezing agent. Calculate the amount of ethylene glycol to be added to 4 kg of water to prevent it from freezing at $-6°C$. K_f for H_2O is 1.85 K mol^{-1} kg. (804.3 g)

325. From the following data for the reaction between A and B:

[A], mol L^{-1}	[B], mol L^{-1}	Initial rate, mol L^{-1} s^{-1}, at	
		300 K	320 K
2.5×10^{-4}	3.0×10^{-5}	5.0×10^{-4}	2.0×10^{-3}
5.0×10^{-4}	6.0×10^{-5}	4.0×10^{-3}	—
1.0×10^{-3}	6.0×10^{-5}	1.6×10^{-2}	—

Calculate

(i) the order of the reaction with respect to A and with respect to B,
(ii) the rate constant at 300 K,
(iii) the energy of activation, and
(iv) the pre-exponential factor. **(IIT 1994)**

[(i) 2, 1 (ii) 2.67×10^8 (iii) 55.3 kJ (iv) 1.145×10^{18}]

326. Show that in case of a first-order reaction, the time required for 93.75% of the reaction to take place is four times that required for half of the reaction.

327. What will be the initial rate of reaction if its rate constant is 10^{-3} min^{-1} and the concentration of the reactant is 0.2 mol dm^{-3}? How much of the reactant will be converted into products in 200 minutes? (2.0×10^{-4} mol dm^{-3} min^{-1}, 18%)

328. In a first-order reaction $A \rightarrow P$, if it takes 20 minutes to bring about decomposition of 30% of the initial substance, calculate the time to decompose (i) 60% of it, and (ii) all of it. (51.3 min, infinite)

329. What is the pH of a 1.0×10^{-8} M solution of NaOH? (7.021)

330. What is the pH of a 0.1 M solution of Na_2SO_4? For H_2SO_4, $K_{a_2} = 1 \times 10^{-2}$. (7.51)

331. A 50-mL aliquot of 0.01 M solution of HCOOH was titrated with 0.1 M NaOH. Predict the pH of the solution

(a) at the beginning of the reaction,
(b) at the half-equivalence point,
(c) at the equivalence point, and
(d) after 10 mL of the base has been added.

$K_a(HCOOH) = 1.772 \times 10^{-4}$ (2.88, 3.75, 7.86, 11.92)

332. The equilibrium constant of the reaction: $2D_2O = D_3O^+ + OD^-$ (D is deuterium) is 1.35×10^{-15} at 25°C. Calculate pD of the heavy water at 25°C. (7.435)

333. How much water should be added to 10 g of acetic acid to give a hydrogen-ion concentration equal to 1×10^{-3} M? $K_a(CH_3COOH) = 1.8 \times 10^{-5}$. (3.0 litres)

334. For the equilibrium: $NH_4HS(s) = NH_3(g) + H_2S(g)$; $K_p = 0.05$ atm^2 at 20°C. If

0.06 mole of solid NH_4HS is placed in a 2.40-litre flask at 20°C, calculate the percentage of the solid decomposed under equilibrium. (37.18%)

335. The progress of the reaction $A \rightleftharpoons nB$, with time is presented in the figure. Determine

(i) the value of n

(ii) the equilibrium constant K

(iii) the initial rate of conversion of A **(IIT 1994)** (2, 1.2, 0.1)

[**Hint:**

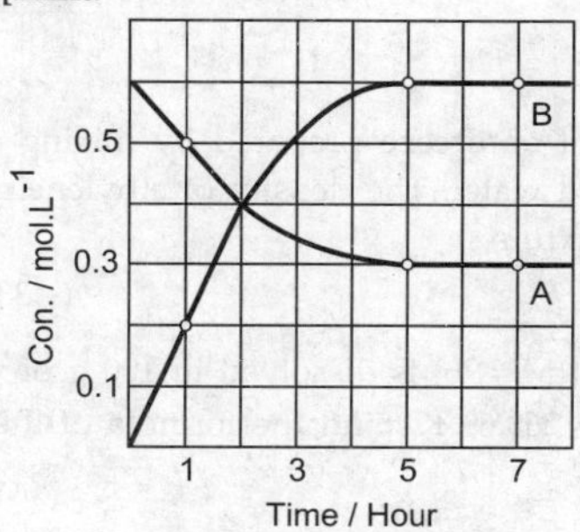

At equilibrium:

Rate (forward) = Rate (backward)

$$\frac{0.6-0.3}{5} = \frac{0.6}{5} \times \frac{1}{n}$$

For the eqb.: $A \rightleftharpoons 2B$, $K = \frac{0.6^2}{0.3}$.

Initial rate = change in concentration of A in the 1st hour]

336. Calculate the per cent dissociation of $H_2S(g)$ if 0.1 mole of H_2S is kept in a 0.4-litre vessel at 1000 K.

For the reaction

$$2H_2S(g) \rightleftharpoons 2H_2(g) + S_2(g),$$

the value of K_c is 1.0×10^{-6}. (2%)

337. Find the amount of time expressed in units of $t_{1/2}$ at which $A/A_0 = 0.125$.

[**Hint:** $N = 2^{-n} N_0$; $n \equiv$ no. of $t_{1/2}$] (3)

338. A radioisotope has a half-life of 900 seconds. Calculate the fraction of the original isotope which will remain behind after four half-life periods. (1/16)

339. What is the minimum half-life of an isotope needed so that not more than 0.1% of the nuclei undergo decay during a 3-hour laboratory period? (88 days)

340. Isotopes of oxygen with mass number less than 16 undergo β^+ emission. Assuming an equimolar mixture of ^{14}O and ^{15}O, find the ratio of the nuclides at the end of one hour. $t_{1/2}$ for $^{14}O = 71$ s, $t_{1/2}$ for $^{15}O = 124$ s. At what time will the ratio of ^{14}O nuclei to ^{15}O nuclei be equal to 0.25? (3.02×10^{-7}, 332 s)

341. What volume of concentrated HCl solution ($d = 1.18$) containing 36% HCl by weight is required to produce 6.55 litres of a solution with pH 1.85? (7.88 mL)

342. An ammonia solution is 9.9% ammonia by mass and has a density of 0.99 g/mL. Calculate the pH of the solution. $K_b(NH_4OH) = 1.7 \times 10^{-5}$. (12.0)

343. How much (volume) of 0.001 M HCl (aq) should be added to 10 mL of 0.001 M NaOH to change its pH by one unit? (8.18 mL)

344. A current of 1 amp is passed through one litre of 1.0 M HCl solution for one day.

Find the pH of the solution after electrolysis. Assume no change in volume. (0.98)

345. In how many litres of water, 10 g of CH_3COOH should be dissolved to give $[H^+] = 10^{-3}$? $K_a(CH_3COOH) = 1.8 \times 10^{-5}$? (3.0 litres)

346. Find the pH of 0.4833% HCl (aq) solution. (0.824)

347. Calculate the pH at the equivalence point when a solution of 0.1 M CH_3COOH is titrated with a solution of 0.1 M NaOH.
$K_a(CH_3COOH) = 1.8 \times 10^{-5}$. (8.66)

348. The coolant usually contains a solution of antifreeze prepared by mixing equal volumes of ethylene glycol, $C_2H_4(OH)_2$ and water. The density of ethylene glycol is 1.113 g/mL. Calculate the f.p. of the mixture.
K_f for H_2O is 1.86. (– 33.4°C)

349. A 7.64-g sample of the salt MF_x (at. wt. of M = 96) is dissolved in 100 g of water and the f.p. of the solution is found to be 268.69 K. Find the formula of the salt, assuming ideal behaviour. $K_f(H_2O) = 1.8$ (MF_5)

350. The f.p. of nitrobenzene is 3°C. When 1.2 g of chloroform (mol. wt. = 120) is dissolved in 100 g of nitrobenzene, the f.p. of the solution is 2.3°C. When 0.6 g of acetic acid is dissolved in 100 g of nitrobenzene, the f.p. of the solution is 2.64°C. Calculate the molecular weight of acetic acid. What conclusion can be drawn from it? (116.6, dimer)

351. By dissolving 0.517 g of nitrogen sulphide in 18.25 g of chloroform, the b.p. raised by 0.6°C. Nirogen sulphide contains 30.5% S. Find the molecular weight and molecular formula of nitrogen sulphide. (184, N_2S_4)

352. Consider the equilibrium: $LiCl \cdot 3NH_3(s) \rightleftharpoons LiCl \cdot NH_3(s) + 2NH_3(g)$ with $K_p = 9\ atm^2$ at 40°C. A 5-litre flask contains 0.1 mole of $LiCl \cdot NH_3$. How many moles of NH_3 should be added to the flask at this temperature to drive the backward reaction practically to completion? (0.7837)

353. Given that

$$S + S^{2-} \rightleftharpoons S_2^{2-} \quad \cdots \quad K_1 = 1.7$$

$$S + S_2^{2-} \rightleftharpoons S_3^{2-} \quad \cdots \quad K_2 = 3.1$$

Calculate the equilibrium constant for

$$2S + S^{2-} \rightleftharpoons S_3^{2-}$$

(5.27)

354. For the equilibrium: $2NO(g) + O_2(g) \rightleftharpoons 2NO_2(g)$, $K_c = 6.45 \times 10^5$.

(i) At what O_2 concentration is the NO_2 concentration equal to the NO concentration?

(ii) At what O_2 concentration is the NO_2 concentration 100 times the NO concentration? (1.55×10^{-6}, 1.55×10^{-2})

355. For the gas reaction: $3H_2 + N_2 = 2NH_3$, the partial pressures of H_2 and N_2 are 0.4

and 0.8 atm respectively. The total pressure of the entire system is 2.8 atm. What will be the value of K_p if all the concentrations are given in terms of atmosphere? (50)

356. What is the value of K_a such that K_a for an acid is equal to K_b for its conjugate base at 25°C? (1×10^{-7})

357. A solution contains 0.1 M H_2S and 0.3 M HCl. Calculate the concentration of S^{2-} and HS^- ions in the solution. For H_2S, $K_{a_1} = 1 \times 10^{-7}$, $K_{a_2} = 1.3 \times 10^{-13}$. $(3.34 \times 10^{-8}$ M, 1.447×10^{-20} M)

358. Calculate the molarity of an aqueous solution of ammonia of pH 9.3. K_b for ammonia is 1.8×10^{-5} and $K_w = 1 \times 10^{-14}$. $(1.95 \times 10^{-5}$ M)

359. 0.98% (by wt.) H_2SO_4 is 96% ionized. Find its pH. (0.71)

360. What volume of N/20 CH_3COONa should be mixed to 250 mL N/10 CH_3COOH to get a solution of pH = 5. $K_a(CH_3COOH) = 1.8 \times 10^{-5}$ (900 mL)

361. Calculate pH of the following mixtures:

(i) 0.4 litre of 0.1 M NaOH and 0.2 litre of 0.05 M C_2H_5COOH

(ii) 0.2 litre of 0.1 M NaOH and 0.4 litre of 0.1 M C_2H_5COOH

(iii) 0.4 litre of 0.1 M NaOH and 0.4 litre of 0.1 M C_2H_5COOH

$K_a(C_2H_5COOH) = 5.6 \times 10^{-6}$, $K_w = 1 \times 10^{-14}$ (12.67, 5.25, 7.97)

362. COF_2 gas passed over a catalyst at 1000°C comes to equilibrium:

$$2\,COF_2(g) \rightleftharpoons CO_2(g) + CF_4(g)$$

Analysis of the equilibrium mixture (after quick cooling to freeze the equilibrium) shows that 500 mL of equilibrium mixture (NTP) contains 300 mL (NTP) of COF_2 and CO_2. Taking the total pressure to be 10 atm, calculate K_p. (4.0)

[**Hint:** At eqb.: $\underset{100\text{ mL}}{2COF_2} \rightleftharpoons \underset{200\text{ mL}}{CO_2} + \underset{200\text{ mL}}{CF_4}$]

363. When 0.5 mole of H_2 and 0.5 mole of I_2 react in a 10-litre evacuated vessel at 450°C, HI is formed. K_c for $H_2 + I_2 \rightleftharpoons 2HI$ is 50.

(i) Calculate K_p.

(ii) Calculate moles of I_2 which are in excess. (50, 0.11 mole)

364. The value of K_c for $2HF(g) = H_2(g) + F_2(g)$ is 1.0×10^{-3} at a particular temperature. At a certain time, the concentrations of HF, H_2 and F_2 were found to be 0.5, 1×10^{-3} and 4×10^{-3} mole/litre respectively. Is the reaction at equilibrium? If not, what would be the direction to attain equilibrium? (No, towards right)

365. Find out the number of waves made by a Bohr electron in one complete revolution in its third orbit. **(IIT 1994)** (3)

[**Hint:** $\frac{nh}{2\pi} = mvr; \quad \frac{2\pi r}{h/mv} = \frac{2\pi r}{\lambda} = n$]

366. The composition of a sample of wustite is $Fe_{0.93}O_{1.00}$. What percentage of iron is present in the form of Fe (III)? **(IIT 1994)** (15.05%)

[**Hint:** Let the moles of Fe in FeO and Fe_2O_3 be n_1 and n_2 respectively. Thus,

$$\frac{\text{moles of Fe}}{\text{moles of O}} = \frac{n_1 + n_2}{n_1 + \frac{3n_2}{2}} = 0.93]$$

367. A is a binary compound of a univalent metal. 1.422 g of A reacts completely with 0.321 g of sulphur in an evacuated and sealed tube to give 1.743 g of a white crystalline solid B, that forms a hydrated salt C, with $Al_2(SO_4)_3$. Identify A, B and C. **(IIT 1994)** (KO_2, K_2SO_4, $K_2SO_4 \cdot Al_2(SO_4)_3 \cdot 24H_2O$)

368. A 1.345-g sample of a compound of barium and oxygen was dissolved in hydrochloric acid to give a solution of barium ion, which was then precipitated with an excess of potassium chromate to give 2.012 g of barium chromate, $BaCrO_4$. What is the formula of the compound? (BaO_2)

369. Haemoglobin is the oxygen-carrying molecule of red blood cells, consisting of a protein and a nonprotein substance. The nonprotein substance is called haeme. A sample of haeme weighing 35.2 mg contains 3.19 mg of iron. If a haeme molecule contains one atom of iron, what is the molecular weight of haeme? (616 amu)

370. The wavenumber of the first line in the Balmer series of hydrogen is 15200 cm^{-1}. What is the wavenumber of the first line in the Balmer series of Be^{3+}?

(2.43×10^5 cm^{-1})

[**Hint:** $\left(\frac{1}{\lambda}\right)_{Be^{3+}} = z^2\left(\frac{1}{\lambda}\right)_H$]

371. Sufficient NaCN was added to 0.015 M $AgNO_3$ to give a solution that was initially 0.1 M in CN^-. What is the concentration of Ag+ in this solution after $Ag(CN)_2^-$ forms? The formation constant K_f for the complex ion $Ag(CN)_2^-$ is 5.6×10^{18}.

$$Ag^+ + 2CN^- \rightleftharpoons Ag(CN)_2^-$$

(5.5×10^{-19} M)

372. In an Arrhenius's equation, $k = Ae^{-E/RT}$, A may be termed as the rate constant at .... . **(IIT 1997)** (infinite temp.)

373. When Fe (s) is dissolved in aqueous acid in a closed vessel, the work done is ... . **(IIT 1997)** (zero)

374. A liquid which is permanently supercooled is frequently called a ...... **(IIT 1997)** (glass)

375. Enthalpy is an ... property. **(IIT 1997)** (extensive)

376. A quantity of 0.25 M NaOH is added to a solution containing 0.15 mole of acetic acid. The final volume of the solution is 375 mL and the pH of the solution is 4.45.

(a) What is the molar concentration of sodium acetate?

(b) How many mL of NaOH were added to the original solution?

(c) What was the original concentration of the acetic acid?

[(a) 0.048 M (b) 1.9×10^2 mL (c) 0.81 M]

377. What are the concentration and percentage of Ag^+ ion remaining after Ag_2CrO_4 precipitates when 25 mL of 0.10 M $AgNO_3$ is added to 25 mL of 0.10 M K_2CrO_4? K_{sp} $(Ag_2CrO_4) = 1.1 \times 10^{-12}$. ($6.6 \times 10^{-6}$ M, 0.013%)

378. The pH of a white-vinegar solution is 2.45. This vinegar is an aqueous solution of acetic acid with a density of 1.09 g/mL. What is the mass percentage of acetic acid in the solution? (4.1%)

379. A chemist needs a buffer with pH 4.35. How many mL of pure acetic acid (density = 1.049 g/mL) must be added to 465 mL of 0.0941 M NaOH solution to obtain such a buffer? (9.1 mL)

380. Calculate the pH of a solution which has a hydronium-ion concentration of 6×10^{-8} M. (7.22)

381. Calculate the per cent error in the hydronium-ion concentration made by neglecting the ionisation of water in a 1×10^{-6} M NaOH solution. (1%)

382. Calculate $[CH_3COOH]/[CH_3COO^-]$ in a buffer solution whose pH is 7.0. Explain how it is possible to have any acid in a neutral solution.

(5.6×10^{-3}, possible when some base is present)

383. Calculate the molar solubility of AgCl in 1.0 M NH_3. K_{sp} (AgCl) = 1.8×10^{-10}, K_f $(Ag(NH_3)_2^+) = 1.7 \times 10^7$. (0.050 M)

384. An acid solution of a $KReO_4$ sample containing 26·83 mg of combined rhenium was reduced by passage through a column of granulated zinc. The effluent solution, including the washings from the column, was then titrated with 0.1 N $KMnO_4$. 11.45 mL of standard permanganate was required for the reoxidation of all the rhenium to the perrhenate ion, ReO_4^-. Assuming that Re was the only element reduced, what is the oxidation state to which Re was reduced by the zinc column? (Re = 186.2) (–1 oxd. state)

385. A diver quickly ascends to the surface of the water from a depth of 4.08 m without exhaling out the air in her lungs. By what factor would the volume of her lungs increase by the time she reaches the surface? Assume constant temperature and ideal gas behaviour. The density of sea water is 1.03 g/cc and g = 980.67 cm s^{-2}. (1.4 times)

386. The density of dry air at 1 atm and 34.4°C is 1.15 g/L. Calculate the composition of air (% by weight) assuming only N_2 and O_2 to be present and ideal gas behaviour. (N_2 72.4%)

[**Hint:** First calculate mol. wt. of air by using $p = \frac{dRT}{M}$.]

387. At what pH will 1×10^{-3} M solution of an indicator with $K_b = 1 \times 10^{-10}$ change colour? (4)

[**Hint:** The indicator changes colour when the conjugates are of equal concentration.]

388. Which has greater molarity in water, AgCl or $Mg(OH)_2$? Can the relative solubilities be predicted on the basis of the values of K_{sp} alone? [Mg(OH)2, No]

389. An electrochemical cell is made by placing a zinc electrode in 1.0 litre of 0.2 M $ZnSO_4$ solution and a copper electrode in 1.0 litre of 0.015 M $CuCl_2$ solution.
(a) What is the initial voltage of this cell when it is properly constructed?
(b) Calculate the final concentration of Cu^{2+} in this cell if it is allowed to produce an average current of 1 amp for 225 seconds. Given that $E^0_{cell} = 1.1$ V.
[(a) 1.07 V (b) 0.014 M]

390. Dinitrogen pentoxide, N_2O_5, undergoes first-order decomposition in chloroform solvent to yield NO_2 and O_2. The rate constant at 45°C is 6.2×10^{-4} min^{-1}. Could the volume of O_2 obtained from the reaction of 1 mole of N_2O_5 at 45°C and 780 mmHg after 20 hours, be calculated? (Insufficient information)

391. What would you expect to be the general temperature and pressure conditions for an optimum yield of nitric oxide, NO, by the oxidation of ammonia?
$4NH_3 (g) + 5O_2 (g) \rightleftharpoons 4NO (g) + 6H_2O (g) ; \Delta H < 0$
(Low temperature and low pressure)

392. At 850°C and 1-atm pressure, a gaseous mixture of CO and CO_2 in equilibrium with solid carbon is 90.55% CO by mass.
$C (s) + CO_2 (g) \rightleftharpoons 2CO (g)$
Calculate K_c for this reaction at 850°C. (0.153)
[**Hint:** For gasses: mole ratio = pressure ratio]

393. The following equilibrium exists in a closed system at 25°C:
$NH_4HS (s) \rightleftharpoons NH_3 (g) + H_2S (g)$
(a) When a sample of pure NH_4HS (s) is placed in an evacuated vessel and allowed to reach equilibrium at 25°C, the total pressure is 0.66 atm. Find the value of K_p.
(b) To this system, sufficient H_2S (g) is injected until the pressure of H_2S is three times that of the ammonia at equilibrium. What are the partial pressures of NH_3 and H_2S?
(c) In a different experiment, 0.75 atm of NH_3 and 0.5 atm of H_2S are introduced into a 1-litre vessel at 25°C. How many moles of NH_4HS are present when equilibrium is established? [(a) 0.1089 (b) 0.19, 0.57 (c) 0.5672 g]

394. Equal volumes of 1.0 M Na_2CO_3 and 1.0 M HCl are mixed. Calculate $[CO_3^{2-}]$ at equilibrium. K_1 and K_2 for H_2CO_3 are 4.5×10^{-7} and 4.7×10^{-11} respectively.
(5×10^{-3} M)
[**Hint:** Solution is 0.50 M $NaHCO_3$ (plus 0.5 M NaCl). Now see Example 71, Chapter 16.]

395. Calculate pH of a 0.1 M Na_2HPO_4 solution. K_1, K_2 and K_3 for H_3PO_4 are 7.1×10^{-3}, 6.3×10^{-8} and 4.5×10^{-13} respectively. Which approximation is necessary for the calculation? (10.10)
[**Hint:** As K_3 for H_3PO_4 is very low compared to K_1, assume no acidic ionisation. Consider only the equilibrium,
$HPO_4^{2-} + H_2O \rightleftharpoons H_2PO_4^- + OH^-$]

396. A tenfold increase in pressure on the reaction,
$N_2 (g) + 3H_2 (g) \rightleftharpoons 2NH_3 (g)$

at equilibrium results in ...... in K_p. **(IIT 1996)** (No change)

397. The reaction

$$Sb_2S_3\ (s) + 3H_2\ (g) \rightleftharpoons 2Sb\ (s) + 3H_2S\ (g)$$

was studied by analysing the equilibrium mixture for the amount of H_2S produced. A vessel whose volume was 2.5 litre was filled with 0.01 mole of Sb_2S_3 and 0.01 mole of H_2S. After the mixture came to equilibrium in the closed vessel at 440°C, the gaseous mixture was removed and the H_2S was dissolved in water. Sufficient Pb^{2+} ions were added to react completely with the H_2S to precipitate PbS. If 1.029 g of PbS was obtained, what is the value of K_c at 440°C? (0.430)

398. An aqueous solution containing 288 g of a nonvolatile compound having composition $C_nH_{2n}O_n$ in 90 g of water boils at 101.24°C at 1-atm pressure. What is the molecular formula of the compound? K_b = 0.512°C/*m*. ($C_{44}H_{88}O_{44}$)

399. Although AgCl is insoluble in water, it readily dissolves upon the addition of ammonia

$$AgCl\ (s) + 2NH_3\ (aq) \rightleftharpoons Ag(NH_3)_2^+\ (aq) + Cl^-\ (aq)$$

(a) What is the equilibrium constant for this dissolving process?

(b) Ammonia is added to a solution containing excess AgCl (s). The final volume is 1 litre and the resulting equilibrium concentration of NH_3 is 0.80 M. Calculate the number of moles of AgCl dissolved, the molar concentration of $Ag(NH_3)_2^+$ and the number of moles of NH_3 added to the original solution.
K_{sp} (AgCl) = 1.8×10^{-10} and K_f [$Ag(NH_3)_2^+$] = 1.7×10^7.

[(a) 3.1×10^{-3} (b) 0.045 mole, 0.045 mole, 0.89 mole]

400. From the dissociation constants K_a and K_b for an acid and its conjugate base, show that $K_a \cdot K_b = K_w$.

[Hint: See text, Chapter 16.]

401. The tallest trees known are the redwoods in California. Assuming the height of a redwood to be 105 m, estimate the osmotic pressure required to push water up from the roots to the treetop. (1.029×10^6 Pa)

402. A sample of impure cuprite, Cu_2O, contains 66.6% copper. Calculate the percentage of pure Cu_2O in the sample. (75%)

403. How much Ag^+ would remain in solution after mixing equal volumes of 0.08 M $AgNO_3$ and 0.08 M HOCN? K_a (HOCN) = 3.3×10^{-4}, K_{sp} (AgOCN) = 2.3×10^{-7}. (5×10^{-3} M)

[Hint: See Example 66, Chapter 16.]

404. The following equilibrium was studied by analysing the equilibrium mixture for the amount of HCl produced.

$$LaCl_3\ (s) + H_2O\ (g) \rightleftharpoons LaOCl\ (s) + 2HCl\ (g)$$

A vessel whose volume was 1.25 litre was filled with 0.0125 mole of lanthanum (III) chloride and 0.025 mole of water. After the mixture came to equilibrium in a closed vessel at 619°C, the gaseous mixture was removed and dissolved in more water. Sufficient silver (I) ion was added to precipitate the chloride ions completely as silver chloride. If 3.59 g of AgCl was obtained, what is the value of K_c at 619°C? (0.04)

405. A solution is 0.10 M Co^{2+} and 0.10 M Hg^{2+}. Calculate the range of pH in which only one of the metal sulphides precipitates when the solution is saturated in H_2S. K_{sp} (CoS) $= 4 \times 10^{-21}$ and K_{sp} (HgS) $= 1.6 \times 10^{-52}$. (pH less than 0.8)

406. A standard electrochemical cell is made by dipping an Ag electrode into a 1.0 M Ag^+ solution and a Cd electrode into a 1.0 M Cd^{2+} solution.

(a) What is the spontaneous chemical reaction and what is the maximum potential produced by the cell?

(b) What would be the effect on the potential of this cell if Na_2S were added to the Cd^{2+} half cell and CdS were precipitated? Why?

(c) What would be the effect on the potential of the cell if the size of the silver electrode was doubled?

See E^0 values from the table if required.

[(a) $Cd(s) + 2Ag^+ = 2Ag(s) + Cd^{2+}$; 1.20 V
(b) It would increase (c) No effect]

407. A sample of impure ore contains 42.34% Zn. Calculate the percentage of pure ZnS in the sample. (67.10%)

408. A peroxidase enzyme isolated from human red blood cells was found to contain 0.29% selenium. What is the minimum molecular weight of the enzyme? (Se = 78.96) (2.7×10^4)

409. In an experiment to measure the charge on an electron, the following values of charge were found on oil droplets: $-1.6 \times 10^{-19}, -2.4 \times 10^{-19}, -4.0 \times 10^{-19}$ (in coulomb). What values of electronic charge would be indicated by these results?

[**Hint:** Find the largest common factor.] (-0.8×10^{-19} C)

410. A diatomic molecule has a dipole moment of 1.2 D. If its bond distance is 1.0 Å, what fraction of an electronic charge, e, exists on each atom? (1 D = 10^{-18} esu cm and $e = 4.8 \times 10^{-10}$ esu) (25%)

411. How many grams of Cu will be replaced from 2 litres of 1.5 M $CuSO_4$ solution by 40 g of Al? (141 g)

412. At the top of a mountain a thermometer reads 0°C and a barometer reads 710 mmHg. At the bottom of the mountain the temperature is 30°C and the pressure is 760 mmHg. Compare the density of the air at the top with that at the bottom. (1.04 : 1)

413. What is the pH of a 0.50 M aqueous NaCN solution? pK_b of CN^- is 4.70. **(IIT 1996)** (11.5)

414. In the reaction $I^- + I_2 \longrightarrow I_3^-$, the lewis acid is .... **(IIT 1997)** (I_2)

415. A monoatomic ion has a charge of +2. The nucleus of the ion has a mass number of 62. The number of neutrons in the nucleus is 1.21 times that of protons. How many electrons are in the ion? What is the atomic number of the element? (26, 28)

416. Potassium superoxide, KO_2, is used in rebreathing gas masks to generate oxygen.

$$4KO_2\,(s) + 2H_2O\,(l) \longrightarrow 4KOH\,(s) + 3O_2\,(g)$$

If a reaction vessel contains 0.15 mole of KO_2 and 0.10 mole of H_2O, what is the limiting reactant? How many moles of oxygen can be produced?

(KO_2, 0.1125 mole)

417. A 0.288 g sample of an unknown monoprotic organic acid is dissolved in water and titrated with a 0.115 M sodium hydroxide solution. After the addition of 17.54 mL of base, a pH of 4.92 is recorded. The equivalence point is reached when a total of 33.83 mL of NaOH is added.

(a) What is the molar mass of the organic acid?

(b) What is the K_a value for the acid? The K_a value could have been determined very easily if a pH measurement had been made after the addition of 16.92 mL of NaOH. Why?

[(a) 74, (b) 1.3×10^{-5}]

418. How many grams of NaCl can be added to 785 mL of 0.0015 M $AgNO_3$ before a precipitate forms? K_{sp} (AgCl) $= 1.8 \times 10^{-10}$. (5.5×10^{-6} g)

419. Metallic Ba has a body-centred cubic structure (all atoms at the lattice points) and a density of 3.51 g/cc. Assume Ba atoms to be spheres. The spheres in a body-centred array occupy 68.0% of the total space. Find the atomic radius of Ba.

(3.14×10^{-8} cm)

420. In a reaction:

$$2ClO_2\,(aq) + 2I^-\,(aq) \longrightarrow 2ClO_2^-\,(aq) + I_2\,(aq)$$

the order of the reaction with respect to ClO_2 was determined by starting with a large excess of I^-, so that its concentration was essentially constant. Then

$$\text{Rate} = k\,[ClO_2]^m\,[I^-]^n = k'\,[ClO_2]^m,$$

where $k' = k\,[I^-]^n$. Determine the order of the reaction and also k' from the following data:

Time (s)	$[ClO_2]$ (mol/L)
0.00	4.77×10^{-4}
1.00	4.31×10^{-4}
2.00	3.91×10^{-4}
3.00	3.53×10^{-4}

(First, 0.101 s^{-1})

421. A 0.1 M solution of an acid (density = 1.01 g/cc) is 4.5% ionised. Calculate the f.p. of the solution. The molecular weight of the acid is 300. $K_f = 1.86$. (–0.199°C)

422. How much heat is required to change 10 g of ice at 0°C to steam at 100°C?

$$\left[\begin{aligned}\Delta H\,(\text{total}) &= \Delta H_{\text{fusion}} + \Delta H_{\text{heating}} + \Delta H_{\text{vap.}}\\ &= 10(80 + 1 \times 100 + 540)\text{ cal}\end{aligned}\right]$$

423. After 11.2 g of carbon reacts with oxygen originally occupying 21.2 litres at 18°C and 750 mmHg, the cooled gases are passed through 3 litres of 2.50 M NaOH solution. Determine the concentration of NaOH remaining in solution which is not converted to Na_2CO_3. (1.95 M)

[**Note**: CO does not react with NaOH under these conditions.]

424. From the kinetic theory of gases, predict the effect on the pressure of a gas inside

a cubic box of side l by reducing the size so that each side measures $l/2$.
(8 times increase)

425. The equilibrium equations and K_a values for three reaction systems are given below:

$$H_2C_2O_4 + H_2O \rightleftharpoons H_3O^+ + HC_2O_4^-; \quad K_a = 5.6 \times 10^{-2}$$
$$H_3PO_4 + H_2O \rightleftharpoons H_3O^+ + H_2PO_4^-; \quad K_a = 6.9 \times 10^{-3}$$
$$HCOOH + H_2O \rightleftharpoons H_3O^+ + HCOO^-; \quad K_a = 1.7 \times 10^{-4}$$

(a) Which conjugate pair would be best for preparing a buffer with a pH of 2.88?

(b) How would you prepare 50 mL of a buffer with a pH of 2.88 assuming that you had available 0.1 M solution of each pair?

[(a) H_3PO_4 and $H_2PO_4^-$ (b) 8 mL H_3PO_4 and 42 mL $H_2PO_4^-$]

426. Tartaric acid is a weak diprotic acid with $K_1 = 1 \times 10^{-3}$ and $K_2 = 4.6 \times 10^{-5}$.

(a) Letting the symbol H_2A represent tartaric acid, write the chemical equations that represent K_1 and K_2. Write the chemical equation that represents $K_1 \times K_2$.

(b) Qualitatively describe the relative concentrations of H_2A, HA^-, A^{2-} and H_3O^+ in a solution that is about 0.5 M in tartaric acid.

(c) Calculate the pH of a 0.025 M tartaric acid solution and the equilibrium concentration of H_2A.

(d) What is the A^{2-} concentration?

[(a) $H_2A + H_2O \rightleftharpoons H_3O^+ + HA^-$
$HA^- + H_2O \rightleftharpoons H_3O^+ + A^{2-}$
$H_2A + 2H_2O \rightleftharpoons 2H_3O^+ + A^{2-}$
(b) $[H_2A] >> [H_3O^+] = [HA^-] >> [A^{2-}]$
(c) pH = 2.34; $[H_2A]$ = 0.0205 M
(d) 4.6×10^{-5} M]

427. Tritium, ^3_1H, is a radioactive nucleus of hydrogen. It is used in luminous watch dials. Tritium decays by beta emission with a half-life of 12.3 years. What is the decay constant (in s^{-1})? What is the activity (in Ci) of a sample containing 2.5 μg of tritium? The atomic mass of tritium is 3.02 amu. (1.79×10^{-9} /s, 0.024 Ci)

428. The equilibrium equations and K_a values for three reaction systems are given below.

$$NH_4^+ + H_2O \rightleftharpoons H_3O^+ + NH_3; \quad K_a = 5.6 \times 10^{-2}$$
$$H_2CO_3 + H_2O \rightleftharpoons H_3O^+ + HCO_3^-; \quad K_a = 4.3 \times 10^{-7}$$
$$H_2PO_4^- + H_2O \rightleftharpoons H_3O^+ + HPO_4^{2-}; \quad K_a = 6.2 \times 10^{-8}$$

(a) Which conjugate pair would be the best for preparing a buffer with a pH of 6.96? Why?

(b) How would you prepare 100 mL of a buffer with a pH of 6.96 assuming that you had available 0.10 M solutions of each pair?

[(a) H_2CO_3 and HCO_3^- (b) H_2CO_3 – 20·4 mL, HCO_3^- – 79·6 mL]

429. How many grams of silver could be plated out on a serving tray by electrolysis of a solution containing silver in +1 oxidation state for a period of 8.0 hours at a current of 8.46 amp? What is the area of the tray if the thickness of the silver plating is 0.00254 cm? Density of silver is 10.5 g/cm^3.

(IIT 1997) (271.65 g, $1.02 \times 10^4\ cm^2$)

430. Assuming that 50% of the heat is useful, how many kg of water at 15°C can be heated to 95°C by burning 200 litres of methane at NTP? $\Delta H_{combustion}$ (CH_4) = 211 kcal/mole, sp. heat of water = 1 kcal/kg K. (11.8 kg)

431. The rate of reaction:

$$CH_3C(S)NH_2\ (aq) + H_2O \longrightarrow H_2S\ (aq) + CH_3C(O)NH_2\ (aq)$$

is given by the rate law:

$$\text{Rate} = k[H_3O^+]\ [CH_3C(S)NH_2]$$

Consider 1 litre of solution that is 0.20 M in $CH_3C(S)NH_2$ and 0.15 M in HCl at 25°C.

(a) For each of the following changes, state whether the rate of reaction increases, decreases or remains the same.

(i) A 4 g sample of NaOH is added to the solution

(ii) 500 mL of water is added to the solution

(iii) The 0.15 M HCl solution is replaced by a 0.15 M acetic acid solution.

(b) State whether the value of k will increase, decrease or remain the same.

(i) A catalyst is added to the solution

(ii) The reaction is carried out at 15°C instead of 25°C

[(a) (i), (ii) and (iii) decreases (b) (i) increases (ii) decreases]

432. What per cent of a sample of nitrogen must be allowed to escape if its temperature, pressure and volume are to be changed from 220°C, 3 atm and 1.65 litre to 110°C, 0.7 atm and 1 litre respectively? (81.8%)

433. The vapour pressure of water at 20°C is 17.5 mmHg. Calculate mass of water per litre of air at 20°C and 45% relative humidity. (7.8 mg)

[**Hint:** Relative humidity is the ratio of the partial pressure of water in air at a given temperature to the vapour pressure of water at that temperature.]

434. A gas is composed of 30.4% N and 69.6% O. Its density is 11.1 g/L at –20°C and 2.5 atm. What are the empirical and molecular formula of the gas? (NO_2, N_2O_4)

435. The active ingredients of an antacid tablet contained only magnesium hydroxide and aluminium hydroxide. Complete neutralisation of the sample of the active ingredients required 48.5 mL of 0.187 M hydrochloric acid. The chloride salts from the neutralisation were obtained by evaporation of the filtrate from the titration; they weighed 0.42 g. What was the percentage by mass of magnesium hydroxide in the active ingredients of the antacid tablet? (61.7%)

436. In a reaction:

$$CH_3COOCH_3\ (aq) + OH^-\ (aq) \rightarrow CH_3COO^-\ (aq) + CH_3OH\ (aq)$$

The overall order of the reaction was determined by starting with methyl acetate and hydroxide ion at the same concentrations, so

$$[CH_3COOCH_3] = [OH^-] = x$$

Then rate = $k\,[CH_3COOCH_3]^m\,[OH^-]^n = kx^{m+n}$.

Determine the overall order and the value of rate constant, k, from the following data:

Time (min)	$[CH_3COOCH_3]$ (mole/litre)
0.00	0.01
3.00	0.0074
4.00	0.00683
5.00	0.00634

(Second, 11.61 L mol^{-1} s^{-1})

437. A sample of natural gas is 85.2% methane, CH_4, and 14.8% ethane, C_2H_6, by mass. What is the density of this mixture at 18°C and 748 mmHg? (0.71 g/L)

438. If the rms speed of the NH_3 molecule is found to be 0.51 km/s, what is the temperature? (–95°C)

439. The disintegration of ^{239}Pu is accompanied by the loss of 5.24 MeV/dis. The half-life of ^{239}Pu is 24400 years. Calculate the energy released per day from 1 g sample of ^{239}Pu in MeV. (1.03×10^{15} MeV/d)

440. Calculate the effective neutron capture radius of a nucleus having a cross section of 1.0 barn. (5.6×10^{-13} cm)

[**Hint:** 1 barn = 10^{-24} cm^2 and area of circle = πr^2]

441. New industrial plants for acetic acid react liquid methanol with carbon monoxide in the presence of a catalyst.

$$CH_3OH + CO \rightarrow CH_3COOH$$

In the experiment, 15 g of methanol and 10 g of carbon monoxide were placed in a reaction vessel. What is the theoretical yield of acetic acid? If the actual yield is 19.1 g, what is the percentage yield? (21.4 g, 89.1%)

442. A 1.0-mg sample of technetium-99 has an activity of 1.7×10^{-5} Ci decaying by β-emission. What is the decay constant for $^{99}_{43}Tc$? (1.0×10^{-13}/s)

443. The nuclide ^{227}Ac undergoes β emission (98.6%) or α emission (1.4%) with a half-life of 21.6 years. Determine λ(α) and λ(β). (4.5×10^{-4} yr^{-1}, 0.0317 yr^{-1})

444. Crystals of AgBr can be removed from black-and-white photographic film by reacting the AgBr with sodium thiosulphate.

$$AgBr\,(s) + 2S_2O_3^{2-}\,(aq) \rightleftharpoons [Ag(S_2O_3)_2]^{3-}\,(aq) + Br^-\,(aq)$$

(a) What is the equilibrium constant for this dissolving process?

(b) In order to dissolve 2.5 g of AgBr in a 1-litre solution, how many moles of $Na_2S_2O_3$ must be added? K_{sp} (AgBr) = 5×10^{-13}, $K_f\,[Ag(S_2O_3)_2]^{3-} = 2.9 \times 10^{13}$

[(a) 14.5 (b) 0.03 mole]

445. One of the hazards of nuclear explosion is the generation of ^{90}Sr and its subsequent incorporation in bones. The nuclide has a half-life of 28.1 years. Suppose one microgram was absorbed by a newborn child, how much ^{90}Sr will remain in his bones after 20 years? **(IIT 1995)** (0.061 μg)

446. A 0.239 g sample of unknown organic base is dissolved in water and titrated with a 0.135 M HCl solution. After the addition of 18.35 mL of acid, a pH of 10.73 is recorded. The equivalence point is reached when a total of 39.24 mL of HCl is added. The base and acid combine in a 1 : 1 ratio.

(a) What is the molar mass of the organic base?

(b) What is the K_b value for the base? The K_b value could have been determined very easily if a pH measurement had been made after the addition of 19.62 mL of HCl. Why? [(a) 45.12 (b) 4.72×10^{-4}]

447. A solution is 1.5×10^{-4} M Zn^{2+} and 0.20 M HSO_4^-. The solution also contains Na_2SO_4. What should be the minimum molarity of Na_2SO_4 to prevent precipitation of ZnS when the solution is saturated with H_2S (0.1 M H_2S)? K_{sp} (ZnS) = 1.1×10^{-21}. (0.18 M)

448. A 0.50-g mixture of Cu_2O and CuO contains 0.425 g of Cu. What is the mass of CuO in the mixture? (0.21 g)

449. An alloy of iron (54.7%), nickel (45.0%) and manganese (0.3%) has a density of 8.17 g/cc. How many iron atoms are there in a block of alloy measuring 10.0 cm × 20.0 cm × 15.0 cm? (1.45×10^{26})

450. (a) Calculate the equilibrium constant for the following reaction at 25°C.

$$Sn\ (s) + Pb^{2+}\ (aq) \rightarrow Sn^{2+}\ (aq) + Pb\ (s)$$

The standard emf of the corresponding voltaic cell is 0.01 V.

(b) If an excess of tin metal is added to 1.0 M Pb^{2+}, what is the concentration of Pb^{2+} at equilibrium? [(a) 2.2 (b) 0.3 M]

451. (a) Calculate the equilibrium constant for the following reaction at 25°C.

$$Ag^+\ (aq) + Fe^{2+}\ (aq) \rightarrow Ag\ (s) + Fe^{3+}\ (aq)$$

The standard emf of the corresponding voltaic cell is 0.03 V.

(b) When equal volumes of 1.0 M solutions of Ag^+ and Fe^{2+} are mixed, what is the equilibrium concentration of Fe^{2+}? [(a) 3.218 (b) 0.268 V]

452. The thermochemical equation for the dissociation of hydrogen gas into atoms may be written as

$$H_2 \rightarrow 2H\ ; \quad \Delta H = 436\ kJ.$$

What is the ratio of the energy yield on combustion of hydrogen atoms to steam to the yield on combustion of an equal mass of hydrogen molecules to steam? $\Delta H_{combustion}$ for H_2 = −241.81 kJ. (2.80)

453. A mixture of N_2 and Ne contains equal moles of each gas and has a total mass of 10.0 g. What is the density of this gas mixture at 500 K and 10 atm? (5.88 g/L)

454. Calculate the hydronium-ion concentration and the sulphide ion concentration of a 0.1 M H_2S solution, $K_1 = 1 \times 10^{-7}$ and $K_2 = 1 \times 10^{-14}$ (1×10^{-4}, 1×10^{-14})

[Hint: $[H^+]$ is mainly due to the first step of ionisation while $[S^{2-}]$ is due to the second step of ionisation.]

455. Calculate $[SO_4^{2-}]$ in 0.15 M H_2SO_4 solution if the first step of ionisation is complete and the second step $K_2 = 1.02 \times 10^{-2}$. (8.9×10^{-3})

456. What is the limiting value of the time required for the radioactive daughter to reach its maximum activity as the value of $t_{1/2}$ (parent)/ $t_{1/2}$ (daughter) increases? (∞)

457. Under standard conditions for all concentrations, the following reaction is spontaneous at 25°C.

$$O_2\,(g) + 4H^+\,(aq) + 4Br^-\,(aq) = 2H_2O\,(l) + 2Br_2\,(l)$$

If $[H^+]$ is decreased so that the pH = 3.6, what value will E_{cell} have, and will the reaction be spontaneous at this $[H^+]$? (–0.05 V, No)

458. An electrode is prepared by dipping an Ag strip into a solution saturated with silver thiocyanate, AgSCN, and containing 0.10 M SCN^-. The emf of the voltaic cell constructed by connecting this electrode as the cathode to the standard hydrogen half cell as the anode is 0.45 V. What is K_{sp} of AgSCN? (1×10^{-7})

459. An ideal gas with density of 3.0 g/L has pressure of 675 mmHg at 25°C. What is the rms speed of the molecules of this gas? $(3.0 \times 10^2$ m/s)

460. Determine the emf of the following cell:

$$Pb \mid PbSO_4\,(s),\ SO_4^{2-}\,(1.0\,M) \mid\mid H^+\,(1.0\,M) \mid H_2\,(1.0\,atm) \mid Pt$$

The anode is essentially a lead electrode, $Pb \mid Pb^{2+}$ (aq). However, the anode solution is saturated with $PbSO_4$, so that Pb^{2+} ion concentration is determined by the solubility product of $PbSO_4$ $(= 1.7 \times 10^{-8})$. See E^0 values from the table if required. (0.36 V)

461. Under standard conditions for all concentrations, the following reaction is spontaneous at 25°C.

$$O_2\,(g) + 4H^+\,(aq) + 4Br^-\,(aq) \rightarrow 2H_2O\,(l) + 2Br_2\,(l);\ E^0 = 0{\cdot}17\text{ V}$$

If $[H^+]$ is adjusted by adding a buffer of 0.10 M NaOCN and 0.10 M HOCN $(K_a = 3.5 \times 10^{-4})$, what value will E_{cell} have, and will the reaction be spontaneous at this $[H^+]$? (–0.042 V, No)

462. How many moles of NH_3 must be added to 1 litre of 0.75 M $AgNO_3$ in order to reduce the Ag^+ concentration to 5×10^{-8} M? $K_d\,[Ag(NH_3)_2^+] = 1 \times 10^{-8}$ (1.9 mole)

463. What fraction of a mole of iron metal will be produced by passage of 4 amp of current through 1 litre of 0.1 M Fe^{3+} solution for 1 hour? (0.025 mole)

464. Metallic Mg has a hexagonal close-packed structure and a density of 1.74 g/cc. Assume Mg atoms to be spheres with radius r. Because Mg has a close-packed structure, 74.1% of the space is occupied by atoms. Calculate the volume of each atom and then find the atomic radius r. $(1.72 \times 10^{-23}$ cc, 1.6×10^{-10} m)

465. A solution is 0.10 M in Na_2SO_4. When 50 mL of 0.1 M $Ba(NO_3)_2$ is added to 50 mL of this solution, what fraction of the sulphate ion is not precipitated? $K_{sp}\,(BaSO_4) = 1.1 \times 10^{-10}$. (2.1×10^{-4})

466. A metallic element crystallises into a lattice containing a sequence of layers

ABABAB.... Any packing of spheres leaves out voids in the lattice. What percentage by volume of this lattice is empty space? **(IIT 1996)** (25.94%)

[**Hint:** The empty space in h.c.p. or c.c.p. arrangement is same as for f.c.c. See Chapter 19.]

467. A sample of $^{14}CO_2$ was to be mixed with ordinary CO_2 for biological tracer experiment. In order that 10 cc (NTP) of the diluted gas should have 10^4 disintegrations per minute, how many microcuries of radioactive carbon are needed to prepare 60 litres of the diluted gas? (27 μCi)

468. Which state of the triply ionised beryllium, Be^{3+}, has the same radius as that of the ground state of hydrogen atom? (Second)

469. At what temperature would the average translational kinetic energy of gaseous hydrogen molecules equal the energy required to dissociate the molecules into atoms, i.e., 104 kcal per mole? (34900 K)

470. Calculate the pH of a 0.005 M Na_2S solution. K_1 and K_2 for H_2S are 1×10^{-7} and 1×10^{-14} respectively. (11.70)

[**Hint:** The first step of hydrolysis, i.e., $S^{2-} + H_2O \rightleftharpoons HS^- + OH^-$ is predominant and hence K_2 value is used in the calculations.]

471. A voltaic cell whose cell reaction is

$$2Fe^{3+}\,(aq) + Zn\,(s) = 2Fe^{2+}\,(aq) + Zn^{2+}\,(aq)$$

has an e.m.f. of 0.72 V. What is the maximum electrical work that can be obtained from the cell per mole of Fe (II) ion? (69 kJ)

472. The dipole moment of HBr is 2.6×10^{-30} C m, and the interatomic spacing is 1.41 Å. What is the per cent ionic character of HBr? (11.5%)

473. How much AgBr would dissolve in 1 litre of 0.40 M NH_3? K_{sp} (AgBr) = 5×10^{-13}, K_d $[Ag(NH_3)_2^+] = 1 \times 10^{-8}$. ($2.83 \times 10^{-3}$ M)

474. An electron beam can undergo diffraction by crystals. Through what potential should a beam of electrons be accelerated so that its wavelength becomes equal to 1.54 Å? See values of m_e and h from the table. **(IIT 1997)** (63.5 V)

[**Hint:** See Example 28, Chapter 11.]

475. Calculate the wavenumber for the shortest wavelength transition in the Balmer series of atomic hydrogen. **(IIT 1996)** (27419 cm^{-1})

476. What is the solubility of CaF_2 in a buffer solution containing 0.45 M HCOOH and 0.20 M HCOONa? K_{sp} (CaF_2) = 3.4×10^{-11}, K (HF) = 6.8×10^{-4}, K (HCOOH) = 1.7×10^{-4}. (2.75×10^{-4} M)

[**Hint:** $2[Ca^{2+}] = [F^-] + [HF]$]

477. What is the solubility of MgF_2 in a buffer solution containing 0.45 M CH_3COOH and 0.20 M CH_3COONa? K_{sp} (MgF_2) = 6.5×10^{-9}, K (HF) = 6.8×10^{-4}, K (CH_3COOH) = 1.7×10^{-5}. (1.22×10^{-3} M)

478. The reaction

$$LaCl_3\,(s) + H_2O\,(g) \rightleftharpoons LaClO\,(s) + 2HCl\,(g)$$

is taking place in a closed container at a constant temperature. After the equilibrium is reached, more water vapour is added and the reaction shifted to a new equilibrium state at which the concentration of water vapour is found to be doubled. Calculate the factor by which the concentration of HCl is increased at the second equilibrium state. ($\sqrt{2}$)

479. ^{227}Ac has a half-life of 22 years with respect to radioactive decay. The decay follows two parallel paths, one leading to ^{227}Th and the other to ^{223}Fr. The percentage yields of these two daughter nuclides are 2.0 and 98.0 respectively. What are the decay constants (λ) for each of the separate paths?

(IIT 1996) (6.302×10^{-4} yr^{-1}, 3.088×10^{-2} yr^{-1})

[**Hint:** See Example 66, Chapter 11.]

480. A space capsule is filled with neon gas at 1 atm and 290 K. The gas effuses through a pinhole into outer space at such a rate that the pressure drops by 0.3 mm/second. If the capsule were filled with 30% He, 20% O_2 and 50% N_2 (mole %) at a total pressure of 1 atm and a temperature of 290 K, calculate the rate of pressure drop. (0.29 mm/s)

[**Hint:** Use Equation 10, Chapter 12.]

481. Find the equivalent weight of Br_2 in each of the following reactions:

(a) $Br_2 + 2e = 2Br^-$ or $5Br_2 + 10\,e = 10Br^-$

(b) $Br_2 + 12OH^- = 2BrO_3^- + 6H_2O + 10\,e$

(c) $6Br_2 + 12OH^- = 10Br^- + 2BrO_3^- + 6H_2O$

Equation (c) is the sum of equations (a) and (b).

What is the relationship between the answer to (c) and the answers to (a) and (b)?

(Br = 80.0) [(a) 80.0 (b) 16 (c) 96]

[**Note:** The equivalent weight of Br_2 in the overall reaction is the sum of that of the two half-reactions (for a species which disproportionates).]

482. Find the equivalent weight of $KMnO_4$ in the reaction:

$$Mn^{2+} + MnO_4^- + H_2O \longrightarrow MnO_2 + H^+ \text{ (unbalanced)}$$

What mass in g of $MnSO_4$ is oxidised by 1.25 g of $KMnO_4$? (52.7, 1.79 g)

[**Hint:** Eq. of $MnSO_4$ = Eq. of $KMnO_4$.]

483. A sample of radioactive material has an apparently constant activity of 2000 dis/min. By chemical means, the material is separated into two fractions, one of which has an initial activity of 1000 dis/min. The other fraction decays with a 24-hour half-life. Estimate the total activity in both samples 48 hours after the separation. Explain your estimate. (2000)

[**Hint:** The total activity when the samples are separated will be the same as the total activity when they are mixed, i.e., the mixing makes no difference to the activity.]

484. The time required for 10% completion of a first-order reaction at 298 K is equal to that required for its 25% completion at 308 K. If the pre-exponential factor for the reaction is 3.56×10^9 s^{-1}, calculate its rate constant at 318 K and also the energy of activation. **(IIT 1997)** (18.42 kcal, 9.46×10^{-4} s^{-1})

[Hint: $k_2/k_1 = \frac{2.303}{t}\log\frac{4}{3} \Big/ \frac{2.303}{t}\log\frac{10}{9}$; now use eqn. (12), Chapter 17 to calculate E and then Arrhenius's equation to calculate k (318 K).**]**

485. For a gaseous reaction 2B → A, the equilibrium constant K_p is ...... to/than K_c.
(IIT 1997) (less)

486. When an aqueous solution of sodium fluoride is electrolysed, the gas liberated at the anode is .... **(IIT 1997)** (O_2)

487. When 10 mL of ethanol of density 0.7893 g/L is mixed with 20 mL of water of density 0.9971 g/L at 25°C, the final solution has a density of 0.9571 g/L. Calculate the percentage change in total volume on mixing. (3.05%)

488. In a Cu-voltameter, mass deposited in 30 seconds is m gram. If the time-current graph is as shown in the figure, calculate the electrochemical equivalent of Cu.

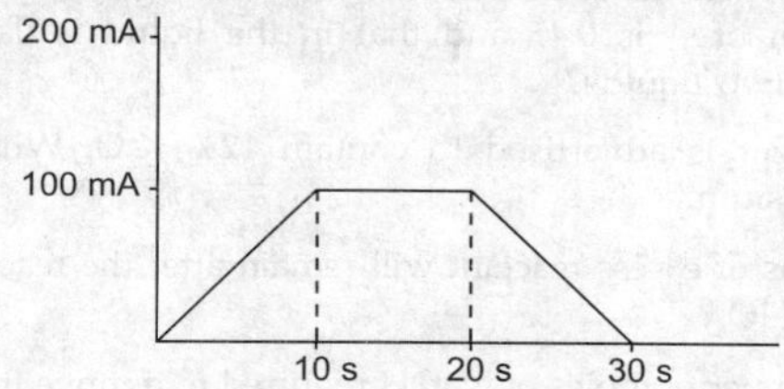

$[\frac{m}{2}]$

489. A litre of CO_2 at 15°C and 1-atm pressure dissolves in 1 litre of water at the same temperature when the pressure of CO_2 is 1 atm. Calculate the molal concentration of CO_2 in a solution over which the partial pressure of CO_2 is 150 mmHg. (0.0083 m)

[Hint: $m_{CO_2} \propto p_{CO_2}$**]**

490. The voltage of the cell,

$$Pb\ (s)\,|\,PbSO_4\ (s)\,|\,NaHSO_4\ (0.6\ M)\ ||\ Pb^{2+}\ (2.5 \times 10^{-5})\,|\,Pb\ (s)$$

is 0.061 V. Calculate $K_2 = \frac{[H^+][SO_4^{2-}]}{[HSO_4^-]}$.

Given $PbSO_4\ (s) + 2e = Pb\ (s) + SO_4^{2-}$; $E^0 = -0.356$ V

$Pb^{2+} + 2e = Pb\ (s)$; $E^0 = -0.126$ V (9.7×10^{-3})

491. 0.75 g of solid benzoic acid was placed in a 0.5-litre pressurised reaction vessel filled with O_2 at 10-atm pressure and 25°C. To the extent of availability of O_2, the acid burned to give CO_2 and H_2O. What were the final mole fractions of CO_2 and H_2O vapour in the resulting gas mixture brought to the initial temperature? The vapour pressure of water at 25°C is 23.8 torr. Neglect the volume occupied by nonaqueous substances and the solubility of CO_2 in H_2O.
(CO_2 0.213, H_2O 0.0033)

[Hint: Benzoic acid is the limiting reactant.**]**

492. A silent electric discharge was passed through 100 mL of air when 95 mL of ozonised air was formed. The ozonised air took 48.7 seconds to diffuse through a very small hole. If 100 mL of air diffused through the same hole under the identical conditions, it took 50 seconds. Find the molecular weight of ozone assuming air to contain 79% N_2 and 21% O_2. (48)

[**Hint:** Composition of 95 mL of ozonised air is N_2 79 mL, O_3 10 mL and O_2 6 mL.]

493. An element forms two oxides, the per cent composition in them A : O = $x:y$ in the first oxide and $y:x$ in the second oxide. If the equivalent weight of A in the first oxide is 10.33, what is the equivalent weight of A in the second oxide? (6.19)

494. For the nonequilibrium process A + B → Products, the rate is first-order w.r.t. A and second-order w.r.t. B. If one mole each of A and B were introduced into a 1-litre vessel, and the initial rate were 1×10^{-2} mol/litre s, calculate the rate when half the reactants have been turned into products. (1.2×10^{-3} mol/L s)

495. A solution of the two liquids A and B obeys Raoult's law. At a certain temperature, it is found that when the total pressure above the given solution is 400 mmHg, the mole fraction of A is 0.45 and that in the liquid is 0.65, what are vapour pressures of the two liquids? (277 mm, 629 mm)

496. A certain fertilizer is advertised to contain 12% K_2O. What percentage of the fertilizer is potassium? (9.96%)

497. How many grams of excess reactant will remain after the reaction of 12.5 g of CaO and 75.0 g of $HClO_4$? (30.4 g)

498. Calculate the number of moles of NaOH required to remove the SO_2 from 10 metric tons of atmosphere if the SO_2 is 0.1% by mass. 1 metric ton = 1.0×10^6 g.

$2NaOH + SO_2 \rightarrow Na_2SO_3 + H_2O$ (312 mole)

499. How many sandwiches, each containing 1 slice of cheese and 2 slices of bread, can you make with 30 slices of bread and 20 slices of cheese? Which is in the limiting quantity? (15, bread)

500. In certain areas where coal is cheap, artificial gas is produced for household use by the 'water gas' reaction

$$C\,(s) + H_2O\,(g) \xrightarrow{600°C} H_2\,(g) + CO\,(g)$$

Assuming that coke is 100% carbon, calculate the maximum heat obtainable at 298 K from the combustion of 1.0 kg of coke and compare this value to the maximum heat obtainable at 298 K from burning the water gas produced from 1.0 kg of coke.

Heat of combustion of C, H_2 and CO are −94.1, −68.4 and −68.0 kcal/mole respectively. (More energy is obtainable from the water gas, 11366 kcal)

501. Calculate the concentration of all the ions in solution if 1 mole of HCl and 2 moles of NaCl are dissolved in sufficient water to make 6 litres of a single solution. (0.17 M H^+, 0.50 M Cl^-, 0.33 M Na^+)

502. Calculate the final concentration of all ions in solution after 2 litres of 1.3 M $Ba(OH)_2$ is treated with 3 litres of 2.0 M HCl. (0.16 M H^+, 0.52 M Ba^{2+}, 1.2 M Cl^-)

503. What is the meaning of a positive sign for (a) a cell potential, and (b) a half-cell potential? [(a) The reaction can proceed as written. (b) Nothing]

504. A solution of silver benzoate has a pH of 8.63. K_a (C_6H_5COOH) = 6.5×10^{-5}. Calculate the value of K_{sp} for silver benzoate. (1.4×10^{-2})

[**Hint:** Use $pH = \frac{1}{2}(pK_w + pK_a + \log C)$.]

505. The rate constant for the first-order decomposition of a certain reaction is described by the equation

$$\log k\,(s^{-1}) = 14.34 - \frac{1.25 \times 10^4\,K}{T}$$

(i) What is the energy of activation for this reaction?

(ii) At what temperature will its half-life period be 256 minutes?

(239.34 kJ mole^{-1}, 669K)

[**Hint:** See Example 51, Chapter 17.]

506. A compound of vanadium has a magnetic moment of 1.73 BM. Work out the electronic configuration of the vanadium ion in the compound. (V = 23)

$\left[\mu = \sqrt{n(n+2)}\text{ BM, for } \mu = 1.73\text{ BM}, n = 1, \; V^{4+}: 1s^2\,2s^2\,2p^6\,3s^2\,3p^6\,3d^1\right]$

507. The decomposition of N_2O into N_2 and O in the presence of gaseous argon follows second-order kinetics, with

$$k = (5.0 \times 10^{11}\text{ L mol}^{-1}\text{ s}^{-1})\,e^{-29000k/T}$$

What is the energy of activation of this reaction? (241 kJ/mol)

[**Hint:** Compare the given equation with Arrhenius's equation.]

508. K_p for the reaction

$$N_2O_4\,(g) = 2NO_2\,(g)$$

is 0.66 at 46°C. Calculate the per cent dissociation of N_2O_4 at 46°C and a total pressure of 380 mm. What are the partial pressures of N_2O_4 and NO_2 at equilibrium? (0.168 atm, 0.332 atm)

509. An excess of liquid mercury is added to an acidified solution of 1×10^{-3} M Fe^{3+}. It is found that 5% of Fe^{3+} remains at equilibrium at 25°C. Calculate $E^0_{Hg_2^{2+}, Hg}$, assuming that the only reaction that occurs is

$$2Hg + 2Fe^{3+} \longrightarrow Hg_2^{2+} + 2Fe^{2+}$$

Given $E^0_{Fe^{3+}, Fe^{2+}} = 0.77$ V. (0.792 V)

[**Hint:** See Example 28, Chapter 18.]

510. Calculate the pH of an aqueous solution of 1.0 M ammonium formate assuming complete dissociation. pK_a of formic acid = 3.8 and pK_b of ammonia = 4.8. (6.5)

511. At 380°C, the half-life period for the first-order decomposition of H_2O_2 is 360 minutes. The energy of activation of the reaction is 200 kJ mol^{-1}. Calculate the time required for 75% decomposition at 450°C. (20.34 min)

512. $^{224}_{88}Ra$ having $t_{1/2}$ = 3.64d emits an α particle to form $^{220}_{86}Rn$, which has $t_{1/2}$ = 54.5 s. Given that the molar volume of radon under these conditions is

35.2 dm^3, what volume of radon is in secular equilibrium with 1 g of radium?
(2.72×10^{-8} m^3)

513. Find the ratio of the mass needed to generate 1 μCi of $^{226}_{88}Ra$ ($t_{1/2}$ = 1622 yr) to that for 1 μCi of $^{222}_{86}Rn$ ($t_{1/2}$ = 3.825 d). (1.58×10^5)

514. The composition of the equilibrium mixture $Cl_2 \rightleftharpoons 2Cl$, which is attained at 1200°C, is determined by measuring the rate of effusion through a pinhole. It is observed that at 1.80 mmHg pressure, the mixture effuses 1.16 times as fast as krypton effuses under the same conditions. Calculate the fraction of chlorine molecules dissociated into atoms. (At. wt. of Kr = 84) **(IIT 1995)** (0.14)

(See Example 33, Chapter 12.)

515. A 5-cc solution of H_2O_2 liberates 0.508 g of iodine from an acidified KI solution. Calculate the strength of H_2O_2 solution in terms of volume strength at STP.
(IIT 1995) (0.8 V)

[Hint: See examples 25 and 29, Chapter 7.]

516. A 20-cc mixture of CO, CH_4 and He gases is exploded by an electric discharge at room temperature with excess of oxygen. The volume contraction is found to be 13 cc. A further contraction of 14 cc occurs when the residual gas is treated with KOH solution. Find out the composition of the gaseous mixture in terms of volume percentage. **(IIT 1995)** (50% 20%, 30%)

[Hint: See Example 19, Chapter 3.]

517. What is the pH of a 0.50 M aqueous NaCN solution? pK_b of CN^- is 4.70.
(IIT 1996) (11.5)

518. A unit cell of sodium chloride has four formula units. The edge length of the unit cell is 0.564 nm. What is the density of sodium chloride?
(IIT 1997) (2.16×10^3 kg m^{-3})

[Hint: See Example 9, Chapter 20.]

519. Electrolysis of a solution of $MnSO_4$ in aqueous sulphuric acid is a method for the preparation of MnO_2 as per the reaction

$$Mn^{2+} (aq) + 2H_2O \longrightarrow MnO_2 (s) + 2H^+ (aq) + H_2 (g)$$

Passing a current of 27 A for 24 hours gives one kg of MnO_2. What is the value of current efficiency? Write the reactions taking place at the cathode and at the anode. **(IIT 1997)** (95.09%)

520. How many equivalents are there in 98 g of H_2SO_4 in the following reaction?

$$8H^+ + H_2SO_4 + 4Zn \longrightarrow H_2S + 4Zn^{2+} + 4H_2O$$

Also, find out the equivalent weight of H_2SO_4 in this reaction. (8 eq., 12.25)

521. What is the molarity of 4N H_2SO_4 in the following reaction?

$$8H^+ + H_2SO_4 + 8e^- \longrightarrow H_2S + 4H_2$$ (0.5 M)

522. Show that the ratio $t_{\frac{1}{2}}/t_{\frac{3}{4}}$ for an *n*th-order reaction is a function of *n* alone.

523. An aqueous solution contains 10% ammonia by mass and has a density of 0.99 g/cc.

Calculate the hydroxyl and hydrogen-ion concentration in this solution. $K_a\,(NH_4^+) = 5\times10^{-10}$ M. $(1.08\times10^{-2},\ 9.28\times10^{-13}$ M)

524. 0.15 mole of pyridinium chloride has been added into 500 cc of 0.2 M pyridine solution. Calculate the pH and hydroxyl-ion concentration in this resulting solution assuming no change in volume. K_b for pyridine $= 1.5\times10^{-9}$ M. $(5,\ 10^{-9}$ M)

525. Calculate the amount of ice that will separate out on cooling a solution containing 50 g of ethylene glycol in 200 g water to −9.3°C. K_f for water = 1.86 K · mol^{-1} · kg (38.71 g)

[Hint: See Example 41, Chapter 13.]

526. Calculate the equilibrium constant for the reaction

$$Fe^{2+} + Ce^{4+} \rightleftharpoons Fe^{3+} + Ce^{3+}$$

Given $E^0_{Ce^{4+}/Ce^{3+}} = 1.44$ V; $E^0_{Fe^{3+}/Fe^{2+}} = 0.68$ V. **(IIT 1997)** (7.6×10^{12})

[Hint: See Example 12, Chapter 18.]

527. Chromium metal crystallizes with a body-centred cubic lattice. The length of the unit-cell edge is found to be 287 pm. Calculate the atomic radius. What would be the density of chromium in g/cm^3? **(IIT 1997)**

[Hint: See Example 8, Chapter 20.] (124.27 pm, 7.32 g/cm^3)

528. Anhydrous $AlCl_3$ is covalent. From the data given below, predict whether it would remain covalent or become ionic in aqueous solution. (Ionization energy for Al = 5137 kJ mole^{-1}; $\Delta H_{hydration}$ for Al^{3+} = −4665 kJ mole^{-1}; $\Delta H_{hydration}$ for Cl^- = −381 kJ mole^{-1}). **(IIT 1997)**

[Hint: Total energy evolved due to hydration = −4665 − 3(−381) = −5808 kJ/mole. As this released energy is greater than ionization energy (5137 kJ/mole) of Al, $AlCl_3$ can be ionic in aqueous solution.]

529. Write a balanced equation for the reaction of ^{14}N with α particle.

$({}^{14}_{7}N + {}^{4}_{2}He \longrightarrow {}^{17}_{8}O + {}^{1}_{1}H)$

530. Isotopes of oxygen with mass number less than 16 undergo β^+ emission. Assuming an equimolar mixture of ^{14}O and ^{15}O, find the ratio of the nuclides at the end of one hour. Given that $t_{1/2}\,(^{14}O) = 71$ s and $t_{1/2}\,(^{15}O) = 124$ s. At what time will the above said ratio be equal to 0.25? $(3.29\times10^{-7}$, 332 s)

[Hint: Use Equation 25, Chapter 11.]

531. What is the minimum half-life of an isotope needed so that not more than 0.1% of the nuclei undergo decay during a 3.0-hour laboratory period? (88 days)

532. Calculate the percentage of hydrolysis in 0.003 M aqueous solution of NaOCN. K_a for HOCN $= 3.33\times10^{-4}$ M. (0.01%)

[Hint: $OCN^- + H_2O \rightleftharpoons HOCN + OH^-$]

533. A sample of ^{238}U (half-life $= 4.5\times10^9$ yr) ore is found to contain 23.8 g of ^{238}U and 20.6 g of ^{206}Pb. Calculate the age of the ore. $(4.489\times10^9$ years)

[Hint: See Example 48, Chapter 11.]

534. An aqueous solution of aniline of concentration 0.24 M is prepared. What concentration of sodium hydroxide is needed in this solution so that anilinium ion concentration remains at 1×10^{-8} M? K_a for $C_6H_5NH_3^+ = 2.4 \times 10^{-5}$ M. (1×10^{-2} M)

535. 20% of N_2O_4 molecules are dissociated in a sample of gas at 27°C and 760 torr. Calculate the density of the equilibrium mixture. (3.12 g/L)

536. The average molar mass of the vapour above solid NH_4Cl is nearly 26.75 g mole^{-1}. What is the composition (by wt.) of this vapour? (NH_3 31.8%, HCl 68.2%)

537. When 12 g of carbon reacted with oxygen to form CO and CO_2 at 25°C and constant pressure, 75 kcal of heat was liberated and no carbon remained. Calculate the mass of oxygen which reacted. Given,

$$C + O_2 \longrightarrow CO_2\ ; \quad \Delta H = -94.05 \text{ kcal/mole}$$

$$C + \frac{1}{2}O_2 \longrightarrow CO\ ; \quad \Delta H = -26.41 \text{ kcal/mole}$$

(27.5 g)

538. A proposed mechanism for the catalysed decomposition of aqueous H_2O_2 is

$$H_2O_2 + I^- \xrightarrow{k_1} H_2O + IO^-$$

$$H_2O_2 + IO^- \xrightarrow{k_2} H_2O + O_2 + I^-,$$

where $k_2 >> k_1$. Derive the rate law for the reaction. $\left(-\frac{d[H_2O_2]}{dt} = k_1\,[H_2O_2]\,[I^-]\right)$

539. Prepare a reaction-coordinate diagram for the reaction

$$2H_2O_2 \longrightarrow 2H_2O_2 + O_2\ ; \quad \Delta H = -191.78 \text{ kJ}$$

in which energy of activation for uncatalysed and catalysed reactions are 75.3 kJ mole^{-1} and 56.6 kJ mole^{-1} respectively at 298 K.

(a) What is the ratio of the rate constant for the forward catalysed reaction to that for the forward uncatalysed reaction?

(b) By what factor will the rate constant for the reverse catalysed reaction increase compared to that for the reverse uncatalysed reaction?

(c) Hence prove that the catalyst increases both the forward and reverse reactions by the same factor. [(a) 1970 (b) 1970]

540. A weak base BOH was titrated against a strong acid. The pH at one-fourth equivalence point was 9.24. Enough strong base was now added (6 m.e.) to completely convert the salt. The total volume was 50 mL. Find the pH at this point.

[**Hint:** $(14 - 9.24) = pK_b + \log\frac{(1/4)}{(3/4)}$; cal K_b (11.2)

6 m.e. of the strong base, added, is used to convert the salt to the weak base. Thus before the addition of the strong base, m.e. of the salt and the base were 6 and 18 respectively. As 6 m.e. of the strong base shall combine with the same number of m.e. of the salt to produce 6 m.e. of BOH, total m.e. of BOH = 6 + 18 = 24 and thus molarity $= \frac{24}{50}$ M. Now using K_b value, calculate the pH.]

541. An aqueous solution containing 0.10 g $KClO_3$ (formula weight = 214.0) was treated with an excess of KI solution. The solution was acidified with HCl. The liberated I_2 consumed 45.0 mL of thiosulphate solution to decolourise the blue starch-iodine complex. Calculate the molarity of the sodium thiosulphate solution.

(IIT 1998) (0.062 M)

[**Hint:** $KIO_3 + 5\ KI = 3K_2O + 3I_2$]

542. Calculate the equilibrium constant for the reaction $2Fe^{3+} + 3I^- \rightleftharpoons 2Fe^{2+} + I_3^-$. The standard reduction potentials in acidic-medium conditions are 0.77 and 0.54 V respectively for Fe^{3+}/Fe^{2+} and I_3^-/I^- couples. **(IIT 1998)** (6.07×10^7)

[**Hint:** Apply Equation 2, Chapter 18, $n = 2$]

543. From the following data, calculate the enthalpy change for the combustion of cyclopropane at 298 K. The enthalpy of formation of CO_2(g), H_2O(l) and propene (g) are −393.5, −285.8 and 20.42 kJ mol^{-1} respectively. The enthalpy of isomerisation of cyclopropane to propene is −33.0 kJ mol^{-1}. **(IIT 1998)** (−2091.32 kJ)

544. The degree of dissociation is 0.4 at 400 K and 1 atm for the gaseous reaction

$$PCl_5 \rightleftharpoons PCl_3 + Cl_2.$$

Assuming ideal behaviour of all the gases, calculate the density of equilibrium mixture at 400 K and 1 atm. **(IIT 1998)** (4.535 g/L)

[**Hint:** Apply $p = \frac{dRT}{M_{mix}}$]

545. Given: $Ag(NH_3)_2^+ \rightleftharpoons Ag^+ + 2NH_3$, $K_c = 6.2 \times 10^{-8}$ and K_{sp} of $AgCl = 1.8 \times 10^{-10}$ at 298 K. Calculate the concentration of the complex in 1.0 M aqueous ammonia.

(IIT 1998) (0.054 M)

[**Hint:** Let the concentration of the complex, $[Ag(NH_3)_2]Cl$, be x mole per litre in 1 M NH_3 and concentration of Ag^+ be y mole/litre

$$Ag(NH_3)_2^+ \rightleftharpoons Ag^+ + 2NH_3;\ K_c = 6.2 \times 10^{-8}$$

At eqb.: x $\quad y \quad 2y + 1 \approx 1$

$$AgCl \rightleftharpoons Ag^+ + Cl^-;\ K_{sp} = 1.8 \times 10^{-10}$$

At eqb.: $y \quad x$

546. A solution of a nonvolatile solute in water freezes at −0.30°C. The vapour pressure of pure water at 298 K is 23.51 mmHg and K_f for water is 1.86 degree/molal. Calculate the vapour pressure of this solution at 298 K. **(IIT 1998)** (23.44 mm)

547. For the reaction, $N_2O_5(g) \rightleftharpoons 2NO_2(g) + 0.5\ O_2(g)$, calculate the mole fraction of $N_2O_5(g)$ decomposed at a constant volume and temperature, if the initial pressure is 600 mmHg and the pressure at any time is 960 mmHg. Assume ideal gas behaviour. **(IIT 1998)** (0.25)

[**Hint:** If 'p' mm of N_2O_5 decomposes then $600 - p + 2p + \frac{p}{2} = 960$]

548. Find the solubility product of a saturated solution of Ag_2CrO_4 in water at 298 K if the emf of the cell

$$Ag \mid Ag^+ \text{ (satd. } Ag_2CrO_4 \text{ soln.)} \parallel Ag^+(0.1\text{ M}) \mid Ag$$

is 0.164 V at 298 K. **(IIT 1998)** (2.44×10^{-12})

549. What will be the resultant pH when 200 mL of an aqueous solution of

HCl(pH = 2.0) is mixed with 300 mL of an aqueous solution of NaOH (pH = 12)?
(IIT 1998) (11.3)

550. The rate constant of a reaction is $1.5 \times 10^7\ s^{-1}$ at 50°C and $4.5 \times 10^7\ s^{-1}$ at 100°C. Evaluate the Arrhenius parameters A and E_a.
(IIT 1998) (2.19×10^4 J/mol, $5.4 \times 10^{10}\ s^{-1}$)

[**Hint:** Apply Arrhenius's equation.]

551. How many millilitres of 0.5 M H_2SO_4 are needed to dissolve 0.5 g of copper (II) carbonate?
(IIT 1999) (8.09 mL)

552. Nitrobenzene is formed as the major product along with a minor product in the reaction of benzene with a hot mixture of HNO_3 and H_2SO_4. The minor product consists of C: 42.86%, H: 2.40%, N: 16.67% and O: 38.07%. (i) Calculate the empirical formula of the minor product, (ii) when 5.5 g of the minor product is dissolved in 45 g of benzene, the b.p. of the solution is 1.84°C higher than that of pure benzene. Calculate the molecular weight of the minor product and determine its molecular and structural formula. $K_b(C_6H_6) = 2.53$ K kg mol^{-1}.
(IIT 1999) ($C_3H_2NO_2$, 168, *m*-dinitrobenzene)

553. A plant virus is found to consist of uniform cylindrical particles of 150 Å in diameter and 5000 Å long. The specific volume of the virus is 0.75 cm^3/g. If the virus is considered to be a single particle, find its molecular weight.
(IIT 1999) (7.0939×10^7)

[**Hints:** Mol wt. = Mass of 1 molecule × Av. constant.]

554. When 3.06 g of solid NH_4HS is introduced into a two-litre evacuated flask at 27°C, 30% of the solid decomposes into gaseous NH_3 and H_2S (i) Calculate K_c and K_p for the reaction at 27°C. (ii) What would happen to the equilibrium when more solid NH_4HS is introduced into the flask?
(IIT 1999) [(i) 8.1×10^{-5} mole/L, 0.049 atm^{-2}, (ii) No effect]

555. One mole of nitrogen gas at 0.8 atm takes 38 s to diffuse through a pinhole, whereas one mole of an unknown compound of xenon with fluorine at 1.6 atm takes 57 s to diffuse through the same hole. Calculate the molecular formula of the compound. (Xe = 131, F = 19)
(IIT 1999) (XeF_6)

556. The pressure exerted by 12 g of an ideal gas at temperature t°C in a vessel of volume V litre is one atm. When the temperature is increased by 10 degrees at the same volume, the pressure increases by 10%. Calculate the temperature t and volume V. (Mol wt. of the gas = 120)
(IIT 1999) (100 K, 0.821 L)

[**Hint:** Apply $pV = nRT$ twice]

557. A cell, $Ag \mid Ag^+ \mid\mid Cu^{2+} \mid Cu$, initially contains 1 M Ag^+ and 1 M Cu^{2+} ions. Calculate the change in the cell potential after the passage of 9.65 A of current for 1 h.
(IIT 1999) (0.1355 V)

[**Hint:** See examples 31 and 32, Chapter 18.]

558. The solubility of $Pb(OH)_2$ in water is 6.7×10^{-6} M. Calculate the solubility of $Pb(OH)_2$ in a buffer solution of pH = 8.
(IIT 1999) (1.2×10^{-3} M)

559. Estimate the average S-F bond energy in SF_6. The standard heat of formation values of $SF_6(g)$, S(g) and F(g) are: −1100, 275 and 80 kJ mol^{-1} respectively.
(IIT 1999) (309.6 kJ mol^{-1})

560. The rate constant for an isomerisation reaction $A \rightarrow B$ is 4.5×10^{-3} min^{-1}. If the initial concentration of A is 1 M, calculate the rate after 1 h.
[**Hint:** See Example 3, Chapter 17.] **(IIT 1999)** (3.43×10^{-3} M/min)

561. A metal cystallises into two cubic phases, f.c.c and b.c.c. whose unit-cell lengths are 3.5 and 3.0 Å. Calculate the ratio of densities of f.c.c. and b.c.c.
[**Hint:** Apply Equation (1), Chapter 20.] **(IIT 1999)** (1.259)

562. ${}^{238}_{92}U$ is radioactive and it emits α and β particles to form ${}^{206}_{82}Pb$. Calculate the number of α and β particles emitted in this conversion. An ore of ${}^{238}_{92}U$ is found to contain ${}^{238}_{92}U$ and ${}^{206}_{82}Pb$ in the weight ratio of 1 : 0.1. The half-life period ${}^{238}_{92}U$ is 4.5×10^9 years. Calculate the age of the ore. **(IIT 2000)** (8, 6, 7.098×10^8 years)
[**Hint:** See Example 36 and 40, Chapter 11.]

563. The average concentration of SO_2 in the atmosphere over a city on a certain day is 10 ppm, when the average temperature is 298 K. Given that the solubility of SO_2 in water at 298 K is 1.3653 mole lit^{-1} and the pK_a of H_2SO_3 is 1.92, estimate the pH of rain on that day. **(IIIT 2000)** (0.913)

[**Hint:** $[SO_2] = [H_2SO_3] = 1.3653$ M, $H_2SO_3 \rightleftharpoons H^+ + HSO_3^-; K_a = 1.2 \times 10^{-2}$]

564. Calculate the pressure exerted by one mole of CO_2 gas at 273 K if the van der Waals constant $a = 3.593$ dm^6 atm mol^{-2}. Assume that the volume occupied by CO_2 molecules is negligible. **(IIT 2000)** (0.9922 atm)

[**Hint:** Apply $\left(p + \frac{a}{V^2}\right) V = RT$; $V = 22.4$ dm^3 (suppose)]

565. The figures given below show the location of atoms in three crystallographic planes in an f.c.c. lattice. Draw the unit cell for the corresponding structure and identify these planes in your diagram. **(IIT 2000)**

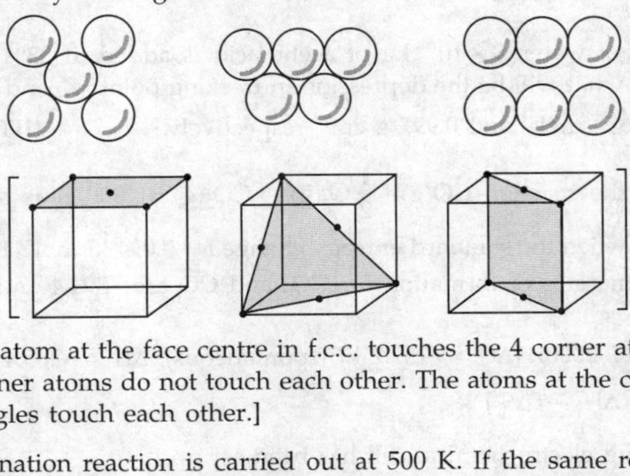

[**Hint:** The atom at the face centre in f.c.c. touches the 4 corner atoms on that face but the corner atoms do not touch each other. The atoms at the centre of the faces at right angles touch each other.]

566. A hydrogenation reaction is carried out at 500 K. If the same reaction is carried out in the presence of a catalyst at the same rate, the temperature required is 400 K.

Calculate the activation energy of the reaction if the catalyst lowers the activation barrier by 20 kJ mol^{-1}. **(IIT 2000)** (100 kJ mol^{-1})

[**Hint:** From Arrhenius's equation, under the given condition, we have,

$-\frac{E}{RT_1} = -\frac{E-20}{RT_2}$]

567. Copper sulphate solution (250 mL) was electrolysed using a platinum anode and copper cathode. A constant current of 2 mA was passed for 16 minutes. It was found that after electrolysis the absorbance of the solution was reduced to 50% of its original value. Calculate the concentration of copper sulphate in the solution to begin with. (7.958×10^{-5} M) **(IIT 2000)**

[**Hint:** Initial mole of $CuSO_4$ per 250 mL

$= 2 \times$ mol of $CuSO_4$ lost

$= 2 \times$ mol of Cu deposited]

568. Calculate the energy required to excite 1 litre of H_2 gas at 1 atm and 298 K to the first excited state of atomic hydrogen. The energy for the dissociation of H—H bonds is 436 kJ mol^{-1}. Rydberg constant for H = 109679 cm^{-1}, $h = 6.626 \times 10^{-34}$ Js and $C = 3 \times 10^8$ m s^{-1}. **(IIT 2000)** (98.19 kJ)

[**Hint:** Mole of H $= 2 \times$ mole $H_2 = 2 \times \frac{pV}{RT}$

Energy to excite 1 H atom $= h\nu = \frac{hc}{\lambda} = hcR\left(\frac{1}{n_1} - \frac{1}{n_2}\right)$.

Total energy = energy to break H—H bonds + energy to excite H atom]

569. A sample of argon gas at 1-atm pressure and 27°C expands reversibly and adiabatically from 1.25 dm^3 to 2.50 dm^3. Calculate the enthalpy change in this process. $C_{V,\mathrm{m}}$ for argon is 12.48 JK^{-1} mol^{-1}. **(IIT 2000)** (–117.6 J)

[**Hint:** $T_1 = 300$ K, cal. T_2 using $\frac{T_2}{T_1} = \left(\frac{V_1}{V_2}\right)^{R/C_V}$ then

$\Delta H = nC_p(T_2 - T_1) = n(C_V + R)(T_2 - T_1)$]

570. To 500 cm^3 of water, 3×10^{-3} kg of acetic acid is added. If 23% of acetic acid is dissociated, what will be the depression in freezing point? K_f and density of water are 1.86 K kg^{-1} mol^{-1} and 0.997 g cm^{-3} respectively. **(IIT 2000)** (0.228 K)

571. Show that the reaction $CO(g) + \frac{1}{2}O_2(g) \rightarrow CO_2(g)$ at 300 K is spontaneous and exothermic, when the standard entropy change is –0.094 kJ mol^{-1} K^{-1}. The standard Gibbs free energies of formation for CO_2 and CO are –394.4 and –137.2 kJ mol^{-1} respectively.

(IIT 2000) [$\Delta G^0 = -257.2$ kJ (spontaneous), $\Delta H^0 = -285.4$ kJ (exothermic)]

[**Hint:** $\Delta G^0 = \Delta H^0 - T\Delta S^0$]

572. The following electrochemical cell has been set up.

Pt(1) | Fe^{3+}, Fe^{2+}($a = 1$) || Ce^{4+}, Ce^{3+}($a = 1$) | Pt(2)

$E°(Fe^{3+}/Fe^{2+}) = 0.77$ V, $E°(Ce^{4+}/Ce^{3+}) = 1.61$ V

If an ammeter is connected between the two platinum electrodes, predict the direction of flow of current. Will the current increase or decrease with time?

(IIT 2000) (Right to left, decrease)

573. Diborane is a potential rocket fuel which undergoes combustion according to the reaction,

$$B_2H_6(g) + 3O_2(g) \rightarrow B_2O_3(s) + 3H_2O(g).$$

From the following data, calculate the enthalpy change for the combustion of diborane.

(i) $2B(s) + \frac{3}{2}O_2(g) \rightarrow B_2O_3(s); \Delta H = -1273$ kJ mol^{-1}

(ii) $H_2(g) + \frac{1}{2}O_2(g) \rightarrow H_2O(l); \Delta H = -286$ kJ mol^{-1}

(iii) $H_2O(l) \rightarrow H_2O(g); \Delta H = 44$ kJ mol^{-1}

(iv) $2B(s) + 3H_2(g) \rightarrow B_2H_6(g); \Delta H = 36$ kJ mol^{-1} **(IIT 2000)** (-2035 kJ mol^{-1})

[Hint: Apply inspection method: (i) + 3 (ii) + 3 (iii) − (iv)]

574. The following solutions were mixed: 500 mL of 0.01 M $AgNO_3$ and 500 mL of a solution that was both 0.01 M in NaCl and 0.01 M in NaBr. Calculate $[Ag^+]$, $[Cl^-]$ and $[Br^-]$ in the equilibrium solution.

$K_{sp}(AgCl) = 1.0 \times 10^{-10}$, $K_{sp}(AgBr) = 5 \times 10^{-13}$ (2.0×10^{-8} M, 0.005 M, 2.5×10^{-5} M)

575. 1.1 g $CH_3(CH_2)_nCOOH$ was burnt in excess of air and the resultant gases ($CO_2 + H_2O$) were passed through a solution of NaOH. The resulting solution is divided into two equal parts. One part required 75 m.e. of HCl for neutralisation using phenolphthalein as indicator. The other part required 100 m.e. of HCl using methyl orange as indicator. Find n. (2)

576. The Mn_3O_4 formed on strong heating of a sample of $MnSO_4.4H_2O$ was dissolved in 100 cm^3 of 0.1 N $FeSO_4$ containing dilute H_2SO_4. The resulting solution reacted completely with 50 cm^3 of $KMnO_4$ solution. 25 cm^3 of this $KMnO_4$ solution required 30 cm^3 of 0.1 N $FeSO_4$ solution for complete reaction. Calculate the amount of $MnSO_4.4H_2O$ in the sample. (1.338 g)

[Hint: See Example 55, Chapter 7.]

577. Determine the number of moles of AgI which may be dissolved in 1.0 litre of 1.0 M CN^- solution. K_{sp} for AgI and K_f for $[Ag(CN)_2]^-$ are 1.2×10^{-17} M^2 and 7.1×10^{19} M^{-2} respectively. (0.49 mole)

[Hint: $AgI \rightleftharpoons Ag^+ + I^-; K_{sp} = 1.2 \times 10^{-17}$

$Ag^+ + 2CN^- \rightleftharpoons Ag(CN)_2^-; K_f = 7.1 \times 10^{19}$

$\therefore$ $AgI + 2CN^- \rightleftharpoons Ag(CN)_2^- + I^-; K = K_{sp} \cdot K_f$

$(1 - 2x)$ x x (x-solubility of AgI)]

578. x g of a nonelectrolytic compound (molar mass = 200) are dissolved in 1.0 L of 0.05 M NaCl aqueous solution. The osmotic pressure of this solution is found to be 4.92 atm at 27°C. Calculate the value of x. Assume complete dissociation of NaCl and ideal behaviour of this solution. (19.9 g)

[Hint: $OP = \left(\frac{x}{200} + 2 \times 0.05\right) \times 0.0821 \times 300$]

579. 0.16 g of N_2H_4 are dissolved in water and the total volume made upto 500 mL. Calculate the percentage of N_2H_4 that has reacted with water in this solution. $K_b(N_2H_4) = 4.0 \times 10^{-6}$ M. (2%)

[Hint: $N_2H_4 + H_2O \rightarrow N_2H_5OH \rightleftharpoons N_2H_5^+ + OH^-$]

580. Determine the value of ΔE and ΔH for the reversible isothermal evaporation of 90.0 g of water at 100°C. Assume that water vapour behaves as an ideal gas and heat of evaporation of water is 540 cal g^{-1}. (44.87 kcal, 48.6 kcal)

581. 12.0 g of an impure sample of arsenious oxide was dissolved in water containing 7.5 g of sodium bicarbonate and the resulting solution was diluted to 250 mL. 25 mL of this solution was completely oxidised by 22.4 mL of a solution of iodine. 25 mL of this iodine solution reacted with same volume of a solution containing 24.8 g of $Na_2S_2O_3 \cdot 5H_2O$ in one litre. Calculate the percentage of arsenious oxide in the sample. (As = 75) (9.24%)

582. Two buffers, (X) and (Y) of pH 4.0 and 6.0 respectively, are prepared from acid HA and the salt NaA. Both the buffers are 0.50 M in HA. What would be the pH of the solution obtained by mixing equal volumes of the two buffers? $K_a(HA) = 1.0 \times 10^{-5}$. (5.70)

[Hint: Calculate [salt] in X and Y using Henderson equation. Find $[\text{salt}]_{mix}$ and again apply the same equation]

583. Calculate the value of $\log K_p$ for the reaction: $N_2(g) + 3H_2(g) \rightleftharpoons 2NH_3(g)$ at 25°C. The standard enthalpy of formation of $NH_3(g)$ is −46 kJ and standard entropies of N_2, H_2 and NH_3 gases are 191, 130, 192 JK^{-1} mol^{-1} respectively.
[Hint: Apply $\Delta G° = \Delta H° - T\Delta S°$ and $\Delta G° = -2.303RT \log K_p$] (5.845)

584. Determine the concentration of NH_3 solution, one litre of which can dissolve 0.10 mole AgCl. K_{sp} of AgCl and K_f of $Ag(NH_3)_2^+$ are 1.0×10^{-10} M^2 and 1.6×10^7 M^{-2} respectively. (2.7 M)

[Hint: 0.1 M Ag^+ combines with 0.2 M NH_3 to produce 0.1 M complex which then dissociates.

$AgCl \rightleftharpoons Ag^+ + Cl^-$; $K_{sp} = 1.0 \times 10^{-10}$
0.1 $\quad x \quad$ 0.1 M

$Ag(NH_3)_2^+ \rightleftharpoons Ag^+ + 2NH_3$; $K_d = \frac{1}{1.6 \times 10^7}$
$(0.1 - x) \quad x \quad (2x + y)$
$\approx 0.1 \quad\quad \approx y$

Total $[NH_3] = y + 0.2$].

585. At 817°C, K_p for the reaction between $CO_2(g)$ and excess hot graphite(s) is 10 atm
(a) What are equilibrium concentrations of the gases at 817°C and a total pressure of 5 atm.

(b) At what total pressure, the gas contains 5% CO_2 by volume?

[(a) 0.0167, 0.041 mole/L
(b) 0.554 atm]

586. 1.4 g of acetone dissolved in 100 g of benzene gave a solution which freezes at 277.12 K. Pure benzene freezes at 278.4 K. 2.8 g of a solid (A) dissolved in 100 g of benzene gave a solution which froze at 277.76 K. Calculate the molecular weight of (A). (232)

587. The rate law of the reaction is given as $2A + B \rightarrow$ product

$$\text{Rate} = k[A]^2[B]$$

$[A]_{Initial}$	$[B]_{Initial}$	$t_{1/2}(s)$
3.0×10^{-4} M	4.0×10^{-5} M	60
3.0×10^{-4} M	6.0×10^{-5} M	x

Find x. (60 s)

[**Hint:** In a reaction with more than one reactant, $t_{1/2}$ of the limiting reactant is $t_{1/2}$ of the reaction.]

588. A 0.025-g sample of a compound that is composed of B and H, has a molecular mass of about 28 amu and burns spontaneously when exposed to air, producing 0.063 g of B_2O_3. Find the molecular formula of the compound. (B_2H_6)

589. Excited hydrogen atoms with very large radii have been detected. How large is an H atom with an electron characterised by a quantum number of 106? How many times larger is that than the radius of an H atom in its ground state?

[11236 times larger (106^2)]

590. One molecule of haemoglobin will combine with four molecules of oxygen. If 1.0 g of haemoglobin combines with 1.53 mL of O_2 at body temperature (37°C) and a pressure of 743 torr, what is the molar mass of haemoglobin? (6.8×10^4 g mol^{-1})

591. What is the half-life for the decomposition of NOCl when the concentration of NOCl is 0.15 M? The rate constant for the reaction is 8.0×10^{-8} L mol^{-1} s^{-1}. (8.3×10^7 s)

592. If both the functional groups of salicylic acid, HOC_6H_4COOH, ionise in water, with $K_a = 1 \times 10^{-3}$ for the –COOH group and 4.2×10^{-13} for the –OH group, calculate pH of the saturated solution of the acid (solubility = 1.8 g/L)? (2.45)

593. The density of trifluoroacetic acid vapour was determined at 117°C and 470 mm and found to be 2.784 g/L. Calculate K_c for (17.09)

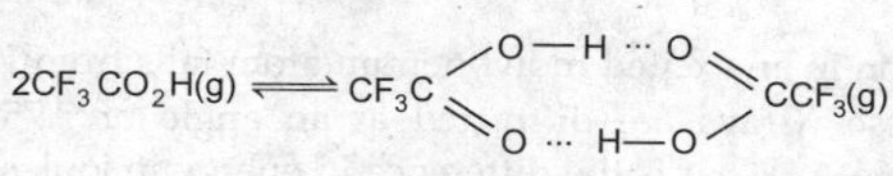

594. The following equilibria exist simultaneously in a vessel.

$$2NO_2(g) \rightleftharpoons N_2O_4(g);\ K_{p_1} = 6.8\ \text{atm}^{-1}$$

and $$NO(g) + NO_2(g) \rightleftharpoons N_2O_3(g);\ K_{p_2}\ \text{(say)}$$

If initially only NO and NO_2 are present in 1 : 2 mole ratio and total pressure at equilibrium is 5.05 atm and the partial pressure of N_2O_4 is 1.7 atm, calculate the equilibrium partial pressure of NO and K_{p_2}. (1.05 atm, 3.43 atm^{-1})

[Hint: See Example 51, Chapter 15]

595. In a reaction of the type

$$A(s) + 2B(g) \rightleftharpoons C(g) + D(g),$$

the equilibrium concentrations of A, B, C and D are 1, 2, 5 and 6 mol/litre respectively. Argon is then introduced at equilibrium at constant volume. Calculate the concentrations of A, B, C and D at the new equilibrium position.

(Eqb. concs. do not change)

596. The rate law of the reaction given below is given as $2A + B \rightarrow$ product

$$\text{Rate} = k[A]^2\,[B]$$

$[A]_{Initial}$	$[B]_{Initial}$	$t_{1/2}$ (s)
5.0×10^{-6} M	3.0×10^{-4} M	400
10.0×10^{-6} M	3.0×10^{-4} M	x

Find x. (200 s)

[Hint: In a reaction with more than one reactant, $t_{1/2}$ of the limiting reactant gives the $t_{1/2}$ of the reaction.]

597. (a) Nitric acid is prepared from ammonia in a three-step process.

(i) $4NH_3 + 5O_2 \rightarrow 4NO + 6H_2O$ (fast)

(ii) $2NO + O_2 \rightarrow 2NO_2$ (slow)

(iii) $3NO_2 + H_2O \rightarrow 2HNO_3 + NO$ (fast)

Calculate how much HNO_3 can be produced from 10^5 kg of ammonia assuming 100% efficiency in each of the reactions.

(b) If equation (ii) is second-order in NO and first-order in O_2, calculate the rate of formation of HNO_3 when oxygen concentration is 0.50 M and the nitric oxide concentration is 0.75 M. $k = 5.8 \times 10^2\ L^6\ mol^{-2}\ s^{-1}$.

[(a) 2.47×10^5 kg (b) 1.63×10^{-6}]

598. The reaction of WCl_6 with Al at about 400°C gives black crystals of a compound that contains only tungsten and chlorine. A sample of this compound, when reduced with hydrogen gives 0.2232 g of tungsten metal and hydrogen chloride, which is absorbed in water. Titration of the hydrochloric acid thus produced required 46.2 mL of 0.1051-M NaOH to reach the end point. What is the empirical formula of the black tungsten chloride? (WCl_4)

599. When an electron in an excited molybdenum atom falls from L to K shell, an X-ray is emitted. These X-rays are diffracted at an angle of 7.75° by planes with a separation of 2.64 Å. What is the difference in energy in joules between the K shell and the L shell in Mo assuming a first-order diffraction? (2.79×10^{-15} J)

600. A bottle of milk stored at 300 K sours in 36 hours. When stored in a refrigerator at 275 K it sours in 360 hours. Calculate the energy of activation of the reaction involved in the souring process. (63.18 kJ/mol)

601. The rate constant for the first-order decomposition at 45°C of N_2O_5, dissolved in chloroform, is 6.2×10^{-4} min^{-1}. $2N_2O_5 \rightarrow 4NO_2 + O_2$

(a) What is the rate of decomposition when $[N_2O_5] = 0.40$ M?

(b) What are the rates of formation of NO_2 and of O_2 when $[N_2O_5] = 0.40$ M?

[(a) 2.5×10^{-4} (b) 5×10^{-4}, 1.2×10^{-4}]

602. A balloon weighing 50 kg has a radius of 10 m. What will be its payload if it is filled with He at 1 atm and 25°C. Density of air = 1.22 kg m^{-3}. Also calculate its payload if H_2 is filled in place of He. (4893.8 kg, 4715.4 kg)

[**Hint:** $pV = \frac{w}{M} RT$; w – wt. of He in g

$$\frac{\text{Wt. of balloon} + w'}{\text{Volume of balloon}} = \text{density of air}$$

w'–wt. of He for the balloon to just lift from the ground.

Payload = $(w' - w)$]

603. Calculate the coefficient of viscosity of CO_2 at 300 K on the basis of kinetic theory of gases. Molecular diameter for $CO_2 = 3.64 \times 10^{-10}$ m. (2.37×10^{-4} poise)

604. Show that the excluded volume is four times the actual volume of a molecule.

[**Hint:** Excluded volume per two molecules $= \frac{4}{3}\pi(2r)^3$]

605. A gas dimerizes to a small extent as

$$2A(g) = A_2(g)$$

Show that to a first approximation

$$\frac{pV}{RT} = 1 - \frac{K_c}{V}.$$

606. A vessel contains three gases A, B and C in the equilibrium

$$A \rightleftharpoons 2B + C$$

At equilibrium, the concentration of A was 3 M and that of B was 4 M. On doubling the volume of the vessel, the new equilibrium concentration of B was 3 M. Calculate K_c and the initial equilibrium concentration of C. (28.8, 5.4)

607. The average velocity of the molecules of a gas is 400 m/s. Calculate its rms velocity at the same temperature. (434.26 m/ s)

608. The wavelength of high energy transition of H-atoms is 91.2 nm. Calculate the corresponding wavelength of He atoms. (22.8 nm)

609. You are given marbles of diameter 10 mm. They are to be placed such that their centres lie in a square bound by four lines, each of length 40 mm. What will be the arrangements of marbles in a plane so that maximum number of marbles can be placed inside the area? Draw the diagram and derive expressions for the number of molecules per unit area.

[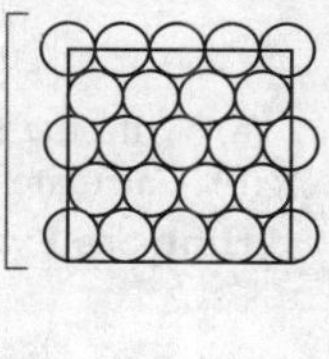]

610. 1 g of charcoal adsorbs 100 mL of 0.5 M CH_3COOH to form a monolayer, and thereby the molarity of CH_3COOH reduces to 0.49. Calculate the surface area of the charcoal adsorbed by each molecule of acetic acid. Surface area of charcoal $= 3.01 \times 10^2\ m^2/g$. ($5 \times 10^{-19}\ m^2/$ molecule)

[**Hint:** No. of molecules of CH_3COOH adsorbed $= 0.001 \times$ Av. constant]

611. Two students use the same stock solution of $ZnSO_4$ and a solution of $CuSO_4$. The emf of one cell is 0.03 V higher than the other. The concentration of $CuSO_4$ in the cell with higher emf value is 0.5 M. Find out the concentration of $CuSO_4$ in the other cell $\left(\frac{2.303\ RT}{F} = 0.06\right)$. (0.05 M)

612. Match the following if the molecular weights of X, Y and Z are the same.

	Boiling point	K_b
X	100	0.68
Y	27	0.53
Z	253	0.98

(Same as given)

613. The C_V value of He is always $3R/2$ but the C_V value of H_2 is $3R/2$ at low temperatures and $5R/2$ at moderate temperatures and more than $5R/2$ at higher temperatures. Explain. [Rotational and vibrational degrees of freedom for diatomic molecules contribute towards C_V at high temperature.]

614. At 298 K, the inversion of sucrose proceeds with constant half-life of 500 min at pH = 5 and with half-life of 50 min at pH = 4 for any concentration of sucrose. If the rate law for the reaction is given by

$$-\frac{d\,[\text{sucrose}]}{dt} = k\,[\text{sucrose}]^x\,[H^+]^y$$

find x and y. (1, 2)

615. Calculate the equilibrium pressure for the conversion of graphite to diamond at 25°C. The densities of graphite and diamond may be taken to be 2.25 and 3.51 g/cc respectively, independent of pressure. The change in ΔG with pressure is $-2900\ J\ mol^{-1}$. (1.52×10^9 Pa)

[**Hint:** $\Delta G = -S\Delta T + V\Delta p;\ \int_1^2 d\,\Delta G = \int_{p_1}^{p_2} \Delta V\,dp,\ p_2 = \frac{\Delta G_2 - \Delta G_1}{\Delta V} + p_1$]

616. For the given reaction: $A + B \rightarrow$ products, the following data were given.

Initial concentration (mol/L) [A]	Initial concentration (mol/L) [B]	Initial rate ($mol\ L^{-1}\ s^{-1}$)
0.1	0.1	0.05
0.2	0.1	0.1
0.1	0.2	0.05

(a) Write the rate equation.

(b) Calculate the rate constant. (Rate $= k[A][B]^0, k = 0.5\ s^{-1}$)

[**Hint:** See Example 27, Chapter 17.]

617. 100 mL of a liquid is contained in an isolated container at a pressure of 1 bar. The pressure is steeply increased to 100 bar. The volume of the liquid is decreased by 1 mL at this constant pressure. Find ΔH and ΔU. $(0, 10^{-2}\ \text{kJ})$

[Hint: $\Delta H = \Delta U + p\Delta V,\ \delta$

$\Delta H = 0, \Delta U = -p\Delta V$

$1\ \text{atm} \cdot \text{mL} = 10^{-4}\ \text{kJ}$]

618. The crystal AB (rock-salt structure) has molecular weight $6.023y$ amu, where y is an arbitrary number in amu. If the minimum distance between the cation and anion is $y^{1/3}$ nm and the observed density is $20\ \text{kg/m}^3$, find (a) density in kg/m^3, and (b) type of defect. $(5\ \text{kg/ m}^3$, metal excess)

[Hint: $\rho = \dfrac{zM}{NV};\ z = 4,\ V = (2\,y^{1/3} \times 10^{-9})^3\ \text{m}^3$

$$M = 6.023\,y \times 10^{-3}\ \text{kg/mol}$$

Observed density ($20\ \text{kg/m}^3$) is greater than calculated density.]

619. (a) The Schrödinger equation for the hydrogen atom is

$$\psi_{2S} = \frac{1}{4(2\pi)^{1/2}}\left(\frac{1}{a_0}\right)^{3/2}\left(2 - \frac{r}{a_0}\right)e^{-r/2a_0},$$

where a_0 is Bohr's radius. If the radial node in 2 s is at r then find r in terms of a_0.

(b) A baseball having a mass of 100 g moves with velocity 100 m/s. Find out the value of the wavelength of the baseball.

(c) ${}^{234}_{92}X \xrightarrow[-6\beta]{-7\alpha} Y$. Find out the atomic number and mass number of Y and identify it. [(a) $2a_0$ (b) 6.626×10^{-25} m (c) ${}^{206}_{84}\text{Po}$]

[Hint: (a) $\psi^2 = 0$ at node $\therefore \left(2 - \dfrac{r}{a_0}\right)$ has to be zero.

(b) Apply $\lambda = \dfrac{h}{mv}$ ·]

620. (a) In the following equilibrium

$$N_2O_4\,(g) \rightleftharpoons 2NO_2\,(g)$$

When 5 mole of each are taken, the temperature is kept at 298 K. The total pressure was found to be 20 bar. Given: $\Delta G_f^0(N_2O_4) = 100$ kJ and $\Delta G_f^0(NO_2) = 50$ kJ,

(i) find ΔG of the reaction, and

(ii) the direction of the reaction in which the equilibrium shifts.

(b) A graph is plotted for a real gas which follows van der Waals equation with pV_m taken on the y-axis and p on the x-axis. Find the intercept and the slope of the line. V_m is the molar volume.

$\left(\begin{array}{l}\text{(a) } +56.03\ \text{L atm, reverse} \\ \text{(b) } RT \text{ and } -\dfrac{a}{RT}\end{array}\right)$

[**Hint:** (a) $\underset{10\text{ bar}}{N_2O_4(g)} \rightleftharpoons \underset{10\text{ bar}}{2NO_2(g)}$ ... initial(1 bar ≈ 1 atm)

Reaction quotient $(Q) = \dfrac{10^2}{10} = 10$ bar

$\Delta G^0 = 2 \times 50 - 100 = 0$

Apply $\Delta G = \Delta G^0 + 2.303\, RT \log Q$.

(b) To determine the intercept and the slope, the pressure has to be very low tending to zero and in such a condition, the volume would be sufficiently large, and b in van der Waals equation could be neglected.

$$\left(p + \frac{a}{V^2}\right) V = RT$$

$$pV = -\frac{a}{RT} \cdot p + RT]$$

621. (a) 1.22 g C_6H_5COOH is added to two solvents.

(i) In 100 g CH_3COCH_3 : $\Delta T_b = 0.17$, $K_b = 1.7$ kg K mol^{-1}

(ii) In 100 g C_6H_6 : $\Delta T_b = 0.13$, $K_b = 2.6$ kg K mol^{-1}

Find out the molecular weight of C_6H_5COOH in both the solvents and interpret the result.

(b) If 0.1 M HA is titrated with 0.1 M NaOH, calculate the pH at the end point. $K_a(\text{HA}) = 5 \times 10^{-6}$ and $\alpha << 1$.

((a) (i) 122 (ii) 244, C_6H_5COOH dimerises in C_6H_6
(b) 9)

[**Hint:** (a) Apply $\Delta T_b = K_b \cdot m$ in both cases.

(b) Apply $\text{pH} = \frac{1}{2}\{\text{p}K_w + \text{p}K_a + \log a\}$, $a = \frac{0.1}{2}$ M]

622. Find the equilibrium constant for the reaction

$$Cu^{2+} + In^{2+} \rightleftharpoons Cu^+ + In^{3+}$$

given that, $E^0_{Cu^{2+}, Cu^+} = 0.15$ V

$E^0_{In^{2+}, In^+} = -0.4$ V

$E^0_{In^{3+}, In^+} = -0.42$ V (10^{10})

[**Hint:** See Example 11, Chapter 18]

623. Fill in the blanks.

(a) ${}^{235}_{92}U + {}^{1}_{0}n \rightarrow {}^{137}_{52}A + {}^{97}_{40}B + \ldots$

(b) ${}^{82}_{34}Se \rightarrow 2\ {}^{0}_{-1}e + \ldots$ [(a) $2\,{}^{1}_{0}n$ (b) ${}^{82}_{36}Kr$]

624. Calculate the amount of calcium oxide required when it reacts with 852 g of P_4O_{10}. (1008 g)

[**Hint:** $CaO + P_4O_{10} \rightarrow Ca_3(PO_4)_2$

Apply POAC for Ca and P atoms.]

625. An element crystallizes in f.c.c. lattice having edge length 400 pm. Calculate the maximum diameter of the atom which can be placed in the interstitial site without distorting the structure. (117.1 pm)

[**Hint:** See p. 745 and Eqn. 4 on p. 740.

$r = 0.414\,R$ and $R = \frac{\sqrt{2}a}{4}$]

626. 20% surface sites have adsorbed N_2. On heating, N_2 gas evolved from sites and was collected at 0.001 atm and 298 K in a container of volume 2.46 cm^3. The density of surface sites is $6.023 \times 10^{14}/cm^2$. Density of surface area is 1000 cm^2. Find out the number of surface sites occupied per molecule of N_2. (2)

[**Hint:** No. of surface sites per molecule of N_2

$$= \frac{\text{no. of surface sites used to adsorb } N_2 \text{ on a surface area of } 1000 \text{ cm}^2}{\text{total no. of adsorbed molecules}}]$$

627. For the reaction: $2X\,(g) \rightarrow 3Y\,(g) + 2Z\,(g)$

Time (min)	0	100	200
p_x (mmHg)	800	400	200

Assuming ideal gas condition, calculate

(a) order of the reaction

(b) rate constant

(c) time taken for 75% completion of reaction

(d) total pressure when $p_X = 700$ mm

((a) 1, (b) 3.46×10^{-3} min^{-1} (c) 200 min (d) 950 mm)

[**Hint:** (a) $\because t_{1/2}$ is constant, order is one. (b) $k = \frac{0.6932}{2t_{1/2}}$

(b) Time for 75% completion of the reaction $= 2t_{1/2}$

(c) $\underset{(800-p)\text{mm}}{\overset{800 \text{ mm}}{2X}} \rightarrow \underset{\frac{3p}{2}\text{ mm}}{3Y} + \underset{p \text{ mm}}{2Z}$; $800 - p = 700$]

628. (a) Calculate the velocity of electrons in the first Bohr orbit of hydrogen atom. $r = a_0 = 0.53 \times 10^{-10}$ m.

(b) Find the de Broglie wavelength of the electron in the first orbit.

(c) Find the orbital angular momentum of the 2p orbital in terms of $h/2\pi$ units.

((a) 2.18×10^6 m/s (b) 3.3×10^{-10} m (c) $\sqrt{2}\,\frac{h}{2\pi}$)

[**Hint:** (a) $mvr = \frac{nh}{2\pi}$; $n = 1$

(b) $\lambda = \frac{h}{mv}$

(c) For 2p, $l = 1$; $L = \sqrt{l(l+1)} \cdot \frac{h}{2\pi}$ (p. 250)]

629. (a) Calculate ΔG^0 for the following reaction.

$$Ag^+\,(aq) + Cl^-\,(aq) \rightarrow AgCl\,(s)$$

Given: $\Delta G_f^0\,(AgCl) = -109$ kJ/mol

$$\Delta G_f^0 (Cl^-) = -129 \text{ kJ/mol}$$
$$\Delta G_f^0 (Ag^+) = 77 \text{ kJ/mol}$$

Represent the above reaction in the form of a cell. Also calculate E^0 of the cell and find log K_{sp} of AgCl.

(b) 6.539×10^{-2} g of metallic Zn (65.39 amu) was added to 100 mL of saturated solution of AgCl. Calculate log $([Zn^{2+}]/[Ag^+]^2)$.

Given: $Ag^+ + e = Ag;\ E^0 = 0.80$ V

$Zn^{2+} + 2e = Zn;\ E^0 = -0.76$ V

Also find how many moles of Ag will be formed.

(a) (Ag | AgCl, Cl^- Ag^+ | Ag) 0.59 V, −10
(b) 52.79, 10^{-6} mol

[**Hint:** (a) For $Ag^+ + Cl^- = AgCl;\ \Delta G^0 = -109 - (-129 + 77)$

$\Delta G^0 = -nFE^0$ and $\Delta G^0 = -2.303\, RT \log K$

Calculate K and then apply $K_{sp}(AgCl) = \dfrac{1}{K}$

(b) $2Ag^+ + Zn = 2Ag + Zn^{2+};\ E^0_{cell} = 0.80 - (-0.76) = 1.56$ V

Apply $E^0_{cell} = \dfrac{0.0591}{2} \log ([Zn^{2+}]/[Ag^+]^2)$

Again, $K_{sp}(AgCl) = [Ag^+][Cl^-] = S^2$

$S = \sqrt{10^{-10}} = 10^{-5}$ mol/L

$\therefore$ mol of Ag^+ per 100 mL $= 10^{-6}$

and mol of Zn $= \dfrac{6.539 \times 10^{-2}}{65.39} = 10^{-3}$

As Ag^+ is the limiting reactant and the cell reaction is feasible in the forward direction, mole of Ag formed $= 10^{-6}$.]

630. The electrode potential of the half-cell Ag | AgCl | Cl^- (1 M) is 0.22 V. What should be the electrode potential ($E_{Ag^+, Ag}$) of the same half-cell if represented as Ag | Ag^+, AgCl, Cl^- (1 M)? (same)

631. The dissociation constant of a substituted benzoic acid at 25°C is 1.0×10^{-4}. Calculate pH of a 0.01 M solution of its sodium salt. **(IIT 2009)** (8)

632. At 400 K, the root mean square (rms) speed of a gas X (molecular weight = 40) is equal to the most probable speed of a gas Y at 60 K. Calculate the molecular weight of the gas Y. **(IIT 2009)** (4)

[**Hint:** $\sqrt{\dfrac{3RT_X}{M_X}} = \sqrt{\dfrac{2RT_Y}{M_Y}}$]

633. Calculate the total number of α and β particles emitted in the nuclear reaction,

$${}^{238}_{92}U \rightarrow {}^{214}_{82}Pb$$

[**Hint:** See Ex. 44, p. 279] **(IIT 2009)** (8 : 6α + 2β)

634. Calculate the oxidation number of Mn in the product of alkaline oxidative fusion of MnO_2.

[**Hint:** $4KOH + 2MnO_2 + O_2 \rightarrow 2K_2MnO_4 + 2H_2O$] **(IIT 2009)** (6)

635. Calculate the coordination number of Al in the crystalline state of $AlCl_3$.
(IIT 2009) (6)

[**Hint:** Cl^- ions form space lattice with Al^{3+} ion occupying octahedral voids in $AlCl_3$ crystal.]

636. In a constant volume calorimeter, 3.5 g of a gas with molecular weight 28 was burnt in excess of oxygen at 298.0 K. The temperature of the calorimeter was found to increase from 298.0 K to 298.45 K due to combustion process. Given that the heat capacity of the calorimeter is 2.5 kJ K^{-1}, calculate the numerical value for the enthalpy of combustion of the gas in kJ mol^{-1}. **(IIT 2009)** (9 kJ mol^{-1})

[**Hint:** Heat of combustion of 3.5 g of the gas = Heat taken by the calorimeter
= 2.5 (298.45 – 298.0) kJ]

637. A student performs a titration with different burettes and finds titre values of 25.2 mL, 25.25 mL and 25.0 mL. What is the number of significant figures in the average titre value? **(IIT 2010)** (3)

[**Hint:** All digits are significant except zeros at the beginning of the number and possibly terminal zeros.]

638. The concentration of R in the reaction $R \rightarrow P$ was measured as a function of time and the following data is

[R] (molar)	1.0	0.75	0.40	0.10
t (min)	0.0	0.05	0.12	0.18

Find the order of reaction. **(IIT 2010)** (0)

[**Hint:** Apply $k = \frac{x}{t} = \frac{a-(a-x)}{t}$]

639. Calculate the number of neutrons emitted when ${}^{235}_{92}U$ undergoes controlled nuclear fission to ${}^{142}_{54}Xe$ and ${}^{90}_{38}Sr$. **(IIT 2010)** (3)

[**Hint:** ${}^{235}_{92}U \rightarrow {}^{142}_{54}Xe + {}^{90}_{38}Sr + 3\,{}^{1}_{0}n$]

❑

CHAPTER TWENTY-TWO

MISCELLANEOUS OBJECTIVE QUESTIONS

1. The degree of dissociation of 100 mL of pure water at 25°C is
   (a) 1.8×10^{-16} (b) 1×10^{-14} (c) 1.8×10^{-9} (d) 1.0

2. The pH of 10^{-8} M NaOH aqueous solution at 25°C is
   (a) 7.02 (b) 7.0 (c) 6.89 (d) 6.0

3. The effect of temperature on the equilibrium constant is given by
   (a) $\log K_{p_2} - \log K_{p_1} = \dfrac{H}{2.303\,R} \cdot \dfrac{(T_2 - T_1)}{T_1 T_2}$
   (b) $\log k_2 - \log k_1 = \dfrac{E}{2.303\,R} \cdot \dfrac{(T_2 - T_1)}{T_1 T_2}$
   (c) $\log K_{p_2} - \log K_{p_1} = \dfrac{\Delta H}{2.303} \cdot \dfrac{(T_2 - T_1)}{T_1 T_2}$
   (d) none of these

4. The rate constant of a first-order reaction of the type $2A \rightarrow P$ is 1.5×10^{-4} s^{-1}. The half-life period of the reaction is
   (a) 2.31×10^{3} s (b) 4.62×10^{3} s (c) 9.24×10^{3} s (d) 1.5×10^{-4} s

5. For the following reaction:
   Initial concentration: 10 mol/L   2 mol/L
   $$2A + B \rightarrow \text{product}$$
   $t_{1/2}$ of the overall reaction is the time when
   (a) half of A changes to product
   (b) half of B changes to product
   (c) half of each of A and B changes to product
   (d) 6 moles of A and B changes to product

6. From the following half cells a galvanic cell is made,
   $A^{2+} + 2e = A$ ... $E_1^0 = 0.8$ V
   $B = B^{3+} + 3e$ ... $E_2^0 = -0.3$ V
   E^0 cell is
   (a) $E_1^0 - E_2^0$ (b) $E_1^0 + E_2^0$ (c) $3E_1^0 - 2E_2^0$ (d) $3E_1^0 + 2E_2^0$

7. An inert gas is added to the following equilibrium,
   $$A(s) + 2B(g) \rightleftharpoons 3C(g)$$
   at constant pressure. The equilibrium
   (a) is not affected (b) shifts to right
   (c) shifts to left (d) may shift right and left both

8. The rate of the process: $Cu \rightarrow Ni + {}^{0}_{+1}e$
   (a) changes with the change in pressure

(b) changes with the change in temperature
(c) changes with the change in pressure and temperature
(d) is independent of pressure and temperature

9. In a successive radioactive disintegration

$$\underset{N_1}{A} \xrightarrow{\lambda_1} \underset{N_2}{B} \xrightarrow{\lambda_2} C \quad \text{... no. of nuclides after time } t.$$

in which the parent has a longer but not much longer half-life than the daughter. Which of the following expressions is correct?

(a) $\frac{N_1}{N_2} = \frac{\lambda_2}{\lambda_1}$ (b) $\frac{N_1}{N_2} = \frac{\lambda_2 - \lambda_1}{\lambda_1}$ (c) $\frac{N_1}{N_2} = \frac{\lambda_1}{\lambda_2}$ (d) $\frac{N_1}{N_2} = \frac{\lambda_2 - \lambda_1}{\lambda_2}$

10. Which of the following curves represents a second-order reaction? ('*a*' is the concentration of the reactant)

(a) 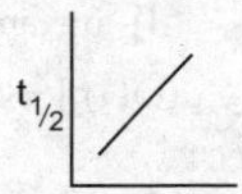

(b) 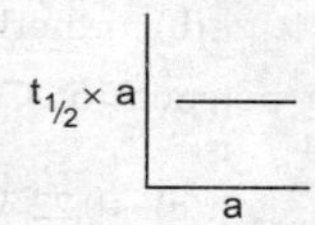

(c) 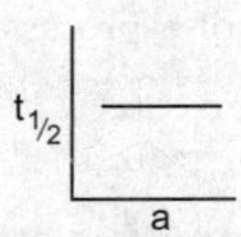

(d) 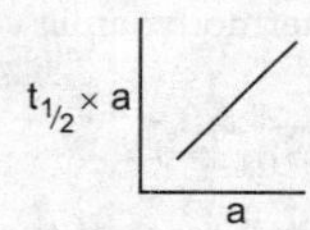

11. For the equilibrium:

$$\begin{array}{cccl} 0.5\text{ M} & 0 & 0 & \text{... Initial concentration} \\ PCl_5(g) \rightleftharpoons & PCl_3(g) + & Cl_2(g) & \\ 0.3\text{ M} & 0.2\text{ M} & 0.2\text{ M} & \text{... Equilibrium concentration} \end{array}$$

the degree of dissociation of PCl_5 is

(a) 0.2 (b) 0.01 (c) 0.1 (d) 0.4

12. The degree of hydrolysis of a salt of weak acid and strong base is ≈ 0.5. The equation to be used to calculate the accurate value of the degree of hydrolysis (h) is

(a) $h = \sqrt{\frac{K_w}{K_a \cdot K_b}}$ (b) $h = \sqrt{\frac{K_w}{K_a \cdot C}}$

(c) $h = \sqrt{\frac{K_w}{K_b \cdot C}}$ (d) none of these

13. For $CuSO_4 \cdot 5H_2O$, which is the correct mole relationship?

(a) $9 \times$ mole of Cu = mole of O
(b) $5 \times$ mole of Cu = mole of O
(c) $9 \times$ mole of Cu = mole of O_2
(d) Mole of Cu = $5 \times$ mole of O

14. The shaded plane in the simple cubic crystal is designated by

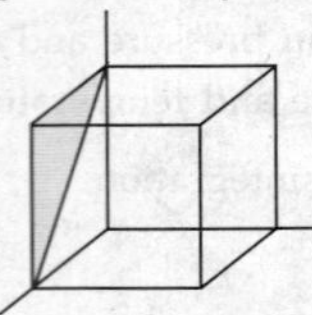

(a) (101) (b) (100) (c) (111) (d) (010)

15. If a reaction is represented by

$2A + 3B \rightarrow$ product

(a) the order of the reaction w.r.t. A and B may be 2 and 3 respectively
(b) the molecularity may be 5
(c) both order and molecularity are same
(d) the choices (a), (b) and (c) are wrong

16. The crystal with $a \neq b \neq c; \alpha = \beta = \gamma = 90°$ is

(a) cubic (b) orthorhombic (c) tetragonal (d) hexagonal

17. The reduction potential of $Cu^{2+} | Cu$ at pH = 14, if $K_{sp}[Cu(OH)_2] = 1 \times 10^{-19}$ and $E^0_{Cu^{2+}, Cu} = +0.34$ V, is

(a) −0.22 V (b) +0.22 V (c) 0.059 V (d) −0.059 V

18. What is the thermodynamic condition for the feasibility of a process?

(a) $(\Delta S)_{sys} > 0$ (b) $(\Delta G)_{sys} > 0$
(c) $(\Delta S)_{sys} + (\Delta S)_{surr} > 0$ (d) $(\Delta G)_{sys} + (\Delta G)_{surr} < 0$

19. If m_0 is the initial mass of the nuclei, the mass of undecayed radioactive isotope at the end of the nth half-life is

(a) $2^n m_0$ (b) $2^{-n} m_0$ (c) $n^{-2} m_0$ (d) $n^2 m_0$

20. Dry air is bubbled successively through (i) a solution, (ii) its solvent, and (iii) through $CaCl_2$. The lowering of vapour pressure of the solvent due to the addition of solute is equal to

(a) $p^0 \times \dfrac{\text{loss in wt. of solvent}}{\text{gain in wt. of } CaCl_2}$ (b) $\dfrac{\text{loss in wt. of solvent}}{\text{gain in wt. of } CaCl_2}$

(c) $\dfrac{\text{loss in wt. of solution}}{\text{gain in wt. of } CaCl_2}$ (d) $\dfrac{\text{loss in wt. of solute}}{\text{gain in wt. of } CaCl_2}$

21. The number of 2-fold axis of symmetry in a cubic crystal is

(a) 10 (b) 13 (c) 23 (d) 9

22. The nuclide lying below the stability belt in n-p graph does not disintegrate by

(a) α-emission (b) ${}^{0}_{-1}\beta$-emission
(c) ${}^{0}_{+1}\beta$-emission (d) K-electron capture

23. The osmotic pressure of glucose solution (400 mm), on dilution, decreased to 100 mm. The extent of dilution is

(a) two times (b) four times (c) one-fourth times (d) eight times

24. The molal depression constant for water is 1.86. The depression constant for 100 g of H_2O is

(a) 1.86 (b) 18.6 (c) 0.186 (d) 186

25. Urea is added to 2 litres of water to such an extent that $\Delta T_b / K_b$ becomes equal to 1/100. The weight of urea added is

(a) 0.6 g (b) 6.0 g (c) 12 g (d) 1.2 g

26. The equilibrium partial pressure of N_2, H_2 and NH_3 are 4, 4 and 8 atm respectively. The value of K_p for the Haber's process in atm^{-1} is

(a) 1/4 (b) 1/2 (c) 4 (d) 2

27. 5 moles of a gas occupy 100 litres at NTP. The compressibility factor of the gas is

(a) < 1 (b) > 1 (c) 1 (d) 0

28. A gas absorbs 400 J of heat and expands by 2×10^{-2} m^3 against a constant pressure 1×10^5 Pa. The internal energy of the gas is

(a) 200 J (b) −200 J
(c) 600 J (d) cannot be calculated

29. 4 g of He is expanded from 1 litre to 10 litres isothermally. The change in entropy is

(a) 19.15 J/K (b) 76.6 J/K (c) −19.15 J/K (d) −76.6 J/K

30. 1.5 moles of electrons are passed through 1 litre of 1 M Fe^{3+} solution. The number of moles of Fe produced is

(a) 0.25 (b) 0.50 (c) 1.0 (d) 1.5

31. In which case, a reaction is impossible at any temperature?

(a) $\Delta H > 0, \Delta S > 0$ (b) $\Delta H > 0, \Delta S < 0$
(c) $\Delta H < 0, \Delta S < 0$ (d) In all cases

32. The molecular weight of NaCl (degree of diss. $= x$) determined by the osmotic pressure method, is found to be different from its actual molecular weight (M). Which of the following relationships is correct?

(a) Obs. mol. wt. $= (1 + x) M$ (b) $M = (1 + x) \times$ obs. mol. wt.
(c) Obs. mol. wt. $= x \times M$ (d) $M = x \times$ obs. mol. wt.

33. Which of the following equations gives the combined form of the first and second laws of thermodynamics?

(a) $q = \Delta E - w$ (b) $\Delta E = T\Delta S - p\Delta V$
(c) $\Delta S = \frac{q}{T}$ (d) $\Delta H = \Delta E + \Delta n_g RT$

34. For a real gas at a given temperature, which of the following facts is not correct?

(a) pV may increase with the increase in pressure
(b) pV may decrease with the increase in pressure
(c) pV will not change with the change in pressure
(d) p/V will increase with the increase in pressure

35. The heat capacity of air is 20 J/K/mol. The amount of heat, required to heat the room through 1°, assuming the amount of air in the room to be 29 kg, is

(a) 20 kJ (b) −20 J (c) 200 J (d) −200 kJ

36. The molarity of 4 N H_2SO_4 in the following reaction is

$$8H^+ + H_2SO_4 + 8e^- \rightarrow H_2S + 4H_2O$$

(a) 2 M (b) 8 M (c) 32 M (d) 0.5 M

37. Phenol associates in benzene to produce double molecules. To what degree phenol associates if van't Hoff factor is 0.54?
(a) 0.54 (b) 0.92 (c) 0.98 (d) 0.46

38. A mixture of He(4) and Ne(20) in a 5-litre flask at 300 K and 1 atm weighs 4 g. The mole % of He is
(a) 2 (b) 0.02 (c) 20 (d) 4

39. The pH of the solution when 0.2 mole of HCl is added to one litre of a solution containing 0.1 M CH_3COOH and 0.1 M CH_3COO^- is
(a) 0.7 (b) 4.57 (c) 3.8 (d) 1.0

40. 100 mL of an $O_2 - O_3$ mixture was passed through turpentine, and 20 mL was reduced. If 100 mL of such a mixture is heated, the increase in volume will be
(a) 110 mL (b) 10 mL (c) 90 mL (d) 100 mL

41. Which of the relationships is wrong?
(a) rms speed $\propto \sqrt{p}$ (b) Diffusion rate $\propto p$
(c) Diffusion rate $\propto T$ (d) rms speed $\propto \sqrt{T}$

42. The heats evolved and absorbed when 100 g each of $CuSO_4$ and $CuSO_4 \cdot 5H_2O$ are dissolved in water are 41.7 and 4.7 kJ/mole respectively. The heat of hydration of $CuSO_4$ is (in kJ/mole)
(a) −66.59 (b) +11.72 (c) −78.2 (d) +78.2

43. The energy of the second Bohr orbit of the hydrogen atom is −3.41 eV. The energy of the third Bohr orbit of the He^+ ion will be
(a) −30.69 eV (b) −13.64 eV (c) −7.67 eV (d) −6.06 eV

44. It requires 40 mL of 0.5 M Ce^{4+} to titrate 10 mL of 1.0 M Sn^{2+} to Sn^{4+}. The oxidation state of Ce in the reduction product is
(a) +2 (b) +3 (c) +6 (d) none

45. From the following reaction sequences

$$2A = B + C$$
$$5B + D = 2E + F$$
$$E + G = 4H + J$$

calculate moles of H produced by 10 moles of A
(a) 20 (b) 10 (c) 5 (d) 8

46. If T_1 and T_2 are the temperatures of the heat source and sink respectively, the efficiency of a heat engine may be expressed as
(a) $\frac{T_2 - T_1}{T_2}$ (b) $\frac{T_1}{T_2}$ (c) $\frac{T_1 - T_2}{T_1}$ (d) $\frac{T_2}{T_1}$

47. Which of the following, when mixed, will give a solution with pH greater than 7?
(a) 0.1 M HCl + 0.2 M NaCl
(b) 100 mL of 0.2 M H_2SO_4 + 100 mL of 0.3 M NaOH
(c) 100 mL of 0.1 M $C_2H_4O_2$ + 100 mL of 0.1 M KOH
(d) 100 mL of 0.1 M HCl + 100 mL of 0.1 M $NaC_2H_3O_2$

48. In the reaction:

$$\overset{3}{A} + \overset{2}{2B} + \overset{1}{+3C} \longrightarrow D \quad \text{... initial no. of moles}$$

the limiting reactant may be

(a) A (b) B (c) C (d) D

49. An element forms two oxides, the weight-ratio composition in them is $A:O = x:y$ in the first oxide and $y:x$ in the second oxide. If the equivalent weight of A in the first oxide is 10.33, the equivalent weight of A in the second oxide is

(a) 6.2 (b) 10.33 (c) x/y (d) y/x

50. For the cell: $A \mid A^{m+} \mid\mid B^{n+} \mid B$; $E_{cell} = -1.1$ V

(a) right electrode is cathode
(b) the cell shall not operate
(c) left electrode is cathode
(d) electrons flow from left to right in the external circuit

51. Which of the following concepts is (are) wrong?

(a) If for, $A \rightarrow B$ $\Delta H = +q$ cal
then for, $B \rightarrow A$ $\Delta H = -q$ cal

(b) If for, $A \rightleftharpoons B$ eqb. constant $= K$
then for, $B \rightleftharpoons A$ eqb. constant $= \frac{1}{K}$

(c) If for, $A \rightarrow B$ rate constant $= k$
then for, $B \rightarrow A$ rate constant $\neq k$

(d) If for, $A \rightarrow B$ energy of activation $= E$
then for, $B \rightarrow A$ energy of activation $= -E$

52. The wrong statement is

(a) heat of neutralisation is always negative
(b) resonance energy is always negative
(c) heat of atomisation is always positive
(d) heat of combustion is always negative

53. Choose the correct answers for a given amount of hydrogen.

(a) Mole of H_2 = mole of H (b) Eq. of H_2 = eq. of H
(c) Mole of H_2 = eq. of H_2 (d) Mole of H = eq. of H

54. 5 moles of H_2SO_4 contain

(a) 5 eq. of H (b) 5 moles of S
(c) 20 eq. of O (d) 10 eq. each of H, S and O

55. In the figure representing variation of the compressibility factor Z of a real gas with pressure

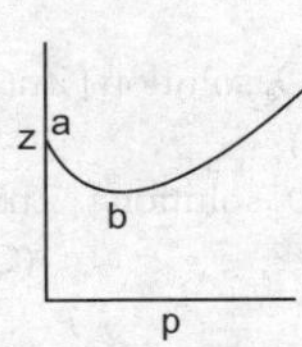

(a) *ab* shows that the gas is more compressible than an ideal gas

(b) *bc* shows that the gas is less compressible than an ideal gas

(c) *ab* may be explained by $p\overline{V} = RT - \frac{a}{\overline{V}}$

(d) *bc* may be explained by $p\overline{V} = RT + pb$

56. Which of the following galvanic cells has no liquid-junction potential?

(a) Zn – Hg | $ZnSO_4$ solution | Zn – Hg
(C_1) (C_2)

(b) Cu | $CuSO_4$ solution ⋮ $CuSO_4$ solution | Cu
(C_1) (C_2)

(c) Ag | AgCl | HCl solution ⋮ HCl solution| AgCl | Ag
(C_1) (C_2)

(d) Zn | $ZnSO_4$ solution | | $ZnSO_4$ solution | Zn
(C_1) (C_2)

57. In the Nernst equation, we substitute

$$\frac{2.303RT}{F} = 0.0591$$

where,

(a) $R = 8.314$ J, $T = 273$ K and $F = 96500$ C

(b) $R = 0.0821$ L. atm, $T = 298$ K and $F = 96500$ C

(c) $R = 8.314$ J, $T = 298$ K and $F = 96500$ C

(d) $R = 2.0$ cal, $T = 298$ K and $F = 1$ C

58. A gas at $T_2°$ is condensed to liquid following the path XYAB. The liquid appears at the point

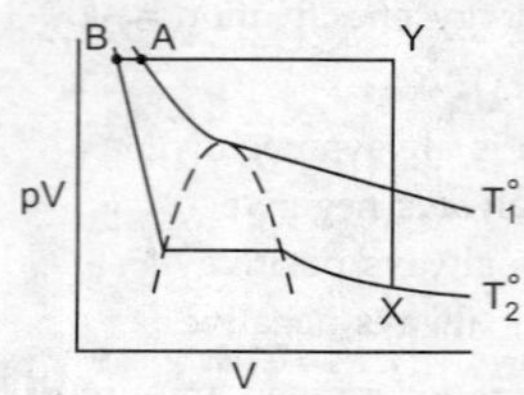

(a) Y at $T_2°$ (b) A at $T_2°$ (c) B at $T_1°$ (d) A at $T_1°$

59. For the electrode process,

$$H^+ + e = \frac{1}{2}H_2;\ E_{H^+, H_2} = x \text{ volt}$$

then for $2H^+ + 2e = H_2;\ E_{2H^+, H_2}$ is equal to

(a) x volt (b) $2x$ volt (c) $\frac{x}{2}$ volt (d) 0 volt

60. For the cell:

Cu(10 g) | $CuSO_4$ solution | | $ZnSO_4$ solution | Zn(10 g), $E_{cell} = E$ V
(C_1) (C_2)

E_{cell} for the cell: Cu(20 g) | $CuSO_4$ solution | | $ZnSO_4$ solution | Zn(20 g), is
(C_1) (C_2)

(a) E volt (b) $2E$ volt (c) $\frac{E}{2}$ volt (d) 1.1 volt

61. In an electrolysis process, 10 electrons could deposit x kg of a univalent metal M. The atomic weight of M is (N is Avogadro constant.)

(a) $x \times N$ (b) $x \times N \times 10^3$ (c) $x \times N \times 10^2$ (d) $x \times 10^3$

62. If a gas gets half compressed, compared to an ideal gas, the compressibility factor Z is equal to

(a) 1 (b) 2 (c) ½ (d) none of these

63. The temperature at which the second virial coefficient of a real gas is zero is called

(a) critical temperature (b) eutectic point
(c) boiling point (d) Boyle temperature

64. A gaseous mixture of 2 moles of A, 3 moles of B, 5 moles of C and 10 moles of D is contained in a vessel. Assuming that the gases are ideal and the partial pressure of C is 1.5 atm, the total pressure is

(a) 3 atm (b) 6 atm (c) 9 atm (d) 15 atm

65. The maximum efficiency of a steam engine operating between 100°C and 25°C is

(a) 20% (b) 22.2% (c) 25% (d) 30%

66. The entropy change accompanying the evaporation of 1 mole of water at 100°C, assuming that the latent heat of vaporisation of water is 540 cal g^{-1}, in cal/K/mole is

(a) 20 (b) 25 (c) 26.06 (d) 30

67. The compressibility factor of a van der Waals gas at the critical point is equal to

(a) 0 (b) 1 (c) 0.375 (d) any value

68. The equivalent volume of CO_2 in the following reaction is

$$2NaHCO_3 \rightarrow Na_2CO_3 + H_2O + CO_2$$

(a) 22.4 litres (b) 11.2 litres (c) 5.6 litres (d) 22 litres

69. The equivalent volume of CO_2 in the following reaction is

$$NaHCO_3 + HCl \rightarrow NaCl + H_2O + CO_2$$

(a) 22.4 litres (b) 11.2 litres (c) 5.6 litres (d) 44 litres

70. One coulomb is the charge of

(a) 1 mole of electrons (b) $\frac{1}{96500}$ mole of electrons
(c) 96500 moles of electrons (d) none of these

71. For a zero-order reaction, with the initial reactant concentration a, the time for completion of the reaction is

(a) k/a (b) a/k (c) $2k/a$ (d) $a/2k$

72. Which statement(s) about the behaviour of a real gas is (are) wrong?

(a) A gas exerts more pressure compared to an ideal gas
(b) A gas can be compressed more compared to an ideal gas
(c) A gas with high value of van der Waals constant, a, is highly liquefiable
(d) H_2 and He are less compressible compared to an ideal gas at ordinary temperature

73. The figure shows the effect of pressure on the compressibility factor, Z, of a gas.

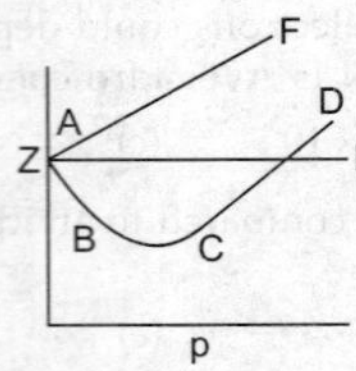

The wrong conclusion(s) is (are),
(a) the curves AE and BC can be explained by $pV = RT$
(b) the curves AF and CD can be explained by $pV = RT + Pb$
(c) the curve AB can be explained by $pV = RT - \frac{a}{V}$
(d) all the three curves AF, AE and AD show the real gas behaviour

74. Equivalent weights of CO_2 in the following reactions

$$2NaHCO_3 = Na_2CO_3 + H_2O + CO_2$$

and $$NaHCO_3 + HCl = NaCl + H_2O + CO_2$$

are respectively,
(a) 22 & 44 (b) 44 & 22 (c) 44 & 44 (d) 22 & 22

75. According to Faraday's laws of electrolysis, the discharge of one electrochemical equivalent of ions should involve
(a) 1 mole of electrons (b) 96500 moles of electrons
(c) $\frac{1}{96500}$ mole of electrons (d) none of these

76. In which of the following cases a gas is more compressible than the ideal gas?
(a) $p\overline{V} > RT$ (b) $p\overline{V} = RT$ (c) $p\overline{V} < RT$

77. In the electrolysis of aq. $CuSO_4$ solution with Pt electrodes, using 1 F of electricity, which of the following processes shall occur?
(a) 1 eq. of Cu is deposited at the cathode
(b) 1 eq. of O_2 is liberated at the anode
(c) 1 eq. of OH^- is discharged at the anode
(d) 1 eq. of H_2SO_4 is produced

78. On electrolysis of 500 mL of an aqueous solution of NaCl, NaOH is produced, the normality of which is found to be N/2. The number of faradays used is
(a) 0.25 (b) 250 (c) 500 (d) 1000

79. 10% of a reactant decomposes in 1 hour, 20% in 2 hours and 30% in 3 hours. The order of the reaction is
(a) 0 (b) 1 (c) 2 (d) 3

80. One way of writing the equation of state for real gases is

$$p\overline{V} = RT\left[1 + \frac{B}{\overline{V}} + \ldots\right]$$

The constant B is equal to
(a) $\left(a - \frac{b}{RT}\right)$ (b) $\left(a + \frac{b}{RT}\right)$ (c) $\left(b + \frac{a}{RT}\right)$ (d) $\left(b - \frac{a}{RT}\right)$

[Hint: Example 66, Chapter 12]

81. Equal number of moles of A and B are allowed to react with each other till it reaches equilibrium.

$$2A + B \rightleftharpoons C + D$$

The value of K_c for this equilibrium can never be

(a) >1 (b) <1 (c) $=1$ (d) ∞

82. In $Ca(OH)_2$, solution (aqueous), the molar concentration of OH^- is found to be x, the solubility of $Ca(OH)_2$ in moles/litre is

(a) $2x$ (b) x (c) $x/2$ (d) $4x$

83. For a reaction of the order of 0.5, when the concentration of the reactant is doubled, the rate

(a) doubles (b) increases four times
(c) decreases four times (d) increases $\sqrt{2}$ times

84. In the decomposition of H_2O_2 at a given temperature T, the energy of activation decreases from E_1 to E_2 by the use of a catalyst. How many times does the rate of the catalysed reaction increase?

(a) $(E_1 - E_2)$ (b) $\frac{E_2}{E_1}$ (c) $\frac{E_1 - E_2}{R}$ (d) $e^{(E_1 - E_2)/RT}$

85. A gas can be condensed to liquid through the paths I, II and III, as shown in the figure. The path(s) through which the gas changes to liquid abruptly is (are)

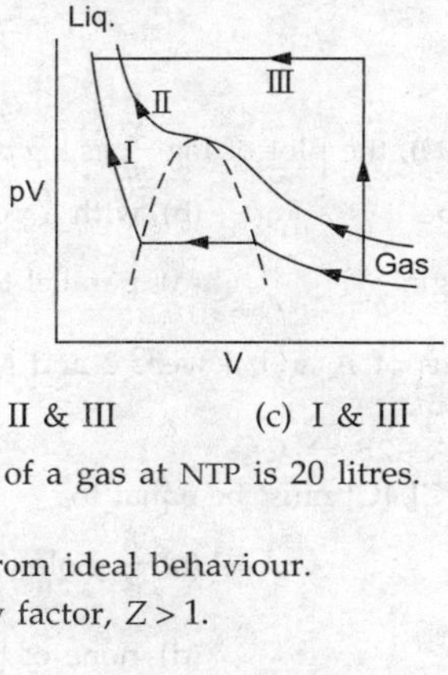

(a) I & II (b) II & III (c) I & III (d) II

86. The volume of 1 mole of a gas at NTP is 20 litres.

(a) The gas is ideal.
(b) The gas deviates from ideal behaviour.
(c) The compressibility factor, $Z > 1$.
(d) $Z < 1$.

87. 10 mL of a solution containing Na_2CO_3 and $NaHCO_3$ is titrated by HCl using phenolphthalein and then methyl orange (added after first end point). The first and second end points were found after adding 10 mL and 15 mL of N/10 HCl respectively. The ratio of m.e. of Na_2CO_3 and $NaHCO_3$ in the solution is

(a) 2/1 (b) 1/2 (c) 5/1 (d) 1/5

88. The initial concentrations of X and Y were 2 and 4 moles/litre respectively for the following equilibrium

$$X + 2Y \rightleftharpoons Z.$$

Which of the following relationships among equilibrium concentrations of X, Y and Z is (are) not feasible?

(a) $[X] = [Z]$ (b) $[X] < [Y]$ (c) $[X] > [Y]$ (d) $[Y] >=< [Z]$

89. For an ideal gas

(a) $\left(\frac{\partial V}{\partial T}\right)_p = 0$ (b) $\left(\frac{\partial p}{\partial T}\right)_V = 0$ (c) $\left(\frac{\partial E}{\partial V}\right)_T = 0$ (d) $\left(\frac{\partial E}{\partial T}\right)_V = 0$

90. x g of a gas is absorbed per m grams of a solid. A plot x/m vs pressure p at two different temperatures T_1 and T_2 shows

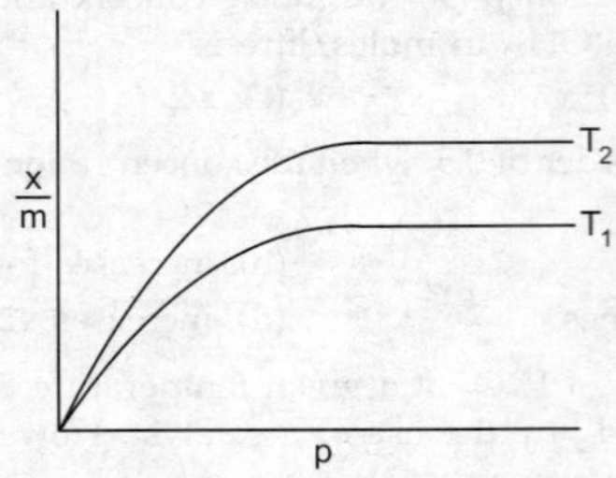

What we conclude is

(a) $\frac{x}{m} \propto p^n$, where $n = 0$ at low pressure

(b) $\frac{x}{m} \propto p^n$, where $n = 1$ at high pressure

(c) $T_1 > T_2$

(d) $T_2 > T_1$

91. In the above problem (90), the plot of $\log \frac{x}{m}$ vs $\log p$ should be a straight line

(a) with a negative slope (b) with a positive slope

(c) passing through origin (d) parallel to $\frac{x}{m}$-axis

92. The initial concentrations of A and B were 2 and 4 moles/litre respectively for the following equilibrium.

$$A + 2B \rightleftharpoons C$$

If at equilibrium, [B] = [C], [C] must be equal to

(a) 2 moles/litre (b) $\frac{8}{3}$ moles/litre

(c) $\frac{4}{3}$ moles/litre (d) none of these

93. The number of molecules of the sweetener saccharin, which can be prepared

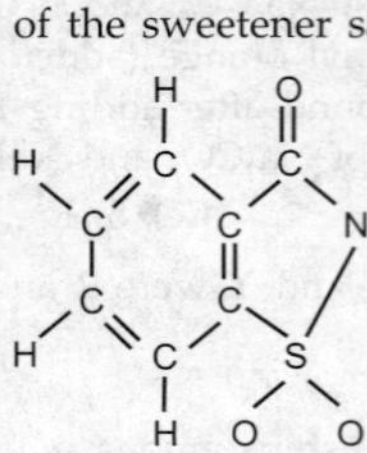

from 30 C-atoms, 25 H-atoms, 12 O-atoms, 8 S-atoms and 14 N-atoms is

(a) 7 (b) 5 (c) 8 (d) 4

94. The equilibrium partial pressures of A, B and C are 1, 2 and 4 atm respectively in the equilibrium

$$A + B \rightleftharpoons C$$

The value of the reaction quotient just after reducing the volume of the container to half is

(a) 1 (b) 2 (c) 4 (d) none of these

95. For a system at equilibrium the system acquires a state of

(a) maximum entropy (b) minimum entropy
(c) zero entropy (d) optimum entropy

96. 14 g of an element X combines with 16 g of oxygen. The element X could have an at. wt. of

(a) 14 and its oxide the formula X_2O
(b) 7 and its oxide the formula X_2O
(c) 7 and its oxide the formula XO
(d) 14 and its oxide the formula XO_2

97. 2 volumes of a diatomic gas of an element combine with 5 volumes of O_2 to give 2 volumes of its oxide. The equivalent weight of the element is 2.8, the atomic weight of the element is

(a) 32 (b) 14 (c) 35.5 (d) 1

98. The vapour density of a volatile chloride of a metal is 55.5. When 40 g of the metal is dissolved in the acid, 22.4 litres of H_2 is liberated at NTP. The valency of the metal is

(a) 1 (b) 2 (c) 3 (d) 4

99. 0.09 g of a trivalent metal (M) combines with 56 mL of O_2 at NTP. The vapour density of its chloride is 133.5. The molecular formula of the chloride is

(a) MCl_2 (b) M_2Cl_4 (c) MCl_3 (d) M_2Cl_6

100. 6.0 g of a solid is heated, the residue left behind weighs 2 g and 1120 cc of a gas measured at 0°C and 1 atm is evolved. The molecular weight of the gas is

(a) 80 (b) 40 (c) 20 (d) 60

101. A compound is found to contain 5.47% nitrogen by weight. The minimum value which can be assigned to its molecular weight is

(a) 256 (b) 100 (c) 130 (d) 65

102. One volume of a diatomic gas (mol. wt. = 28) combines with three volumes of another diatomic gas (mol. wt. = 2) and forms 2 volumes of a gaseous compound. The molecular weight of the gas is

(a) 16 (b) 17 (c) 30 (d) 32

103. 2000 mL of air on ozonisation formed 1915 mL of ozonised air. The volume of ozone formed is

(a) 85 mL (b) $\frac{2}{3} \times 85$ mL (c) 49.5 mL (d) 170 mL

104. The volume of oxygen needed for complete combustion of 20 cc of a gaseous hydrocarbon C_xH_y is

(a) 20 cc (b) $\left(x + \frac{y}{4}\right)$ cc (c) $20\left(x + \frac{y}{4}\right)$ cc (d) $\frac{1}{2}\left(x + \frac{y}{4}\right)$ cc

105. At high temperature, S_4N_4 decomposes into sulphur vapour and nitrogen. If 1 mL of S_4N_4 is decomposed and 2.5 mL of mixture are obtained, the formula of sulphur vapour is

(a) S_2 (b) S_4 (c) S_6 (d) S_8

106. The approximate atomic weight of an element is 26.89. If its equivalent weight is 8.9, the exact atomic weight of the element would be

(a) 26.89 (b) 8.9 (c) 26.70 (d) 17.8

107. The oxide of an element possesses the formula M_2O_3. If the equivalent weight of the metal is 9, the molecular weight of the oxide will be

(a) 54 (b) 102 (c) 120 (d) 200

108. A container contains a certain gas of mass m at high pressure. A little amount of the gas has been allowed to escape from the container and after some time, the pressure of the gas becomes half and its absolute temperature two-third. The mass of the gas escaped is

(a) $\frac{2m}{3}$ (b) $\frac{m}{2}$ (c) $\frac{m}{4}$ (d) $\frac{m}{6}$

109. Two vessels A and B contain the same gas. If the pressure, volume and absolute temperature of the gas in A are two times as compared to that in B, and if the mass of the gas in B is x grams, the mass of the gas in A will be

(a) $4x$ g (b) $\frac{x}{2}$ g (c) $2x$ g (d) x g

110. The vapour density of undecomposed N_2O_4 is 46. When heated, the vapour density decreases to 24.5 due to its dissociation to NO_2. The per cent dissociation of N_2O_4 at the final temperature is

(a) 88 (b) 60 (c) 40 (d) 70

111. The energy of an electron in the first Bohr orbit of H-atom is −13.6 eV. The possible energy value(s) of the excited state(s) for electrons in Bohr orbits of hydrogen is (are)

(a) −3.4 eV (b) −4.2 eV (c) −6.8 eV (d) +6.8 eV

(IIT 1998)

112. Which of the following statement(s) is (are) correct?

(a) The coordination number of each type of ion in CsCl crystal is 8.
(b) The metal that crystallises in b.c.c. structure has a coordination number of 12.
(c) A unit cell of an ionic crystal shares some of its ions with other unit cells.
(d) The length of the unit cell in NaCl is 552 pm.
($r_{Na^+} = 95$ pm, $r_{Cl^-} = 181$ pm) **(IIT 1998)**

113. The standard reduction potential values of three metallic cations of X, Y and Z are 0.52, −3.03 and −1.18 V respectively. The order of reducing power of the corresponding metals is

(a) Y > Z > X (b) X > Y > Z
(c) Z > Y > X (d) Z > X > Y **(IIT 1998)**

114. Decrease in atomic number is observed during

(a) α-emission (b) β-emission
(c) positron emission (d) electron capture **(IIT 1998)**

115. For a first-order reaction

(a) the degree of dissociation $= (1 - e^{-kt})$

(b) a plot of reciprocal of concentration of the reactant vs time gives a straight line

(c) the time taken for the completion of 75% reaction is thrice the $t_{1/2}$ of the reaction

(d) the pre-exponential factor in the Arrhenius's equation has the dimension of time, T^{-1} **(IIT 1998)**

116. According to Graham's law at a given temperature the ratio of the rates of diffusion r_A/r_B of gases A and B is given by

(a) $\left(\frac{p_A}{p_B}\right)\left(\frac{M_A}{M_B}\right)^{1/2}$ (b) $\left(\frac{M_A}{M_B}\right)\left(\frac{p_A}{p_B}\right)^{1/2}$

(c) $\left(\frac{p_A}{p_B}\right)\left(\frac{M_B}{M_A}\right)^{1/2}$ (d) $\left(\frac{M_A}{M_B}\right)\left(\frac{p_B}{p_A}\right)^{1/2}$ **(IIT 1998)**

117. For the reaction: $CO(g) + H_2O(g) \rightleftharpoons CO_2(g) + H_2(g)$ at a given temperature, the equilibrium amount of CO_2 can be increased by

(a) adding a suitable catalyst

(b) adding an inert gas

(c) decreasing the volume of the container

(d) increasing the amount of CO(g) **(IIT 1998)**

118. Which of the following statement(s) is (are) correct?

(a) the pH of 1×10^{-8} M solution of HCl is 8.

(b) The conjugate base of $H_2PO_4^-$ is HPO_4^{2-}.

(c) Autoprotolysis constant of water increases with temperature.

(d) When a solution of a weak monoprotic acid is titrated against a strong base, at half-neutralisation point, $pH = 1/2\ pK_a$. **(IIT 1998)**

119. The number of neutrons accompanying the formation of ${}^{139}_{54}Xe$ and ${}^{94}_{38}Sr$ from the absorption of a slow neutron by ${}^{235}_{92}U$, followed by nuclear fission, is

(a) 0 (b) 2 (c) 1 (d) 3 **(IIT 1999)**

120. A gas will approach ideal behaviour at

(a) low temperature and low pressure

(b) low temperature and high pressure

(c) high temperature and low pressure

(d) high temperature and high pressure **(IIT 1999)**

121. The normality of 0.3 M phosphorus acid (H_3PO_3) is

(a) 0.1 (b) 0.9 (c) 0.3 (d) 0.6 **(IIT 1999)**

122. The coordination number of a metal crystallising in a hexagonal close-packed structure is

(a) 12 (b) 4 (c) 8 (d) 6

123. A gas X at 1 atm is bubbled through a solution containing a mixture of 1 M Y^- and 1 M Z^- at 25°C. If the reduction potential of $Z > Y > X$, then

(a) Y will oxidise X and not Z (b) Y will oxidise Z and not X

(c) Y will oxidise both X and Z (d) Y will reduce both X and Z **(IIT 1999)**

124. The pH of a 0.1 M solution of the following salts increases in the water,
(a) $NaCl < NH_4Cl < NaCN < HCl$
(b) $HCl < NH_4Cl < NaCl < NaCN$
(c) $NaCN < NH_4Cl < NaCl < HCl$
(d) $HCl < NaCl < NaCN < NH_4Cl$ **(IIT 1999)**

125. For the chemical reaction $3X(g) + Y(g) \rightleftharpoons X_3Y(g)$ the amount of X_3Y at equilibrium is affected by
(a) temperature and pressure
(b) temperature only
(c) pressure only
(d) temperature, pressure and catalyst **(IIT 1999)**

126. One mole of calcium phosphide on reaction with excess water gives
(a) one mole of phosphine
(b) two moles of phosphoric acid
(c) two moles of phosphine
(d) one mole of phosphorus pentoxide **(IIT 1999)**

127. The oxidation numbers of S in S_8, S_2F_2 and H_2S respectively are
(a) 0, +1 & −2 (b) +2, +1 & −2 (c) 0, +1 & +2 (d) −2, +1 & −2

128. Which of the following statement(s) is (are) correct?
(a) A plot of log K_p vs $\frac{1}{T}$ is linear.
(b) A plot of log [X] vs time is linear for a first-order reaction, $X \rightarrow p$.
(c) A plot of log p vs $\frac{1}{T}$ is linear at constant volume.
(d) A plot of p vs $\frac{1}{V}$ is linear at constant temperature.

129. In depression-of-freezing-point experiment, it is found that the
(a) vapour pressure of the solution is less than that of pure solvent
(b) vapour pressure of the solution is more than that of pure solvent
(c) only solute molecules solidify at the freezing point
(d) only solvent molecules solidify at the freezing point **(IIT 1999)**

130. A buffer solution can be prepared from a mixture of
(a) sodium acetate and acetic acid in water
(b) sodium acetate and hydrochloric acid in water
(c) ammonia and ammonium chloride in water
(d) ammonia and sodium hydroxide in water

131. The ΔH_f° for $CO_2(g)$, $CO(g)$ and $H_2O(g)$ are −393.5, −110.5 and −241.8 kJ mol^{-1} respectively. The standard enthalpy change (in kJ) for the reaction,

$$CO_2(g) + H_2(g) \rightarrow CO(g) + H_2O(g), \text{ is}$$

(a) 524.1 (b) 41.2 (c) −262.5 (d) −41.2 **(IIT 2000)**

132. The rate constant for the reaction $2N_2O_5 \rightarrow 4NO_2 + O_2$ is $3.0 \times 10^{-5}\ s^{-1}$. If the rate is 2.40×10^{-5} mol $lit^{-1}\ s^{-1}$ then the concentration of N_2O_5 (in mol L^{-1}) is

(a) 1.4 (b) 1.2 (c) 0.04 (d) 0.8 **(IIT 2000)**

133. For the electrochemical cell, $M | M^+ || X^- | X$, $E^0(M^+/M) = 0.44$ V and $E^0(X/X^-) = 0.33$ V. From this data, one can deduce that

(a) $M + X \rightarrow M^+ + X^-$ is the spontaneous reaction

(b) $M^+ + X^- \rightarrow M + X$ is the spontaneous reaction

(c) $E_{cell} = 0.77$ V

(d) $E_{cell} = -0.77$ V **(IIT 2000)**

134. For the reversible reaction $N_2(g) + 3H_2(g) \rightleftharpoons 2NH_3(g)$ at 500°C, the value of K_p is 1.44×10^{-5} when partial pressure is measured in atm. The corresponding value of K_c, with concentration in mol L^{-1} is

(a) $1.44 \times 10^{-5}/(0.082 \times 500)^{-2}$ (b) $1.44 \times 10^{-5}/(8.314 \times 773)^{-2}$

(c) $1.44 \times 10^{-5}/(0.082 \times 773)^{2}$ (d) $1.44 \times 10^{-5}/(0.082 \times 773)^{-2}$

(IIT 2000)

135. The rms velocity of hydrogen is $\sqrt{7}$ times the rms velocity of nitrogen. If T is the temperature of the gas,

(a) $T(H_2) = T(N_2)$ (b) $T(H_2) > T(N_2)$

(c) $T(H_2) < T(N_2)$ (d) $T(H_2) = \sqrt{7}\, T(N_2)$ **(IIT 2000)**

136. The compressibility of a gas is less than unity at STP. Therefore,

(a) $V_m > 22.4$ L (b) $V_m < 22.4$ L (c) $V_m = 22.4$ L (d) $V_m = 44.8$ L

(IIT 2000)

137. At 100°C and 1 atm, if the density of liquid water is 1.0 g cm^{-3} and that of water vapour is 0.0006 g cm^{-3}, the volume occupied by water molecules in 1 litre of steam at that temperature is

(a) 6 cm^3 (b) 60 cm^3 (c) 0.6 cm^3 (d) 0.06 cm^3

(IIT 2000)

138. The number of nodal planes in a p_x orbital is

(a) 1 (b) 2 (c) 3 (d) 0 **(IIT 2000)**

139. When two reactants A and B are mixed to give products C and D, the reaction quotient, Q, at the initial stages of the reaction

(a) is zero (b) decreases with time

(c) is independent of time (d) increases with time

140. Which of the following mixtures will have pH equal to 2.0 if the normalities of HCl and NaOH are 0.1 N?

(a) 50 mL HCl + 50 mL NaOH (b) 55 mL HCl + 45 mL NaOH

(c) 75 mL HCl + 25 mL NaOH (d) 60 mL HCl + 140 mL NaOH

141. The equilibrium constant depends only on temperature because

(a) the energy of activation of forward reaction is different from that of the backward reaction

(b) for a certain change in temperature, the rate of the forward reaction and that of the backward reaction do not change by the same amount

(c) rate constants of the forward and the backward reactions change to different extents
(d) every reaction is caused by heating

142. K_p and K_c are related by $K_p = K_c(RT)^{\Delta n}$. Under what practical condition/s, $K_p = K_c$?
(a) $RT = 1$ (b) $T = 12K$ (c) $\Delta n = 0$ (d) $R \propto \frac{1}{T}$

143. The equilibrium constant shall not depend on temperature if (*E*-energy of activation)
(a) $E_f = E_b$ (b) $E_f > E_b$ (c) $E_b > E_f$ (d) $(\Delta H)_{\text{reaction}} = 0$

144. The pK_a of an indicator is 4. Its pH range is
(a) 1–3 (b) 3–5 (c) 5–8 (d) 8–12

145. The heat of fusion is 334.7 J g^{-1}. The entropy change in J K^{-1} kg^{-1} in melting of 1 g of ice at 0°C is
(a) 1.226 (b) 0.293 (c) 293 (d) 1226

146. Which statement is not correct about internal energy?
(a) Internal energy is the sum of kinetic energy and potential energy of the particles making up the system.
(b) Internal energy of the system can be measured.
(c) Internal energy is a state function.
(d) ΔU is the heat of reaction at constant volume.

147. At what temperature does an aqueous solution containing 3×10^{23} molecules of a nonelectrolyte substance in 250 g of water freeze? ($K_f = 1.86$).
(a) 269.28 K (b) 271.14 K (c) 271 K (d) 276.72 K

148. A plot between time and the amount of reactant consumed is found to be a straight line passing through origin for a reaction. The order of the reaction is
(a) 0 (b) 1 (c) 2 (d) 3

149. The unit of rate constant is the same as that of the rate of the reaction of order
(a) 0 (b) 1 (c) 2 (d) 3

150. The time required to decompose half of the reactant for an *n*th-order reaction is inversely proportional to (*a* is the initial concentration)
(a) a^n (b) a^{n-2} (c) a^{n-1} (d) a^{n+1}

151. For a chemical change $A \rightarrow B$, it is found that the rate doubles when the concentration of A is increased 4 times. The order in A is
(a) 2 (b) 1 (c) 0 (d) $\frac{1}{2}$

152. A metallic element exists as a cubic lattice. Each edge of the unit cell is 2.88 Å and the density of the metal is 7.2 g cm^{-3}. The number of unit cells in 100 g of the metal is
(a) 5.82×10^{23} (b) 6.023×10^{23} (c) 8.52×10^{23} (d) none of these

153. The vapour pressure of a solvent decreased by 10 mmHg when a nonvolatile solute was added to the solvent. The mole fraction of the solute in the solution is 0.2. What would be the mole fraction of the solvent if the decrease in the vapour pressure is to be 20 mmHg?

(a) 0.8 (b) 0.6 (c) 0.4 (d) 0.2

154. Schottky defect in crystals is observed when
(a) an ion leaves its normal site and occupies an interstitial site
(b) unequal number of cations and anions are missing from the lattice
(c) density of the crystal is increased
(d) equal number of cations and anions are missing from the lattice

155. The face diagonal length of f.c.c. cubic cell is 508 pm. If the radius of the cation is 110 pm, the radius of the anion is
(a) 288 pm (b) 398 pm (c) 144 pm (d) 618 pm

156. Haemoglobin contains 0.334% of Fe by weight. The molecular weight of haemoglobin is approximately 67200. The number of Fe atoms present in one molecule of haemoglobin is
(a) 1 (b) 6 (c) 4 (d) 2

157. Given that

$$C + O_2 = CO_2; \quad \Delta H = -x \text{ kJ}$$
$$2CO + O_2 = 2CO_2; \quad \Delta H = -y \text{ kJ}$$

The enthalpy of formation of CO will be
(a) $(y - 2x)$ (b) $(2x - y)/2$ (c) $(y - 2x)/2$ (d) $(2x - y)$

158. E^0 for some half reactions are given below:

$$Sn^{4+} + 2e = Sn^{2+} \quad ; E^0 = 0.15 \text{ V}$$
$$2Hg^{2+} + 2e = Hg_2^{2+} \quad ; E^0 = 0.92 \text{ V}$$
$$PbO_2 + 4H^+ + 2e = Pb^{2+} + 2H_2O \quad ; E^0 = 1.45 \text{ V}$$

Based on the given data which statement is correct?
(a) Sn^{4+} is a stronger oxidising agent than Pb^{4+}.
(b) Sn^{2+} is a stronger reducing agent than Hg_2^{2+}.
(c) Hg^{2+} is a stronger oxidising agent than Pb^{4+}.
(d) Pb^{2+} is a stronger reducing agent than Sn^{2+}.

159. An element (atomic mass = 100 g/mole) having b.c.c. structure has unit cell edge 4.00 Å. The density in g/cc of the element is
(a) 10.376 (b) 5.188 (c) 7.289 (d) 2.144

160. Solid ammonium carbamate dissociates as follows:

$$NH_2COONH_4(s) \rightleftharpoons 2NH_3(g) + CO_2(g)$$

At equilibrium, total pressure is found to be 0.3 atm at a given temperature. The value of K_p is
(a) 0.3 atm^3 (b) 0.108 atm^3
(c) 4.0×10^{-3} atm^3 (d) 0.158 atm

161. The dimensions of pressure are the same as that of
(a) force per unit volume (b) energy per unit volume
(c) force (d) energy
[**Hint:** Force – kg m s^{-2}, energy – kg m^2s^{-2}, pressure – kg $m^{-1}s^{-2}$]

162. The number of atoms in 100 g of an f.c.c. crystal, with density 10 g/cc and cell edge equal to 100 pm, is equal to

(a) 3×10^{25} (b) 4×10^{25} (c) 1×10^{25} (d) 2×10^{25}

[**Hint:** No. of atoms per g $= \frac{N}{M}$: Apply Equation 1, Chapter 20]

163. Consider the following reactions occurring in an automobile

$$C_8H_{18}(g) + \frac{25}{2}O_2(g) \rightarrow 8CO_2(g) + 9H_2O(g)$$

The sign of ΔH, ΔS and ΔG would be respectively,

(a) +, −, + (b) −, +, − (c) −, +, + (d) +, +, −

164. 8 moles of SO_2 and 4 moles of O_2 are mixed in a closed vessel. The reaction proceeds at constant temperature. By the moment when equilibrium sets in, 80% of the initial amount of SO_2 enters into the reaction. The equilibrium pressure would be

(a) 2.17 atm (b) 8 atm

(c) 12 atm (d) cannot be calculated

165. For the adsorption of a gas on a solid, the plot of $\log \frac{x}{m}$ vs $\log p$(x-axis) is linear with slope equal to

(a) k (b) $\log K$ (c) n (d) $\frac{1}{n}$

166. Select the rate law corresponding to the data for the following reaction.

$$A + B \rightarrow C$$

$[A]_{Initial}$	$[B]_{Initial}$	Initial Rate
0.012	0.035	0.10
0.024	0.070	0.80
0.024	0.035	0.10
0.012	0.070	0.80

(a) Rate $= k[B]^3$ (b) Rate $= k[B]^4$

(c) Rate $= k[A][B]^3$ (d) Rate $= k[A]^2[B]^2$

167. Under what conditions will a pure sample of an ideal gas not only exhibit a pressure of 2 atm but also a concentration of 2 mole/litre?

(a) at STP (b) when $V = 22.4$ L

(c) at 12 K (d) impossible under any condition

168. At NTP, 0.5 mole of H_2 and 1.0 mole of He gas

(a) have equal kinetic energy (b) have equal molecular velocity

(c) occupy equal volume (d) have equal diffusion rate

169. In which of the following solvents will AgBr have the highest solubility?

(a) 10^{-3} M HBr (b) 10^{-3} M NaBr

(c) 10^{-3} M NH_4OH (d) pure water

170. Based on the following thermochemical equation:

$$H_2O(g) + C(s) = CO(g) + H_2(g) \quad ; \Delta H = 131 \text{ kJ}$$

$$CO(g) + \frac{1}{2}O_2(g) = CO_2(g) \quad ; \Delta H = -282 \text{ kJ}$$

$$H_2(g) + \frac{1}{2}O_2(g) = H_2O(g) \quad ; \Delta H = -242 \text{ kJ}$$

$$C(s) + O_2(g) = CO_2(g) \quad ; \quad \Delta H = -x \text{ kJ}$$

The value of x will be

(a) 393 kJ (b) −393 kJ (c) 655 kJ (d) −655 kJ

171. The equivalent mass of NO_3^- in the following reaction is

$$NO_3^- \rightarrow NH_4^+$$

(a) 7.75 (b) 31 (c) 62 (d) 15

172. The molarity of 0.56 L of water vapour at STP is

(a) 0.089 M (b) 0.045 M (c) 0.45 M (d) 0.22 M

173. If the compressibility factor, Z, of a gas increases with the increase in pressure (usually at very high pressure),

(a) $\frac{a}{V^2}$ is negligible in comparison to p

(b) b is negligible in comparison to V

(c) both are negligible

(d) none of them is negligible

174. For an acid, it is always true that lower the

(a) pH value, higher the acid strength

(b) molarity, higher the acid strength

(c) K_a value, higher the acid strength

(d) pK_a value, higher the acid strength

175. The hydrogen-ion concentration in weak acid HA of dissociation constant K_a and concentration C is nearly equal to

(a) $\sqrt{\frac{K_a}{C}}$ (b) $\frac{C}{K_a}$ (c) $K_a \cdot C$ (d) $\sqrt{K_a \cdot C}$

176. For pure water at an unknown temperature, the pH should always be equal to

(a) 7 (b) >7 (c) <7 (d) any of them

177. A first-order reaction is carried out with an initial concentration of 10 mol/L and 80% of the reactant changed into the product. Now if the same reaction is carried out with an initial concentration of 5 mol/L, the percentage of the reactant changing to the product will be

(a) 40 (b) 80 (c) 160 (d) 20

178. The buffering action of an acidic buffer is maximum when its pH is equal to

(a) 5 (b) 7 (c) 1 (d) pK_a

179. van der Waals equation: $\left(p + \frac{a}{4V^2}\right)\left(V - \frac{b}{2}\right) = \frac{RT}{2}$ is valid for

(a) 1 mole of an ideal or real gas (b) 2 moles of an ideal or real gas

(c) $\frac{1}{2}$ mole of an ideal or real gas (d) $\frac{1}{2}$ mole of real gas only

180. When an ideal gas filled in a closed vessel is heated through 1°C, its pressure increases by 0.4%, the initial temperature of the gas was

(a) 250 K (b) 2500 K (c) 250°C (d) 25°C

181. Which of the following relations is correct for the mole fraction of the solute (x)? (m and M are the molality of the solution and molar mass of the solvent respectively.)

(a) $x = \frac{mM}{1000 + mM}$ (b) $x = \frac{mM}{1000 - mM}$

(c) $x = \frac{1 + mM}{mM}$ (d) $x = \frac{1 - mM}{mM}$

182. The correct statement(s) of Charles's law is (are),

(a) pressure remaining constant, all gases expand (or contract) by the same fraction of its volume at 0°C for 1° rise (or fall) in temperature.

(b) The volume of a given amount of any gas at constant pressure increases linearly with temperature at any scale, i.e., °C or K.

(c) In the plot V vs T at constant pressure, a straight line of positive slope is obtained. This line passes through the origin only if the temperature is in the Kelvin scale.

(d) For any number of gases of the same volume and at the same temperature, the slopes of the above plot (c) would be same for all these gas samples.

183. The volume of 0.1 M HCl required to neutralise completely 2 g of an equimolar mixture of Na_2CO_3 and $NaHCO_3$ is

(a) 318.76 mL (b) 325 mL (c) 215 mL (d) 225 mL

184. The diameter of molecule B is half that of molecule A. The ratio of mean free path (λ_A/λ_B) will be

(a) $\frac{1}{2}$ (b) $\frac{1}{4}$ (c) $\frac{2}{1}$ (d) $\frac{4}{1}$

185. Which of the following is not a state function?

(a) $(q + w)$ (b) $\frac{q}{T}$ (c) $(E + pV)$ (d) $\frac{q}{w}$

186. Two moles of an ideal monoatomic gas are allowed to expand adiabatically and reversibly from 300 K to 200 K. The work done by the system is ($C_V = 12.5$ J K^{-1} mol^{-1})

(a) +2.5 kJ (b) −2.5 kJ (c) +6.25 kJ (d) −50 kJ

187. The work required to raise a mass of 0.5 kg through a height of 1 m is

(a) 4.9 J (b) 49 J (c) 490 J (d) 4900 J

188. Entropy change for an adiabatic reversible process is

(a) positive (b) zero (c) negative (d) infinite

189. Three engines A, B and C take steam at 130°C and reject it at 20°C, 40°C and 50°C respectively. The most efficient engine would be

(a) A (b) B (c) C (d) all same

190. One mole of an ideal gas is expanded isothermally from 1 dm^3 to 10 dm^3 at 300 K. ΔG will be equal to

(a) 5.744 kJ (b) 57.44 J (c) 574.4 J (d) −5744 J

191. Which of the following relations gives the mathematical expression for the combined first and second law of thermodynamics?

(a) $\Delta G = V\Delta p - S\Delta T$ (b) $\Delta A = -p\Delta V - S\Delta T$

(c) $\Delta U = T\Delta S - p\Delta V$ (d) $\Delta H = T\Delta S + V\Delta p$

192. For the expression: $dG = Vdp - SdT$, which of the following is correct?

(a) $\left(\frac{\partial G}{\partial T}\right)_p = V$ (b) $\left(\frac{\partial G}{\partial p}\right)_T = V$

(c) $\left(\frac{\partial G}{\partial T}\right)_S = V$ (d) $\left(\frac{\partial G}{\partial p}\right)_T = -S$

193. Which of the following will increase with the increase in temperature?

(a) Surface tension (b) Viscosity
(c) Molality (d) Vapour pressure

194. One mole of an ideal gas expands reversibly and isothermally at 300 K from 5 dm^3 to 50 dm^3. The work done by the gas for the process is equal to

(a) −1.382 kcal (b) +1.382 kcal (c) −1381.8 kcal (d) +1382 kcal

195. The equilibrium constant for the reaction

$$A + B \rightleftharpoons C + D$$

is 10. $\Delta G°$ for the reaction at 300 K is

(a) −0.6 kcal (b) −116 kcal (c) −691 kcal (d) −1.382 kcal

196. For the reaction

$$3Br_2 + 6OH^- = 5Br^- + BrO_3^- + 3H_2O$$

Equivalent weight of Br_2 (mol. wt. M) is

(a) $\frac{M}{2}$ (b) $\frac{M}{10}$ (c) $\left(\frac{M}{2} + \frac{M}{10}\right)$ (d) $\frac{M}{6}$

[**Hint:** As Br_2 disproportionates (simultaneous oxidation and reduction) its equivalent weight is the sum of equivalent weights of the two half reactions.]

Note: Some other reactions of this type are given below. If the molecular weight of the underlined substance is M, the equivalent weights of the underlined substances are as follows:

$\underline{3I_2} + 6OH^- = 5I^- + IO_3^- + 3H_2O$; eq. wt. $= \left(\frac{M}{2} + \frac{M}{10}\right)$

$\underline{P_4} + 3OH^- + 3H_2O = 3H_2PO_2^- + PH_3$; eq. wt. $= \left(\frac{M}{4} + \frac{M}{12}\right)$

$\underline{3KClO_3} + 3H_2SO_4 = 3KHSO_4 + HClO_4 + 2ClO_2 + H_2O$;

eq. wt. $= \left(M + \frac{M}{2}\right)$

197. The equivalent weight of Cu_2S (mol. wt. = M) in the following reaction is

$$Cu_2S + MnO_4^- = Cu^{2+} + SO_2 + Mn^{2+}$$

(a) $\frac{M}{2}$ (b) $\frac{M}{6}$ (c) $\frac{M}{8}$ (d) $\frac{M}{4}$

[**Hint:** Both Cu and S in Cu_2S undergo oxidation.]

Note: Some other reactions of this type are given below. The equivalent weight of the underlined substances (mol. wt. = M) are as follows:

$\underline{Cu_2S} + 4SO_4^{2-} + 2H^+ = 2Cu^{2+} + 5SO_3^{2-} + H_2O$; eq. wt. $= \frac{M}{8}$

$\underline{As_2S_3} + 7NaClO_3 + 12NaOH = 2Na_3AsO_4 + 7NaClO + 3Na_2SO_4$;

eq. wt. $= \frac{M}{28}$

198. The equivalent weight of HNO_3 (mol. wt. = 63) in the following reaction is

$$3Cu + 8HNO_3 = 3Cu(NO_3)_2 + 2NO + 4H_2O$$

(a) $\frac{4 \times 63}{3}$ (b) $\frac{63}{5}$ (c) $\frac{63}{3}$ (d) $\frac{63}{8}$

[**Hint:** Out of 8 N atoms in HNO_3, only 2 N atoms undergo reduction to NO. Further, 6 moles of electrons are involved with 8 moles of HNO_3. $E = M + \frac{M}{3} = \frac{8}{6} M; M = 63$]

199. The equivalent weight of H_2SO_4 in the following reaction is

$$Na_2Cr_2O_7 + 3SO_2 + H_2SO_4 = 3Na_2SO_4 + Cr_2(SO_4)_3 + H_2O$$

(a) 98 (b) $\frac{98}{6}$ (c) $\frac{98}{2}$ (d) $\frac{98}{8}$

[**Hint:** 6 moles of electrons are involved with 1 mole of H_2SO_4 in this redox reaction.]

200. The largest diagonal in a cubic crystal of edge length a is

(a) $\sqrt{3a}$ (b) $\sqrt{2a}$ (c) $\sqrt{3}a$ (d) $\sqrt{2}a$

201. The decreasing order of density of different crystals is

(a) f.c.c. > b.c.c. > simple cube (b) b.c.c. > f.c.c. > simple cube

(c) simple cube > f.c.c. > b.c.c. (d) h.c.p. > f.c.c. > simple cube

202. The empty space left between the spheres in close-packed structure is called voids. The decreasing order of the size of voids is

(a) cubic > octahedral > tetrahedral > trigonal

(b) octahedral > tetrahedral > trigonal > cubic

(c) tetrahedral > trigonal > cubic > octahedral

(d) trigonal > cubic > octahedral > tetrahedral

203. One unit cell of NaCl contains

(a) 1 Na^+ and 6 Cl^- (b) 6 Na^+ and 1 Cl^-

(c) 4 Na^+ and 4 Cl^- (d) 6 Na^+ and 6 Cl^-

204. In a close-packed structure, if r is the radius of the spherical void and R is the radius of the spheres forming voids, the critical r/R value for the tetrahedral void is equal to

(a) 0.155 (b) 0.225 (c) 0.414 (d) 0.732

205. A reaction A + B → D + E takes place as

$$A + B \xrightarrow{k_1} 2C$$

$$C + B \xrightarrow{k_2} 2D$$

$$C + A \xrightarrow{k_3} 2E$$

The rate of disappearance of C is given by

(a) $-\dfrac{d[C]}{dt} = k_2[B][C] + k_3[A][C] - k_1[A][B]$

(b) $-\dfrac{d[C]}{dt} = k_2[B][C] + k_3[A][C]$

(c) $-\dfrac{d[C]}{dt} = k_2[D] + k_3[E] - k_1[C]$

(d) $-\dfrac{d[C]}{dt} = k_2 \cdot k_3[D][E]$

206. In the NaCl crystal, the coordination number of each ion is 6. How can the CN change to 8?

(a) By increasing temperature
(b) By keeping it for some time
(c) By increasing pressure
(d) CN does not change by any way

[**Note:** In CsCl : 8 : 8 CN changes to 6 : 6 on increasing temperature]

207. The pK_a for acetyl salicylic acid (aspirin) is 3.5. The pH of gastric juice in the human stomach is about 2–3 and the pH in the small intestine is about 8. Aspirin will be

(a) un-ionised in the small intestine and in the stomach
(b) completely ionised in the small intestine and in the stomach
(c) ionised in the stomach and almost un-ionised in the small intestine
(d) ionised in the small intestine and almost un-ionised in the stomach

208. The degree of dissociation of an electrolyte does not depend on

(a) nature of electrolyte (b) catalyst
(c) dilution (d) temperature

209. From the given plot between Z and p for a real gas, the correct relation is

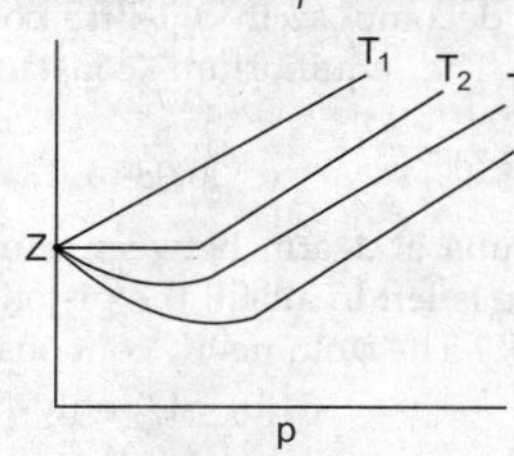

(a) $T_1 = \dfrac{2a}{Rb}$ (b) $T_2 = \dfrac{a}{Rb}$ (c) $T_3 = \dfrac{a}{Rb}$ (d) $T_4 = \dfrac{2a}{Rb}$

210. The van der Waals equation for 1 mole of gas is

$$\left(p + \frac{a}{V^2}\right)(V - b) = RT$$

where,

(a) p and V are ideal pressure and ideal volume respectively

(b) $(V-b)$ is the real volume

(c) $\left(p+\frac{a}{V^2}\right)$ is the ideal pressure

(d) $\left(p+\frac{a}{V^2}\right)$ is the real pressure

211. A mixture of H_2 and O_2 taken in a bulb in 2 : 1 mole ratio is allowed to diffuse through a fine hole. The composition of the gases (mole ratio) coming out initially is 8 : 1. The composition of the gases after some time may be

(a) 8 : 1 (b) 9 : 1 (c) 7 : 1 (d) 4 : 1

212. For a dilute solution,
lowering of vapour pressure $\propto$ mole fraction of the solute
or, lowering of vapour pressure = $K \times$ mole fraction of the solute, where K is

(a) a constant for the solute (b) a constant for the solvent
(c) a constant for the solution (d) vapour pressure of the solvent

213. For ideal binary solution,

$$p = x_A \cdot p_A^0 + x_B \cdot p_B^0$$

This equation reflects

(a) Boyle's law (b) Charles's law
(c) Dalton's law of partial pressure (d) none of these

214. Dry air is passed through a solution and then through its solvent and finally through $CaCl_2$. The ratio of weight loss in the solvent to the weight gain in $CaCl_2$ gives

(a) relative lowering of vapour pressure
(b) lowering of vapour pressure
(c) mole fraction of the solute
(d) mole fraction of the solvent

215. The relative lowering of vapour pressure of an aqueous solution of urea is 0.018. If K_b for H_2O is 0.54° m^{-1}, the elevation in boiling point will be

(a) 0.54°C (b) 0.18°C (c) 0.54 K (d) 0.18 K

216. 1 mole of A is allowed to decompose in a 1-litre container. 0.4 mole of B was produced at equilibrium. The equilibrium constant K_c for the equilibrium $A(g) \rightleftharpoons 2B(g) + C(g)$, is

(a) 0.04 (b) 0.053 (c) 0.08 (d) 0.106

217. V_1 mL of SO_2, filled in a tube at 1 atm between a movable frinctionless piston and a porous plug. The tube is left in air till the piston stops moving. The volume V_1 is found to increase to V_2. The volume V_2 contains

(a) air only (b) SO_2 only
(c) both air and SO_2 (d) neither air nor SO_2

218. The equilibrium: $PCl_5(g) \rightleftharpoons PCl_3(g) + Cl_2(g)$, may be shifted to the right if

(a) an inert gas is added at constant volume
(b) an inert gas is added at constant pressure
(c) pressure is increased
(d) PCl_3 and Cl_2 are added

219. Two flasks A and B of equal volume containing 1 mole and 2 moles of O_3

respectively, are heated to the same temperature. When the reaction $2O_3 = 3O_2$ practically stops, then both the flasks shall have

(a) the same ratio: $[O_2]/[O_3]$
(b) the same ratio: $[O_2]^{3/2}/[O_3]$
(c) only O_2
(d) the same time to reach equilibrium

220. For a 0.1 M aqueous solution of a weak acid, HA ($K_a = 10^{-9}$), the pH is approximately equal to

(a) 9 (b) 3 (c) 11 (d) 10

221. If 720 litres of a gas were collected over water at 25°C and 720 mm then the volume of the dry gas at the same temperature and pressure is (aq. tension of water at 25°C = 23.8 mm)

(a) 696.2 litres (b) 360 litres (c) 743.8 litres (d) 1440 litres

222. Which of the following values of heat of formation indicates that the product is the least stable?

(a) –94 kcal (b) –231.6 kcal (c) +21.4 kcal (d) +64.8 kcal

223. In a first-order reaction of the type: $A(g) \rightarrow 2B(g)$, the initial and final pressures are p_1 and p respectively. The rate constant can be expressed by

(a) $k = \frac{1}{t} \ln \frac{p_1}{2p_1 - p}$
(b) $k = \frac{1}{t} \ln \frac{p_1}{p_1 - p}$
(c) $k = \frac{1}{t} \ln \frac{p_1}{p - p_1}$
(d) $k = \frac{1}{t} \ln \frac{p_1}{p}$

224. The pH of a 0.1 M solution of a weak acid HA is found to be 2 at a temperature T. The osmotic pressure of the acid solution would be equal to

(a) 0.11 RT (b) 0.22 RT (c) 2 RT (d) RT

225. For a first-order reaction of the type

$$nA \rightarrow \text{product}$$

where initial concentration of A is 'a' moles/litre. The correct expression for the rate constant or half-life is

(a) $k = \frac{2.303}{nt} \log \frac{a}{(a - x)}$
(b) $k = \frac{2.303}{t} \log \frac{a}{a - x}$
(c) $t_{1/2} = \frac{0.6932}{k}$
(d) $t_{1/2} = \frac{0.6932}{nk}$

226. Which of the following give(s) the strength of the acid or base?

(a) pH
(b) Normality
(c) Degree of dissociation
(d) Dissociation constant

227. Choose the correct statement about the equilibrium

$$A(s) + 2B(g) \rightleftharpoons C(g) + D(g)$$

(a) change in temp. changes both the state of eqb. and the eqb. constant
(b) change in pressure changes both the state of eqb. and the eqb. constant
(c) change in temperature changes eqb. constant but not the state of eqb.
(d) change in pressure does not affect either eqb. constant or state of eqb.

228. Choose the wrong statement(s).

(a) If the eqb. constant changes, the state of eqb. has to change.
(b) If the state of eqb. changes, the eqb. constant has to change.

(c) The acid with a lower pK_a value has higher acid strength than the acid with a higher value of pK_a.
(d) Two acids of the same pH are always of equal acid strength

229. x and y moles/litre of A and B respectively were allowed to react $A + 2B \rightleftharpoons \frac{1}{2}C$. At equilibrium, the concentrations of A, B and C were found to be 4, 2 and 2 moles/litre respectively. x and y are
(a) 6 & 4 (b) 8 & 10 (c) 6 & 10 (d) 8 & 4

230. 116 g of A_3B_4 has 1.5 moles of A. Molecular weight of A_3B_4 is
(a) 164 (b) 232
(c) 77 (d) cannot be calculated

231. 0.8 g of silver salt of a dibasic organic acid on ignition yielded 0.54 g of metallic silver. Molecular weight of the acid is (Ag = 108)
(a) 106 (b) 108 (c) 320 (d) 85
[**Hint:** See Example 12, Chapter 5]

232. 0.80 g of chloroplatinate of a monoacid organic base on ignition gave 0.25 g of Pt. The molecular weight of the base is (Pt = 195)
(a) 624 (b) 214 (c) 107 (d) 312
[**Hint:** See Example 14, Chapter 5]

233. One volume of a gaseous organic compound of C, H and N on combustion produced 2 volumes of CO_2, 3.5 volumes of H_2O vapour and 0.5 volume of N_2 under identical conditions of temperature and pressure. The molecular formula of the compound is
(a) C_2H_7N (b) $C_4H_{14}N_2$ (c) $C_2H_7N_2$ (d) $C_4H_{14}N_4$

234. In a gravimetric determination of phosphorus, 0.248 g of an organic compound was strongly heated in a Carius tube with concentrated HNO_3. Phosphoric acid so produced was precipitated as $MgNH_4PO_4$ which on ignition yielded 0.444 g of $Mg_2P_2O_7$. The percentage of phosphorus in the compound is
(a) 2.5 (b) 5.0 (c) 7.5 (d) 50

235. The empirical formula of a substance whose composition includes H, C, O and N in the mass ratio 1 : 3 : 4 : 7 is
(a) $HC_3O_4N_7$ (b) H_4CON_2 (c) $HC_4O_4N_2$ (d) none of these

236. 500 cc of a hydrocarbon gas burnt in excess of oxygen yielded 2500 cc of CO_2 and 3 litres of H_2O vapour, measured under identical conditions of temperature and pressure. The formula of the hydrocarbon is
(a) C_5H_{12} (b) C_5H_{10} (c) C_4H_{10} (d) none of these

237. The minimum weight of H_2 filled in a balloon of weight w_1 kg to just lift it from the ground is w_2 kg. If w_3 kg of H_2 is filled into this balloon, the payload of the balloon will be (the volume of the balloon is supposed to be same with w_2 or w_3 kg of H_2)
(a) $w_2 - w_3$ (b) $(w_2 - w_3 - w_1)$ (c) $(w_3 - w_2)$ (d) $(w_3 - w_2 - w_1)$

238. The degree of dissociation (α) can be calculated using the formula,

$$\alpha = \frac{dt - d_0}{(n-1)d_0}$$

This formula is applicable at
(a) constant volume and constant temperature
(b) constant volume and variable temperature
(c) constant pressure and constant temperature
(d) all the above conditions

239. For a reaction a graph plotted between log $\left(\frac{dx}{dt}\right)$ and log $(a-x)$ along y and x axes respectively shows a straight line with a positive slope of 45°. The order of the reaction is
(a) 0 (b) 1 (c) 2 (d) 3

240. The pH of an aqueous solution of HCl is 2. Its osmotic pressure at a temperature T would be equal to
(a) 0.01 RT (b) 0.02 RT (c) RT (d) 2 RT

241. The K_p value for a homogeneous gaseous reaction is found to vary with the change in pressure. The correct conclusion is
(a) the gases behave ideally
(b) the gases deviate from ideal behaviour
(c) the gases in a reaction do not behave ideally
(d) K_p variation with pressure is a consequence of experimental error

242. For a first-order reaction, $A \xrightarrow{k} B$, the degree of dissociation is equal to
(a) e^{-kt} (b) $1-e^{-kt}$ (c) e^{kt} (d) $1+e^{-kt}$

[**Hint:** For $\underset{(1-x)}{\overset{1\text{ M}}{A}} \longrightarrow \underset{x}{B}$ integrate $\frac{dx}{dt}=k(1-x)$]

243. 1 mole of N_2 and 3 moles of H_2 filled in a one-litre bulb were allowed to react. When the reaction attained equilibrium, two-thirds of N_2 converted to NH_3 ($N_2+3H_2 \rightleftharpoons 2NH_3$). If a hole is then made in the bulb, the mole ratio of the gases N_2, H_2 and NH_3 effusing out initially would be respectively.
(a) 1 : 3 : 4 (b) $\sqrt{28}:\sqrt{2}:\sqrt{17}$
(c) $\frac{1}{\sqrt{28}}:\frac{1}{\sqrt{2}}:\frac{1}{\sqrt{17}}$ (d) $\frac{1}{\sqrt{28}}:\frac{3}{\sqrt{2}}:\frac{4}{\sqrt{17}}$

244. In the above problem (243) what would be to the state of equilibrium during the course of effusion?
(a) Remains same (b) Shifts right
(c) Shifts left (d) First shifts right and then left

245. The reaction A + 2B → product, is first order in A and second order in B. If $t_{1/2}$ w.r.t. A is same as the $t_{1/2}$ of the reaction,
(a) $[A]_{\text{Initial}} > [B]_{\text{Initial}}$ (b) $[A]_{\text{Initial}} = [B]_{\text{Initial}}$
(c) A is the limiting reactant (d) B is the limiting reactant

246. For which of the following reactions, the degree of dissociation cannot be calculated applying the formula: $\alpha=\frac{dt-d_0}{(n-1)d_0}$
(a) $PCl_5 \rightleftharpoons PCl_3+Cl_2$ (b) $2NH_3 \rightleftharpoons N_2+3H_2$
(c) $2HI \rightleftharpoons H_2+I_2$ (d) $2KClO_3 \rightleftharpoons 2KCl+3O_2$

247. In an f.c.c. unit cell, atoms are numbered as shown below. The atoms not touching each other are

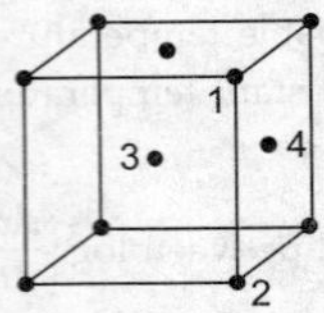

(a) 3 & 4 (b) 1 & 2 (c) 1 & 3 (d) 2 & 4

248. In a b.c.c. unit cell, atoms are numbered as shown below. The atoms touching each other are

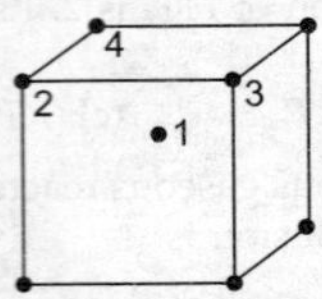

(a) 1 & 2 (b) 2 & 3 (c) 3 & 4 (d) 1 & 4

249. The coordination number of h.c.p. or c.c.p. structure is 12, that is, one atom touches 12 other atoms. The 12 atoms lie as

(a) 8 atoms are on the same plane, 2 above and 2 below the plane
(b) 6 atoms are on the same plane, 3 above and 3 below the plane
(c) 4 atoms are on the same plane, 4 above and 4 below the plane
(d) 2 atoms are on the same plane, 5 above and 5 below the plane

250. In a simple cubic crystal, the corner atom touches

(a) 3 other atoms (b) 4 other atoms
(c) 6 other atoms (d) 8 other atoms

251. In a simple cubic crystal, each atom is shared by

(a) 2 unit cells (b) 4 unit cells (c) 6 unit cells (d) 8 unit cells

252. In an f.c.c. crystal, which of the following shaded planes contains the following arrangement of atoms?

(a) 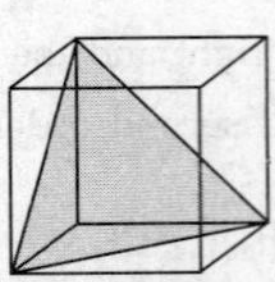 (b) 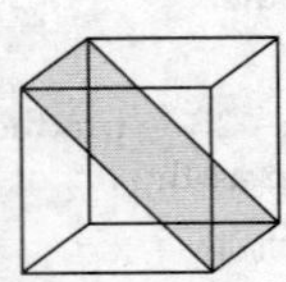

(c) 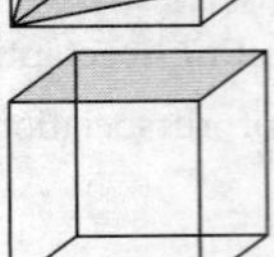 (d) 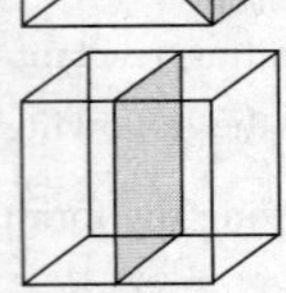

253. In which of the following aqueous solutions, the degree of dissociation of water is maximum?

(a) NH_4Cl solution (b) CH_3COONa solution
(c) CH_3COONH_4 solution (d) Same in all

254. When a drop of a concentrated HCl solution is added to one litre of pure water at 25°C, the pH drops from 7 to about 4. When the second drop of the same HCl solution is added, the pH further drops to about
(a) 3.7 (b) 2 (c) 1 (d) 0

255. The quantum numbers $+\frac{1}{2}$ and $-\frac{1}{2}$ for the electron spin represent
(a) rotation of the electron in clockwise and anticlockwise directions respectively
(b) rotation of the electron in anticlockwise and clockwise directions respectively
(c) magnetic moment of the electron pointing up and down respectively
(d) two quantum mechanical spin states which have no classical analogue
(IIT 2001)

256. Which one of the following statements is false?
(a) Work is a state function.
(b) Temperature is a state function.
(c) Change in the state is completely defined when the initial and final states are specified.
(d) Work appears at the boundary of the system. **(IIT 2001)**

257. In thermodynamics, a process is called reversible when
(a) surroundings and system change into each other
(b) there is no boundary between system and surroundings
(c) the surroundings are always in equilibrium with the system
(d) the system changes into the surroundings spontaneously **(IIT 2001)**

258. In a solid 'AB' having the NaCl structure, A atoms occupy the corners of the cubic unit cell. If all the face-centred atoms along one of the axes are removed then the resultant stoichiometry of the solid is
(a) AB_2 (b) A_2B (c) A_4B_3 (d) A_3B_4 **(IIT 2001)**

259. An aqueous solution of 6.3 g oxalic acid dihydrate is made up to 250 mL. The volume of 0.1 N NaOH required to completely neutralize 10 mL of this solution is
(a) 40 mL (b) 20 mL (c) 10 mL (d) 4 mL **(IIT 2001)**

260. In the standardization of $Na_2S_2O_3$ using $K_2Cr_2O_7$ by iodometry, the equivalent weight of $K_2Cr_2O_7$ is
(a) (molecular weight)/2 (b) (molecular weight)/6
(c) (molecular weight)/3 (d) same as molecular weight **(IIT 2001)**

261. If '*I*' is the intensity of absorbed light and '*C*' is the concentration of AB for the photochemical process $AB + h\nu \rightarrow AB^*$, the rate of formation of AB^* is directly proportional to
(a) *C* (b) *I* (c) I^2 (d) *C.I* **(IIT 2001)**

262. The correct order of equivalent conductance at infinite dilution of LiCl, NaCl and KCl is
(a) LiCl > NaCl > KCl (b) KCl > NaCl > LiCl
(c) NaCl > KCl > LiCl (d) LiCl > KCl > NaCl **(IIT 2001)**

263. At constant temperature, the equilibrium constant (K_p) for the decomposition reaction $N_2O_4 \rightleftharpoons 2NO_2$ is expressed by $K_p = (4x^2p)/(1-x^2)$, where p = pressure, x = extent of decomposition. Which one of the following statements is true?

(a) K_p increases with increase of p.

(b) K_p increases with increase of x.

(c) K_p increases with decrease of x.

(d) K_p remains constant with change in p and x. **(IIT 2001)**

264. The root-mean-square velocity of an ideal gas at constant pressure varies with density (d) as

(a) d^2 (b) d (c) $\sqrt{d}$ (d) $\frac{1}{\sqrt{d}}$ **(IIT 2001)**

265. The wavelength associated with a golf ball weighing 200 g and moving at a speed of 5 m/h is of the order (Planck constant, $h = 6.626 \times 10^{-34}$ J s)

(a) 10^{-10} m (b) 10^{-20} m (c) 10^{-30} m (d) 10^{-40} m **(IIT 2001)**

266. The set with correct order of acidity is

(a) $HClO < HClO_2 < HClO_3 < HClO_4$

(b) $HClO_4 < HClO_3 < HClO_2 < HClO$

(c) $HClO < HClO_4 < HClO_3 < HClO_2$

(d) $HClO_4 < HClO_2 < HClO_3 < HClO$ **(IIT 2001)**

267. The reaction $3ClO^-_{(aq)} \rightarrow ClO^-_{3(aq)} + 2Cl^-_{(aq)}$ is an example of

(a) oxidation reaction (b) reduction reaction

(c) disproportionation (d) decomposition reaction **(IIT 2001)**

268. Saturated solution of KNO_3 is used to make 'salt-bridge' because

(a) velocity of K^+ is greater than that of NO_3^-

(b) velocity of NO_3^- is greater than that of K^+

(c) velocities of both K^+ and NO_3^- are nearly the same

(d) KNO_3 is highly soluble in water **(IIT 2001)**

269. For a sparingly soluble salt A_pB_q, the relationship of its solubility product (L_s) with its solubility (S) is

(a) $L_s = S^{p+q} \cdot p^p \cdot q^q$ (b) $L_s = S^{p+q} \cdot p^q \cdot q^p$

(c) $L_s = S^{pq} \cdot p^p \cdot q^q$ (d) $L_s = S^{pq} \cdot (pq)^{p+q}$ **(IIT 2001)**

The questions below (270 to 279) consist of an *'assertion'* in column 1 and the *'reason'* in column 2. Use the following key to choose the appropriate answer.

(a) If both the *assertion* and *reason* are correct, and the *reason* is the correct explanation of the *assertion*.

(b) If both the *assertion* and *reason* are correct but the *reason* is not the correct explanation of the *assertion*.

(c) If *assertion* is correct but *reason* is incorrect.

(d) If *assertion* is incorrect but *reason* is correct.

	Assertion (column 1)	*Reason* (column 2)
270.	The pressure of a fixed amount of an ideal gas is proportional to its temperature.	Frequency of collisions and their impact both increase in proportion to the square root of temperature. **(IIT 2001)**
271.	The heat absorbed during the isothermal expansion of an ideal gas against vacuum is zero.	The volume occupied by the molecules of an ideal gas is zero. **(IIT 2001)**
272.	In any ionic solid [MX] with Schottky defects, the number of positive and negative ions are same.	Equal number of cation and anion vacancies are present. **(IIT 2001)**
273.	When H_2S is passed into an aqueous solution of Zn^{2+} in the presence of NH_4OH, ZnS gets precipited.	Common-ion effect supresses the dissociation of weak electrolytes.
274.	The time for completion of a first-order reaction is infinite.	The time for completion of any definite fraction of the reaction is constant for first-order reaction.
275.	The addition of an inert gas at constant volume may affect the state of equilibrium: $PCl_5(g) = PCl_3(g) + Cl_2(g)$	The addition of inert gas at constant volume increases both the number of molecules and pressure to the same extent.
276.	$Fe^{3+} + e = Fe^{2+}$; $E^0 = +0.77$ V $Fe^{2+} + 2e = Fe$; $E^0 = -0.44$ V $\therefore$ $Fe^{3+} + 3e = Fe$; $E^0 = -0.04$ V	E^0 is an extensive property.
277.	van der Waals equation describes the behaviour of real gases.	The kinetic theory postulates of negligible volume of gaseous molecules and intermolecular forces of attraction do not stand correct at high pressure and low temperature.
278.	The nuclide ${}^{30}_{13}Al$ is less stable than ${}^{40}_{20}Ca$.	Nuclides having odd number of protons and neutrons are generally unstable. **(IIT 1998)**
279.	The value of van der Waals constant '*a*' is larger for ammonia than for nitrogen.	Hydrogen bonding is present in ammonia. **(IIT 1998)**

280. For which of the following types of reactions does $t_{1/2}$ go on decreasing with the progress of the reaction?

(a) Zero order (b) First order (c) Second order (d) Third order

281. When a reaction progresses, which of the following facts is not true?

(a) $t_{1/2}$ remains constant throughout in first-order reactions.

(b) $t_{1/2}$ goes on decreasing in zero-order reactions.

(c) $t_{1/2}$ goes on decreasing in first-order reactions.

(d) $t_{1/2}$ goes on increasing in second-order reactions.

282. A hypothetical reaction $A_2 + B_2 \rightarrow 2AB$ follows the following mechanism.

$$A_2 \rightarrow A + A \quad \text{(fast)}$$

$$A + B_2 \rightarrow AB + B \quad \text{(slow)}$$
$$A + B \rightarrow AB \quad \text{(fast)}$$

The order of the overall reaction is

(a) 0 (b) 1 (c) 2 (d) 3/2

283. The internal energy change when a system goes from the state A to B is 40kJ/mol. If the system goes from A to B by a reversible path and returns to the state A by an irreversible path, what would be the net change in internal energy?

(a) > 40 kJ (b) < 40 kJ (c) Zero (d) 40 kJ

284. If at 298 K the bond energies of C—H, C—C, C=C and H—H bonds are respectively 414, 347, 615 and 435 kJ mol^{-1}, the value of enthalpy change for the reaction given below at 298 K will be

$$H_2C{=}CH_2\,(g) + H_2\,(g) \rightarrow H_3C{-}CH_3\,(g)$$

(a) –250 KJ (b) +125 KJ (c) –125 KJ (d) +250 kJ

285. One mole of the complex compound $Co(NH_3)_5Cl_3$, gives 3 moles of ions on dissolution in water. One mole of the same complex reacts with two moles of $AgNO_3$ solution to yield two moles of AgCl (s). The structure of the complex is

(a) $[Co(NH_3)_3\,Cl_3]\cdot 2NH_3$ (b) $[Co(NH_3)_4\,Cl_2]Cl\cdot NH_3$

(c) $[Co(NH_3)_4\,Cl]Cl_2\cdot NH_3$ (d) $[Co(NH_3)_5\,Cl]Cl_2$

286. How many unit cells are present in a cube-shaped ideal crystal of NaCl of mass 1.00 g?

(a) 5.14×10^{21} (b) 1.28×10^{21} (c) 1.71×10^{21} (d) 2.57×10^{21}

287. In an irreversible process taking place at constant temperature and constant pressure and in which only *p*–*V* work is being done, the change in Gibbs free energy (dG) and the change in entropy (dS), satisfy the criteria

(a) $(dS)_{V,E} > 0, (dG)_{T,p} < 0$ (b) $(dS)_{V,E} = 0, (dG)_{T,p} = 0$

(c) $(dS)_{V,E} = 0, (dG)_{T,p} > 0$ (d) $(dS)_{V,E} < 0, (dG)_{T,p} < 0$

288. What volume of hydrogen gas at 273 K and 1 atm pressure will be consumed in obtaining 21.6 g of elemental boron (atomic mass = 10.8) from the reduction of BCl_3 by hydrogen?

(a) 67.2 L (b) 44.8 L (c) 22.4 L (d) 89.6 L

289. Standard reduction electrode potentials of three metals A, B and C are respectively + 0.5 V, –3.0 V and –1.2 V. The reducing powers of these metals are

(a) A > B > C (b) C > B > A (c) A > C > B (d) B > C > A

290. If liquids A and B form an ideal solution,

(a) the entropy of mixing is zero

(b) the free energy of mixing is zero

(c) the free energy as well as the entropy of mixing are each zero

(d) the enthalpy of mixing is zero

291. The rate law from a reaction between the substances A and B is given by

$$\text{rate} = k\,[A]^n\,[B]^m$$

On doubling the concentration of A and halving the concentration of B, the ratio of the new rate to the earlier rate of the reaction will be as

(a) $(m+n)$ (b) $(n-m)$ (c) $2^{(n-m)}$ (d) $\frac{1}{2^{(m+n)}}$

292. The effect of a tenfold pressure increase on the equilibrium composition of the reaction

$$N_2\,(g) + 3H_2\,(g) \rightleftharpoons 2NH_3\,(g) \quad \text{is}$$

(a) 100-fold increase in K_x (b) 10-fold increase in K_x
(c) no change in K_x (d) 10-fold decrease in K_x

293. The addition of one of the reactants in a reaction at equilibrium
(a) always shifts the equilibrium towards product at constant volume but not always at constant pressure
(b) always shifts the equilibrium towards product at constant pressure but not always at constant volume
(c) always shifts the equilibrium towards product either at constant pressure or at constant volume
(d) does not shift the equilibrium either at constant pressure or at constant volume

294. A crystal of formula AB_3 has A ions at the cube corners and B ions at the edge centres. The coordination numbers of A and B are respectively
(a) 6 and 2 (b) 2 and 6 (c) 6 and 6 (d) 8 and 8

295. The rate of a first-order reaction is 1.5×10^{-2} mol L^{-1} at 0.5 M concentration of the reactant. The half-life of the reaction is
(a) 0.383 min (b) 23.1 min (c) 8.73 min (d) 7.53 min

296. The standard emf of a galvanic cell involving cell reaction with $n = 2$ is found to be 0.295 V at 25°C. The equilibrium constant of the reaction would be
(a) 2.0×10^{11} (b) 4.0×10^{12} (c) 1.0×10^{2} (d) 1.0×10^{10}

297. The rapid change of pH near the stoichiometric point of an acid–base titration is the basis of indicator detection. The pH of the solution is related to ratio of the concentration of the conjugate acid (HIn) and base (In^-) forms of the indicator by the expression

(a) $\log \dfrac{[In^-]}{[HIn]} = pK_{In} - pH$ (b) $\log \dfrac{[HIn]}{[In]} = pK_{In} - pH$

(c) $\log \dfrac{[HIn]}{[In^-]} = pH - pK_{In}$ (d) $\log \dfrac{[In^-]}{[HIn]} = pH - pK_{In}$

298. The frequency of radiation emitted when the electron falls from $n = 4$ to $n = 1$ in a hydrogen atom will be (given ionization energy of $H = 2.18 \times 10^{-18}$ J atom^{-1} and $h = 6.625 \times 10^{-34}$ J s
(a) 1.54×10^{15} s^{-1} (b) 1.03×10^{15} s^{-1} (c) 3.08×10^{15} s^{-1} (d) 2.00×10^{15} s^{-1}

[Hint: $\Delta E = h\nu = E_1\left(\dfrac{1}{n_1^2} - \dfrac{1}{n_2^2}\right)$**]**

299. The work done during the expansion of a gas from a volume of 4 dm^3 to 6 dm^3 against a constant external pressure of 3 atm is (1 L atm = 101.32 J)
(a) –6 J (b) –608 J (c) +304 J (d) –304 (J)

300. The emf of the cell,

$$Zn \mid Zn^{2+}\ (a = 0.1\ M) \parallel Fe^{2+}\ (a = 0.01\ M) \mid Fe$$

is 0.2905 V. The equilibrium constant of the cell reaction is
(a) $10^{0.32/0.0591}$ (b) $10^{0.32/0.0295}$ (c) $10^{0.26/0.0295}$ (d) $e^{0.32/0.295}$

301. HX is a weak acid ($K_a = 10^{-5}$). It forms a salt NaX (0.1 M) on reacting with caustic soda. The degree of hydrolysis of NaX is

(a) 0.01% (b) 0.0001% (c) 0.1% (d) 0.5%

302. Spontaneous adsorption of a gas on solid surface is an exothermic process because

(a) ΔH increases for system (b) ΔS increases for gas

(c) ΔS decreases for gas (b) ΔG increases for gas

303. For a monoatomic gas, kinetic energy is equal to E, its relation with rms velocity is

(a) $C = \sqrt{\frac{2E}{M}}$ (b) $C = \sqrt{\frac{3E}{2M}}$ (c) $C = \sqrt{\frac{E}{2M}}$ (d) $C = \sqrt{\frac{E}{3M}}$

304. The pair of compounds having metals in their highest oxidation state is

(a) MnO_2, $FeCl_3$ (b) $[MnO_4]^-$, CrO_2Cl_2

(c) $[Fe(CN)_6]^{3-}$, $[Co(CN)_3]$ (d) $[NiCl_4]^{2-}$, $[CoCl_4]^-$

305. The spin magnetic moment of the compound $Hg[Co(SCN)_4]$ is

(a) $\sqrt{3}$ (b) $\sqrt{15}$ (c) $\sqrt{24}$ (d) $\sqrt{8}$

[**Hint:** $\mu = \sqrt{n(n+2)}$ BM]

306. Which hydrogenlike species will have the same radius as that of the Bohr orbit of the hydrogen atom?

(a) $n = 2$, Li^{2+} (b) $n = 2$, Be^{3+} (c) $n = 2$, He^+ (d) $n = 3$, Li^{2+}

[**Hint:** $r = \frac{n^2h^2}{4\pi^2 m z e^2}$]

307. 0.004 M Na_2SO_4 is isotonic with 0.01 M glucose. Degree of dissociation of Na_2SO_4 is

(a) 75% (b) 50% (c) 25% (d) 85%

[**Hint:** $(1 + 2\alpha)0.004 = 0.01$]

308. $\Delta H_{vap} = 30$ kJ/mol and $\Delta S_{vap} = 75$ J mol^{-1} K^{-1}. Find the temperature of the vapour at 1 atm.

(a) 400 K (b) 350 K (c) 298 K (d) 250 K

[**Hint:** See p. 429: $T = \frac{\Delta H}{\Delta S}$]

309. 2 mol of an ideal gas expands isothermally and reversibly from 1 litre to 10 litres at 300 K. What is the enthalpy change?

(a) 4.98 kJ (b) 11.47 kJ (c) −11.47 kJ (d) 0 kJ

310. (A) follows a first-order reaction: (A) → Product, and the concentration of A changes from 0.1 M to 0.025 M in 40 minutes. Find the rate of reaction of A when the concentration of A is 0.01 M.

(a) 3.47×10^{-4} M min^{-1} (b) 3.47×10^{-5} M min^{-1}

(c) 1.73×10^{-4} M min^{-1} (d) 1.73×10^{-5} M min^{-1}

311. When I^- is oxidised by MnO_4^- in an alkaline medium, I^- converts into

(a) IO_3^- (b) I_2 (c) IO_4^- (d) IO^-

312. Which of the following will not be oxidised by O_3?

(a) KI (b) $FeSO_4$ (c) $KMnO_4$ (d) K_2MnO_4

[**Hint:** In $KMnO_4$, Mn is in maximum possible oxidation state.]

313. Which of the following f.c.c. structures contains cations in alternate voids?

(a) NaCl (b) ZnS (c) Na_2O (d) CaF_2

[**Hint:** See p. 744]

314. The elevation in boiling point when 13.44 g of freshly prepared $CuCl_2$ are added to one kilogram of water is ($K_b = 0.52$ kg K mol^{-1}, mol. wt. of $CuCl_2 = 134.4$)

(a) 0.05 (b) 0.1 (c) 0.156 (d) 0.21

315. The half-cell reactions of rusting of iron are

$$2H^+ + \frac{1}{2}O_2 + 2e \rightarrow H_2O; \quad E^0 = +1.23 \text{ V}$$

$$Fe^{2+} + 2e \rightarrow Fe; \quad E^0 = -0.44 \text{ V}$$

ΔG^0 (in kJ) for the reaction is

(a) –76 (b) –322 (c) –122 (d) –176

[**Hint:** $Fe + 2H^+ + \frac{1}{2}O_2 \rightarrow Fe^{2+} + H_2O;\ E^0_{cell} = 1.67$ V]

316. The number of radial nodes in 3s and 2p respectively are

(a) 2 and 0 (b) 1 and 2 (c) 0 and 2 (d) 2 and 1

[**Hint:** See p. 250, Chapter 11]

317. 0.1 mole of $CH_3 \cdot NH_2$ ($K_b = 5 \times 10^{-4}$) is added to 0.08 mole of HCl and the solution is diluted to one litre. The resulting $[H^+]$ is

(a) 1.6×10^{-11} (b) 8×10^{-11} (c) 5×10^{-5} (d) 8×10^{-2}

[**Hint:** The resulting solution is a buffer solution.]

318. If He and CH_4 are allowed to diffuse out of the container under similar conditions of temperature and pressure then the ratio of the rate of diffusion of He to CH_4 is

(a) 2.0 (b) 1.0 (c) 0.5 (d) 4.0

319. Which of the following is correct for lyophilic sols?

(a) They are irreversible.

(b) They are formed by inorganic substances.

(c) They are readily coagulated by addition of electrolytes.

(d) They are self-stabilized.

320. Which of the following statements is incorrect about the order of a reaction?

(a) The order of a reaction is determined experimentally.

(b) It is the sum of the powers of concentration terms in the rate law expression.

(c) It does not necessarily depend on stoichiometric coefficients.

(d) The order of a reaction can not have a fractional value.

321. One mole of a monoatomic ideal gas expands adiabatically at initial temperature T against a constant external pressure of 1 atm from one litre to two litres. Find out the final temperature. (R = 0.0821 L atm K^{-1} mol^{-1})

(a) T (b) $\frac{T}{(2)^{5/3-1}}$ (c) $T-\frac{2}{3\times 0.0821}$ (d) $T+\frac{2}{3\times 0.0821}$

[**Hint:** In adiabatic expansion, $q=0$

$\therefore\ \Delta U = q + w = w = -p_{ext}(V_2 - V_1)$

Also $\Delta U = n\,C_V(T_2 - T_1)$ (Eqn. 6, Chapter 14)]

322. van der Waals equation may be reduced to the following form:

$$Z = 1 + \left(b - \frac{a}{RT}\right)\frac{1}{\overline{V}} + \left(\frac{b}{\overline{V}}\right)^2 + \ldots.$$

This equation shows that (tick the incorrect one).

(a) if $b > a/RT$, then the size effect dominates the behaviour of the gas and the slope of Z versus P is positive.

(b) if $b < a/RT$, then the effect of the attractive forces dominates the behaviour of the gas and the initial slope of Z versus P is negative.

(c) if $b = a/RT$, T is then the Boyle temperature and Z is independent of P.

(d) the ideal behaviour of the gas vanishes not only as the P approaches zero but also as T approaches ∞.

[**Hint:** Substitute $\overline{V} = RT/P$ in the given equation, see Ex. 66, p. 349]

323. Which of the following statements is incorrect for the chemical reaction

$$H_2O_2 + O_3 = H_2O + 2O_2?$$

(a) H_2O_2 is oxidised to O_2 (b) H_2O_2 is reduced to H_2O

(c) O_3 simply decomposes to O_2 (d) O_3 is reudced to O_2

[**Hint:** $\underset{-1}{H_2O_2} + \underset{0}{O_3} = \underset{-2}{H_2O} + \underset{0}{2O_2}$]

324. In the following fuel cell reaction

$$C_5H_{12} + 8O_2 = 5CO_2 + 6H_2O$$

the number of moles of electrons involved per mole of C_5H_{12} is

(a) 32 (b) 12 (c) 20 (d) 8

[**Hint:** $\underset{\substack{(-12)\\ \text{(for 5 C)}}}{C_5H_{12}} + 8O_2 = \underset{\substack{(+4\times 5)\\ \text{(for 5 C)}}}{5CO_2} + 6H_2O$]

325. In Bohr theory, if E_n, K_n, V_n and r_n denote total energy, kinetic energy, potential energy and radius of nth orbit respectively, which of the following is (are) correct?

(a) $\frac{V_n}{K_n} = -2$ (b) If $r_n \propto E_n^x$, then $x = -1$

(c) Angular momentum in lowest orbit = 0

(d) If $\frac{1}{r_n} \propto Z^y$, then $y = 1$

[**Hint:** $K_n = \frac{e^2}{8\pi\varepsilon_0 r_n}$, $V_n = -\frac{e^2}{4\pi\varepsilon_0 r_n}$

$E_n = -\frac{e^2}{8\pi\varepsilon_0 r_n}$, $r_n = \frac{\varepsilon_0 n^2 h^2}{\pi m z e^2}$

and $L = \sqrt{l(l+1}\cdot\frac{h}{2\pi}$]

326. A monoatomic gas undergoes a process in which the ratio P to V at any instant is constant and equals to 1. What is molar heat capacity of the gas?

(a) $\frac{4R}{2}$ (b) $\frac{3R}{2}$

(c) $\frac{5R}{2}$ (d) 0

[**Hint:** On heating the gas both P and V increase in such a way that their ratio is maintained at 1.]

327. The direct conversion of A to B is difficult, hence it is carried out by the followed path.

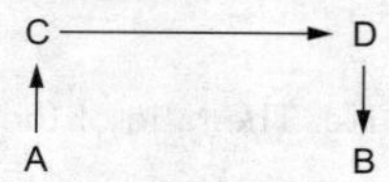

Given: $\Delta S_{A\to c} = 50$ eu, $\Delta S_{c\to D} = 30$ eu, $\Delta S_{B\to D} = 20$ eu. The value of $\Delta S_{A\to B}$ is

(a) +100 eu (b) +60 eu (c) –100 eu (d) –60 eu

328. The species present in solution when CO_2 is dissolved in water are

(a) CO_2, H_2CO_3, HCO_3^-, CO_3^{2-} (b) H_2CO_3, CO_3^{2-}

(c) CO_3^{2-}, HCO_3^- (d) CO_2, H_2CO_3

329. The given graph represents the variation of compressibility factor (Z) versus P for three real gases A, B and C. Identify the only incorrect statement.

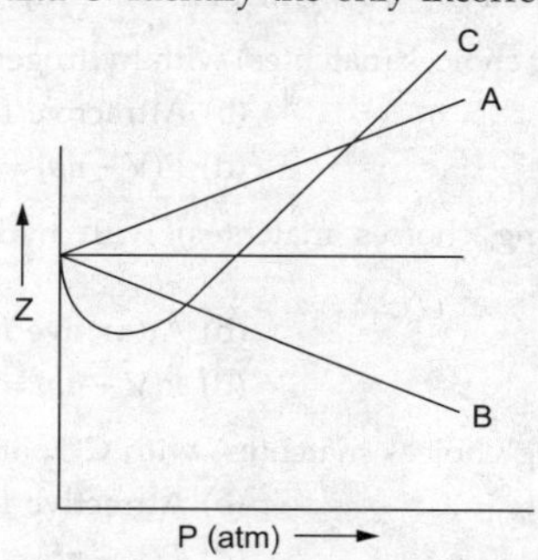

(a) For the gas A, $a = 0$ and its dependance on P is linear at all pressures.

(b) For the gas B, $b = 0$ and its dependance on P is linear at all pressures.

(c) For the gas C, which is typical real gas for which neither 'a' nor 'b' $= 0$. By knowing the minima and the point of intersection, with $Z = 1$, a and b can be calculated.

(d) At high pressure, the slope is positive for all gases.

330. A nuclear explosion has taken place leading to increasing in concentration of C^{14} in nearby areas. C^{14} concentration is C_1 in nearby areas and C_2 in areas far away. If the age of the fossil is determined to be T_1 and T_2 at the places respectively then

(a) The age of the fossil will increase where explosion has taken place and $T_1 - T_2 = \frac{1}{\lambda} \ln \frac{C_1}{C_2}$

(b) The age of the fossil will decrease at the place where explosion has taken place and $T_1 - T_2 = \frac{1}{\lambda} \ln \frac{C_1}{C_2}$

(c) The age of the fossil will be same

(d) $\frac{T_1}{T_2} = \frac{C_1}{C_2}$

$\left[\textbf{Hint: } \lambda = \frac{1}{T_1} \ln \frac{C_1}{X} = \frac{1}{T_2} \ln \frac{C_2}{X}\right]$

331. A positron is emitted from $^{23}_{11}Na$. The ratio of the atomic mass and atomic number of the resulting nuclide is

(a) $\frac{22}{10}$ (b) $\frac{22}{11}$ (c) $\frac{23}{10}$ (d) $\frac{23}{12}$ **(IIT 2007)**

332. The edge length of unit cell of a metal having molecular weight 75 g/mol is 5Å which crystallises in cubic lattice. If the density is 2 g/c.c., then the radius of the metal atom in pm is

(a) 216 (b) 320 (c) 432 (d) 108

[**Hint:** First calculate $Z(=2)$ which means the structure is b.c.c. for which $r = \sqrt{3/4} \cdot a$;

$1 \text{ Å} = 10^{-10}$ m and $1 \text{ m} = 10^{12}$ pm; $1 \text{ Å} = 10^2$ pm]

333. Which of the following choices match(es) with hydrogen gas at 200 atm and 273 K?

(a) $Z \neq 1$ (b) Attractive forces are dominant

(c) $PV = nRT$ (d) $P(V - nb) = nRT$ **(IIT 2007)**

334. Which of the following choices match(es) with hydrogen gas at $P \approx 200$ and $T = 273$ K?

(a) $Z \neq 1$ (b) Attractive forces are dominant

(c) $PV = nRT$ (d) $P(V - nb) = nRT$ **(IIT 2007)**

335. Which of the following choices match(es) with CO_2 at $P = 1$ atm and $T = 273$ K?

(a) $Z \neq 1$ (b) Attractive forces are dominant

(c) $PV = nRT$ (d) $P(V - nb) = nRT$ **(IIT 2007)**

336. Which of the following choices match(es) with the real gas with very large molar volume?

(a) $Z \neq 1$ (b) Attractive forces are dominant

(c) $PV = nRT$ (d) $P(V - nb) = nRT$ **(IIT 2007)**

337. Which of the following characteristic features match(es) with the crystal systems: simple cubic and face centred cubic?

(a) Both have the cell parameters $a = b = c$ and $\alpha = \beta = \gamma$

(b) Both belong to different crystal systems

(c) Both have only two crystallographic angles of 90°

(d) Both belong to same crystal system **(IIT 2007)**

338. Which of the following characteristic features match(es) with the crystal systems: cubic and rhombohedral?

(a) Both have the cell parameters $a = b = c$ and $\alpha = \beta = \gamma$

(b) Both belong to different crystal systems

(c) Both have only two crystallographic angles of 90°

(d) Both belong to same crystal system **(IIT 2007)**

339. Which of the following characteristic features match(es) with the crystal systems: cubic and tetragonal?

(a) Both have the cell parameters; $a = b = c$ and $\alpha = \beta = \gamma$

(b) Both belong to different crystal systems

(c) Both have only two crystallographic angles of 90°

(d) Both belong to same crystal system **(IIT 2007)**

340. Which of the following characteristic features match(es) with the crystal systems: hexagonal and monoclinic?

(a) Both have the cell parameters; $a = b = c$ and $\alpha = \beta = \gamma$

(b) Both belong to different crystal systems

(c) Both have only two crystallographic angles of 90°

(d) Both belong to same crystal system **(IIT 2007)**

341. Consider a reaction $aG + bH \rightarrow$ products. When concentration of both the reactants G and H is doubled, the rate increases by eight times. However, when concentration of G is doubled keeping the concentration of H fixed, the rate is doubled. The overall order of the reaction is

(a) 0 (b) 1 (c) 2 (d) 3 **(IIT 2007)**

[**Hint:** See Ex. 15, p. 624]

342. When 20 g of naphthoic acid ($C_{11}H_8O_2$) is dissolved in 50 g of benzene ($K_f = 1.72$ K kg mol^{-1}), a freezing point depression of 2 K is observed. The van't Hoff factor (i) is

(a) 0.5 (b) 1 (c) 2 (d) 3 **(IIT 2007)**

343. The value of $\log_{10} K$ for a reaction $A \rightleftharpoons B$ is

(Given: $\Delta r H^\circ_{298\text{ K}} = -54.07$ kJ mol^{-1}, $\Delta r S^\circ_{298\text{ K}} = 10$ J K^{-1} mol^{-1}, $R = 8.314$ J K^{-1} mol^{-1}; $2.303 \times 8.314 \times 298 = 5705$)

(a) 5 (b) 10 (c) 95 (d) 100 **(IIT 2007)**

A 4.0 M aqueous solution of NaCl is prepared and 500 mL of this solution is electrolysed. This leads to the evolution of chlorine gas at one of the electrodes. (Na = 23, Hg = 200, 1 F = 96500 C)

The following examples (Q. Nos. 344 to 346) illustrate a typical case, involving chemical/electrochemical reaction, which requires a clear understanding of the mole concept.

344. The total number of moles of chlorine gas evolved is

(a) 0.5 (b) 1.0 (c) 2.0 (d) 3.0 **(IIT 2007)**

345. If the cathode is a Hg electrode, the maximum weight (g) of amalgam formed from this solution is

(a) 200 (b) 225 (c) 400 (d) 446 **(IIT 2007)**

[**Hint:** Maximum moles of Na deposited is 2 moles and the amalgam is (Na–Hg).]

346. The total charge (coulombs) required for complete electrolysis is

(a) 24125 (b) 48250 (c) 96500 (d) 193000 **(IIT 2007)**

Given below are a set of half-cell reactions (acidic medium) along with their *E*° values (with respect to normal hydrogen electrode). Using this data, obtain the correct explanation to questions 347 and 348.

$I_2 + 2e \rightarrow 2I^- \ldots$ $E^\circ = 0.54$ V

$Cl_2 + 2e \rightarrow 2Cl^- \ldots$ $E^\circ = 1.36$ V

$Mn^{3+} + e \rightarrow Mn^{2+} \ldots$ $E^\circ = 1.50$ V

$Fe^{3+} + e \rightarrow Fe^{2+} \ldots$ $E^\circ = 0.77$ V

$O_2 + 4H^+ + 4e \rightarrow H_2O \ldots$ $E^\circ = 1.23$ V

347. Among the following, identity the correct statement.

(a) Chloride ion is oxidised by O_2 (b) Fe^{2+} is oxidised by iodine

(c) Iodide ion is oxidised by chlorine (d) Mn^{2+} is oxidised by chlorine **(IIT 2007)**

[Hint: E°_{cell} should be +ve for the cell reaction]

348. While Fe^{3+} is stable, Mn^{3+} is not stable in acidic solution because

(a) O_2 oxidises Mn^{2+} to Mn^{3+}

(b) O_2 oxidises both Mn^{2+} to Mn^{3+} and Fe^{2+} to Fe^{3+}

(c) Fe^{3+} oxidises H_2O to O_2

(d) Mn^{3+} oxidises H_2O to O_2 **(IIT 2007)**

349. For the process: $H_2O(l)$ (1 bar, 373 K) $\rightarrow H_2O(g)$ (1 bar, 373 K), the correct set of the thermodynamic parameters is

(a) $\Delta G = 0$, $\Delta S = +ve$ (b) $\Delta G = 0$, $\Delta S = -ve$

(c) $\Delta G = +ve$, $\Delta S = 0$ (d) $\Delta G = -ve$, $\Delta S = +ve$ **(IIT 2007)**

350. Among the following, the surfactant that will form micelles in aqueous solution at the lowest molar concentration at ambient conditions is

(a) $CH_3(CH_2)_{15}N^+(CH_3)_3Br^-$ (b) $CH_3(CH_2)_{11}OSO_3^- Na^+$

(c) $CH_3(CH_2)_6COO^- Na^+$ (d) $CH_3(CH_2)_{11}N^+(CH_3)_3 Br^-$ **(IIT 2007)**

[Hint: During micelle formation, number of ions decreases as a result of which conductance reduces. The formation of micelles thus also reduces the osmotic pressure thereby increasing the average molar mass of the hydrocarbon chain of the surfactant.]

In the following reasoning-type questions from 351 to 353 each question has 4 choices (a), (b), (c) and (d), out of which only one is correct. (Tick the correct one.)

351. Statement 1: For every chemical reaction at equilibrium, standard Gibbs energy of reaction is zero; and

Statement 2: At constant temperature and pressure, chemical reactions are spontaneous in the direction of decreasing Gibbs energy.

(a) Statement 1 is true, statement 2 is true; statement 2 is correct explanation for statement 1.

(b) Statement 1 is true, statement 2 is true; statement 2 is not a correct explanation for statement 1.

(c) Statement 1 is true, statement 2 is false

(d) Statement 1 is false, statement 2 is true **(IIT 2008)**

[**Hint:** At equilibrium, $\Delta G = 0$, but ΔG° may or may not be zero]

352. **Statement 1:** The plot of atomic number (y-axis) versus number of neutrons (x-axis) for stable nuclei shows a curvature towards x-axis from the line of 45° slope as the atomic number is increased; and

Statement 2: Proton–Proton electrostatic repulsions begin to overcome attractive forces involving protons and neutrons in heavier nuclides.

(a) Statement 1 is true, statement 2 is true; statement 2 is correct explanation for statement 1

(b) Statement 1 is true, statement 2 is true; statement 2 is not a correct explanation for statement 1

(c) Statement 1 is true, statement 2 is false

(d) Statement 1 is false, statement 2 is true **(IIT 2008)**

353. **Statement 1:** Micelles are formed by surfactant molecules above the critical micellar concentration (CMC), because

Statement 2: The conductivity of a solution having surfactant molecules decreases sharply at the CMC.

(a) Statement 1 is true, statement 2 is true; statement 2 is a correct explanation for statement 1

(b) Statement 1 is true, statement 2 is true; statement 2 is not a correct explanation for statement 1

(c) Statement 1 is true, statement 2 is false

(d) Statement 1 is false, statement 2 is true **(IIT 2007)**

354. 2.5 mL of 2/5 M weak monoacidic base ($K_b = 1 \times 10^{-12}$ at 25°C) is titrated with 2/15 M HCl in water at 25°C. The concentration of H^+ at equivalence point is ($K_w = 1 \times 10^{-14}$ at 25°C)

(a) 3.7×10^{-13} M (b) 3.2×10^{-7} M

(c) 3.2×10^{-2} M (d) 2.7×10^{-2} M **(IIT 2008)**

[**Hint:** The base is very weak, the cations of the salt formed at the equivalence point shall undergo hydrolysis to a greater extent and so the degree of hydrolysis, h cannot be neglected in comparison to 1. Now apply

$$Kh = \frac{Ch^2}{1-h} = \frac{K_w}{K_b};\ C = 0.1\,\text{M},\ [H^+] = Ch]$$

355. Under the same reaction conditions, initial concentration of 1.386 mol dm^{-3} of a substance becomes half in 40 seconds and 20 seconds through first order and zero order kinetics, respectively. Ratio (k_1/k_0) of the rate constants for first order (k_1) and zero order (k_0) of the reactions is

(a) 0.5 mol^{-1} dm^3 (b) 1.0 mol^{-1} dm^{-3}

(c) 1.5 mol^{-1} dm^{-3} (d) 2.0 mol^{-1} dm^3 **(IIT 2008)**

356. A gas described by van der Waals equation (multiple correct answers type)

(a) behaves similar to an ideal gas in the limit of large molar volumes.

(b) behaves similar to an ideal gas in the limit of large pressures.

(c) is characterised by van der Waals coefficients that are dependent on the identity of the gas but are independent of the temperature.

(d) has the pressure that is lower than the pressure exerted by the same gas behaving ideally. **(IIT 2008)**

A solution M is prepared by mixing ethanol and water. The mole fraction of ethanol in the mixture is 0.9. Given

K_f (water) = 1.86 K kg mol^{-1},
K_f (ethanol) = 2.0 K kg mol^{-1},
K_b (water) = 0.52 K kg mol^{-1}
K_b (ethanol) = 1.2 K kg mol^{-1},
Standard f.p. of water = 273 K
Standard f.p. of ethanol = 155.7 K
Standard b.p. of water = 373 K
Standard b.p. of ethanol = 351.5 K
Vapour pressure of pure water = 32.8 mmHg
Vapour pressure of pure ethanol = 40 mmHg
Molecular weight of water = 18 g mol^{-1}
Molecular weight of ethanol = 46 g mol^{-1}

In answering the following questions (357 to 359), consider the solutions to be ideal dilute solutions and solutes to be nonvolatile and non-dissociative.

357. The f.p. of the solution M is
(a) 268.7 K (b) 268.5 K (c) 234.2 K (d) 150.9 K **(IIT 2008)**
[**Hint:** Solute is water and solvent is ethanol]

358. The vapour pressure of the solution M is
(a) 39.3 mmHg (b) 36.0 mmHg
(c) 29.5 mmHg (d) 28.8 mmHg

359. Water is added to the solution M such that the mole fraction of water in the solution becomes 0.9. The b.p. of this solution is
(a) 380.4 K (b) 376.2 K (c) 375.5 K (d) 354.7 K **(IIT 2008)**
[**Hint:** Solute is ethanol and solvent is water]

360. The solubility product (K_{sp}) of the salts of types MX, MX_2 and M_3X at temperature 'T' are 4.0×10^{-8}, 3.2×10^{-14} and 2.7×10^{-15} respectively. Solubilities (mol dm^{-3}) of the salts at temperature 'T' are in the order
(a) $MX > MX_2 > M_3X$ (b) $M_3X > MX_2 > MX$
(c) $MX_2 > M_3X > MX$ (d) $MX > M_3X > MX_2$ **(IIT 2008)**

361. Electrolysis of dilute aqueous NaCl solution was carried out by passing 10 milli ampere current. The time required to liberate 0.01 mol of H_2 gas at the cathode is (1 F = 96500 C mol^{-1})
(a) 9.65×10^4 sec (b) 19.3×10^4 sec
(c) 28.95×10^4 sec (d) 38.6×10^4 sec **(IIT 2008)**
[**Hint:** Number of equivalents of hydrogen = 0.02]

362. How many atoms (spheres) are contained within the hexagonal prism shown below in h.c.p. space lattice?

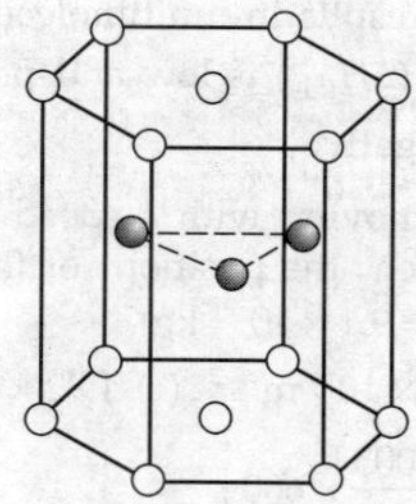

(a) 2 (b) 6 (c) 12 (d) 17 **(IIT 2008)**

[Note: One corner atom is shared by six other prisms and one hexagon base central atom is shared by two prisms. So number of atoms in one hexagonal prism $= (12 \times \frac{1}{6} + 2 \times \frac{1}{2} + 3)$]

363. The cell content (Z) of the unit cell of a h.c.p. space lattice is

(a) 2 (b) 6 (c) 12 (d) 17

[Hint: h.c.p. is the only one of the four (simple cubic, b.c.c., h.c.p. and c.c.p.) that has noncubic unit cell. The unit cell is rectangular ($z = 2$ as in b.c.c. as indicated below by dark lines.]

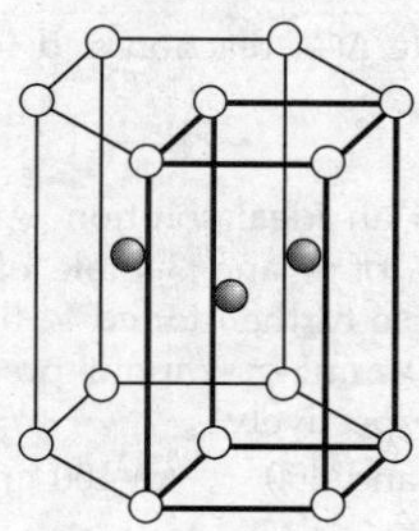

364. The volume of the hexagonal prism is (radius of every sphere is r)

(a) $24\sqrt{2}\,r^3$ (b) $16\sqrt{2}\,r^3$ (c) $12\sqrt{2}\,r^3$ (d) $\frac{64}{3\sqrt{3}}\,r^3$ **(IIT 2008)**

[Hint: Calculate base area and height from the figures:]

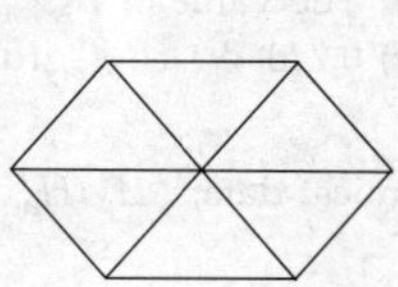

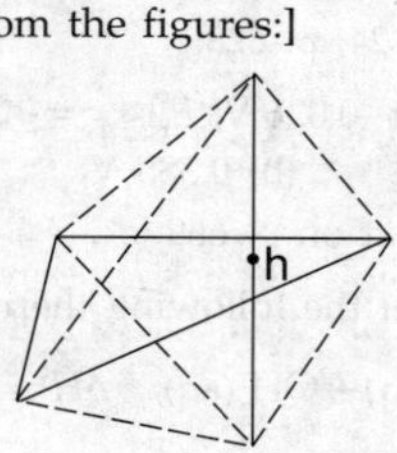

365. The empty space in the h.c.p. structure is

(a) 74% (b) 47.6% (c) 32% (d) 26% **(IIT 2008)**

[Hint: See p. 742]

366. Which of the following statements is incorrect regarding physical adsorption?

(a) It occurs because of van der Waals forces.

(b) More easily liquefiable gases are adsorbed readily.

(c) Under high pressure it results in multimolecular layer on adsorbent surface.

(d) Enthalpy of adsorption (ΔH_{ads}) is low and positive.

[Hint: ΔH_{ads} is low but negative]

367. In an atom, an electron is moving with a speed of 600 m/s with an accuracy of 0.005%. Certainty with which the position of the electron can be located is, ($h = 6.6 \times 10^{-34}$ kg m^2 s^{-1}, $m_e = 9.1 \times 10^{-31}$ kg)

(a) 1.52×10^{-4} m (b) 5.10×10^{-3} m (c) 1.92×10^{-3} m (d) 3.84×10^{-3} m

$$\left[\textbf{Hint: } \Delta x = \frac{h}{4\pi\, m\, \Delta v};\ \Delta v = \frac{0.005}{100} \times 600\right]$$

368. In a fuel cell methanol is used as fuel and oxygen gas is used as an oxidiser. The reaction is

$$CH_3OH(l) + \frac{3}{2}O_2(g) \rightarrow CO_2(g) + 2H_2O(l)$$

At 298 K standard Gibbs energies of formation for $CH_3OH(l)$, $H_2O(l)$ and $CO_2(g)$ are −166.2, −237.2 and −394.4 kJ mol^{-1} respectively. If standard enthalpy of combustion of methanol is −726 kJ mol^{-1}, efficiency of the fuel cell will be

(a) 80% (b) 87% (c) 90% (d) 97%

[Hint: $\Delta G_f^\circ(O_2) = 0$. Calculate ΔG_f° reaction and then apply: Efficiency $\frac{\Delta G^\circ}{\Delta H^\circ} \times 100$.

See Ex. 38, p. 438.]

369. Two liquids X and Y form an ideal solution. At 300 K, vapour pressure of the solution containing 1 mole of X and 3 mole of Y is 550 mmHg. At the same temperature, if 1 mole of Y is further added to this solution, the vapour pressure of the solution increases 10 mmHg. Vapour pressure of X and Y (in mmHg) in their pure states will be, respectively.

(a) 200 and 300 (b) 300 and 400 (c) 400 and 600 (d) 500 and 600

[Hint: See Ex. 23, p. 384.]

370. The half-life period of a first order chemical reaction is 6.93 minutes. The time required for the completion of 99% of the chemical reaction will be

(a) 230.3 minutes (b) 23.03 minutes (c) 46.06 minutes (d) 460.6 minutes

[Hint: See Ex. 24, p. 628.]

371. Given $E^\circ_{Fe^{3+}/Fe} = -0.036$ V, $E^\circ_{Fe^{2+}/Fe} = -0.439$ V. The value of $E^\circ_{Fe^{3+}/Fe^{2+}}$ will be

(a) −0.072 V (b) 0.385 V (c) 0.770 V (d) −0.270 V

[Hint: Read text on p. 669.]

372. On the basis of the following thermochemical data: $\Delta H_f^\circ(H^+_{aq}) = 0$,

$$H_2O(l) \rightarrow H^+(aq) + OH^-(aq);\quad \Delta H^\circ = 57.32 \text{ kJ}$$

$$H_2(g) + \frac{1}{2}O_2(g) \rightarrow H_2O(l);\quad \Delta H^\circ = -286.20 \text{ kJ}.$$

The value of enthalpy of formation of OH^- ions at 25°C is

(a) −22.88 kJ (b) −228.88 kJ (c) +228.88 kJ (d) −343.52 kJ

[Hint: Add two equations]

373. Copper crystalizes in f.c.c. with unit cell length of 361 pm. What is the radius of copper atom?

(a) 108 pm (b) 127 pm (c) 157 pm (d) 181 pm

[**Hint:** Use equation 4, p. 740.]

374. Solid $Ba(NO_3)_2$ is gradually dissolved in a 1.0×10^{-4} M Na_2CO_3 solution. At what concentration of Ba^{2+} will a precipitate begin to form? (K_{sp} for $BaCO_3 = 5.1 \times 10^{-9}$)

(a) 4.1×10^{-5} M (b) 5.1×10^{-5} M (c) 8.1×10^{-8} M (d) 8.1×10^{-7} M

375. Calculate the wavelength in nanometer associated with a proton moving at 1.0×10^3 m s^{-1}, (mass of proton $= 1.67 \times 10^{-27}$ kg and $h = 6.63 \times 10^{-34}$ J s)

(a) 0.032 nm (b) 0.40 nm (c) 2.5 nm (d) 14.0 nm

$$\left[\textbf{Hint: } \lambda = \frac{h}{mv}\right]$$

376. A binary liquid solution is prepared by mixing *n*-heptane and ethanol. Which one of the following statements is correct regarding the behaviour of the solution?

(a) The solution formed is an ideal solution.

(b) The solution is non-ideal, showing positive deviation from Raoult's law.

(c) The solution is non-ideal, showing negative deviation from Raoult's law.

(d) *n*-heptane shows positive deviation while ethanol shows negative deviation from Raoult's law.

377. The standard enthalpy of formation of NH_3 is -46.0 kJ mol^{-1}. If enthalpy of formation of H_2 from its atoms is 436 kJ mol^{-1} and that of N_2 is -712 kJ mol^{-1}, the average bond enthalpy of N—H bond in NH_3 is

(a) -964 kJ mol^{-1} (b) $+352$ kJ mol^{-1} (c) $+1056$ kJ mol^{-1} (d) -1102 kJ mol^{-1}

[**Hint:** See Ex. 54, p. 458.]

378. The time for half-life period of a certain reaction A→products, is 1 hour. When the initial concentration of the reactant, A is 2.0 mol L^{-1}, how much time does it take for its concentration to come from 0.50 to 0.25 mol L^{-1} if it is a zero-order reaction?

(a) 4 h (b) 0.5 h (c) 0.25 h (d) 1 h

[**Hint:** For zero-order reaction, $t_{1/2} \propto a$]

379. A solution containing 2.675 g of $CoCl_3 \cdot 6NH_3$ (molar mass = 267.5 g mol^{-1}) is passed through a cation exchanger. The chloride ions obtained in solution were treated with excess of $AgNO_3$ to give 4.31 g of AgCl (molar mass = 143.5 g mol^{-1}). The formula of the complex is (Ag = 108 u)

(a) $[Co(NH_3)_6]Cl_3$ (b) $[CoCl_2(NH_3)_4]Cl$

(c) $[CoCl_3(NH_3)_3]$ (d) $[CoCl(NH_3)_5]Cl_2$

[**Hint:** Mol of Cl^- = mol of AgCl $= \frac{4.31}{143.5} = 0.03$

$\because$ 0.01 mol of $CoCl_3 \cdot 6NH_3$ produces 0.03 mol of Cl^-

$\therefore$ 1 mole of $CoCl_3 \cdot 6NH_3$ contains 3 Cl^- anions]

380. Consider the reaction:

$$Cl_2(aq) + H_2S(aq) \rightarrow S(s) + 2H^+(aq) + 2Cl^-(aq)$$

The rate equation for this reaction is

Rate = $k[Cl_2][H_2S]$.

Which of these mechanisms is/are consistent with this rate equation?

(A) $Cl_2 + H_2S \rightarrow H^+ + Cl^- + Cl^+ + HS^-$ (slow)

$Cl^+ + HS^- \rightarrow H^+ + Cl^- + S$ (fast)

(B) $H_2S \rightleftharpoons H^+ + HS^-$ (fast equilibrium)

$Cl_2 + HS^- \rightarrow 2Cl^- + H^+ + S$ (slow)

(a) B only (b) Both A and B

(c) Neither A nor B (d) A only

381. If 10^{-4} dm^3 of water is introduced into a 1.0 dm^3 flask at 300 K, how many moles of water are in the vapour phase when equilibrium is established?
(**Given:** Vapour pressure of H_2O at 300 K is 3170 Pa, $R = 8.314\ J\ K^{-1}\ mol^{-1}$)

(a) 5.56×10^{-3} mol (b) 1.53×10^{-2} mol

(c) 4.46×10^{-2} mol (d) 1.27×10^{-3} mol

[**Hint:** Use $pV = nRT$]

382. If sodium sulphate is considered to be completely dissociated into cations and anions in aqueous solution, the change in freezing point of water (ΔT_f), when 0.01 mol of sodium sulphate is dissolved in 1 kg of water, is ($K_f = 1.86$ K kg mol^{-1})

(a) 0.0372 K (b) 0.0558 K (c) 0.0744 K (d) 0.0186 K

[**Hint:** $\Delta T_f = i \cdot K_f \cdot m;\ i = 3$]

383. 29.5 mg of an organic compound containing nitrogen was digested according to Kjeldahl method and the evolved ammonia was absorbed in 20 mL of 0.1 M HCl solution. The excess of the acid required 15 mL of 0.1 M NaOH solution for complete neutralisation. The percentage of nitrogen in the compound is

(a) 59.0 (b) 47.4 (c) 23.7 (d) 29.5

[**Hint:** See Ex. 4, p. 202.]

384. The energy required to break one mole of Cl—Cl bonds in Cl_2 is 242 kJ mol^{-1}. The longest wavelength of light capable of breaking a single Cl—Cl bond is
($C = 3 \times 10^8$ m s^{-1} and $N = 6.02 \times 10^{23}$ mol^{-1})

(a) 594 nm (b) 640 nm (c) 700 nm (d) 494 nm

[**Hint:** E (for one Cl—Cl bond) $= \dfrac{242 \times 10^3\ J}{6.02 \times 10^{23}};\ E = \dfrac{hc}{\lambda}$]

385. Ionisation energy of He^+ is 19.6×10^{-18} J atom^{-1}. The energy of the first stationary state ($n = 1$) of Li^{2+} is

(a) 4.41×10^{-16} J atom^{-1} (b) -4.41×10^{-17} J atom^{-1}

(c) -2.2×10^{-15} J atom^{-1} (d) 8.82×10^{-17} J atom^{-1}

[**Hint:** See Ex. 12, p. 264.]

386. On mixing, heptane and octane form an ideal solution. At 373 K, the vapour pressures of the two liquid components (heptane and octane) are 105 kPa and 45 kPa respectively. Vapour pressure of the solution obtained by mixing 25.0 g of heptane and 35 g of octane will be (molar mass of heptane =100 and octane = 114 g mol^{-1})

(a) 72.0 kPa (b) 36.1 kPa (c) 96.2 kPa (d) 144.5 kPa

[**Hint:** Use equation 3 on p. 370.]

387. In an aqueous solution the ionisation constants for carbonic acid are $K_1 = 4.2 \times 10^{-7}$ and $K_2 = 4.8 \times 10^{-11}$. Select the correct statement for a saturated 0.034 M solution of the carbonic acid.

(a) The concentration of CO_3^{2-} is 0.034 M

(b) The concentration of CO_3^{2-} is greater than that of HCO_3^-

(c) The concentration of H^+ and HCO_3^- are approximately equal

(d) The concentration of H^+ is double that of CO_3^{2-}

[**Hint:** $K_1 >> K_2$, i.e., second dissociation is practically negligible]

388. The edge length of face-centred cubic cell of an ionic substance is 508 pm. If the radius of the cation is 110 pm, the radius of the anion is

(a) 288 pm (b) 398 pm (c) 618 pm (d) 144 pm

[**Hint:** For ionic f.c.c. cell: Edge length $= 2(r_+ + r_-)$]

389. Solubility product of AgBr is 5.0×10^{-13}. The quantity of KBr (M = 120) to be added to 1 litre of 0.05 M solution of $AgNO_3$ to start the precipitation of AgBr is

(a) 1.2×10^{-10} g (b) 1.2×10^{-9} g (c) 6.2×10^{-5} g (d) 5.0×10^{-8} g

[**Hint:** Wt. in g = no. of moles × mol. wt.]

390. The Gibbs energy for decomposition of Al_2O_3 at 500°C is as follows:

$$\frac{2}{3}Al_2O_3 \rightarrow \frac{4}{3}Al + O_2; \qquad \Delta G = +966 \text{ kJ}$$

The potential difference needed for electrolytic reduction of Al_2O_3 at 500°C is at least

(a) 4.5 V (b) 3.0 V (c) 2.5 V (d) 5.0 V

[**Hint:** $\underset{+6}{Al_2O_3} \rightarrow 2Al + \underset{0}{\frac{3}{2}O_2}$; $\Delta G = \frac{3}{2} \times 966$ kJ mol^{-1}, $n = 6$. Then apply $\Delta G = -nFE$]

391. At 25°C, the solubility product of $Mg(OH)_2$ is 1.0×10^{-11}. At what pH, will Mg^{2+} ions start precipitating in the form of $Mg(OH)_2$ from a solution of 0.001 M Mg^{2+} ions?

(a) 9 (b) 10 (c) 11 (d) 8

[**Hint:** Calculate $[OH^-]$ from $K_{sp} = [Mg^{2+}][OH^-]^2$]

392. The correct order of $E^\circ_{M^{2+}/M}$ values with negative sign for the four successive elements Cr, Mn, Fe and Co is

(a) Mn > Cr > Fe > Co (b) Cr > Fe > Mn > Co

(c) Fe > Mn > Cr > Co (d) Cr > Mn > Fe > Co

[**Hint:** $E^\circ_{Mn^{2+}, Mn} = -1.18$ V, $E^\circ_{Cr^{2+}, Cr} = -0.91$ V

$E^\circ_{Fe^{2+}, Fe} = -0.44$ V, $E^\circ_{Co^{2+}, Co} = -0.28$ V]

393. For a particular reversible reaction at temperature T, ΔH and ΔS were found to be both +ve. If T_e is the temperature at equilibrium, the reaction would be spontaneous when

(a) $T_e > T$ (b) $T > T_e$ (c) T_e is 5 times T (d) $T = T_e$

[**Hint:** ΔG should be negative. $\Delta G = \Delta H - T_e\Delta S = 0$]

394. The term that corrects for the attractive forces present in a real gas in the van der Waals equation is

(a) nb (b) $\frac{an^2}{V^2}$ (c) $-\frac{an^2}{V^2}$ (d) $-nb$ **(IIT 2009)**

395. Given that the abundances of isotopes ^{54}Fe, ^{56}Fe and ^{57}Fe are 5% , 90% and 5% respectively, the atomic mass of Fe is

(a) 55.85 (b) 55.95 (c) 55.75 (d) 56.05 **(IIT 2009)**

[**Hint:** See Ex. 25, p. 10.]

396. The Henry's law constant for the solubility of N_2 gas in water at 298 K is 1.0×10^5 atm. The mole fraction of N_2 in air is 0.8. The number of moles of N_2 from air dissolved in 10 moles of water at 298 K and 5 atm pressure is

(a) 4.0×10^{-4} (b) 4.0×10^{-5}

(c) 5.0×10^{-4} (d) 4.0×10^{-6} **(IIT 2009)**

[**Hint:** $\underset{\text{partial pressure}}{p_{N_2}} = \underset{\text{Henry's constant atm}}{k} \times \underset{\text{mole fraction of N2}}{x_{N_2}} = k \times \frac{\text{mol } N_2 \text{ in solution}}{\underbrace{\text{mol } (N_2+O_2)}_{\text{neglect}} + \text{mol } H_2O}$]

397. Among the electrolytes Na_2SO_4, $CaCl_2$, $Al_2(SO_4)_3$ and NH_4Cl, the most effective coagulating agent for Sb_2S_3 sol is

(a) Na_2SO_4 (b) $CaCl_2$ (c) $Al_2(SO_4)_3$ (d) NH_4Cl **(IIT 2009)**

[**Hint:** SbS_3 sol is negatively charged. Apply Hardy–Schultze rule.]

398. The correct statement(s) regarding defects in solids is (are)

(a) Frankel defect is usually favoured by a very small difference in the sizes of cations and anions

(b) Frankel defect is a dislocation defect

(c) Trapping of an electron in the lattice leads to the formation of F-center

(d) Schottky defects have no effect on the physical properties of solids. **(IIT 2009)**

[**Hint:** F-center is created when an anion is lost from the lattice and vacancy is filled by trapping of an electron]

399. For a first order reaction $A \rightarrow P$, the temperature (T) dependent rate constant (k) was found to follow the equation

$$\log k = -(2000)\frac{1}{T} + 6.0.$$

The pre-exponential factor A and the activation energy E_a, respectively, are

(a) $1.0\times10^6\ s^{-1}$ and $9.2\ kJ\ mol^{-1}$ (b) $6.0\times s^{-1}$ and $16.6\ kJ\ mol^{-1}$

(c) $1.0\times10^6\ s^{-1}$ and $16.6\ kJ\ mol^{-1}$ (d) $1.0\times10^6\ s^{-1}$ and $38.3\ kJ\ mol^{-1}$

(IIT 2009)

[**Hint:** Compare the given equation with $\log k = -\frac{E}{2.303\,R}\cdot\frac{1}{T} + \log A$ (p. 617)]

400. The spin magnetic moment value (in Bohr magneton unit) of $Cr(CO)_6$ is

(a) 0 (b) 2.84 (c) 4.90 (d) 5.92 **(IIT 2009)**

[**Hint:** No. of unpaired e in $Cr(CO)_6 = 0$. Use $\mu = \sqrt{n(n+2)}$ BM]

401. For the reduction of NO_3^- ion in an aqueous solution, $E°$ is +0.96 V. Values of $E°$ for some metals ions are given below:

$V^{2+}(aq) + 2e^- \rightarrow V$; $E° = -1.19$ V

$Fe^{3+}(aq) + 3e^- \rightarrow Fe$; $E° = -0.04$ V

$Au^{3+}(aq) + 3e^- \rightarrow Au$; $E° = +1.40$ V

$Hg^{2+}(aq) + 2e^- \rightarrow Hg$; $E° = +0.86$ V

The pair(s) of metals that is (are) oxidised by NO_3^- in aqueous solution is (are)

(a) V and Hg (b) Hg and Fe (c) Fe and Au (d) Fe and V

(IIT 2009)

[**Hint:** An electrode with higher value of std. red. pot. in a cell undergoes reduction]

402. Among the following, the state function(s) is (are)

(a) internal energy (b) irreversible exp. work

(c) reversible exp. work (d) molar enthalpy **(IIT 2009)**

403. Plots showing the variation of the rate constant (k) with temperature (T) are given below. The plot that follows Arrhenius equation is

(a)

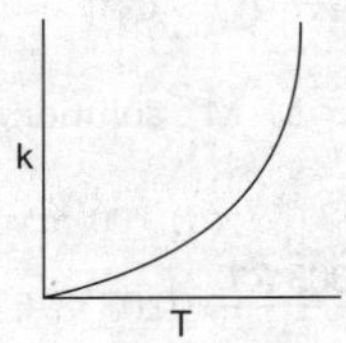

(b)

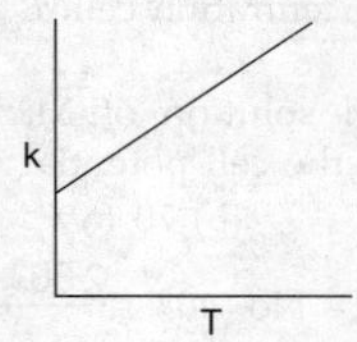

(c)

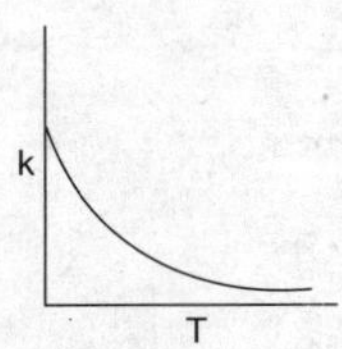

(d)

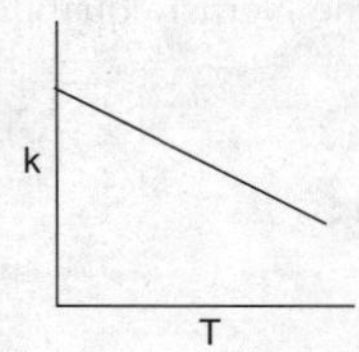

(IIT 2010)

[**Hint:** See Q. 50, p. 659.]

404. The species which by definition has zero standard molar enthalpy of formation at 298 K is

(a) $Br_2(g)$ (b) $Cl_2(g)$ (c) $H_2O(g)$ (d) $CH_4(g)$ **(IIT 2010)**

[**Note:** $\Delta H_f^\circ(Br_2(l)) = 0$ but $\Delta H_f^\circ(Br_2(g)) \neq 0$]

405. The bond energy (in kcal mol^{-1}) of C—C bond is approximately

(a) 1 (b) 10 (c) 100 (d) 1000 **(IIT 2010)**

[See Table on p. 423.]

406. Aqueous solutions of HNO_3, KOH, CH_3COOH and CH_3COONa of identical concentration are provided. The pair(s) of solutions which form a buffer upon mixing is (are)

(a) HNO_3 and CH_3COOH (b) KOH and CH_3COONa

(c) HNO_3 and CH_3COONa (d) CH_3COOH and CH_3COONa **(IIT 2010)**

[**Hint:** Mixture of a weak acid and its salt gives a buffer. Option (c) may also be correct if only HNO_3 is present in limiting amount.]

407. Among the following, the intensive property is (are)

(a) molar conductivity (b) electromotive force

(c) resistance (d) heat capacity **(IIT 2010)**

Paragraph for Q. Nos. 408 and 409

The concentration of K^+ inside a biological cell is at least 20 times higher than the outside. The resulting potential difference across the cell is important in several processes such as transmission of nerve impulses and maintaining the ion balance. A simple model for such a concentration cell involving a metal M is:

$$M(s) \mid M^+(aq; 0.05\ M) \parallel M^+(aq, 1\ M) \mid M(s)$$

For the above galvanic cell the magnitude of the cell potential $|E_{cell}| = 70$ mV.

408. For the above cell

(a) $E_{cell} < 0;\ \Delta G > 0$ (b) $E_{cell} > 0;\ \Delta G < 0$

(c) $E_{cell} < 0;\ \Delta G^\circ > 0$ (d) $E_{cell} > 0;\ \Delta G^\circ > 0$ **(IIT 2010)**

[**Hint:** For concentration cell, $E_{cell} = \frac{2.303\ RT}{nF} \log \frac{C_2}{C_1} = +ve\ \ (C_2 > C_1)$]

409. If the 0.05 M solution of M^+ is replaced by 0.0025 M M^+ solution, then the magnitude of the cell potential would be

(a) 35 mV (b) 70 mV (c) 140 mV (d) 700 mV **(IIT 2010)**

[**Hint:** From Q. No. 408 : $\frac{2.303\ RT}{F} \log \frac{1}{0.05} = 0.07;\ \frac{2.303\ RT}{F} = 0.0538$. Then apply again the Nernst equation.]

Answers

1-c, **2**-a, **3**-d, **4**-a, **5**-b, **6**-b, **7**-b, **8**-d, **9**-b, **10**-b, **11**-d, **12**-d, **13**-a, **14**-d, **15**-d, **16**-b, **17**-a, **18**-c, **19**-b, **20**-a, **21**-a, **22**-b, **23**-b, **24**-b, **25**-d, **26**-b, **27**-a, **28**-d, **29**-a, **30**-a, **31**-b, **32**-b, **33**-b, **34**-c, **35**-a, **36**-d, **37**-b, **38**-a, **39**-d, **40**-b, **41**-a, **42**-c, **43**-d, **44**-b, **45**-d, **46**-c, **47**-c, **48**-c, **49**-a, **50**-c, **51**-d, **52**-d, **53**-b & d, **54**-b & d, **55**-All correct, **56**-a & d, **57**-c, **58**-d, **59**-a, **60**-a, **61**-c, **62**-b, **63**-d, **64**-b, **65**-a, **66**-c, **67**-c, **68**-b, **69**-a, **70**-b, **71**-b, **72**-a & b, **73**-d, **74**-a, **75**-c, **76**-c, **77**-All correct, **78**-a, **79**-a, **80**-d, **81**-c & d, **82**-c, **83**-d, **84**-d, **85**-b & d, **86**-b & d, **87**-a, **88**-c, **89**-c, **90**-c, **91**-b, **92**-c, **93**-d, **94**-a, **95**-a, **96**-b, **97**-b, **98**-b, **99**-d, **100**-a, **101**-a, **102**-b, **103**-d, **104**-c, **105**-d, **106**-c, **107**-b, **108**-c, **109**-c, **110**-a, **111**-a, **112**-a & c, **113**-a, **114**-a, c & d, **115**-a & d, **116**-c, **117**-d, **118**-c, **119**-d, **120**-c, **121**-d, **122**-a, **123**-a, **124**-b, **125**-a, **126**-c, **127**-a, **128**-a, b & d, **129**-a & d, **130**-a, b & c, **131**-b, **132**-d, **133**-b, **134**-d, **135**-c, **136**-b, **137**-c, **138**-a, **139**-d, **140**-b, **141**-a, b & c, **142**-c, **143**-a & d, **144**-b, **145**-d, **146**-b, **147**-a, **148**-a, **149**-a, **150**-c, **151**-d, **152**-a, **153**-b, **154**-d, **155**-c, **156**-c, **157**-c, **158**-d, **159**-b, **160**-c & d, **161**-b, **162**-b, **163**-b, **164**-a, **165**-c, **166**-a, **167**-c, **168**-d, **169**-c, **170**-a, **171**-a, **172**-b, **173**-a, **174**-d, **175**-d, **176**-d, **177**-b, **178**-d, **179**-c, **180**-a, **181**-a, **182**-All correct, **183**-a, **184**-b, **185**-d, **186**-b, **187**-a, **188**-b, **189**-a, **190**-d, **191**-c, **192**-b, **193**-d, **194**-a, **195**-d, **196**-c, **197**-c, **198**-a, **199**-b, **200**-c, **201**-a, **202**-a, **203**-c, **204**-b, **205**-a, **206**-c, **207**-d, **208**-b, **209**-b, **210**-c, **211**-c & d, **212**-b & d, **213**-c, **214**-a & c, **215**-a & c, **216**-a, **217**-a, **218**-b, **219**-b, **220**-a, **221**-a, **222**-b, **223**-a, **224**-a, **225**-a & d, **226**-c & d, **227**-a & d, **228**-b & d, **229**-b, **230**-b, **231**-a, **232**-c, **233**-a, **234**-d, **235**-b, **236**-a, **237**-a, **238**-c, **239**-b, **240**-b, **241**-b, **242**-b, **243**-d, **244**-c, **245**-c, **246**-c, **247**-b, **248**-a & d, **249**-b, **250**-c, **251**-d, **252**-a, **253**-c, **254**-a, **255**-d, **256**-a, **257**-c, **258**-d, **259**-a, **260**-b, **261**-b, **262**-b, **263**-d, **264**-d, **265**-c, **266**-a, **267**-c, **268**-c, **269**-a, **270**-a, **271**-b, **272**-a, **273**-b, **274**-a, **275**-d, **276**-c, **277**-a, **278**-a, **279**-a, **280**-a, **281**-c, **282**-d, **283**-c, **284**-c, **285**-d, **286**-d, **287**-a, **288**-a, **289**-d, **290**-d, **291**-c, **292**-a, **293**-a, **294**-a, **295**-b, **296**-d, **297**-b, **298**-c, **299**-b, **300**-b, **301**-a, **302**-c, **303**-a, **304**-b, **305**-b, **306**-b, **307**-a, **308**-a, **309**-d, **310**-a, **311**-a, **312**-c, **313**-b, **314**-c, **315**-b, **316**-a, **317**-b, **318**-a, **319**-d, **320**-d, **321**-c, **322**-d, **323**-d, **324**-a, **325**-All correct, **326**-a, **327**-b, **328**-a, **329**-b, **330**-a, **331**-c, **332**-a, **333**-a & d, **334**-c, **335**-a & b, **336**-c, **337**-a & d, **338**-a & b, **339**-b, **340**-b & c, **341**-d, **342**-a, **343**-b, **344**-b, **345**-d, **346**-d, **347**-c, **348**-d, **349**-a, **350**-a, **351**-d, **352**-a, **353**-b, **354**-d, **355**-a, **356**-a, c & d, **357**-d, **358**-a, **359**-b, **360**-d, **361**-b, **362**-b, **363**-a, **364**-a, **365**-d, **366**-d, **367**-c, **368**-d, **369**-c, **370**-c, **371**-c, **372**-b, **373**-b, **374**-b, **375**-b, **376**-b, **377**-b, **378**-c, **379**-a, **380**-d, **381**-d, **382**-b, **383**-c, **384**-d, **385**-b, **386**-a, **387**-c, **388**-d, **389**-b, **390**-c, **391**-b, **392**-a, **393**-b, **394**-b, **395**-b, **396**-a, **397**-c, **398**-b & c, **399**-d, **400**-a, **401**-a, b & d, **402**-a & d, **403**-a, **404**-b, **405**-c, **406**-c & d, **407**-a & b, **408**-b, **409**-c.

❑

ATOMIC WEIGHTS OF ELEMENTS

Element	*Symbol*	*Atomic number*	*Atomic weight*
Aluminium	Al	13	26·98
Antimony	Sb	51	121·80
Arsenic	As	33	74·92
Barium	Ba	56	137·30
Beryllium	Be	4	9·012
Bismuth	Bi	83	209·00
Boron	B	5	10·81
Bromine	Br	35	79·90
Cadmium	Cd	48	112·04
Calcium	Ca	20	40·08
Carbon	C	6	12·00
Chlorine	Cl	17	35·45
Chromium	Cr	24	52·00
Cobalt	Co	27	58·93
Copper	Cu	29	63·55
Europium	Eu	63	152·00
Fluorine	F	9	19·00
Gallium	Ga	31	69·72
Germanium	Ge	32	72·59
Gold	Au	79	197·00
Hydrogen	H	1	1·008
Iodine	I	53	126·9
Iron	Fe	26	55·85
Krypton	Kr	36	83·80
Lead	Pb	82	207·20
Lithium	Li	3	6·94
Magnesium	Mg	12	24·31
Manganese	Mn	25	54·94
Mercury	Hg	80	200·60
Molybdenum	Mo	42	95·94
Neon	Ne	10	20·18
Nickel	Ni	28	58·70
Nitrogen	N	7	14·01
Oxygen	O	8	16·00
Palladium	Pd	46	106·40

Element	*Symbol*	*Atomic number*	*Atomic weight*
Phosphorus	P	15	30·97
Platinum	Pt	78	195·10
Plutonium	Pu	94	244·00
Polonium	Po	84	209·00
Potassium	K	19	39·10
Radium	Ra	88	226·00
Radon	Rn	86	222·00
Selenium	Se	34	78·96
Silicon	Si	14	28·09
Silver	Ag	47	107·90
Sodium	Na	11	22·99
Strontium	Sr	38	87·62
Sulphur	S	16	32·06
Tin	Sn	50	118·70
Tungsten	W	74	183·90
Uranium	U	92	238·00
Vanadium	V	23	50·94
Zinc	Zn	30	65·38
Zirconium	Zr	40	91·22

❑

LOGARITHMS

	0	1	2	3	4	5	6	7	8	9	Mean Differences								
											1	2	3	4	5	6	7	8	9
10	0000	0043	0086	0128	0170	0212	0253	0294	0334	0374	4	8	12	17	21	25	29	33	37
11	0414	0453	0492	0531	0569	0607	0645	0682	0719	0755	4	8	11	15	19	23	26	30	34
12	0792	0828	0864	0899	0934	0969	1004	1038	1072	1106	3	7	10	14	17	21	24	28	31
13	1139	1173	1206	1239	1271	1303	1335	1367	1399	1430	3	6	10	13	16	19	23	26	29
14	1461	1492	1523	1553	1584	1614	1644	1673	1703	1732	3	6	9	12	15	18	21	24	27
15	1761	1790	1818	1847	1875	1903	1931	1959	1987	2014	3	6	8	11	14	17	20	22	25
16	2041	2068	2095	2122	2148	2175	2201	2227	2253	2279	3	5	8	11	13	16	18	21	24
17	2304	2330	2355	2380	2405	2430	2455	2480	2504	2529	2	5	7	10	12	15	17	20	22
18	2553	2577	2601	2625	2648	2672	2695	2718	2742	2765	2	5	7	9	12	14	16	19	21
19	2788	2810	2833	2856	2878	2900	2923	2945	2967	2989	2	4	7	9	11	13	16	18	20
20	3010	3032	3054	3075	3096	3118	3139	3160	3181	3201	2	4	6	8	11	13	15	17	19
21	3222	3243	3263	3284	3304	3324	3345	3365	3385	3404	2	4	6	8	10	12	14	16	18
22	3424	3444	3464	3483	3502	3522	3541	3360	3579	3598	2	4	6	8	10	12	14	15	17
23	3617	3636	3655	3674	3692	3711	3729	3747	3766	3784	2	4	6	7	9	11	13	15	17
24	3802	3820	3838	3856	3874	3892	3909	3927	3945	3962	2	4	5	7	9	11	12	14	16
25	3979	3997	4014	4031	4048	4065	4082	4099	4116	4133	2	3	5	7	9	10	12	14	15
26	4150	4166	4183	4200	4216	4232	4249	4265	4281	4298	2	3	5	7	8	10	11	13	15
27	4314	4330	4346	4362	4378	4393	4409	4425	4440	4456	2	3	5	6	8	9	11	13	14
28	4472	4487	4502	4518	4533	4548	4564	4579	4594	4609	2	3	5	6	8	9	11	12	14
29	4624	4639	4654	4669	4683	4698	4713	4728	4742	4757	1	3	4	6	7	9	10	12	13
30	4771	4786	4800	4814	4829	4843	4857	4871	4886	4900	1	3	4	6	7	9	10	11	13
31	4914	4028	4942	4955	4969	4983	4997	5011	5024	5038	1	3	4	6	7	8	10	11	12
32	5051	5065	5079	5092	5105	5119	5132	5145	5159	5172	1	3	4	5	7	8	9	11	12
33	5185	5198	5211	5224	5237	5250	5263	5276	5289	5302	1	3	4	5	6	8	9	10	12
34	5315	5328	5340	5353	5366	5378	5391	5403	5416	5428	1	3	4	5	6	8	9	10	11
35	5441	5453	5465	5478	5490	5502	5514	5527	5539	5551	1	2	4	5	6	7	9	10	11
36	5563	5575	5587	5599	5611	5623	5635	5647	5658	5670	1	2	4	5	6	7	8	10	11
37	5682	5694	5705	5717	5729	5740	5752	5763	5775	5786	1	2	3	5	6	7	8	9	10
38	5798	5809	5821	5832	5843	5855	5866	5877	5888	5899	1	2	3	5	6	7	8	9	10
39	5911	5922	5933	5944	5955	5966	5977	5988	5999	6010	1	2	3	4	5	7	8	9	10
40	6021	6031	6042	6053	6064	6075	6085	6096	6107	6117	1	2	3	4	5	6	8	9	10
41	6128	6138	6149	6160	6170	6180	6191	6201	6212	6222	1	2	3	4	5	6	7	8	9
42	6232	6243	6253	6263	6274	6284	6294	6304	6314	6325	1	2	3	4	5	6	7	8	9
43	6335	6345	6355	6365	6375	6385	6395	6405	6415	6425	1	2	3	4	5	6	7	8	9
44	6435	6444	6454	6464	6474	6484	6493	6503	6513	6522	1	2	3	4	5	6	7	8	9
45	6532	6542	6551	6561	6571	6580	6590	6599	6609	6618	1	2	3	4	5	6	7	8	9
46	6628	6637	6646	6656	6665	6675	6684	6693	6702	6712	1	2	3	4	5	6	7	7	8
47	6721	6730	6739	6749	6758	6767	6776	6785	6794	6803	1	2	3	4	5	5	6	7	8
48	6812	6821	6830	6839	6848	6857	6866	6875	6884	6893	1	2	3	4	4	5	6	7	8
49	6902	6911	6920	6928	6937	6946	6955	6964	6972	6981	1	2	3	4	4	5	6	7	8
50	6990	6998	7007	7016	7024	7033	7042	7050	7059	7067	1	2	3	3	4	5	6	7	8
51	7076	7084	7093	7101	7110	7118	7126	7135	7143	7152	1	2	3	3	4	5	6	7	8
52	7160	7168	7177	7185	7193	7202	7210	7218	7226	7235	1	2	2	3	4	5	6	7	7
53	7243	7251	7259	7267	7275	7284	7292	7300	7308	7316	1	2	2	3	4	5	6	6	7
54	7324	7332	7340	7348	7356	7364	7372	7380	7388	7396	1	2	2	3	4	5	6	6	7

LOGARITHMS

	0	1	2	3	4	5	6	7	8	9	Mean Differences								
											1	2	3	4	5	6	7	8	9
55	7404	7412	7419	7427	7435	7443	7451	7459	7466	7474	1	2	2	3	4	5	5	6	7
56	7482	7490	7497	7505	7513	7520	7528	7536	7543	7551	1	2	2	3	4	5	5	6	7
57	7559	7566	7574	7582	7589	7597	7604	7612	7619	7627	1	2	2	3	4	5	5	6	7
58	7634	7642	7649	7657	7664	7672	7679	7686	7694	7701	1	1	2	3	4	4	5	6	7
59	7709	7716	7723	7731	7738	7745	7752	7760	7767	7774	1	1	2	3	4	4	5	6	7
60	7782	7789	7796	7803	7810	7818	7825	7832	7839	7846	1	1	2	3	4	4	5	6	6
61	7853	7860	7868	7875	7882	7889	7896	7903	7910	7917	1	1	2	3	4	4	5	6	6
62	7924	7931	7938	7945	7952	7959	7966	7973	7980	7987	1	1	2	3	3	4	5	6	6
63	7993	8000	8007	8014	8021	8028	8035	8041	8048	8055	1	1	2	3	3	4	5	5	6
64	8062	8069	8075	8082	8089	8096	8102	8109	8116	8122	1	1	2	3	3	4	5	5	6
65	8129	8136	8142	8149	8156	8162	8169	8176	8182	8189	1	1	2	3	3	4	5	5	6
66	8195	8202	8209	8215	8222	8228	8335	8241	8248	8254	1	1	2	3	3	4	5	5	6
67	8261	8267	8274	8280	8287	8293	8299	8306	8312	8319	1	1	2	3	3	4	5	5	6
68	8325	8331	8338	8344	8351	8357	8363	8370	8376	8382	1	1	2	3	3	4	4	5	6
69	8388	8395	8401	8407	8414	8420	8426	8432	8439	8445	1	1	2	2	3	4	4	5	6
70	8451	8457	8463	8470	8476	8482	8488	8494	8500	8506	1	1	2	2	3	4	4	5	6
71	8513	8519	8525	8531	8537	8543	8549	8555	8561	8567	1	1	2	2	3	4	4	5	5
72	8573	8579	8585	8591	8597	8603	8609	8615	8621	8627	1	1	2	2	3	4	4	5	5
73	8633	8639	8645	8651	8657	8663	8669	8675	8681	8686	1	1	2	2	3	4	4	5	5
74	8692	8698	8704	8710	8716	8722	8727	8733	8739	8745	1	1	2	2	3	4	4	5	5
75	8751	8756	8762	8768	8774	8779	8785	8791	8797	8802	1	1	2	2	3	3	4	5	5
76	8808	8814	8820	8825	8831	8837	8842	8848	8854	8859	1	1	2	2	3	3	4	5	5
77	8865	8871	8876	8882	8887	8893	8899	8904	8910	8915	1	1	2	2	3	3	4	4	5
78	8921	8927	8932	8938	8943	8949	8954	8960	8965	8971	1	1	2	2	3	3	4	4	5
79	8976	8982	8987	8993	8998	9004	9009	9015	9020	9025	1	1	2	2	3	3	4	4	5
80	9031	9036	9043	9047	9053	9058	9063	9069	9074	9079	1	1	2	2	3	3	4	4	5
81	9085	9090	9096	9101	9106	9112	9117	9122	9128	9133	1	1	2	2	3	3	4	4	5
82	9138	9143	9149	9154	9159	9165	9170	9175	9180	9186	1	1	2	2	3	3	4	4	5
83	9191	9196	9201	9206	9212	9217	9222	9227	9232	9238	1	1	2	2	3	3	4	4	5
84	9243	9248	9253	9258	9263	9269	9274	9279	9284	9289	1	1	2	2	3	3	4	4	5
85	9294	9299	9304	9309	9315	9320	9325	9330	9335	9340	1	1	2	2	3	3	4	4	5
86	9345	9350	9355	9360	9365	9370	9375	9380	9385	9390	1	1	2	2	3	3	4	4	5
87	9395	9400	9405	9410	9415	9420	9425	9430	9435	9440	0	1	1	2	2	3	3	4	4
88	9445	9450	9455	9460	9465	9469	9474	9479	9484	9489	0	1	1	2	2	3	3	4	4
89	9494	9499	9504	9509	9513	9518	9523	9528	9533	9538	0	1	1	2	2	3	3	4	4
90	9542	9547	9552	9557	9562	9566	9571	9576	9581	9586	0	1	1	2	2	3	3	4	4
91	9590	9595	9600	9605	9609	9614	9619	9624	9628	9633	0	1	1	2	2	3	3	4	4
92	9638	9643	9647	9652	9657	9661	9666	9671	9675	9680	0	1	1	2	2	3	3	4	4
93	9685	9689	9694	9699	9703	9708	9713	9717	9722	9727	0	1	1	2	2	3	3	4	4
94	9731	9736	9741	9745	9750	9754	9759	9763	9768	9773	0	1	1	2	2	3	3	4	4
95	9777	9782	9786	9791	9795	9800	9805	9809	9814	9818	0	1	1	2	2	3	3	4	4
96	9823	9827	9832	9836	9841	9845	9850	9854	9859	9863	0	1	1	2	2	3	3	4	4
97	9868	9872	9877	9881	9886	9890	9894	9899	9903	9908	0	1	1	2	2	3	3	4	4
98	9912	9917	9921	9926	9930	9934	9939	9943	9948	9952	0	1	1	2	2	3	3	4	4
99	9956	9961	9965	9969	9974	9978	9983	9987	9991	9996	0	1	1	2	2	3	3	3	4

ANTILOGARITHMS

	0	1	2	3	4	5	6	7	8	9	Mean Differences								
											1	2	3	4	5	6	7	8	9
.00	1000	1002	1005	1007	1009	1012	1014	1016	1019	1021	0	0	1	1	1	1	2	2	2
.01	1023	1026	1028	1030	1033	1035	1038	1040	1042	1045	0	0	1	1	1	1	2	2	2
.02	1047	1050	1052	1054	1057	1059	1062	1064	1067	1069	0	0	1	1	1	1	2	2	2
.03	1072	1074	1076	1079	1081	1084	1086	1089	1091	1094	0	0	1	1	1	1	2	2	2
.04	1096	1099	1102	1104	1107	1109	1112	1114	1117	1119	0	1	1	1	1	2	2	2	2
.05	1122	1125	1127	1130	1132	1135	1138	1140	1143	1146	0	1	1	1	1	2	2	2	2
.06	1148	1151	1153	1156	1159	1161	1164	1167	1169	1172	0	1	1	1	1	2	2	2	2
.07	1175	1178	1180	1183	1186	1189	1191	1194	1197	1199	0	1	1	1	1	2	2	2	2
.08	1202	1205	1208	1211	1213	1216	1219	1222	1225	1227	0	1	1	1	1	2	2	2	3
.09	1230	1233	1236	1239	1242	1245	1247	1250	1253	1256	0	1	1	1	1	2	2	2	3
.10	1259	1262	1265	1268	1271	1274	1276	1279	1282	1285	0	1	1	1	1	2	2	2	3
.11	1288	1291	1294	1297	1300	1393	1306	1309	1312	1315	0	1	1	1	2	2	2	2	3
.12	1318	1321	1324	1327	1330	1334	1337	1340	1343	1346	0	1	1	1	2	2	2	2	3
.13	1349	1352	1355	1358	1361	1365	1368	1371	1374	1377	0	1	1	1	2	2	2	3	3
.14	1380	1384	1387	1390	1393	1396	1400	1403	1406	1409	0	1	1	1	2	2	2	3	3
.15	1413	1416	1419	1422	1426	1429	1432	1435	1439	1442	0	1	1	1	2	2	2	3	3
.16	1445	1449	1452	1455	1459	1462	1466	1469	1472	1476	0	1	1	1	2	2	2	3	3
.17	1479	1483	1486	1489	1493	1496	1500	1503	1507	1510	0	1	1	1	2	2	2	3	3
.18	1514	1517	1521	1524	1528	1531	1535	1538	1542	1545	0	1	1	1	2	2	2	3	3
.19	1549	1552	1556	1560	1563	1567	1570	1574	1578	1581	0	1	1	1	2	2	3	3	3
.20	1585	1589	1592	1596	1600	1603	1607	1611	1614	1618	0	1	1	1	2	2	3	3	3
.21	1622	1626	1629	1633	1637	1641	1644	1648	1652	1656	0	1	1	2	2	2	3	3	3
.22	1660	1663	1667	1671	1675	1679	1683	1687	1690	1694	0	1	1	2	2	2	3	3	3
.23	1698	1702	1706	1710	1714	1718	1722	1726	1730	1734	0	1	1	2	2	2	3	3	4
.24	1738	1742	1746	1750	1754	1758	1762	1766	1770	1774	0	1	1	2	2	2	3	3	4
.25	1778	1782	1786	1791	1795	1799	1803	1807	1811	1816	0	1	1	2	2	2	3	3	4
.26	1820	1824	1828	1832	1837	1841	1845	1849	1854	1858	0	1	1	2	2	3	3	3	4
.27	1862	1866	1871	1875	1879	1884	1888	1892	1897	1901	0	1	1	2	2	3	3	3	4
.28	1905	1910	1914	1919	1923	1928	1932	1936	1941	1945	0	1	1	2	2	3	3	4	4
.29	1950	1954	1959	1963	1968	1972	1977	1982	1986	1991	0	1	1	2	2	3	3	4	4
.30	1995	2000	2004	2009	2014	2018	2023	2028	2032	2037	0	1	1	2	2	3	3	4	4
.31	2042	2046	2051	2056	2061	2065	2070	2075	2080	2084	0	1	1	2	2	3	3	4	4
.32	2089	2094	2099	2104	2109	2113	2118	2123	2128	2133	0	1	1	2	2	3	3	4	4
.33	2138	2143	2148	2153	2158	2163	2168	2173	2178	2183	0	1	1	2	2	3	3	4	4
.34	2188	2193	2198	2203	2208	2213	2218	2223	2228	2234	1	1	2	2	3	3	4	4	5
.35	2239	2244	2249	2254	2259	2265	2270	2275	2280	2286	1	1	2	2	3	3	4	4	5
.36	2291	2296	2301	2307	2312	2317	2323	2328	2333	2339	1	1	2	2	3	3	4	4	5
.37	2344	2350	2355	2360	2366	2371	2377	2382	2388	2393	1	1	2	2	3	3	4	4	5
.38	2399	2404	2410	2415	2421	2427	2432	2438	2443	2449	1	1	2	2	3	3	4	4	5
.39	2455	2460	2466	2472	2477	2483	2489	2495	2500	2506	1	1	2	2	3	3	4	5	5
.40	2512	2518	2523	2529	2535	2541	2547	2553	2559	2564	1	1	2	2	3	4	4	5	5
.41	2570	2576	2582	2588	2594	2600	2606	2612	2618	2624	1	1	2	2	3	4	4	5	5
.42	2630	2636	2642	2649	2655	2661	2667	2673	2679	2685	1	1	2	2	3	4	4	5	6
.43	2692	2698	2704	2710	2716	2723	2729	2735	2742	2748	1	1	2	3	3	4	4	5	6
.44	2754	2761	2767	2773	2780	2786	2793	2799	2805	2812	1	1	2	3	3	4	4	5	6
.45	2818	2825	2831	2838	2844	2851	2858	2864	2871	2877	1	1	2	3	3	4	5	5	6
.46	2884	2891	2897	2904	2911	2917	2924	2931	2938	2944	1	1	2	3	3	4	5	5	6
.47	2951	2958	2965	2972	2979	2985	2992	2999	3006	3013	1	1	2	3	3	4	5	5	6
.48	3020	3027	3034	3041	3048	3055	3062	3069	3076	3083	1	1	2	3	4	4	5	6	6
.49	3090	3097	3105	3112	3119	3126	3133	3141	3148	3155	1	1	2	3	4	4	5	6	6

ANTILOGARITHMS

	0	1	2	3	4	5	6	7	8	9	Mean Differences								
											1	2	3	4	5	6	7	8	9
.50	3162	3170	3177	3184	3192	3199	3206	3214	3221	3228	1	1	2	3	4	4	5	6	7
.51	3236	3243	3251	3258	3266	3273	3281	3289	3296	3304	1	2	2	3	4	5	5	6	7
.52	3311	3319	3327	3334	3342	3350	3357	3365	3373	3381	1	2	2	3	4	5	5	6	7
.53	3388	3396	3404	3412	3420	3428	3436	3443	3451	3459	1	2	2	3	4	5	6	6	7
.54	3467	3475	3483	3491	3499	3508	3516	3524	3532	3540	1	2	2	3	4	5	6	6	7
.55	3548	3556	3565	3573	3581	3589	3597	3606	3614	3622	1	2	2	3	4	5	6	7	7
.56	3631	3639	3648	3656	3664	3673	3681	3690	3698	3707	1	2	3	3	4	5	6	7	8
.57	3715	3724	3733	3741	3750	3758	3767	3776	3784	3793	1	2	3	3	4	5	6	7	8
.58	3802	3811	3819	3828	3837	3846	3855	3864	3873	3882	1	2	3	4	4	5	6	7	8
.59	3890	3899	3908	3917	3926	3936	3945	3954	3963	3972	1	2	3	4	5	5	6	7	8
.60	3981	3990	3999	4009	4018	4027	4036	4046	4055	4064	1	2	3	4	5	6	6	7	8
.61	4074	4083	4093	4102	4111	4121	4130	4140	4150	4159	1	2	3	4	5	6	7	8	9
.62	4169	4178	4188	4198	4207	4217	4227	4236	4246	4256	1	2	3	4	5	6	7	8	9
.63	4266	4276	4285	4295	4305	4315	4325	4335	4345	4355	1	2	3	4	5	6	7	8	9
.64	4365	4375	4385	4395	4406	4416	4426	4436	4446	4457	1	2	3	4	5	6	7	8	9
.65	4467	4477	4487	4498	4508	4519	4529	4539	4550	4560	1	2	3	4	5	6	7	8	9
.66	4571	4581	4592	4603	4613	4624	4634	4645	4656	4667	1	2	3	4	5	6	7	9	10
.67	4677	4688	4699	4710	4721	4732	4742	4753	4764	4775	1	2	3	4	5	7	8	9	10
.68	4786	4797	4808	4819	4831	4842	4853	4864	4875	4887	1	2	3	4	6	7	8	9	10
.69	4898	4909	4920	4932	4943	4955	4966	4977	4989	5000	1	2	3	5	6	7	8	9	10
.70	5012	5023	5035	5047	5058	5070	5082	5093	5105	5117	1	2	4	5	6	7	8	9	11
.71	5129	5140	5152	5164	5176	5188	5200	5212	5224	5236	1	2	4	5	6	7	8	10	11
.72	5248	5260	5272	5284	5297	5309	5321	5333	5346	5358	1	2	4	5	6	7	9	10	11
.73	5370	5383	5395	5408	5420	5433	5445	5458	5470	5483	1	3	4	5	6	8	9	10	11
.74	5495	5508	5521	5534	5546	5559	5572	5585	5598	5610	1	3	4	5	6	8	9	10	12
.75	5623	5636	5649	5662	5675	5689	5702	5715	5728	5741	1	3	4	5	7	8	9	10	12
.76	5754	5768	5781	5794	5808	5821	5834	5848	5861	5875	1	3	4	5	7	8	9	11	12
.77	5888	5902	5916	5929	5943	5957	5970	5984	5998	6012	1	3	4	5	7	8	10	11	12
.78	6026	6039	6053	6067	6081	6095	6109	6124	6138	6152	1	3	4	6	7	8	10	11	13
.79	6166	6180	6194	6209	6223	6237	6252	6266	6281	6295	1	3	4	6	7	9	10	11	13
.80	6310	6324	6339	6353	6368	6383	6397	6412	6427	6442	1	3	4	6	7	9	10	12	13
.81	6457	6471	6486	6501	6516	6531	6546	6561	6577	6592	2	3	5	6	8	9	11	12	14
.82	6607	6622	6637	6653	6668	6683	6699	6714	6730	6745	2	3	5	6	8	9	11	12	14
.83	6761	6776	6792	6808	6823	6839	6855	6871	6887	6902	2	3	5	6	8	9	11	13	14
.84	6918	6934	6950	6966	6982	6998	7015	7031	7047	7063	2	3	5	6	8	10	11	13	15
.85	7079	7096	7112	7129	7145	7161	7178	7194	7211	7228	2	3	5	7	8	10	12	13	15
.86	7244	7261	7278	7295	7311	7328	7345	7362	7379	7396	2	3	5	7	8	10	12	13	15
.87	7413	7430	7447	7464	7482	7499	7516	7534	7551	7568	2	3	5	7	9	10	12	14	16
.88	7586	7603	7621	7638	7656	7674	7691	7709	7727	7745	2	4	5	7	9	11	12	14	16
.89	7762	7780	7798	7816	7834	7852	7870	7889	7907	7925	2	4	5	7	9	11	13	14	16
.90	7943	7962	7980	7998	8017	8035	8054	8072	8091	8110	2	4	6	7	9	11	13	15	17
.91	8128	8147	8166	8185	8204	8222	8241	8260	8279	8299	2	4	6	8	9	11	13	15	17
.92	8318	8337	8356	8375	8395	8414	8433	8453	8472	8402	2	4	6	8	10	12	14	15	17
.93	8511	8531	8551	8570	8590	8610	8630	8650	8670	8690	2	4	6	8	10	12	14	16	18
.94	8710	8730	8750	8770	8790	8810	8831	8851	8872	8892	2	4	6	8	10	12	14	16	18
.95	8913	8933	8954	8974	8995	9016	9036	9057	9078	9099	2	4	6	8	10	12	15	17	19
.96	9120	9141	9162	9183	9204	9226	9247	9268	9290	9311	2	4	6	8	11	13	15	17	19
.97	9333	9354	9376	9397	9419	9441	9462	9484	9506	9528	2	4	7	9	11	13	15	17	20
.98	9550	9572	9594	9616	9638	9661	9683	9705	9727	9750	2	4	7	9	11	13	16	18	20
.99	9772	9795	9817	9840	9863	9886	9908	9931	9954	9977	2	5	7	9	11	14	16	18	20